Coasts and Estuaries
The Future

Coasts and Estuaries
The Future

Edited by

Eric Wolanski

John W. Day

Michael Elliott

Ramesh Ramachandran

ELSEVIER

Elsevier
Radarweg 29, PO Box 211, 1000 AE Amsterdam, Netherlands
The Boulevard, Langford Lane, Kidlington, Oxford OX5 1GB, United Kingdom
50 Hampshire Street, 5th Floor, Cambridge, MA 02139, United States

Notices
Knowledge and best practice in this field are constantly changing. As new research and experience broaden our understanding, changes in research methods, professional practices, or medical treatment may become necessary.

Practitioners and researchers must always rely on their own experience and knowledge in evaluating and using any information, methods, compounds, or experiments described herein. In using such information or methods they should be mindful of their own safety and the safety of others, including parties for whom they have a professional responsibility.

To the fullest extent of the law, neither the Publisher nor the authors, contributors, or editors, assume any liability for any injury and/or damage to persons or property as a matter of products liability, negligence or otherwise, or from any use or operation of any methods, products, instructions, or ideas contained in the material herein.

Library of Congress Cataloging-in-Publication Data
A catalog record for this book is available from the Library of Congress

British Library Cataloguing-in-Publication Data
A catalogue record for this book is available from the British Library

ISBN: 978-0-12-814003-1

For information on all Elsevier publications visit our
website at https://www.elsevier.com/books-and-journals

Working together
to grow libraries in
developing countries

www.elsevier.com • www.bookaid.org

Publisher: Candice Janco
Acquisition Editor: Louisa Hutchins
Editorial Project Manager: Lindsay Lawrence
Production Project Manager: Omer Mukthar
Cover Designer: Matthew Limbert

Typeset by SPi Global, India

Dedication

We dedicate this book to our grandchildren: Oliver, Grace, and Harry Wolanski; Olly, Dylan, and Mycah Elliott; and Daisy and Sunny Day; and to Ramachandran's children Gowtham and Niveda Ramesh; we hope that they will enjoy healthy estuaries and coastal waters by 2050 and beyond and we hope that these will remain healthy to entrust to their children.

Contents

1. A Synthesis: What Is the Future for Coasts, Estuaries, Deltas and Other Transitional Habitats in 2050 and Beyond?

Michael Elliott, John W. Day,
Ramesh Ramachandran, Eric Wolanski

Section A
Estuaries

2. An Assessment of Saltwater Intrusion in the Changjiang (Yangtze) River Estuary, China

Maotian Li, Zhongyuan Chen

Section B
Deltas

8. Arctic Deltas and Estuaries: A Canadian Perspective

Donald L. Forbes

9. Delta Winners and Losers in the Anthropocene

John W. Day, Ramesh Ramachandran, Liviu Giosan, James Syvitski, G. Paul Kemp

10. Mississippi Delta Restoration and Protection: Shifting Baselines, Diminishing Resilience, and Growing Nonsustainability

John W. Day, Craig Colten, G. Paul Kemp

Section F
Coral Reefs

26. Successful Management of Coral Reef-Watershed Networks

Robert H. Richmond, Yimnang Golbuu, Austin J. Shelton III

27. Challenges and Opportunities in the Management of Coral Islands of Lakshadweep, India

Purvaja, R., Yogeswari, S., Debasis, T., Hariharan, G., Raghuraman, R., Muruganandam, R., Ramesh Ramachandran

28. The Future of the Great Barrier Reef: The Water Quality Imperative

Brodie, J., Grech, A., Pressey, B., Day, J., Dale, A.P., Morrison, T., Wenger, A.

Section G
Over-Arching Topics

33. Temperate Estuaries: Their Ecology Under Future Environmental Changes

Ducrotoy J.-P., Michael Elliott, Cutts N.D., Franco A., Little S., Mazik K., Wilkinson M.

34. Plastic Pollution in the Coastal Environment: Current Challenges and Future Solutions

K. Critchell, A. Bauer-Civiello, C. Benham, K. Berry, L. Eagle, M. Hamann, K. Hussey, T. Ridgway

35. Changing Hydrology: A UK Perspective

Peter E. Robins, Matt J. Lewis

Section H
Management of Change

36. Global Change Impacts on the Future of Coastal Systems: Perverse Interactions Among Climate Change, Ecosystem Degradation, Energy Scarcity, and Population

John W. Day, John M. Rybczyk

Contributors

Numbers in parentheses indicate the pages on which the authors' contributions begin.

Waqar Ahmed (213), National Institute of Oceanography, Karachi, Pakistan

A. Bauer-Civiello (595), College of Science and Engineering, James Cook University, Townsville, QLD, Australia

C. Benham (595), College of Science and Engineering, James Cook University, Townsville, QLD, Australia

K. Berry (595), College of Science and Engineering, James Cook University, Townsville, QLD, Australia

Morris Bidjerano (293), School of Public Policy and Administration, Walden University, Greenville, SC, United States

Sophie Blackburn (661), Department of Geography, King's College London, London, United Kingdom

Erik Bonsdorff (343), Åbo Akademi University, Turku, Finland

J. Brodie (477), ARC Centre of Excellence for Coral Reef Studies, James Cook University, Townsville, QLD, Australia

Nguyen Ba Cao (321), Vietnam Academy of Water Resources, Hanoi, Vietnam

Zhongyuan Chen (31, 233), State Key Laboratory of Estuarine and Coastal Research, East China Normal University, Shanghai, People's Republic of China

Peter Clift (213), Louisiana State University, Baton Rouge, LA, United States

Craig Colten (167), Department of Geography and Anthropology, Louisiana State University, Baton Rouge, LA, United States

K. Critchell (595), College of Science and Engineering, James Cook University, Townsville; Marine Spatial Ecology Lab, University of Queensland, Brisbane, QLD, Australia

N.D. Cutts (577), Institute of Estuarine and Coastal Studies, University of Hull, Hull, United Kingdom

Christopher F. D'Elia (293), College of the Coast and Environment, Louisiana State University, Baton Rouge, LA, United States

A.P. Dale (477), The Cairns Institute, James Cook University, Cairns, QLD, Australia

Moslem Daliri (57), Department of Fisheries, Faculty of Marine and Atmospheric Sciences and Technologies, University of Hormozgan, Bandar Abbas, Iran

Steve E. Davis (277), Everglades Foundation, Palmetto Bay, FL, United States

John W. Day (1, 149, 167, 237, 377, 621), Department of Oceanography and Coastal Sciences, College of the Coast and Environment, Louisiana State University, Baton Rouge, LA, United States

J. Day (477), ARC Centre of Excellence for Coral Reef Studies, James Cook University, Townsville, QLD, Australia

T. Debasis (461), National Centre for Sustainable Coastal Management, Ministry of Environment, Forest and Climate Change, Government of India, Anna University Campus, Chennai, India

Omar Defeo (45), UNDECIMAR, Faculty of Sciences, University of the Republic, Montevideo, Uruguay, Montevideo, Uruguay

Mustafa Dihkan (363), Department of Geomatics, Faculty of Engineering, Karadeniz Technical University, Çamburnu, Trabzon

Salif Diop (311), Cheikh Anta Diop University, Dakar-Fann, Senegal

Sabine R. Dittmann (523), College of Science & Engineering, Flinders University, Adelaide, SA, Australia

Tom Dreschel (277), Everglades Systems Assessment Section, South Florida Water Management District, West Palm Beach, FL, United States

J.-P. Ducrotoy (577), Institute of Estuarine and Coastal Studies, University of Hull, Hull, United Kingdom

Ryan J.K. Dunn (69), Ocean Science & Technology, RPS, Gold Coast, QLD, Australia

L. Eagle (595), College of Business, Law and Governance, James Cook University, Townsville, QLD, Australia

Michael Elliott (1, 577), Institute of Estuarine and Coastal Studies, University of Hull, Hull, United Kingdom

Muzaffer Feyzioğlu (363), Department of Marine Science and Technology, Faculty of Marine Sciences, Karadeniz Technical University, Çamburnu, Trabzon

Donald L. Forbes (123), Geological Survey of Canada, Natural Resources Canada, Bedford Institute of Oceanography, Dartmouth, NS, Canada; Department of Geography, Memorial University of Newfoundland, St. John's, NL; Department of Earth Sciences, Dalhousie University, Halifax, NS, Canada

A. Franco (577), Institute of Estuarine and Coastal Studies, University of Hull, Hull, United Kingdom

D. Ganguly (187), National Centre for Sustainable Coastal Management, Ministry of Environment, Forest and Climate Change, Government of India, Anna University Campus, Chennai, India

Javier García-Alonso (45), Departament of Ecology, CURE, University of the Republic, Maldonado, Uruguay

Chris L. Gillies (427), The Nature Conservancy, Carlton, VIC; James Cook University, Townsville, QLD, Australia

Liviu Giosan (149, 213), Department of Geology and Geophysics, Woods Hole Oceanographic Institution, Woods Hole, MA, United States

Bernhard Glaeser (641), Freie Universität; German Society for Human Ecology (DGH), Berlin, Germany

Yimnang Golbuu (445), Palau International Coral Reef Center, Koror, Palau

A. Grech (477), ARC Centre of Excellence for Coral Reef Studies, James Cook University, Townsville, QLD, Australia

Abdulaziz Güneroğlu (363), Department of Marine Ecology, Faculty of Marine Sciences, Karadeniz Technical University, Çamburnu, Trabzon

C.S. Hallett (103), Centre for Sustainable Aquatic Ecosystems, Harry Butler Institute, Murdoch University, Perth, WA, Australia

M. Hamann (595), College of Science and Engineering, James Cook University, Townsville, QLD, Australia

Boze Hancock (427), The Nature Conservancy, Graduate School of Oceanography, University of Rhode Island, Narragansett, RI, United States

G. Hariharan (461), National Centre for Sustainable Coastal Management, Ministry of Environment, Forest and Climate Change, Government of India, Anna University Campus, Chennai, India

Anna-Stiina Heiskanen (343), Finnish Environment Institute, Helsinki, Finland

K. Hennig (103), Department of Water and Environmental Regulation, Perth, WA, Australia

Claudia Teutli Hernández (377), Center for Research and Advanced Studies of the National Polytechnic Institute, Merida Campus, Mexico

Jorge A. Herrera-Silveira (377), Center for Research and Advanced Studies of the National Polytechnic Institute, Merida Campus, Mexico

M.R. Hipsey (103), Aquatic Ecodynamics, UWA School of Agriculture and Environment, The University of Western Australia, Perth, WA, Australia

Steeg D. Hoeksema (523), Centre for Sustainable Aquatic Ecosystems, Harry Butler Institute, Murdoch University, Murdoch; Department of Biodiversity, Conservation and Attractions, Bentley Delivery Centre, Western Australia, Australia

Jianyin Huang (69), Natural and Built Environments Research Centre, School of Natural and Built Environments; Future Industries Institute, University of South Australia, Adelaide, SA, Australia

P. Huang (103), Aquatic Ecodynamics, UWA School of Agriculture and Environment, The University of Western Australia, Perth, WA, Australia

Austin Humphries (427), Department of Fisheries, Animal and Veterinary Science, University of Rhode Island, Kingston; Graduate School of Oceanography, University of Rhode Island, Narragansett, RI, United States

K. Hussey (595), Centre for Policy Futures, Faculty of Humanities and Social Sciences, The University of Queensland, St Lucia, QLD, Australia

Carles Ibáñez (237), Aquatic Ecosystems Program, IRTA, San Carles de la Rapita, Catalonia, Spain

Asif Inam (213), National Institute of Oceanography, Karachi, Pakistan

Marko Joas (343), Åbo Akademi University, Turku, Finland

Ehsan Kamrani (57), Department of Fisheries, Faculty of Marine and Atmospheric Sciences and Technologies, University of Hormozgan, Bandar Abbas, Iran

Coura Kane (311), Cheikh Anta Diop University, Dakar-Fann, Senegal

G. Paul Kemp (149, 167, 377), Department of Oceanography and Coastal Sciences, College of the Coast and Environment, Louisiana State University, Baton Rouge, LA, United States

Samina Kidwai (213), National Institute of Oceanography, Karachi, Pakistan

K.L. Kilminster (103), Department of Water and Environmental Regulation, Perth, WA, Australia

Brian A. King (69), Ocean Science & Technology, RPS, Gold Coast, QLD, Australia

R. Kirby (413), Ravensrodd Consultants Ltd., Liverpool, United Kingdom

Cheikh Tidiane Koulibaly (311), Cheikh Anta Diop University, Dakar-Fann, Senegal; University of Ibadan, Ibadan, Nigeria

Ahana Lakshmi (187), National Centre for Sustainable Coastal Management, Ministry of Environment, Forest and Climate Change, Government of India, Anna University Campus, Chennai, India

Janet M. Lanyon (87), School of Biological Sciences, The University of Queensland, St. Lucia, QLD, Australia

Ana L. Lara-Domínguez (377), Institutue of Ecology, Veracruz, Mexico

Diego Lercari (45), UNDECIMAR, Faculty of Sciences, University of the Republic, Montevideo, Uruguay, Montevideo, Uruguay

Matt J. Lewis (611), School of Ocean Sciences, Marine Centre Wales, Bangor University, Menai Bridge, United Kingdom

Maotian Li (31), State Key Laboratory of Estuarine and Coastal Research, East China Normal University, Shanghai, People's Republic of China; Institute of Eco-Chongming Shanghai, China

S. Little (577), School of Animal, Rural and Environmental Sciences, Nottingham Trent University, Nottinghamshire, United Kingdom

Amy Lauren Lovecraft (671), Center for Arctic Policy Studies, University of Alaska Fairbanks, Fairbanks, AK, United States

Concepción Marcos (253), Department of Ecology and Hydrology, Regional Campus of International Excellence "Mare Nostrum", University of Murcia, Murcia, Spain

César Marques (661), National School of Statistical Science—Brazilian Institute of Geography and Statistics (ENCE/IBGE), Rio de Janeiro, Brazil

Osamu Matsuda (401), Graduate School of Biosphere Sciences, Hiroshima University, Higashihiroshima, Japan

K. Mazik (577), Institute of Estuarine and Coastal Studies, University of Hull, Hull, United Kingdom

John F. Meeder (277), Sea Level Solutions Center and Southeast Environmental Research Center, Florida International University, Miami, FL, United States

Chanda L. Meek (671), Department of Political Science, University of Alaska Fairbanks, Fairbanks, AK, United States

Ian Michael McLeod (427), TropWATER, Centre for Tropical Water and Aquatic Ecosystem Research, James Cook University, Townsville, QLD, Australia.

T. Morrison (477), ARC Centre of Excellence for Coral Reef Studies, James Cook University, Townsville, QLD, Australia

R. Muruganandam (461), National Centre for Sustainable Coastal Management, Ministry of Environment, Forest and Climate Change, Government of India, Anna University Campus, Chennai, India

Alice Newton (253), NILU-IMPACT, Kjeller, Norway; CIMA-Centre for Marine and Environmental Research, Gambelas Campus, University of Algarve, Faro, Portugal

Nguyen Huu Nhan (321), Vietnam Academy of Water Resources, Hanoi, Vietnam

Awa Niang (311), Cheikh Anta Diop University, Dakar-Fann, Senegal

Sara Morales Ojeda (377), Center for Research and Advanced Studies of the National Polytechnic Institute, Merida Campus, Mexico

Maria-Lourdes D. Palomares (569), Institute for the Oceans and Fisheries, University of British Columbia, Vancouver, BC, Canada

Daniel Pauly (569), Institute for the Oceans and Fisheries, University of British Columbia, Vancouver, BC, Canada

Mark Pelling (661), Department of Geography, King's College London, London, United Kingdom

Angel Pérez-Ruzafa (253), Department of Ecology and Hydrology, Regional Campus of International Excellence "Mare Nostrum", University of Murcia, Murcia, Spain

Isabel M. Pérez-Ruzafa (253), Department of Plant Biology I, Complutense University of Madrid, Madrid, Spain

Didier Pont (237), Institute of Hydrobiology and Aquatic Ecosystem Management (IHG), University of Natural Resources and Life Sciences, Vienna, Austria

Ian C. Potter (523), Centre for Sustainable Aquatic Ecosystems, Harry Butler Institute; School of Veterinary and Life Sciences, Murdoch University, Murdoch, Western Australia, Australia

B. Pressey (477), ARC Centre of Excellence for Coral Reef Studies, James Cook University, Townsville, QLD, Australia

R. Purvaja (187, 461), National Centre for Sustainable Coastal Management, Ministry of Environment, Forest and Climate Change, Government of India, Anna University Campus, Chennai, India

R. Raghuraman (461), National Centre for Sustainable Coastal Management, Ministry of Environment, Forest and Climate Change, Government of India, Anna University Campus, Chennai, India

Ramesh Ramachandran (1, 149, 187, 461), National Centre for Sustainable Coastal Management, Ministry of Environment, Forest and Climate Change, Government of India, Anna University Campus, Chennai, India

Robert H. Richmond (445), Kewalo Marine Laboratory, University of Hawaii at Manoa, Honolulu, HI, United States

T. Ridgway (595), Global Change Institute, The University of Queensland, St Lucia, QLD, Australia

R.S. Robin (187), National Centre for Sustainable Coastal Management, Ministry of Environment, Forest and Climate Change, Government of India, Anna University Campus, Chennai, India

Peter E. Robins (611), School of Ocean Sciences, Marine Centre Wales, Bangor University, Menai Bridge, United Kingdom

Pablo L. Ruiz (277), South Florida Caribbean Network, National Park Service, Palmetto Bay, FL, United States

John M. Rybczyk (621), Department of Environmental Science, Western Washington University, Bellingham, WA, United States

Bonthu S.R. (187), National Centre for Sustainable Coastal Management, Ministry of Environment, Forest and Climate Change, Government of India, Anna University Campus, Chennai, India

Osman Samsun (363), Faculty of Fisheries, Sinop University, Sinop, Turkey

Swati Mohan Sappal (187), National Centre for Sustainable Coastal Management, Ministry of Environment, Forest and Climate Change, Government of India, Anna University Campus, Chennai, India

Francesco Scarton (237), SELC Societá Cooperativa, Venezia, Italy

Peter Scheren (311), WWF Regional Office for Africa, Nairobi, Kenya

Nickolai Shalovenkov (547), The Centre for Ecological Studies, Russia

Moslem Sharifinia (57), Iranian National Institute for Oceanography and Atmospheric Science (INIOAS), Gulf of Oman and Indian Ocean Research Center, Marine Biology Division, Chabahar, Iran

Austin J. Shelton III (445), Center for Island Sustainability and Sea Grant Program, University of Guam, Mangilao, Guam

Fred H. Sklar (277), Everglades Systems Assessment Section, South Florida Water Management District, West Palm Beach, FL, United States

Mary Divya Suganya (187), National Centre for Sustainable Coastal Management, Ministry of Environment, Forest and Climate Change, Government of India, Anna University Campus, Chennai, India

James Syvitski (149), Community Surface Dynamics Modeling System, University of Colorado, Boulder, CO, United States

Syed Mohsin Tabrez (213), National Institute of Oceanography, Karachi, Pakistan

Peter R. Teasdale (69), Natural and Built Environments Research Centre, School of Natural and Built Environments; Future Industries Institute, University of South Australia, Adelaide, SA, Australia

Tiffany G. Troxler (277), Sea Level Solutions Center and Southeast Environmental Research Center, Florida International University, Miami, FL, United States

James R. Tweedley (523), Centre for Sustainable Aquatic Ecosystems, Harry Butler Institute; School of Veterinary and Life Sciences, Murdoch University, Murdoch, Western Australia, Australia

F.J. Valesini (103), Centre for Sustainable Aquatic Ecosystems, Harry Butler Institute, Murdoch University, Perth, WA, Australia

Nathan J. Waltham (69), Centre for Tropical Water and Aquatic Ecosystem Research (TropWATER), Division of Tropical Environments and Societies, James Cook University, Douglas, QLD, Australia

A. Wenger (477), School of Earth and Environmental Sciences, University of Queensland, St. Lucia, QLD, Australia

Timothy B. Wheeler (293), Bay Journal, Seven Valleys, PA, United States

Alan K. Whitfield (523), South African Institute for Aquatic Biodiversity, Grahamstown, South Africa

M. Wilkinson (577), Institute of Life and Earth Sciences, Heriot-Watt University, Edinburgh, United Kingdom

Kim Withers (523), Department of Life Sciences, Texas A&M University, Corpus Christi, TX, United States

Eric Wolanski (1, 503), TropWATER and College of Marine & Environmental Sciences, James Cook University and Australian Institute of Marine Science, Townsville, QLD, Australia

Tetsuo Yanagi (401), International EMECS Center, Kobe, Japan

Alejandro Yáñez-Arancibia (377), Institutue of Ecology, Veracruz, Mexico

S. Yogeswari (461), National Centre for Sustainable Coastal Management, Ministry of Environment, Forest and Climate Change, Government of India, Anna University Campus, Chennai, India

Jing Zhang (213), State Key Laboratory in Estuarine and Coastal Research, Shanghai, China

Philine S.E. zu Ermgassen (427), School of GeoSciences, University of Edinburgh, Edinburgh, United Kingdom

About the Editors

Professor Eric Wolanski

Eric Wolanski is an estuarine oceanographer and ecohydrologist at James Cook University and the Australian Institute of Marine Science. His research interests range from the oceanography of coral reefs, mangroves, and muddy estuaries to the interaction between physical and biological processes determining ecosystem health in tropical waters. He has over 400 scientific publications, including 12 books, and technical reports. Eric is a fellow of the Australian Academy of Technological Sciences and Engineering, the Institution of Engineers Australia (ret.), and l'Académie Royale des Sciences d'Outre-Mer. He was awarded a Doctorate Honoris Causa by the catholic University of Louvain, another Doctorate Honoris Causa by the University of Hull, a Queensland Information Technology and Telecommunications Award for Excellence, and a Lifetime Achievement Award by the Estuarine & Coastal Sciences Association. Eric is an Editor-in-Chief of *Wetlands Ecology and Management*, *Treatise on Estuarine and Coastal Science*, the Honorary Editor of *Estuarine, Coastal and Shelf Science*, and a member of the editorial board of four other journals. He is also a member of the Scientific and Policy Committee of Japan's EMECS (focusing on the Seto Inland Sea) and the European Union DANUBIUS-PP Scientific and Technical Advisory Board, which is a pan-European distributed research infrastructure dedicated to interdisciplinary studies of large river–sea systems throughout Europe.

Professor John Day

John Day is distinguished professor emeritus in the Department of Oceanography and Coastal Sciences at Louisiana State University. He has over 400 publications focusing on the ecology and management of coastal and wetland ecosystems, with emphasis on the Mississippi delta, as well as, among many, coastal ecosystems in Mexico and the impacts of climate change on wetlands in Venice Lagoon and in the Po, Rhone, and Ebro deltas in the Mediterranean. John is the coeditor of 14 books including *Estuarine Ecology, Ecological Modeling in Theory and Practice, The Ecology of the Barataria Basin, An Estuarine Profile, Ecology of Coastal Ecosystems in the Southern Mexico: The Terminos Lagoon Region, Ecosystem Based Management of the Gulf of Mexico, America's Most Sustainable Cities and Regions—Surviving the 21st Century Megatrends*. John served as chair of the Science and Engineering Special Team on restoration of the Mississippi delta, on the Scientific Steering Committee of the Future Earth Coasts program, and a National Research Council panel on urban sustainability.

Professor Michael Elliott

Michael Elliott is the professor of Estuarine and Coastal Sciences at the University of Hull, United Kingdom. He is a marine biologist with a wide experience and interests and his teaching, research, advisory, and consultancy work includes estuarine and marine ecology, policy, governance, and management. Mike has published widely, coauthoring/coediting 18 books/proceedings and >270 scientific publications. This includes coauthoring *The Estuarine Ecosystem: Ecology, Threats and Management*, *Ecology of Marine Sediments: Science to Management*, and *Estuarine Ecohydrology: An Introduction*' and as a volume editor and contributor to the *Treatise on Estuarine & Coastal Science*. He has advised on many environmental matters for academia, industry, government, and statutory bodies worldwide. Mike is a past-President of the international Estuarine & Coastal Sciences Association (ECSA) and is an Editor-in-Chief of the international journal *Estuarine, Coastal & Shelf Science*; he has been adjunct professor and held research positions at Murdoch University (Perth), Klaipeda University (Lithuania), the University of Palermo (Italy), and the South African Institute for Aquatic Biodiversity, Grahamstown. He was awarded Laureate of the Honorary Winberg Medal of the Russian Hydrobiological Academic Society in 2014.

Professor Ramesh Ramachandran

Ramesh Ramachandran is director of the National Centre for Sustainable Coastal Management at the Ministry of Environment, Forest and Climate Change, Government of India. His expertise includes coastal/marine biogeochemistry, conservation of coastal/marine biodiversity, and Integrated Coastal Zone Management. He has over 135 research publications and over 100 technical reports. Among the several awards Professor Ramesh has received are the University Grants Commission UGC-Swami Pranavananda Saraswathi Award in Environmental Sciences and Ecology for the Year 2007 (awarded in February 2010). He was the chair of the Scientific Steering Committee of LOICZ (currently renamed as Future Earth Coasts), member of the Scientific Steering Committee of the Monsoon Asia Integrated Regional Study, chairman of the International Working Group on Coastal Systems on the Role of Science in International Waters Projects of UNEP-GEF, as well as being affiliated with the Bay of Bengal Large Marine Ecosystem Programme of the FAO. He is currently the chair of the Global Partnership in Nutrient Management (GPNM) of UNEP.

Preface: Why This Book?

Coastal ecosystems are at the nexus of the Anthropocene, with enormous environmental issues, and inhabited by nearly half of the human population. These coastal systems and the surrounding human societies form coastal social-ecological systems that increasingly face enormous environmental issues from multiple pressures, which threaten their ecological and economical sustainability. The pressures are derived from hazards which then become risks where they impact the society and where, in some cases, human responses exacerbate the risks. There is only one big idea in managing these systems— how to maintain and protect the natural ecological structure and functioning and yet at the same time allow them to deliver ecosystem services which produce societal goods and benefits. The pressures include basically all human activities within the river catchments such as changes to land use and hydrology in the river catchment, and directly on coastal ecosystems from land claim, coastal sand mining, harbor dredging, pollution and eutrophication, overexploitation such as overfishing and extraction of groundwater, gas and petroleum extraction. In addition, coastal zones are impacted by climate change— this is not just the 'usual' culprits of sea level rise, ocean acidification, and increased temperature but also, just as important, changes in the rainfall-runoff of the river catchments, stronger coastal storms, and the changes to species distributions, including the influx of invasive species.

The problems faced by half of the humanity worldwide living near coasts are truly a worldwide challenge as well as an opportunity for science to study commonality and differences and provide solutions. During the five decades of monitoring the degradation of estuaries and coastal waters in the 20th century, coastal scientists studied the problems and issues arising along the coasts worldwide. Now, in the 21st century, the scientists need to use their science to help find solutions to these problems through science-informed management and innovation. The issues to solve are complex because they involve large areas, many users, and sociopolitical-environmental mosaics.

This book provides a typology of the human interaction with estuaries and coastal waters worldwide as a comprehensive description of what works and what does not work for estuaries and coastal waters worldwide and what remediation measures are possible and likely to succeed within limits. This is the first time that such a worldwide approach to estuarine and coastal sustainability has been initiated.

Thus the book addresses these real-life issues in order to learn from each other, by having a series of chapters written by the leading local experts detailing case studies from estuaries and coastal waters worldwide in the full range of natural variability and human pressures. The study sites are located in all the continents, except for the Antarctic, and several oceanic islands. This is followed by a series of chapters written by scientific leaders worldwide synthesizing the problems and offering solutions for specific issues graded within the framework of the socioeconomic-environmental mosaic. These include coastal fisheries, climate change, biophysical limits and energy costs, coastal megacities, evolving human-nature interactions, remediation measures for a number of worldwide issues such as mud and metal legacy as well as plastic pollution, integrated coastal management, and international water conflicts affecting estuaries, deltas, and coastal waters.

We wish to thank Jaclyn Truesdell and Lindsay Lawrence at Elsevier for their help in producing this book.

Eric Wolanski
John Day
Michael Elliott
Ramesh Ramachandran

Chapter 1

A Synthesis: What Is the Future for Coasts, Estuaries, Deltas and Other Transitional Habitats in 2050 and Beyond?

Michael Elliott[*], John W. Day[†], Ramesh Ramachandran[‡], Eric Wolanski[§]

[*]Institute of Estuarine and Coastal Studies, University of Hull, Hull, United Kingdom, [†]Department of Oceanography and Coastal Sciences, College of the Coast and Environment, Louisiana State University, Baton Rouge, LA, United States, [‡]National Centre for Sustainable Coastal Management, Ministry of Environment, Forest and Climate Change, Government of India, Anna University Campus, Chennai, India, [§]TropWATER and College of Marine & Environmental Sciences, James Cook University and Australian Institute of Marine Science, Townsville, QLD, Australia

1 INTRODUCTION

Estuaries, deltas, coastal lagoons, and fjords are transitional waters and contain ecosystems between riverine and coastal marine ecosystems. They are sites of important connectivity and intense gradients that make them among the world's most productive ecosystems. The coastal and transitional areas considered here are only a small fraction of the marine and brackish areas worldwide (~5%) but produce approximately half of the global fish catch per year (Palomares and Pauly, 2019). Although up to half of this catch is from small-scale fisheries (artisanal, subsistence, and recreational), it gains less attention worldwide than the larger-scale and open ocean industrial fisheries. At the same time, they develop ecological communities with an important diversity and complex mechanisms of self-regulation (Pérez-Ruzafa et al., 2019) and they provide significant ecosystem services and societal goods and benefits (Van den Belt and Costanza, 2011; Wolanski and Elliott, 2015). These and associated coastal ecosystems, including the open coast, enclosed and semi-enclosed seas, and special systems such as polar and coral environments, will be subject to change in the coming decades, and those changes will have to be either managed or accommodated by society.

Coastal and transitional ecosystems have been, are being, and will continue to be adversely affected by global climate change in many ways. These changes include increasing temperatures and sea levels, either reduced, increased, or at least subject to more erratic rainfall and freshwater discharges, especially in temperate areas, and the likelihood of more frequent or more severe droughts and storms (Day and Rybczyk, 2019). The changes to biogeographical regimes are likely with the movement of organism distributions toward higher latitudes. Higher sea levels, perhaps up to 1.5 m higher in the next century, will both increase saline intrusion and water levels into transitional areas, thereby changing vegetation and foodweb structure and perhaps causing loss of wetlands that have been important in producing ecosystem services and delivering societal goods and benefits. Growing resource scarcity, especially of energy, will limit our ability to handle these evolving problems effectively (Day et al., 2016, 2018; Wiegman et al., 2017; Day and Rybczyk, 2019).

In addition to climate changes, coastal and transitional ecosystems are increasingly subject to other types of environmental degradation, not least from increased industrialization, urbanization (urban development), and agricultural and aquacultural expansion. There is an increasing occurrence of non-native species with more vectors and migration routes becoming available; for example, the loss of polar ice may open up migration routes. There is increasing habitat loss and fragmentation not the least of which is because of land use changes for perhaps short-term economic gain but with long-term environmental and societal consequences; for example, the loss of mangroves for shrimp ponds ultimately reduces the resilience of coasts to hazards and storm events (Elliott et al., 2015, 2016; Day and Rybczyk, 2019). Because of this, there is the need for a holistic approach which incorporates the catchment-river-estuary continuum of ecosystems as well as the adjacent coastal and marine areas.

This synthesis, which is based largely on the chapters in this volume, aims to show that an eventual reduction in land and water resources, and perhaps increases in energy use in restoration and alleviation schemes, have long-term consequences (Day and Rybczyk, 2019). The loss of these resources and the increase in arid areas, the reduction in deltas and wetlands,

Coasts and Estuaries. https://doi.org/10.1016/B978-0-12-814003-1.00001-0

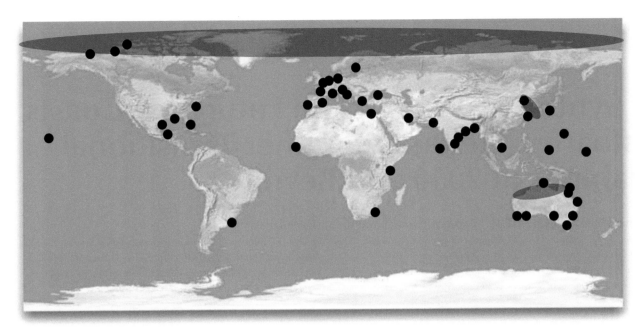

FIG. 1 Location map of the main study sites discussed in the text.

and the loss of resilience and resistance to natural hazards may all exceed biophysical limits. Therefore, using a set of case studies covering a large geographical area (Fig. 1), in this chapter we focus on the need for a holistic approach to create sustainable management of the coastal areas. We emphasise that there is the need for a good and appropriate use of the best-available science linked to that management. Such science will help to indicate the causes and consequences of the problems as well as the solutions to them.

2 SETTING THE SCENE: THE DAPSI(W)R(M) FRAMEWORK

All environments are affected by change, both natural and anthropogenic. To determine the causes and consequences of change, there is an increasing need for risk assessment and risk management frameworks that center on the human uses and abuses of the environment. One such framework originated as the DPSIR approach, but it more recently has been refined into DAPSI(W)R(M) (Drivers, Activities, Pressures, State change, Impacts (on human Welfare), Responses (using management Measures), (Patricio et al., 2016; Elliott et al., 2017). This relates to the acceptance that society has basic demands, termed *Drivers*, from the environment, such as the need for food, for shelter, well-being, and security, which require current *Activities* in an area. These activities in turn create *Pressures* in an area, which are termed *endogenic managed pressures* (for example, the need to go fishing for food and to build sea defenses for shelter and security as the mechanisms of change; again, for example, fishing involves scraping nets over the bed whereas building sea defenses may influence hydrographic processes and sediment-scouring in an area). The pressures are the mechanisms of both *State change* on the natural system (the loss of biota or the interference with normal hydrographic processes) and *Impacts* (on human *Welfare*). For example, the latter may be a loss of fish for human food or the reduced resilience of an area to storm events that results in the loss of human assets and livelihoods. The *State changes* and the *Impacts* on human *Welfare* then require *Responses*, which are management *Measures* to prevent the adverse effects. Hence the *Responses* should operate on the *Drivers*, *Activities* and *Pressures*.

The *Responses* (including management *Measures*) can fall within what has been termed the *10-tenets*: that for the management to be successful and sustainable, our actions have to be ecologically sustainable, technologically feasible, economically viable, socially desirable/tolerable, legally permissible, administratively achievable, politically expedient, ethically defensible (morally correct), culturally inclusive, and effectively communicable (Barnard and Elliott, 2015). It is emphasized that whereas only one of these relates to ecological well-being, the remainder are all society-based. As noted earlier, growing pressures arising from climate change, resource scarcity, environmental degradation, and growing population will challenge the ability to manage these problems effectively.

The Endogenic Managed Pressures emanate from inside an area in which management can address both the causes and the consequences of the pressures (Elliott, 2011). For example, future demographic changes to estuaries and coastal areas will increase urbanization and industrialization, which are likely to result in pollution loadings, eutrophication, and the discharge of ballast water. In contrast, global climate change is regarded as an Exogenic Unmanaged Pressure, in which the cause emanates from outside the area of concern but the consequences have to be managed within an area. For example, sea-level rise is caused by global changes in greenhouse gases and by isostatic events, so either cause needs global action or it cannot be subject to any action to stop it, but the consequences sea-level rise, such as increased flooding and erosion, require actions (responses) inside the management area (Ducrotoy and Elliott, 2008).

The pressures mentioned earlier constitute hazards, both natural and anthropogenic, which, if they affect human assets and livelihoods, become risks (Cormier et al., 2013; Elliott et al., 2014) (Table 1). As shown by many of the contributions referenced here, to determine the future trajectories of change, it is important to use the best-available science to know what

TABLE 1 Hazards and Risks

Hazard	Natural or Anthropogenic?	Examples
A. Surface hydrological hazards	Natural, but exacerbated by human activities	High-tide, spring-tide, and equinoctial flooding; flash flooding, ENSO/NAO patterns; flow delivery repercussions of catchment modifications (land use increasing sediment loading, dams decreasing peak flows, and sediment loadings, etc.)
B. Surface physiographic removal by natural processes—chronic/long-term	Natural, but exacerbated by human activities	Gradual erosion of soft cliffs by slumping; estuary bank erosion by prevailing currents
C. Surface physiographic removal by human actions—chronic/long-term	Anthropogenic	Land claim, removal of wetlands for urban and agricultural area
D. Surface physiographic removal—acute/short-term	Natural	Cliff failure, undercutting of hard cliffs, and intermittent erosion
E. Climatological hazards—acute/short-term	Natural, but exacerbated by human activities	Storm surges, cyclones, tropical storms, hurricanes, offshore surges, fluvial and pluvial flooding
F. Climatological hazards—chronic/long-term	Natural, but exacerbated by human activities, or anthropogenic	Ocean acidification, sea-level rise, storminess, ingress of seawater/saline intrusion
G. Tectonic hazards—acute/short-term	Natural	Tsunamis, seismic slippages, earthquakes
H. Tectonic hazards—chronic/long-term	Natural	Isostatic rebound, subsidence
I. Anthropogenic microbial biohazards	Anthropogenic	Sewage pathogens
J. Anthropogenic macrobial biohazards	Anthropogenic	Alien, introduced and invasive species, GMOs, bloom-forming species
K. Anthropogenic introduced technological hazards	Anthropogenic	Failures or mismanagement of infrastructure, coastal defenses, catchment impedance structures (dams, weirs)
L. Anthropogenic extractive technological hazards	Anthropogenic	Removal of space, removal of biological populations (fish, shellfish, etc.); seabed extraction and oil/gas/coal extraction leading to subsidence
M. Anthropogenic acute chemical hazards	Anthropogenic	Pollution from one-off spillages, oil spills
N. Anthropogenic chronic chemical hazards	Anthropogenic	Diffuse pollution, ocean acidification, litter/garbage, nutrients from land run-off, constant land-based discharges, aerial inputs
O. Anthropogenic acute geopolitical hazards	Anthropogenic	Terrorism attacks leading to damage on infrastructure
P. Anthropogenic chronic geopolitical hazards	Anthropogenic	Wars created by shortage of resources (e.g., land, water, minerals)

Modified from Elliott, M., Cutts, N.D., Trono, A., 2014. A typology of marine and estuarine hazards and risks as vectors of change: a review for vulnerable coasts and their management. Ocean Coast Manag. 93, 88–99.

the areas are like now and what they were like in the past and to determine what hazards and risks were in the past, are occurring now, and will occur in the future. Hence the management of transitional and coastal areas becomes an exercise in risk assessment and management, that is, determining which risks are real and need to be addressed becomes the main challenge both now and with future developments. However, in considering the future for the coast and associated habitats, it is of note than the effects of a natural hazard can become much greater through human actions (see below), and biophysical constraints will result in increasing challenges to the ability to solve growing problems.

3 CURRENT STATUS OF ESTUARINE AND COASTAL ECOSYSTEMS

3.1 Estuaries

While many estuaries globally have changed throughout geological times and in many cases show their ephemeral nature, recent anthropogenic changes have received the most attention and indeed potentially indicate the trajectory of future changes (Duarte et al., 2015). As an example, Li and Chen (2019) show the value of a long-term dataset (e.g., half a century) for the changing hydrographical parameters in the Changjiang River estuary in China resulting from a combination of three hazards: water transfer projects, building of infrastructure (the Three-Gorges Dam), and sea-level rise. The increasing population in the catchment exacerbated the effects of each of these. By assessing the pattern between saline intrusion and river discharge levels, and calculating threshold values, they showed that fresh water will not be available each decade after 2040 during the annual low-flow periods. This emphasizes the need for early planning for management and remedial measures, possibly starting decades in advance.

In some cases, several adjacent estuaries are just components of a delta but they need to be considered as a single system. The so-called Delta of the rivers Rhine, Meuse, and Scheldt in the southwest part of the Netherlands is formed by the estuaries of these three rivers (Professor Patrick Meire, University of Antwerp, pers. comm.). In the last two millennia, human-induced changes have been greater than natural changes, especially with navigation and agriculture becoming the dominant drivers. However, the repercussions of such changes were then exposed due to the natural pressure of the storm surges of 1953 and 1976. The many lives lost following the first of these and the fears in the 1970s led to the so-called Delta plan in the Netherlands (Nienhuis and Smaal, 1994) and the Sigma plan in Flanders (Belgium). Excessive engineering of the estuaries, including closures and diversions, both increased public safety but at the same time led to unforeseen consequences such as changes to the water quality (stratification, anoxia, eutrophication) and ecological features (loss of commercial shellfish beds). This is now leading to further engineering and adaptive management and confirms the adage that once the systems begin to be engineered, then interventions have to continue, otherwise the systems revert to an unwanted state (for human uses). These estuaries are now central the European economic prosperity, through the ports of Antwerp and Rotterdam, and so their managers have to reconcile the natural functioning of the estuaries with the need for access by larger and larger vessels. Because of this, innovative engineering is aiming to work with nature-based solutions such as controlled inundation areas with reduced tides, managed retreat, adapted dredging, and dredging-disposal strategies to cope with human demands while ensuring natural functioning (Professor Patrick Meire, University of Antwerp, pers. comm.).

The trajectories of the change of estuaries and other transitional waters relate strongly to the features of the catchment, but also there are site-specific differences depending on the size of the estuary. For example, García-Alonso et al. (2019) studied the Río de la Plata, one of the largest estuaries in the world, and the second largest in South-America, with a catchment (of two main tributaries, the Paraná and the Uruguay Rivers) extending into five countries: Argentina, Bolivia, Brazil, Paraguay, and Uruguay. Its catchment has pressures from arable and pastoral agriculture and forestry as well as the infrastructure of hydroelectric dams, whereas the pressures in the de la Plata are from marine and river vessels, fisheries, and coastal urbanization, including tourism, and polluting discharges. Abstraction upstream for drinking water for Buenos Aires has changed the freshwater inputs, and hence the salinity balance, whereas downstream, the estuary supports cultural ecosystem services for recreation and tourism. These show that pressures equally affect neo-tropical areas and cross national boundaries. With increasing populations in these South American countries, it is expected that all of these pressures will increase.

In some areas, authorities have focused on increasing the resilience of the system in protecting the coastal and marine biodiversity, for example along the coastline of the Middle East and the northern Persian Gulf (Sharifinia et al., 2019). The Iranian coastal waters support internationally important habitats including coral reefs, mangrove forests, and seagrasses, with sandy beaches, rocky shores, and estuaries also being important. These, importantly, deliver ecological services and also support recreational, economic, and cultural activities and societal benefits. Despite this, the Persian Gulf is strongly affected by human activities, has suffered from the effects of conflicts and various human pressures even though the population density is not as high as in other areas. In contrast to many other areas, sea-level rise does not appear to be a high-priority concern given the abundance of unoccupied land and the ability to move infrastructure landward. Management here aims to increase the ability of the ecosystem to absorb natural and anthropogenic disturbances, but this area is continuing to experience potential risks because of geopolitical hazards (Table 1) with uncertain effects on the ecosystem. This area, especially the delta of the Tigris-Euphrates River, is also likely to be affected by growing water scarcity (Day et al., 2019a).

Urban estuaries, especially those in arid areas such as Australia, face severe problems of both water quality and quantity. Dunn et al. (2019), using examples from three cities in Australia, show the way in which the important provision of environmental, social, cultural, and economic ecosystem services in estuaries are often negatively affected by urbanization in the catchment and along the coast; they also show some practical and proven remediation measures. In these measures, stormwater management is a priority—for example, for new suburbs to divert their stormwater to ponds in public parks—whereas multi-solution plans cover water-sensitive urban design, removal of polluted sediment, restoration of vegetation along creeks and riverbanks, debris traps, and monitoring/research. They rightly put a high value on education and on actively canvasing the community for their views and their input to the solutions.

Although the understanding and management of estuaries has to acknowledge the bottom-up approach by looking after the hydrology, ecohydrology, and fundamental lower-level ecological processes (Wolanski and Elliott, 2015), it is often the charismatic megafauna that gets the attention of the public, the policy-makers, and the politicians. In a good news story, which is a rare occurrence for the world's estuaries, Lanyon (2019) illustrates the success of preserving charismatic and iconic marine megafauna (turtles, dolphins, and dugongs, as well as migratory habitats for whales and shorebirds) in Moreton Bay, Australia, despite increasing estuarine urbanization. This is one of the few sites in the world where proactive management has succeeded in ensuring the survival of this high-profile and mobile marine wildlife next to a large city. This was made possible by integrating protection legislation that incorporates water quality and critical habitats protected by Marine Protected Areas (MPA), with regulating activities such as shipping, dredging, and fishing, and enforcing the legislation. Most areas worldwide have appropriate environmental protection legislation but this is not always successfully implemented and enforced. Monitoring also has to be used, as in Moreton Bay, to determine whether the management is effective and is leading to adaptive management. This is necessary, because although MPAs are effective at protecting the static features and seabed, they may be less effective for highly mobile species. In the case of the high-profile megafauna, the health of individuals and their populations similarly receives a large amount of attention and may be the end point of stressors further down the system (Lanyon, 2019).

The Moreton Bay success story contrasts with the disastrous evolution of the tidal flats along the coast of the Yellow Sea coastline in North and South Koreas and China; here the extent of the tidal flats has decreased by 50%–80% over the past 50 years because of land claim, and the decline continues at a rate of about 1.2% year^{-1} (Murray et al., 2012, 2014, 2015). As a result, the migratory waterbird populations that use the East Asian-Australasian Flyway, from Japan to Australia, are decreasing at a rate of about 5%–9% year^{-1} (MacKinnon et al., 2012; Clemens et al., 2016). As detailed by Fang (Box 1), efforts are underway in China to at least preserve some of the remaining tidal flats.

In the case of some estuarine areas, the management of future conditions to regain a natural system will not be possible given that the physiography has been significantly altered by engineering from the original state. Hence the management is

---

BOX 1 China's Mudflats

There are few data published in China at the national scale about the country's coastal mudflats. The total area of China's coastal mudflats in the 1980s was 20,293 km^2 according to *China Statistical Yearbook* (1989). However, a more recent analysis indicated an area of 13,104 km^2 of China's coastal natural wetlands (including tide zone/shallow beach, marine marshes/mangrove, estuarine water, estuarine delta, and lagoons) in 1978, 11,463 km^2 in 1990, 9108 km^2 in 2000, and 7890 km^2 in 2008 (Niu et al., 2012). Another similar study showed that the area of coastal natural wetland was 14,674.3 km^2 in 2000 and 14,318.5 km^2 in 2010 (Hou et al., 2016). There are thus wide discrepancies in the absolute estimates, but all studies point to a rapid loss of mudflats: for example, a 39.8% loss from 1978 after three decades of rapid coastal development in China (Niu et al., 2012). A historical satellite image analysis of the Yellow Sea coastal region (~4000 km coastline of China, North Korea, and South Korea) also shows that 65.3% (70.2% on China's part) of tidal flats that existed in the 1950s had disappeared by the late 2000s (Murray et al., 2014). The rate of loss of China's mudflats (wetlands) is decreasing in recent years; it was 13.4% from 2000 to 2008 (Niu et al., 2012; Murray et al., 2014).

This loss of coastal mudflats resulting from the sea enclosing and from land reclamation, coupled with other anthropogenic threats such as land-based pollution and climate-change effects, has led to ecosystem degradation at different scales, from local water quality degradation (Ma et al., 2017) to a regional decline of shorebirds migrating through the East Asian–Australasian Flyway (MacKinnon et al., 2012; Murray et al., 2015).

Continued

BOX 1 China's Mudflats—cont'd

Coastal wetland administration in China was fragmented among several departments, which include Development and Reform, Oceanic Administration, Land Resources, Water Resources, Agriculture, Fisheries, Forestry, Tourism, Industry and Information, and Environmental Protection, etc., and each department has different priorities. However, a strong political will to protect coastal wetlands has been displayed in China in recent years. The Chinese central government has developed a national protection goal for wetlands, which lasts until 2020, including a series of polices such as the Wetland Protection and Restoration Plan in 2016, the 13th Five-Year National Wetland Protection Implementation Plan in 2017, Management Regulation on Coastline Protection and Use in 2016, and Guiding Opinions on Strengthening Coastal Wetland Management and Protection in 2016—as well as the reshuffling of the central government in 2018 to address ambitiously the problem of the fragmentation of wetland management.

With the efforts mentioned, it is believed that the coastal mudflats in China will likely enter a period of stabilization. However, a long process will be needed that depends on the continuous implementation of those good governmental policies and the support of the whole society.

Qinhua Fang[*,†]

*Coastal and Ocean Management Institute, Xiamen University, Xiamen, China
†Fujian Provincial Key Laboratory for Coastal Ecology and Environmental Studies, Xiamen University, Xiamen, China

References

Hou, X., Xu, X., Wu, T., Li, X., 2016. Change characteristics and scenario analysis of coastal wetlands in China. Wetland Sci. 14 (5), 597–606 (in Chinese).

Ma, D.Q., Zhang, L., Fang, Q., Jiang, Y., Elliott, M., 2017. The cumulative effects assessment of a coastal ecological restoration project in China: an integrated perspective. Mar. Pollu. Bull. 118, 254–260.

MacKinnon, J., Verkuil, Y.I., Murray, N., 2012. IUCN situation analysis on East and Southeast Asian intertidal habitats, with particular reference to the Yellow Sea (including the Bohai Sea). Occasional Paper of the IUCN Species Survival Commission No. 47. IUCN. Gland, Switzerland and Cambridge, UK. 70 pp.

Murray, N.J., Clemens, R.S., Phinn, S.R., Possingham, H.P., Fuller, R.A., 2014. Tracking the rapid loss of tidal wetlands in the Yellow Sea. Front. Ecol. Environ. 12 (5), 267–272.

Murray, N.J., Ma, Z., Fuller, R.A., 2015. Tidal flats of the Yellow Sea: a review of ecosystem status and anthropogenic threats. Austral Ecol. 40, 472–481.

Niu, Z.G., Zhang, H.Y., Wang, X.W., et al., 2012. Mapping wetland changes in China between 1978 and 2008. Chin. Sci. Bull. 57, 2813–2823.

of (and for) a non-natural system. Such an example is the Peel-Harvey Estuary in Western Australia (Valesini et al., 2019). Despite the natural features and ecological importance of the estuary, which is part of the Ramsar-listed Peel-Yalgorup wetland system, and a large emphasis to maintain the cultural heritage of the indigenous peoples, it is subject to the growing pressures from a rapidly expanding urban area (the City of Mandurah) and the associated pressures. Severe water-quality problems resulting from nutrient inputs created eutrophic conditions in the 1970s and 1980s. The adopted remedy involved changing the hydrodynamics through engineering a second mouth into the system to increase tidal flushing of this microtidal system, and it was hoped that this would improve the ecology (Elliott et al., 2016). However, improvement in land use did not occur with the engineering works, so urban and agricultural changes accelerated; as a result, the system improved for a while before degrading again. In the future, as the area is subject to a rapidly drying climate, the hazards and risks of a lack of water will become increasingly apparent. Ecological health improvements will have to take priority over economic growth and exploitation, otherwise the quality of urban life will degrade because of a degraded environment. It is of note that a large, nationally funded research project is looking for holistic solutions to the management of the Peel-Harvey and its catchment.

The ecology of macrotidal temperate estuaries has been well studied, especially their highly dynamic nature that is greatly affected by temporal and spatial anthropogenic pressures (Elliott and Whitfield, 2011). However, there is still a need to determine how these areas will be influenced by current hazards and risks such as increasing climate variability (Ducrotoy et al., 2019). If the estuarine physical and biological parameters are close to vital thresholds related to the tolerances of organisms, they will be susceptible to changing temperatures, climate variability, or accelerated sea-level rise (Solan and Whiteley, 2016). Hence, research is required to determine the tolerance responses by organisms to changing physical-chemical conditions and the way in which such changes are superimposed over the likely tidal regime, wave hydrodynamics, and morphosedimentary evolutions (Solan and Whiteley, 2016; Ducrotoy et al., 2019). There is the continuing need for information about the anthropogenic impacts on estuarine communities and habitats resulting from the complex forces on the environment that climate change causes, including repercussions for biogeochemical cycles, ocean acidification, and dissolved oxygen and nutrients levels. Such repercussions will then need to be incorporated into conservation and resource management in the transitional and coastal areas.

The case studies in this volume have shown that lessons, such as those from engineered systems and a legacy of stressors (Valesini et al., 2019), can be learned. However, this also shows that after we start engineering systems, it has to continue, as otherwise the system may revert to something incompatible with the wishes of society. Whether we have the ability to engineer greater resilience into systems to cope with future pressures remains to be seen. Management has to cope with these moving baselines caused by climate change and so it requires adaptability—fortunately we now have many examples of management responses to past degradation (see below).

3.2 Deltas

Globally, deltas occur across the full range of coastal environments, ranging from tropical to Arctic and from the areas with high precipitation to very arid areas, and they are largest in areas with high river discharge and broad, shallow continental shelves. Deltas encompass some of the most productive and economically important ecosystems, and their wetlands are important for the connectivity with coastal fisheries (Day et al., 2019a). Their contribution to ecosystem services and societal goods and benefits is well recognized, but they are under threat from many pressures, hazards, and risks. Changes to their catchments, affecting sediment and water supply, increased urbanization and industrialization, changing hydrogeomorphology, and large-scale engineering projects to protect assets and human lives are all affecting previously stable cycles in hydrodynamics and sediment balance. Deltas are particularly susceptible to climate change. All of these aspects affect the long-term geomorphological, ecological, and economic sustainability of deltas.

Given previous data, Day et al. (2019a) consider that there will be "winners and losers" among deltas as a result of local, regional, and global change. They present a continuum of delta types based on their susceptibility to climate change and other human pressures. Those in arid areas are most susceptible and may disappear or at least function differently from the present through reduced water input. Those in highly urbanized areas, particularly in temperate regions, but increasingly in tropical and subtropical areas, will increasingly be subject to engineering modifications such as flood and erosion prevention to protect human assets and lives. In contrast, tropical deltas in sparsely populated areas and with high freshwater discharges may, for the time being, not be at as great a risk as their counterparts elsewhere. Despite this, these deltas may lose wetlands through increased flooding.

High latitude (polar) coasts, deltas, and estuaries may be more fragile ecosystems than their temperate counterparts, and they have not been afforded the same attention and degree of study despite having many similar characteristics (Forbes, 2019). However, they have additional polar characteristics of ice cover and ice movement, often low precipitation, very large seasonal changes, and sparse vegetation, especially tree cover. As shown in Canadian systems, the MacKenzie Delta may or may not be increasing, depending on sediment supply, which may be low because of previous glacial effects or may be pulsed because of spates of meltwater. The impacts of a sea-level rise become more subtle as the areas are affected by global sea-level rise but at the same time may experience isostatic rebound because of the loss of ice since the last ice age (Snoeijs-Leijonmalm and Andren, 2017); as such, relative sea levels are falling in these areas as new land emerges (Ducrotoy and Elliott, 2008).

Climate change is disproportionately likely to affect high latitude deltas and estuaries through retreating ice cover, exposed areas, and increased coastal erosion (Forbes, 2019). The important ecosystem services currently provided by the ice cover, such as climate modification, ice habitat for polar bears, sunlight reflection, etc. (Moline et al., 2008), will be reduced. At the same time, navigation routes may be opened, producing economic benefits but also increasing the pressures through oil spills, habitat change, and increased vectors for non-indigenous species. Monitoring these high-latitude deltas and estuaries may therefore give the most important indications of the effects of climate change.

Given its size and economic and ecological importance, perhaps one of the most studied deltas worldwide is the Mississippi (Day et al., 2019b). In being modified by natural and human pressures since human colonization, the delta has changed in shape, dynamics, and sustainability, and it is an excellent example of the adage that after we start managing systems we then have to keep doing it; otherwise they revert to an undesirable state for humans. Although the dynamic hydrographic and sediment cycles are increasingly understood, their increasing unpredictability due to climate change, will affect the fundamental interactions between the catchment, the deltaic plain, and the open sea.

Deltas function because of the intricate balance between the natural and anthropogenic changes to their hydrology, sediment supply, relative sea-level rise, and land use. Anthropogenic changes have distorted the natural balance and will increasingly do so in the coming decades unless radical management plans are implemented. Day et al. (2019b) emphasize the importance for the Mississippi Delta of state funding of $50B in a 50-year Coastal Master Plan (CMP) and adaptive management scheme to achieve a "win-win situation" for human safety through reduced flood risk and the protection of ecologically healthy and self-sustaining wetlands. Although hard engineering in some areas, such as levees and floodwalls, is inescapable, working with nature through soft engineering (diverting the river into the deltaic plain to nourish wetlands and maintain the barrier islands) is needed to increase resilience to hurricanes in populated areas.

The Mississippi Coastal Master Plan shows the importance of policy underpinned by good and adequate science, modeling, and understanding (Day et al., 2019b). The sequence of interventions by engineering and ecoengineering is critical but even so the local populations have to be prepared for unexpected events including evacuations. They will even have to accept that some areas cannot be defended and may become uninhabitable. Managed retreat will have to be accepted but with the overall aim to protect the natural and human system over large parts of the delta. This is becoming the policy elsewhere where, for example, the managed retreat of populations from eroding coasts has been accepted as a legitimate management measure in some areas such as the low-lying, soft, North Sea coasts (Elliott et al., 2016).

Socioecological challenges in wealthy countries for deltas such as the Mississippi become magnified in poorer and developing countries (Ramesh et al., 2019), and the human impacts become greater even if the overall economic consequences are not as high. An example of this is the extensive Ganges-Brahmaputra-Meghna (GBM) Delta covering large areas of Bangladesh and West Bengal and supporting the Sundarbans wetlands. The huge population in both the delta and the catchment not only creates many pressures on the system but also shows the consequences of natural and anthropogenic hazards and risks in the area. It exemplifies a classic dilemma: the area is highly fertile and productive because of inundations by riverine flooding, but this means the populations want to exploit those ecosystem services and gather the societal goods and benefits and are consequently at risk. Water abstraction, monsoonal rains, cyclones, and rising sea levels all compound problems with land subsidence and in turn all act to exacerbate the seasonal variations in flooding and sediment patterns. These features and the effects of coastal erosion are likely to become more accentuated with global climate change.

In the GBM Delta, the anthropogenic pressures of agriculture and fisheries, together with increased urbanization, industrialization, and the associated pollution loading, can exacerbate the impact of natural hazards (Ramesh et al., 2019). For example, clearance of natural vegetation such as mangroves would remove protection against natural hazards such as cyclonic storm surges. This delta, like all deltas, requires a strategy for sustainability and integrated management, including land planning and maintaining environmental freshwater flows. However, again, as with many areas in developing countries and with international (transboundary) areas, a lack of data and weak sociopolitical will, as well as geopolitical uncertainties, create impediments to that integrated management.

The Indus Delta, its catchment, river, and coast show many of the same challenges of the GBM Delta (Kidwai et al., 2019). Its complex geopolitical position (crossing Pakistan, India, Afghanistan, and China) and its development have resulted in morphological changes, both natural and anthropogenic. Notably, it has only one functional channel (the Khobar Creek), as freshwater and sediment flows have been reduced by the construction of large dams and other man-made structures. Abstraction for agriculture and the arid conditions create water scarcity and so also increase saline penetration into the deltaic and riverine systems. These pressures, compounded by unstable politics in the area, increase the difficulty of reaching agreements and determining a catchment management plan. In these areas, not only are future environmental changes affected by global warming and infrastructure (such as dams) required by an increasing population, but there is also the large potential for conflicts triggered by scarce water resources (Kidwai et al., 2019).

The downstream effects of catchment-level engineering are seen particularly in large deltaic systems that are responding to hydrographical changes as in the Mekong River Delta (Nhan and Cao, 2019); this delta is now rapidly shrinking and becoming increasingly environmentally degraded. The Mekong catchment has suffered from poorly planned water management and hydropower schemes in China as well as sediment starvation and increased nutrient inflows, all of which are expected to increase in the coming decades. Water is retained behind the dams in China, thus reducing sediment loads, decreasing flooding events, disturbing fish migration and seasonal cropping cycles, and increasing saline intrusion in the Vietnamese Mekong Delta. In turn, the future additional and increasing pressures from mineral mining of the bed, bed subsidence, and removal of protective mangroves will all reduce the resistance and resilience to climate change. Furthermore, the increased erosion of the river banks, deltas, and the coasts not only reduce the ability of the area to cope with climate change but also make the consequences much greater, especially for the populations in the Delta. For the Mekong, all of these adverse aspects are predicted to increase, so an international integrated management approach in the riparian countries (principally China, Cambodia, Laos, and Vietnam) will be needed to solve these issues. Notably, there is no organization to manage this, as China is not a partner of the Mekong River Commission and it is argued that the Commission has limited powers for such management. These problems, coupled with poor construction methods, were sadly illustrated with catastrophic consequences by the failure of the Xepian Xe Nam Noy Dam in Laos on a tributary of the Mekong in July 2018.

The Nile Delta has also experienced serious environmental degradation, including severe water pollution, since the completion of the Aswan High Dam in 1964. This enabled control of the water flows and year-long intensive irrigation in the delta while its environmental health was ignored, particularly sediment starvation (sediment is needed to replenish the system), and the floods needed to flush the system were suppressed (Chen, 2019). Each person receives on average only 660 m^3 of water per year, out of which about 86% is used for irrigation. This water allocation is among the world's lowest and will lead to water shortages by around the late 2020s because Egypt's population is rapidly growing (projected to double by the late 2060s) and because the Great Ethiopian Renaissance Dam (GERD) on the Blue Nile, once built, will further decrease the water supply (Day et al., 2019a). Climate change effects that increase climate variability and water scarcity will compound existing challenges, resulting in large repercussions for the countries in the Nile River catchment and the Egyptian Nile Delta. An international integrated management approach is needed to solve these issues and prevent conflicts; there is no organization to manage this, as the Nile Basin Initiative is a partnership between countries, which, as with the Mekong River Commission, have little power and ability to manage the catchment successfully.

In contrast to the large deltas in developing and arid countries mentioned previously, the Ebro, Rhône, and Po are temperate deltas that are set in a developed, European context. They have all been highly modified by human actions but have been the sites of well-developed management schemes with restoration of some of the natural features (Day et al., 2019c). The Ebro Delta in NE Spain has lost freshwater and sediment inputs, partly due to dam building, and more than half of its wetlands have been converted to rice cultivation. Before the construction of those dams, the sedimentation and accretion in the delta were maintained often by sediment deposition during river floods and were delivered by the rice-field irrigation network. Given the economic value of the dams and the acceptance that they have created a new equilibrium that will not be reversed, the future maintenance of the delta requires coordinated action to allow freshwater and sediment inputs irrespective of the operation of the dams. However, as a note of caution, future political instability in the region, through the dispute between Spain and Catalonia, could compromise the integrated management of the catchment and delta and thus jeopardize the ecosystem services and ecological well-being provided by the system.

In contrast to the Ebro, the Rhône River in France has retained most of its discharge and suspended sediment concentration, but the river has been severely constrained by dikes (Day et al., 2019c). Although very large flooding events still occur, sea-level rise will continue to have an effect unless there is sediment nourishment, either naturally or by human actions, to keep pace with rising sea levels. The delta shows that managed river diversions can be used to balance the sediment, but with future climatic and hydrographic changes, these actions increasingly will be required. Such features are also shown by the Po Delta in Italy, which relies on a set of five functioning distributaries to deliver water to the edges of the delta. As large areas of the Po Delta are up to 5 m below sea level, it is unlikely that these can be maintained in the next century unless they receive sediment nourishment or are converted to shallow lagoons.

The southern Gulf of Mexico is characterized by coastal ecosystems with high freshwater input, extensive wetlands and coastal lagoons, productive fisheries, and human settlements whose economy is largely based on the rich natural resources of the area (Herrera-Silveira et al., 2019). The Grijalva-Usumacinta River and Delta region has high riverine input and extensive wetlands, and the Laguna de Terminos, the largest lagoon-mangrove system in Mesoamerica, has high habitat diversity. Primary producers have their peak production at different times of the year, leading to overall sustained high productivity throughout the year. There is a high-diversity migratory nekton community that uses the lagoon habitats at times when they are most productive, ensuring overall high secondary production that supports a multistock fishery. All of these features have to be incorporated into the management scheme, and the area has been designated as a natural protected area to ensure sustainable management. The coastal zone of the Yucatan Peninsula is fed by the extensive groundwater system that supports a unique system of karstic freshwater lakes, brackish lagoons, estuarine coastal lagoons, and reef lagoons with extensive mangrove swamps and submerged aquatic vegetation. The main source of inorganic nutrients is groundwater and, in some parts of the coast, elevated levels can cause eutrophication symptoms including harmful algal blooms. Rapid human development has led to extensive habitat degradation and water-quality problems, so ecosystem-based management is required to restore the area and to improve environmental quality. Some global climate models suggest that this already-arid region will experience strong decreases in precipitation and this may become the most serious problem for this region. As a way ahead, Mexico has established 17 natural protected areas in the southern Gulf of Mexico and Yucatan to enhance the sustainable management in the region.

3.3 Wetlands, Lagoons, and Catchments

As with deltaic and estuarine areas, it is important to determine for coastal lagoons the environmental characteristics, their variability and complexity, their role in connectivity between the catchment and the sea, and their delivery of ecosystem services and societal goods and benefits. For example, for the Venice Lagoon (Day et al., 2019c), sediment input into marshes is necessary if they are to survive sea-level rise. Subsidence of the area through aquifer water abstraction and sea-level rise has dictated the need for a hard engineering solution such as construction of the MOSE flood protection scheme that consists of rows of mobile gates installed at the inlets to the Lagoon to isolate the Lagoon from the Adriatic Sea during *Acqua alta* high tides. However, operation of these gates will partly decrease resuspended sediment input to marshes, which may have to be offset by nourishing marshes with dredged sediment and diverted river water.

Together with their other characteristics in common with estuaries, the lagoon ecological features create unique ecosystems that support ecosystem services and deliver societal goods and benefits. Although differing in size, lagoons influence the genetic diversity of adjacent marine populations and contribute to the genetic diversity of those adjacent populations (Pérez-Ruzafa et al., 2019). Despite this, lagoons are also subject to increasing anthropogenic pressures that will further increase in the coming decades with increasing urbanization, population increase (especially in tourist areas), and climate change. As with all areas, the successful and sustainable management of lagoons depends on knowledge of both the structure and the functioning of the systems.

As the *raison d'être* for management, it is important to determine the value of the transitional and coastal habitats for their ecosystem services and societal goods and benefits (see Turner and Schaafsma, 2015, for further definitions and explanation). The well-studied freshwater and coastal ecosystems of the Florida Everglades (Sklar et al., 2019) serve as a good example of these features, and their high importance for recreation and tourism relies heavily on their good water and environmental quality. This leads to other benefits such as their importance for fisheries, their importance for flood and erosion mitigation, and their ability to sequester carbon and other nutrients, thereby playing a role in ameliorating global change. With global change expected throughout the coming decades, the Everglades will show responses common to many of the habitats described here, not the least of which is changes to ecosystem service provision and societal goods and benefits delivery.

The ecological functioning of all the transitional systems discussed here relies on a successful physical-chemical functioning, of which salinity is often a dominant forcing factor. At the extreme, hypersalinity is the usually the result of restricted physical connectivity with the open sea, hence producing closed or temporarily closed systems that receive insufficient fresh water to maintain the marine connection (Whitfield and Elliott, 2011; Tweedley et al., 2019). Such a restriction in turn increases the effects of evaporation and hence becomes the driving force behind the structure and functioning of the biota, given that few species can tolerate salinities >100 (see also Solan and Whiteley, 2016). The salinities and hydrography in turn will be greatly affected by land use changes but will also drive changes in land use, such as the change to types of agriculture present in an area. These changes may have the greatest repercussions in hypersaline environments (i.e., those with salinities >40) (Tweedley et al., 2019). Hypersaline estuaries, lagoons, and marine embayments, which can vary in size from <0.1 to >10,000 km^2, occur worldwide but principally in tropical, arid, and warm temperate climates with low and/or highly seasonal rainfall. It is of note that those environments are likely to be most affected by global temperature and rainfall changes and thus will require a particular set of management measures and adaptation.

All of the habitats considered here will be altered with global changes (discussed below) depending not only on the prevailing human pressures but also on their geological history. For example, the Florida Everglades have peat areas at which the water depths, and hence the degree of inundation, will change according to sea level, rainfall, compaction, and subsidence (Sklar et al., 2019). These changes will affect not only the communities but also the ability of the peat to store and release carbon and other nutrients. With gradual change, and as long as the habitats are not constrained at the landward margin (thus avoiding "coastal squeeze"), the Everglades coastal habitats will gradually migrate inland although the saline water could migrate into the freshwater sawgrass marshes. In turn, as shown in the Everglades, an increased salinization and inundation change to the vegetation could lead to the collapse of the peat (a reduction by 0.5 m in elevation over several decades) or even expose the peat, thereby leading to a release of the stored carbon.

The movement inland of coastal habitats, such as mangroves in tropical and subtropical areas and saltmarshes in temperate areas, is a natural mechanism for adapting to increasing sea levels (Sklar et al., 2019). However, this first depends on an unimpeded ability to migrate in contrast to most developed areas in which there are constraints to landward movement (such as high shore barriers, levees, and dikes), so "coastal squeeze" occurs. Secondly, that habitat migration relies on the ability of the vegetation to migrate at the appropriate rate—for example, in the Florida Everglades, evidence suggests that mangroves could not develop in line with the increased inundation, so an open-water mangrove-free area was created.

Many of the habitats considered here have undergone many decades of the combined pressures from increasing urbanization and industrialization, polluting inputs (especially nutrients and organic matter), change to land use following intensification of agriculture and uniform culture, and, increasingly, climate change. One of the best studied of such cases is the Chesapeake Bay in the eastern United States (D'Elia et al., 2019), which shows impressive improvements such as reducing anoxic bottom waters and restoring submersed aquatic vegetation (SAV), in itself also a proxy of improved conditions, especially for water quality. Although the Chesapeake Bay is a high-profile area, being close to the site of the federal US government as well as to many adjacent states, it shows the value of coordinated action supported by good science linked to the policy.

Perhaps the most important conundrum in tackling the effects of global climate change on coastal ecosystems is that the greatest impacts are likely in those regions of the globe that are most ill-prepared and can least afford to carry out remedial measures. This is exemplified by African catchments and coastal areas where the hugely increasing population, demand for water and land, and increasing industrialization, urbanization, agriculture, and aquaculture are creating the problems (Niang et al., 2019). Although climate change-related impacts, such as reduced rainfall, more frequent droughts, and more erratic weather patterns, will exacerbate the problems, authorities may still continue to overexploit resources and increase economic growth irrespective of the effects on the environment and its resilience to adapt to climate change.

Infrastructure and engineering in the catchment eventually affects the functioning of the downstream transitional waters and coastal ecosystems. Again, this is especially so in developing countries where there is a high demand for catchment water for irrigation and hydroelectric schemes. For example, the Senegal River in Senegal and the Pagane River in Tanzania have suffered from large-scale human interventions, such as dams and irrigation schemes, as well as climate change reductions in freshwater availability (Niang et al., 2019). Despite this, during the planning process, the downstream effects of such catchment modifications are usually not considered by the developers nor by the governments. Both the local people and the estuarine environments then have to adapt to the new conditions. All of these examples argue for the spatial and temporal footprints of these engineering works to be defined, especially the far-field effects as well as the near-field effects, and also argue for the integrated management to take the view that long-term good environmental management will also have economic benefits.

3.4 Enclosed, Semienclosed, and Open Coasts

Although transitional waters often show the greatest and most immediate effects of environmental change, enclosed and semienclosed seas often do not have the hydrographic properties, especially the flushing characteristic, to ameliorate the effects of catchment pressures. For example, the Baltic Sea has changed markedly in the past century as a result of human pressures, particularly eutrophication caused by elevated nutrient inputs and a lowered assimilative capacity because of its enclosed nature (Heiskanen et al., 2019). The resulting impaired ecosystem status and negative impacts on human welfare have led to governance responses by coordinated actions via the HELCOM regional seas program (Scharin et al., 2016). The Baltic Sea has been subjected to, and has responded to, geopolitical changes as well as changes in the economic fortunes of the countries in the catchment, not the least being the removal of the "Iron Curtain" and membership of the European Union in many states; the latter led to the adoption of new environmental management regimes (Heiskanen et al., 2019). Despite the large changes throughout the 20th century in the Baltic Sea, it is expected that greater responses will be needed to cope with climate change, including more adaptive governance and the adoption of the circular economy to lead to sustainable solutions.

As with the Baltic Sea, the Black Sea has suffered environmental degradation during the 20th century from eutrophication, the proliferation of nonindigenous species, overfishing, and marine litter, and the degradation has become more severe since the 1960s (Güneroğlu et al., 2019). International action, notably through the Black Sea (Bucharest) Commission, has begun to alleviate the problems, and the area is increasingly becoming "Europeanized" with the gradual adoption of the European Union Environmental Directives. For example, the ability to meet Good Environmental Status (GES) for the Marine Strategy Framework Directive (MSFD) and Good Ecological and Chemical Status for the Water Framework Directive (WFD) requires concerted action. The countries in the Black Sea catchment are now recognizing the need for action on the causes and consequences of increasing sea levels and temperatures, for actions with 30- to 50-year projections, and for associated remedial measures (Güneroğlu et al., 2019). In the future, it is expected that the increasing implementation of environmental legislation in line with the European Union will bring large benefits, especially for coordinate catchment and marine management.

Climate change leading to a greater unpredictability of climate, often resulting in reduced rainfall and lower freshwater inputs to coastal areas, will pose major challenges for the saline balance, the residence time, and the ecological structure and functioning of these areas and their associated transitional environments. This is particularly likely in arid areas such as the southern Gulf of Mexico and Yucatan Peninsula, including the Grijalva-Usumacinta River and delta (Herrera-Silveira et al., 2019). This area has large expanses of wetlands and coastal lagoons formed from a large riverine input, which support productive fisheries; they also provide the natural resources that support urban and rural areas dependent on the ecosystem services and societal goods and benefits.

The resistance and resilience of coasts, that is, the respective ability to withstand and recover from natural and anthropogenic risks and hazards, depend on maintaining the physical features. For example, the Yucatan Peninsula's coastline relies on a large groundwater system that supports a unique system of karstic freshwater lakes, and of brackish, estuarine coastal and reef lagoons with extensive mangrove swamps and submersed aquatic vegetation (SAV) (Herrera-Silveira et al., 2019). These vegetated habitats have been adversely affected by the elevated nutrients in groundwater, which reportedly caused eutrophication signs and symptoms including harmful algal blooms. Increased developmental pressures and deteriorating water quality are likely to exacerbate environmental degradation in the coming years. However, there are encouraging management initiatives; for example, the Laguna de Terminos, the largest lagoon-mangrove system in Meso-America, has long been shown to have a high biodiversity that sustains a high annual productivity based on a multistock fishery (Herrera-Silveira et al., 2019). Protection of this crucial ecosystem to ensure its functioning and its continued ability to provide ecosystem services and societal goods and benefits is being achieved by sustainable management reliant on its designation as a natural protected area.

3.5 Coral Reefs

The survival of coral reefs is intimately linked with changes to the coast, the adjoining transitional environments, and the river catchments; poor catchment practices, such as poor land use generating erosion, changes to freshwater run-off, polluting discharges, and overfishing, affect coral reefs at distances away from the coast depending on the hydrographic patterns (Birkeland, 2015). Hence the management of coral reefs requires catchment-scale measures and needs to integrate the physical-chemical, biological cultural, and socioeconomic aspects. In addition, global warming and ocean acidification leading to coral bleaching are of major concern for the future (Hughes et al., 2017; Richmond et al., 2019).

Although many case studies indicate the problems in these areas, it is also beneficial to provide good news stories; for example, Richmond et al. (2019) highlight the success story of coral reef management in some Pacific Island states. In those islands, the community, traditional leaders, and national politicians have learned from their history of 2000 years of continuous settlement in a small island that the catchment, the estuaries, and the coral reef form one ecosystem that must be managed as such (Koshiba et al., 2014). Thus there are strict regulations readily accepted by the population about managing and using the land and the reefs and, where degradation has occurred, implementing practical restoration measures such as creating sediment traps, removing invasive algae, catchment modifications of agricultural run-off, and erosion control. Such an effective and sustainable management relies on stakeholder engagement and communication with all parties, and both natural and social sciences are required to achieve an adequate understanding for management. The financial, human, and institutional resources are not yet adequate to address fully the challenges facing coral reefs in small Pacific Island states, but the results so far are encouraging.

Perhaps the site that most epitomizes the concerns regarding past, present, and future threats to coral reefs is Australia's Great Barrier Reef (GBR). The GBR ecological degradation includes the loss of corals and charismatic animals such as the dugong, due to poor water quality associated with land runoff and climate change (Wolanski et al., 2017; Hughes et al., 2018; Brodie et al., 2019). Inputs of sediment, pesticides, and nutrients from farmland (especially sugarcane and cattle grazing) in the catchment have long adversely affected the GBR. Even though management controls are now purported to exist, the improvements have been slow and small (Brodie et al., 2019). Although management targets have been set, meeting these is especially difficult and will require a greater implementation of existing governance mechanisms as well as much higher funding. The timelines for targets are not sufficiently ambitious, and there is also the need to consider changes at decadal and century scales (Brodie et al., 2019).

The environment in India is quite degraded, as shown in Ramesh et al. (2019), but there are still a few sites with healthy environments such as the Lakshadweep group of islands that are relatively little impacted by human activities. The management of India's coral islands creates specific challenges and, as with most areas, requires a holistic and multisectoral approach that involves all uses and users (Purvaja et al., 2019). Such islands have minimal land available for development, and sea-level rise becomes a major driving factor because of the low-lying nature of the islands. As with most areas, the main aim is to ensure economic well-being with ecological health and the conservation of scarce resources. Such islands have a high fisheries resource potential and, perhaps most important, the development of the tourism industry. However, it is paradoxical that the features that make the islands more desirable for tourists (the beaches) are most at risk from an influx of tourism, which brings the required infrastructure. The fragile coral ecosystem is therefore also at risk from the joint pressures of tourism and climate change (Purvaja et al., 2019). Hence an integrated management plan has been developed to control land-based pollution, the supply of drinking water, fisheries, the number of tourists, and biodiversity protection so as to preserve the reef system for the long term. In the longer term, climate change impacts will also need to be addressed.

TABLE 2 Examples of Moving Baselines in Coastal and Transitional Water Bodies

Water and hydrographic parameters varying on tidal, diurnal, lunar, seasonal, and decadal scales
Erosion-deposition cycles varying on tidal, weekly, lunar, equinoctial, seasonal, annual cycles
Physiography and shape varying through natural and anthropogenic activity on decadal, century, millennial, and geological timescales
Relative sea level varying on decadal to geological scales
Population size and community structure depending on seasonal, annual, and decadal cycles

4   QUANTIFYING CHANGES: THE NEED TO ACCOMMODATE MOVING BASELINES

As shown here, all areas are changing and will change as a result of natural and anthropogenic factors including climate change (discussed later). Assessments must detect that change and, in the case of the anthropogenic aspects, determine the causes, consequences, and management actions. This involves measuring the change of interest (that is, the "signal") against the background inherent variability (the "noise"), hence defining the signal-to-noise ratio. That change also needs to be determined against baselines, which, in coastal and transitional areas, are moving for many reasons (Table 2).

5   CHANGES TO STRESSORS: THE INPUT OF PHYSICAL, CHEMICAL, AND BIOLOGICAL POLLUTANTS AND THE EXTRACTION OF BIOLOGICAL RESOURCES

5.1   Dredging

Several of the case studies mentioned earlier describe water pollution of estuarine and coastal waters, mainly by sediment from dredging navigation channels and also by chemicals, including excess nutrients that cause eutrophication, and metals. For example, Kirby (2019) takes the view that anthropogenic modifications of estuarine hydrodynamics, especially of the bed conditions, is "exploitation" rather than "management." Estuaries such as the Scheldt, Weser, Ems, and Loire have been "managed" by deepening to ensure the passage of increasing deeper-draughted vessels and the resulting reduction in bed-friction in deeper areas has increased the tidal range and flood velocity. This creates greater upstream pumping of sediment and results in non-natural hyperturbid areas that essentially bury the ecology in areas of deposition (De Jonge et al., 2014). Despite this, the estuarine biota is naturally adapted to withstand such sediment changes (Elliott and Quintino, 2018).

5.2   Legacy Pollution

Chemical pollution has been tackled, with moderate success, in the developed world at least, with treatment controls of point-source discharges and the reduction of wastewater-contained contaminants (McLusky and Elliott, 2004); hence this is less likely to be of major concern in the coming decades. Despite this, there remains "legacy pollution" of the sediment, which is extremely difficult to remedy and where the accepted management policy is to leave the material in situ so that the contaminants are not biologically available. The common method is to store and so sequester the polluted sediment in enclosures for long-term storage, for example the Slufter system in the Rhine delta and Port of Rotterdam. However, as an alternative, Kirby (2019) suggests in situ bioremediation as a new method to solve the problem instead of storing the polluted sediment. Legacy metal pollution is a serious, widespread problem worldwide; the example of the Tamar Estuary, Australia, is shown in Box 2 (Sheehan et al.).

Despite these examples, physically diffuse pollutants such as plastics have increased and gained recognition in the public profile, both as microplastics (particles < 5 mm) and ocean litter (Critchell et al., 2019). The now worldwide and common presence of marine microplastics has adverse impacts on many marine species, and especially the high-profile megafauna such as marine birds, turtles, cetaceans, and fish. As with all types of contamination (the anthropogenic presence of low levels of undesirable materials in the environment) and pollution (the ability of contaminants to cause biological harm), there is a pre-eminent need to determine the sources, sinks, mechanisms of transport, uptake routes, consequences, and management measures for plastics. Critchell et al. (2019) discuss these aspects for microplastics

BOX 2 Industrial Pollution Legacy of Muddy Estuaries: A Case Study in Australia

Worldwide, estuaries have undergone change as a result of catchment land-use changes, bringing an alteration in hydrology and sediment yield, and industrial development with contaminant releases. As a combination of both impacts, sediments have been deposited in estuaries over the last century as mudbanks and wetlands, which may harbor harmful contaminants that were previously deposited. Timeframes vary between continents, according to the developmental history of agricultural intensification and industrialization, and the current generation that is considering sustainable futures is struggling to understand and manage this legacy.

Estuarine sediments worldwide have recorded historical heavy metal contaminants, such the Baltic Sea coast (Vaalgamaa and Conley, 2008), with an increase from 1950 toward the present, and most muddy estuaries in Northern Europe (Kirby, this book; Day et al., this book). Also in the United States, the Galveston Bay estuary system (Al Mukaimi et al., 2018) includes core profiles of trace metals (Ni, Zn) showing substantial inputs starting in 1905, peak concentrations between 1960 and the 1970s, and levels mostly decreasing thereafter. Far more recent records emerge from China, for example a moderate enrichment of Cd of ≤3.5× background values is evident since the early 1990s (Xia et al., 2011).

In Australia, the Tamar estuary in northern Tasmania extends 70 km from the major South and North Esk Rivers confluence to enter the Bass Strait (Fig. 1). At the head of the estuary is Launceston, one of the oldest cities in Australia, founded in 1804. The Tamar estuary has since been affected by catchment land-use changes, mining, and industrial development. Contaminant releases occurred from Launceston and the North Esk River (with a history of mining of metals), and the records show elevated levels of a range of heavy metals above background levels (Ellison and Sheehan, 2014), and recent water-quality annual monitoring reports show continuing problems (TEER, 2016). We obtained core samples from four estuarine sites, from close to Launceston and also further north in the estuary, and we used *aqua regia* digestion to assess for the presence and biological availability of major trace metals such as Cd, Cu, Fe, Ni, Pb, and Zn.

High levels of contaminant concentrations were found in sediments of the upper estuary extending to lower estuary locations close to where the dredged historical sediments were released in the period 1950 to 1970. Cd, Cu, Pb, and Zn concentrations were higher at greater depths relative to the surface at all four sites, exemplified by Cd results shown in Fig. B.1, cores with a post-1950s

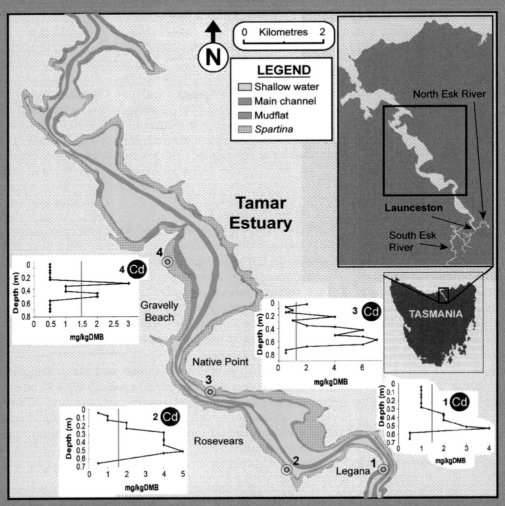

FIG. B.1. The Tamar estuary, showing locations of the study sites. Cadmium profiles in sediment show the ISQG-Low trigger value for sediment (ANZECC and ARMCANZ, 2000) as mentioned earlier, and which biological effects may be expected to occur.

BOX 2 **Industrial Pollution Legacy of Muddy Estuaries: A Case Study in Australia—cont'd**

chronology provided by records of invasive species introductions and macrofossil analysis. Contaminant levels occur at higher levels than ISQG-Low guidelines for several heavy metals at depth, which is of concern if the sediments were to become eroded. Estuarine sediment erosion has occurred in the Tamar in recent decades (Sheehan and Ellison, 2014), and further erosion is likely with global sea-level rise. This indicates, with widespread records of relic contaminants in global estuaries, an issue of worldwide concern for future sustainable management.

Matthew R. Sheehan*, Andrew J. Seen[†], Joanna C. Ellison*
*School of Technology, Environments and Design, University of Tasmania, Launceston, TAS, Australia
[†]School of Physical Sciences, University of Tasmania, Launceston, TAS, Australia

Acknowledgment
The Australian Research Council Linkage Grant LP0214145 funded this study.

References

Al Mukaimi, M.E., Kaiser, K., Williams, J.R., Dellapenna, T.M., Louchouarn, P., Santschi, P.H., 2018. Centennial record of anthropogenic impacts in Galveston Bay: evidence from trace metals (Hg, Pb, Ni, Zn) and lignin oxidation products. Environ. Pollut. 237, 887–899.

ANZECC/ARMCANZ, 2000. Australian and New Zealand Guidelines for Fresh and Marine Water Quality. Australian and New Zealand Environment and Conservation Council, Agriculture and Resource Management Council of Australia and New Zealand, Canberra, ACT.

Ellison, J.C., Sheehan, M.R., 2014. Past, present and futures of the Tamar estuary, Tasmania. In: Wolanski, E. (Ed.), Estuaries of Australia in 2050 and Beyond. Springer, The Netherlands, pp. 69–89.

Sheehan, M.R., Ellison, J.C., 2014. Intertidal morphology change following *Spartina anglica* introduction, Tamar Estuary, Tasmania. Estuar. Coast. Shelf Sci. 149, 24–37.

TEER (Tamar Estuary and Esk Rivers Program), 2016. Tamar Estuary 2016 Report Card. Monitoring Period December 2014–November 2015. NRM North, Launceston. Available from: https://www.nrmnorth.org.au/reports-and-manuals. (Accessed May 14, 2018).

and ocean litter, arguably of major concern and one of the most difficult types of diffuse pollution to control. Global changes in industrialization and urbanization, especially of transitional and coastal areas, and socioeconomic changes in the coming decades are expected to increase the amount of ocean litter and microplastics. Hence there is the need to investigate current global initiatives to see how and where actions need be taken to reduce the plastic load and use, including the circular economy approach, and to show that successful and sustainable solutions require behavioral, social, and economic changes.

5.3 Invasive Species

The presence of invasive species, which may have an impact on indigenous communities, has been termed *biological pollution* and climate change will result in conditions becoming more favorable for some species and less favorable for others (Elliott et al., 2015). This will change species distributions and allow some to migrate into and out of areas depending on the means of transport (the vector) and the prevailing conditions in relation to the organism tolerances (Olenin et al., 2011). In turn, greater urbanization and industrialization and the opening of various polar sea routes for navigation may increase the ease of species moving around the world. The current dominant navigation corridors will be even more at risk of invasive species movements; an example is the Mediterranean Sea and its connection to other waterways via the Suez Canal (Galil et al., 2014). In turn, the Mediterranean is the main conduit for invasive species into the Black Sea (Shalovenkov, 2019). Consequently, the invasive species in the Black Sea differ with area, with the hydrographic conditions (principally salinity and temperature and the resulting density-dependent fronts) and the different levels of marine traffic. In addition, the climate changes are superimposed over the large-scale climate features, such as the Atlantic Multidecadal Oscillation (AMO) index and their influences on species supply and movements. The connectivity of enclosed and semienclosed coastal seas, and the connectivity in estuaries and lagoons from the marine and catchment areas, will influence the supply and rate of colonization by nonindigenous species (NIS). Some areas, such as the Mediterranean, may thus act as a "holding area" for NIS, pending movement by the diffusion of vectors (such as ballast water) into the Black Sea (Shalovenkov, 2019). These consequences of climate change will have legal and management repercussions for estuaries and coastal areas worldwide (Saul et al., 2016; Shalovenkov, 2019) and are expected to continue over the coming decades.

6 ADDITIONAL FUTURE THREATS AND CHALLENGES

The earlier chapters describe the threats, the successes, and the failures to sustainably manage estuaries and coastal waters worldwide. The future will not just be "business as usual" along these scenarios but rather it includes accelerated threats and challenges from several developments. The DAPSI(W)R(M) problem-solving framework is, of course, anthropocentric and revolves around the drivers of basic human needs, such as the increasing need for space, water, energy, goods, security, and food, and hence management needs the ability to protect and deliver these sustainably. The framework is required to guide management actions but also to check that they are effective.

6.1 Increasing Globalization and Human Population Growth

The global human population increased from ~1 billion in 1800 to ~7.6 billion in 2018, and it is likely to reach 9.8 billion by 2050 and 11.2 billion by 2100 (http://www.un.org/en/development/desa/population/publications/index.shtml). The populations moving to the coast have long been increasing, especially in developing countries (Fig. 2). In addition to that coastal migration, the population growth is not uniform worldwide, and many nations with the fastest population growth have low standards of living (https://assets.prb.org/pdf14/2014-world-population-data-sheet_eng.pdf); it is therefore likely that in those countries the degradation of estuaries and coastal waters will increase in the coming decades. Of greatest concern is that much of that population growth will be in urban and semirural centers on low-lying and vulnerable coasts (Barbier, 2015) and that these are most vulnerable to risks and hazards.

There will also be the demand for goods and services and the accompanying globalization needed to deliver these. As an example of increasing globalization, China's Belt & Road Initiative has been agreed upon by the Constitution of the Chinese Communist Party. It will involve some 7000 infrastructure and mining, fossil fuel, logging, and other extractive-industry projects spanning much of the world, including the South Pacific, Southeast Asia, Indochina, East Asia, Central Asia, the Middle East, Europe, Africa, South America, Central America, and the Caribbean region. Hence it will affect hundreds of catchments (and thus their runoff to estuaries), estuaries, and coastal waters worldwide. It has the potential to be the biggest environmental challenge of our time. Its proponents assert that it will be "green," "sustainable," and "circular," but this is questionable, as the evidence so far suggests that there are no clear mechanisms in place to ensure environmental sustainability (https://www.bfm.my/damming-the-tapanuli-apes-to-death; http://theconversation.com/the-belt-and-road-initiative-chinas-vision-for-globalisation-beijing-style-77705).

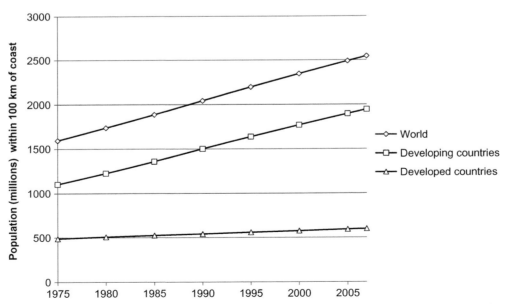

FIG. 2 Population within 100 km of the coast, 1975–2007. *(From Barbier, E.B.,2015. Climate change impacts on rural poverty in low-elevation coastal zones. Estuar. Coast Shelf Sci. 165:A1–A13; United Nations Environment Programme (UNEP). 2014b. The UNEP Environmental Data Explorer, as compiled from UNEP/DEWA/GRID-Geneva. UNEP, Geneva, http://geodata.grid.unep.ch.)*

6.2 Climate Change

Climate change is predicted to lead to global warming, increased atmospheric CO_2, changes to interannual weather patterns [NAO (North Atlantic Oscillation) and ENSO (El Niño Southern Oscillation)], changes to rainfall, sea-level rise, and ocean acidification; in turn these physical changes will have ecological repercussions and then societal effects (Elliott et al., 2015). Other likely changes may involve more variable weather patterns including more severe storms, increased evaporation from higher temperatures, and changes to the water balance and catchment run-off. All these changes to the environments have been summarized as a set of flow diagrams showing the causes and consequences of change; the changes to the physics (hydrology, physiography, etc.) lead to ecosystem changes and eventual changes to the ecosystem services, and in turn this leads to changes in the delivery of societal goods and benefits (Elliott et al., 2015). For instance, changing rainfall, a rise in the mean sea level, and increased climate variability, combined with increased urbanization, more dams and catchment infrastructure, more intensified agriculture, and increased diffuse inputs, will each exacerbate the present problems in estuaries and coastal waters (Fig. 3A–C).

Similar links from physical changes to ecological changes can be drawn for other effects of climate change; changes to the physical–chemical conditions linked to the environmental tolerances of organisms will change faunal and floral distributions including the migration of species to higher latitudes (Elliott et al., 2015). There are likely to be other mechanisms of species change such as the loss of polar ice cover thereby opening navigation routes and so increasing the vectors for the transport of nonindigenous species. All of these changes to distributions are likely to displace native species, although it is questionable whether that would be sufficient to alter the functioning of systems; that is, although the species complement in an area may change, the functioning may remain the same.

Certain habitats may change more than others, for example, climate change is predicted to exacerbate hypersalinity in lagoons in arid areas because such lagoons are shallow (usually <2 m deep) and are often poorly flushed, being microtidal (tidal range <2 m) (Pérez-Ruzafa et al., 2019; Tweedley et al., 2019). All of these changes will eventually affect the generation of ecosystem services and then the produced societal goods and benefits. Hence, when planning for the future changes, it is necessary to consider the effects on resource availability. For instance, inshore and transitional waters fisheries may become particularly affected by climate change in the coming decades in addition to being impacted from artisanal and commercial fisheries and by increasing urbanization and industrialization of the coastal fringe (Palomares and Pauly, 2019).

The central cause of the climate changes is increased atmospheric CO_2 and associated warming of land and sea surfaces. In addition to warmed conditions, and the repercussions for faunal and floral distributions, ocean acidification is predicted to increase under the business-as-usual CO_2 emission scenario, with pH decreasing from 8.1 to 7.8 by 2100 (IPCC, 2013). This will particularly affect many marine organisms, especially those secreting calcium carbonate structures; thus coral reefs, oysters, mussels, shrimps, and lobsters will not be able to form sufficiently strong shells, tests, and exoskeletons in acidified waters. On the other hand, noncalcareous taxa may benefit from an increased dissolved CO_2 (Connell et al., 2013), so that the ecological structure of estuarine and marine ecosystems may change. The ecological and societal effects of ocean acidification in estuaries and coastal waters either may thus be profound (Gattuso and Hansson, 2011) or, alternatively, may just adapt to a new equilibrium. As emphasized throughout this synthesis, the highly variable transitional environments and their biota, with an ability for "environmental homeostasis" (Elliott and Quintino, 2018, and references therein), can tolerate and adapt to wide-ranging environmental changes.

7 TOOLS AND APPROACHES FOR THE MANAGEMENT OF NEW CHANGES

7.1 Monitoring to Inform Management

The management of the ecosystems covered in this synthesis can only be achieved with adequate management targets and a clear vision of management goals, and both the targets and vision need adequate monitoring and the production of quantitative information. Similarly, the effectiveness of management measures can only be determined by monitoring the changes to the systems following management actions. All the areas covered in this synthesis require ecosystem-based management (EBM) in order to effect environmental quality improvement and to restore systems where possible, and the EBM needs to be underpinned by models and data. For example, global climate models are predicting large decreases in precipitation and changes in the water balance, and this is more severe in some arid areas (e.g., the Yucatan Peninsula; see Herrera-Silveira et al., 2019). Tools such as the database and approach of Giosan et al. (2014) will become increasingly important to provide the information on which to base future management of sensitive areas and to determine whether existing management measures have been effective.

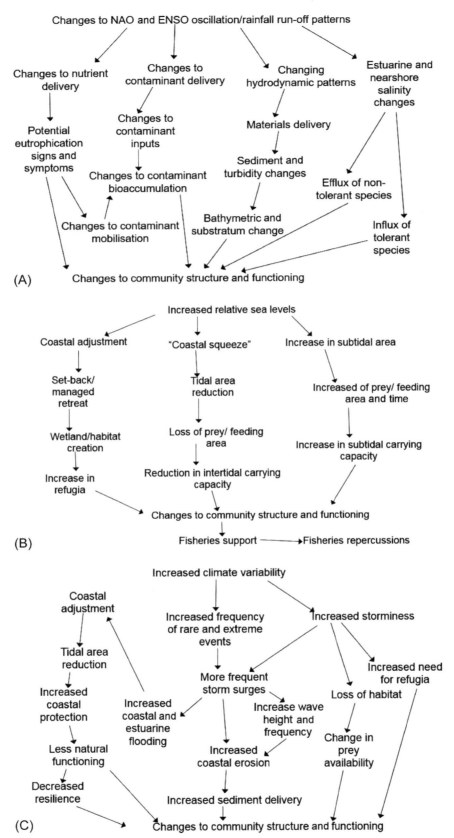

FIG. 3 Examples of the linked sequences of changes due to climate change. (A) Land-based discharges and run-off due to regional climate perturbations leading to ecosystem effects; (B) Estuarine hydrodynamic changes due to increased relative sea levels leading to ecosystem effects; (C) Coastal hydro-dynamic changes due to increased climate variability leading to ecosystem effects. *(Modified from Elliott M, Borja Á, McQuatters-Gollop A, Mazik K, et al., 2015. Force majeure: will climate change affect our ability to attain good environmental status for marine biodiversity? Mar. Pollut. Bull. 95:7-27; https://doi.org/10.1016/j.marpolbul.2015.03.015.)*

7.2 Environmental Impact Modeling to Guide Management

Considering future changes and implementing effective management requires a predictive capability, especially using modeling, either in qualitative (conceptual) terms or as a combination of deterministic or empirical modeling. There is, therefore, a practical need for models to assess not only the impact on estuarine ecosystems of development proposals throughout the catchment, within the transitional environments, and also at sea, but also the effectiveness of remedial measures (Wolanski, 2019). Given the importance of connectivity between the catchment and the open sea, such models must account for that connectivity and, especially, the relationship between the natural physical–ecological functioning and the societal system.

It is axiomatic that numerical models are only as good as our understanding and the availability of data to support the empirical assessment. For example, there is the need for linked models and end-to-end models where the latter covers the bottom-up processes and the top-down responses, especially from the societal system (Wolanski, 2019). Hydrodynamic models, sediment models, and ecological models need to be linked to understand the ecosystem and hence to model environmental impacts. Whereas the hydrodynamic models are adequate in general terms, they are still inadequate at predicting precise features such as the effect of fronts as barriers to the movement of organisms, especially the young stages. The sediment models have difficulty in coping with the precise transport mechanisms for certain sediment types as well as with the influence of ecosystem engineers on modifying the substrata. Such feedback mechanisms, including the role of micro- and macroorganisms in flocculation, erosion, and deposition processes, are still poorly understood and yet are critical for determining the way in which the natural system will respond in the future to climate change and to other human pressures. The ecological models often have to oversimplify the natural system and are thus restricted to the essential processes (Wolanski, 2019). This becomes even more difficult when attempting to predict the effect of stochastic processes such as those attributable to climate change. Based on the result of several case studies, ecohydrological modeling gives encouraging results but it needs further improvements to include societal parameters; such end-to-end models will be vital for understanding the repercussions of global change (Wolanski and Elliott, 2015).

7.3 Community Involvement and Culture

Management of transitional and coastal waters against future change requires the input of natural and social sciences. Communication and integration across fields, partnerships, and involvement with all stakeholder groups, with researchers, community groups, and nongovernmental organizations (NGOs) have achieved success through targeted outreach and education programs. This is particularly successful where there is a culture responding to the needs to better manage the environment, for example, in Pacific Island states and in Moreton Bay (Richmond et al., 2019; Lanyon, 2019).

Similarly, as an example, Japanese environmental conservation, restoration, and management policies have existed since the 1950s in full cooperation with the community. They initially focused on hazardous substances and water pollution measures (Matsuda and Yanagi, 2019). This then led to the adoption of the total pollution load control (TPLC) approach to reduce nutrients and organic matter to prevent the signs and symptoms of eutrophication such as red tides. As in many places, pollution control is now succeeding but is giving way to holistic policies such as habitat-centered conservation that focuses on biodiversity, enhancing productivity, and balanced nutrient fluxes. In addition, again as shown in Japan, there is the adoption of regional management measures to ensure sustainable coastal communities and use, such as the national research project, "Development of Coastal Management Methods to Achieve a Sustainable Coastal Sea," started in 2014 and lasting until 2018. Matsuda and Yanagi (2019) emphasize the philosophy of *Satoumi* to ensure that the vision of "a clean, productive and prosperous coastal sea" is achieved.

All these examples demonstrate the importance of culture and society and hence the need to emphasize these in the DAPSI(W)R(M) framework and the Responses (using Measures) (see below). There is not only the need for formal governance involving policies, politics, administration, and legislation, but also the input from informal governance involving key stakeholders. Such an integrated approach may lead to sustainable and successful management, especially with changing environments, as is demonstrated in the United States for Chesapeake Bay and the Everglades (Delia et al., 2019; Sklar et al., 2019).

7.4 National Planning

Given the previous data, there are lessons to be learned from the case studies; for example, the Grijalva-Usumacinta River and Delta (Mexico) supports the CENTLA Biosphere Reserve, whose management plan has identified natural environmental units. These encompass two central zones that are highly protected and surrounded by several buffer zones in which economic activities are allowed (Herrera-Silveira et al., 2019). Faced with such environmental degradation and the likely effects of climate change, Mexico has designated many natural protected areas as management units. As another example, management measures in areas such as the Florida Everglades require the restoration of freshwater flow conditions, the protection of habitats such as underlying peat, and the prevention of vegetation change such as the migration of saline-tolerant mangrove habitats (Sklar et al., 2019).

Globally, there are now excellent examples of science-based policy-making and implementation of management measures in which local, regional, and national initiatives have been combined. For example, the Chesapeake Bay system in the eastern United States has shown that different stakeholders have different but complementary roles leading to the success of the scientifically based "US EPA Chesapeake Bay Program" (D'Elia et al., 2019). Despite this, it now remains to be seen whether the integrated and widely accepted program can cope with continuing pressures of population growth and land-use change over which are superimposed accelerating sea-level rise and other adverse consequences of climate change. Although studies have shown what the Bay was like before the emergence of such anthropogenic pressures, it is not likely that such conditions will be regained given the population surrounding this part of the United States. Therefore, the management will have to accommodate the current baseline as a new equilibrium.

7.5 International Water Governance

When the river and its estuary fall within one level of regional governance, coordinated management may be more easily achieved than for transboundary and international areas. Indeed, Nhan and Cao (2019) suggest that the problems of the Mekong Delta are the result of changes as far away as the Chinese part of the international catchment, so solutions require internationally coordinated actions. In particular, operating the large dams in China in the Upper Mekong basin (the Lancang basin) should accommodate the needs of the lower Mekong population, and not ignore them. Similarly for the Nile, the operators of the Great Ethiopian Renaissance Dam in Ethiopia should keep the wishes and needs of the Nile Delta population in mind. Neither for the Nile nor for the Mekong is there an obvious mechanism to ensure an equitable and fair solution to transboundary water problems. In contrast to these water conflicts, the Argentinean–Uruguayan Binational Commission, in operation since the 1970s, regulates the estuarine activities and has prioritized the deteriorating coast, overfishing, and pollution to be tackled in the Rio de la Plata (García-Alonso et al., 2019). The Commission's aim is to protect the estuarine health and deliver ecosystem services, although effective implementation is slow and limited because of the overdrawing of fresh water for Buenos Aires and because of pollution.

7.6 Coastal Cities

The increasing urbanization and industrialization of coasts and transitional environments has led to an increased knowledge of both large urban systems and of the coast; yet there has been little work on the effects of large-scale urbanization on the coast (Blackburn et al., 2019). The large and increasing policy responses will have to build on good and appropriate science in natural and social research, but in essence relate to hydrological characteristics and to environmental health risk analysis and management (Cormier et al., 2013). The increasing number of hazards and risks affecting human assets (Table 1) are increasing vulnerability and reducing resistance to and resilience in the face of climate change. Indeed, in many cases, human actions can exacerbate the effects of natural hazards, such as removing protective vegetation (Wolanski and Elliott, 2015). Tackling these challenges in increasingly urbanized areas will require innovative and inclusive approaches that take a long-term view for coastal–urban futures.

7.7 Examples of Arguably the Earth's Last Pristine Catchment to Estuarine Ecosystems

As emphasized earlier, managing coastal and transitional environments requires a standardized framework, in essence to determine the causes of change and their solutions. Whereas such approaches are well-developed in temperate areas (such as North America, Europe, and Australia), they need to be adapted for developing areas as well as those areas that, because of climate change, will be increasingly exploited, such as the Arctic coastal systems (Lovecraft and Meek, 2019). These areas are still sparsely populated and include little permanent infrastructure, and yet global warming may make the environmental changes even more severe than at lower latitudes. The areas are subject to different biophysical, sectoral, governance, and cultural conditions and constraints. However, the environmental management complexity and the relatively lesser-studied conditions of the Pan-Arctic coasts will increase the challenges against impending climate change.

Australia's tropical coast (Wolanski, Box 3) and the Arctic (Forbes, 2019; Lovecraft and Meek, 2019) are the only places worldwide with numerous pristine or nearly pristine catchments–river–estuaries ecosystems. These ecosystems are a global example of what pristine rivers and estuaries look like. These pristine ecosystems should not be used and abused; they need to be preserved whole for the benefit of future generations to see and understand what these systems were like and how they functioned before the Anthropocene and to enjoy their natural beauty.

BOX 3 What Is the Future for Australia's Tropical Estuaries?

Australia still has relatively pristine catchment–estuary systems in the tropics west of Cape York (Fig. B.2A). This is a vast area; the distance between Cape York and Broome is ~2300 km. The total population is a few hundred thousand people and there is only

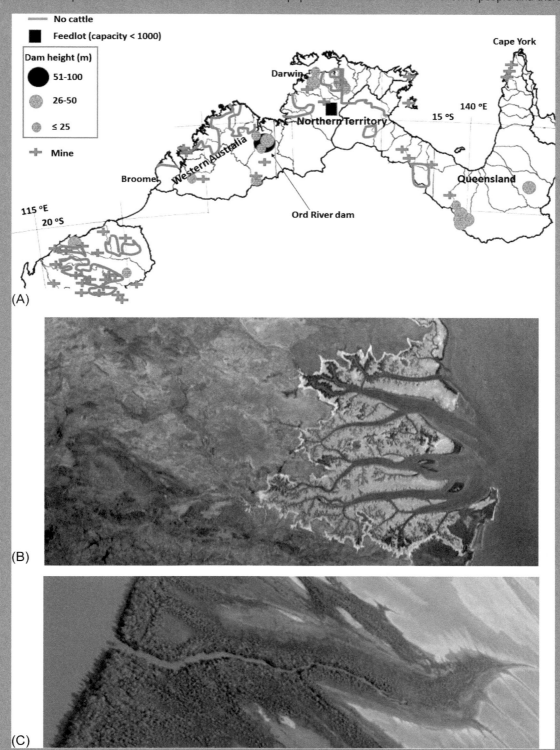

FIG. B.2. (A) Map of tropical coastal Australia west of Cape York showing the river catchments, the areas with no cattle, the dams, the mines, Darwin and Broome, and the feedlot. (B) Satellite photograph of a pristine catchment–estuary system. (C) Aerial photograph of a pristine estuary, mangroves, and mudflats.

Continued

BOX 3 What Is the Future for Australia's Tropical Estuaries?—cont'd

one large dam (the Ord River dam). There are several mines, but their pollution footprint is strictly controlled and there is only one large city (Darwin) and only one large irrigated area (from the Ord River dam). Large areas have no cattle, whereas the average cattle density elsewhere is <0.1 cattle/ha, and there is only one large feedlot, and in some rivers fishing is minimal.

It is difficult to find any other pristine tropical catchment–estuary systems left worldwide. They are truly biodiversity hotspots in a world where such hotspots are increasingly rare, and they are an Australian gift to the world showing what pristine ecosystems look like. With the marked monsoonal and interannual variability of the rainfall and sporadic cyclones, these estuaries vary constantly. Scientific research is scant there compared to that in the rest of Australia, and it is largely restricted to specific studies such as geomorphology, oceanography, and an inventory of vegetation, fish, and birds (e.g., see Woodroffe and Mulrennan, 1963; Woodroffe et al., 1989; Wolanski et al., 2006; Williams, 2014; Leseberg and Campbell, 2015; http://www.nespnorthern.edu.au/). There are very few studies on the biophysical processes, with rare exceptions such as Ridd et al. (1988) and Cobb et al. (2007). This vast area is a gold mine in waiting for natural scientists! However developments are just around the corner, and the science has not been done to assess the impact of these developments.

With the rapid human population growth worldwide and the economic imperative dominating human behavior, there is an urgent need for society to preserve and cherish the few remaining relatively pristine catchment–estuary systems left in the world. There are only two such large areas left, namely the Arctic (Lovecraft and Meek, 2019) and the Australian tropics west of Cape York. Science alone cannot provide the political will for such conservation action, but aesthetic appreciation and spiritual values can. Australian tropical pristine catchment–estuary systems offer spectacular landscapes to enjoy (Fig. B.2B and C) together with abundant wildlife including birds, charismatic animals, and fish (Rains, 2016). The beauty of nature is vital for maintaining our quality of life. The human need for beauty is a prime factor in our call for preserving some of the last pristine catchment–estuaries worldwide. As Dorst (1965) wrote (translated from French): *"Nature will only be preserved if man loves it a little, simply because it is beautiful, and because we need beauty whatever is the form we are sensitive to as a result of our culture and intellectual formation. Indeed, this is an integral part of the human soul."* Indeed this human need for beauty is what drove visionaries to preserve exceptionally beautiful sites like Yellowstone in the United States and Serengeti in Tanzania! It is hoped that this appeal to beauty can also be successful in tropical coastal Australia!

Eric Wolanski
James Cook University and the Australian Institute of Marine Science, Townsville, QLD, Australia

References

Cobb, S.M., Saynor, M.J., Eliot, M., Eliot, I., Hall, R. 2007. Saltwater intrusion and mangrove encroachment of coastal wetlands in the Alligator Rivers Region, Northern Territory, Australia. Supervising Scientist Report 191, Supervising Scientist, Darwin NT.

Dorst, J., 1965. Avant Que Nature Meure. In: Delachaux and Niestle. Neuchatel, Switzerland.

Leseberg, N., Campbell, N., 2015. Birds and Animals of Australia Top End. Princeton University Press, Princeton, p. 272.

Lovecraft, A.M., Meek, C.L., 2019. Arctic Coastal Systems: An Overview of Pressures, Pitfalls, and Potentials. This book.

Rains, N., 2016. Tropical Australia. Explore Australia Publishing Pty, p. 192.

Ridd, P., Sandstrom, M., Wolanski, E., 1988. Outwelling from tropical tidal salt flats. Estuar. Coast. Shelf Sci. 26, 243–253.

Williams, D., 2014. Recent, rapid evolution of the lower Mary River estuary and flood plains (pp. 277–287). In: Wolanski, E. (Ed.), Estuaries of Australia in 2050 and Beyond. Springer, Dordrecht. 202 pp.

Wolanski, E., Williams, D., Hanert, E., 2006. The sediment trapping efficiency of the macro-tidal Daly Estuary, tropical Australia. Estuar. Coast. Shelf Sci. 69, 291–298.

Woodroffe, C.D., Mulrennan, M.E., 1963. Geomorphology of the lower Mary River Plains, Northern Territory. Australian National University North Australian Research Unit and the Conservation Commission of the Northern Territory, Darwin.

Woodroffe, C., Chappell, J., Thom, B.G., Wallensky, E., 1989. Depositional model of a macrotidal estuary and floodplain, South Alligator River, Northern Australia. Sedimentology 36, 737–756.

8 CHANGES TO STRESSORS: RESPONSES TO INCREASING COASTAL POPULATIONS, THEIR ENVIRONMENT, AND INFRASTRUCTURE

Much of this synthesis indicates that increasing population and the migration of populations into coastal and transitional environments pose the greatest threats. The increased populations will have a high need for land, food, water, and energy, and the high cost of energy may make this impossible in the long term, especially if the situation is aggravated by climate change (Day et al., 2019a). They will require public safety from flooding and erosion, and will increase the modification of the water bodies, the land mass, and the catchments. Indeed, the three dominant priorities will need to be ranked in the order of public safety, economic benefits, and ecological well-being, with the latter two being equal or, in many countries, the economic imperative will have precedence over the ecology.

The production of infrastructure such as dams will be needed for water provision and hydropower to service the increasing population together with the resultant downstream knock-on effects (see the case of the Mekong and the Nile in this synthesis). The high-density population in the transitional and coastal areas will result in changed inhabited areas, which can no longer be returned to anything resembling natural areas, and so the modified system will be the new baseline. It is a paradox that the system will have to accommodate the population pressures and the population will have to accommodate the nature changes and, through feedback loops, the changes that are exacerbated by increased populations. Although some areas such as Manhattan, New York, will have such high population pressure that it can no longer allow the coast to function naturally, so it will only rely on hard engineering for protection, other areas, such as eroding and sinking North Sea coasts, will increase the controls on inhabitation of the vulnerable coasts. Increasingly, as shown by Day et al. (2019a), governments will increasingly balance the costs and benefits of living in vulnerable areas, particularly deltas, and whether populations can be protected. Further, as the case studies of Egypt's Nile Delta and Vietnam's Mekong Delta demonstrate, a method to reach a fair share of water resources is needed between upstream countries (e.g., Ethiopia and China, respectively) and downstream countries (Egypt and Vietnam, respectively) to avoid conflicts as well as to benefit all the riparian countries and the environment.

9   SUSTAINABLE SOLUTIONS

This synthesis has identified a series of problems and their causes, which are likely to occur in transitional and coastal ecosystems both now and in the coming decades (Table 3, columns 1 and 2). Solving these problems is extremely challenging but not beyond human ingenuity, and indeed, and as shown in this book, there are now many global initiatives to reverse and remediate problems (Table 3, column 3). Many of these solutions involve ecoengineering or at least working with nature

TABLE 3 Causes of Environmental Change and Their Solutions in Coastal and Transitional Areas

What?	Cause	Solution
Land-claim	Wetland removal/dike construction	Restocking with vegetation, reconnection, resculpting, managed realignment, habitat creation
Dissolved oxygen minima and water quality barriers	Waste discharges	Reduction/treatment of inputs, reoxygenation, bubbling
Biogenic reef loss (e.g., bivalve beds)	Siltation, overharvesting	Adaptation, flushing, regulation, restocking
Eutrophication	Poor flushing, excess nutrients	Nutrient controls and land management, reconnection of waters, regulation through treatment (e.g., urban stormwater management)
Biota kills	Toxin input, WQ problems	Regulation, industry removal
Coral reef loss	Siltation, direct damage, bleaching	Fishing and run-off controls, restoration
Loss of fish and shellfish	Overharvesting, climate change, hydrodynamic barriers	Restocking, adaptation, regulation
Salinity change	Upstream abstraction, impediments to flow (e.g., dams)	Removal, reconnection
Loss of seagrass	Smothering, nutrient excess, disease, hydrographic change	Reduction, removal, reconnection, replanting
Flow loss or regulation	Diversion, abstraction, structures (e.g., dams, hydropower schemes)	Reconnection, reallocation, removal of infrastructure (weirs, dams), flow management (e.g., generate artificial freshets from dams)
Seabed extraction	Aggregate removal, loss of sediment fraction	Reseeding, regulation, reallocation
Taxonomic changes	Nonindigenous species influx	Removal, eradication, prevention

(Wolanski and Elliott, 2015; Elliott et al., 2016) and they may involve either engineering the habitat (Type A ecoengineering) or introducing or reintroducing species (Type B ecoengineering) (see below). There are some successes in restoration and there is an increasing body of literature showing the resilience of estuarine ecosystems (Duarte et al., 2015; Wolanski and Elliott, 2015; Elliott et al., 2016). Hence, in addition to restoring the physical conditions and the shape (physiography) of estuaries, both to remediate against previous degradation and to increase the resilience to systems in the event of climate change, it may be necessary to actively or passively restore populations of dominant habitat-forming ecosystem engineers (Gilby et al., 2018). For example, restoring biogenic reefs and especially those of bivalves will enhance the production of ecosystem services and the delivery of societal goods and benefits, possibly even at large scales; however, this is on the condition that the original pressures causing the decline or removal have been addressed (McLeod et al., 2019). As shown earlier, management of estuarine and transitional habitats is only successful if accompanied by catchment-wide improvements and also the management of all other associated habitats such as seagrasses, mangroves, etc.

Restoring estuarine and coastal systems is considered successful if the restored areas become self-sustaining, that is, with positive feedback such as an improved habitat (e.g., more bivalves), which encourage the development of greater habitats. If successful, this not only delivers benefits for improving the resilience of areas to coastal climate change (such as erosion protection, energy attenuation, and flood defense) but also increases food security and a bioremediation method for water quality by filtering the waters (McLeod et al., 2019).

The solutions to the predicted changes in coastal and transitional waters described in this synthesis and its literature, and summarized in Table 3, are encapsulated in the Responses (as Measures) part of the DAPSI(W)R(M) framework. They include human actions for prevention, mitigation, and compensation initiatives, which, to be successful and sustainable, are required to cover many aspects. These are the so-called 10-tenets shown in Table 4; among others, these include the legal, economic, and administrative instruments, the sociocultural aspects, ecological modifications, and suitable techniques and technologies (Table 4). It is emphasized that only by accommodating all these aspects will there be anything approaching an ability to tolerate and adapt to the changes proposed over the coming century. However, it is acknowledged here that different sociopolitical systems include different priorities. For example, as shown by the Chinese Belt and Road initiative, there is an imperative for economic growth that is perhaps more important than environmental well-being. This may be regarded

TABLE 4 The 10-Tenets Framework for Examples of Management Solutions

Tenet	Examples of Measures
Ecologically sustainable	Achieving a healthy ecology, suitable fish stocks, carbon sequestration areas; delivering ecosystem services
Technologically feasible	Building treatment works; remediating habitats; flow regulation measures from dams
Economically viable	Sufficient funding for the measure, acceptable cost–benefit analysis/ratio, willingness-to-pay by society; tolerable insurance costs for damage
Socially desirable/tolerable	Stakeholder agreement based on consultation; accommodating what the community will tolerate; acknowledging that society may tolerate less than it desires; delivering societal goods and benefits
Legally permissible	Compliance with laws and regulations; licence compliance for waste disposal; requirement for environmental impact assessments
Administratively achievable	Agreement from administrative and statutory bodies such as an Environmental Protection Agency, nature conservation body, etc.
Politically expedient	Agreement with the manifesto commitments of the ruling party; emphasis provided by the national leader or the political ruling party
Ethically defensible (morally correct)	Ensuring funding mechanisms are not a burden on future generations, and that discounting mechanisms are not a liability; ensuring that specific groups of society are not disadvantaged
Culturally inclusive	Protection of culturally and aesthetically important areas, no interference of indigenous human population areas, achieving equitable solutions
Effectively communicable	Agreement by consultation, advertised decision-making; stakeholder consultation and conflict resolution

Modified from Elliott, M., Burdon, D., Atkins, J.P., Borja, A., Cormier, R., de Jonge, V.N., Turner, R.K., 2017. "And DPSIR begat DAPSI(W)R(M)!"—a unifying framework for marine environmental management. Mar. Pollut. Bull. 118(1–2) 27–40. https://doi.org/10.1016/j.marpolbul.2017.03.049, and references therein.

as analogous to the environmental degradation shown by polluted estuaries and catchments and excessive land-claim in the United Kingdom during the 1800s in the Industrial Revolution (McLusky and Elliott, 2004). Fortunately, most countries eventually realize that a good environment is also good for economic prosperity and the quality of life.

Future challenges are emerging, such as projected climate change, so that the solutions will have to evolve with them (Robins and Lewis, 2019). Temperate areas are likely to experience seasonal changes toward wetter winters and drier summers, so there will be repercussions for transitional and coastal environments at the receiving end of those catchment systems (Robins et al., 2016). The repercussions for the biota in those systems can be gained from increasing evidence (e.g., Little et al., 2017; Ducrotoy et al., 2019) but they are still uncertain, especially given the ability of the dynamics of highly variable systems to absorb change; furthermore, the ability to detect change due to non-natural features and climate change in highly variable systems continues to be a challenge (Elliott and Quintino, 2018). Robins and Lewis (2019) consider that, despite the difficulties in understanding how estuaries will change with climate change, hydrological drivers may be as important, if not more so, than other climate drivers such as sea-level change or morphological change. As shown by Pérez-Ruzafa et al. (2019), to tackle these challenges we need appropriate science and research and international collaboration across different types of water bodies so that we can test our assumptions and paradigms and determine site-specific and generic features.

Environmental management has changed in the last two decades from a sectoral approach to management, where each activity or ecological component was managed separately, toward integrated, holistic, and sustainable management approaches that incorporate conflict resolution, hazard and risk assessment, and the linking of natural and sociological systems (Glaeser, 2019). However, it may not be possible to achieve all of these, so compromises will have to be made on a case-to-case basis; limited outcomes and likely nonsustainability on a case-by-case basis will have to be accepted based on the biophysical limits of each individual site taking into account the legacy of environmental degradation and population growth. Thus, the reality of biophysical constraints will ultimately limit what is possible to achieve (Day and Rybczyk, 2019).

10 CONCLUSIONS

This synthesis concludes with a set of main points and lessons that arise from it and its source literature:

- The local human culture has a dominant influence in shaping the future of the estuaries; the outcome varies between regions, creating a mosaic of successes and failures.
- The future of coasts and transitional environments has to confront the huge joint challenges of increasing population demands, including trade globalization, loss of resources such as energy and space, and climate change.
- Interface coastal and transitional environments are arguably subject to more and more severe risks and hazards than environments either in the open sea, or in the catchment, or on land.
- High-risk coastal environments have many ecosystem services and societal goods and benefits, and therefore the increasing population is increasingly migrating to exploit them.
- Human populations are moving to the coast and transitional environments, but the increasing populations degrade the environments, which in many areas cannot accommodate the increases; this feature applies to both migrants for residence and to tourism.
- The repercussions of human actions, and even those actions planned to tackle damage, are poorly understood, so near and far field spatial and temporal effect footprints are poorly defined.
- Coastal and transitional environments thus have a finite and decreasing assimilative capacity to absorb adverse effects and reduced carrying capacity to accommodate the increasing drivers, activities, and pressures.
- After society engineers the coast, transitional waters, and catchment, it then has to keep doing so, otherwise the areas revert to situations not desired by society or in conflict with societal demands.
- Catchment modifications, such as dams, appear to be planned without thought for the downstream consequences, and these features are compounded in catchments crossing international boundaries.
- There is a "vicious circle" in that the pressure of increasing urbanization on the coast forces the authorities to spend more and more money to protect human safety; arguably, this is also to protect areas that perhaps should not be protected, so populations may have to be discouraged from settling in vulnerable areas.
- Managing the ecology for natural and societal benefits requires good and appropriate science and an understanding of and the management of the physical system and the socioecological system.
- Monitoring is required to inform management responses, but climate change is producing moving baselines against which it is increasingly difficult to detect other natural and anthropogenic change.

- Modeling can indicate future changes, but it is only as good as the availability of data and the current scientific and societal understanding; many areas receive insufficient or decreasing funding for gathering those data and that understanding.
- As there are few pristine estuarine and coastal areas left worldwide, the emphasis should be on protecting these areas; in all other areas, current management methods are designed for accommodating either unnatural systems or the new equilibria created by society and its engineering.
- Transnational boundary problems occur with many transitional waterbodies and require international action and cooperation, but this is often not achieved in areas with cultural differences or because of conflicts in many geopolitically unstable areas.
- The greatest effects of future climate and population changes are expected to be in areas with the least ability (e.g., competence, organization, and finance) to tackle those effects.
- The ecology of highly variable environments is adapted to those conditions and arguably may be most able to withstand the future natural and anthropogenic changes; it is society that will have to adapt to the changing future of our coasts and transitional waters.

REFERENCES

Barbier, E.B., 2015. Climate change impacts on rural poverty in low-elevation coastal zones. Estuar. Coast. Shelf Sci. 165, A1–A13.

Barnard, S., Elliott, M., 2015. The 10-tenets of adaptive management and sustainability—applying an holistic framework for understanding and managing the socio-ecological system. Environ. Sci. Policy 51, 181–191.

Birkeland, C., 2015. Coral Reefs in the Anthropocene. Springer, Dordrecht, p. 271.

Blackburn, S., Pelling, M., Marques, C., 2019. Megacities and the coast: global context and scope for transformation. This book.

Brodie, J., Grech, A., Pressey, B., Day, J., Dale, A.P., Morrison, T., Wenger, A., 2019. The future of the Great Barrier Reef: the water quality imperative. This book.

Chen, Z., 2019. A brief overview of ecological degradation of the Nile delta: what we can learn? This book.

Clemens, R.S., Rogers, D.I., Hansen, B.D., et al., 2016. Continental-scale decreases in shorebird populations in Australia. EMU 116, 119–135.

Connell, S.D., Kroeker, K.J., Fabricius, K.E., Kline, B.D., 2013. The other ocean acidification problem: CO_2 as a resource among competitors for ecosystem dominance. Philos. Trans. R. Soc. Lond. B Biol. Sci. 368 (1627), 20120442. https://doi.org/10.1098/rstb.2012.0442.

Cormier, R., Kannen, A., Elliott, M., Hall, P., Davies, I.M., 2013. In: Marine and coastal ecosystem-based risk management handbook. ICES Cooperative Research Report, No. 317, March 2013, International Council for the Exploration of the Sea, Copenhagen. pp. 60. ISBN: 978-87-7472-115-1.

Critchell, K., Bauer-Civiello, A., Benham, C., Berry, K., et al., 2019. Plastic pollution in the coastal environment: current challenges and future solutions. This book.

Day, J.W., Rybczyk, J.M., 2019. Global change impacts on the future of coastal systems: Perverse interactions among climate change, ecosystem degradation, energy scarcity, and population. This book.

Day, J., Hall, C., Roy, E., Moerschbaecher, M., et al., 2016. America's Most Sustainable Cities and Regions—Surviving the 21st Century Megatrends. Springer, New York, p. 348.

Day, J.W., Delia, C., Wiegman, A., Rutherford, J., et al., 2018. The energy pillars of society: perverse interactions of human resource use, the economy, and environmental degradation. Biophys, Eco. Res. Qual. 3, 2. https://doi.org/10.1007/s41247-018-0035-65.

Day, J.W., Ramesh, R., Giosan, L., Syvitski, J., Kemp, P., 2019a. Delta winners and losers in the Anthropocene. This book.

Day, J.W., Colten, C., Kemp, G.P., 2019b. Mississippi Delta restoration and protection: Shifting baselines, diminishing resilience, and growing non-sustainability. This book.

Day, J.W., Ibáñez, C., Pont, D., Scarton, F., 2019c. Status and sustainability of Mediterranean Deltas: The case of the Ebro, Rhône, and Po Deltas and Venice Lagoon. This book.

De Jonge, V.N., Schttelaars, H.M., van Beusekon, J.E.E., Talke, S.E., de Swart, H.E., 2014. The influence of channel deepening on estuarine turbidity levels and dynamics, as exemplified by the Ems Estuary. Estuar. Coast. Shelf Sci. 139, 46–59.

D'Elia, C., Bidjerano, M., Wheeler, T.B., 2019. Population growth, nutrient enrichment, and science-based policy in the Chesapeake Bay Watershed. This book.

Duarte, C.M., Borja, A., Carstensen, J., Elliott, M., Krause-Jensen, D., Marbà, N., 2015. Paradigms in the recovery of estuarine and coastal ecosystems. Estuar. Coasts 38 (4), 1202–1212. https://doi.org/10.1007/s12237-013-9750-9.

Ducrotoy, J.-P., Elliott, M., 2008. The science and management of the North Sea and the Baltic Sea: natural history, present threats and future challenges. Mar. Pollut. Bull. 57, 8–21.

Ducrotoy, J.-P., Elliott, M., Cutts, N.D., Franco, A., Little, S., Mazik, K., Wilkinson, M., 2019. Temperate estuaries: their ecology under future environmental changes. This book.

Dunn, R.J.K., Waltham, N.J., Huang, J., Teasdale, P.R., King, B.A., 2019. Protecting water quality in urban estuaries: Australian case studies. This book.

Elliott, M., 2011. Marine science and management means tackling exogenic unmanaged pressures and endogenic managed pressures—a numbered guide. Mar. Pollut. Bull. 62, 651–655.

Elliott, M., Quintino, V.M., 2018. The Estuarine Quality Paradox Concept. Encyclopaedia of Ecology, second ed. Elsevier, Amsterdam, ISBN: 9780444637680.

Elliott, M., Whitfield, A., 2011. Challenging paradigms in estuarine ecology and management. Estuar. Coast. Shelf Sci. 94, 306–314.

Elliott, M., Cutts, N.D., Trono, A., 2014. A typology of marine and estuarine hazards and risks as vectors of change: a review for vulnerable coasts and their management. Ocean Coast. Manag. 93, 88–99.

Elliott, M., Borja, Á., McQuatters-Gollop, A., Mazik, K., et al., 2015. *Force majeure*: will climate change affect our ability to attain good environmental status for marine biodiversity? Mar. Pollut. Bull. 95, 7–27. https://doi.org/10.1016/j.marpolbul.2015.03.015.

Elliott, M., Mander, L., Mazik, K., Simenstad, C., et al., 2016. Ecoengineering with ecohydrology: successes and failures in estuarine restoration. Estuar. Coast. Shelf Sci. 176, 12–35. https://doi.org/10.1016/j.ecss.2016.04.003.

Elliott, M., Burdon, D., Atkins, J.P., Borja, A., Cormier, R., de Jonge, V.N., Turner, R.K., 2017. *"And DPSIR begat DAPSI(W)R(M)!"*—a unifying framework for marine environmental management. Mar. Pollut. Bull. 118 (1-2), 27–40. https://doi.org/10.1016/j.marpolbul.2017.03.049.

Forbes, D.L., 2019. Arctic deltas and estuaries: a Canadian perspective. This book.

Galil, B.S., Marchini, A., Occhipinti-Ambrogi, A., Minchin, D., et al., 2014. International arrivals: widespread bioinvasions in European Seas. Ethol. Ecol. Evol. 26 (2-3), 152–171. https://doi.org/10.1080/03949370.2014.897651.

García-Alonso, J., Lercari, D., Defeo, O., 2019. Rio de la Plata: A neotropical estuarine system. This book.

Gattuso, J.-P., Hansson, L., 2011. Ocean Acidification. Oxford University Press, Oxford, p. 352.

Gilby, B.L., Olds, A.D., Peterson, C.H., Connolly, R.M., et al., 2018. Maximising the benefits of oyster reef restoration for finfish and their fisheries. Fish Fish. 1–17. https://doi.org/10.1111/faf.12301.

Giosan, L., Syvitski, J., Constantinescu, S., Day, J., 2014. Protect the world's deltas. Nature 516, 31–33. https://doi.org/10.1038/516031a.

Glaeser, B., 2019. Human-nature relations in flux: Two decades of research in coastal and ocean management. This book.

Güneroğlu, A., Samsun, O., Feyzioğlu, M., Dihkan, M., 2019. The Black Sea—the past, present and future status. This book.

Heiskanen, A.-H., Bonsdorff, E., Joas, M., 2019. Baltic Sea: a recovering future from decades of eutrophication. This book.

Herrera-Silveira, J.A., Lara-Domínguez, A.L., Day, J.W., Yáñez-Arancibia, A., et al., 2019. Ecosystem functioning and sustainable management in coastal systems with high freshwater input in the Southern Gulf of Mexico and Yucatan Peninsula. This book.

Hughes, T.P., Barnes, M.L., Bellwood, D.R., Cinner, J.E., et al., 2017. Coral reefs in the Anthropocene. Nature 546, 82–90.

Hughes, T.P., Anderson, K.D., Connolly, S.R., Heron, S.F., et al., 2018. Spatial and temporal patterns of mass bleaching of corals in the Anthropocene. Science 359, 80–83.

IPCC, 2013. In: Stocker, T.F., Qin, D., Plattner, G.-K., Tignor, M., Allen, S.K., Boschung, J., Nauels, A., Xia, Y., Bex, V., Midgley, P.M. (Eds.), Climate change 2013: the physical science basis. Contribution of Working Group I to the Fifth Assessment Report of the Intergovernmental Panel on Climate Change. Cambridge University Press, Cambridge, NY, pp. 1535.

Kidwai, S., Ahmed, W., Tabrez, S.M., Zhang, J., et al., 2019. The Indus Delta Catchment, river, coast and people. This book.

Kirby, R., 2019. Challenges of restoring polluted industrialised muddy NW European Estuaries. This book.

Koshiba, S., Besebes, M., Soaladaob, K., Ngiraingas, M., Isechal, A.L., Victor, S., Golbuu, Y., 2014. 2000 years of sustainable use of watersheds and coral reefs in Pacific Islands: a review for Palau. Estuar. Coast. Shelf Sci. 144, 19–26.

Lanyon, J., 2019. Management of megafauna in estuaries and coastal waters: Moreton Bay as a case study. This book.

Li, M., Chen, Z., 2019. An assessment of saltwater intrusion in the Changjiang (Yangtze) River Estuary. This book.

Little, S., Wood, P.J., Elliott, M., 2017. Quantifying salinity-induced changes on estuarine benthic fauna: the potential implications of climate change. Estuar. Coast. Shelf Sci. 198 (B), 610–625. https://doi.org/10.1016/j.ecss.2016.07.020.

Lovecraft, A.L., Meek, C.L., 2019. Arctic coastal systems: evaluating the DAPSI(W)R(M) framework. This book.

MacKinnon, J., Verkuil, Y., Murray, N., 2012. In: IUCN situation analysis on East and Southeast Asian intertidal habitats, with particular reference to the Yellow Sea (including the Bohai Sea). Occasional Paper of the IUCN Species Survival Commission No. 47. IUCN, Gland, Switzerland. pp. 70.

Matsuda, O., Yanagi, T., 2019. Restoration of estuaries and bays in Japan—what's been done so far and future perspectives. This book.

McLeod, I., et al., 2019. Can bivalve habitat restoration improve degraded estuaries? This book.

McLusky, D.S., Elliott, M., 2004. The Estuarine Ecosystem: Ecology, Threats and Management, third ed. Oxford University Press, Oxford, p. 216.

Moline, M.A., Karnovsky, N.J., Brown, Z., Divoky, G.J., et al., 2008. High latitude changes in ice dynamics and their impact on polar marine ecosystems. Ann. N. Y. Acad. Sci. 1134, 267–319.

Murray, N., Phinn, S., Clemens, R., Roelfsema, C., Fuller, R., 2012. Continual scale mapping of tidal flats across East Asia using the Landsat archive. Remote Sens. (Basel) 4, 3417–3426. https://doi.org/10.3390/rs4113417.

Murray, N., Clemens, R., Phinn, S., Possingham, H., Fuller, R., 2014. Tracking the rapid loss of tidal wetlands in the Yellow Sea. Front. Ecol. Environ. 12, 267–272. https://doi.org/10.1890/130260.

Murray, N., Ma, Z., Fuller, R., 2015. Tidal flats of the Yellow Sea: a review of ecosystem status and anthropogenic threats. Austral Ecol. 40, 472–481. https://doi.org/10.1111/aec.12211.

Nhan, N.H., Cao, N.B., 2019. Damming the Mekong: impacts in Vietnam and solutions. This book.

Nienhuis, P.H., Smaal, A.C. (Eds.), 1994. The Oosterschelde Estuary: A Case-Study of a Changing Ecosystem. Kluwer Academic Publication, Dordrecht, The Netherlands, pp. 597.

Ning, A., Scheren, P., Diop, S., Kane, C., Koulibaly, C.T., 2019. The Senegal and Pagane Rivers: examples of over-used river systems within water stressed environments in Africa. This book.

Olenin, S., Elliott, M., Bysveen, I., Culverhouse, P., et al., 2011. Recommendations on methods for the detection and control of biological pollution in marine coastal waters. Mar. Pollut. Bull. 62 (12), 2598–2604.

Palomares, M.-L.D., Pauly, D., 2019. Coastal Fisheries: the past, present and possible futures. This book.

Patricio, J., Elliott, M., Mazik, K., Papadopoulou, K.-N., Smith, C.J., 2016. DPSIR—Two decades of trying to develop a unifying framework for marine environmental management? Front. Mar. Sci. 3, 177. https://doi.org/10.3389/fmars.2016.00177.

Pérez-Ruzafa, A., Perez-Ruzafa, I.M., Newton, A., Marcos, C., 2019. Coastal lagoons: environmental variability, ecosystem complexity and goods and services uniformity. This book.

Purvaja, R., Yogeswari, S., Tudu, D., Hariharan, G., et al., 2019. Challenges and opportunities in management of Coral Islands of Lakshadweep, India. This book.

Ramesh, R., Lakshmi, A., Mohan, S., Bonthu, S.R., et al., 2019. Integrated management of the Ganges Delta, India. This book.

Richmond, R.H., Golbuu, Y., Shelton, A.J.I.I.I., 2019. Successful management of coral reef-watershed networks. This book.

Robins, P.E., Lewis, M.J., 2019. Changing hydrology: a UK perspective. This book.

Robins, P.E., Skov, M.W., Lewis, M.J., Gimenez, L., et al., 2016. Impact of climate change on UK estuaries: a review of past trends and potential projections. Estuar. Coast. Shelf Sci. 169, 119–135.

Saul, R., Barnes, R., Elliott, M., 2016. Is climate change an unforeseen, irresistible and external factor—a *force majeure* in marine environmental law? Mar. Pollut. Bull. 113 (1–2), 25–35.

Scharin, H., Ericsdotter, S., Elliott, M., Turner, R.K., et al., 2016. Processes for the sustainable stewardship of marine environments. Ecol. Econ. 128, 55–67.

Shalovenkov, N., 2019. Alien species invasion: case study of the Black Sea. This book.

Sharifinia, M., Daliri, M., Kamrani, E., 2019. Estuaries and coastal zones in the Northern Persian Gulf (Iran). This book.

Sklar, F.H., Meeder, J.F., Troxler, T.G., Dreschel, T., et al., 2019. The everglades: at the forefront of transition. This book.

Sloan, M., Whiteley, N., 2016. Stressors in the Marine Environment: Physiological and Ecological Responses: Societal Implications. Oxford University Press, Oxford, pp. 384.

Snoeijs-Leijonmalm, P., Andren, E., 2017. Why is the Baltic Sea so special to live in? In: Snoeijs-Leijonmalm, P., Schubert, H., Radziejewska, T. (Eds.), Biological Oceanography of the Baltic Sea. Springer, Dordrecht, pp. 23–84. (Chapter 2).

Turner, R.K., Schaafsma, M., 2015. Coastal Zones Ecosystem Services: From Science to Values and Decision Making. Springer Ecological Economic Series. Springer International Publication, Switzerland, ISBN: 978-3-319-17213-2.

Tweedley, J.R., Dittmann, S.R., Whitfield, A.K., Withers, K., et al., 2019. Hypersalinity: global distribution, causes and present and future effects on the biota of estuaries and lagoons. This book.

Vaalgamaa, S., Conley, D.J., 2008. Detecting environmental change in estuaries: nutrient and heavy metal distributions in sediment cores in estuaries from the Gulf of Finland. Baltic Sea. Estuar. Coast. Shelf Sci. 76 (1), 45–56.

Valesini, F.J., Hallett, C.S., Hipsey, M.R., Kilminster, K.L., Huang, P., Hennig, K., 2019. Peel-Harvey Estuary, Western Australia. This book.

Van den Belt, M., Costanza, R., 2011. Ecological economics of estuaries and coasts. In: Treatise on Estuarine and Coastal Science. vol. 12. Elsevier, Amsterdam.

Whitfield, A.K., Elliott, M., 2011. Ecosystem and Biotic Classifications of Estuaries and Coasts. Treatise on Estuarine & Coastal Science. vol. 1. Elsevier, Amsterdam, p. 99–124.

Wiegman, R., Day, J., D'Elia, C., Rutherford, J., Morris, J., Roy, E., Lane, R., Dismukes, D., Snyder, B., 2017. Modeling impacts of sea-level rise, oil price, and management strategy on the costs of sustaining Mississippi delta marshes with hydraulic dredging. Sci. Total Environ. https://doi.org/10.1016/j.scitotenv.2017.09.314.

Wolanski, E., 2019. Estuarine ecohydrology modeling: what works and within what limits? . This book.

Wolanski, E., Elliott, M., 2015. Estuarine Ecohydrology: An Introduction. Elsevier, Amsterdam, p. 322.

Wolanski, E., Andutta, F., Deleersnijder, E., Li, Y., Thomas, C.J., 2017. The Gulf of Carpentaria heated Torres Strait and the Northern Great Barrier Reef during the 2016 mass coral bleaching. Estuar. Coast. Shelf Sci. 194, 172–181.

Xia, P., Meng, X., Yin, P., Cao, Z., Wang, X., 2011. Eighty-year sedimentary record of heavy metal inputs in the intertidal sediments from the Nanliu River estuary, Beibu Gulf of South China Sea. Environ. Pollut. 159 (1), 92–99.

Section A

Estuaries

Chapter 2

An Assessment of Saltwater Intrusion in the Changjiang (Yangtze) River Estuary, China

Maotian Li[*†], Zhongyuan Chen[*]

*State Key Laboratory of Estuarine and Coastal Research, East China Normal University, Shanghai, People's Republic of China †Institute of Eco-Chongming, Shanghai, China

1 INTRODUCTION

The Changjiang River is now the main source of freshwater supply to Shanghai, in terms of the large-scale Qingcaosha reservoir, which presently supplies up to 70% of the freshwater to Shanghai. This reservoir was built in the freshwater region of the estuary and is mostly likely to become even more important in the future. Demand is rising as the population (presently >23 million) is growing. Since the local sources, mainly the Taihu basin with 30% freshwater source (Fig. 1B), are heavily polluted, threats relating to water diversions and saltwater intrusions already occur in the Changjiang River estuary in the dry season (Chen et al., 2001; Gong et al., 2013). The ultimate purpose of this study is to understand the nature of the salinity distribution and its persistence through time in the Changjiang River mouth. This is of particular concern given that only 68 days of freshwater supply for Shanghai can be stored in the reservoirs in the Changjiang River estuary.

1.1 Water Transfer Projects

The Changjiang River, China is listed as the fifth largest river in the world in terms of annual discharge (Fig. 1A). Its catchment holds a population exceeding 400 million, accounting for ca. 35% of the total population of China. Moreover, it produces >25% of its gross domestic product at the national scale. The Changjiang River estuary is a large eco-complex, which receives not only a huge amount of freshwater ($960 \times 10^9 \, \mathrm{m^3 \, a^{-1}}$) from the river basin, but also holds ca. 100 million people, including ca. 23 million living in the metropolitan city Shanghai (Fig. 1A).

The Changjiang River flows from west to east roughly along ~31°N into the East China Sea. It is located in the monsoon region with a huge amount of runoff, ca. 35% of the total of China, to irrigate ca. 25% of the arable land of China. In contrast, the annual runoff of Northern China, where the Yellow River basin is located, takes only ca. 17% of the total, while sustaining ca. 62% of the total arable land of China (http://www.360doc.com/content/15/0827/14/7536781_495176991.shtml). Also, the Yellow River flow has at times occasionally ceased or has severely decreased to a trickle in the last few decades due to climate and anthropogenic activities, including over-irrigation and numerous water diversions in the upper basin (Pereira et al., 2007).

Given these severe challenges, an initiative was proposed by administrators and scientists decades ago, calling for the South-North Water transfer, which was to divert water from the Changjiang River basin to supply water to Northern China. This initiative was implemented via three geo-engineering routes, that is, the West-, the Middle-, and the East-Route (Fig. 1A). The capacity of water diversion via the West-Route averages at $145–195 \times 10^8 \, \mathrm{m^3 \, year^{-1}}$, that via the Middle-Route totals $130 \times 10^8 \, \mathrm{m^3 \, year^{-1}}$, and that via the East-Route is $278.6 \times 10^8 \, \mathrm{m^3 \, year^{-1}}$ (sourced from: http://www.nsbd.gov.cn/).

The Changjiang Water Conservancy Committee (2007) also reported that there are hundreds of water diversion projects in the lower Changjiang River basin (Fig. 1B). The total designed capacity of water diversion reaches $788 \times 10^9 \, \mathrm{m^3 \, year^{-1}}$. Saltwater intrusion in the estuary would have been extreme had all water diversion projects been in operation at the same

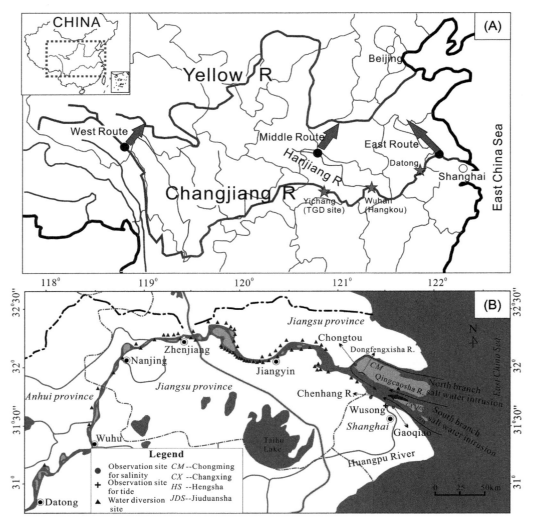

FIG. 1 (A) Map of China; the arrows show the South-North Water Transfer Project: West-Route, Middle-Route, and East-Route; ☆—hydrological gauging station. Datong represents the lower Changjiang below Hankou; Hankou represents the area between Yichang and Hankou, and Yichang represents the upper Changjiang; and the TGD site. (B) The lower Changjiang River estuary showing sites of water diversions, the location of freshwater reservoirs, and the sites where hydrological data were collected. Also indicated in shaded areas is the pattern of saltwater intrusion into the estuary via the South and North Branches. *(Data from Mao, Z.C., Shen, H.T., Yao, Y.D., 1993. Analysis of sources of saltwater intrusion along south bank of south branch of the Changjiang River estuary, Mar. Sci. Bull. 12(1):17–26; Mao, Z.C., 1994. The salt intrusion character of the South Channel in the Changjiang mouth, Shanghai Water Conserv. 4:22–32; Mao, Z.C., Shen, H.T., Xiao, C.Y., 2001. Saltwater intrusion patterns in the Qing Caosha area, Changjiang River estuary, Oceanol. Limnol. Sin. 32(1):58–66; Wu, H., 2006. Saltwater intrusion in the Changjiang River estuary (Ph.D. thesis) East China Normal University, pp. 205; Xu, J.Y., Yuan, J.Z., 1994. The research on salt-water intrusion into the south branch of the Changjiang River estuary, Hydrology 5:1–5; Zhu, J.R., Wu, H., Li, L., Wang, B., 2010. Saltwater intrusion in the Changjiang River estuary in the extremely drought hydrological year 2006, J. East China Normal Univ. (Nat. Sci.) 4:1–4; Tang, J.H., Xu, J.Y., Zhao, S.W., Liu, W.Y., 2011. Research on saltwater intrusion of the south branch on the Changjiang River estuary based on measured data, Res. Environ. Changjiang Basin 20(6):677–685; Li, L., 2012. Spatial-temporal dynamic characteristics of saltwater intrusion in the Changjiang River estuary, (Ph.D. thesis) East China Normal University, 161 pp.)*

time; fortunately for Shanghai they were not. Had these diversions all occurred at the same time, they would have resulted in severe social issues in the river basin arising from water diversion because they would have increased the salinity intrusion in the water reservoirs for Shanghai.

There are two major concerns with regards to freshwater resources: (1) the timing of water diversion. An example is the year 2010 when a drought occurred in the middle Changjiang River basin, while $83.93 \times 10^8 \, m^3$ was diverted from the Danjiangkou Reservoir (Fig. 1A); as a result the freshwater supply for people's livelihood in the middle Changjiang River basin was ca. 3.8% less than during a normal year (http://www.nsbd.gov.cn/). (2) Saltwater intrusion. Saltwater intrusion into the river delta is extremely sensitive to freshwater diversion especially during the dry season (winter). It appears that the saltwater intrusion in recent years has occurred earlier in winter than ever before, and its duration was also longer, as was the case during 2006 (Chen and He, 2009). Considering this case, the Chenhang Reservoir (Fig. 1B) was built as a tidal-flat reclamation along the southern bank of the South Branch of the Changjiang estuary. It is an emergency source of freshwater for Shanghai, with a storage capacity of ca. $9.5 \times 10^6 \, m^3$ (Li et al., 2011a, b); for instance saltwater intrusion lasted about 10 days in February 2009, and resulted in a severe shortage of freshwater for Shanghai (Zheng et al., 2014).

1.2   Three Gorges Dam: Changing Hydrology at Seasonal Scales

The Three Gorges Dam (TGD), which is located in the middle Changjiang River basin at Yichang (Fig. 1A), was built to generate hydroelectricity and assists in flood control. There has been increasing argument about the TGD with regard to its hydrological function and its estuarine ecological and social responses (e.g., Chen and He, 2009; Yang et al., 2015).

Truly, TGD is the biggest dam in the world, but this is only in terms of the size of the dam wall and its electricity generating capacity. In hydrological terms, TGD is quite small, since it can store only 4% of the mean annual flow of the Changjiang River and is therefore not capable of changing significantly the hydrological characteristics of the river at annual time scales (Fig. 2, Insert map; Chen et al., 2014). Also, there are abundant freshwater resources in the Changjiang River catchment; this is evident from the relationship between the annual runoff and annual precipitation recorded at the three major hydrological gauging stations (Yichang, Hankou and Datong; see Fig. 1A for their location), indicating that the impact of the dam on the annual freshwater discharge is quite limited.

Nevertheless, Fig. 2 shows that hydrographic changes at seasonal time scales have occurred in relation to the dam operation since 2011 (Yuan et al., 2016). The TGD began storing water in 2003, which cut down sediment flux dramatically into the coast, but not the water discharge at annual time scales (Yuan et al., 2012; Chen et al., 2014). Once flow data for the Changjiang River in the post TGD period became available, question were raised and discussions were initiated about the significance of the observed changing trends of the monthly and seasonal flows of water and sediment for the estuary. Yang et al. (2015) argued that the seasonal effects of impoundment and release affect short-term discharge and not overall annual flow. Yuan et al. (2012) estimated that the highest 30% of flows are reduced in magnitude while the lowest 40% of flows have increased. The TGD reduces water flows during the high flow season (June to October) but increases the flows during the low flow months (November to May).

1.3   Sea Level Rise

According to the long-term observation reported by the State Ocean Administration (SOA), China, the mean sea level (MSL) has risen >1.0 m since 1980 on the east coast of China, and the sea level rise on the Changjiang coast is 2.1 mm per year on average (Fig. 3).

In summary, the above three natural and anthropogenic forcings are the key factors that have played an integrated role in affecting saltwater intrusion into the Changjiang River estuary. We present below further evidence and conclusions based on our long-term data.

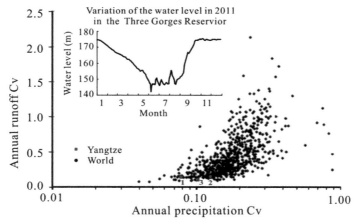

FIG. 2   The lowest rate of runoff Cv versus precipitation Cv at three major hydrological stations of the Changjiang basin: 1—Yichang, 2—Hankou; 3—Datong (see Fig. 1A for their location). Inset map shows the influence of the operational scheme of Three Gorges Dam on the water level in the reservoir.*(Modified from Yuan WH, Li MT, Chen ZY, Yin DW, Wei TY: Responses of runoff-sediment flux and bedform in the middle Yangtze River to the completion of the Three Gorges Dam, J East China Normal Univ (Nat Sci) 2:90–100, 2016 (in Chinese, with English summary).)*

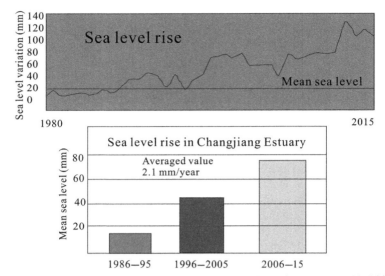

FIG. 3 Sea level rise at the East China Sea coast. *(Data from http://www.coi.gov.cn/gongbao/nrhaipingmian/nr2015/201603/t20160328_33810.html.)*

2 DATA SOURCES AND OBSERVATION

Saltwater intrusion into the Changjiang River estuary is a natural process, but it has been intensified by human activities in the basin in recent decades. To quantify this argument, we now show the variations in discharge, salinity, tidal level, and water diversion in the study area over the past 50 years.

2.1 Discharge Variations

Daily discharge data from Datong (Fig. 1A) were sourced from the hydrology yearbooks of the Changjiang for the periods 1950–87 and 1997–2011 (The Changjiang Water Conservancy Committee, 1950–1986, 1997–2011). Datong is the furthest downstream hydrological gauging station, ca. 600 km upstream of the river mouth. From the daily discharges shown in Fig. 4A the high flow season can be defined as May 1–October 30 and the low flow season as November 1–April 30. We have grouped daily discharge into four types, that is, high flow, normal flow, low flow, and extreme low flows. High flow is the maximum discharge recorded on a particular day of the year in the 51-year record; normal flow is the average discharge on the same day of the year over 51 year; extreme low flow is the minimum discharge on that date over the 51 year; and low flow is 10% more than the extreme low flow (Fig. 4B and C).

In this chapter we focus on the dry season, when strong saltwater intrusion occurs normally in the estuary. The frequency distribution of daily discharges was used to determine the probability of particular threshold flows that will be cited in the following discussion (Fig. 4D and E).

The daily discharge recorded at Datong station (Fig. 4A) averages $25,500 \, \mathrm{m^3 \, s^{-1}}$ and shows no trend through time. The four types of discharge, that is, high flow, normal flow, low flow, and extreme low flow, are all $>15,000 \, \mathrm{m^3 \, s^{-1}}$ during the wet season (Fig. 4B). This is in contrast with the dry season when the extreme low flow discharge is always below $15,000 \, \mathrm{m^3 \, s^{-1}}$, and the low flow and normal flow discharges are below $15,000 \, \mathrm{m^3 \, s^{-1}}$, accounting for 70% and 50% of the dry season, respectively (Fig. 4C).

The frequency distribution of daily discharge in the dry season shows a rapid rise from 0.1% to 1.7% over the range of discharge from 6000 to $10,000 \, \mathrm{m^3 \, s^{-1}}$ (Fig. 4D). At discharges above $10,000 \, \mathrm{m^3 \, s^{-1}}$ the frequency rapidly drops to ca. 0.2% as discharge rises to ca. $30,000 \, \mathrm{m^3 \, s^{-1}}$, and further to nearly 0.1% as discharge increases to ca. $50,000 \, \mathrm{m^3 \, s^{-1}}$. The cumulative frequency plot (Fig. 4E) shows that 50% of flows are less than or equal to $15,000 \, \mathrm{m^3 \, s^{-1}}$ and 20% are less than or equal to $10,000 \, \mathrm{m^3 \, s^{-1}}$.

2.2 Salinity Variations

Daily salinity readings for 769 days when saltwater intrusions occurred between 1979 and 2011 were collected from the three key hydrological gauging stations, that is, Gaoqiao (representing the Qingcaosha Reservoir, 316 days), Chenhang Reservoir (277 days), and Chongtou (representing Dongfengxisha reservoir, 176 days), all located in the Changjiang River estuary (Fig. 1B).

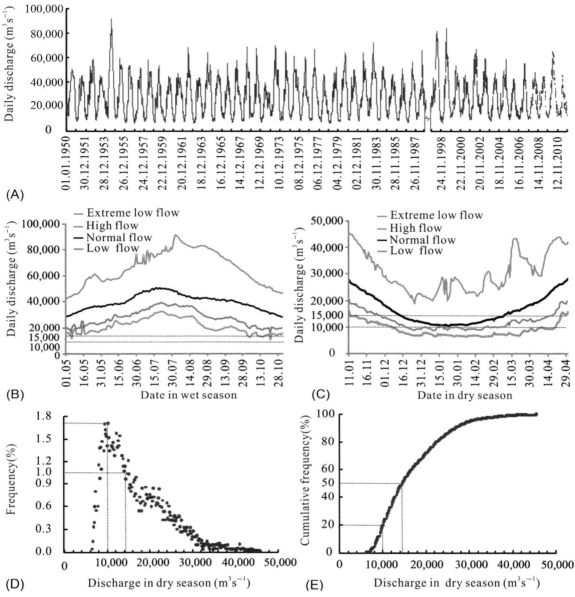

FIG. 4   (A) Daily river discharge at Datong from 1951 to 2011. (B) The range of daily discharge in the wet season. (C) The range of daily discharge in the dry season. (D) Frequency distribution of daily discharge in the dry season. (E) The cumulative frequency of daily discharge in the dry season.

The measured daily salinity is plotted against the discharge measured at Datong in Fig. 5 (subsets A1–A3) with a 7-day lag to account for the flow time between Datong and the river mouth. Salinity values were divided into different sections, each with $150\,mg\,L^{-1}$ range, and their mean value was plotted against the river discharge at Datong (with a 7-day lag) in Fig. 5 (subsets B1–B3), together with the maximum and minimum value for each range. Least squares regressions have been fitted to each of the minimum, mean, and maximum series for each station and these regressions are all significant ($P < .001$). Fig. 5 (subsets C1–C3) shows the duration (in days) of salinity events when the maximum salinity recorded was greater than $250\,mg\,L^{-1}$. As would be expected in this situation, higher salinities are reached as the duration of the event increases.

There is a negative correlation between discharge and salinity during the dry season. Salinity at the three stations decreases from $3400\,mg\,L^{-1}$ to nearly zero, which is associated with an increase in discharge from 6000 to $20,000\,m^3\,s^{-1}$ (Fig. 5A and B). Salinity was frequently greater than $1000\,mg\,L^{-1}$ (1 PSU) when the river discharge fell below the threshold of $15,000\,m^3\,s^{-1}$. Also of note is that the salinity corresponding to a river discharge of $15,000$–$20,000\,m^3\,s^{-1}$ at the

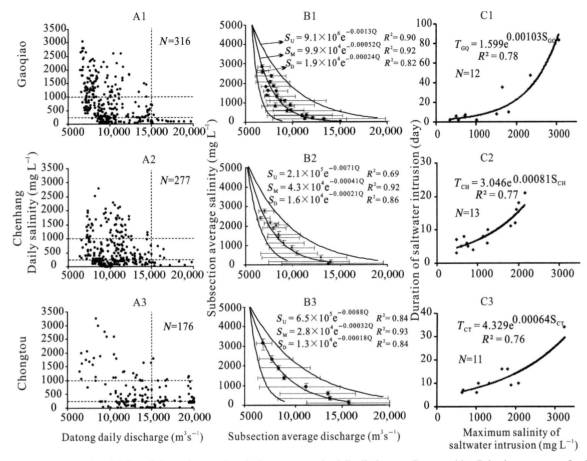

FIG. 5 (A1–A3) Scatter plot of daily salinity at three stations in the estuary vs the daily discharge at Datong with a 7-day lag to account for the travel time of water to the estuary. (B1–B3) The correlation between daily salinity and river discharge (with a 7-day lag as above), showing sectional minima, means, and maxima (see the text for details). (C1–C3) Duration of the salinity intrusion in relation to maximal salinity level during saltwater intrusion events in the estuary.

three stations showed a spatial difference, that is, from upstream to downstream, from 100 to $200\,\mathrm{mg\,L^{-1}}$ at Gaoqiao, to 200–$500\,\mathrm{mg\,L^{-1}}$ at Chenhang, and to 100–$1500\,\mathrm{mg\,L^{-1}}$ at Chongtou (Figs. 1B and 5A1–A3).

Fig. 5B shows that there is a negative relation between mean salinity and river discharge at all three stations and this is also the case for minimum and maximum salinities. As shown in Fig. 5C, high salinity persists for significantly longer durations than low salinities.

2.3 Water Diversion

The annual water diversions in the mid-lower Changjiang River basin for the period 2005–10 were obtained from the Changjiang Water Conservancy Commission (The Water Resources Department of Anhui Province, 2005–2011; The Water Resources Department of Jiangsu Province, 2005–2011; The Water Resources Department of Shanghai, 2005–2011). This includes the total amount of water diverted via the S-N water project and local water diversion projects.

There are 739 water diversion projects in the lower Changjiang basin below Datong as reported by The Changjiang Water Conservancy Committee (2007). The designed capacity for water diversion of these projects in 2000 totals $788 \times 10^9\,\mathrm{m^3}$ $(24{,}979\,\mathrm{m^3\,s^{-1}})$, but the actual monthly water diversion ranges from 2.5 to 6.0 billion $\mathrm{m^3\,a^{-1}}$ $(800$–$2200\,\mathrm{m^3\,s^{-1}})$ (Fig. 6A), and the yearly total is $38.6 \times 10^9\,\mathrm{m^3}$ $(1250\,\mathrm{m^3\,s^{-1}})$ (Fig. 6B).

Water diversions are expected to increase in the future following the pattern of increases in the past. The pattern of monthly water diversions in the Changjiang River estuary in 2000 depicted in Fig. 6A can be referenced for future simulation. Fig. 6B shows the time-series plot of annual water diversions against years (with the year 2000 being assigned the value of 1) and used this to predict the possible diversions out to 2040. Using the diversion levels for 2000, we apportioned

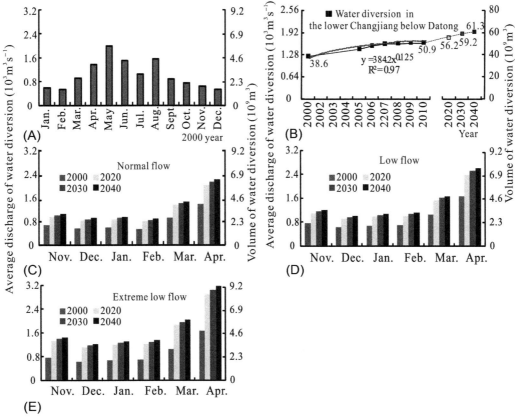

FIG. 6   (A) Monthly water diversions from the lower Changjiang River reach in 2000. (B) Water diversions in 2000 and 2005–10, and predicted water diversions in 2020, 2030, and 2040. (C–E) Prediction of monthly water diversion in the dry season for normal flow, low flow, and extreme low flow, for 2020, 2030, and 2040.

the water diversion of the dry season for 2020–40, based on normal flow conditions. Three scenarios were made for the normal flow, low flow, and extremely low flow types. The future water diversions for the low flow and the extreme low flow type were based on the assumption of 10% and 15% more than that of the normal flow type (Zhang et al., 2003) (Fig. 6D–E).

The actual water diversions over the period 2000–10 has increased gradually from 38.6 billion m^3 to 50.9 billion m^3 (Fig. 6B). Using the regression shown in Fig. 4B, the predicted annual water diversions for 2020, 2030, and 2040 are 56.2, 59.2, and 61.3 × 10^9 m^3, respectively. The future monthly water diversions in the dry season, apportioned according to the pattern in 2000 (normal flow; Fig. 6A), will be more than that of 2000, as presented further.

Normal flow type: The water volume to be diverted in 2020, 2030, and 2040 is 0.6–1.8 billion m^3 (200–600 m^3 s^{-1}), 0.75–2.1 billion m^3 (250–700 m^3 s^{-1}), and 0.9–2.4 billion m^3 (300–800 m^3 s^{-1}), respectively.

Low flow type: The water volume to be diverted in 2020, 2030, and 2040 is 0.9–2.1 billion m^3 (300–700 m^3 s^{-1}), 1.05–2.4 billion m^3 (350–800 m^3 s^{-1}), and 1.2–2.7 billion m^3 (400–900 m^3 s^{-1}), respectively.

Extreme low flow type: The water volume to be diverted in 2020, 2030, and 2040 is 1.2–2.4 billion m^3 (400–800 m^3 s^{-1}), 13.5–2.7 billion m^3 (450–900 m^3 s^{-1}), and 1.5–3 billion m^3 (500–1000 m^3 s^{-1}), respectively (Fig. 6C–E).

2.4   Tide Level

The monthly average high tide levels at Wusong station (Fig. 1B) for the period 1967–85 were collected from the hydrology yearbook of the lower Changjiang (The Changjiang Water Conservancy Committee, 1967–1985) and are plotted in Fig. 7A together with discharge. Fig. 7B shows the relationship between the high tide level at Wusong and the discharge at Datong, separated into the wet and dry season flows. The tidal levels in the dry season were separated into 11 height ranges

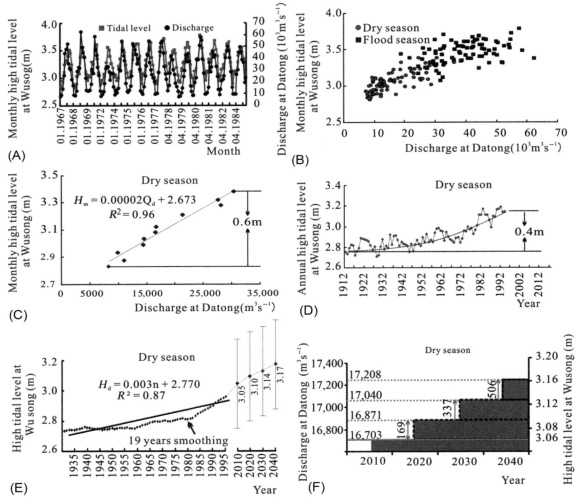

FIG. 7 (A) Time-series plot of the monthly high tide levels at Wusong and the river discharge Datong for the period 1967–84. (B) The relationship between monthly high tide water level and discharge for the period 1967–84. (C) The scatter plot of monthly high tide water level and the monthly river discharge in the dry season (note the 0.6-m water level difference between the high and low flows). (D) Sea level rise as recorded by the high tide levels at Wusong (1912–94). (E) Trend lines of high tide level at Wusong for a linear trend and for a 19-year smoothing of dry season water level (SAROS cycle) and the predicted water levels in 2020, 2030, and 2040. (F) Equivalent river discharge ($m^3 s^{-1}$) decreases as a result of sea level rise and its impact on salinity intrusion for the dry season in 2020, 2030, and 2040 (see details in the text).

at intervals of 5 cm from 2.81 to 3.38 m, and correlated with the dry season discharges (Fig. 7C). This shows that discharge from upstream can influence the high tide level by up to 0.6 m.

The mean annual high tide level at Wusong station for the period 1912–94 was sourced from Chen (1990) and Zheng and Yu (1996). The Wusong datum was used in this study. The tidal levels were referred to the Sheshan datum to remove the effect of land subsidence. The difference in high tide level between the annual and dry season tidal levels was 0.31 m over the period 1967–85. This value was subtracted from the annual average values dataset to obtain tide level fluctuations in the dry seasons of 1912–94 (Fig. 7D).

We used the SAROS cycle of approximately 19 years to smooth the dry season tidal levels (Fig. 7E), and then applied the regression equation of high tide level against the year (1934 = year 1) to predict the future high tide water levels in 2020, 2030, and 2040 (Fig. 7E). Combining the equation that describes the effect of upstream discharge on tidal level ($H_d = 0.00002Q_d + 2.6729$, Fig. 7C) and the equation that describes the change in the high tide levels through time ($H_d = 0.003n + 2.770$, Fig. 7E) allows us to calculate the future monthly discharge into the estuary, which correlates with different water level positions (Fig. 7F). This corresponds to an average discharge 16,703 $m^3 s^{-1}$ for the dry season of 2010 as recorded at Datong.

3   DISCUSSIONS

3.1   Freshwater Sources—Dry Season Shortage

Undoubtedly, there is much freshwater available in the Changjiang River basin as the annual precipitation averages 1042 mm across the whole basin (Finlayson et al., 2013), and the annual discharge flowing into the estuary averages $900 \times 10^9 \, m^3$ (Chen et al., 1988), with a relatively small interannual variability and thus a long-term stability (Fig. 4A). However, the seasonal differences in freshwater discharge are quite substantial (Fig. 4B and C). Obviously, no saltwater intrusion occurs in the wet season, but in the dry season, when discharge falls below $15,000 \, m^3 \, s^{-1}$, this occurs frequently (Mao et al., 2000; Shen et al., 2003).

Therefore, the dry season is the focus of this discussion, concentrating on the three flow types, that is, normal flow, low flow, and extreme low flow (Fig. 4C), with discharge of $<15,000 \, m^3 \, s^{-1}$ for 50%, 70%, and 100% of the time, respectively (Fig. 4C). The high flow type in the dry season is not discussed here, since flow $>15,000 \, m^3 \, s^{-1}$ generates no significant saltwater intrusion (Fig. 4C).

Clearly, flows in the range $8000–15,000 \, m^3 \, s^{-1}$ are critical, as this range includes the highest frequencies of flows in the dry season (Fig. 4D). The cumulative frequency plot (Fig. 4E) indicates that discharge $<15,000 \, m^3 \, s^{-1}$ occurs 50% of the time during the dry season. This range of discharges is therefore important in relation to the upstream water diversion projects and future sea level rise.

3.2   Salinity Distribution in Relation to Freshwater Availability

The spatial differences of salinity at Gaoqiao, Chenhang, and Chongtou reveal significant features in the pattern of saltwater intrusion (Fig. 5A). The higher salinities at river discharges $<15,000 \, m^3 \, s^{-1}$ at Gaoqiao and Chongtou station than at Chenhang indicates that the saltwater intrusion occurs not only directly from the South Branch but also as an intrusion around Chongming Island of brackish water from the North Branch of the estuary into the South Branch (Figs. 1B and 5A) (Shen et al., 2003; Xue et al., 2009). In the range of river discharge $15,000–20,000 \, m^3 \, s^{-1}$, the salinity decreases only slightly from the river mouth at Chongtou landward in the South Branch to Chenhang, and Gaoqiao (Fig. 1B). This demonstrates the importance of the saltwater intrusions via the North Branch in the overall salinity levels in the main river channel, the South Branch, where the Shanghai Municipality sources some 70% of its freshwater supplies (Shen et al., 2003; Mao et al., 2000; Xue et al., 2009).

A salinity level of $250 \, mg \, L^{-1}$ is the limit for freshwater intake to the Shanghai supply system (Shen et al., 2003). It needs also to be noted that Shanghai only has the capacity to store up to 68 days of freshwater supply. The spatial differences of salinity distribution show that at Gaoqiao (Qinchaosha Reservoir), while discharge is $>15,000 \, m^3 \, s^{-1}$, the salinity is below the threshold value of $250 \, mg \, L^{-1}$, while at Chongtou (Dongfengxisha Reservoir) and Chenhang station, salinity $> 250 \, mg \, L^{-1}$ occurs frequently (Fig. 5A1–A3). This is a direct consequence of the saltwater intrusions that comes via the North Branch. Salinity levels $>250 \, mg \, L^{-1}$ occur at all the three stations while the river discharge is $<15,000 \, m^3 \, s^{-1}$ (Fig. 5A1–A3) and this occurs for ca. 90 days during the dry season (Fig. 4E).

The 90 days (50% probability; Fig. 4E) with salinity $>250 \, mg \, L^{-1}$ were summed from discrete events of saltwater intrusions occurring at the river mouth station. However, our analyses show not only the negative correlation of salinity with discharge (Fig. 5B), but also the fact that the higher the salinity, the longer the time duration (Fig. 5C). In particular, the results derived for the three levels (minimum, sectional mean, and maximum; Fig. 5B) suggest that high salinities can persist longer than the 68-day storage limit (see below for further discussion).

3.3   Water Diversion—Present and Future Case

As we know that there are 739 water diversion projects below Datong, in addition to the S-N water transfer project, our investigations have verified that the design capacity of these water diversions, reaching $788 \times 10^9 \, m^3 \, a^{-1}$, is already equal to 82% of the mean annual discharge at Datong ($960 \times 10^9 \, m^3$). In fact, the monthly water diversion in 2000 ranged from $700–1500 \, m^3 \, s^{-1}$ ($1.8–3.9 \times 10^9 \, m^3$) (February–April) to $2200 \, m^3 \, s^{-1}$ ($5.7 \times 10^9 \, m^3$) (May) (Fig. 6A). Also, the annual water diversion in 2000 was $1250 \, m^3 \, s^{-1}$ ($38.6 \times 10^9 \, m^3$), which had increased to $1650 \, m^3 \, s^{-1}$ ($50.9 \times 10^9 \, m^3$) in 2010 (Fig. 6B). Assuming this increase continues at its historic rate, then it will reach $1850 \, m^3 \, s^{-1}$ ($56.2 \times 10^9 \, m^3$), $1920 \, m^3 \, s^{-1}$ ($59.2 \times 10^9 \, m^3$), and $2000 \, m^3 \, s^{-1}$ ($61.3 \times 10^9 \, m^3$) in 2020, 2030, and 2040, respectively (Fig. 6B).

Water diversion that occurs between May and October (wet season) does not create problems in relation to saltwater intrusions because of the high discharge. However, the quantity of water diverted in the dry season is a potential threat to freshwater supplies in the upper estuary, especially considering there is a 50% probability that the discharge will be small, in the range of 8000–15,000 $m^3 s^{-1}$ (Fig. 4E).

According to our predictions, there could be a further water diversion of ca. 800–2200, 1000–2800, and 1200–3200 $m^3 s^{-1}$ in the dry season of 2020, 2030, and 2040, for the normal flow, low flow, and extreme flow types, respectively (Fig. 6C–E). These values thus need to be subtracted from the monthly average discharge in the dry season in the discussion of future scenarios below about the effective river flow reaching the estuary (Figs. 4C and 8).

3.4 Sea Level Rise—Equivalent to Increase in Salinity and Decrease in Discharge

Fig. 7A and B shows that there is a correlation between high tide level at the mouth and discharge at monthly time scales; while this fluctuation at the river mouth does not represent sea level fluctuation, the long-term trend of the high tide level can be used to estimate the trend in sea level rise, based on the observation of monthly high tidal level to Datong discharge (Fig. 7C, 0.6 m driven up by an increasing discharge from 10,000 to 35,000 $m^3 s^{-1}$). This monthly rate can be added to the long-term (interannual) water level (sea level) at Wusong, leading to an observation that the average high tide level (sea level) of the dry season 1912–95 had risen by 0.4 m (Fig. 7D).

Using SAROS (~19-year cycle) simulated future sea level rise predictions, the peak high tide during the dry season has reached 3.05 m of 2010 (and this is verified by the long-term observation by the State Oceanic Administration, China; http://www.soa.gov.cn), and will reach 3.10 m of 2020, 3.14 m of 2030, and 3.17 m of 2040 (Fig. 7D and E).

This prediction based on a linear regression (Fig. 7E) is more conservative than that if an exponential curve is fitted to the SAROS smoothed data in Fig. 7D. This simulated water level is at annual scale, and to it the additional value of tidal fluctuation at monthly scale must be added, that is, 0.6 m as recorded at Wusong (Fig. 7C and E).

Increase in sea levels in the future will drive up the salinity in the river mouth area, which is basically equivalent to a decrease in the river discharge as shown in Fig. 7F for the periods 2010–20, 2010–30, and 2010–40. To estimate the impact of salinity intrusion in the future as a result of MSL rise, the effective river discharge reaching the estuary was decreased by the values shown in Fig. 7F. The results will be used to discuss the effect of reduced discharge, in terms of an increase in salinity and its duration, in the future scenarios for 2020, 2030, and 2040.

4 FUTURE SCENARIOS

Fig. 8A shows the monthly average discharge for the low flow and extremely low flow months (November–April) for the past 51 year, and those predicted for 2020, 2030, and 2040 after subtraction of the amount lost to future water diversions and the equivalent effects of sea level rise on saltwater intrusion. The results show that: (1) in 2020–40, during low flow the dry season discharge will be reduced by ca. 2100–2600 $m^3 s^{-1}$, this comprises ca. 1930–2094 $m^3 s^{-1}$ due to water diversion averaged for 6 months of the dry season, and 169–506 $m^3 s^{-1}$ due to sea level rise (Fig. 8A1); and (2) during extreme low flow it will be reduced by 2200–2700 $m^3 s^{-1}$ (comprising 2030–2194 $m^3 s^{-1}$ due to water diversion and 169–506 $m^3 s^{-1}$ due to sea level rise) (Fig. 8A2).

These results are based on the assumption that future discharge at Datong will remain at 7000–11,000 and 6500–9000 $m^3 s^{-1}$ for low flow and extreme low flow, respectively, in the dry seasons during the period 2020–40 (Fig. 6A1 and A2).

Fig. 8B and C shows the salinity and time duration simulation (based on the correlation shown in Fig. 6C) for the Qingcaosha Reservoir. Taking the Qingcaosha Reservoir (represented by Gaoqiao station; Fig. 1) as the most important case, we can calculate the duration of high salinity using the minimum and maximum salinity values shown in Fig. 5B1. In the worst case with the lowest discharge in February, taking deviations ranging 20%–30%, 10%–12%, and 4%–5% (Fig. 8B2; the higher the salinity, the less the deviation, see Fig. 5B1), we estimate the longest time that the salinity would exceed 250 mg L^{-1} would be 20–65, 75–90, and 120–128 days for 2020, 2030, and 2040, respectively (Fig. 8C2).

These findings have important implications for China because they show that the water supply for Shanghai is threatened by dams, water diversions schemes, and the sea level rise. They also highlighted that (1) there is a need to continue on-site measurements in the dry season at the Qingcaosha Reservoir, in order to establish a direct correlation between discharge and salinity, and (2) further studies are needed to better assess the impact of future sea level rise on salinity in the river mouth area.

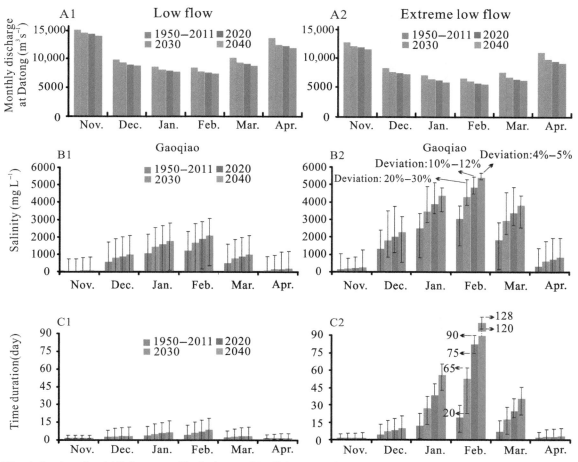

FIG. 8 Historical and predicted (A) future monthly discharges, (B) salinity at Gaoqiao (representing the Qingcaosha reservoir), and (C) duration of saltwater intrusion, for the dry season low flow and extremely low flow types under the scenario of combined sea level rise and water diversion.

5   THE WAY FORWARD

Our study clearly demonstrates that freshwater resources at the Changjiang River estuary are threatened at present and will be increasingly threatened in the near future. This is an important issue for China because the metropolitan city of Shanghai with >23 million people solely depend on these resources. Here we have proposed countermeasures that would be applicable for remediation in a long-term perspective for societal development, especially taking into consideration the local political and cultural background.

- River-basin scale: there is only one river basin, and it needs a better planning and management of water resources than the present administration by different provinces. In particular, while planning water diversion, regional benefits should compromise with each other in the interest of the whole basin.
- Management scale: governmental officers need to work more intensively with physical and social scientists for to stipulate and modify the relevant policies, especially on the basis of better understanding of the needs of people's demands in relation to the capacity of water resources of the basin.
- Scientific scale: scientists (physical and social) should put more efforts for observation and collection of dataset, and based on this, simulation or hydrological processes modeling of freshwater sources at different scales is highly necessary; subjectively, scientists should be more visible and more vocal in our society against any water-abusing bureaucracies.
- Stakeholder scale: all relevant communities should coherently work together toward a long-term target for harmonious development.
- Education scale: to diffuse water-related knowledge with regard to river-basin-estuary environmental protection is fundamental to our society toward sustainable development.

ACKNOWLEDGMENT

This project is financially funded by the National Natural Science Foundation of China (Grant No. 41671007).

REFERENCES

Chen, X.Q., 1990. Sea level changes from 1922 to 1987 in the Changjiang River mouth and its significance. Acta Geograph. Sin. 45 (4), 387–398.

Chen, J.Y., He, Q., 2009. Study of Water Security of Shanghai, a Special Reference to the Drought Year 2006. Ocean Press, Beijing, p. 145. (in Chinese).

Chen, J.Y., Shen, H.T., Yun, C.X., 1988. Processes of Dynamics and Geomophology of the Changjiang Estuary. Shanghai Scientific and Technical Publisher, Shanghai, p. 452. (in Chinese).

Chen, X.Q., Zong, Y.Q., Zhang, E.F., Xu, J.G., Li, S.J., 2001. Human impacts on the Changjiang basin, China, with special reference to the impacts on the dry season water discharges into the sea. Geomorphology 41, 111–123.

Chen, J., Wu, X.D., Finlayson, B.L., Webber, M., Wei, T.Y., Li, M., Chen, Z., 2014. Variability and trend in the hydrology of the Yangtze River, China: annual precipitation and runoff. Hydrology 513, 403–412.

Finlayson, B.L., Barnett, B., Wei, T.Y., Webber, M., Li, M.T., Wang, W.Y., Chen, J., Xu, H., Chen, Z., 2013. The drivers of risk to water security in Shanghai. Reg. Environ. Chang. 13, 329–340. https://doi.org/10.1007/s10113-012-0334-1.

Gong, W.P., Shen, J., Jia, L.W., 2013. Salt intrusion during the dry season in the Huangmaohai Estuary, Pearl River Delta. J. Mar. Syst. 111–112, 235–252.

Li, L., 2012. Spatial-temporal dynamic characteristics of saltwater intrusion in the Changjiang River estuary, PhD thesis, 161 pp., East China Normal University.

Li, L.Y., Zhu, J., Wu, H., 2011a. Distributions of current and chlorinity in the Chenhang Reservoir in the case of brackish water in-taking. In: 2011 International Conference on Remote Sensing, Environment and Transportation Engineering. 24–26 June, Nanjing, China.

Li, M.T., Chen, Z., Yin, D.W., Chen, J., Wang, Z.H., Sun, Q.L., 2011b. Morphodynamic characteristics of the dextral diversion of the Changjiang River mouth, China: tidal and the Coriolis force controls. Earth Surf. Process. Landf. 36, 641–650.

Mao, Z.C., 1994. The salt intrusion character of the South Channel in the Changjiang mouth. Shanghai Water Conserv. 4, 22–32 (in Chinese, with English summary).

Mao, Z.C., Shen, H.T., Yao, Y.D., 1993. Analysis of sources of saltwater intrusion along south bank of south branch of the Changjiang River estuary. Mar. Sci. Bull. 12 (1), 17–26 (in Chinese, with English summary).

Mao, Z.C., Shen, H.T., Xu, P.L., 2000. The pattern of saltwater intruding into the Changjiang River estuary and the utilization of freshwater resources. Acta Geograph. Sin. 55 (2), 243–250. (in Chinese, with English summary).

Mao, Z.C., Shen, H.T., Xiao, C.Y., 2001. Saltwater intrusion patterns in the Qing Caosha area, Changjiang River estuary. Oceanol. Limnol. Sin. 32 (1), 58–66 (in Chinese, with English summary).

Pereira, L.S., Gonçalves, J.M., Dong, B., Mao, Z., Fang, S.X., 2007. Assessing basin irrigation and scheduling strategies for saving irrigation water and controlling salinity in the upper Yellow River Basin, China. Agric. Water Manag. 93 (3), 109–122.

Shen, H.T., Mao, Z.C., Zhu, J., 2003. Saltwater Intrusion in the Changjiang River estuary. China Ocean Press, Beijing, p. 175. (in Chinese, with English summary).

Tang, J.H., Xu, J.Y., Zhao, S.W., Liu, W.Y., 2011. Research on saltwater intrusion of the south branch on the Changjiang River estuary based on measured data. Res. .Environ. Changjiang Basin 20 (6), 677–685 (in Chinese, with English summary).

The Changjiang Water Conservancy Committee, 1950–1986, 1997–2011. Hydrology Year Book of the Changjiang, China (in Chinese, unpublished). p. 31620.

The Changjiang Water Conservancy Committee, 1967–1985. Hydrology year book of the lower Changjiang, China (in Chinese, unpublished). p. 10830.

The Changjiang Water Conservancy Committee, 2007. The Report of Water Resource Assigning Plan in the Lower Changjiang Below Datong in Dry Season (in Chinese). p. 156. http://wenku.baidu.com/view/74b035c60c22590102029db3.htm.

The Water Resources Department of Anhui Province, 2005–2011. The Water Resource Bulletin of Anhui Province (in Chinese). p. 200. http://www.ahsl.gov.cn.

The Water Resources Department of Jiangsu Province, 2005–2011. The Water Resource Bulletin of Jiangsu Province (in Chinese). p. 210. http://szyc.jswater.gov.cn.

The Water Resources Department of Shanghai, 2005–2011. The Water Resource Bulletin of Shanghai (in Chinese). 186. http://www.shanghaiwater.gov.cn.

Wu, H., 2006. Saltwater intrusion in the Changjiang River estuary. Ph.D. thesis, East China Normal University, p. 205.

Xu, J.Y., Yuan, J.Z., 1994. The research on salt-water intrusion into the south branch of the Changjiang River estuary. Hydrology 5, 1–5 (in Chinese, with English summary).

Xue, P., Chen, C., Ding, P., Beardsley, R.C., Lin, H., Ge, J., Kong, Y., 2009. Saltwater intrusion into the Changjiang River: a model-guided mechanism study. J. Geophys. Res. C Oceans 114, C02006https://doi.org/10.1029/2008JC004831.

Yang, S.L., Xu, K.H., Milliman, J.D., Yang, H.F., Wu, C.S., 2015. Decline of Yangtze River water and sediment discharge: impact from natural and anthropogenic changes. Sci. Rep. 5, 12581.

Yuan, W.H., Yin, D.W., Finlayson, F., Chen, Z., 2012. Assessing the potential for change in the middle Changjiang River channel following impoundment of the Three Gorges Dam. Geomorphology 147-148, 27–34.

Yuan, W.H., Li, M.T., Chen, Z.Y., Yin, D.W., Wei, T.Y., 2016. Responses of runoff-sediment flux and bedform in the middle Yangtze River to the completion of the Three Gorges Dam. J. East China Normal Univ. (Nat. Sci.) 2, 90–100 (in Chinese, with English summary).

Zhang, E.F., Chen, X.Q., Wang, X.L., 2003. Water discharge changes of the Changjiang River downstream from Datong during the dry season. J. Geogr. Sci. 13, 355–362 (in Chinese, with English summary).

Zheng, D.W., Yu, N.H., 1996. Research on the long-range forecast for the trend of sea level rising in the area of Shanghai, Annals of Shanghai observatory. Acad. Sinica 17, 36–42 (in Chinese, with English summary).

Zheng, X.Q., Xiao, W.J., Yu, Y., Pan, L.Z., 2014. Statistical prediction study of runoff and tide on Changjiang River saltwater intrusion. Marine Forecast. 31 (4), 18–23 (in Chinese, with English summary).

Zhu, J.R., Wu, H., Li, L., Wang, B., 2010. Saltwater intrusion in the Changjiang River estuary in the extremely drought hydrological year 2006. J. East China Normal Univ. (Nat. Sci.) 4, 1–4 (in Chinese, with English summary).

Chapter 3

Río de la Plata: A Neotropical Estuarine System

Javier García-Alonso[*], Diego Lercari[†], Omar Defeo[†]

*Departament of Ecology, CURE, University of the Republic, Maldonado, Uruguay, †UNDECIMAR, Faculty of Sciences, University of the Republic, Montevideo, Uruguay, Montevideo, Uruguay

1 GENERAL INTRODUCTION

1.1 Geographical and Morphological Features

South America, the Neotropical continent, includes very large basins draining water into the Atlantic Ocean, notably the equatorial Amazonia and the tropical and subtropical Río de la Plata (RdlP) basin, with a very wide estuary that ends at the Atlantic Ocean. The RdlP estuary (35°S) is a highly productive area with subtropical climate, sustaining valuable fisheries for Uruguay and Argentina (Acha et al., 2008). The inner part is the source of drinking water for the city of Buenos Aires but it also receives the discharge of industrial, agriculture, and urbanized areas. The basin covers a huge area of 3,100,000 km^2 (17% of South America), involving entirely Paraguay, most of Uruguay (88%), and part of Brazil and Argentina (Fig. 1). Approximately 150 million people inhabit the basin, with almost 60 cities inhabited by more than 100,000 residents, including the capitals Buenos Aires, Brasilia, Asunción, and Montevideo (Achkar et al., 2016). The RdlP basin comprises a huge area with the Parana and Uruguay rivers as the main tributaries, and it includes 10 different ecological areas or biomes with multiple land uses, mainly agriculture (grain and fruits) and livestock. Other activities include intensive forestry, petroleum extraction, refineries, pulp mills, natural gas extraction and transport, and fruit production (Box 1). More than 20 million inhabitants have access to drinking water from the internal part of the RdlP estuary (AQUASTAT-FAO, 2016).

The main natural forces in the RdlP can be divided into topographic, oceanographic (astronomical and meteorological tides), hydrological (catchment discharge), and meteorological conditions (winds) in the estuary itself. At the western coast of the South Atlantic Ocean, where the RdlP estuary ends, converges the warm and salty Brazil current and the cold and less dense Malvinas current with the presence of a strong and dynamic frontal region (Piedra-Cueva and Fossati, 2007; Simionato et al., 2009). The El Niño Southern Oscillation phenomenon (ENSO) also affects the mixohaline conditions of the RdlP (Nagy, 2006).

In the estuarine region, maximum concentrations of inorganic suspended material occur at the Argentinean (Southwest) coast of RdlP, associated with the higher discharge of sediments from the Paraná River (PNUD-GEF, 2007). Higher turbidity occurs in winter, associated with the main discharge of tributaries and higher wind speeds causing resuspension. The amount of suspended particulate matter declines to a high degree with distance downstream, where the depth of the estuary greatly increases, acting as a barrier for the saline front. Flocculation and precipitation of suspended matter generated by the increase of salinity regulates the turbidity (PNUD-GEF, 2007). This is the natural limit of the inner (continental) part of the RdlP and the start of the brackish (mixture) water (Fig. 2). The external political limit is set by an imaginary line from Punta del Este (Uruguay) to Punta Rasa (Argentina, Brazeiro et al., 2003) although the influence of the freshwater discharge of the RdlP reaches hundreds of kilometers over the continental platform (Simionato et al., 2009).

The RdlP seabed is mainly composed of soft sediments. Sandy silt dominates at the head, silt in the intermediate estuary, and sand at the mouth, following a gradual arrangement of textures (Moreira et al., 2016). Sediment sorting and organic matter content increase toward the estuary mouth, relating to the morphology and hydrodynamics of the estuary. The topology and differential distribution of sediment types at both margins reflect estuarine coastal landscapes, with mudflats on the west coast (Argentina) and sandy beaches at the east (Uruguay) (Psuty and Mizobe, 2006).

Coasts and Estuaries. https://doi.org/10.1016/B978-0-12-814003-1.00003-4

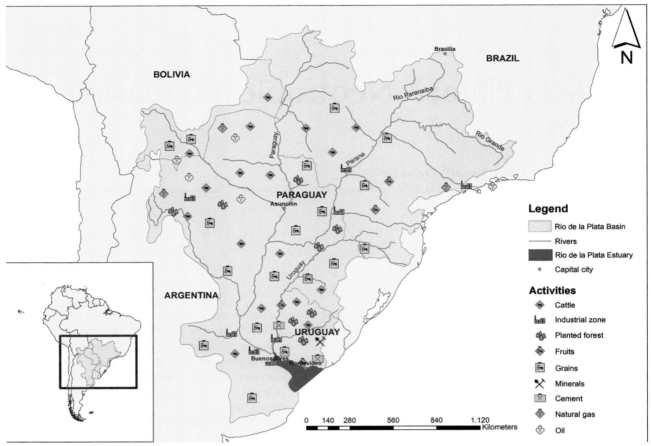

FIG. 1 Map of Río de la Plata basin showing the main tributaries, capitals, and main land uses in the catchment area. *(Modified from Achkar M, Díaz I, Domínguez A, Pesce F: Uruguay, Naturaleza, Sociedad, Economía, Ediciones de la Banda Oriental, 2016, Una visión desde la geografía, 369pp., ISBN 978-9974-1-0980-3.)*

BOX 1

The Río de la Plata (RdlP) estuary receives the water from a basin of 3,100,000 km² (17% of South America), the second in South America after Amazonia. There are two main tributaries, the Paraná and the Uruguay rivers. Five countries administer the basin area: Argentina, Bolivia, Brazil, Paraguay and Uruguay. Several important cities, including capitals, are in the basin such as Sao Paulo, Asuncion, Rosario, Buenos Aires, La Plata, and Montevideo. Agriculture, forestry, and animal farms are the most relevant human drivers at the basin level, with several dams for electricity supply. Several human activities occur in the region, notably marine traffic, fisheries, and coastal urbanization (including tourism expansion and pollution). The RdlP offers direct ecosystem services such as the main drinking-water source for the city of Buenos Aires, and sandy beaches uses for recreation, leisure and tourism (Sathicq et al., 2014). A binational commission from Argentina and Uruguay administers most aspects related to this transitional water system. However, coastal degradation, diffuse, and point-source contamination from agriculture and urban areas (nutrient enrichment, inorganic, and organic chemical pollutants) and overexploited fisheries are threatening the system (CTMFM, 2017). Future scenarios of concern given by global change predictions may increase the risks of environmental and human health deterioration. Continued assessment of the effects of global (including climate) change should be given a high priority in conservation and management planning.

1.2 Biodiversity

Environmental conditions, including nutrient supply from the basin and upwelling events at the external part of the estuary (Simionato et al., 2009), lead to a high primary production and biodiversity at the RdlP. This Neotropical estuary is a sensitive area, containing at the coastline an UNESCO Man and the Biosphere Reserve and a Ramsar site. In addition, the economic and ecological significance of the saline front has resulted in many species of commercial interest.

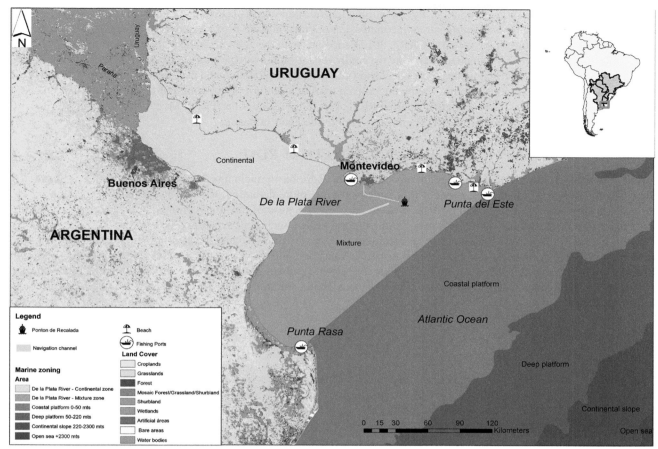

FIG. 2   Map of the Río de la Plata estuary and its main drivers and land uses in surrounding areas. Yellow lines at the estuary indicate navigation channels. Estuary limits are based on Brazeiro et al. (2003).

The biota encompasses environmental seasonal variations at different temporal scales. At the RdlP, phytoplankton blooms, including toxic Cyanobacteria are clearly associated with natural events (Sathicq et al., 2015). The benthic soft bottom communities of the RdlP estuary are composed of more than 350 taxa, mainly polychaetes, molluscs, and crustaceans. Nematodes and bivalves dominate the community in reaching more than 60% of the total abundance, while the community biomass is highly dominated by the bivalves, particularly *Mactra isabelliana* (reported as the main food resource for the fish the white-mouth croaker *Micropogonias furnieri*) (Cortelezzi et al., 2007). The density and biomass of two introduced and alien mollusc species is especially notable. The bivalve *Corbicula fluminea* is very common in soft sediments all along the estuary, while the gastropod *Rapana venosa* is more common on hard bottoms at the external area of the estuary. The macrobenthic abundance and biomass is higher in the intermediate estuarine region, followed by the outer zone (Cortelezzi et al., 2007). Trophic studies (Lercari and Bergamino, 2011) on the role of these two species in the food web of the RdlP suggest several negative effects concerning ecological and conservation issues, but also socioeconomic concerns.

Approximately 90 marine and freshwater species belonging to 81 genera and 54 families comprise the fish community along the RdlP estuary, which is structured in three assemblages: riverine, estuarine, and shelf although an additional group may be recognized outside the estuary on the continental slope. These assemblages differ spatially in their species composition: the inner zone is slightly dominated by the catfish *Paraloricaria vetula*, the estuarine area by the sciaenid *M. furnieri* and the shelf by the carangid *Trachurus lathami*. Bottom salinity and temperature are the main environmental drivers of the assemblage structure. Changes in assemblage structure between areas are gradual, with no sharp boundaries (Jaureguizar et al., 2016).

1.3   Management

In 1969, Argentina, Bolivia, Brazil, Paraguay, and Uruguay signed the first treat aiming to manage the region as a Whole Basin System for regional integration and harmonic development (CIC, 2016). In 1973, Argentina and Uruguay signed the treaty of the RdlP and its maritime front (TRPyFM) aiming for a sustainable management of aquatic resources and environmental conservation. A binational administrative commission (Comisión Administradora del RdlP, CARP) was created to regulate almost all activities in the estuary (CARP, 2012), together with the Joint Technical Commission for the Maritime Front (CTMFM). The TRPyFM aimed to overcome the controversial situation regarding the jurisdiction over the river's waters and also regulated navigation, fishing, bed and subsoil exploration-exploitation and use, pollution prevention, and other issues. Some binational initiatives are now being undertaken to pilot monitoring programs and to gather information about the basic features of the water and sediment status (FREPLATA, 2014). Under the TRPyFM, each country has exclusivity in the use of coastal areas (2 km in the inner region and 7 km in the outer estuary).

2 MAJOR ANTHROPOGENIC DRIVING FORCES AT RDLP

Various and extensive ecosystem services are provided by the RdlP basin, including production services (i.e., agriculture, livestock, and fisheries), regulating services (i.e., energy source, bioremediation of waste by wetlands and floodplains), and cultural services (i.e., leisure and recreation at the beaches). Because of this, the region is threatened by several drivers and associated pressures at different spatial and temporal scales, including expanding coastal urbanization, fisheries, intensive agriculture, livestock, and industries, including the construction of dams for electricity generation. In addition, human driving forces of these pressures increase maritime transport and propagate other pressures (e.g., dredging, invasive species). These pressures could be categorized either as exogenic unmanaged pressures (ExUPs) from a strictly estuarine perspective, or endogenic managed pressures (EMPs) from a whole system integrated perspective (Elliott, 2011; Elliott and Whitfield, 2011).

2.1 Food Supply

2.1.1 Agriculture and Livestock

As the region shows rich soils and the estuary ends in a confluence of two marine currents (the Brazil and Malvinas currents), the catchment lands and the estuary itself have been exploited for food. The catchment area covers different land uses, even though agriculture and animal husbandry dominate the area (Fig. 1). Specifically for land areas near the estuary, Argentina and Uruguay have more than 160,000,000 ha, most of them belonging to the catchment area, with almost 60% for livestock, 20% for grain, 6% for forestry, and few areas for rice and milk production (AQUASTAT-FAO, 2016). Historically, the main activity was extensive animal husbandry for export although currently soybeans have become the main export product of the region.

One of the main pressures associated with several drivers is the release of nutrients to the environment with the subsequent eutrophication processes along the basin and subestuaries of the RdlP. Red tides and cyanobacteria blooms are increasingly documented (Bonilla et al., 2015), with serious risks to environmental and human health (Vidal et al., 2017). The recent exponential increase of intensive agriculture, with soybean as the main product (https://atlas.media.mit.edu/en/profile/country/arg/; http://www.uruguayxxi.gub.uy), was not supported by scientific studies directed to assess environmental impacts in the catchment areas. However, preliminary evidence suggests that almost all the RdlP basin and coastal areas of the estuary are experiencing the adverse consequences of eutrophication.

At the basin level, land denudation and erosion have been increasing in recent decades in relation to land uses such as intensive agriculture (Bonachea et al., 2010). Several pesticides have been detected (De Gerónimo et al., 2014), such as endosulfan, chlorpyrifos, and cypermethrin, particularly along the Paraná subbasin (Etchegoyen et al., 2017). Pollution by agrochemicals and associated with eutrophication processes have also been detected (Nagy, 2006). An intensive culture of grain and oilseeds occurs, mainly maize, wheat, and soybeans (Leguizamón, 2014; Achkar et al., 2016). The compounds used to control pests are mainly herbicides (e.g., glyphosate, dicamba, atrazine) and insecticides (e.g., chlorpyrifos, cypermethrin). Insecticides released into creeks in peri-urban horticultural areas of Buenos Aires have been found at sublethal and lethal concentrations for benthic fauna (Mac Loughlin et al., 2017). Not only do emergent pollutants occur in the system, but also the earlier banned DDT and endosulfan, an insecticide banned in Uruguay in 2012 that occurs at relatively high concentrations near the mouth of the Uruguay river (Williman et al., 2017). Few management actions have been taken to counteract this effect.

2.1.2 Fishing

Fish consumption in Uruguay and Argentina has been historically below 10 kg per year per person (half the per capita fish world consumption, FAO, 2012). Fisheries production by both countries is mainly channeled to foreign markets (mainly to Brazil, the United States and Nigeria) and the remaining is sufficient to meet the local demand.

Fish catches in the estuary are mainly focused on four or five species. The white croaker (*Micropogonias furnieri*) (approx. 40,000 t/yr), striped weakfish (*Cynoscion guatucupa*) (approx. 12000 t/yr), and sharks (50 t/yr) represent more than 90% of catches in the estuary. Other commercially important fish species include: tope shark (*Galeorhinus galeus*), Argentine angelshark (*Squatina argentina*), leatherjack (*Parona signata*), Argentine croaker (*Umbrina canosai*), southern kingfish (*Menticirrhus americanus*), plaice (*Paralichthys* spp.), Brazilian codling (*Urophycis brasiliensis*), white sea catfish (*Netuma barba*), black croaker (*Pogonias cromis*), king weakfish (*Macrodon ancylodon*), and Brazilian menhaden (*Brevoortia aurea*). Freshwater species such as dorado (*Salminus brasiliensis*) are also fished along the estuary, particularly during occasional events of very large freshwater discharge from the catchment basin. The mussel (*Mytilus edulis*) is the most important invertebrate species exploited for human consumption on the Uruguayan coast of the RdlP.

Different fishing boat types and gear types are employed, and both countries (Argentina and Uruguay) have different fleet categorizations. The industrial fleet is composed of almost 30 Uruguayan vessels with a capacity higher than 10 gross register tonnage (GRT) with 30 m length and a fishing autonomy of no more than 7 days. This fleet uses bottom trawls, most of them operating by pair trawling. Argentinean industrial vessels operating in the estuary are categorized in two types: (1) the smaller boats called "Rada" or "Ría" with lengths between 7 and 17 m and up to 350 HP, mostly operated by pair trawling (75 boats in 2015), and (2) "Costeros," with a length close to 17 m and 350 HP, usually using standard bottom trawls (>50 boats in 2015) (CTMFM, 2017).

Artisanal fisheries also operate up to 5.5 km offshore, both in Uruguay (Defeo et al., 2011) and Argentina (Lasta et al., 2001). The Uruguayan fleet is composed of more than 500 small boats (mainly made of wood or fiberglass) of less than 10 GRT with outboard motors and one day autonomy. The Argentinean artisanal fleet operating in the RdlP is estimated as 370 boats (Errazti et al., 2009). This fleet targets the same stocks as the industrial fleet, thereby producing a high spatial overlap and producing negative externalities to the artisanal fleets (Horta and Defeo, 2012). Along the Uruguayan coast, the fleets follow migrations of *M. furnieri* to coastal spawning areas (Defeo et al., 2011) which are associated with the inner salinity front (Jaureguizar et al., 2008). Both fleets employ almost 2000 (artisanal) and 1400 (industrial) fishers in Uruguay (Defeo et al., 2011).

Ecological pressures by fishing activities are affecting fish populations and ecosystems. Time-series analysis of the main fishery indicators for the RdlP and contiguous oceanic zone showed that fishing effort for both fleet types has remained constant or even declined slightly (CTMFM, 2017). However, catches for the two most important species (*C. guatucupa* and *M. furnieri*) have declined (Fig. 3), reaching in Uruguay their lowest values for the last 35 years, concurrently with decreasing export volumes (Gianelli and Defeo, 2017).

Biomass population models showed a pronounced decrease in the total biomass of *M. furnieri* and a slight decrease in biomass *C. guatucupa*. In the former, the decline in total biomass might be associated with an increase in fishing mortality (CTMFM, 2017). Total biomass values are below the biomass at maximum sustainable yield (B_{MSY}). For *C. guatucupa*, total biomass (about 182,000 t) is greater than the B_{MSY} (154,000 t: CTMFM, 2017). Similar decreasing trends were observed for chondrichthyes, including *Mustelus schmitti* (De Wysiecki et al., 2017).

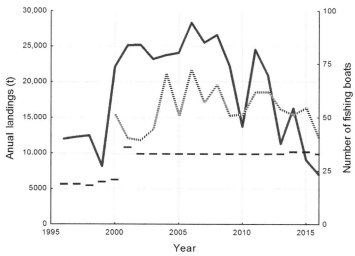

FIG. 3 Variations in fishing effort (right axis, dashed line) and annual landings (left axis) of the main two exploited species in the RdlP estuary, *M. Furnieri* (solid line) and *C. guatucupa* (dotted line). (Modified from CTMFM: El recurso corvina (Micropogonias furnieri) en el área del Tratado del Río de la Plata y su Frente Marítimo, Montevideo, 2017, Diagnóstico poblacional. Documento conjunto DINARA-INIDEP-SSPyA.)

Potential fishing impacts include bycatch, discards, and habitat degradation (e.g., the destruction of the benthic community). Bycatch and discards are also poorly assessed and understood at the level of the RdlP fisheries, and no systematic monitoring exists. A recent work reconstructing fisheries statistics for Uruguay (Lorenzo et al., 2015) has estimated a discard rate of 14% for the period 1960–2000, whereas this rate was set at 9% for 2001–2010. Industrial discards are mainly composed of small organisms of the main target species (*C. guatucupa* and *M. furnieri*), sharks and rajids (Paesch et al., 2014). Of special concern is the entanglement of the estuarine dolphin (franciscana) *Pontoporia blainvillei* by artisanal gear, which has been categorized as a vulnerable species by IUCN. This species has been incidentally caught by gillnet fisheries since the early 1940s (Secchi et al., 2001). A recent study reported that more than 300 franciscana (*P. blainvillei*) were incidentally caught annually by the artisanal Uruguayan fleet (Franco-Trecu et al., 2009). The green turtle *Chelonia mydas* (Endangered) is also incidentally caught. In one particular fishing area (Bajos del Solis) an annual incidental catch of 500–1500 turtles by artisanal fisheries has been estimated (Lezama et al., 2013). The South American sea lion (*Otaria flavescens*) also frequently interacts with Uruguayan artisanal fisheries by preying upon entangled fish such as sciaenids (Szteren and Páez, 2002). Although this interaction is viewed as a major problem to the economic sustainability of artisanal fisheries, recent studies showed only a small percentage of sea lion predation, which therefore should not be considered responsible for the low artisanal landings (De María et al., 2014). Effects of bottom trawling by the industrial fleets damaging the seafloor (Hiddink et al., 2017) may be expected as one of the most important pressures in the RdlP, although scientific evidence is lacking.

2.2 Marine Traffic

The estuary of the RdlP has a fluvial connection with Paraguay-Paraná, Rivers, a multinational fluvial transport system with a total length of 3302 km. This fluvial-maritime system is a principal way to transport products from Argentina, Bolivia, Paraguay, and a part of Brazil to export goods worldwide. This has led to intensive marine traffic and the development of port infrastructure. La Plata-Buenos Aires (Argentina) and Montevideo (Uruguay) are the biggest ports in the RdlP basin, moving jointly almost 20,000 *t* of cargo and 550,000 passengers. The Montevideo port recorded more than 4000 vessel arrivals during 2016 and Buenos Aires almost 1000 arrivals.

Most Uruguayan primary products of agriculture, such as wood, grain, meat, and dairy, are exported from the Montevideo port, which also acts as a regional cargo hub, having a 50% of its containers in transit to/from Argentina, Paraguay, and Bolivia. Most cargo moved from/to the Buenos Aires port is composed by manufactured goods, followed by chemical products, meat, fruits, textiles, fuels, lubricants, and cars. Transport of passengers (by ferries and cruises) is also important for both ports. In 2016, Buenos Aires handled more than 1.5 million passengers and Montevideo more than 700,000.

Traffic can generate environmental impacts through different mechanisms such as space reclamation and dredging together with chemical pollution. The Buenos Aires port is composed of six docks protected by two breakwaters (2500 and 950 m long), reaching more than 7000 m of piers and occupying 92 ha. Most of the space occupied by the Buenos Aires port was land-claimed from the estuary when constructed in 1928 (www.worldportsource.com). The Montevideo port is placed in a deeper natural embayment, enclosed by two breakwaters of 1300 and 900 m. The port is divided into three docks and a refinery terminal reaching a total of 4500 m of piers and a terrestrial area of 110 ha. More than 135 ha had been claimed from the estuary (Gautreau, 2006) to develop additional piers for containers, fishing, and agricultural forestry products (www.worldportsource.com). Both ports need continuous dredging to be fully operative. In addition, the channels in the RdlP are dredged in order to provide access to the Paraná and Uruguay rivers, and then to the entire estuary basin.

The RdlP estuary is freely navigable for ultramarine ships in the external region (>10 m). However, artificial channels are maintained to reach the ports of Buenos Aires and Montevideo and associated ports at the Parana and Uruguay River basins (Fig. 4). The entrance channel (Punta Indio) extends 120 km along the estuary from Montevideo to La Plata where the canal bifurcates into a northern branch (Martín García) and a southern branch (Emilio Mitre—Buenos Aires port access).

The official statistics of maritime shipping in the RdlP estuary shows a rather constant trend. Montevideo has a greater amount of ship movements (arriving plus departing) than Buenos Aires, possibly related to a major activity of fishing fleets. Despite this, there is a decreasing temporal trend in the number of ships in both ports. In terms of containers, Montevideo showed a constant increase, and Buenos Aires a slight decreasing trend (Fig. 5).

Pressures of marine traffic on the RdlP ecosystems are produced by port operations and those particularly exerted by the ships, impacting directly the water body or, indirectly, air, and terrestrial areas. At present, the most common environmental concerns of shipping refer to chemical discharges (oil, wastewater, antifouling paint, ballast water, and marine litter) and physical pollution (underwater noise, ports, fairways and channels, dredging, and shipwrecks).

Dredging is a frequent activity in the estuary. Montevideo dredges more than 60 km of canals including the docks, reaching a total sediment volume of more than 17 million m^3 in a 5-year period. In addition, more than 24 million m^3 were

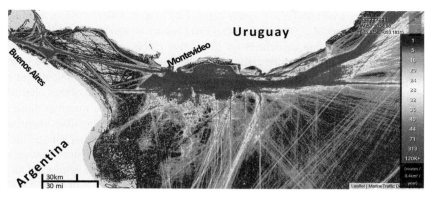

FIG. 4 Marine traffic density map for the Rio de la Plata estuary. A clear canalized pattern is observed with huge activity near Montevideo city. (https://www.marinetraffic.com/en/ais/home/centerx:-55.6/centery:-35.3/zoom:8) Publicly display content with proper attribution online allowed.

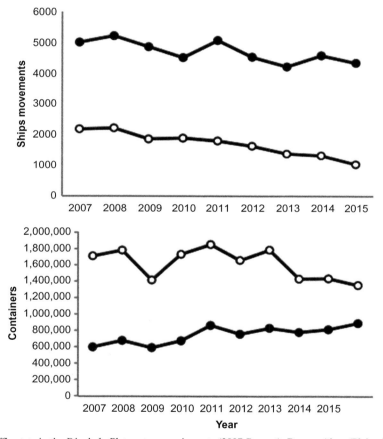

FIG. 5 Indicators of main traffic state in the Río de la Plata estuary main ports (2007-Present). Buenos Aires (Plain circle) and Montevideo (Shaded circle). From official statistics.

recently dredged to construct an access channel and port infrastructure for a re-gasification plant. The Buenos Aires port is artificial, placed in terrain claimed from the estuary. In the 1980s, 30 million m^3 of sediments were dredged, and the sediment removed was used to claim more than $4\,km^2$ from the estuary. At present, this area is included in a reserve and park along the city coast. Buenos Aires dredges more than 2 million m^3 per year, including main docks and access channels. The dredged sediment is disposed in the RdlP at few kilometers from both ports. There have been no specific studies focused on the impacts of dredging.

Chemical and biological pollution is generated by traffic. The Montevideo Bay was described as highly polluted (Danulat et al., 2002). Sediments in the RdlP have been affected by high PAHs (polynuclear aromatic hydrocarbons) concentrations since at least 1989, particularly in the main ports. In Buenos Aires, La Plata, and Montevideo, a range of 2.5–555 µg/g were reported (Colombo et al., 1989; Muniz et al., 2004). The mean PAHs concentration is very high (29.5 ± 63.4 µg/g), exceeding by several orders of magnitude Canadian guidelines for individual PAHs (5.9–111 ng/g for 13 PAHs; sum = 468 ng/g). Disentangling the sources of diverse PAHs in RdlP estuary may be challenging since data are scarce. The main sources of aliphatic and polycyclic aromatic hydrocarbons are petroleum inputs and combustion (Venturini et al., 2015) due to oil transport and refining (Fig. 2), maritime activities, and vehicular emissions. The release of these harmful substances into the estuary is not regulated. Total aliphatic hydrocarbons and PAH levels are in the same range as those determined in heavily polluted areas worldwide. Metals are released into the environment by leaking combustible materials from ships. Lead, a metal associated with fuels, is present even at moderated toxic levels on some beaches of Montevideo city (García-Alonso et al., 2017), although the multiple sources of pollution prevent the determination of the main source. Although now banned under International Maritime Organization (IMO) regulations, tri-butyl-tin (TBT) has been used as an antifouling in ship paints to prevent settlement of sessile organisms that affect passage and navigation. Although TBT concentrations in the RdlP are unknown, imposex in female mollusks was detected, suggesting potential deleterious effects of TBT (Bigatti et al., 2009). Marine debris in the RdlP includes packaging objects, plastics, and fishing gears (Acha et al., 2003). The salinity front of the estuary acts as a barrier accumulating debris, particularly in the inner estuary, where important ecological areas (e.g., spawning and nursery areas) occur (Acha et al., 2003).

Argentina and Uruguay have signed UN treaties and conventions, such as the International Convention for the Prevention of Pollution from Ships (1973) and the Protocol of 1978 (MARPOL 73/78). The 1973 TRPyFM agrees about the common use of the space in the estuary and assigns rights and responsibilities about navigation and infrastructure in the common zone, excluding a 7-mile coastal fringe of national exclusive jurisdiction. The binational commission CARP manages and controls the activities at the estuary, such as navigation (del Castillo-Laborde, 2008). Despite its wide remit and competence, most resolutions of this Commission focus on navigation infrastructure (e.g., dredging of new channels), and there are no apparent resolutions concerning impacts and pressures of the activities.

Uruguayan and Argentine navies are responsible for enforcing international treaties related to maritime traffic and pollution, including protocols for environmental protection and security, and implementing monitoring, control, and surveillance of national and international regulations. This is achieved by long-term strategies (e.g., by planning and training) but also rapid responses to contingencies (e.g., oils spills). Both port administrations are committed to the reduction of environmental impacts, delineating diverse policies toward this direction. National and international rules and certification programs are followed to manage waste, transport of dangerous materials, and to develop environmental impact assessments for new developments. For instance, the port of Buenos Aires is following the Environmental Ship Index strategy, offering discounts in the services for ships with a good environmental score (http://www.environmentalshipindex.org).

3 MINOR DRIVERS: INDUSTRY, URBANIZATION, AND TOURISM

Several industries are located along the coast of the RdlP, particularly near Buenos Aires and Montevideo. In Uruguay, there are pulp mill factories at the inner estuary, while petro-chemicals, cloro-alkali plants, tanneries, refineries, pulp mills, chemical industries, and bitumen producers are found near Montevideo (Danulat et al., 2002; SADI, 2013). In Argentina, several industries are located in the inner RdlP. The subbasins of Riachuelo and Reconquista rivers are the most polluted watercourses in Argentina, receiving untreated disposal points from urban, agricultural, cattle farming and industrial effluents. Approximately 90% of industrial waste produced in the Reconquista river subbasin is discharged directly into watercourses (Salibián, 2006). Since permissive environmental laws exist on industrial discharge and chemical importation in comparison to the EU REACH legislation, toxic chemicals are still used and released by industries to the environment without treatment. For example, carcinogenic surfactants alkylphenols polyethoxylated with endocrine disrupting properties are freely commercialized in the region.

Polluted areas have been detected aling both banks of the estuary using different markers of pollutants such as toxicities (Basílico et al., 2017), metal traces (Tatone et al., 2012, García-Alonso et al., 2017), macrophyte assemblages (Feijoó and Lombardo, 2007), biofilm coating, and metagenomics (Bauer et al., 2007; Piccini and García-Alonso 2015). The population increase and the tourist demand for accommodation have induced the construction of housing and related infrastructure with a lack of sewage treatment, thereby negatively impacting coastal streams and beaches (Arreghini et al., 2007; Basílico et al., 2017; Rigacci et al., 2013). At Buenos Aires, tourism is more than half of the Argentinean tourist GDP (gross domestic product), reaching 11.1 billion US$ (WTTC-Report, 2017). In Uruguay, sandy beaches attract tourists for recreation, with tourism reaching 7% of the country GDP, with the number of visitors (particularly in summer) being the same as the local permanent residents (Achkar et al., 2016). In addition, almost 200 cruise liners arrived at Punta del Este and Montevideo in summer 2014–15 (Achkar et al., 2016), reaching 13% of total tourists (Ministerio de Turismo del Uruguay, MINTUR, 2016). Montevideo is the Uruguayan port where more tourists disembarked (Bellani et al., 2017).

Urbanized coasts are inducing heavy pressures on the environment. Discharges of untreated effluents into the RdlP estuary are a major concern, since this produces a conflict with water for human consumption and recreation. Unknown amounts of biological and chemical pollutants are released into the estuary from the main Buenos Aires sewer at Berazategui (2 million m^3 per day: Cirelli and Ojeda, 2008) and from the sewage discharge at Punta Carretas in Montevideo city (Nagy et al., 2014). At Montevideo, only the presence of fecal coliforms is monitored in beaches used for recreation to reduce health risk by posting sanitary flags when fecal Coliforms exceed 1000 UFC per 100 mL (www.montevideo.guv.uy/ciudadania/desarrollo-ambiental/playas). Chemical pollutants are not routinely measured, including cyanotoxins, although polycyclic chlorinated biphenyls (PCBs) and polybrominated diethyl ethers (PBDEs) were detected in suspended organic matter near the sewage discharge of Buenos Aires. These chemicals bioaccumulate almost 40 times more than organic matter in the streaked prochilod fish *Prochilodus lineatus* (Cappelletti et al., 2015). These pollutants acting additively with other industrial and domestic chemicals (as endocrine disrupting compounds: EDCs) interfere in the normal functioning of hormones. The effects of EDCs with estrogenic properties have been detected in the estuary itself (Valdés et al., 2015) or in tributaries such as the Santa Lucía River (Griffero et al., 2018). Urban beaches of the RdlP accumulate anthropogenic marker elements such as phosphorous, lead, copper, chromium, zinc, and mercury, with an unknown potential health impact in humans and ecosystems (García-Alonso et al., 2017).

Eutrophication processes originating from different pressures are one of the most important environmental impacts. Point source (domestic sewage and industrial discharge) and diffuse (agriculture activities) sources throughout the whole basin and the RdlP estuary increase the concentration of nutrients, mainly near the coast. A lack of basic knowledge on the system and poor monitoring programs do not allow predicting the carrying capacity of the RdlP to cope with the nutrient enrichment by environmental homeostasis.

4 THE FUTURE OF THE RDLP ESTUARY

Estuaries are complex systems where several natural stressors occurring simultaneously may interfere in environmental monitoring programs (Elliott and Quintino, 2007; Elliott and Whitfield, 2011). For instance, integrating the whole pollution-toxicity relationship in estuarine systems is needed for estuaries, since the homeostatic properties are specific for each region (García-Alonso et al., 2011).

Natural climatic factors and external unmanaged pressures from anthropogenic activities at the basin level (e.g., dams, irrigation, intensive agriculture) together with main drivers inside the estuary may affect the status of the RdlP estuary both in the short and long terms. The relatively scarce monitoring, control, and surveillance programs of pressures and impacts in a basin administered by five countries, and the lack of historical data makes it difficult to determine the current potential level of impact and to predict the impact of human activities at the basin level. At the level of the RdlP estuary and associated ecosystems (wetlands and floodplains), there is a substantial lack of environmental management and conservation plans. Since long-term management strategies are lacking, predictions of potential scenarios in the future are unreliable.

Even though several fishery management measures have been proposed by CTMFM and CARP to improve stock conditions (e.g., restrictions in fishing effort, minimum landing sizes, protected areas and time closures, total allowable catches and quotas per year), commercial fish populations are declining, and more efforts are needed in order to maintain sustainable stocks (CTMFM, 2017).

It is expected that maritime traffic will increase in the RdlP in the near future. For example, the port of Montevideo is experiencing a significant capacity increase. New and planned infrastructure includes a multipurpose dock, a new terminal for forest products and bulk solids, and a new fisheries terminal. These developments will result in more than 600 m of docks, and dozens of hectares have been claimed from the estuary. In addition, there are plans to extend port activities (in Montevideo bay) by constructing a new passenger terminal and a new logistic hub with an industrial zone and port (Puntas de Sayago). The port of Buenos Aires is also expanding its capacity. The historic docks will be land filled to attain deeper waters, allowing larger vessels to enter (such as the New Panamax type). This cargo would be operated by two terminals instead of three, as at present, requiring an investment of 850 million US$.

Different drivers and pressures exert impacts that interact with other drivers and ecosystems services of the RdlP. For instance, exogenic unmanaged pressures (ExUPs) from agriculture and domestic sewage generate eutrophication that affects the quality of drinking water at its source or recreation areas. Without an effective and increased regulation of fertilizers, phosphorous-based biocides and sewage treatments, an increase in cyanobacteria and harmful algal blooms may be expected to occur at the basin and coastal areas of the RdlP (Martínez et al., 2017). In addition, urbanization and industrialization may threaten the environmental quality in spawning and nursery areas, exerting unknown negative effects on fisheries. In fact, most sciaenid species spawn in the sub-estuarine systems of the Uruguayan coast. Nutrient enrichment associated with eutrophication of RdlP may have a detrimental impact in some ecosystem services of the estuary (drinking-water source, food supply, and recreation) if the markers become above the managed environmental homeostasis level (Elliott and Quintino, 2007).

Hundreds of sandy beaches at the RdlP are used for human welfare, where recreational activities occur. Monitoring programs of environmental quality need to be updated to actual international standards for marine and transitional waters, such as the reduction of the maximum allowable concentration of fecal coliforms, quantification of *Enterococci* and chemical pollutants, and analyzes for metal pollution (WHO, 2003).

Several binational projects have been developed in the inner part of the RdlP, including the use of water for human consumption, and at brackish areas with sandy beaches for recreational leisure (Sathicq et al., 2014). In addition, international projects are being executed aiming to improve the environmental management of the RdlP and its maritime front (e.g., ECOPLATA and FREPLATA: www.freplata.org). However, the lack of basic studies and monitoring programs precludes estimating basic reference points for assessing ecosystem health status (García-Alonso et al., 2017). Therefore, there is a need to identify and assess drivers and pressures in order to reach sustainable development in these complex social-ecological systems (Norgaard et al., 2009).

Finally, global change (climate change and environmental pollution) is likely to increase eutrophication and toxicity within the RdlP. Eutrophication-like processes are expected under climate change scenarios and are likely to deeply impact the already eutrophic estuaries in South America (Kopprio et al., 2015). Assessment of impacts and attributing causes to global change pressures are challenges for all estuaries (Little et al., 2017), in particular in low-industrialized regions such as the RdlP estuary, where more baseline studies are needed to develop long-term sustainable management programs.

In summary, we have described the main social and ecological issues of the RdlP estuary, with special emphasis on the main drivers and pressures affecting the system, highlighting information gaps, and the lack of long-term management policies and comprehensive plans that should consider the RdlP as a social-ecological system. The inclusion of multisectoral goals and the reinforcement of comprehensive long-term monitoring programs to assess ecosystem status are needed to attain a sustainable use of the RdlP estuary.

REFERENCES

Acha, E.M., Mianzan, H.W., Iribarne, O., Gagliardini, D.A., Lasta, C., Daleo, P., 2003. The role of the Río de la Plata bottom salinity front in accumulating debris. Mar. Pollut. Bull. 46 (2), 197–202.

Acha, E.M., Mianzán, H., Guerrero, R., Carreto, J., Giberto, D., Montoya, N., Carignan, M., 2008. An overview of physical and ecological processes in the Rio de la Plata Estuary. Cont. Shelf Res. 28, 1579–1588.

Achkar, M., Díaz, I., Domínguez, A., Pesce, F., 2016. Uruguay, Naturaleza, Sociedad, Economía. Una visión desde la geografía, Ediciones de la Banda OrientalISBN: 978-9974-1-0980-3369.

AQUASTAT-FAO. 2016. La Plata basin. Regional report. Available from: http://www.fao.org/nr/water/aquastat/basins/la-plata

Arreghini, S., de Cabo, L., Seoane, R., Tomazin, N., Serafini, R., de Iorio, A.F., 2007. A methodological approach to water quality assessment in an ungauged basin, Buenos Aires, Argentina. GeoJournal 70 (4), 281–288.

Basílico, G., Magdaleno, A., Paz, M., Moretton, J., Faggi, A., de Cabo, L., 2017. Sewage pollution: genotoxicity assessment and phytoremediation of nutrients excess with *Hydrocotyle ranunculoides*. Environ. Monit. Assess. 189, 182.

Bauer, D.E., Gómez, N., Hualde, P.R., 2007. Biofilms coating *Schoenoplectus californicus* as indicators of water quality in the Río de la Plata Estuary (Argentina). Environ. Monit. Assess. 133, 309–320.

Bellani, A., Brida, J.G., Lanzilotta, B., 2017. El turismo de cruceros en Uruguay: determinantes socioeconómicos y comportamentales del gasto en los puertos de desembarco. Rev. Eco. del Rosario 20 (1), 71–95.

Bigatti, G., Primost, M.A., Cledón, M., Averbuj, A., Theobald, N., Gerwinski, W., Arntz, W., Morriconi, E., Penchaszadeh, P.E., 2009. Biomonitoring of TBT contamination and imposex incidence along 4700 km of Argentinean shoreline (SW Atlantic: From 38S to 54S). Mar. Pollut. Bull. 58 (5), 695–701.

Bonachea, J., Bruschi, V.M., Hurtado, M.A., Forte, L.M., da Silva, M., Etcheverry, R., Cavallotto, J.L., Dantas, M.F., Pejon, O.J., Zuquette, L.V., Bezerra, M.A.O., Remondo, J., Rivas, V., Gómez-Arozamena, J., Fernández, G., Cendrero, A., 2010. Natural and human forcing in recent geomorphic change; case studies in the Rio de la Plata basin. Sci. Total Environ. 408, 2674–2695.

Bonilla, S., Haakonsson, S., Somma, A., Gravier, A., Britos, A., Vidal, L., De León, L., Brena, B.M., Pírez, M., Piccini, C., Martínez de la Escalera, G., Chalar, G., González-Piana, M., Martigani, F., Aubriot, L. 2015. Cyanobacteria and cyanotoxins in freshwaters of Uruguay. INNOTEC 10, 9-22. ISSN 1688-3691-9.

Brazeiro A., Achkar M., Mianzan H., Gomez-Erache M., Fernandez, V. 2003. Areas prioritarias para la conservacion y manejo de integridad biologica del Rio de la Plata y su Frente Maritimo. FREPLATA PNUD/GEF/RLA99/G31.

Cappelletti, N., Speranza, E., Tatone, L., Astoviza, M., Migoya, M.C., Colombo, J.C., 2015. Bioaccumulation of dioxin-like PCBs and PBDEs by detritus-feeding fish in the Rio de la Plata estuary, Argentina. Environ. Sci. Pollut. Res.Int. 22, 7093–7100.

CARP, Comision Administradora del Río de la Plata. Webpage. 2012. Available from: http://www.comisionriodelaplata.org

CIC. 2016. The intergovernmental coordinating committee of the countries of La Plata Basin. Available from: http://www.cicplata.org/.

Cirelli, A.F., Ojeda, C., 2008. Wastewater management in Greater Buenos Aires, Argentina. Desalination 218, 52–61.

Colombo, J.C., Pelletier, E., Brochu, C., Khalil, M., Catoggio, J.A., 1989. Determination of hydrocarbon sources using n-alkane and polyaromatic hydrocarbon distribution indexes. Case study: Rio de la Plata Estuary, Argentina. Environ. Sci. Technol. 23, 888–894.

Cortelezzi, A., Capítulo, A.R., Boccardi, L., Arocena, R., 2007. Benthic assemblages of a temperate estuarine system in South America: transition from a freshwater to an estuarine zone. J. Mar. Syst. 68 (3), 569–580.

CTMFM, 2017. El recurso corvina (*Micropogonias furnieri*) en el área del Tratado del Río de la Plata y su Frente Marítimo. Diagnóstico poblacional. Documento conjunto DINARA-INIDEP-SSPyA, Montevideo.

Danulat, E., Muniz, P., García-Alonso, J., Yannicelli, B., 2002. First assessment of the highly contaminated Harbour of Montevideo (Uruguay). Mar. Pollut. Bull. 44, 554–565.

De Gerónimo, E., Aparicio, B.C., Bárbaro, S., Portocarrero, R., Jaime, S., Costa, J.L., 2014. Presence of pesticides in surface water from four sub-basins in Argentina. Chemosphere 107, 423–431.

Defeo, O., Puig, P., Horta, S. de Álava, A., 2011. Coastal fisheries of Uruguay. In: Salas, S., Chuenpagdee, R., Charles, A., Seijo, J.C. (Eds.), Coastal Fisheries of Latin America and the Caribbean. FAO Fisheries and Aquaculture Technical Paper No. 544. (Food and Agriculture Organization of the United Nations (FAO), Rome, pp. 357–384.

del Castillo-Laborde, L., 2008. The Río de la Plata and its maritime front legal regime. Brill. ISBN: 978-90-47-43204-3. 428pp. https://doi.org/10.1163/ej.9789004163447.i-4728.

De María, M., Barboza, F.R., Szteren, D., 2014. Predation of South American sea lions (*Otaria flavescens*) on artisanal fisheries in the Rio de la Plata estuary. Fish. Res. 149, 69–73.

De Wysiecki, A., Jaureguizar, A., Cortés, F., 2017. The importance of environmental drivers on the narrownosesmoothhound shark (Mustelus schmitti) yield in a small-scale gillnetfishery along the southern boundary of the Río de la Plata estuarine area. Fish. Res. 186, 345–355.

Elliott, M., Quintino, V., 2007. The estuarine quality paradox, environmental homeostasis and the difficulty of detecting anthropogenic stress in naturally stressed areas. Mar. Pollut. Bull. 54, 640–645.

Elliott, M., Whitfield, A.K., 2011. Challenging paradigms in estuarine ecology and management. Estuar. Coast. Shelf Sci. 94, 306–314.

Elliott, M., 2011. Marine science and management means tackling exogenic unmanaged pressures and endogenic managed pressures – a numbered guide. Mar. Pollut. Bull. 62, 651–655.

Errazti, E., Bertolotti, M.I., Gualdoni, P., 2009. In: Sistema pesquero artesanal de la Provincia de Buenos Aires. Comunicación presentada en XIII Congreso Latinoamericano de Ciencias del Mar y VIII Congreso de Ciencias del Mar, La Habana (Cuba), 26–30 October.

Etchegoyen, M.A., Ronco, A.E., Almada, P., Abelando, M., Marino, D.J., 2017. Occurrence and fate of pesticides in the Argentina stretch of the Paraguay-Paraná basin. Environ. Monit. Assess. 189 (63), 12pp.

FAO, 2012. The State of World Fisheries and Aquaculture. Food and Agriculture Organization of the United Nations, Rome, p. 230.

Feijoó, C.S., Lombardo, R.J., 2007. Baseline water quality and macrophyte assemblages in Pampean streams: a regional approach. Water Res. 41 (7), 1399–1410.

Franco-Trecu, V., Costa, P., Abud, C., Dimitriadis, C., Laporta, P., Passadore, C., Szephegyi, M., 2009. By-catch of franciscana *Pontoporia blainvillei* in Uruguayan artisanal gillnet fisheries: an evaluation after a twelve-year gap in data collection. Lat. Am. J. Aquat. Mamm. 7 (1-2), 11–22.

FREPLATA, 2014. Estrategia para la implementación del Programa Binacional de Monitoreo del Río de la Plata y su Frente Marítimo. Grupo Binacional de Monitoreo. Documento Técnico. Montevideo, FREPLATA.

García-Alonso, J., Greenway, G.M., Munshi, A., Gómez, J.C., Mazik, K., Knight, A.W., Hardege, J.D., Elliott, M., 2011. Biological responses to contaminants in estuaries: disentangling complex relationships. Mar. Environ. Res. 71, 295–303.

García-Alonso, J., Lercari, D., Araujo, B.F., Almeida, M.G., Rezende, C.E., 2017. Total and extractable elemental composition of the intertidal estuarine biofilm of the Río de la Plata: disentangling natural and anthropogenic influences. Estuar. Coast. Shelf Sci. 187, 53–61.

Gautreau, P., 2006. La Bahía de Montevideo: 150 años de modificación de un paisaje costero y subacuático. Bases para la conservación y manejo de la costa Uruguaya. Vida Silvestre, 401–411.

Gianelli, I., Defeo, O., 2017. Uruguayan fisheries under an increasingly globalized scenario: long-term landings and bioeconomic trends. Fish. Res. 190, 53–60.

Griffero, L., Gomes, G., Berazategui, M., Fosalba, C. Teixeira de Mello, F., Rezende, C.E., Bila, D.M., García-Alonso, J. Estrogenicity and cytotoxicity of sediments and water from the drinkwater source-basin of Montevideo city, Uruguay, Ecotoxicol. Environ. Contam. 13, 2018, 15–22.

Hiddink, J.G., Jennings, S., Sciberras, M., Szostek, C.L., Hughes, K.M., Ellis, N., Collie, J.S., 2017. Global analysis of depletion and recovery of seabed biota after bottom trawling disturbance. Proc. Natl. Acad. Sci. 114 (31), 8301–8306.

Horta, S., Defeo, O. 2012. The spatial dynamics of the Whitemouth Croaker artisanal fishery in Uruguay and interdependencies with the industrial fleet. Fish. Res., vol. 125-126, 2012, pp. 121–128. https://doi.org/10.1016/j.fishres.2012.02.007.

Jaureguizar, A.J., Militelli, M.I., Guerrero, R., 2008. Distribution of *Micropogonias furnieri* at different maturity stages along an estuarine gradient and in relation to environmental factors. J. Mar. Biol. Assoc. UK 88, 175–181. https://doi.org/10.1017/s0025315408000167.

Jaureguizar, A.J., Solari, A., Cortés, F., Milessi, A.C., Militelli, M.I., Camiolo, M.D., García, M., 2016. Fish diversity in the Río de la Plata and adjacent waters: an overview of environmental influences on its spatial and temporal structure. J. Fish Biol. 89 (1), 569–600.

Kopprio, G.A., Biancalana, F., Fricke, A., Garzón Cardona, J.E., Martínez, A., Lara, R.J., 2015. Global change effects on biogeochemical processes of Argentinian estuaries: an overview of vulnerabilities and ecohydrological adaptive outlooks. Mar. Pollut. Bull. 91, 554–562.

Lasta, C.A., Ruarte, C.O., Carozza, C.R., 2001. Flota costera argentina: antecedentes y situación actual. In: Bertolotti, M.I., Verazay, G.A., Akselman, R. (Eds.), El Mar Argentino y sus recursos pesqueros. Evolución de la flota pesquera argentina, artes de pesca y dispositivos selectivos, Instituto Nacional de Investigacion y Desarrollo Pesquero 3. Mar del Plata, pp. 89–106. ISBN 987-20245-0-2M.

Leguizamón, A., 2014. Modifying Argentina: GM Soy and socio-environmental change. Geoforum 53, 149–160. https://doi.org/10.1016/j.geoforum.2013.04.001.

Lercari, D., Bergamino, L., 2011. Impacts of two invasive mollusks, *Rapana venosa* (Gastropoda) and *Corbicula fluminea* (Bivalvia), on the food web structure of the Río de la Plata estuary and nearshore oceanic ecosystem. Biol. Invasions 13 (9), 2053–2061.

Lezama, C., Estrades, A., Rivas, F., Viera, N., Fallabrino, A., 2013. In: Green turtle interactions with coastal gillnet fishery of the Rio de la Plata estuary, Uruguay. 33° Annual Symposium on Sea Turtle Biology and Conservation. Baltimore (USA).

Little, S., Spencer, K.L., Schuttelaars, H.M., Millward, G.E., Elliott, M., 2017. Unbounded boundaries and shifting baselines: estuaries and coastal seas in a rapidly changing world. Estuar. Coast. Shelf Sci. 198, 311–319.

Lorenzo, M.I., Defeo, O., Moniri, N.R., Zylich, K., 2015. Fisheries catch statistics for Uruguay. Working Paper Series, vol. 25, pp. 1-6, University of British Columbia.

Mac Loughlin, T.M., Peluso, L., Marino, D.J.G., 2017. Pesticide impact study in the peri-urban horticultural area of Gran La Plata, Argentina. Sci. Total Environ. 598, 572–580.

Martínez, A., Méndez, S., Fabre, A., Ortega, L., 2017. Intensification of marine dinoflagellates blooms in Uruguay. INNOTEC 13, 19–25. ISSN 1688-6593.

Moreira, D., Simionato, C.G., Dragani, W., Cayocca, F., Tejedor, M.L.C., 2016. Characterization of bottom sediments in the Río de la Plata Estuary. J. Coast. Res. 32, 1473–1494.

Muniz, P., Danulat, E., Yannicelli, B., García-Alonso, J., Medina, G., Bícego, M.C., 2004. Assessment of contamination by heavy metals and petroleum hydrocarbons in sediments of Montevideo Harbour (Uruguay). Environ. Int. 29, 1019–1028.

Nagy, G., 2006. Vulnerabilidad de las aguas del Río de La Plata: Cambio de estado trófico y factores físicos. In: Barros, V., Menéndez, A., Nagy, G. (Eds.), El cambio climático en el Río de la Plata. Final report submitted to Assessments of Impacts and Adaptations to Climate Change (AIACC), Project No. LA 32.

Nagy GJ, Muñoz N., Verocai J.E., Bidegain M., de los Santos, B., Seijo, L., García, J.M., Feola, G., Brena, B., Risso, J. 2014. Integrating climate science, monitoring, and management in the Rio de la Plata estuarine front (Uruguay). In: Leal Filho W., Alves F., Caeiro S., Azeiteiro U. (eds) International Perspectives on Climate Change Management. Springer, Cham.

Norgaard, R.B., Kallis, G., Kiparsky, M., 2009. Collectively engaging complex socio-ecological systems: re-envisioning science, governance, and the California Delta. Environ. Sci. Pol. 12 (6), 644–652.

Paesch, L., Norbis, W., Inchausti, P., 2014. Effects of fishing and climate variability on spatio-temporal dynamics of demersal chondrichthyans in the Rio de la Plata, SW Atlantic. Mar. Ecol. Prog. Ser. 508, 187–200.

Piccini, C., García-Alonso, J., 2015. Bacterial diversity patterns of the intertidal biofilm in urban beaches of Río de la Plata. Mar. Pollut. Bull. 91, 476–482.

Piedra-Cueva, I., Fossati, M., 2007. Residual currents and corridor of flow in the Rio de la Plata. Appl. Math. Model. 31, 564–577.

PNUD-GEF, 2007. Protección Ambiental del Río de la Plata y su Frente Marítimo: Prevención y Control de la Contaminación y Restauración de Hábitats. Project RLA 99/G31, FREPLATA, CARP, CTMFM 25.

Psuty, N.P., Mizobe, C. 2006. South America, coastal geomorphology. In Encyclopedia of Coastal Science. Schwartz, M. (Ed.). Springer Science & Business Media. (pp. 905-909). Springer, Netherlands.

Rigacci, L.N., Giorgi, A.D., Vilches, C.S., Ossana, N.A., Salibián, A., 2013. Effect of a reservoir in the water quality of the Reconquista River, Buenos Aires, Argentina. Environ. Monit. Assess. 185 (11), 9161–9168.

SADI, 2013. Solicitud de Desagüe Industrial, Ministerio de Vivienda, Ordenamiento Territorial y Medio Ambiente, Uruguay. Available from: http://www.mvotma.gub.uy/ciudadania/tramites/tramites-medio-ambiente/item/10004360-trámite-de-informe-ambiental-de-operación-iao-vía-web.html

Salibián, A., 2006. Ecotoxicological assessment of the highly polluted Reconquista river of Argentina. In: Ware, G.W. (Ed.), Reviews of Environmental Contamination and Toxicology 185. Springer, New York, pp. 35–65.

Sathicq, M.B., Gómez, N., Andrinolo, D., Sedán, D., Donadelli, J.L., 2014. Temporal distribution of cyanobacteria in the coast of a shallow temperate estuary (Río de la Plata): some implications for its monitoring. Environ. Monit. Assess. https://doi.org/10.1007/s10661-014-3914-3.

Sathicq, M.B., Bauer, D.E., Gómez, N., 2015. Influence of El Niño Southern Oscillation phenomenon on coastal phytoplankton in a mixohaline ecosystem on the southeastern of South America: Río de la Plata estuary. Mar. Pollut. Bull. 98, 26–33.

Secchi, ER., Ott, PH., Crespo, EA., Kinas, PG., Pedraza, SN. Bordino. P. 2001. A first estimate of Franciscana (*Pontoporia blainvillei*) abundance off southern Brazil. J. Cetacean Res. Manag. 3, 95–100.

Simionato, C.G., Meccia, V.L., Dragani, W.C. 2009. On the path of plumes of the Río de la Plata estuary main tributaries and their mixing time scales. Geoacta 34, 87e116.

Szteren, D., Páez, E., 2002. Predation by southern sea lions (*Otaria flavescens*) on artisanal fishing catches in Uruguay. Mar. Freshw. Res. 53, 1161–1167.

Tatone, L.M., Bilos, C., Skorupka, C.N., 2012. Trace metals in settling particles from the sewage impacted Buenos Aires coastal area in the Río de la Plata Estuary, Argentina. Bull. Environ. Contam. Toxicol. 90, 318–322.

Valdés, M.E., Marino, D.J., Wunderlin, D.A., Somoza, G.M., Ronco, A.E., Carriquiriborde, P., 2015. Screening concentration of E1, E2 and EE2 in sewage effluents and surface waters of the "Pampas" region and the "Río de la Plata" estuary (Argentina). Bull. Environ. Contam. Toxicol. 94, 29–33. https://doi.org/10.1007/s00128-014-1417-0.

Venturini, N., Bícego, M.C., Taniguchi, S., Sasaki, S.T., García-Rodríguez, F., Brugnoli, E., Muniz, P., 2015. A multi-molecular marker assessment of organic pollution in shore sediments from the Río de la Plata Estuary, SW Atlantic. Mar. Pollut. Bull. 91, 461–475.

Vidal, F., Sedan, D., D'Agostino, D., Cavalieri, M.L., Mullen, E., Parot Varela, M.M., Flores, C., Caixach, J., Andrinolo, D., 2017. Recreational Exposure during Algal Bloom in Carrasco Beach, Uruguay: a liver failure case report. Toxins 9, 267. https://doi.org/10.3390/toxins9090267.

WHO, 2003. Guidelines for Safe Recreational Water Environments. Volume 1, Coastal and Fresh Waters. World Health Organization, Geneva.

Williman, C., Munitz, M.S., Montti, M.I.T., Medina, M.B., Navarro, A.F., Ronco, A.E., 2017. Pesticide survey in water and suspended solids from the Uruguay River Basin. Argentina Environ. Monit. Assess. 189, 259. https://doi.org/10.1007/s10661-017-5956-9.

WTTC, 2017. WTTC annual economic impact analysis. Available from: https://www.wttc.org/research/economic-research/economic-impact-analysis.

Chapter 4

Estuaries and Coastal Zones in the Northern Persian Gulf (Iran)

Moslem Sharifinia*, Moslem Daliri[†], Ehsan Kamrani[†]

*Iranian National Institute for Oceanography and Atmospheric Science (INIOAS), Gulf of Oman and Indian Ocean Research Center, Marine Biology Division, Chabahar, Iran, [†]Department of Fisheries, Faculty of Marine and Atmospheric Sciences and Technologies, University of Hormozgan, Bandar Abbas, Iran

1 GEOLOGY AND GEOMORPHOLOGY OF THE PERSIAN GULF

The Persian Gulf is a Mediterranean Sea type basin in the western Asia and situated between the Eurasian and the Arabian tectonic plates. The Persian Gulf has also been known as a shallow marginal sea of the Indian ocean that is located between the south side of Iran and the Arabian Peninsula, including the countries like Saudi Arabia, United Arab Emirates, Iraq, Kuwait, Bahrain, Qatar, and Oman (Fig. 1). One of the tectonic influences on the Persian Gulf is raising the southeastern border which makes deeper water and numerous islands along the southern Persian Gulf coast of Iran (Carpenter, 1997; Konyuhov and Maleki, 2006). This inland sea is connected to the Gulf of Oman in the east by the Strait of Hormuz, and its western end is marked by the major river delta of the Shatt al-Arab (called Arvand Roud by some countries), which carries the waters of the Euphrates, Tigris, and Karun rivers. The Hendijan, Hileh, and Mond rivers also flow in from Iran. The Persian Gulf is about 56 km (35 mile) wide at its narrowest, in the Strait of Hormuz. The waters are overall very shallow, with a maximum depth of 100 m (near opening of Strait of Hormuz) and an average depth of 36 m (IHO, 1953; Reynolds, 1993).

The bottom topography is also mostly flat and dominated by soft sediments (Fig. 2). Along the Iranian coast of the Persian Gulf, the seabed is mostly muddy and some coral islands, such as Kish, are also present. The maximum width of the Persian Gulf is 338 km, and the length to its northern coast is nominally 1000 km. The surface area of the Gulf is approximately $2.39 \times 10^5 \, km^2$, and a mean depth of 36 m implies an average volume of $8.63 \times 10^3 \, km^3$.

The Persian Gulf is one of the most important waterways in the world. A review of the world's energy indicates that approximately 60% of the world's marine transport of oil comes from this region and one ship passes the Strait of Hormuz each 6 min (Reynolds, 1993).

2 CLIMATIC CONDITIONS AND RECENT CHANGES

The Persian Gulf is located in the subtropical region, laying almost entirely between 24°10′–30°16′N and 47°46′–56°13′E and connected to the Gulf of Oman by the Strait of Hormuz, where the climate is classified as arid. The combination of high evaporation ($1.4 \, m \, yr^{-1}$) and low precipitation rates (0.03–$0.11 \, m \, yr^{-1}$) (Privett, 1959; Almazroui et al., 2012). Dominating winds, called Shamal winds, in the region are mainly from the N-NW throughout the year, whereby the topography of the region plays a major role in strengthening of this wind with the high terrain along the Iranian Coast (Al Senafi and Anis, 2015). The dominant current pattern in the Persian Gulf is a clockwise movement with less saline water at the surface and more saline water at the bottom (Reynolds, 1993; Carpenter, 1997). Tides, which predominately are semidiurnal, do not have a dominant or long-lasting effect on the current pattern but they only have localized effects. The tidal fluctuation and range is high, and ranges from 1 to 3 m (Carpenter, 1997).

3 HYDROLOGY AND CIRCULATION IN THE PERSIAN GULF

The Persian Gulf includes a large region of shallow water in the northwest part of the Arabian Sea between the coasts of Iran and Arabia. The length and mean depth of the Persian Gulf are about 990 km and 36 m, respectively, and there are several

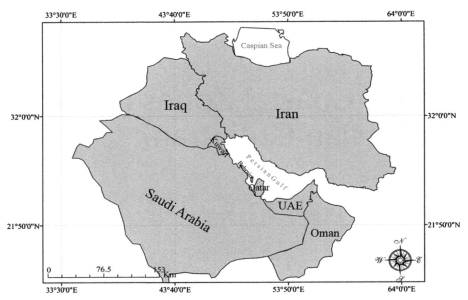

FIG. 1 Location of the Persian Gulf.

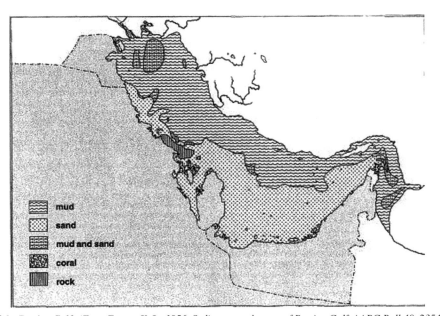

FIG. 2 Bottom types of the Persian Gulf. *(From Emery, K.O., 1956. Sediments and water of Persian Gulf, AAPG Bull 40, 2354–2383).*

places where the depth is greater than 100 m (Emery, 1956). The maximum and minimum (near the strait) width of the Gulf are approximately 338 and 65 km, respectively. The area of the Persian Gulf is approximately 239,000 km² and the volume of water in it is approximately 8630 km³ (Emery, 1956, Kämpf and Sadrinasab, 2006). The waters of the Persian Gulf in the winter season are well mixed whereas in the summer season, there are three layers of water in the Persian Gulf (Strait of Hormuz) (Mubarak and Kubryakov, 2001) (Table 1).

Along the west coast of the United Arab Emirates, three water layers are very visible. Nevertheless, in this shallow-water region, because of more heating and evaporation, the temperature and salinity values are higher than the Strait of Hormuz (26.6°C–28.2°C and 39.5°C–41.2°C, respectively). In the southern coast of the Persian Gulf (near the Qatar and Bahrain coasts), there is a thin surface isohaline layer (2 m thickness) with high temperature and relatively low salinity (26°C–27°C and 38.1°C, respectively). In depths of 18–24 m, the temperature experiences a rapid decrease, down to 22.2°C, while the salinity increases, up to 39.8. In deeper layers (>24 m), temperature and salinity are 20.4°C and 39.6°C,

TABLE 1 The Stratified Nature of the Waters of the Persian Gulf as Shown by the Temperature and Salinity (Salinity as Practical Salinity Units)

Water Layer	Temperature Range (°C)	Salinity Range
Surface layer (0–20 m)	26.7–28	36.8–37.5
Intermediate layer (20–43 m)	21.5–26.7	37.5–37.8
Bottom layer (43–95 m)	19.8–21.5	37.8–40.6

respectively. For the deep-water part of the Persian Gulf (near the west coast of Iran), there is two layers: (1) a surface isohaline layer with a salinity of about 38.6 (with 20 m thickness) and (2) the bottom layer located at depths of 42–65 m (Mubarak and Kubryakov, 2001). In the northern part of the Persian Gulf, there are only two layers as below (Mubarak and Kubryakov, 2001) (Table 2).

Based on an integrated hydrographic data set obtained a year after the Persian Gulf War in 1991, Reynolds (1993) gave a sketch of the Persian Gulf general circulation (Fig. 3) that has been confirmed by Sadrinasab and Kämpf (2004) and Meshkati and Tabibzadeh (2016). The main flow path is counterclockwise. Ocean water with normal salinity enters the Persian Gulf from the Gulf of Oman, flows westward along Iranian coastline and turns southeast to exit the Persian Gulf along the coasts of United Arab Emirates and Oman (Reynolds, 1993). It should be noted that water circulation in the

TABLE 2 The Water Characteristics of the Northern Part of the Persian Gulf

Water Layer	Attributes
Surface layer (0–5 m)	Uniform distribution of temperature (24.8°C) Salinity varies within the range 36–37.2
Bottom layer	Quite smooth variation of temperature (19.4°C–24.8°C) and salinity (37.2–40.3)

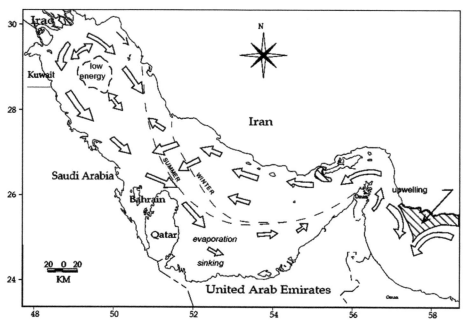

FIG. 3 Sketch of the surface currents and general circulation in the Persian Gulf. (*Figure copied from the online version of the article by Reynolds, R.M., 1993. Physical oceanography of the Gulf, Strait of Hormuz, and the Gulf of Oman—results from the Mt Mitchell expedition. Mar. Pollut. Bull. 27, 35-59*).

Persian Gulf is expected to take from 2 to 5 years (Reynolds, 1993). The spatial distribution of salinity shown by Johns et al. (2003) proposed that during summer the cyclonic (clockwise) circulation extends the full length of the Persian Gulf. Moreover, the joint discharge from the Bahmanshir and Shatt-al-Arab Rivers makes a classical river plume that generally runs along the coasts of Kuwait and Saudi Arabia.

4 BIODIVERSITY OF THE IRANIAN COASTAL WATERS OF THE PERSIAN GULF

Protecting the coastal and marine biodiversity in the Persian Gulf will increase the ability of ecosystem to adjust and improve following natural or anthropogenic disturbances, including the effects of global change in its many forms. Conserving natural coastal and marine biodiversity as a critical part of preserving marine ecosystem functions and services will play an important role in protecting the water quality, safeguarding the shoreline, providing recreational space, and ensuring a sustainable exploitation of fisheries resources (Palumbi et al., 2009).

In coastal and marine environments, there is growing evidence that conserving natural biodiversity can provide a broad spectrum of ecosystem services such as providing food, medicines, recreation, and climate modulation (Duraiappah, 2005; Palumbi et al., 2009). Nevertheless, on a global scale, nearly 60% of these ecosystem services have been degraded (Duraiappah, 2005). Along the Iranian coasts, loss or impairment of biodiversity occurs together with anthropogenic activities (Taherizadeh and Sharifinia, 2015; Kamrani et al., 2016). Efforts to improve national marine spatial development as a part of a national ocean policy will be required to provide a major progress in safeguarding coastal and marine biodiversity.

The most important natural coastal habitats in the Persian Gulf include coral reefs, mangroves, seagrass beds, sandy shores, rocky shores, and estuarine ecosystems. In Iranian coastal waters, coral reefs, mangrove forests, and sea grass habitats play an important role in ecological, recreational, economic, and cultural activities. These habitats deliver food and shelter for various fish and marine species, protect coastal areas from storms, prevent the coast from erosion, improve commercial fishing, and provide an array of recreational activities.

4.1 Coral Reefs

Coral reefs are part of a unique and highly complex system of intertidal and subtidal habitats in the Persian Gulf. The Persian Gulf coral communities, due to living in a harsh environment with respect to salinities, sea temperatures (ranges between 14°C and 34°C), and extreme low tides, can be considered as a unique living laboratory to investigate the effects of anthropogenic activities, climate change, and the potential for adaptation among reefs universally (Coles and Fadlallah, 1991; Sheppard and Loughland, 2002).

The best development of coral growth takes place virtually throughout the offshore parts of the Persian Gulf. Corals in the Iranian waters of the Persian Gulf are mostly restricted to the offshore islands with significant coral development in regions with depths of <10 m. Knowledge of Iranian coral assemblages was reviewed by Rosen (1971) and Harger and Thamrin (1978), with additional information from Rezai and Savari (2004), Wilson et al. (2002), Rezai et al. (2004), Namin et al. (2009), Kavousi et al. (2011), and Jafari et al. (2016). There are few records of coral communities on the Iranian mainland coast (Maghsoudlou et al., 2008; Rezai et al., 2010). Riegl and Purkis (2011) stated that this could be due to unfavorable conditions caused by runoff from the mountainous hinterland. Pleistocene coral reefs are well known from the Kish (Preusser et al., 2003) and Qeshm (Pirazzoli et al., 2004) islands. Of the known coral rich regions, Kish (62 ha), Hendourabi (20 ha), Farur (19 ha), Lavan (18 ha), Larak (16 ha), Sirri (16 ha), and Farurgan (2.5 ha) islands have been surveyed. Some of corals that have been reported from these islands for the first time are *Echinopora gemmacea*, *Leptoria irregularis*, *Montipora incrassata*, gorgonian *Subergorgia suberosa*, *Heteropsammia* sp., *Tubastraea* spp., *Antipathes* sp., *Dendronephthya*, and *Sarcophyton*.

Rezai et al. (2004) stated that due to the deeper water, slightly lower temperatures, and more stable salinities, some better-developed reefs can be found nearer to the Straits of Hormuz. Other coral reefs are known to be in places such as Kharko (266 ha), Nay Band (181 ha), Khark (181 ha), Hormuz (59 ha), Hengam (36 ha), Tonb-e-Bozorg and Tonb-e-Koochak (21 ha), Shidvar (13 ha), and Qeshm and Aboomusa (11 ha) islands; however, these have not been investigated. It should be noted that the best development of coral reefs is in the Kharg, Farur, Farurgan, and Larak Islands (Rezai et al., 2004). Generally, because of extremes of water temperature and salinity that are close to the physiological tolerance limits of many species in the Persian Gulf, the diversity of coral reefs in this ecosystem is usually low (Maghsoudlou et al., 2008) compared to those in the Indian Ocean. Typically, the composition of coral species in this area is Indo-Pacific and hard coral species richness in the entire of the Persian is less than in the Gulf of Oman (which is about 70 species) (Sheppard and Salm, 1988).

4.2   Mangrove Forests

Mangroves forests, the only woody halophytes living in the intertidal region between the sea and the land in the tropical and subtropical coastlines of the world, are normally distributed from mean sea level to highest spring tide (Alongi, 2002). They are a valuable ecological and economic resource and have been heavily used as food, timber, fuel, and medicine. Moreover, mangrove forests deliver nursery grounds and breeding sites for animals, especially fishes, and provide shoreline protection against coastal erosion (Alongi, 2002). They grow in harsh environmental conditions such as areas with high temperatures, extreme tides, high salinities, high sedimentation, and muddy anaerobic soils.

In Iran, mangrove forests comprised about 93.37 km² of Iranian shorelines (between longitude 25°19′ and 27°84′) with the largest area (about 67.5 km²) in the Persian Gulf occurring between the Khamir Port and the northwest side of Qeshm Island, and the smallest area (about 0.01 km²) in the Bardestan estuary (Zahed et al., 2010) (Fig. 4). Moreover, these forests are one the richest ecosystems in terms of biodiversity in the northern part of the Persian Gulf. Many species of plants and animals linked to mangrove forests in Iran have been reported in the Persian Gulf (Zahed et al., 2010). More than 100 taxa of sea birds, including 9 classes and 35 families, have been identified within the Iranian mangrove forests (Behrouzi Rad, 1996). More than 100 species of fishes belonging to 40 families, 10 species of crabs, and 6 species of prawn have been identified. Moreover, 51 genera of phytoplankton and 37 groups of zooplankton, 5 species of sea snake, 2 species of sea turtle, and only 1 taxon of a marine mammal have been reported in this region (Danehkar, 1996; Mashaii, 2006; Mehrabian et al., 2008; Zahed et al., 2010; Kamrani et al., 2016).

In total, there are 69 species in 27 genera, belonging to 20 families mangrove species in the world (Duke, 1992; Kathiresan and Bingham, 2001). *Rhizophora*, *Avicennia*, *Bruguiera*, and *Sonneratia* are the most important of mangrove genera. *Avicennia marina* (family *Avicenniaceae*) and *Rhizophora macrunata* (family *Rhizophoraceae*) are the only two mangrove species that are found within the Iranian mangrove forests. The Persian name of *A. marina* is "Harra" and for *Rh. macrunata* is "Chandal" (Danehkar, 1996, 1998). The dominant mangrove species within the Iranian forests is *A. marina*, while *Rh. macrunata* is found only in the Sirik region (Zahed et al., 2010). Zahed et al. (2010) stated that the dominance of *A. marina* in the Iranian mangrove forests and Persian Gulf could be due to its high tolerance to temperature and salinity variations.

4.3   Phytoplankton Community

In general, studies on plankton communities in subtropical estuaries and coastal ecosystems are less than in temperate ecosystems (Badylak and Phlips, 2004). Nevertheless, there is a growing interest in subtropical and tropical ecosystems, particularly those subject to anthropogenic activities and eutrophication phenomenon (Knoppers et al., 1991; Oliveira and Kjerfve, 1993; Badylak and Phlips, 2004; Royer et al., 2015). Due to the importance of the eutrophication phenomenon and the effect of human activities on coastal ecosystems, the issue of phytoplankton composition has gained increasing importance (Hallegraeff, 1993; Badylak and Phlips, 2004; Sharifinia et al., 2015).

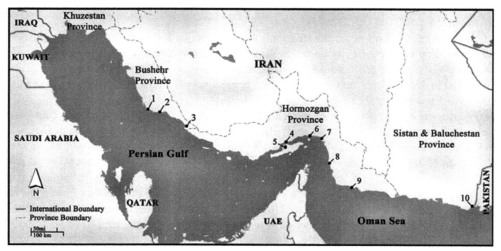

FIG. 4   Location of Mangrove forests in the Persian Gulf (Zahed et al., 2010).

The Iranian coastal waters are physically, chemically, and biologically diverse marine waters that are subject to different type of anthropogenic pressures (Taherizadeh and Sharifinia, 2015; Kamrani et al., 2016; Sharifinia et al., 2018). One of the main concerns about the Iranian coastal waters has been the occurrence of harmful algal blooms (HABs), principally toxic species (Fatemi et al., 2012; Moradi and Kabiri, 2012; Sharifinia et al., 2015). In 2008–2009, fish kills were related to blooms of the toxic dinoflagellate *Cochlodinium polykrikoides* (Sharifinia et al., 2015).

In recent years, the estuarine and coastal waters of the Persian Gulf have been an important subject for the ecological research of marine phytoplankton. Every year algal blooms and red tides occurs in the Persian Gulf due to the flourishing of various species of dinophyceae. One of the main species that cause algal blooms repeatedly throughout the Persian Gulf is *Noctiluca scintillans*. Phytoplankton blooms and red tides are common in Iranian waters of the Persian Gulf. Since the 1980s, the most important species that cause algal blooms include *N. scintillans*, *Trichodesmium* sp., and *Nitzschia* sp. (Fatemi et al., 2012). Regardless of the concerns about phytoplankton blooms in the Iranian coastal waters of the Persian Gulf over the last years, it should be noted that there are some studies of the abundance and composition of phytoplankton communities. The first study of the Persian Gulf phytoplankton communities was conducted by Böhm (1931), which lists the occurrence of 34 Pyrrophyta. Fallahi et al. (2005) investigated the distribution of phytoplankton communities in the Persian Gulf. In total, 244 species of phytoplankton have been identified in this study, of which 124 were diatoms (*Bacillariophyceae*), 113 *Dinophyceae*, 5 *Cyanophyceae*, 1 *Chrysophyceae*, and 1 *Euglenophyta*. The results showed that the density and diversity of phytoplankton over the past few decades has decreased. The density of phytoplankton has increased from east to west of the Persian Gulf and there were two peaks in the summer and winter seasons during the year (Fallahi et al., 2005).

4.4 Macrobenthic Community

In this section, we summarize a description of these studies and their main outcomes in an effort to combine our understanding of the macrobenthic ecology in Iranian estuarine and coastal waters of the Persian Gulf. Macrobenthic assemblages in the Persian Gulf are characterized by high levels of biodiversity but low species richness because of harsh environmental conditions (Price, 2002; Sheppard et al., 2010; Pourjomeh et al., 2014). In recent decades, the Iranian estuarine and coastal waters of the Persian Gulf have been an active area for the ecological study of macrobenthic assemblages (Farsi et al., 2015; Taherizadeh and Sharifinia, 2015; Amini-Yekta et al., 2017; Sharifinia, 2017). These studies are concerned with evaluating the impact of anthropogenic activities on macrobenthic communities or understanding the general relationship between the distribution of these organisms and their physicochemical environment using univariate and multivariate analyses. Overall, these studies contain valuable information about species diversity and density, distribution, seasonality, and the impact of a range of anthropogenic activities such as aquaculture effluents on macrobenthic population dynamics. Moreover, the majority are from waters relatively impacted by anthropogenic activities, and as such, they provide a valuable record of macrobenthic variability in the stressed environment. For example, Taherizadeh and Sharifinia (2015) investigated the changes in macrobenthic community structure using macrobenthic-based indices such as BENTIX. In this study, 125 taxa have been identified and most species belong to the crustacea, polychaeta, or bivalvia. This study concluded that the BENTIX index was a suitable tool to evaluate the status of benthic quality in the estuaries of the Persian Gulf. This index precisely classifies the macrobenthic communities into ecological quality classes and is able to detect spatial changes, due to the differential impact of the pressures.

The effects of environmental changes on the distribution of macrobenthic community in the Bushehr coasts of the Persian Gulf were assessed by Farsi et al. (2015). In this study, 17 taxa of macrobenthic were identified, of which the most dominant taxa were belonging to the *Mollusca*, *Annelida*, and *Arthropoda*. The average abundance and biomass of macrofauna ranged from 450 to 4380 ind m^{-2} and 9 to 165 g m^{-2}, respectively. The highest abundance and biomass have been reported in depths of 10 m and intertidal zone.

Sharifinia (2017) assessed the impact of anthropogenic activities on the estuarine ecosystems using macrofauna as biotic indices (e.g., the AMBI and BENTIX). In this study, 165 taxa of macrobenthic taxa were identified, which including 7 phyla *Annelids* (65 taxa), *Mollusca* (65 taxa), *Arthropods* (28 taxa), *Echinoderms* (3 taxa), *Hydrozoans* (2 taxa), *Nemertea* (1 taxon), and *Nematoda* (1 taxon). For the total abundance, *Polychaeta* showed the richest taxonomic feature with 62 species. This study revealed that the AMBI and BENTIX indices are suitable for evaluating the environmental situation of coastal ecosystems and they enabled the separation of areas less impacted by human activities from areas affected by these activities and could be used as a robust management tool for monitoring programs in coastal areas. In addition, in this study, *Capitella capitata* and *Clymene robusta* were species as resistant to pollution and the taxa *Assiminea* sp. and *Littorina intermedia* were introduced as species sensitive to disturbance and pollution.

5 ANTHROPOGENIC STRESSES IN THE NORTHERN PERSIAN GULF

The Persian Gulf presents stressful environmental conditions, such as extreme heat ($>30°C$) in summer and cold ($<10°C$) in winter, low annual rainfall, high evaporation and salinity, a low entry of fresh water, etc. In addition, the Persian Gulf is also considered as one of the regions most affected by human activity in the world (Halpern et al., 2008). The human activities which mostly have an impact include the following.

5.1 Oil Pollution

The highest concentration of oil and gas extraction and desalination facilities are located in the Persian Gulf region (Van Lavieren et al., 2011). Therefore, the petroleum industry is one of the biggest threats to the marine and coastal environments in the region. Multiple wars in the Persian Gulf have added to the problem of oil pollution. For example, between 1980 and 1988 (Iran-Iraq war) attacks on petroleum fields from both side caused the spillage of 2–4 million barrels into waters of the Persian Gulf. During the Iraq-Kuwait war in 1991, the First Persian Gulf war, Iraq also released 11 million barrels of oil into Kuwait waters which was the largest oil contamination in human history (Sheppard et al., 1992). Once again, during the Second Gulf War in 2003, 6–8 million barrels of oil spilled into the Persian Gulf and the Arabian Sea. In recent years, there have been 21 major oil spills globally, with more than 100 million barrels discharged, in which 7 of them have occurred in the Persian Gulf region.

5.2 Fisheries and Overfishing

Globally, 61.3% of fish stocks are estimated to be fully exploited and 28.8% fished at stock levels that are biologically unsustainable whereas only 9.9% remain underexploited worldwide (FAO, 2014). The revealed catch data, by the Regional Commission for Fisheries (RECOFI), show that most of the Persian Gulf stocks are fully exploited and overfished. In support of this claim, Hosseini et al. (2015) stated that because juvenile commercially important species are mostly fished, increasing overfishing is occurring in the Iranian waters of the Persian Gulf. Thus, some fish species (especially demersal species) are assessed as endangered by the International Union for Conservation of Nature and Natural Resources (IUCN, 2014). Illegal fishing also occurs and contributes to overfishing in Iranian waters of the Persian Gulf (Daliri et al., 2015). Daliri et al. (2016) reviewed the reasons for illegal fishing occurrence in the northern Persian Gulf (Hormozgan province), and designed a model with categories such as culture, management issues, economic conditions, personal skills, and area features. Underreporting of catches is also a serious risk to the marine environment in the Persian Gulf, so that Al-Abdulrazzak and Pauly (2013) and Daliri (2016) argued that the officially catches are mostly underreported in the Persian Gulf region.

6 THE FUTURE CHANGES TO THE PERSIAN GULF

Globally, coastal and estuarine ecosystems have been influenced by anthropogenic pressures such as pollution and habitat degradation and the Persian Gulf is no exception. Some of the environmental issues resulting from anthropogenic activities in the coastal waters of the Persian Gulf include eutrophication, chemical pollution, sedimentation, and changes to fisheries (Taherizadeh and Sharifinia, 2015; Kamrani et al., 2016; Daliri et al., 2017; Sharifinia, 2017; Sharifinia et al., 2018). In general, society is heavily dependent on the coastal and marine habitat and resources for various activities but has used the coastal area as a waste disposal ground (Naden et al., 2016; Sharifinia et al., 2018). Due to the increase in population growth, urbanization, and industrialization, it is expected that living and nonliving coastal resources will continue to be under increasing pressure. Therefore, a comprehensive management plan is needed to protect coastal areas in the future.

The estuaries and coasts of the Persian Gulf face important challenges from different types of natural and human-based issues. In the densely populated coastline of the Bandar-Abbas, Bushehr, and Khuzestan provinces, the risk of threats to the Persian Gulf ecosystem will be increasing by population growth, unmanaged wastewater effluents, and residential urban development. One of the main impacts of urbanization and industrialization on marine ecosystems is the loss of sensitive species and reducing biodiversity. Sharifinia (2017) argued that industrial sewage and aquaculture effluents have been the most important factors in reducing biodiversity in the studied estuaries in the Persian Gulf. These threats (e.g., eutrophication, chemical pollution, and sedimentation) can affect fisheries, tourism, public works, marine transport, and industry.

6.1 Increasing Urban Populations

As Khan (2007) argued, the growth of the exploitation of abundant oil and gas reserves and tourism has caused the Persian Gulf countries mostly to have amongst the highest levels of economic growth rate worldwide. This economic growth has led to population growth particularly in the United Arab Emirates (UAE), Bahrain, and Qatar. For example, the population of Bahrain and Qatar has increased from 661,000 in 2001 to 1 million in 2009 and 770,000 in 2001 to 1.4 million in 2009, respectively. Subsequently, this rapid population increase and developing industries are leading to large-scale and deleterious changes in the coastal areas in the region.

Using seawater desalination systems (SDSs), in order to supply fresh water, could be an irreparable threat for ecosystem health in the Persian Gulf. Hoepner and Lattemann (2003) examined chemical impacts from SDSs in the northern Red Sea and concluded that a seawater desalination plant with daily capacity 1.5 million $m^3 d^{-1}$ discharges amounts equal to 2708 kg chlorine (Cl), 36 kg copper (Cu), and 9478 kg antiscalants (as the pretreatment water additive for reverse osmosis system). Currently, 213 active seawater desalination plants exist in the Persian Gulf (Fig. 5), although 51 others are under planning, construction, or installation. The daily capacity of these plants is 11 million $m^3 d^{-1}$ which is equal to half of global production (Taghavi et al., 2016). As reviewed above, due to excessive evaporation (1.4 m yr^{-1}) and low precipitation rates (0.03–0.11 m yr^{-1}), the Persian Gulf is a saline water body and the rapid increase of seawater desalination facilities leads to critical ecological stresses by increasing water temperature and salinity, decreasing dissolved oxygen, and discharging heavy metals (Fig. 6).

6.2 An Unstable Political Situation

There are persistently high levels of hydrocarbon pollution throughout the waters of the Persian Gulf of which a major proportion is due to the wars that have happened in the region (El-Baz and Makharita, 2016). For example, Iraq began an eight-year war against Iran in 1980. This war first was predominantly on the ground front but, to disrupt Iran's oil exports,

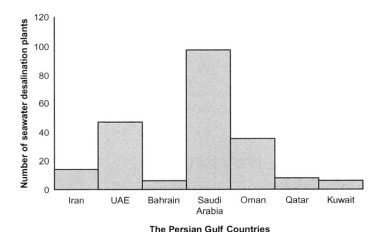

FIG. 5 The number of seawater desalination plants in Arabian countries of the Persian Gulf (Taghavi et al., 2016).

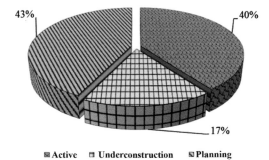

FIG. 6 Relative frequency of seawater desalination plants in Iran (Taghavi et al., 2016).

Iraq attacked the Iranian oil tankers and terminals. Thereafter, the war on the sea front was intensified and the number of marine attacks from 5 and 22 in the first and second years reached 290 in the last 2 years of the war. In 1991, during the Iraq and Kuwait war, Iraq also spilled an estimated 10.8 million barrels of oil in the Persian Gulf coastal areas, where the benthic habitats in particular were severely impacted (Sale et al., 2011). An historical overview indicates that the Persian Gulf has always been influenced by an unstable political situation and the recent raised tensions among the Persian Gulf countries potentially are leading to a new crisis in the region.

6.3 Fisheries and Fish Species Conservation

As stated above, the Persian Gulf fish stocks are fully exploited and overfished, whereas many demersal fish species are protected, listed, or endangered. Fish species such as *Scomberomorus commerson* (Lacepède, 1800) and *Carcharhinus dussumieri* (Müller and Henle, 1839) have been listed as near threatened (NT) and others are assessed as data deficient (DD) (IUCN, 2018). Valinassab et al. (2006) argued that the population composition of demersal fishes in the Persian Gulf has been altered during the recent decades, and some noncommercial species such as *Netuma thalassina* (Rüppell, 1837) and *Saurida tumbil* (Bloch, 1795) are more abundant in catch composition. Decreasing commercial fish abundance, due to a high fishing effort, compared to an increase in the noncommercial fish stocks, constitutes an ecological disruption in the Persian Gulf.

6.4 Impacts of Climate Change on the Persian Gulf Ecosystem

Coastal and estuaries ecosystems are essential and important habitats and nursery grounds for many species that deliver many benefits to society and natural ecosystems (Beck et al., 2001; Gillanders et al., 2003). The Persian Gulf is also important in terms of providing marine resources and essential habitats for different species of aquatic animals. The Persian Gulf, due to its unique features such as high temperatures and high salinities, has recently attracted the attention of the international scientific community to investigate the effects of physicochemical and environmental factors on marine organisms and provide insights into the adaptation to global climate change (Riegl and Purkis, 2012; Bayani, 2016). The phenomenon of global warming and rising sea surface temperature is increasing the probability of coral bleaching, which can lead to the death and destruction of corals and the loss of vital habitats for other species (Pandolfi et al., 2011). Another consequence of climate change is the rise in atmosphere carbon dioxide (CO_2). The rise in CO_2 levels in the atmosphere subsequently causes ocean acidification and makes the ocean water more acidic. Scientists have predicted that the likelihood of such a trend would be greater in the coming decades (Raven et al., 2005; Feely et al., 2009). Such conditions would cause undesirable effects on marine organisms, including plankton, mollusks, and other shellfish but, in particular, corals are more susceptible to CO_2 concentrations, when the reefs are already stressed by anthropogenic activities. Coral reefs in the Persian Gulf are likely to experience considerable losses if concentrations of CO_2 in the atmosphere continue to rise at their current rate. As Bayani (2016) stated in her article "*It is now an opportune moment for Iran and other littoral states to extend their cooperation, raise their environmental concerns, and take concerted action to address the threats to the Persian Gulf's environment, and protect it for us and future generations.*"

REFERENCES

Al Senafi, F., Anis, A., 2015. Shamals and climate variability in the Northern Arabian/Persian Gulf from 1973 to 2012. Int. J. Climatol. 35, 4509–4528.

Al-Abdulrazzak, D., Pauly, D., 2013. Managing fisheries from space: Google Earth improves estimates of distant fish catches. ICES J. Mar. Sci. 71, 450–454.

Almazroui, M., Islam, M.N., Jones, P., Athar, H., Rahman, M.A., 2012. Recent climate change in the Arabian Peninsula: seasonal rainfall and temperature climatology of Saudi Arabia for 1979–2009. Atmos. Res. 111, 29–45.

Alongi, D.M., 2002. Present state and future of the world's mangrove forests. Environ. Conserv. 29, 331–349.

Amini-Yekta, F., Izadi, S., Asgari, M., Aghajan-Pour, F., Shokri, M.-R., 2017. Higher taxa as surrogates for species richness in intertidal habitats of Qeshm Island in the Persian Gulf. Mar. Biodivers. 48, 1–8.

Badylak, S., Phlips, E., 2004. Spatial and temporal patterns of phytoplankton composition in subtropical coastal lagoon, the Indian River Lagoon, Florida, USA. J. Plankton Res. 26, 1229–1247.

Bayani, N., 2016. Ecology and environmental challenges of the Persian Gulf. Iran. Stud. 49, 1047–1063.

Beck, M.W., Heck Jr., K.L., Able, K.W., Childers, D.L., Eggleston, D.B., Gillanders, B.M., Halpern, B., Hays, C.G., Hoshino, K., Minello, T.J., 2001. The identification, conservation, and management of estuarine and marine nurseries for fish and invertebrates: a better understanding of the habitats that serve as nurseries for marine species and the factors that create site-specific variability in nursery quality will improve conservation and management of these areas. Bioscience 51, 633–641.

Behrouzi Rad, B., 1996. The community of birds of Iranian mangrove forests. Environ. Sci. Quatr. J. 8, 70–80.

Böhm, A., 1931. Peridineen aus dem Persichen Golf und dem Golf von Oman. Archivfur Protestenkunde 74, 188–197.

Carpenter, K.E., 1997. Living Marine Resources of Kuwait, Eastern Saudi Arabia, Bahrain, Qatar, and the United Arab Emirates. Food & Agriculture Org, Rome.

Coles, S.L., Fadlallah, Y.H., 1991. Reef coral survival and mortality at low temperatures in the Arabian Gulf: new species-specific lower temperature limits. Coral Reefs 9, 231–237.

Daliri, M., 2016. Illegal, Unreported and Unregulated (IUU) Small-Scale Fishing in the Northern Persian Gulf (Hormozgan Waters). University of Hormozgan, Bandar Abbas, Iran.

Daliri, M., Jentoft, S., Kamrani, E., 2017. Illegal, unreported, and unregulated fisheries in the Hormuz Strait of Iran: how the small-scale fisheries guidelines can help. In: Jentoft, S., Chuenpagdee, R., Barragán-Paladines, M.J., Franz, N. (Eds.), The Small-Scale Fisheries Guidelines: Global Implementation. Springer International Publishing, Cham, pp. 557–572.

Daliri, M., Kamrani, E., Jentoft, S., Paighambari, S.Y., 2016. Why is illegal fishing occurring in the Persian Gulf? A case study from the Hormozgan province of Iran. Ocean Coast. Manag. 120, 127–134.

Daliri, M., Kamrani, E., Paighambari, S.Y., 2015. Illegal shrimp fishing in Hormozgan inshore waters of the Persian Gulf. Egypt. J. Aquatic Res. 41, 345–352.

Danehkar, A., 1996. Iranian mangroves forests. Environ. Sci. Quatr. J. 8, e22.

Danehkar, A., 1998. Marine sensitive areas of Iran. Environ. Sci. Quatr. J. 24, 28–38.

Duke, N.C., 1992. Mangrove floristics and biogeography. In: Robertson, A.I., Alongi, D.M. (Eds.), Tropical Mangrove Ecosystems. Coastal and Estuarine Studies Series, vol. 41. American Geophysical Union, Washington, DC, pp. 63–100.

Duraiappah, A.K., 2005. Ecosystems and Human Well-Being: Biodiversity Synthesis. A Report of the Millennium Ecosystem Assessment, World Resources Inst and Island Press, Washington, DC.

El-Baz, F., Makharita, R., 2016. The Gulf War and the Environment. Routledge, London.

Emery, K.O., 1956. Sediments and water of Persian Gulf. AAPG Bull. 40, 2354–2383.

Fallahi, M., Fatemi, S.M., Seraji, F., 2005. Biodiversity of phytoplankton communities in Iranian coastal waters of the Persian Gulf. In: 6th Marine Science and Technology Conference. Center for Oceanographic and Atmospheric Research, Tehran.

FAO, 2014. The State of World Fisheries and Aquaculture 2006. Food and Agriculture Organization of the United Nations. Rome, Italy.

Farsi, P., Seyfabadi, J., Owfi, F., Aramli, M.S., 2015. Effect of environmental conditions on spatial distribution of macrobenthic community in the Bushehr Coasts of the Persian Gulf. Turk. J. Fish. Aquat. Sci. 15, 869–878.

Fatemi, S., Nabavi, S., Vosoghi, G., Fallahi, M., Mohammadi, M., 2012. The relation between environmental parameters of Hormuzgan coastline in Persian Gulf and occurrence of the first harmful algal bloom of Cochlodinium polykrikoides (Gymnodiniaceae). Iran. J. Fish. Sci. 11, 475–489.

Feely, R.A., Doney, S.C., Cooley, S.R., 2009. Ocean acidification: present conditions and future changes in a high-CO_2 world. Oceanography 22, 36–47.

Gillanders, B.M., Able, K.W., Brown, J.A., Eggleston, D.B., Sheridan, P.F., 2003. Evidence of connectivity between juvenile and adult habitats for mobile marine fauna: an important component of nurseries. Mar. Ecol. Prog. Ser. 247, 281–295.

Hallegraeff, G.M., 1993. A review of harmful algal blooms and their apparent global increase. Phycologia 32, 79–99.

Halpern, B.S., Walbridge, S., Selkoe, K.A., Kappel, C.V., Micheli, F., D'agrosa, C., Bruno, J.F., Casey, K.S., Ebert, C., Fox, H.E., 2008. A global map of human impact on marine ecosystems. Science 319, 948–952.

Harger, J., Thamrin, J., 1978. Rapid survey techniques to determine distribution and structure of coral communities. In: Comparing Coral Reef Survey Methods. Report of a Regional UNESCO/UNEP Workshop. Phuket Marine Biological Center, Thailand, UNESCO, Paris, pp. 83–91.

Hoepner, T., Lattemann, S., 2003. Chemical impacts from seawater desalination plants—a case study of the northern Red Sea. Desalination 152, 133–140.

Hosseini, S.A., Daliri, M., Raeisi, H., Paighambari, S.Y., Kamrani, E., 2015. Destructive effects of small-scale shrimp trawl fisheries on by-catch fish assemblage in Hormozgan coastal waters. Fisheries 68, 61–78.

IHO, 1953. Limits of Oceans and Seas. International Hydrographic Organization, Monaco.

IUCN. 2014. The IUCN red list of threatened species. Available from: http://www. iucn red list. org. Version 2014. (Accessed 12 June 2014).

IUCN, 2018. The IUCN red list of threatened species. Version 2017-3. Available from: https://www.iucnredlist.org. (Accessed 25 April 2018).

Jafari, M.A., Seyfabadi, J., Shokri, M.R., 2016. Internal bioerosion in dead and live hard corals in intertidal zone of Hormuz Island (Persian Gulf). Mar. Pollut. Bull. 105, 586–592.

Johns, W., Yao, F., Olson, D., Josey, S., Grist, J., Smeed, D., 2003. Observations of seasonal exchange through the Straits of Hormuz and the inferred heat and freshwater budgets of the Persian Gulf. J. Geophys. Res. Oceans (C12)108. 1–18.

Kämpf, J., Sadrinasab, M., 2006. The circulation of the Persian Gulf: a numerical study. Ocean Sci. 2, 27–41.

Kamrani, E., Sharifinia, M., Hashemi, S.H., 2016. Analyses of fish community structure changes in three subtropical estuaries from the Iranian coastal waters. Mar. Biodivers. 46, 561–577.

Kathiresan, K., Bingham, B.L., 2001. Biology of mangroves and mangrove ecosystems. Adv. Mar. Biol. 40, 81–251.

Kavousi, J., Seyfabadi, J., Rezai, H., Fenner, D., 2011. Coral reefs and communities of Qeshm Island, the Persian Gulf. Zool. Stud. 50, 276–283.

Khan, N.Y., 2007. Multiple stressors and ecosystem-based management in the Gulf. Aquat. Ecosyst. Health Manag. 10, 259–267.

Knoppers, B., Kjerfve, B., Carmouze, J.-P., 1991. Trophic state and water turn-over time in six choked coastal lagoons in Brazil. Biogeochemistry 14, 149–166.

Konyuhov, A., Maleki, B., 2006. The Persian Gulf Basin: geological history, sedimentary formations, and petroleum potential. Lithol. Miner. Resour. 41, 344–361.

Maghsoudlou, A., Araghi, P.E., Wilson, S., Taylor, O., Medio, D., 2008. Status of coral reefs in the ROPME sea area (The Persian Gulf, Gulf of Oman and Arabian Sea). In: Wilkinson, C. (Ed.), Status of Coral Reefs of the World: 2008. Global Coral Reef Monitoring Network, Townsville, pp. 79–90.

Mashaii, N., 2006. Phytoplankton Abundance and Distribution In Bahoo-Kalat Estuary at Southeast of Iran. Pajouhesh and Sazandegi, Tehran.

Mehrabian, A., Naghinezhad, A., Mostafavi, H., Kiabi, B., Abdoli, A., 2008. Contribution to the Flora and Habitats of Mond Protected Area (Bushehr province). J. Environ. Studies 34, 1–18.

Meshkati, N., Tabibzadeh, M., 2016. An integrated system-oriented model for the interoperability of multiple emergency response agencies in large-scale disasters: implications for the Persian Gulf. Int. J. Disaster Risk Sci. 7, 227–244.

Moradi, M., Kabiri, K., 2012. Red tide detection in the Strait of Hormuz (east of the Persian Gulf) using MODIS fluorescence data. Int. J. Remote Sens. 33, 1015–1028.

Mubarak, W.A.-M., Kubryakov, A., 2001. Hydrological structure of waters of the Persian Gulf according to the data of observations in 1992. Phys. Oceanogr. 11, 459–471.

Naden, P., Bell, V., Carnell, E., Tomlinson, S., Dragosits, U., Chaplow, J., May, L., Tipping, E., 2016. Nutrient fluxes from domestic wastewater: a national-scale historical perspective for the UK 1800–2010. Sci. Total Environ. 572, 1471–1484.

Namin, K.S., Rezai, H., Kabiri, K., Zohari, Z., 2009. Unique coral community in the Persian Gulf. Coral Reefs 28, 27.

Oliveira, A.M., Kjerfve, B., 1993. Environmental responses of a tropical coastal lagoon system to hydrological variability: Mundau-Manguaba, Brazil. Estuar. Coast. Shelf Sci. 37, 575–591.

Palumbi, S.R., Sandifer, P.A., Allan, J.D., Beck, M.W., Fautin, D.G., Fogarty, M.J., Halpern, B.S., Incze, L.S., Leong, J.-A., Norse, E., 2009. Managing for ocean biodiversity to sustain marine ecosystem services. Front. Ecol. Environ. 7, 204–211.

Pandolfi, J.M., Connolly, S.R., Marshall, D.J., Cohen, A.L., 2011. Projecting coral reef futures under global warming and ocean acidification. Science 333, 418–422.

Pirazzoli, P., Reyss, J.-L., Fontugne, M., Haghipour, A., Hilgers, A., Kasper, H., Nazari, H., Preusser, F., Radtke, U., 2004. Quaternary coral-reef terraces from Kish and Qeshm Islands, Persian Gulf: new radiometric ages and tectonic implications. Quat. Int. 120, 15–27.

Pourjomeh, F., Hakim Elahi, M., Rezai, H., Amini, N., 2014. The distribution and abundance of macrobenthic invertebrates in the Hormozgan province, the Persian Gulf. J. Persian Gulf 5, 25–32.

Preusser, F., Radtke, U., Fontugne, M., Haghipour, A., Hilgers, A., Kasper, H., Nazari, H., Pirazzoli, P., 2003. ESR dating of raised coral reefs from Kish Island, Persian Gulf. Quat. Sci. Rev. 22, 1317–1322.

Price, A.R., 2002. Simultaneous "hotspots" and "coldspots" of marine biodiversity and implications for global conservation. Mar. Ecol. Prog. Ser. 241, 23–27.

Privett, D., 1959. Monthly charts of evaporation from the N. Indian Ocean (including the Red Sea and the Persian Gulf). Q. J. R. Meteorol. Soc. 85, 424–428.

Raven, J., Caldeira, K., Elderfield, H., Hoegh-Guldberg, O., Liss, P., Riebesell, U., Shepherd, J., Turley, C., Watson, A., 2005. Ocean Acidification Due to Increasing Atmospheric Carbon Dioxide. The Royal Society, London.

Reynolds, R.M., 1993. Physical oceanography of the Gulf, Strait of Hormuz, and the Gulf of Oman—results from the Mt Mitchell expedition. Mar. Pollut. Bull. 27, 35–59.

Rezai, H., Samimi, K., Kabiri, K., Kamrani, E., Jalili, M., Mokhtari, M., 2010. Distribution and abundance of the corals around Hengam and Farurgan islands, the Persian Gulf. J. Persian Gulf 1, 7–16.

Rezai, H., Savari, A., 2004. Observation on reef fishes in the coastal waters off some Iranian Islands in the Persian Gulf. Zoo. Middle East 31, 67–76.

Rezai, H., Wilson, S., Claereboudt, M., Riegl, B., 2004. Coral reef status in the ROPME sea area: Arabian/Persian Gulf, Gulf of Oman and Arabian Sea. In: Wilkinson, C. (Ed.), Status of Coral Reefs of the World: 2004. 1. Australian Institute of Marine Science, Townsville, pp. 155–170.

Riegl, B., Purkis, S., 2011. Persian/Arabian Gulf coral reefs. In: Hopley, D. (Ed.), Encyclopedia of Modern Coral Reefs: Structure, Form and Process. Springer, Dordrecht, pp. 790–798.

Riegl, B.M., Purkis, S.J., 2012. Coral reefs of the Gulf: adaptation to climatic extremes in the world's hottest sea. In: Coral Reefs of the Gulf. Springer, Dordrecht, pp. 1–4.

Rosen, B.R., 1971. The distribution of reef coral genera in the Indian Ocean. Symp. Zool. Soc. Lond. 28, 263–299.

Royer, S.J., Mahajan, A., Galí, M., Saltzman, E., Simó, R., 2015. Small-scale variability patterns of DMS and phytoplankton in surface waters of the tropical and subtropical Atlantic, Indian, and Pacific Oceans. Geophys. Res. Lett. 42, 475–483.

Sadrinasab, M., Kämpf, J., 2004. Three-dimensional flushing times of the Persian Gulf. Geophys. Res. Lett. 31, . L24301.

Sale, P.F., Feary, D.A., Burt, J.A., Bauman, A.G., Cavalcante, G.H., Drouillard, K.G., Kjerfve, B., Marquis, E., Trick, C.G., Usseglio, P., 2011. The growing need for sustainable ecological management of marine communities of the Persian Gulf. AMBIO: J. Human Environ. 40, 4–17.

Sharifinia, M., 2017. Macrobenthic assemblage distribution modeling and assessment of health/pollution status of Persian Gulf and Oman Sea estuaries (Hormozgan Province) using biotic indices. Ph.D. thesis, University of Hormozgan, Iran119.

Sharifinia, M., Penchah, M.M., Mahmoudifard, A., Gheibi, A., Zare, R., 2015. Monthly variability of chlorophyll-α concentration in Persian Gulf using remote sensing techniques. Sains Malays. 44, 387–397.

Sharifinia, M., Taherizadeh, M., Namin, J.I., Kamrani, E., 2018. Ecological risk assessment of trace metals in the surface sediments of the Persian Gulf and Gulf of Oman: evidence from subtropical estuaries of the Iranian coastal waters. Chemosphere 191, 485–493.

Sheppard, C., Al-Husiani, M., Al-Jamali, F., Al-Yamani, F., Baldwin, R., Bishop, J., Benzoni, F., Dutrieux, E., Dulvy, N.K., Durvasula, S.R.V., 2010. The gulf: a young sea in decline. Mar. Pollut. Bull. 60, 13–38.

Sheppard, C., Loughland, R., 2002. Coral mortality and recovery in response to increasing temperature in the southern Arabian Gulf. Aquat. Ecosyst. Health Manag. 5, 395–402.

Sheppard, C., Price, A., Roberts, C., 1992. Marine Ecology of the Arabian Region: Patterns and Processes in Extreme Tropical Environments. Academic Press, New York. 359 p.

Sheppard, C., Salm, R., 1988. Reef and coral communities of Oman, with a description of a new coral species (Order Scleractinia, genus Acanthastrea). J. Nat. Hist. 22, 263–279.

Taghavi, L., Mohebian, M., Sa'adatian, S., 2016. Impacts of seawater desalination industerial plants on the Persian Gulf ecosystem. Sustain. Dev. Environ. 2, 1–14.

Taherizadeh, M., Sharifinia, M., 2015. Applicability of ecological benthic health evaluation tools to three subtropical estuaries (Azini, Jask and Khalasi) from the Iranian coastal waters. Environ. Earth Sci. 74, 3485–3499.

Valinassab, T., Daryanabard, R., Dehghani, R., Pierce, G., 2006. Abundance of demersal fish resources in the Persian Gulf and Oman Sea. J. Mar. Biol. Assoc. U.K. 86, 1455–1462.

Van Lavieren, H., Burt, J., Feary, D., Cavalcante, G., Marquis, E., Benedetti, L., Trick, C., Kjerfve, B., Sale, P., 2011. Managing the growing impacts of development on fragile coastal and 724 marine ecosystems: Lessons from the Gulf. A policy report UNU-INWEH, Hamilton, ON, Canada.

Wilson, S., Fatemi, S.M., Shokri, M., Claereboudt, M., 2002. Status of coral reefs of the Persian/Arabian Gulf and Arabian Sea region. In: Wilkinson, C. (Ed.), Status of Coral Reefs of the World: 2002. Australian Institute of Marine Science, Townsville, pp. 53–62.

Zahed, M.A., Rouhani, F., Mohajeri, S., Bateni, F., Mohajeri, L., 2010. An overview of Iranian mangrove ecosystems, northern part of the Persian Gulf and Oman Sea. Acta Ecol. Sin. 30, 240–244.

Chapter 5

Protecting Water Quality in Urban Estuaries: Australian Case Studies

Ryan J.K. Dunn*, Nathan J. Waltham†, Jianyin Huang‡,§, Peter R. Teasdale‡,§, Brian A. King*

*Ocean Science & Technology, RPS, Gold Coast, QLD, Australia, †Centre for Tropical Water and Aquatic Ecosystem Research (TropWATER), Division of Tropical Environments and Societies, James Cook University, Douglas, QLD, Australia, ‡Natural and Built Environments Research Centre, School of Natural and Built Environments, University of South Australia, Adelaide, SA, Australia, §Future Industries Institute, University of South Australia, Adelaide, SA, Australia

1 INTRODUCTION

Estuaries provide important environmental, social, cultural, and economic services (Costanza et al., 1997; Woodward and Wui, 2001; Barbier et al., 2011). Such services are well documented and include coastal protection, nutrient supply and cycling, carbon sequestration, filtration of contaminants, and the provision of habitats for flora and fauna, including commercial and recreational fish species (as well as various others listed further). The provision of these services is often negatively impacted by urban development within the catchment and along the shoreline (Table 1), most notably through habitat loss and alteration of hydrological and sedimentation regimes influencing system structure and function (Kennish, 2002; Lee et al., 2006).

Australian estuaries are areas of significant importance for indigenous peoples, having provided settings for subsistence economies, hunting and gathering, freshwater use, and cultural traditions (Tindale, 1974; Woodroffe et al., 1988; Dortch, 1997; Hamilton and Gehrke, 2005; Jackson et al., 2005). Historically, in Australia, estuaries also served as the preferred sites for European settlements owing to multiple benefits offered, including ease of vessel access (especially when colocated with a natural harbor), availability of freshwater and fertile soils suitable for agriculture nearby (Edgar et al., 2000). Since then colonization has continued with pronounced landscape alterations through urbanization and sustained population growth. Today many of these locations are important cities with large industrial, commercial and urban centers, whose footprints have incorporated several adjacent estuaries (Wolanski and Ducrotoy, 2014) and population growth continues to accelerate (Clark and Johnston, 2017). Urban pressures exerted within estuarine settings are strongly related to development and catchment land use, with the intensity of pressures generally correlating with population density (Clark and Johnston, 2017). Present day population densities within Australian coastal and estuarine regions differ significantly between tropical and temperate regions, with tropical regions characterized with low population density while temperate coastal and estuarine settings are characterized by high population density, especially for capital cities which often have near-continuous infrastructure along adjacent coastlines. Continued population growth and expansion of urban and industrial areas ensures that there will be ongoing challenges for the management and long-term sustainability of these environments (Waltham and Sheaves, 2015).

Key to urbanization is the replacement of previously pervious areas with impervious surfaces invariably resulting in significant changes to hydrological and sedimentation regimes, for example, increased peak stormwater runoff, reduced time to peak flows, and deterioration of stormwater quality (Mein and Goyen, 1988; Lee and Bang, 2000; Dietz and Clausen, 2008; Ahiablame et al., 2012). The (water) quality of this runoff is also modified, as loadings of sediment, litter (including dumped items), nutrients, and toxic contaminants is increased resulting from anthropogenic activities common to urban environments (Fig. 1, Lindegarth and Hiskin, 2001; Kennish, 2002; Dunn et al., 2007; Klosterhaus et al., 2013). In addition, urban infrastructure (including wastewater treatment plant discharges, septic tank, and sewage system overflows and leakages) may also alter inputs and dynamics of organic matter, in addition to the biogeochemistry, biodiversity, trophic structures, and ecological function of estuaries (Sewell, 1982; Magnien et al., 1992; Downing et al., 1999; Kennish, 2002; Lotze et al., 2006; Fellman et al., 2011). The intensities of urban-sourced disturbances (drivers, see Table 1) vary across spatial and temporal scales (Lindegarth and Hiskin, 2001; Dafforn et al., 2015), with the potential for complex interactions.

TABLE 1 Key Pressures of Urbanization on Water Quality and Related Issues in Estuaries Environments and Example Protection and Management Approaches

Urbanization Components	Example Driver	Key Pressures	Example Protection and Management Approaches (Including Water Sensitive Urban Design and Stormwater Quality Improvement Devices)
Physical changes to landscape	Clearing riparian vegetation and adjacent habitat including draining wetlands	– Increase in range of flow rates, low flows are diminished and high flows amplified – Change in timing and quantity of flow – Change to tidal flows hydrodynamic characteristics (i.e., flow velocities, tidal prism, patterns) – Physical disturbance to habitat – Increased potential debris and pollutant loads (e.g., total suspended solids, nutrients, trace metals, pesticides, herbicides, hydrocarbons) – Alteration of dynamics and inputs of organic matter supply	Restoration of natural channels and habitats, dry basins, wet vaults, wet ponds, constructed wetlands, bioretention, rainwater harvesting, grass channels, wet swales, vegetated swales and strips, infiltration basins and trenches, porous pavements, soft engineering strategies, pit inserts, grate covers, gross pollutant traps, sediment traps
	Increased impervious surfaces (urban land development)	– Increased runoff peak, runoff volume and reduced time to peak flows – Physical disturbance to habitat (e.g., erosion and scouring) – Increased structural and functional ecological changes	Dry basins, wet vaults, wet ponds, constructed wetlands, bioretention, rainwater harvesting, grass channels, wet swales, vegetated swales and stripes, infiltration basins and trenches, porous pavements, soft engineering strategies, vegetated roofs
Physical changes to foreshore/seascape	Waterway development (e.g., canal developments, weirs, locks)	– Change to tidal flows hydrodynamic characteristics (i.e., flow velocities, tidal prism, patterns) – Alteration to sediment transport processes (e.g., erosion and deposition patterns) – Physical disturbance to habitat – Alteration to biogeochemical processes	Installation and scheduled openings of tidal gates, revegetation of banks, adaptive designs to minimize erosion processes, erosion monitoring
	Dredging	– Physical disturbance to benthic habitat – Increased loading of suspended solids, nutrients, organic matter – Alteration to biogeochemical processes	Flexible dredge plans, silt curtains, compliance regulations and monitoring
Urban land use	Point source and nonpoint source pollution inputs (e.g., effluent discharge, stormwater, runoff)	– Increased input of bacteria, nutrients, trace metals, hydrocarbons, pesticides, sediments, chemicals (e.g., endocrine disruptors) – Deterioration to water quality parameters – Debris and litter inputs	Pit inserts, grate covers, wet vaults, oil/water/grit separators, vortex separators, inert media filters, sorptive media filters, gross pollutant traps, sediment traps, stormwater booms, event based monitoring, community education
Seascape use	Vessel use	– Physical disturbance to benthic and shoreline habitat – Source of trace metals, herbicides and hydrocarbons – Source of debris and litter	Sewage holding tanks, advanced antifouling hull coatings, stakeholder education, environmentally friendly mooring systems, vessel use/speed restrictions
	Fishing/aquaculture	– Physical disturbance to benthic habitat – Source of debris and litter	Stakeholder education, management acts, rubbish minimization initiatives

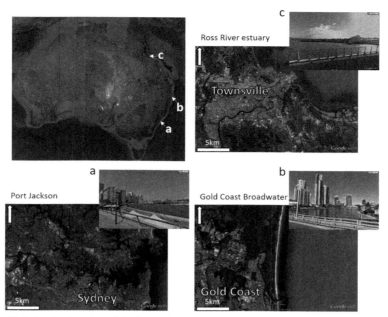

FIG. 1 Location map of Australia indicating location of the featured case study estuaries presented in this chapter; and example images of: (A) Port Jackson; (B) Gold Coast Broadwater; and, (C) Ross River estuary. *(Modified from: Satellite and street view imagery from Google Earth.)*

The protection, remediation, and restoration efforts adopted to minimize the pressures of urbanization can be difficult and often complex (Wolanski, 2014; Cook, 2017). Furthermore, challenges can be exacerbated by lack of community awareness and engagement, maintenance costs, compliance with regulations or lack of compliance effort, governing difficulties, and conflicts between multiple user groups over resource access and use. Nevertheless, various combined protection and management approaches, restoration, monitoring/research, and education initiatives are implemented to best minimize urban pressures concerning water quality within Australian cities. For effective efforts by government authorities, collective initiatives are necessary, which integrates multidisciplinary science and proactive management driven by the high values placed on the estuarine systems. Such efforts include: large and site-scale strategies, including low impact developments or water-sensitive urban design (WSUD) (Wong, 2006; Ahiablame et al., 2012; Healthy Waterways, 2006) and treatment device installations used at site scales in efforts to reduce unwanted constituent loads (Table 1 and Fig. 2). For example, available strategies include: retention/detention basins, constructed stormwater wetlands, grass swales, wet ponds, vegetated buffer strips, green roofs, sand filters, and rainwater tanks (e.g., Stanley, 1996; Moglen et al., 2003; Birch et al., 2004; Hogan and Walbridge, 2006; Collins et al., 2010; Voyde et al., 2010; Ahiablame et al., 2012; Sample et al., 2012). Furthermore, stormwater quality improvement devices (SQIDs) may also include: gross pollutant traps, sediment traps, filtration units, and vortex devices (e.g., Lee et al., 2003; Armitage, 2007; Boving and Neary, 2007; Collins et al., 2010; Sample et al., 2012; Reddy et al., 2014). The success of these devices often installed as a longitudinal series, or "treatment train," targeting source, in-line and end-of-line control, is reliant upon residence time, cost, maintenance requirements, and community acceptance of the practice (Collins et al., 2010; Ahiablame et al., 2012).

The condition, management, and outlook of Australian estuaries vary widely (Wolanski, 2014). Although the causes and sources of stormwater pressures are understood, mitigating the effects provides a great challenge for managers. This is particularly relevant when considering the diversity of Australian estuaries and catchments settings (e.g., land use patterns, population density, estuary size, hydrodynamic regime, sediment, and vegetation type associated waterbodies) merged with associated climatic zones (e.g., rainfall seasonality and intensity), complicating nation-wide solutions fit for all settings and scenarios. The establishment of appropriate management practices is dependent on the stressors and targeted pollutants, influenced by individual physical, climatic, and urban characters.

This chapter presents an account of combined management plans, restoration, monitoring/research, and education efforts used in addressing and managing the issues of urban pressures in (east coast) Australian estuarine settings through case study locations (Fig. 1 and Table 2), focusing on water quality issues. Furthermore, a summary of recommended considerations and measures likely to succeed through management and innovation are also offered.

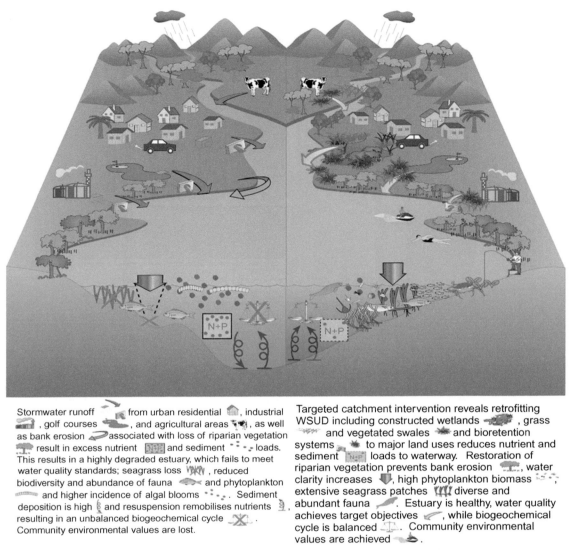

Stormwater runoff ⌁ from urban residential 🏠, industrial 🏭, golf courses ⌁, and agricultural areas 🐄, as well as bank erosion ⌁ associated with loss of riparian vegetation 🌳 result in excess nutrient [N+P] and sediment ⌁ loads. This results in a highly degraded estuary, which fails to meet water quality standards; seagrass loss ⌁, reduced biodiversity and abundance of fauna ⌁ and phytoplankton ⌁ and higher incidence of algal blooms ⌁. Sediment deposition is high ⌁ and resuspension remobilises nutrients ⌁, resulting in an unbalanced biogeochemical cycle ⌁. Community environmental values are lost.

Targeted catchment intervention reveals retrofitting WSUD including constructed wetlands ⌁, grass ⌁ and vegetated swales ⌁ and bioretention systems ⌁ to major land uses reduces nutrient and sediment [N+P] loads to waterway. Restoration of riparian vegetation prevents bank erosion ⌁, water clarity increases ⬇, high phytoplankton biomass ⌁, extensive seagrass patches ⌁ diverse and abundant fauna ⌁. Estuary is healthy, water quality achieves target objectives ⌁, while biogeochemical cycle is balanced ⌁. Community environmental values are achieved ⌁.

FIG. 2 Catchments with urban centers contribute stormwater from a range of different land use areas. Opportunities to intercept stormwater runoff with treatment either using wetlands or engineered structures can improve water quality and downstream condition of rivers and estuaries. *(From Integrated Analysis Network.)*

2 CASE STUDY EXAMPLES

2.1 Port Jackson

Port Jackson (Sydney Estuary) is one of the most famous natural harbors in the world, approximately 30 km long with a total area of 50 km^2 and is the centerpiece of the heavily urbanized city of Sydney (Hedge et al., 2014; Lee and Birch, 2014). The temperate estuarine system offers enormous social, economic, and ecological benefits to a broad range of national and international stakeholders, in addition to providing significant cultural, historical, and spiritual values to the broader Australian community (Birch et al., 2015; Hedge et al., 2014; Lee and Birch, 2014). However, the urbanized system has been altered by historical and ongoing anthropogenic activities, resulting in environmental impacts on biological diversity and ecosystem functioning (Lee and Birch, 2014). Present-day contaminants within Port Jackson from nonpoint and point stormwater and sewer overflows (Hedge et al., 2014), negatively impact water and sediment quality within the estuary with elevated concentrations of nutrients (Birch and Rochford, 2010), heavy metals (Birch and Taylor, 1999), organochlorine residues (Birch and Taylor, 2000), polycyclic aromatic hydrocarbons and polychlorinated biphenyls (Birch, 2017; McCready et al., 2000), and microplastics (Banks et al., 2016; Montoya, 2015) have been reported in the surface sediments of Port Jackson.

TABLE 2 Australian Urban Estuary Case Study Locations

	Port Jackson	Gold Coast Broadwater	Ross River Estuary
Location	33.849°S 151.193°E	27.941°S 153.411°E	19.276°S 146.830°E
Associated urban center	Sydney	Gold Coast	Townsville
Climatic zone[a]	Temperate (uniform rainfall)	Subtropical (summer rainfall)	"Dry" tropical (summer rainfall)
Mean annual rainfall (mm)[b]	1215	1260	1130
Monthly rainfall range (mm)[b]	60	122	286
Estuary classification	Tide dominated; multiple constricted mouth branched lagoon	Tide dominated; unconstricted mouth branched estuary	Tide dominated; tidal flat creek; Constricted mouth, unbranched channel
Tidal range (m)	2.1	2.0	1.7
Condition[a]	Extensively modified	Modified	Modified
Catchment area (km)[a]	589	5872	881
Waterbody/tidal flats area (km)[a]	50	103	7
Greater regional Population (2015)[c]	~1,960,000	~570,000	~238,000
Ecological status[a]	Considerably affected	Moderately affected	Moderately affected
Urban features	Stormwater/increased runoff, port/port works	Entrance modified, canal estates	Port/port works, dam and weir structures

[a] Geoscience Australia: OzCoasts Australian online coastal information, 2017. Available from: http://www.ozcoasts.gov.au/search_data/estuary_search.jsp. Accessed 17 August 2017.

[b] Australian Bureau of Meteorology: Climate data online, 2017. Available from: http://www.bom.gov.au/climate/data/index.shtml?bookmark=200. Accessed 4 March 2017.

[c] Australian Bureau of Statistics: Data by region, 2017. Available from: http://stat.abs.gov.au/itt/r.jsp?databyregion. Accessed 4 April 2017.

As a result of the pressures, various management and mitigation measures have been developed and implemented by Federal, State, and Local Governments, industry, and the community in an effort to balance the requirements and aspirations of the local stakeholders, while addressing the impacts of anthropogenic activities influencing the water quality of the estuary (e.g., Fig. 3, Caton and Harvey, 2015; Davies and Wright, 2014). Table 3 provides examples of such measures instrumental in the protection of Port Jackson, which often act simultaneously and overlap spatially and temporally. The management of the system is weakened by the complex governance frameworks because of "conflicting authorities, unclear or lack of responsibilities allocation, stakeholder involvement and scientific knowledge" (Bruns, 2013).

Sewerage overflow and aging sewage and stormwater infrastructure are responsible for major pollutant inputs into the system, suggesting an urgent need to upgrade stormwater infrastructure, integrate WSUD, and improve the inflows within the urban catchment (Davies and Wright, 2014; Hedge et al., 2014). In response, the New South Wales (NSW) State Government established the *Urban Stormwater Trust*, which required state, regional, and local governments to prepare stormwater management plans (see http://www.environment.nsw.gov.au/stormwater/usp/). The Trust allocated $18.1 million to 81 stormwater projects over a 4-year period (Davis and Birch, 2009), providing opportunities for inter- and intragovernment to cooperate on urban water management. Predominantly, trust funding was allocated for gross pollutant traps, education initiatives, WSUDs, and source control measures. A review by Lee and Birch (2014) note that various field and numerical investigations demonstrate stormwater runoff must be treated before discharged into the estuary waterways in order to reduce contaminant concentrations. This is reportedly attributable to limited flushing within the system, where contaminants supplied via stormwater runoff become entrained down the water column and settle on the estuary bed (Lee and Birch, 2014). The NSW Environment Protection Authority (EPA) established legally binding licenses in 2000, which aimed to reduce the impacts of sewer overflows on the environment and ensure continuous improvement. Between 2007 and 2012, Sydney Water (http://www.sydneywater.com.au/) spent ~$250–$300 million on a *Sewer Fix Program* in an effort

FIG. 3 Example water-sensitive urban design projects for Port Jackson and catchment waterways: (A) Sydney Park wetlands overview; (B) Pitt Street Mall WSUD impervious surface stormwater capture drainage; and, (C–E) example WSUDs providing significant value as both environmental and aesthetic landscape resources. *(From: (A) Google Earth; Courtesy: (B) ACO Pty Ltd (www.heelsafe.com.au/ACO) Drain: Aco Drain News, 2017, Available from: https://www.acodrain.com.au/news/wsud-design-recognition.htm (Accessed 20 November); (C–E) Turf Design Studio & Environmental Partnership for 202020Vision: The directory of good design: Sydney Park, 2017. Available from: http://202020vision.com.au (Accessed 20 November 2017).)*

to replace leaking sewers, upgrade sewage pumping stations, and implement system maintenance improvements. In addition, Sydney Water also created the *Sewer and Stormwater Rehabilitation Program*, which aims to upgrade the wastewater and stormwater pipes across Sydney and minimize leaks from water pipes. This project will continue until 2020.

Great outcomes have been achieved since the commencement of the *Urban Stormwater Program* (see Office of Environment and Heritage, 2011a); the waterways in Greater Metropolitan Region of Sydney (GMRS) demonstrated reduced pollutant inputs, including the collection of 10 tons of rubbish/litter from a single installed pollution trap in Sydney's

TABLE 3 Examples of Major and Relevant Management Instruments Applicable for Port Jackson

Management Tool	Example
Main State Instruments	Assessment Act, 1979; Coastal Management Policy Coastline Management Plans Development Control Plans Estuary Management Plans Environment Protection and Biodiversity Conservation Act 1999 Environmental Planning and Assessment Act, 1979 Fisheries Management Act, 1994 Estate Management Act, 2014 NSW Coastal Protection Act, 1979 NSW Diffuse Source Water Pollution Strategy NSW Estuary Management Policy NSW Protection of the Environment Operations Act, 1997 NSW State Rivers and Estuaries Policy
Other closely relevant legislation	Crown Lands Act, 1989 Land and Environment Court Act, 1991 Protection of the Environment Administration Act, 1991
Supporting instruments or strategies	Coastline Management Manual Estuary Management Manual
Local governments	Local Environmental Plans Local Government Act, 1995
Regional planning and strategies	Sydney Harbour Foreshores Area Development Control Plan, 2005 Sydney Regional Environmental Plan (Sydney Harbour Catchment) 2005 Sydney Harbour Foreshores and Waterways Area Development Control Plan, 2005 Sydney Harbour Catchment Water Quality Improvement Plan

Modified from Caton B, Harvey N: Coastal management in Australia, Adelaide, University of Adelaide Press, 2015.

Centennial Park (Office of Environment and Heritage, 2011b); while additionally more than 2.1 million people across NSW have reportedly changed their behavior to avoid actions that would pollute stormwater, including nearly two out of every five people making an effort to keep sand and soil out of the drain by such action as cleaning gutters and drains, and over 300,000 people reporting to use fewer garden chemicals as a result of education programs; all urban areas in the GMRS have completed stormwater management plans; and, the capability of local Councils to manage urban stormwater has improved (Environment Protection Authority (EPA), 2001).

Local governments have a crucial role to play in stormwater management, and they largely support federal and state government with water conservation and urban water management. However, the management of urban waterways has fallen by-and-large on local government (Davies and Wright, 2014). Although cooperation between local governments exists for Port Jackson through the Sydney Coastal Councils Group, collaboration is insufficient due to lack of a statutory force, different priorities, and limited funding. To assist alleviating funding shortcomings, the *NSW Local Government Act* 2005 and *Local Government Regulation* allowed local Councils to increase the charge of stormwater management service up to an annual rate of AU$25 per household (since July 1, 2006), potentially providing councils with a sustainable long-term funding mechanism. In addition, under the Government's *Urban Sustainability Program* (Office of Environment and Heritage, 2015) grants for stormwater projects were available for government and local Councils of the Port Jackson catchment.

While improvements have been made to address processes affecting water quality and estuary condition/health, continued monitoring of contaminant concentrations within the estuary is required to define the success of management strategies and to better inform future management decisions. To date, extensive research has been conducted in Port Jackson, including water quality and management disciplines. Such research provides a better understanding of the threats and pressures to the marine environment (e.g., Birch et al., 2008; Yu, 2008; Beck and Birch, 2012; Mayer-Pinto et al., 2015). For example, in 2009 collaboration between Macquarie University, University of New South Wales, University of Sydney, and University of Technology Sydney led to the formation of the Sydney Institute of Marine Science and its associated Marine Discovery center in 2015 (see http://sims.org.au/community/sims-discovery-centre; Banks et al., 2016). Research efforts are focused on developing management and policy efforts targeted at addressing point pollution sources (Birch, 2011; Davies and Wright, 2014). In addition, the Sydney Harbour Research Program (SHRP) was launched in 2011 (see http://sims.org.au/research/long-term-projects/sydney-harbour-research-program). This program promotes collaboration of scientists from multiple disciplines to expand the understanding of diversity and dynamics of Port Jackson and the influence of human activity, assisting policy makers, industries, and the public to manage the system. The implementation of this project assists with high-quality management of both the science and outreach components.

Implementing policy, legislation, and management strategies in addressing the issues of urban estuaries concerning water quality is assisted in part through community involvement and education of water-quality issues/catchment processes and potential influences arising from urbanization of catchments and foreshore regions (Fig. 4). Currently, multiple local volunteer groups undertake works within the catchment (e.g., https://www.urgdiveclub.org.au, http://www.underwatersydney.org, https://www.northsydney.nsw.gov.au/Waste_Environment/Get_Involved/HarbourCare), which focus on monitoring the marine environment within the harbor (Dalton and Smith, 2009) in addition to national initiative groups (e.g., https://reeflifesurvey.com/reef-life-survey/about-rls). The initiatives of these groups include the contribution of large amounts of data valuable for management and monitoring purposes (Koss, 2010). Public education within Sydney has

Pollution at home and in your street

Typical activities that can cause stormwater pollution are:

- **car washing on the street:** using detergent and allowing it to run down the street drain.

- **fixing your car on the street:** letting oil or other substances flow into the street drain.

- **disposing of garden waste:** letting leaves or garden clippings accumulate in gutters or driveways where they can end up in the street drain.

- **dropping litter:** dropping litter where it will be swept into the street drains next time it rains.

- **cleaning paint brushes:** letting the contaminated water flow into the street drain.

- **hosing the footpath:** letting the water carry dirt, soil or other waste into the street drains.

- **not picking up dog droppings:** left dog droppings will be carried into the stormwater system next time it rains. (Imagine the cumulative effect of all the dogs in your neighbourhood.)

(A)

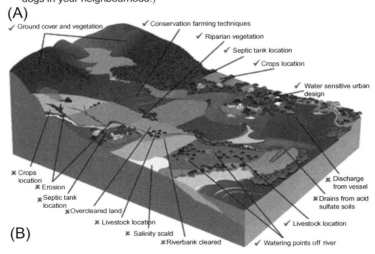

(B)

FIG. 4 Example community education material: (A) literature regarding home and street activities influencing stormwater pollution for residential home occupiers; and (B) catchment processes and example management practices. *((A) This text 'Pollution at home and in your street' (October 2013) is an extract from the Office of Environment and Heritage webpage on Stormwater - what causes it, is reproduced with permission of the NSW Office of Environment and Heritage. (B) This diagram "Examples of good and poor diffuse source water pollution management" (August 2011) is from the Office of Environment and Heritage webpage on Diffuse source water pollution, is reproduced with permission of the NSW Office of Environment and Heritage.)*

proven successful through the (NSW) Government established "*Urban Stormwater Education Program* (USEP)," part of the Government's 3-year, $60 million *Urban Stormwater Program.* The USEP aimed to improve the public's knowledge and change their behavior to prevent stormwater pollution and develop stronger partnerships between councils, the community, industries, and other relevant organizations. The program included education initiatives regarding preventative measures with respect to stormwater and water quality through TV, radio, billboard advertising, and alternate media, in addition to course delivery and training materials. Evaluation of the 2000 campaign demonstrated that nine out of ten people surveyed understood what actions they could take to reduce stormwater pollution as a result of the campaign (Environment Protection Authority (EPA), 2001).

In addition, to assist local councils and other organizations with urban stormwater management planning and improving stormwater management practices, the NSW Government has prepared a series of managing urban stormwater documents (e.g., NSW Environment Protection Authority (NSW EPA), 1997, 1998).

2.2 Gold Coast Broadwater (Southern Moreton Bay)

The Gold Coast Broadwater in southern Queensland, a large, shallow lagoon, is a major estuary of the northern Gold Coast City, which has undergone dramatic changes over the past few decades (Dunn et al., 2014). The Broadwater is positioned in one of the fastest growing regions in Australia and consequently will experience further large-scale urban expansion. The Broadwater is the recipient of four principal rivers which includes the largely urbanized Nerang River featuring extensive intertidal residential canal developments, and the Coomera, Pimpama, and Logan-Albert Rivers that although currently less urbanized than the Nerang River has, and continues to, demonstrated rapid urban development, including extensive residential housing and canal development, golf courses, and public amenities. A striking feature of the Gold Coast is the extensive network of residential canal and lake estates, increasing usable waterways and waterfront property development, which accounts for up to 90% of Australia's canal estates (Waltham and Connolly, 2011). Such waterfront developments, in concert with the regions urban infrastructure present numerous point and diffuse sources of contaminants potentially entering the Broadwater (Waltham et al., 2011).

Given the significance of the Broadwater and the importance of maintaining its many ecosystem and recreational functions (Dunn et al., 2014), efforts have been made to better understand the natural and anthropogenic processes and associated contaminant concentrations based on urban development pressures and land use (e.g., Moss and Cox, 1999; Dunn et al., 2007, 2013; Jordan et al., 2008; Davies et al., 2009; Waltham et al., 2011). Although increases in contaminant loads entering the Broadwater have been reported for receiving sediment, water column, and biota (e.g., Moss and Cox, 1999; Waltham, 2002; Teasdale et al., 2009; Dunn et al., 2014), on the whole, the accumulated body of evidence on contaminants within the Broadwater illustrates that there are no major threats of contamination, with perhaps the exception of marina facilities (Dunn et al., 2014). However, episodic event-driven (i.e., rainfall and runoff) deterioration of water quality has been reported (e.g., Moss and Cox, 1999; Waltham, 2002; Teasdale et al., 2009; Dunn et al., 2013) relating to land clearing and urban expansion, suggesting the region is not immune to stormwater pollution risks. For example, following a 2008 rainfall event (90 mm over previous 72 h and 65 mm in preceding 24 h period), total suspended solids (TSS) and dissolved inorganic nitrogen concentrations were observed to increase 400% and 2000%, in comparison to sampled "dry" conditions at the same adjoining Broadwater sample site (Dunn et al., 2013).

The City of Gold Coast (CGC) is responsible for the management of the Gold Coast Broadwater and has been actively involved with local and regional stakeholders (e.g., *Water by Design*, a program of the South East Queensland Healthy Waterways Partnership) to develop integrated catchment management strategies, which are supported by urban stormwater management plans, including WSUD objectives (e.g., Healthy Waterways, 2006, 2007; Gold Coast City Council (GCCC), 2007; City of Gold Coast (CGC), 2017). These catchment management plans and design objectives look to adopt a total management approach, by combining engineering, environment, social, planning, and architecture disciplines, including the preparation of policies and guidelines relating to stormwater treatment and reuse (Alam et al., 2008; Waltham et al., 2014). The CGC, which was one of the first local governments in Queensland (Australia) to implement WSUD practices, has incorporated grass swales, bio-swales, bioretention basins, wetlands, and gross pollutants traps as a statutory requirement under the city's planning framework (Gold Coast City Council (GCCC), 2007) as part of sustainable development practices for the city for all future urban development (Fig. 5). Investigations regarding the effectiveness of adding WSUD infrastructure (mainly SQIDs) to the existing urban stormwater network at several locations (Waltham et al., 2014) have shown improvements to the quality of stormwater runoff, while also extending available habitat for local species (Alam et al., 2008). Although no current policies are in place to retrofit remaining urban areas with such features, given this would require large financial expenditure (Wolanski and Ducrotoy, 2014; Water by Design, 2010), such investment may ultimately avoid greater financial and infrastructure burden in the future (Water by Design, 2010). Regional stormwater design objectives for urban

FIG. 5 Example projects, managed design and treatment devices for the Gold Coast Broadwater and catchment waterways: (A) aquatic plant harvesting; (B, C) photographs illustrating before (1997) and after (2000) conditions along a waterway following restoration efforts; (D) water-sensitive urban design infrastructure for stormwater treatment within a residential setting; and, (E) a stormwater quality improvement structure within a residential/parkland setting. *(From Waltham NJ: Gold Coast waterway management images (2003–2009), 2017a, [photographs] (Nathan Waltham's own private collection).)*

development are contained in the *Urban Stormwater Quality Planning Guidelines 2010* DERM (2010, as amended) with objectives specified for both the construction and operational phases of development in accordance with landscape features and the regional location of proposed development (https://www.ehp.qld.gov.au/water/policy/urban-stormwater.html).

In addition, restoration projects have also focused on weed removal, foreshore stabilization works, revegetation of cleared areas, and weed eradication programs (Fig. 5). These programs over the past decade have achieved major successes (e.g., see Gold Coast Catchment Association Inc. (http://www.goldcoastcatchments.org); Gold Coast Management Groups (http://www.goldcoast.qld.gov.au/environment/catchment-management-groups-576.html); Gold Coast Water Watch (https://www.natura-pacific.com/waterwatch-qld); Griffith Center for Coastal Management (https://www2.griffith.edu.au/cities-research-institute/griffith-centre-coastal-management/community-engagement/coasted/coasted-resources), and SEQ Catchments (http://www.seqcatchments.com.au/our-programs-water.html)). Although development intervention along the Broadwater foreshores has led to land reclamation increasing recreational and development purpose areas, efforts have been made to incorporate WSUD and SQID concepts and infrastructure. For example, as part of the Southport Broadwater Parklands development a 1.2 ha constructed mangrove wetland habitat area was constructed (GCCC and GCCM; Henkelmann, 2012), serving as a stormwater treatment device, while also providing ecological benefits and aiding in community awareness and education.

Because of the potential for anthropogenic disturbances diminishing the various values associated with the Broadwater, long-term monitoring has occurred as part of the Healthy Land and Water Reporting program (Healthy Land and Water, 2017), including water quality indicators that provide an understanding of the ecosystem health and response to land use activities (Healthy Land and Water, 2017). The Healthy Land and Water reporting program, which commenced in 2000 provides an annual measure of the environmental pressures facing catchments throughout South East Queensland, providing an environmental condition grade (e.g., A–F) and the level of social and economic benefits (e.g., 1–5 stars) the waterways provide to local communities (Healthy Land and Water, 2017). In addition, comprehensive monitoring of the various coastal zone waters potentially impacted by the release of reclaimed water into the Gold Coast Seaway has been done to assist in the development of a decision support system leading to optimization of the release parameters (Kirkpatrick and Hughes, 2009). Furthermore, in addition to more conventional water and sediment sampling approaches, time-integrated in situ sampling and both routine and experimental biomonitoring approaches have also been conducted to investigate trace metals and metalloids, pesticides, polychlorinated biphenyls (PCBs), polyaromatic hydrocarbons (PAHs), and indicators of bacterial contamination for monitoring and research interests (e.g., Mortimer and Cox, 1998; Dunn et al., 2007; Jordan et al., 2008; Waltham et al., 2011).

Community education and involvement/volunteer programs such as catchment groups, community and school education programs, tree planting initiatives, landholder, financial support and training, and industry support and relations are implemented and supported within the Broadwater region (e.g., see https://goldcoastcatchments.org, http://www.goldcoastwaterwatch.com.au, http://www.gcwa.qld.gov.au/environment, http://www.goldcoast.qld.gov.au/community/environmental-volunteering-3336.html; https://www.natura-pacific.com/waterways-field-guide-teaching-resource, Fig. 6). A recent educational initiative within the Broadwater has been the *"Tackle Bin Project"* launched in 2017, a marine debris reduction initiative by the Gold Coast Marine Debris Network funded by private and government bodies. The initiative provides disposal bins for unwanted fishing line in an effort to reduce marine debris within the Gold Coast Broadwater, while alerting fishers of the consequences of discarded fishing tackle for water quality and wildlife through signage. In addition, an educational interactive initiative provided by the Queensland Government Department of Environment and Heritage Protection (WetlandInfo (Department of Environment and Heritage Protection), 2016a) provides another example of community education, where internet-based map journals provide extents and values of catchments and commentary regarding key features which influence water flow, including geology, topography, rainfall and runoff, natural features, human modifications, and land uses (Fig. 7). Such programs are very important to protect, achieve, and maintain the community waterway values. Collectively, initiatives undertaken appear successful to date and demonstrate that future management requires a multidisciplinary and proactive approach driven by the high community values placed on the Broadwater and its associated waterways.

FIG. 6 Example photos of (A) local council and state government authority public notice of stormwater drainage upgrades within the Broadwater catchment and (B and C) educational signage on stormwater drains promoting community awareness regarding the receiving waterways. *(From Dunn RJK: Gold Coast waterway management images (2017), 2017, [photographs] (Ryan Dunn's own private collection).)*

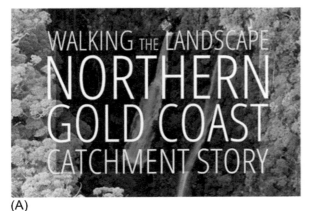

(A)

(B)

Water Flow

Water flows across the landscape into streams and eventually into the main watercourses of the Pimpama River and Coomera River.

Click to see animation

The remaining water either sinks into the ground where it supports a variety of terrestrial and groundwater dependent ecosystems or is used for other purposes.

The upper reaches of the catchment have relatively steep slopes which create the potential for increased runoff which may lead to flooding in areas where the floodplain has restricted channels and gullies.

The restricted channels and gullies eventually flatten out to form waterways that meander across the floodplain. They pass through alluvial areas which store and release water, prolonging the time streams flow.

(C)

FIG. 7 Example computer screenshots of the internet-based educational interactive map journal of the Northern Gold Coast catchment story as part of a series prepared for catchments of Southeast Queensland by Queensland Wetland Program: (A) journal title page; (B) significance of map journal; and (C) animation slide. *(From WetlandInfo (Department of Environment and Heritage Protection): Northern gold coast catchment story, 2016. Available from: http://qgsp.maps.arcgis.com/apps/MapJournal/index.html?appid=e05ca7ca37fe4609a2ac0b3f35dce3c7. Accessed 4 November 2017.)*

2.3 Ross River Estuary

The Ross River estuary is situated in the dry tropics region, within the limits of northern Australia's largest urban center, Townsville (North Queensland), where the population is expected to increase, along with urban and industrial expansion in the next few decades. The city is situated adjacent to Cleveland Bay, a large shallow coastal bay with fringing natural vegetation including seagrass, mangroves, and saltmarsh, which provides habitats for a range of important marine mammals, fish, and sea turtles (NQ Dry Tropics, 2016). During and after the wet season (November to April) rainfall flow can deliver high sediment and nutrient loads to the bay resulting in persistent turbid conditions, which has been shown to reduce seagrass area (Petus et al., 2014). The river system flows through the city, where a series of weirs and dam (Ross River Dam, which provides drinking water for the city) result in controlled delivery of freshwater to the estuary. The freshwater section of Ross River has extensive invasive aquatic weeds which require substantial investment from the local government authority to remove and to ensure water quality conditions do not contribute to summer fish kills (NQ Dry Tropics, 2016; Dubuc et al., 2017). Water quality in the freshwater region is influenced by runoff from the surrounding catchment, which includes urban residential, light industry and commercial facilities, in addition to some forested and protected areas. Although the estuary has been highly modified and includes a mix of urban engineering structures (e.g., rock walls, jetties, pylons, concrete retaining walls), interspersed with natural vegetation including mangroves and saltmarsh (Waltham and Sheaves, 2015), the system provides important ecological and economic services (Sheaves et al., 2012). Overall, the estuary generally experiences reduced water quality owing to the delivery of urban stormwater runoff, characterized by high nutrient, heavy metals, and sediment concentrations, particularly following rainfall (NQ Dry Tropics, 2016). Best estimates of the pollutant loads of Ross River include: 80 k tons/year of TSS, 140 tons/year of total phosphorus (TP), 50 tons/year of dissolved inorganic nitrogen (DIN), and 690 tons/year of total nitrogen (TN) (Kroon et al., 2012).

Recognizing the expansion of the city, the need to protect coastal waterways and wetland ecosystems, and preserve their various service functions, the local government authority (Townsville City Council, TCC) in partnership with regional stakeholders have prepared urban stormwater quality improvement guidelines. These guidelines set out design criteria and guidance on how to approach the treatment of urban stormwater as part of proposed development projects. In addition, the *Burdekin Region Water Quality Improvement Plan* (2016) has been developed, which aims to support and guide decision-making and investment around protection of the local ecosystems as they relate to water quality and the receiving waters of the Great Barrier Reef (GBR) (NQ Dry Tropics, 2016). Further examples of protection plans in effect within the Ross River estuary include *The Reef Water Quality Protection Plan* (2013), which is a Commonwealth and Queensland government collaboration that quantifies pollutant loads within regional subcatchments as well as the relative contribution of different land uses. Establishing such criteria ensures the benefits of improving and reducing the impact of urban stormwater quality reaching adjacent sensitive waterways, but also ensures that this local government authority is not burgeoned with on-going, expensive, maintenance problems long after completion of development projects. Within the region stormwater discharge from development areas must reach >80% reduction in the mean annual load of TSS, >65% reduction in the mean annual load of TP, >40% reduction in the mean annual load of TN, and >90% reduction in the mean annual load of gross pollutants (Townsville City Council (TCC), 2011). In addition to achieving these load reductions, the policy outlines that WSUD plans, accounting for the regionally specific rainfall patterns and intensity, soil and vegetation types, etc., look to:

- protect ecologically and economically valuable freshwater, estuarine and marine aquatic ecosystems, including the GBR;
- reduce demand on potable water use during construction, and also during the initial years of maintenance under the control of the developer;
- protect water quality of surface and ground waters;
- protect natural features and ecological processes;
- maintain natural hydrologic behavior of catchments; and
- minimize wastewater generation and discharge to the natural environment.

The City of Townsville WSUD guidelines (see Townsville City Council (TCC), 2011) comprise (bioretention) swales, sediment and bioretention basins, infiltration measures, constructed stormwater wetlands, sand filters, and aquifer storage and recovery and are based on the Healthy Waterways Partnership WSUD Technical Design Guidelines for South East Queensland. However, worked examples have been specifically rewritten to address adaptations required for effective operation in the coastal dry tropics climate. The guidelines outline concepts, best management practices, application advice, and site planning. They also identify the appropriate design flows for use in sizing stormwater management measures (e.g., sediment basins and swales), design adaptations for constructed wetlands and bioretention systems required in response to the seasonal rainfall patterns, and any other requirements to sustain vegetation or to meet best practice stormwater treatment objectives. The guidelines also recognize the suitable designs for stormwater management measures (e.g., sediment basins and swales), design adaptations and bioretention systems required in response to seasonal rainfall patterns, and other requirements to meet best practice stormwater treatment objectives. Additional discussion and example case studies regarding the use of WSUD, SQIDs, and stormwater strategies specific to the unique rainfall patterns of the region (differing from other climatic regions) include those prepared by local and regional stakeholders, industry, and consultant groups (e.g., NQ Dry Tropics, 2016; Bligh Tanner Pty Ltd, 2016), including sediment basins, grass filters, wetlands, and gross pollutant traps (Fig. 8). The efficacy of these structures therefore is likely to be within the realm of the design specifications, though data confirming modeling efforts are urgently needed to validate modeling to local conditions (which would also provide confidence to government authorities). Full lifecycle costs of these structures are also needed; without this data the local authority is not able to appropriately allocate funds to clean and maintain these engineering assets. The risk of not maintaining these engineering structures is poor performing water quality treatment, spread of environmental weeds, and overall reduced public amenity for locals living nearby.

FIG. 8 (A) The Strand along the Townsville City Council esplanade attracts tourists and locals to enjoy the natural amenity; (B) stormwater outlet to Ross River estuary where untreated stormwater flows from adjacent urban areas; (C) example of WSUD where a grass filter strip treats water from local carpark; and, (D) public place a high recreational value on Ross River estuary highlighting the need to treat stormwater in the region. *(From Waltham NJ: Townville waterway management images (2013–2016), 2017b, [photographs] (Nathan Waltham's own private collection).)*

In addition, institutional collaborations such as the Reef Urban Stormwater Management Improvement Group (RUSMIG), which aims to improve runoff to the reef, and community education, and involvement/volunteer programs are important initiatives helping to increase awareness and address the issues of urban estuaries concerning water quality within the Ross River estuary and adjacent waters (e.g., http://www.creektocoral.org, https://www.qff.org.au/projects/reef-alliance, http://www.nqdrytropics.com.au/events, https://www.townsville.qld.gov.au/water-waste-and-environment/sustainability/sustainability-tours). The Creek to Coral (C2C; http://www.creektocoral.org) program is a local government initiative, in partnership with the Queensland Department of Environment and Natural Resources, established to maintain and enhance the regional waterways. The program promotes a "whole of catchment" perspective aimed to maintain system services from impacts from urbanization. Its success is built on a partnering approach that integrates all levels of government, the community, sciences, business, and industry. The C2C program also managed the development of a Water Quality Improvement Plan for the Ross River Basin [funded by the Australian Government's former Coastal Catchment Initiative (CCI) program (http://www.creektocoral.org)]. Erosion and sediment control training initiatives provided by the TCC also contributes to the education of stakeholders, and its establishment has been an important initiative given the soil condition in the Townsville area and the proximity to the GBR world heritage area.

3 CONSIDERATIONS AND SUMMARY

Management of urban stormwater is a challenging issue where no single solution is apparent but requires varied approaches. Important to the protection and management of estuaries in response to urban pressures on water quality is an effort to increase revegetation and reduce pollutant loads from point and nonpoint sources. A survey of protection and management initiatives, illustrated through the featured east Australian case study estuaries, highlight the following themes of management practices employed:

- catchment/estuary management policies and plans including SQID initiatives practices;
- low impact and purpose-designed development options including WSUD options;
- restoration programs;
- routine water quality monitoring programs;

- complementary research activities; and
- stakeholder and community education campaigns.

Without any doubt, best management outcomes require implementation of catchment-based management plans that are supported by clear objectives regarding ecosystem services and regional challenges. Management plans and policies should look to clearly define the plan/policy scope as a first step, then identify responsibilities and roles, establish timeframes, define expectations of best-practices regarding SQIDs implementation, define governance boundaries, be inclusive of stakeholder interests, classify legal frameworks, limit resource conflict, outline understanding of key trade-offs, and establish assessment criteria to report the effectiveness of the plan/policy. To improve management of estuarine systems, especially those with major urban areas spanning the catchment (e.g., Port Jackson), consideration should be given to reduce the quantity and complexity of government management plans and policies currently employed (e.g., state and federal governments, and local councils) in order to adopt improved efficiency regarding implementation and enforcement. Such collaboration may provide clearer overall objectives and leadership, while also reducing expended efforts and costing. Most importantly, management plans and policies require adequate funding to be appropriately implemented and disseminated. A range of funding sources and mechanisms should therefore be considered and support future management plans, restoration programs, routine water quality monitoring programs, research endeavors, and stakeholder and community education campaigns. Examples include licensing or taxes related to the use of the catchment and estuary.

Understanding the regional climatological influences and physical settings of estuaries and the associated scale at which urban pressures are influencing water quality is particularly important regarding effective management options (including appropriately selected and scaled WSUD features and adoption of SQIDs, including site selection, suitability of technology, cost-benefit-ratios) to reduce the negative impacts of urbanization on estuary water quality. Although the management of point and nonpoint source pollution has seen improvement within urban estuaries, diffuse stormwater pollutant loads are more difficult to limit, particularly under increasing population growth and catchment development. In addition to addressing water-quality issues downstream, many management solutions should be tackled by implementing a range of management options addressing the catchment-wide activities contributing to downstream concerns.

Research efforts are also an important component when considering the management of Australian estuaries, directly supporting decision-making and investment. Collaborative partnerships between research, stakeholder, and management bodies provide a vital role in the provision of science-based frameworks for estuary management and should therefore be supported and sustained. Research themes such as approaches to reduce point and nonpoint pollutant loads, nutrient cycling and estuary metabolism investigations, are examples of key focus areas that would further enhance system knowledge regarding threats, pressures, and mitigation measures regarding urbanized estuaries. Criteria for prioritizing knowledge acquisition through research efforts include (1) importance of the issue to healthy estuaries, (2) relative importance of new information in addressing this issue, (3) urgency of the information needed, and (4) costliness of research (Thomson et al., 2016). Research and monitoring efforts are also important for the design and assessment phases of WSUD/SQID initiatives. Collaboration between researchers and managing authorities can effectively ensure that adequate information is obtained for performance assessments of such initiatives to support (or redirect as required) further decision-making and investment. Support of multidiscipline research programs, which include stakeholder engagement, is crucial for increasing the understanding of system behavior regarding urban pressures used to develop management and policy efforts, while also providing opportunity for stakeholder education and interest.

Stakeholder education seems to be a successful management means in reducing urban pressures on water quality within Australian estuaries and is key to the current and future protection of these systems through awareness and consequences of urban-related issues. In addition, education and awareness presents an opportunity to bestow a sense of ownership and responsibility for the protection of estuaries that communities treasure and rely on for important environmental, social, cultural, and economic services.

Although implementation of the management practices outlined here has had some positive influence on urban estuaries in Australia, further research and implementation is necessary. The management of these important systems requires continued execution of management themes presented herein, with consideration and improvement to clearer governance, improved compliance with regulations and compliance efforts, risk assessment tools, water-quality guidelines and objectives, and catchment scale ecosystem-ecohydrology-based (Elliott et al., 2016) management approaches partnering engineering, science, and technology. Science through research and monitoring plays key roles in the management of estuaries and water quality concerns through availability and accessibility of data, development of new sampling techniques for improved detection and reporting of baseline, event-driven conditions, and pre- and postremediation efforts used to help determine effectiveness of management initiatives. Research and monitoring that improves our understanding of the persistence, pathways and threats of newly emerging classes of contaminants (e.g., pharmaceuticals, cosmetic products, and plastics) are also important considerations that require attention in Australia.

REFERENCES

Ahiablame, L.M., Engel, B.A., Chaubey, I., 2012. Effectiveness of low impact development practices: literature review and suggestions for future research. Water Air Soil Pollut. 223, 4253–4273.

Alam, K., Hossain, S., Dalrymple, B., 2008. In: WSUD in the Gold Coast. New South Wales and Queensland Stormwater Industry Association Conference, 8–11 July 2008, Gold Coast, Australia.

Armitage, N., 2007. The reduction of urban litter in the stormwater drains of South Africa. Urban Water J. 4, 151–172.

Banks, J., Hedge, L., Hoisington, C., Strain, E., Steinberg, P., Johnston, E., 2016. Sydney Harbour: beautiful, diverse, valuable and pressured. Reg. Stud. Marine Sci. 8, 353–361.

Barbier, E.B., Hacker, D.D., Kennedy, C., Koch, E.W., Stier, A.C., Silliman, B.R., 2011. The value of estuarine and coastal ecosystem services. Ecol. Monogr. 81, 169–193.

Beck, H.J., Birch, G.F., 2012. Metals, nutrients and total suspended solids discharged during different flow conditions in highly urbanised catchments. Environ. Monit. Assess. 184, 637–653.

Birch, G.F., 2011. Contaminated soil and sediments in a highly developed catchment-estuary system (Sydney estuary, Australia): an innovative stormwater remediation strategy. J. Soils Sediments 11, 194–208.

Birch, G.F., 2017. Assessment of human-induced change and biological risk posed by contaminants in estuarine/harbour sediments: Sydney Harbour/estuary (Australia). Marine Poll. Bull. 116, 234–248.

Birch, G.F., Rochford, L., 2010. Stormwater metal loading to a well-mixed/stratified estuary (Sydney Estuary, Australia) and management implications. Environ. Monit. Assess. 169, 531–551.

Birch, G., Taylor, S., 1999. Source of heavy metals in sediments of the Port Jackson estuary, Australia. Sci. Total Environ. 227, 123–138.

Birch, G., Taylor, S., 2000. Distribution and possible sources of organochlorine residues in sediments of a large urban estuary, Port Jackson, Sydney. Aust. J. Earth Sci. 47, 749–756.

Birch, G.F., Matthai, C., Fazeli, M.S., Suh, J.Y., 2004. Efficiency of a constructed wetland in removing contaminants from stormwater. Wetlands 24, 459–466.

Birch, G.F., McCready, S., Long, E.R., Taylor, S.S., Spyrakis, G., 2008. Contaminant chemistry and toxicity of sediments in Sydney Harbour, Australia: spatial extent and chemistry–toxicity relationships. Mar. Ecol. Prog. Ser. 363, 71–87.

Birch, G., Lean, J., Gunns, T., 2015. Growth and decline of shoreline industry in Sydney estuary (Australia) and influence on adjacent estuarine sediments. Environ. Monit. Assess. 187, 314.

Bligh Tanner Pty Ltd, 2016. Townsville Regional Stormwater Strategy Part 1. Report Prepared for Townsville Regional Council, Brisbane.

Boving, T.B., Neary, K., 2007. Attenuation of polycyclic aromatic hydrocarbons from urban stormwater runoff by wood filters. J. Contam. Hydrol. 91, 43–57.

Bruns, A., 2013. The environmental impacts of megacities in the coast. In: Pelling, M., Blackburn, S. (Eds.), Megacities and the Coast. Taylor & Francis Group, New York.

Caton, B., Harvey, N., 2015. Coastal Management in Australia. University of Adelaide Press, Adelaide.

City of Gold Coast (CGC), 2017. Water Sensitive Urban Design. Available from: http://www.goldcoast.qld.gov.au/environment/water-sensitive-urban-design-3924.html. (Accessed July 7, 2017).

Clark, G.F., Johnston, E.L., 2017. Australia State of the Environment 2016: Coasts, Independent Report to the Australian Government Minister for Environment and Energy. Australian Government Department of the Environment and Energy, Canberra.

Collins, K.A., Lawrence, T.J., Stander, E.K., Jontos, R.J., Kaushal, S.S., Newcomer, T.A., Grimm, N.B., Cole Ekberg, M.L., 2010. Opportunities and challenges for managing nitrogen in urban stormwater: a review and synthesis. Ecol. Eng. 36, 1507–1519.

Cook, H.F., 2017. The Protection and Conservation of Water Resources. John Wiley & Sons, West Sussex.

Costanza, R., d'Arge, R., De Groot, R., Farber, S., Grasso, M., Hannon, B., Limburg, K., Naeem, S., O'neill, R.V., Paruelo, J., Raskin, R.G., 1997. The value of the world's ecosystem services and natural capital. Nature 387, 253–260.

Dafforn, K.A., Mayer-Pinto, M., Morris, R.L., Waltham, N.J., 2015. Application of management tools to integrate ecological principles with the design of marine infrastructure. J. Environ. Manag. 158, 61–73.

Dalton, S.J., Smith, S.D., 2009. A Review of Underwater Volunteer Groups in NSW. Report to the Hunter-Central Rivers Catchment Management Authority. National Marine Science Centre, Coffs Harbour.

Davies, P.J., Wright, I.A., 2014. A review of policy, legal, land use and social change in the management of urban water resources in Sydney, Australia: a brief reflection of challenges and lessons from the last 200 years. Land Use Policy 36, 450–460.

Davies, S., Mirfenderesk, H., Tomlinson, R., Szylkarski, S., 2009. Hydrodynamic, water quality and sediment transport modelling of estuarine and coastal waters on the Gold Coast. Australia J. Coast. Res. 56, 937–941.

Davis, B.S., Birch, G.F., 2009. Catchment-wide assessment of the cost-effectiveness of stormwater remediation measures in urban areas. Environ. Sci. Pol. 12, 84–91.

Dietz, M.E., Clausen, J.C., 2008. Stormwater runoff and export changes with development in a traditional and low impact subdivision. J. Environ. Manag. 87, 560–566.

Dortch, C.E., 1997. New perceptions of the chronology and development of Aboriginal fishing in South-Western Australia. World Archaeol. 29, 15–35.

Downing, J.A., McClain, M., Twilley, R., Melack, J.M., Elser, J., Rabalais, N.N., Lewis, W.M., Turner, R.E., Corredor, J., Soto, D., Yanez-Arancibia, A., Kopaska, J.A., Howarth, R.W., 1999. The impact of accelerating land use change on the N-cycle of tropical aquatic ecosystems: current conditions and projected changes. Biogeochemistry 46, 109–148.

Dubuc, A., Waltham, N., Malerba, M., Sheaves, M., 2017. Extreme dissolved oxygen variability in urbanised tropical wetlands: the need for detailed monitoring to protect nursery ground values. Estuar. Coast. Shelf Sci. 198, 163–171.

Dunn, R.J.K., Teasdale, P.R., Warnken, J., Jordan, M.A., Arthur, J.M., 2007. Evaluation of the in situ, time-integrated DGT technique by monitoring changes in heavy metal concentrations in estuarine waters. Environ. Pollut. 148, 213–220.

Dunn, R.J.K., Robertson, D., Teasdale, P.R., Waltham, N.J., Welsh, D.T., 2013. Benthic metabolism and nitrogen dynamics in an urbanised tidal creek: domination of DNRA over denitrification as a nitrate reduction pathway. Estuar. Coast. Shelf Sci. 131, 271–281.

Dunn, R.J.K., Waltham, N.J., Benfer, N.P., King, B.A., Lemckert, C.J., Zigic, S., 2014. Gold Coast Broadwater: Southern Moreton Bay, Southeast Queensland (Australia). In: Wolanski, E. (Ed.), Estuaries of Australia in 2050 and Beyond. Springer, Dordrecht.

Edgar, G.J., Barrett, N.S., Graddon, D.J., Last, P.R., 2000. The conservation significance of estuaries: a classification of Tasmanian estuaries using ecological, physical and demographic attributes as a case study. Biol. Conserv. 92, 383–397.

Elliott, M., Mander, L., Mazik, K., Simenstad, C., Valesini, F., Whitfield, A., Wolanski, E., 2016. Ecoengineering with ecohydrology: successes and failures in estuarine restoration. Estuar. Coast. Shelf Sci. 176, 12–35.

Environment Protection Authority (EPA), 2001. Evaluation of the Urban Stormwater Program: Summary Report—Urban Stormwater Program Evaluation Part A. Environment Protection Authority, Sydney.

Fellman, J.B., Petrone, K.C., Grierson, P.F., 2011. Source, biogeochemical cycling, and fluorescence characteristics of dissolved organic matter in an agro–urban estuary. Limnol. Oceanogr. 56, 243–256.

Gold Coast City Council (GCCC), 2007. Gold Coast planning scheme policies. In: Policy 11: Land Development Guidelines, Section 13 Water Sensitive Urban Design (WSUD) Guidelines, Gold Coast.

Hamilton, S.K., Gehrke, P.C., 2005. Australia's tropical river systems: current scientific understanding and critical knowledge gaps for sustainable management. Mar. Freshw. Res. 56, 243–252.

Healthy Land & Water, 2017. An holistic understanding of our waterways. Available from: http://hlw.org.au/report-card/monitoring-program. (Accessed September 17, 2017).

Healthy Waterways, 2006. Strategy for Water Sensitive Urban Design in SEQ. Final Report to the Department of Environment & Heritage, Brisbane.

Healthy Waterways, 2007. Water Sensitive Urban Design. Developing Design Objectives for Urban Development in South East Queensland, Brisbane.

Hedge, L., Johnston, E., Ahyong, S., Birch, G., Booth, D., Creese, R., Doblin, M., Figueira, W., Gribben, P., Hutchings, P., 2014. Sydney Harbour: A Systematic Review of the Science. The Sydney Institute of Marine Science, Sydney.

Henkelmann, M., 2012. Monitoring the Ecological Health of the Southport Broadwater Parklands Mangrove and Seagrass Habitat. Griffith University, Gold Coast.

Hogan, D.M., Walbridge, M.R., 2006. Best management practices for nutrient and sediment retention in urban stormwater runoff. J. Environ. Qual. 36, 386–395.

Jackson, S., Storrs, M., Morrison, J., 2005. Recognition of aboriginal rights, interests and values in river research and management: perspectives from northern Australia. Ecol. Manag. Restor. 6, 105–110.

Jordan, M.A., Teasdale, P.R., Dunn, R.J.K., Lee, S.Y., 2008. Modelling copper uptake by *Saccostrea glomerata* with diffusive gradients in a thin film measurements. Environ. Chem. 5, 274–280.

Kennish, M.J., 2002. Environmental threats and environmental future of estuaries. Environ. Conserv. 29, 78–107.

Kirkpatrick, S., Hughes, L., 2009. The Gold Coast Seaway *SmartRelease* Study: Monitoring Campaign. Research Report No. 91.3. Griffith Centre for Coastal Management, Griffith University, Gold Coast.

Klosterhaus, S.L., Grace, R., Hamilton, M.C., Yee, D., 2013. Method validation and reconnaissance of pharmaceuticals, personal care products, and alkylphenols in surface waters, sediments, and mussels in an urban estuary. Environ. Int. 54, 92–99.

Koss, R.S., 2010. Volunteer health and emotional wellbeing in marine protected areas. Ocean Coast. Manag. 53, 447–453.

Kroon, F.J., Kuhnert, P.M., Henderson, B.L., Wilkinson, S.N., Kinsey-Henderson, A., Abbott, B., Brodie, J.E., Turner, R.D.R., 2012. River loads of suspended solids, nitrogen, phosphorus and herbicides delivered to the Great Barrier Reef lagoon. Mar. Pollut. Bull. 65, 167–181.

Lee, J.H., Bang, K.W., 2000. Characterization of urban stormwater runoff. Water Res. 34, 1773–1780.

Lee, S.B., Birch, G.F., 2014. Sydney estuary, Australia: geology, anthropogenic development and hydrodynamic processes/attributes. In: Wolanski, E. (Ed.), Estuaries of Australia in 2050 and Beyond. Springer, Dordrecht.

Lee, J., Bang, K., Choi, J., Ketchum, L.H., Cho, Y., 2003. The vortex concentrator for suspended solids treatment. Water Sci. Technol. 47, 7–8.

Lee, S.Y., Dunn, R.J.K., Young, R.A., Connolly, R.M., Dale, P.E.R., Dehayr, R., Lemckert, C.J., McKinnon, S., Powell, B., Teasdale, P.R., Welsh, D.T., 2006. Impact of urbanization on coastal wetland structure and function. Austral Ecol. 31, 149–163.

Lindegarth, M., Hiskin, M., 2001. Patterns of distribution of macro-fauna in different types of estuaries, soft sediment habitats adjacent to urban and non-urban areas. Estuar. Coast. Shelf Sci. 52, 237–247.

Lotze, H.K., Lenihan, H.S., Bourque, B.J., Bradbury, R.H., Cooke, R.G., Kay, M.C., Kidwell, S.M., Kirby, M.X., Peterson, C.H., Jackson, J.B.C., 2006. Depletion, degradation, and recovery potential of estuaries and coastal seas. Science 312, 1806–1809.

Magnien, R.E., Summers, R.M., Sellner, K.G., 1992. External nutrient sources, internal nutrient pools, and phytoplankton production in Chesapeake Bay. Estuaries Coast. 15, 497–516.

Mayer-Pinto, M., Johnston, E., Hutchings, P., Marzinelli, E., Ahyong, S., Birch, G., Booth, D., Creese, R., Doblin, M., Figueira, W., 2015. Sydney Harbour: a review of anthropogenic impacts on the biodiversity and ecosystem function of one of the world's largest natural harbours. Mar. Freshw. Res. 66, 1088–1105.

McCready, S., Slee, D.J., Birch, G.F., Taylor, S.E., 2000. The distribution of polycyclic aromatic hydrocarbons in surficial sediments of Sydney Harbour, Australia. Mar. Poll. Bull. 40, 999–1006.

Mein, R.G., Goyen, A.G., 1988. Urban runoff. Civil Engineering Transactions, CE30, No. 4, Institution of Engineers Australia, Australia.

Moglen, G.E., Gabriel, S.A., Faria, J.A., 2003. A framework for quantitative smart growth in land development. J. Am. Water Resour. Assoc. 39, 947–959.

Montoya, D., 2015. Pollution in Sydney Harbour: Sewage, Toxic Chemicals and Microplastics. NSW Parliamentary Research Service, New South Wales.

Mortimer, M.R., Cox, M., 1998. Contaminants in Oysters and Crabs in a Canal Adjacent to Lae Drive, Runaway Bay, Queensland. Department of Environment and Heritage, Brisbane.

Moss, A., Cox, M., 1999. Southport Broadwater and Adjacent Pacific Ocean: Water Quality Study 1979–1998. Queensland Environmental Protection Agency, Brisbane.

NQ Dry Tropics, 2016. Burdekin Region Water Quality Improvement Plan 2016. NQ Dry Tropics, Townsville.

NSW Environment Protection Authority (NSW EPA), 1997. Managing Urban Stormwater: Council Handbook. Environment Protection Authority, South Sydney.

NSW Environment Protection Authority (NSW EPA), 1998. Managing Urban Stormwater: Source Control. Environment Protection Authority, South Sydney.

Office of Environment and Heritage, 2011a. Stormwater Trust Program. Available from: http://www.environment.nsw.gov.au/stormwater/usp. (Accessed October 3, 2017).

Office of Environment and Heritage, 2011b. Pollution Traps. Available from: http://www.environment.nsw.gov.au/stormwater/whatis/pollutiontrap.htm. (Accessed December 23, 2017).

Office of Environment and Heritage, 2015. Urban Sustainability Program. Available from: http://www.environment.nsw.gov.au/grants/urbansustainability. htm. (Accessed October 3, 2017).

Petus, C., Collier, C., Devlin, M., Rasheed, M., McKenna, S., 2014. Using MODIS data for understanding changes in seagrass meadow health: a case study in the Great Barrier Reef (Australia). Mar. Environ. Res. 98, 68–85.

Reddy, K.R., Xie, T., Dastgheibi, S., 2014. Removal of heavy metals from urban stormwater runoff using different filter materials. J. Environ. Chem. Eng. 2, 282–292.

Sample, D.J., Grizzard, T.J., Sansalone, J., Davis, A.P., Roseen, R.M., Walker, J., 2012. Assessing performance of manufactured treatment devices for the removal of phosphorus from urban stormwater. J. Environ. Manage. 113, 279–291.

Sewell, P.L., 1982. Urban groundwater as a possible nutrient source for an estuarine benthic algal bloom. Estuar. Coast. Shelf Sci. 15, 569–576.

Sheaves, M., Johnston, R., Connolly, R.M., Baker, R., 2012. Importance of estuarine mangroves to juvenile banana prawns. Estuar. Coast. Shelf Sci. 114 (1), 208–219.

Stanley, D.W., 1996. Pollutant removal by a stormwater dry detention pond. Water Environ. Res. 68, 1076–1083.

Teasdale, P.R., Welsh, D.T., Dunn, R.J.K., Robertson, D., 2009. Investigating Nutrient Sources and Processes in the Saltwater Creek Catchment. Griffith Centre for Coastal Management, Gold Coast.

Thomson, C., Kilminster, K., Hallett, C., Valesini, F., Hipsey, M., Trayler, K., Gaughan, D., Summers, R., Syme, G., Seares, P., 2016. Research and Information Priorities for Estuary Management in South-West Western Australia: Consultation Draft. Western Australian Institute of Marine Science (WAMSI), Crawley.

Tindale, N.B., 1974. Aboriginal Tribes of Australia: Their Terrain, Environmental Controls, Distribution, Limits, and Proper Names. Australian National University Press, Canberra.

Townsville City Council (TCC), 2011. Water Sensitive Urban Design for the Coastal Dry Tropics (Townsville): Technical Design Guidelines for Stormwater Management. Townsville City Council, Townsville.

Voyde, E., Fassman, E., Simcock, R., 2010. Hydrology of an extensive living roof under sub-tropical climate conditions in Auckland, New Zealand. J. Hydrol. 394, 384–395.

Waltham, N.J., 2002. Health of the Gold Coast Waterways Report. Gold Coast City Council, Gold Coast.

Waltham, N.J., Connolly, R.M., 2011. Global extent and distribution of artificial, residential waterways in estuaries. Estuar. Coast. Shelf Sci. 94, 192–197.

Waltham, N.J., Sheaves, M., 2015. Expanding coastal urban and industrial seascape in the Great Barrier Reef world heritage area: critical need for coordinated planning and policy. Mar. Policy 57, 78–84.

Waltham, N.J., Teasdale, P.R., Connolly, R.M., 2011. Contaminants in water, sediment and fish from natural and artificial residential waterways in southern Moreton Bay. J. Environ. Monit. 13, 3409–3419.

Waltham, N.J., Barry, M., McAlister, T., Weber, T., Groth, D., 2014. Protecting the green behind the gold: catchment-wide restoration efforts necessary to achieve nutrient and sediment load reduction targets in Gold Coast city, Australia. Environ. Manag. 54, 840–851.

Water by Design, 2010. A Business Case for Best Practice Urban Stormwater Management (Version 1.1). South East Queensland Healthy Waterways Partnership, Brisbane.

WetlandInfo (Department of Environment and Heritage Protection), 2016a. Northern Gold Coast Catchment Story. Available from: https://wetlandinfo. ehp.qld.gov.au/wetlands/ecology/processes-systems/water/catchment-stories/transcript-northern-gold-coast.html. (Accessed November 4, 2017).

Wolanski, E. (Ed.), 2014. Estuaries of Australia in 2050 and Beyond. Springer, Dordrecht.

Wolanski, E., Ducrotoy, J.-P., 2014. Estuaries of Australia in 2050 and beyond—a synthesis. In: Wolanski, E. (Ed.), Estuaries of Australia in 2050 and Beyond. Springer, Dordrecht.

Wong, T.H.F., 2006. An overview of water sensitive urban design practices in Australia. Water Prac. Technol. 1, . wpt2006018.

Woodroffe, C.D., Chappell, J., Thom, B.G., 1988. Shell middens in the context of estuarine development, South Alligator River, Northern Territory. Archaeol. Ocean. 23, 95–103.

Woodward, R.T., Wui, Y.-S., 2001. The economic value of wetland services: a meta-analysis. Ecol. Econ. 37, 257–270.

Yu, X., 2008. Use of low quality water: an integrated approach to urban stormwater management (USM) in the Greater Metropolitan Region of Sydney (GMRS). Int. J. Environ. Stud. 65, 119–137.

Chapter 6

Management of Megafauna in Estuaries and Coastal Waters: Moreton Bay as a Case Study

Janet M. Lanyon

School of Biological Sciences, The University of Queensland, St. Lucia, QLD, Australia

1 INTRODUCTION

Degradation of marine coastal systems is a major global problem (Halpern et al., 2008). Increases in human populations around coastal land margins have brought about declines in water quality (siltation, pollution, marine debris), habitat modification (dredging, mining, construction), overfishing, and disturbance through human activities (vessel traffic, noise, fishing gear). Over the past four decades in particular, combinations of these human impacts on coastal systems, have caused declines in the abundance of marine wildlife populations, with marine mammals, reptiles, and birds particularly affected (WWF (World Wildlife Fund), 2012; McCauley et al., 2015). Almost two decades ago, Jackson et al. (2001) suggested that large marine vertebrates such as marine mammals were functionally or entirely extinct in most coastal ecosystems around the world. Certainly, there have been recent marine mammal species extinctions driven by anthropogenic activities, for example, Caribbean monk seal (*Monachus tropicalis*) in the 1950s (MacPhee and Flemming, 1999) and more recently, the Baiji river dolphin in 2006 (Turvey et al., 2007). Local extirpations have also occurred, for example, Atlantic population of gray whale *Eschrichtius robustus* (Rice et al., 1984) and also severe depletion of local populations, for example, Vaquita porpoise (*Phocoena sinus*) (Rojas-Bracho et al., 2006). For some other marine mammals, current population status is unknown but suspected to be dire, for example, the Okinawan dugongs (Marsh et al., 2011). Threat levels for marine mammals living close to urbanized areas appear to be greater than for terrestrial mammals, and until recently were thought to be driven by different processes, principally accidental mortality and pollution (Schipper et al., 2008). Later evidence suggests that not only habitat degradation but also habitat fragmentation leading to contraction of geographic range and eventual range collapse may be a major contributing threat, for example, for the Baiji dolphin (Turvey et al., 2010). Marine mammals are not the only marine megafauna in trouble due to human impacts in the marine system. All sea turtle species are classified as *vulnerable to extinction*, *endangered*, or *critically endangered* globally (IUCN (International Union for the Conservation of Nature), 2018) and since their migration patterns span international borders/waters, conservation efforts in some countries do not always translate to protection across the species/population range. Migratory shorebirds with reliance on inshore, often near-urban coastlines as foraging grounds are also threatened (Croxall et al., 2012; IUCN (International Union for the Conservation of Nature), 2018).

In this time of deteriorating coastal ecosystems (Dobson et al., 2006), effective management of marine megafauna requires a multifaceted approach that addresses: (1) legislative protection of species, populations; (2) identification and protection of critical habitat, (3) monitoring of population trends and health, and (4) identification and mitigation of threatening processes. It is also imperative that water quality is managed sufficiently that the marine habitats are protected from degradation (Bearzi et al., 2004; Waycott et al., 2009).

2 MORETON BAY: A MEGAFAUNA CASE STUDY

Moreton Bay (Fig. 1) in the state of Queensland (Qld) on the mid-east coast of Australia supports a diverse community of marine megafauna (Chilvers et al., 2005). Resident dolphins and transient whales are found within the bay environs, some of the largest and most significant Australian populations of sea turtles coexist (Limpus et al., 1994), and the bay is

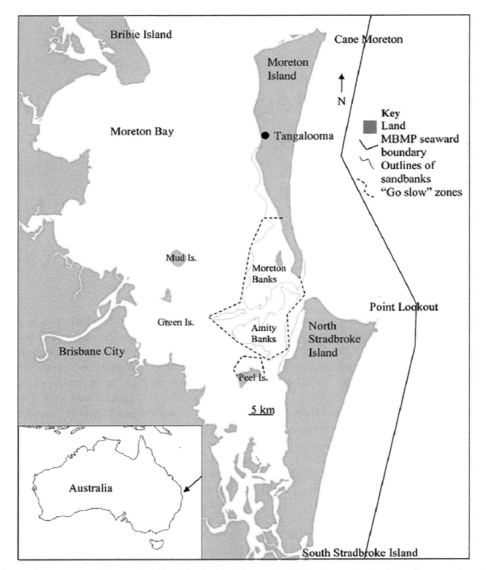

FIG. 1   Moreton Bay on the east coast of Queensland, Australia (see inset), showing the locations of each of the places mentioned in the text: the four barrier islands (Bribie, Moreton, North Stradbroke, and South Stradbroke Islands) and Brisbane city; Mud Island; and the southern Bay island group. The Eastern Bankscomprise the Moreton and Amity Banks. The seaward boundary of the Moreton Bay Marine Park (MBMP) is indicated by the solid line to the east; The *Go Slow for dugongs and turtles* area is indicated by the dotted line. (*Source Chilvers et al., 2005*).

one of the most important migratory shorebird sites in the southern hemisphere (EPA (Environment Protection Agency), 2005; Wilson et al., 2011). In addition, the southern-most population of dugongs in the world is supported by the extensive seagrass beds in these sheltered waters (Lanyon, 2003). Moreton Bay is also adjacent to Australia's third most populous city with a human population of over three million and growing (QOESR (Queensland Office of Economic and Statistical Research), 2011), and contains the busy Port of Brisbane.

Moreton Bay is vulnerable to all of the pressures expected for a capital city situated on a large coastal estuary and providing agricultural, industrial, residential, and shipping services. Boating traffic includes container ships servicing the port, and barge, ferry, and water taxi services across the bay. Shipping-related activities include dredging and underwater dredge dump sites. As of 2016, more than 145,000 vessels including 18,500 jet skis were registered in the greater Brisbane area (MSQ (Maritime Safety Queensland), 2016). Commercial fishing operations in the bay include netting, otter and beam trawling and pot and line fishing, with Moreton Bay's catch comprising 13% of Queensland's catch, making the bay a major seafood source (QFish, 2018). The economic value of recreational fishing in the bay has been estimated at ~$20 million per year (Pascoe et al., 2014). Tourism also brings pressures to the bay; resort and tourist destinations can be found on the bay islands, and increasing ecotourism both within and outside the bay results in over 12 million visitors to the bay each year (QPWS (Queensland Parks and Wildlife Service), 2012).

As "sea-changers" and sun-seekers flock north from the southern states to Brisbane, and as local industry and recreational use of the bay grows, the pressures on the coastal systems and their fauna can only increase. How the marine megafauna communities of the heavily urbanized Moreton Bay will fare into the future is of concern. Below, I examine the perceived threats to marine wildlife within the bay and further species-specific protective measures.

3 MORETON BAY—PHYSICAL CHARACTERISTICS

Moreton Bay lies between 27° and 28° South, and is a large (125 km long, 1500 km^2), shallow, wedge-shaped embayment with a broad (35 km) opening to the north and a narrower navigation channel at its southern end (Fig. 1). This lagoon-like bay is bordered by four large continental sand islands (Bribie, Moreton, North and South Stradbroke Islands) that provide shelter from the oceanic influences of the Coral Sea. The eastern central bay in the lee of these barrier islands contains extensive shallow sandbanks, known to locals as the Eastern Banks. To the north are open deeper waters containing the main shipping channel that cuts across the bay into the Port of Brisbane. To the south is a labyrinth of more than 300 landmasses, from continental islands supporting human settlements to small mangrove islets, creating a vast southern network of shallow channels and banks. The western shoreline of the bay is muddy and mostly developed: lined with the suburbs of Brisbane and dotted with industrial precincts, including the port and the city's airport.

The Moreton Bay catchment is extensive with the western bay receiving water from a number of rivers, the most significant of which, in terms of size and outflow, are the Caboolture, Pine, Brisbane, Logan, Albert, and Pimpana Rivers. These rivers wend their ways through modified bushland in the higher reaches, then rural agricultural followed by residential and industrial sites, and open at intervals along the western shore. The outflow from these rivers includes effluent from riparian agriculture, treated sewage, residential and industrial outflow, and urban litter. Rainfall in this subtropical region is mildly monsoonal with the heaviest falls normally occurring during the austral summer months. At this time, freshwater runoff into the bay increases markedly, and coastal flooding occurs occasionally. Large-scale coastal flooding does occur at times, especially in the cyclone season, and these events produce episodic outflows of large quantities of freshwater into the bay.

Moreton Bay is mostly shallow (max depth ~30 m) with semidiurnal tides and a moderate tidal range (1.5–2 m). There is a complex pattern of water circulation in the bay that changes with freshwater input and tides; the northerly outflow of water out of the bay with the ebb tide is an important removal mechanism for contaminants that originate along the landward shores of the bay (Church, 1979; Gibbes et al., 2014). Whilst the waters of the urban shoreline are generally muddy and turbid, the waters over the Eastern Banks are mostly blue and clear, especially over the incoming and high tide phases. A significant proportion of the Eastern Banks falls within the Moreton Bay Marine Park (see below).

4 MORETON BAY MEGAFAUNA

The confluence of tropical and temperate systems in the Moreton Bay vicinity results in a diverse subtropical fauna including the southernmost populations of several tropical species, that is, sea turtles, dugongs, and Australian humpback dolphins. The extensive wetlands, shallow sandbanks, mudflats, mangrove forests, coral reefs, and myriad of waterways provide a variety of habitats for wildlife within the bay, and the more exposed coastal waters outside the bay provide further habitat for offshore fauna. In particular, vast aggregations of marine megafauna (defined here as marine mammals and reptiles) are found in the eastern (seaward) part of Moreton Bay, over the Eastern Banks sandbanks that have relatively clear waters and support mixed meadows of seagrasses and benthic invertebrates.

4.1 Sea Turtles

Moreton Bay is globally unique in that large populations of sea turtles occur close to a capital city: these are principally loggerhead (*Caretta caretta*) and green turtles (*Chelonia mydas*), but hawksbill (*Eretmochelys imbricata*), flatback (*Natator depressus*), and Pacific ridley turtles (*Lepidochelys olivacea*) also occur (DNPSR (Department of National Parks Sport and Racing), 2018). Each of these turtle species, and also the leatherback turtle (*Dermochelys coriacea*) has sporadically nested on the oceanic sides of the bay's barrier islands.

The loggerhead turtle is listed as *endangered* in both Queensland (Nature Conservation Act, 1992) and nationally (Environment Protection and Biodiversity Conservation Act, 1999). Moreton Bay is of particular significance as one of the most important feeding grounds for the Australian east coast population (Limpus and Limpus, 2001), which in turn is possibly the most significant population in the southern Pacific Ocean (Limpus, 2008). The hundreds of loggerheads that feed on the benthos in Moreton Bay are part of the breeding stock that migrates to the internationally important rookery of Mon Repos near Bundaberg, over 300 km north (Limpus et al., 1992). This population declined by about 86% between the late 1970s and 2000 (Limpus and Limpus, 2003), probably due to migrating turtles becoming entrapped in longline and coastal otter trawl fisheries operating along the coast, and through egg predation by foxes (Limpus and Reimer, 1994; Heppel et al., 1996, Chaloupka and Limpus, 2001). Since then, with mandatory implementation of turtle exclusion devices (TEDs) in the Queensland east coast trawl fishery, a significant (99%) reduction in turtle bycatch has been achieved (Brewer et al., 2006) so that the population may have a chance of recovery.

The green turtle is classified as *endangered* globally (Seminoff, 2004), and as *vulnerable* in both Queensland (*Nature Conservation Act*, 1992) and nationally (*Environment Protection and Biodiversity Conservation Act*, 1999). Green turtles occur in large numbers throughout the shallow areas of Moreton Bay, most conspicuously in the Eastern Banks region (Limpus et al., 1994). The thousands of grazing green turtles range in size from new recruits, fresh from their juvenile-phase Pacific Ocean circumnavigation, to adults of 50 years or older (Limpus et al., 1994). They mostly forage over vast meadows of seagrasses and algae (Brand-Gardner et al., 1999; Read and Limpus, 2002) but may also take significant amounts of animal material (Arthur et al., 2008a). Green turtles are active all year in Moreton Bay, even during the winter months when turtles at similar latitudes elsewhere in the world may enter winter diapause (Hazel et al., 2009). These Moreton Bay green turtles sporadically migrate north to rookeries, joining turtles from other foraging grounds and thus forming part of the southern Great Barrier Reef breeding stock (Limpus et al., 1994).

Threats to sea turtles in Moreton Bay are numerous. With every passing year, more turtles fall victim to boat strike and entanglement. Boat strike in Moreton Bay Marine Park currently accounts for more than 50 turtle deaths each year, and about 50% of the boat strikes reported for Queensland (DNPSR (Department of National Parks Sport and Racing), 2018). Significant numbers of turtles also become entangled and drown in operational and also discarded fishing gear (ghost gear), especially in crab pots which are set for both commercial and recreational fisheries in the bay (DNPSR (Department of National Parks Sport and Racing), 2018). Ingestion of marine debris, including plastics is a growing cause of mortality (Schuyler et al., 2012), whilst harvest by local indigenous people for consumption is sporadic and probably accounts for few mortalities (Limpus, 2008). The effects of contaminants (chemical and terrestrial discharge) and coastal development on the health of sea turtles are more difficult to ascertain. Higher numbers of sick and dead turtles have been reported in heavily urbanized areas such as Moreton Bay compared with nonurban areas (Flint et al., 2009). For example, the incidence of chelonid fibropapilloma (FP) disease (and associated herpesvirus) has been recorded in Queensland since the 1970s and appears to be associated with degrading water quality (Jones et al., 2016).

4.2 Dugongs

The dugong population of Moreton Bay is unique; it has one of the highest densities on the Australian coast, and is the only one in the world situated close to a major city (Lanyon, 2003). Almost a thousand dugongs live year-round in Moreton Bay, with the vast majority found in the relatively clear water habitat of the Eastern Banks (Fig. 2). During the cooler winter months, dugongs make short forays (up to 10–12 km) out of the bay via South Passage into warmer oceanic waters, sometimes on a daily basis, but otherwise are mostly resident over the seagrass banks (Preen, 1993; Lanyon, 2003). Recent genetic studies have confirmed that the dugongs of Moreton Bay comprise a largely separate breeding stock to the dugongs further up the Queensland coast (Seddon et al., 2014; Cope et al., 2015). Although small numbers of dugongs do sometimes travel between Moreton Bay and the next closest population in Hervey Bay, these animals appear to be mostly ranging with only a few individuals outbreeding each year (Cope et al., 2015). The Moreton Bay dugong population appears to have been relatively stable since the 1980s in contrast to other populations along the Queensland coast and elsewhere throughout the Indo-Pacific (Marsh et al., 2011), with minimum population estimates from aerial surveys mostly fluctuating between 600 and 950 (Preen, 1993; Lanyon, 2003; Sobtzick et al., 2015, 2017). These fluctuations are more likely due to survey artifacts rather than to true changes in numbers: there has been no evidence of mass movements or mortality events.

There is some anecdotal evidence to suggest that dugongs were once found throughout Moreton Bay, but that their range has contracted to the eastern bay as the western shoreline has become increasingly urbanized (Preen et al., 1992). As coastal development progressed, nearshore boating traffic increased and seagrass habitat was lost or modified with respect to seagrass community structure (Preen et al., 1992). Dugongs are fussy feeders, preferring to feed on seagrass species that are low in fibre and high in nitrogen and carbohydrates; these are also the pioneer seagrass species, that is, the first to recolonize after disturbance (Lanyon and Sanson, 2006). These preferred seagrasses, for example, *Halophila*

FIG. 2 Aerial photograph of a small herd of dugongs grazing over the shallow seagrass of the Eastern Banks. The north-western shoreline of North Stradbroke Island can be seen in the background. (*Photo credit: Darren Jew*).

sp. and *Halodule* sp. are found in extensive meadows over the Eastern Banks, but are not so common now in the muddy waters of the western and southern bay. Feeding herds as large as 50–100 animals occur over the Eastern Banks at high tide, day after day until local areas of seagrass are depleted, at which time the herds move on to new areas. As the tide ebbs, dugongs move into deeper waters (up to 26 m) where they continue to feed on sparse seagrasses (Preen, 1993; Sprogis, 2008).

Like dugongs elsewhere, the dugongs of Moreton Bay are entirely dependent on healthy abundant seagrasses for their well-being and survival. Reproductive and life history patterns are dependent on availability of seagrass nutrients so that higher quality forage appears to result in faster growth and more frequent reproduction (Marsh and Kwan, 2008). Since Moreton Bay is at the southern limit of the species' range, dugongs are exposed to distinct seasonal shifts in each of water temperature, seagrass abundance, and nutritional quality. These dugongs are consequently slower to mature, have a protracted pubertal period but appear to achieve larger final adult body sizes than dugongs elsewhere (Burgess et al., 2012a,b). Breeding is also slow and the long periods between successive calves, and prolonged lactation (several years) probably have their basis in seagrass availability.

Dugongs are listed as a *vulnerable* species globally (Marsh and Sobtzick, 2015), but it has been suggested that they may be more correctly classified as *critically endangered* in Queensland, due to sharp declines along the coast since the 1960s (Marsh et al., 2011). Threats to dugongs are regionally specific so that those in urban Moreton Bay are different to those elsewhere on the Australian coast. For almost half of the dugongs recovered dead in the Moreton Bay region, cause of death has been indeterminate (Greenland and Limpus, 2007; Woolford et al., 2015a). Of those with cause of death assigned, boat strike, trauma and netting, inflammatory, and infectious disease were most common (Owen et al., 2012). Moreton Bay appears to have the highest incidence of boat strike of anywhere in the dugong's range, probably due to the high coincidence of dugongs and boaters in shallow areas. Fortuitously for the dugongs of Moreton Bay, the sheer distance between the principal foraging area (Eastern Banks) and the developed coastline (~7 km) probably helps to buffer them from nearshore human activities.

4.3 Whales and Dolphins

Moreton Bay has the highest recorded diversity and abundance of resident and transient cetaceans (whales and dolphins) in Australia. Two species of whales pass by the bay each year as part of their annual breeding migration, that is, humpback whales and southern right whales, whilst other whales are sighted or strand sporadically, for example, killer whale (*Orca orcinus*), false killer whale (*Pseudorca crassidens*), sperm whale (*Physeter microcephalus*), pygmy sperm whale (*Kogia breviceps*), Minke whale (*Balaenoptera acutorostrata*), Bryde's whale (*B. edeni*), blue whale (*B. musculus*), Blainville's beaked whales (*Mesoplodon densirostris*), and melon-headed whales (*Peponocephala electra*) (Chilvers et al., 2005; Lanyon et al., 2018). Moreton Bay and environs are also home to three species of coastal dolphin: the offshore and inshore bottlenose dolphins (*Tursiops truncatus* and *T. aduncus,* respectively), and the rarer tropical Australian humpback dolphin *Sousa sahulensis* (Chilvers et al., 2005); with other dolphins visiting on occasion, for example, *Orcaella* sp., *Stenella* sp., *Grampus* sp., and *Delphinus* sp. (DES (Department of Environment and Science), 2015).

4.4   Humpback Whales

The East Australian Area V (130°E–170°W) stock of humpback whales, *Megaptera novaeangliae*, migrates past Moreton Bay each year between May and October: north from their summer feeding grounds in Antarctica to winter breeding grounds of the Great Barrier Reef and Coral Sea, and then south on their return (Paterson and Paterson, 1984; Paterson et al., 1994). Between 96% (Bryden, 1985) and 100% (Brown, 1998) of migrating whales pass within 10 km of Point Lookout on North Stradbroke Island. The proximity of the whale's migration route to Moreton Bay was exploited mid-last century when a whaling station operated out of Moreton Island (Chittleborough, 1965). Since its closure in the 1960s due to collapse of the population, and the complete protection of humpbacks in 1973, this population has recovered from an estimated low of 200–500 animals (IWC, 2015) to a 2015 estimate of almost 26,000 (Noad et al., 2016) which is thought to be close to prewhaling numbers (Lanyon et al., 2018). This recovery rate of 11% per annum, is one of the fastest recovery rates for all of the great whale populations (Noad et al., 2011, 2016) and close to the theoretical limit of 12% for cetaceans (Brandao et al., 2000). The humpback whale is currently listed globally as of *least concern* (Reilly et al., 2008), but is *vulnerable* in Australia at both federal (Environment Protection and Biodiversity Conservation Act (EPBC), Aust.) and state levels (Nature Conservation Act 1992 Qld).

Interestingly, over the past few years as the population continues to increase, humpback whales have started to move into the shallow waters of Moreton Bay proper, with singing adult males recorded in the eastern and northern bay during the whales' northward migration, and mothers and calves encountered over the shallow Eastern Banks and central bay during the southern migration. The purpose of these visits is uncertain. Although there have not yet been recent mortalities of humpbacks recorded within Moreton Bay itself, net entanglements have occurred with increasing frequency along the coasts outside the bay in each of recent years, that is, along the northern Sunshine Coast and southern Gold Coast (Meynecke and Meager, 2016). The carrying capacity of this population is unknown but under current recovery rates, the frequency of encounters between humpback whales and threatening processes such as nets and vessels can only increase.

4.5   Southern Right Whales

Populations of the southern right whale *Eubalaena glacialis* were so severely depleted by commercial whaling around the south coast of Australia in the nineteenth century that by the 1850s, the industry based around this species became economically unviable (Dawbin, 1986). Since the cessation of whaling, recovery of right whales around the Australian coast has been slow. Each year, southern right whales undergo an annual migration from high latitude (Antarctic) feeding grounds to lower latitude (temperate) breeding grounds (Burnell, 2001). The South-West genetic population moves up the west coast of Tasmania and across the Great Australian Bight whilst the South-East population moves east of Tasmania and into the waters of Victoria and New South Wales (Carroll et al., 2011). The sizes of the South-West and South-East populations are thought to number in the hundreds and tens, respectively (Carroll et al., 2011).

In 1998, the first record was made of a southern right whale in Queensland—in the oceanic waters off North Stradbroke Island (Noad, 2000) and in the next year a sighting was made in Moreton Bay (Chilvers, 2000). Since then, small numbers of southern right whales have been sighted each year in the bay, usually near Victoria Point in the central-western bay. Unfortunately, Victoria Point is also the jumping off point for passenger and vehicular ferries servicing the southern bay islands. In 2014, the first recorded mortality of a southern right whale occurred within the bay (Lanyon and Janetzki, 2016). The juvenile whale was struck and killed by a water taxi, whilst a second whale injured at the same time could not be located: it is possible that the dead juvenile was a calf sighted in the bay the previous year. Tragically, this was not the first death of a southern right whale in Queensland; in the previous year, a whale was found dead stranded just north of Moreton Bay, most probably a victim of boat strike (Lanyon and Janetzki, 2016).

The recent annual occurrence of southern right whales in Moreton Bay, so far north of their presumed historical range, suggests a couple of possibilities: Moreton Bay may have been part of this species' original prewhaling range, or this "invasion" may represent a range extension with the bay becoming a new overwintering ground (Lanyon and Janetzki, 2016). Either way, there is a need to protect these whales from ship strike, entanglement, and other anthropogenic stressors to which slow swimming, surface dwelling right whales are prone (Verboom, 2002; Kemper et al., 2008). The southern right whale is currently listed as *endangered* under Australia's EPBC Act 1999. Since the whales visiting Moreton Bay are likely part of the small and vulnerable South-East population; this group's listing as *critically endangered* (Department of Sustainability and Environment, 2013) may be a more appropriate guide.

4.6 Dolphins

The offshore bottlenose dolphin is found in oceanic waters on the seaward side of the barrier islands outside the bay proper and is thought to be one of the largest resident populations of bottlenose dolphins in the world (Chilvers and Corkeron, 2003). Members of this population appear to have very small home ranges, restricted to the Point Lookout area of North Stradbroke Island (Chilvers and Corkeron, 2003). In contrast, the inshore Indo-Pacific bottlenose (Ansmann et al., 2013) and the Australian humpback dolphin are both found within the bay ranging over broader areas, sometimes in mixed species pods (Ansmann et al., 2012a). More is known about the distribution and population sizes of these bay dolphins.

The inshore bottlenose dolphin ranges across the entire bay, however, perhaps surprisingly for a highly mobile species, there are two genetically distinct subpopulations (Ansmann et al., 2012b). The larger subpopulation of ~350–400 dolphins (Ansmann et al., 2013) is found in the northern and central part of the bay (Ansmann et al., 2012b), foraging in the open deeper waters on pelagic fishes (Ansmann et al., 2015). In contrast, the smaller subpopulation (~200 dolphins, Ansmann et al., 2013) is found around islands and channels of the southern bay (Ansmann et al., 2012b), foraging on mostly shallow-water demersal species (Ansmann et al., 2015). There is limited genetic mixing between the two subpopulations, and they appear to be adapted to different feeding niches and are presumably vulnerable to different threats. Certainly, tissue biopsy of live members of the southern subpopulation has shown relatively higher levels of some heavy metal contaminants, for example, lead, in the blubber, perhaps reflecting these dolphins' proximity to land-based sources of pollutants (Ansmann et al., 2015).

How inshore bottlenose dolphins move around within the bay is largely unknown at this time. Some members of the north subpopulation are part of a familial group that regularly visits Tangalooma Resort on Moreton Island, where the dolphins are hand-fed by tourists (Neil and Brieze, 1998). Another form of supplemental feeding to which some of the Moreton Bay dolphins have been exposed is discards from commercial trawling within the bay. In the late 1990s, inshore bottlenose dolphins in the eastern part of Moreton Bay formed two distinct but sympatric communities: one group followed prawn trawlers to forage on trawler discards, whilst the other group did not (Chilvers and Corkeron, 2001). Subsequent to this, trawling practices in the bay altered markedly over five years with a reduced overall catch and a shift from several previously trawled inshore areas to more central and northern offshore areas (Ansmann et al., 2012a). The trawler and nontrawler community structure in the dolphins disappeared with these changes in fishing operations, reverting to a more conventional fission-fusion association pattern (Ansmann et al., 2012a).

The Australian humpback dolphin *S. sahulensis* is a tropical species (Cagnazzi, 2013) and Moreton Bay is the one of the southern-most bay systems recorded with a resident population (Meager et al., 2015). Humpback dolphins are mostly found in the central and north-western areas of the bay along the developed coastline between the Brisbane River and Scarborough and across to Mud Island (Chilvers et al., 2005; Lanyon et al., 2018) (Fig. 1). Small groups of dolphins are also found along the southern shores of Bribie Island and in the central eastern bay near the town of Amity on North Stradbroke Island. They frequent shallow water habitats, including river mouths and are thought to be primarily piscivorous (Cagnazzi, 2013). In Moreton Bay, humpback dolphins frequently form pods with inshore bottlenose dolphins but the nature of these associations is not yet understood.

The number of humpback dolphins in Moreton Bay has been variously estimated at as high as 163 in 1984–86 (95% CI: 108–251) (Corkeron, 1989) to as low as 100 in the 1990s (Hale et al., 1998). More recently, the population has become the focus of a mark-recapture program with an initial population estimate of 132 in the first year of the study, 2014 (95% CI: 62–276) (Meager et al., 2015). The extent of the genetic connectivity between humpback dolphins in Moreton Bay and the nearest resident population 150km north in the Great Sandy Strait is unknown but occasional sightings and strandings along the intervening ocean-exposed coast suggests that some exchange may be occurring (Meager et al., 2015).

The Australian humpback dolphin was recently listed as *vulnerable* in Queensland under the *Nature Conservation Act 1999 [Qld]* due to their generally small discrete populations (Meager et al., 2015). The small size and potentially isolated nature of the Moreton Bay humpback dolphin population as well as its proximity to a large city gives it special status and suggests a particular and unique vulnerability. Anthropogenic threats to humpback dolphins in Moreton Bay are probably similar to bottlenose dolphins and presumably include resource depletion, net entanglement, disturbance by vessel traffic, underwater construction and noise, and pollution (Meager et al., 2015; Meager and Sumpton, 2016). Concentrations of the contaminants polychlorinated biphenyls (PCBs) and DDT (dichlorodiphenyltrichloroethane) and its metabolites in the tissues of Australian humpback dolphins stranded in Moreton Bay have been recorded as above or close to levels where adverse physiological effects may occur (Weijs et al., 2016). As apex predators, dolphins foraging in the most shallow, nearshore habitats of this urbanized environment may be particularly susceptible to bioaccumulation of contaminants.

5 PROTECTIVE MEASURES FOR MORETON BAY MEGAFAUNA

5.1 Protective Legislation

Green turtles, dugongs, and humpback whales of the Moreton Bay region have each been commercially exploited by Europeans in the past (Daley et al., 2008). Bribie Island in the north of Moreton Bay was the base for a cottage industry manufacturing dugong oil products from the 1850s until 1920 (Johnson, 2002), whilst in the 1920s, a green turtle soup industry operating in the Great Barrier Reef region markedly reduced numbers of breeding females in Moreton Bay. Moreton Island was home to a whaling station in the 1950s–60s, targeting passing humpback whales on their breeding migration. Each of the marine mammal and reptile species in Moreton Bay is protected under the Australian Government's *Environment Protection and Biodiversity Conservation Act 1999* (EPBC Act) and various State and Northern Territory legislation. In 1968, Queensland became the first jurisdiction in the world to protect its marine turtles.

Dugongs and turtles have cultural and social significance for Australian Aboriginal and Torres Strait Islander people in coastal and island communities (Johannes and MacFarlane, 1991). Hunting of these species provides not only an important protein source, but also is vital for maintaining social relations, traditional ceremonies, and community co-hesiveness. Under Section 211 of the Native Title Act 1993, traditional owners have the right to take marine resources, including hunting of dugongs and sea turtles for personal, domestic, or noncommercial communal needs and in exercise and enjoyment of their native title rights and interests. Indigenous communities throughout Queensland, including those in Moreton Bay, work in collaboration with government agencies and scientists to develop and implement community-based management for sustainable hunting. This approach is supported by various government-sponsored programs such as *Caring for Our Country [2008]*, and *North Australian Indigenous Land and Sea Management Alliance Ltd. [2001]*. In Moreton Bay, the Quandamooka Yoolooburrabee Aboriginal Corporation represents the interests of the local traditional owners.

The protection and conservation of Moreton Bay's urban marine environments has been high priority for both the Australian Federal and Queensland Governments. In 1993, a large area within Moreton Bay (exceeding 113,000 ha) was recognized as an Australian wetland of international significance under the Ramsar Wetland Convention 1971. Also in 1993, the Queensland Government declared the *Moreton Bay Marine Park* covering some 3400 km^2 of bay habitats. This multiuse marine protected area is divided into zones with differing degrees of protection and allowable activities, from general use zones where all activities are permitted, to Marine National Park Zones (green, no-take) which are areas of highest conservation value and protected from all activities except boating and diving, although other activities may occur with permits. In addition, there are several shallow areas designated as critical habitat for marine turtles and dugongs in which vessels must Go-Slow to avoid collision. In 2009, the *Moreton Bay Marine Park* was rezoned and extended from 0.5% to 16% of the bay under protection, along with introduction of new fishing restrictions to protect important marine and coastal environments (Van De Geer et al., 2013; Kenyon et al., 2018).

Today, most of the protection for megafauna in Moreton Bay is through enforced zoning regulations of the Moreton Bay Marine Park. In particular, the boating *Go-slow* and protection zones over the principal foraging areas of the Eastern Banks aim to protect turtles and dugongs from boat strike and drowning in nets. Day-to-day management of the park (including permitting, patrols, and penalty infringements) is handled by conservation officers and rangers from each of Marine Parks (Queensland Parks and Wildlife Service) and the Queensland Indigenous Land and Sea Ranger Program. Marine Parks also coordinate and regulate to some degree research activities and ecotourism, the latter a burgeoning industry with the Moreton Bay Marine Park.

5.2 Management of Water Quality

Whilst the existing marine park legislation provides an effective tool to manage impacts on Moreton Bay's wildlife, it has no capacity to regulate activities that affect the Moreton Bay catchment. Protecting discrete areas of the bay may in fact be pointless if overall water quality degrades within the catchment and subsequently within the bay. Thankfully, there are stringent overarching Queensland Government regulations governing what enters the Moreton Bay waterways. The regulation of pollution in Queensland waters is a complex matter that involves various legislation and multiple jurisdictions. The *Environmental Protection (EP) Act 1994* (Qld) makes it an offence to cause unlawful environmental harm. More specifically, it prohibits people from unlawfully causing or allowing prescribed contaminants to enter waterways.[1]

---

1 Environmental Protection Act 1994 (Qld), s440ZG.

Prescribed contaminants include substances such as oil, fuel and pesticides, as well as sewage and sediment.[2] The *Waste Reduction and Recycling Act 2011* has further littering and illegal dumping provisions which can overlap with these *EP Act* provisions. Industry regulated by the Department of Environment and Science under the *EP Act* often has permit conditions imposing monitoring and reporting requirements, pre- and postrelease testing and contaminant limits; industry types range from heavy industry (mining), through to waste, sewage, and wastewater management. Similarly, construction and tidal works (e.g., the construction of jetties) will often have imposed conditions that limit what can be released to waters through approvals granted by local government under planning legislation and the *Coastal Protection and Management Act 1995*, respectively. Emissions from ships are regulated by Maritime Safety Queensland under the *Transport Operations (Marine Pollution) Act 1995* (Qld), legislation that imposes international standards on all ships in Queensland coastal waters.[3] Essentially, this legislative framework aims to minimize pollution and protect the marine and coastal environments. When pollution does occur, the whole of government approach in Queensland is "polluter pays" for the cleanup.

Healthy Land and Water is one of a number of Queensland government-funded initiative to improve the health of the catchment and rivers of south-east Queensland including Moreton Bay. Scientists monitor sites throughout the region, provide advice about the health of the region's waterways, guide rehabilitation efforts, and facilitate regional scientific projects. Each year, an annual report card is produced on the condition of local waterways. In addition, this is one group of many to deliver education programs to motivate the community to value and protect waterways. These types of efforts have spawned a shift in awareness and action, evident from the fact that some councils have put in place further voluntary measures to collect rubbish entering our waters such as litter collection bags at the end of stormwater pipes as well as floating stormwater quality improvement devices (SQIDs). At a state government level, there has been the widely anticipated single-use plastic bag ban as of July 2018, which aims to curb the 16 million bags entering our marine environment, and November 2018 will see the commencement of a Queensland container refund scheme. Further, a large number of local conservation, community, and sporting groups conduct Moreton Bay catchment clean-up days, for example, South Brisbane Sailing Club's "Clean up the (Brisbane) River Race 2018" sponsored by the Sea World Research and Rescue Foundation's litter initiatives, *Zero Waste* and *Community Marine Debris* grants.

5.3 Monitoring Population Size and Trends

Monitoring population size and trends is critical for management of threats to marine wildlife. In Moreton Bay, population monitoring programs are species specific.

There are no population estimates for the foraging populations of loggerhead and green turtles in Moreton Bay, since these form part of larger breeding stocks, that is, eastern Australian and southern GBR stocks of these species, respectively. However, the Queensland Government continues to commit to conservation of these turtles through an intensive, long-term (>20 years) tagging and demographic study based in the Moreton Bay feeding ground in association with annual monitoring of key nesting sites for each of these species (Limpus et al., 1992). Monitoring of numbers of breeding females, nests and hatchling success has been conducted for more than 40 years at each of Mon Repos and Heron Island, in association with the *Recovery Plan for Marine Turtles in Australia 2017–27*, and the *Reef 2050 Plan*. This approach is appropriate for the loggerhead and green turtles of Moreton Bay, given their breeding grounds further north within the Great Barrier Reef region. Recognition that sea turtles need protection at each of foraging grounds and breeding grounds and also along migration routes (which may cross international borders) is critical for long-term conservation of this group.

The dugongs of Moreton Bay and elsewhere along the Queensland coast have been monitored by aerial survey at roughly 5-yearly intervals since the 1980s, initially to determine distribution within the bay (Preen, 1993) and then to obtain minimum population estimates and examine population trends (e.g., Marsh et al., 1990, 2011; Lanyon, 2003; Sobtzick et al., 2012, 2015, 2017). Quantitative aerial surveys over 20 years have suggested population sizes ranging from 344 ± 88 (Lawler, 2002) up to 1019 ± 116 (Lanyon, 2003) with most estimates between 500 and 1000. In the absence of evidence of migration or mortality, variation in population estimates is most likely due to variation in survey conditions (season, weather, tidal phase) or dugong behavior (water depth). In addition, a longitudinal capture-mark-recapture program has been established specifically for Moreton Bay (Lanyon et al., 2002) to circumvent errors associated with aerial survey of a cryptic species (Marsh et al., 2018), even in relatively clear waters. This program has been collecting population and life history information for 17 years; such an approach allows determination of minimum population size based on minimum number of gene-tagged individuals, that is, >750 so far, as well as the opportunity to collect life history information (e.g., Burgess et al., 2012a,b, 2013).

---

2 Environmental Protection Regulation 2008 (Qld), schedule 9.
3 These standards are taken from the International Convention for the Prevention of Pollution from Ships (MARPOL).

Dolphins in Moreton Bay have been studied sporadically since the 1980s, usually as graduate student projects (e.g., Corkeron, 1989; Chilvers and Corkeron, 2003; Ansmann et al., 2013). Line transect surveys and short-term capture mark-recapture studies based on photo-identification of individuals have been the main approaches (Lukoschek and Chilvers, 2008). However, because sampling areas, sampling efforts, and methods of analyses have varied, the studies are largely incomparable for determination of population trends (see Ansmann et al., 2013). However, a monitoring program has commenced recently for the Australian humpback dolphin population in the bay, and since this has government backing, the long-term outlook is promising (Meager et al., 2015). Humpback whales, in contrast, have been monitored since the 1970s, through biennial migration counts as they travel past Point Lookout (Bryden, 1985; Bryden et al., 1990; Paterson et al., 1994, 2001, 2004). Since these surveys have used broadly similar counting techniques, population trends have been deduced, and have suggested successful recovery of this species postwhaling (Noad et al., 2016).

5.4 Health Assessment of Wildlife

StrandNet is the marine stranding and mortality database administered by the Queensland government to monitor the mortality of marine wildlife (dugongs, dolphins, whales, sea turtles) along the Queensland coast, including within Moreton Bay Marine Park (QDEHP, 2017). Each year, small numbers of marine mammal and turtle carcasses are recovered and transported to local universities to be necropsied for determination of cause of death; the number of recovered dead sea turtles each year in Moreton Bay is now too high for all to be necropsied. Assigning cause of death is challenging if a carcass is in a state of decomposition, but also if little baseline information is available regarding normal health parameters (Woolford et al., 2015a).

Health assessment of live animals is becoming an increasingly important tool for conservation and management of marine wildlife and their habitats (Bonde et al., 2004; Bossart, 2006), particularly as coastal habitats are impacted by increasingly large human populations. Early detection of health issues, including emerging disease and contaminants, offers managers the opportunity to mitigate threats to health. Health assessment of wildlife in their natural habitat is critical to understanding not only health of individual animals, but also providing an indication of overall health of a population and the ecosystem (Moore, 2008).

An annual health assessment program has been conducted for the dugongs in Moreton Bay for the past decade (since 2008), monitoring health of a small subset of the population through hands-on medical examination (Lanyon et al. 2010). The initial objectives of the program have been achieved, that is, development of clinical blood reference intervals for blood hematology and chemistry against which abnormals may be compared and health issues identified (Lanyon et al., 2015; Woolford et al., 2015b). Recently, this program was instrumental in development and application of dugong-specific antibodies to screen for pathogens. Early results have indicated that dugongs on the seaward side of Moreton Bay have been exposed to a disease-causing pathogen of terrestrial origin, *Toxoplasma gondii*, possibly after coastal flood events (Wong, 2016).

A turtle health program in Moreton Bay has also established clinical reference intervals for blood from normal healthy turtles (Flint et al., 2009, 2010). This program has identified fibropapilloma (FP) virus as a problem for the turtles in Moreton Bay (Arthur et al., 2006), as has been found close to urban centers in other parts of the world (Herbst, 1994). Furthermore, in Moreton Bay, tissue concentrations of lyngbyatoxin A, produced by the alga *Lyngbya majuscula*, have been correlated with the presence of FP lesions in dead green turtles (Arthur et al., 2008b). The incidence of marine debris (Schuyler et al., 2012, 2014) including microplastics and other pollutants (e.g., Hermanussen et al., 2008) in Moreton Bay turtles is also being assessed.

There are currently no programs examining health of cetaceans along the Queensland coast, although an inshore dolphin health assessment program is set to commence in Moreton Bay in late 2019. Although no mass disease outbreaks have occurred in the region, health issues identified from east Australian dolphin carcasses include isolated cases of cetacean morbillivirus (Stone et al., 2012), poxvirus (Fury and Reif, 2012), and *Toxoplasma* (Bowater et al., 2003). It has been suggested that there may be a relationship between declines in water quality (e.g., after flood events) and heightened incidence of disease pathogens in inshore dolphins (e.g., Bowater et al., 2003; Fury and Reif, 2012), including those pathogens that lead to dermal diseases, for example, lobomycosis and tattoo skin disease (GBRMPA (Great Barrier Reef Marine Park Authority), 2012).

6 THE FUTURE

Globally, there is a poor outlook for all biomes, including coastal and estuarine environments (Sala et al., 2000). In particular, urban coastal environments with degrading water quality threaten the well-being and survival of wildlife and their habitats. What the future holds for the marine megafauna of Moreton Bay is uncertain. Fortuitously, the megafauna of Moreton

Bay happens to live in waters of a wealthy developed country with a general societal ethos of protection and conservation. Furthermore, the endangered sea turtles and dugongs are almost all found in the seaward side of the bay, several kilometers from the city shores. However, despite these protective measures (legislation, marine protected areas, geographical distance), Moreton Bay's megafauna is still under considerable threat.

Habitat degradation and loss due to deteriorating water quality has already seen changes to the western (landward) side of Moreton Bay, as sediment and nutrients pour in via the catchment waterways (Abal et al., 2001, 2005; South East Queensland Healthy Waterways Partnership (SEQHWP), 2007), despite the apparently stringent water quality codes. After recent (2011, 2012) severe rainfall events that resulted in coastal flooding in Moreton Bay and elsewhere along the Queensland coast, severe degradation of inshore water quality and seagrass habitats (particularly in Bramble and Deception Bays) ensued (O'Brien et al., 2012). Of concern, the effects of the Brisbane River mud plumes were felt as distant as the Eastern Banks area (Yu et al., 2011, 2014), carrying chemical contaminants and pathogens. The mortality rates for green turtles and dugongs along the Queensland coast increased markedly at this time, as their seagrass food source was destroyed. The megafauna of the bay was largely spared at this time, but with predictions that frequency and severity of weather events are increasing, the long-term effects of future flooding can only be guessed at.

Acknowledgments Many thanks to Kristy Murray (Queensland Parks and Wildlife Service) for supplying information regarding the operation of Moreton Bay Marine Park; Jemma Lanyon (Department of Environment and Science) for information regarding the legislation surrounding regulation of pollution in Queensland waters; and John Kirkwood and Eric Wolanski for review of the chapter.

REFERENCES

Abal, E.G., Dennison, W.C., Greenfield, P.F., 2001. Managing the Brisbane River and Moreton Bay: an integrated research/management program to reduce impacts on an Australian estuary. Water Sci. Technol. 43 (9), 57–70.

Abal, E.G., Dennison, W.C., Bunn, S.E., 2005. Setting. In: Abal, E.G., Bunn, S.E., Dennison, W.C. (Eds.), Healthy Waterways, Healthy Catchments: Making the Connection in South East Queensland. Moreton Bay and Catchments Partnership, Brisbane, pp. 13–34.

Ansmann, I.C., Parra, G.J., Chilvers, B.L., Lanyon, J.M., 2012a. Dolphins restructure social system after reduction of commercial fisheries. Animal Behav. 84, 575–581.

Ansmann, I.C., Parra, G.J., Lanyon, J.M., Seddon, J.M., 2012b. Fine scale genetic population structure in a mobile marine mammal: inshore bottlenose dolphins in Moreton Bay, Australia. Mol. Ecol. 21 (18), 4472–4485.

Ansmann, I.C., Lanyon, J.M., Seddon, J.M., Parra, G.J., 2013. Monitoring dolphins in an urban marine system: total and effective population size estimates of Indo-Pacific bottlenose dolphins in Moreton Bay, Australia. PLoS One 8 (6), e65239.

Ansmann, I.C., Lanyon, J.M., Seddon, J.M., Parra, G.J., 2015. Habitat and resource partitioning among Indo-Pacific bottlenose dolphins in Moreton Bay, Australia. Mar. Mammal Sci. 31, 211–230.

Arthur, K., Shaw, G., Limpus, C., Udy, J., 2006. A review of the potential role of tumour promoting compounds produced by *Lyngbya majuscula* in marine turtle fibropapillomatosis. Afr. J. Mar. Sci. 28, 441–446.

Arthur, K.E., Boyle, M.C., Limpus, C.J., 2008a. Ontogenetic changes in diet and habitat use in green sea turtle (*Chelonia mydas*) life history. Mar. Ecol. Prog. Ser. 362, 303–311.

Arthur, K., Limpus, C., Balazs, G., Capper, A., Udy, J., Shaw, G., Keuper-Bennett, U., Bennett, P., 2008b. The exposure of green turtles (*Chelonia mydas*) to tumour promoting compounds produced by the cyanobacterium *Lyngbya majuscula* and their potential role in the aetiology of fibropapillomatosis. Harmful Algae 7, 114–125.

Bearzi, G., Holcer, D., Notarbartolo di Sciara, G., 2004. The role of historical dolphin takes and habitat degradation in shaping the present status of northern Adriatic cetaceans. Aquat. Conserv. 14, 363–379.

Bonde, R.K., Aguirre, A.A., Powell, J., 2004. Manatees as sentinels of marine ecosystem health: are they the 2000-pound canaries? EcoHealth 1, 255–262.

Bossart, G.D., 2006. Marine mammals as sentinels for ocean and human health. Oceanography 19 (2), 134–137.

Bowater, R.O., Norton, J., Johnson, S., Hill, B., O'Donoghue, P., Prior, H., 2003. Toxoplasmosis in Indo-Pacific humpbacked dolphins (*Sousa chinensis*), from Queensland. Aust. Vet. J. 81, 627–632.

Brandao, A., Butterworth, D.S., Brown, M.R., 2000. Maximum possible humpback whale increase rates as a function of biological parameter values. J. Cetacean Res. Manage. (Suppl.) 2, 192–193.

Brand-Gardner, S.J., Limpus, C.J., Lanyon, J.M., 1999. Diet selection by immature green turtles, *Chelonia mydas*, in subtropical Moreton Bay, south-east Queensland. Aust. J. Zoo 47 (2), 181–191.

Brewer, D., Heales, D., Milton, D., Dell, Q., Fry, G., Venables, B., Jones, P., 2006. The impact of turtle excluder devices and bycatch reduction devices on diverse tropical marine communities in Australia's northern prawn trawl fishery. Fish. Res. 81 (2–3), 176–188.

Brown, M.R., 1998. Population Biology and Migratory Characteristics of East Australian Humpback Whales *Megaptera novaeangliae*. (Unpublished Ph.D. thesis), University of Sydney, p. 322.

Bryden, M.M., 1985. Studies of humpback whales (*Megaptera novaeangliae*), Area V. In: Ling, J.K., Bryden, M.M. (Eds.), Studies of Sea Mammals in South Latitudes. South Australian Museum, Adelaide, pp. 115–123.

Bryden, M.M., Kirkwood, G.P., Slade, R.W., 1990. Humpback whales, Area V. An increase in numbers off Australia's east coast. In: Antarctic Ecosystems. Springer, Berlin, Heidelberg, pp. 271–277.

Burgess, E.A., Lanyon, J.M., Brown, J.L., Blyde, D., Keeley, T., 2012a. Diagnosing pregnancy in free-ranging dugongs using fecal progesterone metabolite concentrations and body morphometrics: a population application. Gen. Comp. Endocrinol. 177, 82–92.

Burgess, E.A., Lanyon, J.M., Keeley, T., 2012b. Testosterone and tusks: maturation and seasonal reproductive patterns of live, free-ranging male dugongs (*Dugong dugon*) in a subtropical population. Reproduction 143, 1–16.

Burgess, E.A., Brown, J.L., Lanyon, J.M., 2013. Sex, scarring and stress: understanding seasonal costs in a cryptic marine mammal. Conserv. Physiol. 1. https://doi.org/10.1093/conphys/cot014.

Burnell, S.R., 2001. Aspects of the reproductive biology, movements and site fidelity of right whales off Australia. J. Cetacean Res. Manage. (Special Issue) 2, 89–102.

Cagnazzi, D., 2013. Review of coastal dolphins in central Queensland, particularly Port Curtis and Port Alma region. Report to the Gladstone Port Corporation.

Carroll, E., Patenaude, N., Alexander, A., Steel, D., Harcourt, R., Childerhouse, S., Baker, C.S., 2011. Population structure and individual movement of southern right whales around New Zealand and Australia. Mar. Ecol. Prog. Ser. 432, 257–268.

Chaloupka, M., Limpus, C., 2001. Trends in the abundance of sea turtles resident in southern Great Barrier Reef waters. Biol. Conserv. 102 (3), 235–249.

Chilvers, B.L., 2000. Southern right whales *Eubalaena australis* (Desmoulins 1822) in Moreton Bay, Queensland. Mem. Queensl. Mus. 45 (2), 576.

Chilvers, B.L., Corkeron, P.J., 2001. Trawling and bottlenose dolphins' social structure. Proc. R. Soc. Lond. B Biol. Sci. 268 (1479), 1901–1905.

Chilvers, B.L., Corkeron, P.J., 2003. Abundance of Indo-Pacific bottlenose dolphins, *Tursiops aduncus*, off point lookout, Queensland, Australia. Marine Mammal Sci. 19, 85–95.

Chilvers, B.L., Lawler, I.R., Macknight, F., Marsh, H., Noad, M., Paterson, R., 2005. Moreton Bay, Queensland, Australia: an example of the co-existence of significant marine mammal populations and large-scale coastal development. Biol. Conserv. 122, 559–571.

Chittleborough, R.G., 1965. Dynamics of two populations of the humpback whale, *Megaptera novaeangliae* (Borowski). Mar. Freshw. Res. 16 (1), 33–128.

Church, J.A., 1979. An Investigation of the Tidal and Residual Circulations and Salinity Distribution in Moreton Bay (PhD thesis). School of Physical Sciences, The University of Queensland, St. Lucia, p. 209.

Cope, R., Lanyon, J.M., Pollett, P.K., Seddon, J.M., 2015. Indirect detection of genetic dispersal (movement and breeding events) through pedigree analysis of dugong populations in southern Queensland, Australia. Biol. Conserv. 181, 91–101.

Corkeron, P.J., 1989. Studies of Inshore Dolphins, *Tursiops* and *Sousa*, in the Moreton Bay Region. . (unpublished PhD thesis), University of Queensland.

Croxall, J.P., Butchart, S.H., Lascelles, B.E.N., Stattersfield, A.J., Sullivan, B.E.N., Symes, A., Taylor, P.H.I.L., 2012. Seabird conservation status, threats and priority actions: a global assessment. Bird Conser. Int. 22 (1), 1–34.

Daley, B., Griggs, P., Marsh, H., 2008. Exploiting marine wildlife in Queensland: the commercial dugong and marine turtle fisheries, 1847–1969. Aust. Econ. Hist. Rev. 48 (3), 227–265.

Dawbin, W.H., 1986. Right whales caught around south eastern Australia and New Zealand during the nineteenth and early twentieth centuries. Rep. Int. Whaling Commission Special Issue 10, 261–267.

Department of Sustainability and Environment (DSE), 2013. Advisory List of Threatened Vertebrate Fauna in Victoria, 2013. Victorian Government Department of Sustainability and Environment, Melbourne. 18 pp.

DES (Department of Environment and Science), 2015. Whales and Dolphins. https://www.npsr.qld.gov.au/parks/moreton-bay/zoning/information-sheets/whales_and_dolphins.html#. (Accessed July 6, 2018).

DNPSR (Department of National Parks Sport and Racing), 2018. Turtles in Moreton Bay Marine Park. https://www.npsr.qld.gov.au/parks/moreton-bay/zoning/information-sheets/turtles.html#turtles_in_moreton_bay_marine (http://4.7.0.18, 04 July 2018). Queensland Government Website. Accessed 4 July 2018.

Dobson, A., Lodge, D., Alder, J., Cumming, G.S., Keymer, J., McGlade, J., Mooney, H., Rusak, J.A., Sala, O., Woletrs, V., Wall, D., Winfree, R., Xenopoulos, M.A., 2006. Habitat loss, trophic collapse and the decline of ecosystem services. Ecology 87, 1915–1924.

EPA (Environment Protection Agency), 2005. Shorebird Management Strategy: Moreton Bay. Environmental Protection Agency (Queensland), Brisbane.

Flint, M., Patterson-Kane, J.C., Limpus, C.J., Work, T.M., Blair, D., Mills, P.C., 2009. Postmortem diagnostic investigation of disease in free-ranging marine turtle populations: a review of common pathologic findings and protocols. J. Vet. Diagn. Investig. 21, 733–759.

Flint, M., Patterson-Kane, J.C., Limpus, C.J., Mills, P.C., 2010. Health surveillance of stranded green turtles in southern Queensland, Australia (2006–2009): an epidemiological analysis of causes of disease and mortality. EcoHealth 7, 135–145.

Fury, C.A., Reif, J.S., 2012. Incidence of poxvirus-like lesions in two estuarine dolphin populations in Australia: links to flood events. Sci. Total Environ. 416, 536–540.

GBRMPA (Great Barrier Reef Marine Park Authority), 2012. Vulnerability Assessment: Indo-Pacific (Inshore) Bottlenose Dolphin.

Gibbes, B., Grinham, A., Neil, D., Olds, A., Maxwell, P., Connolly, R., Udy, J., 2014. Moreton Bay and its estuaries: a sub-tropical system under pressure from rapid population growth. In: Wolanski, E. (Ed.), Estuaries of Australia in 2050 and Beyond. Springer, Dordrecht, pp. 203–222.

Greenland, J., Limpus, C.J., 2007. Marine Wildlife Stranding and Mortality Database Annual Report 2007, I: Dugong. . Queensland Environmental Protection Agency Conservation technical and data report.

Hale, P., Long, S., Tapsall, A., 1998. Distribution and conservation of delphinids in Moreton Bay. In: Tibbetts, I.R., Hall, N.J., Dennison, W.C. (Eds.), Moreton Bay and Catchment. University of Queensland, Brisbane, pp. 477–486.

Halpern, B.S., Walbridge, S., Selkoe, K.A., Kappel, C.V., Micheli, F., D'agrosa, C., Fujita, R., 2008. A global map of human impact on marine ecosystems. Science 319 (5865), 948–952.

Hazel, J., Lawler, I.R., Hamann, M., 2009. Diving at the shallow end: green turtle behaviour in near-shore foraging habitat. J. Exp. Mar. Biol. Ecol. 371 (1), 84–92.

Heppel, S.S., Limpus, C.J., Crouse, D.T., Frazer, N.B., Crowder, L.B., 1996. Population model analysis for the loggerhead sea turtle, *Caretta caretta*, in Queensland. Wildl. Res. 23 (2), 143–161.

Herbst, L.H., 1994. Fibropapillomatosis of marine turtles. Annu. Rev. Fish Dis. 4, 389–425.

Hermanussen, S., Matthews, V., Paepke, O., Limpus, C.J., Gaus, C., 2008. Flame retardants (PBDEs) in marine turtles, dugongs and seafood from Queensland, Australia. Mar. Pollut. Bull. 57, 409–418.

IUCN (International Union for the Conservation of Nature), 2018. The IUCN Red List of Threatened Species. Version 2018-1. www.iucnredlist.org. (Accessed July 10, 2018).

IWC (International Whaling Commission), 2015. Report of the Scientific Committee, Annex H. Report of the sub-committee on other southern hemisphere whale stocks. J. Cetacean Res. Manage. (Suppl.) 16, 196–221.

Jackson, J.B., Kirby, M.X., Berger, W.H., Bjorndal, K.A., Botsford, L.W., Bourque, B.J., Bradbury, R.H., Cooke, R., Erlandson, J., Estes, J.A., Hughes, T.P., 2001. Historical overfishing and the recent collapse of coastal ecosystems. Science 293, 629–637.

Johannes, R.E., MacFarlane, J.W., 1991. Traditional Fishing in the Torres Strait Islands. CSIRO Division of Fisheries, Marine Laboratories, Hobart.

Johnson, M., 2002. A modified form of whaling: the Moreton Bay dugong fishery 1846–1920. In: Johnson, M. (Ed.), Moreton Bay Matters. Brisbane History Group papers 19. 148 pp. (Chapter 4).

Jones, K., Ariel, E., Burgess, G., Read, M., 2016. A review of fibropapillomatosis in green turtles (*Chelonia mydas*). Vet. J. 212, 48–57.

Kemper, C., Coughran, D., Warneke, R., Pirzl, R., Watson, M., Gales, R., Gibbs, S., 2008. Southern right whale (*Eubalaena australis*) mortalities and human interactions in Australia, 1950–2006. J. Cetacean Res. Manag. 10 (1), 1–8.

Kenyon, R.A., Babcock, R.C., Dell, Q., Lawrence, E., Moeseneder, C., Tonks, M.L., 2018. Business as usual for the human use of Moreton Bay following marine park zoning. Mar. Freshw. Res. 69 (2), 277–289.

Lanyon, J.M., 2003. Distribution and abundance of dugongs in Moreton Bay, Queensland, Australia. Wildl. Res. 30 (4), 397–409.

Lanyon, J.M., Janetzki, H., 2016. Mortalities of southern right whales (*Eubalaena australis*) in a subtropical wintering ground, southeast Queensland. Aquat. Mamm. 42 (4), 470–475.

Lanyon, J.M., Sanson, G.D., 2006. Mechanical disruption of seagrass in the digestive tract of the dugong. J. Zool. 270 (2), 277–289.

Lanyon, J.M., Sneath, H.L., Kirkwood, J.M., Slade, R.W., 2002. Establishing a mark-recapture program for dugongs in Moreton Bay, south-east Queensland. Australian Mamm. 24, 51–56.

Lanyon, J.M., Sneath, H.L., Long, T., Bonde, R.K., 2010. Physiological response of wild dugongs (*Dugong dugon*) to out-of-water sampling for health assessment. Aquat. Mamm. 36, 46–58.

Lanyon, J.M., Wong, A., Long, T., Woolford, L., 2015. Serum biochemistry reference intervals of live wild dugongs (*Dugong dugon*) from urban coastal Australia. Clin. Vet. Pathol. 44, 234–242.

Lanyon, J.M., Noad, M., Meager, J., 2018. Ecology of the marine mammals of Moreton Bay (Chapter 5). In: Tibbetts, I.R., Rothlisberg, P.C., Neil, D.T., Homburg, T.A., Brewer, D.T., Arthington, A.H. (Eds.), Moreton Bay Quandamooka & Catchment: Past, Present & Future. In Press.

Lawler, I.R., 2002. Distribution and Abundance of Dugongs and Other Megafauna in Moreton Bay and Hervey Bay Between December 2000 and November 2001. James Cook University, Townsville.

Limpus, C.J., 2008. A biological review of Australian Marine Turtles. 1. Loggerhead Turtle *Caretta caretta* (Linneaus). Queensland Environment Protection Agency. Available from: http://www.epa.qld.gov.au/publications/p02785aa.pdf/A_Biological_Review_Of_Australian_Marine_Turtles_1_Loggerhead_Turtle_emCaretta_Caretta/em_Linnaeus.pdf. (Accessed July 4, 2018).

Limpus, C.J., Limpus, D.J., 2001. The loggerhead turtle, *Caretta caretta*, in Queensland: breeding migrations and fidelity to a warm temperate feeding area. Chelonian Conser. Biol. 4 (1), 142–153.

Limpus, C.J., Limpus, D.J., 2003. Loggerhead turtles in the equatorial and southern Pacific Ocean: a species in decline. In: Bolten, A., Witherington, B. (Eds.), Loggerhead Sea Turtles. Smithsonian Institution Press, Washington, DC, pp. 199–210.

Limpus, C.J., Reimer, D., 1994. In: James, R. (Ed.), The loggerhead turtle, *Caretta caretta*, in Queensland: a population in decline. Proceedings of the Australian Marine Turtle Conservation Workshop, Gold Coast, 14–17 November 1990. Qld Dept Env. & Heritage, Canberra, ANCA, pp. 39–59.

Limpus, C.J., Miller, J.D., Parmenter, C.J., Reimer, D., McLachlan, N., Webb, R., 1992. Migration of green (*Chelonia mydas*) and loggerhead (*Caretta caretta*) turtles to and from eastern Australian rookeries. Wildl. Res. 19 (3), 347–357.

Limpus, C.J., Couper, P.J., Read, M.A., 1994. The green turtle, *Chelonia mydas*, in Queensland: population structure in a warm temperature feeding area. Mem. Queensl. Mus. 35 (1), 139–154.

Lukoschek, V., Chilvers, B.L., 2008. A robust baseline for bottlenose dolphin abundance in coastal Moreton Bay: a large carnivore living in a region of escalating anthropogenic impacts. Wildl. Res. 35 (7), 593–605.

MacPhee, R.D.E., Flemming, C., 1999. Requiem aeternam. In: MacPhee, R.D.E., Sues, H.D. (Eds.), Extinctions in Near Time. Springer, Boston, MA, pp. 333–371.

Marsh, H., Kwan, D., 2008. Temporal variability in the life history and reproductive biology of female dugongs in Torres Strait: the likely role of seagrass dieback. Cont. Shelf Res. 28 (16), 2152–2159.

Marsh, H., Sobtzick, S., 2015. *Dugong dugon*. The IUCN Red List of Threatened Species 2015: e.T6909A43792211. https://doi.org/10.2305/IUCN.UK.2015-4.RLTS.T6909A43792211.en. (Accessed 6 July 2018).

Marsh, H., Saalfeld, W.K., Preen, A.R., 1990. The distribution and abundance of dugongs in southern Queensland waters: implications for management. . Report to Queensland Department of Primary Industries.

Marsh, H., O'Shea, T.J., Reynolds III, J.E., 2011. Ecology and Conservation of the Sirenia: Dugongs and Manatees (No. 18). Cambridge University Press, Cambridge.

Marsh, H., Brooks, L., Hagihara, R., 2018. The challenge of monitoring coastal marine mammals. In: Legge, S., Robinson, N., Lindenmayer, D., Scheele, B., Southwell, D., Wintle, B. (Eds.), Monitoring Threatened Species and Ecological Communities. CSIRO Publishing, Melbourne, pp. 291.

McCauley, D.J., Pinsky, M.L., Palumbi, S.R., Estes, J.A., Joyce, F.H., Warner, R.R., 2015. Marine defaunation: animal loss in the global ocean. Science 347 (6219), 1255641.

Meager, J.J., Sumpton, W.D., 2016. Bycatch and strandings programs as ecological indicators for data-limited cetaceans. Ecol. Indic. 60, 987–995.

Meager, J., Hawkins, E., Gaus, C., 2015. Health and status of Australian humpback dolphins in Moreton Bay. . Queensland. Final report to the Australian Marine Mammal Centre.

Meynecke, J.O., Meager, J.J., 2016. Understanding strandings: 25 years of humpback whale (*Megaptera novaeangliae*) strandings in Queensland, Australia. J. Coast. Res. 75, 897–901.

Moore, S., 2008. Marine mammals as ecosystem sentinels. J. Mammal. 89 (3), 534–540.

MSQ (Maritime Safety Queensland), 2016. Recreational Ship Census June 2016. Department of Transport and Roads. 12 pp. (Accessed 10 July 2018).

Neil, D.T., Brieze, I., 1998. Wild dolphin provisioning at Tangalooma, Moreton Island: an evaluation. In: Tibbetts, I.R., Hall, N.J., Dennison, W.C. (Eds.), Moreton Bay and Catchment. The University of Queensland, Brisbane, pp. 487–500.

Noad, M.J., 2000. A southern right whale *Eubalaena australis* (Desmoulins, 1822) in southern Queensland waters. Mem. Queensl. Mus. 45 (2), 556.

Noad, M.J., Dunlop, R.A., Paton, D., Cato, D.H., 2011. Absolute and relative abundance estimates of Australian east coast humpback whales (*Megaptera novaeangliae*). J. Cetacean Res. Manag., 243–252. (Special Issue 3).

Noad, M.J., Dunlop, R.A., Bennett, L., Kniest, H., 2016. Abundance Estimates of the East Australian Humpback Whale Population (BSE1): 2015 Survey and Update. . IWC document SC/66b/SH/21.

O'Brien, K., Tuazon, D., Grinham, A., Callaghan, D., 2012. Impact of Mud Deposited by 2011 Flood on Marine and Estuarine Habitats in Moreton Bay. Healthy Waterways, Brisbane.

Owen, H., Gillespie, A., Wilkie, I., 2012. Postmortem findings from dugong (*Dugong dugon*) submissions to the University of Queensland: 1997–2010. J. Wildl. Dis. 48 (4), 962–970.

Pascoe, S., Doshi, A., Dell, Q., Tonks, M., Kenyon, R., 2014. Economic value of recreational fishing in Moreton Bay and the potential impact of the marine park rezoning. Tour. Manag. 41, 53–63.

Paterson, R., Paterson, P., 1984. A study of the past and present status of humpback whales in east Australian waters. Biol. Conserv. 29 (4), 321–343.

Paterson, R., Paterson, P., Cato, D.H., 1994. The status of humpback whales *Megaptera novaeangliae* in east Australia thirty years after whaling. Biol. Conserv. 70 (2), 135–142.

Paterson, R.A., Paterson, P., Cato, D.H., 2001. Status of humpback whales, *Megaptera novaeangliae*, in east Australia at the end of the 20th century. Mem. Queensl. Mus. 47 (2), 579.

Paterson, R., Paterson, P., Cato, D.H., 2004. Continued increase in east Australian humpback whales in 2001, 2002. Mem. Queensl. Mus. 49 (2), 712–731.

Preen, A.R., 1993. Interactions Between Dugongs and Seagrasses in a Subtropical Environment (unpublished PhD thesis). James Cook University, Townsville, p. 392.

Preen, A.R., Thompson, J., Corkeron, P.J., 1992. Wildlife and management: dugongs, waders and dolphins. In: Crimp, O.N. (Ed.), Moreton Bay in the Balance. Australian Littoral Society Inc., Brisbane, pp. 61–70.

QDEHP (Qld Department of Environment and Heritage Protection), 2017. Marine Wildlife Strandings Annual Reports. https://www.ehp.qld.gov.au/wildlife/caring-for-wildlife/strandnet-reports.html. (Accessed July 9, 2018).

QFish, 2018. Queensland Commercial Fishery. http://qfish.fisheries.qld.gov.au/query/acdaf4fb-5e46-4755-9249-2c1dd80b7059/table?customise=True. (Accessed July 10, 2018).

QOESR (Queensland Office of Economic and Statistical Research), 2011. Queensland Government Population Projections: Local Government Area Report. QOESR, Brisbane. Available from: http://www.oesr.qld.gov.au/products/publications/qld-govt-pop-proj-lga/index.php. (Accessed July 6, 2018).

QPWS (Queensland Parks and Wildlife Service), 2012. Queensland Parks and Wildlife Service Community Survey 2012. p. 21.

Read, M.A., Limpus, C.J., 2002. The green turtle, *Chelonia mydas*, in Queensland: feeding ecology of immature turtles in Moreton Bay, southeastern Queensland. Mem. Queensl. Mus. 48 (1), 207–214.

Reilly, S.B., Bannister, J.L., Best, P.B., Brown, M., Brownell Jr., R.L., Butterworth, D.S., Clapham, P.J., Cooke, J., Donovan, G.P., Urbán, J., Zerbini, A.N., 2008. *Megaptera novaeangliae*. The IUCN Red List of Threatened Species 2008: e.T13006A3405371. https://doi.org/10.2305/IUCN.UK.2008.RLTS.T13006A3405371.en. (Accessed 8 July 2018).

Rice, D.W., Wolman, A.A., Braham, H.W., 1984. The gray whale, *Eschrichtius robustus*. Mar. Fish. Rev. 46 (4), 7–14.

Rojas-Bracho, L., Reeves, R.R., Jaramillo-Legorreta, A., 2006. Conservation of the vaquita Phocoena sinus. Mammal Rev. 36, 179–216.

Sala, O.E., Chapin, F.S., Armesto, J.J., Berlow, E., Bloomfield, J., Dirzo, R., Leemans, R., 2000. Global biodiversity scenarios for the year 2100. Science 287, 1770–1774.

Schipper, J., Chanson, J.S., Chiozza, F., Cox, N.A., Hoffmann, M., Katariya, V., Lamoreux, J., Rodrigues, A.S., Stuart, S.N., Temple, H.J., Baillie, J., 2008. The status of the world's land and marine mammals: diversity, threat, and knowledge. Science 322, 225–230.

Schuyler, Q., Hardesty, B.D., Wilcox, C., Townsend, K., 2012. To eat or not to eat? Debris selectivity by marine turtles. PLoS One 7 (7), e40884.

Schuyler, Q., Hardesty, B.D., Wilcox, C., Townsend, K., 2014. Global analysis of anthropogenic debris ingestion by sea turtles. Conserv. Biol. 28 (1), 129–139.

Seddon, J.M., Ovenden, J.R., Sneath, H.L., Broderick, D., Dudgeon, C.L., Lanyon, J.M., 2014. Fine scale population structure of dugongs (*Dugong dugon*) implies low gene flow along the southern Queensland coastline. Conserv. Genet. 15, 1381–1392.

Seminoff, J.A., 2004. *Chelonia mydas*. The IUCN Red List of Threatened Species 2004: e.T4615A11037468. https://doi.org/10.2305/IUCN.UK.2004. RLTS.T4615A11037468.en. (Accessed 5 July 2018).

Sobtzick, S., Hagihara, R., Grech, A., Marsh, H., 2012. Aerial survey of the urban coast of Queensland to evaluate the response of the dugong population to the widespread effects of the January 2011 floods and Cyclone Yasi. . Final report to the Australian Marine Mammal Centre, Hobart, Australia.

Sobtzick, S., Hagihara, R., Grech, A., Jones, R., Pollock, K., Marsh, H., 2015. Improving the time series of estimates of dugong abundance and distribution by incorporating revised availability bias corrections. . Final report to the Australian Marine Mammal Centre, Project 13/31.

Sobtzick, S., Cleguer, C., Hagihara, R., Marsh, H., 2017. Distribution and abundance of dugong and large marine turtles in Moreton Bay, Hervey Bay and the southern Great Barrier Reef. A report to the Great Barrier Reef Marine Park Authority, Centre for Tropical Water & Aquatic Ecosystem Research (TropWATER) Publication 17/21, James Cook University, Townsville, p. 91.

South East Queensland Healthy Waterways Partnership (SEQHWP), 2007. South East Queensland Healthy Waterways Strategy. 2007–2012. SEQ Healthy Waterways Partnership, Brisbane.

Sprogis, K.A., 2008. Small Scale Habitat Use and Movements of Dugongs (*Dugong dugon*) With Respect to Tides. Foraging Time and Vessel Traffic on the Eastern Banks are of Moreton Bay. (unpublished Honours thesis) The University of Queensland, Brisbane75.

Stone, B.M., Blyde, D.J., Saliki, J.T., Morton, J.M., 2012. Morbillivirus infection in live stranded, injured, trapped, and captive cetaceans in southeastern Queensland and northern New South Wales, Australia. J. Wildl. Dis. 48, 47–55.

Turvey, S.T., Pitman, R.L., Taylor, B.L., Barlow, J., Akamatsu, T., Barrett, L.A., Wei, Z., 2007. First human-caused extinction of a cetacean species? Biol. Lett. 3 (5), 537–540.

Turvey, S.T., Barrett, L.A., Hart, T., Collen, B., Yujiang, H., Lei, Z., Ding, W., 2010. Spatial and temporal extinction dynamics in a freshwater cetacean. Proc. R. Soc. Lond. B Biol. Sci. 277 (1697), 3139–3147.

Van De Geer, C., Mills, M., Adams, V.M., Pressey, R.L., McPhee, D., 2013. Impacts of the Moreton Bay Marine Park rezoning on commercial fishermen. Mar. Policy 39, 248–256.

Verboom, W.C., 2002. Noise criteria for marine mammals. . Report HAG-RPT-010120 TNO TPD Delft, The Netherlands.

Waycott, M., Duarte, C.M., Carruthers, T.J.B., Orth, R.J., Dennison, W.C., Olyarnik, S., Calladine, A., Fourqurean, J.W., Heck Jr., K.L., Hughes, A.R., Kendrick, G.A., Kenworthy, W.J., Short, F.T., Williams, S.L., 2009. Accelerating loss of seagrasses across the globe threatens coastal ecosystems. PNAS 106, 12377–12381.

Weijs, L., Vijayasarathy, S., Villa, C.A., Neugebauer, F., Meager, J.J., et al., 2016. Screening of organic and metal contaminants in Australian humpback dolphins (*Sousa sahulensis*) inhabiting an urbanised embayment. Chemosphere 151, 253–262.

Wilson, H.B., Kendall, B.E., Fuller, R.A., Milton, D.A., Possingham, H.P., 2011. Analyzing variability and the rate of decline of migratory shorebirds in Moreton Bay, Australia. Conserv. Biol. 25 (4), 758–766.

Wong, A., 2016. Health Surveillance of a Subtropical Wild Dugong Population: Development and Application of Tools for Assessment of Clinical Health and Immunology (unpublished PhD thesis). The University of Queensland, Brisbane, p. 83.

Woolford, L., Franklin, C., Whap, T., Loban, F., Lanyon, J.M., 2015a. Pathological findings in wild harvested dugongs *Dugong dugon* of central Torres Strait, Australia. Dis. Aquat. Org. 113 (2), 89–102.

Woolford, L., Wong, A., Sneath, H.L., Long, T., Boyd, S.P., Lanyon, J.M., 2015b. Hematology of dugongs (*Dugong dugon*) in southern Queensland. Clin. Vet. Pathol. 44, 530–541.

WWF (World Wildlife Fund), 2012. Living Planet Report. WWF International, Gland. http://wwf.panda.org/lpr. (Accessed July 5, 2018).

Yu, Y., Zhang, H., Lemckert, C.J., 2011. The response of the river plume to the flooding in Moreton Bay, Australia. J. Coast. Res. 64, 1214–1218.

Yu, Y., Zhang, H., Lemckert, C.J., 2014. Numerical analysis on the Brisbane River plume in Moreton Bay due to Queensland floods 2010–2011. Environ. Fluid Mech. 14 (1), 1–24.

Chapter 7

Peel-Harvey Estuary, Western Australia

Valesini, F.J.*, Hallett, C.S.*, Hipsey, M.R.[†], Kilminster, K.L.[‡], Huang, P.[†], Hennig, K.[‡]

*Centre for Sustainable Aquatic Ecosystems, Harry Butler Institute, Murdoch University, Perth, WA, Australia, [†]Aquatic Ecodynamics, UWA School of Agriculture and Environment, The University of Western Australia, Perth, WA, Australia, [‡]Department of Water and Environmental Regulation, Perth, WA, Australia

1 OVERVIEW

The Peel-Harvey Estuary is the largest inland water body in the southern half of Western Australia (WA), covering ~130 km^2 (Hodgkin and Hesp, 1998; Brearley, 2005). It lies in the Peel Region south of Perth, the State capital. Although Peel is the smallest of WA's nine regional areas, its coastal city of Mandurah is among the fastest growing in the State (Australian Bureau of Statistics, 2018). The estuary, which is part of the Ramsar-listed Peel-Yalgorup wetland system, is a major natural asset in the region and is intrinsically tied to Peel's cultural heritage, as well as the modern lifestyles and livelihoods of the people it supports (Hale and Butcher, 2007).

As with many estuaries in developed catchments, changes in the Peel-Harvey over recent centuries have been strongly influenced by the societies it has helped develop, particularly since European settlement in the late 1800s (Bradby, 1997). During the 1970s and 1980s, the estuary became nationally and internationally iconic for its hypereutrophication and extreme algal bloom issues, following decades of catchment and estuary modification (McComb and Humphries, 1992). These symptoms were managed by an engineering intervention in the mid-1990s, namely an artificial second channel (the Dawesville Channel, or "Cut") to increase tidal flushing, which led to further major environmental and ecological shifts for this system.

As societal relationships with the Peel-Harvey have shifted from exploitation for socio-economic gain to valuing its ecological health, and as the system faces further challenges from a fast growing urban sector and rapidly drying climate, it is timely to consider current and evolving approaches for better balancing the societal demands and ecosystem integrity of this waterway. The Peel-Harvey also provides an interesting case study for exploring the resilience of estuarine systems and the societies they support, given its particular history of fundamental ecosystem shifts through both chronic decline and an acute engineered "recovery," as well as its current and forecast climate and development stressors.

The overarching objective of this chapter is to outline current and developing initiatives that are helping to grow our understanding of not only the present status of the Peel-Harvey system, but also how it is likely to respond to anticipated changes over the coming decades. These projections will provide a basis for exploring options to help support key societal values under shifting baselines, not only for the Peel-Harvey but also potentially for other urbanized estuaries in temperate microtidal regions worldwide.

Following a brief overview of the Peel-Harvey system (Section 2), we summarize what is known about its past development and subsequent response to provide the context for future projections, drawing from various excellent summaries by historians and scientists (Section 3). We then examine in more detail how aspects of the estuarine environment and its ecology have responded to key pressures in recent decades (Section 4). This is followed by the current status of this socio-ecological system, including relevant management and scientific initiatives (Section 5). Finally, we speculate about how the Peel-Harvey and its interlinked society may change in coming decades, and identify some potential approaches for better maintaining it into the future (Section 6).

2 THE PEEL-HARVEY SYSTEM

The Peel-Harvey Estuary is located along the temperate and microtidal coastline of southwestern Australia (tidal range ~0.5 m; Fig. 1A). The system experiences a Mediterranean-type climate characterized by a strong seasonal pattern of cool wet winters and hot dry summers, with almost all of the annual rainfall occurring during the cooler months of May to

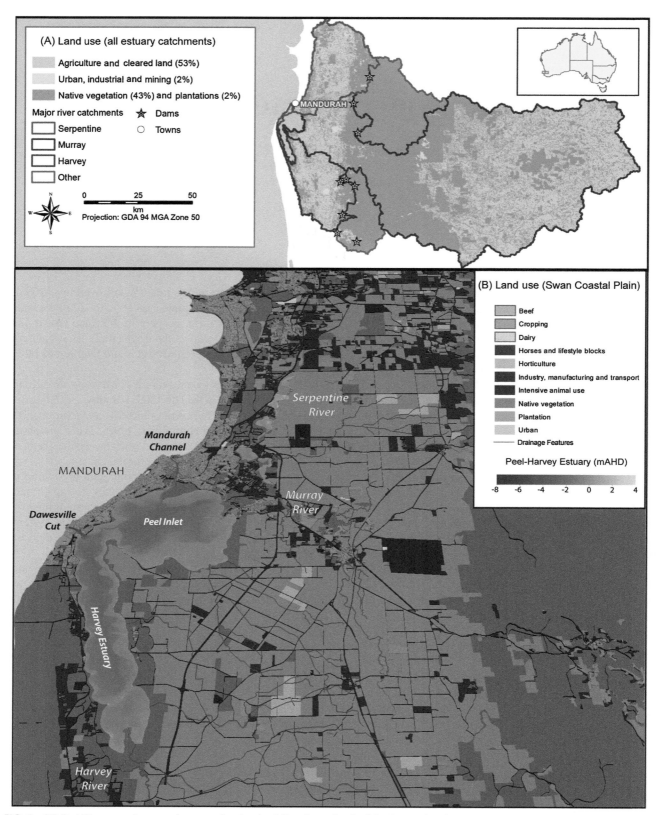

FIG. 1 (A) Peel-Harvey catchment and estuary, showing the delineations of each of the three main subcatchments and all broad land uses that drain to the estuary. Inset shows the location of the system in Australia; (B) Peel-Harvey Estuary and detailed land uses on the Swan Coastal Plain portion of the catchment.

October (see Section 4). It has an unusual geomorphology that reflects its Pleistocene inheritance of both interbarrier and basin coastal depressions on the low-lying Swan Coastal Plain, which is bordered to the east by an escarpment (the Darling Scarp). These depressions were flooded during the Holocene marine transgression, and intercepted river systems flowing off the adjacent uplands (Hodgkin and Hesp, 1998). The estuary initially had a relatively large tidal incursion, which shifted to greater freshwater inputs over time as it silted up and formed a shallow intermittently-open mouth along with river bars and basin sand banks (Bradby, 1997). It currently comprises two large and shallow basins (mainly <2 m deep), namely the roughly circular Peel Inlet and the elongate Harvey Estuary, as well as three rivers, comprising the Murray and Serpentine which discharge into the former basin, and the Harvey which flows into the latter (Fig. 1B). Two narrow and permanently open entrance channels connect the estuary to the sea, one of which is natural (the Mandurah Channel at the northern end of Peel Inlet), and the other of which is artificial (the Dawesville Channel at the northern end of Harvey Estuary). The latter was constructed in 1994 to increase tidal exchange with the ocean and alleviate a host of responses to hypereutrophication (see Sections 3 and 4).

The three main river subcatchments of the Peel-Harvey Estuary cover about 9400 km^2 across both the coastal plain and escarpment (Fig. 1A). Much of the sandy coastal plain has been cleared of its native vegetation and developed mainly for agriculture (dominated by beef cattle grazing), with some industrial areas (e.g., mines, refineries and intensive animal uses) and a fast-growing urban sector along the coastal fringe (Fig. 1B). The upland area, much of which is in the large Murray subcatchment, is dominated by native vegetation and, further inland, by extensive cropping areas with some forestry plantations (Kelsey et al., 2011). Soil types differ throughout the catchment, with the Serpentine and Harvey dominated by sandy soils, while the Murray comprises clays and sands (Hale and Butcher, 2007).

The estuary, together with a series of nearby wetlands and nature reserves, forms the Peel-Yalgorup wetland system, which was designated as a Ramsar site in 1990 and extended in 2001. Its listing reflects its uniqueness and biological diversity in the region, importance to waterbirds, fish and other biotic communities, as well as the bioregionally rare thrombolites it supports (Hale and Butcher, 2007). Like many estuaries, the societal benefits provided by the healthy functioning of this ecosystem are extensive. They include crucial provisioning and supporting ecosystem services such as nutrient cycling, flood control and climate regulation, and also key industries such as fishing and tourism (Section 5). The Peel-Harvey also supports many recreational activities, particularly bird watching, boating and fishing (Hale and Butcher, 2007; Department of Transport, 2011; Johnston et al., 2014, 2017), and is home to many culturally significant Nyoongar sites (Cuthbert et al., 2007).

3 HISTORICAL SOCIO-ECOLOGICAL DEVELOPMENTS

To begin to understand the future of this complex socio-ecological system, it is informative to understand its past development. To this end, a nonexhaustive series of timelines summarizing some key historical environmental changes, the subsequent estuary response and accompanying socioeconomic shifts, is presented in Fig. 2.

From what was a widespread and highly productive wetland system that flooded the coastal plain in a coalescence of rivers, lakes and estuary basins in winter, and which had been used sustainably for many thousands of years by the Binjareb Aboriginals (a dialect group of the Nyoongar people from wider southwestern Australia; Cuthbert et al., 2007), the estuary and its catchment were irrevocably changed following European settlement in 1829. This occurred mainly through systematic land clearing, drain construction (~3000 km of arterial drains) and river training (desnagging, damming and diversion) to develop the area for agriculture (Fig. 2). Much social and economic hardship was encountered for at least the first century, not only from ongoing crop failure and stock malnourishment, but also from conflicts between the settlers and Indigenous people (Bradby, 1997; Brearley, 2005; Cuthbert et al., 2007).

Legislative developments from 1900 and freely available labor during the Great Depression helped facilitate the gradual conversion of the once densely vegetated coastal plain into drained and irrigated land for dairy pasture and horticulture. Agricultural productivity accelerated from the mid-1950s through high commodity prices, new farming technologies, cheap fertilizers, the postwar population boom and growing infrastructure. However, the estuary, which was showing signs of decline as early as the 1890s, developed progressive signs of change over the next century, including salinization, loss of freshwater faunae, fish kills, and macroalgal blooms (Fig. 2). By the 1970s, the wider community was keenly aware of the poor health of the estuary, as extensive macroalgal growths carpeted Peel Inlet and later blooms of the toxic cyanobacterium *Nodularia spumigena* plagued the Harvey Estuary (McComb and Humphries, 1992). This sparked a shift in attitude toward recognizing the value of a healthy estuary, and agitated development of new management and community groups (Bradby, 1997), coordinated estuarine research (Hodgkin et al., 1981), and politicization of the issues. In concert with State elections in the early to mid-1980s, rising public frustration at feeling left out of decision-making and the slow response in "fixing" the problems, large-scale engineering projects were conceived. That which gained most momentum was the Dawesville

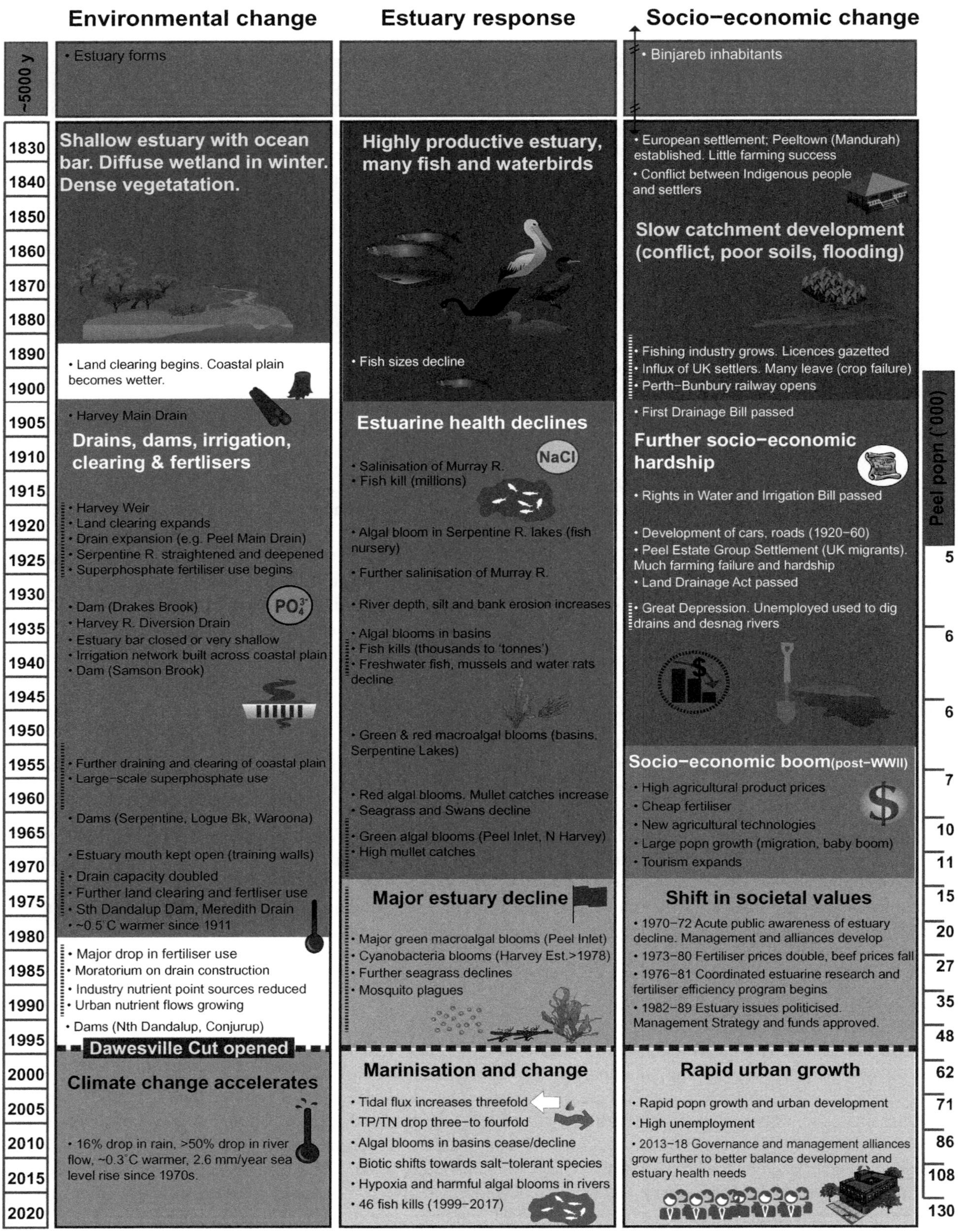

Environmental change

~5000 y
- Estuary forms

1830–1900
Shallow estuary with ocean bar. Diffuse wetland in winter. Dense vegetatation.

- Land clearing begins. Coastal plain becomes wetter.

1905
- Harvey Main Drain

Drains, dams, irrigation, clearing & fertlisers

- Harvey Weir
- Land clearing expands
- Drain expansion (e.g. Peel Main Drain)
- Serpentine R. straightened and deepened
- Superphosphate fertiliser use begins

PO_4^{3-}

- Dam (Drakes Brook)
- Harvey R. Diversion Drain
- Estuary bar closed or very shallow
- Irrigation network built across coastal plain
- Dam (Samson Brook)

- Further draining and clearing of coastal plain
- Large-scale superphosphate use

- Dams (Serpentine, Logue Bk, Waroona)

- Estuary mouth kept open (training walls)
- Drain capacity doubled
- Further land clearing and fertiliser use
- Sth Dandalup Dam, Meredith Drain
- ~0.5°C warmer since 1911

- Major drop in fertiliser use
- Moratorium on drain construction
- Industry nutrient point sources reduced
- Urban nutrient flows growing
- Dams (Nth Dandalup, Conjurup)

Dawesville Cut opened

Climate change accelerates

- 16% drop in rain, >50% drop in river flow, ~0.3°C warmer, 2.6 mm/year sea level rise since 1970s.

Estuary response

Highly productive estuary, many fish and waterbirds

- Fish sizes decline

Estuarine health declines

- Salinisation of Murray R. **NaCl**
- Fish kill (millions)

- Algal bloom in Serpentine R. lakes (fish nursery)
- Further salinisation of Murray R.

- River depth, silt and bank erosion increases

- Algal blooms in basins
- Fish kills (thousands to 'tonnes')
- Freshwater fish, mussels and water rats decline

- Green & red macroalgal blooms (basins, Serpentine Lakes)

- Red algal blooms. Mullet catches increase
- Seagrass and Swans decline
- Green algal blooms (Peel Inlet, N Harvey)
- High mullet catches

Major estuary decline 🚩

- Major green macroalgal blooms (Peel Inlet)
- Cyanobacteria blooms (Harvey Est.>1978)
- Further seagrass declines
- Mosquito plagues

Marinisation and change

- Tidal flux increases threefold
- TP/TN drop three-to fourfold
- Algal blooms in basins cease/decline
- Biotic shifts towards salt-tolerant species
- Hypoxia and harmful algal blooms in rivers
- 46 fish kills (1999–2017)

Socio-economic change

- Binjareb inhabitants

- European settlement; Peeltown (Mandurah) established. Little farming success
- Conflict between Indigenous people and settlers

Slow catchment development (conflict, poor soils, flooding)

- Fishing industry grows. Licences gazetted
- Influx of UK settlers. Many leave (crop failure)
- Perth–Bunbury railway opens

- First Drainage Bill passed

Further socio-economic hardship

- Rights in Water and Irrigation Bill passed

- Development of cars, roads (1920–60)
- Peel Estate Group Settlement (UK migrants). Much farming failure and hardship
- Land Drainage Act passed

- Great Depression. Unemployed used to dig drains and desnag rivers

Socio-economic boom (post-WWII)

- High agricultural product prices
- Cheap fertiliser
- New agricultural technologies
- Large popn growth (migration, baby boom)
- Tourism expands

Shift in societal values

- 1970–72 Acute public awareness of estuary decline. Management and alliances develop
- 1973–80 Fertiliser prices double, beef prices fall
- 1976–81 Coordinated estuarine research and fertiliser efficiency program begins
- 1982–89 Estuary issues politicised. Management Strategy and funds approved.

Rapid urban growth

- Rapid popn growth and urban development
- High unemployment
- 2013–18 Governance and management alliances grow further to better balance development and estuary health needs

Peel popn ('000)

Year	Popn
1925	5
1935	6
1945	6
1955	7
1960	10
1965	11
1975	15
1980	20
1985	27
1990	35
1995	48
2000	62
2005	71
2010	86
2015	108
2020	130

FIG. 2 Summaries of key environmental, estuary response and socioeconomic changes in the Peel-Harvey system from the early 1800s to present. Note that these timelines are nonexhaustive. Information sources include Hodgkin et al. (1981), Bradby (1997), McComb and Humphries (1992), Brearley (2005), Cuthbert et al. (2007), Silberstein et al. (2012), Australian Bureau of Statistics, 2018 and DWER unpubl. data, with further details given in the text.

Cut, a 2.5 km long, 200 m wide and 4.5–6.5 m deep channel connecting the northern Harvey Estuary with the sea to increase tidal flushing. After several years of stalling over feasibility, costs, funding and leadership, construction commenced in 1992 and the Cut opened in 1994 (Bradby, 1997; Brearley, 2005).

As well as the environmental, ecological and societal shifts brought about by the Cut (Section 4), two other key changes also occurred across this system from the mid-1970s/early 1980s. The first was the reduction in rainfall and rising temperatures across southwestern Australia, which underwent a step change around this time and have exacerbated since (Silberstein et al., 2012; Fig. 2; Section 4). The second was the rapidly growing Peel population, particularly in the urban center of Mandurah (Fig. 2; Australian Bureau of Statistics, 2018), and the new challenges this would, and will, bring for the region (Sections 5 and 6).

4 ESTUARY RESPONSES OVER RECENT DECADES

4.1 Hydrology and Water Quality

Unlike typical "funnel-shaped" estuaries common in other regions of the world, water circulation in the microtidal Peel-Harvey Estuary is a product of the complex interaction of freshwater inputs from the three main rivers, tidal exchange through the two entrance channels and wind forces across the large water surface. Nearly 90% of the annual rainfall and 95% of the river flow occurs during the cooler months of May to October (Fig. 3C and D). Average monthly inflows in winter have historically been of the order of 70–80 GL/month prior to 2006 (approximately distributed as 45, 20 and 10 GL from the Murray, Harvey and Serpentine, respectively), which is about half the estuary volume (~150 GL). Combined with tidal exchange, these flows led to a relatively low water retention time in the estuary during winter, while the low summer flows led to a high retention time in the warmer months, which was also associated with higher nutrient retention and phytoplankton biomass (Fig. 3G).

Prior to the Dawesville Cut being created, horizontal salinity gradients typically occurred north-south along the length of the Harvey basin, and west-east across the Peel Inlet to the deltas of the Serpentine and Murray rivers. Due to the seasonal rainfall cycle, these gradients showed opposite patterns in summer and winter. During summer, marine salinities (~35) prevailed close to the natural channel and hypersaline conditions (up to 50) developed towards the southern Harvey Estuary and western Peel Inlet. In winter, however, fresh conditions occurred throughout the tidal rivers and near their mouths, while marine salinities prevailed close to the channel. The Peel Inlet was generally more saline than the Harvey Estuary in winter due to greater tidal influence, while this was reversed in summer due to the higher evaporation rates and more limited tidal exchange in the Harvey Estuary (Fig. 3F). A recently developed hydrodynamic model of the estuary (see Section 5) has revealed the complexity of the circulation within the lagoons and provided greater insights into the water retention time across the basins and rivers (Huang et al., 2017). When used to characterize pre-Cut conditions, this model showed that the average age of water in the system during winter was ~45 days, although water retention in the Harvey Estuary could be substantially higher due to poor mixing in its southern reaches. In summer, the average water age increased to ~68 days, with some areas in the tidal rivers exceeding 100 days.

The hydrology of the estuary was fundamentally changed in the mid-1990s by the Cut, which increased the mean daily tidal range from 17% to 48% of the ocean tidal range in the Peel Inlet, and from 15% to 55% in the Harvey Estuary. This was estimated to increase water exchange with the ocean by about 3 times (D.A. Lord and Associates, Water and Rivers Commission, 1998). Although marked salinity gradients still developed post-Cut, winter salinities in both basins increased noticeably (Fig. 3F), and summer hypersalinity became more prevalent along the inland margins. While retention time decreased substantially (to 22 days in winter and 50 days in summer), periods of stratification increased in the main lagoons, and the propagation of the salt wedge into the tidal rivers became more significant (Huang et al., 2017).

In addition to the hydrological changes brought about by the Dawesville Cut, southwestern Australia has been experiencing a warming and drying climate over the last century, and particularly in recent decades (Petrone et al., 2010; Silberstein et al., 2012; Smith and Power, 2014). From the 1970s, rainfall has decreased by 16% and stream flows have declined by more than 50%, a trend which has appeared to accelerate since the 2000s (Silberstein et al. 2012). Climate data from the Peel region show that since the 1970s, air temperature has increased by 0.3°C (not shown) and average annual precipitation has declined from 871 to 670 mm (Fig. 3A). These reductions in rainfall have led to amplified reductions in runoff, given the low runoff:rainfall ratio in the surrounding catchments. Flow through the largest river subcatchment (Murray) has declined by nearly 50% (Fig. 3B), and the seasonal peak of river flow entering the estuary is now lower and arriving later (Fig. 3D). Flow reductions in the Serpentine and Harvey rivers have been occurring at an even faster rate than in the Murray (Ruibal-Conti, 2014). In addition, sea level rise in the broader region has amplified considerably since the 1970s. The past 100-year record at the nearby Fremantle tidal gauge station (~50 km north of the Peel-Harvey) has shown a mean

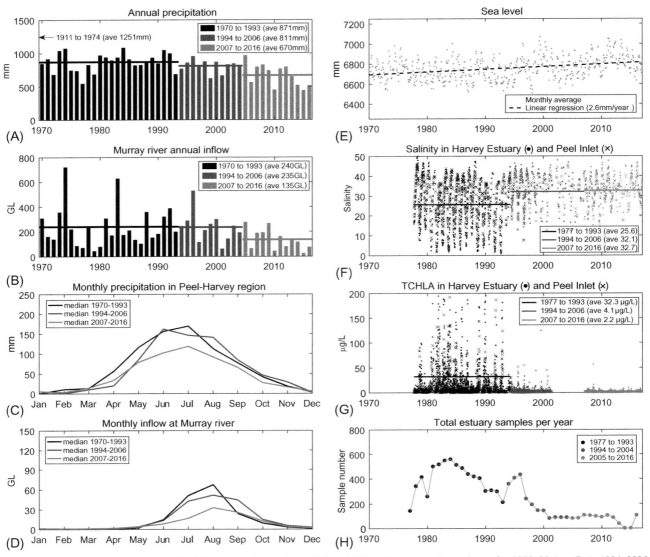

FIG. 3 (A) Annual precipitation in the Peel-Harvey catchment from 1970 to 2016, with averages (ave) shown for 1970–93 (pre-Cut), 1994–2006 (post-Cut) and 2007–16 (recent) periods. *Arrow* indicates the average annual precipitation in southwestern Australia throughout the earlier part of the century (1911–74; Silberstein et al., 2012); (B) Annual inflow into the Murray River from 1970 to 2016, with period averages as per (A); (C) Median monthly precipitation in the Peel region for the same periods as in (A); (D) Median monthly inflow into the Murray River for the same periods as in (A); (E) Average monthly sea level at the Fremantle gauge station, 1970–2016; (F) Salinity at the centers of Harvey Estuary and Peel Inlet from 1977 to 2016, with period averages shown for 1977–93 ("pre-Cut") and the remainder as per (A); (G) Total chlorophyll-a (TCHLA) at the centers of Harvey Estuary and Peel Inlet in the same periods as in (F); and (H) Total number of water-quality samples taken from the Peel-Harvey Estuary in 1977–93 (pre-Cut), 1994–2004 (post-Cut) and 2005–16 (recent).

rising rate of 1.5 mm/year (Kuhn et al., 2011), but if estimated from 1970 onward, is a far greater 2.6 mm/year (Fig. 3E). It is noteworthy, however, that while reduced river inflows in tandem with increasing sea level have certainly increased salt intrusion into the estuary over recent decades, these climate-related impacts have been less substantial than those brought about by the Cut (Fig. 3F).

Given the significant changes in the hydrology of the system, there has been an equally fundamental shift in water quality from the 1970s to present. Water quality throughout the estuary (including physical parameters, nutrient concentrations and/ or phytoplankton communities) and rivers (physical parameters, flow and nutrient loads) has been monitored since 1977 by the Department of Water and Environmental Regulation (DWER) and its preceding agencies, as well as other organizations (e.g., Hale and Paling 1999b). These water quality programs have varied in sampling intensity over this period (Fig. 3H), with the most intense programs operating in the 1980s when both micro- and macro-algal blooms were commonplace.

Over the monitoring period, the estuary concentrations of total phosphorus (TP) and total nitrogen (TN) have shown a strong correlation with the amount of nutrient input from the rivers. However, the sensitivity of this relationship has changed in response to the Cut, and more recently due to reduced river inflows linked with climate change (Fig. 4). Prior to the Cut, mean nutrient concentrations were very high, with annual averages exceeding 0.15 mg/L for TP and 2.0 mg/L for TN (with some samples exceeding 0.5 and 10.0 mg/L, respectively), coinciding with high chlorophyll-a concentrations of >200 μg/L in late spring and early summer (Fig. 3G). The system at this time tended toward P limitation (Ruibal-Conti, 2014), although blooms of the N-fixing cyanobacteria *N. spumigena* became a frequent occurrence from the late 1970s (McComb and Humphries, 1992). Despite a concerted effort by managers, farmers, industry groups and researchers to reduce nutrient flows to the estuary from the 1980s, as well as economic drivers such as high fertilizer prices from the 1970s (Bradby 1997; Fig. 2), nutrient concentrations remained high at > 0.10 mg/L for TP and > 1.2 mg/L for TN.

Nutrient concentrations across the estuary fell rapidly following the opening of the Cut in 1994, with annual averages of ~0.05 mg/L for TP and 0.5 mg/L for TN. Chlorophyll-a levels decreased to generally <10 μg/L in the basins, assisted greatly by the increased salinity which impeded *N. spumigena* growth, given the salinity tolerance of this species is <20 (Zhu et al., 2016). Some blooms of this cyanobacterium continued to occur in the Serpentine River for a short period after the opening of the Cut (D.A. Lord and Associates, Water and Rivers Commission, 1998), but in general chlorophyll-a levels have continued to decline until present (Fig. 3G). Total phosphorus concentrations have continued to decrease, with the lowest values recorded from 2005 to 2016, whereas TN had a much larger initial reduction post-Cut, but has increased slightly since 2005 (Fig. 4). Salinity stratification has increased on the western side of the basins, although oxygen depletion in their bottom waters has been less severe (Ruibal-Conti, 2014). However, hypoxia in the tidal rivers has been increasing in more recent years, impacted by the greater intrusion and persistence of the salt wedge.

4.2 Sediment Condition

The condition and biogeochemical functioning of sediments in the Peel-Harvey have been of long-standing concern, given the widespread catchment clearance and cultural eutrophication that has impacted this system since 1900 (Fig. 2). The first intensive investigation of Peel-Harvey sediments, undertaken in the late 1970s, indicated that phosphorus concentrations in the surface layers were >20 times those in the water column, and that the sediment may exceed river flow as the major source of dissolved phosphorus and nitrogen (Gabrielson, 1981; McComb et al., 1998). Further surveys of the surficial sediments throughout the basins in 1978–89 revealed that concentrations of organic matter, TP and TN were consistently higher in the finer-grained sediments of the central Harvey Estuary than in the Peel Inlet (Gabrielson and Lukatelich, 1985; McComb et al., 1998).

Increased estuary flushing following the Cut was widely anticipated to gradually reduce sediment nutrient concentrations and immobilize phosphorus into biologically unavailable forms (Peel Inlet Management Authority, 1994). Yet, a subsequent study in 1998 found little change in sediment nutrient or organic matter concentrations from pre- to post-Cut periods, although there was some evidence to suggest a decline in the proportion of biologically available phosphorus (Hale and Paling, 1999a). The hydrological changes brought about by this engineering intervention appear, however, to have

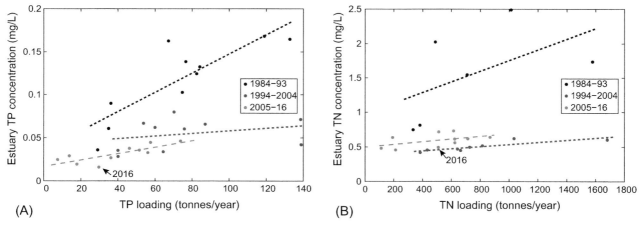

FIG. 4 (A) Total phosphorus (TP) loading from all rivers vs average TP concentration in the estuary (based on all available station data) in the pre-Cut period (1984–93), post-Cut period (1994–2004) and recent years (2005–16); (B) As for (A), except for total nitrogen (TN). Note that only those years where adequate river and estuary nutrient concentration data were available were included in the analysis.

altered the spatial distribution of sediment nutrients and organic matter throughout the basins. Nutrient concentrations near the Cut have decreased since its opening, while the organic and nutrient content of sediments at more distant sites, such as near the Harvey River mouth, have increased (Hale and Paling, 1999a).

Over the last decade or so, concern has been growing over the extent and potential impact of fine sulfidic sediments in the estuary, sometimes called monosulfidic black oozes (MBOs; Smith and Melville, 2004). These fluid black muds are prominent in deeper, low-energy environments such as navigation channels (Kraal et al., 2013a), where finer-grained silts, clays and organic matter settle out and become enriched under anoxic conditions. The risks associated with MBOs include their potential to act as sources of nutrients (e.g., dissolved inorganic phosphorus and ammonium), metals and hydrogen sulfide to the water column (Kraal et al., 2013a, b). The resuspension of MBOs by dredging, which is commonplace in the Peel-Harvey to maintain boating safety, can also lead to deoxygenation and acidification of the water column (Sullivan et al., 2002). However, relatively recent work around a large dredging event in the system suggested that such impacts may be spatially and temporally limited (Morgan et al., 2012).

4.3 Submerged and Fringing Vegetation

The large growths of macroalgae that proliferated in the estuary from around the 1960s (Fig. 2) were a clear biotic reflection of the hypereutrophication of the system. These algal growths were initially dominated by *Willeella brachyclados* (previously *Cladophora montagneana*; Wynne, 2016), particularly in the deeper waters of Peel Inlet and the northern Harvey Estuary, where it formed beds that later accumulated along the shorelines (McComb et al., 1981). A pronounced ecological shift occurred in 1979, after which *Chaetomorpha linum* and to a lesser extent *Enteromorpha/Ulva* spp. became dominant (Lavery et al., 1991), reaching a mean biomass of 10,000–20,000 tonnes dry weight in 1985–91 (Wilson et al., 1999). In contrast, the mean biomass of seagrass (mainly *Ruppia megacarpa* and *Halophila ovalis*) over the same period was only ~2000 tonnes (Wilson et al., 1999), which had been contracting from the late 1950s, most likely due to algal smothering (Hodgkin et al., 1981; Bradby, 1997).

The conditions created by the Cut decreased the biomass of macroalgae in Peel Inlet (Wilson et al., 1999; Pedretti et al., 2011), while that of both macroalgae and seagrass in the Harvey Estuary increased by 90% from 1996 to 2009 (Pedretti et al., 2011), reflecting mainly *Ulva* spp. growths near the Harvey River mouth, *Zostera* spp. and *Caulerpa* spp. in the northern Harvey Estuary and *R. megacarpa* throughout both basins. Macrophyte diversity has also increased since the Cut, probably reflecting reduced nutrient concentrations, increased salinities and/or greater light availability. Macroalgal biomass continues to be greatest near the river mouths, revealing the ongoing influence of elevated nutrients in those areas of the estuary (Pedretti et al., 2011).

The Cut also had notable impacts on the riparian vegetation fringing the estuary. Greater inundation of low lying areas has led to an inland extension of salt marsh (*Sarcocornia quinqueflora* and *Suaeda australis*), and elevated salinities have also caused the death of those species with greater freshwater requirements, including the rush *Juncus kraussii* and swamp paperbark *Melaleuca rhaphiophylla* (Calvert, 2002).

4.4 Benthic Macroinvertebrates

The estuary supports diverse benthic macroinvertebrate assemblages, ranging from commercially and recreationally important crab and prawn species, to infaunal communities. In the 1930s, the then Department of Fisheries and Wildlife noted enormous quantities of prawns and Blue swimmer crabs (*Portunus armatus*) in the estuary, the latter being so abundant that licensed fishers moved to have them declared vermin in 1950 (Bradby, 1997). *Portunus armatus* is a relatively short-lived and fast-growing species whose juveniles recruit from marine waters into the estuary over summer, with the ovigerous females then leaving the system in autumn/winter of the following year to spawn (Potter et al., 1983a; de Lestang et al., 2003). The increased salinities, tidal flushing and ocean-estuary access brought about by the Cut have notably influenced the biology of this species in the Peel-Harvey. In the years following the opening of the Cut (1995–98), crabs increased in density and became more evenly distributed throughout the estuary, ovigerous females left earlier, their 0+ recruits entered the system earlier, and crabs grew faster in their first year of life compared to the early 1980s (Potter et al., 1983a; de Lestang et al., 2003). In more recent years (2007–11), crab densities have generally increased further, although have shown notable interannual variability that may be related to the timing of winter river flows (Johnston et al., 2014).

Abundant prawn species in the estuary include the Western king prawn (*Penaeus latisulcatus*), which also spawns in the marine environment and uses the estuary as a nursery (Potter et al., 1991), and the Western school prawn (*Metapenaeus dalli*), which completes its life cycle in the estuary and spawns in its upper reaches (Potter et al., 1989). The abundances of these two species have shown opposing longer-term trends, with the formerly dominant school prawn declining from the

1980s and the king prawn becoming more abundant (Bradby, 1997). These trends probably reflect the influence of increasing salinities brought about by the Cut, as well as declining freshwater flows since the 1970s and the consequent effects on the estuary sediments.

Infaunal communities in the estuary basins have also changed notably over the last few decades, including marked pre-to post-Cut shifts. Mollusc species which were highly dominant in the late 1970s (*Arthritica semen* and *Hydrococcus brazieri*) were markedly reduced or absent in the mid-1980s, potentially due to the combined effects of macro- and micro-algal blooms at that time. Further sampling in 2000 showed that densities of both of these species had increased, but to far lower levels than recorded in the 1970s (Whisson et al., 2004). Wildsmith et al. (2009) also found marked shifts across the whole infaunal assemblage from the mid-1980s to mid-2000s, including declines in mean density and diversity (taxonomic distinctness), as well as major compositional shifts that reflected losses of more environmentally sensitive crustacean and molluscan taxa and increases in more tolerant annelid taxa. These findings were attributed to a general deterioration in the condition of the benthic environment (Wildsmith et al., 2009).

4.5 Fish

The structure of the fish communities in the Peel-Harvey Estuary has been studied extensively since the late 1970s, particularly in the shallows of the basins (e.g., Potter et al., 1983b, 2016; Loneragan et al., 1986, 1987; Young and Potter, 2003a, b). Most recently, Potter et al. (2016) compared the fish faunae in these waters during 1980–81, when massive macroalgal growths dominated; 1996–97, following the opening of the Cut; and 2008–10, when accelerated climate change influences were also apparent. The increases in tidal exchange and salinities and decreases in macroalgae following the Cut were accompanied by increases in fish species with marine affinities and declines in those associated with macrophytes. Fish faunal composition also showed less spatial differentiation throughout the basins as environmental conditions became more homogeneous, but increased temporal variability. Marked seasonal cyclicity became established (Young and Potter, 2003a), reflecting both the recruitment patterns of marine species and their responses to the more pronounced changes in basin salinities following winter freshwater flows. Ongoing declines in river flows, combined with increased tidal exchange, have since further elevated marine influences on the estuary, with the numbers of fish species in the estuary basins, and particularly of marine stragglers, increasing progressively from the early 1980s to late 2000s (Potter et al., 2016). Interestingly, macrophyte-associated species have also increased in prevalence in the latest period, most likely reflecting the increased macroalgae in the Harvey Estuary and the re-establishment of seagrass communities close to the Cut.

In contrast to the shallows of the estuary basins, there is a pronounced lack of understanding of the fish communities in the tidal rivers of the Peel-Harvey and in the deeper waters throughout the estuary, neither of which have been studied for >30 years. A current research project is helping to address this knowledge gap, as well as proposing a robust fish monitoring regime and fish-based index of ecosystem health for the estuary (see Section 5).

Fish kills have occurred regularly in the system over recent decades, with 46 verified events since 1999 (DWER, unpubl. data). Major events of >10,000 fish mortalities occurred in 2003, 2004, 2005 and 2017, and those exceeding 1000 fish occurred on six other occasions. Most of these fish kills have occurred in the tidal rivers, with 21 in the Murray and 17 in the Serpentine. While it is often difficult to confidently attribute the cause of such events, 17 were likely to be related to hypoxia associated with phytoplankton blooms, 5 were linked with hypoxia following unseasonal rainfall, while a further 5 were attributed to other causes such as stranding, disease and bycatch.

5 CURRENT SOCIO-ECOLOGICAL CHARACTERISTICS

5.1 Catchment Land Use

Just over half (53%) of the land that drains to the Peel-Harvey Estuary has been cleared for agriculture, with a further 2% developed for urban or industrial uses (Fig. 1A). Developed land on the Swan Coastal Plain (Fig. 1B) has greater potential to adversely impact the water quality in the estuary than that on the Darling Scarp, given that its land uses are generally more nutrient intensive and it experiences higher rainfall. As soil type varies across these catchments, nutrient retention and runoff yield also differs (Kelsey et al., 2011).

Current modeling of the catchment to estimate patterns, sources and loads of nutrients exported to the estuary shows that beef cattle grazing, which covers 42% of the landuse on the coastal plain, is the largest contributor to nitrogen and phosphorus loads (Kelsey et al., 2011; DWER, unpubl. data; Fig. 1B). However, far more intensive animal uses such as free-range poultry, piggeries and feedlots, which comprise <1% of the coastal catchment area, can also be considerable sources of nutrient export. Nutrient application in these localities is more than 8 times higher in nitrogen and 4 times higher in phosphorus than in beef grazing areas. Furthermore, horticulture, which also comprises ~1% of the catchment, is the most phosphorus-intensive land use across the system, with application rates that are ~10 times higher than those for beef farming.

Comparison of land use maps from 2005/2006 (Kelsey et al., 2011) and 2015 (DWER, unpubl. data) show that over the past decade, the area of dairy farming has reduced and that for urban development and horticulture has increased. While these changes are relatively small, they may have significantly increased nutrient export to the estuary given their nutrient use and runoff characteristics.

5.2   Estuary Condition

Based on routine water quality monitoring at 12 sites in the estuary basins and tidal portions of the Murray and Serpentine Rivers in 2016–17 (http://wir.dwer.wa.gov.au), the condition of the basins was considered satisfactory, whereas the rivers had distinct water quality problems. Peel Inlet and Harvey Estuary typically experienced marine salinities, only a few significant salinity stratification events and nutrient and chlorophyll-a concentrations generally below the Australian and New Zealand Environment and Conservation Council (ANZECC) guidelines. However, stratification and low dissolved oxygen conditions occurred frequently in the bottom waters of the lower Murray River (<2 mg/L in ~45% of samples), and chlorophyll-a concentrations in both the Murray and Serpentine during autumn were more than 4 and 9 times the ANZECC guidelines, respectively (12.6 and 28.1 μg/L). Harmful algae were present throughout the estuary, but were the most speciose and abundant in the Serpentine River (18 species, predominantly cyanophytes). The ichthyotoxic dinophyte *Karlodinium* sp. was also linked with three of four fish kill events that occurred in the Murray River in 2016–17, one of which led to the death of >10,000 fish. Total nitrogen concentrations in the surface waters of the Serpentine (1.69 mg/L) and Murray rivers (0.97 mg/L) were, respectively, >3.5 and >2 times greater than in the basins. Total phosphorus concentrations were also far higher in the Serpentine (0.10 and 0.077 mg/L in surface and bottom waters, respectively) than in all other zones of the Peel-Harvey, and were also higher than those in five other southwestern Australian estuaries monitored in 2016–17 (DWER, unpubl. data).

Regular quantitative monitoring and reporting on the condition of key ecological aspects of the estuary, including macrophytes and fauna such as invertebrates, fish and birds, is a major knowledge gap for the Peel-Harvey and many other estuaries throughout southwestern Australia (Hallett et al., 2016a, b; Thomson et al., 2017). However, a current collaborative research project (see Section 5.4) is helping to address some of these gaps for the Peel-Harvey through developing (i) robust and easily interpretable indices of estuarine health based on benthic invertebrate and fish faunae, and (ii) proposed monitoring regimes for these fauna and the macrophyte assemblages. Potential environmental impacts of regular dredging operations throughout the estuary are also an ongoing concern, and while there has been some research in this area (e.g., Morgan et al., 2012), many questions regarding longer-term effects remain unanswered.

5.3   Socioeconomic System

The Peel region has a population of ~130,340, comprising 5.3% of the WA population (2016 Australian census), and has grown by 121% since the previous census in 2011 (Australian Bureau of Statistics, 2018). The urban city of Mandurah is among the fastest growing regional centers State wide and has increased by ~145% from 2006 to 2016 (Australian Bureau of Statistics, 2018). Peel residents identify happiness, health, and wellbeing, training/education options and lifestyle as their primary reasons for living in the region, and rank among the highest for these attitudinal descriptors throughout regional WA (Government of Western Australia, 2013).

Based on a 2012–13 assessment (Peel Development Commission, 2018), by far the main industry supporting the Peel economy in terms of production value is mining and mineral processing ($3.2 billion). Peel is the third largest mineral producing region in WA (alumina, gold and copper) and has the second largest bauxite mine globally and one of Australia's largest producing gold mines. Retail ($1.1 billion), building construction, tourism and agriculture ($419–125 million) also have relatively high production values in the region. The largest employers over a comparable period were the construction (mainly heavy and civil engineering), manufacturing (mainly metal and metal products) and retail industries (13.7%–11.1%), followed by health care and mining (7.8%–9%; 2011 Australian census data; Australian Bureau of Statistics, 2018; Peel Development Commission, 2018). These industries were also key employers during the most recent (2016) census, but with a general increase in people working in mining and a decrease in those in manufacturing (Australian Bureau of Statistics, 2018). The median unemployment rate in the Peel is higher than that for the State (8.7% vs 7.8%), with the highest rates recorded in Mandurah (11%; Australian Bureau of Statistics, 2018). Confidence regarding employment prospects is also lower in the Peel than any other regional area State wide (Government of Western Australia, 2017).

Industries more directly supported by the estuary itself include fishing, tourism, parts of the manufacturing and retail industries associated with boating, and real estate. There are commercial and recreational fisheries for Blue swimmer crabs (*P. armatus*) and various finfish species, mainly Sea mullet *Mugil cephalus*, Yellowfin whiting *Sillago schomburgkii*, Tailor *Pomatomus saltatrix*, estuarine cobbler *Cnidoglanis macrocephalus*, and Perth herring *Nematalosa vlaminghi*. Crabs are iconic in the system and are the highest recreationally caught species in the Peel-Harvey and other estuarine and coastal waters in southwestern Australia (Johnston et al., 2014, 2017). Recreational crab catches in the estuary are often far higher (i.e., up to 85% higher) than the commercial catch, the latter of which has ranged from ~0 to 105 tonnes over the last two decades (Johnston et al., 2014, 2017). The Peel-Harvey supports the majority of commercial landings of several of the above finfish species across the whole of Australia's West Coast Bioregion (Smith and Holtz, 2017). Yet, >75% of the value of Peel's commercial fishing industry ($5.7 million in 2011–12; Peel Development Commission, 2018) reflects catches of the highly valuable Western rock lobster *Panulirus cygnus* in marine waters outside the estuary. There is a poor understanding of the economic value of the recreational fishery in the Peel-Harvey, and indeed in most estuaries across the southwest. Moreover, disentangling which aspects of the tourism, manufacturing, retail, and real-estate industries in the Peel are due to the estuary itself is difficult from higher-level census assessments. However, it is relevant that tourist activities related to estuarine wildlife (e.g., birds, dolphins and fish) are highly popular, and that the Peel has one of the highest rates of boat ownership in WA (~10% of the population in 2009, and predicted to rise to ~12% by 2031; Department of Transport, 2011). Moreover, real-estate prices near the estuary are notably higher and are increasing at a faster rate than those for land further away (Economics Consulting Services, 2010).

5.4 Management and Science

Management and governance strategies and the information needed to support them are evolving for the Peel-Harvey system. More recent initiatives are taking further steps toward holistic understanding of catchment-estuarine function and response, as well as building institutional relationships and frameworks that recognize the intrinsic links between the societal and ecological health of this ecosystem. However, there are inherent challenges in managing for sustainability and resilience due to the poor fit of different operational scales of management institutions and ecosystems (Folke et al., 2007; Benson and Garmestani, 2011) and the inherently multijurisdictional nature of estuarine management from national to local levels.

For much of its past, water-quality management in the Peel-Harvey Estuary has focused around a "control and command" model, with phosphorus reduction targets being the primary instrument for improving estuary health. A target of <75 tonnes/pa of total phosphorus entering the estuary was gazetted in 1992, instigated by an overarching *Management Strategy for the Peel Inlet and Harvey Estuary System* (Kinhill Engineers, Western Australian Department of Agriculture, Western Australian Department of Marine and Harbours, 1988) that set an environmental regulation framework with legally binding conditions under which the Dawesville Cut could proceed (Bradby, 1997). As well as nutrient reduction targets for the estuary, a catchment management plan to reduce nutrient export was also among those conditions. Unfortunately, a lack of political and institutional will and effective implementation mechanisms prevented these objectives being met following the opening of the Cut in the mid-1990s. In 2003, the WA Environmental Protection Authority highlighted that a catchment management plan was still needed to reduce the flow of nutrients into the estuary given its fragile condition (Environmental Protection Authority, 2003). This need was again highlighted in a Water Quality Improvement Plan for the Peel-Harvey which was released in 2008 (Environmental Protection Authority, 2008), and was primarily aimed at phosphorus reduction targets.

While reducing nutrient delivery is still a key element of current water quality and catchment management approaches for the Peel-Harvey, the effectiveness of an ongoing target-based approach has been questioned given its limited ability to account for the complex dynamics of estuarine systems. More recent strategies are further growing cross-agency collaborations and socio-ecological knowledge to better support a sustainable future for the estuary. Importantly, a Peel-Harvey Estuary Management Committee, comprising leaders from key environmental, industry-based, land planning and other agencies, was established in 2014 to coordinate governance and oversee management and scientific efforts in the system. One major State-Commonwealth initiative overseen by this committee is the *Strategic Assessment of the Perth and Peel Regions* (released in 2015 as the *Draft Perth and Peel Green Growth Plan for 3.5 million*; Government of Western Australia, 2015), which aims to set regional land planning and development directions to accommodate a projected 70% population increase by 2050. It takes an integrative approach to balancing protection of environmentally significant sites with land development needs, through streamlining environmental approvals in areas where development can be considered, clarifying the environmental offset obligations, and implementing a conservation program for environmentally valuable areas (including the Peel-Harvey Estuary). A further key management program is the *Regional Estuaries Initiative*

(REI; 2016–20), led by DWER, which aims to build regional capability, promote whole-of-industry engagement, and strengthen resource and knowledge networks across six "at-risk" estuaries in southwestern Australia. These estuaries include the Peel-Harvey, where this program is also implementing drainage intervention projects and trialling innovative nutrient-binding treatments. Lastly, a new *Estuary Protection Plan* led by DWER is also being developed for the Peel-Harvey. This plan will again focus on water-quality improvement, but will be supported, for the first time, by a coupled catchment-estuary response model being developed in a current research project (see below).

Although a full account of current ecosystem management approaches for the Peel-Harvey is beyond the scope of this chapter, one other notable development is the 2016 certification by the Marine Stewardship Council of the sustainability of the Blue swimmer crab and Sea mullet fisheries (Johnston et al. 2015). Attaining this global standard reflects stock sustainability, minimal environmental impact from fishing operations, and sound fishery management. The Peel-Harvey crab fishery is also the first in the world to achieve dual certification for a combined recreational and commercial fishery.

A recent prioritization of current science needs for southwestern Australian estuaries, representing a consensus view among managers and scientists (Thomson et al., 2017), showed that an overarching need for the Peel-Harvey and other systems was a decision support tool for optimizing trade-offs between catchment development aspirations and estuarine health. One current research project contributing to this need is an Australian Research Council Linkage Project led by Murdoch University and partnering with eight other research, management and community agencies (LP150100451; 2016–19). It aims to identify Peel catchment land use strategies that best balance regional socioeconomic goals with minimal impacts on estuarine health, and combines four main components: (1) a coupled catchment-estuary response model, supported by detailed nutrient "source-to-fate" tracing; (2) estuarine ecological health indices; (3) estuarine ecosystem services indices; and (4) socioeconomic health assessments. The overarching predictive platform that will be produced will enable exploration of socioeconomic and estuarine health trade-offs under future land planning and climate scenarios.

6 LOOKING FORWARD

Toward a future horizon of 2050, two major changes are anticipated for the Peel-Harvey system, as well as for many other temperate urbanized estuaries globally. The first is a major expansion of its population, and the second is significant further drying and warming of the regional climate. Some anticipated projections for this estuarine system, as well as some potential adaptation strategies, are outlined below.

6.1 Population Growth

The population of the Peel region is expected to increase more than threefold by 2050 (Peel Development Commission, 2015). Much of this growth is anticipated to be in the urban sector, with necessary underpinning of industrial, commercial and infrastructure developments, and supply of basic raw materials. The key challenges include balancing this socioeconomic development with minimal impacts on environmentally significant areas, containing urban sprawl, and maintaining livability, housing affordability, and access to facilities. It is anticipated that the final plan of the *Strategic Assessment of the Perth and Peel Regions* (Government of Western Australia, 2015) will provide the overarching management and implementation strategies for meeting these challenges. Some proposed development approaches include balancing increases in suburban infill with those in greenfield developments, and using alternative building methods to increase housing density and diversity. Given that the current population density in the Perth-Peel region is low relative to other national and international cities (i.e., about a quarter of that in Sydney and less than a fifth of that in London; Western Australian Planning Commission, 2015), there is considerable scope to partly accommodate the anticipated population growth through higher density housing. One new tool for assisting with sustainable urban development decisions on the sandy Swan Coastal Plain is *Urban Nutrient Decision Outcomes* (UNDO). Designed for use by urban developers and managers, this web-based tool (http://kumina.water.wa.gov.au/undo) evaluates nutrient reduction approaches for urban developments, and has considerable application in helping to minimize downstream impacts on the health of the Peel-Harvey Estuary.

6.2 Climate and Hydrology Predictions

Further drying of the southwestern Australian climate has been consistently predicted by numerous assessments of global and regional scale models. Silberstein et al. (2012) projected a uniform decline in rainfall based on an ensemble of 15 global climate models (GCMs) under typical Intergovernmental Panel on Climate Change (IPCC) global warming scenarios. Smith and Power (2014) further reported median rainfall declines of about 25% and a dramatic 72% reduction in inflows by the end of the 21st century from 38 GCMs. More recently, a regionally downscaled model by Firth et al. (2017) predicted a

10–20 mm/month reduction in winter rainfall during 2030–59 compared to 1970–99, potentially extending the downward trend that has already been observed (Fig. 3C and D).

Sea level projections specific to the region suggest a rise of 20–84 cm above current levels before the end of this century (Kuhn et al., 2011), with mean predictions varying depending on the likely representative concentration pathway (RCP). For RCP 4.5 and RCP 8.5, the mean predictions are 46 and 61 cm, respectively (Hope et al., 2015). Given that the estuary basins have a mean depth of only 1 m, the projected sea level rise will play a substantial role in changing the hydrology of the system. The increasing potential for extreme sea level events to impact the estuary has also been identified, with estimates that a 50 cm rise in mean sea level will result in a 100- to 1000-fold increase in the frequency of such events (Braganza et al., 2014).

The combined effects of reduced river flow and rising sea level on the Peel-Harvey Estuary will clearly result in a greater intrusion of marine water into the system, as has already been observed over the last decade (Fig. 3F). Numerical modeling experiments for the system predict that, if climate trends continue as projected by Silberstein et al. (2012) and Kuhn et al. (2011), winter salinities will increase by ~9 by 2058 (Fig. 5A and B) and hypersaline conditions in summer will worsen (Fig. 5E; Huang et al., 2017). In addition, water residence time during winter will increase by about 25 days,

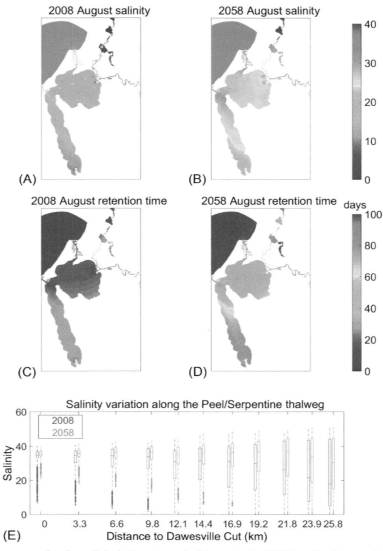

FIG. 5 (A) Modeled monthly averages of surface salinity in the wet month of August during 2008, based on the numerical model by Huang et al. (2017). Gray dots indicate the locations of the transect in (E); (B) As for (A), except in August 2058; (C) Modeled monthly averages of surface water retention time in August 2008; (D) As for (C), except in August 2058; and (E) Box-whisker plot of annual salinity variability along an estuary transect from the Cut to the middle Serpentine River, comparing the 2008 and 2058 model simulations. The 2058 scenario was run using the same settings as for 2008, but assuming a river flow decline of 66% and sea level rise of 0.2 m (refer to Huang et al., 2017 for details).

bringing it closer to pre-Cut conditions (Fig. 5C and D). Greater salt intrusion and evaporation caused by increasing air temperatures is expected to push the salt wedge further upstream and reduce oxygen solubility, thereby increasing the frequency and/or severity of hypoxic events in the bottom waters of the tidal rivers, and increasing the risk of algal blooms. Changes to the flow and salt distribution, plus increasing frequency of extreme sea level events, will also alter patterns of sedimentation and shoreline erosion. This is expected to have a host of ecological effects (see Section 6.3), as well as impacts on coastal infrastructure (e.g., harbors, canals, boat ramps etc.). These effects are increasingly being considered in future environmental management plans for the region, however, improved ongoing monitoring and decadal-scale forecasts are needed for adequate adaptation.

6.3 Ecological Responses

The ecology of the Peel-Harvey is expected to undergo further shifts as a result of the above climate and hydrological changes, which is likely to include both positive and negative responses. For example, more marine conditions are expected to encourage greater use of the estuary basins and tidal rivers by marine biota, while elevated water temperatures will likely accelerate growth rates of endothermic species and perhaps increase the prevalence of tropical species (Hallett et al., 2017). For some of the recreationally and commercially targeted marine fauna such as Blue swimmer crabs, such changes would also increase their vulnerability to fishing pressure, requiring new adaptive management approaches. While rising sea levels could expand the amount of available estuarine habitat for subtidal biota, the opposite is likely to be true for less salt-tolerant riparian vegetation and wading birds.

The most severe ecological consequences are likely to occur in the tidal rivers. A combination of increasing salt-wedge penetration, elevated water temperatures and a potentially greater influence of summer flow events (Hallett et al., 2017), will increase the frequency, persistence and/or severity of water column stratification, hypoxia and algal blooms. Such water quality conditions have often been linked with fish kill events in the Peel-Harvey over the last couple of decades, as well as having well-documented negative effects on the invertebrate and fish communities of other estuaries in southwestern Australia (Hallett et al., 2016c; Tweedley et al., 2016).

6.4 Strategies to Mitigate Impacts of a Drying Trend

The prospect of further declines in freshwater inflows, and the associated risks of higher salinities, water temperatures and retention times, necessitates new strategies to mitigate the various negative impacts on estuarine water quality and biota. Given that estuarine responses to climate change effects may surpass those due to ongoing eutrophication (and will likely act in concert), "adapting scientists" will be required to work with "adapting managers" (Hobday et al., 2015; Alderman and Hobday, 2017) to better deal with shifting environmental stressors. This will primarily require managers to "do more with less", namely achieve better or at least acceptable environmental outcomes despite less water delivery into the estuary.

Defining the necessary water requirements for the Peel-Harvey Estuary, analogous to ecological water requirements that are more routinely applied in river management (Davies et al., 2014), is a necessary first step to guide planning and decision-making. This will facilitate more strategic delivery of available water to the system to avoid crossing threshold conditions (e.g., for salinity or oxygen) that are known to be deleterious to key habitats or biota. One example of such an approach is that being implemented in the Coorong Estuary in South Australia, in which salinities are kept below the maximum tolerated by *Ruppia tuberosa* for successful flowering and reproduction, given the importance of this seagrass species to the food web of that system (Kim et al., 2015). In the Peel-Harvey context, meeting necessary water requirements may involve approaches to hold water in the catchment for longer, and allowing release over the warmer months to prevent prolonged extreme conditions. However, the use of environmental flows for estuaries is not current policy in Western Australia, and the current hydrologic drainage system is poorly suited for this function; complications such as stratification-induced hypoxia, for example, may present a negative risk. Improved infrastructure designed for this purpose would need to be considered, such as modifications to surface storage areas and/or enhanced infiltration initiatives to increase local areas of groundwater recharge and storage.

Other strategies for improving water-use efficiency in the catchment could also help mitigate the downstream effects of flow declines. Improving irrigation and crop/pasture water-use efficiency will reduce the demands on water abstractions from groundwater systems that contribute to summer base flow and groundwater-dependent environments. Closed-cycle water management approaches for intensive horticulture are also being considered, whereby water is reused at the plot/farm cycle to reduce wastage and demand for local water allocation, although the practicalities for achieving this are yet to be realized. Furthermore, given the increasing reliance on desalination for potable water in the Perth-Peel region, opportunities for "new" water inputs to the estuary via treated wastewater could help offset declines in inputs from rainfall. Elsewhere on the Swan Coastal Plain, recent managed aquifer recharge trials of treated wastewater have proved promising for replenishing groundwater, though further research and planning on the best approaches and locations is required.

Given the potentially long timescales needed to implement a rigorous environmental flow regime for the estuary, other complementary measures to help mitigate negative impacts of a drying climate could include local nutrient interventions (e.g., soil amendments, revegetation and water-sensitive urban design technologies), and in situ actions such as artificial oxygenation (e.g., Huang et al., 2018). Managers and scientists will need to be innovative and prepared to trial solutions with imperfect understanding, perhaps a more risky approach than has traditionally been supported. Tools such as coupled ecosystem-response models that can quantitatively elucidate the feedbacks between social and ecological processes and are able to integrate across the various challenges facing the system, will be highly valuable in supporting this process and engaging the wider community (Hipsey et al., 2015; Qiu et al., 2018).

7 CONCLUDING REMARKS

The Peel-Harvey shares many historical legacies and future challenges with other temperate Mediterranean estuaries whose catchments were vastly developed for agriculture and are now rapidly urbanizing, as well as experiencing warming and drying climates. In a global context, however, the Peel-Harvey system has several characteristics that set it apart as a case study. These include its natural predisposition to degradation, given its largely sandy catchment that readily leaches nutrients, wide shallow receiving basins and limited flushing potential. The latter not only reflects the microtidal conditions of the region, but also the highly restricted tidal flushing of the estuary prior to its natural channel being kept permanently open and construction of the Dawesville Cut, and now its diminishing river flushing. Several other microtidal estuaries in southern Australia share these characteristics to varying extents (Warwick et al., 2018). Unparalleled by many other systems, however, is the Peel-Harvey's dramatic trajectory of decline through hypereutrophication, reflected most visibly by massive algal blooms sustained over ~30 years, and the fundamental shifts in its hydrology following the remedial intervention of the Cut. Nearly 25 years on, the environmental, ecological and societal effects of that ecoengineering initiative are still being unraveled, with the added complexity of understanding the synergistic impacts of subsequent catchment development and climate change. Nevertheless, ecosystem transformation is hardly new for the Peel-Harvey, and its ongoing function and adaptation is in part a testament to its resilience.

Societal values and management approaches toward this system have also evolved over past decades, and will need to continue to adapt under anticipated future scenarios. Focus has sharpened on the pressing need to not only maintain the ecosystem health of the estuary, but also to accommodate the housing, infrastructure, food and water requirements of a burgeoning urban population. Understanding and balancing these socio-ecological tradeoffs is clearly also a challenge in other urbanizing catchment-estuarine systems worldwide. Various studies present a range of modeling approaches for exploring such tradeoffs and evaluating the power of different management strategies to sustain key ecosystem services under future development and climate scenarios (e.g., Cloern et al., 2011; Pinto et al., 2014; Shoyama and Yamagata, 2014; Paolisso et al., 2015; Qiu et al., 2018). Similar approaches are currently being developed for the Peel-Harvey, supported by collaborative efforts across the management, science and community sectors. Also key is understanding the societal barriers to engaging with management strategies and better incorporating human behavior into these approaches (e.g., Paolisso et al., 2015; Sheaves et al., 2016), which are also currently being explored for the Peel-Harvey and other southwestern Australian estuaries.

Baselines for the Peel-Harvey have clearly shifted with regard to its inherent functioning and key stressors, and the system today is thus very different from that of decades to a century ago. Focus needs to remain on sustaining a well-functioning estuary that is able to deliver key societal values and accepting further regime shifts, rather than attempting to replicate conditions of the past that match personal perceptions of "good" or "right". Recognition of socio-ecological tradeoffs must also be built into management approaches and societal expectations as we learn to do more with less. It is hoped that, given the demonstrated resilience and adaptation of the Peel-Harvey system, the relatively recent lessons from its past decline and expensive remediation, and the vast community investment in the health of this waterway, sufficient change can be invoked to sustain it well into the future.

ACKNOWLEDGMENTS

The authors gratefully acknowledge the many scientists, managers and community members who have contributed to the knowledge outlined in this chapter. They also thank Malcolm Robb (DWER) for his contributions to the manuscript and Brendan Busch (University of Western Australia) for his assistance with figures. They further acknowledge the funding support provided by the Australian Research Council Linkage Program (LP150100451).

REFERENCES

Alderman, R., Hobday, A.J., 2017. Developing a climate adaptation strategy for vulnerable seabirds based on prioritisation of intervention options. Deep Sea Res. Part II Top. Stud. Oceanogr. 140, 290–297.

Australian Bureau of Statistics, 2018. Census. Available from: http://www.abs.gov.au/websitedbs/D3310114.nsf/home/census?opendocument&ref=topBar (Accessed January 2018).

Benson, M.H., Garmestani, A.S., 2011. Can we manage for resilience? The integration of resilience thinking into natural resource management in the United States. Environ. Manag. 48, 392–399.

Bradby, K., 1997. Peel-Harvey: The Decline and Rescue of an Ecosystem. Greening the Catchment Taskforce (Inc.), Mandurah, WA, ISBN: 0730980413223.

Braganza, K., Hennessy, K., Alexander, L., Trewin, B., 2014. Changes in extreme weather. In: Christoff, P. (Ed.), Four Degrees of Global Warming: Australia in a Hot World. Routledge, London, pp. 33–59.

Brearley, A., 2005. Ernest Hodgkin's Swanland: Estuaries and Coastal Lagoons of South-Western Australia. University of Western Australia Press, Perth, p. 550.

Calvert, T., 2002. Assessment of foreshore vegetation changes in the Peel-Harvey Estuary since the opening of the Dawesville Channel: with focus on *Juncus kraussii, Melaleuca rhaphiophylla* and *M. cuticularis*. Unpublished Honours thesis, Murdoch University, Perth, WA, p. 174.

Cloern, J.E., Knowles, N., Brown, L.R., Cayan, D., Dettinger, M.D., Morgan, T.L., Schoellhamer, D.H., Stacey, M.T., van der Wegen, M., Wagner, R.W., Jassby, A.D., 2011. Projected evolution of California's San Francisco Bay-Delta-River system in a century of climate change. PLoS One 6, e24465. https://doi.org/10.1371/journal.pone.0024465.

Cuthbert, B., Cuthbert, D., Dortch, J., 2007. An indigenous heritage management plan for the eastern foreshores of the Peel and Harvey Inlets. Phase 5 of the Peel Cultural Landscape Project. Unpublished Technical Report. Dortch and Cuthbert, Fremantle, Western Australia, 48 pp.

D.A. Lord and Associates, Water and Rivers Commission, 1998. Dawesville Channel Monitoring Programme. Technical Review Report. Water and Rivers Commission. Perth, Western Australia. ISBN 0730974391, 201 pp.

Davies, P.M., Naiman, R.J., Warfe, D.M., Pettit, N.E., Arthington, A.H., Bunn, S.E., 2014. Flow–ecology relationships: closing the loop on effective environmental flows. Mar. Freshw. Res. 65, 133–141.

de Lestang, S., Hall, N., Potter, I.C., 2003. Influence of a deep artificial entrance channel on the biological characteristics of the blue swimmer crab *Portunus pelagicus* in a large microtidal estuary. J. Exp. Mar. Biol. Ecol. 295, 41–61.

Department of Transport, 2011. Peel region recreational boating facilities study 2010. Technical Report 449. Unpublished Technical Report. WA Department of Transport, Perth, Western Australia, 80 pp.

Economics Consulting Services, 2010. Peel estuary and land values near waterways; an economic valuation. Unpublished Technical Report. Economics Consulting Services, Perth, Western Australia, 13 pp.

Environmental Protection Authority, 2003. Peel Inlet and Harvey Estuary System Management Strategy: Progress and Compliance by the Proponents with the Environmental Conditions Set by the Minister for the Environment in 1989, 1991 and 1993 Advice of the Environmental Protection Authority. Environmental Protection Authority, Perth, WA, p. 353. ISBN 07307 6725 6.

Environmental Protection Authority, 2008. Water Quality Improvement Plan for the Rivers and Estuary of the Peel-Harvey System—Phosphorus Management. Environmental Protection Authority, Perth, WA, ISBN: 0-7309-7293-3, p. 74.

Firth, R., Kala, J., Lyons, T.J., Andrys, J., 2017. An analysis of regional climate simulations for Western Australia's wine regions—model evaluation and future climate projections. J. Appl. Meteorol. Climatol. 56, 2113–2138.

Folke, C., Pritchard, L., Berkes, F., Colding, J., Svedin, U., 2007. The problem of fit between ecosystems and institutions: ten years later. Ecol. Soc. 12, 30.

Gabrielson, J.O., 1981. The sediment contribution to nutrient cycling in the Peel-Harvey estuarine system: a technical report to Department of Conservation and Environment, EPA Bulletin No. 96. Unpublished Technical Report. Environmental Protection Authority, Perth, Western Australia, 131 pp.

Gabrielson, J.O., Lukatelich, R.J., 1985. Wind-related resuspension of sediments in the Peel-Harvey estuarine system. Estuar. Coast. Shelf Sci. 20, 135–145.

Government of Western Australia, 2013. Living in the regions 2013: A survey of attitudes and perceptions about living in regional Western Australia. Government of Western Australia, p. 96. ISSN 2203-2401.

Government of Western Australia, 2015. Perth and Peel Green Growth Plan for 3.5 million; Summary. Unpublished Technical Report. Government of Western Australia, Perth, Western Australia, 8 pp.

Government of Western Australia, 2017. Living in the Regions 2016: Insights Report. Government of Western Australia, p. 26. ISSN 2203-2401.

Hale, J., Butcher, R., 2007. Ecological character description of the Peel-Yalgorup Ramsar site. A Report to the Department of Environment and Conservation and the Peel-Harvey Catchment Council. Unpublished Technical Report. Perth, Western Australia.

Hale, J., Paling, E., 1999a. Sediment nutrient concentrations of the Peel Harvey Estuary: comparisons before and after the opening of the Dawesville Channel. Unpublished Technical Report. Marine and Freshwater Laboratories, Murdoch University. Perth, Western Australia, 25 pp.

Hale, J., Paling, E., 1999b. Water Quality of the Peel-Harvey Estuary: comparisons before and after the opening of the Dawesville Channel (July 1985 to June 1999). Unpublished Technical Report. Marine and Freshwater Laboratories, Murdoch University, Perth, Western Australia, 40 pp.

Hallett, C.S., Valesini, F.J., Scanes, P., Crawford, C., Gillanders, B., Pope, A., Udy, J., Fortune, J., Townsend, S., Barton, J., Ye, Q., Ross, D.J., Martin, K., Glasby, T., Maxwell, P., 2016a. A review of Australian approaches for monitoring, assessing and reporting estuarine condition: II state and territory programs. Environ. Sci. Pol. 66, 270–281.

Hallett, C.S., Valesini, F.J., Elliott, M., 2016b. A review of Australian approaches for monitoring, assessing and reporting estuarine condition: III evaluation against international best practice and recommendations for the future. Environ. Sci. Pol. 66, 282–291.

Hallett, C.S., Valesini, F.J., Clarke, K.R., Hoeksema, S.D., 2016c. Effects of a harmful algal bloom on the community ecology, movements and spatial distributions of fishes in a microtidal estuary. Hydrobiologia 763, 267–284.

Hallett, C.S., Hobday, A., Tweedley, J.R., Thompson, P., McMahon, K., Valesini, F.J., 2017. Observed and predicted impacts of climate change on the estuaries of south-western Australia, a Mediterranean climate region. Reg. Environ. Chang. https://doi.org/10.1007/s10113-017-1264-8.

Hipsey, M.R., Hamilton, D.P., Hanson, P.C., Carey, C.C., Coletti, J.Z., Read, J.S., Ibelings, B.W., Valesini, F.J., Brookes, J.D., 2015. Predicting the resilience and recovery of aquatic systems: a framework for model evolution within environmental observatories. Water Resour. Res. 51, 7023–7043.

Hobday, A., Chambers, L., Arnould, J., 2015. Prioritizing climate change adaptation options for iconic marine species. Biodivers. Conserv. 24, 3449–3468.

Hodgkin, E.P., Hesp, P., 1998. Estuaries to salt lakes: Holocene transformation of the estuarine ecosystems of south-western Australia. Mar. Freshw. Res. 49, 183–201.

Hodgkin, E.P., Birch, P.B., Black, R.E., Humphries, R.B., 1981. The Peel-Harvey Estuarine System Study (1976–1980): A Report to the Estuarine & Marine Advisory Committee, December 1980. Department of Conservation and Environment, Perth, WA, ISBN: 0724467300, p. 72.

Hope, P., Abbs, D., Bhend, J., Chiew, F., Church, J., Ekström, M., Kirono, D., Lenton, A., Lucas, C., McInnes, K., Moise, A., Monselesan, D., Mpelasoka, F., Timbal, B., Webb, L., Whetton, P., 2015. Southern and South-Western Flatlands Cluster Report, Climate Change in Australia Projections for Australia's Natural Resource Management Regions: Cluster Reports. CSIRO and Bureau of Meteorology, Australia, ISBN: 978-1-4863-0429-5, p. 58.

Huang, P., Hipsey, M.R., Pritchard, D., 2017. In: Hydrologic evolution due to an artificial channel and climate change in the Peel-Harvey Estuary. E-proceedings of the 37th IAHR World Congress.

Huang, P., Kilminster, K., Larsen, S., Hipsey, M.R., 2018. Assessing artificial oxygenation in a riverine salt-wedge estuary with a three-dimensional finite-volume model. Ecol. Eng. accepted pending revision.

Johnston, D., Chandrapavan, A., Wise, B., Caputi, N., 2014. Assessment of Blue Swimmer Crab Recruitment and Breeding Stock Levels in the Peel-Harvey Estuary and Status of the Mandurah to Bunbury Developing Crab Fishery. Fisheries Research Report No. 258. Department of Fisheries, Western Australia, ISBN: 978-1-921845-80-2, p. 148.

Johnston, D.J., Smith, K.A., Brown, J.I., Travaille, K.L., Crowe, F., Oliver, R.K., Fisher, E.A., 2015. Western Australian Marine Stewardship Council report series No. 3. In: West Coast Estuarine Managed Fishery (Area 2: Peel-Harvey Estuary) and Peel-Harvey Estuary Blue Swimmer Crab Recreational Fishery. Department of Fisheries, Western Australia, ISBN: 978-1-877098-08-6, p. 284.

Johnston, D., Marks, R., O'Malley, J., 2017. West coast Blue swimmer crab resource status report 2016. In: Fletcher, W.J., Mumme, M.D., Webster, F.J. (Eds.), Status Reports of the Fisheries and Aquatic Resources of Western Australia 2015/16: The State of the Fisheries. Department of Fisheries, Western Australia, pp. 44–49. ISSN 2200-7857.

Kelsey, P., Hall, J., Kretschmer, P., Quinton, B., Shakya, D., 2011. Hydrological and nutrient modelling of the Peel-Harvey catchment. In: Water Science Technical Series, Report no 33. Department of Water, Western Australia, ISBN: 978-1-921789-05-2, p. 234.

Kim, D., Aldridge, K.T., Ganf, G.G., Brookes, J.D., 2015. Physicochemical influences on *Ruppia tuberosa* abundance and distribution mediated through life cycle stages. Inland Waters 5, 451–460.

Kinhill Engineers, Western Australian Department of Agriculture, Western Australian Department of Marine and Harbours, 1988. Peel Inlet and Harvey Estuary Management Strategy: Environmental Review and Management Programme, Stage 2. Kinhill Engineers, Western Australia, ISBN: 0949397164, p. 182.

Kraal, P., Burton, E.D., Bush, R.T., 2013a. Iron monosulfide accumulation and pyrite formation in eutrophic estuarine sediments. Geochim. Cosmochim. Acta 122, 75–88.

Kraal, P., Burton, E.D., Rose, A.L., Cheetham, M.D., Bush, R.T., Sullivan, L.A., 2013b. Decoupling between water column oxygenation and benthic phosphate dynamics in a shallow eutrophic estuary. Environ. Sci. Technol. 47, 3114–3121.

Kuhn, M., Tuladhar, D., Corner, R., 2011. Visualising the spatial extent of predicted coastal zone inundation due to sea level rise in south-west Western Australia. Ocean Coast. Manag. 54, 796–806.

Lavery, P.S., Lukatelich, R.J., McComb, A.J., 1991. Changes in the biomass and species composition of macroalgae in a eutrophic estuary. Estuar. Coast. Shelf Sci. 33, 1–22.

Loneragan, N.R., Potter, I.C., Lenanton, R.C.J., Caputi, N., 1986. Spatial and seasonal differences in the fish fauna in the shallows of a large Australian estuary. Mar. Biol. 92, 575–586.

Loneragan, N.R., Potter, I.C., Lenanton, R.C.J., Caputi, N., 1987. Influence of environmental variables on the fish fauna of the deeper waters of a large Australian estuary. Mar. Biol. 94, 631–641.

McComb, A.J., Humphries, R., 1992. Loss of nutrients from catchments and their ecological impacts in the Peel-Harvey estuarine system, Western Australia. Estuaries 15, 529–537.

McComb, A.J., Atkins, R.I., Birch, I.B., Gordon, D.M., Lukatelich, R.J., 1981. Eutrophication in the Peel-Harvey estuarine system. Western Australia. In: Neilson, B.J., Cronin, L.E. (Eds.), Estuaries and Nutrients. Humana Press, New Jersey, pp. 323–342.

McComb, A.J., Qiu, S., Lukatelich, R.J., McAuli, T.F., 1998. Spatial and temporal heterogeneity of sediment phosphorus in the Peel-Harvey estuarine system. Estuar. Coast. Shelf Sci. 47, 561–577.

Morgan, B., Rate, A.W., Burton, E.D., 2012. Water chemistry and nutrient release during the resuspension of FeS-rich sediments in a eutrophic estuarine system. Sci. Total Environ. 432, 47–56.

Paolisso, M., Trombley, J., Hood, R.R., Sellner, K.G., 2015. Environmental models and public stakeholders in the Chesapeake bay watershed. Estuaries Coast. 38, S97–S113.

Pedretti, Y.M., Kobryn, H.T., Sommerville, E.F., Wienczugow, K., 2011. Snapshot survey of the distribution and abundance of seagrass and macroalgae in the Peel-Harvey Estuary from November/December 2009. Unpublished Technical Report. Marine and Freshwater Laboratories, Murdoch University, Perth, Western Australia, 89 pp.

Peel Development Commission, 2015. Peel Regional Investment Blueprint. Unpublished Technical Report. Peel Development Commission, Perth, Western Australia, 142 pp.

Peel Development Commission, 2018. Available from: http://www.peel.wa.gov.au/wp-content/uploads/2014/10/PEEL-ECONOMY.pdf. (Accessed January 2018).

Peel Inlet Management Authority, 1994. Dawesville channel: environmental impacts and their management. Working Paper. Report 50. Unpublished Technical Report. Peel Inlet Management Authority, Perth, Western Australia.

Petrone, K.C., Hughes, J.D., Van Niel, T.G., Silberstein, R.P., 2010. Streamflow decline in southwestern Australia, 1950–2008. Geophys. Res. Lett. 37, L11401. https://doi.org/10.1029/2010GL043102.

Pinto, R., da Conceição Cunha, M., Roseta-Palma, C., Marques, J.C., 2014. Mainstreaming sustainable decision-making for ecosystems: integrating ecological and socio-economic targets within a decision support system. Environ. Process 1, 7–19.

Potter, I.C., Chrystal, P.J., Loneragan, N.R., 1983a. The biology of the blue manna crab *Portunus pelagicus* in an Australian estuary. Mar. Biol. 78, 75–85.

Potter, I.C., Loneragan, N.R., Lenanton, R.C.J., Chrystal, P.J., Grant, C.J., 1983b. Abundance, distribution and age structure of fish populations in a Western Australian estuary. J. Zool. (Lond.) 200, 21–50.

Potter, I.C., Baronie, F.M., Manning, R.J.G., Loneragan, N.L., 1989. Reproductive biology and growth of the Western School Prawn, *Metapenaeus dalli*, in a large Western Australian estuary. Aust. J. Mar. Freshwat. Res. 40, 327–340.

Potter, I.C., Manning, R.J.G., Loneragan, N.L., 1991. Size, movements, distribution and gonadal stage of the western king prawn (*Penaeus latisulcatus*) in a temperate estuary and local marine waters. J. Zool. 223, 419–445.

Potter, I.C., Veale, L., Tweedley, J.R., Clarke, K.R., 2016. Decadal changes in the ichthyofauna of a eutrophic estuary following a remedial engineering modification and subsequent environmental shifts. Estuar. Coast. Shelf Sci. 181, 345–363.

Qiu, J., Carpenter, S.R., Booth, E.G., Motew, M., Zipper, S.C., Kucharik, C.J., Chen, X., Loheide II, S.P., Seifert, J., Turner, M.G., 2018. Scenarios reveal pathways to sustain future ecosystem services in an agricultural landscape. Ecol. Appl. 28, 119–134.

Ruibal-Conti, A., 2014. Shifts in natural nutrient flux pathways in a catchment-estuarine system and their implications for eutrophication of coastal waters. Unpublished PhD thesis, The University of Western Australia, Perth, WA, p. 163.

Sheaves, M., Sporne, I., Dichmont, C.M., Bustamante, R., Dale, P., Deng, R., Dutra, L.X.C., van Putten, I., Savina-Rollan, M., Swinbourne, A., 2016. Principles for operationalizing climate change adaptation strategies to support the resilience of estuarine and coastal ecosystems: an Australian perspective. Mar. Policy 68, 229–240.

Shoyama, K., Yamagata, Y., 2014. Predicting and-use change for biodiversity conservation and climate change mitigation and its effect on ecosystem services in a watershed in Japan. Ecosyst. Serv. 8, 25–34.

Silberstein, R.P., Aryal, S.K., Durrant, J., Pearcey, M., Braccia, M., Charles, S.P., Bonieck, L., Hodgson, G.A., Bari, M.A., Viney, N.R., McFarlane, D.J., 2012. Climate change and runoff in south-western Australia. J. Hydrol. 475, 441–455.

Smith, K., Holtz, M., 2017. West coast nearshore and estuarine finfish resource status report 2016. In: Fletcher, W.J., Mumme, M.D., Webster, F.J. (Eds.), Status Reports of the Fisheries and Aquatic Resources of Western Australia 2015/16: The State of the Fisheries. Department of Fisheries, Western Australia, pp. 44–49. ISSN 2200-7857.

Smith, J., Melville, M.D., 2004. Iron monosulfide formation and oxidation in drain-bottom sediments of an acid sulfate soil environment. Appl. Geochem. 19, 1837–1853.

Smith, I., Power, S., 2014. Past and future changes to inflows into Perth (Western Australia) dams. J. Hydrol. Reg. Stud. 2, 84–96.

Sullivan, L.A., Bush, R.T., Fyfe, D., 2002. Acid sulfate soil drain ooze: distribution, behaviour and implications for acidification and deoxygenation of waterways. In: Lin, C., Melville, M.D., Sullivan, L.A. (Eds.), Acid Sulfate Soils in Australia and China. Science Press, Beijing, pp. 91–99.

Thomson, C., Kilminster, K., Hallett, C., Valesini, F., Hipsey, M., Trayler, K., Gaughan, D., Summers, R., Syme, G., Seares, P., 2017. Research and information priorities for estuary management in southwest Western Australia. Unpublished Technical Report. Western Australian Marine Science Institution, Perth, Western Australia, 87 pp.

Tweedley, J.R., Hallett, C.S., Warwick, R.M., Clarke, K.R., Potter, I.C., 2016. The hypoxia that developed in a microtidal estuary following an extreme storm produced dramatic changes in the benthos. Mar. Freshw. Res. 67, 327–341.

Warwick, R.M., Tweedley, J.R., Potter, I.C., 2018. Microtidal estuaries warrant special management measures that recognise their critical vulnerability to pollution and climate change. Mar. Pollut. Bull. 135, 41–46.

Western Australian Planning Commission, 2015. Draft Perth and Peel @ 3.5 million. Unpublished Technical Report. Western Australian Planning Commission, Perth, Western Australia, 75 pp.

Whisson, C.S., Wells, F.E., Rose, T., 2004. The benthic invertebrate fauna of the Peel-Harvey estuary of south-western Australia after completion of the Dawesville Channel. Rec. West. Aust. Mus. 22, 81–90.

Wildsmith, M.D., Rose, T.H., Potter, I.C., Warwick, R.M., Clarke, K.R., Valesini, F.J., 2009. Changes in the benthic macroinvertebrate fauna of a large microtidal estuary following extreme modifications aimed at reducing eutrophication. Mar. Pollut. Bull. 58, 1250–1262.

Wilson, C., Hale, J., Paling, E. I., 1999. Macrophyte abundance, distribution and composition of the Peel-Harvey Estuary: Comparisons before and after the opening of the Dawesville Channel (July 1985 to June 1999), Report Number MAFRA 99/5. Unpublished Technical Report. Marine and Freshwater Research Laboratory, Perth, Western Australia, 64 pp.

Wynne, M.J., 2016. The proposal of *Willeella brachyclados* (Montagne) M.J.Wynne comb. nov. (Ulvophyceae). Notulae Algarum 18, ISSN 2009-8987.

Young, G.C., Potter, I.C., 2003a. Induction of annual cyclical changes in the ichthyofauna of a large microtidal estuary following an artificial and permanent increase in tidal flow. J. Fish Biol. 63, 1306–1330.

Young, G.C., Potter, I.C., 2003b. Influence of an artificial entrance channel on the ichthyofauna of a large estuary. Mar. Biol. 142, 1181–1194.

Zhu, Y., Hipsey, M.R., McCowan, A., Beardall, J., Cook, P.L., 2016. The role of bioirrigation in sediment phosphorus dynamics and blooms of toxic cyanobacteria in a temperate lagoon. Environ. Model. Softw. 86, 277–304.

Section B

Deltas

Chapter 8

Arctic Deltas and Estuaries: A Canadian Perspective

Donald L. Forbes

Geological Survey of Canada, Natural Resources Canada, Bedford Institute of Oceanography, Dartmouth, NS, Canada

Department of Geography, Memorial University of Newfoundland, St. John's, NL, Canada

Department of Earth Sciences, Dalhousie University, Halifax, NS, Canada

1   INTRODUCTION

The mouths of rivers at marine coastlines are among the most complex and biologically productive environmental systems on Earth. These transition zones between fluvial and marine environments involve the mixing of fresh and saline water, with associated biogeochemical effects, and the deposition of entrained particulate sediments. The details of the resulting circulation and sedimentary response (primarily deltaic and basin-margin landforms of varying scale and complexity) depend on the freshwater discharge, water properties, seabed morphology and accommodation space of the receiving basin, the effects of land subsidence or uplift on relative sea level, tidal range, and wave energy, among other factors.

Arctic deltas and estuaries are subject to very strong seasonal effects, with important variability in temperature, solar irradiance, ice cover, water storage in snow, nutrient loading, primary production, and a host of other factors, including positive and negative effects on human activities. Ice plays a strong role in regulating flooding, fluvial discharge, dissolved and particulate concentrations, air-sea energy and gas exchange, mixing, and productivity (Carmack and Macdonald, 2002, 2008; Emmerton et al., 2007, 2008).

This contribution focuses on examples of high-latitude deltaic and estuarine systems from northern Canada, highlighting a range of distinctive Arctic characteristics (Fig. 1). In this context, the ecosystem services from deltas and estuaries are important components of food security for indigenous communities, where reliance on "country food" remains culturally, nutritionally, and economically important (e.g., Usher 2002; Ford et al., 2006, 2016). Arctic deltas and estuaries provide essential habitat for migratory birds, fish, and marine mammals, such as beluga (*Delphinapterus leucas*) and narwhal (*Monodon monoceros*). Although the human population of Arctic regions is low compared to many other parts of the global coast, Inuit communities are coastal and indigenous northern residents remain highly dependent on marine living resources (Ford et al., 2016). Few population centers in Canada are directly located on Arctic deltas: Kugluktuk (2016 population: 1057) is located on the Coppermine Delta in Nunavut; and the much larger Mackenzie Delta in the Northwest Territories hosts the communities of Aklavik (population: 590) and Inuvik (population: 3140) (Statistics Canada, 2017), with extensive conventional and subsistence infrastructure. With mixed wage and subsistence economies (Usher, 2002; Dana et al., 2008), northern majority-indigenous communities make extensive use of deltaic and estuarine resources. Traditional ecological and cultural knowledge is a rich source of information on these important coastal systems and a key factor in the resilience of indigenous residents (Carmack and Macdonald, 2008; Kokelj et al., 2012; Ford et al., 2016). This chapter focuses on biophysical aspects of Arctic deltas and estuaries, but it is important to recognize their role in northern social-ecological systems for the maintenance of human nutrition, health, and well-being.

The nutrient status of estuarine and coastal waters is a function of inputs from land and the sea and dependent on biogeochemical processes operating in these transitional environments. The long-term stability of many northern deltaic systems in a changing climate, with loss of glacial ice mass, reduction of sea ice, warming and thaw of permafrost, and changing sea levels, is likely to be variable and dependent on the interplay of these factors with local conditions, including accommodation space and meltwater supply (Bendixen and Kroon, 2017). In some cases, the outcome may be positive (Bendixen et al., 2017), and in others cause for concern (Day et al., 2016).

Coasts and Estuaries. https://doi.org/10.1016/B978-0-12-814003-1.00008-3
123

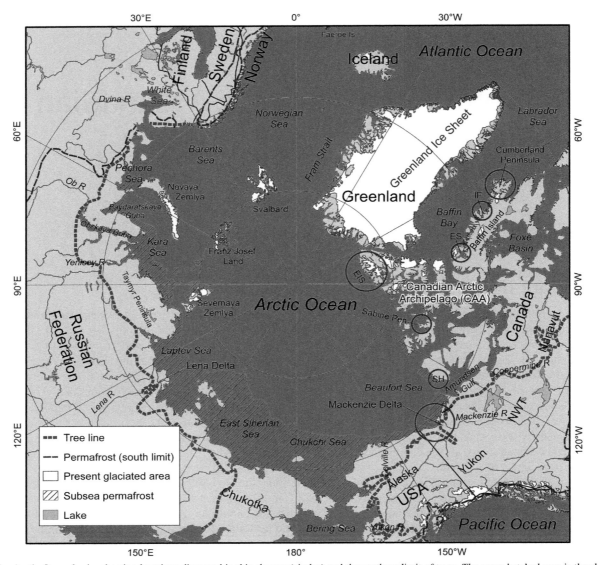

FIG. 1   Arctic Ocean basin, showing locations discussed in this chapter (circles) and the northern limit of trees. The cross-hatched area is the shallow shelf region in which subsea permafrost is known or presumed to occur, and the thinner broken line is the southern limit of permafrost (Brown et al., 1997). *EIS*, Ellesmere Island ice-shelf complex; *ES*, Eclipse Sound between Baffin and Bylot islands; *IF*, Itirbilung Fiord, Baffin Island; *SH*, Sachs Harbour, Banks Island.

1.1   Sediment Balance and Delta Stability

Deltaic emergence or submergence depends on the balance between sediment supply and relative sea level. The latter is the mean water level with respect to the land surface, as measured by tide gauges. In general, a larger sediment supply promotes deltaic progradation (seaward growth) and aggradation (vertical accumulation). Falling relative sea level, resulting from a climatically driven reduction in ocean volume, land uplift, or uplift exceeding sea-level rise, promotes emergence (marine regression). Under this so-called "forced regression," common to many formerly glaciated regions, delta aggradation may be replaced by channel incision into earlier deposits, with most sediment contributing to delta-front advance. Rising relative sea level resulting from climate-driven sea-level rise, land subsidence, or sea-level rise outpacing uplift, promotes submergence and delta-front retreat (marine transgression). The transgression may be mitigated, arrested, or reversed if there is sufficient sediment supply to support aggradation to keep pace with the rise in relative sea level. Adapting Syvitski et al. (2009), this balance can be represented as

$$\Delta z = \Delta s - \Delta c \pm \Delta \eta \pm \Delta g \tag{1}$$

where Δz is the change in level of the delta plain, Δs is delta plain aggradation (sediment supply), Δc is compaction (both natural and anthropogenic), $\Delta \eta$ is the change in sea level, and Δg is vertical crustal displacement (predominantly isostatic). Aggradation rates, $\Delta s/\Delta t$, vary widely, both spatially across individual deltas and between deltas as a function of climate, geology, runoff and sediment supply, and temporally with climate change or human intervention; typical rates on larger deltas are 1–50 mm/a (Syvitski et al., 2009). Compaction and subsidence, $\Delta c/\Delta t$, occur naturally as auto-compaction of aggraded sediments (through dewatering, organic oxidation, and tighter grain packing) or through deeper fluid migration and subsidence. These components typically account for less than 3 mm/a downward displacement, comparable to recent rates of global mean sea-level rise, $\Delta \eta/\Delta t$ (Church et al., 2013). Human-induced compaction through pumping of subsurface fluids can be an order of magnitude (or more) greater and a major contributor to flooding in large coastal cities globally, but a negligible concern in most Arctic deltas. Crustal motion (uplift or subsidence) due to glacial-, hydro-, or sedimentary isostatic adjustment is typically negative in the vicinity of major deltas (−5 to 0 mm/a) but can be as high as 10 mm/a in some Arctic regions of ongoing glacial-isostatic adjustment (James et al., 2014). The long-term survival of a delta or infill of an estuary depends in addition on the relative importance of tidal and littoral (wave-driven) sediment transport. Various classifications have been developed to account for the relative influence of river, tide, and wave processes and typical patterns of deposition associated with each (e.g., Boyd et al., 1992).

1.2 Arctic Estuaries

The Arctic Ocean is the smallest ocean basin, containing about 1% of global seawater volume, but receiving ~11% of global river discharge to the oceans (McClelland et al., 2012). It has an extensive, broad, and shallow shelf on the eastern (Siberian) side of the basin, underlying almost a third of the ocean surface area (Fig. 1). The complex network of inter-island channels of the Canadian Arctic Archipelago (CAA) borders the western side of the Arctic Ocean, mediating flow from the Arctic to the Atlantic through the Northwest Passages and Nares Strait. Arctic-Atlantic exchange also occurs in Fram Strait and the Norwegian Sea. Pacific inflow of low-salinity water through Bering Strait is one of three major sources of freshwater to the Arctic Ocean, the others being continental runoff and net precipitation over the ocean (McClelland et al., 2012). The stratification is typified by a low-salinity surface mixed layer (as much as 50 m deep) over cold saline water, underlain by warmer salty Atlantic water down to about 1600 m, with deep water beneath (Carmack, 2000).

The Arctic Ocean has numerous marginal seas and estuaries of varying scale. Arctic estuaries range from large embayments, such as the Gulf of Ob, Baydaratskaya Bay (Ogorodov et al., 2013), or Eclipse Sound (Figs. 1 and 2A), to small-scale river mouths, lagoons, and breached thaw-lake basins (Fig. 2B) (Forbes et al., 1994, 2014; Hill and Solomon, 1999). Some of the most iconic Arctic estuaries are the fjords of Greenland, Svalbard, Norway, Novaya Zemlya, and the eastern Canadian Arctic (Fig. 2C). These deep glacial troughs transecting mountain ranges can have as much as 2500 m total relief above and below water in extreme cases (Syvitski et al., 1987).

The largest Arctic rivers (Figs. 1 and 2D) exert a strong influence on the properties and circulation of shelf waters, including salinity, suspended sediment, organic carbon, nutrients, plankton, and overall productivity. Although the Mackenzie River plume in the Beaufort Sea can extend to the west under appropriate wind forcing, Coriolis acceleration in northern high latitudes diverts the discharge primarily to the right and thus the Mackenzie influence is most effective over the central and eastern Canadian shelf. This has been described as "a great estuary" (Carmack and Macdonald, 2002) covering an area about 530 km long and 120 km wide. At the largest scale, some have asserted that the Arctic Ocean as a whole can be considered an estuarine system (McClelland et al., 2012).

2 ENVIRONMENTAL FORCING

Arctic deltaic systems conform to Eq. (1), but the high-latitude setting influences the sediment supply (Δs) and compaction (Δc). Proximity to former and present ice sheets and ice caps has a profound influence on vertical crustal motion ($\Delta g/\Delta t$) (Fig. 3) and local sea-level trends ($\Delta \eta/\Delta t$), while also enhancing the supply of proglacial and paraglacial sediment (Fig. 2A) (Church, 1972; Syvitski and Hein, 1991; Bendixen et al., 2017).

2.1 Crustal Motion

The legacy of glacial isostatic adjustment (GIA) compounds the Δg term in Eq. (1), adding positive values (uplift) as high as 10 mm/a or more in areas within the margins of LGM continental ice sheets or negative values (subsidence) as low as −3 mm/a in areas of former glacial forebulges around the margins of the LGM ice (Fig. 3) (Peltier, 2004).

FIG. 2 Arctic estuaries and deltas. (A) Paraglacial (left) and proglacial (right) deltas and suspended sediment plumes emanating from Bylot Island, Eclipse Sound; note glaciers in distance (August 2009). (B) Breached thermokarst-lake estuaries, south shore of Kugmallit Bay near Tuktoyaktuk; note pingos (ice-cored conical mounds), nearshore shoal, and thin transgressive barrier beach (August 1992). (C) Itirbilung Fiord, central Baffin Island, Nunavut (cf. Fig. 8): note sediment plumes from head of fjord and proglacial Nuuksatuguluk River in lower right (July 2008). (D) East Channel, Mackenzie Delta, with Caribou Hills (edge of trough) in background; lake-delta fed by levee-breach in foreground; note small stands of spruce at this location not far from tree line (May 2007). (E) Terraced finger deltas, west coast of Sabine Peninsula, Melville Island, with Chads Point delta in distance and Hecla and Griper Bay in upper left (July 1971). (F) Foreset beds in terrace bank of river in foreground of (E), with ground ice, thermal niche formation, and block failure (July 1971). All photos Geological Survey of Canada (DLF).

Rapid glacial-isostatic uplift exceeding the rate of regional sea-level rise imposes a falling relative sea level (James et al., 2014) and an advancing shoreline irrespective of sediment supply ("forced regression"). Deltas may still advance in such settings if sediment supply is sufficient and the receiving basin not too deep, while the river becomes incised within and upstream of the rising delta terraces. Where Δg is negative (subsidence), the delta may be vulnerable to inundation, as appears to be the case for the outer Mackenzie Delta (Hill, 1996; Forbes et al., 2015a). In the extreme, some deltas are drowned and preserved as submarine terraces (see below).

Fig. 3 demonstrates that large areas of the circumpolar coast are subject to residual GIA uplift, in some places augmented by elastic rebound from recent ice mass loss. The coasts from northwest Alaska to the northwestern islands of the CAA, as well as the easternmost fringe of the CAA and parts of Greenland are currently experiencing isostatic subsidence. Elastic response, deeper tectonic processes, salt dome growth, or other factors may also affect vertical motion. In this chapter, we consider examples of Arctic delta systems developed in regions of uplift with forced regression (Fig. 2E) and contrasting small and large transgressive systems on the submergent coasts of eastern Baffin Island and the western Canadian Arctic (Fig. 2B and D).

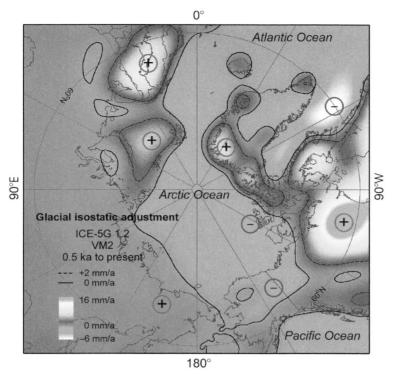

FIG. 3 Vertical crustal motion Δg (mean last 500 years) representing glacial-isostatic adjustment from the Last Glacial Maximum, as output from ICE-5G 1.2/VM2 model (Peltier, 2004). A more advanced model ICE-6G is now available (Peltier et al., 2015), but the broad patterns remain very similar; "+" and "−" symbols highlight areas of uplift and subsidence, respectively.

2.2   Ice in Arctic Deltas

Ice is the dominant environmental driver distinguishing high-latitude coasts, including deltas and estuaries, from lower latitude shorelines (Forbes and Taylor, 1994; Walker, 1998; Forbes and Hansom, 2011). It occurs in a wide variety of forms: (1) permafrost and ground ice (Fig. 2F); (2) river, lake, and sea ice (Figs. 4 and 5); (3) glacial ice (Fig. 2A and C); and (4) snow (water storage and discharge regulation). The many ways in which multiple types of these four forms of ice influence hydrological, hydraulic, geotechnical, geomorphic, oceanographic, and biogeochemical processes lead to distinctive Arctic coastal landforms, including deltas and estuaries (Forbes and Hansom, 2011).

2.2.1   Permafrost and Ground Ice

Most of the Arctic coast lies within the zones of spatially continuous or discontinuous permafrost (defined as material maintaining temperatures <0°C for ≥2 years) (Fig. 1). This has profound effects on delta development and stability, compaction, channel migration, lake formation, and other issues. Ice bonding of delta alluvium in fine-grained deltas reduces the compressibility of the sediment and has been presumed to constrain near-surface compaction, although preliminary findings in the Mackenzie Delta have led us to question this axiom (Forbes et al., 2015a). At the same time, near-surface excess ground ice may be concentrated near the base of the seasonal thaw layer (the "active layer") and in ice-wedge polygons that develop in thermal contraction cracks (Kokelj and Burn, 2005; Bode et al., 2008; Burn and Kokelj, 2009). With climate warming and deepening of the active layer, degradation of near-surface ice may contribute to overbank elevation loss, adding to other components of subsidence, and a risk of delta inundation. Bank erosion along delta distributary channels occurs by thermal-mechanical niche incision leading to failure by block collapse (e.g., Fig. 2F) (Walker, 1998).

Lakes deeper than 2 m do not freeze to the bottom in winter. Thaw zones ("taliks") develop beneath them, which may not penetrate to the base of permafrost, but raise the temperature below and around them. In these circumstances beneath larger lakes (>100 m diameter), unfrozen water content may be as high as 20% and almost half the pore space may be unfrozen, allowing upward migration of fluids from beneath thick permafrost, despite subzero temperatures (Taylor et al., 2008). Permafrost may also form or be preserved in areas of the shallow seabed where bottomfast ice (BFI) forms in winter

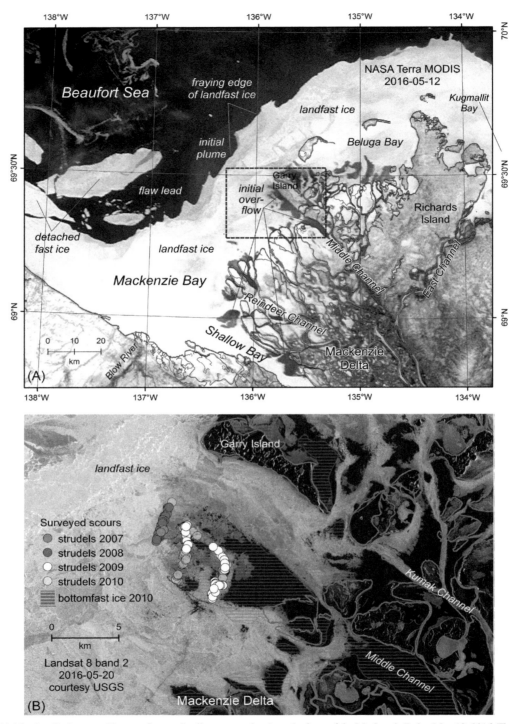

FIG. 4 (A) Initial freshet discharge and ice overflow at distributary mouths along the front of the Mackenzie Delta, May 12, 2016. Flood water concentrated over areas of bottomfast ice with some expansion to floating ice. Note landfast ice, 20–60 km wide, fraying along the outer edge at the flaw lead, and the weak river plume emerging from beneath landfast ice. (B) Overflow and partially drained ice a week later (May 20, 2016), also showing extent of bottomfast ice in spring 2010 and locations of strudel-scour pits surveyed after breakup in 2007–10. *(Image data from (A) NASA Worldview application (https://worldview.earthdata.nasa.gov/) operated by the NASA/Goddard Space Flight Center, Earth Science Data and Information System (ESDIS) project; (B) Landsat-8 imagery courtesy of the U.S. Geological Survey.)*

FIG. 5 Ice in Arctic deltas. (A) Initial overflow of clear freshet water over snow-covered bottomfast and floating ice seaward of Middle Channel, Mackenzie Delta (May 17, 2007). (B) Pressure ridge in landfast ice off front of Mackenzie Delta, with facture and melt pools along landward edge (June 1, 2009). (C) Typical radial flow to strudel drain hole along edge of bottomfast ice seaward of Middle Channel, Mackenzie Delta (May 18, 2010), cf. Fig. 4B. (D) Curtain of water >0.5 m high along ice fracture at edge of bottomfast ice, with two strudel drain holes initiated along the crack, same date and near location of previous photo. (E) Overbank flooding of turbid water along small distributary channel in the outer Mackenzie Delta (June 1, 2009). (F) Ice ride-up and pile-up along cutbank of Middle Channel with flooded stand of alder, near tree line, northern Mackenzie Delta (June 2, 2009). All photos: Geological Survey of Canada (DLF).

(Solomon et al., 2008). Furthermore, the transgressive nature of the Mackenzie Delta is such that permafrost in the now-inundated parts of the delta plain seaward of the delta front may be preserved at depth (Taylor et al., 1996). Large areas of the shallow continental shelf in the Beaufort, Chukchi, East Siberian, and Kara seas, where what is now seabed was exposed in glacial times of depressed sea level, preserve permafrost at depth (Fig. 1) (Mackay, 1972; Rachold et al., 2007; Overduin et al., 2007). The degradation of subsea permafrost influences the extensive release of methane from the seabed in shallow shelf waters (Shakhova et al., 2015).

2.2.2 Snow, Ice, and Arctic Runoff Hydrology

Winter and longer-term storage of precipitation as snow and ice, with its subsequent release as spring snow-melt or summer ice-melt drives the distinctive nival and proglacial runoff regimes of Arctic rivers (Church, 1972, 1974). Nival (snowmelt-dominated) regimes are characterized by spring snowmelt peak flow, occasional precipitation runoff events in summer, and very low flow or complete freezing and shutdown in winter. Large, north-flowing rivers such as the Mackenzie or the Lena, with basin areas of the order of $10^6 km^2$, have peak flows in spring (May–June) produced by widespread snowmelt, beginning earlier in the south and progressing north with breakup (Lesack et al., 2013; Federova et al., 2015). Shallow channels may freeze to the bottom (Forbes, 1979; Walker, 1998), leading in estuarine reaches to high salinity (through salt rejection on freezing) in ice-isolated pools (Forbes et al., 1994), as can also occur along estuarine shores (Ogorodov et al., 2013). With the spring thaw, discharge rises rapidly, flooding BFI, lifting floating ice confined in channels, flooding overbank, and flushing the estuary (Figs. 4 and 5).

Partially glaciated basins with proglacial runoff regimes are characterized by ice-melt discharge peaking in summer (Fig. 2A) and, in small basins, reflecting diurnal melt rates (Church, 1972). In some cases, these basins may also be subject to glacial outburst floods (jökulhlaups) of modest to catastrophic proportions (Church, 1972; Syvitski and Hein, 1991; Smith et al., 2006); such floods can overflow large areas of braided outwash plains (also known as sandurs). Proglacial basins draining large ice bodies such as the Greenland Ice Sheet may see increased discharge with climate warming (Bendixen et al., 2017), while basins with smaller glaciers or thin ice caps may see a decline as the surface area of ice diminishes.

2.2.3 Sea Ice

In coastal waters, winter ice may be bottomfast, landfast, or drifting free, depending on the depth, distance from shore, and wind conditions at freeze-up. BFI may freeze to the bottom and enable deep freezing, growing, or preserving permafrost in shallow water. It is also the locus of initial spring flooding in the mouths of delta distributaries (Figs. 4A and 5A) (Walker, 1998; Solomon et al., 2008; Whalen et al., 2010).

Most parts of the Arctic coast are bounded by wide expanses of landfast ice in winter. Fig. 4A shows the wide landfast ice zone off the Mackenzie Delta in early spring, at a time when easterly winds have driven the mobile pack away from the ice edge, forming a flaw lead, and the ice is shedding very large floes (note scale). Under other conditions, when the pack is pressed against the landfast ice-forming pressure ridges, with keels extending to the bottom, the resulting inverted ice dam can hold back winter freshwater discharge, forming a $70 km^3$ pool of fresh or brackish water over $12,000 km^2$ of the inner shelf (Macdonald and Carmack, 1991; Macdonald et al., 1995; Carmack and Macdonald, 2002). Fractures and pressure ridges can also form within the landfast ice (Fig. 5B). As both a barrier and a crack for drainage, these can limit the seaward expansion of over-ice flow in spring (note also linear distribution of 2007–08 strudel drainage in Fig. 4B).

The Arctic Ocean and marginal seas are seasonally (and in some areas perennially) ice covered (Fig. 2E). The basin-wide seasonal minimum ice cover has been decreasing at an accelerated pace for the past 20 years (Comiso et al., 2008; Richter-Menge et al., 2017). At the same time, there has been a dramatic decline in the total areal coverage of thick, hard, and multiyear ice (Comiso, 2002; Barnhart et al., 2016). As the proportion of multiyear ice declines, the ice taking its place each winter is thinner, more prone to summer melt or breakup, and less likely to survive to become thick multiyear ice.

The length of the open-water season and the over-water fetch for wave generation are progressively increasing in coastal waters of the CAA (St-Hilaire-Gravel et al., 2012; Forbes et al., 2018). The "melt season" and open-water season across the Canadian Arctic are lengthening at rates of several days per decade (Howell et al., 2009; Stroeve et al., 2014; Ford et al., 2016), attributable both to earlier melt and later freeze-up. Even within Parry Channel, historically ice-plugged, Barnhart et al. (2016) show that the annual number of open-water days has departed from a 20th-century mean near zero and begun to climb, with some recent years exceeding 50 days.

From 1971 to 1997, the annual minimum ice cover in Canadian Arctic waters was variable but stable with a near-constant mean (Fig. 6). Beginning in 1998, however, a pronounced downward trend in total and multiyear ice has been apparent. The record 1998 year was succeeded by a record low of multiyear ice in 2008, and then by subsequent records in 2011 and 2012 (Fig. 6). Besides opening up navigation and promoting export of ice through the CAA, these reductions in multiyear ice may have played a role in the decline of Arctic ice shelves (Copland et al., 2017). Extensive open water, unheard of historically, is now occurring regularly along the north coast of Ellesmere Island (Fig. 7).

2.2.4 Glacial Ice and Ice Shelves

Tidewater glaciers are widespread in the eastern Canadian Arctic (Syvitski et al., 1987; Forbes, 2011) and represent efficient systems for delivery of large volumes of sediment to the marine environment. Calving ice fronts can be biologically productive sites by virtue of enhanced dissolved oxygen content and upwelling driven by several mechanisms carrying saline water, sediment, nutrients, and plankton to the surface and promoting the retention of open water at the ice face (Dunbar, 1951; Syvitski et al., 1987). Dynamic ice loss through calving at tidewater ice fronts has increased and may represent up to 40% of total ice mass loss for some High Arctic ice caps (Gardner et al., 2011). Loss of multiyear sea-ice buttressing for both tidewater glaciers and ice shelves may be among the most important causes of increased iceberg

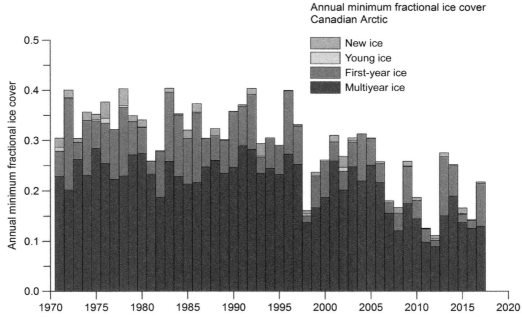

FIG. 6 Annual minimum fractional ice cover by class for Canadian Arctic waters including Baffin Bay, the Beaufort Sea, and the Canada Basin east of 141°W but excluding Hudson Bay, with a total area of 2,775,567 km². Note low minima years 1981, 1998, and 2012. *(Courtesy Canadian Ice Service, http://iceweb1.cis.ec.gc.ca/IceGraph/page1.xhtml?lang=en. (Accessed 7 February 2018).)*

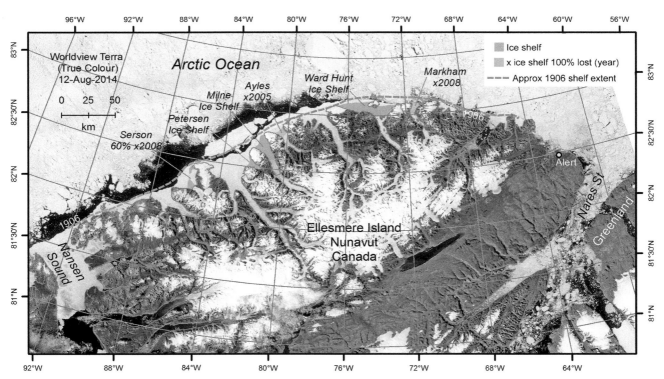

FIG. 7 Terra true-color reflectance image of northern Ellesmere Island (August 12, 2014), showing approximate area covered by ice shelves in 1906 and remaining extent as of late 2008, with two that were lost completely in 2005 (Ayles Ice Shelf) and 2008 (Markham Ice Shelf); note extensive open water along the coast, which previously was hemmed in perennially by thick multiyear ice. *(Image data from NASA Worldview application (https://worldview. earthdata.nasa.gov/) operated by the NASA/Goddard Space Flight Center, Earth Science Data and Information System (ESDIS) project.)*

production (Herdes et al., 2012), with negative impacts on marine transportation, the only means of bulk supply to most Canadian Arctic communities.

Reductions in multiyear ice are implicated in the dramatic loss of ice shelves from northern Ellesmere Island (Copland et al., 2017). The Ellesmere ice shelves are formed primarily from accumulations of very old, thick, landfast, multiyear ice augmented by snowfall and are typically 30–100 m thick (Jeffries, 1992). These features formed about 5.5 ka (3.0 ka in Clements Markham Inlet), as demonstrated by the cessation of driftwood accumulation in inner fjords, a change attributed to ice shelf growth across the fjord mouths (England et al., 2008). At the start of the 20th century, the Ellesmere ice shelf complex formed a continuous band along the north coast of Ellesmere Island, covering an area of about 9000 km^2 (Fig. 7). The ice shelves formed epishelf lakes, a very rare type of estuary (Mueller et al., 2003, 2006, 2008; Veillette et al., 2008). These are pools of fresh water, dammed by the fjord-mouth ice shelves, lying over dense seawater with virtually no mixing, draining seaward along the base of the ice. The depth of the surface freshwater layer in Disraeli Fiord behind the Ward Hunt Ice Shelf (Fig. 7) was equivalent to the ice-shelf draft (Vincent et al., 2001). The epishelf lakes host unusual communities of mixed marine and freshwater organisms (van Hove et al., 2001; Mueller et al., 2003). In addition, the ice shelves support unique ecosystems with well-developed communities of cold-tolerant microbes (Vincent et al., 2000; Mueller et al., 2006).

Between 1906 and 2015, 94% of ice shelf area was lost, most of it by intermittent calving during the first 60 years of the 20th century (Mueller et al., 2017). The total area is reduced to <500 km^2 in isolated small ice-shelf remnants and the epishelf lakes and associated mixed communities are being lost. Three of the six ice shelves remaining in 2005 have now completely disappeared (Copland et al., 2007; Mueller et al., 2008, 2017).

2.2.5 Vegetation

A striking feature of Arctic coastal systems is the near-absence of trees and generally low or sparse vegetation (Walker et al., 2005). The northern limit of trees approaches within <100 km of the Arctic coast in the Mackenzie Delta and the lower Coppermine (Fig. 1). In the proximal Mackenzie Delta south of the tree line, trees play a key role in anchoring sediment and promoting the development and stability of levees up to 8 m high. Levee height declines to 1–3 m beyond the tree line and to almost imperceptible at the delta front (Hill et al., 2001). Intertidal marshes are limited in extent, with supratidal marsh (inundated in synoptic high-water events) more common in some regions. A circumpolar salt-tolerant species indicative of intermittent flooding is the hardy dwarf grass *Puccinellia phryganodes* (Martini et al., 2009).

3 ARCTIC ESTUARIES AND DELTAS

3.1 Fjords

Fjords are quintessential estuarine systems of the Arctic coast, though not restricted to the polar regions (Syvitski et al., 1987). Fjords are glacially overdeepened valleys, typically with U-shaped cross-sectional profiles, having relatively wide and flat basin floors and steep (often near-vertical) valley sides (Fig. 2C). Most fjords have shallow sills at the mouth and often at intermediate distances: while some sills are rock-cored, most represent ice terminal positions at pauses or re-advances in the glacial recession process. The sills typically separate intervening deep basins and the outer basin from the continental shelf, although large former ice stream systems have extended depressions well out onto the shelf.

Focusing on representative fjords of the Baffin Island coast and excluding the largest bays and sounds, there are 227 fjords with axial length greater than 5 km on the island (Dowdeswell and Andrews, 1985), with the following range of dimensions

- surface area: 3–953 km^2
- length: 5–39 km
- mouth width: 1–39 km
- mid-length width: 0.4–8 km
- maximum adjacent elevation: 229–1905 m
- maximum depth: 950 m

The density (number of fjords within 50 km) ranges from 0 to 12. A sample of 29 fjords with recent multibeam bathymetry (Hughes Clarke et al., 2015), found basin depths ranging from 154 to 885 m and sill depths from 64 to 448 m (Johnathan Carter, pers. comm., 2017). These values represent seafloor depths; the rock floor of the fjords (deepest glacial scour) is often much deeper: 20–200 m deeper in a sample of 10 Baffin Island fjords (Syvitski et al., 1987).

Itirbilung Fiord (official place names in Canada use "fiord" spelling), on the central east coast of Baffin Island (Fig. 1), is a well-described example (Syvitski and Hein, 1991). It is 55 km long, with a drainage area of 2192 km^2, partially glaciated

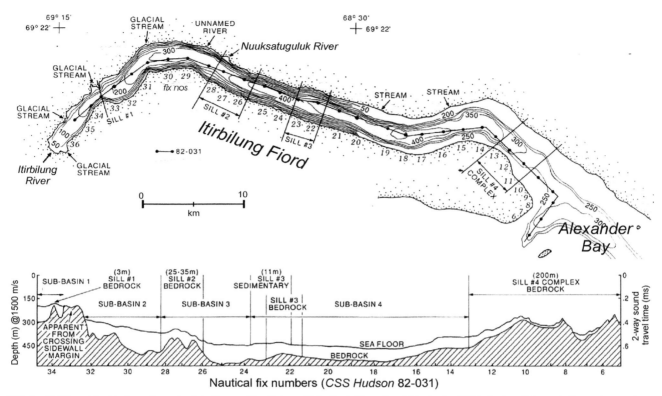

FIG. 8 Itirbilung Fiord bathymetry and sub-bottom profile. Note multiple fjord sills and variable depth of fill in basins. Ship-track fix numbers on lower profile *x*-axis relate to italic numbers below fjord in upper panel. Fjord-head sandur is just out of profile frame at left and sill complex at right is fjord mouth in Alexander Bay. Elevations of adjacent mountain plateau peaks in the central fjord are ≥1500 m (Fig. 2C). *(Modified from Syvitski, J.P.M., Hein, F.J., 1991. Sedimentology of an Arctic Basin: Itirbilung Fiord, Baffin Island, Northwest Territories. Geological Survey of Canada, Paper 91-11, https:// doi.org/10.4095/132684.)*

(Figs. 2C and 8). The fjord is divided by sills into four basins, two of which have sediment fill >150 m thick (Fig. 8). The postglacial sediment fill is impounded behind sill 1 at <200 m depth, slopes seaward from a high of 300 m in basin 2, rising over sill 2 (Fig. 8). The seabed extends down to 450 m in basin 4 and the basin-fill sediments decrease in thickness up the inner side of sill 4. Side-entry glaciers, some of which were tidewater in the 1950s (Fig. 17 in Syvitski and Hein, 1991), have now receded up their valleys (Fig. 2C). The steep walls of many fjords, vertical or even overhanging in places, represent a potential rockfall and tsunami hazard (Forbes et al., 2018). Sidewall slope failure has been documented in Itirbilung Fiord (Syvitski and Hein, 1991).

3.2 Proglacial Deltas in Fjords

Proglacial deltas and fan deltas are widespread features of glaciated terrain and particularly typical of ice-free fjord-head and side-entry settings in the Arctic (Syvitski et al., 1987; Syvitski and Hein, 1991; Lønne and Nemec, 2004; Forbes, 2011). Raised proglacial deltas are frequently the most prominent indicators of the postglacial marine limit (e.g., Dyke, 1979; Nixon et al., 2014). Proglacial fjord-head deltas commonly develop under conditions of early rapid uplift following ice recession and multiple terrace levels can develop (Church, 1972). Sediment supply declines rapidly when receding ice uncovers lake traps or withdraws from the basin. In these circumstances, deltas may become stranded as isolated raised terraces when meltwater runoff ceases or drainage is rerouted as the ice margin pulls back.

The delta at the head of Itirbilung Fiord is one of the largest on Baffin Island (40 km^2) and feeds sediment to a steep (15°–25°) delta front and an extended prodelta slope declining toward the basin (Syvitski and Hein, 1991). The delta advances by coarse-grained (sand and gravel) gravity flows (slipface deposits making up the bulk of the foreset facies), high-density, turbid, supercritical underflows (Deitrich et al., 2016; Normandeau et al., 2016), and turbidity currents.

Channel-confined and unconfined bedforms are typical over a wide depth range on active prodelta slopes (Fig. 9B–E). Chute channels on the foreset slopes converge to form fewer and deeper prodelta channels, which may merge into a single large channel with levees running some distance down-fjord. The channel in Southwind Fiord (Fig. 9E) on the outer

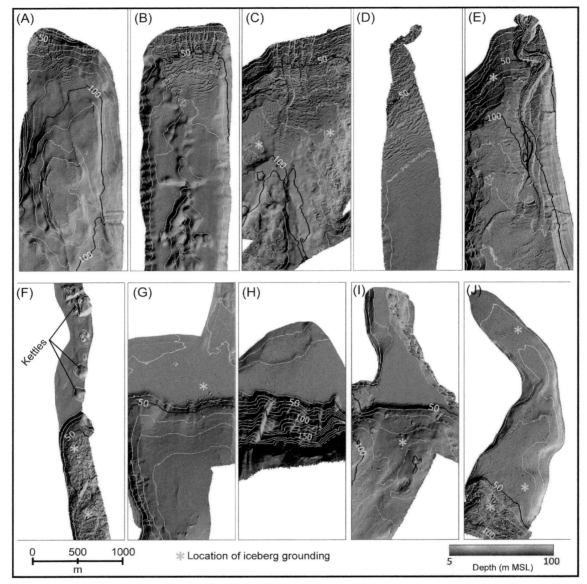

FIG. 9   Shaded-relief bathymetry of active deltas (A–E) and relict submerged deltas (F–J) in fjords of the eastern Cumberland Peninsula, Baffin Island (Fig. 1). Active fjord-head depocenters show increasing development of prodelta channel incision and bedform development from (A) western inner tributary to Totnes Road; to (B) Pangnirtung Fiord; (C) Touak Fiord; (D) Mermaid Fiord; and (E) Southwind Fiord (deeply incised channel with bedforms). Relict (drowned) delta terraces include (F) Kangiqtugaapiruluk (formerly Kangert Fiord), (G and H) Boas Fiord; (I) unnamed tributary to Durban Harbor; and (J) unnamed tributary to Clephane Fiord. Stars mark locations of iceberg grounding. Pits in Kangiqtugaapiruluk are kettles, demonstrating that the delta was coeval with recession of the last glacial ice in the fjord (Cowan, 2015). *(Modified from Figs. 4 and 6 in Hughes Clarke, J.E., Muggah, J., Renoud, W., Bell, T., Forbes, D.L., Cowan, B., Kennedy, J., 2015. Reconnaissance seabed mapping around Hall and Cumberland peninsulas, Nunavut: opening up southeastern Baffin Island to nearshore geological investigations. In: Summary of Activities 2014. Canada-Nunavut Geoscience Office, Iqaluit, NU, 133–144. cngo.ca/app/uploads/Summary-of-Activities-2014-P15.pdf, in which locations can be found.)*

Cumberland Peninsula is 150-m wide at 750 m from the delta lip, with in-channel bedforms (wavelength ~25–45 m) and 20-m high levees (Cowan, 2015). An important sediment delivery process is aeolian: Syvitski and Hein (1991) estimated wind transport into the fjord to be 0.36 Mt/a, three times their estimate of the fluvial transport.

The fjord-head delta in Itirbilung Fiord still bears the scars of a very large Little Ice Age jökulhlaup, which removed raised delta terraces and deposited gravel waves and boulders across part of the upper sandur. How much sediment was transported into the fjord at that time is uncertain: Syvitski and Hein (1991) documented a 10-m thick mass transport deposit with a volume of $2 \times 10^7 \, m^3$, located 7–9 km down-fjord, a feature they hypothesize may have been triggered by the jökulhlaup event, although an earthquake cannot be discounted in this region. Quite apart from the impact of a catastrophic discharge event, the slide would likely have generated a tsunami.

Relict proglacial fjord-head and side-entry deltas are preserved below present sea level on the eastern Cumberland Peninsula (Fig. 1), an area that experienced a postglacial lowstand of sea level (Miller and Dyke, 1974; Dyke, 1979; Cowan, 2015; Hughes Clarke et al., 2015). Examples of drowned deltas in the region include the proglacial (kettled) delta terrace at 19 m depth at the head of Kangiqtugaapiruluk (formerly Kangert Fiord) (Fig. 9F), and relict delta terraces in Boas and several other fjords (Fig. 9G–J). These deltas apparently lost their sediment supply following ice recession. The former sand and gravel sandur surfaces are now covered with 2 m or less of hemipelagic mud. In Boas Fiord, the subsequent transgression back-flooded the original fjord-head sandur surface, which was reactivated up-valley, likely at the end of the Little Ice Age. The modern sandur sits atop the lowstand delta 3.8 km up-valley, with a prodelta wedge that extends ~17% of the distance to the old delta lip.

3.3 Incised Terraced Deltas on Low-Relief Coasts

Terraced finger deltas on small rivers of Sabine Peninsula, Melville Island, provide an example of deltaic progradation outpacing forced regression (Forbes et al., 1986). Chads Point (76°12′N, 109°54′W) on the western shore of the peninsula (Fig. 2E) and Invincible Point (76°16′N, 108°02′W) on the eastern shore (Forbes et al., 2018) are finger deltas of this kind. The associated rivers drain low-relief (150 m) drainage basins of 161 and 54 km², respectively. The surficial material is sparsely vegetated, weathered, poorly cemented to unlithified Mesozoic sand and shale (Harrison et al., 2015), mantled by a thin and discontinuous veneer of postglacial marine mud and locally thicker deltaic deposits up to the marine limit at 64 m elevation, dating no later than 11.6 ka (Nixon et al., 2014).

As seen at Chads Point (Fig. 10), numerous small rivers in the region have deposited long fingers, typically 1.0–1.5 km wide, projecting up to 4.5 km beyond the adjacent coastline, at which the terraces can be 12–16 m high, implying a postglacial deltaic package of equivalent thickness. The terraced delta facies include marine silty clay and bottomset rippled silt and muddy sand, overlain by sand and pebbly sand foresets typically 4–6 m thick (Fig. 2F), with a thin topset veneer of braided channel sand and minor gravel. Permafrost has developed in the finger deltas, evidenced by ice-wedge polygons on the terraces (Fig. 10) and massive ice at the base of some exposed sections (Fig. 2F). Most of the terraces are rimmed with subtle ridges attributed to shoreward ice-push, most prominently on the exposed western side at Chads Point (Fig. 10). Because of the consequent drainage pattern, several finger deltas have coalesced; in one case, two have grown together to enclose a lake (Fig. 10 inset).

These deltas, which have outrun the shoreline advance resulting from forced regression, are stable because of incision and self-confinement between ice-bonded banks and, crucially, because of the near-permanent ice cover and the absence of waves. They are potentially threatened by climate change if the multiyear ice currently surrounding Sabine Peninsula becomes mobile and open water comes to be more prevalent in the future.

The postglacial emergence is reasonably well defined and appears to have been anomalously linear since 9 ka (Forbes et al., 1986; Nixon et al., 2014). From an elevation of 30 m about 9 ka, RSL fell to 20 m ~6 ka (the highest terrace in Fig. 10), 4 m ~2 ka, and 2 m ~1000 years ago. Using the RSL curve to date the terraces enables us to unravel the full Holocene history of sediment delivery from these basins. Progradation rates were typically 0.5–0.7 km/ka over the past 2000 years, a decrease from rates as high as 1.6 km/ka in the early to mid-Holocene. This may be attributable in part to increasing accommodation space as the deltas extended seaward, but bathymetric data to test this hypothesis are sorely lacking for these ice-locked waters.

3.4 Breached-Lake Estuaries of the Arctic Coastal Plain

Extensive areas of the Arctic coastal plain are low-lying and characterized by large concentrations of lakes (Hill et al., 1994). The latter are primarily thermokarst in origin, resulting from differential melt of excess ground ice (Murton, 2001; Grosse et al., 2013). With ongoing postglacial marine transgression, the coast has migrated inland, successively breaching these shallow lake basins (Ruz et al., 1992; Hill et al., 1994; Forbes et al., 2014) and forming an intricate network of shallow estuarine embayments (Figs. 2B and 11). These typically have very shallow sills on which winter ice can freeze to the bottom, creating closed systems in which very high salinity can develop through the process of brine exclusion on freezing (Forbes et al., 1994).

The incursion of marine water into breached-lake estuaries produces sedimentary successions of lacustrine deposits overlain by brackish marine sediments, with varying proportions of fluvial influence (Hill and Solomon, 1999; Forbes et al., 2014). Sedimentation rates can be quite high in basins receiving Mackenzie plume water with high-suspended

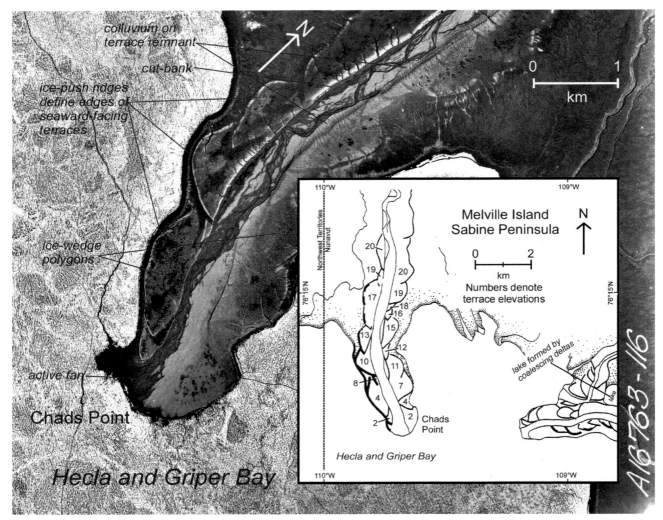

FIG. 10 Terraced finger delta at Chads Point, west coast of Sabine Peninsula, Melville Island, detail from 1959 airphoto A16763-116 ©HM the Queen in Right of Canada (Natural Resources Canada). Inset: Three terraced deltas, two of which have merged to enclose a small lake; delta at lower right is shown in Fig. 2E and F. Numbers on Chads Point delta are terrace elevations in meters (Forbes et al., 1986). Note subtle ice-push ridges on edges of terraces facing outward to Hecla and Griper Bay.

sediment concentrations, possibly augmented by a contribution from coastal erosion. Forbes et al. (1994) reported rates ranging from 3 to 17 mm/a in the breached lake embayments of northern Richards Island. In one case, they concluded from the depth of the fresh-to-marine transition in core 4 (Fig. 11 inset) that the basin was breached between 1919 and 1958. The sedimentation rate in a nearby unbreached lake (core 5 in Fig. 11 inset) was ~0.3 mm/a, an order of magnitude lower.

A shallow-silled (seasonally isolated), hypersaline, marine basin has been described from the south coast of Melville Island, where a switch from emergence to marine transgression has recently occurred and the seasonally isolated tidal system is key to the development of high salinity (Dugan and Lamoureux, 2011). In another case, year-round hypersaline conditions were discovered serendipitously in a linked pair of near-circular, shallow-silled, thermokarst lakes (500–900 m diameter, 11–19 m deep) on the coastal plain of southern Banks Island, near Sachs Harbour (Fig. 1). One basin, breached by coastal erosion, and lake expansion, has a narrow tidal inlet with a sill depth <1.2 m. The second basin is linked to the first through basin expansion, with a 2.5-m deep channel between the two. Seawater enters both basins during the ice-free season, but the tidal range is low (0.2–0.4 m) and the dense, anoxic brines remain undisturbed at depth. Progressive surface freezing in winter cuts off tidal circulation at the outer sill and has, since the as-yet undated breaching, increased stratification and salinity to the point of saturation for mirabilite ($Na_2SO_4 \cdot 10H_2O$), resulting in crystalline deposits up to 1 m thick on the basin floors (Grasby et al., 2013).

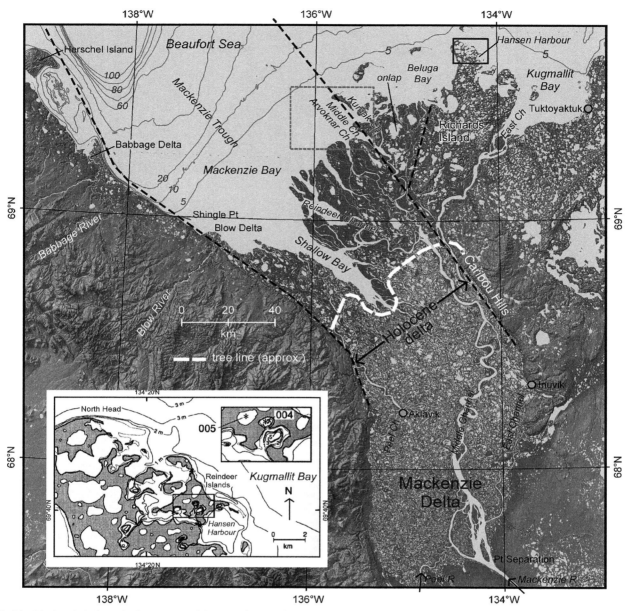

FIG. 11 Mackenzie Delta and adjacent parts of the Canadian Beaufort Sea coast, Yukon and Northwest Territories. Black broken lines denote margins of the trough in which the Holocene delta has developed and the area of onlap over older Pleistocene deposits of Richards Island. Isobaths (in meters) outline Mackenzie Trough and broad prodelta fan morphology in Mackenzie and Kugmallit bays. Broken white line is the approximate northern limit of trees in the delta, farther north than on adjacent uplands. Box outlined by gray broken line shows the extent of Fig. 4B. Box delimited by solid back line is area of inset. *(Topographic data from Natural Resources Canada; inset reproduced from Forbes, D.L., Solomon, S.M., Hamilton, T.S., 1994. Morphology and sedimentary processes of microtidal embayments, Beaufort Sea coast, western Arctic Canada. In: Viggósson G. (Ed.), Proceedings, Hornafjörður International Coastal Symposium, Höfn. Icelandic Harbour Authority, Kópavogur, 363–372 courtesy of Icelandic Road and Coastal Administration.)*

3.5 Small Transgressive Deltas

Deltas and estuaries on submergent coasts face rising relative sea level, which counteracts deltaic progradation. In many cases, sediment supply is insufficient to hold the line; over the long term, the delta retreats up-valley, sometimes with associated estuarine development. This is the case in the Babbage River delta and its associated spit-fronted lagoonal estuary, located along the Yukon coast west of the Mackenzie Delta (Fig. 11).

This delta has a drainage basin area of ~5000 km^2 on the coastal plain, pediment, and mountains to the south. Covering about 40 km^2, the system comprises the lagoon (34%), intertidal flats (16%), distributary channels (7%), and supratidal

FIG. 12 Small and large transgressive deltas. (A) Outer Babbage Delta, showing lake-studded vegetated flood plain and encroaching tidal flats, with lagoonal estuary and enclosing spit in left distance (August 1974). (B) Driftwood sourced from Mackenzie River deposited at storm-surge limit over ice-wedge polygons in valley headward of Babbage Delta (August 1974). (C) Channels and lakes in central Mackenzie Delta during breakup flooding (June 2009). (D) Mackenzie Delta lakes and channels, looking north toward Aklavik on large bend in Peel Channel, showing spruce on levees and willow on point bars (August 2007). (E) Storm-deposited silt-sand sheet on subtle levee (<0.2 m relief) with ice wedges and stranded logs along erosional front of Mackenzie Delta; wave-trimmed overbank silt with high particulate carbon content at left and low herbaceous cover on delta plain at right (August 2007). (F) Small pockmarks attributed to methane release on channel bar at bifurcation of Kumak Channel from Middle Channel (August 2007). All photos: Geological Survey of Canada (DLF).

flats (42%). These last include both the delta plain and the supratidal marsh rimming parts of the lagoon basin (Fig. 12A). Sedimentation rates over the past two millennia, based on radiocarbon dates in cores, were <0.5 mm/a in both intertidal and delta plain settings (Forbes et al., 1994). Over the long term, the sea is gradually flooding the Babbage valley (Fig. 12B) and the modern delta is a thin veneer over transgressed alluvial deposits such as occur today in the lower valley upstream of the tidal limit (Forbes, 1983). As this transgression proceeds, massive ice at depth in the alluvial deposits may be subject to differential thaw, giving rise to thermokarst topography and a proliferation of ponds and lakes in the delta (Forbes, 2004).

3.6 Large Transgressive Deltas

The Mackenzie Delta, the second largest delta (after the Lena) on the Arctic Ocean, covers an area of about 13,000 km². The river catchment has a total area of 1.8×10^6 km² and delivers 284 km³ of water annually to the Beaufort Sea from headwaters in British Columbia, Alberta, Yukon, and the Northwest Territories. The annual sediment load is estimated at 124 Mt (Carson et al., 1998; Holmes et al., 2002). The modern delta overlies a long-term (70 Ma) deltaic depocenter with

~12–16 km of Late Cretaceous to Holocene sediments beneath the central Beaufort Shelf (Dixon et al., 1992). The delta occupies a near-rectangular, glacially scoured, structural depression ~200 km long and 60–80 km wide (Fig. 11). This depression extends seaward as the shelf-crossing Mackenzie Trough.

The Mackenzie River enters the delta in the southeast corner and multiple bifurcations give rise to innumerable major and minor distributary channels (Fig. 11), the largest of which initially are Middle Channel and the narrower East Channel, the main navigation channel flowing past Inuvik. Peel Channel (fed from the Peel River entering in the southwest), flows north past Aklavik (Fig. 12D). Under summer low-flow conditions, the hydraulic gradient is ~5×10^{-5} (m/m) in the central delta and declines to 1×10^{-5} toward the delta front (Hopkinson et al., 2011). Most of the delta channels empty into Mackenzie Bay, but one major distributary (East Channel) cuts to the northeast into Kugmallit Bay (Fig. 11). Shallow Bay is a large, shallow, elongate embayment in the delta front; its origin is unclear. Linear channel reaches in places suggest the possibility of shallow faulting.

Richards Island and smaller islands north of the delta have greater relief and consist primarily of Pleistocene fluvial, aeolian, and ice-contact deposits, with up to 700 m of permafrost (Todd and Dallimore, 1998). As sea level has risen through the Holocene, the delta has filled the trough and spilled over onto the Pleistocene higher ground on western Richards Island (marked "onlap" in Fig. 11). Within the trough, permafrost extends to about 60 m depth in the Holocene delta.

The tree line cuts across the delta near the head of Shallow Bay (Fig. 11). It appears as a broad zone of transition from tall spruce woodland occupying levees in the upper delta (Fig. 12C and D) to low shrub willow (on levees) and sedge wetlands in the outer delta (Fig. 5E). This transition is clearly visible in Fig. 11 as a change from the high lake concentration in the upper and central delta to fewer but larger lakes in the outer delta. The almost imperceptible levees on the outermost delta front have almost no willow, in part because they are repeatedly rebuilt as the shore retreats and sediment is redeposited over the backshore vegetation (Fig. 12E).

Emmerton et al. (2007) enumerated 49,046 lakes on the Mackenzie Delta, with a combined area of 3331 km² (25% of the total delta extent). Other components of the delta plain include 1744 km² of channels (13%), 1614 km² of wetlands (12%), and 6446 km² of dry floodplain (49%). Sediment and nutrient supply to delta lakes depends on connectivity to the channel network, which is determined by the lake channel or sill elevation and water level in the delta. The water level and sediment transport capacity are functions of sea level at the delta front, discharge into the delta, and the hydraulic gradient through the delta distributary network (a function of the boundary roughness of the channels and any overhead ice cover). Sediment supply to overbank surfaces is largely accomplished by spring flooding, and on the outermost delta, by storm-surge flooding.

Radiocarbon ages show that the Holocene delta had prograded as far as Inuvik by 6.9 ka (Johnston and Brown, 1965) and advanced to Unipkat, 25 km short of the modern delta front by 4.5 ka (Taylor et al., 1996; Hill et al., 2001). Hill (1996) described the mid- to late-Holocene sediment body forming the modern delta surface as a prograding delta wedge (and shallow prodelta platform) which advanced some distance seaward of the present delta front. The associated low-angle outer delta clinoforms show an upward reduction in gradient, consistent with rising base level. The retreat of the delta front at a mean rate (1972–2000) of 1.77 ± 0.07 m/a (locally >10 m/a) (Solomon, 2005) reflects the regional transgression and the inability of Mackenzie River discharge to counteract the effects of rising sea level along a wider delta front, as it has onlapped the eastern margin of the trough. Other factors such as delta subsidence, increased sediment trapping in the delta, unquantified sedimentation in Shallow Bay (but low accommodation volume), or greater wave exposure may also be at play.

Large parts of the delta plain are flooded annually during the spring freshet and breakup (Emmerton et al., 2007; Goulding et al., 2009). In addition, the outer delta plain, within 20–30 km of the delta front, is subject to intermittent marine flooding through the open-water season by positive storm surges of up to 2 m or more (Manson and Solomon, 2007; Pisaric et al., 2011).

3.6.1 Spring Breakup Flooding

The annual snowmelt and breakup flood event is vital to lake water renewal and nutrient inputs (Marsh and Hey, 1989). Annual peak flooding, which by 2011 was occurring on average 8 days earlier than in 1964 (Lesack et al., 2013), spreads up to 47% of annual Mackenzie River breakup discharge across the delta plain in late May and early June at a time of 24 h/day solar irradiance (Emmerton et al., 2007). The peak water level and recession rate in any given breakup season and the varying sill elevations of the lakes (Crasto et al., 2015) determine the number of lakes overtopped and the length of connectivity: some lakes are connected on a continuous basis, others annually, and still others only once in several years. The highly varying sill elevations and durations of intermixing with river flood water result in variable water properties, particularly in high-closure lakes with restricted connection to the channel network.

Temporary storage of such a high proportion of river discharge in the delta has an important influence on nutrient and particulate delivery to the Beaufort Sea. Using a two-source mixing model for river discharge after recovery of flood-diverted water, corroborated by field sampling, Emmerton et al. (2008) found that:

- particulate levels were 10%–18% lower (due to sedimentation in the delta) with 75%–280% enrichment in organic content (higher ratios of POC, PN, PP to TSS);
- dissolved organic content was higher by 15% (DOC), 62% (DON), and 239% (DOP); and
- dissolved inorganic nutrients, except ammonium (+10%), were reduced by 14% (NO_3^- and SRP) and 5% (SRSi),

leading to nutrient enrichment relative to phosphorus of 79% for carbon (TOC:TP) and 77% for nitrogen (TN:TP). Although sample data generally supported these results, they also revealed a more complex system due to variability in flood-water interaction with lake waters and flood-plain vegetation.

The first pulse of higher spring flow appears as clear water flooding over BFI (Fig. 5A). The locus of first flow varies from one year to another, depending on distributary channel cross-sectional area beneath ice and channel roughness. Within a day or two, water spreads out over and beneath the ice across a broad front in Mackenzie Bay (Fig. 4). Because the BFI initially has lower elevation than the floating ice, it is more rapidly flooded, but the overflow also expands over floating ice and through sub-ice channels. Subaqueous channels extend some distance seaward of the delta front in northern Shallow Bay, off Arvoknar and Kumak channels, and northeastward from the mouth of East Channel in Kugmallit Bay. As the floating ice begins to lift with rising water levels, floodwater drains though cracks that enlarge to circular drain holes (Fig. 5C and D), located preferentially along the margins of the BFI, which we can map using synthetic aperture radar imagery (Solomon et al., 2008). The vertical drainage jets produce circular scour holes in the nearshore seabed, a process called "strudel scour" (Reimnitz et al., 1974). Postflood shallow multibeam mapping from 2007 to 2010 documented 111 drain-related scours off Middle and Kumak channels west of Garry Island (Fig. 4B). Located in water depths of 0.8–1.6 m along the margins of BFI where drainage features had been observed, these had diameters of 20–30 m and subseabed relief ranging from 0.8 to 2.6 m, with one outlier 22 m in diameter and 6 m deep, with a flat floor 8 m across, in surrounding depths of 1.5 m (Whalen et al., 2010).

3.6.2 Storm-Surge Flooding

Marine flooding can affect large areas of the outer delta and appears to be increasing in frequency (Kokelj et al., 2012; Vermaire et al., 2013). Major surges occurred in 1944, 1963, 1970, 1993, 1999, and 2000, among numerous lesser events (Manson and Solomon, 2007). Surges can damage delta vegetation, alter lake-water habitat, destroy nests in the Kendall Island Bird Sanctuary and more widely across the outer delta, affect water quality in delta channels, erode the delta front, and pose a serious safety hazard (two people died in the 1970 storm).

An exceptional event in September 1999 raised water levels in the outer delta to 2.5 m (CGG05 datum) in an area with a spring tidal range <0.5 m. Brackish water spread across the outer delta causing mortality of willow, alder, and other shrub and herbaceous vegetation over at least 132 km^2 of the delta (Kokelj et al., 2012). A decade later, some partial recovery was underway, although high areas less exposed to spring breakup flooding retained high soil salinity. Although alder growth was suppressed for 8 years following the 1944 event (Pisaric et al., 2011), other major and lesser surge events have not produced such a dramatic impact as the 1999 event.

For much of the year, the Mackenzie River plume dominates the water column off the delta, so that water flooding over the delta in a surge event may be relatively fresh. However, if river levels are low and a surge is preceded by easterly winds, which drive upwelling of deep water from Mackenzie Trough, as in September 1999, the surface water driven onto the delta under subsequent northwesterly winds may be quite saline. In the 1999 event, the water had sufficient salinity to contaminate the soil and convert outer-delta lakes from fresh to saline.

3.6.3 Potential for Delta Inundation

Seasonal flooding and storm surges cause temporary flooding. Here we consider the potential for permanent inundation in the Mackenzie Delta. The balance at the delta front is a function of rising relative sea level and land subsidence versus sediment accretion and possible local permafrost accretion. The latter is primarily occurring in areas of recent progradation such as the Olivier Islands near the outlet of Reindeer Channel in eastern Shallow Bay (Jenner and Hill, 1998). Elsewhere, active layer deepening, involving deeper seasonal thaw and melting of excess near-surface ice may be adding to deeper-seated subsidence. Sedimentation rates are poorly defined but generally low (<5 mm/a) in preliminary results using SET devices, except along levees, where rapid sand deposition can occur during flood events (Forbes et al., 2015a). In lakes sampled on the outer delta, the 1999 surge deposit was ~25–28 mm below the 2010 surface, 11 years after the event (Vermaire et al., 2013), giving a sedimentation rate of 2.4 ± 0.2 mm/a. In more extensive lake sampling throughout the delta, Marsh et al. (1999) reported sedimentation rates in outer delta lakes of 1.46 mm/a (based on their mean bulk density of 931 kg/m^3).

Preliminary analysis of trends from several years of GPS monitoring indicates subsidence rates between 1.7 and 6.4 mm/a with standard errors in the range of 0.3–0.9 mm/a at a subset of stations across the delta (Forbes et al., 2015a). These are with respect to a station on limestone bedrock at Inuvik, where GIA motion is -0.48 ± 0.58 mm/a, while at Tuktoyaktuk, which is closer, the motion is -1.04 ± 0.55 mm/a (James et al., 2014). Realistic scenarios of projected sea-level rise, subsidence, and sedimentation rates applied to a LiDAR digital elevation model for a representative part of the outer delta indicated permanent inundation of 49% of the area by 2085 (vs 31% in 2000) under an RCP8.5 median sea-level rise scenario, assuming subsidence of 4.2 mm/a and sedimentation of 5 mm/a (Forbes et al., 2015a). Adopting the 95% upper limit projection for RCP8.5, the area under water at mean water level in 2100 would be 73%. As of 2010, this area already had extensive wetland and degraded ice-wedge polygons typically indicative of gradual inundation.

4 DISCUSSION

4.1 Arctic Tidewater Ice Fronts and Ice Shelves

Tidewater ice fronts of marine terminating glaciers generate large-scale circulation involving buoyancy of meltwater (as subglacial discharge or direct melt) and entrainment of deep, saline, nutrient-rich water to the surface (Carroll et al., 2017). Under appropriate circumstances, these are foraging hotspots for narwhal (Laidre et al., 2016), sea birds (Urbanski et al., 2017), and other organisms. Arctic fjord-mouth ice shelves impound freshwater "epishelf" lakes with depths comparable to the ice shelf draft, and harbor distinctive mixed communities of freshwater and marine species (e.g., Vincent et al., 2001). With continued losses of ice mass from northern ice sheets and ice caps, the number of marine-terminating glaciers can be expected to shrink, as observed with the loss of tidewater side-entry glaciers in Itirbilung Fiord, Baffin Island. Meanwhile, the rapid loss of Ellesmere Island ice shelves threatens the near-term survival of the remaining epishelf lakes and associated microbial and planktonic ecosystems that depend on them.

4.2 Arctic Deltas

River and sea ice influence deltaic and estuarine processes in several ways. Sea ice is central to the formation and survival of the finger deltas in the northwest CAA, protecting them from wave action while contributing to terrace formation through ice push (Fig. 10). In these and other small basins, surface runoff ceases in winter when all precipitation arrives as snow and channels freeze to the bottom. The runoff season in high Arctic rivers is a small faction of the year, amounting ~40 days in southern Melville Island, where projections suggest that this could increase to 70 days with doubling of total annual runoff and maximum discharge, and an increase by up to six times in the suspended sediment yield by 2100 (Lewis and Lamoureux, 2010). These changes in flow duration and sediment delivery would clearly benefit Melville Island deltas

First spring flow in small nival (snowmelt-dominated) catchments is typically over ice (Forbes, 1979) and the presence of ice complicates reliable discharge measurement during the all-important breakup season. In estuarine reaches, as in the Babbage Delta, hypersalinity can develop in pools isolated by BFI on bars and shoals, but arrival of the snowmelt freshet in spring brings a sudden flushing and precipitous drop in salinity, as outflow is continuous for some weeks in this microtidal setting. Salinity only begins to build back slowly later in the summer (Forbes et al., 1994).

Ice effects in larger delta systems such as the Mackenzie include ice jamming, which raises water levels rapidly (Goulding et al., 2009), contributing to the flooding and water renewal in high-sill lakes, but also posing a safety hazard with flooding of the community of Aklavik and many small fishing and trapping camps throughout the delta (Forbes et al., 2015b). As in smaller rivers, BFI occurs in some of the largest channels and the circumstances of freeze-up determine the flow capacity of various distributaries in spring, directing first flow to various channels from year to year and undoubtedly affecting the growth directions of the delta. Confinement of prodelta flow to sub-ice channels and flow over BFI may create a bypass zone with little sedimentation at the delta front (Hill et al., 2001). At the same time, pressure ridges on the inner shelf may confine and direct freshwater inflow under the ice through the winter (Macdonald et al., 1995; Carmack and Macdonald, 2002).

4.3 Sediment Supply

As seen in Eq. (1), sedimentation is critical to the creation and maintenance of deltas and is a dominant feature of estuarine systems. In cases of self-confinement, such as the Arctic finger deltas, a component of sediment supply comes from cannibalization, but building out progressively beyond the coast increases the accommodation space and sediment required to advance the delta front.

In areas of rising relative sea level, sedimentation is the key to maintaining a positive balance in Δz. In the Mackenzie Delta, with sediment accretion rates on the delta plain and lakes amounting for the most part to <5 mm/a, this appears inadequate to maintain the balance in the face of surface subsidence rates of the same magnitude, compounded by accelerated sea-level rise.

Glacier-fed proglacial deltas and estuaries are highly sensitive to sediment supply, forming quickly during ice recession. Where a portion of the basin remains glaciated, as in Itirbilung Fiord, proglacial sedimentation can persist for thousands of years. In other cases, where the basin becomes unglaciated or flow from receding glaciers is diverted to other valleys, the tap is mostly turned off. The resulting starved system is left as a terrace above sea level (in areas of glacial-isostatic uplift) or stranded on the seabed (in areas of rising sea level) (Fig. 9).

4.4 Vulnerability to Environmental Change

Polar amplification of climate warming is closely linked to rapidly diminishing Arctic sea ice cover (Screen and Simmonds, 2010). Within Canadian waters, inter-island plugs of multiyear ice have opened up, raising the possibility of more open water, even in the northernmost parts of the archipelago currently protected by perennial ice cover (Barnhart et al., 2016; Pope et al., 2017). This would enable the development of surface waves and longshore sediment transport, which would radically alter the stability of nonresistant shores in the CAA, including the distinctive finger deltas described in this chapter (Forbes et al., 2018).

More open water combined with rising relative sea levels and possibly enhanced storm activity can be expected to increase already high rates of frontal retreat on the Mackenzie Delta. Coupled with warming temperatures promoting deeper seasonal thaw (Burn and Kokelj, 2009), the stability of the delta front in most sectors (possibly excluding eastern Shallow Bay) is likely to be compromised.

A major issue for the Mackenzie Delta is the sediment balance (Eq. 1). The discovery of modest rates of surface subsidence in a permafrost delta with ice-bonded sediment was unexpected. This may be explicable in the context of taliks (thaw zones) penetrating permafrost or promoting high concentrations of unfrozen water to the base of permafrost (Taylor et al., 2008). There is evidence for fluid escape to the surface (Forbes et al., 2015a), including persistent thermogenic methane emission over decades (Bowen et al., 2008). The widespread occurrence of methane gas pockmarks and seeps in delta ponds near Middle Channel and in the channel itself (Fig. 12F) raise the question of whether the strudel scour pits could be seeps. The deepest scour (4.5 m below seabed) had a diameter similar to that of the large, active, delta-lake pockmark described by Bowen et al. (2008). However, it had a flat floor and was less than half as deep a year later (Whalen et al., 2010). The smaller seeps observed in ponds and the fluid escape structures in the channel bar are much smaller in diameter than the strudel-scour depressions. The latter can be correlated to observed drainage activity shortly before (Fig. 5C and D) along the SAR-mapped boundaries of BFI (Fig. 4B). The scour depressions are often infilled within a single open-water season, whereas the gas seeps (as observed with repeat mapping) persist in the same locations. Nevertheless, because the strudel scours occur along the trough boundary, which appears to influence the seep locations, we cannot rule out the possibility of pockmarks occurring within the strudel field. At the same time, the evidence for the release of large volumes of methane and ethane from below permafrost in the outer delta (Bowen et al., 2008) supports the hypothesis that deep fluid escape may contribute to the observed near-surface subsidence.

5 CONCLUSIONS

Polar deltas and estuaries are distinctive coastal systems with numerous features not found in other regions of the world. Apart from the Mackenzie (basin area $1.8 \times 10^6 \, km^2$), the Nelson ($1.1 \times 10^6 \, km^2$, flowing into southern Hudson Bay), most drainage basins in the Canadian Arctic are more modest and their deltas and estuaries are correspondingly small. Some of the largest and deepest estuaries are fjords. The fjord region extends from Nunatsiavut (northern Labrador) north into Nunavut, primarily on Baffin, Bylot, Devon, Ellesmere, and Axel Heiberg islands. The sizes, catchment areas, and proportion glaciated of rivers flowing into the fjords cover a wide range, but generally fjord-head deltas are laterally confined and relatively small (Itirbilung at 40 km^2 is one of the largest).

Large, well-vegetated, systems such as the Mackenzie, in which a high proportion of annual discharge is retained for temporary storage during the breakup flood, exert a significant influence on carbon and nutrient delivery to the coastal ocean. In contrast, other systems such as the many small, gravel-dominated, proglacial deltas, are unvegetated and nonretentive systems and promote rapid flow-through of water and sediment with little transformation.

Among the world's rarest estuarine systems are the ice shelves and epishelf lakes of northern Ellesmere Island. These have been breaking up over the past century and more dramatically over the past 20 years. The remaining ice shelves cover only about 5% of the area occupied in 1906 and the epishelf lakes have all but disappeared.

Among the dominant high-latitude effects on Arctic deltas and estuaries (some not exclusive to the polar regions) are:

- extreme seasonality;
- snow (seasonal water storage);
- daily solar irradiance ranging from 0 to 24 h;
- permafrost and ground ice;
- river and sea ice;
- glaciers, ice caps, and ice shelves;
- sparse vegetation including almost total absence of trees; and
- low human impact on natural ecosystems.

Arctic deltas and estuaries of the circumpolar coast provide essential ecosystem services and are responding, in some cases rapidly, to global climate trends exacerbated by polar amplification. The Arctic ice shelves and many deltas can be considered sentinel systems for detection and tracking of climate change and its impacts at high latitudes.

ACKNOWLEDGMENTS

This work, based on experience in numerous collaborative projects over the past 50 years, is a contribution to Future Earth Coasts. Supporting partners and agencies include the Geological Survey of Canada, Canadian Geodetic Survey, and Polar Continental Shelf Project (Natural Resources Canada); the Canadian Hydrographic Service and Canadian Coast Guard (Fisheries and Oceans Canada); the Water Survey of Canada and Canadian Ice Service (Environment and Climate Change Canada); the Ocean Mapping Group (University of New Brunswick); ArcticNet Network of Centres of Excellence; Natural Sciences and Engineering Research Council of Canada; Aurora Research Institute; Inuvialuit Regional Corporation; Government of Nunavut; Canada-Nunavut Geoscience Office; Nunavut Research Institute; community partners, industry partners, and many colleagues, students, pilots, ship masters and crews, indigenous elders and others. Thanks to Eric Wolanski, Nicole Couture, and another reviewer for helpful comments on a draft of this chapter. This is contribution no. 20180012 of Natural Resources Canada. Canadian Crown Copyright reserved.

REFERENCES

Barnhart, K.R., Miller, C.R., Overeem, I., Kay, J.E., 2016. Mapping the future expansion of Arctic open water. Nat. Clim. Chang. 6, 280–285.

Bendixen, M., Kroon, A., 2017. Conceptualizing delta forms and processes in Arctic coastal environments. Earth Surf. Process. Landf. 42, 1227–1237.

Bendixen, M., Iversen, L.L., Bjørk, A.A., Elberling, B., Westergaard-Nielsen, A., Overeem, I., Barnhart, K.R., et al., 2017. Delta progradation in Greenland driven by increasing glacial mass loss. Nature 550, 101–104.

Bode, J.A., Moorman, B.J., Stevens, C.W., Solomon, S.M., 2008. Estimation of ice wedge volume in the Big Lake area, Mackenzie Delta, NWT, Canada. In: Kane, D.L., Hinkel, K.M. (Eds.), Proceedings of the Ninth International Conference on Permafrost. University of Alaska, Fairbanks, vol. 1. pp. 131–136.

Bowen, R.G., Dallimore, S.R., Coté, M.M., Wright, J.F., Lorenson, T.D., 2008. Geomorphology and gas release from pockmark features in the Mackenzie Delta, Northwest Territories, Canada. In: Kane, D.L., Hinkel, K.M. (Eds.), Proceedings of the Ninth International Conference on Permafrost. University of Alaska, Fairbanks, vol. 1. pp. 171–176.

Boyd, R., Dalrymple, R., Zaitlin, B.A., 1992. Classification of clastic coastal depositional environments. Sediment. Geol. 80, 139–150.

Brown, J., Ferrians, O.J., Jr., Heginbottom, J.A., Melnikov, E.S., 1997. *Circum-Arctic Map of Permafrost and Ground-Ice Conditions*. International Permafrost Association, US Geological Survey, Map CP-45, scale 1:10 000 000. pubs.usgs.gov/cp/45/plate-1.pdf.

Burn, C.R., Kokelj, S.V., 2009. The environment and permafrost of the Mackenzie Delta area. Permafr. Periglac. Process. 20, 83–105.

Carmack, E.C., 2000. The freshwater budget of the Arctic Ocean: sources, storage and sinks. In: Lewis, E.L., et al. (Eds.), The Freshwater Budget of the Arctic Ocean. Kluwer, Dordrecht, pp. 91–126.

Carmack, E.C., Macdonald, R.W., 2002. Oceanography of the Canadian shelf of the Beaufort Sea: a setting for marine life. Arctic 55, 29–45.

Carmack, E.C., Macdonald, R.W., 2008. Water and ice-related phenomena in the coastal region of the Beaufort Sea: some parallels between native experience and western science. Arctic 61 (3), 265–280.

Carroll, D., Sutherland, D.A., Shroyer, E.L., Nash, J.D., Catania, G.A., Stearns, L.A., 2017. Subglacial discharge-driven renewal of tidewater glacier fjords. J. Geophys. Res. 122 (8), 6611–6629.

Carson, M.A., Jasper, J.N., Conly, F.M., 1998. Magnitude and sources of sediment input to the Mackenzie Delta, Northwest Territories, 1974–94. Arctic 51, 116–124.

Church, M., 1972. Baffin Island Sandurs: A Study of Arctic Fluvial Processes. Geological Survey of Canada, Bulletin 216, https://doi.org/10.4095/100928.

Church, M., 1974. Hydrology and permafrost with reference to northern North America. In: Permafrost Hydrology. Canadian National Committee, International Hydrological Decade, Environment Canada, Ottawa, pp. 7–20.

Church, J.A., Clark, P.U., Cazenave, A., Gregory, J.M., Jevrejeva, S., Levermann, A., Merrifield, M.A., et al., 2013. Sea level change. In: Stocker, T.F., Qin, D., Plattner, G.K., Tignor, M., Allen, S.K., Boschung, J., Nauels, A., et al. (Eds.), Climate Change 2013: The Physical Science Basis. Contribution of Working Group I to the Fifth Assessment Report of the Intergovernmental Panel on Climate Change. Cambridge University Press, Cambridge and New York, pp. 1137–1216.

Comiso, J., 2002. A rapidly declining perennial sea ice cover in the Arctic. Geophys. Res. Lett. 29 (20), 1956. https://doi.org/10.1029/2002GL015650.

Comiso, J.C., Parkinson, C.L., Gersten, R., Stock, L., 2008. Accelerated decline in the Arctic sea ice cover. Geophys. Res. Lett. 35, L01703, https://doi.org/10.1029/2007GL031972.

Copland, L., Mueller, D.R., Weir, L., 2007. Rapid loss of the Ayles ice shelf, Ellesmere Island, Canada. Geophys. Res. Lett. 34, L21501, https://doi.org/10.102/2007GL031809.

Copland, L., Mortimer, C., White, A., Richer McCallum, M., Mueller, D., 2017. Factors contributing to recent Arctic ice shelf losses. In: Copland, L., Mueller, D. (Eds.), Arctic Ice Shelves and Ice Islands. Springer Polar Sciences, Springer, Dordrecht, pp. 263–285.

Cowan, B., 2015. Shorelines Beneath the Sea: Geomorphology and Characterization of the Postglacial Sea-Level Lowstand, Cumberland Peninsula, Baffin Island, Nunavut (M.Sc. thesis). Memorial University of Newfoundland, St. John's.

Crasto, N., Hopkinson, C., Forbes, D.L., Lesack, L., Marsh, P., Spooner, I., van der Sanden, J.J., 2015. A LiDAR-based decision-tree classification of open water surfaces in an Arctic delta. Remote Sens. Environ. 164, 90–102.

Dana, L.-P., Meis-Mason, A., Anderson, R.B., 2008. Oil and gas and the Inuvialuit people of the Western Arctic. J. Enterpris. Commun.: People Places Global Eco. 2 (2), 151–157.

Day, J.W., Agboola, J., Chen, Z., D'Elia, C., Forbes, D.L., Giosan, L., Kemp, P., et al., 2016. Approaches to defining deltaic sustainability in the 21st century. Estuar. Coast. Shelf Sci. 183 (Part B), 275–291.

Deitrich, P., Ghienne, J.-F., Normandeau, A., Lajeunesse, P., 2016. Upslope-migrating bedforms in a proglacial sandur delta: cyclic steps from river-derived underflows? J. Sediment. Res. 86 (2), 113–123.

Dixon, J., Dietrich, J.R., MacNeil, D.H., 1992. Upper Cretaceous to Pleistocene Sequence Stratigraphy of the Beaufort-Mackenzie and Banks Island Areas, Northwest Territories, Canada. Geological Survey of Canada, Bulletin 407, https://doi.org/10.4095/133237.

Dowdeswell, E.K., Andrews, J.T., 1985. The fiords of Baffin Island: description and classification. In: Andrews, J.T. (Ed.), Quaternary Environments: Eastern Canadian Arctic. Baffin Bay and West GreenlandAllen & Unwin, Boston, pp. 93–123.

Dugan, H.A., Lamoureux, S.F., 2011. The chemical development of a hypersaline coastal basin in the High Arctic. Limnol. Oceanogr. 56, 495–507.

Dunbar, M.J., 1951. Eastern Arctic waters. Fish. Res. Board Bull. 88, 1–31.

Dyke, A.S., 1979. Glacial and sea-level history of southwestern Cumberland Peninsula, Baffin Island, N.W.T., Canada. Arct. Alp. Res. 11 (2), 179–202.

Emmerton, C.A., Lesack, L.F.W., Marsh, P., 2007. Lake abundance, potential water storage, and habitat distribution in the Mackenzie River delta, western Canadian Arctic. Water Resour. Res. 43, W05419, https://doi.org/10.1029/2006WR005139.

Emmerton, C.A., Lesack, L.F.W., Vincent, W.F., 2008. Mackenzie River nutrient delivery to the Arctic Ocean and effects of the Mackenzie Delta during open water conditions. Glob. Biogeochem. Cycles 22, GB1024, https://doi.org/10.1029/2006GB002856.

England, J.H., Lakeman, T.R., Lemmen, D.S., Bednarski, J.M., Stewart, T.G., Evans, D.J.A., 2008. A millennial-scale record of Arctic Ocean sea ice variability and the demise of the Ellesmere Island ice shelves. Geophys. Res. Lett. 35, L19502, https://doi.org/10.1029/2008GL034470.

Federova, I., Cheverova, A., Bolshiyanov, D., Makarov, A., Boike, J., Heim, B., Morgenstern, A., et al., 2015. Lena Delta hydrology and geochemistry: long-term hydrological data and recent field observations. Biogeosciences 12, 345–363.

Forbes, D.L., 1979. In: Bottomfast ice in northern rivers: hydraulic effects and hydrometric implications. *Proceedings, Canadian Hydrology Symposium 79*. Associate Committee on Hydrology, National Research Council of Canada, Ottawa. pp. 175–184.

Forbes, D.L., 1983. Morphology and sedimentology of a sinuous gravel-bed channel system: lower Babbage River, Yukon coastal plain, Canada. In: Collinson, J.D., Lewin, J. (Eds.), Modern and Ancient Fluvial Systems. International Association of Sedimentologists, Special Publication 6, pp. 195–206.

Forbes, D.L., 2004. Ice-bonded sediments and massive ground ice in a transgressive barrier-lagoon and delta complex, Yukon Coast of Beaufort Sea, western Arctic Canada. EOS Trans. Am. Geophys. Union 85, 47. Fall Meeting Supplement, abstract and e-poster, C13A-0267.

Forbes, D.L., 2011. Glaciated coasts. In: Wolanski, E., McLusky, D. (Eds.), Treatise on Estuarine and Coastal Science. vol. 3. Academic Press, Waltham, pp. 223–243.

Forbes, D.L., Hansom, J.D., 2011. Polar coasts. In: Wolanski, E., McLusky, D. (Eds.), Treatise on Estuarine and Coastal Science. vol. 3. Academic Press, Waltham, pp. 245–283.

Forbes, D.L., Taylor, R.B., 1994. Ice in the shore-zone and the geomorphology of cold coasts. Prog. Phys. Geogr. 18 (1), 59–89.

Forbes, D.L., Taylor, R.B., Frobel, D., 1986. Coastal Studies in the Western Arctic Archipelago (Melville, Mackenzie King, Lougheed and Nearby Islands). Geological Survey of Canada, Open File 1409, https://doi.org/10.4095/130198.

Forbes, D.L., Solomon, S.M., Hamilton, T.S., 1994. Morphology and sedimentary processes of microtidal embayments, Beaufort Sea coast, western Arctic Canada. In: Viggósson, G. (Ed.), Proceedings, Hornafjörður International Coastal Symposium, Höfn. Icelandic Harbour Authority, Kópavogur. pp. 363–372.

Forbes, D.L., Manson, G.K., Whalen, D.J.R., Couture, N.J., Hill, P.R., 2014. Coastal products of marine transgression in cold-temperate and high-latitude coastal-plain settings: Gulf of St Lawrence and Beaufort Sea. In: Martini, I.P., Wanless, H.R. (Eds.), Sedimentary Coastal Zones from High to Low Latitudes: Similarities and Differences. Geological Society, London, Special Publications, vol. 388. pp. 131–163.

Forbes, D.L., Craymer, M.R., Whalen, D.J.R., 2015a. Subsidence and inundation of a large Arctic permafrost delta. In: *CANQUA 2015*, Canadian Quaternary Association, Memorial University of Newfoundland, St. John's. www.researchgate.net/publication/323599628.

Forbes, D.L., Whalen, D.J.R., Fraser, P.R., 2015b. Sharing remote and local information for tracking spring breakup in the Mackenzie Delta and Beaufort Sea. In: American Geophysical Union, Fall Meeting, San Francisco, CA, December 2015, abstract and e-poster PA13A–2178.

Forbes, D.L., Bell, T., Manson, G.K., Couture, N.J., Cowan, B., Deering, R.L., Hatcher, S.V., et al., 2018. Coastal environments and drivers. In: Bell, T., Brown, T. (Eds.), From Science to Policy in the Eastern Canadian Arctic: An Integrated Regional Impact Study (IRIS) of Climate Change and Modernization. ArcticNet, Québec, pp. 211–249.

Ford, J.D., Smit, B., Wandel, J., MacDonald, J., 2006. Vulnerability to climate change in Igloolik, Nunavut: what we can learn from the past and present. Polar Record 42 (221), 127–138.

Ford, J.D., Bell, T., Couture, N.J., 2016. Perspectives on Canada's north coast region. In: Lemmen, D.S., Warren, F.J., James, T.S., Mercer Clarke, C.S.L. (Eds.), Canada's Marine Coasts in a Changing Climate. Government of Canada, Ottawa, pp. 153–206.

Gardner, A.S., Moholdt, M., Wouters, B., Wolken, G.J., Burgess, D.O., Sharp, M., Cogley, J.G., et al., 2011. Sharply increased mass loss from glaciers and ice caps in the Canadian Arctic Archipelago. Nature 473, 357–360.

Goulding, H., Prowse, T., Beltaos, S., 2009. Spatial and temporal patterns of break-up and ice-jam flooding in the Mackenzie Delta, NWT. Hydrol. Process. 23 (18), 2654–2670.

Grasby, S.E., Smith, I.R., Bell, T., Forbes, D.L., 2013. Cryogenic formation of brine and sedimentary mirabilite in submergent coastal lake basins, Canadian Arctic. Geochim. Cosmochim. Acta 110, 12–28.

Grosse, G., Jones, B., Arp, C., 2013. Thermokarst lakes, drainage, and drained basins. In: Shroder, J.F. (Ed.), Treatise on Geomorphology. Elsevier, Amsterdam, pp. 325–353.

Harrison, J.C., Lynds, T., Ford, A., 2015. Geology, Tectonic Assemblage Map of Byam Martin Channel Area, Melville and Surrounding islands, Nunavut—Northwest Territories. Geological Survey of Canada, Canadian Geoscience Map 32 (second ed., preliminary), scale1:500 000, https://doi.org/10.4095/296220.

Herdes, E., Copland, L., Danielson, B., Sharp, M., 2012. Relationships between iceberg plumes and sea-ice conditions on northeast Devon Ice Cap, Nunavut, Canada. Ann. Glaciol. 53 (60), 1–9.

Hill, P.R., 1996. Late Quaternary sequence stratigraphy of the Mackenzie Delta. Can. J. Earth Sci. 33 (7), 1064–1074.

Hill, P.R., Solomon, S.M., 1999. Geomorphologic and sedimentary evolution of a transgressive thermokarst coast, Mackenzie Delta region, Canadian Beaufort Sea. J. Coast. Res. 15, 1011–1029.

Hill, P.R., Barnes, P.W., Héquette, A., Ruz, M.-H., 1994. Arctic coastal plain shorelines. In: Carter, R.W.G., Woodroffe, C.D. (Eds.), Coastal Evolution: Late Quaternary Shoreline Morphodynamics. Cambridge University Press, Cambridge, pp. 341–372.

Hill, P.R., Lewis, C.P., Desmarais, S., Kauppaymuthoo, V., Rais, H., 2001. The Mackenzie Delta: sedimentary processes and facies of a high-latitude, fine-grained delta. Sedimentology 48, 1047–1078.

Holmes, R.M., McClelland, J.W., Peterson, B.J., Shiklomanov, I.A., Shiklomanov, A.I., Zhulidov, A.V., Gordeev, V.V., Bobrovitskaya, N.N., 2002. A circumpolar perspective on fluvial sediment flux to the Arctic Ocean. Glob. Biogeochem. Cycles 16 (4), 1098. https://doi.org/10.1029/2001GB001849.

Hopkinson, C., Crasto, N., Marsh, P., Forbes, D., Lesack, L., 2011. Investigating the spatial distribution of water levels in the Mackenzie Delta using airborne LiDAR. Hydrol. Process. 25 (19), 2995–3011.

Howell, S.E.L., Duguay, C.R., Markus, T., 2009. Sea ice conditions and melt season duration variability within the Canadian Arctic Archipelago: 1979–2008. Geophys. Res. Lett. 36, L10502, https://doi.org/10.1029/2009GL037681.

Hughes Clarke, J.E., Muggah, J., Renoud, W., Bell, T., Forbes, D.L., Cowan, B., Kennedy, J., 2015. Reconnaissance seabed mapping around Hall and Cumberland peninsulas, Nunavut: opening up southeastern Baffin Island to nearshore geological investigations. In: *Summary of Activities 2014*. Canada-Nunavut Geoscience Office, Iqaluit, NU, 133-144. cngo.ca/app/uploads/Summary-of-Activities-2014-P15.pdf.

James, T.S., Henton, J.A., Leonard, L.J., Darlington, A., Forbes, D.L., Craymer, M., 2014. Relative Sea-level Projections in Canada and the Adjacent Mainland United States. Geological Survey of Canada, Open File 7737, https://doi.org/10.4095/295574.

Jeffries, M.O., 1992. Arctic ice shelves and ice islands: origin, growth and disintegration, physical characteristics, structural-stratigraphic variability, and dynamics. Rev. Geophys. 30, 245–267.

Jenner, K.A., Hill, P.R., 1998. Recent Arctic deltaic sedimentation, Olivier Islands, Mackenzie Delta, Northwest Territories, Canada. Sedimentology 45, 987–1004.

Johnston, G.H., Brown, R.J.E., 1965. Stratigraphy of the Mackenzie River delta, Northwest Territories, Canada. Geol. Soc. Am. Bull. 76 (1), 103–112.

Kokelj, S.V., Burn, C.R., 2005. Near-surface ground ice in sediments of the Mackenzie Delta, Northwest Territories, Canada. Permafr. Periglac. Process. 16, 291–303.

Kokelj, S.V., Lantz, T.C., Solomon, S., Pisaric, M.F.J., Keith, D., Morse, P., Thienpont, J.R., Smol, J.P., Esagok, D., 2012. Using multiple sources of knowledge to investigate northern environmental change: regional ecological impacts of a storm surge in the outer Mackenzie Delta, N.W.T. Arctic 65 (3), 257–272.

Laidre, K.L., Moon, T., Hauser, D.D.W., McGovern, R., Heide-Jørgensen, M.P., Dietz, R., Hudson, B., 2016. Use of glacial fronts by narwhals (*Monodon monoceros*) in West Greenland. Biol. Lett. 12 (10), 20160457.

Lesack, L.F.W., Marsh, P., Hicks, F.E., Forbes, D.L., 2013. Timing, duration, and magnitude of peak annual water levels during ice breakup in the Mackenzie Delta and the role of river discharge. Water Resour. Res. 49, 8234–8249.

Lewis, T., Lamoureux, S.F., 2010. Twenty-first century discharge and sediment yield predictions in a small high Arctic watershed. Glob. Planet. Chang. 71, 27–41.

Lønne, I., Nemec, W., 2004. High-Arctic fan delta recording deglaciation and environmental disequilibrium. Sedimentology 51, 553–589.

Macdonald, R.W., Carmack, E.C., 1991. The role of large-scale under-ice topography in separating estuary and ocean on an Arctic shelf. Atmosphere-Ocean 29, 37–53.

Macdonald, R.W., Paton, D.W., Carmack, E.C., Omstedt, A., 1995. The freshwater budget and under-ice spreading of Mackenzie River water in the Canadian Beaufort Sea based on salinity and $^{18}O/^{16}O$ measurements in water and ice. J. Geophys. Res. 100, 895–919.

Mackay, J.R., 1972. Offshore permafrost and ground ice, southern Beaufort Sea, Canada. Can. J. Earth Sci. 9, 1550–1561.

Manson, G.K., Solomon, S.M., 2007. Past and future forcing of Beaufort Sea coastal change. Atmosphere-Ocean 45 (2), 107–122.

Marsh, P., Hey, M., 1989. The flooding hydrology of Mackenzie Delta lakes near Inuvik, N.W.T., Canada. Arctic 42 (1), 41–49.

Marsh, P., Lesack, L.F.W., Roberts, A., 1999. Lake sedimentation in the Mackenzie Delta, NWT. Hydrol. Process. 13, 2519–2536.

Martini, I.P., Jefferies, R.L., Morrison, R.I.G., Abraham, K.F., 2009. Polar coastal wetlands: development, structure, and land use. In: Perillo, G., Wolanski, E., Cahoon, D., Brinson, M. (Eds.), Coastal Wetlands: An Integrated Ecosystems Approach. Elsevier, Amsterdam, pp. 119–155.

McClelland, J.W., Holmes, R.M., Dunton, K.H., Macdonald, R.W., 2012. The Arctic Ocean estuary. Estuar. Coasts 35, 353–368.

Miller, G.H., Dyke, A.S., 1974. Proposed extent of late Wisconsin Laurentide ice on Baffin Island. Geology 2, 125–130.

Mueller, D.R., Vincent, W.F., Jeffries, M.O., 2003. Break-up of the largest Arctic ice shelf and associated loss of an epishelf lake. Geophys. Res. Lett. 30 (20), 2031. https://doi.org/10.1029/2003GL017931.

Mueller, D.R., Vincent, W.F., Jeffries, M.O., 2006. Environmental gradients, fragmented habitats, and microbiota of a northern ice shelf cryoecosystem, Ellesmere Island, Canada. Arct. Antarct. Alp. Res. 38 (4), 593–607.

Mueller, D.R., Copland, L., Hamilton, A., Stern, D., 2008. Examining Arctic ice shelves prior to the 2008 breakup. EOS Trans. Am. Geophys. Union 89 (49), 502–503.

Mueller, D., Copland, L., Jeffries, M.O., 2017. Changes in Canadian Arctic ice shelf extent since 1906. In: Copland, L., Mueller, D. (Eds.), Arctic Ice Shelves and Ice Islands. Springer Polar Sciences, Springer, Dordrecht, pp. 263–285.

Murton, J.B., 2001. Thermokarst sediments and sedimentary structures, Tuktoyaktuk coastlands, western Arctic Canada. Glob. Planet. Chang. 28 (1–4), 175–192.

Nixon, F.C., England, J.H., Lajeunesse, P., Hanson, M.A., 2014. Deciphering patterns of postglacial sea level at the junction of the Laurentide and Innuitian ice sheets, western Canadian High Arctic. Quat. Sci. Rev. 91, 165–183.

Normandeau, A., Lajeunesse, P., Poiré, A.G., Francus, P., 2016. Morphological expression of bedforms formed by supercritical sediment density flows on four fjord-lake deltas of the southeastern Canadian Shield (eastern Canada). Sedimentology 63 (7), 2106–2129.

Ogorodov, S., Arkhipov, V., Kokin, O., Marchenko, A., Overduin, P., Forbes, D., 2013. Ice effect on coast and seabed in Baydaratskaya Bay, Kara Sea. Geography Environ. Sustain. 6 (3), 21–37.

Overduin, P.P., Hubberten, H.-W., Rachold, V., Romanovskii, N., Grigoriev, M., Kasymskaya, M., 2007. The evolution and degradation of coastal and offshore permafrost in the Laptev and East Siberian Seas during the last climatic cycle. In: Harff, J., Hay, W.W., Tetzlaff, D.M. (Eds.), Coastline Changes: Interrelation of Climate and Geological Processes. Geological Society of America, Special Paper, vol. 426. pp. 97–111.

Peltier, W.R., 2004. Global glacial isostasy and the surface of the ice-age Earth: the ICE-5G (VM2) model and GRACE. Annu. Rev. Earth Planet. Sci. 32, 111–149.

Peltier, W.R., Argus, D.F., Drummond, R., 2015. Space geodesy constrains ice age terminal deglaciation: the global ICE-6G_C (VM5a) model. J. Geophys. Res. 120 (1), 450–487.

Pisaric, M.F.J., Thienpont, J.R., Kokelj, S.V., Nesbitt, H., Lantz, T.C., Solomon, S., Smol, J.P., 2011. Impacts of a recent storm surge on an Arctic delta ecosystem examined in the context of the last millennium. Proc. Am. Acad. Sci. 108 (22), 8960–8965.

Pope, S., Copland, L., Alt, B., 2017. Recent changes in sea ice plugs along the northern Canadian Arctic Archipelago. In: Copland, L., Mueller, D. (Eds.), Arctic Ice Shelves and Ice Islands. Springer Polar Sciences, Springer, Dordrecht, pp. 317–342.

Rachold, V., Bolshiyanov, D.Y., Grigoriev, M.N., Hubberten, H.-W., Junker, R., Kunitsky, V.N., Merker, F., et al., 2007. Nearshore Arctic permafrost in transition. EOS Trans. Am. Geophys. Union 88 (13), 149–150.

Reimnitz, E., Rodeick, C.A., Wolf, S.C., 1974. Strudel scour: a unique Arctic marine geologic phenomenon. J. Sediment. Petrol. 44 (2), 409–420.

Richter-Menge, J., Overland, J.E., Mathis, J.T., Osborne, E. (Eds.), 2017. Arctic Report Card 2017. National Oceanic and Atmospheric Administration, Silver Springs, MD. www.arctic.noaa.gov/Report-Card.

Ruz, M.-H., Héquette, A., Hill, P.R., 1992. A model of coastal evolution in a transgressed thermokarst topography. Mar. Geol. 106, 251–278.

Screen, J.A., Simmonds, I., 2010. The central role of diminishing sea ice in recent Arctic temperature amplification. Nature 464, 1334–1337.

Shakhova, N., Semiletov, I., Sergienko, V., Lobkovsky, L., Yusupov, V., Salyuk, A., Salomatin, A., et al., 2015. The East Siberian Arctic Shelf: towards further assessment of permafrost-related methane fluxes and the role of sea ice. Phil. Trans. R. Soc. A 373, 20140451.

Smith, L.C., Sheng, Y., Magilligan, F.J., Smith, N.D., Gomez, B., Mertes, L.A.K., Krabill, W.B., Garvin, J.B., 2006. Geomorphic impact and rapid subsequent recovery from the 1996 Skeiðarársandur jökulhlaup, Iceland, measured with multi-year airborne lidar. Geomorphology 75 (102), 65–75.

Solomon, S.M., 2005. Spatial and temporal variability of shoreline change in the Beaufort-Mackenzie region, Northwest Territories, Canada. Geo-Mar. Lett. 25, 127–137.

Solomon, S.M., Forbes, D.L., Fraser, P., Moorman, B.G., Stevens, C.W., Whalen, D., 2008. In: Nearshore geohazards in the southern Beaufort Sea, Canada. Paper IPC2008-64349. *Proceedings of the 7th International Pipeline Conference*, Calgary, Alberta. American Society of Mechanical Engineers, New York, vol. 4. pp. 281–290.

Statistics Canada. 2017. *Census Profile, 2016 Census*. Statistics Canada, Ottawa. www12.statcan.gc.ca/census-recensement/2016/dp-pd/prof/index.cfm?Lang=E.

St-Hilaire-Gravel, D., Forbes, D.L., Bell, T., 2012. Multitemporal analysis of a gravel-dominated coastline in the central Canadian Arctic Archipelago. J. Coast. Res. 28 (2), 421–441.

Stroeve, J.C., Markus, T., Boisvert, L., Miller, J., Barrett, A., 2014. Changes in Arctic melt season and implications for sea ice loss. Geophys. Res. Lett. 41, 1216–1225.

Syvitski, J.P.M., Hein, F.J., 1991. Sedimentology of an Arctic Basin: Itirbilung Fiord, Baffin Island, Northwest Territories. Geological Survey of Canada, Paper 91-11, https://doi.org/10.4095/132684.

Syvitski, J.P.M., Burrell, D.C., Skei, J.M., 1987. Fjords: Processes and Products. Springer-Verlag, New York.

Syvitski, J.P.M., Kettner, A.J., Overeem, I., Hutton, E.W.H., Hannon, M.T., Brakenridge, G.R., Day, J., Vörösmarty, C., Saito, Y., Giosan, L., Nicholls, R.J., 2009. Sinking deltas due to human activities. Nat. Geosci. 2, 681–686.

Taylor, A.E., Dallimore, S.R., Outcalt, S.I., 1996. Late Quaternary history of the Mackenzie-Beaufort region, Arctic Canada, from modelling of permafrost temperatures. 1. The onshore-offshore transition. Can. J. Earth Sci. 33 (1), 52–61.

Taylor, A.E., Dallimore, S.R., Wright, J.F., 2008. Thermal impact of Holocene lakes on a permafrost landscape, Mackenzie Delta, Canada. In: Kane, D.L., Hinkel, K.M. (Eds.), *Proceedings of the Ninth International Conference on Permafrost*. University of Alaska, Fairbanks, vol. 2. pp. 1757–1762.

Todd, B.J., Dallimore, S.R., 1998. Electromagnetic and geological transect across permafrost terrain, Mackenzie River delta, Canada. Geophysics 63 (6), 1914–1924.

Urbanski, J.A., Stempniewicz, L., Weslawski, J.M., Draganska-Deja, K., Wochna, A., Goc, M., Iliszko, L., 2017. Subglacial discharges create fluctuating foraging hotspots for sea birds in tidewater glacier bays. Sci. Rep. 7, 43999. https://doi.org/10.1038/srep43999.

Usher, P.J., 2002. Inuvialuit use of the Beaufort Sea and its resources, 1960–2000. Arctic 55 (suppl. 1), 18–28.

van Hove, P., Swadling, K., Gibson, J.A.E., Belzile, C., Vincent, W.F., 2001. Farthest north lake and fiord populations of calanoid copepods in the Canadian High Arctic. Polar Biol. 24, 303–307.

Veillette, J., Mueller, D.R., Antoniades, D., Vincent, W.F., 2008. Arctic epishelf lakes as sentinel ecosystems: past, present, and future. J. Geophys. Res. 113, G04014, https://doi.org/10.1029/2008JG000730.

Vermaire, J.C., Pisaric, M.F.J., Thienpont, J.R., Courtney Mustaphi, C.J., Kokelj, S.V., Smol, J.P., 2013. Arctic climate warming and sea ice declines lead to increased storm surge activity. Geophys. Res. Lett. 40 (7), 1386–1390.

Vincent, W.F., Gibson, J.A., Pienitz, R., Villeneuve, V., Broady, P.A., Hamilton, P.B., Howard-Williams, C., 2000. Ice shelf microbial ecosystems in the High Arctic and implications for life on snowball Earth. Naturwissenschaften 87, 137–141.

Vincent, W.F., Gibson, J.A.E., Jeffries, M.O., 2001. Ice shelf collapse, climate change, and habitat loss in the Canadian High Arctic. Polar Record 37 (201), 133–142.

Walker, H.J., 1998. Arctic deltas. J. Coast. Res. 14 (3), 718–738.

Walker, D.A., Raynolds, M.K., Daniëls, F.J.A., Einarsson, E., Elvebakk, A., Gould, W.A., Katenin, A.E., et al., 2005. The circumpolar Arctic vegetation map. J. Veg. Sci. 16, 267–282.

Whalen, D., Solomon, S., Forbes, D.L., Couture, N., Lintern, G., Lavergne, J.C., Craymer, M., 2010. In: Environmental impacts and geohazards in the Mackenzie Delta and shallow Beaufort Sea. Yellowknife Geoscience Forum, November 2010. www.researchgate.net/publication/323599725.

Chapter 9

Delta Winners and Losers in the Anthropocene

John W. Day*, Ramesh Ramachandran[†], Liviu Giosan[‡], James Syvitski[§], G. Paul Kemp*

*Department of Oceanography and Coastal Sciences, College of the Coast and Environment, Louisiana State University, Baton Rouge, LA, United States, [†]National Centre for Sustainable Coastal Management, Ministry of Environment, Forest and Climate Change, Government of India, Anna University, Chennai, India, [‡]Department of Geology and Geophysics, Woods Hole Oceanographic Institution, Woods Hole, MA, United States, [§]Community Surface Dynamics Modeling System, University of Colorado, Boulder, CO, United States

1 INTRODUCTION

Coastal systems are highly productive with great ecological, social, and economic value and the present book reaffirms this paradigm. They are also most threatened by human activities and climate change. This is especially the case for deltas, which are the most productive and economically important global ecosystems, associated with some of the largest coastal marine fisheries and the majority of global coastal wetlands. They are often regions of intense economic activities including agriculture, navigation and trade, fisheries, forestry, fossil energy production, and manufacturing. Because of the ecological richness, deltas support the highest values of ecosystem goods and services (EGS) in the world (Day et al., 1997, 2007a,b; Costanza et al., 1997; Batker et al., 2014; Kuenzer and Renaud, 2012; Vörösmarty et al., 2009; Syvitski et al., 2009; Chen and Saito, 2011). Many large cities are located in or adjacent to deltas such as Yangtze, Ganges, Yellow, Mekong, Rhine, Nile, and Mississippi, and they are important sites for maritime trade (e.g., Blackburn et al., 2019). One in fourteen people globally live in and around deltas. As demonstrated in many chapters of this book, deltas have been tremendously altered by human activities (Syvitski et al., 2009; Vörösmarty et al., 2009, Renaud et al., 2013; Day et al., 2016) and are more sensitive to global climate change than most other coastal systems due to large areas of near sea level wetlands and often high rates of subsidence (Day and Rybczyk, 2019).

In this chapter, we synthesize information presented on deltas in this book and elsewhere and discuss how individual deltas will fare given the given global megatrends of the 21st century. In doing so, we use the conceptual framework developed by Day et al. (2016) to define deltaic sustainability and to ask "which deltas will be winners and which will be losers?"

Deltas worldwide are becoming highly stressed and degraded systems (Day et al., 2007a,b, 2014, 2016; Syvitski et al., 2009; Vörösmarty et al., 2009; Renaud et al., 2013; Giosan et al., 2014; Tessler et al., 2015; Ibañez, 2015). Most medium and large deltas face a reduction in area due to reduced sediment input and sea-level rise (Giosan et al., 2014). Large areas in deltas have been "reclaimed" for agriculture, aquaculture, urban growth, and industry (Kuenzer et al., 2014). Deltas often receive high levels of pollutants (Chen et al., 2010a,b), as is the case of excessive nitrogen levels (Rabalais and Turner, 2001). Deltas are highly vulnerable to sea-level rise because of their low relief combined with high rates of natural subsidence, often exacerbated by peat oxidation and extraction of subsurface ground water, natural gas, and petroleum (Sestini, 1992; Morton et al., 2005; Chen et al., 2008; Day et al., 2011b; Wang et al., 2012; Kuenzer et al., 2014; Higgins et al., 2013, 2014; Ibañez, 2015). Both climate change and growing energy costs will limit options for sustainable management and flood protection because many restoration approaches are energy intensive and climate change will make restoration more challenging (Day et al., 2005, 2007a,b, 2014, 2016; Giosan et al., 2014; Tessler et al., 2015; Day and Rybczyk, 2019).

2 A FRAMEWORK FOR UNDERSTANDING THE DEVELOPMENT, FUNCTIONING, AND SUSTAINABILITY OF DELTAS AND THE ROLE OF ENERGETIC FORCING EVENTS IN THE FUNCTIONING OF DELTAS

All current deltas formed over the past several thousand years after sea level stabilization following the end of the last glacial epoch. This period had three characteristics that enhanced the development of deltas—relatively stable sea level,

Coasts and Estuaries. https://doi.org/10.1016/B978-0-12-814003-1.00009-5

149

predictable and regular input from drainage basins, and open deltaic systems with a high degree of interaction among drainage system, river channels, the deltaic plain, and the coastal ocean. The size of deltas was impacted by the area and geomorphology of drainage basins, river discharge, the area and slope of the continental shelf, and the energy regime of the coastal ocean. Human activity has changed all of this.

Delta sustainability must be considered within the context of global biophysical and socioeconomic constraints (e.g., Burger et al., 2012). This means that continuous flows of energy and materials are needed to maintain the highly organized, far-from-equilibrium nature of deltas, without which deltas quickly become destabilized. Past human activities have shown this to be the case (Syvitski et al., 2009; Vörösmarty et al., 2009; Day et al., 2014; Giosan et al., 2014). Deltas are embedded in the larger biogeosphere that includes climate, drainage basins, and the oceans. When exchanges with the larger systems are disrupted, deltaic ecosystems degrade (e.g., Snedaker, 1984; Milliman et al., 1984, 1989; Syvitski and Saito, 2007; Syvitski, 2008; Syvitski et al., 2005a,b, 2009; Vörösmarty et al., 2009; Day et al., 1997, 2011a,b, 2014; Kuenzer et al., 2013; Giosan et al., 2014; Ibañez, 2015).

The functioning and sustainability of deltas depend on regular and episodic, external and internal, inputs of energy and materials that produce benefits over different spatial and temporal scales (Odum et al., 1995; Day et al., 1997, 2007a,b, 2016). These scales range from daily tides to development of new delta lobes (Table 1). Infrequent events such as channel switching, crevasses, large river floods, and strong storms largely control sediment delivery and impact geomorphology. For example, two large floods of the Rhone River that broke dikes formed large depositional splays in the delta (Pont et al., 2017). More frequent events maintain salinity gradients and regulate biogeochemical processes.

3 PERSPECTIVES ON DELTA SUSTAINABILITY

Deltaic sustainability refers to the persistence through time of the structure and function of deltaic systems, especially as related to human and climate forcings. Because deltas are such open systems, their functioning is significantly impacted by broader global processes at the scale of the drainage basin (e.g., freshwater and sediment input), ocean (e.g., sea-level rise), and atmosphere (e.g., climate warming) (Day et al., 2016).

TABLE 1 A Hierarchy of Forcing or Pulsing Events Affecting the Formation and Sustainability of Deltas

Event	Time Scale	Impact
Deltaic lobes	100s to >1000 years	Deltaic lobe development
Crevasses	10s–100s years	Natural levee development Minor lobe development
Sea level rise	10s years	Delta regeneration by flooding the estuarine flood plain
Major river floods	20–100 years	Channel switching Major sediment deposition
Major storms	5–20 years	Moderate deposition Enhanced production
Average river floods	Annual	Enhanced deposition Freshening (lower salinity) Nutrient input Enhanced 1° and 2° production
Normal storm events (frontal passages)	Weekly	Enhanced deposition Organism transport Net sediment and chemical transport
Tides	Daily	Drainage/marsh production Low net transport

Modified from Day, J., Martin, J., Cardoch, L., Templet, P., 1997. System, functioning as a basis for sustainable management of deltaic ecosystems. Coast. Manag. 25, 115–154; Day, J.W., Boesch, D., Clairain, E., Kemp, P., Laska, S., Mitsch, W., Orth, K, Mashriqui, H., Reed, D., Shabman, L., Simenstad, L, Streever, B., Twilley, R., Watson, W.J., Whigham, D. 2007. Restoration of the Mississippi Delta: lessons from hurricanes Katrina and Rita. Science 315, 1679–1684; Day, J., Gunn, J., Folan, W., Yáñez-Arancibia, A., 2007. Emergence of complex societies after sea level stabilization. Eos 88, 169-170; Day, J., Agboola, J., Chen, Z., D'Elia, C., Forbes, D., Giosan, L., Kemp, P., Kuenzer, C., Lane, R., Ramachandran, R., Syvitski, J., Yanez, A., 2016. Approaches to defining deltaic sustainability in the 21st century. Estuar. Coast. Shelf Sci. 183(B), 275–291. https://doi.org/10.1016/j.ecss.2016.06.018; Williams, D., 2013. The lower Mary River and flood plains. In: Wolanski, E. (Ed.), Estuaries of Australia in 2050 and Beyond. Springer, Dordrecht (Chapter 16).

Deltaic sustainability can be considered from geomorphic, ecological, and economic perspectives (Day et al., 1997, 2016). A delta is geomorphically sustainable if the long-term net change in surface elevation is greater than relative sea-level rise (RSLR) and if total delta area does not decrease over time. Giosan et al. (2014) reported that almost all large deltas $>10,000\,km^2$ and most intermediate deltas from 1000 to $10,000\,km^2$ do not have sufficient historic mineral sediment input to offset projected sea-level rise of 1 m by 2100. We revisit the database of Giosan et al. taking into consideration the deltas discussed in this paper and the acceleration of sea-level rise. A delta is ecologically sustainable if the long-term change in net primary productivity (NPP) of delta plants is greater than or equal to zero. This is important because primary productivity supports coastal ecosystems by providing food and habitat, supporting organic soil formation and carbon sequestration, improving water quality, and buffering storm surges. A delta is economically sustainable if the net value of economic activity generated in the delta is greater than economic subsidies imported into the delta. Deltas have very high values of EGS (Costanza et al., 1997, 2017; Kuenzer and Vo Quoc, 2013; Batker et al., 2014).

Deltas that are more vulnerable are likely to be less sustainable. Wolters and Kuenzer (2015) defined resilience as the "degree to which a system and its components are able to anticipate, absorb, accommodate, or recover from perturbations or stress". Kemp et al. (2016) reported that large river plumes of low-salinity water on the shelf in the Southern and Northern Gulf of Mexico promote resilience to climate change. Thus, vulnerability is related to a reduction in the forcings.

Deltaic sustainability depends on interactive and hierarchical interactions among geomorphic, ecological, and economic sustainability. Economic sustainability is dependent on ecological sustainability that delivers the EGS that sustain economic activity. Ecological sustainability is thus dependent on and contributes to geomorphic sustainability, for example, wetland maintenance being dependent on both mineral sediment input and organic soil formation (Giosan et al., 2014; Day et al., 2011b). Human activity has had far ranging impacts on deltaic sustainability from the global biosphere-geosphere, to the drainage basin, to the delta (Syvitski et al., 2009; Vörösmarty et al., 2009).

4 IMPACT OF CLIMATE CHANGE AND RESOURCE SCARCITY ON DELTAS

Day and Rybczyk (2019) discuss the impacts of climate change and energy scarcity on coastal systems. They showed that growing impacts and decreasing energy availability and higher energy prices will combine to limit options for restoration of deltas and complicate human response to climate change (Day et al., 2007a,b, 2014, 2016; Tessler et al., 2015; Wiegman et al., 2017). Climate impacts affect deltas through rising temperatures, sea-level rise, changes in river discharge, stronger tropical cyclones and other storms, and more extreme weather events (IPCC (Intergovernmental Panel on Climate Change), 2013; Meehl et al., 2007; FitzGerald et al., 2008; Pfeffer et al., 2008, Vermeer and Rahmstorf, 2009; Koop et al., 2016, Emanuel, 2005; Webster et al., 2005; Hoyos et al., 2006; Goldenberg et al., 2001; Kaufmann et al., 2011; Mei et al., 2015; Min et al., 2011; Pall et al., 2011; Royal Society, 2014; Horton et al., 2014; Deconto and Pollard, 2016). Increasing energy costs will lead to higher costs for energy-intensive activities (Tessler et al., 2015; Wiegman et al., 2017).

Much delta restoration and management, especially in rich countries, is highly energy intensive including dredging, maintenance of navigation channels, building and maintaining dikes, transporting dredged sediments in pipelines, and building and maintaining large water control structures. Wiegman et al. (2017) reported that interactions of energy costs and sea-level rise may increase the cost of marsh creation using dredged sediments in the Mississippi delta by as much as an order of magnitude.

5 CLASSIFICATION OF DELTA TYPES IN RELATIONSHIP TO SUSTAINABILITY

In terms of future potential for sustainability, Day et al. (2016) classified deltas into several generalized types; each representing a particular set of conditions that impact deltaic sustainability (Table 2). Running from less to more sustainable the delta types are as follows:

- deltas in arid environments;
- deltas with highly energy-intensive management and flood defense systems;
- deltas with significant areas below sea level;
- Arctic deltas;
- tropical and subtropical deltas with growing human impact;
- deltas with relatively low energy management;
- tropical deltas with relatively low human impact and high freshwater input.

Deltas in arid environments are among the most threatened deltas, with many already in advanced state of deterioration such that they are no longer functioning deltas. Human demand for freshwater diminishes delta sustainability and climate

TABLE 2 Factors Affecting Delta Sustainability

	Aridity in Drainage Basin	Aridity in the Delta Plain	Salinity in Delta Plain	Freshwater Input From Drainage Basin	Reduced Sediment Input From Drainage Basin (%)	Subsidence Rate (Natural, Induced)	Cyclones	Erosion of Delta Fringe Due to Wave Attack	Human Impact	Energy Intensive Management	Population Density in and Adjacent to Delta Plain
Nile	X	X	X	Low	>90	Moderate	Low		Very high	Low	Very high
Ebro	X			Low	>90	Low	Low		Very high	Moderate	Low
Indus	X	X	X	Low	>90		High		Very high	Low	High
Mississippi					>50	High	High		Very high	High	Moderate
Yangtze				Reduced	>50	High	High		Very high	Moderate	High
Rhone					<20	Low	Low		Moderate	Moderate	Low
Rhine				Reduced		High	Low		Very high	High	High
Ganges				Reduced		High	High		Very high	Low	Very high
Po				Reduced	>50	High	Low		High	Moderate	Moderate
Mackenzie					<20		Low		Low	Low	Low
Mekong				Reduced	>90	High	High		Very high	Low	High
Grijalva-Usumacinta					<20	Moderate	Moderate		Low	Low	Low
Paraná					<20	?	Low		Low	Low	Low
Danube						Low	Low		Moderate	Moderate	Low
Mahanadi							High		Moderate	Moderate	Moderate
Godavari-Krishna				Reduced	>90	Moderate	High	High	High	Moderate	High
Cauvery		X	X	Reduced	>90		Moderate		Moderate	Low	Moderate

change will lead to greater competition for water. Deltas with highly energy-intensive management will become less sustainable as energy costs increase and climate impacts grow. Deltas with significant areas below sea level will also be less sustainable due to the combination of climate change (e.g., sea-level rise, stronger storms) and growing cost of management due to rising energy costs and resource scarcity (Tessler et al., 2015; Wiegman et al., 2017). Arctic deltas will be strongly impacted by climate change due to rapid rise in temperature, loss of sea ice, and accelerating coastal erosion as well as enhanced subsidence due to permafrost thaw. Many tropical and subtropical deltas with growing human impact will have challenged sustainability due to human impact (Syvitski et al., 2009; Tessler et al., 2015). Issues include reduction in river input, retention of sediments in reservoirs, coastal erosion, wetland loss, and hydrologic alteration (Renaud et al., 2013). Deltas with relatively low energy management are potentially more sustainable but active management will require significant changes in the way people live in and use these deltas. Inner tropical deltas with relatively low human impact will have a lower risk of becoming progressively less sustainable due to projected increases in precipitation in the inner tropics (IPCC, 2013). Large tropical deltas may decrease in size but functioning coastal ecosystems can remain due to high freshwater input.

6   DELTA WINNERS AND LOSERS—SUSTAINABILITY OF INDIVIDUAL DELTAS

Giosan et al. (2014) reported that large deltas will shrink as the mineral sediment input cannot compensate for the sea-level rise (Fig. 1). The sustainability of individual deltas will depend on a variety of factors including rates of subsidence,

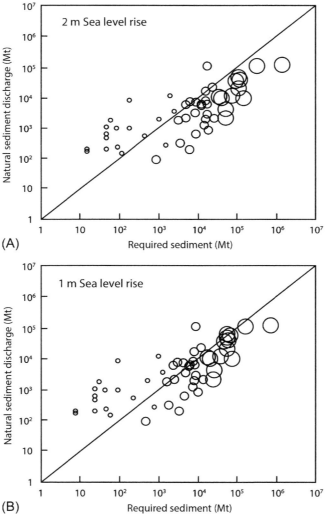

FIG. 1   Sustainability of deltas based on historic sediment availability with a 1-m sea-level rise (B) and a 2-m sea-level rise (A). Large circles represent deltas >10,000 km², intermediate circles represent deltas of 1000–10,000 km², and small circles represent deltas <1000 km². Deltas falling below the diagonal line are less sustainable. *(Modified from Giosan, L., Syvitski, J., Constantinescu, S.D., Day, J., 2014. Protect the world's deltas. Nature 516, 31–33.)*

quantity of freshwater and mineral sediment input from the basin, and degree of connectivity between the river and deltaic plain. Functioning deltaic ecosystems may be maintained with high freshwater input but there will be less wetlands and more open water. Giosan (2017) termed this the estuarization of deltas. Even though large tropical deltas are more sustainable from a biophysical perspective, human communities will have to adjust.

Based on the foregoing discussion, the following criteria were used to classify the sustainability of individual deltas:

- increasing aridity in the drainage basin due to climate and/or human forcings;
- increasing aridity in the delta plain due to climate and/or human forcings;
- significant hypersalinity in the delta plain;
- reduced freshwater and sediment input from the drainage basin;
- reduced input of freshwater and sediments to the delta plain due to dams, dikes, and other restrictions;
- large areas of delta plain below sea level;
- high subsidence rates in the delta plain due to natural and human-induced forcings;
- high rates of relative sea-level rise;
- high probability of strong tropical cyclone impacts (outside of subarctic and arctic regions);
- high erosion on the delta fringe due to wave attack;
- high human impact in the delta plain;
- the degree of energy-intensive management;
- high population density in and adjacent to the delta plain.

7 SUSTAINABILITY OF INDIVIDUAL DELTAS

In this section, we summarize deltas addressed in other chapters of this book and use the database of Giosan et al. (2014) who judged the sustainability of deltas based on historic sediment inputs under natural conditions and projected sea-level rise. They reported that almost all large deltas >10,000 km^2 and most intermediate deltas from 1000 to 10,000 km^2 do not have sufficient mineral sediment input to offset projected sea-level rise of 1 m by 2100, even in a "no dam" world (Fig. 1). But we are not in a no dam world. As noted throughout this book, human impacts have seriously degraded coastal systems worldwide.

The diagonal 1:1 line in the figure separates delta sustainability above the line and nonsustainability below. Nonsustainability implies that the wetland area of deltas will decline. Fig. 1A presents delta sustainability with a 2-m sea-level rise clearly showing the importance of rising water levels for the preservation of deltas. These diagrams consider the availability of sediment as the ultimate cause for delta sustainability and as such they do not include the role of wetlands in contributing toward vertical growth of deltaic wetlands. But in the end, deltas cannot survive rising water levels without sediment input. Sediments form the matrix within which wetlands grow. Next, we consider the future of deltas considered in this book within the context of the sustainability diagram by highlighting individual deltas. The deltas are discussed beginning in Asia and moving west through Europe and Africa to the Americas.

8 ASIAN DELTAS

8.1 Yangtze (Changjiang)

The Changjiang River is now the primary freshwater source to Shanghai. Saltwater intrusion in the Changjiang River estuary is exacerbated by water transfer projects, the Three Gorges dam in the upper basin, and sea-level rise in the estuary (Li and Chen, 2019). The area has a high population density and part of the delta is reclaimed and located below sea level. The historical relationships between discharge, salinity, and the duration of saltwater intrusion were established quantitatively and demonstrate that (1) the lower the river discharge, the higher the salinity and the longer the residence time of saltwater in the estuary, and (2) the threshold value of the river discharge for salinity intrusion impacting the freshwater supply is 15,000 m^3/s discharge. Measures for remediation on a long-term perspective have to be at multiple scales. Under natural conditions, the Yangtze River would be nearly sustainable in terms of sediment (Fig. 2). But given the wide range of human impacts and climate change, the delta is moving toward nonsustainable.

8.2 Mekong

The Mekong River delta, the third largest delta in the world, is not only shrinking but also suffering from extensive environmental degradation (Nhan and Cao, 2019). These changes are due to several factors including ill-planned water

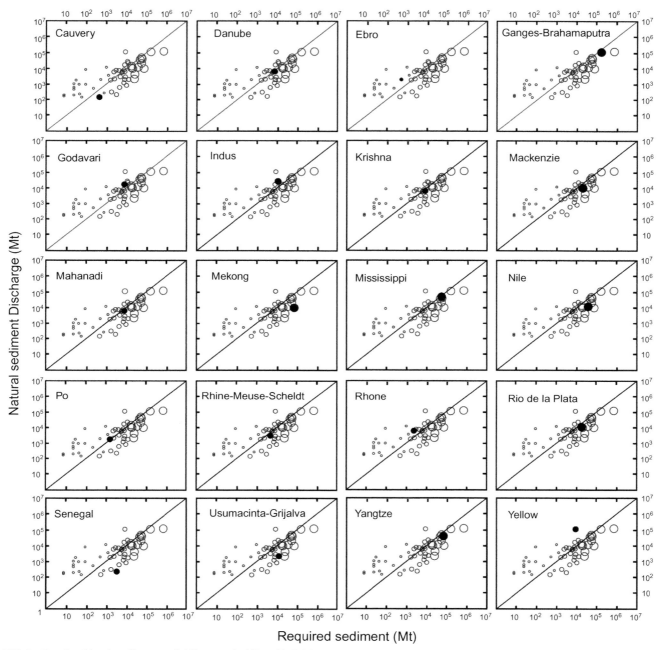

FIG. 2 Based on historic sediment availability, sustainability of individual deltas with 1 m sea level rise included in Wolanski et al. (2019). Several deltas are not discussed in this book but are included for reference (Danube—see e.g., Giosan et al., 2013). Black dots represent the delta referenced in each box. (*Modified from Giosan, L., Syvitski, J., Constantinescu, SD., Day, J., 2014. Protect the world's deltas. Nature 516, 31–33.*)

management schemes, hydropower dams in the river basin, sediment starvation, increased nutrient inflows, salt water intrusion, in combination with other human activities that include infrastructural extension, riverbed mining, delta subsidence, climate change and sea-level rise, degradation of coastal mangrove belt, and gaps in governance in the whole Mekong basin (Wolanski and Nguyen, 2005; Nilsson and Reidy, 2005; Hoa et al., 2007; Syvitski et al., 2009; Kuenzer et al., 2013; Leinenkugel et al., 2014; Nhan and Cao, 2019). Much of the degradation in the Vietnamese Mekong delta is due to hydropower dams in the entire Mekong river basin, particularly the big dams in China in the Upper Mekong basin have resulted in the reduction in sediment load by 50%–60%. The flood discharges have also decreased, low flow events are now common, the hydrological seasonal regime has shifted resulting in earlier and more severe salinity intrusion into the delta as well as flooding from storms at sea now not blocked by the river discharge. Further the river bed is on the average deeper by 1.3 m, to which riverbed mining also contributes. There has been a recent increase in the erosion of river banks (400 locations) and coasts (66% of the foreshore is eroding) and their occurrence is increasing. The sediment sustainability diagram shows that even under natural conditions, the current Mekong delta is unsustainable with a 1-m sea-level rise (Fig. 2). Given the ongoing human and climate impacts, the delta seems headed for collapse.

8.3 Ganges

The Ganges-Brahmaputra-Meghna (GBM) delta is the world's largest delta and extends across two countries. Two-thirds of the delta is located in Bangladesh, and the rest in the Indian state of West Bengal. This delta is a particularly fertile region supporting one of the most densely populated regions in the world with over 150 million people. The enormous size of the Ganges basin and the GBM delta, and the wide variation in physiography mean that both changes in upstream areas and coast-related phenomena have tremendous impacts on the delta. The challenges related to the upstream effects include disparities in water flow and sediment transport due to seasonal variations as well as withdrawal for irrigation and other purposes (Ramesh et al., 2019). There is also distinct variation in the water flows within the Indian parts of the delta resulting in a hypersaline central sector while the eastern and western regions are hyposaline impacting ecosystem functioning as evidenced by changing species diversity within the delta. Coastal effects include impacts due to flooding and inundation caused by monsoonal rainfall, cyclones, and rising sea levels which may be intensified by delta subsidence; and loss of land due to coastal erosion. Large areas of the delta are polderized with ground elevations below the high tide level. Embankments can increase the effective sea-level rise because of increased tidal range in estuary channels constricted by embankments (Pethick and Orford, 2013). Anthropogenic impacts are cross-cutting, caused by rapidly rising population whose mainstay is agriculture and fisheries. New challenges are due to industrialization and the heavy pollution load carried by the rivers. The various threats to the delta are closely interrelated having a cascading effect. Naturally, the combined sediment input from Ganges and Brahmaputra could have been sufficient to offset a significant part of 1 m of sea-level rise (Fig. 2) but the combined impacts of climate change and human impact, especially on the Ganges, will push the delta further from sustainability (Ramesh et al., 2019).

8.4 Other Indian Deltas

Important deltas of Peninsular India include the Mahanadi, Godavari-Krishna, and Cauvery along the east coast (see Ramesh et al., 2019).

8.5 Mahanadi

The 860-km Mahanadi River along with the Brahmani, a major seasonal river and the smaller Baitarani, form a large delta in the eastern state of Odisha. The delta plain extends over an area of 0.9×10^4 km^2. The delta plain is a major rice-growing region in India and population density is extremely high. The country's longest major earthen dam (Hirakud) has been built across the Mahanadi creating a huge artificial reservoir. Its water-holding capacity has been reduced by 28% due to siltation. While the Hirakud dam reduced flooding of the Mahanadi in the upper reaches, it has reduced the flushing capacity of the Chilika lagoon due to clogging and siltation of the mouth. Other rivers have been embanked resulting in control of overflow to associated wetlands resulting in a reduction in breeding area for migratory fish, lack of fluvial-derived nutrients affecting the fishery in both wetlands and lagoon. Additional threats are due to the extensive shrimp farms in the area. The average rainfall is 1572 mm and more than 70% of this is during the southwest monsoon (June-September). Numerous lakes and bays are present on the delta plain, many of them the remnants of former river courses. Monsoon floods and cyclones are the main hazards in the delta. In the last 15 years, there have been severe floods especially in the delta areas. Deforestation in the upper reaches has resulted in heavy erosion which has reduced the capacity of the river to carry the heavier inflows during monsoon. Seasonal cyclones affect the coast on a regular basis. Wave energy is quite high along the delta front and well-developed beaches and barrier islands are present along the coast. Erosion is more dominant across the lower reach of the delta. Mangrove is the most common type of vegetation along the seaward edges of the delta and there have been extensive efforts in mangrove afforestation as their protective role during cyclones has been clearly proved. The delta is potentially sustainable if engineered structures are controlled and the river systems are allowed to follow natural courses. Historic sediment loads would have been barely sufficient to support the sustainability of Mahanadi delta (Fig. 2).

8.6 Godavari-Krishna

The second largest river in India, the Godavari River originates in the state of Maharashtra and flows 1465 km eastwards before draining into the Bay of Bengal in Andhra Pradesh. The Godavari drainage basin extends over nearly 10% of the total geographical area of the country. The River Krishna rises in the Western Ghats, flows about 1400 km eastwards before draining into the Bay of Bengal. The Godavari delta is wave dominated with one of the largest sediment deliveries in the world. The delta prograded during the Holocene but pronounced shoreline erosion led to a net negative growth of the delta during recent decades due to sediment retention by upstream dams (Rao et al., 2012; Malini and Rao, 2004). The sediment load is believed to have decreased from 145.26 million tons in 1971–79 to 56.76 million tons during 1990–98 (Kallepalli et al., 2017). Studies indicated a net loss of 42.1 km^2 between 1977 and 2008 along the 330 km long K-G delta and a net erosion of 57.6 km^2 along the K-G delta coast is expected by 2050. Both the Godavari and Krishna deltas have large tracts of mangroves (e.g., the Coringa region) which are under high stress because of extensive aquaculture, industrialization, and the associated human impacts. The combined Krishna-Godavari delta, like all deltas along the east coast, is important for rice production and there are likely to be high levels of fertilizer and pesticide residues in the delta systems though no systematic assessment is available. The basin faces water scarcity in many areas leading to water conflicts while floods are also common during the monsoon. The delta is also affected regularly by cyclones that originate in the Bay of Bengal. An additional threat is that the offshore basin is a proven petroliferous basin. The Ravva oil and gas field located in the shallow offshore area of the Krishna-Godavari basin produces 8% of the domestic oil production in India. While the prodigious natural sediment load of the Godavari would have been enough to sustain the delta under a 1-m sea-level rise (Fig. 2), the sediment deficit for the Krishna is larger.

8.7 Cauvery

A much smaller delta than the others described above, nevertheless this is an important delta in southern India especially for rice cultivation and food security in the region. However, upstream dams and interstate disputes have resulted in water scarcity in the delta region. High groundwater extraction compounds existing water scarcity issues. The delta has shrunk by 20% due to anthropogenic factors such as diversion of land for nonagricultural purposes, as well as factors linked to climate change. The tail-end regulators in many of the rivers and channels are in poor condition allowing seawater intrusion. Increase in mangrove cover has been noted but this is linked with increased saline intrusion as there is no freshwater flow in the rivers.

8.8 Indus

The River Indus is one of the oldest documented rivers formed by the collision of the Indian and Eurasian Plates prior to 45 million years ago. It runs 2900 km through four countries (Pakistan, India, Afghanistan, and China). Within Pakistan, the River Indus runs the entire length of the country and basin forms 65% of the total country. Many smaller rivers join the course. The river courses have changed over geological times and the delta has shifted westward. Today there is only one functional channel (Khobar creek). The freshwater water and sediment flows have been reduced with the construction of large dams and man-made structures and water is consumed to support the agriculture production in the catchment area, nevertheless, the water use efficiency is very low (Syvitski et al., 2013). The scarcity of freshwater during the dry season results in the sea moving upstream reaching several km up the river, thus affecting the position of the estuary and the riverine, deltaic ecosystem. Indus delta has experienced widespread salinization and most mangroves have died. Under natural conditions, the Indus delta would be sustainable in terms of sediment supply with a 1-m sea-level rise (Fig. 2). But changes throughout the drainage basin are pushing the delta toward non-sustainability and collapse (Kidwai et al., 2019).

Other deltas impacted by arid conditions include the Tigris-Euphrates, Nile, and Colorado, all of which have very low potential for sustainability (e.g., for the Colorado see Carriquiry et al., 2011; Gerlak et al., 2013). Most of these deltas are already in an advanced state of deterioration and from a geomorphic and ecological point of view, there is little if any functioning deltaic system.

9 EUROPEAN AND AFRICAN DELTAS

9.1 Mediterranean Deltas

Mediterranean deltas, including the Ebro, Rhone, Po, and Nile, encompass a range of environmental settings and management challenges. All of these systems have been strongly modified by human activities, however, each system has a unique combination of impacts that informs management and restoration approaches (see Chen, 2019; Day et al., 2019a). The Ebro faces a massive reduction in freshwater and sediment input and over 65% of wetlands in the delta plain have been converted to rice fields. The sustainability diagram shows that under natural conditions of sediment input, the Ebro could survive a 1-m sea-level rise (Fig. 2). Sustainable management must include higher freshwater inflow and mobilization of sediments from reservoirs and transport to the delta, without which the delta will move below the sustainability line and deteriorate despite ongoing wetland restoration in the delta plain. Although the majority of the delta is rice fields, the irrigation network can be used to distribute sediments over much of the delta (Ibáñez et al., 2010). The ongoing conflict between the Spanish government and Catalonia demonstrates how political issues can compromise sustainable management and ecological restoration (see Day et al., 2019a).

The Rhône River has not had a significant decrease in discharge or suspended sediment concentration and the main problem is that almost all river input to the delta plain has been eliminated by dikes but very large floods with high sediment concentrations still occur regularly. Two large floods that flooded extensive parts of the delta demonstrate that river water and sediment can be delivered to large areas of the delta (Pont et al., 2017). Under natural sediment discharge, the Rhone River would be sustainable with 1 m of sea-level rise (Fig. 2). An aggressive program to reintroduce Rhone River water back into the delta would maintain the delta in this century. Both the Rhone and Ebro have relatively low subsidence rates so less vertical accretion is needed to offset sea-level rise (Pont et al., 2002; Day et al., 2011a,b).

There has been a strong decrease in its sediment load in the Po River but the river has at least five functioning distributaries that distribute water and sediment widely on the deltaic wetland fringe. However, the delta has already subsided by over 1 m due to water and gas mining with much of the surrounding delta plain already at or below sea level, protected only by the strength of engineered levees and coastal barriers (Syvitski, 2008). A program to remobilize sediments from the drainage basin would maintain most of the current wetlands in the delta. Over time sediment from the Po River could be used to fill in some of these areas whereas others could be converted to shallow lagoons. Unfortunately the sediment load of the Po has continued to decrease over the last century and river channel mining continues this sediment reduction vector (Syvitski and Kettner, 2007).

9.2 The Nile

The fan-shaped 20,000 km^2 subaerial Nile delta in Egypt is fed solely by the water and sediment from the African Plateau and Ethiopian Highlands where rainfall is ~1500–2000 mm/a, whereas the rainfall is only 50–100 mm/a in the lower Nile basin, including its delta. But human impacts have changed input to the delta (Milliman et al., 1989; Chen et al., 2010b; Stanley and Clemente, 2017; Chen, 2019). Particularly after the completion of the High Aswan Dam in 1964 and the ongoing construction of the Grant Ethiopian Renaissance Dam (GERD) on the Blue Nile, the Nile delta is increasingly suffering from serious environmental degradation because of reduced sediment and freshwater and increasing nutrients which have negatively impacted the eco-health of the Nile delta coast. Egypt initially benefited by overextraction of water from Lake Victoria when a new hydroelectric dam was built in Uganda in the early 2000s at the outlet of the lake that forms the White Nile River, but once the lake level was lowered this benefit stopped (Kiwango and Wolanski, 2008). The pre-GERD Nile flow now supplies 97% of Egypt's present water needs with only 660 m^3 per person per year. About 86% of that water is for irrigation and industrial use, so this is one of the world's lowest annual per person water shares. Egypt's population is expected to double in the next 50 years, and this is will lead to countrywide freshwater shortages as early as 2025 after the GERD dam is completed. Some form of arbitration is clearly needed to prevent a conflict over water between Egypt, Sudan, and Ethiopia (https://www.npr.org/2018/02/27/589240174/in-africa-war-over-water-looms-as-ethiopia-nears-completion-of-nile-river-dam).

Projections are for increasing salinization and subsidence in the delta. Even under natural levels of sediment, the Nile delta would not maintain its area with a 1-m sea-level rise (Fig. 2; Stanley, 1988; Stanley and Warne, 1993; Stanley and Clemente., 2017). The Nile is so fundamentally changed that sustainable management is likely impossible given the trajectories of major 21st century environmental and socioeconomic trends.

9.3 Senegal and Pangani Deltas

The Senegal River in Senegal and the Pangani River in Tanzania are examples of the many river systems in Africa that have been heavily affected by both human interventions, such as dams and large-scale irrigation schemes, and climate change (Niang et al., 2019). The 116-year average (1900–2016) rainfall in Senegal estuary is about 375 mm/year but appears to have reduced and is close to 200 mm/year in the last 50 years. Located in the Sahel zone, the Senegal River estuary in the North of Senegal primarily houses disadvantaged and low-income communities. Weak floods trigger severe salt intrusion in the lower river ranges and upstream migration of species. The low groundwater recharge has resulted in compromised food security for local communities. Coastal erosion on the immediate north of the Pangani River mouth, and changes in sediment deposition patterns in the estuary itself have been reported as one of the major environmental issues of concern. Overall, while dams and associated reservoirs have remediated drought conditions, they have stressed the hydrological regime. The strong relation between rainfall and river runoff constitutes a growing constraint in such river basins. Thus, the Senegal delta is being impacted by declining precipitation in the drainage basin and decreases in river flow. This combined with human impacts is leading to salinization of the delta. Even under natural conditions of sediment input, the Senegal delta is unsustainable with respect to 1 m of sea-level rise. Growing climate impacts due to drying will further compromise delta sustainability.

9.4 Danube

The Danube delta is both a puzzling and a hopeful case. While sediment discharge has been reduced by at least 70% due to constructions of dams on tributaries and the main trunk, channel dredging for fishing has kept the delta plain connected to the river (Giosan et al. 2013). This favored sediment trapping in the delta that combined with the low subsidence (Giosan et al. 2006b) and restoration activities (Schneider et al., 2008) keeps most of the delta as viable as in historic conditions (Fig. 2). If management of sediment resources is not wise in the future, Danube delta could easily move into the unsustainable category due to the enormous decrease in sediment load in the river. Furthermore, under a severely unidirectional wave climate, some of the largest rates of net alongshore sediment transport rates in the world make erosion a chronic problem at the coast of the delta (Giosan et al., 1999). Navigation dredging has made the delta a conduit for marine invasive species in Europe (Shalovenkov, 2019).

9.5 Rhine-Meuse-Scheldt

Most of the Rhine delta is below sea level, is highly developed and densely populated, and has been for centuries. The landscape below sea level developed during a time of relatively stable climate. The land was first drained with windmills beginning in the 16th century and then with fossil fuel-powered pumps. Most of the delta since then has been converted to fast lands for agriculture, industry, and urban development, and is no longer a functioning delta (Knights, 1979). During the 20th century, there was a massive increase in energy-intensive infrastructure with high maintenance costs. As the century progresses and climate impacts grow more severe and energy costs escalate, the risk of not maintaining the current system will grow (Tessler et al., 2015; Meire, P., n.d.). There is no real chance of returning in a significant way to a more natural system for the Rhine.

10 AMERICAN DELTAS

10.1 Mississippi Delta

The Mississippi River Delta (MRD) is one of the largest of global deltas. Within the last century, the delta has been profoundly altered by humans with respect to hydrology, sediment supply, sea-level rise, ecology, and land use that directly affect sustainability, especially in the context of global change forcings (see Day et al., 2019b). The MRD has lost over 25% of coastal wetlands since the 1930s and is in the process of a physical, ecological, and societal collapse. Human activities have reduced sediment input from the basin, isolated much of the deltaic plain from the river due to levees, pervasively altered delta plain hydrology due to over 15,000 km of canals, enhanced subsidence due to oil and gas production and drainage, and introduced a number of extremely damaging invasive species. There is presently a large-scale coastal master plan to reduce flood risk and restore deltaic wetlands to a more self-sustaining and healthy condition. This involves both hard structures for flood protection (levees, floodwalls) and wetlands sustained by river diversions, marsh nourishment, and barrier island restoration. The large degree of control envisioned in the Master Plan may not be possible to maintain and thus society in the delta will have to learn again to live with a much more dynamic system. But climate change and resource scarcity, especially energy costs, will make achieving these goals ever more challenging and expensive. Large areas, especially around New Orleans, are below sea level and susceptible to both river and hurricane flooding. Repeated evacuations followed by more or less managed retreat will also continue to be necessary for much of the population. The CMP is ecological engineering on a grand scale, but to be successful it must operate in consonance with complex social processes. This will mean living in a much more open system, accepting natural and social limitations, and utilizing the resources of the river more fully. The sustainability diagram shows that the Mississippi delta is slightly below the sustainable line for a 1-m sea-level rise given historic sediment inputs (Fig. 2). The conditions in the current delta have shifted the delta toward lower sustainability. Maximum use of riverine sediment resources using river diversions (Day et al., 2019b; Day and Erdman, 2018), mobilization of fine sediments from the drainage basin (Kemp et al., 2016), and restoration of a more natural hydrology could shift the delta toward greater sustainability and a functional coastal ecosystem. But the delta will shrink considerably and retreat inland will be necessary for a considerable portion of the coastal population.

10.2   Usumacinta-Grijalva Delta

The Grijalva-Usumacinta delta (GUD) region of Mexico has high riverine input and extensive wetlands. The management plan of the CENTLA Biosphere Reserve in the GUD was established to protect this important region. The management plan is based on the identification of natural environmental units that are grouped into two nucleus zones with high levels of protection and a number of buffer zones designed to allow some economic activities (Herrera-Silveira et al., 2019). Laguna de Terminos, the largest lagoon-mangrove system in Meso-America, has a long history of scientific study. The system has a high habitat diversity. Primary producers have peak production at different times of the year leading to overall sustained high productivity throughout the year. There is a high diversity migratory nekton community that uses the lagoon habitats at times when they are most productive ensuring overall high secondary production that supports a multistock fishery. The area has been designated as a natural protected area to ensure sustainable management. Mexico has established 17 natural protected areas in the Southern Gulf of Mexico and the adjacent groundwater fed Yucatan Peninsula to enhance sustainable management in the region. Much of the area is still relatively natural with abundant freshwater resources. Some global climate models suggest that this region will experience strong decreases in precipitation and this may become the most serious problem for this region. The sediment sustainability diagram shows that the current delta is not sustainable with 1-m sea-level rise (Fig. 2) but drying due to climate change will worsen this situation.

10.3   Rio de la Plata and the Parana Delta

The drainage basin of the Parana River system is about 3.1 million km^2 or about the size of the Mississippi basin. The drainage basin includes tropical and subtropical areas. With a discharge of about 17,000 m^3/s, it is one of the largest rivers in the world. The Parana delta covers about 14,000 km^2. Precipitation in the delta ranges between 1000 and 1400 mm/year. Precipitation over the basin comes from tropical areas. With natural sediment inputs the delta is slightly below the sustainability line for 1 m of sea-level rise (Fig. 2). Future sustainability of the delta depends on climate impacts on precipitation and sediment trapping by dams in the basin (Garcia-Alonso et al., 2019).

Other tropical American deltas include the Amazon and Orinoco. These deltas in the inner tropics are still relatively natural without large-scale interventions in the delta plain that separate the wetlands from the river. They have high freshwater input that is projected to increase with climate change. These deltas may shrink due to sea-level rise but will remain as functional deltaic systems.

10.4   Mackenzie Delta

Arctic deltas are strongly impacted by sea ice by protection from wave action and contributing to terrace formation. With climate warming, the runoff season will increase significantly leading to a large increase in sediment yield. Ice effects in larger delta systems such as the Mackenzie include ice jamming, which raises water levels rapidly, contributing not only to the flooding and water renewal in high-sill lakes, but also posing a safety hazard with flooding of communities. As in smaller rivers, bottomfast ice occurs in some of the largest channels and the circumstances of freeze-up determine the flow capacity of various distributaries in spring, directing first flow to various channels from year to year and undoubtedly affecting the growth directions of the delta. Confinement of prodelta flow to subice channels and flow over bottomfast ice may create bypass zones with little sedimentation at the delta front. At the same time, pressure ridges on the inner shelf may confine and direct freshwater inflow under the ice through the winter. Sedimentation is critical for the creation and maintenance of Arctic deltas. In cases of self-confinement, such as the Arctic finger deltas, a component of sediment supply comes from cannibalization, but building out progressively further beyond the coast increases the accommodation space and sediment required to advance the delta front. In areas with relative sea-level rise, sedimentation is the key to maintaining a positive elevation balance. In the Mackenzie Delta, sediment accretion rates on the delta plain appear inadequate to maintain the balance in the face of surface subsidence rates of the same magnitude, compounded by accelerated sea-level rise (Forbes, 2011, 2019; Forbes and Hansom, 2011).

More open water combined with rising relative sea levels and possibly enhanced storm activity can be expected to increase already high rates of frontal retreat on the Mackenzie delta. Coupled with warming temperatures promoting deeper seasonal thaw, the stability of the delta front in most sectors is likely to be compromised. In the Mackenzie delta the presence of unfrozen water in the permafrost and fluid escape combined with widespread methane gas pockmarks and seeps supports the hypothesis that deep fluid escape may contribute to the observed near-surface subsidence. These observations are supported by the sustainability diagram showing that the current delta is not sustainable with 1-m sea-level rise. The ongoing changes in the Mackenzie due to climate change will move the delta toward greater non-sustainability. Other important Arctic deltas include the Yukon and Lena. The Yukon discharges into the Bering Sea via North Sound and is sheltered by the Seward Peninsula thus it is not exposed to sustained wave attack. It also flows mostly east to west so the climate impacts are more evenly distributed over the basin. The Lena is a forced-regression delta where isostatic rebound has resulted in a relative decrease in sea level but it will be exposed to increasing wave attack due to loss of Arctic sea ice.

11   RANKING SUSTAINABILITY

Here we summarize the information on deltaic sustainability and classify a representative group of deltas in terms of relative sustainability. We classify the different deltas into four levels of sustainability (Fig. 3, Table 2) based on quantitative and qualitative data discussed herein and from the literature. The rankings are based on the current status of individual deltas, the degree of human impact, and the ability to deal with limitations that will be imposed by climate change and growing energy scarcity as they affect the different measures of sustainability discussed above. This approach based on the functional characteristics of deltas and biophysically based sustainability criteria should contribute to the development of more quantitative indices based on metrics such as those we have discussed in this paper and also in the works of Syvitski et al. (2009), Renaud et al. (2013), and Tessler et al. (2015).

These rankings are in general agreement with analyses by Syvitski et al. (2009), Renaud et al. (2013), Tessler et al. (2015), and Day et al. (2016). All of the analyses agree that large deltas in the inner tropics with high freshwater input and low human impact, such as the Amazon and Orinoco, are most sustainable. Large areas of wetlands are likely to be lost in these deltas but functioning deltaic coastal ecosystems with large expanses of open water will survive. An interesting but unanswered question at this point is how will the biophysical functioning of these deltas change in systems with ongoing estuarization that are dominated by large expanses of shallow open water with relatively small areas of wetlands. At the other end of the scale, deltas in arid regions with diminishing freshwater input and high human impact are least sustainable, such as the Colorado, Indus, Nile, and Tigris, which to a considerable degree are no longer functioning deltas.

12   CONCLUSIONS

Sustainability of deltaic systems can be defined and determined in terms of the persistence through time of the structure and function of deltaic systems. Achieving sustainable deltas will be very difficult in the face of climate change, sea-level rise, reduced inputs from drainage basins, high human impacts, and growing resource scarcity. We classified deltas from different regions as having high, moderate, and potential for sustainability and not likely sustainable based on the current status of these deltas, the degree of human impact, how energy-intensive delta management is, and the degree to which they are impacted by climate change. Deltaic sustainability is being increasingly compromised by human activity, climate change, and energy costs. These factors will reduce options for achieving deltaic sustainability and make them more challenging.

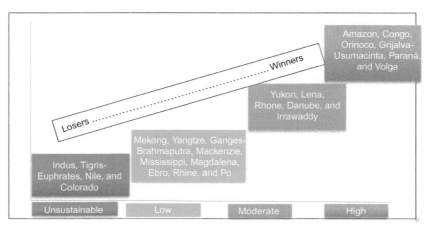

FIG. 3   Ranking of deltas discussed in this paper with other deltas.

Indices of sustainability based on measures of vulnerability and resilience should reflect system functioning and provision of ecosystem services from both biophysical and human perspectives.

ACKNOWLEDGMENTS

We thank Eric Wolanski and Mike Elliott for helpful comments.

REFERENCES

Batker, D., de la Torre, I., Costanza, R., Swedeen, P., Day, J., Boumans, R., Bagstad, K., 2014. The value of ecosystem goods and services of the Mississippi delta. In: Day, J., Kemp, P., Freeman, A., Muth, D. (Eds.), The Once and Future Delta. Springer, New York, pp. 155–174.

Blackburn, S., Pelling, M., Marques, C., 2019. Megacities and the coast: global context and scope for transformation. In: Wolanski, E., Day, J., Elliot, M., Ramachandran, R. (Eds.), Coasts and Estuaries—The Future. Elsevier, Amsterdam.

Burger, J.R., Alle, C., Brown, J., Burnside, W., Davidson, A., Fristoe, T., Hamilton, M., Mercado, N., Nekola, J., Okie, J., Zuo, W., 2012. The macroecology of sustainability. PLoS Biol. 10 (6), E1001345.

Carriquiry, J.D., Villaescusa, J.A., Camacho-Ibar, V., Daesslé, L.W., Castro-Castro, P.G., 2011. The effects of damming on the materials flux in the Colorado River delta. Environ. Earth Sci. 62 (7), 1407–1418.

Chen, Z., 2019. A brief overview of ecological degradation of the Nile delta: what we can learn? In: Wolanski, E., Day, J., Elliot, M., Ramachandran, R. (Eds.), Coasts and Estuaries—The Future. Elsevier, Amsterdam.

Chen, Z., Saito, Y., 2011. The megadeltas of Asia: interlinkage of land and sea and human development—virtual special issue. Earth Surf. Process. Landf. 1096–9837.

Chen, Z., Zong, Y., Wang, Z., Chen, J., Wang, H., 2008. Migration patterns of Neolithic settlements on the abandoned Yellow and Yangtze River deltas of China. Quat. Res. 70 (2), 301–314.

Chen, Z., Wang, Z., Finlayson, B., Chen, J., Yin, D., 2010a. Implications of flow control by the three Gorges Dam on sediment and channel dynamics of the middle Yangtze (Changjiang) River, China. Geology 38 (11), 1043–1046.

Chen, Z., Salem, A., Xu, Z., Zhang, W., 2010b. Ecological implications of heavy metal concentrations in the sediments of Burullus Lagoon of Nile delta, Egypt. Estuar. Coast. Shelf Sci. 86, 491–498.

Costanza, R., d'Arge, R., de Groot, R., Farber, S., Grasso, M., Hannon, B., Limburg, K., Naeem, S., O'Neill, R., Paruelo, J., Raskin, R., Sutton, P., van den Belt, M., 1997. The value of the world's ecosystem services and natural capital. Nature 387, 253–260.

Costanza, R., de Groot, R., Braat, L., Kubiszewski, I., Fioramonti, L., Sutton, P., Farber, S., Grasso, M., 2017. Twenty years of ecosystem services: how far have we come and far do we still have to go. Ecosys. Serv. 28, 1–16.

Day, J., Erdman, J. (Eds.), 2018. Mississippi delta restoration – pathways to a sustainable future. Springer, New York. 261 p.

Day, J.W., Rybczyk, J.M., 2019. Global change impacts on the future of coastal systems: perverse interactions among climate change, ecosystem degradation, energy scarcity, and population. In: Wolanski, E., Day, J., Elliot, M., Ramachandran, R. (Eds.), Coasts and Estuaries—The Future. Elsevier, Amsterdam.

Day, J., Martin, J., Cardoch, L., Templet, P., 1997. System functioning as a basis for sustainable management of deltaic ecosystems. Coast. Manag. 25, 115–154.

Day, J., Barras, J., Clairain, E., Johnston, J., Justic, J., Kemp, P., Ko, J., Lane, R., Mitsch, W., Steyer, G., Templet, P., Yañez-Arancibia, A., 2005. Implications of global climatic change and energy cost and availability for the restoration of the Mississippi Delta. Ecol. Eng. 24, 251–263.

Day, J.W., Boesch, D., Clairain, E., Kemp, P., Laska, S., Mitsch, W., Orth, K., Mashriqui, H., Reed, D., Shabman, L., Simenstad, L., Streever, B., Twilley, R., Watson, W.J., Whigham, D., 2007a. Restoration of the Mississippi Delta: lessons from hurricanes Katrina and Rita. Science 315, 1679–1684.

Day, J., Gunn, J., Folan, W., Yáñez-Arancibia, A., 2007b. Emergence of complex societies after sea level stabilization. Eos 88, 169–170.

Day, J., Kemp, P., Reed, D., Cahoon, D., Boumans, R., Suhayda, J., Gambrell, R., 2011a. Vegetation death and rapid loss of surface elevation in two contrasting Mississippi delta salt marshes: the role of sedimentation, autocompaction and sea-level rise. Ecol. Eng. 37, 229–240.

Day, J.W., Ibanez, C., Scarton, F., Pont, D., Hensel, P., Day, J.N., Lane, R., 2011b. Sustainability of mediterranean deltaic and lagoon wetlands with sea-level rise: the importance of river input. Estuar. Coasts 34, 483–493.

Day, J., Moerschbaecher, M., Pimentel, D., Hall, C., Yáñez-Arancibia, A., 2014. Sustainability and place: how megatrends of the 21st century will impact humans and nature at the landscape level. Ecol. Eng. 65, 33–48.

Day, J., Agboola, J., Chen, Z., D'Elia, C., Forbes, D., Giosan, L., Kemp, P., Kuenzer, C., Lane, R., Ramachandran, R., Syvitski, J., Yanez, A., 2016. Approaches to defining deltaic sustainability in the 21st century. Estuar. Coast. Shelf Sci. 183 (B), 275–291. https://doi.org/10.1016/j.ecss.2016.06.018.

Day, J.W., Ibáñez, C., Pont, D., Scarton, F., 2019a. Status and sustainability of Mediterranean Deltas: the case of the Ebro, Rhône, and Po Deltas and Venice Lagoon. In: Wolanski, E., Day, J., Elliot, M., Ramachandran, R. (Eds.), Coasts and Estuaries—The Future. Elsevier, Amsterdam.

Day, J.W., Colten, C., Kemp, G.P., 2019b. Mississippi Delta restoration and protection: shifting baselines, diminishing resilience, and growing non-sustainability. In: Wolanski, E., Day, J., Elliot, M., Ramachandran, R. (Eds.), Coasts and Estuaries—The Future. Elsevier, Amsterdam.

Deconto, R., Pollard, D., 2016. Contribution of Antarctica to past and future sea-level rise. Nature 531, 591–597.

Emanuel, K., 2005. Increasing destructiveness of tropical cyclones over the last 30 years. Nature 436, 686.

FitzGerald, D., Fenster, M., Argow, B., Buynevich, I., 2008. Coastal impacts due to sea-level rise. Annu. Rev. Earth Planet. Sci. 36, 601–647.

Forbes, D.L. (Ed.), 2011. State of the Arctic Coast 2010: Scientific Review and Outlook. Helmholtz-Zentrum, Geesthacht, pp. 168. International Arctic Science Committee, Land-Ocean Interactions in the Coastal Zone, Arctic Monitoring and Assessment Program, International Permafrost Association.

Forbes, D.L., 2019. Arctic deltas and estuaries: a Canadian perspective. In: Wolanski, E., Day, J., Elliot, M., Ramachandran, R. (Eds.), Coasts and Estuaries—The Future. Elsevier, Amsterdam.

Forbes, D.L., Hansom, J., 2011. Polar coasts. In: Wolanski, E., McLusky, D.S. (Eds.), Treatise on Estuarine and Coastal Science. 3. Academic Press, Waltham, pp. 245–283.

Garcia-Alonso, J., Lercari, D., Defeo, O., 2019. Rio de la Plata: a neotropical estuarine system. In: Wolanski, E., Day, J., Elliot, M., Ramachandran, R. (Eds.), Coasts and Estuaries—The Future. Elsevier, Amsterdam.

Gerlak, A.K., Zamora-Arroyo, F., Kahler, H.P., 2013. A delta in repair: restoration, binational cooperation, and the future of the Colorado River Delta. Environ.: Sci. Policy Sustain. Develop. 55 (3), 29–40.

Giosan, L., 2017. In: Survival of deltas under anthropogenic global changes. American Geophysical Union, Fall Meeting 2017, abstract #U32A-02.

Giosan, L., Bokuniewicz, H.J., Panin, N., Postolache, I., 1999. Longshore sediment transport pattern along the Romanian Danube delta coast. J. Coastal Res. 15 (4), 859–871.

Giosan, L., Donnelly, J., Vespremeanu, E., Constantinescu, S., Filip, F., Ovejanu, I., Vespremeanu-Stroe, A., Duller, G., 2006b. Young Danube Delta documents stable black sea level since middle Holocene: morphodynamic, paleogeographic and archaeological implications. Geology 34, 757–760.

Giosan, L., Constantinescu, S., Filip, F., Bing, D., 2013. Maintenance of large deltas through channelization: nature vs. humans in the Danube delta. Anthropocene 1, 35–45.

Giosan, L., Syvitski, J., Constantinescu, S.D., Day, J., 2014. Protect the world's deltas. Nature 516, 31–33.

Goldenberg, S., Landsea, C., Mestas-Nunez, A., Gray, W., 2001. The recent increase in Atlantic hurricane activity: causes and implications. Science 293, 474–479.

Herrera-Silveira, J.A., Lara-Domínguez, A.L., Day, J.W., Yáñez-Arancibia, A., Ojeda, S.M., Hernández, C.T., Kemp, G.P., 2019. Ecosystem functioning and sustainable management in coastal systems with high freshwater input in the Southern Gulf of Mexico and Yucatan Peninsula. In: Wolanski, E., Day, J., Elliot, M., Ramachandran, R. (Eds.), Coasts and Estuaries—The Future. Elsevier, Amsterdam.

Higgins, S., Overeem, I., Tanaka, A., Syvitski, J., 2013. Land subsidence at aquaculture facilities in the Yellow River delta, China. Geophys. Res. Lett. 40, 3898–3902.

Higgins, S., Overeem, I., Steckler, M.S., Syvitski, J.P.M., Akhter, S.H., 2014. InSAR measurements of compaction and subsidence in the Ganges-Brahmaputra Delta, Bangladesh. J. Geophys. Res. Earth Surf. 119, 1768–1781.

Hoa, L., Nguyen, H., Wolanski, E., Tran, T., Haruyama, S., 2007. The combined impact on the flooding in Vietnam's Mekong River delta of local man-made structures, sea level rise, and dams upstream in the river catchment. Estuar. Coast. Shelf Sci. 71, 110–116.

Horton, B.P., Rahmstorf, S., Engelhart, S., Kemp, A., 2014. Expert assessment of sea-level rise by AD 2100 and AD 2300. Quat. Sci. Rev. 84, 1–6.

Hoyos, C., Agudelo, P., Webster, P., Curry, J., 2006. Deconvolution of the factors contributing to the increase in global hurricane intensity. Science 312, 94–97.

Ibañez, C., 2015. Sustainable management of deltas under relative sea level-rise: looking at the past to cope with future conditions (Rhine, Ebre, Mississippi). In: Yáñez-Arancibia, A. (Ed.), Adaptacion y Mitigacion hacia Agendas Siglo XXI. AGT Editorial S.A., INECOL, Mexico, pp. 103–120.

Ibáñez, C., Sharpe, P., Day, J.W., Day, J.N., Prat, N., 2010. Vertical accretion and relative sea level rise in the Ebro delta wetlands. Wetlands 30, 979–988.

IPCC (Intergovernmental Panel on Climate Change), 2013. *Climate Change 2013: The Physical Science Basis, Contribution of Working Group 1 to the Fifth Assessment Report of the Intergovernmental Panel on Climate Change* Cambridge, UK. 1535.

Kallepalli, A., Kakani, N.R., James, D.B., Richardson, M., 2017. Digital shoreline analysis system-based change detection along the highly eroding Krishna–Godavari delta front. J. Appl. Remote. Sens. 11 (3). 036018-1.

Kaufmann, R.F., Kauppi, H., Mann, M., Stock, J., 2011. Reconciling anthropogenic climate change with observed temperature 1998–2008. Proc. Natl. Acad. Sci. U S A. 108 (29), 11790–11793. https://doi.org/10.1073/pnas.1102467108.

Kemp, G.P., Day, J., Yáñez-Arancibia, A., Peyronnin, N., 2016. Can continental shelf river plumes in the northern and southern Gulf of Mexico promote ecological resilience in a time of climate change? Water 8, 83. https://doi.org/10.3390/w8030083.

Kidwai, S., Ahmed, W., Tabrez, S.M., Zhang, J., Giosan, L., Clift, P., Inam, A., 2019. The Indus delta-catchment, river, coast and people. In: Wolanski, E., Day, J., Elliot, M., Ramachandran, R. (Eds.), Coasts and Estuaries—The Future. Elsevier, Amsterdam.

Kiwango, Y., Wolanski, E., 2008. Papyrus wetlands, nutrients balance, fisheries collapse, food security, and Lake Victoria level decline in 2000–2006. Wetl. Ecol. Manag. 16, 89–96.

Knights, B., 1979. Reclamation in the Netherlands. In: Knights, B., Philipps, A. (Eds.), Estuarine and Coastal Land Reclamation and Water Storage. Saxon House, London.

Koop, R., Kemp, A., Bitterman, K., Horton, B., Donnelly, J., Gehrels, W., Hay, C., Mitrovica, J., Morrow, E., Rahmstorf, S., 2016. Temperature-driven global sea-level variability in the common era. Proc. Natl. Acad. Sci. https://doi.org/10.1073/pnas.1517056113.

Kuenzer, C., Renaud, F., 2012. Climate change and environmental change in river deltas globally. In: Renaud, F., Kuenzer, C. (Eds.), The Mekong Delta System—Interdisciplinary Analyses of a River Delta. Springer, Netherlands, ISBN: 978-94-007-3961-1pp. 7–48. https://doi.org/10.1007/978-94-007-3962-8.

Kuenzer, C., Vo Quoc, T., 2013. Assessing the ecosystem services value of can Gio Mangrove biosphere reserve: combining earth-observation- and household-survey-based analyses. Appl. Geogr. 45, 167–184.

Kuenzer, C., Campbell, I., Roch, M., Leinenkugel, P., Quoc, V., T., and Dech, S., 2013. Understanding the impacts of hydropower developments in the context of upstream-downstream relations in the Mekong river basin. Sustain. Sci. 8 (4), 565–584. https://doi.org/10.1007/s11625-012-0195-z.

Kuenzer, C., Ottinger, M., Lie, G., Sun, B., Dech, S., 2014. Earth observation-based coastal zone monitoring of the Yellow River delta: dynamics in China's second largest oil producing region over four decades. Appl. Geogr. 55, 72–107.

Leinenkugel, P., Wolters, M., Oppelt, N., Kuenzer, C., 2014. Tree cover and forest cover dynamics in the Mekong basin from 2001 to 2011. Remote Sens. Environ. 158, 376–392.

Li, M., Chen, Z., 2019. An assessment of saltwater intrusion in the Changjiang (Yangtze) River Estuary, China—implications for future water availability for Shanghai. In: Wolanski, E., Day, J., Elliot, M., Ramachandran, R. (Eds.), Coasts and Estuaries—The Future. Elsevier, Amsterdam.

Malini, B.H., Rao, K.N., 2004. Coastal erosion and habitat loss along the Godavari delta front—a fallout of dam construction? Curr. Sci. 87, 1232–1236.

Meehl, G.A., et al., 2007. Global climate projections. In: Solomon, S., Qin, D., Manning, M., Chen, Z., Marquis, M., Avery, K.B., Tignor, M., Miller, H.L. (Eds.), Climate Change 2007: The Physical Science Basis. Contribution of Working Group I to the Fourth Assessment Report of the Intergovernmental Panel on Climate Change. Cambridge University Press, Cambridge/New York, NY.

Mei, W., Xie, S., Primeau, F., McWilliams, J., Pasquero, C., 2015. Northwestern Pacific typhoon intensity controlled by changes in ocean temperatures. Sci. Adv. https://doi.org/10.1126/sciadv.1500014.

Meire, P., n.d. Pers. Comm., University of Antwerp.

Milliman, J.D., Quraishee, G., Beg, M., 1984. Sediment discharge from the Indus River to the ocean: past, present and future. In: Haq, B., Milliman, J. (Eds.), Marine Geology and Oceanography of Arabian Sea and Coastal Pakistan. Van Nostrand Reinhold, New York, pp. 65–70.

Milliman, J.D., Broadus, J., Gable, F., 1989. Environmental and economic implications of rising sea level and subsiding deltas: the Nile and Bengal examples. Ambio 18, 340–345.

Min, S.-K., et al., 2011. Human contribution to more-intense precipitation extremes. Nature 470, 378–381.

Morton, R., Bernier, J., Barras, J., Ferina, N., 2005. Rapid subsidence and historical wetland loss in the Mississippi delta plain: likely causes and future implications. U.S. Geological Survey, Open-File Report 2005-1216.

Nhan, N.H., Cao, N.B., 2019. Damming the Mekong: impacts in Vietnam and solutions. In: Wolanski, E., Day, J., Elliot, M., Ramachandran, R. (Eds.), Coasts and Estuaries—The Future. Elsevier, Amsterdam.

Niang, A., Scheren, P., Diop, S., Kane, C., Koulibaly, C.T., 2019. The Senegal and Pangani Rivers: examples of over-used river systems with water stressed environments in Africa. In: Wolanski, E., Day, J., Elliot, M., Ramachandran, R. (Eds.), Coasts and Estuaries—The Future. Elsevier, Amsterdam.

Nilsson, C., Reidy, D.M., Revenga, C., 2005. Fragmentation and flow regulation of the world's large river systems. Science 308, 405–408.

Odum, W.E., Odum, E.P., Odum, H.T., 1995. Nature's pulsing paradigm. Estuaries 18, 547–555.

Pall, P., et al., 2011. Anthropogenic greenhouse gas contribution to flood risk in England and Wales in autumn 2000. Nature 470, 382–385.

Pethick, J., Orford, J.D., 2013. Rapid rise in effective sea-level in southwest Bangladesh: its causes and contemporary rates. Glob. Planet. Chang. 111, 237–245. ISSN 0921-8181. https://doi.org/10.1016/j.gloplacha.2013.09.019.

Pfeffer, W., Harper, J., O'Neel, S., 2008. Kinematic constraints on glacier contributions to 21st century sea-level rise. Science 321, 1340–1343.

Pont, D., Day, J., Hensel, P., Franquet, E., Torre, F., Rioual, P., Ibáñez, C., Coulet, E., 2002. Response scenarios for the deltaic plain of the Rhône in the face of an acceleration in the rate of sealevel rise, with a special attention for *Salicornia*-type environments. Estuaries 25, 337–358.

Pont, D.J., Day, J., Ibáñez, C., 2017. The impact of two large floods (1993–1994) on sediment deposition in the Rhone delta: implications for sustainable management. Sci. Total Environ. https://doi.org/10.1016/j.scitotenv.2017.07.155.

Rabalais, N.N., Turner, R.E., 2001. Hypoxia in the northern Gulf of Mexico: description, causes and change. In: Rabalais, N., Turner, R. (Eds.), Coastal Hypoxia: Consequences for Living Resources and Ecosystems. American Geophysical Union, Washington, DC, pp. 1–36.

Ramesh, R., Lakshmi, A., Mohan, S.S., Bonthu, S.R., Mary Divya Suganya, G., Ganguly, D., Robin, R.S., Purvaja, R., 2019. Integrated management of the Ganges delta, India. In: Wolanski, E., Day, J., Elliot, M., Ramachandran, R. (Eds.), Coasts and Estuaries—The Future. Elsevier, Amsterdam.

Rao, K.N., Saito, Y., Nagakumar, K.C.V., Demudu, G., Basavaiah, N., Rajawat, A.S., Tokanai, F., Kato, K., Nakashima, R., 2012. Holocene environmental changes of the Godavari delta, east coast of India, inferred from sediment core analyses and AMS 14C dating. Geomorphology 175–176, 163–175.

Renaud, F.G., Syvitski, J.P.M., Sebesvari, Z., et al., 2013. Tipping from the Holocene to the anthropocene: how threatened are major world deltas? Curr. Opin. Environ. Sustain. 5, 644–654. https://doi.org/10.1016/j. cosust.2013.11.007.

Royal Society, 2014. Resilience to extreme weather. 122 p. https://royalsociety.org/topics-policy/projects/resilience-extreme-weather/

Schneider, E., Tudor, M., Staras, M. (Eds.), 2008. Ecological restoration in the danube delta biosphere reserve/Romania. Evolution of Babina Polder after Restoration Works. WWF Auen Institute/Danube Delta National Institute, Germany, pp. 81.

Sestini, G., 1992. Implications of climatic changes for the Po delta and the Venice lagoon. In: Jeftic, L., Milliman, J., Sestini, G. (Eds.), Climate Change and the Mediterranean. E. Arnold, London, pp. 429–495.

Shalovenkov, N., 2019. Alien species invasion: case study of the Black Sea. In: Wolanski, E., Day, J., Elliot, M., Ramachandran, R. (Eds.), Coasts and Estuaries—The Future. Elsevier, Amsterdam.

Snedaker, S., 1984. Mangroves: a summary of knowledge with emphasis on Pakistan. In: Haq, B., Milliman, J. (Eds.), Marine Geology and Oceanography of Arabian Sea and Coastal Pakistan. Van Nostrand Reinhold, New York, pp. 255–262.

Stanley, D., 1988. Subsidence in the northeastern Nile delta: rapid rates, possible causes, and consequences. Science 240, 497–500.

Stanley, J.D., Clemente, P.L., 2017. Increased land subsidence and sea-level rise are submerging Egypt's Nile delta coastal margin. GSA Today 27, 4–11. https://phys.org/news/2017-03-looming-crisis-decreased-fresh-water-egypt.html#jCp.

Stanley, D., Warne, A., 1993. Nile delta: recent geological evolution and human impacts. Science 260, 628–634.

Syvitski, J.P.M., 2008. Deltas at risk. Sustain. Sci. 3, 23–32.

Syvitski, J.P.M., Kettner, A., 2007. On the flux of water and sediment into the Northern Adriatic. Cont. Shelf Res. 27, 296–308.

Syvitski, J.P.M., Saito, Y., 2007. Morphodynamics of deltas under the influence of Humans. Glob. Planet. Chang. 57, 261–282.

Syvitski, J.P.M., Kettner, A., Correggiari, A., Nelson, B., 2005a. Distributary channels and their impact on sediment dispersal. Mar. Geol. 222-223, 75–94.

Syvitski, J.P.M., Vörösmarty, C., Kettner, A., Green, P., 2005b. Impact of humans on the flux of terrestrial sediment to the global coastal ocean. Science 308, 376–380.

Syvitski, J., Kettner, A., Overeem, I., Hutton, E., Hannon, M., Brakenridge, G., Day, J., Vörösmarty, C., Saito, Y., Giosan, L., Nichols, R., 2009. Sinking deltas due to human activities. Nat. Geosci. 2, 681–686.

Syvitski, J.P.M., Kettner, A., Overeem, I., Giosan, L., Brakenridge, R., Hannon, M., Bilham, R., 2013. Anthropocene metamorphosis of the Indus delta and lower floodplain. Anthropocene 3, 24–35.

Tessler, Z.D., Vörösmarty, C., Grossberg, M., Gladkova, I., Aizenman, H., Syvitski, J., Foufoula-Georgiou, E., 2015. Profiling risk and sustainability in coastal deltas of the world. Science 349, 638–643.

Vermeer, M., Rahmstorf, S., 2009. Global sea level linked to global temperature. Proc. Natl. Acad. Sci. 106, 21527–21532.

Vörösmarty, C., Syvitski, J., Day, J., Sherbinin, A., Giosan, L., Paola, C., 2009. Battling to save the world's river deltas. Bull. At. Sci. 65, 31–43.

Wang, J., Gao, W., Xu, S.Y., Yu, L.Z., 2012. Evaluation of the combined risk of sea level rise, land subsidence, and storm surges on the coastal areas of Shanghai, China. Clim. Chang. 115, 537–558.

Webster, J., Holland, G.J., Curry, J.A., Chang, H.-R., 2005. Changes in tropical cyclone number, duration, and intensity in a warming environment. Science 309, 1844.

Wolanski, E., Nguyen, H.N., 2005. Oceanography of the Mekong River Estuary. In: Chen, Z., Saito, Y., Goodbred, S.L. (Eds.), Mega-Deltas of Asia-Geological Evolution and Human Impact. China Ocean Press, Beijing, pp. 113–115. 268 pp.

Wolanski, E., Day, J., Elliott, M., Ramesh, R. (Eds.), 2019. Coasts and Estuaries—The Future. Elsevier. this book.

Wolters, R., Kuenzer, C., 2015. Vulnerability assessments of coastal river deltas—categorization and review. J. Coast. Conserv. 19, 345–368.

Wiegman, R., Day, J., D'Elia, C., Rutherford, J. Morris, J., Roy, E., Lane, R., Dismukes, D., Snyder, B., 2017. Modeling impacts of sea-level rise, oil price, and management strategy on the costs of sustaining Mississippi delta marshes with hydraulic dredging. Sci. Total Environ. https://doi.org/doi.org/10.1016/j.scitotenv.2017.09.314.

Chapter 10

Mississippi Delta Restoration and Protection: Shifting Baselines, Diminishing Resilience, and Growing Nonsustainability

John W. Day*, Craig Colten[†], G. Paul Kemp*

**Department of Oceanography and Coastal Sciences, College of the Coast and Environment, Louisiana State University, Baton Rouge, LA, United States, [†]Department of Geography and Anthropology, Louisiana State University, Baton Rouge, LA, United States*

1 INTRODUCTION

The Mississippi delta is arguably the most intensively studied delta in the world. Beginning in the 1930s, a series of foundational studies laid much of the framework for modern deltaic science. The works of Russell (1936), Russell et al. (1936), Fisk (1944, 1952, 1954), Fisk and McFarlan (1955), Kolb and van Lopik (1958), Saucier (1963), and Frazier (1967) provide a robust appreciation for the forms and processes important to deltas. This knowledge base was developed during a period of relatively static sea level, and now is being reevaluated in view of the "new normal," an escalation in sea-level rise forced by anthropogenetic climate change (Blum and Roberts, 2009, 2012).

The coast is where most economic activity takes place in Louisiana. Businesses are put at increased flood risk if wetlands that now protect levees by attenuating storm surge and waves continue to disappear (Barnes et al., 2015; Day et al., 2000). At-risk areas include major import-export commodity ports on the Mississippi, Calcasieu, and Sabine Rivers that are linked to internationally significant petrochemical manufacturing facilities by highway, waterway, and rail infrastructure, in addition to coastal cities housing more than 2 million (e.g., New Orleans, Houma, Lake Charles, Morgan City, the north shore of Lake Pontchartrain). Batker et al. (2014) estimated that the delta ecosystem provides between $12 and 47 billion annually in ecosystem goods and services. The replacement cost of capital stock that will be damaged over 50 years if measures included in the CMP are not built is estimated to be $2.1–3.5 billion. An additional loss of $5.8–7.4 billion is projected from reduced economic activity associated with storm related impacts and abandonment of infrastructure (e.g., exodus of businesses, schools, churches, and residents) (Barnes et al., 2015). These numbers illustrate the cost of replacing the essential social fabric of the region at risk of catastrophic flooding (Glavovic, 2014). It is this existential societal threat—made real in the Hurricane Katrina and Rita catastrophes of 2005—that has led the people and leaders of Louisiana to embark on the Coastal Master Plan (CMP) to restore and protect coastal Louisiana ecosystems, communities, and infrastructure (CPRA, 2007, 2012, 2017).

We briefly describe the geologic, ecologic, and socioeconomic factors that have been important to both the building and loss of the MRD as both a landform and a place for people to live and work. We point out significant biophysical, economic, and social constraints and irreversible trends that will affect the likelihood that CMP-approved projects will be constructed and prove effective, and where there are more sustainable paths forward. Finally, we describe an alternative approach that we believe has a greater chance to enhance deltaic sustainability.

2 DEVELOPMENT OF THE DELTA

The Mississippi delta is made up of several interdistributary hydrologic basins separated by active or abandoned river distributary ridges (Roberts, 1997; Day et al., 2000; Blum and Roberts, 2012, Fig. 1). There are two physiographic provenances. The river dominated Deltaic Plain is on the east and is where active deltaic lobe formation occurred, while the Chenier Plain to the west was created by ocean waves moving muddy river sediments temporarily stored on the nearshore bottom into shore-attached mudflats soon colonized by marsh vegetation. Shore-parallel beach ridges composed of shells

Coasts and Estuaries. https://doi.org/10.1016/B978-0-12-814003-1.00010-1

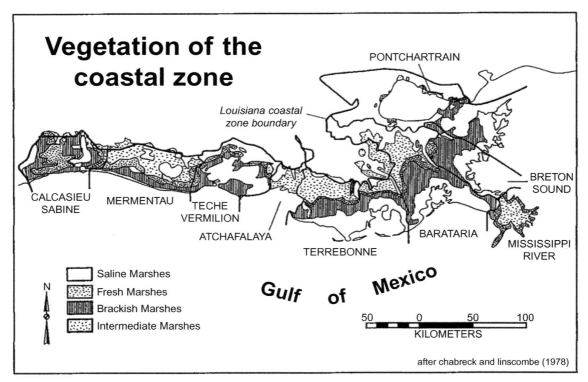

FIG. 1 Interdistributary basins and vegetation zones of the Mississippi delta. The Calcasieu-Sabine and Mermentau basins make up the Chenier Plain while the eastern seven basins form the deltaic plain. *(From Day, J., Shaffer, G., Britsch, L., Reed, D., Hawes, S., Cahoon, D., 2000. Pattern and process of land loss in the Mississippi delta: a spatial and temporal analysis of wetland habitat change. Estuaries 23, 425–438.)*

eroded out of marsh/mudflats alternate with wetlands in a strand plain that has developed over the last 4 ka. This coast is building seaward with progressively younger beach ridges approaching the active shoreline (Roberts, 1997).

The first inner shelf delta lobe (Sale-Cypremort) formed after infilling of the low-stand Mississippi River trench is now completely submerged, but significant parts of all subsequent subdeltas have been incorporated into the modern deltaic plain (Fig. 2). The deltaic surface encompassed about 25,000 km^2 of wetlands, inshore water bodies, and low relief uplands when it reached its greatest expanse in the 19th century (Roberts, 1997; Day et al., 2007, 2014; Hijma et al., 2018; Day and Erdman, 2018). These wetlands form a series of vegetation zones (Fig. 1) that are determined primarily by salinity, ranging inland from the coast, and including saline, brackish and freshwater emergent herbaceous wetlands, and freshwater forested wetlands.

After sea-level rise slowed at the end of the last glacial age and finally stabilized at near present sea level about 5000 years ago, the Mississippi delta began forming this vast deltaic wetland complex that by the 18th century encompassed about 25,000 km^2 (Roberts, 1997; Roberts et al., 2015; Boesch et al., 1994; Day et al., 1995, 2000, 2007; Hijma et al., 2018, Fig. 2). A hierarchy of energetic processes distributed water and materials through the delta from both the watershed and Gulf and exchanged the biotic products of the estuarine ecosystems with the coastal ocean. River sediment was deposited at active distributary mouths or, laterally through overbank flooding and crevasses (breaks in the natural levees).

Delta lobes formed as river channels elongated and bifurcated and were ultimately abandoned through a process of upstream avulsion in favor of shorter routes to sea level (Fig. 2). MRD subdeltas have lifespans of thousands of years and cover 100s–1000s of kilometers. Crevasses functioned at a smaller scale during high water for a few years to as long as a century to build sediment splays adjacent to the natural levee with areas of 10s–100s of km^2 (Saucier, 1963; Welder, 1959; Davis, 2000; Roberts, 1997, Fig. 3).

Most abandoned distributaries from earlier lobe building episodes nonetheless flowed with river water during major floods to maintain a skeletal framework of interconnected distributary channels and natural levee ridges not only within the youngest lobes but also in older, largely abandoned sub-deltas, while barrier islands, and beach ridges formed after lobe abandonment at the marine boundaries (Penland et al., 1988). These ridge features protect interior wetlands from hurricane surge and waves and salinity intrusion, but they have also been important to both precontact and postcolonial societies as preferred sites for human habitation and cultivation (e.g., Day et al., 2007; Xu et al., 2016; Colten and Day, 2018).

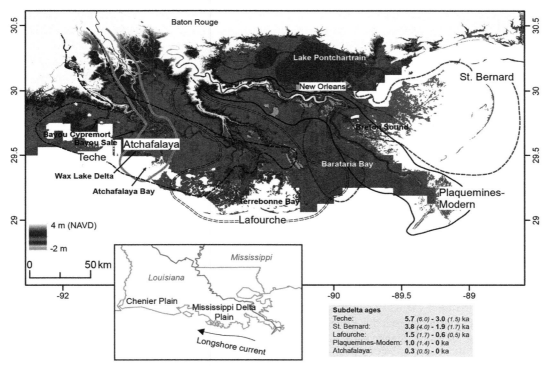

FIG. 2 Delta lobes of the Mississippi delta with times of active growth. *(Adapted from Hijma, M., Shen, Z., Törnqvist, T., Mauz, B., 2018. Late Holocene evolution of a coupled, mud-dominated delta plain-chenier plain systems, coastal Louisiana, USA. Earth Surf Dynam. 5, 689–710.)*

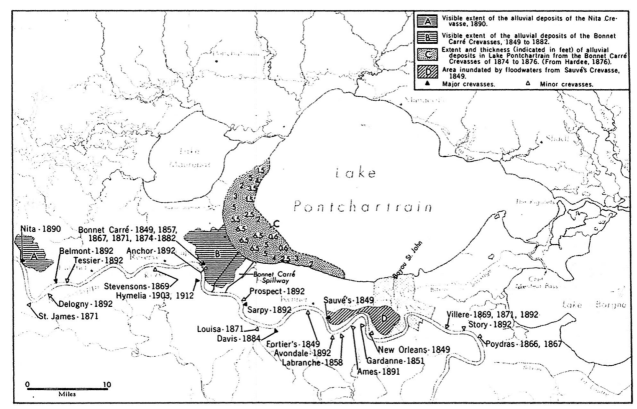

FIG. 3 Location of some of the crevasses along the Mississippi River from 1849 to 1927 including areas affected by three major crevasses (A, B, C, and D). Note that the Bonnet Carré crevasse filled in areas of western Lake Pontchartrain by more than 2 m (6.5 ft). *(From Saucier, R.T., 1963. Recent Geomorphic History of the Pontchartrain Basin. Coastal Studies Institute, Report 9. Louisiana State University, Baton Rouge, Louisiana. 114 p.)*

Human-built levees, some dating back to the 1720s, inhibited crevassing locally, but did not eliminate the process. Saucier (1963) and Davis (1993) documented hundreds of crevasses on the lower Mississippi after the arrival of Europeans. The splays overlap the natural levees to form a nearly continuous band that laterally expands both higher land and new wetlands in adjacent water bodies like Lake Pontchartrain (Fig. 3, Saucier, 1963; Davis, 2000; Allison and Meselhe, 2010; Shen et al., 2015; Day et al., 2016c). Crevassing continued even after federal investment began in a more robust artificial levee system. Colten (2017) notes that a levee breach below New Orleans in 1922 augmented the flanks of the natural levee with about 2 m of new "high" land. Also, Day et al. (2016b) report that an artificial crevasse intentionally opened at Caernarvon for 3 months during the great flood of 1927 had a peak discharge of nearly 10,000 $m^3 s^{-1}$ and deposited a layer of river sediment up to 40 cm thick over about 130 km^2.

Self-sustaining deltaic wetlands must build vertically at a rate sufficient to stay within the tidal frame and offset naturally high rates of RSLR (Baumann et al., 1984; Day et al., 2011). Most of this is due to subsidence caused by compaction and dewatering of underlying soft sediments in the upper 10 m of the sediment column (Jankowski et al., 2017). While RSLR varies spatially over short distances across the delta depending on local subsidence, mean values for the past decade from the nearly 300 stations that make up the Coastwide Reference Monitoring System (CRMS) (Steyer et al., 2003) are 13.2 to 9.5 mm-yr^{-1} for the Deltaic and Chenier Plains, respectively. RSLR includes a eustatic (global) component independently measured by satellite altimetry in the Northern Gulf of Mexico at 2 mm-yr^{-1} for the same period. Local subsidence on average makes up 85% of RSLR in the Deltaic Plain, and 79% in the Chenier Plain (Jankowski et al., 2017). Furthermore, Jankowski et al. (2017) report that at least 60% of subsidence occurs in the upper 5–10 m of the sediment/soil column.

Near the river mouths unconsolidated coarse sediment is deposited on shallow bay bottoms and in the nearshore where it is susceptible to resuspension by waves and tides. Deposition and permanent capture of these resuspended sediments on marsh surfaces takes place during storms that generate large waves along with sequential water level set-up and set-down (Day et al., 2011). Surface elevation change (SEC), and vertical accretion (VA) are marsh properties that have been systematically measured at the CRMS stations. SEC occurs as a consequence of both in situ organic soil formation and the capture and incorporation of mineral sediment inputs.

Ultimately, the river is directly or indirectly the source of all mineral sediment reaching delta wetlands. Easily suspended, fine-grained mineral sediments (silts and clays) increase marsh soil strength and bulk density while nutrients also delivered by the river enhance plant growth (Day et al., 2011). Introduction of fresh river water into adjacent deltaic estuaries displaces saltwater seaward, while dissolved iron from the continental basin precipitates toxic sulfides that otherwise build up in brackish soils and stress wetland plants (DeLaune and Pezeshki, 1988; Delaune and Pezeshki, 2003). Despite a 30-cm tide range, water level set-up and set-down of a meter or more occurs several times each winter during cold front passages, and less frequently during hurricanes in the summer and fall (Baumann et al., 1984; Perez et al., 2000).

Globally, the late Holocene has been a good time to both grow and live in a delta (Day et al., 2016a; Colten and Day, 2018). Despite the coexistence of areas of wetland building and loss, the net size of the MRD increased over all but the last century, when humans intervened in a big way (Jankowski et al., 2017; Day and Erdman, 2018). Stable sea-level concentrated land-building inputs from the continental basin (freshwater, sediments, nutrients) while also limiting the inland reach of waves and storm surge. This allowed for maturation of productive estuarine ecosystems and barrier island tracts well adapted to deltaic dynamics (Giosan et al., 2014). Because flooding by the river and by hurricanes occurs in different seasons in Louisiana, and because the availability of migratory food sources is also predictable, native peoples found higher ridges of the delta to be good sites for villages and temporary camps (Colten and Day, 2018).

Condrey et al. (2014) used journals and charts of 16th century Spanish explorers to describe what they have called the "last natural delta" of the Mississippi as it existed prior to European settlement. The delta is described in early accounts as fringed with treacherous oyster reefs and barrier islands that fronted all of the five most recent delta complexes of the Mississippi River and extended entirely across the deltaic plain. Ship crews marveled that they could refill their water casks from plumes of fresh water that extended far into the Gulf during the spring flood and nourished a vast offshore oyster reef. Condrey et al. (2014) suggest that the shoreline of the coast was advancing seaward at this time in many places. However, this changed quickly after the arrival of European colonists a couple of centuries later.

3 DETERIORATION OF THE DELTA

Human impact on the MRD was rather minor during the 18th and 19th centuries, as the delta continued to grow. Completion of mainline levees along the river, and the invasion of steam dredges and the oil and gas industry, however, reversed the long-term trend, and brought about a collapse of a quarter of the deltaic plain wetland inventory (about 5000 km^2) in less than 80 years from 1930 to 2010 (Couvillion et al., 2011). In 2005, the weakened coast was devastated by Hurricanes Katrina and Rita leading to 1800 deaths and about $200 billion in damage and reconstruction costs (Day et al., 2007).

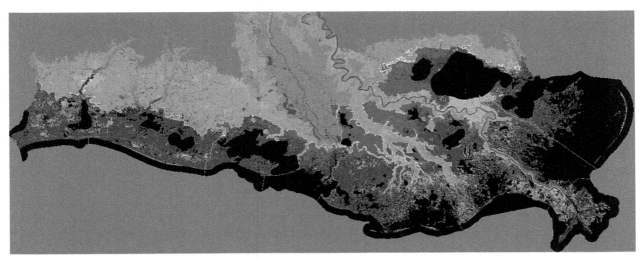

FIG. 4 Land area change in coastal Louisiana from 1932 to 2010. Red and yellow areas have high land loss rates. Note that land loss is low in the central coast and in the northeastern flank of the delta. *(Source: Couvillion, B., Barras, J., Steyer, G., Sleavin, W., Fishcher, M., Beck, H., et al., 2011. Land area change in coastal Louisiana from 1932 to 2010. U.S. Geological Survey, Scientific Investigations Map 3164, scale 1:265,000, 12 p. pamphlet. The map can be downloaded at https://pubs.usgs.gov/sim/3381/sim3381.pdf for detialed examination of specific areas of change. See also https://pubs.er.usgs.gov/ publication/sim3381.)*

The devastating effects of these hurricanes were seen as a predictable result of two centuries of mismanagement of the Mississippi River and MRD, and galvanized attention to impacts of global climate change on deltas worldwide.

Couvillion et al. (2011) produced a highly detailed map of coastal land loss from 1932 to 2010 (Fig. 4). Various shades of red and yellow show areas of land loss. The map can be downloaded for examination of specific areas of change. A casual glance at this map immediately conveys two impressions. First, wetland loss has been pervasive across the coast, but especially near the mouth of the Mississippi River, in the Barataria and Terrebonne basins, and west into the Chenier Plain. Second, two areas stand out as much less affected by land loss. One is the central coast that is nourished by the Atchafalaya River that discharges into shallow bays and wetlands over a wide arc along the central Louisiana coast. This input constitutes 30% of the combined discharge of the Red and Mississippi Rivers. The other zone of lower land-loss is on the northeastern flank of the delta in the seaward reaches of the Pontchartrain estuary.

Beginning in the late 19th century, local appeals for more flood protection ultimately led to federal investment in more effective levees that largely severed the river from its deltaic distributaries and the adjacent wetlands (Reuss, 1998; Camillo and Pearcy, 2004).

A 'levees-only' policy that tightly controlled the Mississippi was adopted by the US Army Corps of Engineers (USACEs) in the 1880s to maximize navigation and flood control benefits from the river. This federal program proved to be a failure, driving up flow lines and leading to more disastrous levee failures and flooding. After the 1927 Mississippi River flood catastrophe, the USACE recognized that during large floods the river needed access to more offline floodplain storage and an increase in the number and capacity of emergency flood outlets in the MRD. This led to construction of the Bonnet Carré and Atchafalaya Basin floodways which have trapped a vast amount of sediment that would have otherwise reached coastal wetlands. But little sediment reaches the delta plain directly from the lower Mississippi River channel past New Orleans until the last 100 km upstream of the birdsfoot delta. Now that some sediment entering the Atchafalaya River bypasses the Atchafalaya Basin, more is reaching the central coast where two new subdelta lobes have been building since the 1970s. The Atchafalaya River continues to drive wetland building and nourishment in its bayhead deltas and adjacent marshes, and also prevent loss of marshes around the receiving bays (Day et al., 2011; Blum and Roberts, 2012; Rosen and Jun Xu, 2013; DeLaune et al., 2016; Twilley et al., 2016).

4 GLOBAL CHANGE CONSTRAINTS ON COASTAL PROTECTION AND RESTORATION

Global change forcings, especially climate change and energy scarcity, along with socioeconomic policy decisions can be expected to limit options of what is possible for coastal protection and restoration (e.g., Day et al., 2016a). Day and Rybczyk (2019, this volume) review global change drivers that will impact coastal systems. Climate change drivers are already affecting the coast and their impact will grow more severe during the 21st century. Sea level is projected to increase by 2 m or more by 2100 or shortly thereafter (DeConto and Pollard, 2016), the frequency of category four and five hurricanes is anticipated to increase, extreme weather events including heavy precipitation and droughts may become more common, and the peak discharge of the Mississippi River is projected to increase by up to 60% (Tao et al., 2014; Min et al., 2011; Pall et al., 2011; Prein et al., 2016).

Environmental migration has been underway in the coastal region for decades (Hemmerling, 2018). Intense storms prompt pulses of relocation from highly exposed communities and younger residents are gradually departing coastal communities. Numerous small communities have disappeared since the 19th century and New Orleans lost about 20% of its population after Hurricane Katrina (2005). The departures drain the social capital of the region while leaving behind more vulnerable residents (Colten, 2017). Public policy has neglected both the costs and the potential benefits of relocation.

Energy scarcity will likely become an important factor affecting delta management (Day et al., 2000; Tessler et al., 2015; Wiegman et al., 2017; Rutherford et al., 2018a, b). Recent analyses suggest that world oil production will peak in 2–3 decades (Maggio and Cacciola, 2012) implying that demand will consistently be greater than supply and that the cost of energy will increase significantly in coming decades. The planning horizon for coastal protection and restoration in Louisiana is 50–100 years, thus energy scarcity in conjunction with climate change will likely constrain our ability to manage the coast. Levee construction, barrier island restoration, and long-distance conveyance of sediments are energy intensive both in construction and maintenance (Tessler et al., 2015; Wiegman et al., 2017). Only engineered river diversions have relatively low long-term maintenance costs. We now consider how these global change drivers and public policies will impact coastal protection and restoration.

5   COASTAL PROTECTION AND RESTORATION

The State of Louisiana is now in the midst of a $50 billion, 50-year program of coastal protection and restoration called the Louisiana's comprehensive master plan (CMP) for a sustainable coast (CPRA, 2017; Wiegman et al., 2018). The vast majority of funding has yet to be identified but work has begun with the $2 billion currently in hand. The goals of the CMP are twofold, namely to protect human infrastructure in the delta with conventional flood control structures (levees, floodwalls, and pumps), and to restore natural processes to create wetlands and sustain the deltas estuarine ecosystems, all with the goal of promoting economic development and cultural traditions in the coastal zone.

The plan does not include specific projects for protecting social capital and restoring coastal cultures. The CMP has "collective goals of reducing economic losses to homes and business from storm surge-based flooding, promoting sustainable ecosystems, providing habitats for a variety of commercial and recreational activities coast wide, strengthening communities, and supporting businesses and industry" (LACMP, 2017). The 2017 CMP contains 13 structural protection projects and 23 nonstructural risk-reduction projects (elevate and flood proof buildings and help property owners prepare for flooding or move out of areas of high flood risk). These components complement a federal flood control system along the Mississippi River, the Mississippi River and Tributaries Project (MR&T), to prevent riverine flooding. The historical reliance on levees was a major factor in creating the land loss crisis and led to dramatic loss of property and life during Katrina. Restoration projects involve a number of activities such as barrier island restoration, shoreline protection and stabilization, hydrologic restoration, ridge restoration, and oyster barrier reefs. But the two most important restoration activities are river diversions and marsh creation for land building in the coastal zone. Marsh creation uses dredged sediments that are pumped in pipelines, often for long distances, to build wetlands (Wiegman et al., 2017). Diversions reintroduce river water into the coastal zone to create and restore wetlands (Day et al., 2014; Rutherford et al., 2018a, b); these will be much more effective if the reduction in fine sediment transport in the river is restored (Kemp et al., 2016). As with levees, past diversions, designed for other purposes, have contributed to ecological disruptions in the delta region. For example, fresh water flushing through the Bonnet Carré Spillway seriously disrupted local fisheries in the short term (Day et al., 2016c; Colten, 2017).

Dredging sediment for marsh creation builds land quickly, but due to relative sea-level rise the resulting marshes require periodic re-nourishment to be sustainable (Wiegman et al., 2017). After construction, diversions have minimal recurring costs but build land gradually (Day et al., 2016) and will likely last for a century or more (e.g., Bonnet Carré Spillway, Day et al., 2012). While it is clear that river diversions will be a necessary and vital aspect of coastal restoration, much uncertainty still persists about their effectiveness, potential damages, and their economic benefit in the short term (Day et al., 2016c). It is possible that they could be a detriment to the economy in the short term, as fisheries struggle to adapt to changes in estuaries such as salinity reduction brought about by fluvial inputs (Day et al., 2016c).

The CMP will spend about 50% of total funding on restoration and 50% on coastal protection. The 2017 CMP proposes $17.1 billion on marsh creation and $5.1 billion on sediment diversions (Table 1). In the 2012 CMP, marsh creation projects had an average cost of about $360,000 per hectare over a 50-year time span, while sediment diversions had an average cost

TABLE 1 2017 Louisiana Coastal Master Plan Funding Allocation by Project Type

Class	Project Type	Funding ($billions)	Percent of Funds	Prime Mover
Restoration	(Total)	25	50	N/A
	Barrier Island	1.5	3	Hydraulic Dredge, Bulldozer
	Hydrologic	0.4	1	Pump or Gravity[a]
	Marsh Creation	17.1	34	Hydraulic Dredge, Bulldozer
	Ridges	0.1	0	Excavator, Dragline or Bucket Dredge
	Sediment Diversion	5.1	10	Gravity[a]
	Shoreline Protection	0.2	0	Barge, Crane or N/A[b]
Risk reduction	(Total)	25	50	N/A
	Structural (Levees)	18.8	38	Excavator, Dragline or Bucket Dredge
	Nonstructural	6.1	12	Various
Total		50	100	N/A

[a]Various machinery is required to build the control structures; after which the displacement of water or sediment is controlled by gravity (and pumps in some cases for hydrological restoration).

[b]Oyster reefs have various methods of creation; Rock armor shorelines and jetties require barges and cranes.

Adapted from CPRA, 2017. Louisiana's Comprehensive Master Plan for a Sustainable Coast. 2017 Coastal Master Plan. Louisiana Coastal Protection and Restoration Authority, Baton Rouge by Wiegman, A., Rutherford, J., Day, J., 2018. Mississippi Delta Restoration Pathways to a Sustainable Future. Springer, Cham, Switzerland, pp. 93–111.

(including engineering, operation, and maintenance) of about $45,000 per hectare. The CMP has not compared the cost of diversions and marsh creation to the costs of supporting community relocation to reduce risk.

Davis et al. (2015) estimated that the actual cost to restore Louisiana's coastline and flood protection infrastructure would be about $90 billion, greatly exceeding the $50 billion estimate. The key environmental drivers included in the 2017 CMP are eustatic sea-level rise, subsidence, tropical storm intensity, tropical storm frequency, precipitation, and evapotranspiration. But the plan does not include increases in peak discharge of the Mississippi River (Tao et al., 2014) or extreme weather events (Min et al., 2011; Pall et al., 2011; Prein et al., 2016).

In addition, the CMP does not fully consider the full range of climate change impacts (more strong hurricanes, and more extreme weather events, drought, and increases in peak discharge of the Mississippi River) nor does it factor in the cost of economic impacts due to increasing cost of energy. It also dismisses relocation as a core option and as an unavoidable eventual cost. Large flood control projects, marsh creation, and river diversions are examples of projects that require a vast amount of energy and capital during implementation (Day et al., 2005, 2014; Tessler et al., 2015; Wiegman et al., 2017). Thus there are significant financial limitations on the CMP that could be exacerbated by fluctuations in energy prices. It is plausible that changes in the global economy, public policy, regional and global climate, and energy availability will render significant components of the CMP unaffordable in a few decades (Day and Rybczyk, 2019).

The CMP uses sophisticated modeling to project future states of the coast given restoration scenarios compared to a future without action. The models do not fully consider how humans can adapt to ongoing and future changes. Fig. 5 presents model results for the coastline and vegetation types of the 2017 CMP 50 years in the future compared to no action for a high sea-level rise scenario. Almost all saline wetlands are predicted to be lost for both the future without action and the future with action (2017 Master Plan projects). The models project that marshes will be formed by diversions south of New Orelans along the river.

6 COASTAL PROTECTION AND RESTORATION IN A CLIMATE-CHALLENGED, ENERGY-SCARCE FUTURE

The information presented thus far indicates that global change forcings and public policies will seriously impact the efficacy of the CMP. Here we address these issues and develop ideas about a more sustainable path forward.

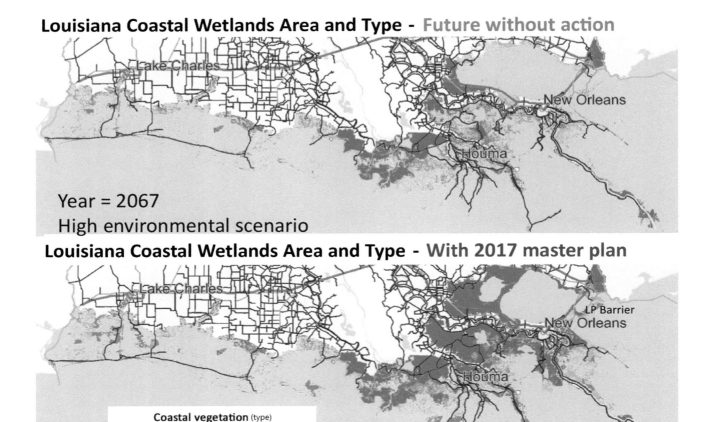

FIG. 5 A projection of the Louisiana coastline and vegetation types 50 years in the future with no action and with the 2017 Coastal Master Plan. The scenario, shown here uses a high environmental scenario, which does not include the highest estimates for eustatic sea-level rise, subsidence, and extreme weather events. In this scenario, almost all saline marshes disappear, and it is unlikely that the New Orleans metropolitan area or most natural levees would survive without protective wetlands. *(Modified from Coastal Protection and Restoration Authority, http://cims.coastal.louisiana.gov/masterplan/.)*

6.1 Flood and Storm Protection

Flood protection in southern Louisiana must confront flooding from three distinct threats including hurricane surge with accompanying waves, Mississippi River flood events, and localized extremely heavy rainfall. These different types of flood threats require different types of protection strategies accompanied by follow through on effective public policies.

6.2 Hurricane Surge and Waves

Since Hurricanes Katrina and Rita, providing adequate hurricane protection has been a central issue in Louisiana. This form of protection is absolutely necessary to continue to have viable social and economic systems in south Louisiana. Hurricane protection must take into consideration climate change and resource scarcity. Current plans include enhanced protection for New Orleans and long linear levees for much of the rest of the coast. For the New Orleans region, hurricane protection includes higher and stronger levees, structures at the entrance to Lake Pontchartrain, enhanced pumping stations for conveying rainfall in New Orleans to Lake Pontchartrain, closure of the Mississippi River—Gulf Outlet (MRGO), and elimination of the funnel associated with MRGO with a surge barrier (Shaffer et al., 2009, see Fig. 5). The CMP includes barriers across the entrance to Lake Pontchartrain, across the mid-Barataria Basin and the northeastern part of the Terrebonne Basin (see Fig. 5). These structures will have gates to prevent hurricane flooding as well as control salinity so that fresh water marshes and swamps proposed in the CMP can be maintained landward of the barriers. With rising sea level, gravity drainage will become less and less effective and at some point all of these areas may have to put under pump if they are to continue to function. This area is far larger than what exists now. The costs of pumping such large areas are likely to be unaffordable.

Hurricane protection must be integrated into coastal restoration. It also will be affected by global climate change and energy scarcity as discussed above. Climate change will require larger and stronger levees and energy scarcity is likely to make currently planned protection much more expensive, possibly prohibitively so.

6.3 The Mississippi River

After the catastrophic 1927 flood, the US government revamped its flood control system for the lower Mississippi River valley from Cairo IL to the mouth of the river—the Mississippi River and Tributaries project or MR&T. It shifted the emphasis from "levees only" to "levees and outlets" (Camillo and Pearcy, 2004; Reuss, 1998)—an acknowledgement that eliminating the natural distributaries reduced the river's ability to handle major floods and that levees alone did not offer a permanent fix. The revised system relies principally on levees but also incorporated a number of flood outlets or spillways and very large flood control and river management structures. The MR&T system is designed for the project flood with a peak discharge of about $85,000\,\mathrm{m^3\,s^{-1}}$ (3 million $\mathrm{ft^3\,s^{-1}}$). Peak discharge of the Mississippi River at Vicksburg was $64,000\,\mathrm{m^3\,s^{-1}}$ (2,278,000 cfs) in 1927 and $65,000\,\mathrm{m^3\,s^{-1}}$ (2,310,000 cfs) in 2011.

The Bonnet Carré Spillway is one of two flood relief outlets and is situated about 40 km upstream of New Orleans. It is designed to lower river levels in the city by allowing water to flow from the river to Lake Pontchartrain (Day et al., 2012). It was completed in 1933 after the flood of 1927 and has been opened 12 times beginning in 1937 with the last opening in 2018. Eight of the 12 openings occurred in the second half of its operational history indicating that large floods are becoming more common.

Tao et al. (2014) modeled the interactive effects of climate change, land use, and river management on discharge of the Mississippi River and projected that river discharge may increase by 10%–60% during this century. Assuming that peak discharge would increase by the same amounts, we plotted the peak discharge of the 2011 flood and added 10%–60%. An increase of 60% over the peak discharge for the 2011 flood would result in a peak discharge of $104,000\,\mathrm{m^3\,s^{-1}}$ (3.70 million $\mathrm{ft^3\,s^{-1}}$) exceeding the project flood by about 20% (Fig. 6). Peak flow increases of this magnitude may compromise the MR&T flood control system on the Mississippi River (Kemp et al., 2014).

Such large floods will likely increasingly threaten the functioning of the MR&T flood control system, as high discharges threaten both river levees and flood control structures. Many factors can compromise levees and lead to collapse including cavities due to rotting logs and burrowing animals, soil weakness, sand boils, and scour at the base of the levee. As John Barry notes, "the biggest danger is pressure, constant unrelenting pressure." (Barry, 1997, p. 191). Structures such as the Old River control structure, which regulates Mississippi River flow into the Atchafalaya and prevents capture of most of the flow by the Atchafalaya, could also be threatened. This structure came close to catastrophic failure due to undermining during the 1973 food (Belt, 1975, see also Kesel et al., 1974; Barnett, 2017) and may fail in the future due to a combination of a reduction of the capacity of the river channel and increased pressure during high water could threaten the control structure during a very large flood (Wang and Xu, 2015, 2016, 2018; Joshi and Xu, 2017; Xu, 2017). Given projections for an increase in the peak discharge of the river as discussed above, the potential for failure of levees and structures

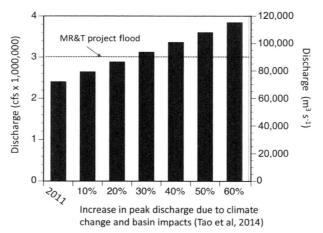

FIG. 6 Potential increase in peak Mississippi River discharge due to climate change and land use changes in the basin. MR&T is the Mississippi River and Tributaries project flood. *(Based on model projections from Tao, B., Tian, H., Ren, W., Yang, J., Yang, Q., He, R., et al., 2014. Increasing Mississippi river discharge throughout the 21st century influenced by changes in climate, land use, and atmospheric CO_2. Geophys. Res. Lett. 41(14), 4978–4986.)*

on the lower Mississippi is increasing. Increasing costs of energy as discussed above would make the cost of sustaining the MR&T flood control system much higher, perhaps prohibitively so (see Wiegman et al., 2017).

6.4 Flooding Due to Extremely Heavy Rainfall

Evidence indicates that the intensification of extreme weather events will occur in a warming climate (Coumou and Rahmstorf, 2012; IPCC, 2014). Warmer air holds more moisture, and heavy precipitation events and flooding are expected to increase in intensity with climate change (Groisman et al., 2005; Min et al., 2011; Pall et al., 2011; Prein et al., 2016). Several recent examples have occurred in the north central Gulf of Mexico. In August, 2016, nearly a meter of rain in 3 days that led to extensive flooding east of the Mississippi River in the Baton Rouge area with over 10,000 houses flooded. The 2016 storm was associated with a near stationary low-pressure area just offshore in the Gulf. In 2017, Hurricane Harvey stalled over Houston and dumped up to 1.3 m over several days. Such intense storms are occurring with a greater frequency. Because local rivers in Louisiana are low relief coastal plain rivers, the river slope is critical in regulating the quantity and rate at which water can be conveyed from large rain storms. A rising sea level will decrease the river slope and greatly exacerbate flooding of low relief areas near the coast.

6.5 A Truly Sustainable New Orleans

New Orleans is a special case for sustainable restoration planning. It is the largest metropolitan area in Louisiana and the region is economically important for the state in terms of port activity, tourism, education, and commerce. The local economy includes petrochemicals, oil and gas activity, shipping, and important fisheries. The city is also one of the most threatened areas in the nation. Katrina led to catastrophistic flooding and loss of life in the metropolitan area and the region has not fully recovered (Kates et al., 2006; Glavovic, 2014). After the storm, there was an expectation that the region would undergo fundamental changes to make the area more resilient to hurricane flooding. But almost all residential and commercial buildings were repaired in place and very few structures were raised above flood levels (Colten, 2015). Because about half the city is below sea level and future climate forcings are projected to be worse (accelerated sea-level rise, more frequent stronger hurricanes, higher peak river discharge, and more extreme weather events like the August 2016 flooding and Hurricane Harvey's impact on Houston), the continued existence of the New Orleans region will likely become increasingly untenable. If the current approach to flood protection continues, it is plausible that New Orleans may fail catastrophically before 2100. Clearly, there is a need for new approaches to integrating resilience into the metropolitan area's long-term planning that will contribute to its long-term sustainability. Of paramount importance is that the city does not remain below sea level.

Erdman et al. (2018a) proposed a bold new vision to make New Orleans truly resilient and sustainable in the 21st century and beyond. They argued that New Orleans provides a case study for all of south Louisiana, as well as cities around the world, especially those that are below sea level, that are increasingly threatened by sea-level rise and strong storms as to how settlement can continue in such a precarious location. They presented a bold design proposal for elevating the city of New Orleans as an adaptive course of action (Fig. 7). Their two-part strategy begins by reinforcing the lake edge along Lake Pontchartrain using infill to extend the higher, buildable ground. This higher ground would be fronted by a cypress swamp and urban edge. The second part of the strategy aims to build a series of leveed polders that they call a "marais" based on the French term. These polders could then be developed by filling them with river sediment, raising structures inside the polder, or managing them for aquatic and wetland systems that could also serve as flood reservoirs. The goal is to provide additional protection at the lake front and to raise all living areas to at least 5 m above sea level. The design proposal further develops edge and fill tactics to complete elevation of the city, in whole or part. Erdman et al. (2018b) review different strategies for raising structures and making them more resilient, and Colten (2018) reviews historical examples where urban land has been raised to avoid flooding threat or to enhance sewage removal.

7 MOVING FORWARD ON COASTAL PROTECTION

The following are suggestions for enhancing coastal protection in conjunction with restoration. These planned protection and restoration activities may help to enhance hurricane protection for New Orleans given that hurricane protection would be much more difficult without restored wetlands. In general, raising buildings above the 500-year flood level should be encouraged, or even mandated, in threatened areas, along with revisions to the National Flood Insurance Program to discourage rebuilding in place. The preservation of freshwater forested wetlands in the coastal zone is especially critical to hurricane surge and wave reduction (Shaffer et al., 2009, 2018). These actions along with raised elevations will result in a more secure New Orleans. Most protection envisions long, linear levee systems. In some cases, it would seem better to build

FIG. 7 Conceptual plan for raising New Orleans above hurricane flood levels. The city is fronted by a cypress swamp and an area formed by dredging from the lake that is about 5-m above sea level. Behind that is a series of polders that can be filled for development, contain raised structures or flooded to create natural habitat and flood basins. *(From Erdman, J., Williams, E., James, C., Coakley, G., 2018a. Raising buildings: the resilience of elevated structures. In: Day, J., Erdman, J. (Eds.), Mississippi Delta Restoration Pathways to a Sustainable Future. Springer, Cham, Switzerland. pp. 143–170; Erdman, J. James, C., Coakley, G., Williams, E., 2018b. In: Day, J., Erdman, J. (Eds.), Mississippi Delta Restoration Pathways to a Sustainable Future. Springer, Cham, Switzerland. pp. 171–200.)*

ring levee systems to protect specific areas. This would also eliminate some of the problem of having wetlands isolated behind levees. Wetlands should be used to help protect levees rather than being threatened by them. In some cases, a leaky levee approach is being advocated, where water control structures will allow water exchange except in times of storms. Much information on such management brings this idea into question (e.g., Wiegman et al., 2017). The massive failure of semi-impoundment management areas in the Chenier Plain during Hurricane Rita illustrates what can go wrong with this management approach. All of this suggests that any plan for the use of leaky levees should be very carefully considered and backed up by scientific study and modeling.

7.1 Managed Retreat

Over the last several decades, many residents left the lower coast for both economic and safety reasons (Bailey et al., 2014; Hemmerling, 2017; Glavovic, 2014; Colten et al., 2018). After hurricanes Katrina and Rita, this retreat greatly increased. Colten and Day (2018) and Colten et al. (2012) reported that human communities developed lifestyles adapted to a dynamic coast and that enhanced resilience. But the baselines that existed prior to the 20th century have changed and 21st century megatrends portend conditions never experienced by either coastal ecosystems or human communities. This suggests that sustainability and resilience of both natural and human systems will depend on new visions that take into consideration a future that is increasingly outside of the range of conditions that existed when these systems developed. Hemmerling (2018) concluded that successive shocks due to flooding, climate change, and economic change may lead to erosion of community resilience. These considerations suggest that moving people out of harm's way is prudent. The plan to move the Native American community of Isle de Jean Charles to higher ground serves as an example of proactive managed retreat (Louisiana Office of Community Development, 2017). There needs to be full integration of currently disconnected state planning for biophysical restoration and community safety. In addition, planners need to incorporate true community participation in the earliest stages of planning—not just to allow the opportunity for comments on fully developed plans.

There is a pressing need to determine what role this retreat will play in the economy of the coast and to include this in planning. There is a need to think about how to maintain the economy and health of the coast in a scenario of managed retreat. Natural resource activities like fishing will continue to be economically important, and perhaps more so in the future. Perhaps fishing and other resource harvest activities could move toward more of a seasonal utilization of the coast. It is important to maintain sustainable fishing communities. Perhaps a number of areas along the coast could be developed as safe fishing communities protected by strong ring levees. These areas would protect boats and other infrastructure from hurricanes and also provide temporary housing, processing, and other facilities to support the fishing industry. There is a need to give much more thought to ways that natural resource-based activities can take place without being threatened by storms.

7.2 Delta Restoration

The CMP proposes to use a variety of management approaches to restore the coast but nearly 90% of restoration funding will go to river diversions ($5.1 billion, ~20%) and marsh management ($17.1 billion, ~68%) (Table 1). Marsh creation uses dredged sediments that are pumped in pipelines, often for long distances, to build wetlands. Diversions reintroduce river water into the coastal zone to create and restore wetlands. Here, we discuss these and other approaches to delta restoration.

7.3 River Diversions—Crevasses, Large Diversions, Reactivated Distributaries

Engineered sediment diversions, which divert sediment and nutrient laden freshwater from the MR to adjacent wetlands, have been identified as critical tools in restoring the Mississippi River delta plain (MRDP) (Day et al., 2007, 2016b, 2018; Kim et al., 2009; Allison and Meselhe, 2010; Paola et al., 2010; CPRA, 2007, 2012; Wang et al., 2014; Esposito et al., 2017). Large river diversions are very expensive, costing over a billion dollars each (CPRA, 2017). After construction, however, diversions have minimal recurring costs (Day et al., 2016c). One problem is that the immediate outfall area may tend to fill in as is the case for the Bonnet Carré Spillway (Day et al., 2012). One option is to build a conveyance channel like the Wax Lake Outlet that is self-scouring. Diversion structures will likely last for more than a century. For example, the Bonnet Carré Spillway structure will be 100 years old in 15 years and will likely last many more years (Day et al., 2012, 2016).

Under natural conditions, river input to the delta plain was dramatically greater than now. Large crevasses occurred every few years with flows reaching $10,000 \, m^3/s$ and discharging tens of millions of tons of sediment/yr (Saucier, 1963; Davis, 2000; Shen et al., 2015; Day et al., 2012, 2016c). Crevasses built up the natural levee and formed large crevasse splays in adjacent wetlands. Some distributaries functioned year round but most were active seasonally. Sustainable restoration, especially with climate change, will require much greater river input than is currently envisioned and should seek to mimic the past functioning of the river. Rutherford et al. (2018a, b) showed that a very large diversion ($\sim 7000 \, m^3 \, s^{-1}$) into the Maurepas Swamp on the east side of the river between Baton Rouge and New Orleans would build and maintain much more land than currently proposed diversions for this area.

Although the first large diversions will be constructed below New Orleans and discharge to the Barataria and Breton Sound basins (CPRA, 2017), sediment retention decreases the closer a diversion is to the river mouth. If diversions are located in the more inland parts of the delta, sediment retention will increase because of shallower water and the high trapping efficiency of forested wetlands; diversions into open waters have sediment retention efficiencies of 5%–30% (e.g., Xu et al., 2016; Blum and Roberts, 2009). Esposito et al. (2017) reported that diversions into settings that are still vegetated have sediment retention efficiencies greater than 75%. A large diversion into the Maurepas Swamp would yield several benefits. It would sustain this deteriorating 57,000 ha swamp that provides a variety of ecosystem goods and services, especially hurricane protection, for the critical New Orleans to Baton Rouge corridor (Shaffer et al., 2009, 2018; Rutherford et al., 2018a, b). It could also serve as an alternative flood outlet to the Bonnet Carré Spillway that is rapidly filling in (Day et al., 2012). An intriguing idea is that a large Maurepas diversion may relieve pressure on the Old River control structure and decrease the potential that a larger portion of Mississippi River flow will be captured by the Atchafalaya River. The Maurepas Swamp is approximately the same distance from the Old River complex as the mouth of the Atchafalaya so that the distance advantage of the Atchafalaya would be reduced and make stream capture less likely. Nienhuis et al. (2018) used Delft3D modeling to investigate the influence of vegetation and soil compaction on the evolution of a natural levee breech and showed that crevasse splays tend to heal because aggradation reduces the water slope. A way around this is to create a bifurcation at the point of the Maurepas diversion like the lower Atchafalaya and Wax Lake Outlet. This would maintain open the lower Mississippi channel past New Orleans. Control at that point could deliver sediment to the lower river below New Orleans. The birdsfoot would be abandoned with a new sub delta between New Orleans and the birdsfoot delta at Head of Passes. The Maurepas lobe would then build out into Lake Maurepas and then into Lake Pontchartrain to enhance wetlands to help protect New Orleans.

Another location for an advantageous diversion is into the Biloxi Marsh complex on the eastern flank of the deltaic plain. Because of the low subsidence of the Biloxi Marsh region, it has a greater potential for survival than wetlands in the center of the deltaic plain where subsidence rates are much higher (Nienhuis et al., 2017). The Biloxi Marsh wetlands play a critical role for hurricane protection on the southeastern flank of New Orleans. If these marshes were to disappear, the threat for hurricane flooding would increase dramatically.

7.4 Marsh Creation

Dredging sediment for marsh creation builds land quickly, but requires periodic re-nourishment to be sustainable (Wiegman et al., 2017). Wiegman et al. (2017) modeled the impact of increasing energy cost and sea-level rise on the cost of marsh creation and reported that the cost of creating 1 ha of marsh could increase to over $1 million by 2100 for the worst-case

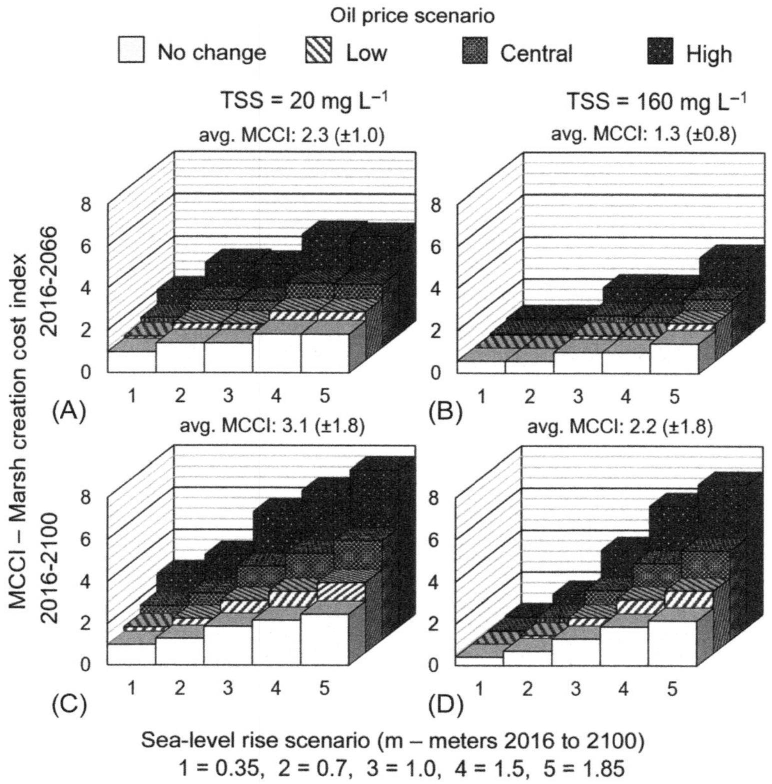

FIG. 8 The interactive effects of sea-level rise and fuel prices on the cost of marsh creation using dredged sediments. *(From Wiegman, R., Day, J., D'Elia, C., Rutherford, J., Morris, J., Roy, E., Lane, R., Dismukes, D., Snyder, B., 2017. Modeling impacts of sea-level rise, oil price, and management strategy on the costs of sustaining Mississippi delta marshes with hydraulic dredging. Sci. Total Environ. https://doi.org/10.1016/j.scitotenv.2017.09.314.)*

scenario for sea-level rise and energy price (Fig. 8). Increasing suspended sediment concentrations from river diversions raised marsh lifespan and decreased long-term dredging costs. Regardless of management scenario, sustaining created marsh suffered declining returns on investment due to the convergence of energy and climate trends. They concluded that marsh creation will likely become unaffordable during the second half of the 21st century. The model results suggest that most created marsh will succumb to rising sea level by a century from now but marsh creation can extend the life of coastal wetlands while plans are put in place to more fully utilize the resources to develop a more sustainable delta. Careful consideration should be given to tradeoffs between building marsh that is destined to disappear and other competing demands for scarce resources such as more diversions or coastal retreat.

7.5   Abandonment of Lower Delta

It is now generally accepted that the lower birdsfoot delta will have to be largely abandoned for effective coast-wide restoration. The lower delta is an anomaly compared to earlier delta lobes because it has grown to the edge of the continental shelf where most of the sediments reaching the birdsfoot delta bypasses wetlands. It is imperative that much more of the sediment reaching the river mouth is retained in the birdsfoot delta complex or further upstream. An extremely important and difficult problem is maintaining a functioning navigation channel while the lower delta undergoes abandonment. Relocation of residents and businesses in the lower delta presents an issue to policy makers who have been reluctant to address and is not a principal component of the CMP.

7.6   The Atchafalaya River Delta Region—An Underused Resource

The Atchafalaya River carries one-third of the total Mississippi flow, but this resource is not optimally used. Two new delta lobes are forming at the mouth of the river and wetland loss is low in a broad area of the central coast. Much of the sediments still flows past the emerging delta but a considerable amount of this sediment is reworked back into the headland marshes of the new delta lobes during frontal storm passages (Roberts et al., 2015). More of the discharge needs to be moved both east and west for successful restoration (see Twilley et al., 2016).

Most studies of the impact of Mississippi River water show positive impacts on coastal wetlands. The Atchafalaya and Wax Lake delta complex are part of the beginning stages of a major new deltaic lobe development fed by the Atchafalaya River distributary (Roberts, 1997; Roberts et al., 2015; DeLaune et al., 2013). Subaqueous deposition occurred in Atchafalaya Bay for much of the 20th century, first via the lower Atchafalaya River and then beginning in 1941 via the dredged Wax Lake Outlet. Both delta lobes first became subaerial during the large flood in 1973. The Atchafalaya delta lobe has a dredged navigation channel running through it, but the Wax Lake lobe has developed naturally at the end of the 26 km long artificial conveyance channel. The delta lobes have grown at about $3\,km^2\,yr^{-1}$ with areas of 160 and $100\,km^2$, respectively, by 2015 (Roberts et al., 2015).

Atchafalaya River discharge has led to mineral sediment deposition in wetlands in a broad arc from Fourleague Bay to the east to Vermillion Bay to the west (e.g., Day et al., 2011) and these wetlands have the lowest rates of land loss in the Mississippi delta. Twilley et al. (2016) analyzed wetland loss in this area compared it to wetland loss in the Terrebonne Bay complex to the east of Atchafalaya Bay that is isolated from riverine input. They measured the change in position of the 50% land:water ratio isopleth in both basins from the 1930s to the present. The 50% land:water line retreated an average of 17,000 m in the Terrebonne region with no river input compared to only 35 m in the Atchafalaya region, demonstrating the dramatic impact of river discharge. It is important to note that the Atchafalaya delta lobe complex developed in open waters of Atchafalaya, Cote Blanche, and Vermillion Bays. By contrast, the largest planned river diversions are from the Mississippi River below Baton Rouge and discharge to coastal basins with significant amounts of wetlands. Thus, these receiving basins have a much lower accommodation space and much higher friction and are expected to retain a higher percent of introduced sediments (Blum and Roberts, 2009). They are also the sites for extensive oyster cultivation and habitat for shrimp and other commercial marine life.

7.7 The Chenier Plain—Potential for Sustainable Management

The Chenier Plain was formed by downdrift sediments from the Mississippi River, especially when the river had significant discharge to the western part of the deltaic plain as it is happening now from the Atchafalaya River (Hijma et al., 2018). Coarse sediments were moved up on the shore face during storms and formed beach ridges. As new shore face accretion of fine sediments formed new marshes seaward of the beaches, the beach ridges were located inland so that the Chenier Plain has a complex series of stranded beach/dune systems called Cheniers, for the oaks growing on them. These ridges play an important role in protecting the Chenier Plain wetlands from Hurricanes. In a natural state, the Chenier ridges resulted in semi-enclosed areas that had connections with the Gulf of Mexico. This has changed with large areas of semi-impounded marshes for freshwater storage and waterfowl management.

Hurricane Rita (2005) brought into sharp focus the limitations of current management of the Chenier Plain. The combination of the salt-water intrusion via the Calcasieu Ship Channel as well as hurricane surge and the large freshwater impoundments set the stage for failure with the death of large areas of freshwater vegetation as levees were overtopped (Barras, 2009; Morton and Barras, 2011). A large body of scientific studies has shown that structural marsh management using impoundments and water control structures generally does not work (Cahoon, 1994; Boumans and Day, 1994). And the Chenier Plain is marsh management on a grand scale. Climate change and energy scarcity will make continuation of current management practices less and less tenable. Several approaches should characterize future Chenier Plain management; minimize salt-water intrusion via the Calcasieu, maximize input of resuspended sediments from the near-shore Gulf, and manage for transition to a more open internal system with brackish marshes in areas likely to be affected by future hurricanes. Hijma et al. (2018) stress that significant discharge of Mississippi River water to the western portion of the delta plain is necessary to sustain the Chenier Palin. This will demand foregrounding both restoration and public safety, rather than continuation of the traditional focus on shipping needs.

7.8 Coastal Forested Wetlands—A Vanishing Resource

Following intensive harvesting in the early 20th century, nearly a million hectares of coastal freshwater forested wetlands regenerated in the Atchafalaya, Terrebonne, Barataria, and Pontchartrain basins. Regeneration of the forests occurred during drought years when natural dry downs allowed seedlings to survive. This was important because seedlings need a dry period when they are not flooded for survival to take place. The forests in the latter three basins are now almost all completely cut off from river input and mostly unsustainable. Most are flooded continuously and are not regenerating (Shaffer et al., 2009, 2016; Keim et al., 2006; Day et al., 2012; Conner et al., 2014). Considerable areas are threatened by salt-water intrusion. Most of these forests will disappear in the 21st century without large-scale river input (Chambers et al., 2005; Shaffer et al., 2016). These forests play a particularly important role in hurricane protection because the three-dimensional structure reduces waves as well as surge (Shaffer et al., 2009). These forests also lead to water quality improvement and sequester large amounts of carbon (Shaffer et al., 2016; Lane et al., 2017; Hunter et al., 2018). Most forests are privately owned and there is considerable pressure to harvest them. But once cut, most will not regenerate because of permanent flooding. Strong public-private cooperation is necessary for the survival of these ecosystems. More sustainable management of these valuable forested wetlands should include maximizing sediment input (Rutherford et al., 2018a, b) and optimal use of freshwater resources to combat salinity intrusion (Shaffer et al., 2018).

7.9   Wasted Freshwater Resources

Freshwater wetlands make up a significant part of the Mississippi delta (see Fig. 3). These systems will become increasingly more threatened by increasing salinity due to sea-level rise and more frequent and intense droughts (Day and Rybczyk, 2019). The drought-induced brown marsh of 2000–01, which killed thousands of acres of wetlands, is a harbinger for the future (Alber et al., 2008; McKee et al., 2004). Thus, it will become necessary to fully utilize all freshwater resources. Currently, most point and nonpoint source runoff flows directly into waterbodies, bypassing wetlands and often causing water quality deterioration. Shaffer et al. (2018) identified several sources of freshwater to the coastal zone and outlined approaches to optimally use these freshwater resources to combat salt water intrusion and simultaneously improve water quality. Sources of freshwater to the coastal zone include the Mississippi River via diversions, smaller local rivers, direct rainfall on coastal wetlands, nonpoint source runoff, once-through, noncontact industrial cooling water, and treated municipal effluent. Nonpoint source runoff, which is often channelized directly into water bodies, should be rerouted to wetlands. Cooling water from the industrial complexes is a potential source of freshwater, especially along the Mississippi River below Baton Rouge (Hyfield et al., 2007). Treated, disinfected, nontoxic municipal effluent can be a continuous base flow of freshwater that has been demonstrated to restore and maintain coastal wetlands (Hunter et al., 2018). In addition to combating salt-water intrusion, such use of freshwater resources results in energy savings, wetland restoration, and enhanced carbon sequestration.

7.10   Restoration of Basin Inputs

Coastal restoration will be much more difficult unless sediment input from the basin is increased. There is a need to remobilize sediments trapped behind dams and move them to the delta, especially fine sediments. Sand transport to the delta remains sufficient to build wetlands in shallow, sheltered coastal bays fed by engineered diversions. Allison et al. (2012) reported that much of the sand fraction of the river was sequestered in overbank storage and channel bed aggradation inside the flood control levees. Nittrouer and Viparelli (2014) concluded that sand supply to the head of the Mississippi delta was unlikely to decrease for centuries. But suspended mud (silt and clay) flux to the coast has dropped from a mean of $390\,Mt\,yr^{-1}$ in the early 1950s, to $100\,Mt\,yr^{-1}$ since 1970 (Meade and Moody, 2010). Unlike sand that is deposited near where it leaves the river, fine-grained sediments can be transported deeper into receiving estuarine basins and play a critical role in sustaining existing wetlands. Practically all of this now-absent mud once flowed from the Missouri River Basin prior to the construction of nearly 100 dams in the Missouri basin. About $100\,Mt\,yr^{-1}$ is currently trapped by the large main-stem Upper Missouri River dams completed by 1953. The remaining $200\,Mt\,yr^{-1}$ is trapped in impoundments built on tributaries to the Lower Missouri in the 1950s and 1960s. In contrast, to the large dams on the upper Missouri, sediment bypassing on the lower river impoundments is part of river management. Sediment flux during the post-dam high discharge years of 1973, 1993, and 2011 approached predam levels when tributaries to the Lower Missouri contributed to flood flows. These lower Missouri river tributaries drain a vast, arid part of the Great Plains, while those entering from the east bank traverse the lowlands of the Mississippi floodplain. Both provinces are dominated by highly erodible loess soils. Reducing the continued decline in Mississippi River fine-grained sediment flux is very important now that river diversions are being built for coastal wetland restoration in the deltaic plain. Kemp et al. (2016) concluded that tributary dam bypassing in the Lower Missouri basin could increase mud supply to the MRD by $100–200\,Mt\,yr^{-1}$ within 1–2 decades. Such measures to restore the Mississippi delta are compatible with objectives of the Missouri River Restoration and Platte River Recovery Programs to restore riparian habitat for endangered species (Kemp et al., 2016).

8   SHIFTING BASELINES AND DIMINISHING RESILIENCE

Colten and Day (2018) discussed how the Mississippi delta and human occupation of the delta evolved at the same time in the context of a predictable baseline in a dynamic deltaic ecosystem. For centuries, despite the highly dynamic coastal system, both natural ecosystems and human communities flourished and coexisted in a sustainable relationship. The riverine system sustained the natural system and human communities developed lifestyles adapted to a dynamic coast which enhanced resilience. But the baselines that existed prior to the 19th century have changed and 21st century megatrends portend conditions not experienced by either coastal ecosystems or human communities over the past. This includes Native American communities over the past several millennia and colonial and postcolonial communities up to about the end of the 19th century. The megatrends include increasingly severe climate change impacts, most notably sea-level rise, stronger hurricanes, more extreme weather events, increased peak river discharge, and growing resource scarcity. These circumstances suggest that sustainability and resilience of both natural systems and human communities will depend on new visions that take into consideration a future that is increasingly outside of range of conditions that existed when these systems developed.

9 COMPREHENSIVE PLANNING—THE IMPORTANCE OF GLOBAL CHANGE

There is a pressing need to carry out planning for coastal restoration and protection to a much greater extent within a comprehensive plan that takes into consideration 21st century global change megatrends including climate change, energy scarcity, ecosystem degradation, economic constraints, and the local cultural context. There is a need to recognize the extremely high ecosystem services and their role in the future ecological, economic, and social health of the state and how these goods and services can be sustained. There is a need to really consider what is possible, what is not, and what it will take to have a sustainable system in the Mississippi delta. There needs to be an acceptance that the delta will shrink considerably (Chamberlain et al., 2018), that there will be significant population shifts accompanying managed retreat, and that the river will have to be used to the maximum extent possible. As this century progresses, much of what is being planned for coastal protection and restoration will become more expensive, perhaps prohibitively so (Tessler et al., 2015; Wiegman et al., 2017). Society will have to adapt to a much more dynamic river and delta system that cannot be controlled as it has been for the last century. The energies of nature must play a much more important role in delta management. This is ecological engineering on a grand scale that must operate in synchronization with complex social processes. This will mean living in a much more open system, accepting natural and social limitations, and utilizing the resources of the river more fully.

ACKNOWLEDGMENTS

The authors would like to thank Dr. Torbjörn Törnqvist for his helpful comments on this chapter. Also, Dr. Marc Hijma for kindly providing the image for Fig. 2.

REFERENCES

Alber, M., Swenson, E.M., Adamowicz, S.C., Mendelssohn, I.A., 2008. Salt Marsh Dieback: an overview of recent events in the US. Estuar. Coast. Shelf Sci. 80, 1–11.

Allison, M.A., Meselhe, E.A., 2010. The use of large water and sediment diversions in the lower Mississippi River (Louisiana) for coastal restoration. J. Hydrol. 387 (3), 346–360.

Allison, M., Demas, C., Ebersole, B., Kleiss, B., Little, C., Meselhe, E., Powell, N., Pratt, T., Bosburg, B., 2012. A water and sediment budget for the lower Mississippi-Atchafalaya River in flood years 2008–2010: implications for sediment discharge to the oceans and coastal restoration in Louisiana. J. Hydrol. 432–433, 84–97.

Bailey, C., Gramling, R., Laska, S., 2014. Complexies of resilience: adaptation and change within human communities of coastal Louisiana. In: Day, J., Kemp, G., Freeman, A., Muth, D., (Eds.), Perspectives on the Restoration of the Mississippi Delta. Springer, New York, pp. 125–140.

Barnes, S., Bond, C., Burger, N., Anania, K., Strong, A., Weilant, S., et al., 2015. Economic Evaluation of Coastal Land Loss in Louisiana. Louisiana State University and the Rand Corporation. Available from: http://coastal.la.gov/economic-evaluation-of-land-loss-in-louisiana/.

Barnett, J.F., 2017. Beyond Control: The Mississippi River's New Channel to the Gulf of Mexico. University Press of Mississippi, Oxford.

Barras, J.A., 2009. Land Area Change and Overview of Major Hurricane Impacts in Coastal Louisiana, 2004–08: US Department of the Interior. US Geological Survey Scientific Investigations Map 3080, 6 p http://pubs.usgs.gov/sim/3080/.

Barry, J., 1997. Rising Tide: The Great Mississippi River Flood of 1927 and How it Changed America. Simon Schuster, New York.

Batker, D., de la Torre, I., Costanza, R., Day, J., Swedeen, P., Boumans, R., Bagstad, a.K., 2014. In: Day, J., Kemp, P., Freeman, A., Muth, D. (Eds.), Perspectives on the Restoration of the Mississippi Delta. Springer, New York, pp. 155–174.

Baumann, R., Day, J., Miller, C., 1984. Mississippi deltaic wetland survival: sedimentation vs coastal submergence. Science 224, 1093–1095.

Belt, C., 1975. The 1973 flood and man's constriction of the Mississippi River. Science 189, 681–684.

Blum, M.D., Roberts, H.H., 2009. Drowning of the Mississippi Delta due to insufficient sediment supply and global sea-level rise. Nat. Geosci. 2 (7), 488–491.

Blum, M.D., Roberts, H.H., 2012. The Mississippi delta region: past, present, and future. Annu. Rev. Earth Planet. Sci. 40, 655–683.

Boesch, D.F., Josselyn, M.N., Mehta, A.J., Morris, J.T., Nuttle, W.K., Simenstad, C.A., Swift, D.J.P., 1994. Scientific assessment of coastal wetland loss, restoration, and management in Louisiana. J. Coastal Res. . Special Issue No. 20.

Boumans, R.M., Day, J., 1994. Effects of two Louisiana marsh management plans on water and materials flux and short-term sedimentation. Wetlands 14, 247–261.

Cahoon, D.R., 1994. Recent accretion in two managed marsh impoundments in coastal Louisiana. Ecological Applications. 4, 166–176.

Camillo, C.A., Pearcy, M.T., 2004. Upon Their Shoulders: A History of the Mississippi River Commission. Mississippi River Commission, Vicksburg.

Chamberlain, E., Tornqvist, T., Sheen, Z., Mauz, B., Sallinga, J., 2018. Anatomy of Mississippi Delta growth and its implication for coastal restoration. Sci. Adv. 4, eaar4740.

Chambers, J., Conner, W., Day, J., Faulkner, S., Gardiner, E., Hughes, M., Keim, R., King, S., McLeod, K., Miller, C., Nynam, J., Shaffer, G., 2005. In: Conservation, protection and utilization of louisiana's coastal wetland forests. Final report to the Governor of Louisiana from the Coastal Wetland Forests Conservation and Use Science Working Group. 102 pp.

Colten, C.E., Rooksby, O., Sobott, J.-K., 2015. Historic city with a poor memory. In: The Katrina Effect. Bloomsbury Publishing, London, pp. 305–330.

Colten, C.E., 2017. Environmental management in coastal Louisiana. J. Coast. Res. 33 (3), 699–711.

Colten, C.E., 2018. Raising urban land: historical perspectives on adaptation. In: Day, J., Erdman, J. (Eds.), Mississippi Delta Restoration Pathways to a Sustainable Future. Springer, Cham, Switzerland, pp. 135–142.

Colten, C.E., Day, J., 2018. Resilience of natural systems and human communities in the Mississippi delta: moving beyond adaptability due to shifting baselines. In: Mossop, E. (Ed.), Coastal Resilience. Taylor & Francis Group, CRC Press, Boca Raton, FL.

Colten, C., Hay, J., Giancarlo, A., 2012. Community resilience and oil spills in coastal Louisiana. Ecol. Soc. 17 (3), https://doi.org/10.5751/ ES-05047-170305.

Colten, C.E., Simms, J.R.Z., Grismore, A.A., Hemmerling, S.A., 2018. Social Justice and Mobility in Louisiana, USA. Reg. Environ. Change 18 (2), 371–383.

Condrey, R.E., Hoffman, P.E., Evers, D.E., 2014. The last naturally active delta complexes of the Mississippi River (LNDM): discovery and implications. In: Day, J.W., Kemp, G.P., Freemen, A.M., Muth, D.P. (Eds.), Perspectives on the Restoration of the Mississippi Delta. Springer, Dordrecht, pp. 33–50.

Conner, W.H., Duberstein, J.A., Day, J.W., Hutchinson, S., 2014. Impacts of changing hydrology and hurricanes on forest structure and growth along a flooding/elevation gradient in a south Louisiana forested wetland from 1986 to 2009. Wetlands XX, 1–12.

Coumou, D., Rahmstorf, S., 2012. A decade of weather extremes. Nat. Clim. Chang. 2 (7), 491–496.

Couvillion, B., Barras, J., Steyer, G., Sleavin, W., Fishcher, M., Beck, H., Trahan, N., Griffin, B., Heckman, D., 2011. Land area change in coastal Louisiana from 1932 to 2010. U.S. Geological Survey, Scientific Investigations Map 3164, scale 1:265,000, 12 p. pamphlet. https://pubs.usgs.gov/ sim/3381/sim3381.pdf.

CPRA, 2007. Louisiana's Comprehensive Master Plan for a Sustainable Coast. Coastal Protection and Restoration Authority of Louisiana (CPRA), Baton Rouge.

CPRA, 2012. Louisiana's comprehensive master plan for a sustainable coast. In: 2012 Coastal Master Plan. CPRA, Baton Rouge.

CPRA, 2017. Louisiana's comprehensive master plan for a sustainable coast. In: 2017 Coastal Master Plan. Louisiana Coastal Protection and Restoration Authority, Baton Rouge.

Davis, D.W., 1993. In: Crevasses on the lower course of the Mississippi River. Coastal Zone'93. Proceedings of the Eighth Symposium on Coastal and Ocean Management. American Society of Civil Engineers, pp. 360–378.

Davis, D.W., 2000. Historical perspective on crevasses, levees, and the Mississippi River. In: Colten, C.E. (Ed.), Transforming New Orleans and Its Environs. University of Pittsburgh Press, Pittsburgh, pp. 84–106.

Davis M, Vorhoff H, Boyer D (2015) Financing the future. Turning coastal restoration and protection plans into realities: the cost of comprehensive coastal restoration and protection. second in an occasional series. An issue paper of the Tulane Institute on water resources law and policy.

Day, J., Erdman, J. (Eds.), 2018. Mississippi Delta Restoration Pathways to a Sustainable Future. Springer, Cham, Switzerland.

Day, J., Rybczyk, J., 2019. Global change impacts on the future of coastal systems: perverse interactions among climate change, ecosystem degradation, energy scarcity, and population. In: Wolanski, E., et al. (Eds.), Future of Coasts and Estuaries. Springer, Glam, Switzerland.

Day, J., Pont, D., Hensel, P., Ibañez, C., 1995. Impacts of sea-level rise on deltas in the Gulf of Mexico and the Mediterranean: the importance of pulsing events to sustainability. Estuaries 18 (4), 636–647.

Day, J., Shaffer, G., Britsch, L., Reed, D., Hawes, S., Cahoon, D., 2000. Pattern and process of land loss in the Mississippi delta: a spatial and temporal analysis of wetland habitat change. Estuaries 23, 425–438.

Day, J., Barras, J., Clairain, E., Johnston, J., Justix, D., Kemp, P., Ko, J.-Y., Lane, R., Mitsch, W., Steyer, G., Templet, P., Yanez, A., 2005. Implications of global climatic change and energy cost and availability for the restoration of the Mississippi Delta. Ecol. Eng. 24, 253–265.

Day, J., Boesch, D., Clairain, E., Kemp, P., Laska, S., Mitsch, W., Orth, K., Mashriqui, H., Reed, D., Shabman, L., Simenstad, C., Streever, B., Twilley, R., Watson, C., Wells, J., Whigham, D., 2007. Restoration of the Mississippi Delta: Lessons from Hurricanes Katrina and Rita. Science 315, 1679–1684.

Day, J.W., Kemp, G.P., Reed, D.J., Cahoon, D.R., Boumans, R.M., Suhayda, J.M., et al., 2011. Vegetation death and rapid loss of surface elevation in two contrasting Mississippi delta salt marshes: the role of sedimentation, autocompaction and sea-level rise. Ecol. Eng. 37, 229–240. https://doi. org/10.1016/j.ecoleng.2010.11.021.

Day, J., Hunter, R., Keim, R.F., DeLaune, R., Shaffer, G., Evers, E., Reed, D., Brantley, C., Kemp, P., Day, J., Hunter, M., 2012. Ecological response of forested wetlands with and without large-scale Mississippi River input: implications for management. Ecol. Eng. 46, 57–67.

Day, J., Kemp, P., Freeman, A., Muth, D., (Eds.), 2014. Perspectives on the Restoration of the Mississippi Delta: The Once and Future Delta. Springer, New York. 194 p.

Day, J., Agboola, J., Chen, Z., D'Elia, C., Forbes, D., Giosan, L., Kemp, P., Kuenzer, C., Lane, R., Ramachandran, R., Syvitski, J., Yanez, A., 2016a. Approaches to defining deltaic sustainability in the 21st century. Estuar. Coast. Shelf Sci. https://doi.org/10.1016/j.ecss.2016.06.018.

Day, J., Cable, J., Lane, R., Kemp, G., 2016b. Sediment deposition at the Caernarvon crevasse during the great Mississippi flood of 1927: implications for coastal restoration. Water 8 (38), https://doi.org/10.3390/w8020038.

Day, J., Lane, R., D'Elia, C., Wiegman, A., Rutherford, J., Shaffer, G., Brantley, C., Kemp, G., 2016c. Large infrequently operated river diversions for Mississippi delta restoration. Estuar. Coast. Shelf Sci. https://doi.org/10.1016/j.ecss.2016.05.001.

DeConto, R.M., Pollard, D., 2016. Contribution of Antarctica to past and future sea-level rise. Nature 531 (7596), 591–597.

DeLaune, R.D., Pezeshki, S.R., 1988. Relationship of mineral nutrients to growth of Spartina alterniflora in Louisiana salt marshes. Northeast Gulf Sci. 10, 195–204.

Delaune, R.D., Pezeshki, S.R., 2003. The role of soil organic carbon in maintaining surface elevation in rapidly subsiding U.S. Gulf of Mexico coastal marshes. Water Air Soil Pollut. 3, 167–179.

DeLaune, R.D., Kongchum, M., White, J.R., Jugsujinda, A., 2013. Freshwater diversions as an ecosystem management tool for maintaining soil organic matter accretion in coastal marshes. Catena 107, 139–144.

DeLaune, R.D., Sasser, C.E., Evers-Hebert, E., White, J.R., Roberts, H.H., 2016. Influence of the Wax Lake Delta sediment diversion on aboveground plant productivity and carbon storage in deltaic island and mainland coastal marshes. Estuar. Coast. Shelf Sci. 177, 83–89.

Erdman, J., Williams, E., James, C., Coakley, G., 2018a. Raising buildings: the resilience of elevated structures. In: Day, J., Erdman, J. (Eds.), Mississippi Delta Restoration Pathways to a Sustainable Future. Springer, Cham, Switzerland, pp. 143–170.

Erdman, J., James, C., Coakley, G., Williams, E., 2018b. In: Day, J., Erdman, J. (Eds.), Mississippi Delta Restoration Pathways to a Sustainable Future. Springer, Cham, Switzerland, pp. 171–200.

Esposito, C., Shen, Z., Törnqvist, T., Marshak, J., White, C., 2017. Efficient retention of mud drives land building on the Mississippi delta plain. Earth Surf. Dynam. 5, 387–397. https://doi.org/10.5194/esurf-5-387-2017.

Fisk, H.N., 1944. Geological Investigation of the Alluvial Valley of the Lower Mississippi River. Illus. U.S. Army Corps of Engineers, Washington, DC. 78 p.

Fisk, H.N., 1952. Geological Investigations of the Atchafalaya Basin and the Problem of Mississippi River Diversion. vol. 1. U.S. Army Corps of Engineers, Mississippi River Commission, Vicksburg, MS. 145 p.

Fisk, H.N., McFarlan Jr., E., 1955. Late quaternary deltaic deposits of the Mississippi River. Geol. Soc. Am., 279–302. Special Paper No. 62.

Frazier, D.E., 1967. Recent deltaic deposits of the Mississippi River: their development and chronology. Gulf Coast Assoc. Geol. Soc Trans. 17, 287–315.

Giosan, L., Syvitski, J., Constantinescu, S., Day, J., 2014. Protect the world's deltas. Nature 516, 31–33.

Glavovic, B.C., 2014. Waves of adversity, layers of resilience: floods, hurricanes, oil spills, and climate change in the Mississippi Delta. In: Glavovic, B.C., Smith, G.P. (Eds.), Adapting to Climate Change. Springer Science, Dordrecht, pp. 369–403.

Groisman, P., Knight, R., Easterling, D., Karl, T., Hegerl, G., Razuvaev, V., 2005. Trends in intense precipitation in the climate record. J Clim 18 (9), 1326–1350.

Hemmerling, S.A., 2017. A Louisiana Coastal Atlas. Louisiana State University Press, Baton Rouge.

Hemmerling, S.A., 2018. Eroding communities and diverting populations: historical population dynamics in coastal Louisiana. In: Day, J., Erdman, J. (Eds.), Mississippi Delta Restoration Pathways to a Sustainable Future. Springer, Cham, Switzerland, pp. 201–230.

Hijma, M., Shen, Z., Törnqvist, T., Mauz, B., 2018. Late Holocene evolution of a coupled, mud-dominated delta plain-chenier plain systems, coastal Louisiana, USA. Earth Surf. Dynam. 5, 689–710. https://doi.org/10.5194/esurf-5-689-2017.

Hunter, R., Day, J.W., Lane, R., Shaffer, G., Day, J.N., Conner, W., Rybczyk, J., Mistich, J., Ko, J., 2018. Using natural wetlands for municipal effluent assimilation: a half-century of experience for the Mississippi delta and surrounding environs. In: Nagabhatla, N., Metcalfe, C. (Eds.), Multifunctional Wetlands, Environmental Contamination Remediation and Management. Springer, New York, pp. 15–81.

Hyfield, E., Day, J., Mendelssohn, I., Kemp, P., 2007. A feasibility analysis of discharging non-contact, once-through industrial cooling water to forested wetlands for coastal restoration in Louisiana. Ecol. Eng. 29, 1–7.

IPCC, R.K., Pachauri, Allen, M.R., Barros, V.R., Broome, J., Cramer, W., Christ, R., Church, J.A., Clarke, L., Dahe, Q., Dasgupta, P., Dubash, N.K., 2014. In: Climate change 2014: synthesis report. Contribution of Working Groups I, II and III to the Fifth Assessment Report of the Intergovernmental Panel on Climate Change.

Jankowski, K., Tornqvist, T., Fernandes, A., 2017. Vulnerability of Louisiana's coastal wetlands to present-day relative sea-level rise. Nat. Commun. 8, 14792. https://doi.org/10.1038/ncomms14792.

Joshi, S., Xu, J., 2017. Bedload and suspended load transport in the 140-km reach downstream of the Mississippi River Avulsion to the Atchafalaya River. Water 9, 716. https://doi.org/10.3390/w9090716.

Kates, R.W., Colten, C.E., Laska, S., Leatherman, S.P., 2006. Reconstruction of New Orleans after Hurricane Katrina: a research perspective. Proc. Natl. Acad. Sci. 103 (40), 14653–14660.

Keim, R., Chambers, J., Huges, M., Nyman, A., Conner, W., Day, J., Faulkner, S., Gardiner, E., King, S., McLeod, K., Shaffer, G., 2006. Ecological consequences of changing hydrological conditions in wetland forests of coastal Louisiana. In: Xu, S., Singh, V. (Eds.), Coastal Hydrology and Processes. Water Resources Publications, Highlands Ranch, CO, pp. 383–396.

Kemp, G.P., Willson, C.S., Rogers, J.D., Westphal, K.A., 2014. Adapting to change in the lowermost Mississippi River: implications for navigation, flood control and restoration of the Delta ecosystem. In: Day Jr., J.W., Kemp, G.P., Freeman, A.M., Muth, D.P. (Eds.), Perspectives on the Restoration of the Mississippi Delta: The Once and Future Delta. Springer, Dordrecht, pp. 51–84.

Kemp, P., Day, J., Rogers, D., Giosan, L., Peyronnin, N., 2016. Enhancing mud supply to the Mississippi River delta: dam bypassing and coastal restoration. Estuar. Coast. Shelf Sci. https://doi.org/10.1016/j.ecss.2016.07.008.

Kesel, R., Dunne, K.C., McDonald, R.C., Allison, K.R., Spicer, B.E., 1974. Lateral erosion and overbank deposition on the Mississippi River in Louisiana caused by 1973 flooding. Geology 2, 461–464.

Kim, W., Mohrig, D., Twilley, R., Paola, C., Parker, G., 2009. Is it feasible to build new land in the Mississippi River delta? Eos Trans. Am. Geophys. Union 90 (42), 373–374.

Kolb, C., van Lopik, J., 1958. Geology of the Mississippi River Deltaic Plain, Southeastern Louisiana. Technical Report No. 3-483. U.S. Army Engineer Waterways Experiment Station, Vicksburg, MS.

Lane, R., Mack, S., Day, J., Kempka, R., Brady, L., 2017. Carbon sequestration at a forested wetland receiving treated sewage effluent. Wetlands https://doi.org/10.1007/s13157-017-0920-6.

Louisiana Office of Community Development. 2017. LASAFE. http://lasafe.la.gov/.

Maggio, G., Cacciola, G., 2012. When will oil, natural gas, and coal peak? Fuel 98, 111–123.

McKee, K.L., Mendelssohn, I.A., Materne, M.D., 2004. Acute salt marsh dieback in the Mississippi river deltaic plain: a drought-induced phenomenon? Glob. Ecol. Biogeogr. 13, 65–73.

Meade, R.H., Moody, J.A., 2010. Causes for the decline of suspended-sediment discharge in the Mississippi River system, 1940–2007. Hydrol. Process. 24 (1), 35–49.

Min, S.K., Zhang, X., Zwiers, F.W., Hegerl, G.C., 2011. Human contribution to more-intense precipitation extremes. Nature 470 (7334), 378–381.

Morton, R.A., Barras, J.A., 2011. Hurricane impacts on coastal wetlands: a half-century record of storm-generated features from Southern Louisiana. J. Coastal Res. 275, 27–43. https://doi.org/10.2112/JCOASTRES-D-10-00185.1.

Nienhuis, J., Törnqvist, T., Jankowski, K., Fernandes, A., Keogh, M., 2017. A new subsidence map for coastal Louisiana. GSA Today (Geologi. Soc. Am.) 27, https://doi.org/10.1130/GSATG337GW.1.

Nienhuis, J., Törnqvist, T., Esposito, C., 2018. Crevasse splays versus avulsions: a recipe for land building with levee breaches. Geophys. Res. Lett. https://doi.org/10.1029/2018GL077933.

Nittrouer, J., Viparelli, E., 2014. Sand as a stable and sustainable resource for nourishing the Mississippi river delta. Nat. Geosci. https://doi.org/10.1038/NGEO2142.

Pall, P., Aina, T., Stone, D.A., Stott, P.A., Nozawa, T., Hilberts, A.G., Lohmann, D., Allen, M.R., 2011. Anthropogenic greenhouse gas contribution to flood risk in England and Wales in autumn 2000. Nature 470 (7334), 382–385.

Paola, C., Twilley, R.R., Edmonds, D.A., Kim, W., Mohrig, D., Parker, G., Viparelli, E., Voller, V.R., 2010. Natural processes in delta restoration: application to the Mississippi delta. Annu. Rev. Mar. Sci. 3, 67–91.

Penland, S., Boyd, R., Suter, J., 1988. Transgressive depositional systems of the Mississippi delta plain: a model for barrier shoreline and shelf sand development. J. Sediment. Petrol. 58, 932–949.

Perez, B., Day, J., Rouse, L., Shaw, R., Wang, M., 2000. Influence of Atchafalaya River discharge and winter frontal passage on suspended sediment concentration and flux in Fourleague Bay, Louisiana. Estuar. Coast. Shelf Sci. 50, 271–290.

Prein, A.F., Rasmussen, R.M., Ikeda, K., Liu, C., Clark, M.P., Holland, G.J., 2016. The future intensification of hourly precipitation extreme. Nat. Clim. Chang. 7 (1), 48–52.

Reuss, M., 1998. Designing the Bayous: The Control of Water in the Atchafalaya Basin 1800–1995. U.S. Army Corps of Engineers Office of History, Alexandria, VA.

Roberts, H.H., 1997. Dynamic changes of the Holocene Mississippi river delta plain: the delta cycle. J. Coast. Res. 13, 605–627.

Roberts, H.H., DeLaune, R.D., White, J.R., Li, C., Sasser, C.E., Braud, D., Khalil, S., 2015. Floods and cold front passages: impacts on coastal marshes in a river diversion setting (Wax Lake Delta Area, Louisiana). J. Coast. Res. 31, 1057–1068.

Rosen, T., Jun Xu, Y., 2013. Recent decadal growth of the Atchafalaya River Delta complex: effects of variable riverine sediment input and vegetation succession. Geomorphology 194, 108–120.

Russell, R., 1936. Physiogaphy of lower Mississippi River Delta. Louisiana Dept. of Conservation, Louisiana Geological Survey. Geol. Bull. 8, 3–199.

Russell, R.J., Howe, H.V., McGuirt, J.H., Dohm, C.F., Hadley Jr., W., Kiffen, F.B., Brown, A., 1936. In: Lower Mississippi River Delta: reports on the geology of Plaquemines and St. Bernard Parishes. Louisiana Geological Survey, Geological Bulletin 13, Louisiana Department of Conservation, New Orleans, LA.

Rutherford, J., Day, J., D'Elia, C., Wiegman, A., Willson, C., Caffey, R., Shaffer, G., Batker, D., 2018a. Evaluating trade-offs of a large, infrequent sediment diversion for restoration of a forested wetland in the Mississippi delta. Estuar. Coast. Shelf Sci. https://doi.org/10.1016/j.ecss.2018.01.016.

Rutherford, et al., 2018b. In: Day, J., Erdman, J. (Eds.), Mississippi Delta Restoration Pathways to a Sustainable Future. Springer, Cham, Switzerland, pp. 201–230.

Saucier, R.T., 1963. Recent Geomorphic History of the Pontchartrain Basin. Louisiana State University Press, Baton Rouge.

Shaffer, G.P., Wood, W.B., Hoeppner, S.S., Perkins, T.E., Zoller, J., Kandalepas, D., 2009. Degradation of Baldcypress-water tupelo swamp to marsh and open water in southeastern Louisiana, USA: an irreversible trajectory? J. Coast. Res. 54, 152–165.

Shaffer, G., Day, D., Kandalepas, D., Wood, W., Hunter, R., Lane, R., Hillmann, E., 2016. Decline of the Maurepas swamp, Pontchartrain Basin, Louisiana and approaches to restoration. Water. 7, https://doi.org/10.3390/w70x000x.

Shaffer, G., Day, J., Lane, R., 2018. Optimum use of freshwater to restore Baldcypress-Water Tupelo swamps and freshwater marshes and protect against salt water intrusion: a case study of the Lake Pontchartrain Basin. In: Day, J., Erdman, J. (Eds.), Mississippi Delta Restoration Pathways to a Sustainable Future. Springer, Cham, Switzerland, pp. 61–76.

Shen, Z., Törnqvist, T.E., Mauz, B., Chamberlain, E.L., Nijhuis, A.G., Sandoval, L., 2015. Episodic overbank deposition as a dominant mechanism of floodplain and delta-plain aggradation. Geology 43, 875–878.

Steyer, G.D., et al., 2003. A proposed coast-wide reference monitoring system for evaluating wetland restoration trajectories in Louisiana. Environ. Monit. Assess. 81, 107–117.

Tao, B., Tian, H., Ren, W., Yang, J., Yang, Q., He, R., Cai, W., Lohrenz, S., 2014. Increasing Mississippi river discharge throughout the 21st century influenced by changes in climate, land use, and atmospheric CO_2. Geophys. Res. Lett. 41 (14), 4978–4986.

Tessler, Z.D., Vörösmarty, C., Grossberg, M., Gladkova, J., H. Aizenman, Syvitski, J., Foufoula-Georgiou, E., 2015. Profiling risk and sustainability in coastal deltas of the world. Science 349 (6248), 638–643.

Twilley, R., Bentley, S., Chen, Q., Edmonds, D., Hagen, S., Lam, N., Willson, C., Xu, K., Braud, D., Peele, R., McCall, A., 2016. Co-evolution of wetland landscapes, flooding, and human settlement in the Mississippi River Delta Plain. Sustain. Sci. 11, 711–731.

Wang, B., Xu, J., 2015. Sediment trapping by emerged channel bars in the lowermost Mississippi River during a major flood. Water 7, 6079–6096.

Wang, B., Xu, J., 2016. Long-term geomorphic response to flow regulation in a 10-km reach downstream of the Mississippi-Atchafalaya diverion. J. Hydrol.: Reg. Stud. 8, 10–25.

Wang, B., Xu, J., 2018. Dynamics of 30 large channel bars in the lower Mississippi River in response to river engineering from 1985 to 2015. Geomorphology 300, 31–44.

Wang, H., Steyer, G.D., Couvillion, B.R., Rybczyk, J.M., Beck, H.J., Sleavin, W.J., Meselhe, E.A., Allison, M.A., Boustany, R.G., Fischenich, C.J., Rivera-Monroy, V.H., 2014. Forecasting landscape effects of Mississippi River diversions on elevation and accretion in Louisiana deltaic wetlands under future environmental uncertainty scenarios. Estuar. Coast. Shelf Sci. 138, 57–68.

Welder, F.A., 1959. Processes of deltaic sedimentation in the lower Mississippi River. In: Louisiana State University, Coastal Studies Institute Technical Report. vol. 12, pp. 1–90.

Wiegman, R., Day, J., D'Elia, C., Rutherford, J., Morris, J., Roy, E., Lane, R., Dismukes, D., Snyder, B., 2017. Modeling impacts of sea-level rise, oil price, and management strategy on the costs of sustaining Mississippi delta marshes with hydraulic dredging. Sci. Total Environ. https://doi.org/10.1016/j.scitotenv.2017.09.314.

Wiegman, A., Rutherford, J., Day, J., 2018. The costs and sustainability of ongoing efforts to restore and protect Louisiana's coast. In: Day, J., Erdman, J. (Eds.), Mississippi Delta Restoration Pathways to a Sustainable Future. Springer, Cham, Switzerland, pp. 93–111.

Xu, J. 2017. What would happen if the Mississippi River changed its course to the Atcdhafalaya (Poster). American Geophysical Union Meeting, 12 December 2017. New Orleans, LA.

Xu, K., Bentley, S.J., Robichaux, P., Sha, X., Yang, H., 2016. Implications of texture and erodibility for sediment retention in receiving basins of coastal Louisiana diversions. Water 8, 26.

Chapter 11

Integrated Management of the Ganges Delta, India

Ramesh Ramachandran, Ahana Lakshmi, Swati Mohan Sappal, Bonthu S.R., Mary Divya Suganya, D. Ganguly, R.S. Robin, R. Purvaja

National Centre for Sustainable Coastal Management, Ministry of Environment, Forest and Climate Change, Government of India, Anna University Campus, Chennai, India

1 INTRODUCTION

The Ganges-Brahmaputra River delta began developing ca. 11,000 year B.P. when rising seas resulted in the flooding of the Bengal basin because of which the river's discharge was trapped on the inner margin (Goodbred Jr and Kuehl, 2000). The plume formed by the sediments transported from the Himalayas into the Bay of Bengal was found to cover the entire floor of the Bay of Bengal (Curray and Moore, 1971). The delta currently ranks first (along with the Amazon) in sediment transport, which is of the order of 1×10^9 t/year (Milliman and Syvitski, 1992). The Ganges-Brahmaputra-Meghna (GBM) delta spans two countries, India and Bangladesh and is a particularly fertile region supporting one of the most densely populated regions in the world. However, the GBM, like all other coastal river deltas, is a hotspot of global climate change impacts (Day et al., 2016; Hagenlocher et al., 2018). In this chapter, the important challenges faced by the Indian part of the delta, for which data are available, are examined by focusing on the present trend and strategies for integrated management of the delta under changing natural and human-induced conditions are explored.

1.1 The Ganges and the GBM Delta

The River Ganges originates in the Gangotri glacier in the Garhwal Himalayas at an elevation of 7010 m as the Bhagirathi River. At Devprayag, River Alaknanda joins Bhagirathi and the combined stream is known as the Ganges. From its origin, the river flows for 2525 km before reaching the Bay of Bengal at Ganga Sagar in the state of West Bengal, India (Fig. 1).

1.1.1 Ganges Basin

The Ganges basin stretches over an area of 1,086,000 km^2 across India, China (Tibet), Nepal, and Bangladesh. In India, it covers 11 states namely Uttarakhand, Himachal Pradesh, Uttar Pradesh, Delhi, Haryana, Rajasthan, Madhya Pradesh, Chhattisgarh, Bihar, Jharkhand, and West Bengal, draining an area of 8,61,452 km^2, which is nearly 26% of the total geographical area of the country (Table 1). The basin is bounded by the Himalayas on the north, by the Aravalli on the west, by the Vindhyas and Chota Nagpur plateau on the south, and by the Brahmaputra ridge on the east. The Ganges and its tributaries have formed a large flat and fertile plain in north India whereas in the eastern part of the basin, migration of the tributaries has resulted in conspicuous back-swamp and meander bolt deposits. These sedimentological features play a dominant role in the hydrodynamics of the region (WRIS, 2018).

1.1.2 River Flow

The Ganges joins the Brahmaputra from the east, and along with Meghna, forms the GBM delta. Together, these three rivers along with their tributaries drain a catchment area of 1.72×10^6 million km^2, on the southern side of the Himalayas, stretching across India (64.02%), China (17.69%), Nepal (8.57%), Bangladesh (7%), and Bhutan (2.73%) (Table 1). The GBM dispersal system, sustained with south Asian monsoon and high Himalayan sources, carries 1.8–2.4×10^9 tons sediment yr^{-1} (Goodbred and Nicholls, 2004) and the deposition of silt occurs within a vast area of about 115,000 km^2 (Coleman, 1969; Goodbred and Nicholls, 2004; Nicholls and Goodbred, 2005; Woodroffe et al., 2006).

Coasts and Estuaries. https://doi.org/10.1016/B978-0-12-814003-1.00011-3

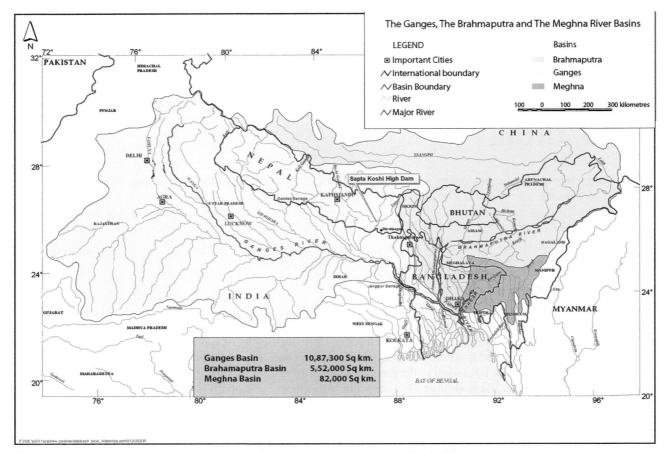

FIG. 1 Ganges-Brahmaputra-Meghna River Basin Map. *(Source Joint Rivers Commission, Bangladesh).*

TABLE 1 Country Catchment Areas in the Ganges-Brahmaputra-Meghna River Basin

Rivers	Total Catchment Area (km²)	Catchment Area (km²)				
		India	Nepal	Bhutan	China	Bangladesh
Brahmaputra	552,000	195,000	–	47,000	270,900	39,100
Ganges	1,087,300	861,452	147,480	–	33,520	46,300
Meghna	82,000	47,000	–	–	–	35,000
GBM total	1,721,300	1,102,000	147,480	47,000	304,420	120,400
(%)	(100%)	(64.02%)	(8.57%)	(2.73%)	(17.69%)	(7%)

Data from Joint Rivers Commission, Bangladesh.

The Ganges delta receives an average annual rainfall between 1520 and 2540 mm and experiences typical hot tropical climate with strong cyclonic storms, both during pre-monsoon (March to May) and post-monsoon (September to October) seasons. The Brahmaputra and Ganges Rivers receive water from both monsoonal rainfall and melting of ice from the Himalayan glaciers. For the period 1993–2011, Papa et al. (2012) estimated the mean aggregate discharge as ~32,000 m³ s⁻¹; whereas the annual maximum monthly discharge had a mean value of ~82,000 m³ s⁻¹ for the same time period.

1.1.3 Sediment Discharge

The GBM river system delivers 30% of the world's total load of river sediment (Milliman and Meade, 1983) to the Bay of Bengal. This is despite the fact that in the case of the Ganges and Brahmaputra, 55% of their combined annual sediment load is retained by their delta, with only 36% reaching the shelf and 9% reaching the deep sea (Syvitski, 2003). According to Subramanian and Ramanathan (1996) the sediment load for the Ganges River varies from 403 to 660×10^6 tons year^{-1} and consists predominantly of coarse silt to sand-size particles. About 88% of the annual sediment load is transported during the monsoon compared to 76% of the annual water discharge (Subramanian, 1996).

1.1.4 GBM Delta

The GBM delta is the world's largest and extends across two countries. Two-thirds of the delta is located in Bangladesh, and the rest in the state of West Bengal in India. The GBM delta extends from the Hooghly River on the west to the Meghna River on the east; extending approximately 18,000 km^2 along and across the Bhagirathi-Hooghly River (the distributary of Ganges). The width of the delta is approximately 350 km across the Bay of Bengal coast and the surface area is over 100,000 km^2. This delta is particularly fertile supporting one of the most densely populated regions in the world. The major land use categories in the GBM delta include irrigated croplands, rainfed and/or mosaic croplands as well as mosaic vegetation, mangrove vegetation, shrubland, broadleaved evergreen forests, and developed (urban) areas (Brown and Nicholls, 2015).

It is estimated that at least 630 million people live in the GBM river basin. Two megacities, Kolkata (India) (population about 14.3 million) and Dhaka (Bangladesh) (population about 18.2 million) are located within the GBM delta. Population density in the GBM river basin varies; it is very low in China and Bhutan whereas it is 432 inhabitants/km^2 in India and more than double that (1013 inhabitants/km^2) in Bangladesh.

1.1.5 The Ganges Delta

In India, the Ganges delta is located within the state of West Bengal and the delta begins at the Farakka Barrage (CWC and IMD, 2015). About 40 km downstream of Farakka, the river splits into two arms; the left arm, Padma, turns eastward and enters Bangladesh and the right arm, Bhagirathi, flows southward. After Nabadwip, it is known as Hooghly, and enters the Bay of Bengal about 150 km downstream of Kolkata. The Hooghly River, which is the lower tidal stretch of the Bhagirathi River, flows past the megacity of Kolkata and forms a vast mangrove-enriched estuarine delta before entering the Bay of Bengal. This is the Indian part of the Sunderbans covering an area of about 9630 km^2 (Ray et al., 2011). Important tributaries entering the Ganges in West Bengal include Ajay, Dwarka, Damodar, Rupnarayan, and Haldi (CWC, 2014). The Ganges delta in West Bengal covers nine districts, draining an area of 71,485 km^2. On the coast, the Sundarban mangrove forests rise up to 2.1 m above mean sea level, providing stabilization and enabling accretion of new land.

1.2 Challenges in the Ganges Delta

Worldwide, delta sustainability is increasingly being challenged because of the multiple stresses that they are being subjected to, including rising populations, intensive agricultural activities, engineering projects that change water and sediment delivery, sea-level rise (SLR) and flooding from rivers and intense tropical storms, groundwater and hydrocarbon extraction, delta subsidence and submergence, and coastal erosion (Syvitski, 2008; Tessler et al., 2015; Brondizio et al., 2016; Renaud et al., 2016; Day et al., 2016). Recent research has indicated that the GBM delta is most prone to multiple natural hazards (i.e., flooding, droughts, salinity intrusion, cyclones, and storm surges), with the highest exposure of social-ecological systems among the major deltas of the globe (Hagenlocher et al., 2018). Because of the enormous size of the Ganges basin and the GBM delta, and the wide variation in physiography (from the high Himalayas through plains to the low-lying delta), changes in the upstream areas can have a tremendous impact on the delta. Simultaneously, the deltaic coast is under stress from rising seas and intensification of hazards of hydro-meteorological origin as well as human impacts. Hence, the major challenges to the sustainability of the Ganges delta can be broadly classified into *Upstream Effects* and *Coastal Effects* (Fig. 2).

2 UPSTREAM EFFECTS

2.1 Water Flows

Rainfall, subsurface flows, and snowmelt from glaciers are the main sources of water in River Ganges and its tributaries. As the Ganges flows from its source toward the coast, it is joined by a large number of rivers including the Yamuna, the Ramganga, the Ghaghra, the Gandak, the Burhi Gandak, the Kosi, the Mahananda, and the Sone. From December to May are the months of low flow in the Ganges. There is considerable variability in water flows as seasonal rainfall in the

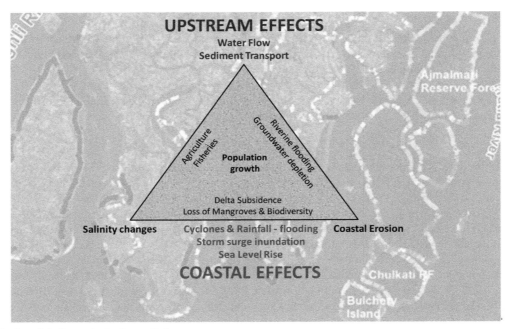

FIG. 2 Classification of challenges in the Ganges Delta.

catchment plays a major role in the amount of water flowing through the delta's river systems. This means that during the rainy season, the Ganges becomes a large single channel while during the dry months it reverts to a sandy braided course. The Central Water Commission (CWC) has divided the Ganges basin into the Upper Ganges, Yamuna, and Lower Ganges basins.

The mean annual available water resources of Ganges basin are 509.52 billion m^3 (BCM), while it is 192.60 BCM in the Lower Ganges basin which extends from the eastern margin of the Punjab in the west to the Bangladesh border in the east (Fig. 3). The Lower Ganges basin (288,283 km^2) is divided into 10 subbasins which include the combined delta (50,525 km^2) as a single subbasin.

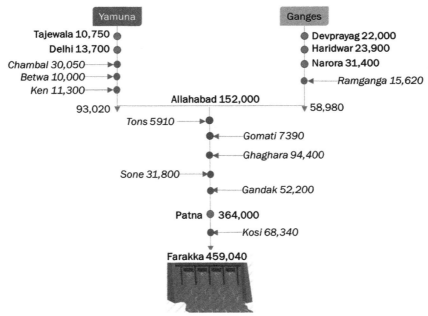

FIG. 3 Ganges and its major tributaries. Numbers indicate average annual flows in 10^6 m^3. *(Modified from Jain SK. Impact of retreat of Gangotri glacier on the flow of Ganges River. Curr. Sci. 95(8): 1012–1014, 2008.)*

The barrage at Farakka in Murshidabad district of West Bengal was constructed in 1974 with the primary purpose of diverting adequate quantity of water from the Ganges via a 40-km feeder canal which discharges into the Bhagirathi-Hooghly River for flushing out the sediment deposition from the Kolkata harbor without the need of regular mechanical dredging. The inflows at Farakka are about $459 \times 10^9 \, \text{m}^3$ (Jain, 2008) whereas observed discharge at Farakka between 1985–86 and 2014–15 was between 200 and $400 \times 10^9 \, \text{m}^3$ (CWC, 2017b). Measurements at Farakka indicate that while the flow remains low through most of the year, high flows occur during the wet season (Jul.-Oct.) including several peaks of very high discharge (Singh, 2008). The average bed slope from Farakka to Nabadwip (deltaic nontidal plain) and from Nabadwip to the outfall (deltaic tidal plain) is 1:23,000 to 1:24,000 (Jain, 2008); as a result intense rainfall during the monsoon causes flooding because of the slow movement of water. During dry seasons, there is hydrological drought.

Most of the deltaic plains are confined to the southeast part of the North and South 24 Parganas districts covering about 60% of the coastal area of West Bengal (Chakrabarty, 1995). These low-lying marshy lands with elevation below the high tide mark get submerged under brackish water during high tides. The Ganges delta is tide dominated with a macro tidal range (>4 m). The districts of North and South 24 Parganas are traversed by a number of moribund rivers, which are primarily spill channels of the Hooghly River. Six major estuarine rivers, namely Muri Ganga, Saptamukhi, Thakuran, Matla, Gosaba, and Herobhanga discharge to the Bay of Bengal and are interconnected with each other through numerous creeks and small rivers creating about 102 islands of which 54 islands are inhabited while the rest are still part of the Sundarbans mangrove ecosystem. These tidal estuarine rivers carry seawater from the Bay of Bengal during high tide and inundate the mangrove forests at regular intervals. River Hooghly in the west is the main river carrying freshwater from upstream reaches of lower Ganges delta into the Indian part of Sundarbans as most of the other estuarine rivers have lost their earlier connections with River Ganges over time (Morgan and McIntire, 1959).

2.2 Sediment Transport

Collision tectonics of the Himalayas control the Ganges system and are responsible for the formation of the vast Ganges plain, the world's largest delta and the world's largest submarine fan—the Bengal Fan (Singh, 2008). The basin area of the Ganges in the Himalayas is undergoing intense erosion which annually contributes a huge amount of sediment via tributaries into the main river; a large portion is carried as bed load. The Ganges plain acts both as a sink and source for sediment in transport to the delta and the submarine fan through initial deposition as channel bars and subsequent downstream movement especially during the wet season. According to Goodbred and Kuehl (2000) the sediment discharge of the Ganges-Brahmaputra system during 7–11,000 years BP was twice that deposited in the following 7000 years.

Throughout Pleistocene times, the site of active deltaic sedimentation has switched. Today, the Ganges merges with the Brahmaputra, and the site of active sedimentation lies to the east, where large bell-shaped distributaries can be discerned. The major area of abandoned deltaic plain lies to the west and is the site of one of the largest mangrove regions in the world, the Sunderbans. The abandoned delta is approximately 1.6 times the size of the active delta plain. Numerous abandoned channel scars dominate the surface morphology of the abandoned delta plain. These scars are apparently remnants of former courses of the Ganges River and many of its distributaries. Most of the scars indicate that a meandering channel was dominant, now extensively modified by humans. Channel scars are of similar size to channels presently active along the Ganges and its distributaries. Many of these former riverine channels are now tidally dominated (Coleman et al., 2008).

Wasson (2003) developed a suspended sediment budget for the Ganges-Brahmaputra catchment according to which of the $794 \times 10^6 \, \text{tons yr}^{-1}$ transported in the rivers of the Ganges catchment, over 80% comes from the High Himalayas. According to him, the rivers of the Ganges plain catchment appear to be aggrading, thereby exacerbating the annual overbank flood. Aggradation may be because of enhanced sediment delivery to the rivers due to land use, rainfall change, or neotectonics in the Himalayas. It could also be caused by neotectonics on the Plain, warping the riverbed. According to Akter et al. (2016) in the last five decades, this delta has prograded at a rate of $17 \, \text{km}^2 \, \text{yr}^{-1}$, whereas most large deltas elsewhere in the world have suffered from sediment starvation.

2.3 Impacts of Changes in Water and Sediment Transport Regimes on the Ganges Delta

The Ganges River system is considered as one of the most engineered with almost 800 projects including irrigation, dams, hydroelectric projects, lift systems, and barrages (WRIS, 2018). Overall, because of the reduction in flows due to dams and consequently, less water reaching the delta, there are periods of drought within the delta while during the monsoonal heavy rains, there is flooding not only because of excess storage being diverted but also because of embankments that have been built to channel flows which often result in flooding as they are unable to contain flows (Jain, 2008).

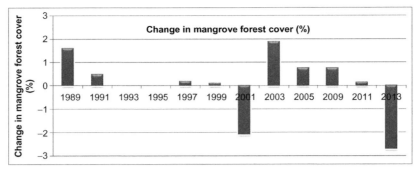

FIG. 4 Biennial change in Sundarban mangrove forest cover. *(Data from FSI, 2013; ISRO, 2015.)*

In the Ganges delta, there is a distinct variation in the water flows within the delta. The altered flow regimes have resulted in the variation in salinity in the different sectors of the delta. Here, salinity is mainly regulated by natural factors like siltation, tilting/subsidence of the delta, rising sea levels as well as by anthropogenic factors such as barrage discharge, run-off from the adjacent landmasses, many polders, etc. A study conducted by Trivedi et al. (2016) on the surface water salinity in the Ganges delta during 1984–2013 reveals that salinity has decreased by 0.63 and 0.86 yr^{-1} in the western and eastern sectors of the delta respectively, whereas in the central sector, it has increased by 1.09 yr^{-1}. The western region of the delta was found to be hyposaline owing to the freshwater discharge from the Farakka barrage. The decadal average discharge (1999–2008) from the barrage into the western part of the delta was found to be $3.7 \pm 1.15 \times 10^3 \, m^3 \, s^{-1}$ with maximum discharge during the monsoon season. Similarly, hyposaline conditions in the eastern sector of the delta were caused because of fresh water carried by several creeks and channels from the Harinbhanga Estuary of Bangladesh's Padma Meghna River basin. On the contrary, the central sector experiences complete obstruction of fresh water due to heavy siltation and minimal fresh water discharge in the Bidyadhari and Matla Rivers (Cole and Vaidyaraman, 1966; Chaudhuri and Choudhury, 1994; Mitra, 2013). As a result, the central sector is hypersaline with tidal inflow of seawater being the only source of water (Trivedi et al., 2016).

As a consequence of salinity changes, enhanced saline intrusion northward is evident in the coastal stretches of the Ganges delta which has been found to adversely impact mangroves and other associated flora and fauna. Gradual disappearance of fresh water species like *Heritiera fomes* and *Nypa fruticans* is a confirmatory test of such salinity variation (Gopal and Chauhan, 2006). The biomass of mangroves and species composition of phytoplankton and fish are also influenced by salinity fluctuation. Fig. 4 shows the declining trends in mangrove forest cover in the Sunderbans over the last 25 years.

A study by Raha et al. (2012) indicates that the growth of dominant mangrove flora is more in the western sector of Ganges delta than the central sector due to changes in the salinity regimes, which also affects species composition and distribution. Increased salinity caused reduced growth in *Sonneratia apetala* whereas salinity did not influence the growth of *Avicennia alba* and *Excoecaria agallocha* due to their salt excretion capability and thus better adaptation to hypersaline conditions. Similar observations were made for fish production and diversity in the deltaic region where changing salinities have changed the fish assemblage and distribution (Raha et al., 2012). Fresh water fish (such as *Tenualosa ilisha*), due to the hypersaline conditions in the central sector, have changed their course and breeding grounds to the western sector; while trash fish which can survive in the stressful conditions are abundant in the central sector.

3 COASTAL EFFECTS

3.1 Floods and Inundation

The state of West Bengal is vulnerable to multiple natural disasters. Floods and cyclones occur on an annual basis inflicting large losses of life and property. Such floods may be caused by tropical cyclones or by monsoonal rains, especially in the catchment areas as well as in the delta. Opening of sluice gates of dams during intense rainfall may result in flash floods. About 42% of the total geographic area of West Bengal state and 69% of its net cropped area is prone to floods. The floods actually bring sediment that has not only shaped the delta but also has allowed high agricultural production with relatively less input. However, the high population density as well as the increase in engineered structures has resulted in an increase in the number of flood-related disasters.

Flooding episodes due to severe cyclonic storms over the Bay of Bengal also pose severe hazards to lives and livelihoods. A 26% increase has been documented over the last 120 years in the frequency of cyclones over the Bay of Bengal, intensifying in the post-monsoon period. Cyclones bring strong wind, heavy rainfall, and flooding, resulting in severe coastal erosion and embankment failure. From 1999 to 2005, while there were a number of cyclonic depressions, only three materialized into severe and super-cyclonic storms. However, in the next 4 years, seven such cyclonic storms were generated from a similar number of cyclonic depressions in the Northern part of the Bay of Bengal. This was found to be closely related to the increase in the sea surface temperatures (Singh, 2007).

Track data were obtained for 56 cyclonic depressions over the GBM over a 25-year period (1990–2015) based on historical cyclone records from the India Meteorological Department (IMD). Out of these, 42 were deep depressions and 14 were severe tropical cyclones with intensity greater than 34 knots. Some of these cyclones had a major impact on the deltaic regions of India and Bangladesh. Documentary evidence is available to indicate extensive changes in geomorphological features due to the storm surges and heavy floods (Singh, 2007).

3.2 SLR and Delta Subsidence

A major challenge for the Ganges delta is related to the global issue of rising sea levels. The Ganges delta is experiencing SLR at an average rate of 3.14 mm yr^{-1} near Sagar Island (Fig. 5) in the western deltaic region (Hazra et al., 2002; WWF-India, 2010) and may rise up to 3.5 mm yr^{-1} in the next few decades (Hazra et al., 2002). The increase in eustatic SLR combined with the intense land subsidence due to the compaction of unconsolidated deltaic deposits can further aggravate relative SLR as much as 10–20 mm yr^{-1} in the seaward sectors of the Ganges delta (Allison, 1998). This is a serious cause of concern for the ecosystem dynamics and sustainability in the Ganges delta.

3.3 Coastal Erosion

The Ganges delta is an extremely dynamic ecosystem and the process of erosion and accretion occurs almost simultaneously in different parts of the deltaic lobe. Studies have reported that the islands of the western delta are gradually eroding which may be attributed to sediment run-off, water flow, and current patterns regulated mostly by the Farakka barrage. By contrast, the islands of the central part are expanding owing to the absence of any head-on discharge, siltation on the riverbed, and land subsidence. The net effect, however, is inclined toward coastal erosion (Fig. 6) in the lower stretches of the delta (Ganguly et al., 2006; Raha et al., 2012).

Along the east coast of India, the terrain has resulted in the formation of many large deltas. It is useful to compare the status of shoreline change of the Ganges delta with other large deltas. Here we compare the change in the shoreline of the Mahanadi, Godavari, and Ganges deltas. Shoreline change patterns between 1972 and 2015 were analyzed after extraction from appropriate remote sensing imagery. Fig. 7 shows the comparison of the shoreline in the three cases. Coastal embankments are seen only in the case of the Ganges delta.

In the Mahanadi and Godavari deltas, erosion and accretion are more or less balanced, while in the case of the Ganges, erosion is greater than accretion indicating that the trend is net loss of land. In the Ganges delta, the shoreline change rate range varies from −98 to 165 m^{-1}. High erosion areas are largely in Dalhousie, Bhangaduni, Bulchery Island, Chulkati Reserve Forest, east of Ajmalmari Reserve Forest, west of Matla Reserve Forest, and west of Gosaba River.

3.4 Impacts of Coastal Effects on the Ganges Delta

Flooding is a regular occurrence in the Ganges delta. Riverine floods are beneficial as they bring in fresh silt that supports crop production without additional inputs. However, dams and other engineered structures have resulted in the reduction

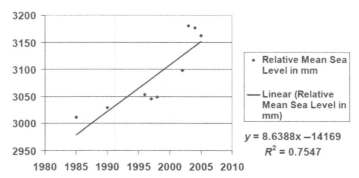

FIG. 5 Relative mean sea level at Sagar Island, Ganges Delta. *(Data from Survey of India, unpublished data.)*

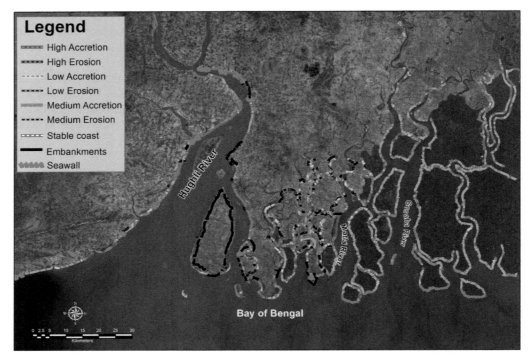

FIG. 6 Shoreline changes along the Hooghly, Matla, and other rivers of Ganges delta in comparison with other deltaic systems along the east coast of India.

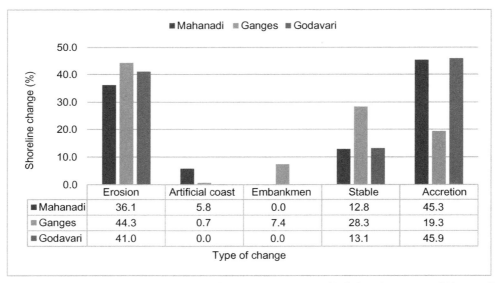

	Erosion	Artificial coast	Embankmen	Stable	Accretion
■Mahanadi	36.1	5.8	0.0	12.8	45.3
■Ganges	44.3	0.7	7.4	28.3	19.3
■Godavari	41.0	0.0	0.0	13.1	45.9

FIG. 7 Comparison of erosion/accretion characteristics in major deltas along the east coast of India based on average of 38 years (1972–2010).

in water flow and consequently the delta is increasingly sediment starved as well (Allison, 1998). Dams also get silted up rapidly because of the high sediment levels in the water reducing their storage capacity. Thus, during periods of heavy rainfall, large quantities of water may have to be released to prevent dam bursts. For example, in 2013 and 2018, there was widespread flooding in many districts of West Bengal due to heavy rainfall and water release from dams of the Damodar Valley Corporation just upstream of where the rivers enter West Bengal. The release also coincided with high tide (SANDRP, 2015).

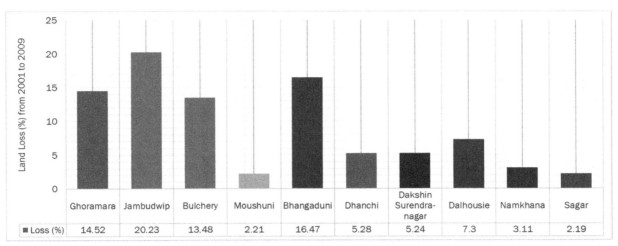

FIG. 8 Land loss in the most vulnerable islands of the Ganges Delta.

Cyclonic storms can cause extensive damage across large swathes of land because of the strong winds, heavy rain, and storm surge. Cyclonic storms and storm surges as well as extensive coastal erosion are key factors in contributing to water salinity especially in North and South 24 Parganas districts apart from increase of salinity in the estuaries of the Sunderbans due to reduced upland discharges. During Cyclone Aila (2009), damage to sluices and breaching of embankments by the storm surge resulted in large-scale flooding as well as intrusion of salt water into agricultural fields (CWC, 2017b). In Bangladesh, large areas remained under water for 2 years until the embankments were repaired, but it is possible that the tidal influx deposited sediment equivalent to decades' worth of normal sedimentation (Auerbach et al., 2015). Fig. 8 provides the extent of land loss in some of the most vulnerable islands in the Indian Sunderbans.

3.5   Anthropogenic Challenges

3.5.1   Population Pressures

The population supported by the GBM delta is large. In the Indian part of the delta, which falls in the state of West Bengal, the population in the nine districts has increased from 9.2 million in 1901 to 16.3 million in 1951 and 57.08 in 2011—a five-fold increase in a century; while the population density (persons per square km) has increased from 3722 in 1991 to 4165 in 2011 (Census of India). Fig. 9 shows how the population in the delta districts has increased since 1901 while Fig. 10 shows the increase of population in the delta districts paralleling the increase in the population of West Bengal state.

In the case of Kolkata, a megacity, the population density went up from 23,783 persons per km^2 in 1991 to 24,760 in 2001 but decreased marginally to 24,348 a decade later. The population in 1991 was 4.39 million in 1991, rising to 4.57 million in 2001 and 4.5 million in 2011. Within the districts, urban populations have been increasing faster than the rural population owing to mass migration, greater employment opportunities, and improved infrastructure facilities in urban centers. This has put greater stress on available resources including access to water which serves as a good indicator for well-being. According to the 2011 census, the sources of potable water include tap water (treated/ untreated source); wells (covered/uncovered); hand pumps, tube wells/boreholes; springs; rivers/canal; tank/pond/lake, and other sources. Hand pumps are the most common sources of potable water except in Kolkata where tap water accounts for over 85% of water supplied. In the context of arsenic contamination of groundwater in certain blocks, as well as saline intrusion into groundwater especially in coastal areas, the need to ensure protected water supply is important for well-being. Chronic arsenic toxicity is caused by drinking arsenic contaminated groundwater and results in a variety of systemic manifestations. Arsenic contamination has been reported extensively from all districts except Burdwan (SOES, 2006; Mazumder and Dasgupta, 2011).

3.5.2   Land Use Change

The primary land use in the delta districts is for agriculture. The large number of tanks, ponds, and other water bodies are used for inland fish culture/capture. The change in the different land use classes between 2003 and 2013 shows the decadal change in land cover for the same time period. Land cover in the delta indicates a reduction in the area under mudflats and marsh vegetation and an increase in the area under industry and aquaculture (Figs. 11A and B).

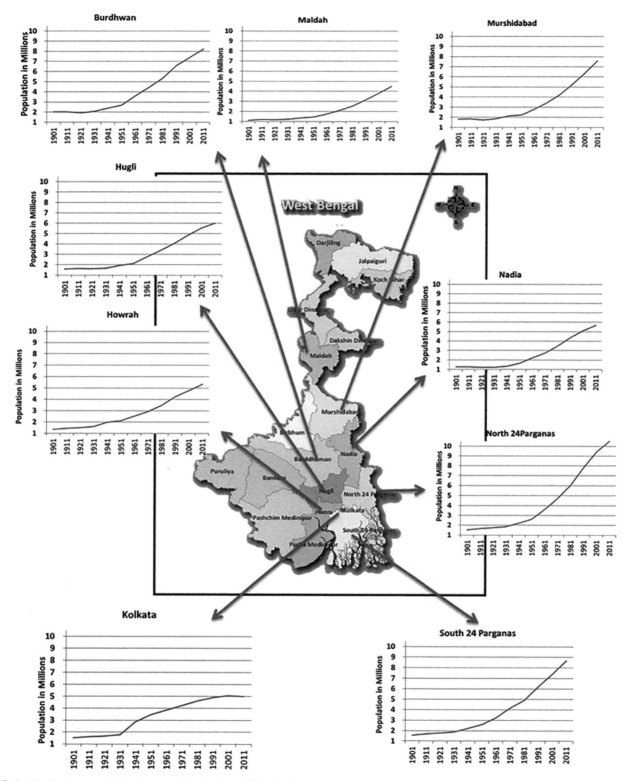

FIG. 9 Population growth in the Ganges Delta Districts (1901–2011).

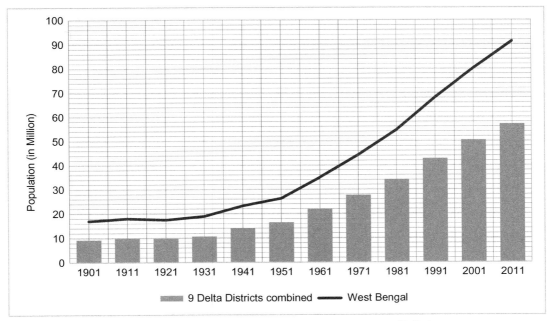

FIG. 10 Population growth in the Ganges Delta districts vs state of West Bengal. *(Data from Census of India.)*

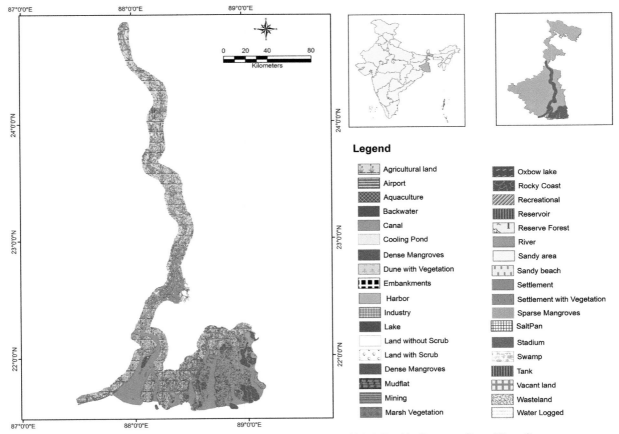

FIG. 11 (A) and (B) Land use and land cover pattern of the Ganges Delta-2003 and 2013 (Farakka Barrage to Bay of Bengal).

(Continued)

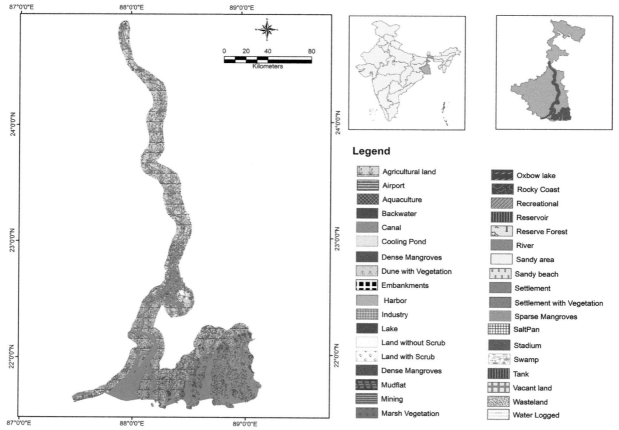

FIG. 11, CONT'D

3.5.3 Livelihoods

West Bengal is predominantly an agrarian state. Comprising of only 2.7% of India's geographical area, it supports nearly 8% of its population. There are 7.123 million farm families of whom 96% are small and marginal farmers. The net cropped area is 5.205 million ha which comprises 68% of the geographical area and 92% of arable land. The rapid growth in West Bengal's agricultural production beginning in the early 1980s can be attributed to the adoption of high-yielding variety of seeds and chemical-based farming practices. Food grains and oilseeds have shown the highest increase in yield rates over the years and have grown by over 75% and 100%, respectively, between 1980 and 2008. The area under production of pulses (lentils) has shown a decline but yield levels have managed to improve over the years. Between 1990 and 2014, there was a steady increase in the consumption of fertilizers (Nitrogen fertilizers dominate); however, the total consumption of fertilizers in the delta districts (Fig. 12) is just over half of the total consumption in the state and in fact has come down, albeit marginally, from almost 58% in 1991 to 55% in 2014 (GoWB, 2015).

The decline in area and increase in yield due to fertilizer-based farming practices could have adverse impacts on the pollution load in the delta region. Cropping intensity has decreased in most of the districts. The agricultural sustainability in the Ganges delta is also vulnerable to the changing climate. Rice is the main food crop cultivated in the deltaic region. Three crop cycles are generally observed, namely rabi (starting from November-February to March-June), kharif (beginning of the Southwest Monsoon and harvested in the autumn months), and autumn or pre-kharif (from March-May to June-October).

Rice cultivation needs differential water supply especially during the sowing time which is largely governed by the seasonal rainfall in this region. Reliable water resource is a prerequisite for rice cultivation, and any change in the pattern and potential magnitude of precipitation in the delta may affect the crop yields. On comparing the rice yield with the seasonal rainfall in the 24 Parganas districts between 1998 and 2010, it was found that there is a strong dependency of crop productivity on precipitation in this region and any alterations in the precipitation patterns affects the agricultural productivity in the delta region. In the inland districts of the Ganges delta, freshwater fisheries dominate. The state has a 158 km

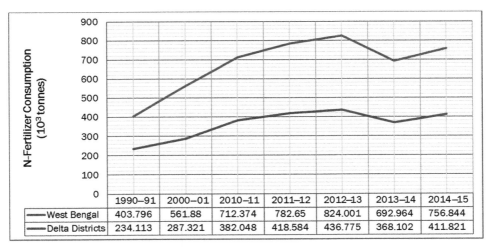

FIG. 12 N-Fertilizer consumption in the delta districts.

	1990–91	2000–01	2010–11	2011–12	2012–13	2013–14	2014–15
West Bengal	403.796	561.88	712.374	782.65	824.001	692.964	756.844
Delta Districts	234.113	287.321	382.048	418.584	436.775	368.102	411.821

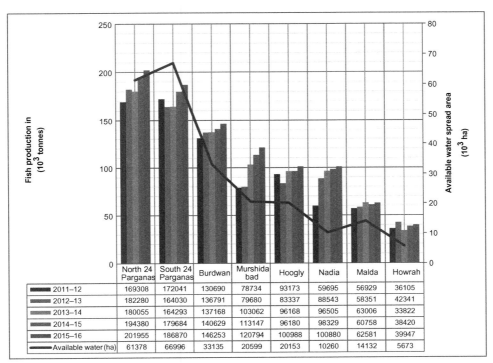

	North 24 Parganas	South 24 Parganas	Burdwan	Murshidabad	Hoogly	Nadia	Malda	Howrah
2011–12	169308	172041	130690	78734	93173	59695	56929	36105
2012–13	182280	164030	136791	79680	83337	88543	58351	42341
2013–14	180055	164293	137168	103062	96168	96505	63006	33822
2014–15	194380	179684	140629	113147	96180	98329	60758	38420
2015–16	201955	186870	146253	120794	100988	100880	62581	39947
Available water (ha)	61378	66996	33135	20599	20153	10260	14132	5673

FIG. 13 Inland fish production vs available water spread in the delta districts. *(Data from GoWB. Handbook of Fisheries Statistics 2015–16. Department of Fisheries, Directorate of Fisheries, Government of West Bengal, 2016.)*

coastline. The estimated annual fish production in 2015–16 was 1.671 million MT (GoWB, 2016). Fig. 13 provides data on fish production over the last 5 years in the delta districts. The production more or less corresponds to the water spread area available in the district (GoWB, 2016).

South 24 Parganas is the only district with a coastline. The fish production reported from this district appears to be leveling off in the last 2 years after showing a steep increase between 2012–13 and 2013–14. Analysis of fish catch data indicates declining trends of most fishes in the marine fisheries realm. Hilsa is an important fish in the West Bengal fishery. After seeing declining catch from both inland and marine fisheries, a sharp increase from marine catch may be noted in 2010–11 after which the catch data once again have fallen to their 2007 values while inland catch shows a continuing declining trend (Fig. 14). This may be attributed to the changing hydrology of the river due to construction of dams and weirs that obstruct the passage upstream as the Hilsa is anadromous, as well as changing water regime. In the post Farakka period, between

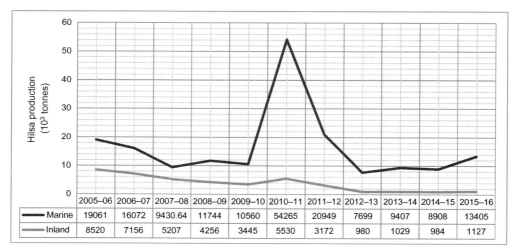

FIG. 14 Hilsa production in West Bengal. *(Data from GoWB. Handbook of Fisheries Statistics 2015–16. Department of Fisheries, Directorate of Fisheries, Government of West Bengal, 2016.)*

1975 and 2010, migratory Hilsa recorded at Patna (upstream of the Farakka barrage) declined even more dramatically from 234.7 to $1.38 \, \mathrm{kg \, km^{-2}}$ (Pathak and Tyagi, 2010).

3.5.4 Pollution

The Ganges River basin is one of the most fertile agriculture belts in the Indian peninsula, making it a focus point for massive human population growth and tremendous urbanization and industrialization (NRCD, 2009). The river basin is the main support system for the industries developed along its stretches and as many as 764 grossly polluting industries have been inventoried throughout the Ganges basin (CPCB, 2013) that utilize the river water and also directly discharge their wastewater into the river.

The industrial units are classified into chemical factories, which mainly include fertilizer, petrochemical, pesticides, and pharmaceuticals; distilleries; dairy, food, and beverage industries; pulp and paper industries; sugar industries; tanneries; textile, bleaching and dyeing units; and other sectors like cement industries, slaughter houses, ordinance factories, packaging and printing units, paint industries, electronics and electrical, thermal plants, electroplating and metallurgical factories, automobile industries, etc. The wastewater generation by these industries is nearly 45% in terms of total water consumption.

In the context of the deltaic region of West Bengal, a large number of industries are situated on the banks of the Hooghly estuary, namely paper, textiles, chemicals, pharmaceuticals, plastics, shellac, food, leather, jute, pesticides, etc. (UNEP, 1982). Though the water consumption of the industries here is comparatively less than the other states in the Ganges basin, their wastewater generation with respect to water consumption is considerably higher and, hence, alarming (CPCB, 2013). The Hooghly River receives 87 million liters per day (MLD) wastewater from 22 grossly polluting industries and as much as 70% of total wastewater generated comes from the chemical industries, followed by pulp and paper industries which discharge around 20% of total wastewater. Thus, large quantities of toxic and hazardous effluents are released into the deltaic ecosystems through these industrial units. In addition, about 360 outfalls on both sides of the river continuously discharge community sewage, virtually without any treatment (Mukherjee et al., 1993).

The area supports large numbers of shrimp (*Penaeus monodon*) aquaculture farms often found coexisting with agricultural lands in the Ganges delta complex. Agriculture-based nutrient pollution is a well-established fact in the Ganges basin and the deltaic region (Banerjee et al., 2014). The region has seen exceptional growth in the agricultural sector in the recent decades, with an increase in the usage of nitrogenous fertilizers and consumption rate as high as $4961 \, \mathrm{kg\text{-}N \, km^2 \, yr^{-1}}$ for the Ganges basin (Chanda et al., 2001). A study by Banerjee et al. (2014) and Swaney et al. (2015) concluded that the net anthropogenic nitrogen input into the basin was considerably high and it was mainly contributed by high agricultural fertilizer inputs. The huge nutrient load emanating from these activities along with the nutrient load from several nonpoint sources (such as discharges from fishing vessels and trawlers and run-off from adjacent landmasses) is leading to ecosystem collapse and problems like harmful algal blooms, eutrophication, hypoxia, and loss of biodiversity at the local scale.

In addition to this, the Ganges delta houses one of the largest urban centers, the megacity Kolkata, with a population of about 14.5 million (Census of India, 2011). The densely populated city depends on the Ganges water for domestic needs. Also, the city of Howrah and the Haldia industrial belt on the bank of the Hooghly add to the anthropogenic stress on the

Ganges delta. Apart from the localized pollution in the deltaic region, the river also brings substantial inputs of pollutants from the upper stretches all along its 2525 km course from Himalayas (CPCB, 2013).

3.5.5 Anthropogenic Impacts on the Delta

The population of the Ganges delta has been rising steeply over time, especially in the 24 Parganas districts (north and south) as the delta has been able to support intensive agriculture. However, rising populations also mean increasing demand for land for settlements and agriculture as well as for resources, especially potable water supply. While the reach of protected water supply is high in urban areas, in rural pockets, dependence on wells is high; however, many of these are in areas of high arsenic contamination and this can have far reaching impacts on the health profile of the population in the delta as given in detail at Annex 1.

The sewage generated by Class-I cities along the Ganges River in West Bengal is 1311.3 MLD whereas the treatment capacity is 548.4 MLD which means that more than half the sewage generated reaches the river systems in the delta. Pollution assessment by the CPCB (CPCB, 2013) indicates that West Bengal generates half of the total sewage generated by Class-I cities in the four Ganges basin states—Uttarakhand, Uttar Pradesh, Bihar, and West Bengal. Within West Bengal, the megacity of Kolkata generates 47% of the total sewage followed by Howrah (10%).This indicates that a significant amount of nutrients enter the river and coastal waters resulting in coastal pollution. The East Kolkata Wetlands (EKWs), a Ramsar Site, stretching over 12.5 km^2 of the north and South 24 Parganas districts, nurture the world's largest wastewater fed aquaculture system which comprises 254 sewage fed fisheries, agricultural land, garbage farming fields, and some built up area (Kundu et al., 2008). The EKW offer a solution that can be implemented on a larger, delta-wide scale to reduce pollution of water bodies as they function as waste stabilization ponds with the slow-moving canal system functioning as the anaerobic and facultative ponds while the fishponds are the maturation ponds. The EKW are under threat from growing urbanization and land reclamation.

The construction of dams, weirs, and other structures along the river at many locations has resulted in not only altering the hydrology of the river but also fish diversity and thus, the profile of the fisheries sector. The construction of the Farakka barrage on the Ganges River resulted in a major stock change in the fishery at Lalgola center about 45 km below the Farakka barrage. From Hilsa being the main fishery (92%), its contribution dropped to only 16.8% post-construction of Farakka; with the niche being replaced by other species including major carps and large catfishes. Also, change in the salinity regime of the Hooghly estuary resulted in a sharp increase in estuarine fishery from 9482 tons (1966–75, pre-Farakka period) to 62,000 tons (1999–2000) (Pathak and Tyagi, 2010). This has not only a bearing on the economic status of fishermen but also resulted in a change in the consumption choice and preferences through the delta. Today, a lot of the fish is imported from other locations including Bangladesh and Myanmar (GoWB, 2016) as Hilsa is an important constituent of the diet of the delta communities.

3.6 Integrated Management of the Ganges delta

3.6.1 The Interrelated Cascading Impacts

It is clear from the above discussion that the delta is under threat from multiple causes and many of these are closely interrelated, with cascading and cumulative impacts (Fig. 15). Reduced river flow can have multiple first-level impacts—such as on agriculture, potable water supplies, saline intrusion and coastal erosion, and linking the social and ecological systems very closely. Reduced water flow in the river leads to reduction in crop production because agriculture becomes highly rain-dependent reducing the number of crops as availability of groundwater for an additional crop shrinks. This reduction in production results in reduced incomes for farmers and higher prices of food resulting in food security issues on the one hand, and heightened poverty and forced migration, on the other, setting off yet another chain of events.

Reduced water flow also means reduced flow of enriched silty sediment. This may raise the demand for chemical fertilizers on the one hand, resulting in coastal erosion as there is insufficient sediment supply at the coast. This may also result in increased penetration of salt water with additional impacts due to rising sea levels as well as storm surges during the regular cyclones, affecting, among other things, diversity of coastal mangroves, which can have far-reaching implications on the structure of the entire Sunderbans ecosystem.

3.6.2 Delta Sustainability and Integrated Management

Considering the intricate connections between the various delta components, the way forward for delta sustainability is an integrated approach, linking the various aspects including water and land management, conservation of ecosystems, ensuring that livelihoods are sustained, managing pollution without straining the delta's water network and building a culture of disaster preparedness. Fig. 16 provides a few strategies under each major class of action that are required for ensuring delta sustainability.

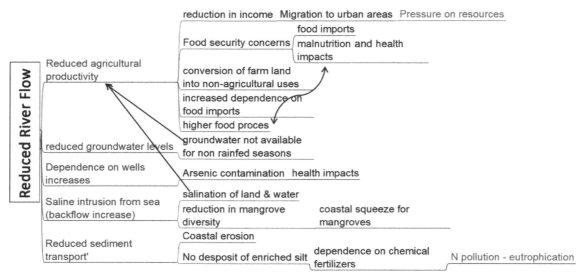

FIG. 15 Example of cascading multiple impacts due to reduced river flow.

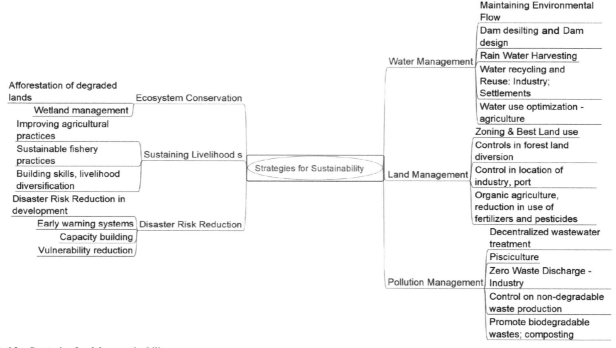

FIG. 16 Strategies for delta sustainability.

Foremost among these is the maintenance of an environmental flow in the river, essential for maintenance of the aquatic ecosystem (Jain and Kumar, 2014). However, the necessary data are not yet available. Even here, investigations have been confined to the upper reaches of the river while the requirements of the delta have been largely ignored. Even within the delta, as mentioned earlier, variation in the salinity regimes indicates how the central section of the delta is becoming hypersaline while the fringes are hyposaline. Thus, in addition to ensuring environmental flow in the major rivers, it is essential to examine and determine environmental flows for all branches of rivers within the delta. Wetlands and the flood plain of rivers are essential for storing water during peak flow, releasing the water during off-season as well as serving as breeding and nursery grounds for many riverine fish.

Many of these are silted up and have lost their functionality and hence appropriate steps including planting of native species on riverbanks to control water (and silt) flow are needed. These will ensure not only regulation of water flows but also restoration of riverine fisheries (Pathak and Tyagi, 2010).

Water use within the delta also needs to be effectively managed by promoting techniques such as decentralized rainwater harvesting that are eco-friendly and cost-effective especially in areas where there is seasonal high fluctuation in water flow. There must be a reduction in over-dependence on ground water, with ground water being utilized within the rechargeable limit to avoid delta subsidence.

Land in the delta needs effective management, especially in the coastal areas where zoning is essential. Effective implementation of the Coastal Regulation Zone Notification 2011 that zones the coastal areas and regulates activities in the CRZ will help in conserving critical coastal ecosystems especially mangroves. Reclamation by landfilling, wholly or partly, of wetlands must be prohibited because such ecosystems (engineered or natural) provide services including flood control during floods and sewage treatment (EKWs) that reduce human stress on the environment/delta.

Water and land management together play an important role in the state of the ecosystems, especially the mangroves that cover a large area of the delta. In the last several decades, large areas of the Sundarban mangroves have been converted into paddy fields and more recently, into shrimp farms (Gopal and Chauhan, 2006). The Sundarbans have been declared as Biosphere Reserve and there are several sanctuaries/national parks/protected areas in the Indian part of the Sunderbans under varying degrees of protection (Singh, 2003). Changes due to upstream controls in water flow have impacted the biodiversity of the Sunderbans. Additional threats are from reclamation and encroachments into the Sunderbans as a huge population (~2.5 million) is reportedly dependent on resources from the forests (Gopal and Chauhan, 2006).

Using satellite data, Giri et al. (2007) noted that between 1973 and 2000, though the loss in total mangrove area was not significant in the Sunderbans, the change matrix showed that turnover due to erosion, aggradation, reforestation, and deforestation was much greater than net change. This means there is a constant stress on the ecosystem structure and functioning and potentially increased vulnerability, as well as reduced functioning in terms of services such as nutrient cycling, carbon sequestration, and shoreline protection. Recent studies (Ray et al., 2011) estimated the overall carbon storage in the Sundarbans mangrove forest reservoir to be 21.13 Tg C and in the soil reservoir (30 cm) 5.49 Tg C. The Sundarbans store 0.41% of the total carbon storage in the Indian forests (6621 Tg C) but their annual increase exhibits faster turnover than the tropical forests. The study also found that the overall annual increase in mean carbon stock across the Sundarbans mangrove forest could be partly due to an increase in resource availability favoring lighter-wooded mangrove species such as *Aegialitis rotundifolia* and *Ceriops*, now dominating in the eastern part rather than the heavier wood species such as *Avicennia* that dominate in the western Sunderbans.

Pollution management through the entire delta has to be implemented with decentralized wastewater management especially for domestic wastes with emphasis on treating at source and recycling and reuse wherever possible. Pisciculture as carried out in the EKWs can be promoted in other suitable locations through the delta which will serve the double purpose of retrieving nutrients and enhancing food security and livelihoods for the local communities.

Sustaining agricultural livelihoods is crucial as agriculture is the primary occupation of the delta inhabitants. However, dependency of crop yield on precipitation and temperature should be reduced. Instead, adaptive management strategies need to be taken up by farmers in choosing what crops to grow, during what time, what is the best suited land/soil type, and how to grow the crops (e.g., using system of rice intensification—SRI), for improving rice yields. Other strategies could include use of indigenous varieties requiring less water and promotion of organic agriculture with use of natural manure rather than chemical fertilizers. The latter may help reduce nutrient runoff into the waterways, consequently reducing eutrophication of coastal waters. Along with agriculture, other traditional resource-based livelihoods such as fisheries and honey collection from mangroves need to be sustained through better practices including market linkages.

The main constraints are with reference to data availability especially with regard to water and sediment flows in the Ganges and the requirements of coordination when actions that can have large-scale impacts on the delta are planned. Very often, political decisions supposed to be for the good of the larger community are taken but the complexities of the ground situation are seldom taken into account resulting in unanticipated adverse impacts. From a broader perspective, sustainable delta management should be based insofar as possible on natural system functioning (Day et al., 2016).

4 CONCLUSIONS

The Ganges delta is the world's largest and spans two countries. We have examined the major challenges to the delta in India that can be primarily construed to be related to the upstream activities that control the flow of the river and sediments. The other set of challenges are due to the coast-related phenomena including rising sea levels and increasing intensity of storms. Anthropogenic challenges are cross cutting across the deltaic landscape. We have also examined how an integrated approach is required to address the various challenges as they can have a cascading impact on the structure and function of the Ganges delta. Appropriate actions are urgently required as the delta is already tending toward a collapsed system.

ANNEX 1

Facts and Figures and Pollution Status Along the Coast of West Bengal with Particular Reference to the Ganges Delta

No.	Coastal Information	
1	Length of coast	200 km
2	Number of coastal districts	4
3	Cities along the coast	6
(a)	Major (Class 1)	Haldia
		Howrah
		Kolkata
		Bangaon
		Basirhat
4	Population in coastal districts	28.17 million
5	Rivers/estuaries	Ganges
		Hooghly
6	Dams	4
7	Number of ports	2
(a)	*Major*	1
(b)	*Minor*	1
8	Number of industries located on the coast	57

Major sources of pollution from the Ganges delta include:

(a) Sewage outfall (treated/ untreated)
(b) Solid wastes
(c) Industries
(d) Port-oil pollution
(e) Thermal power plants
(f) Agriculture
(g) Aquaculture
(h) Water quality
(i) Tourism

(a) Sewage Outfall (Treated/Untreated)

Total Sewage Generated Million liters per day (MLD)	Sewage from Coastal Districts	Coastal Districts	No. of STPs (Coastal Districts)	Capacity of STPs In MLD	Quantity Treated	Total Untreated Sewage
2345.21	250.048	East Midnapur	–	16.3	6.7	10.35
		South 24 Parganas	2	8.25	–	–
		North 24 Parganas	9	114.46	–	–

(b) Solid Waste

State	District	Quantity of Solid Waste Generated (TPD)	Quantity of MSW Collected	Quantity of MSW Treated (TPD)	Number of Landfills	Landfilling Capacity (m²)
381.3 TPD	East Midnapur	13.2	–	–	–	–
	South Parganas	7	–	–	–	–
	North Parganas	361.1	–	–	–	–

(c) Industries

Coastal Districts	Small	Medium	Large	Category (Medium + large)			
				R	O	G	W
Purba Midnapur	–	–	12	9	2	0	1
South-24 Parganas	–	7		2	1	2	2
North-24 Parganas	–	9	29	14	10	3	11

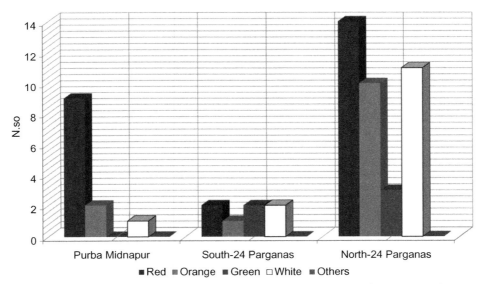

ANNEX FIG. 1 District-wise categorization of Industries in the Ganges Delta, West Bengal.

(d) Port—Oil Pollution: No oil spill has been reported along the coast of West Bengal. There are two ports and one mooring point located along the coast of west Bengal.

(e) Thermal Power plant

Nuclear power plants	**0**
Thermal power plants	**16**

(f) Agriculture
- Total cropping area: 5.5 Mha—3.9% of total cropping area of India
- Total agricultural production: 19.65 million tons
- Fertilizer consumption: 2.82 million metric tons and consists of 5.72% of total fertilizer consumption in India
- Pesticide consumption: 3800 tons

(g) Aquaculture
Estimated potential of West Bengal is 405,000 ha of which area developed consists of 50,405 ha with a production of 52,581 tons, all of which are concentrated in the lower part of the Ganges delta

(h) Water Quality Index

River/Estuary	O$_2$ (mg L^{-1})	BOD (mg L^{-1})	Turbidity (NTU)	Chlorophyll (µg L^{-1})	DIN (µmol L^{-1})	DIP (µmol L^{-1})
Hooghly	7.2	1.1	22.7	2.0	23.6	1.0

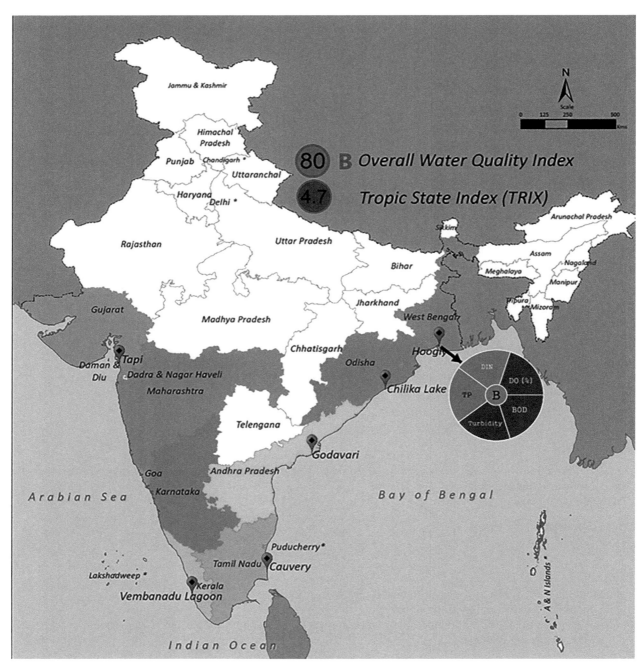

ANNEX FIG. 2   Water Quality Index of the River Hooghly.

Trophic Index for Ganges Delta and coastal waters of West Bengal

Nutrients may stimulate primary production under sufficient light and temperature regimes which may cause eutrophication when released in favorable conditions. The extent to which this process has influenced is reflected in Trophic State Index (TRIX). TRIX is an estimate of nitrogen and phosphorus, and chlorophyll pigments in a water body. It is defined by a linear combination of the logarithm of chlorophyll a, absolute percent deviation of O_2 from saturation, DIN, and IP.

$$TRIX = (log10(Chl\ a\ x \times OD\% \times DIN \times IP) = k)/m$$

TRIX Value	State Water Quality	Level of Eutrophication
0<TRIX≤4	High	Low
4<TRIX≤5	Good	Medium
5<TRIX≤6	Moderate	High
6<TRIX≤10	Poor and degraded	Elevated

Eutrophication index calculated for West Bengal coastal waters was 4.7 indicating good seawater qualities with medium eutrophication level. The observed trophic index is indicative of high potential for algal blooms'

Harmful Algal Blooms

Only one incidence of algal bloom has been reported off Kolkata in 2001 due to *Trichodesmium erythraeum*

State/Place	Latitude	Longitude	Year	Species
Off Kolkata	21.3377	88.1036	April 25, 2001	*Trichodesmium erythraeum*

(i) Tourism

Total number of beaches	14
Pilgrimage sites along the coast	0

(j) Ecologically Sensitive Areas in the Ganges Delta

Ecologically Sensitive Area		Area (km²)
	Mangrove	202.81
	Coral	0.00

Ecologically Sensitive Area		Area (km²)
	Seagrass	0.00
	Salt marsh	0.00
	Horseshoe crab habitat	41.05
	Turtle nesting sites	2.87
	Bird nesting sites	Sunderbans
	Sand dune	2.69
	Mudflat	216.35
	Archaeological and heritage site	0.05

(k) Pollution Hotspots along the coast of West Bengal and Ganges Delta:
The following pollution stretches have been identified in the Ganges Delta and the coast of West Bengal

- Digha
- Diamond Harbor
- Sapthamukhi

REFERENCES

Akter, J., Sarker, M.H., Popescu, I., Roelvink, D., 2016. Evolution of the Bengal Delta and its prevailing processes. J. Coast. Res. 32 (5), 1212–1226. ISSN 0749-0208.

Allison, M.A., 1998. Historical changes in the Ganges–Brahmaputra delta. J. Coast. Res. 14 (4), 1269–1275.

Auerbach, L.W., Goodbred Jr., S.L., Mondal, D.R., Wilson, C.A., Ahmed, K.R., Roy, K., Steckler, M.S., Small, C., Gilligan, J.M., Ackerly, B.A., 2015. Flood risk of natural and embanked landscapes on the Ganges–Brahmaputra tidal delta plain. Nat. Clim. Change 5, 153–157.

Banerjee, K., Purvaja, R., Ray, A.K., Ramesh, R., 2014. Nutrient conservation dynamics in Ganges Delta using LOICZ and NANI biogeochemical approach. In: Fast Track Paper on Module- Modeling Approaches in Deltas: Hydro-eco-geomorphologic Aspects and Links to Ecosystem Services and Human Dimension, Report 2013-2014.

Brondizio, E.S., Foufoula-Georgiou, E., Szabo, S., Vogt, N., Sebesvari, Z., Renaud, F.G., Newton, A., Anthony, E., Mansur, A.V., Matthews, Z., Hetrick, S., Costa, S.M., Tessler, Z., Tejedor, A., Longjas, A., Dearing, J.A., 2016. Catalyzing action towards the sustainability of deltas. Curr. Opin. Environ. Sustain. 19, 182–194. https://doi.org/10.1016/j.cosust.2016.05.001.

Brown, S., Nicholls, R.J., 2015. Subsidence and human influences in mega deltas: the case of the Ganges–Brahmaputra–Meghna. Sci. Total Environ. 527–528 (2015), 362–374.

Chakrabarty, P., 1995. Subarnarekha delta-a geomorphic appraisal. Indian J. Earth Sci. 22, 125–134.

Chanda, T.K., Sundaram, K.P., Dubey, A.C., Sati, K., Robertson, C., 2001. Fertiliser Statistics 2000-2001. Fertiliser Association of India, New Delhi. pp. I-128-164.

Chaudhuri, A.B., Choudhury, A., 1994. Mangroves of the Sundarbans, Vol 1. India. IUCN-The World Conservation Union.

Cole, C.P., Vaidyaraman, P.P., 1966. In: Salinity distribution and effect of fresh water flows in the Hooghly River. Proceedings of Tenth Conference on Coastal Engineering, Tokyo, Japan, September. pp. 1312–1434.

Coleman, J.M., 1969. Brahmaputra River: channel processes and sedimentation. Sediment. Geol. 3, 129–239.

Coleman, J.M., Huh, O.K., Braud Jr., D.W., 2008. Wetland loss in world deltas. J. Coast. Res. 24 (1A, Supplement (January 2008)), 1–14.

CPCB, 2013. Pollution Assessment: River Ganges. Central Pollution Control Board, Ministry of Environment and Forests. Govt. of India, New Delhi, p. 197.

Curray, J.R., Moore, D.G., 1971. Growth of the Bengal Deep-Sea Fan and denudation in the Himalayas. Geol. Soc. Am. Bull. 82, 563–572.

CWC, 2014. Ganges Basin. Central Water Commission and National Remote Sensing Centre. Ministry of Water Resources and ISRO, Department of Space, Government of India.

CWC, 2017b. A Report on Problems of Salination of Land in Coastal Areas of India and Suitable Protection Measures. Hydrological Studies Organization, Central Water Commission, Government of India, Ministry of Water Resources, River Development and Ganges Rejuvenation New Delhi, 2017.

CWC and IMD, 2015. PMP Atlas for Ganges River Basin Including Yamuna. Final Report.

Day, J.W., Agboola, J., Chen, Z., D'Elia, C., Forbes, D.L., Giosan, L., Kemp, P., Kuenzer, C., Lane, R.R., Ramachandran, R., Syvitski, J., Yañez-Arancibia, A., 2016. Approaches to defining deltaic sustainability in the 21st century. Estuar. Coast. Shelf Sci. 183, 275–291.

Ganguly, D., Mukhopadhyay, A., Pandey, R.K., Mitra, D., 2006. Geomorphological study of Sundarban deltaic estuary. J. Indian Soc. Remote Sens. 34 (4), 431–435.

Giri, C., Pengra, B., Zhu, Z., Singh, A., Tieszen, L.L., 2007. Monitoring mangrove forest dynamics of the Sundarbans in Bangladesh and India using multi-temporal satellite data from 1973 to 2000. Estuar. Coast. Shelf Sci. 73, 91–100.

Goodbred Jr., S.L., Kuehl, S.A., 2000. Enormous Ganges-Brahmaputra sediment discharge during strengthened early Holocene monsoon. Geology 28 (12), 1083–1086.

Goodbred, S.L., Nicholls, R., 2004. In: Towards integrated assessment of the Ganges Brahmaputra Delta. Proceedings of the 5th International conference on Asian marine geology, and 1st Annual Meeting of the IGCP475 Delta and APN Mega Delta, 13th February 2004.

Gopal, B., Chauhan, M., 2006. Biodiversity and its conservation in the Sundarban mangrove ecosystem. Aquat. Sci. 68, 338–354.

GoWB, 2015. Statistical Abstract, West Bengal, 2015. Bureau of Applied Economics and Statistics, Government of West Bengal.

GoWB, 2016. Handbook of Fisheries Statistics 2015–16. Department of Fisheries, Directorate of Fisheries, Government of West Bengal.

Hagenlocher, M., Renaud, F.B., Haas, S., Sebsevari, Z., 2018. Vulnerability and risk of deltaic social-ecological systems exposed to multiple hazards. Sci. Total Environ. 631–632, 71–80.

Hazra, S., Ghosh, T., Dasgupta, R., Gautam, S., 2002. Sea level and associated changes in the Sundarbans. Sci. Cult. 68 (9–12), 309–321.

Jain, S.K., 2008. Impact of retreat of Gangotri glacier on the flow of Ganges River. Curr. Sci. 95 (8), 1012–1014.

Jain, S.K., Kumar, P., 2014. Environmental flows in India: towards sustainable water management. Hydrol. Sci. J. 59 (3–4), 751–769. https://doi.org/10.1080/02626667.2014.896996.

Kundu, N., Pai, M., Saha, S., 2008. In: Sengupta, M., Dalwani, R. (Eds.), East Kolkata wetlands: a resource recovery system through productive activities. Proceedings of Taal2007: The 12th World Lake Conference. pp. 868–881.

Mazumder, D.G., Dasgupta, U.B., 2011. Chronic arsenic toxicity: studies in West Bengal, India. Kaohsiung J. Med. Sci. 27, 360–370.

Milliman, J.D., Meade, R.H., 1983. Worldwide delivery of river sediment to the oceans. J. Geol. 91 (1), 1–21.

Milliman, J.D., Syvitski, J.P.M., 1992. Geomorphic/tectonic control of sediment discharge to the ocean: The importance of small mountainous rivers. J. Geol. 100, 525–544.

Mitra, A., 2013. Sensitivity of Mangrove Ecosystem to Changing Climate. Springer, India. https://doi.org/10.1007/978-81-322-1509-7.

Morgan, J.P., McIntire, W.G., 1959. Quaternary Geology of Bengal basin, East Pakistan and India. Bull. Geolog. Soc. Am. 70, 319–342.

Mukherjee, D., Chattopadhyay, M., Lahiri, S.C., 1993. Water quality of the River Ganges (The Ganges) and some of its physico-chemical properties. Environmentalist 13 (3), 199–210.

Nicholls, R.J., Goodbred Jr., S.L., 2005. Towards an Integrated Assessment of the Ganges-Brahmaputra Delta. In: Chen, Z., Saito, Y., Goodbred, S.L. Jr. (Eds.), Mega-Deltas of Asia: Geological Evolution and Human Impact. 2005. China Ocean Press, Beijing, China, pp. 168–181.

NRCD, 2009. Status Paper on River Ganges: State of Environment and Water Quality. National River Conservation Directorate, Ministry of Environment and Forests, Government of India.

Papa, F., Bala, S.K., Pandey, R.K., Durand, F., Gopalakrishna, V.V., Rahman, A., Rossow, W.B., 2012. Ganges-Brahmaputra river discharge from Jason-2 radar altimetry: an update to the long-term satellite-derived estimates of continental freshwater forcing flux into the Bay of Bengal. J. Geophys. Res. 117, https://doi.org/10.1029/2012JC008158. C11021.

Pathak, V., Tyagi, R.K., 2010. Riverine ecology and fisheries visa`-vis hydrodynamic alterations: impacts and remedial measures. . CIFRI, Bulletin No. 161.

Raha, A., Das, S., Banerjee, K., Mitra, A., 2012. Climate change impacts on Indian Sunderbans: a time series analysis (1924–2008). Biodivers. Conserv. 21 (5), 1289–1307.

Ray, R., Ganguly, D., Chowdhury, C., Dey, M., Das, S., Dutta, M.K., Mandal, S.K., Majumder, N., De, T.K., Mukhopadhyay, S.K., Jana, T.K., 2011. Carbon sequestration and annual increase of carbon stock in a mangrove forest. Atmos. Environ. 45 (28), 5016–5024. https://doi.org/10.1016/j.atmosenv.2011.04.074.

Renaud, F.G., Szabo, S., Matthews, Z., 2016. Sustain. Sci. 11, 519. https://doi.org/10.1007/s11625-016-0380-6.

SANDRP, 2015. Damodar Valley Dams role in W Bengal Floods—DVC Dams could have helped reduce the floods, they increased it. South Asia Network on Dams, Rivers and People. Available from: https://sandrp.wordpress.com/2015/08/05/damodar-valley-dams-role-in-w-bengal-floods-dvc-dams-could-have-helped-reduce-the-floods-they-increased-it/. (Accessed 26 March 2018).

Singh, H.S., 2003. Marine protected areas in India. Indian J. Mar. Sci. 32, 226–233.

Singh, I.B., 2008. The Ganges River. In: Gupta, A. (Ed.), Large Rivers: Geomorphology and Management. John Wiley and Sons, UK.

Singh, O.P., 2007. Long-term trends in the frequency of severe cyclones of Bay of Bengal: observations and simulations. Mausam 58, 59–66.

SOES, 2006. Groundwater arsenic contamination in West Bengal—India. School of Environmental Sciences, Jadavpur University, Kolkata. Available from: http://www.soesju.org/arsenic/arsenicContents.htm?f=profile.htm. (Accessed 15 March 2018).

Subramanian, V., 1996. In: The sediment load of Indian rivers—an update. Erosion and Sediment Yield: global and regional perspectives. Proceedings of the Exeter Symposium, July 1996). IAHS Publ. no. 236.

Subramanian, V., Ramanathan, A.L., 1996. Nature of sediment load in the Ganges-Brahmaputra River Systems in India. In: Milliman, J.D., Haq, B.U. (Eds.), Sea-Level Rise and Coastal Subsidence. Coastal Systems and Continental Margins. vol. 2. Springer, Dordrecht.

Swaney, D.P., Hong, B., Selvam, A.P., Howarth, R.W., Ramesh, R., Purvaja, R., 2015. Net anthropogenic nitrogen inputs and nitrogen fluxes from Indian watersheds: an initial assessment. J. Mar. Syst. 141, 45–58.

Syvitski, J.P.M., 2003. Supply and flux of sediment along hydrological pathways: research for the 21st century. Glob. Planet. Chang. 39 (1/2), 1–11.

Syvitski, J.P.M., 2008. Deltas at risk. Sustain. Sci. 3, 23–32. https://doi.org/10.1007/s11625-008-0043-3.

Tessler, Z.D., Vörösmarty, C.J., Grossberg, M., Gladkova, I., Aizenman, H., Syvitski, J.P.M., Foufoula-Georgiou, E., 2015. Profiling risk and sustainability in coastal deltas of the world. Science 349 (6248), 638–643.

Trivedi, S., Zaman, S., Chaudhuri, T.R., Pramanick, P., Fazli, P., Amin, G., Mitra, A., 2016. Inter-annual variation of salinity in Indian Sundarbans. Indian J. Mar. Sci. 45 (3), 410–415.

UNEP, 1982. Pollution and the Marine Environment in the Indian Ocean. UNEP Regional Seas Reports and Studies 13. Switzerland, Geneva.

Wasson, R.J., 2003. A sediment budget for the Ganges–Brahmaputra catchment. Curr. Sci. 84 (8), 1041–1047.

Woodroffe, C.D., Nicholls, R.J., Saito, Y., Chen, Z., Goodbred, S.L., 2006. Landscape variability and the response of Asian megadeltas to environmental change. In: Harvey, N. (Ed.), Global Change and Integrated Coastal Management, Volume 10. Springer, Dordrecht, The Netherlands, pp. 277–314.

WRIS. Ganges. Water Resources Information System of India-WRIS Wiki. Available from: http://www.india-wris.nrsc.gov.in/wrpinfo/index.php?title=Ganges#River_System. (Accessed 31 January 2018).

WWF-India, 2010. Sunderbans: Future Imperfect. Climate Adaptation Report.

Chapter 12

The Indus Delta—Catchment, River, Coast, and People

Samina Kidwai*, Waqar Ahmed*, Syed Mohsin Tabrez*, Jing Zhang[†], Liviu Giosan[‡], Peter Clift[§], Asif Inam*

*National Institute of Oceanography, Karachi, Pakistan, [†]State Key Laboratory in Estuarine and Coastal Research, Shanghai, China, [‡]Department of Geology and Geophysics, Woods Hole Oceanographic Institution, Woods Hole, MA, United States, [§]Louisiana State University, Baton Rouge, LA, United States

1 ORIGIN OF THE RIVER

The River Indus basin spans four countries (Pakistan, India, Afghanistan, and China) for a total length of 2900 km in an area that is 30% arid (Wong et al., 2007). The river originates near Lake Manasarovar to the north of the Himalayan range on the Kailash Parbat Mountain in China at an elevation of 5500 m, and it flows through India and reaches Pakistan, where it flows through the entire length of Pakistan. The River Indus basin forms 65% of the total area of the country and forms the main stream of fresh water into which smaller rivers join course (Fig. 1). The riverbanks have altered over geological times, and the delta has shown evidence of shifting westward (Kazmi, 1984; Giosan et al., 2006; Overeem and Syvitski, 2009).

The Indus is considered as one of the oldest documented rivers. The River Indus was formed when the Indian and Eurasian Plates collided more than 45 million years ago (Clift et al., 2001). The earliest Indus is older than the uplift that formed the Greater Himalaya during the Early Middle Miocene period, around 20–25 million years ago (Searle and Owen, 1999), and the river has followed a similar course along the Indus-Tsanpo Suture Zone in southern Tibet and Ladakh since then. Two parallel west-flowing streams were in existence during the Eocene, one north and the other south of the Himalaya (Qayyum et al., 2001). These rivers jointly formed the Katawaz Delta at the western margin of the Katawaz Ocean, an embayment of the larger Tethys Ocean. The northern stream they recognized as the palaeo-Indus.

The main stream of the River Indus has not shown much deviation from its past course in spite of tectonic events such as the uplift of the Sulaiman Range west of Punjab that displaced the main stream about 100 km toward the east since the Early Eocene. Subsequent growth of the Sulaiman Range must have pushed the course of the Indus southward by 200–300 km (Clift, 2002). Najman et al. (2003) interpret that 18 million years ago the palaeo-Indus first followed its modern course, cutting south through the Himalaya and into the foreland basin, and the river has been flowing in approximately the same location since (Clift, 2002). Isotopic data show that the source of the sediment reaching the Arabian Sea changed sharply about 15 million years ago (Clift and Blusztajn, 2005).

Before 1870, the river flowed with 17 branches into the Arabian Sea (Syvitski et al., 2014). During the colonial times, irrigation canals and flood levees were constructed and all but one branch was blocked; today there is only one functional channel (Khobar Creek). During the dry season, the sea moves upstream, reaching several kilometers up the river, thus affecting the position of the estuary and the riverine, deltaic ecosystem (Kidwai et al., 2016).

The fresh water and sediment flow to the Indus Delta has generally decreased over the years (Fig. 2), which is attributed to the upstream infrastructure for agriculture, damming, groundwater extraction, and climatic variability (Wong et al., 2007). This decrease in the fresh water and sediment flow results in the shrinking and subsidence of the delta, sea intrusion, and loss of the deltaic ecosystem.

2 GEOMORPHOLOGY AND HYDROLOGY

2.1 Catchment, Fluvial, Estuary/Delta, Coastal Offshore

The Indus catchment, including tributaries, is spread in Pakistan as well as adjoining countries (Table 1). The catchment area of River Indus within Pakistan is ~56% of the total, or 52,9135 km^2 of its total 97,0469 km^2. Hydrologically, Pakistan

Coasts and Estuaries. https://doi.org/10.1016/B978-0-12-814003-1.00012-5

213

FIG. 1 River Indus, origin, and catchment to the coast.

is divided into two main catchment regions—the Indus basin and the dry area of Balochistan. Excluding the tributaries, the Indus River basin constitutes 55,3413 km^2 (NDMA-UNDP, 2010).

2.2 Water and Sediment Discharge Downstream

The Indus River discharge has decreased dramatically since the construction of channels, dams, and barrages over the past several decades (Fig. 2), which has resulted in a reduction in sediment reaching the sea and in pronounced erosion in some areas of the Indus Delta, and consequently in the extent of the mangrove forests (Quraishee, 1988). Although the Indus Water Apportionment Accord (WAA) recognized that the annual minimal flow below the Kotri Barrage should be 5 million cusec (10 million acre per foot). But a small proportion to almost no fresh water reaches the delta and sea. This is also because the annual rainfall is very low, averaging between 180 and 200 mm.

Pakistan's vast irrigation network comprises three major storage reservoirs: 19 barrages or headworks, 43 main canals with a conveyance length of 57,000, and 89,000 water courses with a running length of more than 1.65 million km (Fig. 3). They feed more than 150,000 km^2 of farmland, affording the country the highest irrigated-to-rainfed agricultural land ratio in the world (Ali et al., 2009). This has greatly reduced the flow to the delta and the sea, both in terms of the total amount of water and its annual distribution. Water flows to the delta started to decline when the Punjab irrigation system was developed in 1890s. But the construction of the Sukkur (1932), Kotri (1955), and Guddu (1962) dams resulted in a drastic reduction of water flow to the delta, a complete shift from agriculture to fisheries, coastal erosion, and the depletion of mangroves.

3 UPSTREAM, LARGE, MANMADE STRUCTURES

3.1 Political Geography

3.1.1 Ancient Civilizations

The Indus valley has been the cradle of ancient civilization. Archaeological findings indicate that the civilization of the Indus valley possibly antedated that of the Euphrates and the Tigris. Excavations reveal that the dwellers of Mohenjo

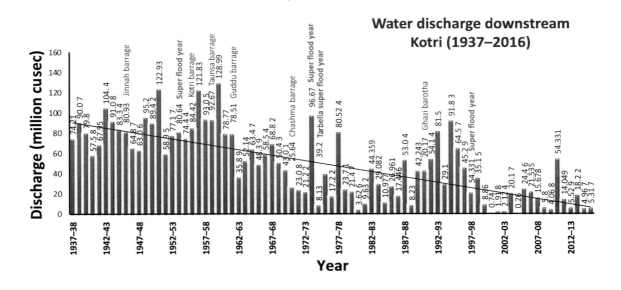

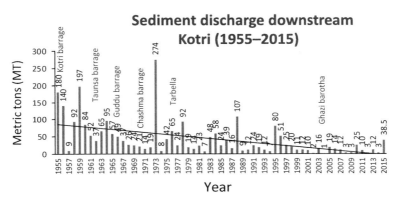

FIG. 2   The downstream water (million cusec), from 1937 to 2016, and sediment discharge (MT, 1 ton = 0.98 metric ton), from 1955 to 2015. Pre-1955 construction is not shown.

TABLE 1 Summary of the Catchment Area of River Indus Within Pakistan

Rivers	Catchment Area (km²)	Main Tributaries in Pakistan
Sutlej	75,369	(8 tributaries, with only 1 in Pakistan) Rohi Nullah, River Beas (largest river)
Ravi	24,960	Deg Nulah
Chenab	41,760	(12 tributaries) Palku Nullah is the largest
Jhelum	39,200	10 tributaries

Daro, Kot Diji, and Harappa settled in this region some 5000 years ago (began 3000 BC, thrived around 2000 BC and completely perished by 1000 BC; Quraeshi, 1974) due to the proximity of water and thrived on the waters provided by the Indus. They were powerful state entities of their time, which were able to construct large cities. The houses in these cities were provided with amenities such as bathrooms, lavatories, drainage, freshwater wells, and tanks—all of which indicate the availability of water. Ancient civilizations in the Indus valley were also based on a successful agriculture (Roberts, 1894; Quraeshi, 1974).

Conquering nomads from Central Asia entered the Indus Valley every five centuries: the Aryans (1500 and 1000 BC), the Persians, Scythians (500–400 BC), Ephthalites (white Huns, 100 BC). Alexander and his army entered through the northern mountain passes in 400 BC, which disturbed the Hindu civilization (the Battle between Alexander and Porus);

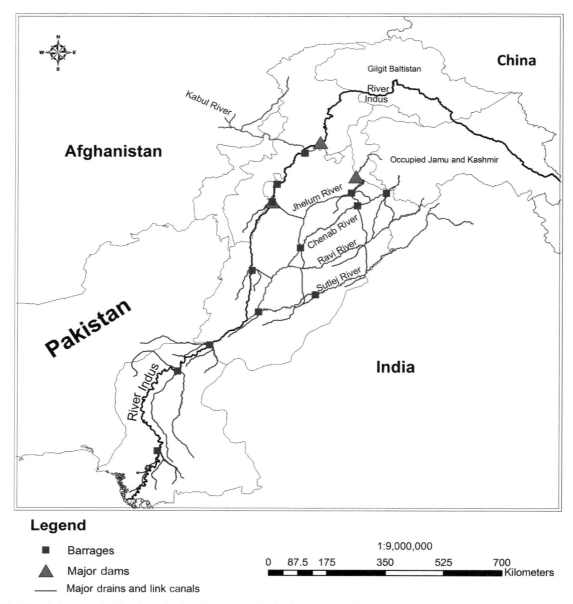

FIG. 3 The River Indus map, showing the major dams, barrages, and other large man-made structures.

later Alexander restored Porus's domain. This is recounted in the story of the Indus and the birth of what the Arab traders called the "Hindoostan" or "Hindia" (which later became the subcontinent of India; Quraeshi, 1974).

The weakening of the monsoons resulted in the extinction of this civilization (Possehl, 1997; Giosan et al., 2012). Nevertheless, until today, the waters of the River Indus and its tributaries are heavily utilized for irrigation in this relatively arid area and the river is a lifeline for the economy and culture of the region (Fahlbusch et al., 2004). Nearly 25% of the modern drainage comprises irrigated cropland (Economic Survey of Pakistan, 2016–17). The communities depend on the river for sustenance in many ways, and the high population density of the Indus basin (169 persons/km^2 in Badin and 65 persons/km^2 in Thatta (District wise Census Results—Census, 2017)) has resulted in major anthropogenic impacts to the river, its delta, and the coastal ecosystem.

3.1.2 The People and the River

The river and its catchment cover a vast area. Pakistan is an agriculture country; more than 70% of the total population is rural and connected to agriculture. The river is the lifeline of all activities and is closely connected to the people (Fig. 4).

FIG. 4 Communities living in the catchment areas of the River Indus. (A) Coastal areas near Badin. (B) Kharo Chan fishing boats. (C) Dhandhari-Gharo. (D) Typical businesses—betel leaf shacks and small "autaks," "baitak," or community center. (E) Present day Bhambhore and the mouth of the River Indus where Muhammad Bin Qasim's army entered Sindh (Subcontinent of India). (F) Soonda—early graveyard of a tribal war, remains of early carved graves. (G) Banana plantations near Sajawal during the 2015 floods. The river flooded the fields and adjacent areas. (H) Fishing boats. (I) Bamboo shacks on the riverbank. (J) Huts made from coconut matting and bamboos in the creeks of the River Indus Delta. (K) Fishers casting net in the creeks of River Indus Delta. (L) Estuarine nets locally known as "Bulla" illegal un regulated fishing gear in operations in the creeks catching indiscriminately. (M) Thatta's "Shehjehani" mosque built in the Mughal-era old capital of Sindh.

The part forming most of the Indus Delta is the Thatta District. The total population of males is 589,341 and females is 523,853, and the population density of the district is 64.1 persons/km². A total of 89% reside in rural areas. The population density in the coastal sections of the Badin district is 169 persons/km², and 84% is rural (District wise Census Results—Census, 2017).

The productive activities in the coastal areas can be divided into three broad classifications: Fishing and related activities, in which an estimated 90% of the population is involved, agriculture and forestry, in which 8% of the population is involved, and the services sector, which engages about 2% of the population (IUCN, 2003).

A sizeable population of local people live in the coastal belt in close proximity to the sea or to creeks. These coastal communities have long been dependent on the coastal resources to meet their demands of food, fodder, fuel wood, sea-salt, timber for their temporary hutments, and the generation of income and for economic activities. The ecological stability of the coastal areas is directly linked with the livelihood of the coastal security of coastal communities. The total population is around 3 million (2010), 90% of the population of the district is rural, and 93% of the coastal sections of the district reside in rural settings. In spite of the close proximity of large cities, the people of this area belong to the marginalized section of the society with 75% living below the poverty line, which is less than US $1 per capita per year (Baseline survey of coastal areas, Sindh Coastal Area Development Project). Each year as the river breaks its banks, the people suffer with loss of their land, crops, and animals.

3.1.3 Invaders (Early Muslims, British Raj)

The first invader to enter was Alexander the Great, who entered Sindh around 400 BC Subsequent invaders followed (Quraeshi, 1974). The land was rich and prosperous and attracted people to trade. The Indus was famous for its textiles, and salt trade flourished. The Muslims came to the Indian Sindh with the army of Muhammad Bin Qasim, a young general who was sent to India to teach the Hindu Raja Dahir a lesson for the treatment he had rendered to the Muslim women and children traveling onboard a trading ship carrying merchandise to Ceylon (present-day Sri Lanka). The present-day Bhambhore was then the mouth of the River Indus (Fig. 4E), and this is where Muhammad Bin Qasim's forces landed in early 8th century. Thatta was the Capital of this Umayyad kingdom (Quraeshi, 1974; Fig. 4M). This area is about 80 km east of Khobar Creek and the remains of Bhambhore tell tales of a glorious history. Islam flourished and spread to Western India from here through the mystics and saints, right up to present-day Multan.

3.2 Indus Treaty—British Raj and Independence

The 4000-year-old Indus civilization has its roots in irrigated agriculture, and canal irrigation development began in 1859 with the completion of the Upper Bari Doab Canal from Madhopur Headworks on River Ravi (WCD, 2000). Until that time, irrigation was undertaken through a network of inundation canals, which were functional only during periods of high river flow.

In the beginning of the 1900s, it became apparent that the water resources of the individual rivers were not in proportion to the potential irrigable land. The supply from the River Ravi, serving a large area of Bari Doab, was insufficient while Jhelum had a surplus. An innovative solution, the Triple Canal Project, was constructed during 1907–15. The project linked the River Jhelum, River Chenab, and River Ravi, allowing a transfer of surplus Jhelum and Chenab water to the River Ravi. The Triple Canal Project was a landmark in integrated interbasin water resources management and provided the key concept for the resolution of the Indus waters dispute between India and Pakistan in 1960 with the Indus Water Treaty (IWT).

The Sutlej Valley Project, comprising four barrages and two canals, was completed in 1933, resulting in the development of the unregulated flow resources of the Sutlej River, which motivated planning for the Bhakra reservoir. During the same period, the Sukkur barrage and its system of seven canals serving 2.95 million ha in the Lower Indus plain were completed and considered as the first modern hydraulic structure on the downstream River Indus. Haveli and Rangpur from Trimmu Headworks on Chenab in 1939 and Thal Canal from Kalabagh Headworks on Indus were completed in 1947. This comprised the system inherited by Pakistan at the time of its creation in 1947. The Indus Basin Irrigation System (IBIS), spread over the flat plains of the Indus Valley, is the largest contiguous irrigation system in the world (Fig. 2) and is the result of large surface irrigation schemes promoted by the British up to 1947, the time of the independence of the Subcontinent of India from colonizers.

At independence, the irrigation system, conceived originally as a whole, had to be divided between India and Pakistan without considering the irrigated boundaries. This resulted in an international water dispute in 1948, which was finally resolved by the enforcement of the Indus Water Treaty in 1960 under the aegis of the World Bank.

The Kotri, Taunsa, and Guddu barrages were completed on the River Indus to provide controlled irrigation to areas previously served by inundation canals. The Taunsa barrage was completed in 1958 to divert water to two large areas on the left and right banks of the river, making irrigated agriculture possible for about 1.18 million ha of arid landscape in Punjab province (Pakistan). Also, three additional interriver link canals were built before the initiation of the Indus Basin Project (IBP). The last inundation canals were connected to weir-controlled supplies in 1962 with the completion of the Guddu barrage on the River Indus. In pursuance of the Indus Water Treaty, the IBP included the Mangla dam, five barrages, one syphon, and eight interriver link canals, completed during 1960–71, and the Tarbela dam started partial operation in 1975–76. The two main components of IBP were the major storage reservoirs on Jhelum (Mangla) and Indus (Tarbela) to mitigate the effect of diverting the three eastern rivers by India and to increase agricultural production in the IBIS. Consequently, construction of Mangla Dam and Tarbela Dam started in 1968 and were completed by 1974, and they started partial operation in 1975–76 (WCD, 2000).

As a result of these extensive developments on the River Indus and its tributaries, Pakistan now possesses the world's largest contiguous irrigation system. The full control-equipped area of 148,000 km^2 (2008) (36% under Mangla and 64% under Tarbela) encompasses the Indus River and its tributaries. More than 95% of irrigation is located in the River Indus basin. The total area equipped for irrigation throughout Pakistan was estimated at 200,000 km^2. The total water-managed area in Pakistan is around 210,000 km^2. In the Indus basin, irrigated agriculture increased, 36%, 44%, 39%, and 52% for wheat, cotton, rice, and sugarcane, respectively. The overall increase in the cropped area was around 39% (WCD, 2000). However, changes in the timings of the monsoons are already adversely affecting the agriculture yields in Pakistan (Gowdy and Salman, 2011).

There is a lot of political bickering over the construction of dams, and tugs-of-war between provinces are getting nowhere. Plans remain just plans; there have been no major dam-building projects since the 1960s. The country faces water scarcity day by day (storage capacity of 30 days) and with the population growth over 2.5% (District wise Census Results—Census, 2017) and more mouths to feed, this situation is worrisome.

3.3 Transboundary Water Issues

The Indus Water Treaty was signed in 1960 to resolve the water dispute between India and Pakistan, but to date, water from the Indus remains one of the unresolved issues of the two countries. This was protracted from the 1948 water dispute when India unilaterally cut off supplies to Pakistan canals originating from the headworks located on the eastern rivers of River Ravi and River Sutlej, thereby asserting its right to the waters of three eastern rivers (River Ravi, River Beas, and River Sutlej). This seriously disrupted Pakistan's water resources development plans (WCD, 2000). India's plans to construct dams upstream to store water from the River Indus, is a cause for concern for the water-deficient Pakistan. The lack of confidence between the neighbors, and the erratic and moody policies of ever-changing government, serves as a cause of tensions on both sides.

3.4 Distribution—Vested Interests and Resolving Internal Conflict

In addition to the upstream dam constructions that have greatly reduced the flow to the downstream deltaic area and the sea, the feudal-style landowners, backed by economic and political power, have managed to procure and maintain huge public subsidies on irrigation water. Roberts (2017), quoting Shamsul Mulk, former Chairman of the Water and Power Development Authority (WAPDA) in the country, says that the "water policy is simply nonexistent in Pakistan. Policymakers act like 'absentee landlords' over water"; "massive corruption" in the water sector and warnings exist regarding "profiteering from the scarcity of a vital resource sector" (http://www.independent.co.uk/news/world/pakistan-droughts-2025-warning-water-levels-a7949226.html).

In addition to capital subsidies on irrigation projects, there is a massive recurring subsidy on irrigation water covering nearly 60% of the operation and maintenance costs of the country's irrigation systems (Hussain et al., 2011). Consequently, water-use efficiency in the Indus Basin has remained low, ~35% (Hussain et al., 2011). Sound water management strategies are required to minimize water losses. One of the solutions proposed was the lining of the irrigation canals; in the early 2000s this was a proposed national project. This is feasible only in places where the groundwater quality was poor and it was expected that water seepage from the irrigation canals would result in mixing with poor quality groundwater that would be damaging for agriculture use (personal communications with the EXN Kotri Barrage, 2017). In the water management process at the national level, coastal areas are often neglected. Allocations for the coastal areas are determined after accounting for all other needs. Over the past 50 years, all additional requirements for irrigation water have been met through the increased diversion of water upstream, thereby reducing the flows to the delta. While the WAA 1990 guarantees 10 million acre per foot (~5 million cusec) for the delta (Table 2), this allocation is not based on any scientific or need-based assessment and therefore nearly always ignored. There is a mindset that propagates that any water that goes to the sea is

TABLE 2 Provincial Wise Distribution According to the Water Accord 1991 (Million cusec)

Province	Summer Monsoon	Winter	Total
Punjab	18.54	9.44	27.97
Sindh	16.97	7.41	24.38
Khyber Pakhtunkhwa	1.74	1.15	2.89
Civil Canal	0.90	0.60	1.50
Balochistan	1.43	0.51	1.94
Total	38.67	18.51	57.18

a waste. Conflict among the upstream and downstream provinces remains unresolved. The downstream provinces often complain that as the upstream provinces consume their water needs, there is hardly any water left to flow downstream. This is true especially during the dry years, as agriculture downstream suffers and dependence on groundwater increases, thus resulting in aquifers drying out in the coastal areas and compaction of the deltaic areas, further adding to the issues of land subsidence and seawater intrusion. A proposed solution is to revise the Water Accord and bring the figures from absolute to a percentage distribution that will result in a somewhat fairer distribution of water. Building of smaller reservoirs to store rainwater and to allow charging of coastal aquifers is proposed especially as a measure to mitigate land compaction.

In a failure to improve water-use efficiency, additional cultivation of land reduces the flow downstream. The government has not succeeded in establishing a water market, due to the pressure from the agricultural lobby. Since the agricultural lobby dominates the assemblies in the country, these pressures will continue to be ignored and the inefficiencies in the use of water are likely to continue. It is quite evident that, unless appropriate policy measures are taken to increase water-use efficiency, population growth and increased demand for irrigation water will continue to create a shortage in the Indus Delta. Adequate supplies of fresh water and silt to the delta region are critical for the stability of the coastline, the health of mangroves, and the region's biodiversity. Any reduction will restrict the mangroves in performing their vital role as primary producers in the ecosystem.

The perhaps unnecessary development projects supported by the International Financial Institutions (IFIs) have not only generated huge economic waste, but have also caused irreparable damage to the environment and livelihoods, especially in the lower reaches of the River Indus. The flooding caused by alterations in the course of water flows has forced communities to migrate and has pushed people into a vicious cycle of deprivation.

The Left Bank Outfall Drainage (LBOD) project, initiated in 1984, aimed to provide a drainage facility for irrigated agriculture in three districts of Sindh covering about $5160\,km^2$ through the construction of a network of surface drains. It was considered that drainage would improve the productive capacity of farmland and the quality of vegetation, while reducing malaria. Earlier environmental assessments had indicated positive effects for the project, and these have been rendered totally misplaced. The reason for this is that the analysis of the sustainability of the infrastructure projects had failed to take into account the needs and risks facing local communities in the first place. The implementation of the project was disastrous (Talpur et al., 1999). The other side of the story narrates that the LBOD was designed for a drainage capacity of 4600 cusec during the flood seasons, and the much larger volume of water was left in the drain far beyond the capacity of the LBOD, which resulted in the breaches (personal communications with EXN Kotri Barrage, 2017). There is hardly any accountability at both the state and IFI levels.

4 CATCHMENT AREAS

4.1 Agrogeography

Agriculture is the leading sector of Pakistan's economy. Over two-thirds (67%) of the population live in rural areas and depend mostly on agriculture for their livelihood. The majority (60%) of the geographical area of the Sindh province is arid (http://sindhforests.gov.pk/rangeland; 2016). A total of $140,000\,km^2$ is cultivable (http://pakistaneconomist.com/issue2000//issue25/i&e2.htm). The country continually faces scarce surface water supplies, and irrigation water is considered the lifeblood of the economy. Cultivated areas contribute significantly to the national agriculture production (Statistics Bureau of Pakistan, 2016).

Projections suggest that future temperatures in South Asia will increase by 2–3°C between the years 2046 and 2065. Rainfall is unpredictable, with estimates indicating that annual rainfall will increase over time. Pakistan is expected to be one of the most affected of the countries in South Asia by climate change (Stocker et al., 2013). Increases in average temperature may be beneficial for some crops or changes in precipitation may affect importance across crops (Siddiqui et al., 2014; Sultana et al., 2009). In the coastal district of Keti Bundar, vegetables, betel leaf, sugarcane, wheat, fruits (chiku, banana, mango, and watermelon) are grown in the inland area. Floods and draught years impact these plantations, sometimes flooding the fields when the river breaks the banks (Fig. 4).

4.2 Agrarian Economy and Dependence on the River

The drinking water for much of India and Pakistan comes from the Himalayan, Karakoram, and Hindu Kush glaciers that are already beginning to melt from warmer temperatures (Jianchu et al., 2007). Climate models indicate that this melting will accelerate in the coming years, with unknown but severe consequences on drinking water, agricultural irrigation, and human health. Increasing salinity in coastal regions from rising sea levels and more severe storms has already taken a toll on coastal communities (Rees and Collins, 2004). Drinking water is scarce in some of the coastal areas of Thatta and Badin; the aquifers have become saline due to the seawater intrusion.

South Asian economies are heavily dependent on agriculture, which is the economic sector most vulnerable to climate change. The main effects on human wellbeing from the physical effects of climate change include loss of agricultural production. The International Panel of Climate Change (IPCC) Chairman Rajendra Pachuri commented that the decline in the wheat production in India was a result of climate change (Worstall, 2007, quoted in Gowdy and Salman, 2011). The availability of water is crucial to the resilience of Pakistan's agricultural sector. Qureshi et al. (2010) argue that the exploitation of groundwater resources has enabled farmers to increase production levels and also to cushion themselves against some of the low rainfall. Qadir et al. (2014), however, highlight the unsustainability of water management in the Indus Basin by studying the impacts of soil salinity on crop yields, and they argue that agricultural productivity losses due to salt-induced land degradation are a growing concern across in Pakistan. Changes in the timing of monsoons are already having an adverse effect on agriculture in Pakistan, and a decline in the yield is generally observed. The untimely rains in Punjab and Sindh have adversely affected the standing crops (Fig. 4).

4.2.1 Riverine Ecosystems

The natural spawning grounds have been lost for commercially important species in the upper and lower Indus plains. The recorded statistics of inland fisheries, both in Sindh and in Punjab, show an increase in fish production over the years, but this is due to increases in aquaculture rather than to increased catches of the indigenous fish from the river.

4.3 Brackish Lakes—Issues and Solutions

There are 27 freshwater and brackish lakes, nine of which are designated RAMSAR sites of the province of Sindh (Box 1, map and list of RAMSAR sites in Sindh). Coastal communities that live nearby depend heavily on the fishing in these lakes. Several of these coastal lakes (western catchment) constituted complexes and clusters (Fig. 5A, satellite image), maintaining their individual identity and unique ecology. The construction of the Left Bank Outfall Drain (LBOD) and its subsequent consequences has over the years resulted in the decline in the lakes' total covered area (Fig. 5B) and the cyclone 2A (1999) made 52 breaches in the LBOD, resulting in the lakes all becoming one large water body that was hypersaline (higher than 45 PSU) (Fig. 5C).

In 2007, a study was conducted to evaluate the ecology in the brackish lakes around the Thatta, Badin District of Sindh. This study showed that the seawater had intruded upstream and the ecology of the brackish lakes that was once presumably fresh water to brackish now supported a mixed-type ecology. The freshwater species as well as purely marine species were reported (NIO archives, unpublished data). Similar observations were made in the Khobar Creek during the 2013–14 Creek Survey Program when marine species are reported from the river during the dry year (2014) and the seawater had moved the estuary further 35 km upstream (Kidwai et al., 2016). Obviously, this fluctuation in the salinity and the estuary would have an impact on the biological productivity in the deltaic creeks and coastal waters (Kidwai et al., 2016).

5 RIVER INDUS DELTA ECOSYSTEM

The average value of wetlands, lakes, rivers, and estuaries ranged from \$8500 to \$22,800 (USD) ha^{-1} year^{-1} compared to the terrestrial ecosystems ranges, which were from \$90 to \$2000 ha^{-1} year^{-1} (Costanza et al., 1997). If a delta is completely

BOX 1 Wetlands of Sindh

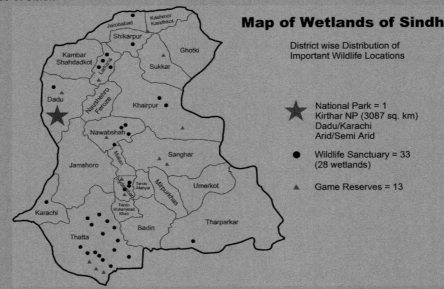

Map of Wetlands of Sindh

District wise Distribution of
Important Wildlife Locations

★ National Park = 1
Kirthar NP (3087 sq. km)
Dadu/Karachi
Arid/Semi Arid

● Wildlife Sanctuary = 33
(28 wetlands)

▲ Game Reserves = 13

The seven sites listed in the RAMSAR list of Wetlands of International Importance (Table B.1).

TABLE B.1 RAMSAR List of Wetlands of International Importance in the Province of Sindh

Name of Site	Location	District	Surface Area (Hectare)	Wetland Type	Recognition RAMSAR Site No. (RS#)
Keenjahar (Kalri) Lake	24° 56'N 68° 03'E	Thatta	13,468[a]	Freshwater lake	1976 RS# 99
Drigh Lake	27° 34'N 68° 06'E	Larkana	164[b] 182[a]	Slightly brackish lake	1976 RS# 100
Haleji Lake	24° 47'N 67° 46'E	Thatta	1704	Artificial freshwater lake	1976 RS# 101
Hub (Hab) Dam	25° 15'N 67° 07'E	Bordered between Sindh and Balochistan	27,000[b] 27,219[a]	Water storage reservoir	2001 RS#1064
Indus Dolphin Reserve	28° 01'N 69° 15'E		125,000[b] 44,200[a]	River	2001 RS#1065
Jubho Lagoon	24° 20'N 68° 40'E		706	Brackish lagoon, mudflats, marshes	2001 RS# 1067
Nurri Lagoon	24° 30'N 68° 47'E		2540	Brackish lagoon, mudflats	2001 RS# 1069

[a] *Scott and Poole (1989).*
[b] *Website.*

Threats to Wetland Habitats

The major threats to wetlands are expressed as percentages of sites by the World Conservation Monitoring Centre (1992), and the activities that contribute toward the loss of the resources are: hunting and allied activities, human settlement, drainage of agriculture, disturbance from recreation, reclamation for urban and industrial development, pollution, catchment degradation, diversion of water, soil erosion and siltation, etc.

Wetlands in Pakistan, Sindh, face a variety of threats, some of which are identified above. Nevertheless, primary among these is the lack of proper management and ignorance of the importance of healthy wetlands. However, the view that the crisis of the aquatic environment is basically an economic issue has been widely recognized by many international fora and organizations. The UNCED held in 1992, to which Pakistan is a signatory, recognized the environment crisis at two levels; at the Global level, resulting from greenhouse effects, warming of the climate/seas, and sea-level rise, and other localized perturbations resulting from indiscriminate destruction of natural wetland resources and the negative feedback from unplanned development processes, leading toward deterioration of the aquatic resources. Despite Pakistan's great reliance on wetlands, these are still being degraded at an alarming rate, more specifically in the Indus Basin. Potable water supply to communities is provided by some of the wetlands. Example, lakes such as Haleji Lake is harnessed for its freshwater resource to supply water to the city of Karachi. Other wetlands, like dams, also provide similar services. The discharge of sewage, effluent, irrigation, and industrial waste into the aquatic ecosystems in Pakistan has become a common phenomenon. The organic sewage load depletes oxygen levels in enclosed water bodies and so reduces the diversity of animal and plant life. As a result, the water quality of deteriorates and ultimately destroys the wetland habitat.

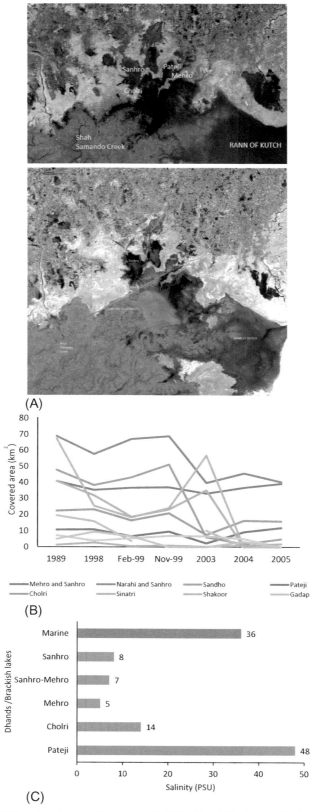

FIG. 5 (A) Precyclone (February 23, 1989) and postcyclone (November 26, 1999) 2A satellite images. (B) Decline in the average area cover of major brackish lakes. (C) Average salinity (PSU) in the major brackish lakes.

converted for human use, the economic value may be higher, but such a conversion will likely make the delta unsustainable in the longer term (Day et al., 2012). The Nile, for example, has been converted to urban development and agriculture. There is no longer a functioning delta and no feasible way to return to one. In the Ebro and Mississippi Deltas, the economic system is closely entwined in and dependent on delta systems that still retain significant natural areas (Day et al., 2012). The River Indus Delta, its creek system, and the mangrove stand are of major economic and ecological significance to the entire coast (Kidwai et al., 2016; Salik et al., 2015).

The mangrove forests serve as major breeding grounds for shrimps, prawns, and other economical species of fish (Bouillon et al., 2008; Reef et al., 2010). A healthy mangrove forest helps to keep this complex ecosystem in its ecological balance and can provide sustainable supply of products and services to meet the growing demands of the coastal communities and the adjacent urban settlements. The socioeconomic conditions of coastal communities are reflected by the health of the ecosystem (Badjeck et al., 2010).

Besides the socioeconomic dependence of the communities that exist around this area, the delta and its ecosystem support a rich biodiversity, is a potential Marine Protected Area, and part of it is a Nature Reserve and RAMSAR site (Box 1). A large number of migrating waterfowl use the area for feeding and breeding in the winter. Approximately 49 species of birds have been reported from the Sindh coastal waters (Ali et al., 2016). The most common are the Gull-billed Tern, Oystercatcher, Sand Plover, Golden Plover, Kentish Plover, Sanderling, Dunlin, Marsh Sandpiper, Curlew, and Whimbrel (WWF-Pakistan, 2007; Ali et al., 2016). If the mangrove go, all this is likely to go too, as well as the fishery on which more than 70% of the population depends (Kidwai et al., 2016).

The mangrove forests in the River Indus Delta once covered the entire delta region, and it was the fifth largest mangrove stand in a semiarid location in the world (Snedaker, 1984). Eight species were reported from the delta during the surveys in 1960–61 (Saifullah, 1982), and four species, including *Avicennia marina, Ceriops tagal, Rhizophora mucronata*, and *Aegiceras corniculata*, existed in the Indus Delta (Meynell and Qureshi, 1993; Qureshi, 1996). It was recently ranked as the fifteenth mangrove stand in the world in terms of size, composed of fewer species (Saifullah, 2017). The rate of degradation of mangrove forests in the delta was estimated to be 6% per year between 1980 and 1995 and only a small percentage are now considered to be healthy (WWF-Pakistan, 2007).

The area of the active delta has consequently shrunk over the years. Studies suggest that mangrove forests cover about 2230–2500 km^2 in the Indus Delta (Mirza et al., 1988; IUCN, 1999–2000). This cover was reduced by half (800 km^2) when WWF Pakistan surveyed the mangroves in 2002 and 980 km^2 is reported also Giri et al. (2015), in Saifullah (2017).

Three species, *Avicennia marina, Ceriops tagal*, and *Aegiceras corniculata*, are now naturally present in the Indus Delta's four open accesses to the sea creeks (Hajamro, Khobar, Dabbo, and Wadi Khuddi), the dominating species is *A. marina* (NIO Pakistan data archives, unpublished data, 2013–17). *Rhizophora macronata* are transplanted on some area of Shah Bundar and Keti Bundar.

The production of fish and shrimp in the Indus Delta and its adjacent coastal waters is about 4000 metric tons of fish per year, 2000 metric tons of shrimp per year, and 2500 metric tons of crabs per year (Hoekstra et al., 1997). This is a considerably high fishing pressure with the present level of exploitation from a relatively smaller area. Most of the fishing for small-sized fish, shrimp, and crabs is carried out within the mangrove ecosystem and in its immediate vicinity in the adjacent open sea.

The findings of the Fishery Resource Appraisal Program (2013–14) from the deltaic creeks and (2009–15) from the shelf areas show a general decline in the fish catch per effort, decline in the size of the fish and shrimp caught, and a decline in the number of species of commercial fin and shellfish (Fanning et al., 2009, 2010, 2015; Kidwai et al., 2016).

6 IN THE LAST 50 YEARS

The bioresources face several threats in the River Indus deltaic ecosystem (Table 3) and some ecological aspects of the River Indus Delta from the 1960s to 2017 have been reported.

7 THE DELTA FACES CLIMATE CHANGE (VARIABILITY IN ARABIAN SEA MONSOON)

Deltaic sustainability becomes a bigger challenge in the face of severe climate impacts. CO_2 concentrations in the atmosphere now exceed 400 ppm, a dramatic increase over levels in the late 19th century that were 280 ppm (IPCC, 2007a,b). These are the highest CO_2 levels of the past three million years, and CO_2 levels are now tracking at the worst-case IPCC scenarios (Friedlingstein et al., 2014). Karl et al. (2015) conducted an analysis of surface temperatures and reported that global temperature trends are higher than reported by the IPCC and that there was no slowdown in the increase in global temperatures in what has been called the global warming "hiatus" (IPCC, 2013a).

TABLE 3 Threats to the Species From the River Indus Delta

Resource		Estimated Potential	Current Exploitation	Remarks/Threats
1	*Mangrove forest*	Mangrove Forest = 680 km²	No recent figures Loss of biodiversity, 4 species not observed	Natural and anthropogenic (N&A)
2	*Fisheries*	Fish = 3000 mt/year Shrimp = 2000 mt/year Crabs = 2500 mt/year	4000 mt/year (approx.) 3000 mt/year (approx.) 500–3000 mt/year (approx.)	Overexploitation, illegal gear, environmental stress
3	*Biodiversity*	NQ	Selective exploitation of commercial/edible species	Habitat loss, coastal development, overexploitation
	Marine plants	Mangroves—4 species Seaweeds—NQ	Mangrove—*Avicennia marina* dominates	Fuel, fodder, building material
	Fish	160 species	NQ	N&A
	Shrimp	15 species	NQ	N&A
	Crabs	20 species	NQ	N&A
	Birds	60 species	NQ	N&A
	Reptiles	13 species	NQ	N&A
	Mammals	6 species	NQ	N&A

A, anthropogenic; N, natural; NQ, not quantified.

Modified and updated from Amjad, S., Rizvi, S.H.N., Memon, G.M., Memon, M.Q., 2002. Indus Mangrove Ecosystem Study to Evaluate the Impact of Coastal Resource Use and Environmental Degradation on the Coastal Communities Living in the Mangrove Ecosystem at Korangi-Gharo Creek System. PNC-UNESCO, NIO (June 2001–April 2002).

Climate change is affecting deltas directly at the level of the entire drainage basin. Basin level impacts include both increases (Mississippi) and decreases (Colorado) in projected precipitation and river discharge. Between 1901 and 2010, global mean sea-level rise averaged 1.7 mm/year compared to 3.2 mm/year between 1993 and 2010 (IPCC, 2013b). High rates of relative sea-level rise are expected to occur as a result of subsidence and accelerated eustatic sea-level rise (Meehl et al., 2007; IPCC, 2013a), more severe storms (Kaufmann et al., 2011), drought (IPCC, 2007a,b), more erratic weather (Min et al., 2011), and other factors.

More than 10 million people were displaced from their coastal residences in 1990 due to storm surges and river flooding; this figure rose to 16–23 million by the 2020s (20 million in 2013; Nicholls et al., 2010; Hinkel et al., 2014).

The regional climate is arid to semiarid with seasonal precipitation and significant variability. Mean annual rainfall is low, ranging from <100 mm over the lower plains to about 500 upstream in Lahore (Pakistan Meteorological Department, 2010). Pakistan has the world's fourth highest rate of water use but is dependent on water from a single source—the Indus River basin in India—and rainfall has been steadily declining (http://www.independent.co.uk/news/world/pakistan-droughts-2025-warning-water-levels-a7949226.html). Rainfall is much higher in the mountains, reaching almost 2000 mm in the frontal Himalayan Ranges. About 60% of precipitation is received during the southwest monsoon (July–September). The summer temperature everywhere in the plains is high, rising above 40°C, resulting in a high evaporation rate. The mean annual evaporation in the upper Indus plain is more than 1500 mm, a figure that rises over 2000 in the lower plains. Decrease in freshwater flows impacts the soil productivity and land use patterns (Dehlavi and Adil, 2012).

In a period of 100 years, 51 severe cyclonic storms have been reported from the Arabian Sea, 4 cyclones landed at Balochistan Coast, and about 15 cyclones landed on Sindh Coast. May through June, and October through November, are favorable periods for the formation of cyclonic storms in the Arabian Sea (Fig. 6).

Floods and cyclones on the Pakistan coast have caused large-scale devastation. The coastal ecosystems are most vulnerable to such erratic weather patterns. Communities living around the River Indus and its delta are marginalized (Dehlavi and Adil, 2012; Naveed and Ali, 2012) and are highly vulnerable to natural calamities (Zaheer et al., 2012). The deltaic ecosystem provides services such as support to fisheries, forestry products, stabilization of shorelines, etc., and livelihoods to coastal communities. Thus this ecosystem has direct socioeconomic implications. The 160-km-long coast of the Sindh adjacent to the Left Bank Outfall Drain (LBOD) in the Thatta and Badin districts received a direct hit from the 1999 big

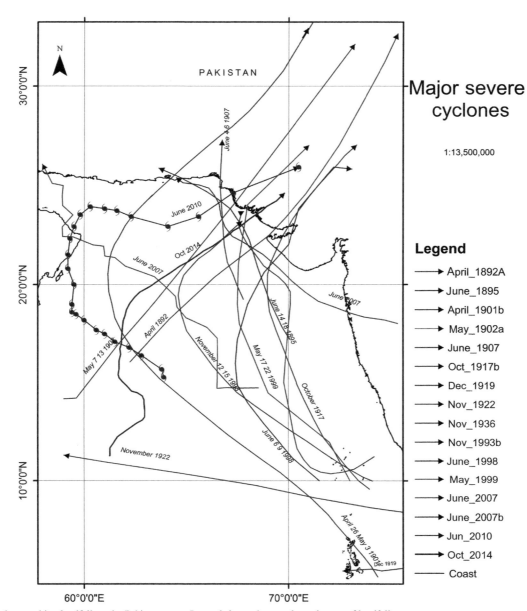

FIG. 6 Cyclone making landfall on the Pakistan coast. Legend shows the months and years of landfall.

cyclone TC02A. The 1999 Cyclone 2A hit the Thatta, Badin district, and caused ecological damage as well as affected the local populations who were forced to relocate, loosing large areas of habitable land that was inundated with sea water, rendering it unfit for agriculture. The cyclone 2A (1999) claimed 118 lives, destroyed 3758 villages and 38,509 houses, and damaged $1158\,km^2$ of agricultural crops (GoS, 2008).

The floods of 2010, designated as "super floods," played havoc with the ecology and the lives of people in Punjab and Sindh including the coastal zones on which the coastal fisher communities heavily depend. The River Indus flows of fresh water and sediment downstream (Fig. 7), and sea intrusion inland makes the delta ecosystem highly dynamic and somewhat complex and therefore unpredictable. After the 2010 flood, it was estimated that the loss in the district of Thatta ranged to approximately US $1840 million and the reconstruction cost was estimated at US $2747 million (World Bank and ADB, 2010). Kamal et al. (2012) describe the issues very openly regarding the politics of floods and the issues surrounding their management.

8   SAVING THE DELTA AND ITS PEOPLE

Deltas are among the most productive and economically important of global ecosystems and are most threatened by human activities (Day et al., 2012; Costanza et al., 1997; Syvitski et al., 2009; Giosan et al., 2014).

FIG. 7   100 km upstream at Sajawal Bridge during the dry season and the flood season each year.

The human population in and around mangrove forests on the coast of Pakistan is estimated to be about 1.2 million. Nearly 900,000 reside in the Indus Delta. The number of households is estimated to be about 140,000 in the Indus Delta. Over 90% of the population is directly or indirectly engaged in fishing. High returns associated with fishing are causing rapid population growth. On average, the population in the coastal areas has been growing at a rate of 6%–8% annually throughout the last 10 years. Migrants from other areas of the country, from Bangladesh, and from Burma, who come mostly to the Indus Delta, have contributed to this growth.

Water use for irrigation accounts for a major proportion of the volume of river abstractions. Over the last 70 years a series of dams, barrages, and irrigation schemes have been built in upstream areas of the Indus, and they are used to feed more than 80% of the country's irrigated farmland.

Land in the area has become unsuitable for agriculture, and potable water sources have become very scarce or have disappeared altogether. In Thatta District, which is located on the mouth of the delta, mangrove areas have suffered heavy destruction; almost a third of the land has been affected by saltwater intrusion and about 12% of cultivable land has been lost. A total of 77% of households solely depend on fishing (Dehlavi and Adil, 2012), in Keti Bundar (District Thatta) alone 44% of households depend on farming, 39% on fishing, and 17% are daily-wage laborers and service providers for fishing and farming (Salik et al., 2015).

Most of the socioeconomic structure, local customs, and traditions in practice by coastal communities in Indus Delta also reflect the integration of their traditional knowledge of coastal resources. The "Shah Jo Risalo," a poetry collection of Shah Abdul Latif Bhittai, describes every aspect of the coastal region, as appears in "Sur Samondi" and "Ghatu" (Shark-Hunters[1]) verses about the concepts of ocean and river and the peoples related to the conservation of resources. This alone reflects the close association of the people to the river and the value it holds in their everyday affairs, general survival, and overall well-being.

9   FISHERS AND FISHERY FROM THE DELTA

The marine landings of the Sindh province go back to 1947, and the record of the fishing vessels (types and numbers) was available from 1955. The total marine catch for that year was compared to the catch per boat. In 1955, the total number of vessels operating in Sindh was about 1000 (1009), comprising no mechanized boats, whereas in 2008, the total numbers of vessels were 17,000 (17,044) of which ~4500 (4475) were mechanized (Anon, 1976–77, 1986, 2012). The total landing in Sindh was 23,910 MT in 1947, the fishing in the EEZ began in 1982 (which explains the increase in fish landings; Fig. 8A). In 2008, the landing recorded was 223,034 MT. Milliman et al. (1984) suggested that with the same stock available to a larger number of boats, the catch available per boat had declined (Fig. 8). Generally, with a 17-fold (mixed-type fishing vessel) increase in the numbers of vessels, a fourfold decline in the fish catch resulted. The declining trend in catch per vessel continues (Fig. 8B).

More than US $100 million per annum are earned through fishery production from these ecosystems (Saifullah, 1997). The fishery in the delta is generally indiscriminate; fish are caught by the local fishermen through the operation of purse-seine nets (locally known as Katra nets) and the installation of large estuarine set-bag nets (locally known as Bulla Gujja) (Fig. 8). Very small, and illegal, mesh sizes (3.75–7.5 mm) are used in the netting of these gears (Waryani et al., 2015), which retain every life form that enters in the range of these fishing gears, including juvenile and undersized fishes (Ahmad and Hasan, 1997).

---

1 Ghatu XXIV (Shark Hunters) extracts: "Where fishers used to seek the fish, the barren sand-dunes lie; Fish-sellers ruined, the river dry; and tax collector gone; You throw the nets in creeks … not so the sharks are ever killed; In search, they into whirlpools got and to fathomlessness … They killed the shark; with happiness now beam fishermen's eyes." Shah-jo-Risalo, Translated by Elsa Kazi, accessed through https://sindhiproverbs.files.wordpress.com/2009/02/shah-jo-risalo3.pdf.

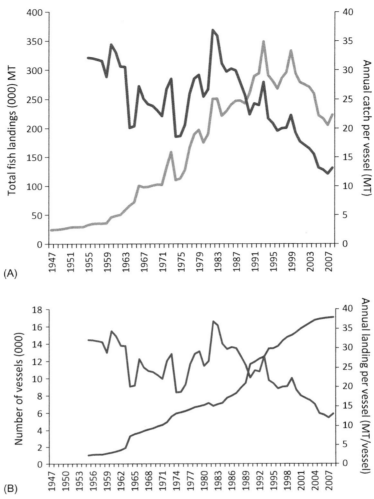

FIG. 8 (A) Total fish landings and landing catch per vessel (catch effort) along the Sindh coast. (B) Number of fishing vessels and landing catch per vessel *(Data from MFD-2010, 1 ton = 0.98 MT.)*

In the postmonsoon season, around 100–200 beach seine and estuarine bag nets regularly operate in the surrounding areas of Keti Bandar (Fig. 8). These nets are illegal, and fishing in the delta during the 2 months of the summer monsoon period is the closed season. Nevertheless, the use of these nets is "common." Local fishers insist that this destructive gear fishing is done by the nonlocal "Bahari" or "outsiders" from the upcountry who are not fishers and are there only for business and profiteering. According to the locals, they have no regard for the resources since they have no roots in the soil. When queried about why the local "bona fide" fishers do not drive these outsiders away, the local fishers inform us that these have the strong backing of influential and rich people from the area (personal conversation with the local fishers).

10 DEPENDENCE ON THE RIVER

The water of the Indus has been used for six millennia, from the Harappan period to the present, through a series of different historical regimes and often in an organized fashion. The last half of the 20th century, however, has seen the transfer to very large-scale management of the water system. Currently, about 60% of the Indus water is estimated to be used for irrigation, supplying water to more than $161,800 \, km^2$, or about 80% of Pakistan's agricultural fields (Iftikhar, 2002).

In some of the major river basins, there is a great demand for water for human use while climate change will lead to further drying in most of these areas. For the Colorado, Indus, Nile, and Tigris-Euphrates rivers, little fresh water regularly enters the sea. Even though the upper Nile basin is wet, the lower basin is dry, and little water reaches the Mediterranean. For the Colorado and Indus Deltas, hypersaline conditions due to freshwater reductions have led to widespread wetland death (Snedaker, 1984; Milliman et al., 1984; Giosan et al., 2006; Syvitski and Brakenridge, 2013).

Over the last decade, concern has grown over the ways in which human activities have altered the mangrove ecosystems of Pakistan. Fresh water scarcity due to upstream diversions of river flows for agriculture, water pollution, overgrazing, cutting for fuel wood and timber, and unsustainable fishing levels are seen as the main factors associated with biodiversity loss in the mangrove forests on the coasts of Pakistan.

Pakistan is an energy- and water-deficient country, the Pakistan Council of Research in Water Resources (PCRWR) made a grim forecast in a report that claimed that the country touched the "water stress line" in 1990 before crossing the "water scarcity line" in 2005 (http://www.independent.co.uk/news/world/pakistan-droughts-2025-warning-water-levels-a7949226.html). The recent flash floods as a result of extreme rains in India and Pakistan have caused plans to gain momentum for new dams and the need to store water and to use this water for energy generation. The Daimer-Bhasha alone could generate 440,000 MW of electricity and could provide a storage capacity of $7,900,000,000 \, m^3$ (https://www.pakistantoday.com.pk/2017/05/13/china-inks-mou-to-invest-in-five-biggest-dams-in-indus-river-cascade). The debate of progress and the national development versus environmental well-being and ecosystems is long and difficult, seemingly one without a win-win solution.

11 WHAT TO SAVE FIRST? WHAT WILL WORK—POLITICAL WILL OR MANAGEMENT STRATEGY?

Ecosystem values are poorly understood, rarely articulated and frequently omitted from decisionmaking

(Emerton and Bos, 2004).

The importance of the delta for maintaining the integrity of the shoreline is without question. Mangroves are important toward maintaining the shoreline; they are productive ecosystems. More than a million acres of the Indus Delta have been lost to the sea (Memon, 2004), and mangrove cover decreased from 2500 km² (Mirza et al., 1988) to about 980 km² (Giri et al., 2015) within the past few decades (Saifullah, 2017). The significance of the flow of fresh water and sediment is critical to maintaining the delta, keeping a check on seawater intrusion, and maintaining a productive and broad shelf off Sindh.

Decision-makers and planners in the water world and in other development and economic sectors have traditionally paid little attention to the economic value of ecosystems. The role of ecosystems in water demand and supply has persistently been undervalued in economic terms, and the economic or financial costs arising from their degradation and loss is often ignored.

There is a general lack of understanding of ecosystems in relation to water, and this is frequently not taken into consideration when decisions are being made. Conventional economic analysis decrees that the "best" or most efficient allocation of resources is one that maximizes economic returns. Inadequate freshwater flows are allocated to downstream ecosystems such as the Indus Delta, because they are not considered as productive water uses when compared to the immediate short-term benefits of irrigated agriculture.

12 STAKEHOLDERS—COMING TOGETHER

Multiple human-induced stressors are also impacting the Indus River Delta in Pakistan (Kidwai et al., 2016). These stressors are the severe decrease of freshwater flow in the Indus River as a result of the extraction of water for irrigation and for industrial and potable water. This is causing hypersalinity, is destroying coastal wetlands on which the coastal fisheries depend, and is impacting the direct exploitation of the delta ecosystems and its coastal fisheries by local delta communities and external users. Proposed remedial measures include the creation of Marine Protected Areas and ecosystem-based management of the Indus River Delta resources.

The community champions need to be educated and the "concepts" of protected areas must be elaborated upon. There seems to be some confusion that, if protected, the access to these resources will be stopped for all and sundry to which they seem to have a birthright. The community thinks that their right to the resource has been usurped and that they are instigated by these community champions to retaliate. This mindset is not likely to benefit anyone, but will destroy the resource forever. The fisher communities are simple folks with limited means and education, and they often play into the hands of the exploiting few who tend to reap individual gain. It is the responsibility of all sensible quarters to help restore confidence between the communities who are the natural custodians and the managers and who are tasked as custodians of the state.

REFERENCES

Ali, G., Hasson, S., Khan, A.M., 2009. Climate Change: Implications and Adaptation of Water Resources in Pakistan, GCISC-RR-13. Global Change Impact Studies Centre (GCISC), Islamabad, Pakistan. Available from: https://www.researchgate.net/publication/274026294_Climate_Change_Implications_and_Adaptation_of_Water_Resources_in_Pakistan.

Ahmad, M.F., Hasan, A., 1997. Effect of Bhoola and Katra fishing nets in the Indus delta Pakistan. Rec. Zool. Surv. Pak. 13, 1–17.

Ali, A., Altaf, M., Khan, M.S.H., 2016. Winter survey of birds at Keti Bunder, district Thatha, Pakistan. Punjab Univ. J. Zool. 31 (2), 203–208.

Amjad, S., Rizvi, S.H.N., Memon, G.M., Memon, M.Q., 2002. Indus Mangrove Ecosystem Study to Evaluate the Impact of Coastal Resource Use and Environmental Degradation on the Coastal Communities Living in the Mangrove Ecosystem at Korangi-Gharo Creek System. PNC-UNESCO, NIO (June 2001–April 2002).

Anon, 1976–77. Final Report "Survey Results of Dr. Fridtjof Nansen" Indian Ocean Fishery and Development Program. Pelagic Fish Assessment Survey North Arabian Sea. Institute of Marine Research, Bergen, p. 26. FAO.

Anon, 1986. Summary of Findings "Dr. Fridtjof Nansen" Surveys of Pakistan Fishery Resources, September 1983–June 1984. . Institute of Marine Research, Bergen Norway UNDP/FAO Programme GLO/82/001.28.

Anon, 2012. Handbook of Fisheries Statistics of Pakistan. Volume 20. Marine Fisheries Department, Government of Pakistan, Fish Harbour, West Wharf, Karachi, p. 145.

Badjeck, M.C., Allison, E.H., Halls, A.S., Dulvy, N.K., 2010. Impacts of climate variability and change on fishery-based livelihoods. Mar. Policy 34 (3), 375–383.

Bouillon, S., et al., 2008. Mangrove production and carbon sinks: a revision of global budget estimates, Global Biogeochem. Cycle 22, GB2013https://doi.org/10.1029/2007GB003052.

Clift, P.D., 2002. A brief history of the Indus River. In: Clift, P.D., Kroon, D., Craig, J., Gaedicke, C. (Eds.), The Tectonics and Climatic Evolution of the Arabian Sea Region. Geological Society of London Special Publication, vol. 195. pp. 237–258.

Clift, P.D., Blusztajn, J., 2005. Reorganization of the western Himalayan river system after five million years ago. Nature 438, 1001–1003. https://doi.org/10.1038/nature04379.

Clift, P.D., Shimizu, N., Layne, G.D., Blusztajn, J.S., Gaedicke, C., Schlüter, H.-U., Clark, M.K., Amjad, S., 2001. Development of the Indus fan and its significance for the erosional history of the Western Himalaya and Karakoram. Geol. Soc. Am. Bull. 113 (8), 1039–1051.

Costanza, R., Ralph d'Arge, R., Groot, R.D., et al., 1997. The value of the world's ecosystem services and natural capital. Nature 387, 253–260.

Day, J., Dudley, N., Hockings, M., Holmes, G., Laffoley, D., et al., 2012. Guidelines for Applying the IUCN Protected Area Management Categories to Marine Protected Areas. IUCN, Gland, Switzerland, p. 36.

Dehlavi, A., Adil, I.H., 2012. Socioeconomic baseline of Pakistani's coastal areas. . World Wide Fund for Nature—Pakistan.

District wise Census Results—Census, 2017. Available from: www.pbs.gov.pk/content/population-census.

Emerton, L., Bos, E., 2004. Value. Counting Ecosystems as an Economic Part of Water Infrastructure. IUCN, Gland and Cambridge, p. 88.

Fahlbusch, H., Schultz, B., Thatte, C.D., 2004. The Indus Basin: History of Irrigation, Drainage and Flood Management, International Commission on Irrigation and Drainage, New Delhi, India.

Fanning, P.L., Kalhoro, M., Hanif, T., Khan, M.W., Ansari, M.A., 2009. Cruise Report—Pakistan Demersal Survey 2009001 29 October–6 November, 2009. Report Prepared for the Fisheries Resources Appraisal in Pakistan Project UTF/PAK/108/PAK, p. 34.

Fanning, L.P., Khan, M.W., Kidwai, S., Macauley, G.J., 2010. Surveys of the Offshore Fisheries Resources of Pakistan—2010. FAO Fisheries and Aquaculture Circular. No. 1065. Karachi, FAO 2011, p. 87.

Fanning, L.P., Khan, M.W., Shafi, D., 2015. Cruise Report—Pakistan Demersal Survey 2015301, 06–26 February, 2015. Report Prepared for the Fisheries Resources Appraisal in Pakistan Project UTF/PAK/108/PAK 157.

Friedlingstein, P., Andrew, R.M., Rogelj, J., Peters, G.P., Canadell, J.G., et al., 2014. Persistent growth of CO2 emissions and implications for reaching climate targets. Nat. Geosci. 7 (10), 709–715.

Giosan, L., Constantinescu, S., Clift, P.D., Tabrez, A.R., Danish, M., Inam, A., 2006. Recent morphodynamics of the Indus Delta shore and shelf. Cont. Shelf Res. 26, 14.

Giosan, L., Clift, P.D., Macklin, M.G., Fuller, D.Q., Constantinescu, S., Durcan, J.A., Stevens, T., Dullerc, G.A.T., Tabrez, A.R., Gangal, K., Adhikari, R., Alizai, A., Filip, F., VanLaninghamj, S., Syvitski, J.P.M., 2012. Fluvial landscapes of the Harappan. Proc Natl Acad Sci USA. www.pnas.org/cgi/doi/10.1073/pnas.1112743109.

Giosan, L., Syvitski, J., Constantinescu, S., Day, J., 2014. Climate change: protect the world's deltas. Nature 516 (7529), 31–33.

Giri, C., Long, J., Abbas, S., Mani Murali, R., Qamer, F.M., Pengra, B., Thau, D., 2015. Distribution and dynamics of Mangrove forests of South Asia. J. Environ. Manag. 148, 101–111.

GoS, 2008. Economic Review, Bureau of Statistics, Government of Sindh, Pakistan.

Gowdy, J.M., Salman, A., 2011. In: Institution and ecosystem functions: the case of Keti Bunder,Pakistan. Ecosystem services economics (ESE). Division of Environmental Policy Implementation Paper No. 10. pp. 21.

Hinkel, J., et al., 2014. Coastal flood damage and adaptation costs under 21st century sea-level rise. Proc. Natl. Acad. Sci. 111, 3292–3297. https://doi.org/10.1073/pnas.1222469111.

Hoekstra, D.A., Mahmood, N., Shah, G.R., Domki, M.A., Ali, Q.M., 1997. Diagnostic study – Indus Delta Mangrove ecosystem. Main subsystem characteristics, problems, potentials, proposed interventions and pilot sites. Sub-project. RRIDM (World Bank/GoS funds). 76 pp.

Hussain, I., Hussain, Z., Sial, M.H., Akram, W., Farhan, M.F., 2011. Water balance, supply and demand and irrigation efficiency of Indus basin. Pak. Econ. Soc. Rev. 49 (1), 13–38.

Iftikhar, U., 2002. In: Valuing the economic costs of environmental degradation due to sea intrusion in the Indus Delta. Sea Intrusion in the Coastal and Riverine Tracts of the Indus Delta—A Case Study (IUCN), IUCN-The World Conservation Union, Pakistan Country Office, Karachi, 48.

IPCC, 2007a. Climate change 2007: impacts, adaptation and vulnerability. In: Contribution of Working Group II to the Fourth Assessment Report of the Intergovernmental Panel on Climate Change. Cambridge University Press, Cambridge, pp. 976.

IPCC, 2007b. Climate change 2007: the physical science basis. In: Contribution of Working Group I to the Fourth Assessment Report of the Intergovernmental Panel on Climate Change. Cambridge University Press, Cambridge and New York, NY, pp. 996.

IPCC, 2013a. Summary for policymakers. In: Stocker, T.F., Qin, D., Plattner, G.-K., Tignor, M., Allen, S.K., Boschung, J., Nauels, A., Xia, Y., Bex, V., Midgley, P.M. (Eds.), Climate Change 2013: The Physical Science Basis. Contribution of Working Group I to the Fifth Assessment Report of the Intergovernmental Panel on Climate Change. Cambridge University Press, Cambridge and New York, NY.

IPCC, 2013b. Climate Change 2013: The Physical Science Basis. In: Stocker, T.F., Qin, D., Plattner, G.-K., Tignor, M., Allen, S.K., Boschung, J., Nauels, A., Xia, Y., Bex, V., Midgley, P.M. (Eds.), Contribution of Working Group I to the Fifth Assessment Report of the Intergovernmental Panel on Climate Change Cambridge University Press, Cambridge and New York, NY. pp. 1535. https://doi.org/10.1017/CBO9781107415324.

IUCN, 1999–2000. Socio-economic data on the Indus Delta. Un-published data. From ICZM Plan for Pakistan (in Qureshi, M. T., 2011).

IUCN, 2003. Case Studies in Wetland Valuation #5: May 2003 Indus Delta, Pakistan: economic costs of reduction in freshwater flows. "Integrating Wetland Economic Values into River Basin Management", Water and Nature Initiative of IUCN – The World Conservation Union, DFID, UK. https://cmsdata.iucn.org/downloads/casestudy05indus.pdf.

Jianchu, X., Shrestha, A., Vaidya, R., Eriksson, M., Hewitt, K., 2007. The Melting Himalayas: regional challenges and local impacts of climate change on mountain ecosystems and livelihoods. ICIMOD Technical Paper, Kathmandu, Nepal, p. 14.

Kamal, S., Amir, P., Mohtadullah, K., 2012. Development of Integrated River Basin Management (IRMA) for Indus Basin. Challenges and Opportunities. WWF Pakistan Publication, p. 80.

Karl, T.R., Arguez, A., Huang, B., Lawrimore, J.H., Mcmahon, J.R., Menne, M.J., Peterson, T.C., Russell, S.V., Zhang, H.M., 2015. Possible artifacts of data biases in the recent global surface warming hiatus. Science, 1469–1472. https://doi.org/10.1126/science.aaa5632.

Kaufmann, R.K., Kauppi, H., Mann, M.L., Stock, J.H., 2011. Reconciling anthropogenic climate change with observed temperature 1998–2008. Proc. Nat. Acad. Sci. 108 (29), 11790–11793. https://doi.org/10.1073/pnas.1102467108.

Kazmi, A.H., 1984. Geology of the Indus Delta. In: Haq, B.U., Milliman, J.D. (Eds.), Marine Geology and Oceanography of Arabian Sea and Coastal Pakistan. Van Nostrand Reinhold Co. Scientific and Academic Editions, NIO, pp. 71–84.

Kidwai, S., Fanning, P.L., Ahmed, W., Tabrez, S.M., Zhang, J., Khan, M.W., 2016. Practicality of Marine Protected Areas – Can there be solutions for the River Indus delta. Special Issue: Future Coasts- Estuaries, Coastal and Shelf Science. Elsevier. https://doi.org/10.1016/j.ecss.2016.09.01.

Meehl, G.A., Stocker, T.F., Collins, W.D., Friedlingstein, A.T., Gaye, A.T., Gregory, J.M., Raper, S.C., 2007. Global Climate Projections. In: Climate Change 2007: The Physical Science Basis. Contribution of Working Group I to the Fourth Assessment Report of the Intergovernmental Panel on Climate Change. Cambridge University Press, Cambridge.

Memon, A.A., 2004. In: Evaluation of impacts on the lower Indus River Basin due to upstream water storage and diversion. Proceedings, World Water & Environmental Resources Congress 2004, American Society of Civil Engineers, Environmental and Water Resources Institute, Salt Lake City, Utah, June 27–July 1, 2004. pp. 1–11.

Meynell, P.J., Qureshi, M.T., 1993. In: Moser, M., Van Vessen, J. (Eds.), Sustainable management of the Mangrove ecosystem in the Indus Delta. Proc. Int. Symp. on Wetland and Waterfowl Conservation in South and West Asia. Karachi. pp. 22–126.

Milliman, J.D., Quraishee, G.S., Beg, M.A.A., 1984. Sediment discharge from the Indus River to the ocean: past, present and future. In: Haq, B.U., Milliman, J.D. (Eds.), Marine Geology and Oceanography of Arabian Sea and Coastal Pakistan. Van Nostrand Reinhold, New York, pp. 65–70.

Min, S., Xuebin Zhang, X., Zwiers, F.W., Gabriele, F.W., Hegerl, C., G.C., 2011. Human contribution to more-intense precipitation extremes. Nature 470, 378–381.

Mirza, M.I., Hassan, M.Z., Akhtar, S., Ali, J., Sanjirani, M.A., 1988. Remote sensing survey of mangrove forest along the coast of Baluchistan. In: Thompson, M.F., Tirmizi, N.M. (Eds.), Marine Science of the Arabian Sea. AIBS, Washington, DC, pp. 339–348.

Najman, Y., Garzanti, E., Pringle, M., Bickle, M., Stix, J., Khan, I., 2003. Early-middle Miocene paleodrainage and tectonics in the Pakistan Himalaya. Geol. Soc. Am. Bull. 115, 1265–1277.

Naveed, A., Ali, N., 2012. Clustered Deprivation: District Profile of Poverty in Pakistan. A Publication of the Sustainable Development Policy Institute (SDPI), ISBN: 978-969-8344-17-7.

NDMA-UNDP, 2010. Pakistan Indus River System. 5-Day Training Course On "Flood Mitigation". 15 February 2010, Islamabad.

Nicholls, R.J., Marinova, N., Jason, A., Lowe, A., Brown, S., Vellinga, P., de Gusmão, D., Hinkel, J., Tol, R.S.J., 2010. Sea-level rise and its possible impacts given a "beyond 4°C world" in the twenty-first century. Phil. Trans. R. Soc. A 369, 161–181. https://doi.org/10.1098/rsta.2010.0291. (2011).

Overeem, I., Syvitski, J.P.M., 2009. Dynamics and vulnerability of delta systems. In: LOICZ Reports & Studies No. 35. GKSS Research Center, Geesthacht, pp. 54. ISSN: 1383 4304.

Pakistan Economic Survey 2016–17.

PMD (Pakistan Meteorological Department), 2010. Rainfall Statement July 2010.

Possehl, G., 1997. Climate and the eclipse of the ancient cities of the Indus. In: Dalfes, H.N., Kukla, G., Weiss, H. (Eds.), Third Millennium BC Climate Change and Old World Collapse, NATO ASI Ser. 1, vol. 49. Springer, New York, pp. 193–244.

Qadir, M., Quillérou, E., Nangia, V., Murtaza, G., Singh, M., Thomas, R.J., Drechsel, P., Noble, A.D., 2014. Economics of salt-induced land degradation and restoration. UN Sus. Dev. J. 38 (I4), 282–295. https://doi.org/10.1111/1477-8947.12054.

Qayyum, M., Niem, A.R., Lawrence, R.D., 2001. Detrital modes and provenance of the Paleogene Khojak Formation in Pakistan; implications for early Himalayan Orogeny and unroofing. Geol. Soc. Am. Bull. 113, 320–332.

Quraeshi, S., 1974. Legacy of the Indus. A Discovery of Pakistan. John Weatherhill, Inc., Tokyo, p. 223.

Quraishee, G.S., 1988. Global warming and rise in sea level in the South Asian seas region. In: The Implication of Climatic Changes and the Impact of Rise in Sea Level in the South Asian Seas Region. pp. 1–21.

Qureshi, M.T., 1996. Restoration of Mangroves in Pakistan. In: Field, C.D. (Ed.), Restoration of Mangrove Ecosystems. International Society for Mangrove Ecosystem, Okinawa, pp. 126–142.

Qureshi, A.S., Gill, M.A., Sarwar, A., 2010. Sustainable groundwater management in Pakistan: challenges and opportunities. Irrig. Drain. 59, 107–116.

Reef, R., Feller, I.C., Lovelock, C.E., 2010. Nutrition of Mangroves. Tree Physiol. 30, 1148–1160.

Rees, G., Collins, D., 2004. SAGARMATHA: Snow and Glacier Aspects of Water Resources Management in the Himalayas. DFID Project R7980—An Assessment of the Potential Impacts of Deglaciation on the Water Resources of the Himalaya, Centre for Ecology and Hydrology, Oxfordshire, UK.

Roberts, M., 1894. The Indus Delta Country. A Memoir Chiefly on Its Ancient Geography and History. Kagan, Paul, Trench, Truber & Co., Ltd., London153.

Roberts, R., 2017. Pakistan could face mass droughts by 2025 as water level nears 'absolute scarcity'. https://www.independent.co.uk/news/world/pakistan-droughts-2025-warning-water-levels-a7949226.html.

Saifullah, S.M., 1982. Mangrove Ecosystem of Pakistan, The Third Research on Mangroves in Middle East, Japan Cooperative Center for the Middle East, Publ., No 137 Tokyo, Japan.

Saifullah, S.M., 1997. Management of the Indus Delta Mangroves. In: Haq, B.U., Haq, S.M., Stel, J.M. (Eds.), Coastal Zone Management Imperative for Maritime Developing Nations Kullenberg. Kluwer Academic publishers, The Netherlands, pp. 336–346.

Saifullah, S.M., 2017. The effect of global warming (climate change) on mangroves of Indus Delta with relevance to other prevailing anthropogenic stresses: a critical review. Eur. Acad. Res. 4, 2110–2138.

Salik, K.M., Jahangir, S., Zahdi, W., Hasson, S., 2015. Climate change vulnerability and adaptation options for the coastal communities of Pakistan. Ocean Coast. Manag. 112, 61–73.

Scott, D.A., Poole, C.M., 1989. A Status Overview of Asian Wetlands, Based on "A Directory of Asian Wetlands". Asian Wetland Bureau, Malaysia.

Searle, M.P., Owen, L.A., 1999. The evolution of the Indus River in relation to topographic uplift, climate and geology of western Tibet, the Trans-Himalayan and High-Himalayan Range. In: Meadows, A., Meadows, P.S. (Eds.), The Indus River, Biodiversity, Resources, Humankind. Linnaean Society of London, Oxford University Press, pp. 210–230.

Siddiqui, R., Samad, G., Nasir, M., Jalil, H.H., 2014. The Impact of Climate Change on Major Agricultural Crops: Evidence from Punjab, Pakistan. PIDE Working Paper.

Snedaker, S.C., 1984. Mangroves: a summary of knowledge with emphasis on Pakistan. In: Haq, B.U., Milliman, J.D. (Eds.), Marine Geology and Oceanography of Arabian Sea and Coastal Pakistan. Van Nostrand Reinhold Co., New York, pp. 255–262.

Statistics Bureau of Pakistan, 2016. http://www.finance.gov.pk/survey/chapters_17/overview_2016-17.pdf.

Stocker, T.F., et al., 2013. Technical summary. In: Stocker, T.F., Qin, D., Plattner, G.-K., Tignor, M., Allen, S.K., Boschung, J., Nauels, A., Xia, Y., Bex, V., Midgley, P.M. (Eds.), Climate Change 2013: The Physical Science Basis. Contribution of Working Group I to the Fifth Assessment Report of the Intergovernmental Panel on Climate Change. Cambridge University Press, Cambridge, United Kingdom and New York, NY, pp. 33–115. https://doi.org/10.1017/CBO9781107415324.005.

Sultana, H., Ali, N., Mohsin, I., Khan, A.M., 2009. Vulnerability and adaptability of wheat production in different climatic zones of Pakistan under climate change scenarios. Clim. Chang. 94, 123–142.

Syvitski, J.P.M., Brakenridge, G.R., 2013. Causation and Avoidance of Catastrophic Flooding Along the Indus River. GSA Today, Pakistan.

Syvitski, J.P.M., et al., 2009. Sinking deltas due to human activities. Nat. Geosci. 2, 681–686.

Syvitski, J.P.M., Kettner, A.J., Overeem, I., Giosan, L., Brakenridge, M., Billam Roger, M.H., 2014. Metamorphosis of the Indus Delta and Lower Floodplain, Anthropocene. https://doi.org/10.1016/j.ancene.

Talpur, M., Ercelan, A., Nauman, M., 1999. The World Bank in Pakistan: See No Suffering, Hear No Cries, Speak No Truth. PILER Publication. Asia Pacific Network. http://www.realityofaid.org/wp-content/uploads/2013/02/The-World-Bank-in-Pakistan.pdf.

Waryani, B., Siddiqui, G., Ayub, Z., Khan, S.H., 2015. Occurrence and temporal variation in the size-frequency distribution of 2 bloom-forming jelly-fishes, *Catostylus perezi* (L. Agassiz, 1862) and *Rhizostoma pulmo* (Cuvier, 1800), in the Indus Delta along the coast of Sindh. Pakistan. Turkish J. Zool. 39 (1), 95–102. https://doi.org/10.3906/zoo-1401-13.

WCD, 2000. Asianics Agro-Dev. International (Pvt) Ltd. *Tarbela Dam and related aspects of the Indus River Basin, Pakistan*, A WCD case study prepared as an input to the World Commission on Dams, Cape Town. www.dams.org.

Wong, C.M., Williams, C.E., Pittock, J., Collier, U., Schelle, P., 2007. World's Top 10 Rivers at Risk. WWF International, Gland.

World Bank, 2010. The Economics of Adaptation to Climate Change (EACC): Synthesis Report. The World Bank Group, Washington, DC.

Worstall, T., 2007. Indian wheat yields. The Broadsheet. September 17. www.timworstall.com.

WWF-Pakistan, 2007. Detailed Ecological Assessment of Fauna, Including Limnology Studies at Keti Bundar 2007–2008. Indus for all Programs.

Zaheer, K., et al., 2012. Community Based Vulnerability Assessment: Kharo Chan, Keti Bunder and Jiwani. World Wide Fund for Nature, Pakistan.

Chapter 13

A Brief Overview of Ecological Degradation of the Nile Delta: What We Can Learn

Zhongyuan Chen

State Key Laboratory of Estuarine and Coastal Research, East China Normal University, Shanghai, People's Republic of China

1 INTRODUCTION

The fan-shaped 20,000 km^2 (exclusive of its subaqueous part) Nile Delta in Egypt (Fig. 1) is solely fed by the water and sediment from the African Plateau and the Ethiopian Highland, where rainfall is ~1500–2000 mm a^{-1}, whereas the rainfall is only 50–100 mm a^{-1} in the lower Nile basin, including its delta coast. The two branches, that is, the Rosette and the Damietta, flow into the delta where there are four lagoons; from east to west, the lagoons are Manzala, Burullus, Idku, and Maryut. The Manzala lagoon is the largest, covering a water-surface area ~1000 km^2, and all lagoons have a water depth ~2.0 m on average.

The Nile Delta not only serves as an agricultural breadbasket for nearly 100 million Egyptians, but it also includes one of the earliest rises of human civilization on the world's coasts (Butzer and Freeman, 1976; Said, 1981).

However, as humans have moved into industrial epoch since the last century, particularly after the completion of the High Aswan Dam in 1964, the Nile Delta has been suffering from serious environmental degradation as is highlighted in the following section.

1.1 Human Impact on the River Basin: Reducing Sediment and Fresh Water, but Increasing Nutrients

The Nile River transported ~150 × 10^6 t^{-1} a^{-1} sediment on average to the delta coast before 1964; however, a large part of it (~60–90 × 10^6 t^{-1} a^{-1}) was trapped in the Aswan reservoir after damming (Fig. 2A), leaving little that can reach the delta coast (Fanos et al., 2001). The formerly turbid estuarine waters became clear, and this has profoundly altered the ecology. Another severe challenge is the freshwater discharge of the Nile; before damming, the delta benefited from seasonal floods from the upper Nile basin. This, together with water diversion for the purpose of agricultural irrigation below the dam site, including the delta, has minimized the freshwater discharge into the Nile coast (Fig. 2A).

Damming did not reduce the transport of nutrients to the delta coast (Fig. 2B). The use of N and P fertilizer in the Nile basin has increased since the 1960s (Nixon, 2004), but P declined after the 1990s, whereas the N flux was just slightly affected by the dam and then was significantly increased to the end of 1990s. The fertilizer application has directly driven up the concentration of N and P in the Nile estuarine water. But, the high P flux that occurred before the High Aswan dam probably indicates multiple sources of P, including soil erosion. It appears that before the dam, the Nile Delta waters were N-limited and they became P-limited after the dam (Ludwig et al., 2009).

1.2 Delta-Estuarine Responses

The altered riverine sediment and freshwater discharge, and the increasing nutrients from irrigation fields in Egypt, have negatively impacted the eco-health of the Nile Delta coast, as is summarized as follows from eco-biological indicators collected since the 1960s.

Coasts and Estuaries. https://doi.org/10.1016/B978-0-12-814003-1.00013-7

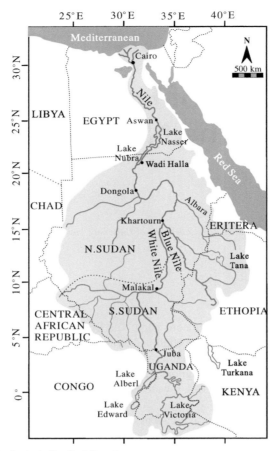

FIG. 1   Geographic site of the Nile River basin, including its delta-estuary.

Immediately after the Aswan Dam was built in 1964, fish landing off the western Nile coast decreased in numbers rapidly, and this trend was followed by a gradual recovery in the 1980s and a significant recovery in 1990s (Fig. 2C). This pattern definitely fit the increasing nutrient delivery to the estuary, here manifested by the parallel increasing N and fish catches in the coastal lagoons (Fig. 2D) (Oczkowski et al., 2008, 2009). The improved fishing technology in more recent times can be an additional attributor.

Dissolved silicate (DSi) into the delta coast (measured below the dam) decreased after the dam (Fig. 2E) (Oczkowski et al., 2008, 2009) because the dam not only blocks the sediment flux, but also increases the time of water residence in reservoirs, thereby promoting the uptake of Si by algae in the reservoir (Li et al., 2007). The primary production in the estuarine waters followed the same trend as that of DSi shortly after damming in 1960s and 1970s, however PPR increased dramatically after the 1980s, suggesting a change from silicate algae to nonsilicate algae (Fig. 2E). In turn, this led to algae-generated hypoxia events (Fig. 2F) (Diaz and Rosenberg, 2008).

2   WHAT WE CAN LEARN?

The environmental degradation of the Nile Delta is typical of many of the world's deltas, such as the Mississippi and the Yangtze (Changjiang). The geographic setting of each delta and its watershed results in different spatiotemporal patterns of distribution of hydrological, ecological, and biological indicators for each delta; however, the key biophysical processes of degradation are similar in all deltas and similarly affect their ecology and food chain dynamics, and this in turn impacts mankind. By knowing this, we must work together to generate a better and sustainable future for deltas and mankind.

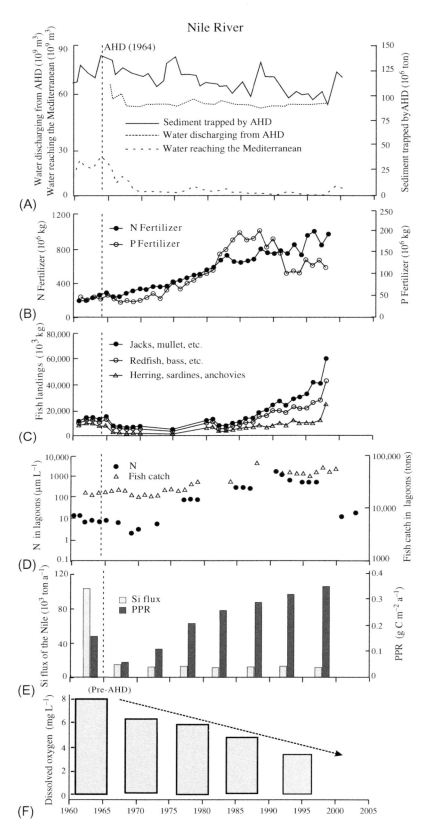

FIG. 2 (A) Long-term variations in sediment load trapped by the Aswan High Dam (AHD) in 1960–2000. The water discharged from the AHD and that finally reaching the Mediterranean Sea are also indicated. (B) Dramatic increase of fertilizer use for agriculture in Egypt after the AHD (Nixon, 2004). (C) Variations of fish landing off the Nile Delta showed the greatest impact of the AHD (Nixon, 2004); fish landing plummeted after the AHD in 1964 and stayed low until the numbers rose again after 1980. (D) Inorganic nitrogen concentrations and the fish catch of the four Nile lagoons from 1960 to 2005. (E) Long-term 5-year averaged variations of Si fluxes by the Nile River into the Mediterranean Sea and the Primary Production sustained by the River (PPR) in the Nile Estuary. (F) Mean decadal dissolved oxygen concentrations in the Nile lagoons. *((A) Modified from Fanos, A.M., Khafagy, A.A., El Kady, M.M. 2001. Variations of the Nile River discharge and sediment regime during the period from 1800 to 2000 and its effects on the Nile Delta coast. In: Eighth International Symposium on River Sedimentation, Cairo; (D) Modified from Oczkowski, A.J., Nixon, S.W., 2008. Increasing nutrient concentrations and the rise and fall of a coastal fishery: a review of data from the Nile Delta, Egypt, Estuar. Coast. Shelf Sci. 77:309–319; Modified from Oczkowski, A.J., 2009. Fertilizing the Land, Lagoons and Sea: A First Look at Human Impacts on the Nile Delta fishery, Egypt (Ph.D. dissertation). University of Rhode Island; (E) From Ludwig, W., Dumont, E., Meybeck, M., Heussner, S., 2009. River discharges of water and nutrients to the Mediterranean and Black Sea: major drivers for ecosystem changes during past and future decades? Prog. Oceanogr. 80:199–217; (F) Modified from Oczkowski, A.J., 2009. Fertilizing the Land, Lagoons and Sea: A First Look at Human Impacts on the Nile Delta fishery, Egypt (Ph.D. dissertation). University of Rhode Island.)*

ACKNOWLEDGMENTS

The author is deeply indebted to Dr. Jiawei Gu and Dr. Hao Xu, who exerted great effort in the collection of dataset and drawings. China National and Natural Science Foundation financially supported this study (Grant No. 41620104004).

REFERENCES

Butzer, K.W., Freeman, L.G., 1976. Early Hydraulic Civilization in Egypt. University of Chicago Press, Chicago.

Diaz, R.J., Rosenberg, R., 2008. Spreading dead zones and consequences for marine ecosystems. Science 321, 926–929. https://doi.org/10.1126/science.1156401.

Fanos, A.M., Khafagy, A.A., El Kady, M.M., 2001. In: Variations of the Nile River discharge and sediment regime during the period from 1800 to 2000 and its effects on the Nile Delta coast. Eighth International Symposium on River Sedimentation, Cairo.

Li, M.T., Xu, K.Q., Watanabe, M., Chen, Z., 2007. Long-term variations in dissolved silicate, nitrogen, and phosphorus flux from the Yangtze River into the East China Sea and impacts on estuarine ecosystem. Estuar. Coast. Shelf Sci. 71, 3–12.

Ludwig, W., Dumont, E., Meybeck, M., Heussner, S., 2009. River discharges of water and nutrients to the Mediterranean and Black Sea: major drivers for ecosystem changes during past and future decades? Prog. Oceanogr. 80, 199–217.

Nixon, S.W., 2004. The artificial Nile. Am. Sci. 92, 158–165.

Oczkowski, A.J., 2009. Fertilizing the Land, Lagoons and Sea: A First Look at Human Impacts on the Nile Delta Fishery, Egypt. (Ph.D. dissertation). University of Rhode Island.

Oczkowski, A.J., Nixon, S.W., 2008. Increasing nutrient concentrations and the rise and fall of a coastal fishery: a review of data from the Nile Delta, Egypt. Estuar. Coast. Shelf Sci. 77, 309–319.

Said, R., 1981. The Geological Evolution of the River Nile. Springer-Verlag, New York, p. 151.

Chapter 14

Status and Sustainability of Mediterranean Deltas: The Case of the Ebro, Rhône, and Po Deltas and Venice Lagoon

John W. Day*, Carles Ibáñez†, Didier Pont‡, Francesco Scarton§

*Department of Oceanography and Coastal Sciences, College of the Coast and Environment, Louisiana State University, Baton Rouge, LA, United States, †Aquatic Ecosystems Program, IRTA, San Carles de la Rapita, Catalonia, Spain, ‡Institute of Hydrobiology and Aquatic Ecosystem Management (IHG), University of Natural Resources and Life Sciences, Vienna, Austria, §SELC Societá Cooperativa, Venezia, Italy

1 INTRODUCTION

The Mediterranean Sea is a semi-enclosed high salinity regional sea whose only outlet to the Atlantic is the Strait of Gibraltar. The drainage basin of the Mediterranean stretches from the slopes of the Alps in temperate east central France to tropical rainforest south of the equator in the Nile drainage. The northern coast of the Mediterranean is the southern shore of Europe. This is an area of high population density, a first world standard of living, and relatively high freshwater input including three of the largest rivers in southern Europe: the Ebro, Rhône, and Po. There are also important coastal lagoons including the Venice Lagoon. By contrast, the southern coast of the Mediterranean is semiarid to arid with one of the largest deserts of the world—the Sahara. With the exception of the Nile, there is almost no riverine input. With a few exceptions such as in the Nile delta, population density is low and the living standards are much lower than Europe.

Continental runoff to the Mediterranean Sea is low in general terms, except in the Adriatic Sea and the Gulf of Lion (mostly due to the Po and Rhône Rivers, respectively). Originally, the highest discharge was from the Nile River, but nowadays it has been greatly reduced due to human intervention. The Mediterranean River mouth estuaries are highly stratified (salt-wedge) due to the low tidal range, and their hydrology is controlled by river discharge (Ibáñez et al., 1997). With only a few exceptions, in the Mediterranean basin all estuaries are part of deltaic systems.

The conspicuous presence of deltas in the Mediterranean is due to low tidal energy and high sediment river discharge due to heavy rains and abrupt relief inland. The Mediterranean Sea is characterized by very weak astronomical tides (20–30 cm) in most of the areas. However, the area near the Gibraltar Strait has higher tides due to the proximity of the Atlantic Ocean. The northern Adriatic Sea also has higher tides (up to 1 m) due to its geomorphic features. Meteorological (barometric) tides can be much higher than astronomical tides, so they play an important role in the ecology of coastal Mediterranean marshes. For instance, the monthly maximum surge height due to meteorological tides is about 1 m in the Ebro delta. Minimum sea level is usually recorded in winter due to atmospheric high pressures.

There is a very large cultural diversity and several of the oldest civilizations began in the Mediterranean basin. Because of this, humans have had significant and wide ranging environmental impacts. Climate change is projected to significantly impact the area in terms of warming and drying, which will be exacerbated by human development (see Day and Rybczyk, 2018, this volume). There has been massive change in the basin especially as related to water management in terms of dams, water diversions, and channelization. The Mediterranean climate is strongly seasonal with wet winters and dry summers and extreme rainfall events are relatively common. There are well-known climate events such as the Mistral in France and the Sirocco in the Adriatic, but no super storms such as tropical cyclones and extratropical storms such as those which occur in the Atlantic and Pacific.

The Mediterranean wetlands have been largely reduced and changed by human activity for centuries, but especially during the last 50 years. Between 1942 and 1984, more than 30,000 ha (about 40%) of wetlands were lost in the Rhône delta, France. Over the last century, Tunisia lost 28% of coastal wetlands, more than 60% in Spain and Greece, and more than 70% in Italy (Ibáñez et al., 2002). The largest wetland areas remaining in the Mediterranean are associated with the main deltaic areas: the Ebro (Spain), Rhône (France), Po (Italy), and Nile (Egypt). There are also important marsh areas surrounding coastal-barrier lagoons, such as those existing in the coast of Languedoc (southwest France) and the lagoon of Venice (Italy).

Large portions of the Mediterranean coastal marshes are occasionally or seasonally flooded by seawater, most of the time due to marine storms, but they are little or not directly influenced by tides. This fact plus the low rainfall leads to salt marshes dominated by succulent halophytes. In the Mediterranean, a considerable portion of salt marshes has been transformed into brackish marshes due to freshwater runoff caused by human activities, especially in those marshes surrounding coastal lagoons, but there are also natural brackish marshes, most of them dominated by reed beds. There are some cases where freshwater marshes are affected by sea-level changes, but there are virtually no tidal freshwater marshes in the Mediterranean [see Ibáñez et al., 2002 for a detailed characterization of the Mediterranean coastal marshes].

In this chapter, in addition to describing the general features of the Mediterranean deltas and their wetlands, we focus on three extensively studied characteristic coastal systems—the Ebro delta in Spain, the Rhône delta in France, and the Po delta and adjacent Venice Lagoon in Italy (Fig. 1). Our objective is to describe the environmental conditions in these four riverine-influenced systems and consider how global change in the 21st century will impact these coastal wetland areas.

The systems discussed in the paper comprise a broad range of coastal wetland habitats in the northwestern Mediterranean (Fig. 1). These include freshwater and low-salinity tidal reed-bed marshes at the mouths of the three large rivers and the

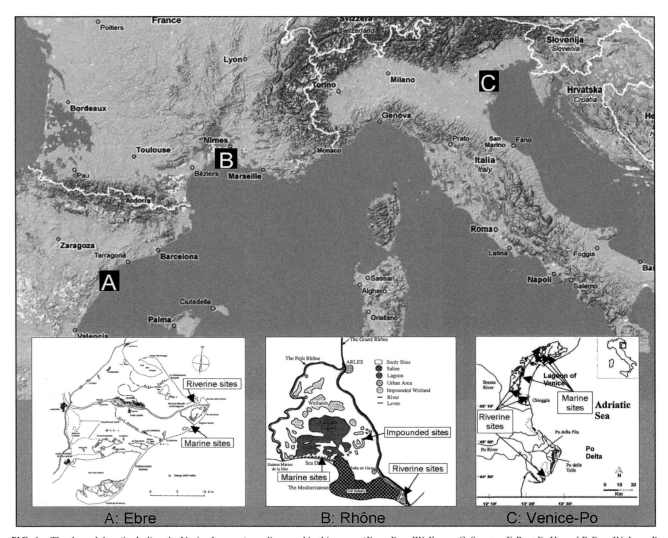

FIG. 1 The three deltas (including the Venice Lagoon) are discussed in this paper.(*From Day JW, Ibanez C, Scarton F, Pont D, Hensel P, Day JN, Lane R: Sustainability of Mediterranean deltaic and lagoon wetlands with sea-level rise: the importance of river input. Estuar. Coasts 34: 483-493. DOI: https://doi.org/10.1007/s12237-011-9390-x, 2011, used by permission.*)

Dese River that drains into the Venice Lagoon; there are marine sites with low freshwater influence, estuarine tidal marshes, and impounded freshwater and saltwater marshes. Detailed descriptions of the marshes are provided elsewhere (Rhone— Hensel et al., 1998, 1999; Pont et al., 2002a, 2017; Ebro—Ibañez et al., 1997, 2010; Curcó et al., 2002; Benito et al., 2014; Po and Venice Lagoon—Scarton et al., 1998, 2002; Day et al., 1999; Solidoro et al., 2010).

2 THE EBRO DELTA

The Ebro delta is the most important delta in Spain. About 65% of the area of the delta is rice fields, while natural areas, including coastal lagoons and wetlands, cover about $80 \, km^2$ (Fig. 2). The river length is $910 \, km$ and the drainage basin is about $85,000 \, km^2$. The Ebro delta has a surface area of about $330 \, km^2$ and contains some of the most important wetland areas in the western Mediterranean. These marshes include a combination of fresh, brackish, and saline wetlands that serve as habitat for waterfowl and fisheries. These natural areas support important economic activities associated with tourism, hunting, fishing, and aquaculture. For more information on the Ebro delta features see Ibáñez and Caiola (2016).

Impacts on the Ebro delta have occurred due to changes at two scales: the delta plain and at the level of the Ebro River basin. Human activities in the drainage basin have had a very notable and large impact, especially in recent decades. Sediment discharge has been reduced by about 99% due to the construction of dams (Sánchez-Arcilla et al., 1996; Ibáñez et al., 1996a, 1997; Rovira et al., 2015). This has led to coastal retreat because of wave erosion and elevation loss in the delta due to a lack sediment input to the delta plain, thus subsidence and sea-level rise (SLR) are no longer offset by new sediments coming from the Ebro River. To prevent excessive waterlogging of wetlands, vertical accretion needs to keep pace with the local combined effects of eustacy and subsidence. The sediment deficit in the delta created by the dams, coupled with land subsidence, accelerated SLR, and the low elevation of the delta plain, puts the delta and its wetlands at major risk for submergence, salt-water intrusion, and coastal erosion (Ibáñez and Prat, 2003; Genua-Olmedo et al., 2016). From the perspective of the drainage basin, the most important management action for the sustainable management of the delta

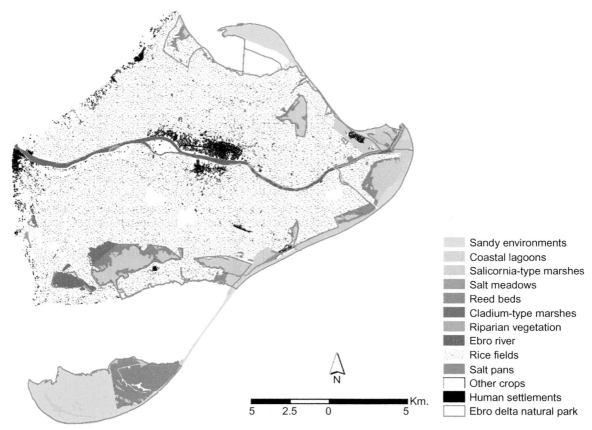

FIG. 2 Map of the Ebro delta showing the distribution of land uses and habitat types in the Ebro delta.*(From Benito X, Trobajo R, Ibáñez C.: Modelling habitat distribution of Mediterranean coastal wetlands: the Ebro Delta as case study. Wetlands, 34(4), 775–785, 2014.)*

is the restoration of the sediment flux in the river, by the remobilization of sediments trapped in reservoirs and increasing freshwater discharge to the delta via controlled floods (Ibáñez et al., 1997; Rovira and Ibànez, 2007).

Prior to the construction of large dams on the lower Ebro in the 1960s, very large floods with high sediment concentrations occurred every few decades. The last very large flood occurred in 1937; it produced the last major change in the position of the river mouth, leading to the creation of the Garxal lagoon. The largest direct impact on the delta was the conversion of over 65% of the wetland area to rice fields, mostly from 1860 to 1960. While this was a very large ecological impact, the extensive irrigation system prior to the construction of the Ribarroja-Mequinenza dams led to high accretion rates in the rice fields that was sufficient to offset SLR and subsidence. Since the construction of the large dams, high sediment delivery no longer occurs via the irrigation network and the fringes of the delta are falling below sea level (Ibáñez et al., 1997). Estimates of relative sea-level rise (RSLR) rates indicate mean rates ranging from 2.08 mm yr^{-1} over 132 years (1965–1833) to 6.26 mm yr^{-1} over 31 years (1965–1934) (Ibáñez et al., 1996b). Recent measures by satellite (unpublished data) confirm this range of subsidence, with maximum values of 5–6 mm yr^{-1} close to the river mouth.

There are significant areas of marshes remaining in the delta. Salt marshes occur along the backshore of the outer coast, especially around the main Ebro River mouth (Garxal) and the secondary mouth (Migjorn), and around some coastal lagoons (Fig. 2). Salt marsh communities are dominated by *Arthrocnemum macrostachyum*, which have low cover (10%–20%) and lower vegetation height, and *Sarcocornia fruticosa* with nearly 100% cover and greater height. The *Sarcocornia* salt marshes are at the lowest elevation. Brackish marshes occur in areas strongly influenced by the Ebro River or by drainage water of the rice fields, and receive periodic influx of fresh water, nutrients, and sediments. They are mostly located around the fresher coastal lagoons and dominated by *Phragmites australis*. Freshwater marshes have almost disappeared and are located close to the inner border of the delta and are dominated by *Cladium mariscus*.

The rate of accretion in the brackish marshes, which are impacted by river input, is higher than in salt marshes (Ibáñez et al., 2010). Because of variations in sediment input, subsidence, and organic soil formation, some of the marshes have the potential to accommodate SLR while others do not (Fig. 3). The factor leading to high elevation gain in the brackish marshes is organic soil formation because of the very low levels of sediments in the river due to dams.

Restoration of wetlands in the Ebro delta has used several different methodological approaches aimed not only at restoration of wetlands and shallow lagoons with submerged aquatic vegetation but also water quality improvement. Re-naturalizing hydrology is central to restoration in deltas. Wetland restoration is essentially taking place in abandoned rice

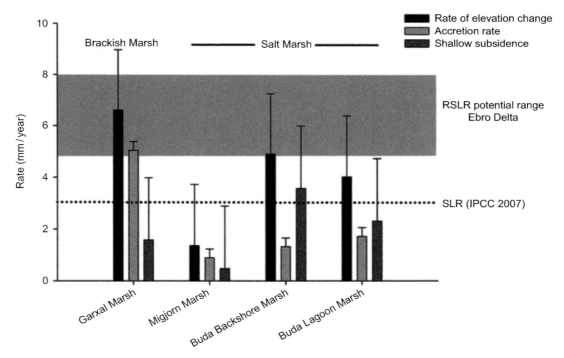

FIG. 3 Comparison of mean elevation change, vertical accretion, and shallow subsidence for the different marsh sites. The dashed line is the IPCC 2007 projection for global sea-level rise (SLR) rates, and the shaded area represents relative sea-level rise (RSLR) rate projections estimated by Ibáñez et al. (1996b) for the Ebro delta.*(From Ibáñez C, James P, Day JW, Day JN, Prat N. Vertical accretion and relative sea level rise in the Ebro delta wetlands (Catalonia, Spain). Wetlands 30: 979–988, 2010.)*

fields to recover wetland habitat (marshes and lagoons), reduce nutrient levels in water draining from rice fields, and to improve habitat for protected bird species (Forès, 1992; Comín et al., 2001; Forès et al., 2002). Typical vegetation in these restored wetlands includes *P. australis, Scirpus maritimus, Typha latifolia,* and *Scirpus lacustris*. The restored wetlands and shallow water bodies provide habitat for endangered species such as the fish the Spanish toothcarp (*Aphanius iberus*) (Prado et al., 2017). Artificial islands have been constructed for a number of endangered bird species to breed, including slender-billed gull (*Croicocephalus genei*), Audouins's gull (*Larus audouinii*), little tern (*Sterna albifrons*), and gull-billed tern (*Gelochelidon nilotica*).

3 THE RHÔNE DELTA

The Rhône is one of the most important rivers in Europe with a drainage basin of 96,000 km^2 and a mean annual flow of about 1700 m^3 (Welcomme, 1985, Fig. 4). The Rhône discharge is characterized by low water levels at the end of the summer and high discharge typically from October through February. There is high interannual variability of mean monthly discharge because it is strongly influenced by the geological and climatic heterogeneity of the catchment, from a continental-oceanic climate in the northern Saone plain to an Alpine climate in the Jura and Alps mountains and a Mediterranean climate in the south (Vivian, 1989).

Major human influences have affected both the Rhône River and its tributaries since the Middle Ages including: (a) channelization to improve navigation, (b) dike construction to protect against floods during the 19th century, and (c)

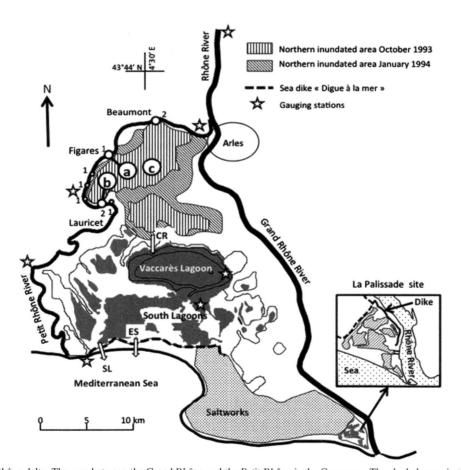

FIG. 4 Map of the Rhône delta. The area between the Grand Rhône and the Petit Rhône is the Camargue. The shaded areas in the northern Camargue show the areas flooded during the floods of October 1993 and January 1994, and the open circles are the locations of levee breaks during the October and January floods, respectively. Large circles are major breaks and small circles are minor breaks. CR is the Canal du Rousty where water flowed from the upper Camargue to the Vaccarès Lagoon. ES is a pumping station where water was pumped to the Mediterranean. SL is the connection between the south lagoons and the sea (Pertuis de la Fourcade). The inset is la Palissade where flooding led to up to 10 cm of sediment deposition.*(From Pont D, Day J, Ibanez C. The impact of two large floods (1993–1994) on sediment deposition in the Rhone delta: implications for sustainable management. Sci. Total Environ. 609: 251–262, 2017, used by permission.)*

hydroelectric development during the second half of the 20th. Nevertheless, the suspended solid discharge of the Rhône River remains significant (Pont et al., 2002b) with a mean annual value of 7.4 million tons ranging annually from 1.2 to 19.7 million tons.

The Rhône discharges to the Mediterranean through two tributaries that are still active today—the Grand Rhône and Petit Rhône that carry about 90% and 10% of the mean annual discharge, respectively. The delta has a total area of 1450 km^2 and the zone between the two branches is called the Isle de Camargue (850 km^2, Fig. 4). The main evolutionary stages since 7000 BP have been described by several authors who have indicated the combined effects of SLR (L'Homer et al., 1981) and climatic changes on the Rhône River discharge and delta development (Probst, 1989). As the main distributaries (i.e., branches of the river that do not return to the main stream after leaving it) have often shifted in the past, the delta has a Mediterranean shoreline of about 50 km along the Golfe du Lyon.

The northern Camargue consists mainly of old inactive river distributary channels (elevation up to 2–3 m) in association with low-salinity marshes (soil elevation from 0 to 0.5 m) where water depth is usually approximately 0.5 m. The southern part of the Camargue consists mainly of a complex pattern of former coastal barrier ridges, abandoned distributary ridges (elevation less than 2 m), and brackish lagoons (soil elevation from −0.8 to 0.1 m), of which the Vaccarès lagoon is the largest with a mean area of 64 km^2. It receives most of the runoff from rainfall (pluvial inputs) and a part of agricultural drainage waters (mainly due to rice cultivation) entering in the Camargue.

After the large floods of 1840 (9640 m^3/s) and 1856 (11,640 m^3/s), the existing dikes were increased in size to protect the Camargue against flooding and to allow agricultural development. The dikes are 96 km long and have an elevation higher than the 100-year flood level (9.7–2.2 m high from North to South). During the same period, a dike (the Digue a la Mer) was also built along the seashore to reduce seawater intrusion during storms. From 1869 to the present, Rhône freshwater enters most of the Camargue only by pumping stations and is distributed by a network of channels for irrigated agriculture, mainly rice. The total volume of irrigation input is of the same order of magnitude as the mean annual rainfall input (509 × 10^6 m^3 per year). Most of the drainage water is returned to the river or the sea by pumping.

Only a small area located near the mouth of the Grand Rhône (La Palissade, about 800 ha) remains open to natural overflow from the river. The vegetation in shallow impoundments semi-isolated from the Rhône influence is dominated by halophytic vegetation (*Arthrocnemum* and *Juncus* spp.) reflecting highly saline soil conditions. The vegetation in areas connected to the Rhône (*S. maritimus* and *Phragmites communis*) shows the importance of regular inputs of the river in mitigating soil salinity levels.

In recent decades, a number of studies have shown that wetland productivity as well as accretion is higher in areas receiving the Rhône River water. Hensel et al. (1998) measured short-term sedimentation rates over periods of a few weeks in different areas of the Rhône delta. The highest sedimentation rates occurred at La Palissade, the natural area near the mouth of the river with a high river water input. Marine and impounded sites without river influence had significantly lower rates of short-term sedimentation. Similarly, vertical accretion and vertical elevation gain in wetlands were more than 10 times higher in areas of the Rhône delta impacted by regular floods of the Rhône compared to marine and impounded sites (Hensel et al., 1999, Fig. 5). The net primary production of wetlands, especially *Salicornia*-type marshes, was much higher in areas affected by the Rhône discharge (Ibañez et al., 1999). *Arthrocnemum* and *Typha* marshes affected by the Rhône water had a much higher productivity than grazed marshes in the same area and productivity was low in impounded marshes without any connection to the Mediterranean Sea. Pont et al. (2002b) concluded that the Rhône delta wetlands with riverine influence were most likely to survive predicted SLR while marshes isolated from the river would not survive. The wetlands in the lower part of the Rhône delta are currently falling below sea level and Pont et al. (2002a) concluded they will disappear if accretion is not enhanced.

The results of this study and others in the Rhône delta indicate that riverine input can enhance accretion. River water will also reduce salinity stress and improve wetland productivity. Day et al. (2016b) concluded that deltas with areas below sea level will not be sustainable due to rising sea levels and resource constraints (see also Tessler et al., 2015).

Pont et al. (2017) reported the effects of two large floods that occurred in October 1993 and January 1994, with peak discharges of 9800 and 10,980 m^3/s and total suspended solid transport of 10.7 × 10^6 and 9.7 × 10^6 tons, respectively (Fig. 4). Both floods led to multiple levee breaches in the Northern part of the delta resulting in the introduction of large volumes of water and sediments which led to the formation of crevasse splays (the fluvial deposits formed after the breach) near the river channel and thus to two extensive depositional fans that covered over 10,000 ha in the northern part of the Camargue. In the Northern inundated area, accretion ranged from 70 mm near the breaches to 4 mm 6–8 km away. In the Palissade, the total deposition from both floods was as high as 10 cm. The Rhône delta is facing an uncertain future with projected SLR. The results of this study show that large introductions of river water can help to sustain the delta in the face of accelerated SLR. Controlled introductions of river water using riverside closable structures, as is being done in other deltas (e.g., Day et al., 2007), and in particular in the Mississippi delta (Day et al., 2016a, b; Peyronnin et al., 2017, Rutherford et al., 2018),

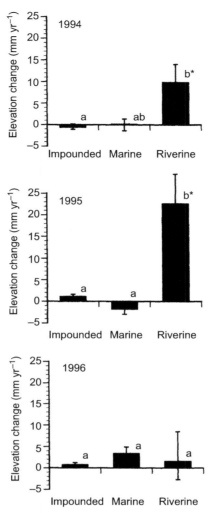

FIG. 5 Vertical accretion and wetland soil elevation change in the Rhône delta. *(From Hensel, P.F., Day Jr., J.W., Pone, D., 1999. Wetland vertical accretion and soil elevation change in the Rhone River Delta, France: the importance of riverine flooding. J. Coast. Res. 15(3), 668–681.)*

could be done in a way that delivers water and sediments to the places where it is needed most and at the same time used to protect important infrastructure.

In summary, the dike breaks led to the formation of crevasse splays near the river channel and to two extensive depositional fans that covered over 10,000 ha in the northern part of the Camargue. The results of such large floods carried out in a controlled manner could lead to sedimentation over large areas of the delta and help to sustain the delta in the face of accelerated SLR. Diversions of river water are currently taking place in the Mississippi delta and could serve as an example for the Rhône delta (Day et al., 2016a, b; Peyronnin et al., 2017; Rutherford et al., 2018).

4 THE PO DELTA AND VENICE LAGOON

The Po is one of the most important rivers discharging to the Mediterranean with a length of 650 km and a mean discharge of about 1500 m^3 s^{-1} (ranging from 275 to 10,000 $m^3 s^{-1}$) m^{-1} and a basin area of about 72,000 km^2 (Zanchettin et al., 2008). The Po delta covers about 61,000 ha (Fig. 1) and discharges to the Adriatic through at least five distributaries. The delta has been created over the past several thousand years as the river successively occupied a number of different river channels (Sestini, 1992). Formerly, most of the deltaic plain was covered by extensive freshwater wetlands, but these were largely claimed for agriculture. Much of the deltaic plain is now 1–4 m below sea level due to subsidence caused primarily by extraction of shallow deposits of natural gas with a high-water content (Sestini, 1992). The fringes of the delta are characterized by beaches and dunes, shallow lagoons, and salt marshes. There are extensive reed swamps (approximately 2500 ha) dominated by *P. australis* bordering the lower ends of the main river channels. The background rate of geological subsidence in the Po delta is 1–3 mm yr (Sestini, 1992; Bondesan et al., 1995; Tosi et al., 2016) but human-induced subsidence has been as high as 5–20 mm yr^{-1} and in a small area in the western Po delta, subsidence was 100 cm between 1958 and 1962 (Bondesan et al., 1995).

During the development of the Po delta, the river migrated widely over the deltaic plain and at times discharged considerably far north of the present main channel, for example, the Po della Pila (Sestini, 1992). During these times, there was a large discharge into the southern part of the Venice Lagoon. The Brenta River also discharged directly into the Venice Lagoon, but it was diverted to the south by the Venetians in 1507 to prevent sedimentation in the lagoon. The Brenta was diverted back into the southern lagoon from 1840 to 1896 to relieve flooding in the agricultural land along the artificial diversion canal. During this period, about 2300 ha of coastal marshes formed in a large fluvial delta (Favero et al., 1988). Thus, the southern part of the Venice Lagoon is considered a deltaic lagoon and has a higher geologic subsidence (1.2 mm yr^{-1}) than in the central lagoon where subsidence is less than 0.5 mm yr^{-1} (Tosi et al., 2016).

Scarton et al. (2002) reported that above (ABG) and belowground (BLG) production of *P. australis* along a distributary channel strong was higher than that of *S. fruticosa* in the southern Venice Lagoon that had a low freshwater input from the Brenta River. Marsh productivity in the Venice Lagoon was higher near sites of freshwater input and at higher elevations (Day et al., 1999; Scarton et al., 1998, 1999).

The Venice Lagoon, the largest Italian lagoon and one of the largest of the Mediterranean, has an area of about 550 km^2 (Fig. 6) and connections to the Adriatic Sea through three large inlets. Over the past five centuries, sediment dynamics of the lagoon have been extensively changed (Gatto and Carbognin, 1981; Sarretta et al., 2010). The Brenta, Sile, and Piave Rivers, which originally discharged into the lagoon, were diverted from the lagoon to the sea beginning in the 16th century. Presently, only a few small rivers (total discharge about 32 m^3 s^{-1}) discharge into the lagoon (Zuliani et al., 2005).

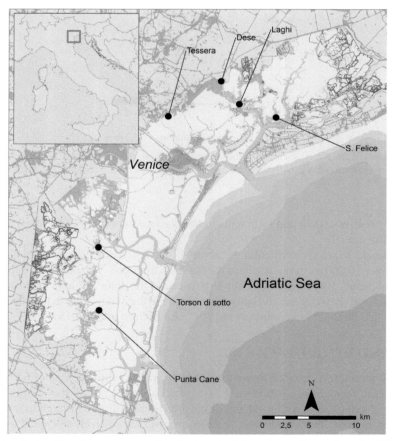

FIG. 6 Venice Lagoon showing the location of measurements of wetland productivity and accretionary dynamics.*(Modified from Day J, Rybczyk J, Scarton F, Rismondo A, Are D, Cecconi G, Soil accretionary dynamics, sea-level rise and the survival of wetlands in Venice Lagoon: a field and modeling approach. Estuar. Coast. Shelf Sci. 49, 607–628, 1999.)*

The import of coarse marine sediments into the lagoon has been greatly reduced because of the construction of long jetties at the inlets at the end of the 19th century. There is a net export of about $0.8 \, M \, m^3$ of sediments from the lagoon (Sarretta et al., 2010).

Most of the lagoon is occupied by a large central waterbody (about $370 \, km^2$) and extensive intertidal salt marshes (about $35 \, km^2$). The mean depth of the lagoon is 1.1 m and the tide range is 0.6–1 m, and extensive tidal flats (about $50 \, km^2$) are exposed at low tide. The subtidal areas are partially vegetated by macroalgae and seagrasses (such as *Zostera marina, Zostera noltei,* and *Cymodocea nodosa*). The dominant salt marsh species include *Limonium serotinum, Puccinellia palustris, S. fruticosa,* and *Spartina maritima* (see Scarton, 2005). The salt marsh area in the lagoon, has decreased from about 12,000 ha at the beginning of the century to about 3500 ha at present due to reclamation (land-claim), erosion, pollution, and natural and human-induced subsidence (Sarretta et al., 2010). An extensive program of saltmarsh creation using sediments dredged from lagoon channels began at the end of the 1980s; these 1300 ha of dredge islands now make a suitable place for breeding waterbirds, with about 4000 pairs counted in recent years (Scarton, 2017).

There have been a number of studies of wetland ecology, accretionary dynamics, and erosion in the Po delta and Venice Lagoon. Short-term sedimentation in the Venice Lagoon averaged $3–7 \, g \, dry \, m^2 \, day^{-1}$ per site with a maximum of $76 \, g \, m^2 \, day^{-1}$ (Fig. 7). The highest values were measured during strong pulsing events, such as storms and river floods that mobilized and transported suspended sediments. Accretion ranged from 2 to $23 \, mm \, yr \, m^2 \, d^{-1}$ and the change in soil elevation ranged from 32 to $13.8 \, mm \, yr^{-1}$. The sites with highest accretion were near a river mouth and in an area where strong wave energy resuspended bottom sediments that were deposited on the marsh surface (Day et al., 1999). The community composition and productivity of marshes in the lagoon are strongly correlated with elevation with respect to mean water level. Species composition varies with elevation and Pignatti (1966) described four plant associations correlated with elevation and other factors: a *Puccinellia-Arthrocnemum* association from 25 to 40 cm msl (mean sea level), a *Limonium-Puccinellia* association from 15 to 30 cm msl, a *Limonium-Spartina* association from 5 to 20 cm msl, and a *Salicornia* spp. association from 5 to 10 cm msl. High-precision elevation measurements for eight saltmarsh species, together with the allochthonous and invasive *Spartina townsendii*, are presented by Scarton et al. (2003). ABG and BLG production at the highest marsh elevations described by Pignatti (1966) were 666 and $1378 \, g \, m^{-2} \, yr^{-1}$, respectively. As marsh elevation is reduced, productivity decreases. For the *Salicornia* community, ABG and BLG productivity are 307 and $100 \, m^{-2} \, yr^{-1}$, respectively (Day et al., 1999).

The potential of coastal marshes to cope with SLR depends on their ability to increase in elevation sufficiently rapidly to keep pace with the water level rise. This depends on a combination of mineral sediment deposition and *in situ* organic soil formation (*sensu* Cahoon et al., 1995). Climate change combined with water level management is likely to make lagoon marshes less sustainable. Since BLG productivity is strongly correlated with marsh elevation, organic soil formation

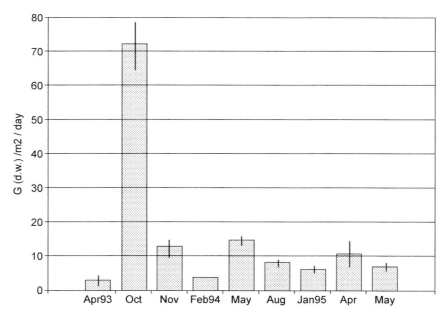

FIG. 7 Short-term sedimentation at a salt marsh site in the Venice Lagoon showing the importance of weather events. A strong storm in October led to very high sediment deposition on the marsh surface.*(From Day, J., Scarton, F., Rismondo, A., Are, D., 1998. Rapid deterioration of a salt marsh in Venice Lagoon, Italy. J. Coast. Res. 14, 583–590).*

decreases as water levels rise relative to marsh elevation. As noted above, almost all riverine input of sediments to the Lagoon has been eliminated and there is a strong net export of sediments to the Adriatic. The highest mineral sediment input to the lagoon marshes occurs during southerly Sirocco winds that elevate water levels and resuspend sediment from open water sediments. The strong waves also lead to elevated erosion rates of exposed marsh shorelines (Day et al., 1998).

The MOSE gates are designed and currently under construction (due for completion in 2022) to prevent flooding of the city of Venice during elevated water levels. This is extremely important for protection of the city. The gates will be closed only when the tide is higher than 1.1 m above sea level. Currently, this happens 3–7 times a year; for the future, 10, 12, and 15 days are estimated with SLR of 10, 30, and 50 cm (Ferrarin et al., 2013). Closing the gates prevents the elevated water levels that lead to flooding of the marshes and high rates of mineral sediment deposition on the marsh surface (Day et al., 1998). As marsh elevation decreases relative to lagoon water levels, BLG productivity will decrease as indicated above. Thus, both processes that lead to high rates of marsh surface elevation gain, mineral sediment input, and organic soil formation, will diminish. This can be offset by the reintroduction of riverine input to the lagoon and by maintenance dredging to maintain marsh elevation.

5 DISCUSSION

The information presented for these Mediterranean systems show that wetlands strongly influenced by river input have the highest productivity and the highest rates of accretion. Day et al. (2011) summarized data on accretionary dynamics for these systems over a 10-year period (Fig. 8). Riverine input led to high average vertical accretion ($10.7\,\mathrm{mm\,yr^{-1}}$) and marsh surface elevation gain ($7.3\,\mathrm{mm\,yr^{-1}}$). By comparison, for non-riverine sites, both accretion ($3.7\,\mathrm{mm\,yr^{-1}}$) and surface elevation change ($3.3\,\mathrm{mm\,yr^{-1}}$) were significantly lower. Impounded habitats in the Rhône Delta had a very low average accretion ($0.8\,\mathrm{mm\,yr^{-1}}$) and elevation change ($1.9\,\mathrm{mm\,yr^{-1}}$).

Accretionary dynamics in deltas are strongly impacted by episodic events such as storm events and river floods (Day et al., 1995, 2007) and human activities have greatly reduced the impact of these events (Day et al., 2007, 2016b; Syvitski et al., 2009; Pont et al., 2002b, 2017; Giosan et al., 2014; Ibáñez et al., 2014; Tessler et al., 2015). The development, functioning, and sustainability of deltas result from external and internal inputs of energy and materials that occur as pulses in a hierarchical manner producing benefits over different spatial and temporal scales (Odum et al., 1995; Day et al., 1997, 2016a, b). Inputs range from daily tides to switching of distributary channels, which occur in the order of hundreds to over a thousand years, and include frontal passages, river floods of varying magnitude, strong storms and associated storm surges, and formation of crevasses. Infrequent events, such as channel switching, crevasse formation, and great river floods control the location and rate of sediment delivery to the delta and thus impact the geomorphology. Pulsed events are especially

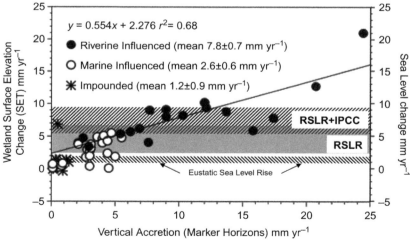

FIG. 8 Wetland vertical accretion vs surface elevation change for the coastal Mediterranean riverine (black circle), marine (white circle), and impounded (black star) sites. Mean surface elevation changes for the marsh types were compared to water level rise due to 20th century eustatic sea-level rise (ESLR); RSLR for the different sites, and RSLR plus the 21st century predicted ESLR from the IPCC (2007) (RSLR + IPCC). To accommodate rising sea level, coastal wetlands must grow at a rate ≥ water level increase, implying that only sites with high sediment input will survive the predicted SLR. *(From Day JW, Ibanez C, Scarton F, Pont D, Hensel P, Day JN, Lane R. Sustainability of Mediterranean deltaic and lagoon wetlands with sea-level rise: the importance of river input. Estuar. Coasts 34: 483-493. https://doi.org/10.1007/s12237-011-9390-x, 2011 used by permission.)*

important considering projections of SLR (e.g., FitzGerald et al., 2008; Deconto and Pollard, 2016). The two last floods in the Rhône delta fit within this framework of pulsing events, specifically lead to crevasse formation during large floods (Pont et al., 2017). In general, however, episodic events have been reduced for all levels in these Mediterranean systems. In order to adapt to climate change and increase the resilience of the Mediterranean Deltas, an integrated management of the river basin and the delta must be implemented through the development and deployment of nature-based solutions based on ecological engineering techniques.

6 SUMMARY AND CONCLUSIONS

The Mediterranean deltaic systems considered in this paper have been strongly modified by a variety of human impacts that affect their sustainability. However, each system has a unique combination of impacts that informs management and restoration approaches although each example has lessons that apply to other deltas both in the Mediterranean and worldwide.

Ebro. There has been a massive reduction of freshwater and sediment delivery to the Ebro delta. Within the delta, over 65% of wetlands have been converted to rice fields. Prior to the construction of large dams, accretion in the delta was maintained by the irrigation network that delivered sediments to the rice fields. There was also high deposition in river mouth areas during large floods. Sustainable management of the delta must include a pulsing regime of freshwater flow and mobilization of sediment from reservoirs to direct these into the delta plain. Thus, the major problem for the Ebro is the reduction of input from the basin. Without increased and better managed sediment and freshwater input to the delta, the delta will deteriorate despite ongoing wetland restoration in the delta plain. The ongoing conflict between the Spanish government and Catalonia demonstrates how political issues can compromise sustainable management and ecological restoration.

Rhône delta. The Rhône River has not had a significant decrease in discharge or suspended sediment concentration. The main problem in the Rhône delta is that almost all river input to the delta plain has been eliminated by dikes. The major freshwater input to the delta is via pumping, mainly for rice. Thus, there is a large amount of river control, but very large floods with high concentrations of sediments still occur regularly. The only area that regularly receives river input is at the mouth of the Grand Rhône. Wetland areas of the lower Camargue are falling below sea level and without more sediment input, this will continue. The two large floods demonstrated how river water and sediment can be delivered to extensive areas of the delta and lead to high rates of accretion. This suggests that managed diversions from the river can be used to sustain the delta. The two functioning distributaries, the Grand Rhône and the Petit Rhône, could be used to deliver water to the entire delta plain.

The Po delta and Venice lagoon: Syvitski and Kettner (2007) suggest that the Po River experienced a strong decrease in its sediment load (17.2–$6.4\,\mathrm{Mt\,yr^{-1}}$) from 1933 to 1987, in contrast to a small increase in water discharge. Other estimates give a sediment discharge of about $15\,\mathrm{M\,m^3\,yr^{-1}}$ (Bever et al., 2009). Furthermore, the river has at least five functioning distributaries that discharge to the Adriatic Sea that leads to river water being distributed widely on the deltaic wetland fringe fronting the sea. A major impact in the Po delta was the formation of the polders that are up to 5 m below sea level. It is likely that these cannot be maintained and will permanently flood in this coming century. Overtime sediment from the Po River could be used to fill in some of these whereas others could be converted to shallow lagoons. Wetlands at the river mouths can be maintained with river sediments (as is true for the other two deltas). However, one questions how much sediment transport for the Po has decreased, given the paucity of data concerning bed and suspended transport. For the Venice Lagoon, there is a need for sediment input to marshes if they are to survive SLR. This would suggest the need for diversion of river water into the lagoon from the Po as well as other small rivers. An ongoing project, financed with EU funds, will study the ecological effects of diverting about $1\,\mathrm{m^3\,s^{-1}}$ from the Sile River into the northern lagoon. Operation of the MOSE flood protection scheme will partly decrease resuspended sediment input to marshes. This can be offset by the reintroduction of river input to the lagoon and maintenance dredging to maintain marsh elevation.

REFERENCES

Benito, X., Trobajo, R., Ibáñez, C., 2014. Modelling habitat distribution of Mediterranean coastal wetlands: the Ebro Delta as case study. Wetlands 34 (4), 775–785.

Bever, A.J., Harris, C.K., Sherwood, C.R., Signell, R.P., 2009. Deposition and flux of sediment from the Po River, Italy: an idealized and wintertime numerical modeling study. Mar. Geol. 260 (1–4), 69–80.

Bondesan, M., Castiglioni, G., Elmi, C., Gabbianelli, G., Marocco, R., Pirazzoli, P., Tomasin, A., 1995. Coastal areas at risk from storm surges and sea-level rise in northeastern Italy. J. Coast. Res. 11, 1354–1379.

Cahoon, D., Reed, D., Day, J., 1995. Estimating shallow subsidence in microtidal salt marshes of the southeastern United States: Kaye and Barghoorn revisited. Mar. Geol. 128, 1–9.

Comín, F.A., Romero, J.A., Hernández, O., Menéndez, M., 2001. Restoration of wetlands from abandoned rice fields for nutrient removal, and biological community and landscape diversity. Restor. Ecol. 9 (2), 201–208.

Curcó, A., Ibanez, C., Day, J., Prat, N., 2002. Net primary production and decomposition of salt marshes of the Ebre delta (Catalonia, Spain). Estuaries 25, 309–324.

Day, J., Scarton, F., Rismondo, A., Are, D., 1998. Rapid deterioration of a salt marsh in Venice Lagoon. Italy. J. Coast. Res. 14, 583–590.

Day, J., Agboola, J., Chen, Z., D'Elia, C., Forbes, D., Giosan, L., Kemp, P., Kuenzer, C., Lane, R., Ramachandran, R., Syvitski, J., Yanez, A., 2016a. Approaches to defining deltaic sustainability in the 21st century. Estuar. Coast. Shelf Sci. https://doi.org/10.1016/j.ecss.2016.06.018.

Day, J., Lane, R., D'Elia, C., Wiegman, A., Rutherford, J., Shaffer, G., Brantley, C., Kemp, G., 2016b. Large infrequently operated river diversions for Mississippi delta restoration. Estuar. Coast. Shelf Sci. 183, 275–291. https://doi.org/10.1016/j.ecss.2016.05.001.

Day, J., Martin, J., Cardoch, L., Templet, P., 1997. System functioning as a basis for sustainable management of deltaic ecosystems. Coast. Manag. 25, 115–154.

Day, J., Pont, D., Hensel, P., Ibañez, C., 1995. Impacts of sea-level rise on deltas in the Gulf of Mexico and the Mediterranean: the importance of pulsing events to sustainability. Estuaries 18 (4), 636–647.

Day, J., Rybczyk, J., 2018. Global change impacts on the future of coastal systems: perverse interactions among climate change, ecosystem degradation, energy scarcity, and population. (this volume).

Day, J., Rybczyk, J., Scarton, F., Rismondo, A., Are, D., Cecconi, G., 1999. Soil accretionary dynamics, sea-level rise and the survival of wetlands in Venice Lagoon: a field and modeling approach. Estuar. Coast. Shelf Sci. 49, 607–628.

Day, J.W., Boesch, D., Clairain, E., Kemp, G., Laska, S., Mitsch, W., Orth, K., Mashriqui, H., Reed, D., Shabman, L., Simenstad, C., Streever, B., Twilley, R., Watson, C., Wells, J., Whigham, D., 2007. Restoration of the Mississippi Delta: lessons from Hurricanes Katrina and Rita. Science 315, 1679–1684.

Day, J.W., Ibanez, C., Scarton, F., Pont, D., Hensel, P., Day, J.N., Lane, R., 2011. Sustainability of Mediterranean deltaic and lagoon wetlands with sea-level rise: the importance of river input. Estuar. Coasts 34, 483–493. https://doi.org/10.1007/s12237-011-9390-x.

Deconto, R., Pollard, D., 2016. Contribution of Antarctica to past and future sea-level rise. Nature 531, 591–597. https://doi.org/10.1038/nature17145. (with supplementary materials).

Favero, V., Parolini, R., Scattolin, M., 1988. Morfologia Storica della Laguna di Venezia. Arsenale Editrice, Venice79.

Ferrarin, C., Ghezzo, M., Umgiesser, G., Tagliapietra, D., Camatti, E., Zaggia, L., Sarretta, A., 2013. Assessing hydrological effects of human interventions on coastal systems: numerical applications to the Venice Lagoon. Hydrol. Earth Sys. Sci. 17 (5), 1733.

FitzGerald, D., Fenster, M., Argow, B., Buynevich, I., 2008. Coastal impacts due to sea-level rise. Annu. Rev. Earth Planet. Sci. 36, 601–647.

Forès, E., 1992. Desecación de la laguna de la Encanyissada: un procedimiento para disminuir los niveles de eutrofia. Butlletí del Parc Nat Delta de l'Ebre 7, 26–31.

Forès, E., Espanya, A., Morales, F., 2002. Regeneración de la laguna costera de La Encanyissada (Delta del Ebro). Una experiencia de biomanipulación. Escosistemas . 2002/2.

Gatto, P., Carbognin, L., 1981. The lagoon of Venice: natural environmental trend and man-induced modification. Hydrol. Sci. Bull. 26, 379–391.

Genua-Olmedo, A., Alcaraz, C., Caiola, N., Ibáñez, C., 2016. Sea level rise impacts on rice production: the Ebro Delta as an example. Sci. Total Environ. 571, 1200–1210.

Giosan, L., Syvitski, J., Constantinescu, S.D., Day, J., 2014. Protect the world's deltas. Nature 516, 31–33.

Hensel, P., Day, J., Pont, D., 1999. Wetland vertical accretion and soil elevation change in the Rhône delta, France: the importance of riverine flooding. J. Coast. Res. 15, 668–681.

Hensel, P., Day Jr., J.W., Pont, D., Day, J.N., 1998. Short term sedimentation dynamics in the Rhône River delta, France: the importance of riverine pulsing. Estuaries 21, 52–65.

Ibáñez, C., Caiola, N., 2016. Ebro Delta (Spain). In: The Wetland Book: II: Distribution, Description and Conservation. pp. 1–9.

Ibáñez, C., Canicio, A., Curcó, A., Day, J.W., Prat, N., 1996b. Evaluation of vertical accretion and subsidence rates. In: MEDDELT Final Report, Ebre Delta Plain Working Group. University of Barcelona, Barcelona, Spain.

Ibáñez, C., Canicio, A., Day, J.W., 1997. Morphologic development, relative sea level rise and sustainable management of water and sediment in the Ebre Delta, Spain. J. Coast. Conserv. 3, 191–202.

Ibáñez, C., Curcó, A., Day Jr., J.W., Prat, N., 2002. Structure and productivity of microtidal Mediterranean coastal marshes. In: Concepts and Controversies in Tidal Marsh Ecology. Springer, Netherlands, pp. 107–136.

Ibáñez, C., Day, J.W., Reyes, E., 2014. The response of deltas to sea-level rise: natural mechanisms and management options to adapt to high-end scenarios. Ecol. Eng. 65, 122–130.

Ibáñez, C., James, P., Day, J.W., Day, J.N., Prat, N., 2010. Vertical accretion and relative sea level rise in the Ebro delta wetlands (Catalonia, Spain). Wetlands 30, 979–988.

Ibañez, C., Pont, D., Prat, N., 1997. Characterization of the Ebre and Rhone estuaries: a basis for defining and classifying salt-wedge estuaries. Limnol. Oceanogr. 42 (1), 89–101.

Ibáñez, C., Prat, N., 2003. The environmental impact of the Spanish hydrological Plan on the lower Ebro river and delta. Water Resour. Dev. 19 (3), 485–500.

Ibáñez, C., Prat, N., Canicio, A., 1996a. Changes in the hydrology and sediment transport produced by large dams on the lower Ebro river and its estuary. Regul. River. 12 (1), 51–62.

Ibañez, C., Day, J., Pont, D., 1999. Primary production and decomposition of wetlands of the Rhone Delta, France: Interactive impacts of human modifications and relative sea level rise. J. Coast. Res. 15, 717–731.

IPCC (Intergovernmental Panel on Climate Change), 2007. Climate Change 2007: The Scientific Basis, Contribution of Working Group I to the Third Assessment Report. Cambridge University Press, Cambridge.

L'Homer, A., Bazile, F., Thommeret, J., Thommeret, Y., 1981. Principales étapes de l'édification du Delta du Rhône de 7000 BP à nos jours; variation du niveau marin. Oceanis 7 (4), 389–408.

Odum, W., Odum, E., Odum, H., 1995. Nature's pulsing paradigm. Estuaries 18, 547–555.

Peyronnin, N., Caffrey, R., Cowan, J., Justic, D., Kolker, A., Laska, S., McCorquodale, A., Melancon, E., Nyman, J., Twilley, R., Visser, J., White, J., Wilkins, J., 2017. Optimizing sediment diversion operations: working group recommendations for integrating complex ecological and social landscape interactions. Water 9, 368. https://doi.org/10.3390/w9060368.

Pignatti, S., 1966. La vegetazione alofila della laguna veneta. Mem. Ist. Ven. Sc. Lett. Arti, Classe di Scienze MM. FF. e NN., Vol. 33.

Pont, D., Day, J., Hensel, P., Franquet, E., Torre, F., Rioual, P., Ibanez, C., Coulet, E., 2002b. Response scenarios for the deltaic plain of the Rhône in the face of an acceleration in the rate of sea level rise, with a special attention for Salicornia-type environments. Estuaries 25, 337–358.

Pont, D., Day, J., Ibanez, C., 2017. The impact of two large floods (1993–1994) on sediment deposition in the Rhone delta: implications for sustainable management. Sci. Total Environ. 609, 251–262.

Pont, D., Simonnet, J.-P., Walter, A., 2002a. Medium-term changes in suspended sediment delivery to the ocean: consequences of catchment heterogeneity and river management (Rhône River, France). Estuar. Coast. Shelf Sci. 54, 1–18.

Prado, P., Alcaraz, C., Jornet, L., Caiola, N., Ibáñez, C., 2017. Effects of enhanced hydrological connectivity on Mediterranean salt marsh fish assemblages with emphasis on the endangered Spanish toothcarp (*Aphanius iberus*. PeerJ 5, e3009. https://doi.org/10.7717/peerj.3009.

Probst, J.L., 1989. Hydroclimatic fluctuations of some European Rivers. In: Petts, G.E., Möller, H., Roux, A.L. (Eds.), Historical Change of Large Alluvial Rivers. John Wiley and Sons, Chichester, UK.

Rovira, A., Ibàñez, C., 2007. Sediment management options for the lower Ebro River and its delta. J. Soils Sediments 7 (5), 285–295.

Rovira, A., Ibáñez, C., Martín-Vide, J.P., 2015. Suspended sediment load at the lowermost Ebro River (Catalonia, Spain). Quat. Int. 388, 188–198.

Rutherford, J., Day, J., D'Elia, C., Wiegman, A., Willson, C., Caffey, R., Shaffer, G., Lane, R., Batker, D., 2018. Evaluating trade-offs of a large, infrequent sediment diversion for restoration of a forested wetland in the Mississippi delta. In: Estuarine, Coastal and Shelf Science. https://doi.org/10.1016/j.ecss.201801.016.

Sánchez-Arcilla, A., Jimenez, J., Stive, M., Ibañez, C., Pratt, N., Day, J., Capobianco, M., 1996. Impacts of sea level rise on the Ebro Delta: a first approach. Ocean Coast. Manag. 30, 197–216.

Sarretta, A., Pillon, S., Molinaroli, E., Guerzoni, S., Fontolan, G., 2010. Sediment budget in the Lagoon of Venice, Italy. Cont. Shelf Res. 30 (8), 934–949. https://doi.org/10.1016/j.csr.2009.07.002.

Scarton, F., 2005. Breeding birds and vegetation monitoring in recreated salt marshes of the Venice lagoon. In: Fletcher, C.A., Spencer, T. (Eds.), Flooding and Environmental Challenges for Venice and Its Lagoon. State of Knowledg. Cambridge University Press, Cambridge, pp. 573–579.

Scarton, F., 2017. Long-term trend of the waterbird community breeding in a heavily man-modified coastal lagoon: the case of the Important Bird Area "Lagoon of Venice". J. Coast. Conserv. 21, 35–45.

Scarton, F., Day, J., Rismondo, A., 1999. Above- and belowground production of *Phragmites australis* in the Po delta. Boll. Museo Civ. St. Nat. Venezia 49, 213–222.

Scarton, F., Day, J.W., Rismondo, A., 2002. Primary production and decomposition of *Sarcocornia fruticosa* (L.) Scott and *Phragmites australis* Trin. Ex Steudel in the Po Delta, Italy. Estuaries 25, 325–336.

Scarton, F., Ghirelli, L., Curiel, D., Rismondo, A., 2003. First Data on Spartina x townsendii in the Lagoon of Venice (Italy). In: Özhan, E. (Ed.), Proceedings of the Sixth International Conference on the Mediterranean Coastal Environment, MEDCOAST 03. 7–11 October 2003, Ravenna, Italy. Vol. 2. pp. 787–792.

Scarton, F., Rismondo, A., Day, J., 1998. Above and belowground production of *Arthrocnemum fruticosum* on a Venice lagoon saltmarsh. Boll. Museo Civ. St. Nat. Venezia 48, 237–245.

Sestini, G., 1992. Implications of climatic changes for the Po delta and Venice lagoon. In: Jeftic, L., Milliman, J., Sestini, G. (Eds.), Climatic Change and the Mediterranean. Edward Arnold, London, pp. 428–494.

Solidoro, C., Bandelj, V., Bernardi, F., Camatti, E., Ciavatta, S., Cossarini, G., Facca, C., Franzoi, P., Libralato, S., Melaku Canu, D., Pastres, R., Pranovi, F., Raicevich, R., Socal, G., Sfriso, A., Sigovini, M., Tagliapietra, D., Torricelli, P., 2010. Response of Venice lagoon ecosystem to natural and anthropogenic pressures over the last 50 years. In: Kennish, M., Paerl, H. (Eds.), Coastal lagoons: critical habitats and environmental change. CRC Press, Taylor and Francis, Boca Raton, FL, USA, pp. 453–511.

Syvitski, J., Kettner, A., Overeem, I., Hutton, E., Hannon, M., Brakenridge, G., Day, J., Vörösmarty, C., Saito, Y., Giosan, L., Nichols, R., 2009. Sinking deltas due to human activities. Nat. Geosci. 2, 681–686.

Syvitski, J.P.M., Kettner, A.J., 2007. On the flux of water and sediment into the Northern Adriatic Sea. Cont. Shelf Res. 27 (3), 296–308.

Tessler, Z., Vörösmarty, C., Grossberg, M., Gladkova, L., Aizenman, H., Syvitski, J., Foufoula-Georgiou, E., 2015. Profiling risk and sustainability in coastal deltas of the world. Science 349, 638–643.

Tosi, L., Da Lio, C., Strozzi, T., & Teatini, P. 2016. Combining L-and X-band SAR interferometry to assess ground displacements in heterogeneous coastal environments: the Po River Delta and Venice Lagoon, Italy. Remote Sens., 8(4), 308: 1-22.

Vivian, H., 1989. Hydrological changes of the Rhône River. In: Petts, G., Möller, H., Roux, A. (Eds.), Historical Change of Large Alluvial Rivers. John Wiley and Sons, Chichester, UK.

Welcomme, R.L., 1985. River fisheries. Tech. Pap. 262, 330.

Zanchettin, D., Traverso, P., Tomasino, M., 2008. Po River discharge: an initial analysis of a 200-year time series. Clim. Change 89 (3-4), 411–433. https://doi.org/10.1007/s10584-008-9395-z.

Zuliani, A., Zaggia, L., Collavini, F., & Zonta, R. 2005. Freshwater discharge from the drainage basin to the Venice Lagoon (Italy). Environ. Int., *31*(7 SPEC. ISS.), 929–938. https://doi.org/10.1016/j.envint.2005.05.004

Section C

Wetlands, Lagoons and Catchments

Chapter 15

Coastal Lagoons: Environmental Variability, Ecosystem Complexity, and Goods and Services Uniformity

Angel Pérez-Ruzafa[*], Isabel M. Pérez-Ruzafa[†], Alice Newton[‡,§], Concepción Marcos[*]

[*]Department of Ecology and Hydrology, Regional Campus of International Excellence "Mare Nostrum", University of Murcia, Murcia, Spain, [†]Department of Plant Biology I, Complutense University of Madrid, Madrid, Spain, [‡]NILU-IMPACT, Kjeller, Norway, [§]CIMA-Centre for Marine and Environmental Research, Gambelas Campus, University of Algarve, Faro, Portugal

1 INTRODUCTION

Despite their important ecological role and the studies conducted in the 1970–80s (UNESCO, 1979, 1980, 1981, 1982, 1986; Barnes, 1980; Lasserre and Postma, 1982; Kapetsky and Lasserre, 1984; Carrada et al., 1988) and 1990–2010 (Pérez-Ruzafa et al., 2011a), our knowledge about coastal lagoons is still limited and fragmented (Esteves et al., 2008; De Wit, 2011), while many of the paradigms on lagoon functioning are changing (Pérez-Ruzafa et al., 2011b, 2013). However, it is clear that their future is compromised by the numerous pressures they face and their extreme sensitivity to them, which is why these natural systems of great socioeconomic and environmental value must be properly managed.

The importance and the ecological and socioeconomic role of coastal lagoons have not always been well understood. For example, wetlands and coastal lagoons have been landfilled for a long time in order to develop agriculture (Doody, 2007), and in past centuries, especially at the end of 19th century, their link with malaria in many parts of the world (Strickland, 1938; Ramasamy and Surendran, 2012; Monfort et al., 2014; Sousa et al., 2014) led to many lagoons being drained.

The ecological functioning of lagoons must be analyzed and evaluated in the context of them representing a specific type of waterbody with their own characteristics (Pérez-Ruzafa et al., 2011a). Their high biological productivity and their biological and geomorphological characteristics make them one of the most-valued habitats for the services they provide. However, at the same time, they are among the most threatened and stressed aquatic habitats due to the pressures they suffer for the same reason. Urbanization, tourism, agriculture and aquaculture, together with pollution and eutrophication, among others, as well as climate change, to which they are especially sensitive, represent a group of threats that will condition the future ecological integrity of these emblematic and valuable systems. Standardizing the way in which we approach their study, the main parameters that control and describe their functioning, the way in which we classify them, and how we use and manage them are an important part of the challenges that researchers and politicians must face in order to guarantee their future sustainability.

2 COASTAL LAGOONS: DEFINITION AND DISTRIBUTION

Coastal lagoons are aquatic ecosystems that share many traits and processes with estuaries and other aquatic ecosystems in the transition between land and sea (Kjerfve, 1994). One of their main characteristics is that they are under the marine influence, but are isolated from the sea by a barrier, land spit, or similar land feature, sometimes with one or more inlets, through which a more or less restricted exchange of water and organisms takes place with the open sea. At the same time, they are characterized by being shallow, with a mean depth rarely greater than 2 m (Pérez-Ruzafa et al., 2011a), and with strong physical-chemical gradients in a constricted space (UNESCO, 1981; Tagliapietra et al., 2009; Pérez-Ruzafa et al., 2011c).

As they are areas separated from open marine conditions by a barrier, in a first approach there are two main systems considered as lagoons: those elongated or irregular stretches of water that lie between the shoreline and some kind of barrier,

Coasts and Estuaries. https://doi.org/10.1016/B978-0-12-814003-1.00015-0

and the more or less circular stretches of water surrounded or protected from the direct wave action by reef barriers or atolls (www.britannica.com). However, although their origin, geomorphological processes, and evolution are different, their biological functioning is conditioned by common features, such as shallowness and a restricted connectivity with the open sea.

Except in coral reef lagoons, the development of barriers that isolate a coastal water body to shape a lagoon is usually linked to the supply, transport, and deposition of sediments resulting from the interaction of river inflows, tides, waves, and coastal currents, in a given tectonic context. Hence, coastal lagoons can occur in tropical (rainy or arid), temperate, and cold coasts (Isla, 2009), from high latitudes such as Alaska or the Baltic sea to the tropics (Ghana, Brazil, India) and equator (Colombia or Kenya), being more common in low-lying coasts. Worldwide, there are differences in the terminology to refer to these environments: coastal lagoons, coastal lakes, semi-enclosed bays, estuaries and rías, lagons, lagunas, lagoas, lagunes, étangs, ponds, albuferas, brednings, caletas, cienagas, marismas, esteros, marsh, marais, marios, stagni, sacca, limans, limnothalassas, zalews, SECS (semienclosed coastal systems), or the more recent terms ICOLLs (intermittently closed and open lakes and lagoons) or RRE, used for regions of restricted exchange (Tett et al., 2003; Tagliapietra et al., 2009; Newton et al., 2014; Cataudella et al., 2015).

Coastal lagoons occupy approximately 13% of the world's coastlines (Barnes, 1980; Nixon, 1982), representing 17.9% of the Africa coastline, 17.6% in North America, 13.8% for Asia, 12.2% for South America, 11.4% for Australia, and 5.3% for Europe (Barnes, 1980; Kjerfve, 1994). Only in Mexico, 123 coastal lagoons along the Caribbean and Pacific coasts (Lankford, 1977) cover an area of 12,500 km^2 (Ortiz-Lozano et al., 2005) and in the Mediterranean region there are about 400 coastal lagoons, covering over 6400 km^2 (Cataudella et al., 2015).

They vary greatly in size, from less than 1 km^2 up to 10,144 km^2 in the case of Lagoa dos Patos in Brazil (Kjerfve, 1994). Most lagoon properties arise from their configuration and geomorphology (Pérez-Ruzafa et al., 2007a). Some are long and narrow, parallel to the coast and separated from the sea by well-defined linear barriers built up in front of the former coastline, while others show a branched configuration, normally where rivers form their estuaries and enclosed by a depositional barrier built up usually by wave action across the river mouths. The largest and usually most complex lagoon systems are found where large bays have been isolated from the sea through deposition barriers (Bird, 1994). In addition, several coastal lagoons may sometimes occur in groups or lagoon complexes that may be interconnected. Examples include Logarou, Tsoukaliou, and Rodias in the Amvrakikos gulf in Greece, Ichkeul and Bizerte in Tunisia, or Lagoa dos Patos, Merín, and Mangueira in Brazil.

However, all coastal lagoons are naturally ephemeral ecosystems that are dynamic, changing shape and size due to natural processes (De Wit, 2011). As mentioned above, barriers enclosing lagoons are typically formed by the longshore drift of sediments derived from the adjacent coastline or supplied by rivers and/or by shoreward drifting of material derived from the sea floor and, once semi-enclosed, the lagoons are re-shaped by erosion and deposition processes around their shores (Bird, 1994), leading normally to the gradual filling of the lagoons and their conversion into marshes.

Barnes (1980) suggested that the typical lifetime of a coastal lagoon is around 1000 years, but many of them exist for longer. During this existence, as in the case of the Mar Menor (Spain), which dates back to its last desiccation 6500 years ago, the lagoons can undergo geomorphological changes, also linked to their greater or lesser degree of connection with the adjacent sea, as a consequence of the fluctuations of the sea level or the action of storms or tsunamis that partially break the sandy barriers that separate them from the sea (Dezileau et al., 2016).

Furthermore, the lifespan of lagoons is related to human manipulation of their morphology, by filling, dredging, and other structural interventions. The increasing number of coastal works and the subsequent alteration of the sediment transport processes in coastal zones affect coast balances and lagoon barriers. Erosion threatens the integrity and consistency of the barriers, or insufficient flushing silts up the channels, interrupting the communication through the inlets between the lagoon and the open sea. Thus, coastal lagoons in their present shape are the result of strong interactions between coastal dynamics and human intervention (Cataudella et al., 2015).

Connections with the open sea are among the main factors that strongly condition the behavior of lagoon ecosystems (Pérez-Ruzafa et al., 2007a). Kjerfve (1994) classified coastal lagoons into three geomorphological types according to their water exchange and degree of isolation. These types represent three points along a spectrum of lagoons variability that reflects the open sea-lagoon dominant force and the time and space scale of hydrological variability: from the least to the most connected these are choked, restricted, and leaky lagoons (Kjerfve, 1994).

Highlighting the importance of the inlets, Pérez-Ruzafa et al. (2007a) showed that the species richness and composition of fish assemblages studied in 40 Atlanto-Mediterranean lagoons were best explained by the degree of lagoon communication with the open sea, expressed as the total area of the transversal section of inlets or its relation with the total lagoon area (the openness parameter). But the importance of the inlets went beyond, and it was noted that the restrictions imposed on connectivity by lagoon barriers lowers the colonization probabilities, which introduces a strong random component in the assemblages that manifests itself in the natural heterogeneity of the ecosystem structure and the response and homeostatic capacity of coastal lagoon ecosystems (Pérez-Ruzafa et al., 2018a). Fishing yields are also improved by geomorphological factors such as a larger lagoon perimeter with shoreline complexity and shallowness, which favor the intensity of physical-chemical gradients and nutrient inputs (Pérez-Ruzafa et al., 2007a; Pérez-Ruzafa and Marcos, 2015).

Other factors, such as the similarity in degree of salinity with the open sea, the size of the lagoon (volume and perimeter) and, to a lesser extent, the trophic status of the water column (limited by minimum phosphate concentrations), as well as other hydrographic factors such as freshwater influence and minimum salinity and temperature, affect the species composition and structure of lagoon communities, but these hydrographic and trophic factors also show a strong dependence on geomorphologic features (Pérez-Ruzafa et al., 2007a).

3 LAGOON FUNCTIONING AND ENVIRONMENTAL VARIABILITY

Coastal lagoons are characterized by the presence of multiple boundaries and transitions, between land, lagoon waters and sea, sediments, water column and atmosphere, and, frequently, between lagoon waters and freshwater inputs. Each boundary involves strong physical and ecological gradients (UNESCO, 1981), which means that lagoons are dynamic systems controlled and subsidized by physical energies, with great spatiotemporal variability. The relatively small volume of the lagoon water mass makes it very sensitive and responds quickly to changes in atmospheric temperature, both daily and seasonal, showing values that are more extreme than the adjacent sea. In the same way, the contributions of fresh water from runoff or rainfall and exchange with the sea determine important changes in the salinity degree of the waters. The direct effect of wind and waves on the bottoms induces extreme changes in turbidity and light penetration. The water column is very sensitive to temperature, oxygen solubility stratification and to change in the respiration/primary productivity balance, facilitating the appearance of hypoxia, anoxia, or dystrophic crises. Hence, most coastal lagoons correspond to the type of coastal ecosystem that is characterized by frequent environmental disturbance and fluctuations (Barnes, 1980; UNESCO, 1980, 1981; Kjerfve, 1994).

Depending on local climatic conditions, the existence or not of flowing rivers, and on the water interchange rates with the open sea, lagoons may have a wide range of salinity. Some lagoons are nearly fresh, such as the Curonian in Lithuania where salinity varies between 0.1 and 7 (Ferrarin et al., 2008) or Lagoa do Peri in Brazil where salinity varies between 0 and 1.5 (Komárková et al., 1999), but in this last case with no direct connection to the sea. Others are hypersaline, like Bardawil in Egypt, with salinity varying between 38.3 and 73 (Krumgalz et al., 1980). It is of note that within the same lagoon salinity can show a very wide range of variation, for example, Anatoliki Klisova in Greece which varies between 0.5 and 42 (Katselis et al., 2003) or in Mundel lagoon in Sri Lanka which ranges between 9 and 109 (Silva et al., 2013).

Numerous works have verified the evolution of different lagoons from hypersaline to hyposaline or freshwater conditions or vice versa (Havinga, 1959; Pérez-Ruzafa et al., 2005a; Ramos Miranda et al., 2005; Dailidienė and Davulienė, 2008; García-Seoane et al., 2016). The relevance of this parameter, which conditions the viability of colonization and settlement of new species and the communities in general, makes it important to follow through any plan for monitoring and sustainable use of lagoon environments. Global climate change studies also point in this direction as the expected increase in the variability and intensity of precipitation events, or intensification of the communication with the open sea due to rising sea levels, are expected to produce increased variability, both spatially and temporally, in salinity concentrations in coastal lagoons (Anthony et al., 2009; Angus, 2017).

Other key environmental variables, such as temperature, oxygen, pH, or turbidity, govern the functioning of coastal lagoons. These key environmental variables are interrelated and changes can act as multistressors. For example, increasing temperature and salinity results in both increased stratification, if the lagoon is not vertically well mixed by wind, and decreased solubility of oxygen, both increasing the probability of hypoxia-anoxia. Furthermore, the redox potential (Eh) affects sediment-water processes, such as adsorption-desorption of phosphate, or the speciation and availability of metals, such as iron or mercury (Viaroli et al., 2008). This can cause the release of "secondary" contaminants due to the "memory effect" and the legacy of past, highlighting the relevance of temporal aspects and processes that have been altered in the past and that are often not considered in new management strategies (O'Higgins et al., 2014). Such boundary effects at the interface between the lagoon and atmosphere or water and sediments are particularly important for fluxes, because of the large surface area to volume ratio of these shallow systems.

Hydromorphology is also one of the main keys to understanding lagoon functioning, since it conditions water movements and will be the primary determinant for the residence time of water within the lagoon (Elliott and Whitfield, 2011). It also conditions sediment and particle movements, the diffusion of pollution and the physical dispersion of eggs and larvae, conditioning the recruitment and the temporal and spatial structure of lagoon and adjacent sea populations; in essence, this encompasses lagoon intra and interconnectivity (Ghezzo et al., 2015; Pérez-Ruzafa et al., 2018b). Derived from this, coastal lagoons are important because they also provide the key to understanding the general dynamics of the nearby waters to which they are connected (Umgiesser et al., 2014), while hydrodynamic modeling is an increasingly important tool for predicting future changes in lagoon ecosystems (Gaertner-Mazouni and De Wit, 2012).

The primary controls acting on coastal lagoon hydromorphology are the climatic conditions, mainly the wind, hydrodynamic setting, topography and substrate materials, and coastal geomorphology (Duck and Figueiredo da Silva, 2012). In a comparison of 10 Mediterranean lagoons, Umgiesser et al. (2014) highlighted the wide hydrodynamic variability of these systems. According to these authors, the main processes controlling water exchange with the open sea are the tidal range and action and the wind setup, while mixing efficiency is controlled by the internal circulation dynamics, which is a function of morphological complexity and wind action in shallow water basins, and stratification in deep ones (Umgiesser et al., 2014). Water residence time can vary from a few days to more than a year, depending on the lagoon and independently of its size and water volume. Lagoons like Venice, with an area of $436 \, km^2$ have a water residence time of only 12 days, while in the Curonian in the Baltic ($1584 \, km^2$) it reaches 152 days, and in the Mar Menor, with only $136 \, km^2$, it is 318 days (Ghezzo et al., 2015). Among the lagoons that Umgiesser et al. (2014) studied, those in the Northern Adriatic, Marano-Grado, and Venice, showed the most active exchange with the open sea, driven by the tidal action. Each day more than half of their basin volume is renewed through the inlets. But most of the lagoons studied by these authors had limited connections with the sea and less than one-tenth of their basin volume was exchanged daily through the inlets (Umgiesser et al., 2014).

The type of substrate is another important source of spatial variability, directly affecting to the biological assemblages. Although many coastal lagoons are dominated by muddy bottoms, many others have a mosaic of natural habitats sandy, muddy, gravelly, or with rocky outcrops and stones of different sizes. In addition, many also have man-made structures, rock, cement, or wood used in the construction of docks and jetties.

4 LAGOON BIOTA AND ECOLOGY

Lagoon functioning is highly dependent on lagoon topography, substrate materials, and coastal geomorphology (Pérez-Ruzafa et al., 2007a; Duck and Figueiredo da Silva, 2012), and this is especially so for the biota. Due to their shallowness, lagoon bottoms are usually well irradiated, and currents and hydrodynamics are closely conditioned by the wind and the bottom topography, winds usually affecting the entire water column and promoting the resuspension of materials and nutrients. Hence, coastal lagoons are also characterized by high levels of biotic production as the photic zone extends to the lagoon floor in most areas, and these ecosystems usually receive substantial amounts of nutrients from the surrounding catchments which stimulate primary production (Kennish, 2016a). The range of annual primary production is large, fluctuating from around 50 to more than $500 \, g \, C \, m^{-2} \, yr^{-1}$ (Knoppers, 1994; Kennish, 2016a), the higher limit being of the same order of magnitude as that of the upwelling areas (Knoppers, 1994). Many lagoons can be considered to lie within the range of eutrophic conditions ($300–500 \, g \, C \, m^{-2} \, yr^{-1}$) or even hypereutrophic conditions ($>500 \, g \, C \, m^{-2} \, yr^{-1}$) (Nixon, 1995; Kennish, 2016a), and in some, like Maryut in Egypt, production can exceed $1800 \, g \, C \, m^{-2} \, yr^{-1}$ (Knoppers, 1994).

Both attributes of coastal lagoons, the existence of frequent environmental fluctuations and high biological productivity, are considered typical of the initial stages of ecological succession (Odum, 1969). In general, species of stressed and physically controlled ecosystems must adapt their behavior and physiology to a broad spectrum of physical fluctuations, and interspecies relationships are not expected to play a decisive role (Sanders, 1968; Pérez-Ruzafa and Marcos, 2015). Therefore, as in anthropogenic stressed communities, and in the early successional stages of an ecosystem (Odum, 1969, 1985), lagoon benthic communities are expected to be characterized by a high abundance of few species and low diversity, providing advantages to r-strategist species (Margalef, 1969; Michel, 1979; Barnes, 1980; Carrada and Fresi, 1988; Kjerfve, 1994; Reizopoulou and Nicolaidou, 2004). Under these naturally fluctuating conditions, the system is expected to be bottom-up controlled and exposed to eutrophication processes and dystrophic crises (Elliott and Quintino, 2007). The result is a status with naturally low biomass/abundance ratio, high production/biomass, and high abundance/species richness ratios, and a trophic system dominated by detritus feeder invertebrates and algae whose growth is conditioned by nutrients (Margalef, 1969; Wilkinson et al., 1995; Elliott and Quintino, 2007). Accordingly, estuarine ecosystems are expected to be composed of a simple and uniform benthic community represented by the euryhaline and eurythermal biocoenosis described by Pérès and Picard (1964) and supported through time by different authors (Augier, 1982; Carrada and Fresi, 1988; European Environment Agency, 2015; Gubbay et al., 2016), habitat lists, and conservation agreements (such as the Regional Seas Conventions OSPAR and Barcelona or the European EUNIS classification).

However, the biota of these ecosystems is well-adapted to this environmental variability and, as we have seen, coastal lagoons are not only extreme and fluctuating environments but are also spatially heterogeneous and can contain a variability of substrates. In fact, more than the range of fluctuations per se, what is really a source of stress is the high frequency or the unpredictability of such fluctuations. Many coastal lagoons also show well-defined temporal patterns, ranging from less than a day, as in the case of tides affecting midlittoral assemblages and exchanges with the open sea that affect the communities in the mouth of the inlets, to monthly or seasonal (Pérez-Ruzafa et al., 2007b, 2008). Hence, far from the assumption of a simple biocoenosis, numerous coastal lagoons boast a variety of communities depending on the type of substrate and vertical zoning and according to horizontal gradients from the inlets to the internal areas (Pérez-Ruzafa et al., 2008, 2011b) (Fig. 1). These communities are well differentiated from those of the open sea. In the Mar Menor, for example, macrophyte species richness and diversity are mainly determined by the vertical gradients in environmental variability, type of substratum, radiation, hydrodynamics, and the stress due to fluctuations in the environmental factors (Pérez-Ruzafa et al., 2008; García-Sánchez et al., 2012; Pérez-Ruzafa and Marcos, 2015).

Moreover, high primary production, together with the input of organic matter and nutrients from adjoining wetlands and their drainage basin, support rich faunal communities that share the lagoon resources with many migratory species that use these environments on a seasonal basis (Kennish, 2016a). Although some naturally stressed lagoons could be bottom-up controlled, with a phytoplankton dominium, coinciding with the functioning of polluted or human eutrophicated lagoons, coastal lagoon ecosystems in pristine conditions can be top-down controlled, with benthic vegetation (either micro or macrophytobenthos) as the main element responsible for ecosystem primary production and with complex trophic webs.

Usually, salinity has been considered the essential parameter to explain lagoon gradients in density, biomass, species richness, or diversity (Por, 1980; Mariani, 2001) and as one of the main factors determining the similarities and differences

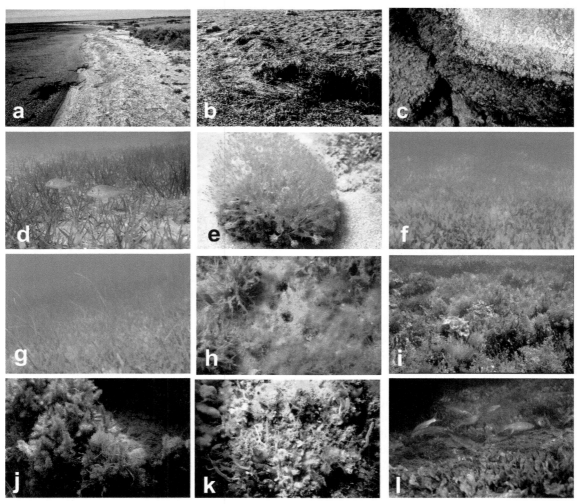

FIG. 1 Main benthic communities of coastal lagoons found in the bottoms of Mar Menor, (A) accumulations of *Cerastoderma* shells and other molluscs in the mid- and supralittoral; (B) *Cymodocea nodosa* and algae washed ashore; (C) communities of the midlittoral rock; (D) *C. nodosa* meadows on sand; (E) facies of *Acetabularia calyculus* in fine calibrated infralittoral sand; (F) *Caulerpa prolifera* meadow on mud; (G) mixed meadow of *C. prolifera* and *C. nodosa* on mud; (H) infralittoral compacted terrigenous red clay; (I) shallow infralittoral rock in a well-illuminated exposed environment, with *Cystoseira* spp.; (J) shallow well-illuminated infralittoral rock without *Cystoseira* spp. and with Rhodomelaceae; (K) sheltered and shaded shallow infralittoral rock; (L) group of gray mullets feeding on an infralittoral rock community colonized by *C. prolifera*.

in lagoon assemblages (Petit, 1953; Aguesse, 1957; Anonymous, 1959; D'Ancona, 1959; Remane in McLusky, 1999). It has therefore been used as a classification system for these environments (Anonymous, 1959; Segerstrale, 1959). However, while salinity undoubtedly plays an important role (Pérez-Ruzafa et al., 2011a), it is not a simple, single factor, and there are differences in salinity with the open sea, lagoon size, the degree of communication with the open sea, wave exposure, habitats distribution, and the trophic status of the water column which seems to be the main parameters that explain lagoon assemblages (Franco et al., 2006a,b; Pérez-Ruzafa et al., 2007a; Maci and Basset, 2009; Manzo et al., 2016; Cavraro et al., 2017). It should also be noted that salinity, as with nutrients and most other parameters, is much affected by human transformations of lagoon systems.

Among these parameters, the configuration of the communities and the functioning of the coastal lagoon ecosystems are largely governed by the degree of isolation (Kjerfve, 1994) and confinement (Guelorget and Perthuisot, 1983; Guelorget et al., 1983), understood as the restrictions imposed by the lagoon barriers to colonization rates, and the spatiotemporal variability existing in them (Pérez-Ruzafa and Marcos, 1993). The reformulation of the confinement concept made by Pérez-Ruzafa and Marcos (1993), replace the original idea involving the renewal of oligoelements of marine origin by a process of colonization of marine species. This is in line with the principles of island biogeography (Carlquist, 1974) and underlines that the settlement of a lagoon is conditioned by the possibilities of colonization of the species in question, but also by their establishment ability. The last is a characteristic of every species but cannot be completely understood except in relation to the ecological conditions provided by the lagoons. Once the species are established and reproduced in the lagoon, competitiveness will act as one of the main forces shaping the definitive assemblages.

Lagoon zonation and the structure of their communities should therefore be considered as the result of the probabilities of colonization of a species and of the balance and energy costs resulting from adaptations, reproduction, population growth rates, and competition with other species, all conditioned by, among the most relevant factors, the hydrodynamics, salinity, water characteristics, substrate, organic matter content, and the range and predictability of environmental fluctuations (Pérez-Ruzafa and Marcos, 1993) (Fig. 2).

Moreover, the low colonization rates due to the restrictions imposed by the communication channels, the barriers represented by the differences in salinity and temperature, and the very randomness of the processes involved, prevent the system from becoming homogenized and from the exclusion of rare species, opening a wide range of possibilities for diversity, adaptation to change and large differences in the species composition between lagoons. This argument also supports the idea that, in lagoon ecosystems, restrictions in the flow of energy and genes will promote structural complexity. The spatiotemporal variability generated by a model such as that described opens a range of possibilities, unpredictable a priori, for the organization of trophic networks that will ensure a stable functioning but with highly variable components (Pérez-Ruzafa, 2015).

Given these features, lagoon ecosystems are characterized by a high heterogeneity where, the main source of variability is probably the random component that restricted connectivity imposes on the colonization process (Pérez-Ruzafa, 2015). In fact, the composition, structure, and distribution of faunal assemblages in coastal lagoons show high spatial and temporal variability, with a high renewal rate of species in any given lagoon. For example, approximately 40% of ichthyoplankton species in the Mar Menor (Quispe, 2014) and of the macrofauna of benthic invertebrates in Venice Lagoon (Sigovini, 2011) changes from 1 year to other.

This temporal and spatial heterogeneity within a lagoon also manifests itself in the heterogeneity and variability observed between different lagoons. The geomorphological characteristics (mainly average depth, degree of opening, the complexity of the coast, the size of communication channels with the open sea and the area of shallow zones), together with the salinity and chlorophyll *a* concentration, explain a large part of this variability between lagoons, their faunal and floral communities, their fish species richness, and their fishing yields (Pérez-Ruzafa et al., 2005b). This variability is exemplified by the fact that of 179 fish species recorded in 40 Atlanto-Mediterranean lagoons, only 98 are present in more than two lagoons (Pérez-Ruzafa et al., 2007b), and of the 944 taxa listed by Basset et al. (2006), 75% are found in fewer than 3 of the 26 lagoons studied. The same occurs with macrophytes: of the 621 species present in Atlanto-Mediterranean estuaries and lagoons, only 45 species (7.3% of the total) appear in more than 10 localities (Pérez-Ruzafa et al., 2011a). In this last work, Pérez-Ruzafa et al. (2011a), after studying the data set of 73 Atlanto-Mediterranean coastal lagoons, found that only the Chlorophyta *Chaetomorpha linum* and *Ulva intestinalis*, and the phanerogams *Zostera noltei*, *Ruppia maritima*, and *Ruppia cirrhosa* are present in more than 50% of these lagoons. In the same study, regarding fish assemblages, only nine species were common to more than 50% of the studied lagoons. These species are *Dicentrarchus labrax*, *Anguilla anguilla*, *Mugil cephalus*, *Chelon ramada*, *Chelon auratus*, *Chelon saliens*, *Chelon labrosus*, *Sparus aurata*, *Solea solea*, and the estuarine *Atherina boyeri* (Pérez-Ruzafa et al., 2011a), all but the last are marine migrants.

In fact, there are no a specific or exclusive lagoon floral or faunal taxa, so assemblages are mainly composed of marine species that are favored by, or at least tolerate, lagoon conditions (Barnes, 1994; Pérez-Ruzafa et al., 2008).

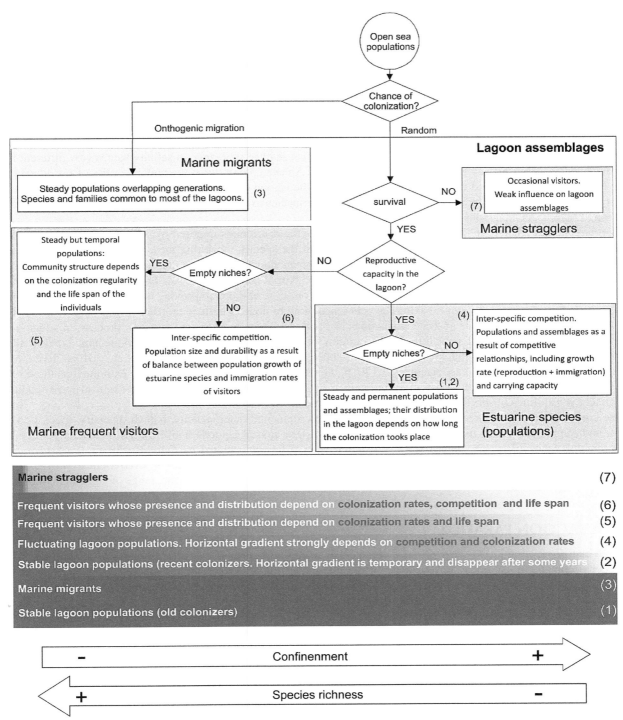

FIG. 2 Conceptual model to explain the composition and structuring of the lagoon assemblages according to the confinement model based on colonization rates in Pérez-Ruzafa and Marcos (1992). (1) Occasional visitors (marine stragglers) are species which colonize the lagoon but cannot survive in lagoon conditions. These species are rare and often limited to the mouth of the inlets, and so have little influence on lagoon assemblages. (2) Species that colonize more or less regularly the lagoon and can survive as juveniles or adults but are unable to reproduce in the lagoon environment. These include the typical marine migrants that support important fisheries and are a characteristic component of lagoon assemblages that are shared by most lagoons and estuaries over a wide geographical range (Pérez-Ruzafa et al., 2007a, 2011a). Other species with no regular migrations can be common but irregularly distributed in space and time. (3) Species which colonize the lagoon (generally after some modification in inlet morphology leading to a change in the hydrographic conditions) and which are able to reproduce in the lagoon environment under the new conditions. These species will establish themselves in the lagoon ecosystems, where their permanence and relative abundance will depend on the results of competition with already established species which are adapted to the variability of the lagoon environment (paralic species). Once established in a stable manner, they give rise to populations adapted to the lagoon conditions, more or less differentiated and containing a high number of apparently exclusive alleles, although they still maintain a certain genetic flow with the open sea populations. In the case of this species, the gradients associated with confinement tend to be temporary and may disappear after several years, when individuals of these species reach the inner areas of the lagoon.

There are genera or families with species that colonize lagoons more frequently than others and, in addition to the selective role played by coastal lagoon environments in local populations, adaptation to these variable and heterogeneous environments is not only a matter of species but of assemblages and communities (Pérez-Ruzafa et al., 2011b, 2013).

Once a species becomes established in a given lagoon, the lagoon environment shapes and conforms locally adapted populations, which gives rise to euryhaline species whose populations reach their maximum development at an optimal site (Cognetti and Maltagliati, 2000; Pérez-Ruzafa et al., 2011b). It is important to emphasize that species with low osmoregulatory capabilities, such as echinoderms, can develop populations adapted to a range of salinity values very different from that of the open sea, but still are less tolerant of this parameter. An example of this is the high mortality of *Holothuria poli* in the Mar Menor that occurred in 2017, when the salinity in the lagoon decreased from 46 to 39 during 3 months due to torrential rainfall. However, Mediterranean populations of this species usually live at salinities between 37 and 39 (Pérez-Ruzafa and Marcos, personal observation).

The randomness inherent in the colonization process due to the restrictions imposed by the separation barrier with the open sea explains the low frequency of occurrence of most of the species inhabiting these environments (Pérez-Ruzafa et al., 2011b). Once a species has colonized a lagoon, the environmental conditions model their genetic and epigenetic adaptations, leading them to optimize their biological strategies (Pérez-Ruzafa et al., 2013). The lagoon selection would not necessarily act on all the biological traits of a species but only on some of them, improving the adaptation of local populations to the lagoon environment, but breaking the coherence of all biological traits in an r/K context (Pérez-Ruzafa et al., 2013). Whether as a consequence of these adaptations or of the randomness of the colonization process, in which a rare allele can be amplified by a bottleneck effect, it is common for lagoon populations to contain unique haplotypes or alleles that are apparently exclusive or that are very rare in the open sea populations (Vergara-Chen et al., 2010a, 2010b; Mejri et al., 2011; Milana et al., 2012; Vasileiadou et al., 2016). Given that lagoon conditions are more extreme than those of the open sea, coastal lagoons become a reservoir of individuals that will be able to face climate change or restrictive conditions with greater possibilities of survival. This aspect deserves more in-depth studies.

The combination of factors such as high productivity, environmental heterogeneity, habitat diversity, species colonization, and migratory species that use this ecosystem means that coastal lagoons house a high biodiversity (European Environment Agency, 2010; De Wit, 2011). For example, the 73 lagoons studied by Pérez-Ruzafa et al. (2011a) in the Atlanto-Mediterranean region contain more than 621 species of macrophytes, 944 species of invertebrates, and 199 of fish, representing a reservoir of singular populations adapted to a wide, but specific for each one, range of lagoon environmental conditions.

5 THE LAGOON PARADOX

The high degree of variability in their physical-chemical characteristics means that estuarine ecosystems must be considered as environmentally naturally stressed areas, while the species living in these environments must be tolerant to such stress, adapting their physiology and behavior to a broad spectrum of fluctuations, both in space and in time. These characteristics of natural stress in estuaries and lagoons are similar to those for anthropogenic stress, what has been called the "Estuarine Quality Paradox" by Elliott and Quintino (2007). This adds to the difficulty of detecting and characterizing the impacts and consequences of human activities in these environments (Dauvin, 2007; Elliott and Quintino, 2007; Dauvin and Ruellet, 2009).

According to the theory of ecological succession (Odum, 1969), highly productive ecosystems such as coastal lagoons must also correspond to early stages of succession, whereby ecosystem constituents are quick growth species, with short life cycles and high fecundity (i.e., r-strategists). This agrees with existing general assumptions on the characteristics of coastal lagoons and the estuarine paradox. But some coastal lagoons, as well as being highly productive, also show structural properties similar to those of mature systems, with complex responses to stress and with sophisticated homeostatic mechanisms typical of climactic stages of ecological succession (Pérez-Ruzafa et al., 2002, 2005b). This homeostatic capability shown by many coastal lagoons has been attributed to their high spatiotemporal variability, the complex interactions that take place in their trophic webs and the unexpected dominance of K-strategies among their inhabitants (Pérez-Ruzafa et al., 2002, 2005c, 2011b, 2013; Pérez-Ruzafa, 2015).

But how can a coastal lagoon show characteristics of early and climax stages of ecosystems succession at the same time? This apparent paradox can be explained by taking into account that while productivity in these environments is enhanced by the strong physicochemical gradients, introducing restrictions to the energy flow would permit the capacity to build physical, hydrological, and biological structures, leading to very complex spatiotemporal heterogeneity and homeostatic behavior. In coastal lagoons, this restriction role is mainly performed by the inlets or communication channels with the open sea, which prevent excessive homogenization of the hydrographic conditions and maintain the differentiation and the presence of singular alleles among the populations; also, at different spatial scales inside a lagoon, the same role is played by benthic vegetation and meadows, stones, and rocky boulders, either natural or introduced by humans to build breakwaters and docks, etc. (Pérez-Ruzafa, 2015).

This emerging complexity differentiates coastal lagoons from estuaries (Pérez-Ruzafa et al., 2011c) and results in the development of the above-mentioned homeostatic mechanism of self-regulation, which also makes it difficult to detect the stress to which these ecosystems may be subjected and the effects of eutrophication or the input of contaminants on them (Pérez-Ruzafa and Marcos, 2015).

6 INFLUENCE OF COASTAL LAGOONS ON THE ADJACENT SEA

Lagoon assemblages are highly dependent on marine colonization. The populations of species settled in coastal lagoons have mainly a marine origin or are regular migrants, the latter are also of marine origin or from fluvial waters, but all playing an essential role in the lagoon trophic networks. Furthermore, up to 40% of the species that inhabit a coastal lagoon may be renewed every year (Sigovini, 2011; Quispe, 2014). However, lagoons exert an even greater influence on the adjacent sea. On the one hand, the populations of regular migrants and occasional colonizers are shaped by the lagoon conditions, while at the same time acting as vectors of the biomass exported from the lagoon to the open sea, helping to dissipate part of the excess energy derived from the high lagoon productivity. On the other hand, the probability of passive colonization in a lagoon is very low (3 larvae/m^2 per million larvae produced, for distances of less than 10 km, in the Mar Menor) (Pérez-Ruzafa et al., 2018b). This represents less than 0.2% of the larval pool in the adjacent open sea, and it is of the same order of magnitude in lagoons with very different water renewal rates, such as the Mar Menor, Curonian, or Venice lagoons (318, 150, and 12 days, respectively) (Ghezzo et al., 2015). Nevertheless, there are still a sufficient number of individuals to maintain genetic flows and stable populations, even in species that do not reproduce in the lagoon. However, the restricted connectivity that this implies and randomness involved act as a bottleneck that can amplify the frequency of alleles that are rare in coastal open sea populations, at the same time as forcing the adaptation of singular populations to lagoon environmental conditions.

Indeed, the probability of the Mar Menor exporting particles to the open sea is much higher that of receiving the same (Pérez-Ruzafa et al., 2018b). This must be taken into account since, in practice, lagoons could play an important role in organizing the genetic structure of the adjacent coastal marine populations, exporting haplotypes and alleles that are rare outside and that have been amplified in the lagoon environment under selective conditions involving a wide range of environmental variations (Pérez-Ruzafa et al., 2011c, 2018b). This could enhance the adaptation capabilities of the species in question to environmental and climatic changes.

Despite all the above, the influence of coastal lagoons on the adjacent coastal sea is not only biological. In a recent study, Ferrarin et al. (2017) identified unexpected effects that lagoons exert on the oceanographic dynamic of the Adriatic Sea. While coastal lagoons represent only 0.002% of the Adriatic Sea volume, the effect of lagoons increases the tidal range by 5% and currents by 10%.

7 ECOSYSTEM SERVICES PROVIDED BY COASTAL LAGOONS: ACTUAL STATUS AND PERSPECTIVES

Coastal lagoons provide a wide spectrum of valuable ecosystem services and societal goods and benefits for human welfare (Table 1). However, both the identification and classification of these services and their assessment are still a complex issue and far from being resolved. There is an important lack of consensus on what constitutes an ecosystem service and this concept can be synonymous with others such as abiotic provisioning, human-related activities, ecological phenomena, or products from ecological systems; moreover, it is not always easy to distinguish between services, goods and benefits (Elliott et al., 2017), which, together with the variety of methodologies and measurements, makes environmental and economic accounting difficult (Haines-Young and Potschin, 2013; La Notte et al., 2017; Newton et al., 2018). Moreover, the literature on goods and services related to coastal lagoons is very scarce (Remoundou et al., 2009; Velasco et al., 2018) and suffers from a great heterogeneity of approaches and methodologies.

Coastal lagoons are among the marine habitats with the highest biological productivity (Nixon, 1982; Alongi, 1998; Kennish and Paerl, 2010) and play an important ecological role by providing a broad set of goods and services (Anthony et al., 2009; Basset et al., 2013; Newton et al., 2018; Velasco et al., 2018). They shelter an important part of global biodiversity (European Environment Agency, 2010; De Wit, 2011), offering a collection of habitat types for many species, while functioning as refugia, nursery areas and feeding grounds for opportunistic marine fishes and wildlife in general (Yañez-Arancibia and Nugent, 1977; Clark, 1998; Vasconcelos et al., 2011). Humans benefit from this productivity,

TABLE 1 Ecosystem Services and Societal Goods provided by coastal lagoons based on a modification of the Common International Classification of Ecosystem Services proposed by the EEA (Haines-Young and Potschin, 2013) and using information from the studies developed in the Mar Menor by the research group of the University of Murcia "Ecology and Management of Coastal Marine Ecosystems" during the last 35 years, and Seeram (2008), Lillebø et al. (2015), Elliott et al. (2017), Newton et al. (2018), and Velasco et al. (2018)

Section/ Category	Division/Type	Human Actions	Lagoon Attribute/ Process	Ecological Service /Societal Good
Provisioning (including nutritional, material, and energetic outputs from living systems, sediments and water)	Nutrition	Fisheries	Biological productivity	Fishes (mainly Sparidae, Mugilidae, Anguillidae, and Moronidae families, which are present in more than 75% of Mediterranean lagoons)
				Molluscs (mainly shellfish, oysters, and clams)
				Crustaceans (mainly shrimps and prawns)
				Export of fishable biomass to the open sea
				Fish feed (wild, farmed, and bait)
		Aquaculture (intensive or extensive exploitation of biological resources)	Biological productivity	Lagoon products (fish, molluscs, and crustaceans) Extraction of algae and macrophytes
			Shelter	Economic benefits for infrastructures and surveillance
			Extreme environmental conditions	Exclusion of parasites and some predators
		Farming in the surrounding areas	Protection from storms and marine intrusion	Farm products
			Organic matter deposition and fertility of riparian lands	
	Non-nutritional biotic materials	Resources extraction	Accumulation of organogenic materials	Construction and ornamental materials
				Fertilizer and biofuels
			Biodiversity	Ornaments and aquaria
		Biotechnology	Chemical and other substances from biota	Medicines and blue biotechnology
				Genetic materials
		Passive	Chemical and other substances from biota	Oxygen production
			Restricted connectivity (inlets) and adaptation to lagoon extreme conditions	Preservation of rare alleles, maintenance of genetic diversity of marine species and selection of genetic adaptations to extreme and variable conditions
			Biological productivity	Nursery for fish and marine and freshwater invertebrates
			Shelter	
			Resources provisioning	Export of production and biomass to the open sea and terrestrial ecosystems (mainly through migratory species and water birds)
			Habitat provisioning	Wildlife refugium (avifauna/insects/small mammals/marine turtles)

TABLE 1 Ecosystem Services and Societal Goods provided by coastal lagoons based on a modification of the Common International Classification of Ecosystem Services proposed by the EEA (Haines-Young and Potschin, 2013) and using information from the studies developed in the Mar Menor by the research group of the University of Murcia "Ecology and Management of Coastal Marine Ecosystems" during the last 35 years, and Seeram (2008), Lillebø et al. (2015), Elliott et al. (2017), Newton et al. (2018), and Velasco et al. (2018)—Cont'd

Section/ Category	Division/Type	Human Actions	Lagoon Attribute/ Process	Ecological Service /Societal Good
	Nonbiological resources	Mining and other resources extraction and management	Biogeochemical process	Water quality
			Freshwater storage	Freshwater for nondrinking purposes
			Evaporation Shallowness Shelter	Salt
			Sedimentation and retention of sediments	Sand for construction or other purposes
			Protection from winds and waves	Transport and harboring
		Passive	Climate regulations	Habitation
	Energy	Biomass culture	Biological productivity	Biomass for energy production (micro and macroalgae)
			Shelter	
		Engineering	Wind, waves and currents (mainly in inlets)	Energy obtained from water, waves, and currents
Regulation and maintenance (covers all the ways in which living organisms can mediate or regulate the physicochemical and biological environment that affect human performance; also covers the mediation of flows of solids, liquids and gases that affect people's performance)	Remediation	Passive / algal cultures / green filters	Biogeochemical processes	Bioremediation
			Accumulation/storage, dilution and filtration (in biota, organic matter, and sediments) of pollutants and chemical elements	Water purification
				Water quality maintenance
				Trap and filtering of heavy metals and other pollutants in inorganic sediments and organic matter
				Nutrient retention / Denitrification
				Sulfur reduction
				CO_2 retention
	Regulation and maintenance of habitats and ecosystem	Passive	Coastal lagoons regulation processes	Hydrological balance
				Improvement in air quality
				Control nutrients and other chemical and mineral particles from agriculture or urban areas
				Reduction of natural hazards
			Lagoon barrier	Flood protection
				Erosion regulation
				Protection against action of the sea
				Shoreline stabilization
				Ecosystem spatiotemporal heterogeneity and complexity
			Habitat diversity	Supports a wide range of biological diversity
				Wildlife refuge
				Shelter for emblematic or protected species

Continued

TABLE 1 Ecosystem Services and Societal Goods provided by coastal lagoons based on a modification of the Common International Classification of Ecosystem Services proposed by the EEA (Haines-Young and Potschin, 2013) and using information from the studies developed in the Mar Menor by the research group of the University of Murcia "Ecology and Management of Coastal Marine Ecosystems" during the last 35 years, and Seeram (2008), Lillebø et al. (2015), Elliott et al. (2017), Newton et al. (2018), and Velasco et al. (2018)—Cont'd

Section/Category	Division/Type	Human Actions	Lagoon Attribute/Process	Ecological Service /Societal Good
	Regulation and maintenance of biological processes	Passive	Lifecycle maintenance Habitat diversity Preservation of rare alleles	Populations regulation Genetic diversity Reproductive and feeding areas
			Extreme environmental conditions	Invasive alien species control Protection against predators and parasites
	Climate regulation	Passive	Primary production	Carbon sequestration
			Organic matter accumulation in sediments	
			Heat balance	Smoothing temperatures
Cultural (all the non-material, and normally non-consumptive, outputs of ecosystems that affect physical and mental states of people)	Physical or experiential use of ecosystems and landscape or seascape	Tourism activities and facilities	Beaches Water quality Shelter Fish and seafood quality	Tourism Social outings
		Environmental protection and nature reserves	Biodiversity Land and seascapes	Ecotourism Bird watching Nature-based aquatic activities Aesthetic experience and well-being Artistic inspiration
		Health care and spa installations	Shelter Shallowness High temperatures and salinity (ionic concentration) Sediment properties	Health care and thalassotherapy
		Nautical facilities and harbors	Winds with small waves Shelter	Nautical sports Military installations for ships and hydroplanes
	Intellectual representations of ecosystems and landscape or seascape	Scientific research	Ecosystem complexity	Scientific knowledge and conditions for new studies
			Semi-enclosed system and well-defined boundaries	
			Intense evaluative and ecological process	
			Shelter and accessibility	
		Cultural research	Ancient history and traditions	Spiritual and symbolic
		Archaeology	Sedimentation process in a shelter area	Archaeological sites and Historical knowledge
			Ancient concentration of human activities	
		Educational actions and technological research	Cultural heritage	Information and knowledge Cultural know-how Lagoon fishing techniques Traditional food Use of wind energy and water (windmills, wells, and irrigation ditches)
			Biodiversity and ecological processes	Environmental education

as lagoons usually support important fisheries, shellfish harvesting, and aquaculture. Frequently these activities coexist with salt extraction, tourism, nautical sports, swimming, and thalassotherapy (the health benefits of the sea) (Pérez-Ruzafa et al., 2009, 2011b). Furthermore, coastal lagoons provide a significant number of environmental services of high value that are less visible. These include flood control, groundwater recharge, the prevention of seawater intrusion, shoreline stabilization and regulation of erosion by offering protection from natural hazards, storm protection, the retention and export of sediment and nutrients, mitigation of climate change, the regulation of water quality and components quantity by biochemical processes, the regulation of carbon, oxygen, nutrients in the water and in the atmosphere and retaining contaminants in the sediments, precluding their incorporation in trophic webs, or as reservoirs of genetic or species biodiversity (Barbier et al., 1997, 2011; Secretaría de la Convención de Ramsar, 2006; Anthony et al., 2009; Pérez-Ruzafa et al., 2011c; Barbier, 2015; Newton et al., 2018; Velasco et al., 2018). In addition, the long history of the human-lagoon relationship has produced a rich and particular cultural heritage and native knowledge of the lagoon ecological processes.

Because of these characteristics (Fig. 3), coastal lagoons are among the habitats that provide high amounts of ecosystem services (De Groot et al., 2012; TEEB, 2012; Russi et al., 2013). However, the perception of these benefits is not always well recognized and may depend on the cultural context or stakeholder characteristics (Newton et al., 2018; Velasco et al., 2018). Significant differences were detected in a survey conducted among a broad representation of Mar Menor lagoon stakeholders, the responses depending on the profession of the respondent (e.g., fishermen's views differ), gender, education, and origin (Velasco et al., 2018). Therefore, these factors should be taken into account in any lagoon assessment, decision-making, or management processes. In the same study, regarding the ecosystem services and societal goods and benefits offered by the lagoon, the most valued were tourism, followed by landscape and its role as refuge and nursery for different species, with average scores of 6.96, 6.80, and 6.56 out of 10, respectively. The least valued services were resources provisioning and farming, with scores of 3.81 and 4.14 (Velasco et al., 2018). This partly coincides with the results obtained by Newton et al. (2018), in which the provision of food and tourism recreation, together with transport and habitation and their role as a wildlife refuges, were considered by scientists as the most important services provided by most of the 32 coastal lagoons studied worldwide.

Fisheries probably represent the most extensive exploitation of biological resources in coastal lagoons. The main commercial species are fish of the Sparidae, Mugilidae, Anguillidae, and Moronidae families (Kapetsky and Lasserre, 1984; Pérez-Ruzafa and Marcos, 2012), which are present in more than 75% of Mediterranean lagoons (Pérez-Ruzafa et al., 2007a, 2011a), although prawns, shrimps, crabs and mussels, oysters, and clams may also be very important locally. According to some estimations, fishing in coastal lagoons accounts for 10% of fish production and 30% of demersal fish

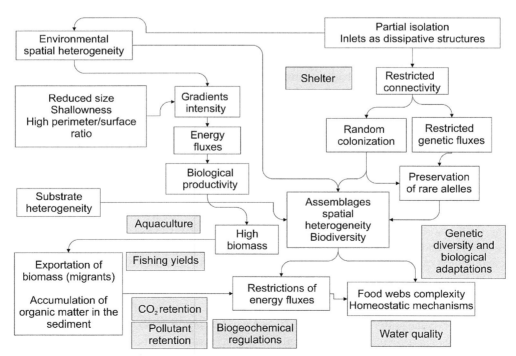

FIG. 3 Main geomorphological features and ecological processes (white boxes) responsible for many of the ecosystem services and societal goods (gray boxes) provided by coastal lagoons.

in the Mediterranean area (Quignard, 1984). The total annual yield of coastal lagoon fisheries in the world was 694,195.9 tons yr^{-1}, with a mean productivity of 137.4 ($\pm$21.6 SE) kg ha^{-1} year^{-1} for the 356 lagoons with available data studied by Pérez-Ruzafa and Marcos (2012). Of this amount, 191,534.3 tons yr^{-1} corresponds to the 169 Mediterranean lagoons included in the mentioned review. For its part, clam fishing activity in Venice, which involves three main species (*Ruditapes decussatus*, *Ruditapes philippinarum*, and *Scrobicularia plana*), produced an estimated annual production of about 4000 tons representing 60% of the Italian national production (Seeram, 2008), while French Mediterranean lagoons produced 4180 tons of mussels in 2010 (Cataudella et al., 2015). Globally, aquaculture produces a mean of 26 tons of molluscs per lagoon and per year (about 10 kg ha^{-1} yr^{-1}), the most productive being French and Italian lagoons with 41 and 26.78 kg ha^{-1} yr^{-1}, respectively (Pérez-Ruzafa and Marcos, 2012).

However, some ecosystem services and societal goods, such as fisheries or tourism, are easier to identify and be assigned a monetary value than others, such as erosion or pollution control. The latter have been neglected because they are difficult to identify, or the lack of available data, or it is difficult to assign an economic value (Barbier et al., 2011). Moreover, the ecological and sociological characteristics of coastal lagoons are dynamic, and in the way in which people use coastal lagoons are expected to change, as will the ecosystem services provided by lagoons. Therefore, it is important to find strategies to incorporate this dynamic nature in the methodologies used to evaluate them.

Lagoon ecosystem services are subject to a great variety of pressures and transformations due to the multiplicity of uses and management actions that coastal lagoons support (Pérez-Ruzafa et al., 2005a, 2009; Velasco et al., 2018). The future of the conservation of these natural systems may depend to a large extent on a socioeconomic approach, where the identification of ecosystem services, their valuation and stakeholder perception may bring users and managers to realize the real value of these ecosystems and allow them to better understand their importance. This approach therefore can be considered a viable strategy to promote the conservation of coastal lagoons, given that the services provided are more readily appreciated when they can be associated with an economic value (Almansa and Martínez-Paz, 2011; TEEB, 2012).

8 THE FUTURE OF COASTAL LAGOONS: MAIN PRESSURES AND IMPACTS ON THE LAGOON SYSTEMS

Coastal lagoons have a long history of human uses that interact with lagoon processes. For example, agricultural development and deforestation have produced an increase in the sedimentation rates and important land and landscape transformations since the middle ages. In some lagoons, such as the Mar Menor, other impacting activities, such as mining, have existed since Roman times and are responsible for introducing large amounts of heavy metals into the lagoon sediments (Simonneau, 1973; Pérez-Ruzafa et al., 1987, 2005a; García and Muñoz-Vera, 2015). In the 19th and early 20th centuries many coastal lagoons were dried to prevent diseases such as malaria or the simple discomfort of mosquitoes, or intentionally isolated from the sea to favor their clogging and to reduce salinity to be used for farming or rice cultivation (Doody, 2007; Monfort et al., 2014). Other activities, such as salt mining, the opening or widening of channels, urban development or, more recently, the construction of leisure harbors, artificial beaches, or tourism infrastructures also constitute a threat of serious impact on these ecosystems (Pérez-Ruzafa and Marcos, 2005; Pérez-Ruzafa et al., 2011c; García-Oliva et al., 2018). Among current threats, the introduction of allochthonous species, usually for aquaculture, can lead to modifications in local biodiversity patterns (Cataudella et al., 2015).

In essence, coastal lagoons suffer multiple pressures from uses and activities that threaten their ecological integrity. As the human population is growing, increasing the demand for resource and the production of wastes, coastal lagoons are expected to be among the ecosystems most at risk of collapsing if social awareness is not improved and management decisions are not taken. This also requires comprehensive knowledge of processes involved in the organization of lagoon ecosystems at the time that decisions are made concerning regulating uses, the management of shorelines, coastal works, or modifications of the inlets communicating with the open sea.

At present, engineering works in coastal lagoons mainly involve land claim and sand filling for coastal uses (e.g., ports, marinas, artificial beaches, and urbanizations) and the dredging of bottoms for sand extraction, the creation of navigation channels or the conditioning of beaches (i.e., beach nourishment). It is also common to close or enlarge the inlets to control exchanges with the open sea, isolating them or increasing communication, for navigation purposes, the control of rising sea levels (such as the Moses barrier infrastructure in the Venice lagoon), or the management of fishing activity.

But there are many other coastal interventions that can directly or indirectly affect lagoons, modifying the different mechanisms that shape them, such as the interaction of tidal and wave energy, hydrology, and riverine inputs or sediment balances. These will affect characteristic features, such as inlets and barrier islands that control the exchange with the adjacent sea, changing the coastal lagoon status from choked to restricted or leaky or vice versa (Kjerfve, 1986), changing their depth, surface area to volume ratios, and, ultimately, producing changes in the natural functioning of these ephemeral and dynamic systems (Williams et al., 2003). The degree of connectivity, with the surrounding catchment area and with the adjacent sea, and the intensity of the physical-chemical gradients are fundamental to the functioning of coastal lagoons. Connectivity can be decreased by damming and building dykes, or increased by opening artificial inlets and dredging, as well as "fixed" by consolidating inlets. Therefore, engineering in coastal lagoons usually changes their natural function (Williams et al., 2003).

9 OUTSTANDING FUTURE THREATS: EUTROPHICATION

One of the main stressors for lagoons, during recent decades and continuing, is nutrient over-enrichment and eutrophication (Sfriso et al., 1992, 2003; Solidoro et al., 2010; Viaroli et al., 2010; Kennish and de Jonge, 2011). Population growth and, consequently, the generation of wastewater, agricultural activities, or the atmospheric deposition of nitrates and ammonium will continue in ever greater quantities (Tagliapietra et al., 2011).

Eutrophication is a process derived from the increase in the supply of organic matter to the ecosystem (Likens, 1972; Nixon, 1995; Gamito et al., 2005), which provokes a general change in biological communities. The process is generally identified with an increase in the entry of inorganic nutrients that favor primary production in the ecosystem (European Environment Agency, 2001), which manifests itself especially in the succession of aquatic vegetation, from perennial benthic macrophytes (mostly seagrasses) to fast growing epiphytes, later to free floating macroalgae and phytoplankton and, finally, to picoplankton and cyanobacteria (Valiela et al., 1997; Sfriso et al., 2003; Viaroli and Christian, 2003; Gamito et al., 2005; Sfriso and Facca, 2007; Viaroli et al., 2008). In turn, the decomposition of the vegetation biomass excess increases oxygen demand, leading to anaerobic processes and dystrophic crises and to profound changes in the lagoon benthic community (Tagliapietra et al., 2011). Such processes have been described in numerous coastal lagoons (Reyes and Merino, 1991; Boynton et al., 1996, Taylor et al., 1999; Newton et al., 2003; Solidoro et al., 2010; Kralj et al., 2016; Leruste et al., 2016; Pérez-Ruzafa et al., 2018a), and even modeled (Giusti et al., 2010; Zhu et al., 2017).

Some lagoons have shown homeostatic and self-regulatory mechanisms with a capacity to delay the effects of eutrophication for decades (Pérez-Ruzafa et al., 2002). These mechanisms are based on top-down control over phytoplankton exerted by ichthyoplankton and jellyfish (Pérez-Ruzafa et al., 2002, 2005c) or by filtering bivalves and other benthic organisms, as it has been described in several coastal bays (Heck and Valentine, 2007; Newell et al., 2007; Lonsdale et al., 2009). In addition to the complexity of the lagoon trophic web, the spatiotemporal heterogeneity of lagoon communities plays an essential role in capacity of lagoon ecosystems to withstand eutrophication (Pérez-Ruzafa et al., 2005c, 2011b). However, every system has a limit and if the input of nutrients is continuous and other sources of stress act at the same time, the systems can collapse, losing resilience and reaching a point of no return so that they reach to a different, generally simpler, ecological state (Pimm, 1984; Elliott et al., 2007; Tett et al., 2007; Pérez-Ruzafa et al., 2018a).

In fact, eutrophication, which is increasing worldwide and to which the great majority of coastal ecosystems are exposed (Nixon, 1995; Kennish, 2009, 2016b; Kennish and Paerl, 2010), besides resulting from contamination by nutrients or an increase of organic matter supply, is a fundamental change in the energetic base, which may propagate through the system in various ways and produce a great variety of changes (Nixon, 2009). But it is not a state, it is a process (Nixon, 2009; Ferreira et al., 2011), a fact that cannot be ignored if we are to understand the evolution and functioning of the "new" ecosystem, and the management actions that need to be adopted.

When eutrophication is chronic, its effects include states of anoxia, toxic phytoplankton blooms, the massive mortality of benthic organisms, a reduction in biodiversity, major changes in the distribution of species, a decline of fisheries, imbalance in the food webs, the alteration of biogeochemical cycles and a reduction of ecosystem services due to the loss of water quality, and simplification of the ecosystem structure and regulatory processes (Kennish and de Jonge, 2011; Kennish, 2016b). Once the process is triggered, it is difficult to slow down. The progressive release of the nutrients accumulated in the sediment, the impoverishment in species and simplification of the trophic network mean that the eutrophic state can be prolonged for a long time, even though the entry of nutrients may cease (Nienhuis, 1992).

9.1 Global Climate Change: Consequences for Coastal Lagoons

Different aspects of climate change, including increasing sea surface temperature (SST), sea-level rise (SLR), and changes in rainfall patterns, threaten the ecological functioning of coastal lagoons and therefore the valuable ecosystem services provided by them (Newton et al., 2018). The evolution of the world's coastlines has been and continue to be much influenced by changing levels of land and sea (Newton and Icely, 2008), and lagoon systems are especially susceptible to these modifications in their formation and dynamics. Global climate change is therefore particularly important, and lagoons are "sentinel" systems (Eisenreich, 2005; Brito et al., 2012) that must be studied in a special way in this particular context.

Sea level, temperature, rainfall, evaporation and hydric balance, and storminess are expected to change significantly with global climate change and to impact coastal lagoons directly (Anthony et al., 2009; Angus, 2017). Changes in air temperature strongly influence the water temperature of slow-moving and shallow water bodies (Turner, 2003). Water temperature in turn influences dissolved oxygen concentrations, the physiology of lagoon organisms and oxygen demand, species ranges of distribution and patterns of migration, as well as the timing of lagoon processes (Anthony et al., 2009). Moreover, beyond this, the barriers that isolate the lagoons will probably be modified, often breaking or disappearing, although it is conceivable that the redistributed sediment may form new barriers or possibly new lagoons (Angus, 2017).

At present, the biological and ecological consequences of climate change for lagoon organisms and ecosystems are more speculative than supported by evidence. Some authors suggest that lagoon biota will have fewer competitive options than their close marine counterparts in a future climate change scenario and that the specialist lagoon biota will face the need to find equivalent places or become extinct in the event that the new environmental conditions become completely unfavorable (Angus, 2017). However, some of these assertions do not take into account that many of the classical assumptions are being reconsidered (Pérez-Ruzafa et al. 2011a) and that the real functioning and genotypic or epigenetic adaptations of coastal lagoon biota suggest other possibilities, including a better capability to face climate change conditions. Several aspects of lagoon biology and ecology, discussed above, suggest that coastal lagoon communities may be more heterogeneous than expected (Pérez-Ruzafa et al., 2004, 2005a, 2007a, 2008), showing complex autoregulation mechanisms based on their spatial and temporal heterogeneity (Pérez-Ruzafa et al., 2005a) and therefore might more easily adapt to pressures and show greater stability than marine communities. Firstly, the populations that inhabit coastal lagoons have been shaped to survive in a more fluctuating and extreme environment than that of the open sea and have adapted to the conditions expected with climate change. Furthermore, the fact that adaptation takes place at population, not species level, means that the species with populations inhabiting lagoons, as a whole, show greater genetic diversity with a capacity to adapt to a broader spectrum of environmental conditions. Secondly, the main constituents of lagoon communities, and common to most coastal lagoons, are migratory species, which are usually adapted to survive in a wide range of environmental conditions. Finally, colonization from the open sea is a crucial process in conforming lagoon assemblages. This process has a high random component and leads to the high inter and intra-lagoon spatial and temporal heterogeneity observed and high renewal rates of species. The fact that approximately 40% of benthic and pelagic species change from 1 year to the next endows coastal lagoon food webs with a greater resilience than if most of the components were fixed, since some elements lost as a consequence of the effects of climate change could be replaced by others in a process that is common in these systems.

However, in a study on the hydrological response of 10 Mediterranean lagoons, Ferrarin et al. (2014) warned that numerical simulations foresee a general loss of such intra-lagoon and inter-lagoon variability in their physical properties. This could be an example of the process that may affect many coastal environments in the future, summarized as a homogenization of the physical characteristics, with a tendency toward adopting features of the open sea (Ferrarin et al., 2014). Hence, the resilience and homeostatic responses of coastal lagoons may decrease under the pressures of climate change.

The consequences of global climate change are undoubtedly a topic of maximum interest in the future of coastal lagoons and for their management. The study of ecological descriptors and the most robust and adequate indicators is fundamental to increase our knowledge and decide mitigation steps (Pitacco et al., 2018). In this sense, some authors, such as Pérez-Ruzafa et al. (2011a) or Brito et al. (2012), have suggested that lagoons are of exceptional value in the context of climate change studies as sentinel systems, and the importance of sharing information and comparing different systems to understand cause-effect relationships and, hence, to draw up conceptual models that anticipate its effects.

Despite all of this, in a scenario of sea level rise as a consequence of climate change, the flooding of extensive areas of low coast would be expected, leading to an expansion of lagoon areas and the formation of new ones. However, this would be valid for unpopulated coasts or countries of low economic development. In inhabited areas, the expected tendency will be to develop defenses and engineering works that impede such processes (Mimura, 2013).

10 FINAL REMARKS

The future of coastal lagoons involves solving several current and future challenges and taking adequate management measures to mitigate the most important of these challenges faced by these ecosystems. One problem is the lack of definition that in some cases weighs on these systems, which leads to management policies and measures that cannot be extrapolated and standardized.

As an example, the European Water Framework Directive (WFD), although it establishes a well differentiated typology of water bodies on the basis of scientific and biological criteria, encounters some problems in the case of coastal lagoons. The fact that European coastal lagoons can be classified as transitional waters or as coastal waters, depending on the entry or not of freshwater, leads to different reference conditions or intercalibration parameters and, moreover, to different management decisions and policies for each of them (Pérez-Ruzafa et al., 2011a; Newton et al., 2014). The present day

WFD definition of transitional waters and any new classification proposal should replace the freshwater influence criterion with another based on common features, such as geomorphological characteristics or relative isolation (Kjerfve, 1994; Pérez-Ruzafa et al., 2011a).

The management of coastal lagoons is not easy, not only due to gaps in our knowledge, but also because of the multiplicity of stakeholders, the complexity of the social and administrative system and the distribution of responsibilities and competences that depend on countries and regions (Pérez-Ruzafa and Marcos, 2008). Management plans must take into account and harmonize agriculture, aquaculture, fisheries, tourism, urban areas, military uses, salt and other mining activities, industrial areas, conservation aspects (many under international agreements such as the Ramsar Convention), and cultural heritage status. All of this occurs many times in a wide territory that includes the drainage basin and the entire environment involved. As an example, in the case of Mar Menor lagoon in Spain, which has recently suffered an eutrophication crisis, it is necessary to coordinate three administration levels and an undefined number of departments, as groundwater management depends on the Hydrographic Confederation of the Segura River, while any management of the coast and submerged beaches depends on the General Directorate of Sustainability of the Coast and Sea, both under the Spanish Ministry of Agriculture and Fisheries, Food, and Environment. In addition, the different departments of the Autonomous Regional Administration have responsibilities concerning tourism, agriculture, fishing, mining, industry and environment, and the local Municipalities have responsibilities for the beaches above sea level and urban development and management. The administrative structure and the division of powers at all levels can change every time if there is a change of government. The problem is more complex in the case of lagoons that are shared by more than one country under different laws and regulations, such as the Vistula and Curonian lagoons, shared by Poland and the Russian Federation and Lithuania and the Russian Federation, respectively.

It is emphasized here that among the most important current threats to coastal lagoons, we can highlight habitat loss and modifications, coastal works as a consequence of urban and tourist development, transformations and pressures exerted by certain uses such as agriculture or industry developed in their drainage basins, together with the consequences that these uses have on the ecosystem, which cause processes of eutrophication or pollution, and the consequences of global climatic change when dealing with these ecosystems especially sensitive to the changes in sea level or water salinity.

Faced with this situation, coastal lagoons sustainability must necessarily involve several steps:

- It is necessary to correctly identify the parameters that condition the functioning of these ecosystems and their spatio-temporal variability, in a biogeographic context beyond the national or local. This will identify clear typologies, their functioning and comparable guidelines that will allow the design of solid and comparable management policies and measures that guarantee sustainability.
- Management must be based on scientific knowledge, understanding the spatiotemporal scales of the processes and clearly differentiate between naturally stressed situations from those alterations of anthropic origin. Coastal lagoons have a homeostatic capacity, based on the restricted connectivity they maintain with the open sea and the heterogeneity of their populations, which are still poorly understood. In many cases, actions taken in communication channels have had unforeseen consequences that irreversibly affect these systems. To know in depth the functioning of connectivity, hydrodynamic processes, and how they condition the lagoon communities are key aspects in their understanding.
- There are also key aspects in managing these systems that must address in a comprehensive manner the study of dystrophic processes: the structure of trophic networks in systems that are highly stable but of great variability in terms of their components, and the role played by invasive species in modifying these networks.
- But all this, without a doubt, requires the need to consider the socioeconomic framework that surrounds the coastal lagoon, and the opinion and sensitivity of the stakeholders. The development of management tools for coastal lagoons needs interdisciplinary research and decisions in a framework of active interactions with the end-users (Aliaume et al., 2007).
- From a broader point of view, the consequences of climate change in coastal lagoons are of great importance in the immediate future. On the one hand, this is because they are especially sensitive ecosystems that can act as sentinels, but, on the other, because they are a reservoir of genetic adaptations to stressed environments that can cope more effectively with the changes that may occur for this cause.
- The important biodiversity that lagoons harbor, their importance in the functioning of coastal ecosystems and the large number of ecosystem services and benefits they produce, justify the need to adequately protect coastal lagoons, guaranteeing relevant management measures, and their sustainability.
- Trying to manage a coastal lagoon ecosystem involves integrating into a complex conceptual model all the uses and actions that take place in it, their effects and the ecological processes that are being affected and modified (Pérez-Ruzafa et al., 2018a). But when we speak about coastal lagoons it must be borne in mind that we include a high diversity of environments with great differences in size, morphology, trophic status, and salinity, all characteristics which condition their biological assemblages structure, species composition, fishing yield and, in brief, ecosystem functioning (Pérez-Ruzafa et al., 2007a).

This means that when we compare the functioning or the effect of human activities in different lagoons, or we want to develop a realistic management plan, we must control these broad sources of variability by selecting lagoon environments with similar characteristics, in the same way that we must be able to control the natural variability of a certain lagoon to differentiate it from others affected by nonnatural factors. For this reason, it is fundamental to have an integrated conceptual model of ecological functioning of the lagoon that being studied, identifying in a very specific way its key and limiting factors.

Furthermore, when we manage these ecosystems, the problem arises of knowing the normal ranges of variation and adaptation for lagoon species, beyond which there is a real harm to the organisms and to the whole ecosystem, and also to differentiate the consequences of human impacts from the natural conditions in such habitats (Elliott and Quintino, 2007; Pérez-Ruzafa and Marcos, 2015). In this framework, to differentiate the anthropogenic effects from the inherent natural variability of these systems, we need to develop experimental designs that include sufficient replicates at the appropriate spatiotemporal scales (Pérez-Ruzafa et al., 2007b). Furthermore, for that, it is necessary to increase our present knowledge concerning the organization, functioning, complexity, and homeostatic regulations that have coastal lagoons. In addition, to reach this target there is a need to share information and to know and understand the lessons learned from similar lagoons around the world establishing a network of observatories and common data bases, and the inclusion in international research programs of topics to cope with the above-mentioned problems and gaps in the knowledge of lagoon ecology (Sorensen, 1993; Pérez-Ruzafa et al., 2011c). This would allow us to establish general guidelines for cause-effect relationships and, hence, to draw up conceptual and numerical models, assign priorities for managerial actions and define future research lines.

ACKNOWLEDGMENTS

This chapter has been written during the realization of the project ConnectMar CTM 2014-56458-R founded by the Spanish Ministry of Economy and Competitiveness, and contains some of the ideas developed in it. Alice Newton's work was supported by NILU IMPACT project 118009. Special acknowledgements are due to Eurolag and Future Earth Coasts.

REFERENCES

Aguesse, P., 1957. La classification des eaux poikilohalines, sa difficulté en Camargue. Nouvelle tentative de classification. Vie et Milieu 8, 341–365.

Aliaume, C., Do Chi, T., Viaroli, P., Zaldivar, J.M., 2007. Coastal lagoons of Southern Europe: Recent changes and future scenarios. Trans. Water. Monograph. 1, 1–12.

Almansa, C., Martínez-Paz, J.M., 2011. Intergenerational equity and dual discounting. Environ. Dev. Econ. 16, 685–707.

Alongi, D.M., 1998. Coastal Ecosystem Processes. CRC Press, Boca Raton, FL.

Angus, S., 2017. Scottish saline lagoons: impacts and challenges of climate change. Estuar. Coast. Shelf Sci. 198, 626–635.

Anonymous, 1959. Final resolution. The Venice System for the Classification of Marine Waters according to salinity. Arch. Oceanogr. Limnol. 11 (suppl), 243–245.

Anthony, A., Atwood, J., August, P., Byron, C., Cobb, S., Foster, C., Fry, C., Gold, A., Hagos, K., Heffner, L., Kellogg, D.Q., Lellis-Dibble, K., Opaluch, J.J., Oviatt, C., Pfeiffer-Herbert, A., Rohr, N., Smith, L., Smythe, T., Swift, J., Vinhateiro, N., 2009. Coastal lagoons and climate change: ecological and social ramifications in U.S. Atlantic and Gulf coast ecosystems. Ecol. Soc. 14 (1), 8. June 2009 [online]. Available from: http://www.ecologyandsociety.org/vol14/iss1/art8/.

Augier, H., 1982. Inventory and classification of marine benthic biocenoses of the Mediterranean. Council of Europe, European Committee for the Conservation of Nature and Natural Resource, Strasbourg, Nature and Environment Series 25.

Barbier, E.B., 2015. Valuing the storm protection service of estuarine and coastal ecosystems. Ecosyst. Service. 11, 32–38.

Barbier, E.B., Acreman, M.C., Knowler, D., 1997. Economic Valuation of Wetlands: A Guide for Policy Makers and Planners. Ramsar Convention Bureau. Gland, Switzerland.

Barbier, E.B., Hacker, S.D., Kennedy, C., Koch, E.W., Stier, A.C., Silliman, B.R., 2011. The value of estuarine and coastal ecosystem services. Ecol. Monogr. 81 (2), 169–193.

Barnes, R.S.K., 1980. Coastal Lagoons. The Natural History of a Neglected Habitat. Cambridge University Press, Cambridge.

Barnes, R.S.K., 1994. A critical appraisal of the application of Guelorget and Perthuisot's concepts of the paralic ecosystem and confinement to macrotidal Europe. Estuar. Coast. Shelf Sci. 38 (1), 41–48.

Basset, A., Galuppo, N., Sabetta, L., 2006. Environmental heterogeneity and benthic macroinvertebrate guilds in Italian lagoons. Transit. Water Bull. 1, 48–63.

Basset, A., Elliott, M., West, R.J., Wilson, J.G., 2013. Estuarine and lagoon biodiversity and their natural goods and services. Estuar. Coast. Shelf Sci. 132, 1–4.

Bird, E.C.F., 1994. Physical setting and geomorphology of coastal lagoons. In: Kjerfve, B. (Ed.), Coastal Lagoon Processes. Elsevier Science, Amsterdam, pp. 9–39.

Boynton, W.R., Murray, L., Hagy, J.D., Stokes, C., Kemp, W.M., 1996. A comparative analysis of eutrophication patterns in a temperate coastal lagoon. Estuar. Coasts 19 (2), 408–421.

Brito, A.C., Newton, A., Tett, P., Fernandes, T.F., 2012. How will shallow coastal lagoons respond to climate change? A modelling investigation. Estuar. Coast. Shelf Sci. 112, 98–104.

Carlquist, S., 1974. Island Biology. Columbia University Press, New York.

Carrada, G.C., Fresi, E., 1988. Le lagune salmastre costiere. Alcune riflessioni sui problemi e metodi. In: Carrada, G.C., Cicogna, F., Fresi, E. (Eds.), Le lagune costiere: ricerca e gestione. CLEM, Massa Lábrense, Nápoles, pp. 36–56.

Carrada, G.C., Cicogna, F., Fresi, E. (Eds.), 1988. Le lagune costiere: ricerca e gestione. CLEM, Napoli.

Cataudella, S., Crosetti, D., Massa, F. (Eds.), 2015. Mediterranean coastal lagoons: sustainable management and interactions among aquaculture, capture fisheries and the environment. Studies and Reviews, General Fisheries Commission for the Mediterranean 95. FAO, Rome.

Cavraro, F., Zucchetta, M., Malavasi, S., Franzoi, P., 2017. Small creeks in a big lagoon: the importance of marginal habitats for fish populations. Ecol. Eng. 99, 228–237.

Clark, J.R., 1998. Coastal Seas: The Conservation Challenge. Blackwell Science, Oxford.

Cognetti, G., Maltagliati, F., 2000. Biodiversity and adaptive mechanisms in Brackish Water Fauna. Mar. Pollut. Bull. 40 (1), 7–14.

D'Ancona, U., 1959. The classification of brackish waters with reference to the North Adriatic lagoons. Arch. Oceanogr. Limnol. 11 (Suppl), 93–109.

Dailidienė, I., Davulienė, L., 2008. Salinity trend and variation in the Baltic Sea near the Lithuanian coast and in the Curonian Lagoon in 1984–2005. J. Mar. Syst. 74, S20–S29.

Dauvin, J.C., 2007. Paradox of estuarine quality: benthic indicators and índices, consensus or debate for the future. Mar. Pollut. Bull. 55, 271–281.

Dauvin, J.C., Ruellet, T., 2009. The estuarine quality paradox: is it possible to define an ecological quality status for specific modified and naturally stressed estuarine ecosystems? Mar. Pollut. Bull. 59, 38–47.

De Groot, R., Brander, L., Van der Ploeg, S., Costanza, R., Bernard, F., Braat, L., Christi, M., Crossman, N., Ghermandi, A., Hein, L., Hussain, S., Kumar, P., McVittie, A., Portela, R., Rodríguez, L.C., Ten Brink, P., Van Beukering, P., 2012. Global estimates of the value of ecosystems and their services in monetary units. Ecosyst. Service. 1, 50–61.

De Wit, R., 2011. Biodiversity of coastal lagoon ecosystems and their vulnerability to global change. In: Grillo, O., Venora, G. (Eds.), Ecosystems Biodiversity. InTech, Rijeka, Croatia, pp. 29–40.

Dezileau, L., Pérez-Ruzafa, A., Blanchemanche, P., Degeai, J.P., Raji, O., Martinez, P., Marcos, C., Von Grafenstein, U., 2016. Extreme storms during the last 6500 years from lagoonal sedimentary archives in the Mar Menor (SE Spain). Clim. Past 12, 1389–1400.

Doody, J.P., 2007. Saltmarsh Conservation, Management and Restoration. Springer Science & Business Media, Dordrecht.

Duck, R.W., Figueiredo da Silva, J., 2012. Coastal lagoons and their evolution: a hydromorphological perspective. Estuar. Coast. Shelf Sci. 110, 2–14.

Eisenreich, S.J. (Ed.), 2005. Climate change and the European water dimension. EU Report No. 21553 European Commission-Joint Research Centre 253.

Elliott, M., Quintino, V., 2007. The Estuarine quality paradox, environmental homeostasis and the difficulty of detecting anthropogenic stress in naturally stressed areas. Mar. Pollut. Bull. 54 (6), 640–645.

Elliott, M., Whitfield, A.K., 2011. Challenging paradigms in estuarine ecology and management. Estuar. Coast. Shelf Sci. 94, 306–314.

Elliott, M., Burdon, D., Hemingway, K.L., Apitz, S.E., 2007. Estuarine, coastal and marine ecosystem restoration: confusing management and science—a revision of concepts. Estuar. Coast. Shelf Sci. 74, 349–366.

Elliott, M., Burdon, D., Atkins, J.P., Borja, A., Cormier, R., De Jonge, V.N., Turner, R.K., 2017. "And DPSIR begat DAPSI(W)R(M)!"—a unifying framework for marine environmental management. Mar. Pollut. Bull. 118 (1–2), 27–40.

Esteves, F.A., Caliman, A., Santangelo, J.M., Guariento, R.D., Farjalla, V.F., Bozelli, R.L., 2008. Neotropical coastal lagoons: an appraisal of their biodiversity, functioning, threats and conservation management. Braz. J. Biol. 68 (4), 967–981.

European Environment Agency, 2001. Eutrophication in Europe's coastal waters. . Topic Report 7, Copenhagen.

European Environment Agency, 2010. Ecosystem accounting and the cost of biodiversity losses. The case of coastal Mediterranean wetlands. EEA Technical Report 3/2010, Copenhagen.

European Environment Agency, 2015. Conservation status of habitat types and species (Article 17, Habitats Directive 92/43/EEC).

Ferrarin, C., Razinkovas, A., Gulbinskas, S., Umgiesser, G., Bliudziute, L., 2008. Hydraulic regime-based zonation scheme of the Curonian Lagoon. Hydrobiologia 611, 133–146.

Ferrarin, C., Bajo, M., Bellafiore, D., Cucco, A., De Pascalis, F., Ghezzo, M., Umgiesser, G., 2014. Toward homogeneization of Mediterranean lagoons and their loss of hydrodiversity. Geophys. Res. Lett. 41, 5935–5941.

Ferrarin, C., Maicu, F., Umgiesser, G., 2017. The effect of lagoons on Adriatic Sea tidal dynamics. Ocean Model. 119, 57–71.

Ferreira, J.G., Andersen, J.H., Borja, A., Bricker, S.B., Camp, J., Cardoso da Silva, M., Garcés, E., Heiskanen, A.S., Humborg, C., Ignatiades, L., Lancelot, C., Menesguen, A., Tett, P., Hoepffner, N., Claussen, U., 2011. Overview of eutrophication indicators to assess environmental status within the European marine strategy framework directive. Estuar. Coast. Shelf Sci. 93 (2), 117–131.

Franco, A., Franzoi, P., Malavasi, S., Riccato, F., Torricelli, P., Mainardi, D., 2006a. Use of shallow water habitats by fish assemblages in a Mediterranean coastal lagoon. Estuar. Coast. Shelf Sci. 66 (1–2), 67–83.

Franco, A., Franzoi, P., Malavasi, S., Riccato, F., Torricelli, P., 2006b. Fish assemblages in different shallow water habitats of the Venice lagoon. Hydrobiologia 555 (1), 159–174.

Gaertner-Mazouni, N., De Wit, R., 2012. Exploring new issues for coastal lagoons monitoring and management. Estuar. Coast. Shelf Sci. 114, 1–6.

Gamito, S., Gilabert, J., Marcos, C., Pérez-Ruzafa, A., 2005. Effects of changing environmental conditions on lagoon ecology. In: Gönenç, I.E., Wolflin, J.P. (Eds.), Coastal Lagoons: Ecosystem Processes and Modeling for Sustainable Use and Development. CRC Press, Boca Ratón, FL, pp. 93–229.

García, G., Muñoz-Vera, A., 2015. Characterization and evolution of the sediments of a Mediterranean coastal lagoon located next to a former mining area. Mar. Pollut. Bull. 100 (1), 249–263.

García-Sánchez, M., Pérez-Ruzafa, I.M., Marcos, C., Pérez-Ruzafa, A., 2012. Suitability of benthic macrophyte indices (EEI, E-MaQI and BENTHOS) for detecting anthropogenic pressures in a Mediterranean coastal lagoon (Mar Menor, Spain). Ecol. Indic. 19, 48–60.

García-Oliva, M., Pérez-Ruzafa, A., Umgiesser, G., McKiver, W., Ghezzo, M., De Pascalis, F., Marcos, C., 2018. Assessing the hydrodynamic response of the Mar Menor lagoon to dredging inlets interventions through numerical modelling. Water 10 (7), 959. https://doi.org/10.3390/w10070959.

García-Seoane, E., Dolbeth, M., Silva, C.L., Abreu, A., Rebelo, J.E., 2016. Changes in the fish assemblages of a coastal lagoon subjected to gradual salinity increases. Mar. Environ. Res. 122, 178–187.

Ghezzo, M., De Pascalis, F., Umgiesser, G., Zemlys, P., Sigovini, M., Marcos, C., Pérez- Ruzafa, A., 2015. Connectivity in three European coastal lagoons. Estuar. Coasts 38, 1764–1781.

Giusti, E., Marsili-Libellia, S., Renzi, M., Focardi, S., 2010. Assessment of spatial distribution of submerged vegetation in the Orbetello lagoon by means of a mathematical model. Ecol. Model. 221, 1484–1493.

Gubbay, S., Sanders, N., Haynes, T., Janssen, J.A.M., Rodwell, J.R., Nieto, A., García-Criado, M., Beal, S., Borg, J., Kennedy, M., Micu, D., Otero, M., Saunders, G., Calix, M., 2016. European Red List of Habitats. Part 1. Marine habitats. European Union.

Guelorget, O., Perthuisot, J.P., 1983. Le domaine paralique. Expressions géologiques, biologiques et économiques du confinement. Travaux du Laboratoire de Géologie 16, 1–136.

Guelorget, O., Frisoni, G.F., Perthuisot, J.P., 1983. Zonation biologique des milieux lagunaires: definition d´une echelle de confinement dans le domaine paralique méditerranéen. J. Rech. Oceanogr. 8 (1), 15–35.

Haines-Young, R., Potschin, M., 2013. Common International Classification of Ecosystem Services (CICES): Consultation on Version 4. EEA Framework Contract No EEA/IEA/09/003.

Havinga, B., 1959. Artificial transformation of salt and brackish water into fresh water lakes in the Netherlands, and possibilities for biological investigations. Arch. Oceanogr. Limnol. 11 (Suppl), 47–62.

Heck Jr., K.L., Valentine, J.F., 2007. The primacy of top-down effects in shallow benthic ecosystems. Estuar. Coasts 30 (3), 371–381.

Isla, F.I., 2009. Coastal lagoons. In: Isla, F.I., Iribarne, O. (Eds.), Coastal Zones and Estuaries (eBook). UNESCO – EOLSS. Available from: http://www.eolss.net/sample-chapters/c09/E2-06-03-02.pdf.

Kapetsky, J.M., Lasserre, G. (Eds.), 1984. Management of coastal lagoon fisheries. Stud. Rev. GFCM 61 (2), 439–776.

Katselis, G., Koutsikopoulos, C., Dimitriou, E., Rogdakis, Y., 2003. Spatial patterns and temporal trends in the fishery landings of the Messolonghi-Etoliko lagoon system (western Greek coast). Sci. Mar. 67 (4), 501–511.

Kennish, M.J., 2009. Eutrophication of Mid-Atlantic coastal bays. Bull. N. J. Acad. Sci. 54 (3), 1–8.

Kennish, M.J., 2016a. Coastal lagoons. In: Kennish, M.J. (Ed.), Encyclopedia of Estuaries. Springer Publisher, Dordrecht, pp. 140–143.

Kennish, M.J., 2016b. Eutrophication. In: Kennish, M.J. (Ed.), Encyclopedia of Estuaries. Springer Publisher, Dordrecht, pp. 304–311.

Kennish, M.J., de Jonge, V.N., 2011. Chemical introductions to the systems: diffuse and nonpoint source pollution from chemicals (Nutrients: Eutrophication). In: Kennish, M.J., Elliott, M. (Eds.), Treatise on Estuarine and Coastal Science. Human-Induced Problems (Uses and Abuses). vol. 8. Elsevier, Oxford, pp. 113–148.

Kennish, M.J., Paerl, W., 2010. Coastal lagoons. Critical habitats of environmental change. In: Kennish, M.J., Paerl, H.W. (Eds.), Coastal Lagoons. Critical Habitats of Environmental Change. CRC Press, Boca Ratón, FL, pp. 1–15.

Kjerfve, B., 1986. Comparative oceanography of coastal lagoons. In: Wolfe, D.A. (Ed.), Estuarine Variability. Academic Press, New York, pp. 63–81.

Kjerfve, B., 1994. Coastal lagoons. In: Kjerfve, B. (Ed.), Coastal Lagoon Processes. Elsevier Oceanography Series 60, Elsevier, Amsterdam, pp. 1–8.

Knoppers, B., 1994. Aquatic primary production in coastal lagoons. In: Kjerfve, B. (Ed.), Coastal Lagoon Processes. Elsevier Oceanography Series 60, Elsevier, Amsterdam, pp. 243–286.

Komárková, J., Laudares-Silva, R., Cabral Senna, P.A., 1999. Extreme morphology of *Cylindrospermopsis raciborskii* (Nostocales, Cyanobacteria) in the Lagoa do Peri, a freshwater coastal lagoon, Santa Catarina, Brazil. Algol. Stud. 94, 207–222.

Kralj, M., De Vittor, C., Comici, C., Relitti, F., Auriemma, R., Alabiso, G., Del Negro, P., 2016. Recent evolution of the physical-chemical characteristics of a Site of National Interest—the Mar Piccolo of Taranto (Ionian Sea)—and changes over the last 20 years. Environ. Sci. Pollut. Res. 23, 12675–12690.

Krumgalz, B.S., Hornung, H., Oren, O.H., 1980. The study of a natural hypersaline lagoon in a desert area (the Bardawil Lagoon in Northern Sinai). Estuar. Coast. Mar. Sci. 10, 403–415.

La Notte, A., D'Amato, D., Mäkinen, H., Paracchini, M.L., Liquete, C., Egoh, B., Geneletti, D., Crossman, N.D., 2017. Ecosystem services classification: a systems ecology perspective of the cascade framework. Ecol. Indic. 74, 392–402.

Lankford, R., 1977. Coastal lagoons of Mexico: their origin and classification. In: Wiley, M. (Ed.), Estuarine Processes. Academic Press, New York, pp. 182–215.

Lasserre, P., Postma, H. (Eds.), 1982. Les lagunes côtières, Actes du Symposium international sur les lagunes côtières. SCOR/IABO/UNESCO, Bordeaux. Oceanologica Acta (volumen spécial).

Leruste, A., Malet, N., Munaron, D., Derolez, V., Hatey, E., Collos, Y., De Wit, R., Bec, B., 2016. First steps of ecological restoration in Mediterranean lagoons: shifts in phytoplankton communities. Estuar. Coast. Shelf Sci. 180, 190–203.

Likens, G.E., 1972. Eutrophication and aquatic ecosystems. Limnol. Oceanogr. 1, 3. (Special Symposia).

Lillebø, A.I., Spray, C., Alves, F.L., Stålnacke, P., Gooch, G.D., Soares, J.A., Sousa, L.P., Sousa, A.I., Khokhlov, V., Tuchkovenko, Y., Marín, A., Loret, J., Bello, C., Bielecka, M., Rozynski, G., Margonski, P., Chubarenko, B., 2015. European coastal lagoons: and integrated vision for ecosystem services, environmental SWOT analysis and human well-being. In: Lillebø, A.I., Stålnacke, P., Gooch, G.D. (Eds.), Coastal Lagoons in Europe. Integrated Water Resource Strategies. IWA Publishing, London, pp. 187–201.

Lonsdale, D.J., Cerrato, R.M., Holland, R., Mass, A., Holt, L., Schaffner, R.A., Pan, J., Caron, D.A., 2009. Influence of suspension-feeding bivalves on the pelagic food webs of shallow, coastal embayments. Aquat. Biol. 6, 263–279.

Maci, S., Basset, A., 2009. Composition, structural characteristics and temporal patterns of fish assemblages in non-tidal Mediterranean lagoons: a case study. Estuar. Coast. Shelf Sci. 83 (4), 602–612.

Manzo, C., Fabbrocini, A., Roselli, L., D'Adamo, R., 2016. Characterization of the fish assemblage in a Mediterranean coastal lagoon: Lesina Lagoon (central Adriatic Sea). Reg. Stud. Mar. Sci. 8, 192–200.

Margalef, R., 1969. Comunidades planctónicas en lagunas litorales. In: UNAM-UNESCO (Eds.), Lagunas costeras, Un Simposio. Memorias del Simposio Internacional de Lagunas Costeras, México D.F. pp. 545–562.

Mariani, S., 2001. Can spatial distribution of ichthyofauna describe marine influence on coastal lagoons? A central Mediterranean case study. Estuar. Coast. Shelf Sci. 52, 261–267.

McLusky, D.S., 1999. Estuarine benthic ecology: a European perspective. Aust. J. Ecol. 24, 302–311.

Mejri, R., Arculeo, M., Ben Hassine, O.K., Lo Brutto, S., 2011. Genetic architecture of the marbled goby *Pomatoschistus marmoratus* (Perciformes, Gobiidae) in the Mediterranean Sea. Mol. Phylogenet. Evol. 58, 395–403.

Michel, P., 1979. Choix d'un descripteur du milieu et planification écologique application a un écosystème lagunaire méditerranéen. Revue de Biologie et Ecologie méditerranéenne VI (3–4), 239–247.

Milana, V., Franchini, P., Sola, L., Angiulli, E., Rossi, A.R., 2012. Genetic structure in lagoons: the effects of habitat discontinuity and low dispersal ability on populations of *Atherina boyeri*. Mar. Biol. 159, 399–411.

Mimura, N., 2013. Sea-level rise caused by climate change and its implications for society. Proc. Jpn. Acad. Ser. B. Phys. Biol. Sci. 89 (7), 281–301.

Monfort, P., Morand, S., Lafaye, M., 2014. Microbiological coastal risks and monitoring systems. In: Monaco, A., Prouzet, P. (Eds.), Vulnerability of Coastal Ecosystems and Adaptation. ISTE and Wiley, London and Hoboken, pp. 95–129.

Newell, R.I.E., Kemp, W.M., Hagy, J.D., Cerco, C.F., Testa, J.M., Boynton, W.R., 2007. Top-down control of phytoplankton by oysters in Chesapeake Bay, USA: comment on Pomeroy et al. (2006). Mar. Ecol. Prog. Ser. 341, 293–298.

Newton, A., Icely, J.D., 2008. Land ocean interactions in the coastal zone, LOICZ: lessons from Banda Aceh, Atlantis, and Canute. Estuar. Coast. Shelf Sci. 77, 181–184.

Newton, A., Icely, J.D., Falcao, M., Nobre, A., Nunes, J.P., Ferreira, J.G., Vale, C., 2003. Evaluation of eutrophication in the Ria Formosa coastal lagoon, Portugal. Cont. Shelf Res. 23, 1945–1961.

Newton, A., Icely, J., Cristina, S., Brito, A., Cardoso, A.C., Colijn, F., Dalla Riva, S., Gertz, F., Hansen, J.W., Holmer, M., Ivanova, K., Leppäkoski, E., Melaku Canu, D., Mocenni, C., Mudge, S., Murray, N., Pejrup, M., Razinkovas, A., Reizopoulou, S., Pérez-Ruzafa, A., Schernewski, G., Schubert, H., Carr, L., Solidoro, C., Viaroli, P., Zaldivar, J.M., 2014. An overview of ecological status, vulnerability and future perspectives of European large shallow, semi-enclosed coastal systems, lagoons and transitional waters. Estuar. Coast. Shelf Sci. 140, 95–122.

Newton, A., Brito, A.C., Icely, J.D., Delorez, V., Clara, I., Angus, S., Schernewski, G., Inácio, M., Lillebø, A.I., Sousa, A.I., Béjaoui, B., Solidoro, C., Tosic, M., Cañedo-Argüelles, M., Yamamuro, M., Reizopoulou, S., Tseng, H., Canu, D., Roselli, L., Maanan, M., Cristina, S., Ruiz-Fernández, A.C., Lima, R., Kjerfve, B., Rubio-Cisneros, N., Pérez-Ruzafa, A., Marcos, C., Pastres, R., Pranovi, F., Snoussi, M., Turpie, J., Tuchkovenko, Y., Dyack, B., Brookes, J., Povilankas, R., Khokhlov, V., 2018. Assessing, quantifying and valuing the ecosystem services of coastal lagoons. J. Nat. Conserv. 44, 50–65.

Nienhuis, P.H., 1992. Eutrophication, water management, and the functioning of Dutch estuaries and coastal lagoons. Estuaries 15 (4), 538–548.

Nixon, S.W., 1982. Nutrient dynamics, primary production and fisheries yields of lagoons. Oceanol. Acta 5, 357–371.

Nixon, S.W., 1995. Coastal marine eutrophication: a definition, social causes and future concerns. Ophelia 41, 199–219.

Nixon, S.W., 2009. Eutrophication and the macroscope. Hydrobiologia 629, 5–19.

O'Higgins, T., Tett, P., Farmer, A., Cooper, P., Dolch, T., Friedrich, J., Goulding, I., Hunt, A., Icely, J., Murciano, C., Newton, A., Psuty, I., Raux, P., Roth, E., 2014. Temporal constraints on ecosystem management: definitions and examples from Europe's regional seas. Ecol. Soc. 19 (4), https://doi.org/10.5751/ES-06507-190446. Art. 46.

Odum, E.P., 1969. The strategy of ecosystem development. Science 164 (3877), 262–270.

Odum, E.P., 1985. Trends expected in stressed ecosystems. Bioscience 35 (7), 419–422.

Ortiz-Lozano, L., Granados-Barba, A., Solis-Weiss, V., Garcia-Salgado, M.A., 2005. Environmental evaluation and development problems of the Mexican Coastal Zone. Ocean Coast. Manag. 48, 161–176.

Pérès, J.M., Picard, J., 1964. Nouveau manuel de bionomie benthique de la Mer Mediterranee. Travaux de la station marine d'Endoume Bulletin 31 (47), 1–137.

Pérez-Ruzafa, A., 2015. El papel de la conectividad restringida en la construcción de los ecosistemas marinos semiaislados: el ejemplo de las lagunas costeras y los archipiélagos. Rev. Acad. Canar. Cienc. 27, 411–456.

Pérez-Ruzafa, A., Marcos, C., 1992. Colonization rates and dispersal as essential parameters in the confinement theory to explain the structure and horizontal zonation of lagoon benthic assemblages. Rapp. Comm. Int. Mer Médit. 33, 100.

Pérez-Ruzafa, A., Marcos, C., 1993. La teoría del confinamiento como modelo para explicar la estructura y zonación horizontal de las comunidades bentónicas en las lagunas costeras. Publ. Espec. Inst. Esp. Oceanogr. 11, 347–358.

Pérez-Ruzafa, A., Marcos, C., 2005. Pressures on Mediterranean coastal lagoons as a consequence of human activities. In: Fletcher, C., Spencer, T., Da Mosto, J., Campostrini, P. (Eds.), Flooding and Environmental Challenges for Venice and Its Lagoon: State of Knowledge. Cambridge University Press, Cambridge, pp. 545–555.

Pérez-Ruzafa, A., Marcos, C., 2008. Coastal lagoons in the context of water management in Spain and Europe. In: Gönenç, I.E., Vadineau, A., Wolflin, J.P., Russo, R.C. (Eds.), Sustainable Use and Development of Watersheds. NATO Science for Peace and Security Series, Springer, Dordrecht, pp. 299–321.

Pérez-Ruzafa, A., Marcos, C., 2012. Fisheries in coastal lagoons: an assumed but poorly researched aspect of the ecology and functioning of coastal lagoons. Estuar. Coast. Shelf Sci. 110, 15–31.

Pérez-Ruzafa, A., Marcos, C., 2015. Monitoring heterogeneous and quick-changing environments: coping with spatial and temporal scales of variability in coastal lagoons and transitional waters. In: Sebastiá, M.T. (Ed.), Coastal Ecosystems. Experiences and Recommendations for Environmental Monitoring Programs. Nova Science Publications, New York, pp. 89–116.

Pérez-Ruzafa, A., Marcos, C., Pérez-Ruzafa, I.M., Ros, J.D., 1987. Evolución de las características ambientales y de los poblamientos del Mar Menor. An. Biol. 12, 53–65.

Pérez-Ruzafa, A., Gilabert, J., Gutiérrez, J.M., Fernández, A.I., Marcos, C., Sabah, S., 2002. Evidence of a planktonic food web response to changes in nutrient input dynamics in the Mar Menor coastal lagoon, Spain. Hydrobiologia 475 (476), 359–369.

Pérez-Ruzafa, A., Quispe-Becerra, J.I., García-Charton, J.A., Marcos, C., 2004. Composition, structure and distribution of the ichthyoplankton in a Mediterranean coastal lagoon. J. Fish Biol. 64, 202–218.

Pérez-Ruzafa, A., Marcos, C., Gilabert, J., 2005a. The ecology of the Mar Menor coastal lagoon: a fast-changing ecosystem under human pressure. In: Gönenç, I.E., Wolflin, J.P. (Eds.), Coastal Lagoons: Ecosystem Processes and Modeling for Sustainable Use and Development. CRC Press, Boca Ratón, FL, pp. 392–422.

Pérez-Ruzafa, A., Mompeán, M.C., Marcos, C., 2005b. To what extent ecological information can be explained by geomorphological characteristics of coastal lagoons? In: Lasserre, P., Viaroli, P., Campostrini, P. (Eds.), Lagoons and Coastal Wetlands in the Global Change Context: Impacts and Management Issues. ICAM Dossier N° 3,UNESCO, Venice, pp. 62–69.

Pérez-Ruzafa, A., Fernández, A.I., Marcos, C., Gilabert, J., Quispe, J.I., García-Charton, J.A., 2005c. Spatial and temporal variations of hydrological conditions, nutrients and chlorophyll a in a Mediterranean coastal lagoon (Mar Menor, Spain). Hydrobiologia 550, 11–27.

Pérez-Ruzafa, A., Mompeán, M.C., Marcos, C., 2007a. Hydrographic, geomorphologic and fish assemblage relationships in coastal lagoons. Hydrobiologia 577, 107–125.

Pérez-Ruzafa, A., Marcos, C., Pérez-Ruzafa, I.M., Barcala, E., Hegazi, M.I., Quispe, J., 2007b. Detecting changes resulting from human pressure in a naturally quick-changing and heterogeneous environment: spatial and temporal scales of variability in coastal lagoons. Estuar. Coast. Shelf Sci. 75, 175–188.

Pérez-Ruzafa, A., Hegazi, M., Pérez-Ruzafa, I., Marcos, C., 2008. Differences in spatial and seasonal patterns of macrophyte assemblages between a coastal lagoon and the open sea. Mar. Environ. Res. 65, 291–314.

Pérez-Ruzafa, A., Marcos, C., Pérez-Ruzafa, I.M., 2009. 30 años de estudios en la laguna costera del Mar Menor: de la descripción del ecosistema a la comprensión de los procesos y la solución de los problemas ambientales. In: Instituto Euromediterráneo del Agua (Eds.), El Mar Menor. Estado actual del conocimiento científico. Fundación Instituto Euromediterráneo del Agua, Murcia, pp. 17–46.

Pérez-Ruzafa, A., Marcos, C., Pérez-Ruzafa, I., Pérez-Marcos, M., 2011a. Coastal lagoons: "transitional ecosystems" between transitional and coastal waters. J. Coast. Conserv. 15, 369–392.

Pérez-Ruzafa, A., Marcos, C., Pérez-Ruzafa, I.M., 2011b. Recent advances in coastal lagoons ecology: evolving old ideas and assumptions. Transit. Water. Bull. 5, 50–74.

Pérez-Ruzafa, A., Marcos, C., Pérez-Ruzafa, I.M., 2011c. Mediterranean coastal lagoons in an ecosystem and aquatic resources management context. Phys. Chem. Earth 36, 160–166.

Pérez-Ruzafa, A., Marcos, C., Pérez-Marcos, M., Pérez-Ruzafa, I., 2013. Are coastal lagoons physically or biologically controlled ecosystems? Revisiting r vs K strategies in coastal lagoons and estuaries. Estuar. Coast. Shelf Sci. 132, 17–33.

Pérez-Ruzafa, A., Marcos, C., Pérez-Ruzafa, I.M., 2018a. When maintaining ecological integrity and complexity is the best restoring tool: the case of the Mar Menor lagoon. In: Quintana, X., Boix, D., Gascón, S., Sala, J. (Eds.), Management and Restoration of Mediterranean Coastal Lagoons in Europe. Recerca i territori 10. Universidad de Gerona, Gerona, pp. 67–95.

Pérez-Ruzafa, A., De Pascalis, F., Ghezzo, M., Quispe, J.I., Hernández-García, R., Muñoz, I., Vergara, C., Pérez-Ruzafa, I.M., Umgiesser, G., Marcos, C., 2018b. Connectivity between coastal lagoons and sea: asymmetrical effects on assemblages' and population's structure. Estuar. Coast. Shelf Sci. https://doi.org/10.1016/j.ecss.2018.02.031.

Petit, G., 1953. Introduction à l'étude écologique des étangs méditerranéens. Vie Milieu 4 (4), 569–604.

Pimm, S.L., 1984. The complexity and stability of ecosystems. Nature 307, 321e326.

Pitacco, V., Mistri, M., Munari, C., 2018. Long-term variability of macrobenthic community in a shallow coastal lagoon (Valli di Comacchio, northern Adriatic): is community resistant to climate changes? Mar. Environ. Res. 137, 73–87.

Por, F.D., 1980. A classification of hypersaline waters, based on trophic criteria. PSZNI Mar. Ecol. 1, 121–131.

Quignard, J.P., 1984. Les caracteristiques biologiques et environnementales des lagunes en tant que base biologique de l'amenagement des pecheries. In: Kapetsky, J.M., Lasserre, G. (Eds.), Management of Coastal Lagoon Fisheries. FAO Studies and Reviews, GFCM No 61,FAO, Rome, pp. 4–38.

Quispe, J.I., 2014. Dinámica espacio-temporal del ictioplancton del Mar Menor (SE España) y factores ambientales asociados. Tesis Doctoral, Universidad de Murcia.

Ramasamy, R., Surendran, S.N., 2012. Global climate change and its potential impact on disease transmission by salinity-tolerant mosquito vectors in coastal zones. Front. Physiol. 3, 198.

Ramos Miranda, J., Mouillot, D., Flores Hernández, D., Sosa López, A., Do Chi, T., Ayala Pérez, L., 2005. Changes in four complementary facets of fish diversity in a tropical coastal lagoon after 18 years: a functional interpretation. Mar. Ecol. Prog. Ser. 304, 1–13.

Reizopoulou, S., Nicolaidou, A., 2004. Benthic diversity of coastal brackish-water lagoons in western Greece. Aquat. Conserv. Mar. Freshwat. Ecosyst. 14, S93–S102.

Remoundou, K., Koundouri, P., Kontogianni, A., Nunes, P.A.L.D., Skourtos, M., 2009. Valuation of natural marine ecosystems: an economic perspective. Environ. Sci. Pol. 12, 1040–1051.

Reyes, E., Merino, M., 1991. Diel dissolved-oxygen dynamics and eutrophication in a shallow, well-mixed tropical lagoon (Cancun, Mexico). Estuaries 14 (4), 372–381.

Russi, D., ten Brink, P., Farmer, A., Badura, T., Coastes, D., Förster, J., Kumar, R., Davidson, N., 2013. The Economics of Ecosystems and Biodiversity for Water and Wetlands. IEEP, London and Brussels, Ramsar Secretariat, Gland.

Sanders, H.L., 1968. Marine benthic diversity – a comparative study. Amer. Nat. 102 (925), 243–282.

Secretaría de la Convención de Ramsar, 2006. Manual de la Convención de Ramsar: Guía a la Convención sobre los Humedales (Ramsar, Irán, 1971). Secretaría de la Convención de Ramsar, Gland, Suiza.

Seeram, L., 2008. Coastal Lagoons Goods and Services and Human Development. Universidade do Algarve, Faro.

Segerstrale, S.G., 1959. Brackishwater classification. A historical survey. Arch. Oceanogr. Limnol. 11 (Suppl), 7–33.

Sfriso, A., Facca, C., 2007. Distribution and production of macrophytes and phytoplankton in the lagoon of Venice: comparison of actual and past situation. Hydrobiologia 577 (1), 71–85.

Sfriso, A., Pavoni, B., Marcomini, A., Orio, A.A., 1992. Macroalgae, nutrient cycles and pollutants in the lagoon of Venice. Estuaries 15, 517–528.

Sfriso, A., Facca, C., Ghetti, P.F., 2003. Temporal and spatial changes of macroalgae and phytoplankton in a Mediterranean coastal area: the Venice lagoon as a case study. Mar. Environ. Res. 56, 316–336.

Sigovini, M., 2011. Multiscale Dynamics of Zoobenthic Communities and Relationships with Environmental Factors in the Lagoon of Venice. Tesi di Dottorato. Universitá Ca'Foscari di Venezia.

Silva, E.I.L., Katupotha, J., Amarasinghe, O., Manthrithilake, H., Ariyaratna, R., 2013. Lagoons of Sri Lanka: From the Origins to the Present. International Water Management Institute (IWMI), Colombo, Sri Lanka, pp. 122.

Simonneau, J., 1973. Mar Menor: Evolution sédimentologique et géochimique récente en remplissage. Universite of Marseille, Thése.

Solidoro, C., Bandelj, V., Aubry Bernardi, F., Camatti, E., Ciavatta, S., Cossarini, G., Facca, C., Franzoi, P., Libralato, S., Melaku Canu, D., Pastres, R., Pranovi, F., Raicevich, S., Socal, G., Sfriso, A., Sigovini, M., Tagliapietra, D., Torricelli, P., 2010. Response of Venice lagoon ecosystem to natural and anthropogenic pressures over the last 50 years. In: Kennish, M.J., Paerl, H.W. (Eds.), Coastal Lagoons:Critical Habitats of Environmental Change. CRC Press, Boca Ratón, FL, pp. 483–511.

Sorensen, J., 1993. The management of enclosed coastal water bodies: the need for a framework for international information exchange. In: Sorensen, J., Gable, F., Bandarin, F. (Eds.), The Management of Coastal Lagoons and enclosed Bays. American Society of Civil Engineers, New York, pp. 1–17.

Sousa, A., García-Barrón, L., Vetter, M., Morales, J., 2014. The historical distribution of main malaria foci in Spain as related to water bodies. Int. J. Environ. Res. Public Health 11 (8), 7896–7917.

Strickland, C., 1938. Malaria in relation to the coastal lagoons of Bengal and Orissa. Indian Med. Gazette 73 (7), 399–402.

Tagliapietra, D., Sigovini, M., Volpi-Ghirardini, A., 2009. A review of terms and definitions to categorise estuaries, lagoons and associates environments. Mar. Freshw. Res. 60, 497–509.

Tagliapietra, D., Aloui-Bejaoui, N., Bellafiore, D., De Wit, R., Ferrarin, C., Gamito, S., Laserre, P., Magni, P., Mistri, M., Pérez-Ruzafa, A., Pranovi, F., Reizopoulou, S., Rilov, G., Solidoro, C., Tunberg, B., Valiela, I., Viaroli, P., 2011. The Ecological Implications of Climate Change on the Lagoon of Venice. UNESCO Venice Office and ISMAR-CNR, Venice.

Taylor, D.I., Nixon, S.W., Granger, S.L., Buckley, B.A., 1999. Responses of coastal lagoon plant communities to levels of nutrient enrichment: a Mesocosm study. Estuaries 22 (4), 104–1056.

TEEB (The Economics of Ecosystems & Biodiversity), 2012. Why Value the Oceans? A discussion Paper. TEEB Communications, Geneva.

Tett, P., Gilpin, L., Svendsen, H., Erlandsson, C., Larsson, U., Kratzer, S., Fouilland, E., Janzen, C., Lee, J., Grenz, C., Newton, A., Ferreira, J.G., Fernandes, T., Scory, S., 2003. Eutrophication and some European waters of restricted exchange. Cont. Shelf Res. 23, 1635–1671.

Tett, P., Gowen, R., Mills, D., Fernandes, T., Gilpin, L., Huxham, M., Kennington, K., Read, P., Service, M., Wilkinson, M., Malcolm, S., 2007. Defining and detecting undesirable disturbance in the context of eutrophication. Mar. Pollut. Bull. 53, 282e297.

Turner, R.E., 2003. Coastal ecosystems of the Gulf of Mexico and climate change. In: Ning, Z.H., Turner, R.E., Doyle, T., Abdollahi, K.K. (Eds.), Integrated Assessment of the Climate Change Impacts on the Gulf Coast Region. Gulf Coast Climate Change Assessment Council and Louisiana State University Graphic Services, Washington, DC, pp. 85–103.

Umgiesser, G., Ferrarin, C., Cucco, A., De Pascalis, F., Bellafiore, D., Ghezzo, M., Bajo, M., 2014. Comparative hydrodynamics of 10 Mediterranean lagoons by means of numerical modeling. J. Geophys. Res.: Oceans . Online research article https://doi.org/10.1002/2013JC009512.

UNESCO, 1979. Coastal ecosystems of the Southern Mediterranean: lagoons deltas and salt marshes. UNESCO Rep. Mar. Sci. 7, .

UNESCO, 1980. Coastal lagoon survey. UNESCO Tech. Pap. Mar. Sci. 31.

UNESCO, 1981. Coastal lagoon research, present and future. UNESCO Tech. Pap. Mar. Sci. 32, .

UNESCO, 1982. Simposio internacional sobre las lagunas costeras. Documentos Técnicos de la Unesco sobre Ciencias del Mar 43, .

UNESCO, 1986. Méthodologie d'étude des lagunes côtières. Rapports de l'Unesco sur les Sciences de la Mer 36, .

Valiela, I., Mcclelland, J., Hauxwell, J., Behr, P.J., Hersh, D., Foreman, K., 1997. Macroalgal blooms in shallow estuaries: controls and ecophysiological and ecological consequences. Limnol. Oceanogr. 42, 1105–1118.

Vasconcelos, R.P., Reis-Santos, P., Costa, M.J., Cabral, H.N., 2011. Connectivity between estuaries and marine environment: integrating metrics to assess estuarine nursery function. Ecol. Indic. 11, 1123–1133.

Vasileiadou, K., Pavloudi, C., Sarropoulou, E., Fragopoulou, N., Kotoulas, G., Arvanitidis, C., 2016. Unique COI haplotypes in *Hediste diversicolor* populations in lagoons adjoining the Ionian Sea. Aquat. Biol. 25, 7–15.

Velasco, A.M., Pérez-Ruzafa, A., Martínez-Paz, J.M., Marcos, C., 2018. Ecosystem services and main environmental risks in a coastal lagoon (Mar Menor, Murcia, SE Spain): the public perception. J. Nat. Conserv. 43, 180–189.

Vergara-Chen, C., González-Wangüemert, M., Marcos, C., Pérez-Ruzafa, A., 2010a. Genetic diversity and connectivity remain high in *Holothuria polii* (Delle Chiaje 1823) across a coastal lagoon-open environmental gradient. Genetica 138 (8), 895–906.

Vergara-Chen, C., Gonzalez-Wanguemert, M., Marcos, C., Pérez-Ruzafa, A., 2010b. High gene flow promotes the genetic homogeneity of the fish goby *Pomatoschistus marmoratus* (Risso, 1810) from Mar Menor coastal lagoon and adjacent marine waters (Spain). Mar. Ecol.-Evol. Persp. 31 (2), 270–275.

Viaroli, P., Christian, R.R., 2003. Description of trophic status, hyperautotrophy and dystrophy of a coastal lagoon through a potential oxygen production and consumption index—TOSI: Trophic Oxygen Status Index. Ecol. Indic. 3 (4), 237–250.

Viaroli, P., Bartoli, M., Giordani, G., Naldi, M., Orfanidis, S., Zaldivar, J.M., 2008. Community shifts, alternative stable states, biogeochemical controls and feedback in eutrophic coastal lagoons: a brief overview. Aquat. Conserv. Mar. Freshwat. Ecosyst. 18, S105–S117.

Viaroli, P., Azzoni, R., Bartoli, M., Giordani, G., Naldi, M., Nizzoli, D., 2010. Primary productivity, biogeochemical buffers and factors controlling trophic status and ecosystem processes in Mediterranean coastal lagoons: a synthesis. Adv. Oceanogr. Limnol. 1 (2), 271–293.

Wilkinson, M., Telfer, T.C., Grundy, S., 1995. Geographical variations in the distribution of macroalgae in estuaries. Neth. J. Aquat. Ecol. 29, 359–368.

Williams, J.J., O'Connor, B.A., Arens, S.M., Abadie, S., Bell, P., Balouin, Y., Van Boxel, J.H., Do Carmo, A.J., Davidson, M., Ferreira, O., Heron, M., Howa, H., Hughes, Z., Kaczmarek, L.M., Kim, H., Morris, B., Nicholson, J., Pan, S., Salles, P., Silva, A., Smith, J., Soares, C., Vila-Concejo, A., 2003. Tidal inlet function: field evidence and numerical simulation in the India project. J. Coast. Res. 19, 189–211.

Yañez-Arancibia, A., Nugent, R.S., 1977. El papel ecológico de los peces en estuarios y lagunas costeras. Anales del Centro de Ciencias del Mar y Limnología, Universidad Nacional Autónoma de México 4, 107–114.

Zhu, Y., McCowan, A., Cook, P.L.M., 2017. Effects of changes in nutrient loading and composition on hipoxia dynamics and internal nutrient cycling of a stratified coastal lagoon. Biogeosciences 14, 4423–4433.

Chapter 16

The Everglades: At the Forefront of Transition

Fred H. Sklar*, John F. Meeder†, Tiffany G. Troxler†, Tom Dreschel*, Steve E. Davis‡, Pablo L. Ruiz§

*Everglades Systems Assessment Section, South Florida Water Management District, West Palm Beach, FL, United States, †Sea Level Solutions Center and Southeast Environmental Research Center, Florida International University, Miami, FL, United States, ‡Everglades Foundation, Palmetto Bay, FL, United States, §South Florida Caribbean Network, National Park Service, Palmetto Bay, FL, United States

1 INTRODUCTION

Everglades Restoration programs involve large landscapes, water bodies, and human populations spanning the Kissimmee River basin (777,000 ha), Lake Okeechobee (189,000 ha), the Caloosahatchee and St. Lucie Rivers, wildlife management areas (24,500 ha), water conservation areas (WCAs) (343,000 ha), Big Cypress National Preserve (290,000 ha), Everglades National Park (607,028 ha), agricultural areas (216,000 ha), and a population of more than 6 million along the dense urban corridor from Miami-Dade, Broward, and Palm Beach Counties (Fig. 1). As such, these programs are at the forefront of integrating wetland science, hydrologic modeling, environmental engineering, and water management to protect society and preserve the environment—fundamentally one and the same. There is a long natural and anthropogenic history of the Everglades that we bring to your attention because their legacies create hysteretic constraints when managing for future climate(s) and sea-level rise (SLR). The object of this discourse is to understand the interactions between Everglades restoration and the geology, ecology, and socioeconomics of South Florida and the impacts of an accelerated rise in sea level during this "Anthropocene Marine Transgression" (Parkinson et al., 2015; Meeder and Parkinson, 2018) of the recently proposed (newly recognized) geologic epoch; the Anthropocene (Zalasiewicz et al., 2008).

In the Florida Everglades, peat soils critical to maintaining wetland elevation have subsided due to drainage (McVoy et al., 2011) and now may be collapsing in response to rising seas, saltwater intrusion, storm surge, nutrient enrichment, and fire (DeLaune et al., 1994; Wanless et al. 1994; Deegan et al., 2012; Voss et al., 2013). Peat soil is the fundamental building block of many wetlands, but the Everglades is relatively unique due to its karst underpinning and its extremely low phosphorus concentrations (Noe et al., 2001; Sklar et al., 2005). This vast "River of Grass" is a home to many tropical species, protects humans from hurricane damages and generates billions of tourist dollars each year (Alpert and Stronge, 2009; Mather Economic, 2010). The Everglades also helps filter water before it reaches the Biscayne Aquifer, which supplies drinking water to millions of residents and tourists in South Florida and irrigation water for agriculture. But peat soils are fragile—too little freshwater can dry them up, while too much saltwater causes plant stress and sometimes death which may compromise the integrity of the peats causing them to degrade and wash out with storm surge (Vlaswinkel and Wanless, 2012). This has been hypothesized to create open water slurry holes where the peat used to be and has been described colloquially as "collapse" due to the apparent rapid rate of elevation loss (see the concept in Fig. 2; DeLaune et al., 1994; Day et al., 2011). Given the flat, low-sloping landscape of the coastal Everglades, SLR is only expected to make the problem worse.

The Everglades is one of the most famous ecological restoration efforts in the world. When the US Congress approved the Water Resources Development Act (WRDA) of 2000, nested within this WRDA was the largest ($7.8 billion) Federally supported authorization for environmental restoration in US history, the Comprehensive Everglades Restoration Plan (CERP). The CERP was and continues to be a 50–50 cost share between the State of Florida and the US Army Corps of Engineers (USACOE). Separate from the CERP; the USACOE and Everglades National Park (the Park) are building bridges across Tamiami Trail to increase flows to Shark River Slough, and the State of Florida spent nearly $2 billion to build 23,067 ha (57,000 acres) of stormwater treatment areas (STAs) and implement elements of restoration strategies (SFWMD, 2011) to preserve the oligotrophic (<10 ppb of Total Phosphorus (TP)) nature of the Everglades. To move the CERP into a "faster lane," the SFWMD and the USACOE received congressional authorization in 2016 to advance the

Coasts and Estuaries. https://doi.org/10.1016/B978-0-12-814003-1.00016-2
2019 Published by Elsevier Inc.

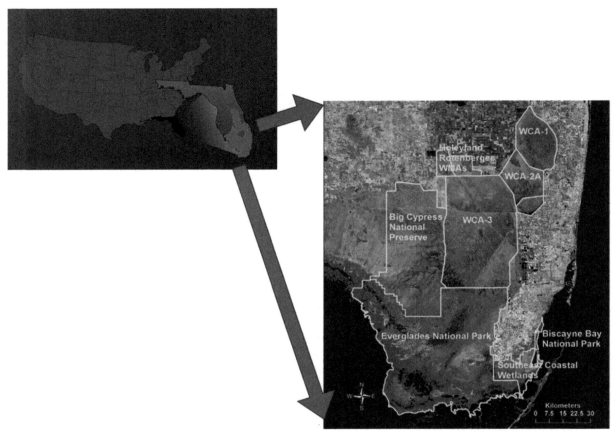

FIG. 1 The southern region of Florida is at the forefront of transition due to its low elevation and a legacy of high peat loss (see Fig. 5). It is a complex suite of parks, wildlife preserves, and water conservation areas (WCAs), bordered on the east by large metropolitan areas such as Miami and Ft. Lauderdale, on the north by the largest contiguous crop of sugarcane in North America, on the west by the Gulf of Mexico, and on the south by Florida Bay and the Atlantic Gulf Stream.

Central Everglades Planning Project (CEPP); a more integrative approach to the CERP planning that combines elements of multiple projects to expedite the storage, treatment, and conveyance of "new" water from Lake Okeechobee to the southern Everglades. The CEPP is now the very heart of Everglades restoration. The state is moving quickly to implement a few of the lower cost CEPP elements (e.g., increasing capacity of a key water control structure) that facilitate delivery of more water from the WCAs into NE Shark River Slough in the Park.

2 THE GEOLOGICAL SETTING

The Tamiami Formation is the oldest strata in South Florida, outcropping in western Monroe, Collier, and Lee Counties, deposited during the mid-Pliocene Pliocene (3 mya) temperature maxima (Miller et al., 2005) when water depth was between 20 and 25 m above present sea level (Raymo et al., 2011; Klaus et al., 2016). This Tamiami strata dips to the east forming the Biscayne Aquifer aqualude (Parker et al., 1955); a surficial aquifer that has become South Florida's primary water source. Sea level during the Pleistocene varied between a high of +120 m and a low of −130 m in respect to present mean sea level and was inversely related to global ice cover. Pleistocene sediments prograded southward forming a series of five wedge-shaped strata (Fort Thompson Formation) each separated by exposure horizons documenting five cycles of deposition behind a contemporaneous fringing coral reef complex, the Key Largo Limestone (Perkins, 1977). The last, Late Pleistocene (80–100 k-year before present; bp) high sea-level stand was ~15 m above present and most of South Florida was inundated resulting in the deposition of the Miami Limestone, the eastern portion forming the Atlantic Coastal Ridge, the "high ground" that supported much of the initial development in South Florida (Hoffmeister et al., 1967; Spratt and Lisiecki, 2016). The low area between the outcropping Tamiami Formation strata to the west and the Atlantic Coastal Ridge to the east forms the Everglades Basin (Parker et al., 1955; Gleason and Stone, 1994).

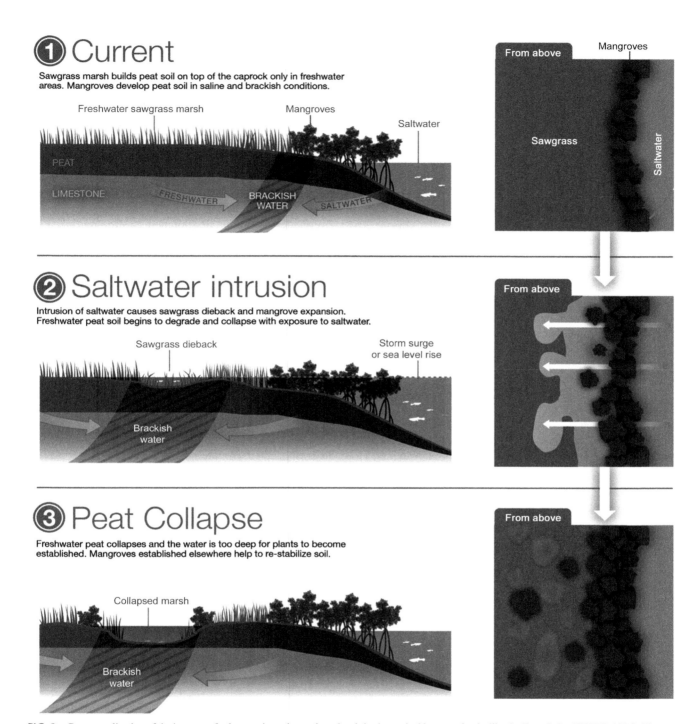

FIG. 2 Conceptualization of the impacts of salt water intrusion and sea-level rise in tropical karst wetlands, like the Everglades (SERES, 2016). The term "peat collapse" is used to describe a relatively dramatic shift in the soil C balance, leading to a net loss of organic C and loss of soil elevation, culminating in a conversion of vegetated freshwater marsh to open water (DeLaune et al., 1994). This process has been documented to varying degrees across the United States (Cahoon et al., 2003; Nyman et al., 2006; Voss et al., 2013) and is a critical concern for the management of freshwater and coastal habitats in Everglades National Park exposed to increasing sea-level rise (CISRERP, 2014).

The lowest elevations during the Late Pleistocene (12 k-year bp) were approximately −130 m below present sea level (Milliman and Emery, 1968; Lambeck et al., 2014). These low sea levels during the Pleistocene exposed the Florida platform to extensive epi-karst processes which: (1) removed most of the Pleistocene deposits exposing Pliocene strata in the Big Cypress Swamp region (Duever et al., 1986), (2) eliminated surface water bodies because of rapid downward percolation (Watts, 1975), (3) dissolved limestone to produce the karst Biscayne Aquifer (Vacher and Mylroie, 2002) and the high concentrations of dissolved calcium important for marl development, (4) produced the micro-topography of the Everglades (Duever et al., 1986; Gleason and Stone, 1994), and (5) served as the foundation upon which the biodiversity, productivity, and peat soils developed (Gleason and Spackman, 1974).

The Holocene (12.7 k-year bp) is used to delineate the beginning of the SLR associated with the last Pleistocene glacier retreat. The rate of SLR was not constant throughout the Holocene; therefore, it is subdivided into three stages based upon different rates of SLR (Scholl et al., 1969; Wanless et al., 1994). Now, we find ourselves in a new, fourth stage called the Anthropocene Marine Transgression, which began with the increased rate of SLR at the end of the Late Holocene (Parkinson et al., 2015; Meeder et al., 2017; Meeder and Parkinson, 2018). Everglades origin and development is strongly linked to these different rates of SLR and sea-level stages (Fig. 3). The rate of SLR during the Early Holocene was >10 mm year^{-1}, too fast for much coastal stabilization or coastal wetland development, and submergence and overstep dominated the coastal environment (Wanless et al., 1994). Only minor, remnant Early Holocene strata are found on the middle and outer shelf areas (Emory et al., 1967; Milliman et al., 1968; Emory and Milliman, 1970). As sea level rose so did the ground water surface producing the Biscayne Aquifer. Because of the very high aquifer porosity, the equilibrium between the two water masses was very fast and freshwater was elevated above sea level because of density differences. When sea level reached −5 m below present 5.5 k-year bp, the rate of SLR decreased to 2–3 mm year^{-1} (Wanless et al., 1994) and freshwater inundation of the Everglades Basin (between 30 and 100 cm above present sea level) first occurred, producing the first short-hydroperiod prairie wetlands (Gleason and Stone, 1994) and their marl sediments (Gleason and Spackman, 1974) over the extent of its Basin.

Some 5 k-year bp this initial marl sedimentation produced a leaky impervious seal separating surface and ground water. This was a critical geological step because with continuing SLR, the hydroperiod across the extended Everglades Basin resulted in the peat deposition we now know as the Everglades (Gleason and Stone, 1994). Along both low-energy south Florida coasts, barrier islands including the 10,000 Islands and mangrove wetlands of the Park began to both develop and retreat with continued SLR (Scholl et al., 1969; Wanless, 1974; Parkinson, 1989). As the rate of SLR further dropped to

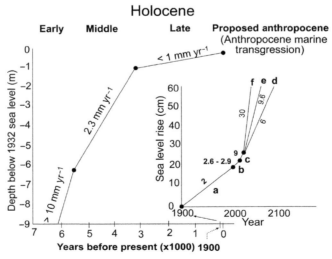

FIG. 3 Four subdivisions of the Holocene. The Early, Middle, and Late subdivisions are based on different rates of SLR and associated different coastal response; Early Holocene characterized by overstep and submergence, Middle Holocene by development of retreating coastal wetlands and barrier islands, and Late Holocene by stable coastal wetlands and barrier islands. The proposed Anthropocene (Smith and Zeder, 2013) has been suggested for the geological time when mankind has altered earth processes. This last stage of the Holocene is presented in a insert because of the great scale difference. All rates of SLR expressed in mm year^{-1}. The rate of sea level has been accelerating since ~1900 as documented: (a) Nicholls and Cazenave (2010); (b) Church and White (2013); and (c) Wdowinski et al. (2016) (regional rate not global). Predictions of future global rate of SLR vary considerably with a range of examples figured: (d) Ezer (2013); (e) IPCC (2013) (high); and (f) Rohling et al. (2013). *(Modified from Wanless, H.R., Parkinson, R.W., Tedesco, L.P., 1994. Sea level control on stability of Everglades wetlands. pp. 199–223. In: Everglades: The Ecosystem and Its Restoration. Davis, S.M., Ogden J.C. (Eds.), St. Lucie Press. Delray Beach, FL.)*

<1 mm year^{-1} about 3 k-year bp (Wanless et al., 1994), barrier islands and coastal wetlands stabilized and coastal peat deposits prograded, continuing to grow vertically or retreating depending upon coastal energy setting (Wanless, 1974). The freshwater sediments of the Everglades Basin expanded both vertically and horizontally to their maximum extent by 1900 (Gleason and Stone, 1994; McVoy et al., 2011). The typical stratigraphic sequence in the Everglades Basin and Big Cypress is limestone bedrock and associated breccia (angular rock embedded in a sandy/marl matrix) overlain by this calcitic marl followed by organic peat reflecting an increasing hydroperiod (Gleason et al., 1974; Gleason and Stone, 1994; Duever et al., 1986). However, extensive peat has been lost since the beginning of the 20th century (Stephens and Johnson, 1951; Johnson, 1974), and periphyton-derived marl is beginning to replace peat in some locations because of decreased hydroperiod (Gleason and Stone, 1994) and salt water encroachment (Vlaswinkel and Wanless, 2012; Meeder et al., 2017).

3   THE ECO-HYDROLOGICAL SETTING

The predrainage Everglades was comprised of a directionally patterned mosaic of tree islands, shrubs, sawgrass ridges, and aquatic sloughs (King, 1917a, b; Parker et al., 1955). To varying degrees, soil cores from freshwater tree islands (Gleason and Stone, 1994; Willard et al. 2006) in the extant Everglades, document a relatively stable 5000-year-old system where all the same elements are still present (Kushlan, 1990). These same soil cores show very rapid community shifts associated with human encroachment and alterations of the basin hydrology (Fig. 4) beginning around 1880 (McVoy et al., 2011).

A comparison of the past and present Everglades landscape is a lesson on the subtle manipulative power of the slow moving, oligotrophic River of Grass. The predrainage landscape (McVoy et al., 2011) was characterized by a major overtopping of the natural spoil banks of Lake Okeechobee during the Florida wet season (approximately, June–October) from seasonal rainfall in the Kissimmee River into a pond apple forest and through a dense sawgrass plain. which created the hydrological velocities and broad sheetflows that many believe are responsible for the distinctive Everglades "corrugation" (Fig. 4A; Sklar and van der Valk, 2003; Larsen and Harvey, 2010, 2011; McVoy et al., 2011; Harvey et al., 2017). Drainage and impoundment associated with the implementation of the Central and Southern Florida (C&SF) Project by the

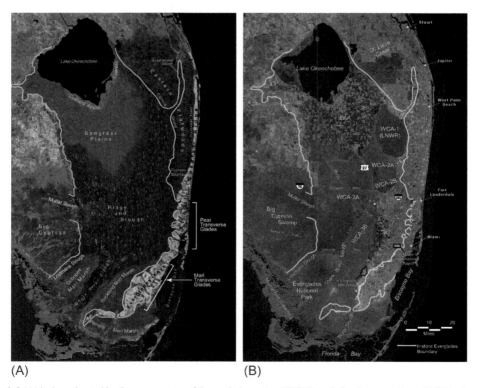

(A)   (B)

FIG. 4   Image on the left (A) is the estimated landscape structure of the predrainage (ca. 1800) Everglades. Image on the right (B) is the present-day landscape structure. EAA—Everglades Agricultural Area. *(Modified from McVoy, C.W., Said, W.P., Obeysekera, J., Van Arman, J.A., Reschel, T.W., 2011. Landscapes and Hydrology of the Predrainage. Everglades University Press of Florida, Gainesville. Available from https://www.researchgate.net/profile/Thomas_Dreschel/publication/257315404_Landscapes_and_Hydrology_of_the_Predrainage_Everglades_Overview/links/02e7e524e97f484b32000000.pdf.)*

USACOE created discontinuous hydrologic subunits (Fig. 4B) each with their own distinctive history of water depths, flows, and ecological impacts. Some areas became drier, others became wetter, and still others first became drier and later became wetter. For example, in WCA-2A, initial over-drainage caused a conversion of predrainage sloughs to shallower wet prairies (Loveless, 1959; Goodrick, 1974), as well as increasing the rate of peat and elevation loss on tree islands. Subsequently, average water elevations have increased some 0.75 m (2.5 ft) under managed conditions between 1955 and 1969. During this period, wet prairies converted back to more closely resemble predrainage aquatic sloughs (Loveless, 1959). However, at the same time, 90% of the tree islands in WCA-2A were lost due to the increasing water elevations and excessive inundation of the subsided and oxidized tree island peats (Dineen, 1972; Worth, 1987; Sklar and van der Valk, 2003).

After completion of the major canals around 1935, there was significant over-drainage, with much of the ridge and slough landscape subject to lower water levels, exposing the soil to microbial oxidation and extensive, long- and deep-burning peat fires. Most of the wetland soils perched atop the limestone of the extant rocky glades oxidized (Craighead, 1966). Vegetation in the central Everglades also changed, with emergent wet prairie species filling in the formerly aquatic, deep-water sloughs (Andrews, 1957; Loveless, 1959). During the 1960s, after construction of the peripheral levees and canals and after considerable peat loss, water levels were raised again to increase storage, creating "high-water" conditions relative to what had been present in the 1950s. All this lead to considerable peat soil loss (Stephens and Johnson, 1951; Johnson, 1974) and shrinkage of wetland area (McVoy et al., 2011). This was dramatically illustrated by Hohner and Dreschel (2015; Fig. 5) who used historical Geographic Information System (GIS) data sets to calculate the difference between original and existing peat volumes and thus, the amount lost over the past approximately 150 years. These predrainage and current volumes were then combined with bulk density data from the USEPA soil monitoring program known

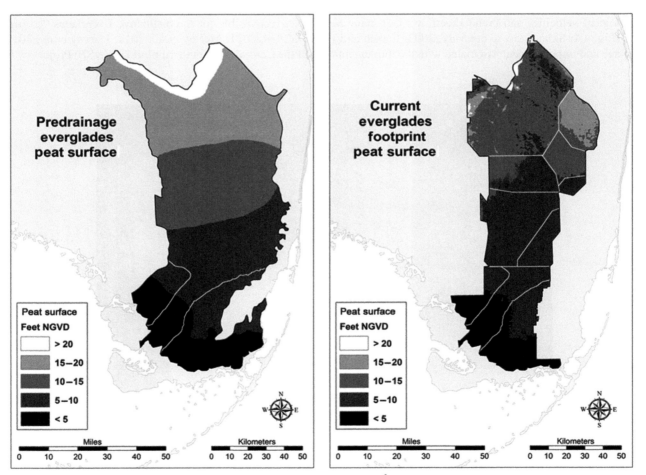

FIG. 5   Left: Predrainage Everglades had a peat depth of 2 m, a peat volume of 20 billion m^3, and a carbon content of about 900 million metric tons. Right: Current Everglades has a peat depth of 0.75 m, a peat volume of 5 billion m^3, and a carbon content of about 200 million metric tons. *(From Hohner, S. M., Dreschel, T.W., 2015. Everglades peats: using historical and recent data to estimate predrainage and current volumes, masses and carbon contents. Mires Peat. 16(1) 1–15. Available from http://www.mires-and-peat.net/pages/volumes/map16/map1601.php.)*

as REMAP (Scheidt et al. 2000) to calculate the corresponding organic masses. They found that the predrainage Everglades had: (1) places where peat depths exceeded 5 m, (2) an average peat depth of 2 m, (3) a peat volume of 20 billion m^3, and (4) a carbon content of about 900 million metric tons. They estimated that the current Everglades has an average peat depth of only 0.75 m, a peat volume of 5 billion m^3, and a carbon content of only 200 million metric tons. When Hohner and Dreschel (2015) assumed optimum hydrological conditions and a maximum historic accretion of 16 cm per 100 years from 1200 YBP to present (McDowell et al.; cited in Davis and Ogden, 1994), they found that it would take about 1600 years to fully restore the peat that has been lost.

In the peat-based portions of the Everglades, water depths can be altered either by changes in water elevation due to rainfall or by changes in elevation of the peat surface due to biophysical processes. Slight changes in the depth ($\pm$10 cm) and period of inundation ($\pm$90 days), over long periods of time (5–10 years), influence the presence and distributions of certain plant species and communities (Armentano et al., 2006; Zweig and Kitchens, 2009). Increased average water elevations in WCA-3A in the early 1980s increased obligate wetland species such as *Sagittaria lancifolia* and slough species such as *Nymphaea odorata* and *Utricularia* spp. (David, 1997), while at the same time reducing the extent and diversity of wet prairies. Zaffke (1983) reported replacement of wet prairies in southern WCA-3A by aquatic sloughs due to extended hydroperiods and ponding due to impoundment. Wood and Tanner (1990) did not find at least 13 species that Loveless (1959) encountered in wet prairies 30 years earlier in the same area. These observations suggest that many wet prairie species only germinate in areas with an annual spring dry period (Goodrick, 1974).

The Everglades is at the "forefront of transition," because its biological resources are primarily linked to changes in the volume, timing, and distribution of freshwater inflows from the WCAs in the North and from tidal exchanges and SLR in the South and Southwest. Everglades National Park, the third largest national park in the contiguous United States, encompasses 5698 km^2 (2200 mi²) of former Ridge and Slough landscape (freshwater sloughs, sawgrass ridges, and tree islands), marl-forming prairies on adjacent higher ground, mangrove forests, and saline tidal flats. Although the Park is a World Heritage site, the International Union for the Conservation of Nature (IUCN), in their 2017 "World Heritage Outlook 2" report, categorized Everglades National Park as the only World Heritage site in the United States as having a *critical* conservation outlook (Osipova et al. 2017).

At the southern end of the Everglades is a vast coastal prairie, known to as the Southeast Saline Everglades (SESE) (Fig. 6), where transition has begun. It consists of short-hydroperiod marl prairies, freshwater marshes dominated by sparse sawgrass, Gulf Coast spikerush, tree islands, short-stature mangrove swamps, coastal hammocks, and mangrove forests. Harshberger (1914) considered this region distinct from the Greater Everglades due to its tidal influence and the presence of low stature red mangroves. Egler (1952) described the SESE as a relic community consisting of seven belts or ecotones paralleling the coast and slowly responding to a lower water table, a decrease in fire frequency, and SLR. Like Harshberger before him, Egler noted that the presence of red mangrove in freshwater dominated marshes but imagined their continued landward migration and colonization of freshwater marshes in response to the absence of fire and salt-water intrusion resulting from a decrease in upstream freshwater availability and SLR. Egler (1952) observed that one of his ecotones, Belt 4, appeared white on the aerial imagery he analyzed. Nearly 50 years later, Ross et al. (2000) documented that the interior boundary of Egler's Belt 4, the "White Zone," had shifted landward by approximately 1 km and that the vegetation composition within this belt had shifted from a sparsely vegetated sawgrass marsh with pockets of Gulf Coast spikerush and emergent short-statured red mangroves, as described by Egler, to a monodominant red mangrove scrub community. Furthermore, Ross et al. (2000) showed that the landward movement of the "White Zone" was least in areas where surface freshwater was not restricted. Today, in response to low freshwater availability, decreased fires, salt water intrusion, and SLR, the location of the "White Zone" is approximately 2.5 km further inland than when Egler first described it (Ruiz et al., 2017).

However, the rate and direction of vegetation change observed within the SESE has been spatially variable. The Red-Mangrove Sawgrass Marsh ecotone (East Taylor Slough), the boundary between the red mangrove dominated "White Zone" and the upstream freshwater marshes, starts out as a thin, less than 1-km wide band, that expands to almost 3 km wide as it approaches US Highway 1 (Fig. 6). In contrast, west of Taylor Slough, this ecotonal unit, is absent (Fig. 6) which may be due to additional inflow of freshwater associated with the removal of the southern spoil bank of the C-111 Canal, an Acceler8 project, thus preventing the expansion of the "White Zone" in this section of the SESE. Once fully operational Everglades restoration (the CERP and CEPP) will rehydrate more of the SESE and potentially slowdown or reverse the expansion of the "White Zone" into graminoid dominated freshwater marshes.

At the most southern end of this River of Grass are the coastal lakes and Florida Bay (Fig. 6). Beginning with the early drainage projects (dredging of the Caloosahatchee River, St. Lucie, New River, and Miami Canals), the Lake Okeechobee water table dropped from between 6.4 m (21 ft) and 7.0 m (23 ft) to 3.0 m (10 ft) (McVoy et al., 2011). Construction of the water conservation basins and the management of the C&SF system resulted in a total reduction of 70% to all south Florida

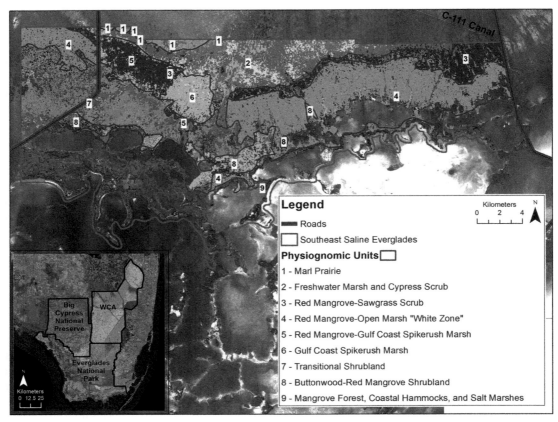

FIG. 6   Vegetation communities of the Southeast Saline Everglades, a region at the forefront of ecological transition.

coasts (Perry, 2004) which was first observed by Parker et al. (1955) to result in salt water encroachment and ground water intrusion. Today, Florida Bay response to SLR is of particular concern, because it has been subject to drastic ecological changes during the past 100 years as a result of declines in freshwater inflows (Boesch et al., 1993; Fourqurean and Robblee, 1999; Day et al., 2013; Madden, 2013). These changes include increased seagrass mortality and algal blooms, and decreased water clarity. It is generally thought that decreased freshwater inflow from the Everglades and resultant increases in salinity, have contributed to these ecological changes.

4   THE ECO-ECONOMIC SETTING

Construction of the C&SF Project physically blocked the flow of water into Shark River Slough, causing sawgrass stands to increase and aquatic slough communities to decline (Alexander and Crook, 1975; Davis and Ogden, 1994; Olmsted and Armentano, 1997) thus, severely impacting populations of fish and crayfish, essential prey for wading birds (Loftus et al., 1990). Restoration is expected to partially remedy this situation. The volume of water flowing through the Everglades will significantly increase with restoration (Fig. 7A), creating an environment that will clearly decrease the number of peat-oxidizing dry-downs (Fig. 7B), restore slough communities, create a larger prey base for target wading birds such as White Ibis (*Eudocimus albus*) and Wood Storks (*Mycteria americana*), facilitate mangrove transgression (with SLR), and improve the seagrass habitats in Florida Bay (CEPP (Central Everglades Planning Project), 2014). The ecological value of restoration is well recognized. However, the socio-ecological and socioeconomic benefits of Everglades restoration are often not appreciated, even though it is also expected to enhance the ability of lawmakers and residents to manage, to a degree, the impacts of SLR. Peat stability maintains elevation and serves as the foundational "platform" for wetlands and all the ecology and ecosystem services (ES) that wetlands support thus, the fate of these peat soils can greatly influence the fishing, recreational, and tourist economy that has emerged around the Florida Everglades.

Increasing pressures from SLR, land use and population growth will influence the number, types, and value of ES expected in the future, especially if the future does not include an active and significant change in the delivery of freshwater to the Everglades. Freshwater and coastal ecosystems provide many socioeconomic benefits including important recreational and tourism opportunities, key fishery habitat, water quality improvements, flood and erosion mitigation, and mitigation

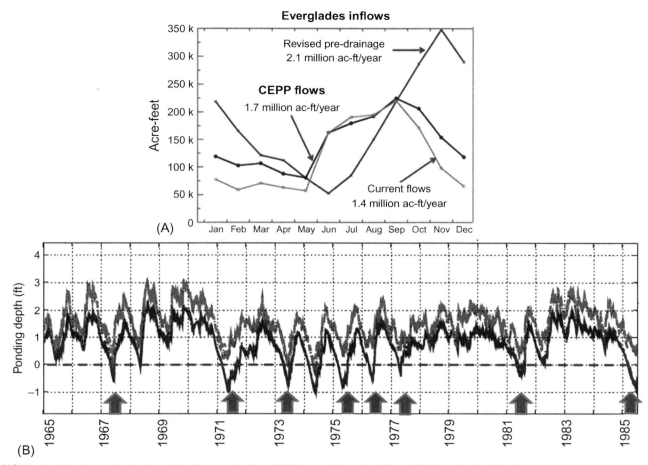

FIG. 7 (A) The Comprehensive Everglades Restoration Plan (CERP) was designed to deliver 300,000 ac-ft of additional water to the extant Everglades, especially during the dry season (November-May). This helps to prevent subsidence and peat oxidation. (B) Multiple restoration scenarios *(light colored lines)* were compared to a future without restoration *(dark colored line)* using the South Florida Water Management Model to hind cast system responses from 1965 to 1986. The restoration scenarios eliminated the degradation of complete dry-drown at the northern tip of Everglades National Park eight times as indicated by the arrows in 1967, 1971, 1973, 1975,1976, 1977, 1981, and 1985. (CEPP (Central Everglades Planning Project), 2014).

of greenhouse gases through carbon storage and sequestration in vegetation (above- and belowground) and soil, each with monetary and nonmonetary value, which can reduce risks to life, property, and economies (Barbier et al., 2011; Mcleod et al., 2011; Lovelock et al., 2017). These ES have been broadly defined as "the benefits people obtain from ecosystems" (MEA, 2005). In 2005, the Millennium Ecosystem Assessment (MEA) highlighted coastal wetland ES including water purification, climate regulation, and mitigation of climate impacts through physical buffering. And Everglades restoration is just that—in addition to being a "replumbing" project, as it is often called, it is indirectly, also about restoring and extending the life of ES.

What are these services? As implied from our review above, the importance of carbon sequestration and the storage of peat cannot be overstated. As reported by Lovelock et al. (2017), "Restoration reinstates the sedimentary biogeochemical conditions and the soil stability in disturbed sites… It also enhances C_{org} storage by increasing the living biomass and its capacity to sequester CO_2 and by trapping organic material that is delivered in tidal flows." Another service is coastal protection. Restored features like coastal wetlands enhance coastal resilience and mitigate the potential impacts of extreme events, SLR, and inundation to coastal communities (Alongi, 2008; Arkema et al., 2014; Sutton-Grier et al., 2015; Vegh et al., n.d.). Coastal risk reduction (storm protection through wave attenuation) and resilience (flood storage) can come in the form of protection, enhancement, and restoration of natural features such as reefs, marshes, mangroves, seagrasses, floodplains, barrier islands, and dunes (Arkema et al., 2014). South Florida coastal mangrove forests are particularly significant due to their ability to attenuate the combined effects of hurricane storm surge and SLR (US Army Corps of Engineers, 2013). There is still a lot to be learned about the coastal risk reductions associated with specific vegetation features (Loder et al. 2009; Zhang et al. 2012; Bricker et al. 2015).

One often hears news accounts linking restoration to the socioeconomic fate of south Florida. To understand the economic stability of south Florida, there have been several economic evaluations focused on Everglades restoration. A 2009 economic study by the Florida Atlantic University (Alpert and Stronge, 2009) estimated that Everglades ES were valued at approximately $82 billion annually, with much of that tied to freshwater marshes or saltwater habitats and associated with services such as water supply and treatment, wildlife habitat, ecotourism, environmental regulation, and land formation. A subsequent report published by the Mather Economics showed that Everglades Restoration generates a significant return on investment ($4 return for every $1 spent on restoration) in areas of water supply, real estate values, wildlife habitat, fishing, open space, and park visitation (Mather Economic, 2010). Much of the improvement to property values was estimated from restoration improvement on water quality in the Caloosahatchee and St. Lucie estuaries through a reduction in discharge of polluted water coming from Lake Okeechobee. These discharges can influence water quality that in turn can significantly alter fishing. Recreational fishing is a highly valued industry throughout the state of Florida, supporting more than 3 million anglers and total expenditures approaching $5 billion annually (American Sportfishing Association (ASA), 2013), much of is associated with estuaries surrounding the Everglades. For example, a 2013 report from the Bonefish and Tarpon Trust estimates that Florida Keys flats fishing alone has an economic impact of $427 million (Fedler, 2013).

5 TRANSITION AWARENESS

As previously mentioned, the Tamiami Formation was deposited when water depths were between 20 and 25 m above present (Miller et al., 2005; Raymo et al., 2011; Klaus et al., 2016). This is roughly the same depth as postulated for a + 4°C climate relative to the preindustrial period (Levermann et al., 2013). Global temperature at present is approximately +1.5°C above preindustrial levels (IPCC, 2013). As recognized in 2013 by the IPCC, the rate of SLR is increasing because of warming oceans and associated ice melt. In the Everglades, the rate of SLR increased to 2 mm year^{-1} around 1930 (Church and White, 2011) initiating retreating coastal wetlands (Meeder and Parkinson, 2018). This increased rate of SLR is associated with: (1) a transgressive stratigraphic sequence (Fig. 8A) (Meeder et al., 2017; Meeder and Parkinson, 2018), (2) shore-line erosion, mangrove retreat, and inundation ponding (Ross et al., 2000, 2002; Meeder and Parkinson, 2018), (3) increased erosion of outer Florida Bay banks (Parkinson and Meeder, 1991), and (4) salt water encroachment in the SESE (Fig. 8B) of up to 70 m year^{-1} (Meeder et al., 2017).

As SLRs and freshwater head continues to be reduced, saltwater inundation into fresh and brackish coastal zones is expected to increase (Pearlstine et al., 2010). Further, SLR and storm surge combined, create nonlinear increases in inundation area that can have significant impacts on the productivity of the plant communities that support wetland elevations (Zhang et al., 2012). In the Everglades, a system that receives very little sediment input as part of accretionary processes, inputs through primary productivity and outputs through decomposition largely govern net peat soil balance and thus the capacity for Everglades peat wetlands to accrete and keep pace with SLR. Key mechanisms that maintain coastal peatland elevation as SLRs are organic matter accumulation via plant production and episodic deposition of mineral sediments (Castañeda-Moya et al., 2010; Smoak et al., 2013). However, there are uncertainties about the fate of coastal peat marshes due to an incomplete mechanistic understanding about how changes in organic matter accumulation and peat stability translate to landscape-scale changes in vegetated wetlands in the face of future SLR. In mangrove and marsh ecosystems that receive little sediment input, root production is identified as the primary driver of vertical peat accretion (Nyman et al., 2006; McKee, 2011; Baustian et al., 2012) and soil C accumulation.

As a flat, low-lying landscape, the conventional thinking is that Everglades coastal habitats (e.g., mangroves) will gradually migrate upslope with increase in sea level as a transgressive transition of saltwater into freshwater sawgrass marshes. Inland transgression of mangroves has been suggested as a means by which subtropical and tropical coastal landscapes will "adapt" to increasing SLR—mangroves will replace inland marshes, stabilizing soils as they transgress (McKee et al., 2007). Historically, rates of vertical soil accretion of mangrove and salt marsh wetlands have kept pace with rates of SLR, especially at SLR rates <1 mm year^{-1} (e.g., McKee, 2011). Landward migration of mangroves and other coastal habitats over the past 50–100 years has been documented for the Everglades (Ross et al., 2000; Gaiser et al., 2006; Fuller and Wang, 2014; Smith et al., 2013), and there is evidence that management of the C&SF system (through reduction in freshwater flow) has accelerated this process in some areas (Davis et al., 2005; Smith et al., 2013).

In several areas of coastal Everglades National Park, freshwater and oligohaline Everglades wetlands are expected to be exposed to increased duration and inundation of seawater, impacting processes that may lead to peat collapse (Fig. 2). In some cases, the freshwater marsh has collapsed by some 0.5 m over a period of a few decades and has converted to an open water, mangrove-free environment (Vlaswinkel and Wanless, 2012). On the surface, these collapsed areas begin as a small pool surrounded by sawgrass with some intermittent mangrove trees. Collectively, they represent a larger area of destabilized peat soil that can coalesce overtime, growing into a larger aggregation of collapsed marsh. Aside from the

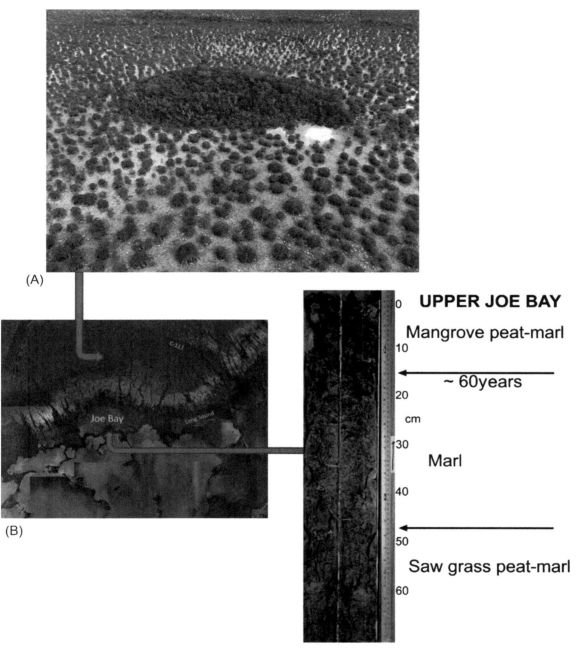

FIG. 8   (A) A core from Upper Joe Bay, approximately 5000 m north of the Florida Bay coastline. Displaying a transgressive stratigraphic sequence of sawgrass peat-marl to marl to mangrove peat-marl documenting salt water encroachment in the SESE. *Arrows* represent breaks in soil types. The date is based upon ^{210}Pb dating of mangrove peat-marl soils producing an accretion rate of 3.2 mm year^{-1} (Meeder et al., 2017). (B) Mangroves retreating from the coast, past a freshwater tree island in the SESE in response to the Anthropocene Marine Transgression. North is to the top, mangrove clumps range between 1 and 3 m in diameter. *(Photo by Mike Ross.)*

mineralized and potentially mobile pool of nutrients associated with peat soil collapse, the collapsed areas are often too deep for emergent plants to become established (personal observation). Therefore, these areas will remain as open water habitats as sea level continues to rise and will ultimately shape the future coastline of south Florida. Therefore, if coastal communities cannot adapt to the salinity changes associated with increasing sea levels, then significant coastal wetland loss may occur, dramatically altering and increasing the vulnerability of the south Florida coastline (Troxler et al., n.d.). Without restoration of freshwater flow to the Everglades, saltwater intrusion-induced peat collapse may be enhanced and landward migration of mangroves into freshwater peat soils stymied (CISRERP, 2014; Chambers et al., 2014).

The southern coastal systems of South Florida will need to adapt to extreme inundation events that will accompany SLR. Dessu et al. (2018) and Park et al. (2017) both illustrate the influence of increasing marine head pressure on freshwater inundation patterns and coastal ecotone salinity. The head difference between freshwater and marine end-members along Shark River Slough explains more than 75% of the salinity variability in Shark River. Although the highest tides of the year occur during the wet season (especially September-November), the estuarine sites have low salinity levels due to high freshwater inflows and high-water levels. The highest salinities are observed in the dry season (especially April-June) and are coincident with the lowest high tides of the year. Based on these observations, increasing freshwater flows during the dry season would be most effective at overcoming the influence of SLR and saltwater intrusion from winter to spring. The CERP projects that increase freshwater flows to Everglades National Park are expected to result in significant dry season benefits over the existing condition.

The present global rate of SLR is estimated at $3.4\,\text{mm year}^{-1}$ (NASA, 2016), but the south Florida regional rate could be as high as $9\,\text{mm year}^{-1}$ that may in part, be due to a slowing of the Florida Gulf Stream (Wdowinski et al., 2016). At rates $>9\,\text{mm year}^{-1}$ coastal response is likely to resemble the Early Holocene. The first evidence of an overstep (i.e., salt water encroachment too fast for all communities to retreat) has already been observed in the Taylor Slough watershed, where scrub mangroves are moving into sawgrass communities (Fig. 8B), bypassing the periphyton marl prairies (Meeder et al., 2017). The biophysics of the Everglades suggest that once sea level reaches 1.3 m higher than current, the transverse glades will become tidal creeks exchanging water with the interior Everglades Basin.

Can we manage for such dramatic transitions? Not if we ignore the transitions that are just ahead. To manage the complex socio-eco-hydro-economic environment of South Florida over the next 50 years, an awareness of the following is needed:

1. In geological terms, the Everglades ecosystem is very young and will continue to evolve and change. SLR is not the demise of the Everglades (unless an ice-free condition develops at the poles; Parkinson et al., 2015).
2. The Everglades developed upon marine strata, sea level controlled the evolution of the Florida Platform and rising sea level is responsible for the origin and genesis of the Everglades ecosystem.
3. Managing for flood control and water supply in the Everglades Basin has had unintended consequences (e.g., reduced water flow, altered flow directions, drowned tree islands, degraded microtopography, salt water encroachment, considerable peat soil loss, and shrinkage of wetland area).
4. Central to the economic value of restoration is slowing the pace of soil oxidation and erosion. Everglades restoration will provide coastal risk reduction and resilience in the form of protection, enhancement, and restoration of natural features such as mangroves, seagrasses, floodplains, swamps, and marshes.
5. The interactions of coastal wetland biology, SLR, and saltwater intrusion in coastal Everglades peatlands will determine wetland elevation in the future, and the capacity for coastal wetlands to provide carbon sequestration, storm surge attenuation, risk reduction, and habitat resilience.

ACKNOWLEDGMENTS

We are grateful to Florida Sea Grant (R/C-S-56) , including cooperative agreements with the South Florida Water Management District, the Everglades Foundation, and Everglades National Park. Additional funding was provided through the National Science Foundation's Florida Coastal Everglades Long Term Ecological Research (FCE LTER) Program (DEB-1237517). This is SERC publication number ## and publication #12 of the Sea Level Solutions Center in the Institute of Water and Environment at the Florida International University. "This is contribution number 893 from the Southeast Environmental Research Center in the Institute of Water & Environment at Florida International University".

REFERENCES

Alexander, T.R., Crook, A.G., 1975. Recent and Long-Term Vegetation Changes and Patterns in South Florida. Part II: Final Report, South Florida Ecological Study, Report to the National Park Service. pp. 1–827.

Alongi, D., 2008. Mangrove forests: resilience, protection from tsunamis, and responses to global climate change. Estuar. Coast. Shelf Sci. 76, 1–13.

Alpert, L., Stronge, W.B., 2009. The Economics of the Everglades Watershed and Estuaries Phase I—Review of Literature and Data Analysis, p. 181. http://www.drivecms.com/uploads/riverofgrasscoalition.com/1022369245The%20Economics%20of%20the%20Everglades%20FINAL%20REPORT.pdf.

American Sportfishing Association (ASA), 2013. Sportfishing in America: An economic force for conservation, p. 12. https://asafishing.org/uploads/2011_ASASportfishing_in_America_Report_January_2013.pdf.

Andrews, R., 1957. Vegetative cover-types of Loxahatchee and their principal components. In: Central and Southern Florida Flood Control Project for Flood Control and Other Purposes. Part 1. Agricultural and Conservation Areas, Supplement 25 – General design memorandum, plan of regulation for Conservation Area 1. US Army Corps of Engineers, Jacksonville, FL. Jacksonville District. Appendix I, 1957, pp. B-25–B-33.

Arkema, K.K., Guannel, G., Verutes, G., Wood, S.A., Guerry, A., Ruckelshaus, M., Kareiva, P., Lacayo, M., Silver, J.M., 2014. Coastal habitats shield people and property from sea-level rise and storms. Nat. Clim. Chang. 3, 913–918.

Armentano, T.V., Say, J.P., Ross, M.S., Jones, D.T., Cooley, H.C., Smith, C.S., 2006. Rapid responses of vegetation to hydrological changes in Taylor Slough, Everglades National Park, Florida, USA. Hydrobiologia 569, 293–309.

Barbier, E., Hacker, S., Kennedy, C., Koch, E., Stier, A., Silliman, B., 2011. The value of estuarine and coastal ecosystem services. Ecol. Monogr. 81, 169–193.

Baustian, J.J., Mendelssohn, I.A., Hester, M.W., 2012. Vegetation's importance in regulating surface elevation in a coastal salt marsh facing elevated rates of sea level rise. Global Change Biol. 18, 3377–3382.

Boesch, D.F., Armstrong, N.E., D'Elia, C.F., Maynard, N.G., Paerl, H.W., Williams, S.L., 1993. Deterioration of the Florida Bay ecosystem—an evaluation of the scientific evidence. Interagency Working Group on Florida Bay Report, 20 pp.

Bricker, J.D., Gibson, S., Takagi, H., Imamura, F., 2015. On the need for larger Manning's roughness coefficients in depth-integrated tsunami inundation models. Coast. Eng. J., 57. https://doi.org/10.1142/S0578563415500059.

Cahoon, D.R., Hensel, P., Rybczyk, J., McKee, K.L., Proffitt, E., Perez, B., 2003. Mass tree mortality leads to mangrove peat collapse at Bay Islands, Honduras after Hurricane Mitch. J. Ecol. 91, 1093–1105.

Castañeda-Moya, E., Twilley, R.R., Rivera-Monroy, V.H., Zhang, K., Davis, S.E., Ross, M., 2010. Spatial patterns of sediment deposition in mangrove forests of the Florida coastal everglades after the passage of Hurricane Wilma. Estuar. Coasts 33, 45–58.

CEPP (Central Everglades Planning Project), 2014. Final Integrated Project Implementation Report and Environmental Impact Statement: Main Report. US Army Corps of Engineers, Jacksonville, FL, p. 430.

Chambers, L.G., Davis, S.E., Troxler, T., Boyer, J.N., Downey-Wall, A., Scinto, L.J., 2014. Biogeochemical effects of simulated sea level rise on carbon loss in an Everglades mangrove peat soil. Hydrobiologia 726, 195–211.

Church, J.A., White, N.J., 2011. Sea-level rise from the late 19th to the early 21st century. Surv. Geophys. 32, 585–602.

CISRERP, 2014. Progress towards restoring the everglades: the fifth biennial review. In: Committee on Independent Scientific Review of Everglades Restoration Progress; Water Science and Technology Board; Board on Environmental Studies and Toxicology; Division on Earth and Life Studies. National Research Council, The National Academies Press, Washington, DC, pp. 320.

Craighead, F.C., 1966. The effects and natural forces on the development and maintenance of the Everglades, Florida. In: National Geographic Society Research Reports, pp. 49–67.

David, P.G., 1997. Changes in plant communities relative to hydrological conditions in the Florida Everglades. Wetlands 16, 15–23.

Davis, S.M., Ogden, J.C., 1994. The Everglades and Its Restoration. CRC Press, Boca Raton, FL, p. 860.

Davis, S.M., Childers, D.L., Lorenz, J.J., Wanless, H.R., Hopkins, T.E., 2005. A conceptual model of ecological interactions in the mangrove estuaries of the Florida Everglades. Wetlands 25 (4), 832–842.

Day, J.W., Kemp, G.P., Reed, D.J., Cahoon, D.R., Boumans, R.M., Suhayda, J.M., Gambrell, R., 2011. Vegetation death and rapid loss of surface elevation in two contrasting Mississippi delta salt marshes: the role of sedimentation, autocompaction and sea-level rise. Ecol. Eng. 37, 229–240.

Day, J.W., Sklar, F., Cable, J., Childers, D., Coronado, C., Davis, S., Kelly, S., Perez, B., Reyes, E., Rudnick, D., Sutula, M., 2013. The salinity transition zone between the southern Everglades and Florida Bay: system functioning and implications for management. In: Day, J.W., Yanez-Arancibia, A. (Eds.), The Gulf of Mexico Origin, Waters and Biota: Ecosystem-Based Management. Texas A&M University Press, College Station, TX, pp. 1–24.

Deegan, D.S., Johnson, R.S., Warren, B.J., Peterson, J.W., Fleeger, Fagherazzi, S., Wollhelm, W.M., 2012. Coastal eutrophication as a driver of salt marsh loss. Nature 490, 388–392.

DeLaune, R.D., Nyman, J.A., Patrick, W.H., 1994. Peat collapse and wetland loss in a rapidly submerging coastal marsh. J. Coast. Res. 10 (4), 1021–1030.

Dessu, S.B., Price, R.M., Troxler, T.G., Kominoski, J.S., 2018. Effects of sea-level rise and freshwater management on long-term water levels and water quality in the Florida Coastal Everglades. J. Environ. Manage. 211, 164–176.

Dineen, J.W., 1972. Life in the tenacious Everglades. Central and South Florida Flood Control District. In Depth Report. Vol. 1, No. 5, 112 pp.

Duever, M.J., Carlson, J.E., Meeder, J.F., Duever, L.C., Gunderson, L.H., Riopelle, L.A., Alexander, T.R., Myers, T.R., Spangler, D.P., 1986. The big cypress national preserve. Nat. Audubon Soc. Res. Rep. 8, 1–455.

Egler, F.E., 1952. Southeast saline Everglades vegetation, Florida, and its management. Vegetation 3, 213–265.

Emory, K.O., Milliman, J.D., 1970. Quaternary sediments of the Atlantic continental shelf of the United States. Quatemaria 12, 3–18.

Emory, K.O., Wigley, R.L., Bartlett, A.S., Rubin, M., Barghoorn, E.S., 1967. Fresh water peat on the continental shelf. Science 158, 1301–1307.

Ezer, T., 2013. Sea level rise, spatially uneven and temporally unsteady: why the U.S. East Coast, the global tide gauge record, and the global altimeter data show different trends. Geophys. Res. Lett. 40, 5439–5444. https://doi.org/10.1002/2013GL057952.

Fedler, T., 2013. Economic Impact of the Florida Keys Flats Fishery. Prepared for the Bonefish & Tarpon Trust, p. 27.

Fourqurean, J.W., Robblee, M.B., 1999. Florida Bay: a history of recent ecological changes. Estuaries 22, 345–357.

Fuller, D.O., Wang, Y., 2014. Recent trends in satellite vegetation index observations indicate decreasing vegetation biomass in the southeastern saline Everglades wetlands. Wetlands 34, 67–77.

Gaiser, E., Zafiris, A., Ruiz, P., Tobias, F., Ross, M., 2006. Tracking rates of ecotone migration due to salt-water encroachment using fossil mollusks in coastal south Florida. Hydrobiologia 569 (237), 257.

Gleason, P.J., Spackman Jr., W., 1974. Calcareous periphyton and water chemistry in the Everglades. In: Gleason, P.J. (Ed.), Environments of South Florida: Present and Past. Memoir 2. Miami Geological Society, Coral Gables, FL, pp. 146–181.

Gleason, P.J., Stone, P.A., 1994. Age, origin, and landscape evolution of the Everglades peatland. In: Davies, S.M., Ogden, J.C. (Eds.), Everglades: The Ecosystem and Its Restoration. St. Lucie Press, Delray Beach, FL, ISBN: 0-9634030-2-8, pp. 149–198.

Gleason, P.J., Cohen, A.D., Smith, W.G., Brooks, H.K., Stone, P.A., Goodrick, R.L. and Spackman, W., 1974. The environmental significance of Holocene sediments from the Everglades and saline tidal plain. In: Environments of South Florida: Present and Past (Memoir 2: 297-341). Miami Geological Society: Coral Gables, FL. Gleason, P.J. 0-9634030-2-8.

Goodrick, R.L., 1974. The wet prairies of northern Everglades. In: Gleason, P.J. (Ed.), Environments of South Florida: Present and Past, Memoir 2. Miami Geological Society, Coral Gables, pp. 287–341.

Harshberger, J.W., 1914. The vegetation of South Florida. Trans. Wagner Free Inst. Sci. Phila. 3, 51–189.

Harvey, J.W., Wetzel, P.R., Lodge, T.E., Engel, V.C., Ross, M.S., 2017. Role of a naturally varying flow regime in Everglades restoration. Restor. Ecol. https://doi.org/10.1111/rec.12558.

Hoffmeister, J.E.K., Stockman, W., Multer, H.G., 1967. Miami Limestone of Florida and its Recent Bahamian counterpart. Geol. Soc. Am. Bull. 78, 175–190.

Hohner, S.M., Dreschel, T.W., 2015. Everglades peats: using historical and recent data to estimate predrainage and current volumes, masses and carbon contents. Mires Peat 16 (1), 1–15. Available from, http://www.mires-and-peat.net/pages/volumes/map16/map1601.php.

IPCC, 2013. Summary for policymakers. In: Stocker, T., Qin, D., et al. (Eds.), Climate Change 2013: The Physical Science Basis. Contribution of Working Group I to the Fifth Assessment Report of the Intergovernmental Panel on Climate Change, Cambridge University Press, Cambridge/New York, NY. Available from, https://www.ipcc.ch/report/ar5/wg1/. (Accessed February 20, 2014).

Johnson, I., 1974. Beyond the Fourth Generation. University Presses of Florida, Gainesville, FL, p. 230.

Klaus, J.S., Meeder, J.F., McNeill, D.F., Woodhead, J.F., Swart, P., 2016. Middle Pliocene coral reef expansion on the west Florida shelf. Global Planet. Change 152, 27–37.

King, J.W., 1917a. Report of Exploration: Examination and Reconnaisance of the Lands of the Tamiami Trail in Dade County, Florida, unpublished manuscript dated Mar. 23, 1917. Jaudon Collection, Box 16. Historical Museum of South Florida, Miami.

King, J.W., 1917b. Map Showing Results of Examination of the Tamiami Trail Lands in Dade County Florida, unpublished blueprint to accompany report dated Mar. 23, 1917, Report item #18), 4 in. = 1 mile (1:15,840), Jaudon Collection, Box 16. Historical Museum of South Florida, Miami.

Kushlan, J.A., 1990. Freshwater marshes, in Ecosystems of Florida. In: Myers, R.L., Ewe, J.J. (Eds.), University of Central Florida Press, Orlando, pp. 324–363.

Lambeck, K., Rouby, H., Purcell, A., Sun, Y., Sambridge, M., 2014. Sea level and global ice volumes from the last glacial Maximum to the Holocene. PNAS 111, 15296–15303.

Larsen, L.G., Harvey, J.W., 2010. How vegetation and sediment transport feedbacks drive landscape change in the Everglades and wetlands worldwide. Am. Nat. 176 (3), E66–E79.

Larsen, L.G., Harvey, J.W., 2011. Modeling of hydrogeological feedbacks predicts distinct classes of landscape pattern, process, and restoration potential in shallow aquatic ecosystems. Geomorphology 126, 279–296.

Levermann, A., Clark, P., Marzeion, B., et al., 2013. The multi-millennial sea-level commitment of global warming. Proc. Natl. Acad. Sci. https://doi.org/10.1073/pnas.1219414110.

Loder, N.M., Irish, J.L., Cialone, M.A., Wamsley, T.V., 2009. Sensitivity of hurricane surge to morphological parameters of coastal wetlands. Estuar. Coast. Shelf Sci. 84, 625–636.

Loftus, W.F., Chapman, J.D., Conrow, R., 1990. Hydroperiod effects on Everglades marsh food webs, with relation to marsh restoration efforts. In: Proc. Conf. Science in the National Park. Vol. 6. Fisheries and Coastal Wetlands Research, U.S. National Park Service and the George Wright Society, Washington, DC, pp. 1–22.

Loveless, C.M., 1959. A study of the vegetation in the Florida everglades. Ecology 40, 1–9.

Lovelock, C.E., et al., 2017. Assessing the risk of carbon dioxide emissions from blue carbon ecosystems. Front. Ecol. Environ. 15 (5), 257–265.

Madden, C., 2013. Use of models in ecvosystem-based management of the southern Evergldaes and Florida Bay. In: Day, J.W., Yanez-Arancibia, A. (Eds.), The Gulf of Mexico Origin, Waters and Biota: Ecosystem-Based Management. Texas A&M University Press, College Station, TX, pp. 25–52.

Mather Economic, 2010. Measuring the Economic Benefits of America's Everglades Restoration: An Economic Evaluation of Ecosystem Services Affiliated with the World's Largest Ecosystem Restoration Project. p. 173. https://www.evergladesfoundation.org/wp-content/uploads/sites/2/2017/12/Report-Measuring-Economic-Benefits-Exec-Summary.pdf.

McKee, K.L., 2011. Biophysical controls on accretion and elevation change in Caribbean mangrove. Estuar. Coast. Shelf Sci. 91, 475–483.

McKee, K.L., Cahoon, D.R., Feller, I.C., 2007. Caribbean mangroves adjust to rising sea level through biotic controls on change in soil elevation. Glob. Ecol. Biogeogr. 16, 545–556.

Mcleod, E., Chmura, G.L., Bouillon, S., Salm, R., Björk, M., Duarte, C.M., Lovelock, C.E., Schlesinger, W.H., Silliman, B.R., 2011. A blueprint for blue carbon: toward an improved understanding of the role of vegetated coastal habitats in sequestering CO_2. Front. Ecol. Environ. 9 (10), 552–560.

McVoy, C.W., Said, W.P., Obeysekera, J., Van Arman, J.A., Reschel, T.W., 2011. Landscapes and hydrology of the predrainage Everglades University Press of Florida, Gainesville. Available from, https://www.researchgate.net/profile/Thomas_Dreschel/publication/257315404_Landscapes_and_Hydrology_of_the_Predrainage_Everglades_Overview/links/02e7e524e97f484b32000000.pdf.

Meeder, J.F., Parkinson, R.W., 2018. SE Saline Everglades transgressive sedimentation in response to historic acceleration in sea-level rise: a viable marker for the base of the Anthropocene? J. Coast. Res. 34 (2), 490–497.

Meeder, J.F., Parkinson, R.W., Ruiz, P., Ross, M.S., 2017. Saltwater encroachment and prediction of future ecosystem response to the Anthropocene Marine Transgression, Southeast Saline Everglades, Florida. Hydrobiologia. https://doi.org/10.1007/s10750-3359-0.

Millennium Ecosystem Assessment, 2005. Ecosystems and Human Well-Being: Wetlands and Water Synthesis. World Resources Institute, Washington, DC.

Miller, K.G., Kominz, M.A., Browning, J.V., Wright, J.D., Mountain, G.S., Katz, M.E., Sugarman, P.J., Cramer, B.S., Christie-Blick, N., Pekar, S.F., 2005. The Phanerozoic record of global sea-level change. Science 310, 1293–1298.

Milliman, J.D., Emery, K.O., 1968. Sea levels during the past 35,000 years. Science 162 (3858), 1121–1123.

Milliman, J.D., Pilkey, O.H., Blackwelder, B.W., 1968. Carbonate Sediments on the Continental Shelf. Duke University, Cape Hatteras to Cape Romano.

NASA, 2016. Global climate change: vital signs of the planet. Available at: http://climate.nasa.gov/vital-signs/sealevel/.

Nicholls, R.J., Cazenave, A., 2010. Sea-level rise and its impact on coastal zones. Science 328, 1517–1520.

Noe, G.B., Childers, D.L., Jones, R.D., 2001. Phosphorus biogeochemistry and the impact of phosphorus enrichment: why is the Everglades so unique? Ecosystems 4, 603–624.

Nyman, J.A., Walters, R.J., DeLaune, R.D., Patrick, W.H., 2006. Marsh vertical accretion via vegetative growth. Estuar. Coast. Shelf Sci. 69, 370–380.

Olmsted, I., Armentano, T.V., 1997. Vegetation of Shark Slough, Everglades National Park. South Florida Natural Resource Center, Technical Report 97-001. National Park Service, Everglades National Park, Homestead, FL, USA.

Osipova, E., Shadie, P., Zwahlen, C., Osti, M., Shi, Y., Kormos, C., Bertzky, B., Murai, M., Van Merm, R., Badman, T., 2017. IUCN World Heritage Outlook 2: a conservation assessment of all natural World Heritage sites. IUCN, Gland, p. 92.

Park, J., Stabenau, E., Redwine, J., Kotun, K., 2017. South Florida's encroachment of the sea and environmental transformation over the 21st century. J. Mar. Sci. Eng. 5, 31.

Parker, G.G., Ferguson, G.E., Love, S.K., 1955. In: Water resources of southeastern Florida with special reference to the geology and ground water of the Miami area. United States Geological Survey Water-Supply Paper 1255. pp. 1–965.

Parkinson, R.W., 1989. Decelerating Holocene SLR and its influence on southwest Florida coastal evolution: a transgressive-regressive stratigraphy. J. Sediment. Petrol. 59, 960–972.

Parkinson, R.W., Meeder, J.F., 1991. Mud-bank destruction and the formation of a transgressive sand sheet, southwest Florida inner shelf. Geol. Soc. Am. Bull. 103, 1543–1551.

Parkinson, R., Harlem, P.W., Meeder, J.F., 2015. Managing the anthropocene marine transgression to the year 2100 and beyond in the State of Florida USA. Clim. Change 128, 83–98. https://doi.org/10.1007/s10584-014-1-1301-2.

Pearlstine, L.G., Pearlstine, E.V., Aumen, N.G., 2010. A review of the ecological consequences and management implications of climate change for the Everglades. J. North Am. Benthology. Soc. 29 (4), 1510–1526.

Perkins, R.D., 1977. Depositional framework of Pleistocene rocks of south Florida. Pt. 2. In: quaternary sedimentation in South Florida. Geol. Soc. Am. Mem. 147, 131–198.

Perry, W., 2004. Elements of south Florida's comprehensive Everglades restorationplan. Ecotoxicology 13, 185–193.

Raymo, M.E., Mitrovica, J.X., O'Leary, M.J., DeConto, R.M., Hearty, P.J., 2011. Departures from eustasy in Pliocene sea-level records. Nat. Geosci. 4, 328–332.

Rohling, E.J., Haigh, I.D., Foster, G.L., Roberts, A.P., Grant, K.M., 2013. A geological perspective on potential future sea-level rise. Sci. Rep. 3, https://doi.org/10.1038/srep03461.

Ross, M.S., Meeder, J.F., Sah, J.P., Ruiz, P.L., Telesnicki, G.J., 2000. The southeast saline Everglades revisited: 50 years of coastal vegetation change. J. Veg. Sci. 11, 101–112.

Ross, M.S., Gaiser, E.E., Meeder, J.F., Lewin, M.T., 2002. Multi-taxon analysis of the "white zone", a common ecotonal feature of South Florida coastal wetlands. In: Porter, J.W., Porter, K.G. (Eds.), The Everglades, Florida Bay and Coral Reefs of the Florida Keys. CRC Press, pp. 205–238.

Ruiz, P.L., Giannini, H.C., Prats, M.C., Perry, C.P., Foguer, M.A., Arteaga Garcia, A., Shamblin, R.B., Whelan, K.R.T., Hernandez, M., 2017. The Everglades National Park and Big Cypress National Preserve Vegetation Mapping Project: Interim report–Southeast Saline Everglades (Region 2), Everglades National Park. Natural Resource Report NPS/SFCN/NRR—2017/1494. Fort Collins, CO., National Park Service.

Scheidt, D., Stober, J., Jones, R., Thornton, K., 2000. South Florida ecosystem assessment: Everglades water management, soil loss, Eutrophication and habitat. In: Report EPA-904-R00-003. Athens: Region 4, Office of Research and Development, USEPA; 2000. pp. 48.

Scholl, D.W., Craighead, F.C., Stuiver, M., 1969. Florida submergence curve revised: its relation to coastal sedimentation rates. Science 163, 562–564.

SERES Project, 2016. Management-Driven Science Synthesis: An Evaluation of Everglades Restoration Trajectories. p. 60. https://www.researchgate.net/publication/308171535_Management-Driven_Science_Synthesis_An_Evaluation_of_Everglades_Restoration_Trajectories.

SFWMD, 2011. Restoration Strategies Regional Water Quality Plan. South Florida Water Management District, West Palm Beach, FL, p. 92.

Sklar, F.H., van der Valk (Eds.), 2003. Tree Islands of the Everglades. Kluwer Academic Publishers, Boston, MA.

Sklar, F.H., Chimney, M.J., Newman, S., McCormick, P., Gawlik, D., Miao, S., McVoy, C., Said, W., Newman, J., Coronado, C., Crozier, G., Korvela, M., Rutchey, K., 2005. The scientific and political underpinnings of Everglades restoration. Front. Ecol. Environ. 3 (3), 161–169.

Smith, B.D., Zeder, A., 2013. The onset of the anthropocene. Anthropocene 4, 8–13.

Smith, T.J., Foster, A.M., Tiling-Range, G., Jones, J.W., 2013. Dynamics of mangrove-marsh ecotones in subtropical coastal wetlands: fire, sea-level rise, and water levels. Fire Ecol. 9 (1), 66–77.

Smoak, J.M., Breithaupt, J.L., Smith, T.J. III, Sanders, C.J., 2013. Sediment accretion and organic carbon burial relative to sea-level rise and storm events in two mangrove forests in Everglades National Park. Catena 104, 58–66.

Spratt, R.M., Lisiecki, L.E., 2016. A late Pleistocene sea level stack. Clim. Past 12, 1079.

Stephens, J.C., Johnson, I., 1951. Subsidence of organic soils in the upper Everglades region of Florida. Soil Sci. Soc. Florida Proc. 11, 191–237.

Sutton-Grier, A., Wowk, K., Bamford, H., 2015. Future of our coasts: the potential for natural and hybrid infrastructure to enhance the resilience of our coastal communities, economies and ecosystems. Environ. Sci. Policy 51, 137–148.

Troxler, TG, G Starr, JN Boyer, JD Fuentes, R Jaffe et al. n.d. Chapter 6: carbon cycling in the Florida coastal everglades social-ecological system across scales. In *The Coastal Everglades: The Dynamics of Socio-Ecological Transformations in the South Florida Landscape*. Childers, DL, Ogden, L, Gaiser, E. (in press), Oxford University Press.

US Army Corps of Engineers, 2013. Coastal Risk Reduction and Resilience. CWTS 2013-3. Directorate of Civil Works. US Army Corps of Engineers, Washington, DC.

Vacher, H.L., Mylroie, J.E., 2002. Eogenetic karst from the perspective of an equivalent porous medium. Carbon. Evap. 17 (2), 182–196.

Vegh, T., T.Troxler, K.Zhang, G. Guannel, E. Castañeda-Moya, A.Sutton-Grier, L.Pendleton, and B.Murray, n.d.. Ecosystem Services and Economic Valuation: Co-benefits of coastal wetlands. Blue Carbon Primer: Science and Policy. CRC Press, Boca Raton, FL (in press).

Vlaswinkel, B.M., Wanless, H.R., 2012. Rapid recycling of organic-rich carbonates during transgression: a complex coastal system in southwest Florida. Perspectives in Carbonate Geology: A Tribute to the Career of Robert Nathan Ginsburg. International Association of Sedimentologists Special Publication 41, 91–112.

Voss, C.M., Christian, R.R., Morris, J.T., 2013. Marsh macrophyte responses to inundation anticipate impacts of sea-level rise and indicate ongoing drowning of North Carolina marshes. Mar. Biol. 160, 181–194.

Wanless, H.R., 1974. Mangrove sedimentation in geological perspective. In: Gleason, P. (Ed.), Environments of South Florida: Present and Past. Miami Geological Society Memoir 2, Coral Gables, FL, pp. 190–200.

Wanless, H.R., Parkinson, R.W., Tedesco, L.P., 1994. Sea level control on stability of Everglades wetlands. In: Davis, S.M., Ogden, J.C. (Eds.), Everglades: The Ecosystem and Its Restoration. St. Lucie Press, Delray Beach, FL, pp. 199–223.

Watts, W.A., 1975. A late Quaternary record of vegetation from Lake Annie, south-central Florida. Geology 3, 344–346.

Wdowinski, S., Bray, R., Kirtman, B.P., Wu, Z., 2016. Increasing flooding hazard in coastal communities due to rising sea level: Case study of Miami Beach, Florida. Ocean Coast. Manag. 126, 1–8.

Willard, D.A., Bernhardt, C.E., Holmes, C.W., Landacre, B., Marot, M., 2006. Response of Everglades tree islands to environmental change. Ecol. Monogr. 76, 565–583.

Wood, J.M., Tanner, G.W., 1990. Graminoid community composition and structure within four Everglades management areas. Wetlands 10, 127–149.

Worth, D., 1987. Environmental responses of water conservation area 2A to reduction schedule and marsh drawdown. Tech. Rep. 87-5. South Florida Water Management District, West Palm Beach.

Zaffke, M., 1983. Plant communities of water conservation area 3A; base-line documentation prior to the operation of S-339 and S-340. Technical memorandum. South Florida Water Management District, West Palm Beach, Florida, USA.

Zalasiewicz, J., Williams, M., Smith, A., Barry, T.L., Coe, A.L., Bown, P.R., Brenchley, P., Cantrill, D., Gale, A., Gibbard, P., Gregory, F.J., Hounslow, M.W., Kerr, A.C., Pearson, P., Knox, R., Powell, J., Waters, C., Marshall, J., Oates, M., Rawson, P., Stone, P., 2008. Are we now living in the Anthropocene? GSA Today 18 (2), 4–8. https://doi.org/10.1130/GSAT01802A.1.

Zhang, K., Liu, H., Li, Y., Xu, H., Shen, J., Rhome, J., Smith, I., T.J., 2012. The role of mangroves in attenuating storm surges. Estuar. Coast. Shelf Sci. 102-103, 11–23.

Zweig, C.L., Kitchens, W.M., 2009. Muliti-state seccession in wetlands: a novel use of state and transition models. Ecology 90 (7), 1900–1909.

Chapter 17

Population Growth, Nutrient Enrichment, and Science-Based Policy in the Chesapeake Bay Watershed

Christopher F. D'Elia*, Morris Bidjerano†, Timothy B. Wheeler‡

*College of the Coast and Environment, Louisiana State University, Baton Rouge, LA, United States, †School of Public Policy and Administration, Walden University, Greenville, SC, United States, ‡Bay Journal, Seven Valleys, PA, United States

1 INTRODUCTION

The Chesapeake Bay is a distinct feature of eastern US geography, easily visible on a US map or from space. With its 166,000 km^2 (64,000 mi^2) watershed, three major rivers and numerous minor tributaries, this bountiful and beautiful estuary is in geological terms a "drowned-river valley." As such, the Bay is a feature of the Holocene, that is, it came into being in the last 12,000 years as a post Ice Age feature formed by riverine erosion and rising sea level. Considering that, natural change due to climate has been a determinant of the state of its ecological systems even before its European colonization occurred four centuries ago.

Over the last seven decades, the Chesapeake Bay watershed's human population has more than doubled. In 1950, an estimated 8.4 million people lived in the watershed in what was predominantly an agricultural economy subsequent to European colonization. By the last several years, this number had jumped to over 18 million and urban and suburban land use has grown (Orth et al., 2017). Since the late 1980s, the watershed's population has increased nearly 10% per decade. The watershed now represents over 5% of the US population. While climate change due to natural causes will always affect the Chesapeake, clearly the expression of the human presence in this system has been a dominant driver of its environmental health for most of the last century. To add to this regional anthropogenic burden, accelerating sea-level rise due to climate change is also looming.

This iconic estuarine system is an important US natural resource in many ways. Given its proximity to the US capital, Washington, DC, the Chesapeake Bay has always garnered special attention from US legislators and policymakers. Accordingly, events in the Chesapeake directly or indirectly influenced by environmental legislation and policymaking in the United States. Thus, events at the local level in the Chesapeake watershed can easily rise to a higher level and end up influencing national policy. The interface between science and policy is a difficult one to bridge successfully, but for the Chesapeake system it is often done. Given the interplay between the effects of climate and local development and population growth in the watershed, today's challenges to manage this resource also pertain to other coastal regions worldwide. In our view, for the Chesapeake Bay, the frequent interaction of scientists with managers and policymakers has made a difference for decades in how policies and legislation came to be. This chapter provides a case study of one of these examples, nutrient management in the Patuxent River (PR) and its basin, a small tributary to the larger Chesapeake Bay system.

In the late 1980s, the Chesapeake Executive Council of the Chesapeake Bay Commission sponsored a report called, *Population Growth and Development in the Chesapeake Bay Watershed to the Year 2020*. We refer to that here simply as the "2020 Report." The Council undertook this effort bearing in mind the rapid growth of the Chesapeake's watershed, and concerned about the environmental expression of the growing human footprint on the land and waters therein. Of particular concern to the panel that produced the report were land use and the generation of human waste. With the 30-year anniversary of that publication fast approaching in 2020, the time has now come to look back and consider what changes have occurred, especially given that the 2020 Report underestimated the population growth rates that have actually occurred. Public and governmental perception of the scientific understanding of the Chesapeake system is also of interest in this chapter.

Coasts and Estuaries. https://doi.org/10.1016/B978-0-12-814003-1.00017-4

293

The 2020 Report projected that the watershed's population would be 16.2 million[1] in 2020, yet an estimated 18.1 million people lived in the Chesapeake Bay watershed in 2016. That is almost 2 million more inhabitants than were anticipated 4 years later. Projections of the watershed's population in 2030 and 2040, respectively, are 20 million and 21.1 million, which we think are underestimates. A critical question looms that remains unanswered and is rarely discussed (D'Elia, 1995): how much population growth can be sustained without sacrificing environmental quality? Put another way, what is the "carrying capacity" of the Chesapeake watershed?

This chapter has three sections. The first considers the scientific considerations attendant to the work of the EPA Chesapeake Bay Program (CBP) that has existed since 1976. We delve into the compelling evidence that supported concerns about its over-enrichment with nutrients and the evidence that first drew the connections between that and declining water quality for which the PR, MD, played an important role in the early stages of the CBP. The second section considers the development of policy and practice in the control of the nutrient elements, N and P, which many scientists consider as major threats to the watershed's environment. Bearing in mind the old adage "think globally but act locally," the actions of local scientists and policy leaders have had a huge impact on larger outcomes not only for the Chesapeake, but also for distant locales around the globe as well. The last section considers the present status of the Chesapeake and its future prospects for its ecological health given concerns about continuing population growth, land use, and climate change.

The three authors each have very different perspectives. The first author lived and worked in the Chesapeake watershed for over 20 years and was involved as a researcher and participant in the inception of the CBP. He was a coprincipal investigator on one of the first major projects commissioned by that program that considered specifically the effects of nutrient enrichment.

The second author, although never a resident of the watershed, is a student of public policy who became interested in the particular case of the Patuxent watershed located between the Washington, DC and Baltimore, MD urban areas. The management of N and P in this watershed has been a major concern since the late 1970s, and the control of N, in particular, has been a controversial feature in it. The history and debate of N control in the Patuxent are of importance to the entire Chesapeake watershed as well as to other coastal areas in the United States. Indeed, what was started in the Patuxent has led to major changes in the management of nutrients nationwide.

The third author has been a journalist with a focus on the environment for over three decades. He has covered the Chesapeake Bay and other environmental issues for most of his career, including nearly 32 years with the Baltimore Sun and Evening Sun. He is a former president of the Society of Environmental Journalists and has won numerous awards, including the 2010 Excellence in Journalism Award from the Renewable Natural Resources Foundation. He now serves as the associate editor of the Bay Journal.

Together, these authors have a perspective of the Chesapeake Bay watershed that spans the entire four-decade existence of the EPA CBP.

2 DESCRIPTION OF THE WATERSHED AND ITS ESTUARY

Many excellent descriptions of the Chesapeake Bay and its watershed are available both online and in the published literature (e.g., Brush, 2017). Accordingly, we provide here only the most salient details that directly pertain to this chapter.

The watershed of the Chesapeake Bay includes New York, Pennsylvania, Delaware, Maryland, West Virginia, Virginia, and District of Columbia. Many US federal agencies have interests and jurisdiction in the Bay and its watershed. Accordingly, its management requires active participation and coordination by a number of jurisdictions, state, and federal. Multiple large tributaries account for over 80% of the freshwater feeding the Bay, adding to the complexity of understanding and managing it: the Susquehanna, Potomac, and James Rivers are the largest tributaries, in declining order of annual discharge. Smaller but significant tributaries include the Rappahannock, York, and PRs, as well as the still smaller Choptank, Pocomoke, and other lower order tributary rivers such as the Appomattox, Pamunkey, and Mattaponi Rivers. Given the large geographic extent of the watershed, depending on the nature of a given meteorological event, discharge may not be synchronous in the different tributaries.

2.1 Environmental History Prior to 1950s

Providing a detailed environmental history of the watershed is beyond the scope of this chapter, but it is worthwhile to consider different stages of deforestation, which has a significant role in increasing nonpoint sources (runoff) from the land and thus, water quality in the Bay and its tributaries. Brush (2017) describes four stages that she has researched over

---

1 This population projection did not include all of the watershed, just MD, VA, PA, and DC. See below for further explanation.

decades of stratigraphic analysis of sediments. In the earliest days before extensive European colonization, "the Pre-Colonial" stage (pre-17th century), the watershed was heavily forested, terrestrial runoff was much less, nutrients from sewage were nonexistent, and the water quality in the Bay was undoubtedly far better than it is now. Early colonists set about immediately at the task of clearing the land for agriculture ("Early Colonial" stage, from the late 17th to early 18th centuries), when less than 20% of the land was deforested. By the middle of the 19th century, much of the forested areas had been converted to agriculture, but by Brush's "Intensive Agriculture" stage (late 19th to early 20th centuries) nearly 80% deforestation had occurred. The effect of this deforestation on Chesapeake Bay watershed was less than it would have been were fertilizer intensively applied. The "Modern Era" (mid-20th century to present) has benefited from early to mid-20th century reforestation. This era is characterized by 40%–60% deforestation with 15% of the land being developed. However, since the end of the 20th century, the watershed has seen accelerating urban and suburban land use with more projected to occur in the future. Fig. 1 illustrates how rapidly both forested and agricultural areas are being lost. See also Orth et al. (2017), Fig. 8.

3 NUTRIENT ENRICHMENT IN THE CHESAPEAKE

In 1977, when the first author joined the faculty of the Chesapeake Biological Laboratory (CBL) of the University of Maryland, considerable discussion had already occurred about secular changes not only in the water quality of the Bay in general, but also in specific of the PR, on whose banks the Laboratory sits. A number of prominent estuarine scientists had worked at the Laboratory since its establishment in the mid-1920s leaving it with an unusual legacy of historic data and published papers in the scientific literature. As one of the oldest estuary-based research facilities in the world, CBL has made many important contributions to our understanding of coastal systems that have led to geographically broader insights as to how to manage them.

Three CBL faculty members were particularly active in the 1970s in the public debate about the causes and extent of changes in water quality of the River, Donald Heinle, Joseph Mihursky, and Robert Ulanowicz. Soon thereafter, Walter Boynton, and Christopher D'Elia joined the faculty and became also involved. Because of the small size of Calvert County, which was home to CBL and was where they all lived and worked, these scientists were well known to local governmental leaders and vice versa. In addition, the Academy of Natural Sciences of Philadelphia had a lab on the PR, and several of their faculty (Sandy Sage, Kent Mountford, and James Sanders) became involved in the policy discourse. Others from other institutions in the Chesapeake Bay community also contributed to the scientific understanding of the nutrient issue in the Bay, although they were less directly involved in PR nutrient enrichment policy issues. These include J.L. Taft and J.J. McCarthy (Johns Hopkins University CBI); W.M. Kemp, J.C. Stevenson, and T.R. Fisher (Horn Point Laboratory, UMCEES); R.J. Orth, K.L. Webb, R.L. Wetzel (VIMS), and others. Indeed, the extensive scientific interest in the nutrient enrichment and related issues of the Chesapeake Bay led to a rapidly expanding literature in the 1970s and 1980s that led the Chesapeake estuary to be one of the most well-researched estuaries in the world at the time. However, with respect to bridging the science-policy divide, in particular, the close relationship of CBL scientists and public officials in Southern Maryland drove what has been an unusually tight connection between scientists and policymakers that we describe in the next section.

Malone et al. (1993) published an excellent review of policy and relevant science that provides more detail than we can provide here. The US Congress authorized the EPA's CBP in 1976. This program came to be largely because of the efforts of former CBL director L. Eugene Cronin who also became the director of the Chesapeake Bay Consortium, which included all academic science institutions on the Bay. A tireless advocate for the Bay and for estuarine science, Cronin had close ties to US Senator Charles McC. "Mac" Mathias, Jr., R-MD, who as a lifelong Marylander had particular affection for the Bay. With Cronin's and other scientists' urging, Mathias himself became more concerned about the changes in the Bay and its watershed during his lifetime. In 1973, Mathias took Russell Train, then Administrator of the US Environmental Protection Agency, on a boat tour of the Chesapeake (Blankenship, 2003). As a much heralded and critical "focusing event," this crystalized the issue of the Bay's decline and galvanized public perception that the Bay and its watershed were in environmental peril.

Nutrient enrichment problems in Chesapeake Bay and its tributaries became a particular target of the nascent CBP, which was established soon thereafter. The loss of submerged aquatic vegetation (SAV) and toxic substances such as herbicides and pesticides also received early attention. Later, the loss of harvestable living resources also became an important topic as well. However, initial doubts existed as to how extensive the nutrient problems were, to what extent they affected not just water quality, but also SAV and other living resources, and how rapidly environmental degradation was occurring. Many believed that the SAV problems resulted not from nutrient enrichment, but rather from agriculturally important herbicides running off the land.

Farmland and forest land loss (2000–2030)

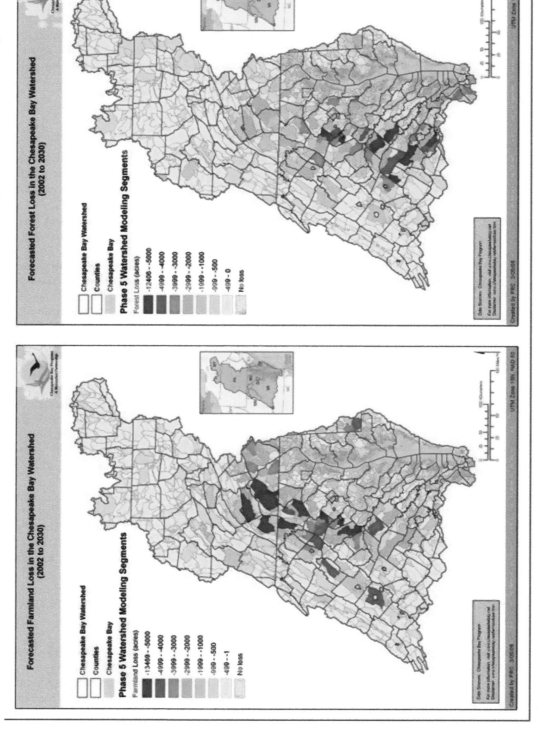

FIG. 1   Past and projected farmland and forest land-use change.

3.1 The "Heinle" Report

By the late 1970s, many scientists and managers had concluded that increased nutrient loadings had their greatest effects in the tributaries nearest the centers of demographic development. The tidal freshwater portions of the Potomac River near Washington, DC showed signs of excessive nutrient enrichment, where point source loadings from municipal sewage treatment plants had noticeable effects early in the 20th century (Cumming, 1916; Cumming et al., 1916). Nonpoint nutrient sources were mostly ignored or put in abeyance for later attention. Although earliest concerns focused on problems of human health and sanitation, clearly the input of untreated sewage to the Potomac caused oxygen depletion in receiving waters. In addition, occasional blooms of cyanobacteria were reported in the upper Potomac estuary as early as 1916 (Cumming et al., 1916). By the mid-1960s, sewage inputs in the tidal freshwater portion of the Potomac sufficiently enriched the water with nutrients that cyanobacterial blooms became a serious problem (Jaworski et al., 1971).

Other tributaries of the Chesapeake also showed signs of nutrient enrichment. The upper Bay near Baltimore, MD and the upper James near Richmond, VA, were over-enriched. Also enriched, but to a lesser extent, were the York, Rappahannock, Patuxent, and Susquehanna Rivers. Growing concern existed at the time that Mathias took Train on the tour of the Bay that human population growth in the Chesapeake watershed had resulted in increased nutrient loadings from point (sewage) and nonpoint (runoff) sources, but some managers such as Maryland Department of Natural Resources (DNRs) Secretary James B. Coulter, refused to acknowledge that fact. Accordingly, in its earliest days, the CBP recognized a need to document the nutrient enrichment phenomenon as quantitatively as possible. Thus, EPA funded a multiinstitution research group headed by CBL's Donald Heinle to undertake that documentation. That study led to the well-known and timely Heinle et al. (1980) report, which was one of the first major products to come out of the CBP, initially a 5-year program that Congress directed EPA to undertake in 1976. While some of the data in that report have been published in the scientific literature, unfortunately, Heinle went to work for a consulting company in Seattle and died at an early age so he never published the data in the peer-reviewed literature. Nonetheless, the compelling evidence of declining water quality presented in the Heinle report did add considerable impetus for EPA and the states to consider what potential remedies existed to control nutrient pollution. This led to a highly circulated "Synthesis Report" that did contain much of those data being released by the CBP in 1982.

Heinle et al. (1980) recognized that of the moderately enriched tributaries on the western shore of the Bay, an exceptionally good data record existed for the PR and estuary, because of the long-term interest of CBL and Benedict scientists in studying the River outside their laboratory doors. This excellent data record, which includes both published and unpublished results, extends back to the mid-1930s and is one of the older and more complete ones for any estuary in the world. Moreover, Heinle discovered laboratory notebooks with unpublished nutrient, oxygen, and Secchi disk data from the 1930s and 1940s in the attic of the CBL Library—this would prove to be extremely helpful in making the case that the River's water quality was declining.

Other factors made the Patuxent interesting for the Heinle group's attention: the Patuxent has a remarkable history of local scientists having close relationships to policymakers and politicians, and its watershed is virtually entirely in one state, Maryland, with a little bit also in the District of Columbia, which simplifies policymaking. Its watershed is located between two major US cities, Washington, DC and Baltimore, MD, and it includes the Maryland state capital, Annapolis. Thus, it has political visibility and impact in a key Chesapeake Bay state as well as at the federal level. Thus the PR is particularly pertinent case study for US policymaking and legislation because of "the strategic interaction of collaboration with conflict" (Bidjerano, 2009). Its watershed has seen very rapid demographic and land-use change, having been primarily agricultural before 1970 with increasing urban/suburban development thereafter. For those reasons, and because the estuary had been undergoing continuing change (dissolved inorganic nutrient levels had increased, and transparency and deep water dissolved oxygen concentrations had decreased), much of the following data analysis and discussion deals with the PR. In the early 1980s became a key battleground in the Chesapeake system over how to undertake and finance nutrient control in the United States. The Patuxent was covered widely in the media.

Background information on nutrient enrichment and its relationship to algal growth is provided below to help underscore why the problem of nutrient enrichment is complex and difficult to assess in light of data gaps in the historical record. We prefer the use of the term "nutrient over-enrichment" to "eutrophication."

3.2 Historical Trends in Nutrient Enrichment in the PR

The following section draws heavily on two EPA CPB reports, one a project final report by Heinle et al. (1980) and the other the "Synthesis Report" chapter by D'Elia (1982). While these reports had widespread circulation in the Chesapeake Bay community, this material was not published in the broader and more accessible scientific literature. Accordingly, this chapter seeks to redress that, albeit nearly four decades later.

In their review of historical trends in nutrient enrichment of Chesapeake Bay, Heinle et al. (1980) concluded that nutrient enrichment problems were greatest in the low salinity areas (less than 8–12 ppt), where summer chlorophyll a concentrations often reach or exceed $60 \mu g \, L^{-1}$. Such areas tend to be those where the tributaries pass through the more populous areas of the Chesapeake watershed, and in fact, were the most rapid land use changes have occurred. Before rapid population growth occurred in the watershed, chlorophyll levels in those areas are unlikely to have exceeded $20 \mu g \, L^{-1}$ chlorophyll a.

Given its excellent historical record, the PR is an excellent proxy for what occurred in the main Bay and western shore tributaries. The River has been increasingly enriched after 1960, although recent point and nonpoint nutrient controls have slowed the rate of enrichment. Similar detrimental effects of nutrient enrichment could also be expected to occur in analogous segments of the Bay system.

Fig. 2 shows the rather striking historical changes that occurred in dissolved inorganic phosphate (DIP) concentrations in surface waters of the Patuxent. Since the ammonium-molybdate methodology for DIP analysis had not changed appreciably over the years, Heinle et al. (1980) considered the data from the 1930s to be valid for comparison with later data. By 1980, maximum concentrations of DIP had clearly increased upstream of the Benedict Bridge (Fig. 3), where salinities are typically less than 9 ppt. Downstream of the bridge, where salinities range from about 8–18 ppt, surface DIP concentrations were significantly lower than those observed upstream, presumably as a result of the dilution of DIP-rich fresh water by less-enriched saline water. Nonetheless, DIP levels in this part of the River seem to have increased sharply between the 1930s and 1980. This increase is most pronounced in the summer. Such a summer phosphate maximum is characteristic of Chesapeake Bay and other estuaries (Taft and Taylor, 1976; Taft et al., 1980). This likely results from surfacing of water rich in phosphate, produced by enhanced rates of benthic regeneration at higher summer temperatures, and by increased phosphate solubility at lower oxygen concentrations below the halocline (Patrick and Khalid, 1974; D'Elia, 1987).

On the basis of the 1968-to-1980 data set which has such high DIP values, phosphorus limitation would appear unlikely anywhere on the PR throughout most of the season when severe oxygen deficits occur (late spring through fall). Light limitation seems likely during this time period (O'Connor et al., 1981). Phosphorus limitation may have been present prior to the late 1960s, when P loadings from sewage treatment plants were considerably lower.

Nitrate plus nitrite-N levels in the Patuxent exhibit the same seasonal cycle of abundance that has been reported for the upper Bay (McCarthy et al., 1977). Moreover, nitrate content of the water increased since the late 1930s (Fig. 4). Most of the increase occurred later than 1965, coinciding with the beginning of extensive development of the PR basin. The source of this nitrate is probably nonpoint. As for DIP, less nitrate is found in the water south of Benedict Bridge, reflecting the dilution of nutrient-rich fresh water by less-enriched saline water from the Atlantic Ocean.

The historical record does not include reliable data on ammonium, the preferred N source for phytoplankton, because of the analytical limitations of the time. Boynton et al. (1980) observed that the regeneration of ammonium by the Patuxent riverbed occurs at some of the highest rates ever recorded anywhere. Ammonium from benthic regeneration accumulates below the halocline and diffuses across that boundary to surface waters. Ammonium is a growth-limiting nutrient for phytoplankton at times in the Chesapeake and its tributaries (D'Elia et al., 1986; Fisher et al., 1992). A suite of factors affectsw

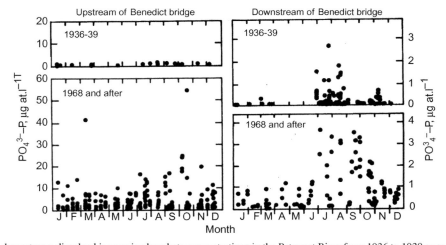

FIG. 2 Upstream and downstream dissolved inorganic phosphate concentrations in the Patuxent River from 1936 to 1939 compared with 1968 to 1980. *(From Heinle, D.R., D'Elia, C.F., Taft, J.L., et al., 1980. Historical review of water quality and climatic data from Chesapeake bay with emphasis on effects of enrichment. U.S. EPA Chesapeake Bay Program Final Report, Grant R306189010. Chesapeake Research Consortium, Inc. Publication No. 84. Annapolis, MD.)*

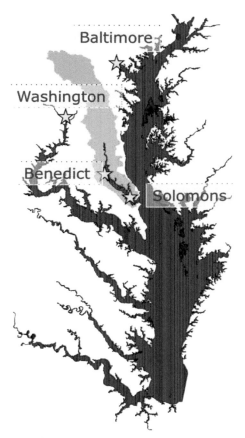

FIG. 3 Map of the Chesapeake Bay and the Patuxent River watershed, showing location of key locations referred to herein.

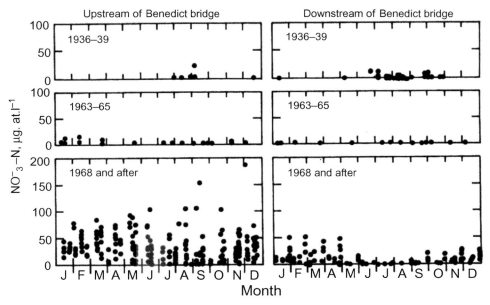

FIG. 4 Nitrate concentrations in the Patuxent River, 1936–1939, 1963–1965, and 1968–1980. *(From Heinle, D.R., D'Elia, C.F., Taft, J.L., et al., 1980. Historical review of water quality and climatic data from Chesapeake bay with emphasis on effects of enrichment. U.S. EPA Chesapeake Bay Program Final Report, Grant R306189010. Chesapeake Research Consortium, Inc. Publication No. 84. Annapolis, MD.)*

benthic nutrient regeneration, especially bottom water oxygen concentration (e.g., Boynton and Kemp, 1985; Kemp et al., 2005). Based on his colleagues and his early studies of sediment oxygen and nutrient fluxes, Boynton (pers. comm.) did not consider the sediment nitrogen reserves to be adequate for more than a few weeks' supply of regenerated ammonium. His subsequent research confirms that (Boynton et al., 2008). Rapid recycling of nitrogen occurs between the water column and the riverbed. A productive system could persist for some time in the absence of added nutrients, so the effects of nutrient controls might not be immediately apparent.

Although the analytical procedure for determination of nitrite has remained essentially the same over the last 50 years, there has been relatively little attention to it, because it rarely achieves significant concentrations in the water column. Observations of periodic accumulations of nitrite in Chesapeake Bay waters (e.g., McCarthy et al., 1977; Webb and D'Elia, 1980) are most common in the late summer and early fall and is probably a consequence of ammonium oxidation (nitrification/ammonia oxidation). Whether this phenomenon occurred historically is unknown, but Webb (1981) suggested that the magnitude of the nitrite accumulation is a function of the degree of nutrient enrichment of the Bay during the summer, and he recommended continued monitoring of the nitrite maximum.

Strong indirect evidence indicates that beginning in the late 1960s, nutrient enrichment stimulated increased phytoplankton production and an accumulation of plant biomass in the lower PR (Heinle et al., 1980). This conclusion is based mainly on Secchi depth data rather than chlorophyll concentration data, because the historical data base for chlorophyll on the Patuxent is less complete (but it does suggest the same trends). Fig. 5 shows Secchi data from July 1937 to July 1978, normalized against surface salinity (to account for variations in river flow). The inability of the Secchi disk to resolve differences in transparency when transparency is low, means that little can be said about historical changes at surface salinities below about eight ppt, which is extremely turbid. Such low salinity regimes are also subject to high levels of turbidity because of inorganic sediment. However, at the greater transparencies found at higher salinities the resolving power of the Secchi disk data is better, particularly at lower flow times of the year such as July. During 1963, water transparencies in the lower estuary were similar to those observed during 1936–40 (Fig. 5). Heinle et al. (1980) felt that increases in phytoplankton and not small particle sediment levels caused decreased Secchi depths in the lower estuary during the summer. Increases in algal standing stocks imply that algal production has increased to a rate greater than that of its consumption, and that a concomitant increase in biological oxygen demand (BOD) has also occurred. This is of concern because, in the lower Patuxent estuary, which is often stratified in the summer, oxygen concentrations are quite low in the earliest data, and they may be driven lower by the settling of organic matter with high BOD produced in surface waters.

One of the more common effects of excessive enrichment is increased variation in diurnal and nocturnal dissolved oxygen concentration in the water column, in response to enhanced rates of photosynthesis and respiration. This represents a particularly serious problem when nighttime consumption of oxygen by respiration becomes great enough to lower oxygen tension to a point where it jeopardizes the viability of aerobic organisms in the community. Maryland State standards

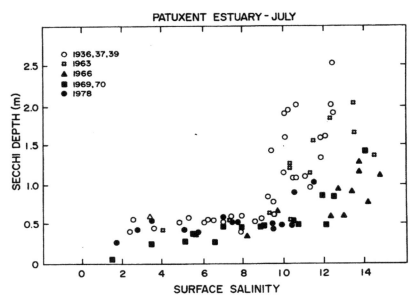

FIG. 5 Secchi Depth in the lower Patuxent Estuary, 1936–78.

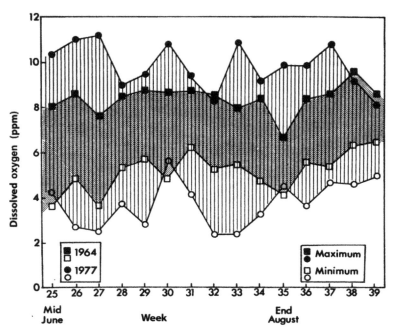

FIG. 6 Comparison of 1964 vs 1977 oxygen concentration maxima and minima at the benedict bridge on the Patuxent River. *(Data are from R. L. Cory of the US Geological Survey.)*

consider this to be ≥4 ppm over a 7-day average, unless naturally lower concentrations exist (Maryland Department of the Environment, 2018). Under such conditions, we observed the nuisance conditions most often associated with excessive nutrient enrichment. Diurnal variations in oxygen concentration in the upper Patuxent appear to have increased substantially by the late 1970s. Cory (1974) and Cory and Nauman (1970) noted evidence for such changes in the Patuxent at Benedict during the period from 1963 to 1969. They observed greater extremes in concentration of dissolved oxygen and a reduced ratio of production in respiration during that period, suggesting that increased levels of heterotrophy are occurring. In a later unpublished study from which Cory made his data available to Heinle et al. (1980), it appears that continued changes have occurred between 1969 and 1977. Fig. 6 shows weekly maximum and minimum concentrations of dissolved oxygen during May through August near the surface at Benedict Bridge. Minimum concentrations observed (about 2 mg O_2) are fortunately transient, but are nonetheless approaching dangerously low values. The increased range of values in 1977 over that of 1964 is clearly evident in Fig. 6.

The greatest ecological concern in the PR rests neither in oxygen concentrations nor in esthetic deterioration by enhanced turbidity in upstream waters but, instead, in the oxygen concentration in the deep waters of the lower estuary. Processes contributing to hypoxia in the Chesapeake are complex and involve both physical and biological processes (Taft et al., 1980; Officer et al., 1984), and they can vary over tidal, seasonal, or annual cycles (D'Elia et al., 1981). In a stratified body of water such as the Patuxent estuary, increased productivity in the surface waters can cause decreased oxygen concentrations in deeper waters as organic matter settles in the water column and decomposes. Sustained oxygen depletion (perhaps by this mechanism) is known to occur naturally in the central part of the Bay (Newcombe and Horne, 1938; Taft et al., 1980). From the evidence available in the early 1980s, the extent of this low-oxygen water was increasing with time.

Nash (1947) observed that the differences between surface and bottom concentrations of dissolved oxygen were greater at times of greater stratification, and he postulated that the degree of stratification was an important determinant of bottom water dissolved oxygen concentration. We now know that stratification strength is a critical consideration in the development of hypoxic waters in many places and must also be considered as a factor in addition to nutrient inputs. Bottom-layer oxygen levels decrease with increasing stratification, because mixing with aerated upper waters is prevented. Similar results have been observed for the mainstream of Chesapeake Bay (Taft et al., 1980) and for the lower York River (Webb and D'Elia, 1980). This greatly complicates the interpretation of nutrient enrichment effects, and it is not surprising that bottom-dissolved-oxygen content in the historical database shows a wide variation within a given year.

4 THE PR CASE AS A DRIVER OF CHESAPEAKE BAY POLICY

The intense scientific debate about the nature of environmental change and management of the PR paved the way for major policy changes elsewhere in the Bay. These include the development of the Maryland Nutrient Control Strategy (http://mda.maryland.gov/resource_conservation/Pages/nutrient_management.aspx), the broader recognition of the large role of nonpoint nutrient sources, and the need to adopt a "Total Maximum Daily Load" (TMDL) approach to nutrient management (Linker et al., 2013).

The Patuxent River runs through the middle of the Baltimore/Washington D.C. corridor, through Southern Maryland, and into the Chesapeake Bay. It is the largest river in Maryland whose watershed is completely within the State. With trout streams and water supply reservoirs in its upper reaches, a tidal freshwater stretch in its middle, and a large estuary in its lower portion, the Patuxent is often considered the Chesapeake Bay in miniature. It has been the proving ground for many of the Chesapeake Bay Program's initiatives.

Patuxent River Commission (2000)

Administratively, the PR basin is overseen by seven counties—Howard, Montgomery, Prince George's and Anne Arundel in the North, and Calvert, Charles, and St. Mary's in the South. Numerous federal and state agencies also have jurisdiction: among them are the EPA, several municipalities and various Maryland and regional planning agencies.

In the late 1970s, the seven counties in the river basin experienced very fast population growth, causing a dramatic increase in sewage treatment plant effluent discharge into the river (Bunker and Hodge, 1982). In 1976, the rapidly developing Howard County applied for federal assistance to upgrade and expand its main wastewater treatment facility in Savage, MD (Boyd, 1980b). In April 1977, EPA approved the plan without requiring an Environmental Impact Statement (EIS), declaring, in accordance with the National Environmental Policy Act provisions, that the expansion would not "significantly" affect the environment (Lemaire, 1982).

In the rural South, however, where generations of watermen had always considered Patuxent their "lifeblood" (Alliance for the Chesapeake Bay, 1988), public concerns were on the rise over the diminished water clarity, disappearing sea grass, and decreasing oyster and fish harvest. Around the same time, as reported earlier in this chapter, the first clear, scientific indicators of deterioration in the condition of the Patuxent began to surface. The time series analysis of historical data from 1930s to the mid-1970s, run by Heinle and D'Elia of the CBL, showed a clear trend that the river was becoming increasingly more turbid and nutrient rich, especially in its lower estuary (Heinle et al., 1980).

As was then called for in the Clean Water Act §208 PR Basin Plan, the "Patuxent River Technical Advisory Group" (PRTAG) had been formed. It was chaired initially by Heinle, with D'Elia taking over after Heinle left Maryland, but most of its members were from the broader community. Its purpose was to advise the state on water quality and pollution control issues for the Patuxent, and the PRTAG strongly endorsed the concerns raised by the scientific community about the secular decline in water quality in the river (Bunker and Hodge, 1982).

State officials, however, such as the Secretary of the Maryland DNRs, James Coulter, remained oblivious of any changes occurring (D'Elia et al., 2003). They argued that any "dead zones" found in the PR (areas of low concentrations of dissolved oxygen) were to be attributed to the natural intrusion of Bay waters into the estuarine part of the river (Coulter, 1977, p. 286).

In light of the position taken by the DNR, the three Southern Maryland counties found themselves compelled to take the initiative. In October 1977, the Board of Commissioners, led by Calvert County Commissioner, C. Bernard ("Bernie") Fowler, who later became a Maryland Senator and a recognized leader for all of the Chesapeake, filed a lawsuit against EPA to halt expansion of the Savage Sewage Treatment until an EIS was issued. Moreover, the initial EPA-approved plan envisioned implementation of phosphorus (P) controls for advanced water treatment. It was in alignment with the State Patuxent River Water Quality Management Plan (PRWQMP) that had been previously released in a draft form in October 1974 and taken to public hearings in January and February 1975. Based on Ryther and Dunstan's (1972) theory and several other scientific studies from that period, however, Heinle and his colleagues at the CBL argued that nitrogen (N), rather than P, was the critical "limiting nutrient" in coastal areas during the summer when hypoxia occurred.

In a subsequent study, D'Elia et al. (1986) confirmed this conclusion, while showing a potential for P limitation during early spring and raising the prospect that a dual nutrient control strategy would be needed. The current consensus, in fact, is that controlling both N and P inputs is preferable to either alone to protect both upstream and coastal waters (Paerl et al., 2014).

In February 1977, the Board of Commissioners of St. Mary's County prepared a "Position Paper" on the Patuxent, in which they questioned the technological soundness of the State plan for the River and the adequacy of resources allocated toward its development. The paper pointed to the omission of nonpoint sources of pollution in the plan, as well as to the need for more studies on the nutrient dynamics of the lower portion of the River, where N was believed to be the greater limiting factor.

In April and May 1977, the Coastal Zone Unit of the Energy and Coastal Zone Administration sponsored a series of public meetings, "Forum on the PR." At the forum, Howard Wilson of the Water Resources Administration (WRA) defended the State focus on P removal, explaining that the decision had been based on "a mathematical model, calibrated to simulate the natural transport and cleansing functions of the Patuxent River" (Maryland Coastal Zone Unit, 1977). Pointing to the dramatic increase of sewage discharge in the PR, Heinle presented in response research data, which, in his opinion, suggested that P was the most likely limiting nutrient in the upper estuary, but that N was the major limiting nutrient in the lower estuary. Accordingly, Heinle felt that "the sewage treatment strategy of P removal would have little effect on the lower estuary" (Maryland Coastal Zone Unit, 1977).

Heinle's suggestion to introduce N control in wastewater treatment, corroborated meanwhile by the findings from another study, conducted independently at the same time Mihursky and Bonyton (1978) of the CBL, was strongly resisted by both the DNR and the EPA, which were at the time committed to P removal nationwide on cost-effectiveness grounds (D'Elia et al., 2003). Thus, while the "Forum on the Patuxent River" was still in session, the WRA submitted to the EPA for approval its latest revised version of the PRWQMP that retained the recommendation of effluent limitations on Total P only.

The response of the Southern Counties was swift. Drawing on Heinle's approach, which was endorsed by the PRTAC, Bernie Fowler, led again the Tri-County Council in a second lawsuit filed in March of 1978 against the EPA, the State of Maryland, and the upper River counties. This suit challenged the adequacy of the PRWQMP, already approved by EPA, which envisioned P control as the sole advanced wastewater treatment method for controlling nutrient enrichment in the Patuxent. The County Commissioners maintained that those water quality standards were outdated, and that N control was also necessary to control over-enrichment of the lower portion of the river (Bunker and Hodge, 1982).

In early 1979, the scientists from the CBL, who were supporting the legal clash of the Tri-County Council Commissioners with the State and EPA on the issue of N removal, submitted written testimonies in court. As reported by Boyd (1980a), in the course of the year, "some of the points Southern Maryland tried to make in those lawsuits were acknowledged by state and federal authorities." In July 1979, EPA publicly admitted to the deficiencies of the existing PRWQMP, which envisioned a "single nutrient" (P) strategy to control overenrichment (Lemaire, 1982) and sent it back to the DNR for revision. Later that year, EPA placed a hold on federal funding for expansion of sewage plants on the PR (Boyd, 1980a).

In July 1980, the US District Court ruled in favor of the Southern Maryland counties and ordered EPA to prepare an EIS on the Savage Sewage Treatment Plant. In October 1980, the US District Court came up again with a favorable ruling on the second lawsuit, brought by the Southern Maryland counties and ordered Maryland and EPA "to prepare a new PRWQMP, including a revised nutrient control strategy, which was scientifically defensible and publicly acceptable" (Hodge, 1987). The State and the US Department of Justice agreed on a court-imposed deadline (January 15, 1982) for submission of the new Nutrient Control Strategy to EPA for approval.

The outcomes of the lawsuits did not go unnoticed in neighboring jurisdictions. A panel discussion, "The Patuxent Litigation: Implications for the Potomac," was promptly convened in the same month, October 1980, with political, administrative, scientific, and citizen representation from Maryland, Pennsylvania, West Virginia, Virginia, and the District of Colombia (Calvert Independent, 1980).

Meanwhile, a newly elected Governor of Maryland, Harry Hughes, came into office in January 1979 and was immediately confronted with the controversy over the PR situation. On December 6, 1979 he toured the lower portion of the river, from Benedict to Solomons, aboard a CBL research boat, accompanied by elected officials from the Southern Maryland counties, agency representatives, scientists, fishermen, and environmental activists and took "a first-hand look at the stress the river was experiencing" (Hodge, 1987). Hailed as a "turning point" on "the long and winding road to clean Patuxent" (Hodge, 1987), the tour convinced the chief executive of the state in the seriousness of the problem and committed him to a campaign to save the river.

Harry Hughes transferred the authority over water treatment decisions from the DNR to a newly created Office of Environmental Programs in the Department of Health and Mental Hygiene. In the spring of 1980, the Governor brought an ambitious environmental lawyer, William Eichbaum, to spearhead the PR action. After an immediate tour of the Bay and sensing the extent of the controversy, Eichbaum decided to come on board with an olive branch and interview all key parties concerned with the policy on the Patuxent. The Tri-County Council for Southern Maryland, the Northern counties, scientists, local officials, citizens' groups, all were given a chance to share with him their point of view in the matter.

Once Eichbaum heard from all stakeholders, he set about finding a way to reconcile the opposing points of view and arrive at an agreed upon nutrient control strategy. He decided to organize a meeting of all key parties involved in the Patuxent conflict (Bunker and Hodge, 1982). Faced with the prospect of losing $29 million in federal funds for wastewater treatment past the court-imposed deadline of January 15, 1982, Eichbaum called the conference "a charette" ("a final, intensive effort to finish a project, especially an architectural design project, before a deadline").

The conference, a 3-day session "of closed discussion and negotiation in a professionally mediated format" (D'Elia et al., 2003), was held from December 2 to 4, 1981 at a secluded convent in Marriottsville, MD and became known as "The Patuxent Charette" or "The Marriottsville Accord" (Rymer, 1981). It was facilitated by a Boston-based consulting firm, Clark-McGlennon Associates (1981), which specialized in environmental conflict resolution, and involved about 40 participants, selected through a series of preliminary interviews conducted by the firm (Horton, 1981). The designers of the conference also formed a Charette Steering Committee, which worked out the procedural rules for the meeting.

The first 2 days of talks at the Charette proved tense. Exposed to presentations of conflicting scientific testimonies, the participants remained in disagreement on the required course of action. Walter Boynton produced a simple nutrient budget for the River, illustrating the increases in nutrient inputs that had occurred in previous decades that focused the group on the dramatic changes that were occurring. The key to the success of the meeting was Maryland Senator Bernie Fowler's common-sense approach that was finally able to get all the participants on board in the third day of the meeting. In reminiscing about his experience as a crabber and fisherman in the 1950s, Bernie Fowler put forward the "reasonable and simply understood goal" of "returning the River to the clarity, quality, and shellfish productivity that it had in the early 1950s" (Boyd, 1980b). Doing that would require a return to a nutrient budget with considerably less N and P inputs.

Eventually, the breakthrough in the negotiations between the opposing camps resulted in an agreed upon action plan that contained specific recommendations, goals for the reduction of the amount of both P and N going into the river and indicators to measure progress toward achievement of each goal. The policies embedded in the adopted nutrient control strategy had to be implemented over the course of 5 years, followed by a 2-year period of intensive monitoring and research on the strategy's effect in water quality improvement. Suboptimal results would lead to modifications to the strategy to ensure attainment of the goals set and, ultimately, of restoring the Patuxent to its condition in the 1950s.

Based on the plan developed at the conference and in compliance with the Federal Court mandate, the Office of Environmental Programs prepared and submitted to the EPA on January 14, 1982, a final Nutrient Control Strategy for the PR Basin. The strategy became the essence of the new PRWQMP, drafted by the PR Commission and opened for public comment in the spring of 1982 (Lemaire, 1982). Subsequently, the plan was signed by Governor Hughes in June 1983 and approved by EPA in October of the same year, making the State's newly redefined commitment to nutrient control official (Ismail, 1984). The plan set strict standards for nutrient discharges and required that the Department of Health and Mental Hygiene use them when issuing sewage treatment permits and authorizing increases in the plants' capacities. In June 1983, Governor Hughes, Virginia Governor Charles Robb and Pennsylvania Governor Richard Thornburg met to discuss the water quality issues in the Chesapeake Bay and coordinate their states' pollution control programs. The Patuxent cleanup effort was brought up as a policy and management model for the entire Bay (Buehler, 1983).

Today, after several revisions and amendments developed over the last two decades, the Patuxent Nutrient Control Strategy, crafted at the landmark 1981 Charette, is still pertinent on the banks of the River and citizens continue to take a special interest in monitoring the progress of the water cleanup (Horton, 2005; Wan, 2006). Maryland Sen. Bernie Fowler, who first got involved in PR policy at the local level as a Calvert County Commissioner, achieved an almost legendary status in the Chesapeake Bay community for his efforts to preserve and protect this great natural resource (Horton, 2005). Little doubt exists that what happened at the local level scaled up to have a much larger impact on Chesapeake Bay-wide and broader USEPA nutrient management policy.

5 THE STATE OF THE BAY: WHAT WAS ACCOMPLISHED SINCE 2020 REPORT WAS PUBLISHED, AND WHAT IS TO BE EXPECTED IN 2020 AND BEYOND?

Long-term monitoring of water quality and living resources in the Chesapeake Bay provides solid evidence of improvement of the ecological health of the system (Zhang et al., 2015), although climatic factors also can contribute to long-term trends (Harding Jr. et al., 2016a, b). While extraordinary accomplishments have been achieved in understanding the scientific and technical issues in the Chesapeake Bay and its watershed, despite recent successes, the future, like the Bay's waters, looks murky. The 2020 Report warned that changing population growth, development, and land use would have enormous detrimental effects on the Bay and possibly overwhelm progress made in nutrient management.

"Today, unmanaged new growth has the potential to erase any progress made in Bay improvements, overwhelming past and current efforts," the report said. "Extensive programs underway to remedy and clean up existing problems hold promise of success. However, success in dealing with existing problems will be temporary, if new growth generates additional quantities of pollutants."

It was a remarkable call to action, taking on some very controversial political topics, including recommendations to change the way states and localities plan for and manage residential and commercial development and transportation. At the core of the issue is the historically dynamic tension that has existed between private property rights and the public good. Economic factors are always at play anytime land use policy is involved.

Since the 2020 Report warning was issued, according to recent reports, the Bay's ecological health has improved (Chesapeake Bay Program, 2018a), even though population growth has exceeded projections. Orth et al. (2017) referred to submersed aquatic vegetation (SAV) as "sentinel species" that reflect broader water quality trends. These authors stated, *"Like estuaries and coastal waters throughout the world, the Chesapeake Bay faces environmental threats from human activities in its watershed and along its shoreline, and many of those threats affect SAV throughout the Bay."* This sentiment is widely shared in the scientific community.

In 2017, SAV covered more than 100,000 acres of the Chesapeake and its tributaries, the greatest abundance seen since the Bay monitoring and restoration effort began in earnest in the early 1980s (Fig. 7) (Bay Journal, 2018). When surveys of SAV, began in 1984, just 40,000 acres could be found. Lefcheck et al. (2018) attribute the significant rebound in this key component of the estuarine ecosystem to "sustained management actions that have reduced nitrogen concentrations in Chesapeake Bay by 23% since 1984." SAV is an excellent proxy for water quality because it is ultimately directly affected by water clarity. If phytoplankton levels in the water are high, due to nutrient enrichment, then SAV growth is negatively impacted because of light limitation (Orth et al., 2017).

So, does that mean the Bay watershed states and localities heeded the 2020 Report's call and took steps that effectively minimized the harmful effects of unmanaged growth? Hardly. As we explain below, efforts to rein in the low-density development patterns that the 2020 Report warned about have been uneven and had only limited success. Unfortunately, nutrient pollution from stormwater runoff continues to grow (Chesapeake Bay Program, 2018b). Instead, the reduction in nutrient concentrations in the Bay stems mainly from a massive investment in upgrading wastewater treatment plants (Chesapeake Bay Program, 2018c), and from air quality improvements driven by federal Clean Air Act regulations (Eshleman and Sabo, 2016). Indeed, thanks to the installation of biological nutrient reduction (BNR) and enhanced nutrient reduction (ENR) technology, the watershed's wastewater treatment plants collectively achieved their 2025 nutrient reduction goals a decade early (Chesapeake Bay Program, 2016).

Neither development was anticipated when the 2020 Report was being written three decades ago. The cost effective technological leap forward in wastewater treatment represented by BNR and ENR (Randall and Cokgor, 2000) was just beginning in the mid-1980s, as were the air emission controls required as a result of the 1980 amendments to the Clean Air Act. The big nutrient reductions from those two sectors, mostly measured since the report's issuance, have masked the meager gains, if any, from curbing sprawl and its attendant stormwater runoff.

Now, though, it is not likely there will be a great deal more progress made in reducing nutrients from wastewater plants—barring some new, as-yet undiscovered, leap forward in treatment technology. Considering that a much greater fraction of the N entering the Bay comes from nonpoint sources, a focus on point sources alone would be insufficient. According to Hagy et al. (2004), *"To eliminate or greatly reduce anoxia will require reducing average annual total nitrogen loading to the Maryland mainstem Bay to 50×10^6 kg yr^{-1}, a reduction of 40% from recent levels."* The atmospheric input of N to estuaries is very significant (Paerl et al., 2002) and needs to be considered. However, in the current political climate in Washington, air pollution regulations are being rolled back rather than strengthened (Washington Post, 2018). The bulk of the nutrient and sediment reductions still needed to achieve the goals set by the bay-wide TMDL are likely going to have come not only from atmospheric sources but also from nonpoint pollution sources, such as runoff from farmland and developed land. With stormwater runoff continuing to increase, the 2020 Report's dire warning could still come true—even if a few decades later than originally anticipated.

Scientists and policy makers have long recognized the harm development can do to water quality and ecosystems. In the second Chesapeake Bay restoration agreement signed in 1987 (Chesapeake Bay Agreement, 1987), watershed states and the federal government acknowledged that "there is a clear correlation between population growth and associated development and environmental degradation in the Chesapeake Bay system." That agreement led to the 2020 Report assessing the future consequences of poorly managed growth, with a series of recommendations for how to prevent them.

In some ways, the 2020 Report understated the threats to the Bay from poorly managed growth, as it only focused on the three largest of the six Bay watershed states, plus the District of Columbia. But those jurisdictions accounted for 83% of the watershed's population, and the vast majority of its growth as well.

At the time of its release in 1988, the report noted that the 1990 population of those four jurisdictions was estimated to be 13.6 million. It projected their population would grow by 20% by 2020, to 16.2 million. By 2016, though, the population for those three states plus the District had already reached about 17.1 million.

And by the year 2020, those three states and the District are projected to have 17.7 million people—nearly 10%, or 1.5 million people more than the report's authors envisioned. If all six watershed states are included, the numbers are even more daunting. The Bay region's 2020 population is projected to hit 18.8 million and keep growing. By 2030, it is forecasted to reach 20.1 million, with another one million people added by 2040.

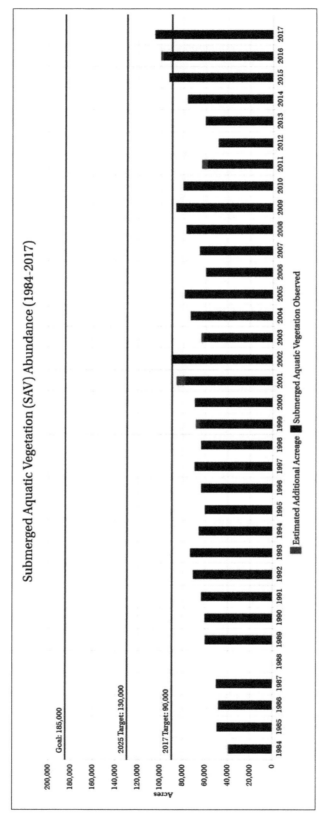

FIG. 7 Submerged aquatic vegetation (SAV) abundance, 1984–2017. *(Courtesy EPA Chesapeake Bay Program.)*

Perhaps even more worrisome to the report's authors than the swelling populace was the conversion of forest and farm lands to houses, stores, and roads in the Bay watershed, which was occurring at a greater rate than the growth in people (Chesapeake Executive Council, 2020; Report, 1988, p. 28). They also warned against the unrelenting spread of pavement across the watershed to accommodate growing motor vehicle traffic.

If we continue to rely on highways and automobiles; and if we continue with the same patterns of growth, it is virtually impossible that the quality of life in the region will get anything but worse.

The report laid out six visions for "an alternative future," in which growth would be managed to minimize environmental harm. Its recommendations included calls for development to be concentrated "in suitable areas" and for sensitive areas, such as wetlands, to be protected. Overall, it called for a "focused and ambitious agenda of cooperative and coordinated planning."

By the report's publication in 1988, a couple steps had already been taken. On the heels of the first Bay cleanup agreement, Maryland had enacted a "Critical Area" law in 1984 to protect water quality by limiting development within 1000 ft of the Bay and its tidal tributaries (Maryland Department of Natural Resources, 2018a). A state commission set criteria for ensuring that near-shore development did not harm the Bay but left the application and enforcement of those criteria largely to local governments. While generally successful at preserving pristine waterfront, the law has been unevenly enforced, particularly its restrictions on disturbing the sensitive 100-ft "buffer" of land along the water (Environmental Law Clinic, 2016).

Virginia lawmakers adopted their own version of waterfront development controls in 1988 (Virginia Department of Environmental Quality, 2018). The Chesapeake Bay Preservation Act aims to improve water quality "by requiring the use of effective land management and land use planning." Like the Maryland law, Virginia's leaves implementation and enforcement up to each Tidewater county, though it is also less restrictive overall.

Maryland lawmakers also enacted measures aimed at preserving nontidal wetlands (1989) and conserving forests (1991). But efforts to impose some statewide order on growth ran into stiff resistance. A 33-member commission appointed by Gov. William Donald Schaefer recommended state legislation modeled on the Year 2020 Report's "visions." But developers and real estate interests teamed up with local officials fearful of losing their traditional control over land use to kill the bills in 1991. Schaefer managed to win passage the following year of the "Economic Growth, Resource Protection and Planning Act," a watered-down measure that articulated a state growth policy but set no standards and left its application entirely up to localities (Tierney, 1994).

Then, in 1997, Maryland's Gov. Parris N. Glendening pushed a "smart growth" law through the General Assembly. Rather than impose top-down regulations on development, the new law aimed to provide a powerful financial incentive to curb suburban sprawl. It spelled out how the state would spend taxpayer dollars on schools, highways, and other infrastructure in already settled communities rather than in "greenfields," farms and forests converted to housing subdivisions.

Smart growth has remained the law of Maryland, though it has been downplayed or even ignored by succeeding pro-business governors. Gov. Martin O'Malley took the antisprawl push one step further in 2012 by winning passage of a measure limiting large-scale rural or suburban development that relies on household septic systems rather than sewered connections to a wastewater treatment plant. That also has been weakened by his successor, Gov. Larry Hogan, who favors economic growth.

Despite the compromises, uneven enforcement and outright backsliding, Maryland has succeeded to a degree in reducing sprawl—though the real estate crash and severe recession that hit around 2008 also played a major role.

Maryland's limited success in addressing sprawl notwithstanding, the insidious ecological harm caused by development has not been effectively countered anywhere in the Bay watershed, and as population continues to increase, more pressure will be put on the ecosystem services that assimilate wastes and protect the Bay. Conversion of forest, farm fields, and other sensitive lands continues, although maybe not at the same headlong rate as before. While we do not have information specifically about changes in land use that have occurred since the 2020 Report was issued in the late 1980s, the USGS does have information over the interval between 1984 and 2013 (Peter Claggett, USGS, pers. comm.). The data are not reassuring (Table 1).

The biggest concern is the loss of forested land, because research shows that forests mitigate runoff better than any other land use. The growth of "impervious" acres is almost as concerning, as pavement and buildings increase runoff by preventing rain from soaking into the ground. Whether it is from increased runoff of nutrients and sediment or other pollutants, a stream's ecological health, as measured by the abundance and diversity of aquatic insects, amphibians, and fish, declines markedly when as little as 5% of its watershed is covered by impervious surfaces (Maryland Department of Natural Resources (2018b).

TABLE 1 Change From 1984–2013 (Percentages Relative to 1984 Conditions and Absolute Change Values Reported in Acres)

Impervious	28.0%	412,210.06
Pervious	32.7%	901,632.23
Forest	−2.3%	558,366.12
Agriculture	−6.1%	547,700.47
Other	−5.9%	207,775.69

The EPA has made clear it expects the Bay watershed states to account for future pollution growth as they work to reduce existing pollution under the TMDL cleanup plan. None of the seven Bay jurisdictions factored projected growth into their initial "watershed implementation plans" spelling out how they intended to reduce nutrient and sediments getting into the Chesapeake. As they prepare new implementation plans in 2018, many of the jurisdictions have indicated they are planning or considering including some kind of offset or trading programs to account for growth and hold the line on Bay water quality by 2025 and beyond. But the details remain to be seen.

In closing, two lines from the 2020 Report ring as true today as they did three decades ago:

"How the land is used is a basic factor in the ecological health of the Chesapeake Bay." This is an absolute truth based on years of in-depth scientific study of nonpoint-source runoff patterns of different land uses. Clear-cutting or even fragmenting forests for agriculture and development of the built environment can have serious effects that go beyond just water quality. Adding impervious surfaces further exacerbates the runoff problem. Building sewerage to increase population density is a two-edged sword. While there are ways to mitigate these effects, they still pale in comparison to what is seen in pristine forests.

"Achieving the future visions will require bold leadership at all levels of government." The challenge inherent in this is huge. Given the multilayered, multijurisdictional nature of environmental management in the Chesapeake watershed, as is true in so many other places worldwide, this has never been and could never be an easy task. The EPA CBP, now over 40 years old can sometimes be plodding and bureaucratic. However, in our view, the CBP has been remarkably successful in bringing different interests together, incorporating cutting edge science into the management of the Bay, and most importantly, maintaining a level of public awareness and civil discourse that is much needed in society today. The success it has had so far faces its sternest test of all in trying to maintain the Bay's health as the watershed's population continues to grow and spread across a landscape that also faces a growing climate change challenge.

Dedication and Acknowledgments This chapter is dedicated to the memories of two men who left us way too early: Donald R. Heinle, PhD played a critical role as a champion of scientifically based policymaking for the Patuxent River and Chesapeake Bay, and T. Wakeman ("Wake") D'Elia grew up in Solomons, MD, on the Patuxent River, which he loved greatly his whole life. The authors thank USEPA's Rich Batiuk and USGS' Peter Claggett of the CBP for their assistance in finding land-use data.

REFERENCES

Alliance for the Chesapeake Bay, 1988. An Abridged History of the Alliance for the Chesapeake Bay and the Chesapeake Bay Restoration Effort: Highlights of Partnership and Progress. Available from: http://tinyurl.com/yd3fxjdk. (Accessed July 14, 2018).

Bay Journal, 2018. Available from: https://www.bayjournal.com/article/bays_underwater_grasses_surge_beyond_100000_acres_for_first_time_in_ages. (Accessed July 14, 2018).

Bidjerano, M., 2009. Collaboration as Paradox: The Case of the Patuxent River, MD Nutrient Control Strategy, Nelson A. Rockefeller College of Public Affairs and Policy Department of Public Administration and Policy. University at Albany, State University of New York, p. 219.

Blankenship, K., 2003. Mathias boat trip in 1973 launched Chesapeake cleanup effort. Bay J. Available from: https://www.bayjournal.com/article/mathias_boat_trip_in_1973_launched_chesapeake_cleanup_effort. (Accessed July 14, 2018).

Boyd, R.H., 1980a. Patuxent River Still a Sewage Ditch; Massive Clean-up Effort Requested. Prince Frederick Recorder. January 2, A-1, A-5.

Boyd, R.H., 1980b. The Life of the Patuxent River. The Washington Star. August 12, Point of View.

Boynton, W.R., Kemp, W.M., 1985. Nutrient regeneration and oxygen consumption by sediments along an estuarine salinity gradient. Mar. Ecol. Prog. Ser. 23, 45–55.

Boynton, W.R., Kemp, W.M., Osborne, C.G., 1980. Nutrient fluxes across the sediment-water interface in the turbid zone of a coastal plain estuary. In: Kennedy, V.S. (Ed.), Estuarine Perspectives. Academic Press, New York, pp. 93–109.

Boynton, W.R., Hagy, J.D., Cornwell, J.C., et al., 2008. Nutrient budgets and management actions in the Patuxent River estuary, Maryland. Estuar. Coasts 31, 623–651. https://doi.org/10.1007/s12237-008-9052-9.

Brush, G., 2017. Decoding the Deep Sediments: The Ecological History of Chesapeake Bay. UM-SG-CP-2017-01 Maryland Sea Grant, College Park, MD, p. 72.

Buehler, I., 1983. Patuxent Cleanup is Model for Bay. Calvert County Recorder. August 5, A-1.

Bunker, S.M., Hodge, G.V., 1982. In: The legal, political and scientific aspects of the Patuxent River nutrient control controversy. Text of a presentation to the Atlantic Estuarine Research Society, April 23, Baltimore, MD.

Chesapeake Bay Agreement, 1987. Available from: http://www.chesapeakebay.net/content/publications/cbp_12510.pdf. (Accessed July 14, 2018).

Chesapeake Bay Program, 2016. Available from: https://www.chesapeakebay.net/news/blog/wastewater_sector_meets_nutrient_goals_of_pollution_diet_a_decade_early. (Accessed July 14, 2018).

Chesapeake Bay Program, 2018a. Available from: https://www.chesapeakebay.net/news/pressrelease/chesapeake_bay_program_notes_continued_progress_in_chesapeake_bay_restorati. (Accessed July 14, 2018).

Chesapeake Bay Program, 2018b. Available from: https://www.chesapeakebay.net/issues/stormwater_runoff. (Accessed July 14, 2018).

Chesapeake Bay Program, 2018c. Available from: https://www.chesapeakebay.net/issues/wastewater. (Accessed July 14, 2018).

Chesapeake Executive Council 2020 Report, 1988. Population growth and development in the Chesapeake Bay watershed to the year 2020. In: The Report of the Year 2020 Panel to the Chesapeake Executive Council. Available from: http://tinyurl.com/yax4pdr4. (Accessed July 14, 2018).

Clark-McGlennon Associates, 1981. Report on the Patuxent River Pre-Charette Technical Meeting, Boston, MA.

Cory, R.L., 1974. Changes in oxygen and primary production of the Patuxent Estuary, Maryland, 1963 through 1969. Chesap. Sci. 15, 78–83.

Cory, R.L., Nauman, J.W., 1970. Temperature and water quality conditions of the Patuxent River Estuary, Maryland, January 1966 through December 1977. Chesap. Sci. 11, 199–209.

Coulter, J.B., 1977. In: Closing statement. Proceedings of the Bi-State Conference on the Chesapeake Bay, April 27–29. Chesapeake Research Consortium Publication No. 61.

Cumming, H.S., 1916. Investigation of the ollution of tidal waters of Maryland and Virginia. U.S. Treasury Dept., Public Health Service, Bull. No. 75, 199.

Cumming, H.S., Purdy, W.C., Ritter, H.P., 1916. Investigation of the Pollution and Sanitary Conditions of the Potomac Watershed. U.S. Treasury Dept. Hygienic Laboratory Bull. No. 104. 239 pp. plus plates.

D'Elia, C.F., 1982. Nutrient enrichment of Chesapeake Bay: an historical perspective. In: Chesapeake Bay Program Technical Studies: A Synthesis. U.S. Government Printing Office, Washington, DC, pp. 509–660. USGPO: 1982.

D'Elia, C.F., 1987. Too much of a good thing. Nutrient enrichment of the Chesapeake Bay. Environment 29, 6–33.

D'Elia, C.F., 1995. Sustainable development of the Chesapeake Bay: a case study. In: Munasinghe, M., Shearer, W. (Eds.), Defining and Measuring Sustainability: The Biogeophysical Foundations. The United Nations University and The World Bank, Washington, DC, pp. 161–176.

D'Elia, C.F., Webb, K.L., Wetzel, R.L., 1981. Time varying hydrodynamics and water quality in an estuary. In: Neilson, B.J., Cronin, L.E. (Eds.), Estuaries and Nutrients. Humana Press, Clifton, NJ, pp. 597–606.

D'Elia, C.F., Boynton, W.R., Sanders, J.G., 1986. Nutrient enrichment studies in a coastal plain estuary: phytoplankton growth in large-scale continuous cultures. Can. J. Fish. Aquat. Sci. 43, 397–406.

D'Elia, C.F., Boynton, W.R., Sanders, J.G., 2003. A watershed perspective on nutrient enrichment, science and policy in the Patuxent River, Maryland, 1960–2000. Estuaries 6, 171–185.

Environmental Law Clinic, 2016. Maryland's critical area protection program: variances and enforcement in selected jurisdictions from 2012 to 2014. Available from: http://www.law.umaryland.edu/programs/environment/documents/CriticalAreaProtectionProgram_2012-14.pdf. (Accessed July 14, 2018).

Eshleman, K.N., Sabo, R.D., 2016. Declining nitrate-N yields in the upper Potomac River basin: what is really driving progress under the Chesapeake Bay restoration? Atmos. Environ. 146, 280–289.

Fisher, T.R., Peele, E.R., Ammerman, J.W., Harding Jr., L.W., 1992. Nutrient limitation of phytoplankton in Chesapeake Bay. Mar. Ecol. Prog. Ser. 82, 51–63.

Hagy, J.D., Boynton, W.R., Keefe, C.W., et al., 2004. Hypoxia in Chesapeake Bay, 1950–2001: long-term change in relation to nutrient loading and river flow. Estuar. Coasts 27, 634–658.

Harding Jr., L.W., Gallegos, C.L., Perry, E.S., et al., 2016a. Long-term trends of nutrients and phytoplankton in Chesapeake Bay. Estuar. Coasts https://doi.org/10.1007/s12237-015-0023-7.

Harding Jr., L.W., Mallonee, M.E., Perry, E.S., et al., 2016b. Variable climatic conditions dominate recent phytoplankton dynamics in Chesapeake Bay. Sci. Rep. 6, 23773.

Heinle, D.R., D'Elia, C.F., Taft, J.L., et al., 1980. Historical review of water quality and climatic data from Chesapeake bay with emphasis on effects of enrichment. In: U.S. EPA Chesapeake Bay Program Final Report, Grant R306189010. Chesapeake Research Consortium, Inc, Annapolis, MD. Publication No. 84.

Hodge, G.V., 1987. The Long and Winding Road to a Clean Patuxent. Tri-County Council for Southern Maryland, Hughesville, MD.

Horton, T., 1981. State Gets Outside Help in Patuxent River Battle. The Baltimore Sun, p. B10. October 16, 1981, Sunday Edition.

Horton, T., 2005. Why Can't We Save the Bay? National Geographic. June.

Ismail, K., 1984. Bay and Patuxent River: Hope for a Clean Future. Calvert County Recorder. January 27, B-4.

Jaworski, N.A., Lear Jr., D.W., Villa Jr., O., 1971. In: Nutrient management in the Potomac Estuary. U.S. Environmental Protection Agency, Middle Atlantic Region, Annapolis Field Office, Tech. Rep. 45. pp. 64.

Kemp, W.M., Boynton, W.R., Adolf, J.E., et al., 2005. Eutrophication of Chesapeake Bay: historical trends and ecological interactions. Mar. Ecol. Prog. Ser. 303, 1–29.

Lefcheck, J.S., Orth, R.J., Dennison, W.C., et al., 2018. Long-term nutrient reductions lead to the unprecedented recovery of a temperate coastal region. Proc. Natl. Acad. Sci. 3658–3662. published ahead of print March 5, 2018. https://doi.org/10.1073/pnas.1715798115.

Lemaire, B., 1982. In: The Legal and Environmental Issues of the Patuxent River Water Quality Management Plan. Report prepared for the Tri-County Council for Southern Maryland, January, 22.

Linker, L.C., Batiuk, R.A., Shenk, G.W., Cerco, C.F., 2013. Development of the Chesapeake Bay watershed total maximum daily load allocation. J. Am. Water Resour. Assoc. 49 (5), 986–1006. https://doi.org/10.1111/jawr.12105.

Malone, T.C., Boynton, W., Horton, T., Stevenson, C., 1993. Nutrient loadings to surface waters: Chesapeake Bay case study. In: Ulman, M.F. (Ed.), Keeping Pace with Science and Engineering. Case Studies in Environmental Regulation. National Academy of Engineering, Washington, DC, pp. 8–38.

Maryland Coastal Zone Unit, 1977. Forum on the Patuxent River: a summary. Maryland Coastal Zone Unit, Annapolis, MD, p. 12 (unpublished).

Maryland Department of Natural Resources, 2018a. Available from: http://dnr.maryland.gov/criticalarea/Pages/background.aspx. (Accessed July 14, 2018).

Maryland Department of Natural Resources, 2018b. Available from: http://dnr.maryland.gov/streams/Pages/streamhealth/How-Impervious-Surface-Impacts-Stream-Health.aspx. (Accessed July 14, 2018).

Maryland Department of the Environment, 2018. Available from: http://mde.maryland.gov/programs/Water/TMDL/WaterQualityStandards/Pages/faqs.aspx.

McCarthy, J.J., Taylor, W.R., Taft, J.L., 1977. Nitrogenous nutrition of the plankton in the Chesapeake Bay, 1. Nutrient availability and phytoplankton preferences. Limnol. Oceanogr. 22, 996–1011.

Mihursky, J.A., Bonyton, W. R., 1978. Review of Patuxent Estuary Data Base. Report prepared for the Maryland Power Plant Silting Program, April. University of Maryland, UNCEES.

Nash, C.B., 1947. Environmental characteristics of a river estuary. J. Marine Res. 6, 147–174.

Newcombe, C.L., Horne, W.A., 1938. Oxygen poor waters of the Chesapeake Bay. Science 88, 80–81.

O'Connor, D.J., Gallagher, T.W., Hallden, J.A., 1981. Water quality analysis of the Patuxent River. Prepared for U.S. Environmental Protection Agency and Maryland Department of Health and Mental Hygiene, Office of Environmental Programs. Prepared by HydroQual, Inc., Mahwah, NJ.

Office of Environmental Programs, 1982. Nutrient Control Strategy for the Patuxent River Basin. Department of Health and Mental Hygiene, State of Maryland.

Officer, C.B., Biggs, R.B., Taft, J.L., et al., 1984. Chesapeake Bay anoxia origin development and significance. Science 223, 22–27.

Orth, R.J., Dennison, W.C., Lefcheck, J.S., et al., 2017. Submersed aquatic vegetation in Chesapeake Bay: sentinel species in a changing world. Bioscience 67, 698–712.

Paerl, H.W., Dennis, R.L., Whittall, D.R., 2002. Atmospheric deposition of nitrogen: implications for nutrient over-enrichment of coastal waters. Estuaries 25, 677–693.

Paerl, H.W., Hall, N.S., Peierls, B.L., Rossignol, K.L., 2014. Evolving paradigms and challenges in estuarine and coastal eutrophication dynamics in a culturally and climatically stressed world. Estuar. Coasts 37, 243–258.

Patrick Jr., W.H., Khalid, R.A., 1974. Phosphorus release and adsorption by soils and sediments: effects of aerobic and anaerobic conditions. Science 186, 53–57.

Patuxent River Commission, 2000. Patuxent River Policy Update Q&A. Available from: http://www.mdp.state.md.us/info/patxattach/PPP-update-Q&A.pdf.

Randall, C., Cokgor, E., 2000. Performance and Economics of BNR Plants in the Chesapeake Bay Watershed, USA.

Rymer, T., 1981. Marriottsville Accord Ends on Hopeful Note. Calvert Independent. December 9, A-6, A-8.

Ryther, J.H., Dunstan, W.M., 1972. Nitrogen, phosphorus and eutrophication in the coastal marine environment. Science 171, 1008–1013.

Taft, J.L., Taylor, W.R., 1976. Phosphorus dynamics in some coastal plain estuaries. In: Wiley, M.L. (Ed.), Estuarine Processes. vol. 1. Academic Press, New York.

Taft, J.L., Taylor, W.R., McCarthy, J.J., 1975. Uptake and release of phosphorus by phytoplankton in the Chesapeake Bay estuary, USA. Mar. Biol. 33, 21–32.

Taft, J.L., Taylor, W.R., Hartwig, E.D., Loftus, R., 1980. Seasonal oxygen depletion in Chesapeake Bay. Estuaries 3, 242–247.

Tierney, P.J., 1994. Bold promises but baby steps: Maryland's growth policy to the year 2020. Univ. Baltimore Law Rev. 23, 461–520.

Virginia Department of Environmental Quality, 2018. Chesapeake Bay Preservation Act. Available from: https://www.deq.virginia.gov/Programs/Water/ChesapeakeBay/ChesapeakeBayPreservationAct.aspx. (Accessed July 14, 2018).

Wan, W., 2006. Caught in Time's Currents: In the Twilight of Life, Md. Man Fears His River Is Too. The Washington Post. July 12, B1.

Washington Post, 2018. Available from: https://www.washingtonpost.com/national/health-science/epa-to-roll-back-car-emissions-standards/2018/04/02/b720f0b6-36a6-11e8-acd5-35eac230e514_story.html?noredirect=on&utm_term=.ae01e4756c93. (Accessed July 14, 2018).

Webb, K.L., 1981. Conceptual models and processes of nutrient cycling in estuaries. In: Neilson, B.J., Cronin, L.E. (Eds.), Estuaries and Nutrients. Humana Press, Clifton, NJ, pp. 25–46.

Webb, K.L., D'Elia, C.F., 1980. Nutrient and oxygen redistribution during a spring-neap tidal cycle in a temperate estuary. Science 207, 983–985.

Zhang, Q., Brady, D.C., Boynton, W.R., Ball, W.P., 2015. Long-term trends of nutrients and sediment from the nontidal Chesapeake watershed: an assessment of progress by river and season. J. Am. Water Resour. Assoc. 51, 1534–1555.

Chapter 18

The Senegal and Pangani Rivers: Examples of Over-Used River Systems Within Water-Stressed Environments in Africa

Awa Niang*, Peter Scheren[†], Salif Diop*, Coura Kane*, Cheikh Tidiane Koulibaly*,[‡]

*Cheikh Anta Diop University, Dakar-Fann, Senegal, [†]WWF Regional Office for Africa, Nairobi, Kenya, [‡]University of Ibadan, Ibadan, Nigeria

1 INTRODUCTION

Climate variability and the resulting effects on river flow dynamics increasingly affect socio-ecological systems in Africa, particularly in the Sahel zone. Human-induced changes to river catchments increase these climate-related sensitivities, resulting in increasing vulnerability of local populations already aggravated by inequalities in access to natural resources and ecological services.

Degradation of ecosystem goods and services is a major challenge in Africa, particularly for countries in sub-Saharan Africa, such as Senegal, where agriculture occupies more than 60% of the active population (Institut de Recherche pour le Développement, 2016). Climatic change, including decreasing rainfall, is pushing the needing populations more and more toward overuse of surface and groundwater, whose renewal is compromised in the more or less long term. Often, in these very precarious and fragile environments, overexploitation of resources has been the response of communities to such changing environments. The examples of the Senegal River (Fig. 1), provided in this chapter, are an illustration of the realities across many river basins in Africa.

In the eastern part of Africa, the situation is similarly complex. This region is particularly troubled by droughts, floods, and "famines" as a result of high hydro-climatic variability, which has often led to regional-scale humanitarian crises. The Pangani River basin (Fig. 1), located between Tanzania and Kenya, does not escape this fate. Despite efforts in developing best practices and sustainable management, the Pangani River is experiencing an advanced degradation of its water resources. This puts local communities, especially those in its estuary, at great environmental and social vulnerability.

Whether in West or East Africa, the resilience of estuarine social-ecological-systems remains a big issue, particularly in the context of emerging climate change impacts.

2 THE SENEGAL RIVER BASIN

2.1 Site Description

The Senegal River estuary in the north of Senegal is a highly vulnerable ecosystem, housing primarily disadvantaged and low-income communities. Located in the Sahel zone where the 1970s drought triggered a series of cyclic crises, undermining livelihood of dwellers and local populations with negative impacts on household economies, the Senegal River estuary has experienced several changes that result both from natural and anthropogenic factors.

The Sahelian climate crisis has particularly affected the northern part of Senegal, manifested in an important decrease in rainfall in Saint-Louis since the early 1970s (Fig. 2). Despite a slight recovery in the early 2000s, a very notable increase in the frequency of occurrence of droughts has been recorded throughout the Sahel region (Fig. 3). The total reduction in annual rainfall as recorded is 35% in the Sahelian part of the Senegal River basin (Albergel et al., 1997), while a rainfall deficit of around 20% was recorded for all stations in the basin (Servat et al., 1999; Bodian, 2010). These recordings confirm the trends observed throughout the Sahel, and have resulted in an advanced degradation of environmental conditions throughout the estuarine zone.

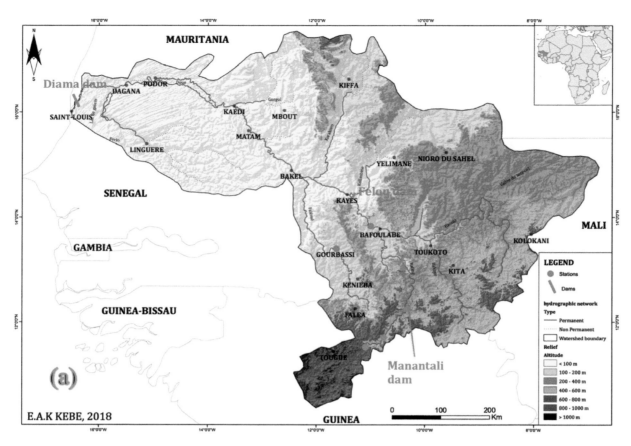

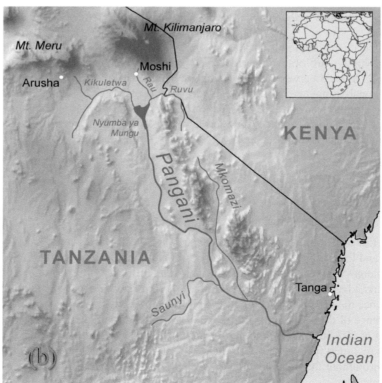

FIG. 1 A general location map: (A) Map of Senegal River Basin, (B) the Pangani River basin, and (C) hydraulic infrastructures. *(Courtesy to the Senegal: Niang, A., 2014. Vulnérabilité de l'environnement et des ressources en eau dans l'estuaire du Sénégal. Dynamique et impacts de la brèche de la Langue de Barbarie entre 2003 et 2013. Thèse de Doctorat Unique de l'Ecole Doctorale « Eau, Qualité et Usages de l'Eau » (EDEQUE), spécialité Hydrologie Continentale, (307 p.) Modified by El Hadj A. K. Kebe (January 2018), and for the Pangani: Kmusser [CC BY-SA 3.0 (https://creativecommons.org/licenses/by-sa/3.0)].)*

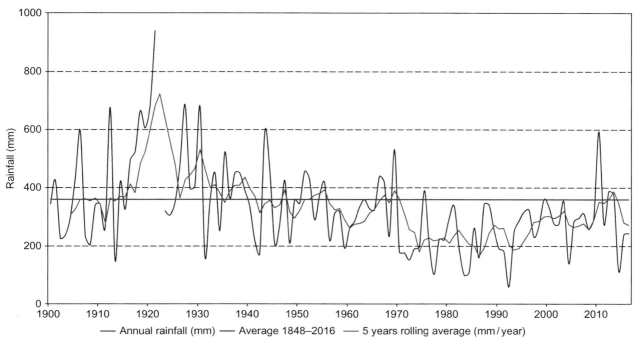

FIG. 2   Annual rainfall in the Senegal River Estuary, Saint-Louis station from 1900 to 2016.

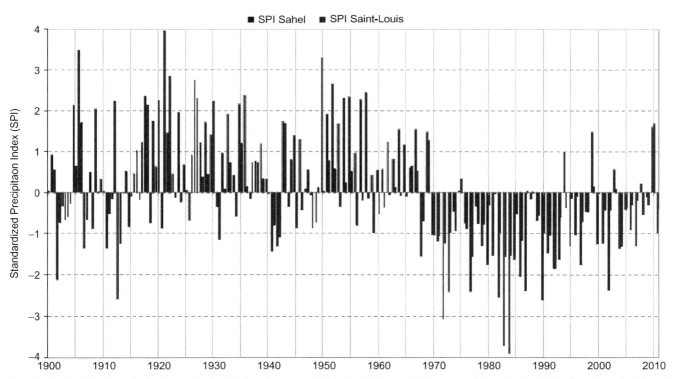

FIG. 3   Standardized Precipitation Index (SPI) of the Sahelian zone and Saint-Louis (Senegal River Estuary) from 1900 to 2010. *(Data from: JISAO and ANACIM-Senegal.)*

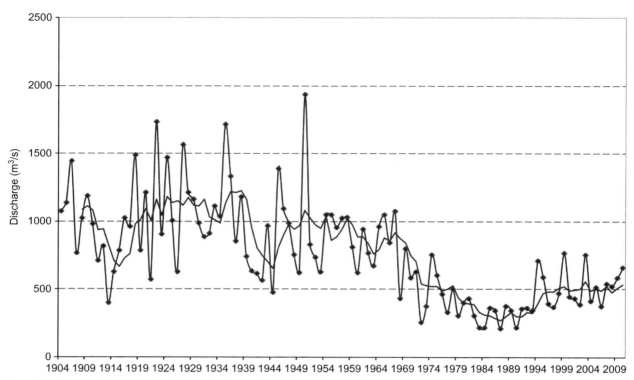

FIG. 4 Annual average discharge at the Bakel station (upper basin) from 1903 to 2010.

In the 1980s, the weakness of the floods in the Senegal River led to more severe salt intrusion; in some years, they were even recorded up to Podor, located 150 km to the river mouth. Data from the Bakel station is a good illustration of this flow reduction in the Senegal River basin (Fig. 4). As groundwater recharge was no longer ensured in these conditions, food security was being compromised for local communities.

Most of the climate analyses and outlooks conducted in West Africa for 2020 and 2050 indicate that current rainfall patterns are likely to continue in the coming years, that is to say, highly variable and irregular, with a high probability of occurrence of extreme events such as droughts or floods (Ardouin, 2004; IPCC, 2007, 2013; Niang, 2014; Mbaye et al., 2015; Tall et al., 2017). In fact, in light of the continuing changing global environment, and the consequent increase in temperature and decrease in rainfall, it is expected that, over the next 5–10 decades, a further decline in river runoff and depletion of groundwater tables is predicted to occur (IPCC, 2013).

2.2 The Damming of the River as a Response to Environmental Degradation

In response to the recurrent series of droughts, during the 1970s, and the resulting reduction of freshwater inflows in the valley and the delta, large dams were established on the Senegal River (Diama dam downstream, Manantali and Felou dams upstream—see Fig. 1). The major objectives of these transnational dams were to produce hydroelectricity, decrease saline intrusion during the dry season, and maintain enough water for domestic freshwater and irrigation purposes. The Diama dam, for example, has created an artificial lake of 235 km^2 surface with a storage capacity of over 250 million m^3, dedicated to the irrigation of 120,000 ha (1200 km^2) of land, as well as for freshwater supply. The Manantali dam is even more important, at a storage capacity of 1200 million m^3. The Felou dam was inaugurated on December 16, 2013; it is located in Mali, 15 km upstream of Kayes; its vocation is hydroelectricity production (431 GWh of capacity and 60 MW of installed power).

Even if those hydraulic works have partly remediated the situation of drought, the changes in the hydrological regime of the Senegal River related to the dams have induced serious impacts on the morphological evolution of the region, as those infrastructures did not take into account factors such as the sedimentology and the ecology of the environment of the Senegal delta.

2.3 The Consequences: Changes in the Hydrological Regime and Morphology, Hyper-Salinization of Lands, Flooding, Changes in Fish Population

In the lower estuary and in the natural region of Gandiolais (see Fig. 1), the management of the dams on the Senegal River has caused serious environmental problems, including freshwater scarcity and recurrent flooding of the city of Saint-Louis. The impending floods in Saint-Louis capital justified the breaching of the "Langue de Barbarie" sandy spit in October 2003. After rapid drainage of waters for the preservation of the city of Saint-Louis, the breach became the new mouth of the Senegal River. With an initial opening of 4 m, the gap rapidly grew and reaches nowadays a width of +6 km, according to the monitoring carried out from LANDSAT satellite imageries (Niang and Kane, 2014; Niang et al., 2015—Fig. 5). The environmental and socioeconomic impacts of these morphological changes are of concern today, putting this socio-ecological system at a critical stage of its evolution.

The accumulation of signs of vulnerability such as hyper-salinization of water and agricultural lands upstream and the rapid morphological changes of the "Langue de Barbarie" spit sand caused by severe erosion along the coast constitutes nowadays a major challenge for the daily activities of the local communities. Fishing communities living along the sand spit of the "Langue de Barbarie" are threatened by sea- level variations; they are experiencing regular storm surges with serious loss of social and economic facilities including their settlements, fishing boats, and gears. This is particularly the case during extreme events with high-level tides. The number of accidents and casualties are now more frequent than before. Moreover, in the ecological zone of the Gandiolais, salinity levels are a standing challenge to economic development. These farmer communities, traditionally specialized in market gardening and livestock breeding, are now facing decreasing income due to salt intrusion into the aquifers and soils that impact their agricultural activities (Koulibaly, 2015).

It also appears that biodiversity has been affected by both the breach and the salinization of surface and groundwater. Certain fish species have been disappearing due to their inability to adapt to the new conditions. On the other hand, fishermen have noted the appearance of new fish species such as *Sardinella*, white carp, cheekfish, or tilapia (Kane, 2010), which has resulted in increased fish landings in Saint-Louis. According to the Regional Fisheries Service of Saint-Louis, the 60,000 tonnes mark was reached in 2008 (Fig. 6), the highest catch on record since 1992 (Seck, 2014).

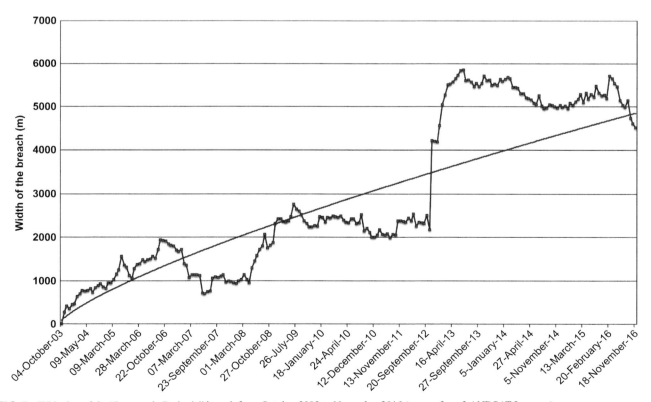

FIG. 5 Widening of the "Langue de Barbarie" breach from October 2003 to November 2016 (survey from LANDSAT Imagery).

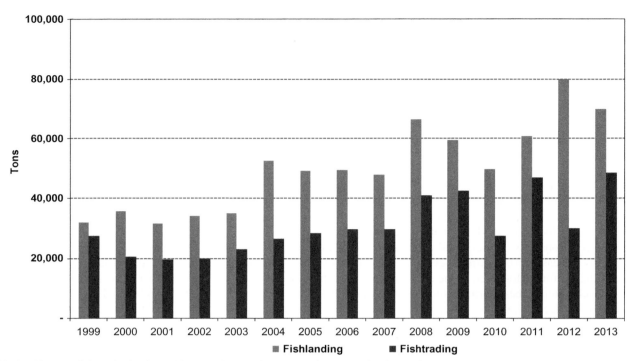

FIG. 6 Change in fish production from 1999 to 2013. *(Data from Regional Division of Fisheries of Saint-Louis.)*

2.4 Adaptation: How People Respond by Relocating and Developing Alternative Economic Activities

Since the opening of the breach in October 2003, the consequent (hyper)salinization of the waters in the lower estuary has been paralyzing agricultural activities of local populations. Due to the salinization of soils, irrigation agriculture has become increasingly difficult in the entire delta of the river, with the total area affected amounting to ~15,000 ha (SAED, 2012; Gning, 2015). To escape such challenges, populations have started to relocate their activities, when and where possible, to areas less affected by salinization processes. Cases have been reported in particular in the lower estuary, specifically in Ndiebene Gandiole and Lahrar (see Fig. 1). One could consider this relocation a way of adapting, although on-going socioeconomic surveys will still have to show in which way this shift affects people's livelihoods.

Traditionally, salt exploitation in Gandiolais area has always been practiced by women but managed by men and placed under the exclusive authority of *Jaaraf* and its *Jambur*. The *Jaaraf* or *Diaraf* is a customary chief of Djolof and *Jambur* constitutes the Assembly of Elders of the *Djolof* Empire. In colonial times, the Gandiolais salt was under the exclusive control of the *Damel of Cayor* who then delegated the administration to his vassal, the *Montel* whose descendants are the *Jaaraf* (Faye and Sambe, 2012). At present, salt exploitation requires official authorization on the basis of a 10-year renewable contract. Each operator has the obligation to pay an annual fee to the local government. After every campaign, the harvest is divided into three parts: one for the farmer and two for the *Jaaraf* (Faye and Sambe, 2012).

Salt extraction is practiced by women in most villages of Gandiolais, particularly in Tassinère, Mouit, and Ndiebene Gandiole, around the Ngaye-Ngaye basin (Fig. 7), tributary of the Gueumbeul basin where the salinity level generally exceeds 35 psu at the beginning of winter season (Corea, 2006). The volumes of salt produced have been steadily increasing over the last 10 years, according to Gandon Local Development Plan (PLD Gandon, 2008), a fact related to the opening of the offloading canal and mainly since the closure of the old Senegal River mouth, which was transformed into a lagoon since 2005 and then evolved almost like a salt swamp. The size of the salt farms is in this respect revealing of the very increase of salt in the estuary.

Salt extraction activity can be seen as an attempt to adapt to changing environmental conditions by developing alternative economic activities. However, the income from salt is not enough to improve the livelihoods of the population, as Gandiolais salt has a low commercial value, despite efforts to improve the product quality through iodization processes.

Despite adaptation efforts of communities through the development of alternative activities such as salt extraction or transfer of agricultural activities to more favored areas, the situation is rather alarming, given the level of impoverishment. There remains, therefore, serious concern about the future of those fragile environments and their communities. Politically,

FIG. 7 Salt mining near Ndiebene Gandiole (lower estuary of the Senegal River). *(From A. Niang on 3rd October 2012.)*

moreover, the situation is complicated by regional interests from the OMVS (Senegal River Basin Management Organization) projects (river navigation, hydroelectricity, and large-scale irrigation schemes) and national policy such as one of the present major *"Plan Sénégal Emergent"* (Senegal Emerging Plan) that aims to develop national irrigation plans for food self-sufficiency, mainly from rice culture. In addition, such a fragile and changing environment is likely to be further weakened by the construction of a maritime-river harbor in Saint-Louis as part of the main Senegal River Navigation Program.

The main risk here is once again, a modification of the dynamics of the estuary and the river mouth, after the major upheaval experienced in October 2003 with the breaching of the Langue de Barbarie. The question is therefore, what will be the future of this region, especially after the construction of the maritime river harbor of Saint-Louis? How will this affect the Gandiolais communities, which have gone from a flourishing market gardening occupation to large-scale salt marshes?

3 THE PANGANI RIVER BASIN

3.1 Site Description

The Pangani River drains a transboundary river basin shared by Kenya and Tanzania (Fig. 1), with its headwaters located in the Kilimanjaro and Meru mountains, fed by cloud-forest precipitation and snowmelt from the glaciers (IUCN, 2007; Hamerlynck et al., 2008). The Pangani is formed through the confluence of Kikuletwa and Ruvu rivers at the Nyumba ya Mungu Dam, after which the river flows across dry plains, through the extensive Kirua swamps, being finally joined by the Mkomazi River at Korogwe and the Luengera River, then traversing the Pangani Falls before entering the Indian Ocean south of the town of Pangani (Akitanda, 2002).

The Nyumba ya Mungu Dam is one of the largest dams in Tanzania; with a total surface area of 15,000 ha (150 km^2), a maximum water depth of 40 m, and a storage capacity of 875 million m^3, the dam supports a storage reservoir for the power plant at Nyumba ya Mungu (8 MW) and two other power plants, Hale (21 MW) and New Pangani Falls (66 MW), located further downstream. Irrigation schemes in the basin can be divided in two categories (Shaghude, 2016): large-scale irrigation systems in the lowland areas, where inadequate rainfall requires the farmers to irrigate their farms to ensure optimum productivity, and small-scale traditional irrigation systems, typically found in the highland areas where irrigation is taken as a safeguard measure to maximize the productivity of the small farms. In particular, the latter are generally characterized by high losses (up to 85%), which combined with high population density (80% of population in the basin resides in the more fertile highlands), results in considerable demand for irrigation water. Population statistics from 1967 to 2012 show that the basin population increased from 2,034,256 people in 1967 to 6,804,733 in 2012 (Shaghude, 2016), resulting in ever-increasing demand.

The current irrigation abstraction systems (both traditional and large scale) are estimated to use at least 400 million m^2 of water per annum, with evaporation losses in the reservoirs estimated at 410 million m^3 per year (Shaghude, 2006), resulting in a total of around 810 million m^3 per annum, or roughly 90% of the total available water flow (900 million m^3). The impacts of high levels of retention and abstraction are further aggravated by climatic change, with a total decrease of at least 250 mm of total annual rainfall at the Mount Kilimanjaro weather station against a total average of ~2000 mm per year over the past century (Shaghude, 2006).

3.2 Consequences: Environmental Degradation

The high water abstraction levels, and consequent strongly reduced river flow, comes with considerable implications. Firstly, increased salt-water intrusion in the lower river ranges has been reported, causing upstream migration of species

(Shaghude, 2004, 2006) and changes in underground water quality and salinization of coastal soils, both with considerable socioeconomic impacts for farmers and fishermen in the lower ranges of the river (Sotthewes, 2008; IUCN, 2009). The study of Pamba et al. (2016) noted that saline water intrusion was highest in December and January (low river flow) when saline water intrusion extended inland 15 km from the river mouth. Furthermore, coastal erosion on the immediate north of the Pangani River mouth, and changes in sediment deposition patterns in the estuary itself have been reported as one of the major environmental issues of concern (Shaghude, 2004, 2006; Pamba et al., 2016). The potential hydrodynamic implications of these morphological changes are expected to have an impact on shipping and port operations, among others.

3.3 Management Strategies

Following the establishment of the Pangani Basin Water Board (PBWB) in 1991, a number of management measures have been put in place. Firstly, a new water right policy was established, under which water users are obliged to hold water rights issued by the PBWB (Mujwahuzi, 2001; Shaghude, 2006). Unfortunately, the policy's introduction did not get full support from the traditional irrigation water users, reportedly as a result of inadequate stakeholder consultation, which led to an increasing number and intensity of illegal water abstractions (Mujwahuzi, 2001; IUCN 2003; Shaghude, 2006). In addition, according to Turpie et al. (2005), the water allocations are not adequately taking into consideration environmental flow demands. As a second measure, furthermore, effort has been put into improving the efficiency of the irrigation systems, in order to reduce the losses from the traditional furrow networks. This involves, among others, the introduction of modern piping systems, which would get rid of both water leakages by infiltration and evaporation along the transport path.

A number of additional management measures are still on the table, including the application of Aquifer Storage and Recovery (ASR) technology, which involves the recharge and storage of water in an artificial aquifer during the rainy season and the recovery of the stored water when it is required; the aquifer essentially functions as a water bank (Pyne, 1995; Shaghude, 2006). Additional measures as suggested by the World Commission on Dams (WCD, 2000) include improved basin and system level management, through afforestation and the promotion of better farming practices. In reality, however, appropriate watershed management is significantly impeded by a number of factors, including poor inter-sectoral coordination at the field level, diverging interests of watershed stakeholders, incompatibility between formal and informal institutions, poor highland-lowland integration, conflicting development interventions, population pressure and migration, and inadequate political support as shown in a recent study by Tuli et al. (2018).

4 CONCLUSION

The cases presented in this chapter, the Senegal and Pangani rivers, are just two examples of the many river systems in Africa that have been affected by human interventions, such as dams and irrigation schemes, designed with the objective to improve the livelihoods of communities through increased access to water for agriculture, energy through hydropower, as well other purposes. While well intended, such interventions often result in important negative consequences further downstream, in particular affecting the vulnerable estuarine environments and their dependent population. As the effects of climate change on rainfall patterns are becoming more obvious, the strong relation between rainfall and river run-off constitutes is indeed a growing constraint in such river basins, a factor that therefore needs to be fully integrated and mainstreamed into the development programs and policies of African countries.

As the two case studies demonstrate, many of the solutions and alternatives that are on the table require trade-offs between upstream water user needs and downstream effects. However, under pressure of an ever-increasing population and related demand for water and land for cultivation and other uses of river basins, solutions that restrain water abstraction or uses otherwise of river catchments tend to encounter a lot of resistance. While such solutions may make sense in a broader eco-economic context, the realities on the ground, and the social and political realities under which decisions are taken, are consequently impeding their effective implementation. What we see in many cases is therefore rather an adaptive response of affected communities migrating or developing alternative livelihoods opportunities.

Yet, it is obvious that the management of Africa's river basins, and the challenges related to this, is increasingly an area of attention. The two river systems presented in this chapter, the Senegal River Basin Organization's projects and the Pangani River Basin Management Project, are examples of efforts underway to establish appropriate management schemes and implement appropriate actions toward a more system-wide management approach. Unfortunately, the current efforts under these schemes are faced with tremendous challenges; it is clear that only through such integrated approaches that appropriate strategies can be developed to secure the long-term ecological future of Africa's river systems.

The Senegal and Pangani Rivers do not constitute an exception; many rivers and other "wet" ecosystems across western and eastern Africa are under stress from natural and anthropogenic conditions (Alhassan and Kwakwa, 2014; Freitas,

2015). The Inner Delta of the Niger is a very good illustration of that situation. Nowadays, the shrinkage of this delta is bound as much by the important water demand for irrigated rice growing as by the reduction of water resources due to modifications of climate conditions (Morand et al., 2014). In East Africa, other estuarine ecosystems like the Wami River in Tanzania and the Tana River in Kenya are also in an alarming situation due to a mix of social and ecological problems linked to climate change, damming and hydropower generation, and over-use of natural resources (Wambura et al., 2015; Kiwango et al., 2015; Sood et al., 2017).

REFERENCES

Akitanda, P., 2002. In: South Kilimanjaro catchment forestry management strategies—perspectives and constraints for integrated water resources management in the Pangani River Basin. Paper presented at the International Conference on Integrated River Basin management, Pangani River Basin, June 2001. Moshi, Tanzania, pp. 13–14.

Albergel, J., Bader, J.-C., Lamagat, J.-P., 1997. In: Rosbjerg, D., Boutayed, N.E., Gustard, A., Kundzewicz, Z.W., Rasmussen, P.F. (Eds.), Flood and drought: application to the Senegal River management. Sustainability of Water Resources Under Increasing Uncertainty: Proceedings of the Rabat Symposium. Wallingford, pp. 509–517.

Alhassan, H., Kwakwa, P.A., 2014. When water is scarce: the perception of water quality and effects on the vulnerable. J. Water Sanit. Hyg. Develop. 4 (1), 43–50.

Ardouin, S., 2004. Variabilité hydro-climatologique et impacts sur les ressources en eau de grands bassins hydrographiques en zone soudano-sahélienne (Logone-Chari, Sénégal, Gambie). Thèse en Sciences de l'eau dans l'environnement continental, Université de Montpellier II, p. 440.

Bodian, A., 2010. Approche par modélisation pluie-débit de la connaissance régionale de la ressource en eau: Application au haut bassin du fleuve Sénégal. Thèse de Doctorat Unique de l'Ecole Doctorale « Eau, Qualité et Usages de l'Eau » (EDEQUE), spécialité Hydrologie Continentale, p. 331.

Corea, M., 2006. Analyse situationnele des ressources en eau dans l'estuaire du fleuve Sénégal: la dynamique de la salinisation dans le bief estuarien. Mémoire de DEA Chaire UNESCO. UCAD / Département de Géographie. 76 p.

Faye, P., Sambe, A.C., 2012. Les relations de genre à propos du sel au Gandiole. Une histoire d'exploitantes exploitées. Negos-GRN Etude de cas n°3, p. 14.

Freitas, A., 2015. Water as a stress factor in sub-Saharan Africa. Population 2005, 47.

Gning, A.A., 2015. Etude et modélisation hydrogéologique des interactions eaux de surface-eaux souterraines dans un contexte d'agriculture irriguée dans le Delta du Fleuve Sénégal. Thèse de Doctorat es Sciences de l'Ingénieur de l'ULg (Belgique) et Hydrologie de l'EDEQUE/UCAD (Sénégal), p. 259.

Hamerlynck, O., Richmond, M.D., Mohamed, A., Mwaitega, S.R., 2008. Fish and Invertebrate Life Histories and Important Fisheries of the Pangani River Basin, Tanzania Final Report. Pangani River Basin Flow Assessment—Specialist Study. IUCN/PBWO, p. 74.

Institut de Recherche pour le Développement (IRD), 2016. Etude prospective en soutien à la programmation européenne conjointe : rapport final [Etat des connaissances scientifiques sur les grands enjeux du développement, scénarios d'évolution]. IRD, Dakar, p. 117. http://www.documentation.ird. fr/hor/fdi:010067188.

IPCC, 2007. Climate Change 2007: Synthesis Report. An Assessment of the Intergovernmental Panel on Climate Change. Fourth Assessment Report, p. 58.

IPCC, 2013. Climate Change 2013: The Physical Science Basis. Working Group I Contribution to the Fifth Assessment Report of the Intergovernmental Panel on Climate Change. Summary for Policy Makers, p. 33.

IUCN, 2003. Eastern African Programme, 2003 The Pangani River Basin: A Situation Analysis. xvi + 104 pp. Available from: https://portals.iucn.org/library/efiles/documents/2003-079.pdf.

IUCN, 2007. In: Socio-economic baseline assessment report: the role of river systems in household livelihoods Pangani River Basin Flow Assessment, Moshi. International Union for Conservation of nature and Pangani Basin Water Board. p. 58. Available from: https://cmsdata.iucn.org/downloads/socio-economic_assessment.pdf.

IUCN Eastern and Southern Africa Programme 2009. In: IUCN (Eds.), The Pangani River Basin Situation Analysis. second ed.. 2009. IUCN Regional Office Nairobi. xii + 82 pp. https://portals.iucn.org/library/efiles/documents/2009-073.pdf.

Kane, C., 2010. Vulnérabilité du système socio-environnemental en domaine sahélien: l'exemple de l'estuaire du fleuve Sénégal. De la perception à la gestion des risques naturels. Thèse de doctorat, Laboratoire Image et Ville, Université de Strasbourg, p. 318.

Kiwango, H., Njau, K., Wolanski, E., 2015. The need to enforce minimum environmental flow requirements in Tanzania to preserve estuaries: case study of the mangrove-fringed Wami River estuary. Ecohydrol. Hydrobiol. 15, 171–181.

Koulibaly, C.T., 2015. Altération des eaux d'irrigation dans la zone du Gandiolais. Mémoire de Master Chaire UNESCO/GIDEL (Gestion Intégrée et Développement Durable du Littoral ouest-africain), Département de Géographie, FLSH/UCAD, p. 122.

Mbaye, M.L., Hagemann, S., Haensler, A., Stacke, T., Gaye, A.T., Afouda, A., 2015. Assessment of climate change impact on water resources in the upper Senegal Basin (West Africa). Am. J. Clim. Chang. 4, 77–93.

Morand, P., Sinaba, F., Niang, A., 2014. Fishers, herders and rice-farmers communities of the Inner Niger Delta facing the huge challenge of adapting to weakened floods: a social-ecological system at risk. In: Tvedt, T., Oestigaard, T. (Eds.), A History of Water, Series III. Vol. 3: Water and Food. pp. 418–436. (Chapter 17), Taurus.

Mujwahuzi, M.R., 2001. Water use conflicts in the Pangani Basin. In: Ngana, J.O. (Ed.), Water Resources Management in the Pangani River Basin: Challenges and Opportunities. Dar es Salaam University Press, Dar es Salaam, pp. 128–137.

Niang, A., 2014. Vulnérabilité de l'environnement et des ressources en eau dans l'estuaire du Sénégal. Dynamique et impacts de la brèche de la Langue de Barbarie entre 2003 et 2013. Thèse de Doctorat Unique de l'Ecole Doctorale « Eau, Qualité et Usages de l'Eau » (EDEQUE), spécialité Hydrologie Continentale, pp. 307.

Niang, A., Kane, A., 2014. Morphological and hydrodynamic changes in the lower estuary of the Senegal River: effects on the environment of the breach of the 'Langue De Barbarie' sand spit in 2003. In: Diop, S., Barusseau, J.-P., Descamps, C. (Eds.), The Land-Ocean Interactions in the Coastal Zone of West and Central Africa. Springer International Publishing, pp. 23–40. ISBN: 978-3-319-06387-4.

Niang, A., Kane, C., Kebe, E.A.K., Kane, A., 2015. Reconstitution de l'évolution de la brèche de la Langue de Barbarie entre 2003 et 2015 à partir d'une série d'images LANDSAT. Revue « Espaces et Sociétés en Mutations », Numéro Spécial en Hommage au Pr Mamadou Moustapha SALL— Décembre 2015, pp. 201–220.

Pamba, S., Shaghude, Y.W., Muzuka, A.N.N., 2016. Hydrodynamic modelling on transport, dispersion and deposition of suspended particulate matter in Pangani estuary, Tanzania. In: Diop, S., Scheren, P.A., Machiwa, J.F. (Eds.), Estuaries: A Lifeline of Ecosystem Services in the Western Indian Ocean, Estuaries of the World. Springer International Publishing, Switzerland, pp. 141–160.

PLD Gandon, 2008. Communauté Rurale de Gandon: Plan Local de Développement (PLD) de la Communauté Rurale de Gandon (2009-2014). Document de synthèse, Novembre 2008, 4 p.

Pyne, R.D.G., 1995. Groundwater Recharge and Wells: A Guide to Aquifer Storage and Recovery in Florida. CRC Press, Gainesville, FL620. 27 p.

SAED, 2012. Bulletin d'information sur la culture irriguée. Saint-Louis. 5 p.

Seck, A., 2014. Les pêcheurs migrants de Guet-Ndar (Saint-Louis du Sénégal): analyse d'une territorialité diverse entre espaces de conflits et espaces de gestion. Thèse de doctorat de Géographieen cotutelle entre l'Université de Liège et de l'Université Cheikh Anta Diop de Dakar, p. 356.

Servat, E., Paturel, J.E., Lubès-Niel, H., Kouamé, B., Masson, J.M., Travaglio, M., Marieu, B., 1999. De différents aspects de la variabilité de la pluvio-métrie en Afrique de l'Ouest et Centrale. Rev. Sci. Eau. 12 (2), 363–387.

Shaghude, Y.W., 2004. Shore Morphology and sediment characteristics south of Pangani River, coastal Tanzania. Western Indian Ocean J. Marine Sci. 3, 93–104.

Shaghude, Y.W., 2006. Review of water resource exploitation and land use pressure in the Pangani river basin. Western Indian Ocean J. Marine Sci. 5, 195–207.

Shaghude, Y.W., 2016. Estuarine Environmental and Socio-economic impacts associated with upland agricultural irrigation and hydropower develop-ments: the case of Rufiji and Pangani estuaries, Tanzania. In: Diop, S., Scheren, P.A., Machiwa, J.F. (Eds.), Estuaries: A Lifeline of Ecosystem Services in the Western Indian Ocean, Estuaries of the World. Springer International Publishing, Switzerland, pp. 169–182.

Sood, A., Muthuwatta, L., Silva, S., McCartney, M., 2017. Understanding the hydrological impacts of climate change in the Tana River Basin. Colombo, Sri Lanka: International Water Management Institute (IWMI) Working Paper 178, p. 40.

Sotthewes, W., 2008. Forcing the salinity distribution in the Pangani estuary. Final Report, Delft University of Technology 81–85.

Tall, M., Sylla, M.B., Diallo, I., Pal, J.S., Faye, A., Mbaye, M.L., Gaye, A.T., 2017. Projected impact of climate change in the hydroclimatology of Senegal with a focus over the Lake of Guiers for the twenty-first century. Theor. Appl. Climatol. 129 (2), 655–665.

Tuli S. Msuya, Makarius C.S. Lalika, 2018. Linking ecohydrology and integrated water resources management: institutional challenges for water man-agement in the Pangani Basin, Tanzania. Ecohydrol. Hydrobiol. 18 (2), 174–191. ISSN 1642–3593. https://doi.org/10.1016/j.ecohyd.2017.10.004.

Turpie, J.K., Ngaga, Y.M., Karanja, K.F., 2005. Preliminary economic assessment of water resources of the Pangani River Basin, Tanzania: Economic values and incentives. Report submitted to IUCN, Eastern Africa Regional Office and Pangani Basin Water Office. Nairobi, p. 10.

Wambura, F.J., Ndomba, P.M., Kongo, V., Tumbo, S.D., 2015. Uncertainty of runoff projections under changing climate in Wami River sub-basin. J. Hydrol. Reg. Stud. 4 (2015), 333–348.

WCD, 2000. Dams and development: a new framework for decision making. The report of the world commission on dams, Nov. 2000. Earthscan Publication Ltd, London and Sterling, VA.

Chapter 19

Damming the Mekong: Impacts in Vietnam and Solutions

Nguyen Huu Nhan, Nguyen Ba Cao
Vietnam Academy of Water Resources, Hanoi, Vietnam

1 INTRODUCTION

The Mekong River drains a catchment of over $800,000\,km^2$ and is the world's 12th longest river (4800 km), the 8th largest water discharge ($470 \times 10^6\,m^3$/yr), and the 10th largest sediment load (160×10^6 tons/yr) (Fig. 1; Meade, 1996). The Mekong River, which is known as the Lancang River in China (Upper Mekong basin, UMB), rises on the Qinghai—Tibet Plateau with a maximal elevation of 5220 m, flows through six countries (China with 16% of its basin, Myanmar with 5% of its basin, Laos with 35% of its basin, Thailand with 18% of its basin, Cambodia with 18% of its basin, and Vietnam with 11% its basin) and empties into the Vietnam East Sea (South China Sea). The Mekong Basin has the world's most diverse river ecosystem (Piman et al., 2013). It is the world's largest inland fishery. Its biodiversity is fundamental to agricultural production and the food security of 90 million people (Lu and Siew, 2006) in the Lower Mekong basin (LMB), including about 18 million people in the Vietnamese Mekong delta (VMD).

Human activities have and will continue to make both positive and negative impacts (directly and indirectly), often severe, on natural resources in most of the world's deltas. Among these activities, hydropower dams are the most critical anthropogenic disturbances to global river systems (Walling, 2006); they generate massive changes in the deltas and to the livelihoods of millions of people and ultimately they threaten delta sustainability (Hoa et al., 2007). This issue is particularly important in the VMD where sustainability is threatened. Therefore, the aim of this chapter is twofold: (1) to present real facts and data and predictions for quantifying human-induced changes in VMD, its estuaries, and coasts due to hydropower dams in the whole Mekong River basin, and (2) to propose solutions to decrease or mitigate their negative effects.

2 HYDROPOWER DAM NETWORK IN THE MEKONG RIVER BASIN

The total hydropower capacity of Mekong River is ~60 GW (MRC, 2016) and about 20 GW is already installed and used. There are more 176 hydropower dams including 39 mainstream dams, 96 planned dams, and 31 dams under construction (https://wle-mekong.cgiar.org/wp-content/uploads/unnamed-11.jpg, 2015) in the entire basin (Figs. 1 and 2). Seven mainstream dams have been commissioned and seven are under construction. In addition, 185 irrigation dams have been completed. Thus, 361 dams will be installed in the entire basin; so, the Mekong River will become one of most dam-fragmented rivers in the world.

The construction of large dams in the Mekong Basin began with China's Manwan Reservoir in 1993; by April 2016, 35 dams had been commissioned for hydropower (>15 MW), for irrigation (reservoirs > 0.5 km²), for water supply, or for mixed purposes (Fig. 1; WLE-Mekong, 2016). A further 226 dams are under construction or planned (WLE-Mekong, 2016). Six mega-dams have been built in the mainstream UMB (Gongguagiao, 900 MW, 2012; Manwan, 1400 MW, 1993, Dochashan, 1350 MW, 2003, Jinhong, 1750 MW, 2009, Xiaowan, 4200 MW, 2010, and Nuozhadu 5850 MW, 2012); they generate ~15,450 MW with a total reservoir capacity of 42 km³. The total accumulated installed capacity has experienced an exponential growth since 2003; the capacity is 4.2 times larger than in 2003 (Allison et al., 2017). Ignoring international concerns from scientists and environmental groups and the protest from the Government of Vietnam, Laos's mainstream Sayaburi dam (reservoir capacity: 1.3 km³) started construction in November 2102.

The impacts of these dams, and, to a lesser extent, of river-bed mining, on fluvial sediment supply and on the future stability of the VMD, including the erosion of the delta's foreshore, has become a particularly important issue in Vietnam, highlighted in recent academic studies and in numerous newspaper reports. The Mekong River Summit has recently taken

Coasts and Estuaries. https://doi.org/10.1016/B978-0-12-814003-1.00019-8

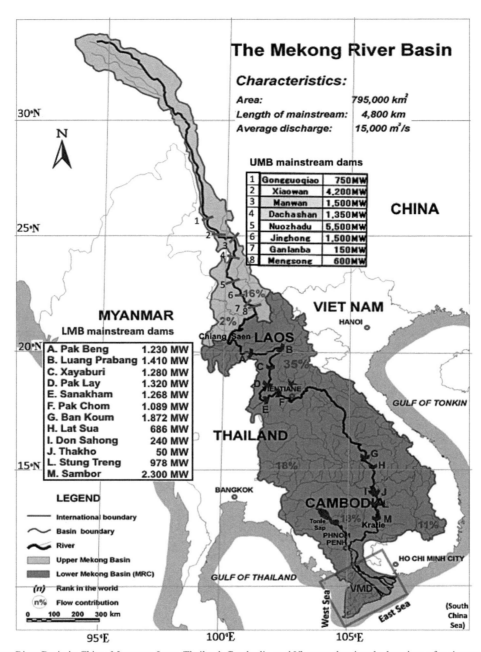

FIG. 1 The Mekong River Basin in China, Myanmar, Laos, Thailand, Cambodia, and Vietnam showing the locations of mainstream hydropower dams in China and the Lower Mekong Basin (Mekong River Commission (MRC), 2010a) and the location of the Vietnamese Mekong River Delta (VMD)—the area of study.

place in Cambodia (March 2018), where environmentalists warned that hydropower dams would destroy the river basin, severely affecting the environment and socioeconomic conditions. This warning is not new. A report by the MRC announced by stated that:

"The amount of sediment is estimated to be reduced by 67%–97% for the period from 2020 to 2040 seriously affecting the wealth of the river, affecting the landform and soil stability in the VMD. The great drought in 2016 leaves a deep wound to the peasants in VMD. The whole of Ben Tre province is severely threatened by saline intrusion. People are thirsty for clean water. Over two hundred thousand hectares of rice, hundreds of thousands of hectares of damaged orchards. These extreme events have relations with dams in UMB. The Mekong will become one of the most heavily affected basins in the world if all planned dams are fully built (Zarfl et al., 2015). The hundreds of dams will unavoidably change the natural

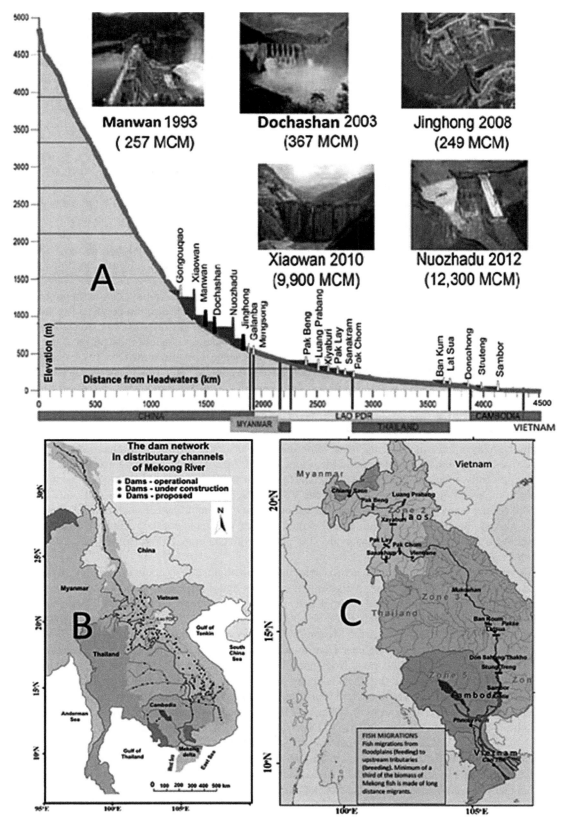

FIG. 2 (A) The dams in the main channel of the Mekong River and the properties of the big operation dams in China (MRC, 2010b); (B) the dam network in distributary channels of the Mekong River (MRC, 2010b); (C) the fish migration routes from the floodplain to upstream distributaries (MRC, 2010b).

flow cycle and sediment yield, further affecting the sediment dispersal pattern in the estuary and along the delta front and ultimately threatening the persistence of the delta landform itself (Bravard et al., 2013; Rubin et al., 2015)."

Dam construction is not the only ongoing human activity in the Mekong Basin and delta region (Brunier et al., 2014). Interactions among multiple factors such as in-channel sand extraction (Bravard et al., 2013), coastal mangrove degradation (Phan et al., 2015), expansion of aquaculture (Nhan, 2016b), climate change (MONRE VN, 2016; MRC, 2009), land subsidence (Schmitt et al., 2017), sea level rise (Rubin et al., 2015), and increasing tidal amplitude due to sea level rise (Nhan, 2016a) will pose additional risks to the delta. Therefore, basin-wide collaborative observation programs are needed to deepen the understanding of the recent evolution of the VMD.

3   THE VIETNAMESE MEKONG DELTA

The coastal tidal regime in the eastern VMD is semi-diurnal with a tidal range ~2–4 m and the mean wave height is 1.25 m. The tidal range decreases toward Ca Mau Cape. The tides are diurnal with amplitudes of 1–2 m in the west coast zone where the mean wave height is 0.65 m. The mangroves are only found between mean sea level and mean high water level. For the past few decades, the VMD has been severely modified and utilized by human activities inside including: dense channel network for irrigation, drainage, and transportation; and a great set of infrastructures including river dykes, sewers, sluices, roads, urbanization, sea dykes, and aquaculture ponds. The natural resources of VMD and the coast mangrove belt have been severely damaged and fragmented.

The VMD (Fig. 3) is the outcome of the Holocene evolution of the delta from the initiation of delta progradation around 8.0 ka BP to the present (Nguyen et al., 2000; Xue et al., 2011) as a result of the interaction between five fundamental factors: (1) the tropical monsoon climate with southwest circulation (known as wet season) from May to October and northeast circulation (known as the dry season) from November to April; (2) the water and material discharge of the Mekong River with huge contrast between wet and dry seasons; (3) the asymmetrical sea impacts from the East Sea (South China Sea) and the West Sea (Gulf of Thailand), in which hydrodynamic and wave power on the East side is over twice stronger than on the West side so that the net southwestward sediment transport induced by the northeast monsoon is stronger than the transport in the southwest monsoon (this asymmetry explains the asymmetrical shape of the VMD; Fig. 3; Xue et al., 2011; Anthony et al., 2013; Tamura et al., 2010; Unverricht et al., 2013); (4) the tectonic activities as well as the nature and structure of geologic and geomorphological formations; and (5) the processes in the VMD and in its coastal zone, where the mangroves and human activities play a key role. During its development, the character of the delta changed from tide-dominated into a wave- and tide-dominated, and its shape and the orientation of the coastline changed through time (Ta et al., 2002). In addition to this, recently, there are some new important factors at work, namely (a) climate change and sea level rise; (b) huge changes in the water and sediment regime of the VMD induced by hydropower dams and other human activities in the entire basin; (c) the many gaps in policies for managing the environment, the ecosystems, and sustainable development; and (d) the almost non-controlled and unlimited exploitation of VMD and coastal recourses, construction of coastal structures, and the destruction of wetlands (Hoa et al., 2007, 2008; Syvitski et al., 2009; Erban et al., 2014; Schmitt et al., 2017; and references within).

The present VMD occupies only 5% of the entire Mekong basin with an area of about 40,000 km^2, with a 700 km-long coastline with multi-distributary rivers forming eight main estuaries and a large flat delta floodplain with low elevation (70% of its area have elevation below 3 m) and a high density of river and canal network (Fig. 3). These estuaries have been relatively stable over the past 2000–3000 years (Ta et al., 2002; Tamura et al., 2010). The tide and saltwater instrusion is impacting on a coastal area of 17,000 km^2 and the Mekong River is flooding on the northern half of the VMD with inundation depth of 1–4 m for 1–6 months. There are near 18 million people (General Statistics Office of Vietnam—GSOV, 2017) living here and most of them earn their livelihood from agricultural activities. The VMD is Vietnam's major production hub for rice, fruits, and aquaculture. However, the VMD is facing major threats due to upstream developments, and hydropower dam impacts are most threatening. Over the past 50 years, the VMD has recorded an average sea level rise of 0.2 m, and it may rise another 0.28–0.33 m by 2050 and by 0.6–1 m by 2100 (Ministry of Natural Resources and Environment of Viet Nam (MONRE) http://www.monre.gov.vn, 2016). MONRE (Ministry of Natural Resource and Environment), 2016 suggested that in 2100, 38.9% of VMD area will be permanently inundated and 35% of the population will be affected.

Moreover, land subsidence (Fig. 4) is expected to be 0.35–1.4 m throughout the delta by 2050 if groundwater pumping continues at present rates (Erban et al., 2014). Local subsidence rates of 2.5 cm/yr outpaced sea level rise by almost an order of magnitude (Minderhoud et al., 2017). Because most of the lower VMD is less than 2 m above sea level, the subsidence, compounded by the SLR, would pose a considerable threat to the sustainability of the delta (Erban et al., 2014). Schmitt et al. (2017) predicted that due to accelerating subsidence and sea level rise, the VMD will almost completely disappear by the end of this century if all the planned dams are built and sediment mining continues at current rates.

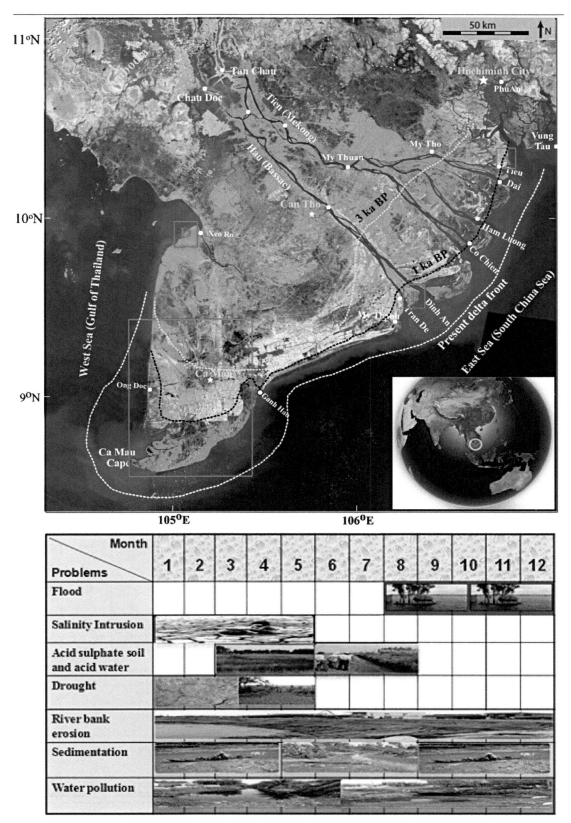

FIG. 3 (Top) Vietnam Mekong Delta (VMD); Locations of past and present delta fronts (Ta et al., 2002); Locations and names of hydro-meteorological stations of the national network; (Bottom) the problems in the VMD during the year.

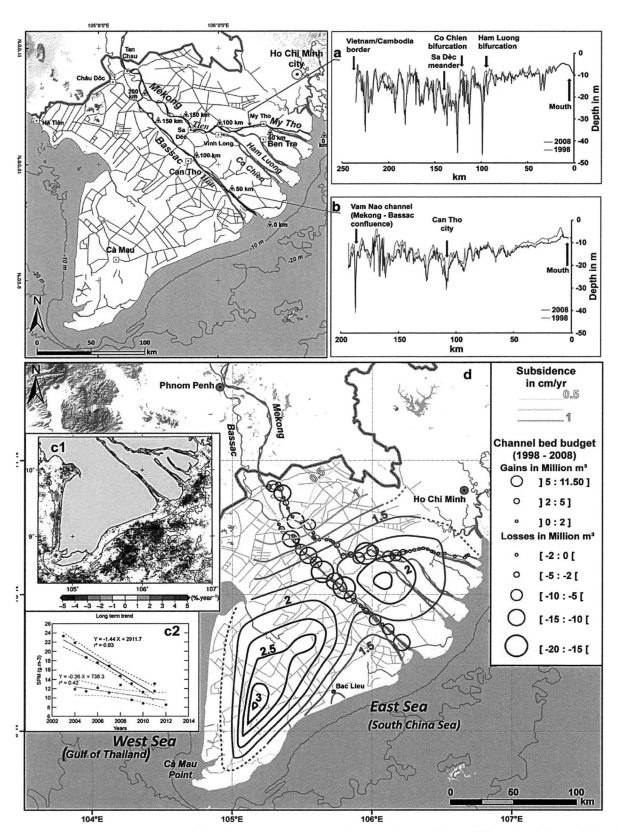

FIG. 4   Recent morphological changes in the VMD and its coast. The riverbed depth changes between 1998 and 2008 as shown in longitudinal profiles of the Mekong branch (A) and Bassac branch (B) for the period 1998–2008 (Brunier et al., 2014); Map (C1) and time-series figure (C2) of significant monotonic trend in % per year of near-surface suspended particulate matter (SPM) in coastal waters off the VMD derived from MERIS satellite data (modified from Anthony et al., 2013); (D) Compaction-based subsidence rates (redrawn from Erban et al., 2014) and the bedload budget changes over 10 years (1998–2008) in the Mekong and Bassac branches of the VMD (modified from Anthony et al., 2013).

For the pre-dam period (before 1993), the most commonly accepted values for the water volume and sediment load to VMD are about $47\,km^3$ and 160 million tons (Milliman and Syvitski, 1992). About 85% of these values occurred during the wet season, and only 15% in the dry season. This seasonal annual cycle generated many problems in VMD (Fig. 3) but the system was resilient; with changes in the river hydrology and sediment fluxes (described further), the VMD is becoming less and less resilient (Schmitt et al., 2017).

The average growth rate of the entire delta in the past 7000 years was ~30 m/yr (Liu et al., 2017), and over the past 3000 years the rate was ~16 m/yr at the Mekong estuarine coast and 26 m/yr in the Ca Mau coast (Ta et al., 2002; Xue et al., 2011). In contrast to this progradation in the Late Holocene, a serious foreshore erosion of the delta has occurred recently, which is largely attributed to significant sediment retention by dams in the Mekong Basin (Tamura et al., 2010; Anthony et al., 2013; Besset et al., 2015), though other factors may also exacerbate the situation (Unverricht et al., 2013).

4 DAM IMPACTS ON THE MEKONG DELTA IN VIETNAM

4.1 The Impact on Water Resource in the Flood Season

Monitoring data show great changes in flood water inflow to the VMD in recent years. The flood season water flows downstream of the Chinese hydropower dams are now lower than those in the dry season. The flood water level in Chang Saen (Fig. 5) is lower than that in the dry season as a result of storage in the hydropower reservoirs. The water level and water flow to the WMD are now also strongly modified, as evidenced by the years of consecutive low floods since 2002 to the present time, with the exception of the high flood in 2011 (Fig. 5). The hydrographs also have changed; in 2014, the high flood peak appeared before the low flood peak, which is contrary to the natural flow regime. Further floods appear to occur now half a month later and the low flood periods are shorter, especially in 2013 and 2015. The results of predictions of total future flood water flow at Kratie (Toan et al., 2016) for the next 100 years as a result of the hydropower dams and based on the historical baseline data are (Toan et al., 2016): (1) The dams increase the number of years with small floods ($P < 76\%$ and $W < 320\,km^3$); this would happen in 21 years in the baseline case, in 36 years with the Chinese dams in UMB only, and 90 years if all planned dams in both UMB and LMB are built; (2) in contrast, the dams decrease the number of years of large flood ($P > 25\%$ and $W > 397\,km^3$); this would happen in 23 years for the baseline case, in 13 years only with the Chinese dams in UMB only, and never (i.e., no large floods) if all planned dams are built; and (3) the dams slightly decrease the number of years with a medium flood ($320 \leq W < 397\,km^3$): there would be 56 years for the baseline case, 51 years with the Chinese dams in UMB only, and 10 years if all planned dams are built. The number of flood years in the next 100 years with maximal water level at Tan Chau exceeding Alert III for the baseline case is 32; it will be only eight with the Chinese dams in UMB only, and one with all planned dams.

The measured hourly data at Tan Chau and Chau Doc-starting cross sections into VMD (Fig. 5E) (their locations are shown in Fig. 3) also show the recent large changes of the flood regime in VMD for 1996–2016 (the period with an increasing impact in the VMD from Chinese dams), namely: (1) after 2002, only one year had a large flood (2011) while 6 years had a small flood (2003, 2004, 2010, 2012, 2015, 2016). These changes in discharges into VMD can be attributed to the Chinese hydropower dams. It must be stressed that these changes induced severe negative effects in VMD. As discussed below, the decreased frequency of large floods generates a deficit in sediment and materials for filling delta land needed to compensate the subsidence, to renovate the soils, and to provide floodwaters for fish migration. The increased frequency of small flood generates a loss of available water for ecosystems and human activities in the VMD and it also increases salinity intrusion in the VMD.

4.2 The Impact on Water Resource in the Dry Season

Because the Lancang River contributes 45% of water to the Mekong basin in the dry season, it is easy to understand that the Chinese dams can make critical changes to the river flow entering the VMD in the dry season. In fact, the severe water shortages and salinity intrusion in 2010, 2015, and 2016 (i.e., after the large Xiaowan and Nuozhadu dams were operating) suggest that. Indeed, the data (Fig. 5) show great nonnatural variations in the water flow into the VMD; the water level decreased abnormally at the start of the dry season and increased unusually slowly at the start of the wet season. The water flow in March and April increased higher than normal in the years of similar hydrological conditions than in the past, and the average water flow during the dry season increased due to the hydropower dams. These effects are most likely due to the regulating hydropower dams in the basin; the early water accumulation in the dams significantly impacts the downstream flow at the end of the dry season and the starting wet season, generating a water shortage in coastal areas at the start of the wet season. In contrast, the late accumulation of water in the dams significantly impacts the downstream flow for the

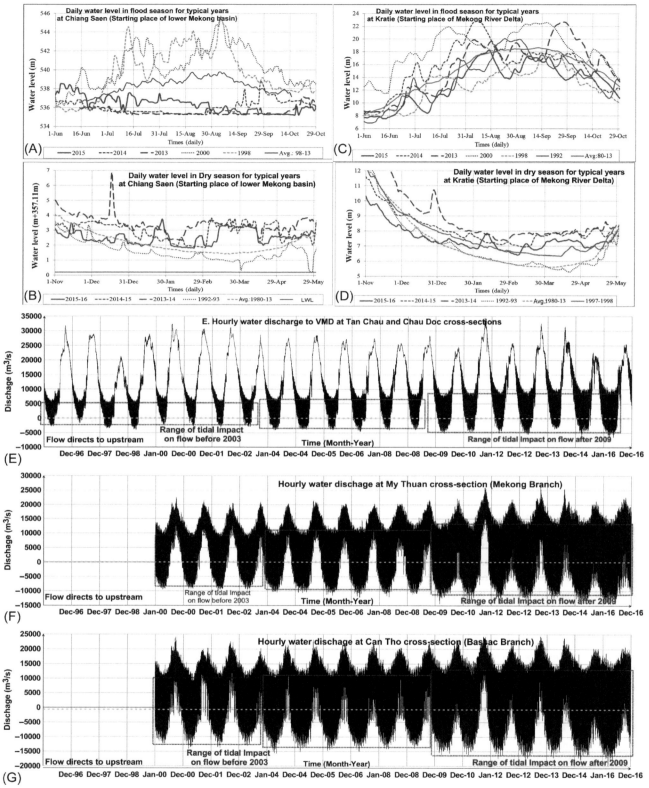

FIG. 5 Changes of the water resource regime in VMD during the past 20 years. Daily water level in flood (A) and dry (B) season for typical years at Chiang Saen, the starting place of the Lower Mekong Basin (From MRC, 2016. Integrated water resources management-based basin development strategy 2016–2020 for the lower Mekong Basin. http://www.mrcmekong.org/assets/Publications/strategies-workprog/MRC-BDP-strategy-complete-final-02.16. pdf.); Daily water level in flood (C) and dry (D) seasons for typical years at Kratie, the starting place of the Mekong River Delta (From MRC, 2016. Integrated water resources management-based basin development strategy 2016–2020 for the lower Mekong Basin. http://www. mrcmekong.org/assets/ Publications/strategies-workprog/MRC-BDP-strategy-complete-final-02.16.pdf.); Hourly water discharge showing the decreasing river discharge and increasing range of sea impact in recent years at Tan Chau and Chau Doc cross-sections (E) in the VMD, and at the My Thuan cross-section (F) and the Can Tho cross-section (G) at the entrance of the strongly tidal region of the lower VMD (From MONRE VN, 2017. *Variation of Water Discharge and Sediment Flux at Lower Mekong River*. National Centre for Hydro-Meteorology. Hanoi. 15p. (In Vietnamese).).

months at the end of the wet season and at the start of the dry reason and induces a water shortage in the VMD coastal areas at these times. These changes are very harmful for agricultural production in the Delta. In fact, for past 10 years, the water shortages and strong salinity intrusions in VMD were more frequent, starting earlier (about 20–35 days) and extending longer than before 1993 (MONRE, 2016; Ministry of agriculture and Rural Development (MARD), www.mard.gov.vn, 2016).

Further, the analysis of the hourly discharge data (Fig. 5) at Tan Chau, Chau Doc, Tan Thuan, and Can Tho (see their locations in Fig. 3) from 1996 to 2016 shows that the range of tidal oscillations of water discharge at the Vietnam-Cambodia border in the dry seasons has increased by 80% after 2009 by comparison with that before 2000; this same increase at Can Tho and Tan Thuan cross sections is 50%–60%. Thus, the sea influence of the hydrological regime of VMD is increasing or, in other words, the Mekong delta is truly shrinking. It is very important to stress that this increase of sea influence in the VMV is in phase with the construction of the dams in the basin; in particular, the largest changes in water discharge into VMD occurred during the massive water accumulation of the big Chinese dams (Xiaowan was fully filled in 2010 and Nuozhadu in 2012). This strongly suggests a direct cause and effect; indirect effects are likely to be the observed river and coastal erosion in the VMD due to the decrease of the sediment load to VMD, other factors being the sea level rise, infrastructural development, and sand mining. Further, for the months from October to January (i.e., in the dry season), when the tidal level is highest for the lower VMD with its mega-city of Ho Chi Minh City, the flood inundation from the sea will be exacerbated by increasing tides due to sea level rise, the sea level rise itself, delta subsidence, urbanization, infrastructures, and the wind surge (Nhan, 2016a, 2016b; Hoa et al., 2007). All these processes put in question the substainable development of VMD.

4.3 Impact on Sediment Resources

4.3.1 River Sediment Load to the Mekong Delta at Kratie

Pre-damming, the Lancang River basin provided about 50% of the sediment load from the entire Mekong River basin (MRC, 2016). The consequences of dams have been estimated in many studies (Bravard et al., 2013; Fan et al., 2015; Grumbine et al., 2012; Ilse and Kim, 2012; International Rivers, 2013; Kondolf et al., 2014, Kuenzer et al., 2012, Li et al., 2017; Nguyen et al., 2015; Kummu et al., 2010); these studies estimated more than half of the sediment of the Upper Mekong River will be trapped when the entire cascade of eight Chinese dams is completed. The estimates of the sediment trapping efficiency when all the planned dams had built range from 51% to 69% (Kummu et al., 2010) to 96% (Kondolf et al., 2014). Pre-damming best estimates of suspended sediment load delivered to the Mekong delta at Kratie are about 160 MT/yr (Syvitski et al., 2009). Milliman and Farnsworth (2011) modified it to 110 MT/yr when including the impact of dams in the basin, while other authors estimated post-dam sediment discharge at Kratie in the range 61–145 MT/yr (Koehnken, 2012; Liu et al., 2017; Lu and Siew, 2006; Wang et al., 2011).

The actual data of monthly suspended sediments concentration (SSC) at Kratie during some typical years (Fig. 6, source: MRC (Mekong River Commission), 2015) showed a large seasonal SSC variability and a huge decrease of SSC in recent years (e.g., 2012) compared to that in 1996 (when there was no large dam in UMB). The small SSC values occurred just as the Nuozhadu dam, the largest dam in UMB, was completed. It clearly suggests the large influence of the Chinese dams on SSC flowing to the VMD. In fact, because of forest clearing and increasing land use and the resulting soil erosion, one would have expected SSC to increase; in fact, SSC decreased, an observation clearly pointing to the Chinese dams as the cause.

4.3.2 River Sediment Load to the VMD

The effect of the Chinese dams on sediment transport in LMB remains poorly studied (Lu and Siew, 2006; Brunier et al., 2014; Wang et al., 2011). However, Vietnam has made serious efforts to study that process in the VMD by monitoring SSC at Tan Chau and Chau Doc stations (Fig. 6; Wang et al., 2011). It is very difficult to quantify the degree that sediment discharge to the VMD has been affected by dams in UMB. Nowacki et al. (2015), working with data for 2 small flood years, suggested that the total suspended loads may be as low as 40 MT/yr. The total sand export from the delta is ~5%–20% of total sediment transport (McLachlan et al., 2017; Li et al., 2017; Ogston et al., 2017).

The measured SSC data for 1987–2002 and for 2008–2017 at key cross sections in VMD are shown in Fig. 6B and C. These data show that the SSC in VMD averaged 64 mg/L from 2013 to 2017, which is 40% the SSC value (158 mg/L) from 1987 to 1993. The step-like decreases of SSC in VMD shown in Fig. 6 coincide with completion of the Chinese dams in UMB. Our estimates of sediment load into the VMD, using the field data and following the methodology of Nowacki et al. (2015), are shown in Table 1. Table 1 shows that the total sediment load to VMD after 2013 was 32% of the pre-dam sediment load. These estimates are in general agreement with Nowacki et al. (2015) results.

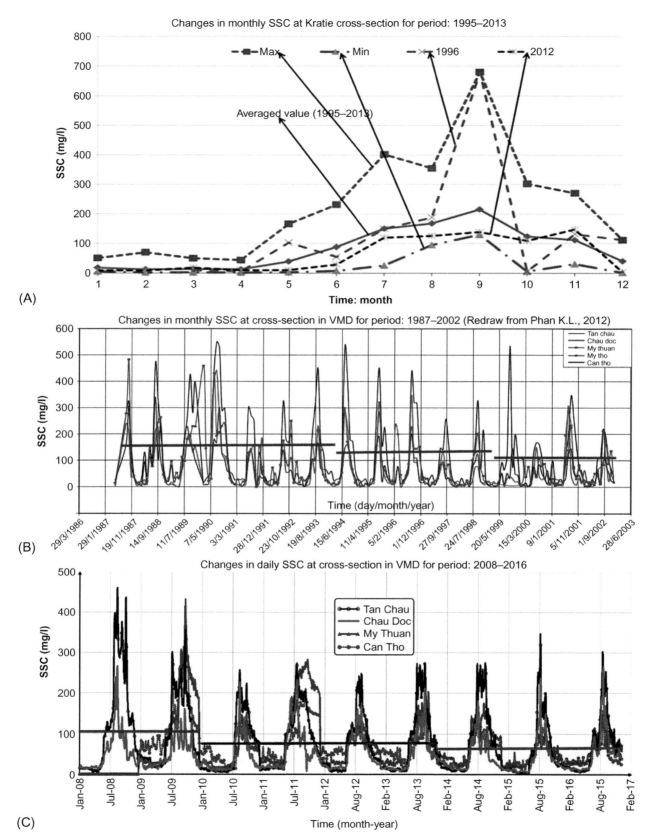

FIG. 6 The recent changes of the sedimentation regime in the VMD, showing a marked decrease of riverine sediment inflow in the VMD: (A) changes during the period 1995–2013 in the monthly suspended sediment concentration (SSC) at the Kratie cross-section, the starting place of the Mekong River Delta (From MRC, 2016. Integrated water resources management-based basin development strategy 2016–2020 for the lower Mekong Basin. http:// www. mrcmekong.org/assets/Publications/strategies-workprog/MRC-BDP-strategy-complete-final-02.16.pdf.); (B) changes in daily SSC at the national hydrometric cross sections in the VMD for the period 1986–2002 (redrawn from Linh, 2012); and (C) changes in daily SSC at the national hydrometric cross sections in the VMD for the period 2008–2016 (From MONRE VN, 2017. *Variation of Water Discharge and Sediment Flux at Lower Mekong River.* National Centre for Hydro-Meteorology. Hanoi. 15p (in Vietnamese).).

TABLE 1 Sediment Loads into the VMD

Sediment Load Component (MT/yr)	Before 1993	2008–2009	2010–2012	After 2013
Suspended sediment load	133	62	42	39
Bed sand load (15% of suspended sediment load)	20	9	6	6
Sediment load by delta soil surface erosion	7	7	7	7
Total sediment load	160	78	55	52

In future, if all planned dams will be constructed, the total sediment load coming into the VMD may further reduce by 24%–66% from the present (2017) conditions (MONRE VN, 2016); in other words, the future total sediment load into the VMD will be ~16–40 MT/yr, that is, 10%–25% of the pre-dam sediment load of 160 MT/yr.

4.4 The Morphological Changes

4.4.1 The Morphological Changes Inside VMD

The severe decrease of sediment load coming into the VMD from 160 MT/yr to only 52 MT/yr by impacts of already constructed Chinese mainstream dams (and in the near future it will be only 16–40 MT after Laos's and Cambodia's dams are constructed) is one of main human's reasons of the VMD shrinking. In addition to this, the massive sand mining removes about 56–57 MT/yr from the river bottom in the whole Mekong delta (Bravard et al., 2013; Brunier et al., 2014). This human activity also plays a very important role in riverbed and riverbank erosion (Figs. 4, 7, and 8). Beside these two main reasons, the morphological changes of the river bottom and riverbank in VMD are also influenced by sea level rise, infrastructural development, environmental degradation, increase of sea impacts, subsidence, etc.

Such data over the past 20 years show an increase of riverbank erosion, and now there are near 400 locations of eroding riverbanks (MARD, 2015). Brunier et al. (2014) estimated the recent morphological changes in the Mekong and Bassac River channels by comparing riverbed topography data in 1998 and 2008; they found net cumulative losses of 200 Mm^3 of bed sand load (Fig. 7A, B, and D) and that over these 10 years the Bassac and Mekong riverbed were on the average 1.3 and 1.44 m deeper, respectively. They attributed these changes to sand mining in the channels and loss of Mekong River sediment load into VMD as a result of the Chinese dams. This sediment starvation is increasing the frequency of riverbank landslides and erosion of the delta (Fig. 8). This sediment starvation issue is exacerbated by compaction-based subsidence (Fig. 4). The highest subsidence rate (>2 cm/yr) occurs in the southwestern sector of the VMD, which essentially comprises easily compressible marsh mud (Erban et al., 2014). The MARD survey reported massive groundwater exploitation especially near the large cities of Can Tho and Ho Chi Minh City that accounted for human-made subsidence of >1.5 cm/yr. Further, the smaller floods as a result of the hydropower dams in the Mekong basin together with local issues in the VMD, such as the construction of dykes and weirs, also decrease the sediment overflow from the channels to the floodplain and thus increase the delta subsidence. For the lower VMD, this subsidence is generating many negative effects for sustainability development, such as increased inundation of urban areas and increasing the erosion rate of the foreshore.

4.4.2 Morphological Changes in the VMD Coastal Zone

The riverine sediment is exported seaward onto the shelf during the wet season where it settles; it is then re-suspended by the wind and the currents transport it to the Ca Mau coastal zone, and only part of it is moved back into the estuaries during the dry season (Wolanski et al., 1996, 1998; Wolanski and Nhan, 2005). The MARD survey data show that since 2005 the erosion rate of the VMD foreshore has increased significantly. Some researchers (Li et al., 2017; Liu et al., 2017) chose 2005 as the turning point of the evolution of the VMD from accretion to erosion, and this may be attributed to the changes in water dynamics and sediment starvation as a result of the Chinese dams; however, these dams are a major but probably not the sole cause of the shrinkage of the VMD.

The decrease of Mekong River sediment load to VMD due to the Chinese dams and several processes inside VMD help explain the recent decrease of near-surface SSC in coastal waters (Fig. 4) that shows a significant monotonic decrease of about −5%/yr between 2003 and 2012. While sediment starvation due to the Chinese dams is one explanation, there are also other factors such as the destruction of mangrove forests for constructing shrimp ponds and sea dykes, though the

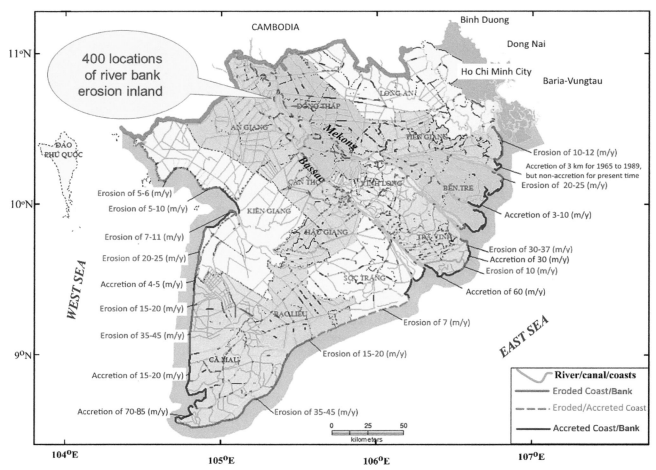

FIG. 7 Map of locations in the VMD showing their erosion/accretion rates (m/yr) of coastal shoreline and riverbank between 1973 and 2013, from the analysis of Landsat images (Nhan, 2016a).

cause-effect relationship is unclear. However, a coincidence or not, the biggest foreshore erosion occurred during the largest building of shrimp ponds in the Ca Mau province (Nhan, 2016b).

Based on results of overlapping Landsat and Spot satellite images from 1973 to 2013 (Besset et al., 2015, Brunier et al., 2014, Kuenzer et al., 2012, Liu et al., 2017, Nhan, 2016b, Li et al., 2017), Fig. 7 shows the average rate of coastline changes of VMD while detailed maps of coastline changes of four areas are shown in Fig. 8. There are 13 eroding segments totally ~400 km in length with an average erosion rate of 5–45 m/yr and seven accreting segments totaling ~300 km in length with an average accretion rate of 4–85 m/yr (Fig. 7). The Ca Mau coast has the largest rate of change (Fig. 8) because of its changing mangrove ecosystem: the mangrove growth is the greatest on the west coast (the Gulf of Thailand), and in contrast on the east coast the mangroves are rapidly disappearing due to both human activities and marine actions (Nhan, 2016b). By starving the delta of sediment, the Chinese dams are an important contributor but not the sole factor of delta erosion.

4.5 The Other Dam Impacts on the VMD

With the construction of the Lancang dam cascade, China controls the quantity of water released to downstream countries, and the lack of an official agreement between all these countries on flow volumes negatively impacts the downstream countries in a number of ways, namely: (1) the high dams are a large obstacle for fish migration (Fig. 2); (2) with less flushing the dissolved oxygen concentration will decrease at depth, water temperature may also decrease at depth, and this may decrease biological productivity and fish stocks, thus reducing fish production in countries downstream of China, and generating a loss of biodiversity and destroying the ecological balance; this directly affects the lives of people who rely on natural freshwater fishing; (3) flow regulation by these dams results in salinity intrusion in the VMD that affect several tens of thousands of ha and therefore impacts the life of hundreds of thousands of families that rely on agricultural cultivation in the coastal zones,

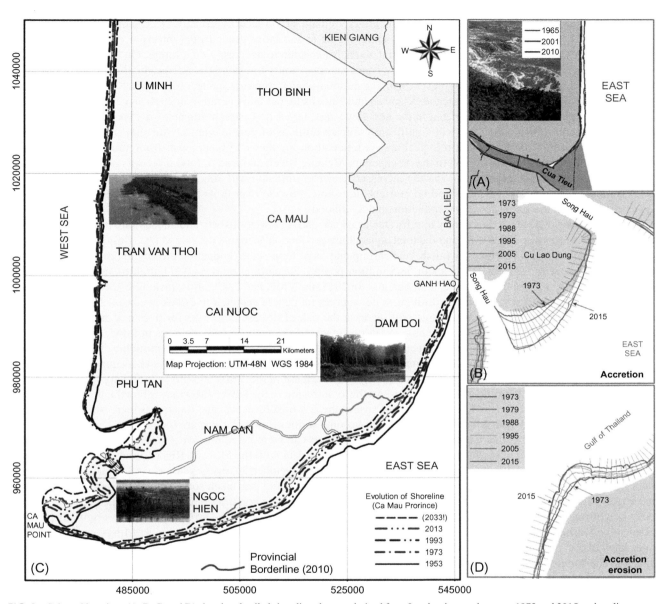

FIG. 8 Selected locations (A, B, C, and D) showing detailed shoreline changes derived from Landsat images between 1973 and 2015, a shoreline survey in 1953, and predictions for 2033. These locations are shown in Fig. 3. Maps for locations B and D are adapted from Liu et al. (2017). Maps for locations B and D are adapted from Nhan (2016b). *(Synthesis report on science and technology results: study mechanisms of formation and development of coastal accretion areas of Ca Mau Peninsula for generating the scientific and technological solutions for their sustainable socio-economic development. The state level project of code: KC08.13/11-15. Available from: http://csdlqg.vista.gov.vn/kq_chitiet_du.asp?id=KQT0112/14/2016~11:23:20~AM. Accessed 12 October 2017).*

and it also causes difficulties for irrigation and affects the schedule of irrigation and cropping in the VMD; (4) the dams reduce the silt deposit in the delta and exacerbate riverbank erosion, causing loss of land and properties of the people along the rivers; and (5) the loss of silt deposit to the land is not only the loss of nutrients for plants, but also undermines the natural accretion that is needed to annually renew the soils and prevent delta subsidence, and this negatively impacts agriculture production.

All this directly impacts Vietnam because the VMD is an important food basket for Vietnam, accounting for over 50% of food production and over 70% of fruit production. By 2015, the VMD had 2760 cultivation farms. The Chinese dams, as well as climate change to a smaller degree, significantly affect the natural water flows in both rainy and dry seasons. The number of years when there is low water flow at what should be the end of the dry season may quadruple compared with the present, and the number of years when there is low water flow at what should be the wet season may double compared with the present; this makes the saline intrusion start early and finish late, which in turn affects the production of both crops of winter-spring and summer-autumn and potentially affect Vietnam's food security strategy. Deep saline intrusion and tidal floods possibly affect the areas of fruit trees due to waterlogging or and saline groundwater intrusion. As a result of salinity intrusion changes, rice cultivation area in the coastal zones is expected to decrease in the future. On the positive side, the reduced frequency of large floods in the future may create an opportunity to extend the rice production in the flood-prone areas, as indeed it already happening at present with the autumn-winter crop increasing.

With the two biggest dams of the cascade, Xiaowan and Nuozhadu, put into operation in 2010 and 2012, and the middle Lancang cascade expected to be completed in the next few years, larger downstream impacts are expected for the hydrology, fisheries, and sedimentation that will significantly impact millions of people who rely directly on the river for their food and livelihoods. Altering the hydrological and sedimentation regimes and blocking fish migration will potentially reduce the quantity and diversity of fish in the downstream Mekong River, and lead to food insecurity and lost livelihoods. Furthermore, the reduction of sediment and the increased seawater intrusion will affect the highly productive agricultural and rice fields in the region, which depend on nutrients transported by the river in its sediment, and therefore create even bigger challenges in the future for food production and livelihoods.

In addition, the VMD leads the country in saltwater, brackish, and freshwater fishery and aquaculture production. The total aquaculture area isover 753,000 ha and the total aquaculture production accounts for 2.45 MT/yr. Besides the household-based fishery model, it also includes rice-fish, rice-shrimp and large farm-scale production models. In 2015, the VMD had 2891 aquaculture farms. The VMD has about 58.5 million poultry, 3.6 million pigs, 34,000 buffaloes, and 689,000 cows, accounting for 10%–17% of their national total quantity. In 2015, the VMD had 7347 farms, including 1560 breeding farms.

Increased salinity intrusion will affect drinking water sources for people and their livestock in coastal areas. In addition, the temperature increase with climate change will increase the risk of disease outbreaks for livestock and poultry. In the context of climate change and sea level rise along with upstream development impact, changes in flooded areas as well as saltwater intrusion may cause less direct impact on the existing aquaculture areas but unusual rainfalls, droughts, and high precipitation will make it more difficult to maintain a reasonable concentration of salinity in brackish aquaculture ponds. In the absence of flushing by large floods, the water quality may decrease in the VMD and this will directly impact fresh and brackish aquaculture. The dams resulting in increased salinity intrusion in the lower VMD may affect the brackish aquaculture areas, and the increasing use of groundwater aggravates the risk of subsidence, and this risk is increased by the lack of silt input to combat subsidence. All these problems will become serious if no solutions are found.

It has been demonstrated that China and Chinese dam builders can be more responsive and responsible when planning and operating dams. China has agreed to share more hydrological data with the Mekong River Commission by extending the hydrological data provision by 30 days, starting on June 1 until October 31, every year, as well as increasing the frequency of the data sharing. The developer of Lancang dams, Hydrolancang, has taken some local environmental and social concerns into consideration in several cases, but the impact on downstream countries is still largely neglected. The Gushui dam's height was reduced due to concerns over inundating a protected area in Tibet. The Guonian dam—originally planned between the Gushui and Wunonglong dams—was canceled because of its potential impacts on the Mingyong Glacier. The water level of the Wunonglong dam was reduced to avoid some impacts, which therefore led to the reduction of installed capacity. The Mengsong dam, originally planned as the last dam on the Lancang, was canceled due to concerns over its negative impact on fish migration. However, as long as the Chinese dam operators ignore the concerns of the countries downstream, large impacts from the Lancang dams are unavoidable in the VMD and mitigation measures need to be implemented even though their effectiveness in practice is unknown.

5 THE CONCEPTUAL SOLUTIONS

5.1 Constraints and Approaches

The principal strategies for generating proposed solutions to protect the VMD and enable sustainable growth are based on the criteria and the priority order as follows:

1. Nonengineering solutions have higher priority than engineering solutions;
2. The solutions to problems caused by human activities have higher priority than those caused by natural estuarine evolution;
3. The solutions to emerging new problems have higher priority than those to old, long-term causes;
4. The solutions for community interests and national sovereignty have higher priority than those for economic and commercial interests.
5. Where engineering solutions are needed, soft engineering solutions (e.g., afforestation, use of environmentally friendly materials…) have priority higher than hard engineering solution (e.g., dykes, embankments …).

The VMD has been shrinking for the past two decades, and our data demonstrate that one of most important causes for this shrinking is the impact of the Chinese operational hydropower dams in the UMB. Both the scientific and the management communities predict that the rate of VMD shrinking will increase markedly in the future as a result of increasing human activities in the whole international Mekong River basin, the main culprits being the completed and planned hydropower dams in China, Laos, and Cambodia. For generating practical solutions that may decrease or mitigate the negative impacts of these dams on the VMD, the proposed remediation measures have to solve fundamental problems related with three factors: human activities, international relations, and the management in planning, operating, and constructing the mainstream hydropower dams. These factors need to be regulated by socioeconomic, political, and scientific people. In practice, when possible nonengineering solutions may be feasible and practical. However, some places in the VMD have critical and emerging crises induced by erosion of riverbank, of foreshore, of mangrove belt, and of sea dykes, together with increasing salinity intrusion and severe river low flow events; for such critical issues, engineering solutions are needed too. Among many types of solutions to problems in the VMD caused by different causes, this chapter focuses on the conceptual solutions to minimize negative, direct, and indirect impacts from hydropower dams.

5.2 The Nonengineering Solutions

The need for effective measures is detailed in the official statements by the Vietnamese Government on these impacts. For example, in his speech at the summit meeting of Mekong country leaders in 2018 in Ho Chi Minh City the Vietnamese PM noted that the Mekong Basin faces major challenges, as the Mekong's water resources are being depleted both in quantity and quality, the amount of sediment is being depleted, and by disturbing the nutrient budget the ecosystem and the environment are seriously degraded. These negative signs are more pronounced and exacerbated in the downstream countries of the Mekong Basin, especially in the VMD where people are frequently exposed to prolonged and severe river low flow events, saline intrusion, riverbank and coastal erosions, and subsidence… all these threaten the livelihoods of more than 18 million people. The PM asked the MRC, the regional coordinator, to focus on fair, rational, and sustainable use of Mekong water resources and related resources.

This fair use of resources is the best way to decrease the negative impact of human activities at both the international and the national scales. The growing vulnerability of resources at the basin scales has significant political, economic, and environmental consequences for many of the world's deltas, the VMD being one of them particularly affected, and this calls for strong coordinated international efforts in terms of research and policy geared toward maintaining or restoring the sustainability of deltas.

The need for a solution is urgent and calls for improvements of the regional Mekong Agreement (signed in 1995 by four countries: Laos, Cambodia, Thailand, and Vietnam) for responding to the recent rapid changes in the Mekong River basin including: (1) it has to include all the countries in the whole basin; (2) the recent situation in the Mekong basin has changed dramatically compared to before 1995, especially with regard to problems of use of water, sediment, fisheries, and other resources; (3) there are severe, negative developments in the Mekong delta induced by hydropower dams in the UMB but the impact of these dams has not been controlled by any international agreement; and (4) the Mekong Agreement is missing several key points about political, social, economic, and environmental issues as well as about processes for response and compensation for losses induced by the illegal use of Mekong River resources. The revised Mekong Agreement needs to define methods to decrease the negative impact of human activities in the whole basin and long-term solutions for sustainability. Therefore, it is important to enhance the monitoring and coordinating role of the MRC in the implementation of commitments by the member states, and its role in developing rational plans in harmony with the Mekong River resources planned by the member countries and to propose joint projects on the sustainable management and use of the Mekong resources.

In the short term, it is urgent to speedily and in real time share research information and databases in the whole Mekong Basin, especially data from hydrological monitoring, forecasting, and warning systems, as well as data about the operation and construction of all dams and land use in the whole basin. Sharing these databases is essential for generating rational and effective measures to minimize negative impacts on VMD and to predict correctly the hydrological processes in VMD; this is vital because the Mekong River hydrology is becoming non-natural mainly as a result of flow regulation by mainstream hydropower dams. This information is urgently needed to enable long-term solutions for the VMD sustainable development. This includes developing an improved policy for food and agricultural production, environmentally friendly use of water, enhancing the development of renewable energy (e.g., wind and solar power in the VMD), mitigating and adapting to the increasing environmental degradation caused by the hydropower dams in the Mekong River basin and by the increasing sea impact, land subsidence, riverbed sand mining, the degradation of the protecting mangrove belt, the SLR, the CC, the infrastructure measures, and the evolution of the VMD and its coastal zone. All these considerations must be integrated in order to sustainably develop the VMD, an important goal in view of its critical geopolitical location in Asia.

All this is complex and this complexity is exacerbated in the international context of globalization and competition, security and many uncertainties, and within the national context of the intense use of resources by millions of people, the low level of labor productivity and education in VMD, and the priority given to practical polices to respond to several problems including increasingly severe river low flow events and salinity intrusion, increasing inundation in the lower VMD, over-exploitation of groundwater, riverbed sand mining, degradation of the protecting mangrove belt, and erosion of riverbanks and foreshore. Thus, in practice, for the VMD the new political social-economic policies yet to be developed must recognize the new man-made hydrological regime and must become practical tools to regulate human activities so that the various interests are balanced and harmonized between all social communities (including the various communities, associations, cooperative farms, farmer households, businesses, etc.), in balance with natural resources with an emphasis on fresh, brackish, and saline waters. Water, energy, health, agriculture, and biodiversity are essential for sustainable human livelihoods, and they are closely related to each other. Economic growth, environmental protection, equity, and social progress are the three pillars of sustainable growth. For all that to be possible in the VMD, it is essential that all projects throughout the whole Mekong Basin, including all hydropower dams in China, Laos, and Cambodia, must be carefully prepared and reviewed objectively, as well as scientifically subjected to scrutiny by the MRC, in order to assure "no regrets" investment and no conflicts between states.

For the VMD, there is a priority need to establish a comprehensive research project on coastal stabilization focusing on: (1) studying and forecasting of coastal evolution for the whole VMD (present and long-term, with particular attention to impacts of upstream developments, sea level rise, and subsidence) as a basis for developing a practical solution to protect the coast; (2) generating solutions suitable for stabilizing different coastal regions (estuaries, East Sea coast, and Gulf of Thailand coast); and (3) implementing measures to protect seriously eroded or seriously eroding coastal areas with high priority for national defense, economy, and population).

In terms of the VMD hydrology and hydrodynamics, it is important to urgently develop a coastal monitoring network focused on monitoring the evolution of coasts and estuaries, hydrological and oceanographic factors (waves, water levels, coastal sediment dynamics, currents, salinity, Mekong River water, and sediment inflows in the VMD...), subsidence, construction works (breakwaters and the increased rate of breakage), the sustainable management of mangrove forests, and also focused on forecasting and early warning. In terms of sustainable management of the coastal zone, this research program must aim at providing solutions for the sustainable management of the coastal zone including: (1) providing practical, useful tools for managing infrastructure and mangroves, and the use of the land, dunes, and mudflat; (2) encouraging community participation in infrastructure management (embankments, marine walls, breakwaters, dykes, soft mud walls); (3) mobilizing the community for infrastructure management and providing remuneration for dyke management by local people; (4) granting rights to local people to sustainably use and manage mangrove forests; (5) ensuring forest protection for all economic sectors, communities, and households; and (6) mobilizing people to restore mangroves (through land lease, capital contribution using coastal land for afforestation, combined with aquaculture and ecotourism).

The solution for coastal land use planning includes: (1) integrating mangrove activities into sectoral policies and plans to exploit mangroves and protect national interests on a sustainable basis; (2) strengthening integrated management among sectors, to effectively manage all uses and development of coastal areas; (3) developing practical experimental models of sustainable management of infrastructure (dykes, embankments to reduce wave energy, soft embankments,...) suitable for different foreshores; and (4) developing practical experimental models of mangrove forest management in combination with livelihood diversification.

Special nonengineering solutions related to the operation of hydropower dams are: (1) in view of the clear adverse effects of these dams on the two main rice-crops in the VMD together with sediment loss, it is necessary to study seasonal alternatives for the water reduction in early wet season and early dry season months; (2) raising community awareness to actively adapt to impacts of these dams in order that people become aware of water environment protection, the proper use of fertilizers and pesticides, and the collection of agricultural waste; (3) improving the adaptive capacity for people in the VMD and providing timely water information to prepare for the future cases that may happen in the Delta, so people can accumulate knowledge, experiences, and experiments to cope with specific situations; (4) develop institutions, policies, appropriate land use planning, and livelihoods solutions to minimize impacting vulnerable groups (poor households, households who have less productive land,...) and at the same time to strengthen the capacity of management by developing operational procedures for building information systems; (5) developing and implementing social solutions to support households through vocational training especially for poor households, training in transferring technical advances, etc., as well as providing clean water and environmental sanitation in rural areas; (6) informing the population about the developing impacts of hydropower dams and their future risks to the VMD; (7) strengthening international cooperation, developing guidelines for operation and regulation of hydropower reservoirs, and operational monitoring at the basin scale with all countries involved; and (8) as the total, final impacts for the VMD are unknown, we need more time for detailed research, survey, and evaluation of possible impacts before any new mainstream hydropower dams can be built on the lower Mekong mainstream.

5.3 Some Engineering Solutions Inland of the VMD

The MARD had proposed some engineering solutions inland of the VMD to address some impacts from the hydropower dams, namely by: (1) considering the needs and the priorities in building flood-control structures to mitigate the increased threats of flooding from the sea in the lower low VMD during low flow events generated by the dams in the UMB; (2) considering the priorities in building salinity control sluices in the Mekong River and the Bassac River in response to saltwater intrusion during low flow events generated by the dams in the UMB; (3) replacing manual sluice gates by readily operational tidal and salinity control sluices, to actively open and close them as needed to facilitate irrigation, water trapping, or drainage, and protection of water quality in the beneficiary areas of the hydraulic systems; (4) linking small-scale hydraulic systems to larger systems to ensure proactive water sources during saline intrusion periods that last longer than in the past; (5) installing medium and small pumping stations for the coastal areas to meet the water requirements for production, irrigation, water supply, and freshwater pumping (e.g., at low ebb tide) when the saline intrusion event intensifies in the future; (6) developing inland hydraulic systems to irrigate large fields in combination with small and medium pump stations to be active in production, irrigation efficiency, control of diseases and to improve the rice quality, contributing to building the VMD rice brand and restructuring the agricultural sector; (7) establishing and building a monitoring system for water, salinity, and turbidity levels in the VMD hydraulic system in order to enhance long-term as well as real-time forecasting to guide the operation of the hydraulic structures; (8) combating riverbank erosion, especially in more important social-economic areas, to protect regions, roads, and dykes; and (9) planning of sea dykes and river dykes must take into account the increasing erosion and where possible dykes should be protected by ecological solutions like mangrove forests instead of concrete.

Generating effective engineering solutions for protecting the foreshore, the mangrove belt, and the sea dykes along the VMD coastal zone is a very difficult problem. Indeed, many experimental models failed (Nhan, 2016b). An analysis of the situation along the VMD coastal zone shows that (1) the mangrove belt is crucially important for protecting the coast, and (2) the sea waves and the range of water level oscillations near the coast are the main cause of erosion of the foreshore, dykes, and the mangrove belt, as the currents are small due to shallow waters (Fig. 5). Thus, for reducing the erosion of the foreshore, mangrove belt, and sea dykes, the engineering measures must reduce the direct impact of sea waves, especially at high tide during storms. Where the protective mangrove belt is eroded, new mangroves must be replanted; where these new mangroves will be promptly eroded again, then underwater wave breakers are needed to protecting the newly replanted mangrove.

Based on these facts, we recommend the following three solutions, sketched in Fig. 9, to protect and stabilize the shoreline. The first solution is for a non-eroding shoreline with a wave height less than 0.6 m and a tidal range less than 0.8 m. Such areas include the east foreshore of the Bassac Estuary and the west coast of Ca Mau Cape (Fig. 8). This solution consists of keeping a natural or planted mangrove belt at least 300 m wide with the only engineering structure being a dyke located inland from Fig. 9A. This sea dyke must be constructed by homogeneous local soil material. The second solution is for a non-severely eroding shoreline with wave height less than 1 m and a tidal range less than 2 m. Such areas include the west coast of VMD (Figs. 7 and 8). This solution consists of replanting a mangrove belt at least 300 m wide and building an underwater wave breaker using wood, bamboo, or concrete with wood or sand-filled tubes; this wave breaker is needed for protecting the new planted mangroves (Fig. 9B). A sea dyke must be built inland of the mangrove belt using homogeneous local soil material. The third solution is for a severely eroding shoreline with a wave height greater than 1 m and a tidal range greater than 2 m. Such areas include sections of the VMD east coast that is eroding at a rate equal to or greater than 10 m/y (Figs. 7 and 8). There are three options in this third solution: (1) if there is no mangrove belt and it cannot be restored, and if wave erosion is directly impacting the sea dyke, then a solid sea dyke or coastal road must be constructed (Fig. 9C); (2) for a foreshore with a severe erosion and large wave height (>2 m), then a solid sea dyke or coastal road must be constructed together with an underwater wave breaker (Fig. 9C); (3) if there is a mangrove belt or it can be restored, in the presence of severe erosion the proposed solution involves three components: (1) replanting a mangrove belt wider than 300 m, (2) building a hard sea dyke or coastal road, and (3) constructing the underwater wave breaker (Fig. 9C).

6 CONCLUSION

The VMD now is in an accelerated shrinking phase with its ecosystem and environment seriously degraded mostly by human activities. The data demonstrate that a major cause of this is the impact of the Chinese operational hydropower dams

Engineering solution for non-eroded coast or wave height < 0.5 m and tidal range < 0.8 m: Keep natural or planted protecting mangrove belt wide > 300 m

(A)

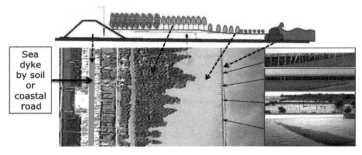

Engineering solution for non-severely eroded coast and wave height < 1 m, tidal range < 2 m: Keep natural or planted protecting mangrove belt wide > 300 m, underwater wave breaking by wood or bamboo or concrete with wood or sand-filled tubes...

(B)

Engineering solution for severely eroded coast and wave height > 1 m, tidal range > 2 m may be 1 of 3 following measures:
1. Build embankment for sea dyke/coastal road;
2. Build embankment for sea dyke/coastal road, underwater wave breaking wall;
3. Keep natural or planted protecting mangrove belt wide >300 m, embankment of sea dyke/coastal road and, underwater wave breaking wall.

(C)

FIG. 9   Three proposed conceptual engineering solutions for protection of the VMD coast for different threat levels.

in the UMB. The VMD shrinking rate will increase markedly in the future as a result of human activities in the whole (i.e., at the transboundary scale) Mekong River basin, especially from the cumulative effect of all the completed and planned hydropower dams in China, Laos, and Cambodia. These negative changes have been exacerbated by other factors such as infrastructural extension, riverbed mining, delta subsidence, climate change and sea level rise, degradation of coastal mangrove belt, and gaps in governance in the whole Mekong Basin including in the VMD. The VMD will never come back to the pre-damming era. However, in view of the severity of the impact in Vietnam, it is necessary to generate optimal, rational, and needed practical solutions to decrease the negative impacts of these dams for the VMD. Nonengineering solutions must be prioritized for practical purposes. A healthy and resilient mangrove belt is a most important condition for stabilizing the VMD foreshore; the power of sea waves and the tidal range are the main factors caused erosion of the foreshore, sea dykes, and coastal mangrove belt and for destabilizing the VMD. Thus, when possible, the first priority is to find measures to reduce the eroding impact of the waves in the presence of a high tidal range using both eco-friendly and engineering solutions. Next comes a number of issues involving remediation measures to address severe low flow events and increased salinity intrusion as a result of the hydropower dams in China, as well as eroding riverbanks, and finally water and socio-economic-political governance and development issues.

While this chapter focuses on remedial solutions that may decrease, but not eliminate, the negative impacts of the dams for the VMD, a long-term solution requires China, and in the future also Cambodia and Laos, to modify the use of their dams to accommodate the concerns of Vietnam.

Acknowledgments The authors express kindest thanks to Prof. Eric Wolanski for his invitation to write this chapter, and for his help in editing the manuscript. Special thanks are also due to Dr. To Quang Toan at the Vietnam Academy of Water Resources for sharing his valuable database.

REFERENCES

Allison, M.A., Nittrouer, C.A., Ogston, A.S., Mullarney, J.C., Nguyen, T.T., 2017. Sedimentation and survival of the VMD: a case study of decreased sediment supply and accelerating rates of relative sea level rise. Oceanography 30 (3), 98–109.

Anthony, E., Dussouillez, P., Goichot, M., Brunier, G., Dolique, F., Nguyen, V., Loisel, H., Mangin, A., Vantrepotte, V. (2013). Erosion of the Mekong delta: the role of human activities. Abstract id. EP34B-08, AGU Fall Meeting, San Francisco.

Besset, M., Brunier, G., Anthony, E. J. (2015). Recent morphodynamic evolution of the coastline of Mekong delta: towards an increased vulnerability. Geophys. Res. Abstracts Vol. 17, EGU2015-5427-1, EGU General Assembly 2015, Vienna.

Bravard, J.P., Goichot, M., Gaillot, S., 2013. Geography of sand and gravel mining in the Lower Mekong River. First survey and impact assessment. EchoGéo. http://echogeo.revues.org/13659.

Brunier, G., Anthony, E.J., Goichot, M., Provansal, M., Dussouillez, P., 2014. Recent morphological changes in the Mekong and Bassac river channels, Mekong delta: the marked impact of river-bed mining and implications for delta destabilisation. Geomorphology 224, 177–191.

Erban, L.E., Gorelick, S.M., Zebker, H.A., 2014. Groundwater extraction, land subsidence, and sea-level rise in the VMD, Vietnam. Environ. Res. Lett. 9, 084010. (6 pp).

Fan, H., He, D., Wang, H., 2015. Environmental consequences of damming the mainstream Lancang-Mekong River: a review. Earth Sci. Rev. 146, 77–91.

Grumbine, R.E., Dore, J., Xu, J., 2012. Mekong hydropower: drivers of change and governance challenges. Front. Ecol. Environ. 10, 91–98.

Hoa, L.T.V., Nhan, N.H., Wolanski, E., Cong, T.T., Shigeko, H., 2007. The combined impact on flooding in Vietnam's Mekong delta of local man-made structures, sea level rise, and dams upstream in the river catchment. Estuar. Coast. Shelf Sci. 71, 110–116.

Hoa, L.T.V., Shigeko, H., Nhan, N.H., Cong, T.T., 2008. Infrastructure effects on floods in the Mekong delta in Vietnam. Hydrol. Process. 22 (3), 1359–1372.

Ilse, P., Kim, G., 2012. The impacts of dams on the fisheries of the Mekong. Available from: https://cgspace.cgiar.org/ Accessed 12 October 2017.

International Rivers, 2013. Lancang River dams: threatening the flow of the lower Mekong. Available from: https://www.internationalrivers.org/sites/default/files/attached-files/ir_lacang_dams_2013_5.pdf. (Accessed December 12, 2017).

Koehnken, L., 2012. Discharge and Sediment Monitoring Program Review, Recommendations and Data Analysis: Part 2—Data Analysis of Preliminary Results. Information and Knowledge Management Programme (IKMP), Mekong River Commission, Phnom Penh. 53 pp.

Kondolf, G.M., Rubin, Z.K., Minear, J.T., 2014. Dams on the Mekong: cumulative sediment starvation. Water Resour. Res. 50, 5158–5169.

Kuenzer, C., Campbell, I., Roch, M., Leinenkugel, P., Tuan, V.Q., Dech, S., 2012. Understanding the impact of hydropower developments in the context of upstream–downstream relations in the Mekong river basin. Sustain. Sci. 8, 565–584.

Kummu, M., Lu, X.X., Wang, J.J., Varis, O., 2010. Basin-wide sediment trapping efficiency of emerging reservoirs along the Mekong. Geomorphology 119, 181–197.

Li, X., Liu, J.P., Saito, Y., Nguyen, V.L., 2017. Recent evolution of the Mekong Delta and the impacts of dams. Earth Sci. Rev. 175, 1–17.

Liu, J.P., DeMaster, D.J., Nguyen, T.T., Saito, Y., Nguyen, V.L., Ta, T.K.O., Li, X., 2017. Stratigraphic formation of the Mekong delta and its recent foreshore changes. Oceanography 30, 72–83.

Lu, X.X., Siew, R.Y., 2006. Water discharge and sediment flux changes over the past decades in the Lower Mekong River: possible impact of the Chinese dams. Hydrol. Earth Syst. Sci. 10, 181–195.

McLachlan, R.L., Ogston, A.S., Allison, M.A., 2017. Implications of tidally varying bed stress and intermittent estuarine stratification on fine-sediment dynamics through the Mekong's tidal river to estuarine reach. Cont. Shelf Res. 147, 27–37.

Meade, R.H., 1996. River-sediment input to major deltas. In: Milliman, J.D., Haq, B.U. (Eds.), Sea-Level Rise and Coastal Subsidence: Causes, Consequences, and Strategies. Kluwer Academic Publishers, Dordrecht, pp. 63–85.

Milliman, J.D., Farnsworth, K.L., 2011. River Discharge to the Coastal Ocean: A Global Synthesis. Cambridge University Press, Cambridge. ISBN: 9780521879873392. https://doi.org/10.1017/CBO9780511781247.

Milliman, J.D., Syvitski, J.P.M., 1992. Geomorphic/tectonic control of sediment discharge to the ocean: the importance of small mountainous rivers. J. Geol. 100 (5), 525–544.

Minderhoud, P.S.J., Erkens, G., Pham, V.H., Bui, V.T., Erban, L., Kooi, H., Stouthamer, E., 2017. Impacts of 25 years of groundwater extraction on subsidence in the Mekong delta, Vietnam. Environ. Res. Lett. 12 (6), 064006.

MONRE VN, 2016. Climate change and sea level rise scenarios for Vietnam. Available from: https://www.preventionweb.net/files/11348_ClimateChangeSeaLevelScenariosforVi.pdf. Accessed 12 October 2017.

MRC, 2009. In: Adaptation to climate change in the countries of the Lower Mekong Basin: regional synthesis report. MRC Technical Paper No. 24. Vientiane, Laos.

MRC, 2010a. State of the Basin report 2010. Available from: http://www.mrcmekong.org/assets/Publications/basinreports/MRC-SOB-report-2010full-report.pdf.

MRC, (2010b). Strategic environmental assessment of hydropower on the Mekong mainstream. Final report (2010). Available from: http://www.mrcmekong.org/assets/Publications/Consultations/SEA-Hydropower/SEA-FR-summary-13oct.pdf. Accessed 20 December 2017.

MRC, (2015). Study on the impacts of mainstream hydropower on the Mekong River. Final Report (2016). Available from: https://www.scientists4mekong.com/wp-content/uploads/2016/04/VMDs-final-project-report-18jan16.pdf. Accessed 12 December 2017.

MRC, 2016. Integrated water resources management-based basin development strategy 2016–2020 for the lower Mekong Basin. http://www.mrcmekong.org/assets/Publications/strategies-workprog/MRC-BDP-strategy-complete-final-02.16.pdf.

Nguyen, L.V., Ta, T.K.O., Tateishi, M., 2000. Late Holocene depositional environments and coastal evolution of the Mekong delta, Vietnam. Southern J. Asian Earth Sci. 18, 427–439.

Nguyen, M.V., Nguyen, D., Hung, N.N., Kummu, M., Merz, B., Apel, H., 2015. Future sediment dynamics in the Mekong Delta floodplains: impacts of hydropower development, climate change and sea level rise. Glob. Planet. Chang. 127, 22–23.

Nhan, N.H., 2016a. Tidal regime deformation by sea level rise along the coast of the VMD. Estuar. Coast. Shelf Sci. 183, 382–391. https://doi.org/10.1016/j.ecss.2016.07.004.

Nhan, N.H., 2016b. Synthesis report on science and technology results: study mechanisms of formation and development of coastal accretion areas of Ca Mau Peninsula for generating the scientific and technological solutions for their sustainable socio-economic development. The state level project of code: KC08.13/11-15. Available from: http://csdlqg.vista.gov.vn/kq_chitiet_du.asp?id=KQT0112/14/2016~11:23:20~AM. (Accessed October 12, 2017).

Nowacki, D.J., Ogston, A.S., Nittrouer, C.A., Fricke, A.T., Van, P.D.T., 2015. Sediment dynamics in the lower Mekong River: transition from tidal river to estuary. J. Geophys. Res C Ocean 120, 6363–6383.

Ogston, A.S., Mead, A., McLachlan, R.L., Nowacki, D.J., Stephens, J.D., 2017. How tidal processes impact the transfer of sediment from source to sink: Mekong River collaborative studies. Oceanography 30, 22–33.

Phan, L.K., van Thiel de Vries, J.F., Stive, M.J.F., 2015. Coastal mangrove squeeze in the Mekong Delta. J. Coast. Res. 31, 233–243.

Piman, T., Lennaerts, T., Southalack, P., 2013. Assessment of hydrological changes in the lower Mekong Basin from basin-wide development scenarios. Hydrol. Process. 27, 2115–2125.

Rubin, Z.K., Kondolf, G.M., Carling, P.A., 2015. Anticipated geomorphic impacts from Mekong basin dam construction. Int. J. River Basin Manag. 13, 105–121.

Schmitt, R.J.P., Rubin, Z., Kondolf, G.M., 2017. Losing ground—scenarios of land loss as consequence of shifting sediment budgets in the Mekong Delta. Geomorphology 294 (Supplement C), 58–69.

Syvitski, J.P.M., Kettner, A.J., Overeem, I., Hutton, E.W.H., Hannon, T., Brakenridge, G.R., Day, J., Vörösmarty, C., Saito, Y., Giosan, L., Nicholls, R.J., 2009. Sinking deltas due to human activities. Nat. Geosci. 2, 681–686.

Ta, T.K.O., Nguyen, V.L., Tateishi, M., Kobayashi, I., Tanabe, S., Saito, Y., 2002. Holocene delta evolution and sediment discharge of the Mekong River, southern Vietnam. Quat. Sci. Rev. 21, 1807–1819.

Tamura, T., Horaguchi, K., Siato, Y., Nguyen, V.L., Tateishi, M., Ta, T.K.O., Nanayama, F., Watanabe, K., 2010. Monsoon-influenced variations in morphology and sediment of a mesotidal beach on the Mekong delta coast. Geomorphology 116, 11–23.

Toan T.Q. et al., 2016. Synthesis report on science and technology results: study for assessing the impacts of hydropower dam ladders on the Mekong downstream mainstream to waterflow, environment and socio-economics in the Mekong Delta and proposing mitigation measures. State level project of code: KC08.13/11-15 (In Vietnamese).

Unverricht, D., Szczuciński, W., Stattegger, K., Jagodziński, R., Le, X.T., Kwong, L.L.W., 2013. Modern sedimentation and morphology of the subaqueous Mekong Delta, southern Vietnam. Glob. Planet. Chang. 110, 223–235.

Walling, D.E., 2006. Human impact on land-ocean sediment transfer by the world's rivers. Geomorphology 79 (3–4), 192–216.

Wang, J.J., Lu, X.X., Kummu, M., 2011. Sediment load estimates and variations in the lower Mekong River. River Res. Appl. 27, 33–46.

WLE-Mekong, 2016. Dam maps. Available from: https://wle-mekong.cgiar.org/maps/.

Wolanski, E., Nhan, N.H., 2005. Oceanography of Mekong River rstuary. pp 113-115 in Chen Z., Saito Y., Goodbred, S.L. (eds.), Mega-Deltas of Asia—Geological Evolution and Human Impact. China Ocean Press, Beijing, 268 pp.

Wolanski, E., Nhan, N.H., Dao, L.T., Huan, N.N., 1996. Fine sediment dynamics in the Mekong river estuary. Estuar. Coast. Shelf Sci. 43, 565–582.

Wolanski, E., Nhan, N.H., Spagnol, S., 1998. Sediment dynamics during low flow conditions in the Mekong River estuary. Vietnam. J. Coast. Res. 14, 472–482.

Xue, Z., Liu, J.P., Ge, Q., 2011. Changes in hydrology and sediment delivery of the Mekong River in the last 50 years: connection to damming, monsoon, and ENSO. Earth Surf. Process. Landf. 36, 296–308.

Zarfl, C., Lumsdon, A., Berlekamp, J., Tydecks, L., Tockner, K., 2015. A global boom in hydropower dam construction. Aquat. Sci. 77, 161–170.

Section D

Enclosed, Semi-enclosed, and Open Coasts

Chapter 20

Baltic Sea: A Recovering Future From Decades of Eutrophication

Anna-Stiina Heiskanen[*], Erik Bonsdorff[†], Marko Joas[†]

[*]Finnish Environment Institute, Helsinki, Finland, [†]Åbo Akademi University, Turku, Finland

1 INTRODUCTION

The Baltic Sea is a sea of change. Since the last glaciation it has undergone several stages from a freshwater lake to a more marine status. Over the last 7000 years, it has gradually developed into its current morphology and hydrodynamics shaped by land lift, climate, and the activities of the growing human population (Andrén et al., 2000; Kotilainen et al., 2014). The specific geological, geomorphological, and hydrodynamical changes of the Baltic Sea have been summarized in Snoeijs-Leijonmalm and Andrén (2017).

The Baltic Sea is the world largest continental brackish water sea. The southern part belongs to the temperate/boreal zone, while the northern part is characterized by almost arctic conditions with ice cover lasting for several months each winter. The harsh conditions combined with low salinity, ranging from almost limnic conditions in the north to over 20 psu in the Danish straits shape the hydrodynamics, species composition, and organism communities that form the ecosystem of the Baltic Sea (Snoeijs-Leijonmalm and Andrén, 2017). The low salinity and large annual variations in temperature restrict the number species that are able to adapt to these conditions, thus leaving open niches for invasive nonnative species to establish viable populations in the entire sea (Bonsdorff, 2006; Ojaveer et al., 2010).

The Baltic Sea has been economically and culturally important for the people living in the coastal areas throughout the postglacial history of the sea, with human settlements following first the retreating ice sheet, and then the retreating coastline as a consequence of land uplift (isostatic rebound) (cf. Snoeijs-Leijonmalm and Andrén, 2017). The Baltic Sea and its coastal regions have provided waterways for exploration, transport and traffic, space and shelter for seafarers and settlements, fish and game for food, materials for energy, building, fertilizers for cultivation, and ornamental material for products such as amber jewelry.

Baltic Sea scenes with open sandy coasts and lagoons, limestone cliffs, rocky, archipelagos, river deltas, moraines, and other coastal environments have been the inspiration for wide variety of arts, music, and literature and have provided the natural and cultural background for recreation and tourism. The value of the sea and its coastline is immense for humans as well as for the biodiversity of the sea. It is only recently, however, that there have been attempts to evaluate the overall benefits of this richness for humans and in economic terms. Thus, its value—both for ecosystem services as well as natural resources—has been and still is very high for human well-being (e.g., Söderqvist et al., 2005; SEPA, 2008; Węsławski et al., 2017).

1.1 Centennial of Changes in the Baltic Sea

The first undisputable signs of pollution started to appear in the early 19th century, near to coastal cities discharging the untreated sewage to the coastal waters. There are reports of algal blooms already from the late 1800s (Finni et al., 2001). Problems with hygiene and odors led to the development of the first wastewater treatment facilities in the coastal cities (Laakkonen and Laurila, 2007), and thus important point sources of nutrient pollution began to be tackled. Local environmental (health) hazards—in cities—were also the bases for early environmental legislation. However, the formative moment for modern, still point-source oriented, environmental policies in the Nordic and other west-European countries emerged in the late 1960s and early 1970s (Joas, 1999).

The change of the Baltic Sea ecosystem has been most tangible since the early 20th century (Zillen et al., 2008; Gustafsson et al., 2012). The increase of human population, intensification of agriculture and forestry, and industrialization of the countries around the Baltic Sea has impacted the marine environment significantly. There are currently 85 million people living in the Baltic Sea catchment, shared by 12 countries, all heavily industrialized. One serious challenge for the management and governance of the Baltic Sea environment is the fact that this land-locked basin has nine countries bordering to the coast, and three more with indirect impacts through runoff from extensive agriculture and heavy industry (Fig. 1).

The socio-ecological system is characterized by differences in cultural, economic, geographical, and ecological features in the different subareas of the Baltic Sea. Those vary from the sparsely inhabited and mostly forested catchment areas in the Northern Baltic Sea, to the densely populated areas around the Baltic proper and in the Southern Baltic Sea, where the catchments are dominated by agricultural, industrial, and urban areas (see HELCOM, 2018, for the latest compilation of spatial distribution of population, land use, and nutrient inputs in the Baltic Sea watershed).

For decades the Baltic Sea has received municipal and industrial waste, containing a cocktail of organic material, toxic chemicals, and pollutants. Even up to the beginning of 21st century some large cities have been without proper wastewater treatment facilities. For instance, the wastewater treatment development in St Petersburg was completed with tertiary treatment in 2005, which has now resulted in improvements in the water quality of the eastern Gulf of Finland (Räike et al., 2016; HELCOM, 2015a), although significant efforts are still needed to tackle remaining municipal and industrial wastewaters from Russia (HELCOM, 2014a). Implementation of the Baltic Sea Action Plan (BSAP) in Russia is conducted through several channels, not the least with support from several diverse multilevel governance and funding institutions, such as transnational city organizations including the Union of the Baltic cities.

Impacts of pesticides and other hazardous chemicals were apparent since the 1970s as the number of sea eagles declined and reproduction of seals was disturbed (HELCOM, 2009a). Toxins were also documented to threaten human health as the levels of toxins in fish and even in human breastmilk exceeded allowed levels for human safety and health (Norén and Meironyté, 2000). Consequently, the Baltic Sea acquired the questionable reputation as the most polluted sea in the world (see Lehtonen et al., 2017; Reusch et al., 2018, for recent reviews of the Baltic Sea pollution history).

The intensifying agriculture and meat production since the Second World War led to the increase of fertilizer use (Gren et al., 2000). Signs of eutrophication increased gradually since the 1950s (Andersen et al., 2017), until the negative impacts of large scale nutrient overenrichment and subsequent eutrophication of the sea became apparent: toxic cyanobacterial blooms, visible even from satellites, regularly covered large sea areas during the vacation season in July and August (Kahru and Elmgren, 2014). Drifting filamentous algae suffocated shallow sandy bottoms and covered perennial algal and seagrass-habitats which normally have important spawning, nursery, and feeding functions for coastal fish and invertebrates (Snoeijs-Leijonmalm et al., 2017).

1.2 Changing Governance Structures

The broad acknowledgment of the environmental problems in the Baltic Sea resulted in increasing international awareness. This development followed a broader international environmental awakening in the 1960s. A more discursive international climate allowed new international conventions and organizations to take place for cooperation in the mid-1970s. The Helsinki Convention, leading to the HELCOM secretariat with a clear scientific role making policy recommendations for members, was signed in 1974 (Jetoo, 2018). An additional circumstance making ambitious environmental policies for the Baltic Sea possible has been the fact that all Nordic countries together with Germany clearly have taken the lead in implementing active and holistic environmental policies, thus together with a number of other forerunners moving the European Union (EU) toward similar actions (Andersen and Liefferink, 1997).

The key change in the governance structures during the last 25 years is the expansion of the EU in the Baltic Sea region. This means that instead of nine riparian and three inland states and their totally separate jurisdictions there is one leading international institution in the region, with clear and binding obligations in environmental governance. Only one coastal state, Russia, is outside the EU. The political change in the 1990s provided a window of opportunity for environmental governance in the Baltic Sea region (Table 1). The EU environmental legislation has introduced the principles of ecosystem-based management as a part of the key directives. EU environmental policies and framework laws mostly agree with most recommendations of the Baltic Marine Environment Protection Commission (Helsinki Commission, HELCOM).

The HELCOM-countries joined forces to tackle some persisting environmental problems through the development of the BSAP in 2007 (HELCOM, 2007, revised and updated in 2013, and continuously adapted to the current environmental needs) to take concrete actions moving toward good environmental and ecological status of the Baltic Sea by 2021. The BSAP sets the overall vision and defines ecological objectives for a healthy Baltic Sea (HELCOM, 2007). Based on empirical and experimental science, and on monitoring and modeling, basin- and country-wise nutrient reduction targets were

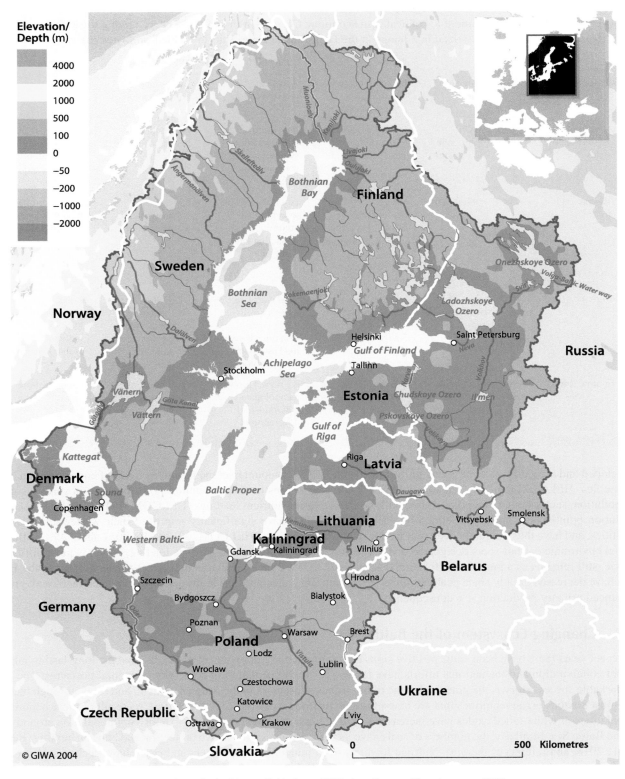

FIG. 1 Map of the Baltic Sea and its' drainage basin. Map available from: GIWA, http://www.grida.no/resources/8300.

TABLE 1 Large Scale Environmental, Political, and Economic Changes in the Baltic Sea Area During the 20th Century. Different Periods are Characterized by Change in the Ecological Regime, Shits in Political and Governance Regimes

Years	<1900	1900–1950	1950–1970	1970–1990	1990–Present
Regime	Preeutrophication	Gradual increase of loading, preeutrophication	Accelerated eutrophication	Large-scale eutrophication established	Remediation starts
Description ecosystem state changes	Baltic Sea in oligotrophic state, exogenic pressures prevail	Enclosed bays close to major urban areas eutrophied	Increasing eutrophication symptoms, anoxic areas increase	Nutrient loads peak, large-scale eutrophication symptoms,	Large areas of anoxic bottoms, expansion of cyanobacterial blooms
Ecosystem regime shifts	No documentation	Little documentation, but some reports indicate shift from clear to turbid waters in coastal bays	From oligrotrophic to large-scale eutrophication impacting whole Baltic Sea	Increase and collapse of cod; regime shift in fish communities	Internal processes maintain and feed eutrophication, some local improvements
Governance shifts	National governance structures	World wars I and II reorganization of national states, Soviet Union expands	Cold war; large differences in governance structures between the Eastern and Western blocks	HELCOM international collaboration start, national governanace structures	Break-up of the Eastern block, EU expands, regional, and global networking
Economic drivers	Agriculture, early industrialization	World wars I and II	Economic growth, technological, and industrial development	From national economics to opening markets	Globalization

developed and agreed by all contracting parties within HELCOM to counteract the negative effects of eutrophication of the Baltic Sea. HELCOM regularly releases pollution-load compilations based on monitoring the changes in various sources of pollution, such as industrial point-source pollution and riverine nutrient loadings (HELCOM, 2015a, 2018). Most of the pollution point-source hot-spots defined by HELCOM in 1992, have significantly improved their wastewater treatment facilities, and have thus gradually been removed from the "blame" lists (HELCOM, 2013d). By 2017 about 25% of the 162 initial environmental hot-spots designated and agreed upon in 1992 remained. Diffuse loading, which is most difficult to tackle, still remains as a major issue for remediation (HELCOM, 2015a), and although direct nutrient inputs to the Baltic Sea have decreased greatly since peaking in the 1980s, nutrient concentrations in the sea have remained high, with grossly enhanced primary production as a consequence.

1.3 Changing Ecosystem of the Baltic Sea

In recent years there have been some positive signs of environmental improvement indicating that the long-lasting international collaboration and the national efforts have led to a lasting remediation of some environmental pressures. The most remarkable success story is the gradual decline of hazardous substances in the water, sediment, and biota, through banning the most noxious organochlorines that are magnifying in the food webs (HELCOM, 2010a). As result of those actions, the populations of white-tailed eagles have increased from the brink of almost extinction to a thriving population around most of the Baltic Sea. Similarly, the numbers of seals have increased as a result of active protection and bans on hunting (Reusch et al., 2018, and references therein). Nutrient loadings from municipal wastewaters and rivers have declined (HELCOM, 2015a), and as a consequence nutrient concentrations in the Baltic Sea have decreased, although not as rapidly as nutrient inputs (HELCOM, 2014b). Nevertheless, the positive signs of decreasing nutrient loads have yet had little improvement in reducing eutrophication (Andersen et al., 2017; HELCOM, 2017a). Moreover, new pollutants (e.g., industrial chemicals, ingredients of personal care products, and pharmaceuticals) are emerging, and banned pollutants are still detected in the environment and in the biota (see Lehtonen et al., 2017, for a review).

The cumulative effects of different human pressures impair the resilience of the ecosystem, particularly in areas where a several human pressures coincide often having additive and synergistic effects (Korpinen et al., 2012). Concurrently, nonmanageable pressures caused by the global climate change (e.g., changes in temperature, salinity, ice-cover, and hydrodynamics) increase the stress and decrease the resilience of species and habitats in such multistressor-impacted areas (BACC, 2015).

Heavy pressure from both coastal and offshore fisheries has caused cascading top-down effects and changes in the fish communities, in turn triggering regime shifts in the ecosystem (Casini et al., 2008). Together with the secondary effects of eutrophication, such as in both deep water and coastal areas of the Baltic Sea suffering persistent and wide-spread hypoxia and anoxia (Carstensen et al., 2014; Conley et al., 2011), these have resulted in profound changes structure of the Baltic Sea food webs, and thus in the goods and services provided by the marine ecosystem for human consumption (Rocha et al., 2015; Reusch et al., 2018). Since the 1970s, the cod stock has declined due to intensive commercial fishing and the concurrent deterioration of spawning conditions which depend on favorable salinity and oxygen levels. Overfishing of cod has cascaded through the food web and led to the dominance by planktivorous sprat which was reflected in the decrease of zooplankton (Casini et al., 2008; Möllmann et al., 2009). Also individual cod are becoming smaller and productivity is decreasing due to the selective removal of large specimens by fishing gear. This enhances the change in the structure of the food web, as the large predatory fish are rare (Svedäng and Hornborg, 2017).

The uncontrolled mixture and combination of human-induced environmental pressures have resulted in regime shift of the Baltic Sea ecosystem, that is, induced shifts in the entire ecosystem from one set of semistable conditions to another, less desirable set of environmental conditions (Yletyinen et al., 2016). Thus it is increasingly difficult to restore the ecosystem to a more pristine status, and when setting environmental and ecological goals, the visions need to take into account that system-wide evolution and succession cannot be returned to a predefined state. Instead management efforts should be targeted to decrease the harmful effects of human pressures, and to improve the resilience of the ecosystem to withstand any further changes in external, nonmanageable stressors. This demands that the ecosystem holds viable populations of organisms at all critical levels, to ensure balanced bottom-up and top-down regulation. This balance has been compromised in the Baltic Sea, as both the primary production and the function of top predators have changed significantly over the past decades (Reusch et al., 2018).

Impaired ecosystems have negative effects not only to the structure and functioning of the ecosystem itself, but indeed on the development of several important economic sectors in the entire Baltic Sea region (Söderqvist and Hasselström, 2008; Ahtiainen and Öhman, 2014).

Globally, there is an increasing awareness and concern about ocean integrated ecosystem health, and in order to synthesize the available information, an ocean health index was launched (Halpern et al., 2015). However, global ocean mapping is based on a very coarse grid and is thus at best indicative rather than specific. Marine spatial planning is also still underdeveloped compared to similar activities on land. In order to get more detailed regional information, efforts have been made also for the Baltic Sea (HELCOM, 2010c; Korpinen et al., 2012), which shows that the importance of adequate marine spatial planning has become obvious, as there is often a conflict of interest between the use of various kinds of ecosystem services (Lester et al., 2013).

In order to further promote the idea of holistic analyses of the entire sea area, Reusch et al. (2018) compiled information for the Baltic Sea covering the long-term development and future projections for basic environmental parameters, anthropogenic stressors including nutrient loads, climatic drivers, oxygen conditions, shipping, pollutants, status of fish stocks, and ecological descriptors, and compared these to human decision systems and aspects of governance. This was also done in comparison with other coastal sea areas in order to illustrate how the long-term experiences (both successes and failures) from the Baltic Sea region could act as a time machine for other regions. In their analysis, Reusch et al. (2018) illustrate both fundamental structural and functional regime shifts of the entire ecosystem, and how negative trends can be halted or even reversed. They also show how international regional legislation and management can lead to gradual recovery regarding both drivers and the organisms of the sea, given that sufficient time is allowed for the process.

1.4 Holistic Framework for management of the Baltic Sea

A holistic approach for marine management is aided by identifying how different drivers of the society are connected to human activities and how the activities produce pressures that impact ecosystems. The so called DAPSI(W)R(M) framework (Elliott et al., 2017) allows us to identify links between **D**rivers, human **A**ctivities, **P**ressures, **S**tate change of the ecosystem, **I**mpacts on the human **W**elfare and finally various **R**esponses (policy, governance, and civil society) using **M**easures to implement the policies and legislation. The integrated DAPSI(W)R(M) scoping framework helps to identify, visualize, and address the linkages between different environmental problems and their drivers and pressures, and how those are and need to be tackled by society to build more holistic management strategies (Atkins et al., 2011; Cooper, 2012). Different drivers (e.g., need for food, transport, living space, etc.) and pressures caused by those that can be tackled at the scale of the system where the framework is being applied, are called endogenic managed pressures (EnMPs). The large-scale drivers, for example, climate change, hydrological changes, etc. that cause pressures such as temperature, salinity and stability anomalies, and which cannot be tackled in the scale of the system are exogenic unmanaged pressures (ExUPs) (Elliott, 2011). The temporal dimension can be used to identify how the drivers and pressures have changed in the past and how those are predicted to change in the future, and how those will impact the rest of the DAPSI(W)R(M) cycle. Concurrently there is also local, subregional, and regional spatial dimension, which can be visualized on maps such as single pressures or pooled as cumulative pressures and impacts (Korpinen et al., 2012).

Different environmental problems can be characterized by their own DAPSI(W)R(M) cycle (Fig. 2). These cycles are interlinked and nested in multiple ways, forming a three-dimensional network which is changing through time (Scharin et al., 2016). The different parts of the DAPSI(W)R(M) framework lead to interlinked and cascading networks (Elliott et al., 2017). One driver can cause changes in many human activities which turn cause several pressures. Different activities can also amplify same pressures. Various pressures result in many cumulative and synergistic effects on the state of the ecosystems. The state change can be linked to a number of ecosystem services to humans which can have consequences on human welfare [I(W); Fig. 2]. Therefore, the ecosystem services connect the changing state of the ecosystem (S) with human well-being through a range of ecosystem services that are necessary for regulation and maintenance of the marine systems or generate benefits for the society (Scharin et al., 2016).

Policy responses and legislation are often multifunctional so that those actually tackle several pressures caused by activities that those are designed to regulate. Such an ecosystem-based approach to management has gradually been applied to all major HELCOM recommendations and especially EU key regulations for the marine environment (Söderström and

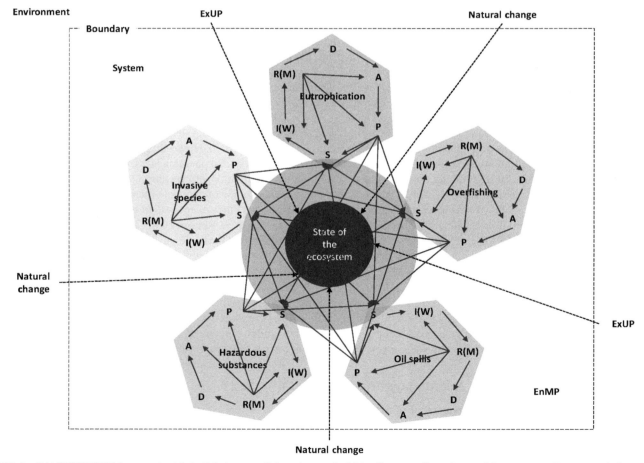

FIG. 2 D(A)PSI(W)R(M) framework with the links between **D**rivers, human **A**ctivities, **P**ressures, **S**tate change of the ecosystem, **I**mpacts on the human **W**elfare and finally various **R**esponses and **M**easures for the major environmental hazards of the Baltic Sea (Scharin et al., 2016). Each of the environmental issue has its own D(A)PSI(W)R(M) cycle, which are also interlinked. Endogenic manageable pressures (EnMP) are within the system to be governed, while the exogenic unmanageable pressures (ExUP) as well as the natural change cannot be influenced and are outside the system to be governed.

Kern, 2017). Although the landscape of policy responses seems to be complicated and impossible to implement in a coherent manner (Boyes and Elliott, 2014), the many national and later the EU regulations in the Baltic Sea regions and particularly how those have been implemented in the past, provide some positive examples with clear impacts on the environment and on human welfare (e.g., Reusch et al., 2018). Nevertheless, some "wicked problems" such as the slow reduction of eutrophication still persist, indicating that the plethora of regulations and policies does not guarantee effective environmental protection in such a large interconnected system, where a lack of scientific knowledge, institutional awareness, and political willingness have resulted in delay and inefficiency in the implementation of regulations (Varjopuro et al., 2014).

These delays can lead to a situation where the ecosystem change becomes irreversible or very difficult to reverse. There are examples of such regime shifts in several marine and coastal areas (Rocha et al., 2015). In the Baltic Sea such regime shifts (Casini et al., 2008; Möllmann et al., 2009) with consequences on human welfare (Blenckner et al., 2015) have been documented. Therefore, the challenge is to be able to recognize the approaching ecosystem tipping points that may cause the system to change to an alternative state with associated risks of repercussions on human welfare and economics. Documented cases and early warning indicators that were recognized to precede regime shifts can be helpful (Yletyinen et al., 2016). We also need to learn from the regulatory failures that did not manage to prevent catastrophic ecosystem changes, in order to identify needs for institutional development toward adaptive governance that is precautionary, agile, and builds resilience of the society and ecosystems.

Demand for food and food production are the major drivers for agriculture, which has long been the major source of diffuse nutrient loading to the Baltic Sea (HELCOM, 2015a). Similarly fisheries, which are driven by food demand, have resulted in major changes in the Baltic Sea food webs (Blenckner et al., 2015). Other major pressures and threats for the Baltic Sea ecosystems are caused by invasive species, marine microplastics and other emerging pollutants, and noise (HELCOM, 2017a). Moreover, some toxic pollutants have still higher concentrations than the maximum allowed in foodstuffs by the EU (dioxins and dioxin-like PCBs in fatty fish) (Elmgren et al., 2015; Scharin et al., 2016; HELCOM, 2017a). However, eutrophication has persisted for a long time (Table 1) and has not been successfully mitigated despite a multitude of responses and actions at national and international level. It is a "wicked problem" that is impacting all habitats and components of the Baltic Sea food web, and causing impairment of human welfare and economics (Ahtiainen et al., 2014). In this chapter, we focus on the long-term changes of the eutrophication in the Baltic Sea and how the different components of the eutrophication DAPSI(W)R(M) framework have changed. Finally, we discuss the future outlooks for mitigation of eutrophication and possibilities to strengthen the sustainable use of the Baltic Sea ecosystems.

BOX: Baltic Sea and the Major Long-term Changes

The Baltic Sea including the Kattegat (area of 393,000 km², volume of 21,631 km³) is a large nontidal brackish water body, characterized by marked climatological and environmental gradients, for example, a strong east-west and north-south salinity gradient, with the most saline water in the south (approximately surface water salinity 8–10 psu) and least saline in the northern bays (approximately 1–3 psu). The Baltic Sea is shallow, with approximately 30% of its area <25 m deep, average depth of 57 m, and a maximum depth of only 459 m. The Baltic Sea drainage basin is about 4 times larger than the area of the sea, with a human population of over 85 million within the drainage area. The total annual fresh water inflow from rivers is about 481 km³, and the residence time of the water mass is about 30–40 years. It is connected to the Kattegat, Skagerrak, and North Sea through the narrow and shallow Danish straits with a threshold-depth of just 17 m (see the review by Snoeijs-Leijonmalm and Andrén, 2017). Low temperatures and low salinity, as well as the harsh winters with ice in the north, impact the biological community species richness which declines with salinity and typically the Baltic Sea is characterized both with freshwater and marine species (cf. Snoeijs-Leijonmalm et al., 2017).

The central part of the Baltic Sea (Baltic Proper) is permanently stratified with a strong halocline which prevents vertical mixing of the water column and ventilation of oxygen-rich waters to the bottom. The deep water renewal and the oxygen levels of the deep areas in the Baltic Proper are influenced by periodic and irregular inflows of oxygen-rich and saline water from the North Sea (Leppäranta and Myrberg, 2009). The geological time series shows that the Holocene climate development in Northwest Europe influenced oxic-anoxic conditions in the deep areas Baltic Sea through the hydrology changes in the catchment, in addition to the inflow of North Sea more saline waters (Zillén et al., 2008). During the past more than 30 years the frequency of inflow events has declined from five to seven major inflows per decade to only one major inflow per decade (Mohrholz et al., 2015). The major inflows are often followed by a period of stagnation during which saline stratification decreases and oxygen deficiency develops in the bottom water. Concurrently anoxic conditions expanded over large areas in the central Baltic and were only temporally relieved by two major inflows in 1993 and 2003 (Fig. 3; Carstensen et al., 2014). The last strong inflow occurred in December 2014 bringing large amounts of saline and oxygenated water into the Baltic Sea (Mohrholz et al., 2015). However, the anthropogenic nutrient loading leading to eutrophication is a major driver increasing area and volume of hypoxic and anoxic deep waters (Carstensen et al., 2014; Kotilainen et al., 2014) and also coastal waters of the Baltic Sea (Conley et al., 2011).

The 20th century has been the warmest in the last 500 years and the annual mean sea-surface temperature has increased by up to 1°C per decade since 1990 (BACC, 2015). Also winters have become milder over the past 100 years: the maximum ice extent has decreased and the ice season has become shorter (BACC, 2015). Over the past century, Baltic Proper deep water temperatures have increased about 2°C and the oxygen depleted bottom water area has increased (Fig. 3) (Carstensen et al., 2014) through decreasing solubility of oxygen and increased oxygen consumption by organic matter.

The back-calculated (estimated) riverine nutrient loads to the Baltic Sea were low until 1900 (Gustafsson et al., 2012). The observed increase of the waterborne nutrient loading started in the 1950s, and peaked in the 1980s, after which both nitrogen (Fig. 3) and phosphorus loads have declined (HELCOM, 2015a). However, the eutrophication status of the Baltic Sea has not improved at same rate (Fig. 3; Andersen et al., 2017).

In coastal areas all over the Baltic Sea, the many local and regional pressures—for example, fish farms, municipal wastewater treatment plants, river estuaries, industries, warm-water outflows from power plants, and coastal structures—create a heavy burden on the marine environment. Most impacted are the south-western sea areas, where bottom trawling, large wind farms, and large-scale extraction of seabed resources are dominant. Riverine pollution is most important in the south (Gulf of Gdansk) and east (Gulf of Finland) (HELCOM, 2010c; Korpinen et al., 2012).

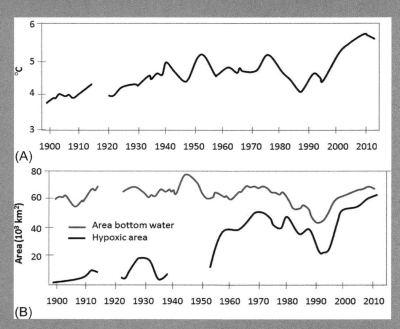

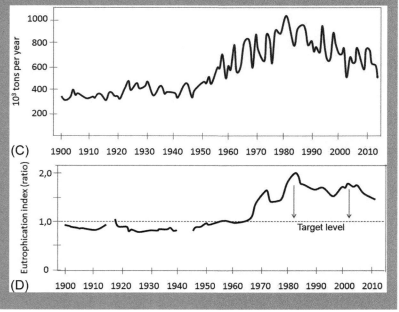

FIG. 3 Long-term trends in Baltic Sea between 1900 and 2010: (A) bottom layer temperature in the Gotland deep, (degrees of Celcius, 0°C), 5-year moving average; (B) areal extent of hypoxia ($<2\,mg\,L^{-1}$) in the Gotland deep with the area of the bottom water below the discontinuity layer (halocline); (C) water-borne inputs of total nitrogen to the Baltic Sea; and (D) integrated assessment of eutrophication in the Baltic Sea presented as eutrophication status index, based on HELCOM Eutrophication Assessment Tool (HEAT 3.0), and combining all (621) individual classifications of eutrophication status into a single assessment, solid line is the 5-year average, dashed line represent the target level of the eutrophication index ("good" ecological status boundary). Below the target line the status is not-affected by eutrophication. *(A) (Modified from Carstensen J, Andersen JH, Gustafsson BG, Conley DJ, 2014. Deoxygenation of the Baltic Sea during the last century. Proc. Natl. Acad. Sci. USA 111, 5628–5633.) (B) (Modified from Carstensen J, Andersen JH, Gustafsson BG, Conley DJ, 2014. Deoxygenation of the Baltic Sea during the last century. Proc. Natl. Acad. Sci. USA 111, 5628–5633.) (C) (Modified from Gustafsson BG, Schenk F, Blenckner T, Eilola K, Meier HEM, Müller-Karulis B, et al., 2012. Reconstructing the development of Baltic Sea eutrophication 1850–2006. Ambio 41, 534–548.) (D) (Modified from Andersen, J.H., Carstensen, J., Conley, D.J., Dromph, K., Fleming-Lehtinen, V., Gustafsson, B.G., Josefson, A.B., Norkko, A., Villnäs, A., Murray, C., 2017. Long-term temporal and spatial trends in eutrophication status of the Baltic Sea. Biol. Rev. 92, 135–149, https://doi.org/10.1111/brv.12221.)*

2 EUTROPHICATION

2.1 Drivers of Eutrophication

The bulk of nutrient inputs to the Baltic Sea originates from anthropogenic activities on land and at sea. The main waterborne sources are agriculture, wastewater treatment plants, industries together with aquaculture, managed forestry, scattered dwellings, storm overflows, and natural background sources and runoff. The main sources contributing to atmospheric inputs are combustion in energy production and industry as well as transport for oxidized nitrogen and agriculture for reduced nitrogen, and emissions from shipping in the Baltic and North Seas. A significant proportion of further atmospheric inputs originate from distant sources outside the Baltic Sea region (HELCOM, 2011, 2015a). Today, eutrophication is also upheld through secondary causes, namely internal loading of primarily phosphorus, hypoxia/anoxia, and shifts from perennial to annual filamentous algal growth, altering, and suffocating large shallow coastal and archipelago areas.

The importance of different drivers and human activities has varied over time. The use of the Baltic Sea has changed from early scattered hunting-fishing settlements to communities focusing on agriculture, industrial development, and finally rapidly increasing urbanization. The different uses have coexisted while the intensity of human activities has varied. The history of eutrophication development of the Baltic Sea is approximately divided into five periods where the driving forces have been different, as the human activities and use of the sea have changed, and concurrently there have been major political and governance changes in the societies around the Baltic Sea (Table 1).

2.2 Urbanization and Wastewaters

Since the early 1970s, there has been a steady increase of the human population living within the Baltic Sea drainage area connected to wastewater treatment facilities, and currently on average approximately 65% of the whole population in the Baltic Sea drainage basin is connected to tertiary treatment and approximately 77% to secondary treatment system (HELCOM, 2011). For example, in the capital of Finland, Helsinki, the first biological wastewater treatment plant was established as early as 1930, and in 1971 tertiary treatment was initiated, which led to a rapid improvement of the water quality in the coastal waters immediately outside the city (Laakkonen and Laurila, 2007), as the effluent pipeline was transferred to the deep waters of the outer coastal fringe. The share of the urban wastewater as a major source for waterborne nutrient loads decreased after the 1970s when Sweden, Denmark, and Finland introduced mandatory tertiary treatment with phosphorus removal to all municipal water treatment facilities. The total point emission of phosphorus decreased to <10% of the previous levels, but still riverine phosphorus transport to the Baltic Sea remained constant in Sweden (Grimwall and Stålnacke, 2001). Since the highest values were recorded between 1975 and 1985, phosphorus loads decreased by more than 90% in Denmark, Finland, Germany, and Sweden. The municipal wastewaters were estimated to account for approximately 70%–90% of the total point source nutrient discharges (except in Sweden and Finland where their share was even less) in 2005 (HELCOM, 2011). The EU directives, such as the Urban Wastewater Treatment Directive (UWWTD; 91/271/EEC), that were implemented by the Baltic Sea countries as they joined the EU, were generally less strict than the HELCOM recommendations (HELCOM, 2006), Nevertheless, the gradual renewal of environmental permits for treatment facilities as well as better technical solutions will also ensure further positive changes in the future.

2.3 Industrial Wastewaters

The HELCOM Joint Comprehensive Environmental Action Program (JCP) approved in 1992, listed a total of 162 environmental "hot spots" of pollution, 50 of which were from various industrial sectors with emission of various types of pollutants and hazardous substances (organic, and inorganic substances, chemicals, nutrients, heavy metals, pesticides, sulfides, nitrates, dust, suspended solids, etc.). The JCP was closed in 2013, when 116 (more than 2/3) of those were considered remediated (HELCOM, 2013d). In 2006, point-source nutrient discharges from industry sector were highest in the Bothnian Bay and Sea, where large pulp and paper and metal industries are located both in Finland and Sweden. In other regions industrial contributions were lower (HELCOM, 2011). Several pollutants have greatly decreased throughout the food chain although the Finnish Food Safety Authority presents some restrictions for consumption of fatty fish such as salmon, trout, and herring from the Baltic Sea which should not be consumed more than twice a month by children, young people, and persons of fertile age (EVIRA, 2017).

3   FOOD PRODUCTION

3.1   Agriculture

World market price fluctuations and trends influence global demand and production for agricultural goods. These drivers also directly influence the potential nutrient loads to the Baltic Sea. The collapse of the Soviet Union in 1990 led to a very large decrease in fertilizer use in Eastern Europe (Grimwall and Stålnacke, 2001). However, due to the time lag in the retention of nitrogen in agricultural soils, no decrease in nitrate transport from major rivers such as the Daugava was observed for a number of years. In Sweden the requirement for phosphorus removal from municipal wastewaters resulted in a shift of the nutrient ratio in municipal wastewaters to more nitrogen in relation to phosphorus. This resulted in a dominance of nitrogen transport downstream of major lakes (Grimwall and Stålnacke, 2001). One of the main reasons for the increased nutrient loads from agriculture to the Baltic Sea was the separation of crop and animal production which started to increase after 1950 in Sweden, Finland, and Denmark (Granstedt, 2000).

Larsson and Granstedt (2010) suggested that changing the conventional farming system toward the low input system of ecological recycling agriculture (ERA), which is a local organic agriculture system based on local and renewable resources would provide one possibility to change into a more sustainable agricultural system and to reduce the source of nutrients from agriculture to the Baltic Sea.

3.2   Aquaculture

Aquaculture and nutrient discharges from fish farms were estimated to be important in Finland and Denmark, although minor in the overall total load to the Baltic Sea (HELCOM, 2011). However, the nutrient loading from fish farms has significant local importance directly impacting primary production of nuisance-algae at the local to regional scales (Nordvarg and Johansson, 2002; Asmala and Saikku, 2010) with highly negative consequences for other users and uses of the coastal environment, including tourism, recreational fishing, and ecosystem health. For Finland, aquaculture accounts for more phosphorus and nitrogen inputs than municipal wastewaters (HELCOM, 2017a, b).

Commercial fisheries also contribute to nutrient release and recycling in the Baltic Sea primarily through two mechanisms: bottom trawling reworking nutrient-rich surface-sediments releasing mainly phosphorus to the water column, and discards of bycatches, which still remain high and largely unreported.

4   NUTRIENT LOADING PRESSURES

After the Second World War, nutrient concentrations in rivers were already high (Grimwall and Stålnacke, 2001), and the highest nutrient loadings to the Baltic Sea were reached in mid-1980s (Gustafsson et al., 2012). It was estimated that since the beginning of the 20th century the total input of phosphorus increased 10 times, and for nitrogen about 4 times (Larsson et al., 1985), leading to rapidly increasing production at all levels in the Baltic Sea trophic web.

A large part of the nutrient loading to the Baltic Sea comes through rivers, but also the subsurface diffuse load can considerable, particularly for nitrogen (Darracq et al., 2008). Similarly, the atmospheric deposition has increased concurrently with waterborne loads (HELCOM, 2011, 2015a, b). Atmospheric N deposition directly into the Baltic was estimated to contribute 22% of the total nitrogen load to the Baltic Sea in 2010 (HELCOM, 2015a, b). Nutrient loading increased until the late 1980s, but stringent measures on land have reversed the trend (Gustafsson et al., 2012; HELCOM, 2011). The reversal is mostly due to reduction in point sources and the introduction of measures to control diffuse loading from fields (HELCOM, 2011). Since the mid-1990s, total normalized nitrogen and phosphorus inputs to the Baltic Sea in 2014 were reduced by 19% and 24%, respectively (HELCOM, 2015a, b). During the latest assessment period of 2012–14, the estimated maximum allowable input (MAI; HELCOM, 2013a) of nitrogen has been achieved in the Kattegat, Danish Straits, and Bothnian Sea and the phosphorus input is achieved only in the Kattegat (Svendsen et al., 2017). However, despite these improvements, the overall eutrophication reduction target of the Baltic Sea has not yet been met (Andersen et al., 2017).

The most important nutrient sources to the Baltic Sea are diffuse losses from agriculture, which have become even more evident during recent decades as point-source inputs have decreased due to improved sewage treatment. Therefore it is unlikely that the eutrophication status of the Baltic Sea will be controlled, if only the nutrient reduction targets from municipal wastewaters are fulfilled (Wulff et al., 2007). Wulff et al. (2014) predicted a massive increase in fertilizer use within the Baltic catchment if the more intensive agricultural practices of Western Europe are also applied in the eastern European/transition countries which may lead to increased nutrient loading, thereby nullifying the gains of improved sewage treatment.

5   EUTROPHICATION STATUS

The negative symptoms of eutrophication started to appear in the Baltic Sea in the mid-1900s (Larsson et al., 1985; Bonsdorff et al., 1997), with algal blooms, increased periods of hypoxia/anoxia in the deep waters, and reduced water quality for human use. Increased nutrient loading to the marine environment led to accelerated algal growth which was not controlled by grazers (Heiskanen et al., 1996). The biomass of phytoplankton accumulates in the water column, leading to reduced light conditions, increased sedimentation of organic matter (Heiskanen, 1998; Blomqvist and Heiskanen, 2001), and oxygen depletion at the sea floor (Conley et al., 2002), which in turn impacted the zoobenthic assemblages, with effects on fish and fisheries (Karlson et al., 2002). Changes in the species composition of phytoplankton and flows of organic material in the pelagic system combined with concurrent changes in the top-down control led to a cascade of ecosystem changes that enhanced nitrogen limitation and rapid regeneration of phosphorus from dead organic material (Heiskanen et al., 1996) and enhanced internal loading of phosphorus from the sediments (Conley et al., 2011; Puttonen et al., 2014). The vicious circle of internal nutrient loading and nitrogen limitation led to an increase of cyanobacterial blooms, anoxic conditions, and nutrient regeneration from sediments (Vahtera et al., 2007). The hypoxic and anoxic conditions have since spread in both the deep open waters and coastal zones of the Baltic Sea (Conley et al., 2011; Carstensen et al., 2014).

Some positive developments in the eutrophication status can be found in the latest assessment period 2011–2015: a marked decrease in nutrient concentrations, improved water clarity in parts of the Baltic Sea, and a decrease in chlorophyll concentrations in some areas. However, the overall assessment indicated that most of the regions of the Baltic Sea are still impacted by eutrophication (HELCOM, 2017a, b). Despite the decrease of nutrient loads to the Baltic Sea, large-scale responses to reduced loading are slow. For instance the average winter concentrations of nutrients have not decreased at the same rate but remained rather stable for approximately 20 years (HELCOM, 2014b). Therefore the recently achieved reductions are not visible in the assessments over the short time frame (Gustafsson et al., 2012; HELCOM, 2013c).

6   EUTROPHICATION IMPACT ON HUMAN WELFARE

The Baltic Sea provides several benefits, ecosystem goods and services, for people living and spending their holidays in the coastal areas, such as recreational opportunities, cultural heritage, enjoyment of scenery, food, materials, and energy, as examples of cultural and provisioning services. Environmental degradation reduces enjoyment and opportunities for marine and coastal recreation or reduces the appreciation of the existence value of the healthy and diverse coastal ecosystems.

Eutrophication affects many intermediate ecosystem services which further impact final ecosystem services such as recreation. These lower carbon sequestration through a decline of eelgrass habitats that enhance carbon sequestration and provide nursery habitat for fish species (Röhr et al., 2016), organic material is suffocating oxidized sediment habitats, and thus reducing both sequestration of phosphorus and denitrification (Arroyo and Bonsdorff, 2016; Griffiths et al., 2017). This also results in the decline of biodiversity and functioning of benthic communities (Korpinen and Bonsdorff, 2015; Weigel et al., 2015; Nordström and Bonsdorff, 2017). However, the impacts of reduced eutrophication are appreciated differently in different parts of the Baltic Sea due to geopolitical, and historic-cultural reasons (Ahtiainen et al., 2013; Kosenius and Ollikainen, 2015).

Nevertheless, the importance of the Baltic Sea for recreation activities is considerable; some 80% of the people living in the Baltic Sea region spend their leisure time by the sea and many (40%–75%) are concerned about the environmental state of the Baltic Sea (Ahtiainen et al., 2013). The long-term benefits of the eutrophication remediation are estimated to be considerable. In willingness-to-pay exercises, most respondents were willing to pay for improvements and the benefits were estimated to be about 4.6 billion € annually for reaching the HELCOM BSAP nutrient reduction goals by 2050 (Ahtiainen et al., 2014). Some of the measures would also benefit drainage areas and to some extent also the North Sea, which would be likely to produce further benefits (Hyytiäinen et al., 2015).

HELCOM has focused previously in the holistic assessments on pressure and status assessments for different themes (eutrophication, biodiversity, and hazardous substances; HELCOM, 2010c) while social and economic assessments were not included (Karlsson et al., 2016). Currently HELCOM has initiated a working group on Economic and Social analysis as a part of the on-going holistic assessment (HELCOM, 2017b). This aims at a coordinated economic assessment of costs and benefits at the Baltic Sea scale which analyzes both the use of the marine waters and the cost of degradation (HELCOM, 2017b). The latter examines the losses in welfare due to the deterioration of the Baltic Sea marine environment illustrating what is at stake if the state does not improve. The estimates of the welfare losses are linked to deterioration of ecological status (due to eutrophication, biodiversity loss, and food web change) and also to recreation as a cultural ecosystem service. A full ecosystem services approach would require identifying and quantifying marine and coastal ecosystem services and their contribution to human welfare. Current knowledge, however, is insufficient to fully apply the ecosystem services approach to the regional cost of degradation analyses. So far the ecosystem service approach has been only applied to evaluate marine and coastal recreation benefits, based on a recent study on recreational use and values in nine countries around the Baltic Sea area (Czajkowski et al., 2015).

7 RESPONSES TO COUNTERACT AND MANAGE EUTROPHICATION

The awakening of Baltic Sea environmental policy took place in late 1960s at a time when it was divided by the Iron Curtain between two political blocks (the eastern bloc with Soviet Union, Poland, and Eastern Germany and the western block with Federal Republic of Germany, Denmark, Sweden, and Finland) (Räsänen and Laakkonen, 2007). The responses to counteract pollution and to protect the Baltic Sea ecosystem led to the signing of the Helsinki Convention for the protection of the Baltic Sea in 1974, despite that Iron Curtain division of the time, and provided a platform for intergovernmental collaboration in the field of environmental protection and monitoring being the first of its kind in the world. The Convention covered all known sources of pollution from land and ships to the entire sea region and developed science-based environmental recommendations for management and protection of the Baltic Sea (Räsänen and Laakkonen, 2007).

The dissolution of the Soviet Union in 1991 and the expansion of the EU introduced new environmental legislation in most of the Baltic Sea countries: Sweden and Finland joined the EU in 1995 as environmental forerunners (Joas, 1997), while Estonia, Latvia, Lithuania, and Poland joined in 2003 as willing followers. Together with the transition of former state socialist countries to a market economy, including downsizing inefficient and highly polluting heavy industry, the global agenda on sustainable development, and the enlargement of the EU, led to many cooperative arrangements for the management and improvement of the Baltic Sea environment (Joas et al., 2008). Since then HELCOM has become a true multilevel and dimensional governing system of the Baltic Sea that allows the contracting parties to plan joint regional actions and environmental targets. However, the historical differences in socioeconomic, political, and administrative systems and the diminishing gap in the living standards between Eastern and Western countries of the Baltic Sea have resulted in different capacities and motivation in implementing the HELCOM recommendations (Tynkkynen et al., 2014).

The main tool of HELCOM to counteract eutrophication and other environmental problems is the BSAP which was established in 2007 (HELCOM, 2007); it represents a comprehensive ecosystem-based approach with a joint vision, objectives, and targets to improve the Baltic Sea (Backer et al., 2010). The eutrophication section of the BSAP sets environmental targets for the marine basins with clear quantitative nutrient reduction goals with the aim of restoring the good ecological status of the Baltic Sea by 2021 (HELCOM, 2007). The BSAP introduces provisional nutrient input reductions targets, which have been allocated to the responsible countries, based on the scientific work and modeling carried out the by the Baltic Nest Institute (HELCOM, 2013a). Development of the BSAP has been attributed to the close cooperation between the Baltic Nest Institute and HELCOM while developing science-based advice that underpins the legitimacy of political claims for regional environmental management (Linke et al., 2016).

Currently there are many multilevel governance institutions (i.e., policies and actors) at national, Baltic Sea and EU level, that all aim to improve the environmental status of the Baltic Sea but which risk to be partly overlapping and contradictory (Boyes and Elliott, 2014; Karlsson et al., 2016). Nevertheless, multilevel governance highlights the adaptive capacity of the political system in the Baltic Sea region, which has developed gradually since early 1970s. New environmental strategies and legislation has been developed together with a growing scientific understanding of the causes and impacts of problems (Elmgren et al., 2015) and for countries such as Finland and Sweden the existing water protection legislation had to be adapted to the requirements of the EU regulations when introduced. The early EU directives (adopted as early as 1991) aimed to reduce nutrient input into water bodies. Those were the UWWTD and the Nitrates Directive (ND), which respectively provide regulations for sewage treatment and farming practices. These were part of the water acquis that countries joining the EU had to adopt. Another instrument for nutrient input reduction is the agri-environmental program of the EU, common agricultural policy (CAP), through which compensation can be paid to farmers who carry out certain water protection measures. Since 2001, airborne emissions of, for example, nitrogen oxides and ammonia have been addressed in the National Emission Ceilings Directive (NECD). More holistic EU regulation aiming to reduce eutrophication problems, such as diffuse nutrient loading that was not sufficiently tackled by former directives, was the Water Framework Directive (WFD), adopted in 2000, which aimed to achieve good ecological status for surface waters (inland and coastal) by 2015.

The EU launched an Integrated Maritime Policy (in 2008) requiring an integrated approach to the management and governance of oceans, seas and coasts to strengthen sustainable economic and environmental development. One of the key instruments of the IMP is the Marine Strategy Framework Directive (MSFD) (adopted in 2008), which aims to achieve or maintain good environmental status in the marine environment by 2020. Together with the Maritime Spatial Planning Directive (MSPD; adopted in 2014) these policies provide a holistic assessment and planning framework for balancing multiple uses of marine environment, space, and resources while protecting the ecosystems. However, the EU member states under certain conditions could extend the target year to 2027 (WFD) and could request exceptions in reaching the target (MSFD) resulting in implementation delay that hampers the protection of the Baltic Sea (Tynkkynen et al., 2014). Other relevant policies such as the WFD and the Habitat and Birds Directives (BHDs) also play a critical role for the environmental status of the European Seas, as do the sector policies such as the Common Fisheries Policy (CFP).

The MSFD brought additional themes (beyond biodiversity, eutrophication, and hazardous substances) to the environmental assessment and management of the Baltic Sea as descriptors for the assessment of the good environmental status (invasive species commercial fish, food webs, sediment integrity, marine litter, hydrological alterations, energy, and sound). Nevertheless, the HELCOM community was methodologically and institutionally prepared to start developing the integrated assessment together with new indicators and assessment tools for the MSFD environmental status assessment (HELCOM, 2013e). The first holistic assessment was carried out prior to the first status assessment of the MSFD (HELCOM, 2010a). The scientific community around the Baltic Sea was also prepared to develop tools that were required in the assessment such as HELCOM Eutrophication Assessment Tool (HEAT; HELCOM, 2009b), HELCOM hazardous assessment tool (CHASE; HELCOM, 2010a) and HELCOM biodiversity assessment tool (BEAT; HELCOM, 2009a) and also applied and developed further tools for the cumulative pressure and impact assessment (HELCOM, 2010b; Korpinen et al., 2012).

The integrated assessment of eutrophication is based on the revised HELCOM eutrophication Assessment Tool (Fleming-Lehtinen et al., 2015) which integrates concentrations of nutrients and chlorophyll a, Secchi depth (water transparency) and oxygen concentration into an overall eutrophication status of the Baltic Sea (HELCOM, 2014b). Based on the scientific work (HELCOM, 2013b), and expert evaluations, targets were derived for these indicators illustrating the concentration levels in the Baltic Sea prior to an apparent eutrophication phase (HELCOM, 2014b).

In addition, the BSAP that was already initiated and the Baltic Nest Institute (that had developed the modeling tools for eutrophication mitigation and assessment of the required nutrient loading reductions) provided the planning and management tools that were globally pioneering (Wulff et al., 2007). Therefore the Baltic Sea countries in having HELCOM as an organization were well-geared to coordinate the joint implementation of MSFD; they had a head start in that implementation and so the Baltic Sea as one of the Regional Seas in the EU could provide a pilot case for the MSFD. However, as pointed out by Gilek et al. (2016), the integration of the EU policies that aim primarily at achieving good environmental status, such as the MSFD, and those aiming to regulate pollutants (e.g., the REACH chemicals regulation) have different institutional arrangements which may lead to institutional tensions and inefficiencies even though primarily those should be supporting each other's policy objectives. Nevertheless, the eutrophication-related EU policies seem to be well in line with HELCOM objectives, particularly with BSAP, which strengthens the role of HELCOM in eutrophication governance (Karlsson et al., 2016).

The major controversy in governance of the Baltic Sea is that the institutional structures linked to sectoral policies (e.g., EU, CAP) have partly different policy objectives than the water and marine EU policies, leading to contradictions and conflicts with the environmental policies in the EU system (Tynkkynen et al., 2014). This is particularly the case as the CAP continues to subsidize intensive agriculture with little consideration of the impacts on the marine environment as there is no effective coordination mechanism with the MSFD in relation to agriculture (Karlsson et al., 2016). It is likely that the earlier national agri-environmental schemes (1995–2006) have not resulted in the desired overall reduction of nutrient loading (Ekholm et al., 2015). However, the current agri-environmental scheme aims to produce multiple environmental benefits related to biodiversity and the improvement of water quality (Anon., 2018). In the future, better multisector and multilevel coordination of policies (WFD, MSFD, and CAP) and objectives adopted by institutions at European, Baltic Sea and national level will be needed (see Karlsson et al., 2016).

The overall importance of the EU is, however, obvious. It gives a direct route to reach almost all Baltic Sea states for new regulations, the harmonizing effect is better than no effect at all, and all major EU directives do acknowledge the need for ecosystem-based management. Finally, the EU has a clear strategy for regional development through the EU Strategy for the Baltic Sea region (EUSBSR). This instrument does not as such provide funding or high-level institutional support, but it does give a solid base for regional development actions, within all common policy sectors.

The key political pressure in the region is the new political tension between Russia, on one hand, and EU member states, on the other. The tensions emerging from the unlawful occupation of Crimean peninsula by Russian troops have cut a deep political cleavage in Europe, postponing initiatives as well as security policy cooperation. The internal tensions in the EU emerging from populistic and nationalistic political movements, for example in Poland, also endanger further policy development in the Baltic Sea region.

8 FUTURE OUTLOOK IN EUTROPHICATION DEVELOPMENT

The recovery of the Baltic Sea from the current eutrophic status is predicted to be slow. Model projections of the future changes indicate that although the nutrient inputs to Baltic Sea have reached the maximum allowable levels, it would still take perhaps half a century or even more to reach the eutrophication status targets set by the HELCOM and BSAP (HELCOM, 2013a). In addition, climate change is expected to slow down improvements, due to a shortening of the snow cover period and increase of river runoff, especially during winter time, leading to increased nutrient leaching from agricultural areas and from natural background in the Baltic Sea catchment (Huttunen et al., 2015). The key drivers for the future development of nutrient loading from agricultural sources are the prices of agricultural products as they affect the level of fertilization and the production intensity and volume. Nevertheless, new innovations such as the introduction of new, high yield cultivars, and other yield-promoting management options such as liming, gypsum treatment, better drainage, and crop protection, may help to reduce nutrient leaching from agricultural soils (Huttunen et al., 2015). Similarly, the regional distribution of manure nutrients and improving agronomic practices could result in a major reduction of nutrient loading and hence the improvement of the Baltic Sea eutrophication status (Hong et al., 2012, 2017). Also, an improved crop yield may also encourage setting land aside from cultivation in order to counteract leaching and improve biodiversity, as also encouraged by the current CAP scheme. Structural changes in agriculture, for instance, by changing the geographical balance between husbandry and crop production, could alleviate nutrient leaching problems, but such changes are difficult to impose as they would necessitate comprehensive national strategies and a major CAP reform (see Karlsson et al., 2016). Moreover, the future changes in human lifestyles as shifts in consumption of animal proteins and population have been suggested to potentially overshadow the climate effects on nutrient runoff for the entire Baltic Sea drainage basin (Hägg et al., 2014).

Sharing the nutrient reduction burden between Baltic Sea countries, by using nutrient trading for an optimal allocation of national obligations for nutrient load reduction has been suggested as they are better targeted to areas where more measures are more cost efficient (e.g., Hautakangas et al., 2014; Tynkkynen et al., 2014; Wulff et al., 2014). Mussel farming has also been suggested as a potential cost-effective means for eutrophication mitigation (Petersen et al., 2014), although it might also introduce further problems of nutrient loading from sediments if not properly managed (Stadmark and Conley, 2011). Currently there are experiments that are testing large-scale mussel farming in the Baltic Sea (Baltic Blue Growth[1]) and pilot projects for testing voluntary nutrient trading, nutrient offsets and joint implementation of nutrient reduction targets, with pilot mussel farming, fishing cyprinids, and gypsum spreading on fields as test cases of cost-effective eutrophication mitigation in Baltic Sea area (NutriTrade[2]). Marine geoengineering has been proposed to alleviate oxygen deficiency in sediments and deep waters and to speed up recovery (Stigebrandt and Gustafsson, 2007; Rydin et al., 2017) but there are doubts about the feasibility and cost-effectiveness in comparison to reduction of land-based nutrient loading (Conley, 2012; Conley et al., 2009).

Modeling projections toward the end of the 21st century imply considerable changes in the Baltic Sea compared to the past 100 years. Future climate conditions may increase runoff and nutrient loading and together with higher water temperature lower oxygen saturation concentrations. This may also cause an increased remineralization of sediment nutrients leading to increased nutrient concentrations in the surface layer and consequently an acceleration of eutrophication in the Baltic Sea. The climate effect may be larger than the anticipated BSAP nutrient reductions (Meier et al., 2012). There is a further risk of regime shifts and possible system collapse due to feedback mechanisms, such as remobilization of nutrients from sediments and changes in the food web components of the system. Therefore further nutrient load reductions and sustainable fishery may be even more important in the future to ensure a healthy marine ecosystem when the stresses from climate change increase (Niiranen et al., 2013; Meier et al., 2014).

9 NEW INNOVATIONS TOWARD SUSTAINABLE BALTIC SEA FUTURE

The concurrent climate and environmental pressures and the rather pessimistic projections of the Baltic Sea future call for a transition into more sustainable solutions for food production, energy, construction, and transport, and new business models that have a potential to foster economic growth while improving the Baltic Sea environment.

The current EU "Blue growth" strategy aims to increase the economic growth based on marine and maritime resources and to support the exploration of new assets based on living and nonliving resources of the marine environment (European Commission, 2012). The focus areas are renewable energies, aquaculture, marine tourism, mineral resources, and biotechnology. The EU Commission also launched "A Sustainable Blue Growth Agenda for the Baltic Sea Region" in 2014 (European Commission, 2014a), which is complementing the existing EUSBSR (European Commission, 2009). This strategy identified short sea shipping, coastal and cruise tourism, offshore wind, shipbuilding, aquaculture, and blue

---

1 Baltic Blue Growth, https://www.submariner-network.eu/projects/balticbluegrowth/about-baltic-blue-growth (website Accessed 28 April 2018), Baltic Sea.
2 NutriTrade Platform; http://nutritradebaltic.eu/ (website Accessed 28 April 2018).

biotechnologies as the most promising marine economic sectors in the Baltic Sea region. It aligns with the HELCOM environmental objectives and provides funding for sustainability innovation projects.

Concurrently, the EU Circular Economy policy (European Commission, 2014b) aims for better recycling of materials and new solutions of producing and saving energy in the society. The scarcity of essential elements (such as mineral phosphorus) and the production of nondegradable waste such as plastics are pushing the move for new solutions to produce food and other goods. The world's first circular economy road map was launched in Finland in 2016 (SITRA, 2016), including nutrient recycling as a part of the focus area on sustainable food systems. The road map supports projects that are piloting new production and business models to reduce the amount of nutrients entering aquatic systems.

The emerging experimentation culture where products, services, and methods are developed swiftly and tested may allow change of direction very quickly and enable transition from linear production systems into a recycling economy (Antikainen et al., 2017). The proliferation and upscaling of viable solutions may result in cutting nutrient emissions and improving resilience of the environment and the society. Examples of such experiments are the new closed circulation concept "industrial ecosystems" where material and waste are circulated and energy produced (e.g., Sybimar[3]) and innovative technical solutions to reduce nutrient pollution (NutriTrade[4]) (SITRA, 2016). Similarly, marine seaweed cultivation to reuse and harvest excess nutrients is reducing eutrophication and producing blue carbon for climate change mitigation (Duarte et al., 2017). The cultivation of macroalgae has been tested also in the Baltic Sea region and it can contribute significantly to nitrogen reduction goals and provide the means for environmental restoration and climate change mitigation (Seghetta et al., 2016), but productivity is varying regionally and heavy metals (such as Cd) may be problem for further use of the produced biomass (Suutari et al., 2017).

The circular economy solutions to curb the growing accumulation of microplastics in marine ecosystems are also calling for societal changes in production, consumption, and discarding of plastics. Recent observations have shown that microplastics are potentially affecting all marine life (Setälä et al., 2014; Auta et al., 2017). HELCOM has initiated a regional action plan against marine litter that should enable concrete measures for the prevention and reduction of marine litter from its main sources, including an overview of the importance of the different sources of microplastics (HELCOM, 2015b). A circular economy experimentation for production, consumption, and recycling of various plastic materials, such as those initiated by the Ellen MacArthur Foundation (NPE, 2017), is crucial for the future health of both humans and the marine environment.

The complexity of the marine ecosystem and its networks of interactions (Rocha et al., 2015; Yletyinen et al., 2016) and the interlinked DAPSI(W)R(M) cycles (Fig. 2) call for adaptive governance and abandoning rigid legislation and institutions (Koontz et al., 2015). The institutions, organizations, and legislation are often path dependent and cannot thus provide solutions for new emerging problems and issues where those are not designed for or where there is no previous experience. In order to build scenarios of the future governance options and pathways of the socio-ecological system, multiple approaches will be needed (Gilek et al., 2011, 2016). The management of environmental, social, economic, and technological uncertainties calls for flexibility in law and policy, which should be still sufficiently predictable to support investments in new economic sectors. To be effective, Baltic Sea adaptive governance should reconcile uncertainties of the dynamic and complex socio-ecological system with predictability and equity, carry wide confidence, support active citizenship, and be informed by science.

ACKNOWLEDGMENTS

The authors thank Professors Mike Elliott and Eric Wolanski for inviting us to write this contribution, and for their valuable comments on the first draft. The authors also thank Dr Harri Kuosa for critical reading of the manuscript. This chapter was produced as part of an interdisciplinary collaboration between the Finnish Environment Institute thematic research area "Sustainable management of the Baltic Sea and freshwater resources" (A-SH), and the Åbo Akademi University Profiling Area "The Sea" under the funding of the Academy of Finland (grant number 311944 / 2017) and the BaltReg-project (grant number 290331 / 2015) funded by the Academy of Finland (EB & MJ). It is also a contribution from the research projects "Adaptive Capacity for Sustainable Blue Growth" (BlueAdapt) funded by the Strategic Research Council at the Academy of Finland (A-SH; grant number 312650/ 2018), and "Nutrient COctails in COAstal zones of the Baltic Sea" (COCOA) funded by the BONUS EEIG (Grant Agreement 2112932-2) and the Academy of Finland (A-SH; grant number 273695/ 2013 & EB; grant number 273732/ 2013). Financial support from the Åbo Akademi University Foundation (EB) is also acknowledged.

---

3 www.sybimar.fi/en
4 http://nutritradebaltic.eu/project-nutritrade/

REFERENCES

Ahtiainen, H., Artell, J., Czajkowski, M., Hasler, B., Hasselström, L., Huhtala, A., 2014. Benefits of meeting nutrient reduction targets for the Baltic Sea—a contingent valuation study in the nine coastal states. J. Environ. Econ. Pol. 3, 278–305. https://doi.org/10.1080/21606544.2014.901923.

Ahtiainen, H., Artell, J., Czajkowski, M., Hasler, B., Hasselström, L., Hyytiäinen, K., Meyerhoff, J., Smart, J., Söderqvist, T., Zimmer, K., Khaleeva, J., Rastrigina, O., Tuhkanen, H., 2013. Public preferences regarding use and condition of the Baltic Sea—an international comparison informing marine policy. Mar. Policy 42, 20–30.

Ahtiainen, H., Öhman, M., 2014. Ecosystem services in the Baltic Sea. Valuation of marine and coastal ecosystem services in the Baltic Sea. Nordic Council of Ministers, Copenhagen, Denmark563. TemaNord 2014 http://norden.diva-portal.org/smash/record.jsf?pid=diva2%3A767673&dswid=-3377.

Andersen, J.H., Carstensen, J., Conley, D.J., Dromph, K., Fleming-Lehtinen, V., Gustafsson, B.G., Josefson, A.B., Norkko, A., Villnäs, A., Murray, C., 2017. Long-term temporal and spatial trends in eutrophication status of the Baltic Sea. Biol. Rev. 92, 135–149. https://doi.org/10.1111/brv.12221.

Andersen, M.S., Liefferink, J.D. (Eds.), 1997. European Environmental Policy: The Pioneers. Issues in Environmental Politics. Manchester University Press, Manchester, pp. 119–160.

Andrén, E., Andrén, T., Kunzendorf, H., 2000. Holocene history of the Baltic Sea as a background for assessing human impact in the sediments of the Gotland Basin. Holocene 10, 687–702.

Anon., 2018. Agriculture and the environment: introduction. Available from: https://ec.europa.eu/agriculture/envir_en. Accessed 30 April 2018.

Antikainen, R., Alhola, K., Jääskeläinen, T., 2017. Experiments as a means towards sustainable societies—lessons learnt and future outlooks from a finnish perspective. J. Clean. Prod. 169, 216–224. https://doi.org/10.1016/j.jclepro.2017.06.184.

Arroyo, N.L., Bonsdorff, E., 2016. The role of drifting algae for marine biodiversity. In: Olafsson, E. (Ed.), Marine Macrophytes as Foundation Species. CRC Press, Taylor & Francis Group, Boca Raton, FL, pp. 285.

Asmala, E., Saikku, L., 2010. Closing a loop: substance flow analysis of nitrogen and phosphorus in the rainbow trout production and domestic consumption system in Finland. Ambio 39, 126–135. https://doi.org/10.1007/s13280-010-0024-5.

Atkins, J., Burdon, D., Elliott, M., Gregory, A.J., 2011. Management of the marine environment: integrating ecosystem services and societal benefits with the DPSIR framework in a systems approach. Mar. Pollut. Bull. 62, 215–226.

Auta, H.S., Emenike, C.U., Fauziah, S.H., 2017. Distribution and importance of microplastics in the marine environment: a review of the sources, fate, effects, and potential solutions. Environ. Int. 102, 165–176. https://doi.org/10.1016/j.envint.2017.02.013.

BACC, 2015. The BACC II Author Team. In: Second Assessment of Climate Change for the Baltic Sea Basin. Springer, https://doi.org/10.1007/978-3-319-16006-1.

Backer, H., Leppänen, J.-M., Brusendorff, A.C., Forsius, K., Stankiewicz, M., Mehtonen, J., Pyhälä, M., Laamanen, M., Paulomäki, H., Vlasov, N., Haaranen, T., 2010. HELCOM Baltic Sea action plan—a regional programme of measures for the marine environment based on the ecosystem approach. Mar. Pollut. Bull. 60, 642–649.

Blenckner, T., Llope, M., Möllmann, C., Voss, R., Quaas, M.F., Casini, M., Lindegren, M., Folke, C., Stenseth, N., 2015. Climate and fishing steer ecosystem regeneration to uncertain economic futures. Proc. R. Soc. B 282, 2014–2809. https://doi.org/10.1098/rspb.2014.2809.

Blomqvist, S., Heiskanen, A.-S., 2001. The challenge of sedimentation in the Baltic Sea. In: Wulff, F., Rahm, L., Larsson, P. (Eds.), A Systems Analysis of the Baltic Sea. Ecological Studies, 148. Springer-Verlag, Berlin Heidelberg, pp. 211–227.

Bonsdorff, E., 2006. Zoobenthic diversity-gradients in the Baltic Sea: continuous postglacial succession in a stressed ecosystem. J. Exp. Mar. Biol. Ecol. 333, 383–391.

Bonsdorff, E., Blomqvist, E.M., Mattila, J., Norkko, A., 1997. Coastal eutrophication: causes, consequences and perspectives in the archipelago areas of the northern Baltic Sea. Estuar. Coast. Shelf Sci. 44, 63–72.

Boyes, S.J., Elliott, M., 2014. Marine legislation—the ultimate "horrendogram": international law, European directives and national implementation. Mar. Pollut. Bull. 86, 39–47.

Carstensen, J., Andersen, J.H., Gustafsson, B.G., Conley, D.J., 2014. Deoxygenation of the Baltic Sea during the last century. Proc. Natl. Acad. Sci. USA 111, 5628–5633. https://doi.org/10.1073/pnas.1323156111pmid:24706804.

Casini, M., Lövgren, J., Hjelm, J., Cardinale, M., Molinero, J.-C., Kornilovs, G., 2008. Multi-level trophic cascades in a heavily exploited open marine ecosystem. Proc. Biol. Sci. 275, 1793–1801.

Conley, D.J., Humborg, C., Rahm, L., Savchuk, O.P., Wulff, F., 2002. Hypoxia in the Baltic Sea and basin-scale changes in phosphorus biogeochemistry. Environ. Sci. Technol. 36, 5315–5320. https://pubs.acs.org/doi/full/10.1021/es025763w.

Conley, D., Bonsdorff, E., Carstensen, J., Destouni, G., Gustafsson, B., Hansson, L.-A., Rabalais, N., Voss, M., Zillén, L., 2009. Tackling hypoxia in the Baltic Sea: is engineering a solution? Environ. Sci. Technol. 43, 3407–3411.

Conley, D.J., 2012. Save the Baltic Sea. Nature 486, 463–464.

Conley, D.J., Carstensen, J., Aigars, J., Axe, P., Bonsdorff, E., Eremina, T., Haahti, B.-M., Humborg, C., Jonsson, P., Kotta, J., Lännergren, C., Larsson, U., Maximov, A., Medina, A.R., Lysiak-Pastuszak, E., Remeikaite-Nikiene, N., Walve, J., Wilhelms, S., Zillén, L., 2011. Hypoxia is increasing in the Coastal Zone of the Baltic Sea. Environ. Sci. Technol. 45, 6777–6783. https://doi.org/10.1021/es201212r.

Cooper, P., 2012. Socio-ecological accounting: DPSWR, a modified DPSIR framework, and its application to marine ecosystems. Ecol. Econ. 94, 106–115. https://doi.org/10.1016/j.ecolecon.2013.07.010.

Czajkowski, M., Ahtiainen, H., Artell, J., Budziński, W., Hasler, B., Hasselström, L., Tuhkanen, H., 2015. Valuing the commons: an international study on the recreational benefits of the Baltic Sea. J. Environ. Manag. 156, 209–217. https://doi.org/10.1016/j.jenvman.2015.03.038.

Darracq, A., Lindgren, G., Destouni, G., 2008. Long-term development of phosphorus and nitrogen loads through the subsurface and surface water systems of drainage basins. Glob. Biogeochem. Cycles 22, https://doi.org/10.1029/2007GB003022. GB3022.

Duarte, C.M., Wu, J., Xiao, X., Bruhn, A., Krause-Jensen, D., 2017. Can seaweed farming play a role in climate change mitigation and adaptation? Front. Mar. Sci. 4, 100. https://doi.org/10.3389/fmars.2017.00100.

Ekholm, P., Rankinen, K., Rita, H., Räike, A., Sjöblom, H., Raateland, A., Vesikko, L., Bernal, J.E.C., Taskinen, A., 2015. Phosphorus and nitrogen fluxes carried by 21 finnish agricultural rivers in 1985–2006. Environ. Monit. Assess. 187, 216. https://doi.org/10.1007/s10661-015-4417-6.

Elliott, M., 2011. Marine science and management means tackling exogenic unmanaged pressures and endogenic managed pressures—a numbered guide. Mar. Pollut. Bull. 62, 651–655.

Elliott, M., Burdon, D., Atkins, J.P., Borja, A., Cormier, R., de Jonge, V.N., Turner, R.K., 2017. "And DPSIR begat DAPSI(W)R(M)!"—A unifying framework for marine environmental management. Mar. Pollut. Bull. 118, 27–40.

Elmgren, R., Blenckner, T., Andersson, A., 2015. Baltic Sea management: successes and failures. Ambio 44, S335–S344. https://doi.org/10.1007/s13280-015-0653-9.

European Commission, 2009. Commission staff working document. European Union Strategy for the Baltic Sea Region ACTION PLAN {COM(2009) 248}.

European Commission, 2012. Communication from the commission to the European parliament, the council, the European economic and social committee and the committee of the regions. Blue Growth opportunities for marine and maritime sustainable growth. . COM(2012) 494 final.

European Commission, 2014a. Commission staff working document. A sustainable Blue Growth agenda for the Baltic Sea region. . SWD(2014) 167 final.

European Commission, 2014b. Communication from the commission to the European parliament, the council, the European economic and social committee and the committee of the regions. Towards a circular economy: a zero waste programme for Europe. . COM/2014/0398 final/2 */.

EVIRA, 2017. Dietary advice on fish consumption. https://www.evira.fi/en/foodstuff/information-on-food/food-hazards/restriction-on-the-use-of-foodstuffs/dietary-advice-on-fish-consumption/.

Finni, T., Laurila, S., Laakkonen, S., 2001. The history of eutrophication in the sea area of Helsinki in the 20th century. Ambio 30, 264–271. https://doi.org/10.1579/0044-7447-30.4.264.

Fleming-Lehtinen, V., Andersen, J.H., Carstensen, J., Łysiak-Pastuszak, E., Murray, C., Pyhälä, M., Laamanen, M., 2015. Recent developments in assessment methodology reveal that the Baltic Sea eutrophication problem is expanding. Ecol. Indic. 48, 380–388.

Gilek, M., Hassler, B., Jönsson, A.-M., Karlsson, M., 2011. Coping with complexity in Baltic Sea risk governance: introduction. Ambio 40, 109–110. https://doi.org/10.1007/s13280-010-0122-4.

Gilek, M., Karlsson, M., Linke, S., Smolarz, K., 2016. Environmental governance of the Baltic Sea: identifying key challenges, research topics and analytical approaches. In: Gilek, M., Karlsson, M., Linke, S., Smolarz, K. (Eds.), Environmental Governance of the Baltic Sea. MARE Publication Series vol. 10. Springer, Dordrecht, pp. 1–20.

Granstedt, A., 2000. Increasing the efficiency of plant nutrient recycling within the agricultural system as a way of reducing the load to the environment—experience from Sweden and Finland. Agric. Ecosyst. Environ. 80, 169–185.

Gren, I.-M., Turner, K., Wulff, F., 2000. Managing a Sea—The Ecological Economics of the Baltic. Earthscan Publications Ltd., London.

Griffiths, J.R., Kadin, M., Nascimento, F.J.A., Tamelander, T., Törnroos, A., Bonaglia, S., Bonsdorff, E., Brüchert, V., Gårdmark, A., Järnström, M., Kotta, J., Lindegren, M., Nordström, M.C., Norkko, A., Olsson, J., Weigel, B., Zydelis, R., Blenckner, T., Niiranen, S., Winder, M., 2017. The importance of benthic-pelagic coupling for marine ecosystem functioning in a changing world. Glob. Chang. Biol. https://doi.org/10.1111/gcb.13642.

Grimwall, A., Stålnacke, P., 2001. Riverine inputs of nutrients to the Baltic Sea. In: Wulff, F., Rahm, L., Larsson, P. (Eds.), A Systems Analysis of the Baltic Sea. Ecological Studies, 148. Springer-Verlag, Berlin Heidelberg, pp. 113–131.

Gustafsson, B.G., Schenk, F., Blenckner, T., Eilola, K., Meier, H.E.M., Müller-Karulis, B., Neumann, T., Ruoho-Airola, T., Savchuk, O.P., Zorita, E., 2012. Reconstructing the development of Baltic Sea eutrophication 1850–2006. Ambio 41, 534–548.

Hägg, H.E., Lyon, S.W., Wällstedt, T., Mörth, C.-M., Claremar, B., Humborg, C., 2014. Future nutrient load scenarios for the Baltic Sea due to climate and lifestyle changes. Ambio 43, 337–351. https://doi.org/10.1007/s13280-013-0416-4.

Halpern, B.S., Longo, C., Lowndes, J.S.S., Best, B.D., Frazier, M., Katona, S.K., Kleisner, K.M., Rosenberg, A.A., Scarborough, C., Selig, E.R., 2015. Patterns and emerging trends in global ocean health. PLoS ONE 10, https://doi.org/10.1371/journal.pone.0117863. e0117863.

Hautakangas, S., Ollikainen, M., Aarnos, K., Rantanen, P., 2014. Nutrient abatement potential and abatement costs of waste water treatment plants in the Baltic Sea region. Ambio 43, 352–360. https://doi.org/10.1007/s13280-013-0435-1.

Heiskanen, A.-S., 1998. Factors governing sedimentation and pelagic nutrient cycles in the northern Baltic Sea. Monograph. Boreal Environ. Res. 8, 1–80.

Heiskanen, A.-S., Tamminen, T., Gundersen, K., 1996. The impact of planktonic food web structure on nutrient retention and loss from a late summer pelagic system in the coastal northern Baltic Sea. Mar. Ecol. Prog. Ser. 145, 195–208.

HELCOM, 2007. HELCOM Baltic Sea action plan. Available from: http://www.helcom.fi/Documents/Baltic%20sea%20action%20plan/BSAP_Final.pdf. (Accessed April 28, 2018).

HELCOM, 2009a. In: Biodiversity in the Baltic Sea—an integrated thematic assessment on biodiversity and nature conservation in the Baltic Sea. Baltic Sea Environment Proceedings No. 116B.

HELCOM, 2009b. In: Eutrophication in the Baltic Sea—an integrated thematic assessment of the effects of nutrient enrichment and eutrophication in the Baltic Sea region. Baltic Sea Environment Proceedings No. 115B.

HELCOM, 2010a. In: Hazardous substances in the Baltic Sea—an integrated thematic assessment of hazardous substances in the Baltic Sea. Baltic Sea Environment Proceedings No. 120B.

HELCOM, 2010b. Towards a tool for quantifying anthropogenic pressures and potential impacts on the Baltic Sea marine environment: a background document on the method, data and testing of the Baltic Sea pressure and impact indices. Baltic Sea Environment Proceedings No. 125.

HELCOM, 2010c. In: Ecosystem health of the Baltic Sea 2003–2007: HELCOM initial holistic assessment. Baltic Sea Environment Proceedings No. 122.

HELCOM, 2011. In: The fifth Baltic sea pollution load compilation (PLC-5). Baltic Sea Environment Proceedings No. 128.

HELCOM, 2013a. Summary report on the development of revised Maximum Allowable Inputs (MAI) and updated Country Allocated Reduction Targets (CART) of the Baltic Sea Action Plan. http://www.helcom.fi/Documents/Ministerial2013/Associated%20documents/Supporting/Summary%20report%20on%20MAI-CART.pdf. (Accessed April 28, 2018).

HELCOM, 2013b. Approaches and methods for eutrophication target setting in the Baltic Sea region. Baltic Sea Environment Proceedings 133. pp. 134.

HELCOM, 2013c. In: Climate change in the Baltic Sea area, HELCOM thematic assessment in 2013. Baltic Sea Environment Proceedings 137. pp. 66.

HELCOM, 2013d. Final report on implementation of hot spots programme under the Baltic Sea joint comprehensive environmental action programme (JCP), 1992–2013. Available from: http://www.helcom.fi/Lists/Publications/Final%20report%20on%20JCP%20efficiency.pdf#search=Joint%20Comprehensive%20Environmental%20Action%20Programme. (Accessed April 29, 2018).

HELCOM, 2013e. In: HELCOM core indicators: final report of the HELCOM CORESET project. Baltic Sea Environment Proceedings No. 136.

HELCOM, 2014a. Towards a healthier Baltic Sea—implementation of the Baltic Sea action plan in Russia. BASE project final report. Recommendations. Available from: http://www.helcom.fi/Lists/Publications/BASE%20Final%20report%20-%20%20chapter%209%20-%20Recommendations.pdf. (Accessed April 29, 2018).

HELCOM, 2014b. In: Eutrophication status of the Baltic Sea 2007–2011—a concise thematic assessment. Baltic Sea Environment Proceedings No. 143.

HELCOM, , 2015a. In: Updated fifth Baltic Sea pollution load compilation (PLC-5.5). Baltic Sea Environment Proceedings No. 145.

HELCOM, 2015b. Regional action plan for Marine litter in the Baltic Sea. 20 pp. http://www.helcom.fi/Lists/Publications/Regional%20Action%20Plan%20for%20Marine%20Litter.pdf. (Accessed April 28, 2018).

HELCOM, 2017a. First version of the "State of the Baltic Sea" report—June 2017—to be updated in 2018. Available from: http://stateofthebalticsea.helcom.fi. (Accessed April 30, 2018).

HELCOM, 2017b. Economic and social analyses in the Baltic Sea region—supplementary report to the first version of the HELCOM "State of the Baltic Sea" report 2017. Available from: http://stateofthebalticsea.helcom.fi/about-helcom-and-theassessment/downloads-and-data/. (Accessed April 28, 2018).

HELCOM, 2018. In: Sources and pathways of nutrients to the Baltic Sea. Baltic Sea Environment Proceedings No. 153.

Hong, B., Swaney, D.P., McCrackin, M., Svanbäck, A., Humborg, C., Gustafsson, B., 2017. Advances in NANI and NAPI accounting for the Baltic drainage basin: spatial and temporal trends and relationships to watershed TN and TP fluxes. Biogeochemistry 133, 245–261. https://doi.org/10.1007/s10533-017-0330-0.

Hong, B., Swaney, D.P., Mörth, C.-M., Smedberg, E., Hägg, H.E., Humborg, C., Howarth, R.W., Bouraoui, F., 2012. Evaluating regional variation of net anthropogenic nitrogen and phosphorus inputs (NANI/NAPI), major drivers, nutrient retention pattern and management implications in the multinational areas of Baltic Sea basin. Ecol. Model. 227, 117–135.

Huttunen, I., Lehtonen, H., Huttunen, M., Piirainen, V., Korppoo, M., Veijalainen, N., Viitasalo, M., Vehviläinen, B., 2015. Effects of climate change and agricultural adaptation on nutrient loading from finnish catchments to the Baltic Sea. Sci. Total Environ. 529, 168–181. https://doi.org/10.1016/j.scitotenv.2015.05.055.

Hyytiäinen, K., Ahlvik, L., Ahtiainen, H., Artell, J., Huhtala, A., Dahlbo, K., 2015. Policy goals for improved water quality in the Baltic Sea: when do the benefits outweigh the costs? Environ. Resour. Econ. 61, 217–241.

Jetoo, S., 2018. Barriers to effective eutrophication governance: a comparison of the Baltic Sea and North American Great Lakes. Water 10, 400. https://doi.org/10.3390/w10040400.

Joas, M., 1997. Finland: from local to global politics. In: Andersen, M.S., Liefferink, J.D. (Eds.), European Environmental Policy: The Pioneers, Series: Issues in Environmental Politics. Manchester University Press, Manchester, pp. 119–160.

Joas, M., 1999. Building up and splitting down—environmental policy organisation in the Nordic Countries. In: Joas, M., Hermanson, A.-S. (Eds.), The Nordic Environments: Comparing Political, Administrative, and Policy Aspects. Ashgate, Aldershot, pp. 133–162.

Joas, M., Jahn, D., Kern, K. (Eds.), 2008. Governing a common sea: environmental policies in the Baltic Sea Region. Earthscan, London, pp. 256.

Kahru, M., Elmgren, R., 2014. Multidecadal time series of satellite-detected accumulations of cyanobacteria in the Baltic Sea. Biogeosciences 11, 3619–3633.

Karlson, K., Rosenberg, R., Bonsdorff, E., 2002. Temporal and spatial large-scale effects of eutrophication and oxygen deficiency on benthic fauna in Scandinavian and Baltic waters—a review. Oceanogr. Mar. Biol. Annu. Rev. 40, 427–489.

Karlsson, M., Gilek, M., Lundberg, C., 2016. Eutrophication and the ecosystem approach to management: a case study of Baltic Sea environmental governance. In: Gilek, M., Karlsson, M., Linke, S., Smolarz, K. (Eds.), Environmental Governance of the Baltic Sea. MARE Publication Series vol. 10. Springer, Dordrecht, pp. 21–44.

Koontz, T.M., Gupta, D., Mudliar, P., Ranjan, P., 2015. Adaptive institutions in social-ecological systems governance: a synthesis framework. Environ. Sci. Pol. 53, 139–151. https://doi.org/10.1016/j.envsci.2015.01.003.

Korpinen, S., Bonsdorff, E., 2015. Eutrophication and hypoxia: impacts of nutrient and organic enrichment. In: Crowe, T.P., Frid, C.L.J. (Eds.), Marine Ecosystems: Human Impacts on Biodiversity, Functioning and Services. Cambridge University Press, Cambridge, pp. 202–243. ISBN 978-1-107-03767-0.

Korpinen, S., Meski, L., Andersen, J.H., Laamanen, M., 2012. Human pressures and their potential impact on the Baltic Sea ecosystem. Ecol. Indic. 15, 105–114. https://doi.org/10.1016/j.ecolind.2011.09.023.

Kosenius, A.-K., Ollikainen, M., 2015. Ecosystem benefits from coastal habitats—a three country choice experiment. Mar. Policy 58, 15–27.

Kotilainen, A.T., Arppe, L., Dobosz, S., Jansen, E., Kabel, K., Karhu, J., Kotilainen, M.M., Kuijpers, A., Lougheed, B.C., Meier, E.H.M., Moros, M., Neumann, T., Porsche, C., Poulsen, N., Rasmussen, P., Ribeiro, S., Risebrobakken, B., Ryabchuk, D., Schimanke, S., Snowball, I., Spiridonov, M., Virtasalo, J.J., Weckström, K., Witkowski, A., Zhamoida, V., 2014. Echoes from the past: a healthy Baltic Sea requires more effort. Ambio 43, 60–68. https://doi.org/10.1007/s13280-013-0477-4.

Laakkonen, S., Laurila, S., 2007. Changing environments or shifting paradigms? Strategic decisions towards water protection in Helsinki 1850–2000. Ambio 36, 212–219.

Larsson, M., Granstedt, A., 2010. Sustainable governance of the agriculture and the Baltic Sea—agricultural reforms, food production and curbed eutrophication. Ecol. Econ. 69, 1943–1951. https://doi.org/10.1016/j.ecolecon.2010.05.003.

Larsson, U., Elmgren, R., Wulff, F., 1985. Eutrophication and the Baltic Sea. Ambio 14, 9–14.

Lehtonen, K.K., Bignert, A., Bradshaw, C., Broeg, K., Schiedek, D., 2017. Chemical pollution and ecotoxicology. In: Snoeijs-Leijonmalm, P., Schubert, H., Radziejewska, T. (Eds.), Biological Oceanography of the Baltic Sea. Springer, Dordrecht, pp. 547–587. https://doi.org/10.1007/978-94-007-0668-2_10. https://link.springer.com/chapter/10.1007/978-94-007-0668-2_10.

Leppäranta, M., Myrberg, K., 2009. Physical Oceanography of the Baltic Sea. Springer, Praxis Publishing Ltd, Chichester.

Lester, S.E., Costello, C., Halpern, B.S., Gaines, S.D., CrowWhite, Barth, J.A., 2013. Evaluating trade offs among ecosystem services to informmarine spatial planning. Mar. Policy 38, 80–89. https://doi.org/10.1016/j.marpol.2012.05.022.

Linke, S., Gilek, M., Karlsson, M., 2016. Science-policy interfaces in Baltic Sea environmental governance: towards regional cooperation and management of uncertainty? In: Gilek, M., Karlsson, M., Linke, S., Smolarz, K. (Eds.), Environmental Governance of the Baltic Sea. Springer, Dordrecht, pp. 173–204.

Meier, H.E.M., Hordoir, R., Andersson, H.C., Dieterich, C., Eilola, K., Gustafsson, B.G., Höglund, A., Schimanke, S., 2012. Modeling the combined impact of changing climate and changing nutrient loads on the Baltic Sea environment in an ensemble of transient simulations for 1961–2099. Clim. Dyn. 39, 2421–2441. https://doi.org/10.1007/s00382-012-1339-y.

Meier, M.H.E., Andersson, H.C., Arheimer, B., Donnelly, C., Eilola, K., Gustafsson, B.G., et al., 2014. Modeling of the Baltic Sea ecosystem to provide scenarios for management. Ambio 43, 37–48. https://doi.org/10.1007/s13280-013-0475-6.

Möllmann, C., Diekmann, R., Muller-Karulis, B., Kornilovs, G., Plikshs, M., Axe, P., 2009. Reorganization of a large marine ecosystem due to atmospheric and anthropogenic pressure: a discontinuous regime shift in the Central Baltic Sea. Glob. Chang. Biol. 15, 1377–1393. https://doi.org/10.1111/j.1365-2486.2008.01814.x.

Mohrholz, V., Naumann, M., Nausch, G., Krüger, S., Gräwe, U., 2015. Fresh oxygen for the Baltic Sea — an exceptional saline inflow after a decade of stagnation. J. Mar. Syst. 148, 152–166.

Niiranen, S., Yletyinen, J., Tomczak, M.T., Blenckner, T., Hjerne, O., MacKenzie, B.R., Müller-Karulis, B., Neumann, T., et al., 2013. Combined effects of global climate change and regional ecosystem drivers on an exploited marine food web. Glob. Chang. Biol. 19, 3327–3342. https://doi.org/10.1111/gcb.12309.

Nordström, M.C., Bonsdorff, E., 2017. Organic enrichment simplifies marine benthic food web structure. Limnol. Oceanogr. https://doi.org/10.1002/lno.10588.

Nordvarg, L., Johansson, T., 2002. The effects of fish farm effluents on the water quality in the Åland archipelago, Baltic Sea. Aquac. Eng. 25, 253–279.

Norén, K., Meironyté, D., 2000. Certain organochlorine and organobromine contaminants in Swedish human milk in perspective of past 20–30 years. Chemosphere 40, 1111–1123.

NPE, 2017. The New Plastics Economy—Catalysing Action. Ellen MacArthur Foundation. https://www.newplasticseconomy.org. (Accessed April 28, 2018).

Ojaveer, H., Jaanus, A., MacKenzie, B.R., Martin, G., Olenin, S., Radziejewska, T., et al., 2010. Status 757 of Biodiversity in the Baltic Sea. PLoS ONE 5, e12467.

Petersen, J.K., Hasler, B., Timmermann, K., Nielsen, P., Bruunshøj Tørring, D., Mørk Larsen, M., Holmer, M., 2014. Mussels as a tool for mitigation of nutrients in the marine environment. Mar. Pollut. Bull. 82, 137–143.

Puttonen, I., Mattila, J., Jonsson, P., Karlsson, O.M., Kohonen, T., Kotilainen, A., Lukkari, K., Malmaeus, J.M., Rydin, M., 2014. Distribution and estimated release of sediment phosphorus in the northern Baltic Sea archipelagos. Estuar. Coast. Shelf Sci. 145, 9–21.

Räike, A., Knuuttila, S., Ekholm, P., Kondratyev, S., Ennet, P., Ulm, R., Oblomkova, N., 2016. Nutrient inputs. In: Raateoja, M., Setälä, O. (Eds.), The Gulf of Finland Assessment. 27. Reports of the Finnish Environment Institute, pp. 89–93. https://helda.helsinki.fi/handle/10138/166296.

Räsänen, T., Laakkonen, S., 2007. Cold War and the environment: the role of Finland in international environmental politics in the Baltic Sea region. Ambio 36, 229–236.

Reusch, T.H.B., Dierking, J., Andersson, H., Bonsdorff, E., Carstensen, J., Casini, M., Czajkowski, M., Hasler, B., Hinsby, K., Hyytiäinen, K., Johannesson, K., Jomaa, S., Jormalainen, V., Kuosa, H., Kurland, S., Laikre, L., MacKenzie, B.R., Margonski, P., Melzner, F., Oesterwind, D., Ojaveer, H., Refsgaard, J.C., Sandström, A., Schwarz, G., Tonderski, K., Winder, M., Zandersen, M., 2018. The Baltic Sea as a time machine for the future coastal ocean. Sci. Adv. 2018, 4. https://doi.org/10.1126/sciadv.aar8195.

Rocha, J., Yletyinen, J., Biggs, R., Blenckner, T., Peterson, G., 2015. Marine regime shifts: drivers and impacts on ecosystems services. Philos. Trans. R. Soc. B 370, 20130273. https://doi.org/10.1098/rstb.2013.0273.

Röhr, E.A., Boström, C., Canal-Vergés, P., Holmer, M., 2016. Blue carbon stocks in Baltic Sea eelgrass (Zostera marina) meadows. Biogeosciences 13, 6139–6153.

Rydin, E., Kumblad, L., Wulff, F., Larsson, P., 2017. Remediation of a Eutrophic Bay in the Baltic Sea. Environ. Sci. Technol. 51, 4559–4566. https://doi.org/10.1021/acs.est.6b06187.

Scharin, H., Ericsdotter, S., Elliott, M., Turner, K.R., Niiranen, S., Blenckner, T., Hyytiäinen, K., Ahlvik, L., Ahtiainen, H., Artell, J., Hasselström, L., Söderqvist, T., Rockström, J., 2016. Processes for the sustainable stewardship of marine environments. Ecol. Econ. 128, 55–67. https://doi.org/10.1016/j.ecolecon.2016.04.010.

Seghetta, M., Tørring, D., Bruhn, A., Thomsen, M., 2016. Bioextraction potential of seaweed in Denmark—an instrument for circular nutrient management. Sci. Total Environ. 563–564, 513–529. https://doi.org/10.1016/j.scitotenv.2016.04.010.

SEPA, 2008. Ecosystem Services Provided by the Baltic Sea and Skagerrak. The Swedish Environmental Protection Agency, 1–193. Stockholm. Report 5873.

Setälä, O., Fleming-Lehtinen, V., Lehtiniemi, M., 2014. Ingestion and transfer of microplastics in the planktonic food web. Environ. Pollut. 185, 77–78.

SITRA, 2016. Leading the cycle—finnish road map to a circular economy 2016–2025. Sitra Stud. 121, 55. https://media.sitra.fi/2017/02/28142644/Selvityksia121.pdf.

Snoeijs-Leijonmalm, P., Andrén, E., 2017. Why is the Baltic Sea so special to live in? In: Snoeijs-Leijonmalm, P., Schubert, H., Radziejewska, T. (Eds.), Biological Oceanography of the Baltic Sea. Springer, Dordrecht, pp. 23–84. https://doi.org/10.1007/978-94-007-0668-2_10. https://link.springer.com/chapter/10.1007/978-94-007-0668-2_10.

Snoeijs-Leijonmalm, P., Schubert, H., Radziejewska, T., 2017. Biological Oceanography of the Baltic Sea. Springer, Dordrecht686. https://doi.org/10.1007/978-94-007-0668-2_10. https://link.springer.com/chapter/10.1007/978-94-007-0668-2_10.

Söderqvist, T., Eggesrt, H., Olsson, B., Soutukorva, Å., 2005. Economic valuation for sustainable development in the Swedish coastal zone. Ambio 34, 169–175.

Söderqvist, T., Hasselström, L., 2008. The economic value of ecosystem services provided by the Baltic Sea and Skagerrak. In: The Swedish Environmental Protection Agency Report 5874. Sweden, Stockholm. https://www.naturvardsverket.se/Documents/publikationer/978-91-620-5874-6.pdf.

Söderström, S., Kern, K., 2017. The ecosystem approach to management in marine environmental governance: institutional interplay in the Baltic Sea region. Environ. Pol. Gov. 27, 619–631. https://doi.org/10.1002/eet.1775.

Stadmark, J., Conley, D.J., 2011. Mussel farming as a nutrient reduction measure in the Baltic Sea: consideration of nutrient biogeochemical cycles. Mar. Pollut. Bull. 62, 1385–1388.

Stigebrandt, A., Gustafsson, B., 2007. Improvement of Baltic proper water quality using large-scale ecological engineering. Ambio 36, 280–286.

Suutari, M., Leskinen, E., Spilling, K., Kostamo, K., Seppälä, J., 2017. Nutrient removal by biomass accumulation on artificial substrata in the northern Baltic Sea. J. Appl. Phycol. 29, 1707–1720. https://doi.org/10.1007/s10811-016-1023-0.

Svedäng, H., Hornborg, S., 2017. Historic changes in length distributions of three Baltic cod (Gadus morhua) stocks: evidence of growth retardation. Ecol. Evol. 7 (16), 6089–6102. https://doi.org/10.1002/ece3.3173.

Svendsen, L., Gustafsson, B.M., Larsen, S.E., Sonesten, L., Knuuttila, S., Kamenetsky, D.F., 2017. Inputs of nitrogen and phosphorus to the Baltic Sea. HELCOM Core Indicator Report. 30 March 2018 http://www.helcom.fi/Core%20Indicators/Nutrient%20inputs%20-%20Core%20indicator%20report-HOLAS%20II%20component%202017.pdf. (Accessed April 28, 2018).

Tynkkynen, N., Schönach, P., Pihlajamäki, M., Nechiporuk, D., 2014. The governance of the mitigation of the Baltic Sea eutrophication: exploring the challenges of the formal governing system. Ambio 43, 105. https://doi.org/10.1007/s13280-013-0481-8.

Vahtera, E., Conley, D.J., Gustafsson, B.G., Kuosa, H., Pitkänen, H., Savchuk, O.P., Tamminen, T., Viitasalo, M., Voss, M., Wasmund, N., Wulff, F., 2007. Internal ecosystem feedbacks enhance nitrogen-fixing cyanobacteria blooms and complicate management in the Baltic Sea. Ambio 36, 186–194. https://doi.org/10.1579/0044-7447(2007)36[186:IEFENC]2.0.CO;2.

Varjopuro, R., Andrulewicz, E., Blenckner, T., Dolch, T., Heiskanen, A.S., Pihlajamäki, M., Steiner Brandt, U., Valman, M., Gee, G., Potts, T., Psuty, I., 2014. Coping with persistent environmental problems: systemic delays in reducing eutrophication of the Baltic Sea. Ecol. Soc. 19 (4), 48. https://doi.org/10.5751/ES-06938-190448.

Weigel, B., Andersson, H.C., Meier, H.E.M., Blenckner, T., Snickars, M., Bonsdorff, E., 2015. Long-term progression and drivers of coastal zoobenthos in a changing system. Mar. Ecol. Prog. Ser. 528, 141–159. https://doi.org/10.3354/meps11279.

Węsławski, J.M., Andrulewicz, E., Boström, C., Horbowy, J., Linkowski, T., Mattila, J., Olenin, S., Piwowarczyk, J., Skóra, K., 2017. Ecosystem goods, services and management. In: Snoeijs-Leijonmalm, P., Schubert, H., Radziejewska, T. (Eds.), Biological Oceanography of the Baltic Sea. Springer, Dordrecht, pp. 609–643. https://doi.org/10.1007/978-94-007-0668-2_10. https://link.springer.com/chapter/10.1007/978-94-007-0668-2_10.

Wulff, F., Savchuk, O.P., Sokolov, A., Humborg, C., Mörth, C.-M., 2007. Management options and effects on a marine ecosystem: assessing the future of the Baltic. Ambio 36, 243–249.

Wulff, F., Humborg, C., Andersen, H.E., Blicher-Mathiesen, G., Mikołaj Czajkowski, M., Elofsson, K., Fonnesbech-Wulff, A., Hasler, B., Bongghi Hong, B., Viesturs Jansons, V., Carl-Magnus Mörth, C.-M., Smart, J.C.R., Smedberg, E., Stålnacke, P., Swaney, D.P., Thodsen, H., Was, A., Zylicz, T., 2014. Reduction of Baltic Sea nutrient inputs and allocation of abatement costs within the Baltic Sea catchment. Ambio 43, 11–25. https://doi.org/10.1007/s13280-013-0484-5.

Yletyinen, J., Bodin, Ö., Weigel, B., Nordström, M.C., Bonsdorff, E., Blenckner, T., 2016. Regime shifts in marine communities: a complex systems perspective on food web dynamics. Proc. R. Soc. B 283, 20152569. https://doi.org/10.1098/rspb.2015.2569.

Zillén, L., Conley, D.J., Andrén, T., Andrén, E., Björck, S., 2008. Past occurrences of hypoxia in the Baltic Sea and the role of climate variability, environmental change and human impact. Earth-Sci. Rev. 91, 77–92. https://doi.org/10.1016/j.earscirev.2008.10.001.

Chapter 21

The Black Sea—The Past, Present, and Future Status

Abdulaziz Güneroğlu*, Osman Samsun†, Muzaffer Feyzioğlu‡, Mustafa Dihkan§

*Department of Marine Ecology, Faculty of Marine Sciences, Karadeniz Technical University, Çamburnu, Trabzon, †Faculty of Fisheries, Sinop University, Sinop, Turkey, ‡Department of Marine Science and Technology, Faculty of Marine Sciences, Karadeniz Technical University, Çamburnu, Trabzon, §Department of Geomatics, Faculty of Engineering, Karadeniz Technical University, Çamburnu, Trabzon

1 INTRODUCTION

The Black Sea is an internal semienclosed basin characterized by brackish and high productivity waters surrounded by six different countries (Turkey, Russian Federation, Ukraine, Romania, Bulgaria, and Georgia). The Black Sea has suffered from unwanted ecological and environmental problems for the last 50 years. Despite an awareness of the issues among bordering countries, there is no satisfactory steps put forward for making positive future projections of this unique ecosystem (O'Higgins et al., 2014). The remedy depends on the synchronization level across the countries bordering the sea and their efforts to overcome the problems. This is because each country has its own regulations and systems and it is difficult to bring them together. The European Union (EU) membership of Romania and Bulgaria is promising. Turkey is also candidate member, which means EU regulations can be applied at least to some extent. It is widely accepted that negative impacts of global climate change on world ecological settings are now observed in many geographic regions such as in the Black Sea ecosystem. One of the most important and measurable effect on world seas and oceans is the elevated trends of mean sea surface temperatures (SSTs). Moreover, changing flood and plume patterns of the rivers, strong stratification of water masses, changing water salinities, and limited vertical exchange of nutrients or oxygen are some other negative impacts mentioned in the literature (Miladinova et al., 2016).

Expected ecosystem-related problems of the Black Sea for the next 50 years can be summarized as unregulated fishing and overfishing, partially mitigated eutrophication, global climate change-related rising mean SSTs, coastal urbanization, and dense housing, solid waste discharges including micro and macroplastics and a high risk from maritime oil transportation. Among all of them, even though related to human-induced activities, eutrophication and high SSTs are demanding and complex problems that cannot be resolved in short time spans and furthermore appropriate solutions for such problems can be considerably expensive. On the other hand, coastal urbanization, solid waste disposal, and overfishing are relatively easy problems to respond to by effective regulations and the application of necessary engineering solutions.

Environmental issues of the Black Sea are very well documented in the literature for the last 50–60 years. The regime shifts observed in the ecological state of the sea were generally attributed to anthropogenic interference. There are mainly two approaches explaining the rationale behind the shifts. The first one is overfishing (especially the top predators) causing the trophic cascade in a top-down ecosystem hierarchy by interrupting the prey-predator relations and the second one is the eutrophication resulted by the massive intrusion of the nutrients into the system and characterized by very intense blooming events and high production rates. However, the double effect of these two approaches is more likely to be responsible for the regime shifts of the Black Sea. However, the impact of the natural climate oscillations in the mitigation of the problem witnessed in the last 20 years has to be explained in more detail in order to be prepared for the next 50 years. The hot agenda of ecosystem-related problems in the Black Sea will probably be an issue of high priority in upcoming decades as a result of persistent anthropogenic activities. The philosophy to follow as a potential solution can be based on the root causes and attempts to formulate the solutions by bringing together science and society. This can only be achieved by consensus among the bordering countries by starting international cooperation that is applicable and legally binding for each nation. Unfortunately, a full recovery of the ecosystem can be more difficult and demanding than expected.

Coasts and Estuaries. https://doi.org/10.1016/B978-0-12-814003-1.00021-6

2 GEOGRAPHIC SETTING AND COASTAL GEOMORPHOLOGY

The Black Sea is the world's largest anoxic sedimentary basin surrounded by a 4400-km long coastline with a relief of approximately 398-m mean elevation. The basin is inhabited by approximately 140 million human population and covers approximately 460,000 km^2 of sea surface. The geomorphologic character of the region is represented by low-relief areas on the western side and steep high-relief zones on the eastern and southeastern parts (Ludwig et al., 2009; Allenbach et al., 2015). The Black Sea has a diverse geomorphology such as plains, mountains, highlands, and forests as a result of its unique geographic location. Boreal climatology dominates the northern sections whereas the southern regions have subtropical character. Generally, the Black Sea faces weather masses from Atlantic, Mid-European, and Siberian origins. Saharan-originated weather systems can also be observed especially during the mid-seasons. The area is the sedimentary basin of Alpin-Himalayan orogeny and situated between the Caucasian, Crimean, and Balkan mountains. However, the northwestern side of the region is under the influence of the Eastern European platform. This explains the roughness and geomorphologic movements of the region as the Black Sea basin is in the middle of two colliding plates with different tectonic formations (Drozdov et al., 1992). The Caucasian and Crimean mountains are located on the northern and northeastern sections while Balkanid-Pontid Mountains extend along the western and northwestern sides (Antonidze, 2010; Kosyan and Velikova, 2016). The compaction of Arabian and South European plaques is considered as the reason for the formation of the sedimentary basin millions of years ago. The sediment accumulation in the middle of the basin proves that the area geologically was possibly a part of ocean or inner sea. The area is an enclosed region formed by two subbasins, one on the west and the other on the east (BSC (Black Sea Commission), 2008).

Lakes, islands, and swamps of the Danube delta characterize the geomorphological setting of the west part. The delta was possibly a bay of the Black Sea 6500 years ago before the transportation of excessive sediment by rivers that shaped the current deltaic formation. The delta moves toward the sea approximately 30 m each year. In the region, the Dobruja plateau exhibits moderate roughness with small rock formations. Beginning from the Istanbul strait, steep and harsh mountain formations extend toward eastern side along the coastal periphery. The southern coast of the sea is fed by many rivers such as the Sakarya, Yesilirmak, Kizilirmak, and Coruh. Associated areas are fertile agricultural lands. Many other small streams are in the form of unregulated creeks transporting waters and nutrients as open drainage to the sea (Zaitsev et al., 2002). Sediment grain sizes transported by rivers differ according to its origin, geomorphological shape, and place. The sediments transported from southern and southeastern sides have generally large grain sizes whereas the northwestern sediment type consists of fine grains due to relatively low slopes. Undoubtedly, the Danube River is the most influential factor in shaping the northwestern deltaic environment (Panin and Jipa, 1998). The western shelves including the Istanbul strait are characterized by very productive waters. This area covers almost 127,000 km^2, that is, around 30% of the total Black Sea surface area and comprises 94% of the total continental shelf of the whole basin (Panin and Jipa, 2002).

The Black Sea has been attributed different characters throughout the history. Sometimes it was regarded as having "unhospitable or dangerous" waters, and sometimes it was given positive attributes. The region has attracted many nations due to its unique geographic locations (Kosyan and Velikova, 2016). The bordering countries of the Black Sea have a remarkable history and the center of many civilizations. It is known that coastal utilization is a common for all countries, as they need the coast for various activities such as transport, settlement, and recreation (Antonidze, 2010). Until the beginning of the 20th century, the Black Sea was a relatively natural and pristine basin with limited human impact. Nevertheless, in recent decades, the situation changed and intense anthropogenic processes caused a degradation of the basin (Kosyan and Velikova, 2016). Linear littoral settlement types appear to be a problem in areas dominated by coastal mountains with limited hinterland. The geomorphological setting and inadequate planning strategies are the major causes of the coastal urbanization in the region (Guneroglu et al., 2013). The vulnerability to inundation and flood risk are some other problems linked to very dense coastal urbanization. The most recent urbanization level around the coastal strips of the Black Sea is shown in Fig. 1.

Many landslides and flooding incidences are experienced each year that result in a loss of life and properties. Furthermore, near coastal and exclusive economic zone (EEZ) areas of the total sea surface there are problems due to the overexploitation of resources and sectoral competition such as tourism and fisheries. Therefore, coastal and open sea interactions, which are vital for the basin, are limited as a result of unregulated usage and degradation. Additional ecological problems associated with institutional and application issues emanate from the mismanagement of coastal resources (Guneroglu et al., 2014). Another human-induced problem related to geomorphology is the coastline change observed along the coastal periphery of the Black Sea. Coastline changes can occur as the result of both natural and anthropogenic factors. Recently, most of the changes reported on the coastal zone of the Black Sea are mainly the result of human activities such as urbanization and inappropriate engineering solutions (Guneroglu, 2015). The importance of coastline change is strongly linked to the economic value of the coastal area. Coastal lands are valuable as they offer productive places for activities such as transport,

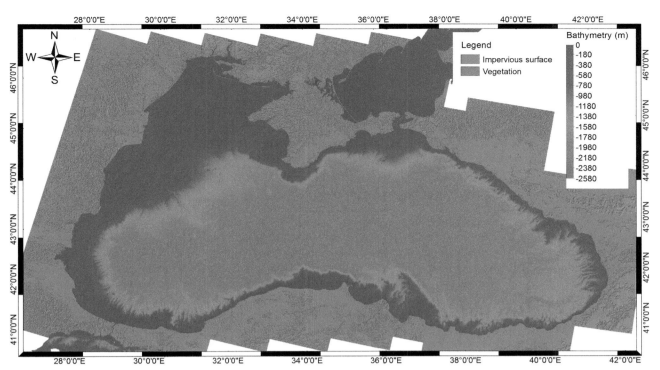

FIG. 1 Landsat 8-LDCM classified mosaic image of the Black Sea basin showing coastal impervious surfaces, vegetation, and bathymetric variability.

tourism, and recreation. Therefore, dynamic changes of the coastline are the expected results of both natural and human-induced causes (Guneroglu, 2015; Kuleli et al., 2011; Karsli et al., 2011). Living very close to the coastline creates some risks and dangers to our daily lives such as storm surges, huge waves, and rising sea levels. Although sea level rise (SLR) is not a big problem today for the Black Sea, if the global climate change continues in its current level for the next decades, it can become a major coastal issue in the near future. Moreover, the populations living on coastal zone of the Black Sea have not yet experienced SLR as the sea is not affected by tides. Therefore, in order to avoid unwanted consequences, a strategic plan about the sea should include a section on SLR. It is reported that the seawaters of Turkey have risen approximately 2.5–2.8 mm/year for the last 85 years. Similarly, on the northern coasts of the Black Sea, the figures reach up to 3.5–4.5 mm/year (Alpar, 2009). Avsar et al. (2015) reported almost the same rates (3.16 ± 0.77 mm/year) for the Black Sea based on data between 1993 and 2014. For example, a project carried out by Stănică and Panin (2009) on the northwestern shelf of the sea revealed that by 2030 the SLR can reach up to 13–14 cm which is very dangerous as it may causes seawater to retreat toward the land by approximately 35–50 m. This can cause a massive destruction in areas especially with narrow coasts in the southern and southeastern parts of the sea. As the most of the investments and urbanization occur on these narrow strips, Kuleli (2010) emphasized that a possible rise by 10 m in Turkish coast can cover 7319 km^2 of the total coastal land of the country. Undoubtedly, Turkish coasts are vulnerable to SLR risk if the necessary precautions are not taken and put into action in the near future. The IPCC climate scenarios must be seriously taken into account and evaluated to avoid possible negative results of SLR in the Black Sea basin.

Regarding the coastal beaches of the Black Sea, there are in total 1228 available beaches of the basin, 2042 km long and occupying an area of 224 km^2. Interestingly, the width of 61% of the beaches is less than 50 m and 47% of them are protected by engineering structures (Allenbach et al., 2015). The adverse effect of SLR may also couple with man-made structures such as building dams in the region. Although dams are necessary for water storage and usage and electricity production, they can limit the sediment supply to the coastal zone and degrade its vulnerability against coastal erosion. A basin-wide action plan can help to mitigate these negative impacts of anthropogenic pressure on the coastal zone of the Black Sea (Tsereteli et al., 2011). The geomorphological setting is also related to land-use type in the region. Most of the coastal regions in the Black Sea are used as agricultural lands especially in places with moderate roughness such as deltaic environments. Agriculture is an important tool for creating revenues as the total population approximates to 140 million in the basin (Ludwig et al., 2009). However, agricultural practices affect both the aquatic and terrestrial ecological formation of the basin. Intense use of fertilizers causes a suite of eutrophication problems in coastal waters leading to a distortion and complex interactions in aquatic food webs. Moreover, the irrigation infrastructure planned to meet the necessary water demand for agriculture alters the terrestrial ecology around the dams or the inland water resources. In accordance with the MSFD and WFD, the bordering countries, as agreed in 1992 Bucharest Convention, are required to stop point or nonpoint (diffuse) nutrient discharges to the aquatic environment by carrying out specific regional projects and monitoring.

Engineering solutions for protecting the coastal zone may also alter the coastal morphology. It is highly recommended that soft engineering techniques should be used in defending the coast instead of hard engineering constructions. Firstly, soft engineering designs are best suited to the natural environment compared to hard structures, they are cheaper than hard solutions, and they can be used to preserve sediment load as well as supply habitats to the aquatic biota. Hard engineering solutions could be used in places where soft engineering techniques tend to fail in defending the coastal zone.

3 ECOLOGICAL STATE AND HEALTH OF THE SEA

The Black Sea, despite various definitions in the literature, is regarded as a semienclosed subbasin of Mediterranean Sea as there are many similarities in terms of biological, geographical, and ecological aspects. Connecting European and Asian continents, the ecology of the Black Sea is determined by geological activities of the region (Zaitsev, 2001). The coastal deep formation is mainly covered by sand dunes and shallow banks on northwestern shelf, whereas rocks and mud banks are found in the northern and southern sections of the coast. The well-known cyclonic (clockwise) circulation pattern of the rim current is also affected by the coastal bathymetry and morphology. The main circulation pattern of the Black Sea can be depicted by two subgyres—one in the west and the other is on the east. The formation of the cyclonic gyres is attributed to a positive freshwater budget and the outflow of many rivers, which creates a quasipermanent buoyancy of surface waters with low salinity above the more saline water masses. Meanders and quasipermanent anticyclonic gyres are formed between the coastal waters and the main circulation pattern and they are very important for triggering mixing and intrusion of the coastal waters to the inner basin. The exchange of these two water masses is quite important for the productivity of the continental shelf. Ekman upwelling zones can be observed in the inner basin as a result of the wind stress curl and frictional convergence effect, which creates a depression zone on the coastal waters. The result is anticyclonic eddies formed by a coupled impact of the wind dynamics and sea topography (Miladinova et al., 2016). The divergence of water masses in the inner basin is balanced by convergence areas formed in the coastal section. A strong vertical and horizontal mixing of water can reach its maxima in the winter season. Some other seasonal upwelling regions can be observed in very near coastal areas of the Black Sea (Daskalov, 1999). A cold intermediate layer (CIL) formation is a characteristic oceanographic process of the Black Sea, which enhances vertical mixing, and nutrient transport. The depth and thickness of the CIL could be changed through heating and cooling mechanisms but it is generally observed between 30 and 100 m depth and most aquatic life activity occurs in the first 50 m depth of the sea that is rich in oxygen, light, and adequate nutrients. Moreover, continental shelves on both narrow and wide coastal margins are responsible for the most productive fisheries of the Black Sea (Miladinova et al., 2016). The other vertical water mass of the sea that begins just below the CIL is the permanent pycnocline. The pycnocline layer limits the water mixing between haline Mediterranean anoxic layer and the CIL. It is well known that most of the aquatic systems are fragile against the excessive anthropogenic nutrient supply. An increasing population and demand for agricultural products triggered usage of fertilizers on coastal lands that can lead to high production rates and algal blooms in the marine environment. The process could even end with eutrophication and very high rates of carbon production, which then consumes the available oxygen in the water column and on the seabed during decomposition (Friedrich et al., 2014).

Annual peak values of bloom dynamics of the Black Sea are typically observed in autumn and winter seasons inside the rim current region and reach its lowest values in summers (Yunev et al., 2002; McQuatters-Gollop et al., 2008). Bloom dynamics and chl-a concentrations show different characteristics in coastal and open waters (BSC (Black Sea Commission), 2008). For instance, high chl-a concentrations can be observed in outflows of Bosporus in summer season. In winter when the prevailing winds are very strong, rim currents limit the exchange of coastal waters with inner regions but under low wind stress, high chl-a concentrations and phytoplankton spread to the entire waters of the basin (BSC (Black Sea Commission), 2008).

The Black Sea is distinguished from the Mediterranean Sea in terms of the phytoplankton and zooplankton assemblages, primary production, and nutrient supply (Zaitsev et al., 2001). Diatoms and dinoflagellates dominate the Black Sea phytoplankton groups. Despite the low plankton diversity, the sea is highly productive with a high abundance of cells per volume. The major taxa observed are Bacillariophyceae, Dinophyceae, Prymneciophyceae, and small flagellates (Moncheva et al., 2012). Some important species of note are *Emiliania huxleyi, Pseudonitzschia pseudodelicatissima, Pseudosolenia calcar-avis, Prorocentrum cordatum, Thalassionema nitzschioides, Ceratium fusus*, and *Gymnodinium* sp.

(Eker-Develi and Kideys, 2003). Furthermore, microflagellates are also remarkably abundant in the Black Sea. A total of 54 classes and 267 genera of dinoflagellates were reported for the basin. Some important species are *Protoperidinium* sp., *Ceratium* sp., *Dinophysis* sp., *Peridinium* sp. *ve Gymnodinium* sp. (Gómez and Boicenco, 2004). Diatoms that are widely reported in the region are *Cerataulina pelagica, Chaetoceros curvisaetus, Chaetoceros socialis, Cylindrotheca closterium*, and *Skeletonema costatum* (Moncheva et al., 2001). The Black Sea phytoplankton structure is strongly linked to environmental conditions and climatic variability. Therefore, the dominant phytoplankton group at any given time that can be observed are diatoms and sometimes switch to dinoflagellates. The causal factors behind this phenomenon are assumed to be the shift of the system toward a small cell community structure during the high primary production periods to make the better use of available sunlight. Some percentage rates of major systematic groups observed before and after 2000 and 2002 are illustrated in Table 1.

The major nanaplankton contributor to total production of the sea is *Emiliania huxleyi* and the first bloom of this species was first reported in 1951 for the Black Sea (Mikaelyan et al., 2005, 2011). Recently, the bloom of this coccolithophorid has been widely observed in coastal waters of the Black Sea with increasing frequency (Mikaelyan et al., 2015). There are different explanations for the primary production mechanism whether it is limited by phosphorus or nitrogen and whether silicate is available in sea column during the photosynthesis. For example, the spring bloom is limited by nitrogen whereas summer bloom is governed by both nitrogen and phosphorus. Moreover, the imbalance between the major nutrients can lead to a domination by bacteria or microzooplankton in the first levels of the food web (Mee et al., 2005). For example, Moncheva et al. (2001) reported that the northwestern shelf was diatom-dominated based on data collected for the period 1999 and 2000. Furthermore, the availability and abundance of different nutrients may enhance the ambient conditions for establishing diatom- or dinoflagellate-dominated ecosystems. Bloom timing and intensity are not only governed by environmental parameters but also they are influenced by sinking, grazing, and a trophic state control mechanism.

The Black Sea trophic structure has been greatly changed during the last 50 years. Considering the primary production and bloom timing between May and October, this period can be partitioned into three sections according to major shifts observed. Preeutrophication (prior to 1980), eutrophication (~1980–90), and posteutrophication (after 2000s) periods are important time spans shaping the Black Sea productivity. The trophic system of the sea evolved in accordance with changing environmental conditions. For instance, one of the most important picoplankton species (*Synechococcus* spp.) abundance has greatly increased in 10 years and reached up to 10-fold compared to rates prior to 2000 (Uysal, 2000, 2001; Feyzioglu et al., 2004, 2015; Kopuz et al., 2012). One of the most severe consequences of Black Sea eutrophication is red tide blooms. Red tides can be initiated by different species in the Black Sea such as dinoflagellates, euglena, or diatoms (Feyzioğlu and Öğüt, 2006). Recently, red tides of *Noctiluca scintillans* have been widely observed by bloom rates reaching up to 6.81×10^9 cells m^{-3} in 2011 on the southeastern coastal waters (Kopuz et al., 2014).

Ecological peculiarities of the Black Sea have shown a unique community structure of the zooplankton assemblage, which then shapes the base of the food web and is responsible for the healthy flow of the energy. Being the key components of the ecosystem, zooplankton are also important in terms of water quality as they are filter-feeding organisms. Oceanographic regular sampling of zooplankton in the Black Sea dates from 1950. Black Sea zooplankton structure is mainly composed of mesozooplankton groups such as Copepods, Cladocera, Chaetognatha, and Oikopleuridae. Advancements in new sampling gears and technologies promoted in situ sampling campaigns by enhancing the scientific quality and assurance of collected data. A micro (<0.5 mm), mesozoo (0.5–10 mm), and macro (>10 mm) classification scheme was applied using these technological gears on board a research vessel (Kovalev et al., 1999). The planktonic assemblage of the Black Sea is affected by the salinity of seawater. Zooplankton structure can ecogeographically be characterized as three origins: Mediterranean,

TABLE 1 The Rates of Some Microplanktonic Groups Observed in the Region in Different Periods

Systematic Groups	Before 2000 (%)	After 2002 (%)
Bacillariophyceae	52.88	44.12
Dinophyceae	33.65	41.18
Euglenophyceae	5.77	5.88
Chloprophyceae	2.88	1.47
Crysophyceae	3.85	5.88
Primnesiophyceae	0.96	1.47

pontic, and fresh water groups. Although they are low in number, those of Mediterranean origins are dominant in the Black Sea. Approximately 50 functional Mediterranean origin species were adapted to Black Sea ecological conditions (Kovalev et al., 1999). The most abundant mesozooplankton species observed are Copepoda, Cladocera, Chaetognatha, Oikopleuridae groups, and some ichthyoplankton such as eggs and larvae. In daylight conditions, some species such as *Calanus euxinus*, *Pseudocalanus elongatus*, and *Sagitta setosa* remain just above the anoxic layer at approximately 60 m depth. The percentage of *Acartia clausii* in the total zooplankton assemblage has greatly increased from 17% to 75% between 1980 and 1990 (Kovalev et al., 1999).

The decreasing trend in biomass of *C. euxinus* at the end of 1990 is followed by the observation of *Oithona similin* and *Oithona nana* in some in situ sampling studies carried out in the Black Sea but more recently *Oithona davisae*, an indo-pacific origin invasive species transported by ballast waters, was observed even in southeastern waters of the sea. It seems that *Oithona davisae* has established within an ecological niche to integrate with the available food web (Kovalev et al., 1999; Gubanova and Altukhov, 2007; Yıldız et al., 2017). Similar fluctuations in the biomass of pelagic fishes and mesozooplankton structure reported in the Black Sea were also observed in the North and the Baltic sea (Niermann et al., 1998). The phenomenon observed in the three seas may have some connections with climatic oscillations (Oguz et al., 2006). Zooplankton biomass intensity was generally observed in the first 50 m depth in coastal waters of the Black Sea. The zooplankton as secondary producers tend to follow the areas with high primary production such as the northwestern shelf, plume areas of important rivers and some near coastal waters of Turkey. A hypoxia problem occurred in northwestern shelf waters in 1973 causing greater eutrophic conditions and leading to long-lasting impacts on the marine ecosystem (Mee et al., 2005). Normally, ecological systems are known for their slow adaptation to changing environmental conditions. This rule is valid for systems with a good and healthy inner ecological balance. If the inner balance of the system is disturbed or broken, then it becomes vulnerable to any environmental stressors and may change very rapidly. This type of sudden deviation from the equilibrium state is a regime shift (Oguz and Gilbert, 2007). A typically changing state continuing for more than a decade is to be regarded as a regime shift in marine ecology. In the Black Sea the period started in 1960 and lasted till 1990. In 1970, the ecological system has evolved from predator fishes to planktivorous pelagic fish groups, which triggered grazing on zooplanktons and caused a high eutrophication state (Daskalov et al., 2007; Pershing et al., 2015). This also coincides with an extreme cooling period of the Black Sea. The cooling period was followed by 20-year warming conditions and experienced a gelatinous carnivorous intrusion at the beginning of 1980 (Oguz and Gilbert, 2007). A gelatinous carnivore *Mnemiopsis leidyi* holds an important place in the zooplankton assemblage of the Black Sea. It was first introduced to the ecosystem in 1980, reached its maximum populations in 1990, and slowly lost its place in the ecosystem after 1994 by biological control from another comb jelly, *Beroe ovata* (Akoglu et al., 2014). The zooplankton community structure and abundance are also influenced by anthropogenic factors such as pollution and overfishing. A similar degradation was also observed in the benthic ecosystem of the Black Sea when the *Rapana* whelk, an East Asian originating-carnivore, was introduced into the system in 1940 (Mee et al., 2005; Oguz et al., 2012). Another benthic indicator affected by the eutrophic state is the red alga "*Phyllophora*." Until the beginning of the 1960s, the biomass of these red algae was abundant in northwestern shelf waters. Due to eutrophic conditions related to hypoxia and limited light problems, this indicator algae disappeared from the system until recent studies that reported sightings of it in field surveys (Minicheva, 2007; Langmead et al., 2009; Capet et al., 2013). Such catastrophic impacts of eutrophication have greatly altered the ecological aquatic life and food chain of the sea. However, there are some signs of recovery of the Black Sea ecosystem and a move to changing the conditions back to the preeutrophication period, which was known as meso-trophic state before 1970. From 1970 to 1980, very high primary and secondary production rates created hypoxic conditions and led to a very large deterioration of benthic life (Akoglu et al., 2014). It seems that precautions and regulations of anthropogenic factors such as agriculture and pollution have been effective at least to some extent in the recovery of the ecological system of the Black Sea. Chl-a satellite images showing the mean annual concentrations from different satellites are depicted in Fig. 2.

The causes of the regime shift encountered in the Black Sea can possibly be attributed to overfishing and the intrusion of *M. leidyi* to the system and new ecological conditions had altered the trophic state from bottom-up to top-down by creating a trophic cascade effect (Akoglu et al., 2014). A trophic cascade is generally observed in benthic life or coastal margins with some exceptions reported in large marine ecosystems (Daskalov et al., 2007). A trophic cascade can occur in both top-down and bottom-up directions in aquatic systems—if the consumers are dominant then the system is top-down, but if primary producers are abundant with high biomass then the system is bottom-up controlled. The ratio between consumers and producers should approach unity for the ecosystems working in balanced conditions (Daskalov, 2002; Daskalov et al., 2007). Top-down systems are generally found in terrestrial ecology and are rarely observed in coastal or freshwater ecosystems (Daskalov, 2002). It should be emphasized that ecological shifts witnessed in the Black Sea are not only caused by anthropogenic factors but also influenced by large-scale climatic oscillations such as the North Atlantic Oscillation (NAO). There are also some known effects of global warming on rising mean SST, which reached up to 4–5°C in the Black Sea and

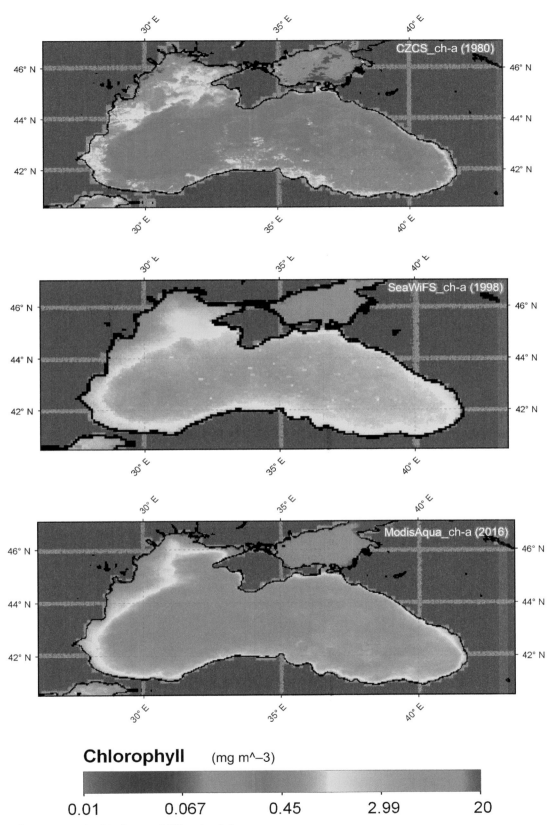

Chlorophyll (mg m^–3)

0.01　0.067　0.45　2.99　20

FIG. 2　Annual mean composite chl-a images of different periods.

approximately 0.3°C in the global oceans according to measurements started in 1950. Increasing trends of SST anomalies are well correlated with the NAO (Oguz et al., 2003). This physical impact has important interactions with oceanographic mixing process of the water column. The CIL formation is directly affected by increasing SST trends. The depth of the CIL is limited to lower values which causes less vertical mixing and inadequate production rates. Moreover, due to a limited nutrient input to the photic zone, the system shifts toward small cell phytoplankton groups for a more efficient use of available light and nutrients in ambient conditions (Oguz et al., 2003).

4 FISHERIES

Fisheries are one of the most important revenue-generating activities in the Black Sea. Although it is not significant in terms of the basin economy, fisheries is a part of the cultural history in the region and dates back several thousand years. Today, industrial, semiindustrial, and small-scale fishing types are widely practiced in the basin. There are approximately 11,000 fishing boats of different sizes and with different gears. The number of the purse seiners totals approximately 455 (FAO, 2016). The largest fishing fleet belongs to Turkey with maximum annual fish yield in the basin. Considering the entire basin fishing effort of all the surrounding countries, registered fishing boat numbers are as follow: Turkey (16,448), Ukraine (135), Russian Federation (33), Georgia (47), Romania (158), and Bulgaria (704). Although the registered boat number is very high in Turkey compared to other countries, approximately 15,000 of total registered boats are less than 12 m in length (FAO, 2016). According to data for 2013, the total catch reported is approximately 1250 kt in the basin. Moreover, 68% of total landings is reported from Turkey.

The number of economically important species is less today than it was during the preeutrophication period. Fishes that comprise the greatest part of the total landings are as follows: European anchovy (*Engraulis encrasicolus*), sprat (*Sprattus sprattus*), Mediterranean horse mackerel (*Trachurus mediterraneus*), Atlantic bonito (*Sarda sarda*), and bluefish (*Pomatomus saltatrix*), turbot (*Psetta maxima*), whiting (*Merlangius merlangus*), picked dogfish (*Squalus acanthias*), striped and red mullets (*Mullus barbatus*, *M. surmuletus*), and four species from the *Mugilidae* family (European Parliament, 2010).

In paralleling the increasing trends of the world population, the Black Sea coastal population has been also increasing in recent decades. Therefore, aquatic resources of the basin fail to meet the necessary high demands of the bordering countries. There are some issues of high priority regarding the fishing regulations in the basin and problems related to fishing must be handled by consensus among all neighboring countries. These problems can be summarized as follows: overfishing problem undoubtedly is the first of all as it is clearly seen in decreasing trends of total catch year by year in the Black Sea. This is governed by stock size, which is related to the recruitment rate of fish stocks (Daskalov, 1999) as well as available food for larvae and eggs in ambient conditions. Nutrient supply by fresh water sources of the basin can also play an important role in developing healthy fish stocks as it is necessary for primary producers. The decreasing trend of anchovy stock size is critically reported to have started in 2005 and linked to enlarged fishing fleet with a too large and unsustainable fishing effort pressure on natural stocks as well as illegal fishing (Raykov et al., 2011). It is well known that overfishing can also alter the food chain and energy flow in the marine environment. Palkovacs (2011) reported that the body size of the same fish captured in the same biogeographic region 100 years later can show a 43% difference in weight and a 69% difference in length. Fish length, weight, and age can be extremely adversely affected by overfishing.

An important threshold showing the success of the fisheries management in the Black Sea could be in achieving Good Environmental Status (GeNS), for the EU Marine Strategy Framework Directive, by taking into account the maximum sustainable yield (MSY) as an indicator. The biggest problem preventing Black Sea fisheries to achieve MSY is the synchronization problem of different institutional arrangements and international environmental policies (Goulding et al., 2014). However, there are also good examples of consensus on environmental policies among bordering countries such as the Bucharest Convention signed in 1992 after recommendations of 1972 Stockholm Conference on environment and development. Following the Bucharest Convention, Black Sea Environmental Program (BSEP) was established with the aim of restoring the Black Sea ecosystem and carrying out specific monitoring programs with financial budgets supported by UNEP, UNDP, and GEF (BSEP (Black Sea Environmental Programme), 1997).

Solution to some other important problems involved in Black Sea fisheries can be summarized as shortening the total length of fishing boats to limit their fishing capacity, switching from industrial type to artisanal and small-scale fishing, controlling by-catch, delineating no-take zones, and imposing necessary fishing quotas. These are necessary and urgent issues that should be handled precisely in order to achieve GeNS and restoring the natural stocks in the basin. Turkey is the first country on the list influenced by the problems stated above. For instance, total anchovy landings in Turkey was around 500 kt in 1960 and decreased to 150 kt within a 30-year period (Akoglu et al., 2014). According to official statistics, the total anchovy catch of Turkey was around 100 kt for 2016 (www.tuik.gov.tr). Recently, the migration behavior of

TABLE 2 Some Fishing Information of Black Sea Countries

Country	Coast Line (km)	Area of EEZ (km²)	Total Fishing Vessel (#)	Average Landings (2000–13) (kt)
Turkey	1400	210.565	5973	307
Russia	475	69.038	33	32
Romania	225	31.108	158	1.26
Ukraine	1628	138.362	135	68.9
Bulgaria	300	34.288	704	7.72
Georgia	310	18.612	47	12.6

(Data from: European Parliament: Directorate general for internal policies, policy department B: structural and cohesion policies, Irina Popescu, 2010, Fisheries in the Black Sea, pp. 1–69; FAO, 2016. The state of Mediterranean and Black Sea fisheries. General fisheries Commission for the Mediterranean, Italy, Rome ; www.tuik.gov.tr; Landings and total fishing vessel information of Turkey represent the average total numbers for Black Sea.)

anchovy has also changed in Southern Black Sea waters. The root causes of the new migration pattern of anchovy must be investigated in detail to determine whether the natural or anthropogenic factors are involved in the problem. The length and mesh size of fishing nets are some additional problems causing depletion of the anchovy and other pelagic fish stocks in the Black Sea. Black Sea fisheries management must be based on scientific monitoring tools and approaches that account for related environmental and ecosystem indicators (Caddy, 2009). The Black Sea fishing figures by bordering countries are summarized in Table 2.

As inferred from Table 2, Turkey is ranked first in terms of fishing activity in the basin. Therefore, any fisheries-related problem can directly influence the socioeconomic well-being of the families living on the Turkish coastal zone. Synchronization with EU fisheries policy could be the best action to be followed by Turkey in order to achieve MSY and GeNS in the region. Necessary regulations and precautions should be implemented without any delay. The power of fishing boat engines, technological fishing gears used on board, fish net length, and mesh size are also the controls that can be used to regulate fishing strategy. Recently, Turkey has also implemented fishing licence revocation for approximately 150 registered fishing boats (purse seiners) by paying a legal compensation amount to reduce fishing pressure on natural stocks.

5 POLLUTION (MARINE LITTER)

Marine litter is one of the most important environmental problems of the Black Sea basin. They can be found in the water column, on the seabed, on the coast, or as floating wastes transported from land. Marine litter could be in the form of nano, micro, or macro-size in marine environment and can accumulate in tissues of aquatic organisms as plastic residuals or cause entanglements of marine fishes. The main pollution source of the Black Sea is land-based solid wastes transported by rivers (Tuncer et al., 1998). Marine litter can be described as all types of solid waste that disturbs the marine environment by creating pollution defined as the damage to the ecology. Apart from their sizes and types, these toxic substances threaten the aquatic living resources and marine environment. The nano and microform of marine litter can be ingested by demersal and pelagic fishes or filter feeding organisms by contaminating their natural diets (UNEP and GRID-Arendal, 2016). Marine litter has been reported from the world seas since 1970 and by an accelerated and alarming increase in 1990s (Corcoran et al., 2009). This might be attributed to a consumption economy and high purchasing power of the people in the developed countries. The problem of marine litter is increasing and more complex in the oceans. Moreover, small gyres of litter are now easily detected by remote sensing satellites around mid-regions of all oceans, especially, the Pacific Ocean (Pichel et al., 2007).

The sources of marine litter pollution in the Black Sea are mainly rivers or creeks distributed all around the coastal periphery of the basin (Guneroglu, 2010; Aytan et al., 2016). A very small part of the total volume might be generated by commercial vessels or fishing boats. In the frame of MSFD to achieve GES in marine environment, 11 descriptors were defined and explained in Annex 1 of the Decision 2010/477/EU by the EU Commission. Descriptor 10 is allocated for marine litter and explained by four subindicators (Galgani et al., 2013). The primary concern of marine litter in the Black Sea was originally started with aesthetic considerations but evolved in course of time and became a problem with high priority. The solution to the problem must be searched for by all bordering countries, as it is a general environmental issue. Increasing public awareness, education, and solid waste management policies can be used for controlling and mitigating the problems.

The best solution can be in stopping the solid waste at its source before reaching the marine environment. Therefore, the initiative by local managers and decision-makers plays a crucial role in recovering the unique Black Sea ecosystem.

6 RECOMMENDATIONS AND CONCLUSIONS

The Black Sea basin is a valuable area with its natural, historical, and ecological assets. Preservation and usage balance have deteriorated because of various factors in the last 50 years. However, setting the successful management of this ecosystem for the next 50 years depends on handling current problems with cautions by using scientific tools and approaches. The essential step in the conservation of the Black Sea ecosystems must focus on sustainability of the basin by tuning usage and protection principles. Considering decreasing the nutrient input to ecosystem, there are some clear signs of recovery of the basin reported in different scientific studies, which should be evaluated precisely before accepting this as a total recovery. Lower algal concentrations and less-frequent bloom events observed in northwestern shelf were also complemented by a decrease in abundance of gelatinous species (McQuatters-Gollop et al., 2008; Shiganova et al., 2008; Oguz and Velikova, 2010). Regular monitoring of the ecosystem is a prerequisite for its ecological protection. In this course, satellite remote sensing and data processing techniques have reached a great potential for the fast and accurate modeling and simulation of the area under investigation. Some basic environmental parameters such as available light, temperature, chlorophyll, and salinity are very easy to obtain and use information such as from satellite images with global coverage. Therefore, almost real-time observation of algal blooms or eutrophication events can be easily carried out by using remote sensing tools. For instance, SST and chl-a data can be embedded into ecosystem models for estimating the ecological state for the next 30 years. The SST data is especially important as it is related to global warming and CIL formation in the Black Sea. Stratification and vertical mixing need to be regularly monitored as they are linked to trophic state of the ecosystem. Although remarkable progress has been achieved in limiting nutrient inputs to the Black Sea environment, scientific projects must be financed to determine point and diffuse sources of current nutrient supply as well as monitoring the state of the coastal productive regions. Another issue regarding anthropogenic nutrient state is the very dense coastal population especially on the northwestern and southern sections of the sea. The correlation between nutrient levels and coastal population settlements can be carried out to determine the hotspots on the coastal environment. There is also a risk of nutrient deficiency in the long term at least at some local coastal regions due to accelerated efforts of dam construction for producing electricity. Sediments with high nutrient content are trapped and deposited in deep layers behind the constructed dams, which limits nutrient concentrations per volume of fresh waters flowing in to the basin.

The final aim in the Black Sea environmental protection must focus on reaching GES as explained in MSFD and WFD documentation. This is also necessary to solve the problems related to fisheries. The state of Black Sea fisheries is definitely alarming and getting worse day by day. Overfishing and inadequate controls of fishing fleet are the main problems. Necessary quotas, fishing area allocation zones, and times must be redefined by consensus of all bordering countries. This can be also be supported by establishing a new Black Sea fisheries coordination body that can make recommendations and suggestions following EU regulations on the fisheries. In this case, Turkey must lead other countries as the country being the highest beneficiary of the Black Sea fisheries. Issuing new fishing licences must be strictly controlled and preferably no new licences of purse seiners should be offered. Engine power and onboard technological gears can be limited to reduce fishing pressure on natural stocks.

Solid waste pollution in the basin is largely due to a mismanaged municipal service that fails in collecting garbage containers on regular schedules. The lack of the service in southern regions of the basin is partly because of scattered settlement patterns. Therefore, unmanaged solid wastes finally reach the sea environment in different shapes and forms. They can float, sink, or suspend in water column as marine litter and enter into food cycle with different pathways. An effective municipal service and monitoring is necessary to stop the marine litter problem in the Black Sea. New policies and regulations are required for designing garbage collection and management at their site of origin. Limiting plastic covers and packaging materials could greatly contribute to reduce marine litter problem. Furthermore, some kind of country marker can be designed to be used in packing materials for determining and controlling marine litter transport. Specially designed solid waste collectors can be also used on predefined sections of rivers or creeks.

The oil and gas industry also has a wide application area in the Black Sea. Oil products transported by large tanker ships create environmental pollution risks in the Black Sea and Bosporus. Alternative routes and an efficient vessel traffic system must be designed in order to avoid any collision or grounding accidents. The environmental pollution risk of maritime transport in the Black Sea cannot be limited with oil spill occurrence, grounding, or collision, as the basin experiences invasive alien species transported by ballast waters and in turn cause catastrophic impacts on the marine food web. The impact of the ctenophore *M. leidyi* on the Black Sea marine ecosystem is such an effect that cannot be overlooked or forgotten. However, more than 100 MT oil was carried via the Bosporus strait each year. Moreover, oil transport pipelines from the Caucasus to European regions traverse the Black Sea deep layers with potential threats to the marine ecosystem. There also some recent initiatives seeking new energy resources in the basin carried out by bordering countries on their EEZs. Therefore, environmental regulations in addition to MARPOL and SOLAS, new alternative solutions and routes such as may help to reduce environmental pollution risk in the southwestern Black Sea and Bosporus regions.

Despite many mitigation policies, it is clear that global warming will continue to be a problem posed by humanity at least for the coming decades. Regarding the coastal environment, SLR could be the most destructive result of the global warming effect in the Black Sea. An expected 10–15 cm rise of seawater could cause billions of dollars of flood or storm surge effects in the region. It is suggested that, starting now, no built-up zones on very near coastal areas have to be defined and planned by responsible decision-makers.

Finally, the future of the Black Sea is in the hands of the countries surrounding the basin and their international cooperation level to achieve a better proposed environmental status for a better and sustainable natural heritage to leave for the next generations. After taking all necessary preventive measures and following internationally recognized environmental arrangements, the Black Sea also needs an element of good luck for totally recovering and reaching the preeutrophication stage as mentioned above in the text. This is because global warming and global climate change impacts are hard to predict nonlinear phenomena.

ACKNOWLEDGMENTS

The authors are grateful to EMODnet Bathymetry Consortium (2016): EMODnet Digital Bathymetry (DTM) for bathymetric data and Global Land Cover Facility GLCF for providing the Landsat satellite data. The CZCS, SeaWiFS and MODIS satellite images were made available by NASA OB.DAAC archives.

REFERENCES

Akoglu, E., Salihoglu, B., Libralato, S., Oguz, T., Solidoro, C., 2014. An indicator-based evaluation of Black Sea food web dynamics during 1960–2000. J. Mar. Syst. 134, 113–125.

Allenbach, K., Garonna, I., Herold, C., Monioudi, I., Giuliani, G., Lehmann, A., Velegrakis, A.F., 2015. Black Sea beaches vulnerability to sea level rise. Environ. Sci. Policy 46, 95–109.

Alpar, B., 2009. Vulnerability of Turkish coasts to accelerated sea-level rise. Geomorphology 107 (1), 58–63.

Antonidze, E., 2010. ICZM in the Black Sea region: experience and perspectives. J. Coast. Conserv. 14 (4), 265–272.

Avsar, N.B., Kutoglu, S.H., Jin, S., Erol, B., 2015. Investigation of sea level change along the Black Sea coast from tide gauge and satellite altimetry. Int. Arch. Photogramm. Remote Sens. Spat. Inf. Sci. 40 (1), 67.

Aytan, U., Valente, A., Senturk, Y., Usta, R., Sahin, F.B.E., Mazlum, R.E., Agirbas, E., 2016. First evaluation of neustonic microplastics in Black Sea waters. Mar. Environ. Res. 119, 22–30.

BSC (Black Sea Commission), 2008. Statee of the environment of the Black Sea (2001–2006/7). Publications of the Commission on the Protection of the Black Sea against Pollution (BSC), 3.

BSEP (Black Sea Environmental Programme) RER/92/G31—RER/93/G31—RER/94/G41—RER/96/006, Final Report, 1997.

Caddy, J.F., 2009. Practical issues in choosing a framework for resource assessment and management of Mediterranean and Black Sea fisheries. Mediterr. Mar. Sci. 10 (1), 83–119.

Capet, A., Beckers, J.M., Grégoire, M., 2013. Drivers, mechanisms and long-term variability of seasonal hypoxia on the Black Sea northwestern shelf-is there any recovery after eutrophication? Biogeosciences 10 (6), 3943.

Corcoran, P.L., Biesinger, M.C., Grifi, M., 2009. Plastics and beaches: a degrading relationship. Mar. Pollut. Bull. 58 (1), 80–84.

Daskalov, G., 1999. Relating fish recruitment to stock biomass and physical environment in the Black Sea using generalized additive models. Fish. Res. 41 (1), 1–23.

Daskalov, G.M., 2002. Overfishing drives a trophic cascade in the Black Sea. Mar. Ecol. Prog. Ser. 225, 53–63.

Daskalov, G.M., Grishin, A.N., Rodionov, S., Mihneva, V., 2007. Trophic cascades triggered by overfishing reveal possible mechanisms of ecosystem regime shifts. Proc. Natl. Acad. Sci. 104 (25), 10518–10523.

Drozdov, V.A., Glezer, O.B., Nefedova, T.G., Shabdurasulov, I.V., 1992. Ecological and geographical characteristics of the coastal zone of the Black Sea. GeoJournal 27 (2), 169–178.

Eker-Develi, E., Kideys, A.E., 2003. Distribution of phytoplankton in the southern Black Sea in summer 1996, spring and autumn 1998. J. Mar. Syst. 39 (3), 203–211.

European Parliament, 2010. Directorate General for Internal Policies, Policy Department B: Structural and Cohesion Policies. Fisheries in the Black Sea, Irina Popescu, pp. 1–69.

FAO, 2016. The State of Mediterranean and Black Sea Fisheries. General Fisheries Commission for the Mediterranean. Rome, Italy.

Feyzioğlu, A.M., Eruz, C., Yıldız, İ., 2015. Geographic variation of Picocyanobacteria Synechococcus spp. along the Anatolian Coast of the Black Sea during the late Autumn of 2013. Turk. J. Fish. Aquat. Sci. 15 (4), 471–475.

Feyzioglu, A.M., Kurt, I., Boran, M., Sivri, N., 2004. Abundance and Distribution of *Synechococcus* spp in the South-Eastern Black Sea During 2001 Summer. Indian J. Marine Sci. 33 (4), 365–368.

Feyzioğlu, A.M., Öğüt, H., 2006. Red tide observations along the eastern Black Sea coast of Turkey. Turk. J. Bot. 30 (5), 375–379.

Friedrich, J., Janssen, F., Aleynik, D., Bange, H.W., Boltacheva, N., Çagatay, M.N., Dale, A.W., Etiope, G., Erdem, Z., Geraga, M., Gilli, A., 2014. Investigating hypoxia in aquatic environments: diverse approaches to addressing a complex phenomenon. Biogeosciences 11, 1215–1259.

Galgani, F., Hanke, G., Werner, S.D.V.L., De Vrees, L., 2013. Marine litter within the European marine strategy framework directive. ICES J. Mar. Sci. 70 (6), 1055–1064.

Gómez, F., Boicenco, L., 2004. An annotated checklist of dinoflagellates in the Black Sea. Hydrobiologia 517 (1–3), 43–59.

Goulding, I., Stobberup, K., O'Higgins, T., 2014. Potential economic impacts of achieving good environmental status in Black Sea fisheries. Ecol. Soc. 19 (3), 1–11.

Gubanova, A.D., Altukhov, D.A., 2007. Establishment of Oithona brevicornis Giesbrecht, 1892 (Copepoda: Cyclopoida) in the Black Sea. Aquat. Invas. 2 (4), 407–410.

Guneroglu, A., 2010. Marine litter transportation and composition in the Coastal Southern Black Sea Region. Sci. Res. Essays 5 (3), 296–303.

Guneroglu, A., 2015. Coastal changes and land use alteration on Northeastern part of Turkey. Ocean Coast. Manag. 118, 225–233.

Guneroglu, A., Karsli, F., Dihkan, M., 2014. Dynamic management of the coasts: marine spatial planning. Proc. Inst. Civil Eng.-Maritime Eng. 167 (3), 144–153.

Guneroglu, N., Acar, C., Dihkan, M., Karsli, F., Guneroglu, A., 2013. Green corridors and fragmentation in South Eastern Black Sea coastal landscape. Ocean Coast. Manag. 83, 67–74.

Karsli, F., Guneroglu, A., Dihkan, M., 2011. Spatio-temporal shoreline changes along the southern Black Sea coastal zone. J. Appl. Remote. Sens. 5 (1), 053545.

Kopuz, U., Feyzioglu, A.M., Agirbas, E., 2012. Picoplankton dynamics during late spring 2010 in the south-eastern Black Sea. Turk. J. Fish. Aquat. Sci. 12 (5), 397–405.

Kopuz, U., Feyzioglu, A.M., Valente, A., 2014. An unusual red-tide event of Noctiluca scintillans (Macartney) in the Southeastern Black Sea. Turk. J. Fish. Aquat. Sci. 14 (1), 261–269.

Kosyan, R.D., Velikova, V.N., 2016. Coastal zone–Terra (and aqua) incognita–Integrated Coastal Zone Management in the Black Sea. Estuar. Coast. Shelf Sci. 169, A1–A16.

Kovalev, A.V., Skryabin, V.A., Zagorodnyaya, Y.A., Bingel, F., Kıdeyş, A.E., Niermann, U., Uysal, Z., 1999. The Black Sea zooplankton: composition, spatial/temporal distribution and history of investigations. Turk. J. Zool. 23 (2), 195–210.

Kuleli, T., 2010. City-based risk assessment of sea level rise using topographic and census data for the Turkish coastal zone. Estuar. Coasts 33 (3), 640–651.

Kuleli, T., Guneroglu, A., Karsli, F., Dihkan, M., 2011. Automatic detection of shoreline change on coastal Ramsar wetlands of Turkey. Ocean Eng. 38 (10), 1141–1149.

Langmead, O., McQuatters-Gollop, A., Mee, L.D., Friedrich, J., Gilbert, A.J., Gomoiu, M.T., Jackson, E.L., Knudsen, S., Minicheva, G., Todorova, V., 2009. Recovery or decline of the northwestern Black Sea: a societal choice revealed by socio-ecological modelling. Ecol. Model. 220 (21), 2927–2939.

Ludwig, W., Dumont, E., Meybeck, M., Heussner, S., 2009. River discharges of water and nutrients to the Mediterranean and Black Sea: major drivers for ecosystem changes during past and future decades? Prog. Oceanogr. 80 (3), 199–217.

McQuatters-Gollop, A., Mee, L.D., Raitsos, D.E., Shapiro, G.I., 2008. Non-linearities, regime shifts and recovery: the recent influence of climate on Black Sea chlorophyll. J. Mar. Syst. 74 (1), 649–658.

Mee, L.D., Friedrich, J., Gomoiu, M.T., 2005. Restoring the Black Sea in times of uncertainty. Oceanography 18 (2), 32–43.

Mikaelyan, A.S., Pautova, L.A., Chasovnikov, V.K., Mosharov, S.A., Silkin, V.A., 2015. Alternation of diatoms and coccolithophores in the north-eastern Black Sea: a response to nutrient changes. Hydrobiologia 755 (1), 89–105.

Mikaelyan, A.S., Pautova, L.A., Pogosyan, S.I., Sukhanova, I.N., 2005. Summer Bloom of Coccolithophorids in the Northeastern Black Sea. Oceanology 45 (Suppl. 1), 127–138.

Mikaelyan, A.S., Silkin, V.A., Pautova, L.A., 2011. Coccolithophorids in the Black Sea: their interannual and long-term changes. Oceanology 51 (1), 39–48.

Miladinova, S., Stips, A., Garcia-Gorriz, E., Moy, D.M., 2016. Changes in the Black Sea physical properties and their effect on the ecosystem. In: Tech. Rep. EUR 28060 EN, Publications Office of the European Union, Luxembourg. https://doi.org/10.2788/69832.

Minicheva, G.G., 2007. Contemporary morpho-functional transformation of seaweed communities of the Zernov phyllophora field (Black Sea). Int. J. Algae 9 (1), 1–21.

Moncheva, S., Gotsis-Skretas, O., Pagou, K., Krastev, A., 2001. Phytoplankton blooms in Black Sea and Mediterranean coastal ecosystems subjected to anthropogenic eutrophication: similarities and differences. Estuar. Coast. Shelf Sci. 53 (3), 281–295.

Moncheva, S., Pantazi, M., Pautova, L., Boicenco, L., Vasiliu, D., Mantzosh, L., 2012. Black Sea Phytoplankton data quality–problems and progress. Turk. J. Fish. Aquat. Sci. 12 (5), 417–422.

Niermann, U., Binge, F., Ergün, G., 1998. Fluctuation of dominant mesozooplankton species in the Black Sea, North Sea and the Baltic Sea: is a general trend recognisable? Turk. J. Zool. 22 (1), 63–82.

Oguz, T., Akoglu, E., Salihoglu, B., 2012. Current state of overfishing and its regional differences in the Black Sea. Ocean Coast. Manag. 58, 47–56.

Oguz, T., Cokacar, T., Malanotte-Rizzoli, P., Ducklow, H.W., 2003. Climatic warming and accompanying changes in the ecological regime of the Black Sea during 1990s. Global Biogeochem. Cycles 17 (3), 1–14.

Oguz, T., Dippner, J.W., Kaymaz, Z., 2006. Climatic regulation of the Black Sea hydro-meteorological and ecological properties at interannual-to-decadal time scales. J. Mar. Syst. 60 (3), 235–254.

Oguz, T., Gilbert, D., 2007. Abrupt transitions of the top-down controlled Black Sea pelagic ecosystem during 1960–2000: evidence for regime-shifts under strong fishery exploitation and nutrient enrichment modulated by climate-induced variations. Deep-Sea Res. I Oceanogr. Res. Pap. 54 (2), 220–242.

Oguz, T., Velikova, V., 2010. Abrupt transition of the northwestern Black Sea shelf ecosystem from a eutrophic to an alternative pristine state. Mar. Ecol. Prog. Ser. 405, 231–242.

O'Higgins, T., Farmer, A., Daskalov, G., Knudsen, S., Mee, L., 2014. Achieving good environmental status in the Black Sea: scale mismatches in environmental management. Ecol. Soc. 19 (3), 1–9.

Palkovacs, E.P., 2011. The overfishing debate: an eco-evolutionary perspective. Trends Ecol. Evol. 26 (12), 616–617.

Panin, N., Jipa, D., 1998. Danube river sediment input and its interaction with the north-western BlackSea: results of EROS-2000 and EROS-21 projects. Geo-Eco-Marina 3, 23–35.

Panin, N., Jipa, D., 2002. Danube River sediment input and its interaction with the north-western Black Sea. Estuar. Coast. Shelf Sci. 54 (3), 551–562.

Pershing, A.J., Mills, K.E., Record, N.R., Stamieszkin, K., Wurtzell, K.V., Byron, C.J., Fitzpatrick, D., Golet, W.J., Koob, E., 2015. Evaluating trophic cascades as drivers of regime shifts in different ocean ecosystems. Philos. Trans. R. Soc. Lond. B Biol. Sci. 370 (1659), 20130265.

Pichel, W.G., Churnside, J.H., Veenstra, T.S., Foley, D.G., Friedman, K.S., Brainard, R.E., Nicoll, J.B., Zheng, Q., Clemente-Colon, P., 2007. Marine debris collects within the North Pacific subtropical convergence zone. Mar. Pollut. Bull. 54 (8), 1207–1211.

Raykov, V., Velikova, V., Lisichkov, K., Kuvendziev, S., 2011. Review of main fisheries indicators in the Black Sea by using diagnostic analysis. Natura Montenegriana. Podgorica 10 (3), 309–312.

Shiganova, T., Musaeva, E., Araskievich, E., Kamburska, L., Stefanova, K., Michneva, V., Polishchuk, L., Timofte, F., Ustun, F., Oguz, T., Khalvashi, M., 2008. The state of zooplankton. In: State of the Environment of the Black Sea (2001–2006/7). 3. pp. 201–246.

Stănică, A., Panin, N., 2009. Present evolution and future predictions for the deltaic coastal zone between the Sulina and Sf. Gheorghe Danube river mouths (Romania). Geomorphology 107 (1), 41–46.

Tsereteli, E., Gobejishvili, R., Bolashvili, N., Geladze, V., Gaprindashvili, G., 2011. Crisis intensification of geoecological situation of the Caucasus Black Sea coast and the strategy of risk reduction. Procedia. Soc. Behav. Sci. 19, 709–715.

Tuncer, G., Karakas, T., Balkas, T.I., Gökçay, C.F., Aygnn, S., Yurteri, C., Tuncel, G., 1998. Land-based sources of pollution along the Black Sea coast of Turkey: concentrations and annual loads to the Black Sea. Mar. Pollut. Bull. 36 (6), 409–423.

UNEP and GRID-Arendal, 2016. Marine Litter Vital Graphics. United Nations Environment Programme and GRID-Arendal, Nairobi and Arendal.

Uysal, Z., 2000. Pigments, size and distribution of Synechococcus spp. in the Black Sea. J. Mar. Syst. 24 (3), 313–326.

Uysal, Z., 2001. Chroococcoid cyanobacteria Synechococcus spp. in the Black Sea: pigments, size, distribution, growth and diurnal variability. J. Plankton Res. 23 (2), 175–190.

Yıldız, İ., Feyzioglu, A.M., Besiktepe, S., 2017. First observation and seasonal dynamics of the new invasive planktonic copepod Oithona davisae Ferrari and Orsi, 1984 along the southern Black Sea (Anatolian Coast). J. Nat. Hist. 51 (3–4), 127–139.

Yunev, O.A., Vedernikov, V.I., Basturk, O., Yilmaz, A., Kideys, A.E., Moncheva, S., Konovalov, S.K., 2002. Long-term variations of surface chlorophyll a and primary production in the open Black Sea. Mar. Ecol. Prog. Ser. 230, 11–28.

Zaitsev, Y., 2001. An introduction to Black Sea Ecology, European Research Office London (Great Britain) Army Engineer Research And Development Center. 174. Defense Technical Information Center.

Zaitsev, Y., Alexandrov, B.G., Berlinsky, N.A., Zenetos, A., 2001. Europe's biodiversity–biogeographical regions and seas, European Environmental Agency (Report). ZooBoTech HB, Sweden.

Zaitsev, Y.P., Alexandrov, B.G., Berlinsky, N.A., Zenetos, A., 2002. Europe's Biodiversity–Biogeographical Regions and Seas: The Black Sea an Oxygen-Poor Sea. EEA (European Environment Agency), Copenhagen.

Chapter 22

Ecosystem Functioning and Sustainable Management in Coastal Systems With High Freshwater Input in the Southern Gulf of Mexico and Yucatan Peninsula

Jorge A. Herrera-Silveira[*], Ana L. Lara-Domínguez[†], John W. Day[‡], Alejandro Yáñez-Arancibia[†,a], Sara Morales Ojeda[*], Claudia Teutli Hernández[*], G. Paul Kemp[‡]

[*]Center for Research and Advanced Studies of the National Polytechnic Institute, Merida Campus, Mexico, [†]Institutue of Ecology, Veracruz, Mexico, [‡]Department of Oceanography and Coastal Sciences, College of the Coast and Environment, Louisiana State University, Baton Rouge, LA, United States

1 INTRODUCTION

Stretching along the southern Gulf of Mexico and Mexican Caribbean for over a thousand kilometers is a remarkable coastal system characterized by high freshwater input, extensive wetlands and coastal lagoons, productive fisheries, and human settlements whose economy is based on the rich natural resources of the area and maritime trade (Fig. 1). The first civilization of Mesoamerica, the Olmecs with the earliest ceremonial center at La Venta, developed in this area and important settlements of the Mayan civilization occurred throughout the region. This zone includes the Alvarado, Coatzacoalcos, and Grijalva-Usumacinta (the second largest riverine input to the Gulf after the Mississippi) Rivers, the large delta of the Grijalva-Usumacinta and the ground water discharge region of the Yucatan Peninsula with coastal lagoons and extensive mangrove forests and beds of submerged aquatic vegetation. Transitional between these two areas is Laguna de Terminos, one of the largest coastal ecosystems in Latin America.

In this chapter, we describe this rich and varied regional system and consider the prospects for its future sustainability in the face of human impacts and substantial global change. Laguna de Terminos, the largest lagoon with extensive mangroves and sea grasses, spans the transition between terrigenous and carbonate provinces. This is one of the best studied tropical coastal systems in the Americas.

Because of its tropical location, this region has moderate seasonal pulses of temperature and light but strong seasonal pulses of precipitation, both river and ground water discharge, and the impacts of a cool season of frontal storms (the *nortes*). The area also has strong near-permanent physical gradients of salinity and a high diversity of estuarine habitats. There are three "seasons" in this region. From June to September-October is the rainy season with frequent afternoon and evening convectional showers associated with the intertropical convergence zone, and occasional hurricanes. From October to March is the period of *nortes* or winter frontal storms. February to May is the dry season when the intertropical convergence zone is south of the equator. During *nortes*, winds are generally from the northwest with speeds often higher than $8\,m\,s^{-1}$. For most of the rest of the year, there is a sea breeze system that is affected by the trades, with predominantly easterly winds with velocities between 4 and $6\,m\,s^{-1}$.

2 THE HIGH RIVER DISCHARGE ZONE AND THE GRIJALVA-USUMACINTA DELTA

The Grijalva-Usumacinta River (GUR) of Mexico and Guatemala is the largest river in Mesoamerica and one of the most significant shared water resources in the Western Hemisphere (Bestermeyer and Alonso, 2000; Yáñez-Arancibia and Day, 2004a, 2006; Yáñez-Arancibia et al., 2009). The GUR drains a watershed of about 7.3 million ha in Mexico and Guatemala,

---

[a] Deceased

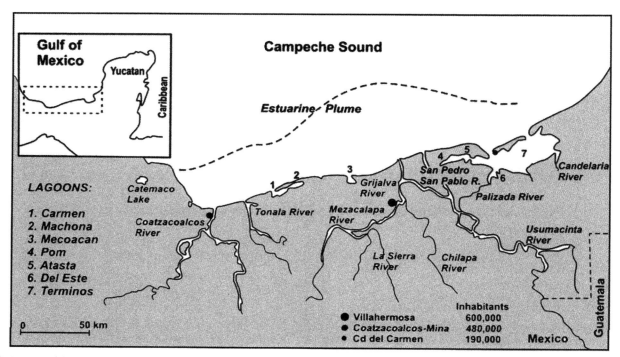

FIG. 1   Map of the southern Gulf of Mexico and Yucatan Peninsula *(insert)*. The southern Gulf has high riverine and groundwater discharge, abundant wetlands, and lagoons. The largest River is the Grijalva-Usumacinta and the largest lagoon is Laguna de Terminos.

one of the largest areas of contiguous tropical forest in the region. About 36% of the land has been altered due to human activities. The delta prairies are an assemblage of the Mezcalapa, Grijalva, and Usumacinta Rivers, and together they constitute a large delta with more than 20,000 km^2 (Yáñez-Arancibia and Day, 2004b). The average discharge of the GUR is 3000–4700 m^3 s^{-1} with peak discharge as high as 9000 m^3 s^{-1}. The highest discharge occurs from September to November and the lowest is in April-May.

One of the important steps in the protection of the UGD was the establishment of the Centla Wetlands Biosphere Reserve (CWBR) in 1992 (INE-SEMARNAT, 2000), one of the largest such preserves in Mesoamerica. In addition to its importance as a natural area, the UGD also has important agricultural activities and contains one of the most important oil and gas production areas of Mexico (Martínez and Gaona, 2005). Therefore, the management plan for CWBR had to take these activities into consideration.

Using the ecosystem approach, Lara-Domínguez et al. (2011, 2013) developed management units that are designed to maintain ecosystem structure and functioning while supporting sustainable use of the biosphere reserve and facilitating oil and gas activities. They defined natural environmental units (NEUs) as a function of the spatial arrangement of a number of environmental variables that characterize the region based on their internal homogeneity, degree of regularity, and their ecological importance (see Fig. 2A). The 29 NEUs included fluvial systems of the coastal plain, natural water courses receiving drainage from local riparian vegetation zones, coastal plain estuaries, tidal channels, dredged channels (mainly for oil and gas), natural levees of active river channels with pasture or agriculture, natural levees of abandoned river channels subject to frequent flooding, crevasses formed by overflow from river channels, major river channels, inland and coastal lagoons, oxbow lakes, interdistributary wetland basins, freshwater swamp and mangrove forests, and abandoned river meanders with riparian vegetation or wetland vegetation.

After the NEUs were identified, the level of disturbance was quantified ranging from low to major disturbance based on types of environmental degradation (Table 1). This information was then used to identify nine environmental management units (EMUs) in order to facilitate better protection, restoration, and conservation of the resources of the CWBR (Fig. 2B). The NEUs were collapsed into similar and geographic proximal areas to form the EMUs and included two nucleus zones and seven buffer zones

The two nucleus zones contain a variety of well-preserved natural habitats of the CWBR, some of which are fragile and in need of protection. There is some agriculture but these areas are not significantly impacted by human activities. Compatible uses are ecotourism, scientific research, environmental education, preservation, and harvest of natural resources. The area is not compatible with new human settlements, oil exploration and exploitation, and agricultural activities. The buffer zones represent a variety of habitat areas and have higher levels of human impact and are in need of restoration and they also serve to protect the nucleus zones.

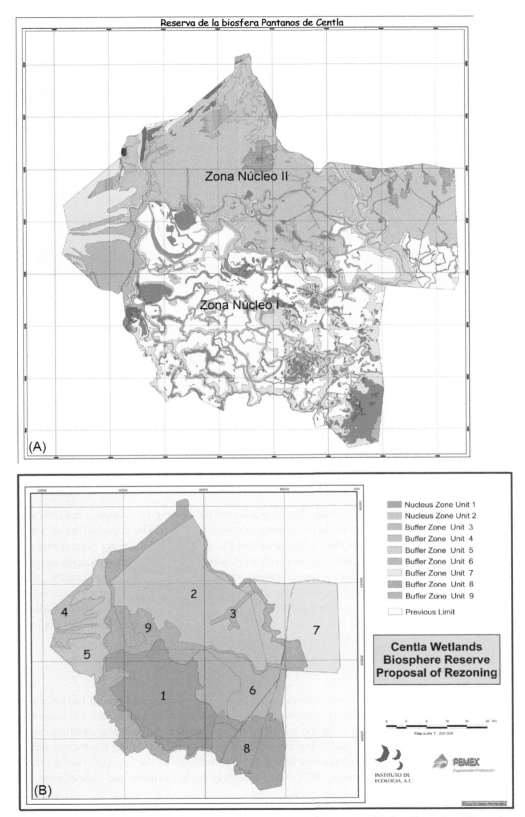

FIG. 2 (A) Natural environmental management units (EMUs) of the CENTLA Biosphere reserve. Nucleus Zones 1 and 2 contain some the best conserved areas in their natural state. The larger areas of *green* and *pale yellow* are different types of wetland ecosystems, *blue areas* are water, and *yellow* and *orange* coastal and palustrine vegetation. The different types of EMUs are listed in the text and presented in detail in Lara-Domínguez et al. (2011, 2013). (B) Environmental management units (EMUs) for the CENTLA Biosphere Reserve. The buffer zones represent geographic zones with similar habitat types, degree of human impact, and types of economic activity. *(Figures modified from Lara-Domínguez et al. (2011).)*

TABLE 1 Diagnosis of Environmental Problems in the Biosphere Reserve Centla Wetlands CWBR

Natural Causes	Impacts
Hydrodynamic changes	Reduction of sediment load Clogging and deposition in channels
Delta subsidence and/or relative sea level rise	Wetlands lost Changes in the original vegetation species composition
Exploration and extraction of petroleum and gas	Fragmentation of ecosystems Sediment retention Water pollution in oil fields
Rural expansion	Wetlands lost due to road construction NO_3 and TP increase Fecal coliform bacteria increase in rivers Introduction of exotic species Deforestation
Agricultural expansion	NO_3 and TP increase Decreased species richness Decreased canopy height
Hunting and fishing Wetland burning Channelization	Sedimentation increase Deposition in channels Alteration of species abundance Pollutants and debris in wetlands

2.1 Environmental Diagnosis

In the last 20 years, important wetland areas have been loss due to natural subsidence processes that occur in the delta. But these processes have been accelerated by external anthropogenic factors such as reduction of sediment input by dikes and dredged canals. These impacts lead to environmental changes such as habitat fragmentation, wetland loss, and increase in open water areas. Factors outside of the reserves such as flood control in Villahermosa and pipelines that link the reserve with other areas also affect the CWBR.

Expansion of human settlements and agriculture has also affected the reserve by changing habitats and reducing water quality. Flood control levees prevent river water from flowing into wetland areas. The introduction of high-yielding grasses for cattle grazing is also affecting the biodiversity of the area and grazing itself is changing community structure.

The diagnosis that we carried out indicated the necessity to modify the land-use classification originally described in the 1992 plan. The purpose of the redefinition of management units is better protection, restoration, and conservation of the resources of the CWBR. In addition, it is to allow more effective management of human activities of the area. Based on ecological, socioeconomic, political, and regulatory criteria, we defined an Ecological Classification Model for the CWBR that identified the nine EMUs that included two nucleus zones and seven buffer zones. More details on the management plan can be found in Lara-Domínguez et al. (2011 and 2013).

3 LAGUNA DE TERMINOS

Laguna de Términos is the largest coastal lagoon-mangrove ecosystem in Mexico and one of the largest in Latin America (Fig. 3). For most of the year, prevailing easterly trade winds cause a net seawater inflow into the eastern Puerto Real inlet and a net outflow of estuarine mixed waters from the western El Carmen inlet. Annual precipitation ranges between 1650 and 1850 mm year^{-1}. The southwestern part of the lagoon receives more than 50% of the freshwater input, primarily from the Palizada River, a distributary of the Usumacinta River. River discharge to the lagoon was historically estimated at 6×10^9 m^3 year^{-1} (Phleger and Ayala-Castanares, 1971). This results in higher salinity and clearer water in the eastern end of the lagoon and more turbid lower salinity in the western end.

Puerto Real inlet is more saline (30–37 psu), warmer (24–28°C), clearer (>60% transparency), and with higher sediment calcium carbonate concentrations (60%–90% $CaCO_3$) than Carmen Inlet (15–25 psu, 22–27°C, 40% transparency, 10%–30% $CaCO_3$). Estero Pargo, a tidal channel on the lagoon side of Carmen Island has conditions intermediate between the two inlets

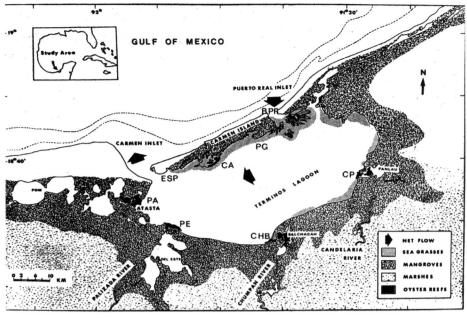

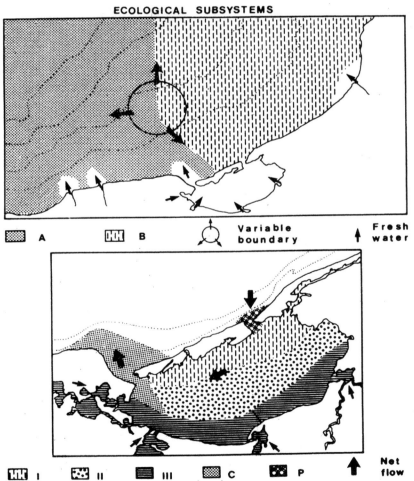

FIG. 3 Top. Laguna de Terminos showing the location of major vegetation habitats and oyster reefs, direction of net flow through the lagoon, and rivers entering the lagoon. Letters indicate sampling locations for nekton discussed in the text. Mid. Location of major subsystems offshore of Terminos Lagoon. Left terrigenous sediments, right carbonate sediments. Bottom. Major habitats of Terminos Lagoon. I Inner litoral of Carmen Island, II Central Basin, III Fluvial-deltaic systems, C Carmen Inlet, P Puerto Real Inlet. *(From Yáñez-Arancibia, A., Day, J.W., Lara-Domínguez, A.L., Sánchez-Gil, P., Villalobos, G.J., Herrera-Silveira, J.A., 2013. Ecosystem functioning, the basis for sustainable management of Terminos Lagoon, Campeche Mexico (Chapter 9). In: Day, J., Yáñez-Arancibia, A. (Eds.), Ecosystem Based Management, vol. 4. The Gulf of Mexico, Origin, Waters, Biota. Texas A&M University Press, College Station, pp.167–199.)*

(26–39 psu, temperatures of 23–32°C and transparency in the water column higher than 80%). In the central lagoon, salinity ranges from about 12 psu in November during high river discharge to 30 psu in May at the end of the dry season in May. Water temperature ranges between 19°C and 30°C. Solar radiation and water transparency are highest at the end of the dry season.

Laguna de Terminos receives a high river input from several rivers, leading to inner estuaries or fluvial lagoon subsystems subject to the bidirectional flow of energy, organic matter, nutrients, and sometimes toxic materials among river mouths, the lagoon and the Gulf (Fig. 3). Such exchanges are highly dynamic and enhance lagoon productivity. These interactions result in seasonal and spatial patterns of water quality. The dominant depositional environments and local habitats (i.e., submerged aquatic vegetation) adjacent to these fluvial lagoon subsystems can influence nutrient concentrations and modify water quality and biota. These differences are more evident during the rainy season when river discharge is highest. N inputs are high with respect to P and may have a strong impact on ecological integrity of Laguna de Terminos in the near future (Medina-Gómez et al., 2015).

3.1 Habitat Diversity

The spatial patterns described above give rise to a high habitat diversity. Because of clearer marine waters, extensive sea grass beds occur in the eastern end of the lagoon and the inner littoral of Carmen Island. In contrast, extensive oyster reefs occur in the western part of the lagoon near the river mouths (Fig. 3). Aquatic habitats include open waters of the nearshore Gulf, brackish waters of the lagoon, and freshwater areas at river mouths. In the nearshore Gulf, terrigenous sediments predominate off the western end of Carmen island while carbonate sediments occur off the eastern end of the island (Fig. 3). There are variety of mangrove and freshwater wetland habitats including all of the mangrove types described by Lugo and Snedaker (1974). Riverine forests occur in areas affected by riverine input. Fringe forests occur along tidal channels and in the inner littoral of Carmen Island and scrub forests are located in a few locations where circulation is restricted. The forests are composed of red (*Rhizophora mangle*), black (*Avicennia germinans*), and white (*Laguncularia racemosa*) mangroves and button bush (*Conocarpus erectus*). Submerged aquatic vegetation dominated by *Thalassia testudinum* is found in clear, high salinity waters while freshwater species such as *Vallisneria americana* and *Cabomba palaeformis* occur in small lagoons associated with areas with freshwater input such as the river mouth systems.

Spatial and temporal patterns of primary productivity have been measured for all major producer groups of the lagoon-wetland ecosystem (Day et al., 1982, 1987, 1988, 1996, and summarized by Rojas-Galaviz et al., 1992, Fig. 4). Mangrove litterfall is higher during the rainy season and early nortes season. Aquatic primary productivity (APP) is highest at the same period when water clarity and nutrients are abundant (Day et al., 1982). APP decreased from $333\,g\,m^{-2}\,year^{-1}$ in Estero Pargo tidal channel (Ley-Lou, 1985), to $222–240\,g\,m^{-2}\,year^{-1}$ in *Thalassia* beds, and $197\,g\,m^{-2}\,year^{-1}$ in the central lagoon. Above ground mangrove NPP was 2475 and $1606\,g\,m^{-2}\,year^{-1}$, respectively, in riverine and fringe forests (Day et al., 1987). Submerged aquatic vegetation biomass was highest during the dry season with high water clarity (Rojas-Galaviz et al., 1992). Above and below ground biomass of Thalassia was 136–275 and $478–1591\,g\,m^{-2}$, respectively (Rojas-Galaviz et al., 1992). Daily photosynthesis rates of SAV were $1.8–12.7\,g\,O_2\,m^{-2}\,day^{-1}$ (Day et al., 1982). Because the different primary producers have peak production periods at different times, there is high year-round productivity but peak productivity shifts seasonally among different locations in the lagoon (Fig. 4).

3.2 Water Quality and Biogeochemistry

Botello and Mandelli (1975) conducted an early study of water quality parameters. Mean salinity was 33.5 psu in May and 21.9 psu in November. PO_4 and NO_2+NO_3 were higher during the rainy season and NH_4 was higher in the dry season. Dissolved oxygen was 147% saturation in November and 99% in May reflecting higher APP during the rainy season (Day et al., 1982, 1987, 1988, 1996). The lagoon was a sink for PO_4 and NO_3. Export from mangroves swamps (Rivera-Monroy et al., 1995b) and denitrification (Rivera-Monroy et al., 1995a) also affect biogeochemistry of the lagoon.

A comparison of earlier nitrogen data with that of Herrera-Silveira et al. (2002a,b) measured 20 years after the earlier studies and found that nitrogen concentrations increased by a factor of five and phosphorus also increased due to land-use changes in the Usumacinta basin. This led to a change in phytoplankton species but nitrogen generally remains the limiting nutrient.

Herrera-Silveria et al. (2002) reported that the highest mean annual chlorophyll-*a* value ($14\,mg\,m^{-3}$) occurred in the inner littoral of Carmen Island during the rainy season. This area is characterized by seagrass beds (*T. testudinum*), macro algae, and drainage from mangroves. All these factors lead to higher phytoplankton production (Day et al., 1982, 1987, 1988; Yáñez-Arancibia and Day, 1982, 1988; Rojas-Galaviz et al., 1992; Rivera-Monroy et al., 1998). The highest mean annual chlorophyll-*a* values were at the end of the *nortes* season (Yáñez-Arancibia and Day, 1982, 1988; Yáñez-Arancibia et al., 1985, 1988, 1993; Soberón-Chávez et al., 1988). Herrera-Silveria et al. reported similar seasonal and spatial patterns of nutrients as in earlier studies.

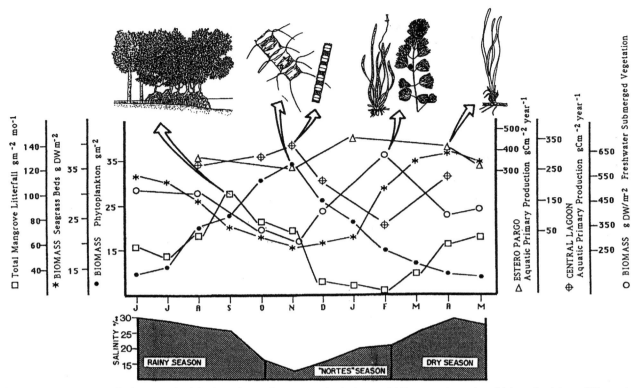

FIG. 4 Seasonal patterns of primary productivity by major producer groups in the lagoon. Different producers have high productivity at different times resulting in high year-round productivity. *(From Rojas-Galaviz, J.L., Yáñez-Arancibia, A., Vera-Herrera, F., Day, J.W., 1992. Estuarine primary producers: the Terminos Lagoon case study. In: Seeliger, U. (Ed.), Coastal Plant Communities in Latin America. Academic Press Inc., New York, pp. 141–154, 392 pp. (Chapter 10).)*

3.3   Consumers

There is a high diversity of consumers including over 250 species of migratory nekton and a rich diversity of benthos and avifauna. Migratory nekton occupies the full range of coastal habitats from tidal freshwater through the lagoon to the nearshore Gulf. Nekton use different habitats in when primary productivity of each habitat is highest (Fig. 5). In doing so, nekton exhibits small-scale migration patterns.

Fishery resources in Campeche Sound and the lagoon are strongly dependent on food availability, high habitat diversity, and the movement of nekton preadults from the lagoon-estuarine system to the sea (Yáñez-Arancibia et al., 1980; Yáñez-Arancibia and Day, 1982; Deegan et al., 1986; Sánchez-Gil et al., 2008). Two examples show the coupling of nekton migration to ecological, biogeochemical, and hydrological process of the system. Sánchez-Gil et al. (2008) showed that two species of flatfishes, *Etropus crossotus* and *Citharichthys spilopterus*, used the estuarine plume and lagoon habitats sequentially using estuarine-related vs estuarine-dependent strategies making optimal use of food resources. They showed that variation in biomass and diversity of fish assemblages using mangrove and seagrass habitats were synchronized with circulation patterns and seasonality of primary production to enhance overall nekton secondary production.

The high productivity and habitat diversity supports a high diversity, multistock fishery resource in Campeche Sound. Seventy-five percent of dominant species are estuarine dependent or estuarine related in the juvenile and preadult stages (Yáñez-Arancibia et al., 1985; Yáñez-Arancibia and Sanchez-Gil, 1986; Sánchez-Gil and Yáñez-Arancibia, 1997).

3.4   Human Impact and Management

Compared to many coastal systems worldwide, Laguna de Terminos has been less impacted by human activities. With the exception of some areas on Carmen Island, mangroves remain largely intact. Water quality has been less impacted in terms of eutrophic status and toxins. But there have been localized wetland impacts and increased nutrient levels. Nutrients increased over a 30-year period and led to increased APP but eutrophication is not yet a problem. The lagoon shifted from nitrogen to phosphorus limitation and mean lagoon chlorophyll-*a* was 3.0 in the 1970s and 5.6 in 2002 suggesting an

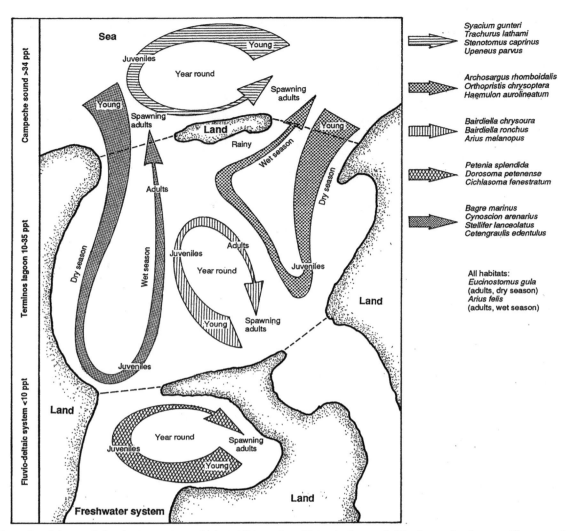

FIG. 5 Examples of characteristic migratory patterns of selected fish species utilizing the Laguna de Términos regional system. The seasonal pattern of migration leads to utilization of different habitats during periods of peak primary productivity. *(From Yáñez-Arancibia, A., Lara-Domínguez, A.L., Rojas-Galaviz, J.L., Sánchez-Gil, P., Day, J.W., Madden, C.J., 1988. Seasonal biomass and diversity of estuarine fishes coupled with tropical habitat heterogeneity (southern Gulf of Mexico). J. Fish Res. 33 (Suppl. A), 191–200 as redrawn by Pauly, D., Yáñez-Arancibia, A., 1994. Fisheries in coastal lagoons. In: Kerfve, B.J. (Ed.), Coastal Lagoons Processes. Elsevier Science Publishers, Amsterdam, pp. 352–372, 577 pp.)*

increase in APP; however, the lagoon has not become eutrophic. The source of nutrients is mainly from the drainage basin. Thus, management for environmental quality should also focus on the drainage basin.

From a comprehensive context, management of Terminos Lagoon should follow an ecosystem-based approach, utilize the energies of nature as much as possible, and involve all stakeholders in development of restoration and management plans (Day and Yáñez-Arancibia, 1982; Currie-Alder, 2013). An important action to achieve protection of the lagoon was the establishment of the Special Area for Protection of Fauna and Flora of Laguna de Términos (CONANP, 2004; Yáñez-Arancibia et al., 1993, 1999b, 2009; CONANP, 2004).

4 YUCATAN GROUNDWATER COASTAL KARSTIC ECOSYSTEMS—ENVIRONMENTAL RISK AND MANAGEMENT OPPORTUNITIES

4.1 Aquatic Coastal Ecosystems in Karstic Settings

Karst landscapes are characterized by the almost complete absence of surface water, rapid infiltration of precipitation, sinkholes, caves, and extensive groundwater flow. Rapid groundwater flow means that karstic aquifers are very susceptible to contamination.

The Yucatán karst aquifer system is one of the most extensive globally. The Yucatán Peninsula is a carbonate platform with an extensive continental shelf. The aquifer system has developed in a nearly horizontal, highly permeable karstic system resulting in high hydrological connectivity between terrestrial and coastal ecosystems and strongly controlled by climate patterns. The aquifer system extends over 165,000 km^2 in México, Guatemala, and Belize. This large groundwater resource maintains highly diverse groundwater-dependent coastal ecosystems. Seawater intrusion affects large parts of the aquifer, which is the only available freshwater resource in the peninsula. There is very little surface water because of high evaporation and rapid infiltration through the soil. Groundwater circulation is consequently high and large volumes move through the subsoil toward the coast. Freshwater discharge is abundant along much of the Yucatán coast (Ward et al., 1985; Herrera-Silveira, 1994; Perry et al., 1995).

Aquatic habitats include karstic freshwater lakes, brackish lagoons, coastal lagoons and reef lagoons, forming the largest and most diverse coastal ecosystems of the Yucatan. The spatiotemporal structure of this complex ecosystem can be used as a model system to study hydrological, biogeochemical, and ecological interactions under the "Transverse Coastal Corridor" conceptual model (Herrera-Silveira and Comín, 2000; Herrera-Silveira et al., 2004; Hernández-Arana et al., 2015).

4.2 Main Features and Ecosystem Characteristics

Factors affecting these coastal ecosystems are variable due to (1) shape and size of coastal lagoons; (2) regional forcing changes significantly with location; (3) tidal range, freshwater input, and human impact are also variable among locations; (4) the tropical climate ranges from arid to humid; and (5) the impact of natural events such as hurricanes varies along the coast (Herrera-Silveira and Comín, 2000; Herrera-Silveira et al., 2013a).

The functional structure of these karstic coastal ecosystems is controlled by regional (i.e., Yucatan Current) and local (i.e., upwelling and groundwater discharges) drivers, as well as pulsing events ranging from high-frequency low-intensity events (i.e., tides) to low-frequency high-intensity events (i.e., hurricanes). In addition, anthropogenic factors interact with all these drivers to determine conditions for individual coastal ecosystems (Herrera-Silveira et al., 2013a) (Fig. 6A and B). The Yucatan coast can be divided into East coast, North coast, and the West coast subregions.

A conceptual framework based on three core concepts was developed to guide ecosystem assessment and management of the Yucatan. Connectivity encompasses biogeochemical, biological, and hydrological interactions. Dynamic land-sea interactions affect system productivity, biodiversity, responses to disturbance, and overall functioning. Ecological stability is the speed and manner with which ecosystems recover after disturbance and their resistance to major disturbances (Herrera-Silveira et al., 1999; Morales-Ojeda et al., 2010).

4.3 Coastal Lagoons

The coastal lagoons of the Yucatan cover about 19,000 km^2. Although they have many features in common, each lagoon responds differently to human activity and environmental forcings. Mean salinity ranges from 10 to 40, reflecting differences in hydrogeomorphology and external forcing (Table 2). Freshwater inputs maintain water chemistry gradients and promote biodiversity and ecological complexity (Herrera-Silveira and Morales-Ojeda, 2010). However, if freshwater sources are contaminated, this can have a negative effect.

The relations between salinity and dissolved inorganic nutrients reflect freshwater input and lagoon circulation. Some lagoons exhibit large salinity gradients associated with submerged groundwater discharge (SGD) inputs. In others, less dilution of high-nutrient SGD by low-nutrient seawater occurs and freshwater inputs rather than mixing with seawater determine nutrient concentrations. Other lagoons are strongly influenced by oceanic controls (Herrera-Silveira, 2006; Herrera-Silveira and Morales-Ojeda, 2010).

Nine coastal lagoons have been characterized in terms of water quality (nutrients, phytoplankton biomass), habitat change (seagrasses and mangrove), and harmful algal species. DIP is generally low due to carbonate soils, while nitrate and silicate are higher due to SGD inputs. Water residence time is variable between and within lagoons, and some lagoons showed symptoms of eutrophication. Inland human activities have quickly contaminated groundwater with nutrients and other pollutants, which affect lagoon water quality and the overall condition of lagoons appears to be worsening due to human activities (Fig. 7).

The variability of inorganic nitrogen species (NH_4^+, NO_3^-) is related to high concentrations in groundwater. The main sources of nitrate are excessive fertilizer use and inadequate manure disposal; the major NH_4^+ source is organic matter decomposition in the aquifer (Aranda-Cirerol et al., 2006; Young et al., 2008; Pacheco et al., 2001). Coastal waters typically exhibit high alkalinity and low DIP due to karstic soils ($<0.1\,\mu M$) where DIP precipitates. Water column DIP is low, limiting coastal productivity. Therefore, lagoons respond rapidly to even small DIP additions (Adame et al., 2015).

The prevalence of HAB species and the frequency and intensity of algal blooms suggest that all the lagoons are at risk if nutrient concentrations increase (Alvarez-Góngora and Herrera-Silveira, 2006; del Merino-Virgilio et al., 2013). Shifts in phytoplankton community structure from diatom dominance to harmful dinoflagellates have been reported in Progreso and Nichupte lagoons, indicating increasing eutrophication (Herrera-Silveira et al., 1999; Herrera-Silveira and Morales-Ojeda, 2010).

The shallowness and high water transparency of the Yucatan coastal lagoons promote high SAV cover (>50%). SAV consists mainly of seagrasses (*T. testudinum*, *Halodule wrightii*, and *Ruppia maritima*) and green and red macroalgae. Temporal changes in SAV composition and abundance depend on hydrological and sediment conditions, and anthropogenic impacts. SAV has not changed significantly over the last decade, suggesting good to fair condition of the lagoons, although lagoons near urban areas or tourism infrastructure are threatened (Herrera-Silveira, 2006). However, all lagoons receive contaminated groundwater, resulting in some level of pollution (Carruthers et al., 2005).

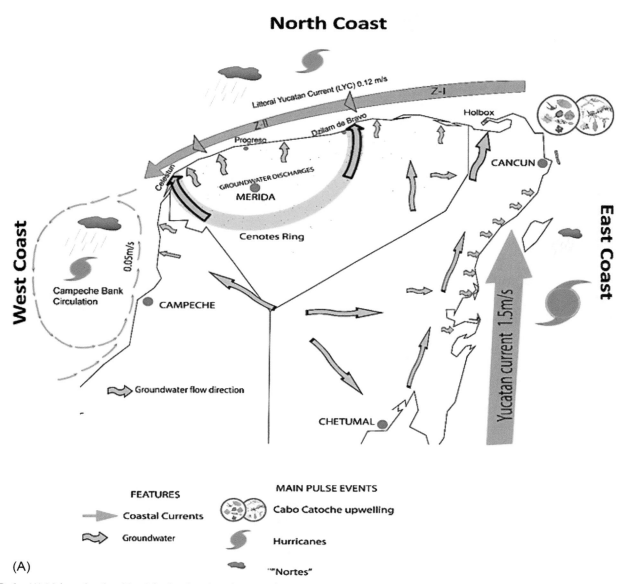

(A)

FIG. 6 (A) Main regional and local forcing functions that control ecosystem processes in the coastal ecosystems of the karstic region of the Yucatán Peninsula. The arrow size is related to the magnitude of each forcing function that acts on the Yucatan Peninsula coast. The cenotes ring is related to the great meteor impact about 65 million years ago. Z-I and Z-II are two distinct zones along the north coast.

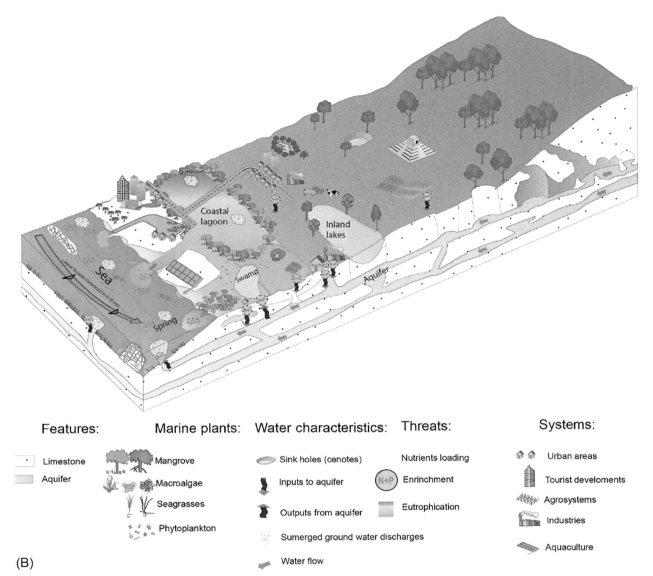

Features:

· Limestone

Aquifer

Marine plants:

Mangrove

Macroalgae

Seagrasses

Phytoplankton

Water characteristics:

Sink holes (cenotes)

Inputs to aquifer

Outputs from aquifer

Sumerged ground water discharges

Water flow

Threats:

Nutrients loading

(N+P) Enrinchment

Eutrophication

Systems:

Urban areas

Tourist develoments

Agrosystems

Industries

Aquaculture

(B)

FIG. 6, CONT'D   (B) Schematic cross-sectional view of the landscapes in the Yucatan coastal zone, showing the major connections among terrestrial and coastal environments, as well as some threats and systems responsible.

Lagoons such as Celestún, Conil, and Chacmochuck have maintained or recovered their original SAV characteristics after hurricane events. Most lagoons have not exhibited such resilience and SAV cover has not rebounded (Asencio, 2008; Arellano-Méndez et al., 2011). The natural variability of the hydrological characteristics of the coastal lagoons is reflected in the salinity gradient. This variable is probably indicative of the resilience of these ecosystems to natural events such as hurricanes. Lagoons with wider salinity ranges (Celestún, Conil) recover faster than those whose average salinities are more stable (Progreso, Nichupte-Bojórquez) (Table 2). On the other hand, trawling and boat traffic can damage SAV (Herrera-Silveira et al., 2000), and SAV in more polluted lagoons has lost some resilience to natural and anthropogenic disturbances.

Differences in mangrove vegetation structure are related to salinity gradients, soil nutrient concentration, and hydroperiod, resulting in high ecotype diversity (Fig. 8). Mangrove degradation is related to human activities that modify the hydroperiod, mainly the time and frequency of flooding, and also due to changes in land use related to boat traffic and tourism. Mangroves in lagoons with low human impact are in good condition, while human impact is affecting others (Herrera-Silveira et al., 2013b). Organic carbon stocks in mangroves of this region are high ($900 \pm 200\,\mathrm{Mg\,C\,h^{-1}}$; Herrera-Silveira et al., 2017) and can contribute to mitigation of climate change through carbon sequestration. Some mangroves can

TABLE 2 Main Features, Activities, and Threats to Coastal Lagoons in the Yucatan Peninsula (SE Mexico)

System	Variable Protection Status	Sal (psu)	A (km²)	Z (m)	T (days)	Activities	Main Drivers	Threat
Celestún	FPNA	5–38	28	1.2	50–300 / 100	– Fisheries – Ecotourism – Mineral extraction	Groundwater	– Loss of SAV coverage – Water level reduction by sediment refill – Eutrophication
Chelem	NP	28–44	14	0.8	300–750 / 400	– Fisheries – Ecotourism – Urban development	Oceanic	– Loss of SAV coverage – Pollution – Eutrophication
Ria Lagartos	FPNA	22–130	91	0.5	450	– Fisheries – Ecotourism	Oceanic	– Pollution from its watershed
Conil	FPNA	32–44	275	1.5	200–350 / 280	– Fisheries – Ecotourism	Oceanic	– Pollution from its watershed
Chacmochuk	NP	28–40	122	1	150–350 / 200	– Fisheries	Oceanic	– Pollution from its watershed – Eutrophication
Nichupte	NP	15–38	41	2.2	200–500 / 300	– Massive tourism	Oceanic and groundwater	– Eutrophication – Loss of SAV coverage
Bojorquez	NP	25–37	3	1.7	360–800 / 400	– Massive tourism	Oceanic and groundwater	– Eutrophication – Pollution – Loss of SAV coverage
Ascension	FPNA	10–40	740	2.5	200–400 / 150	– Fisheries – Ecotourism	Groundwater and oceanic	– Pollution from its watershed
Chetumal	NP	2–20	1098	3	200–400 / 399	– Fisheries – Ecotourism – Urban development	Groundwater and river	– Pollution from its watershed

A, lagoon area; FPNA, federal protected natural area; NP, not protected; Z, mean depth of the lagoon; T, residence time of water, upper-lower, and average.
Variables are listed along the top of the table and systems are listed in the first column.
Modified from Herrera-Silveira, J.A., Morales-Ojeda, S.M., 2010. Subtropical karstic coastal lagoon assessment, SE Mexico. The Yucatan peninsula case. In: Kennish, M.J., Paerl, H.W. (Eds.), Coastal Lagoons: Critical Habitats of Environmental Change. CRC Press, pp. 309–336. ISBN: 978-1-4200883-0-4.

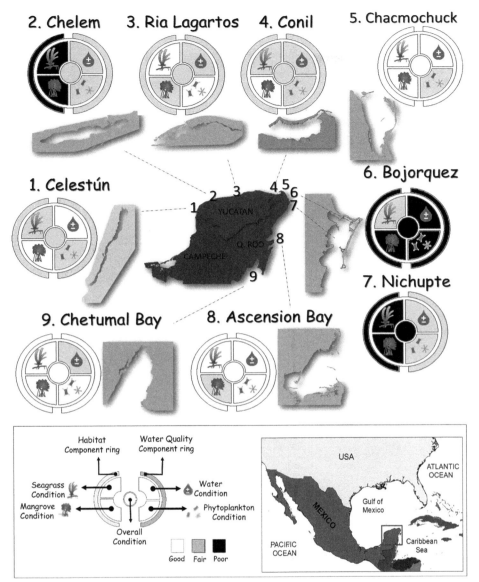

FIG. 7 Environmental condition of coastal lagoons in the Yucatan Peninsula as a result of biological and hydrological characteristics. *White* good, *gray* fair, *black* poor condition.

compensate for sea-level rise due to high accretion while others are sinking, making them more vulnerable to sea-level rise (Herrera-Silveira, 2006; Herrera-Silveira et al., 2013b). Mangroves have high resilience to hurricanes when they are healthy (Celestun, Conil, Chetumal), but remain degraded for long periods if their condition is poor prior to the storms (Progreso, Nichupte-Bojorquez) and restoration is necessary (Teutli-Hernández and Herrera-Silveira, 2016; Comín et al., 2005).

4.4 Nearshore Coastal Systems

The extent of the nearshore ecosystems varies from a few hundred meters to several kilometers depending on SGD plumes and coastal currents in each. SGD is particularly important in the nearshore zone due to the volume and nature of flow pathways. In general, two types of flows occur: focused fracture flow and diffuse flow. The latter contributes the majority of SGD and a large fraction of nutrient loading (Beddows, 2002; Bauer-Gottwein et al., 2011).

In nearshore coastal systems, SGD contributes about 90% of freshwater inputs on the north coast of the Yucatan, while coastal lagoons such as Celestun SGD contributes 5%, surface runoff 4%, and harbors such as Progreso 1%. Although the

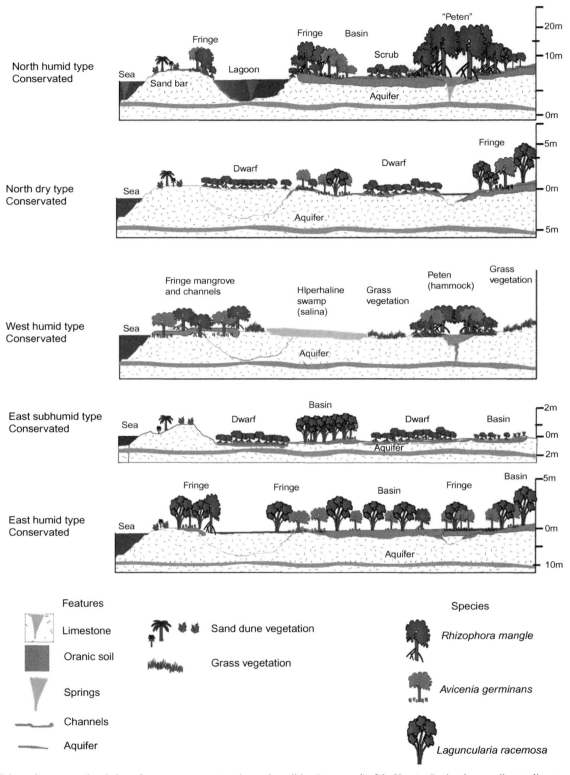

FIG. 8 Schematic cross-sectional view of mangrove ecosystems in good condition (conserved) of the Yucatan Peninsula according to climate conditions.

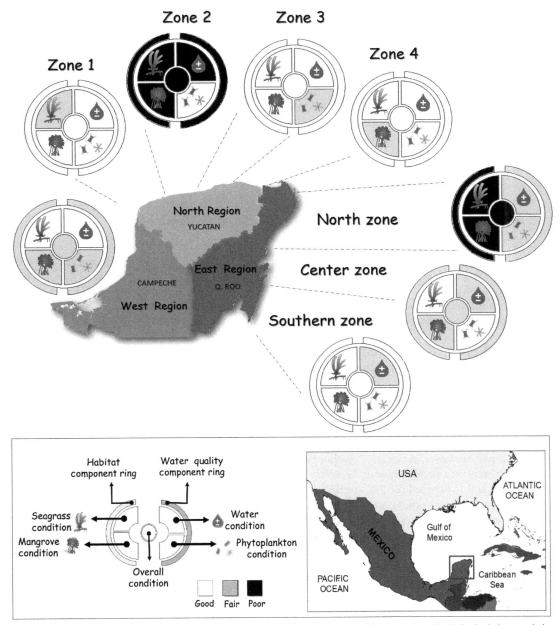

FIG. 9 Environmental condition of nearshore waters the in the Yucatan Peninsula as a result of biological and hydrological characteristics.

main freshwater and nutrient source is groundwater, differences in water quality are related to land use changes, residence time, and weather (Aranda-Cirerol et al., 2006; Herrera-Silveira and Morales-Ojeda, 2009).

Water quality diagnosis based on nutrients, DO, chlorophyll-*a*, HAB species richness, and SAV coverage (Fig. 9) indicate that the south and center zones of the east coast are in good condition for all indicators except nutrients (Null et al., 2014; Hernández-Terrones et al., 2015). Wastewater discharges from agricultural and urban areas of the Rio Hondo basin and Holbox fractures influence water quality. The northern zone shows signs of deterioration for all variables especially nutrients, chlorophyll-*a*, and SAV coverage due to nutrients that favor macroalgae, and cleaning of seagrasses from touristic beaches (Herrera-Silveira et al., 2010a,b; Hernández-Arana et al., 2015). Mangroves in this zone have been cleared for development leading to loss of more than 7000 ha in 25 years (Valderrama et al., 2014). Land use changes, development, and population density are the top contributors to pollution of coastal groundwater systems. Primary symptoms of eutrophication are HABs, and changes in SAV seagrasses and macroalgae (Herrera-Silveira and Morales-Ojeda, 2009; Aranda-Cirerol et al., 2011).

The west coast generally has good water quality conditions. This region has the least land use change due to urban development, although mangroves have been impacted by hydrological changes caused by road construction. This area has the largest extent of SAV (López-Herrera et al., 2015).

Differences that exist along the three coasts are due to changes in the type and intensity of land use change and surface and ground water flow and these are the main drivers of the health of nearshore ecosystems. Nutrient concentrations, however, are higher in areas with greater freshwater influence rather than amount of development (Mutchler et al., 2007), but population density and tourism along the east coast is also important. There is a need to better understand how anthropogenic activity and groundwater delivery in karst systems are linked to land use and pollutant transport that impact local coastal ecosystems. These will have serious consequences if mitigation and adaptation measures are not taken, the most important being restoration of critical ecosystems such as mangroves and seagrasses.

4.5 Land Use and Coastal Ecosystem Risk of the Yucatan Peninsula

Four general types of land use have occurred in the coastal Yucatan over the past 30 years: conservation, semi-intensive urbanization, construction of port facilities, and massive tourist development. The use of natural areas for conservation, especially the northwest coast, has integrated human activities, especially ecotourism, with natural systems that maintain a diversified economy for local people. Massive tourist development on the northeastern and central parts of the east coast, supported by foreign investment also contributed to the socioeconomic status of local people but with more negative impact. The major environmental problem associated with development is loss of habitat, widespread pollution, and loss of mangroves.

The Mexican government has made an effort to implement sustainable management strategies in the coastal ecosystems of the Gulf of Mexico and the Caribbean through establishing protected areas for ecosystems with high biodiversity of species and ecosystems (Table 2). However, many of these efforts have yet to produce good management practices (Chelem, Conil, Nichupte, Bojorquez), while in others (Celestun, Ascensión), a society-nature relationship has developed taking advantage of the environmental services of the ecosystems to yield economic benefits to inhabitants (ecotourism), while conserving or restoring their ecosystems (Teutli-Hernández and Herrera-Silveira, 2016).

If these problems persist, increased environmental degradation will occur. Remediation of these impacts is very difficult in urban developments. However, restoration should be a priority in areas surrounding urban developments. This would yield both environmental and socioeconomic benefits. Massive urbanization has occurred in the northern part of the east coast (Riviera Maya) and is planned for the southern part of this coast (Costa Maya). This type of development generates income and attracts people from other parts of Mexico. However, this type of land use has also caused irreparable loss of habitats and degraded coastal lagoons and coral reefs.

This type of tourist development is not considered sustainable because of the impact on the stability of both ecosystems and socioeconomic conditions. In fact, failures in the Cancun-Tulum tourist corridor can be observed today, as evidenced by tourists going elsewhere, increased crime in recent years including some crimes unheard of before, and a reduction in the quality of life of inhabitants (INEGI, 2011).

Analysis of these experiences suggests that: (1) management and planning for development should include evaluating alternative strategies that include both the larger context as well as the local area; (2) the concepts of ecosystem, ecosystem connectivity, and resilience must be incorporated during the planning processes; and (3) basic physical, chemical, and biological characteristics should be preserved.

Although some of these perspectives to prevent and mitigate environmental problems have been recently incorporated in some areas (e.g., opening culverts in old roads), most of these problems remain unresolved and are often not considered for new activities. A key strategy for tourism development is not to destroy the resource base (Cater and Goodall, 1992). Sustainable use of the coastal zone in the Yucatan Peninsula should be associated with human uses if it incorporates the proposals suggested here. The prediction of impacts and anticipation of scenarios are possible using models. Using ecohydrology as a tool for integrated management can help achieve long lasting and ecologically sustainable solutions. This is the key to the management and conservation of coastal ecosystems.

The functioning of Yucatan coastal ecosystems sustains environmental goods and services. However, the type and intensity of local and regional activities are causing serious threats to coastal ecosystems. We must identify the causes of the problems, their impacts, and consequences to develop specific approaches for their solution through mitigation and restoration. We must apply new eco-technologies appropriate for the natural resources and humans using integrated coastal zone management and an ecosystem approach.

5 MANAGEMENT PERSPECTIVES FOR THE YUCATAN PENINSULA AND SOUTHERN GULF OF MEXICO

System functioning should form the basis for a sustainable management in the southern Gulf of Mexico and Yucatan Peninsula (Day et al., 1997, 2000) and focus on important aspects of the system including wetland preservation, hydrologic restoration, and using resources of freshwater input to enhance coastal ecosystems.

Climate change and increasing energy cost of energy have important implications for restoration (Day et al., 2005). Coastal restoration efforts will have to be more intense to offset the impacts of climate change (Day et al., 2018). Restoration and management of coastal ecosystems in the southern Gulf of Mexico should include (Day and Yáñez-Arancibia, 1988): (1) protect structure and functioning of regional ecosystems, (2) utilize natural energies, (3) carefully plan urban and industrial development in harmony with nature, (4) determine the optimal yield of biotic resources, and (5) monitor changes in resources and habitats.

From a regional perspective, Kemp et al. (2016) suggested that continental shelf river and groundwater plumes in the southern Gulf of Mexico promote ecological resilience in the face of climate change. Riverine and groundwater input sustain coastal ecosystems of the southern Gulf of Mexico and Yucatan Peninsula. The most important rivers are the GUR and adjacent rivers. These systems owe much of their natural resilience to a coastal geomorphology that spreads risk across the coast scape while providing ecosystem connectivity through shelf plumes that connect estuaries. Freshwater input generates large plumes that extend estuarine conditions into the Gulf and strongly influence fisheries over large areas of the adjacent continental shelf. Recent global change models indicate that that precipitation and freshwater input may significantly decline in the 21st century. A great threat to coastal ecosystems in the southern GOM is the severe drying predicted for the entire Mesoamerican "climate change hot-spot" (Fuentes Franco et al., 2015; Imbach et al., 2012). Reconnecting rivers to coastal ecosystems and releasing fine-grained sediments trapped behind dams are ways to counter the impacts of climate change. The UGR and delta are threatened by land use changes that interfere with coastal functioning.

The recognition that multiple threats endanger coastal ecosystems in the southern Gulf of Mexico and the Yucatan has led to successful efforts in Mexico to protect still intact coastal systems by the establishment of natural protected areas. Seventeen protected areas have been established by Mexican authorities at the Federal and State level (Fig. 10, Table 3). These include Biosphere Reserves, Flora and Fauna Protection areas, State Reserves, National Marine Parks, and National Parks. The total area of these coastal protected reserves is about 2.8 million ha in the states of Tabasco, Campeche, Yucatan, and Quintana Roo.

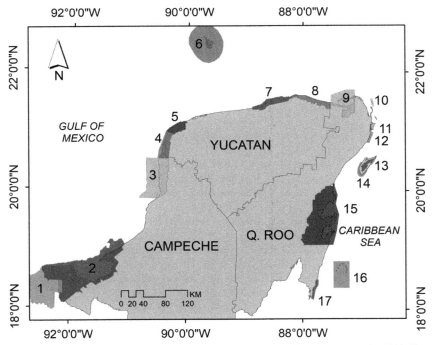

FIG. 10 Map of coastal protected areas in the states of Tabasco, Campeche, Yucatan, and Quintana Roo (see also Table 3).

TABLE 3 Natural Coastal Protected Areas in the Southern Gulf of Mexico and the Yucatan Peninsula

	Name	Category	Type	States	Area (ha)
1	Pantanos de Centla	Federal	RB	Tab, Cam	302,706
2	Laguna de Terminos	Federal	APFF	Cam, Tab	706,147
3	Los Petenes	Federal	RB	Cam	282,857
4	Ría Celestún	Federal	RB	Cam, Yuc	81,482
5	El Palmar	State	SR	Yuc	47,345
6	Arrecife Alacranes	Federal	PMN	Yuc	333,768
7	Reserva Estatal de Dzilam	State	SR	Yuc	68,374
8	Ría Lagartos	Federal	RB	Yuc	60,347
9	Yum Balam	Federal	APFF	QRoo	154,052
10	Isla Contoy	Federal	PN	QRoo	5,126
11	Manglares de Nichupté	Federal	APFF	QRoo	4,257
12	Arrecife de Puerto Morelos	Federal	PN	QRoo	9,066
13	Selvas y Humedales de Cozumel	State	SR	QRoo	19,703
14	Arrecifes de Cozumel	Federal	PMN	QRoo	11,987
15	Sian Ka'an	Federal	RB	QRoo	528,147
16	Banco Chinchorro	Federal	RB	QRoo	144,360
17	Arrecifes de Xcalak	Federal	PN	QRoo	17,949

Type key: RB: BR, biosphere reserve; *APFF*, flora and fauna protection area; *PN*, national park; *PMN*, national marine park, *RB*, biosphere reserve; *SR*, state reserve.
State key: *Cam*, Campeche; *Qroo*, Quintana Roo; *Tab*, Tabasco; *Yuc*, Yucatan.

REFERENCES

Adame, M.F., Fry, B., Gamboa, J.N., Herrera-Silveira, J.A., 2015. Nutrient subsidies delivered by seabirds to mangrove islands. Mar. Ecol. Prog. Ser. 525, 15–24.

Álvarez-Góngora, C., Herrera-Silveira, J.A., 2006. Variations of phytoplankton community structure related to water quality trends in a tropical karstic coastal zone. Mar. Poll. Bull. 52, 48–60.

Aranda-Cirerol, C., Herrera-Silveira, J.A., Comín, F.A., 2006. Nutrient water quality in a tropical coastal zone with groundwater discharge, northwest Yucatán, Mexico. Est. Coast. Shelf Sci. 68, 445–454.

Aranda-Cirerol, N., Comín, F.A., Herrera-Silveira, J.A., 2011. Nitrogen and phosphorus budgets for the Yucatán littoral: an approach for groundwater management. Environ. Monit. Assess. 172, 493–505.

Arellano-Méndez, L.U., Liceaga-Correa Ma, A., Herrera-Silveira, J.A., Hernández-Núñez, H., 2011. Impacto por huracanes en las praderas de *Thalassia testudinum* (Hydrocharitaceae) en el Caribe Mexicano. Rev. Biol. Trop. 59 (1), 385–401.

Asencio, E.J.M., 2008. Cambios en la vegetación sumergida de las lagunas costeras de Celestún, Chelem y Dzilam 2000–2005. (Tesis de Lic.). Inst. Tec. de Conkal.

Bauer-Gottwein, P., Gondwe, B.R.N., Charvet, G., Marín, L.E., Rebolledo-Vieyra, M., Merediz-Alonso, G., 2011. Review: the Yucatán Peninsula karst aquifer, Mexico. Hydrogeol. J. 19, 507–524.

Beddows, P.A., 2002. Where does the sewage go? In: The Karst Groundwater System of the Municipalidad de Solidaridad, Quintana Roo, Mexico. Association for Mexican Cave Studies Activities, Houston, TX, pp. 47–52.

Bestermeyer, B., Alonso, L.E., 2000. A biological assessment of Laguna del Tigre National Park. Higher Usumacinta basin, Peten Guatemala. In: RAP Bulletin of Biological Assessment 16. Conservation International, Washington, DC. 220 pp.

Botello, A., Mandelli, E., 1975. A study of variables related to the water quality of Terminos Lagoon and adjacent coastal areas, Campeche, Mexico. Final Report Project GU 853. ICMyL, UNAM, Mexico, DF92.

Carruthers, T.J.B., van Tussenbroek, B.I., Dennison, W.C., 2005. Influence if submarine springs and wastewater on nutrient dynamics of Caribbean seagrass meadows. Est. Coast. Shelf Sci. 64, 191–199.

Cater, E., Goodall, B., 1992. Must tourism destroy its resource base? In: Mannion, A.M., Bowlby, S.R. (Eds.), Environmental Issues in the 1990s. Wiley, Chichester, pp. 309–323.

Comín, F.A., Menéndez, M., Pedrocchi, C., Moreno, S., Sorando, R., Cabezas, A., García, M., Rosas, M., Moreno, D., González, E., Gallardo, B., Herrera-Silveira, J.A., Ciancarelli, C., 2005. Wetland restoration: integrating scientific-technical, economic, and social perspectives. Ecol. Restor. 23 (3), 181–185.

CONANP, 2004. http://www.semarnat.conanp.gob.mx.

Currie-Alder, B., 2013. The role of participation in ecosystem-based management: insight from the Usumacinta watershed and the Terminos Lagoon, Mexico. In: Day, J.W., Yáñez-Arancibia, A. (Eds.), Gulf of Mexico Origin, Waters and Biota: Ecosystem-Based Management. vol. 4. Texas A & M University Press, College Station, pp. 201–212.

Day, J.W., Yáñez-Arancibia, A., 1982. Coastal lagoons and estuaries: ecosystem approach. Cienc. Interam. 22 (1–2), 11–25.

Day, J.W., Yáñez-Arancibia, A., 1988. Environmental considerations and ecological fundamentals for management of the Terminos Lagoon region, its habitats and fishery resources. In: Yáñez-Arancibia, A., Day, J.W. (Eds.), Ecology of Coastal Ecosystems in the Southern Gulf of Mexico: The Terminos Lagoon Region. Instituto de Ciencias del Mar y Limnologia UNAM, Coastal Ecology Institute LSU, Organization of American States, OAS, UNAM Press Mexico, Washington, DC, pp. 453–482. 518 pp. (in Spanish).

Day, J.W., Day, R.H., Barreiro, M.T., Ley, F., Madden, C.J., 1982. Primary production in Terminos Lagoon, a tropical estuary in the southern Gulf of Mexico. Oceanol. Acta 5 (4), 269–276.

Day, J.W., Conner, W., Ley, F., Day, R.H., Machado, A., 1987. The productivity and composition of mangrove forest, Laguna de Terminos Mexico. Aquat. Bot. 27, 267–284.

Day, J.W., Madden, C.J., Ley, F., Wetzel, R.L., Machado, A., 1988. Aquatic primary productivity in the Terminos Lagoon. In: Yáñez-Arancibia, A., Day, J.W. (Eds.), Ecology of Coastal Ecosystems in the Southern Gulf of Mexico: The Terminos Lagoon Region. UNAM Press, Mexico, pp. 221–236.

Day, J.W., Coronado-Molina, C., Vera-Herrera, F.R., Twilley, R., Rivera-Monroy, V.H., Alvarez-Guillén, H., Day, R., Conner, W., 1996. A 7 year record of above-ground net primary production in a southeastern Mexican mangrove. Aquat. Bot. 55 (1), 39–60.

Day, J.W., Martin, J., Cardoch, L., Templet, P., 1997. System functioning as a basis for sustainable management of deltaic ecosystems. Coast. Manag. 25, 115–154.

Day, J.W., Britsch, L.D., Hawes, S.R., Shaffer, G.P., Reed, D.J., Cahoon, D., 2000. Pattern and process of land loss in the Mississippi delta: a spatial and temporal analysis of wetland habitat change. Estuaries 23 (4), 425–438.

Day, J.W., Barras, J., Clairain, E., Johnston, J., Justic, D., Kemp, P., Ko, J.-Y., Lane, R., Mitsch, W.J., Steyer, G., Templet, P., Yáñez-Arancibia, A., 2005. Implications of global climatic change and energy cost and availability for the restoration of the Mississippi Delta. Ecol. Eng. 24, 253–265.

Day, J.W., D'Elia, C.F., Wiegman, A.R.H., Rutherford, J.S., Hall, C.A.S., Lane, R.R., Dismukes, D.E., 2018. The energy pillars of society: perverse interactions of human resource use, the economy, and environmental degradation. BioPhys. Econ. Res. Qual. 3 (1), 2.

Deegan, L., Day, J., Gosselink, J., Yáñez, A., Soberón, G., Sánchez, P., 1986. Relationships among physical characteristics, vegetation distribution and fisheries yield in Gulf of Mexico Estuaries. In: Wolfe, D. (Ed.), Estuarine Variability. Academic Press, New York, pp. 83–100.

del Merino-Virgilio, F.C., Okolodkov, Y., Aguilar-Trujillo, A.C., Osorio-Moreno, I., Herrera-Silveira, J.A., 2013. In: Botello, A.V., Rendón von Osten, J., Benítez, J.A., Gold-Bouchot, G. (Eds.), Florecimientos algales nocivos en las aguas costeras del norte de Yucatán (2001–2013). UAC, UNAM-ICMyL, CINVESTAV Unidad Mérida, pp. 161–180.

Fuentes Franco, R., Coppola, E., Giorgi, F., Pavia, E.G., Diro, G.T., Graef, F., 2015. Inter-annual variability of precipitation over Southern Mexico and Central America and its relationship to sea surface temperature from a set of future projections from CMIP5 GCMs and RegCM4 CORDEX simulations. Clim. Dyn. 45, 1–16. https://doi.org/10.1007/s00382-014-2258-6.

Hernández-Arana, H.A., Vega-Zepeda, A., Ruíz-Zárate, M.A., Falcon-Álvarez, L.I., López-Adame, H., Herrera-Silveira, J.A., Kaster, J., 2015. Transverse coastal corridor: from freshwater lakes to coral reefs ecosystems. In: Islebe, G.A., et al. (Eds.), Biodiversity and Conservation of the Yucatan Peninsula. Springer International Publishing, New York, pp. 355–376.

Hernández-Terrones, L.M., Null, K.A., Ortega-Camacho, D., Paytan, A., 2015. Water quality assessment in the Mexican Caribbean: impacts on the coastal ecosystem. Cont. Shelf Res. 102, 62–72.

Herrera-Silveira, J.A., 1994. Nutrients from underground water discharges in a coastal lagoon (Celestun, Yucatan, Mexico). Int. Ver. Theor. Angew. Limnol.: Verhandlungen 25 (3), 1398–1401.

Herrera-Silveira, J.A., 2006. Lagunas Costeras De Yucatán (Se, México): Investigación, Diagnóstico Y Manejo. Interciencia 19 (2), 94–108.

Herrera-Silveira, J.A., Comín, F.A., 2000. An introductory account of the types of aquatic ecosystems of Yucatan Peninsula (SE Mexico). In: Munawar, M., Lawrence, S.G., Munawar, I.F., Malley, D.F. (Eds.), Ecovision World Monographs Series. Aquatic Ecosystems of Mexico: Status and Scope, Backhuys Pub., Leiden, pp. 213–227.

Herrera-Silveira, J.A., Morales-Ojeda, S.M., 2009. Evaluation of the health status of a coastal ecosystem in southeast Mexico: assessment of water quality, phytoplankton and submerged aquatic vegetation. Mar. Pollut. Bull. 59, 72–86.

Herrera-Silveira, J.A., Morales-Ojeda, S.M., 2010. Subtropical karstic coastal Lagoon assessment, SE Mexico. The Yucatan Peninsula case. In: Kennish, M.J., Paerl, H.W. (Eds.), Coastal Lagoons: Critical Habitats of Environmental Change. CRC Press, ISBN: 978-1-4200883-0-4, pp. 309–336.

Herrera-Silveira, J.A., Martín, M.B., Díaz-Arce, V., 1999. Variaciones del fitoplancton en cuatro lagunas costeras del Estado de Yucatán, México. Rev. Biol. Trop. 47, 37–46.

Herrera-Silveira, J.A., Zaldivar, A., Ramírez, J., Alonzo, D., 2000. In: Comin, F.A., Herrera-Silveira, J.A., Ramírez, J. (Eds.), Habitat use of the American Flamingo (*Phoenicopterus reber ruber*) in the Celestun Lagoon, Yucatan, Mexico. Proceedings Limnology and Waterfowl, Monitoring, Modeling, and Management, Workshop, Aquatic Birds Working Group. Societas Internationalis Limnologiae, 24–27 November, 1997. Universidad Autónoma de Yucatán, Mérida, Yucatán.

Herrera-Silveira, J.A., Zaldívar, A., Aguayo-González, M., Trejo-Peña, J., Medina-Chan, I., Tapia-González, F., Medina Gómez, I., and O. Vázquez-Montiel. 2002a. Calidad del agua de la Bahía de Chetumal a través de indicadores de su estado trófico. In: F.J. Rosado-May, R. Romero Mayo y A. De Jesús Navarrete (Eds.), Contribuciones de la ciencia al manejo costero integrado de la Bahía de Chetumal y su área de influencia. Universidad de Quintana Roo, Chetumal, Q. Roo, México, pp. 185-196.

Herrera-Silveira, J., Silva, A., Villalobos, A.G.J., Medina, I., Espinal, J., Zaldivar, A., Trejo, J., González, C.A., Ramirez, J., 2002b. Análisis de la calidad ambiental usando indicadores hidrobiológicos y modelo hidrodinámico actualizado de Laguna de Términos, Campeche. CINVESTAV-Merida, EPOMEX-Campeche, UNAM-Mexico DF. Informe Técnico, 187 pp.

Herrera-Silveira, J.A., Comin, F.A., Aranda-Cirerol, N., Troccoli, L., Capurro, L., 2004. Coastal waters quality assessment in the Yucatan Peninsula: management implications. Ocean Coast. Manag. 47, 625–639.

Herrera-Silveira, J.A., Zaldivar, J.A., Ramírez-Ramírez, J., Cortés Balan, O., Borges, G.G., Trejo Sánchez, J., 2010a. Caracterización hidrológica de la Eco-región los Petenes-Celestún-El Palmar. In: Acosta, E., Andrade, M., Duran, R. (Eds.), Plan de conservación Eco-región Petenes-Celestún-El Palmar. Universidad Autónoma de Campeche, Pronatura Península de Yucatán A.C., ISBN: 978-607-7887-08-9, pp. 49–66.

Herrera-Silveira, J.A., Cebrian, J., Hauxwell, J., Ramirez-Ramirez, J., Ralp, P., 2010b. Evidence of negative impacts of ecological tourism on turtlegrass (*Thalassia testudinum*) beds in a marine protected area of the Mexican Caribbean. Aquat. Ecol. 44, 23–31.

Herrera-Silveira, J.A., Comin, F.A., Capurro-Filograsso, L., 2013a. Landscape, land-use, and management in the coastal zone of Yucatan Peninsula. In: Day, J.W., Yáñez-Arancibia, A. (Eds.), Gulf of Mexico: Origin, Waters, and Biota. Volume 4, Ecosystem-Based Management. Harte Research Institute for Gulf of Mexico Studies Series, Texas A&M University—Corpus Christi, Texas A&M University Press, College Station. 460 pp.

Herrera-Silveira, J.A., Teutli-Hernández, C., Zaldívar-Jiménez, A., Pérez-Ceballos, R., Cortés-Balán, O., Osorio-Moreno, I., Ramírez-Ramírez, J., Caamal-Sosa, J., Andueza-Briceño, M.T., Torres, R., Hernández-Aranda, H., 2013b. Programa Regional Para La Caracterización Y El Monitoreo De Ecosistemas De Manglar Del Golfo De México Y El Caribe Mexicano: Inicio De Una Red Multi-Institucional. Península De Yucatán. CINVESTAV-ECOPEY, Noviembre 2013. Informe Final SNIBCONABIO. Proyecto FN009. México, D.F.

Herrera-Silveira, J.A., Camacho, R.A., Medina, G.I., Ramírez-Ramírez, J., López, H.M., Morales, O.S.M., 2017. Síntesis basada en el análisis y diagnóstico documental sobre Carbono Azul en México. PNUD CSP-2016-057. Programa Mexicano del Carbono-CINVESTAV-IPN.

Imbach, P., Molina, L., Locatelli, B., Roupsard, O., Mahe, G., Neilson, R., Corrales, L., Scholze, M., Ciais, P., 2012. Modeling potential equilibrium states of vegetation and terrestrial water cycle of mesoamerica under climate change scenarios. J. Hydrometeorol. 13, 665–680.

INEGI, 2011. Principales resultados del. Censo de Población y Vivienda 2010. Instituto Nacional de Estadística y Geografía, México.

INE-SEMARNAT, 2000. Estadísticas del Medio Ambiente. México 1996-1998. Instituto nacional de ecología. Secretaria de medio ambiente recursos naturales y Pesca, México.

Kemp, G.P., Day, J.W., Yáñez-Arancibia, A., Peyronnin, N.S., 2016. Can continental shelf river plumes in the northern and southern Gulf of Mexico promote ecological resilience in a time of climate change? Water 8 (3), 83.

Lara-Domínguez, A.L., Contreras-Espinosa, F., Castañeda-López, O., Barba-Macías, E., Pérez-Hernández, M.A., 2011. Lagunas costeras y estuarios. La biodiversidad en Veracruz: Estudio de caso. Comisión para el Conocimiento y Uso de la Biodiversidad (CONABIO), Gobierno del Estado de Veracruz, Universidad Veracruzana (UV), Instituto de Ecología AC (INECOL)301–317.

Lara-Domínguez, A.L., Reyes, E., Ortíz-Pérez, M.A., Méndez-Linares, P., Sánchez-Gil, P., Zárate Lomeli, D., Day, J.W., Yáñez-Arancibia, A, Sainz, E, 2013. Ecosystem approach based on environmental units for management of the Centla Wetlands Biosphere Reserve: A critical review for its future protection. Chap.11, pp. 213–223. In: Day, J., A., Yáñez-Arancibia (eds.), Ecosystem Based Management, Vol. 4. The Gulf of Mexico, Origin, Waters, Biota. Texas A&M University Press. College Station.

Ley-Lou, F., 1985. Aquatic Primary Productivity, Nutrient Chemistry, and Oyster Community Ecology in a Mangrove Bordered Tidal Channel, Laguna de Términos, Mexico. (M.S. thesis). Louisiana State University. 59 pp.

López-Herrera, M., Herrera-Silveira, J., Ramírez-Ramírez, J., 2015. Pastos marinos como almacenes de carbono en la Bahía de Campeche. In: Paz, F., Wong, J. (Eds.), Estado Actual del Conocimiento del Ciclo del Carbono y sus Interacciones en México: Síntesis a 2014. Texcoco, Estado de México, México. ISBN: 978-607-96490-2-9, pp. 387–389. 639 p.

Lugo, A.E., Snedaker, S.C., 1974. The ecology of mangroves. Annu. Rev. Ecol. Syst. 5, 39–64.

Martínez, V.G., Gaona, S.O., 2005. Identificación y variación de la vegetación y uso del suelo en la reserva pantanos de Centla, Tabasco (1990–2000) mediante sensores remotos y sistemas de información geográfica. Ra Ximhai 1 (2), 325–346.

Medina-Gómez, I., Villalobos-Zapata, G.J., Herrera-Silveira, J.A., 2015. Spatial and temporal hydrological variations in the inner estuaries of a large coastal lagoon of the Southern Gulf of Mexico. J. Coast. Res. 31 (6), 1429–1438.

Morales-Ojeda, S.M., Herrera-Silveira, J.A., Montero, J., 2010. Terrestrial and oceanic influence on spatial hydrochemistry and trophic status in subtropical marine near-shore waters. Water Res. 44, 5949–5964.

Mutchler, T., Dunton, K.H., Townsend-Small, A., Fredriksen, S., Rasser, M.K., 2007. Isotopic and elemental indicators of nutrient sources and status of coastal habitats in the Caribbean Sea, Yucatan Peninsula, Mexico. Est. Coast. Shelf Sci. 74, 449–457.

Null, K.A., Knee, K., Crook, L., de Sieyes, E.D., Rebolledo-Vieyra, N.R., Hernández-Terrones, M.L., Paytan, A., 2014. Composition and fluxes of submarine groundwater along the Caribbean coast of the Yucatan Peninsula. Cont. Shelf Res. 77, 38–50.

Pacheco, J., Marín, L., Cabrera, A., Steinich, B., Escolero, O., 2001. Nitrate temporal and spatial patterns in 12 water-supply wells, Yucatan, Mexico. Environ. Geol. 40 (6), 708–715.

Perry, E., Marín, L., McClain, J., Velazquez, G., 1995. Ring of Cenotes (sinkholes), northwest Yucatan, Mexico: its hydrogeologic characteristics and possible association with the Chicxulub Impact Crater. Geology 23 (1), 17–20.

Phleger, F.B., Ayala-Castanares, A., 1971. Processes and history of Terminos Lagoon, Mexico. Bull. Am. Assoc. Pet. Geol. 55 (2), 2130–2140.

Rivera-Monroy, V., Twilley, R., Boustany, R., Day, J., Vera-Herrera, F., Ramirez, M., 1995a. Direct denitrification in mangrove sediments in Terminos Lagoon, Mexico. Mar. Ecol. Prog. Ser. 126, 97–109.

Rivera-Monroy, V., Day, J., Twilley, R., Vera-Herrera, F., Coronado-Molino, C., 1995b. Flux of nitrogen and sediment in a fringe mangrove forest in Terminos Lagoon, Mexico. Est. Coast Shelf Sci. 40, 139–160.

Rivera-Monroy, V., Madden, C., Day, J., Twilley, R., Vera-Herrera, F., Alvarez-Guillen, H., 1998. Seasonal coupling of a tropical mangrove forest and an estuarine water column: enhancement of aquatic primary productivity. Hydrobiologia 379, 41–53.

Rojas-Galaviz, J.L., Yáñez-Arancibia, A., Vera-Herrera, F., Day, J.W., 1992. Estuarine primary producers: the Terminos Lagoon case study. In: Seeliger, U. (Ed.), Coastal Plant Communities in Latin America. Academic Press Inc., New York, pp. 141–154. 392 pp. (Chapter 10).

Sánchez-Gil, P., Yáñez-Arancibia, A., 1997. Ecological functional groups and tropical fish resources. In: Flores, D., Sánchez-Gil, P., Seijo, J.C., Arreguiin, F. (Eds.), Análisis y Diagnóstico de los Recursos Pesqueros Críticos del Golfo de México. EPOMEX-UAC México, Serie Cientifica, vol. 7, pp. 357–389. 496 pp. (in Spanish).

Sánchez-Gil, P., Yáñez-Arancibia, A., Tapia, M., Day, J.W., Wilson, C.A., Cowan, J.H., 2008. Ecological and biological strategies of *Etropus crossotus* and *Citharichthys spilopterus* (Pleuronectiformes: Paralichthydae) related to the estuarine plume, Southern Gulf of Mexico. Neth. J. Sea Res. 59, 173–185.

Soberón-Chávez, G., Yáñez-Arancibia, A., Day, J.W., 1988. Fundamentals for a preliminary ecological model of Terminos Lagoon. In: Yáñez-Arancibia, Day, J.W. (Eds.), Ecology of Coastal Ecosystems in the Southern Gulf of Mexico: The Terminos Lagoon Region. UNAM Press, Mexico, pp. 381–414. 518 pp. (in Spanish).

Teutli-Hernández, C., Herrera-Silveira, J.A., 2016. Estrategias de restauración de manglares de México: el caso Yucatán. In: Ceccon, E., Martínez-Garza, C. (Eds.), Experiencias mexicanas en la restauración ecológica de ecosistemas. UNAM-UEM-CONABIO, pp. 459–484.

Valderrama, L., Troche, C., Rodriguez, M.T., Marquez, D., Vázquez, B., Velázquez, S., Vázquez, A., Cruz, M.I., Ressl, R., 2014. Evaluation of mangrove cover changes in Mexico during the 1970–2005 period. Wetlands 34 (4), 747–758.

Ward, W.C., Weidie, A.E., Back, W., 1985. Geology and Hydrogeology of the Yucatan and Quaternary Geology of Northeastern Yucatan Peninsula. New Orleans Geological Society, New Orleans, LA.

Yáñez-Arancibia, A., Day, J.W., 1982. Ecological characterization of Terminos Lagoon, a tropical lagoon-estuarine system in the southern Gulf of Mexico. Oceanol. Acta 5 (4), 431–440.

Yáñez-Arancibia, A., Day, J.W. (Eds.), 1988. Ecology of Coastal Ecosystems in the Southern Gulf of Mexico: The Terminos Lagoon Region. Instituto de Ciencias del Mar y Limnologia UNAM, Coastal Ecology Institute LSU, Organization of American States OAS, UNAM Press Mexico, Washington, DC. 518 pp.

Yáñez-Arancibia, A., Day, J.W., 2004a. Environmental sub-regions in the Gulf of Mexico coastal zone: the ecosystem approach as an integrated management tool. Ocean Coast. Manag. 47 (11–12), 727–757.

Yáñez-Arancibia, A., Day, J.W., 2004b. The Gulf of Mexico: towards an integration of coastal management with large marine ecosystem management. Ocean Coast. Manag. 47 (11-12), 537–564.

Yáñez-Arancibia, A., Day, J.W., 2006. Hydrology, water budget, and residence time in the Terminos Lagoon estuarine system, southern Gulf of Mexico. In: Coastal Hydrology and Processes. Water Resources Publications, Colorado.

Yáñez-Arancibia, A., Sánchez-Gil, P., 1986. The Demersal Fishes of Continental Shelf in the Southern Gulf of Mexico: Environmental Characterization, Ecology and Evaluation of Species, Populations, and Communities. Instituto Ciencias del Mar y Limnología, UNAM, México DF, pp. 1–230. Spec. Publ. 9.

Yáñez-Arancibia, A., Amezcua, F., Day, J.W., 1980. Fish community structure and function in Terminos Lagoon, a tropical estuary in the southern Gulf of Mexico. In: Kennedy, V. (Ed.), Estuarine Perspectives. Academic Press Inc., New York, pp. 465–482. 534 pp.

Yáñez-Arancibia, A., Lara-Domínguez, A.L., Sánchez-Gil, P., Vargas, I., Garcáa-Abad, M.C., Alvarez Guillén, H., Tapia-García, M., Flores, D., Amezcua-Linares, F., 1985. Ecology and evaluation of fish community in coastal ecosystems: estuary-shelf interrelationships in the southern Gulf of Mexico. In: Yáñez-Arancibia, A. (Ed.), Fish Community Ecology in Estuaries and Coastal Lagoons: Towards an Ecosystem Integration. UNAM Press, Mexico DF, pp. 475–498. 654 pp. (Chapter 22).

Yáñez-Arancibia, A., Lara-Domínguez, A.L., Rojas-Galaviz, J.L., Sánchez-Gil, P., Day, J.W., Madden, C.J., 1988. Seasonal biomass and diversity of estuarine fishes coupled with tropical habitat heterogeneity (southern Gulf of Mexico). J. Fish Res. 33 (Suppl. A), 191–200.

Yáñez-Arancibia, A., Lara-Domínguez, A.L., Day, J.W., 1993. Interactions between mangrove and seagrass habitats mediated by estuarine nekton assemblages: coupling of primary and secondary production. Hydrobiologia 264, 112.

Yáñez-Arancibia, A., Lara-Domínguez, A.L., Rojas-Galaviz, J.L., Villalobos, G.J., Zárate Lomeli, D., Sánchez-Gil, P., 1999b. Integrated coastal zone management plan for términos Lagoon, Campeche, Mexico. In: Kumpf, H., Steidingert, K., Sherman, K. (Eds.), The Gulf of Mexico Large Marine Ecosystem: Assessment, Sustainability, and Management. Blackwell Science, Inc., Malden, MA, pp. 565–592. 704 pp. (Chapter 33).

Yáñez-Arancibia, A., Day, J., Alder, B., 2009. Functioning of the Grijalva-Usumacinta river delta. In: Ocean Yearbook. vol. 23. Challenge for Coastal Management, Mexico, pp. 473–501.

Young, M.B., Gonneea, M.E., Fong, D.A., Moore, W.S., Herrera-Silveira, J.A., Paytan, A., 2008. Characterizing sources of groundwater to a tropical coastal lagoon in a karstic area using radium isotopes and water chemistry. Mar. Chem. 109, 377–394.

Section E

Restoration of Estuaries

Chapter 23

Restoration of Estuaries and Bays in Japan—What's Been Done So Far, and Future Perspectives

Osamu Matsuda*, Tetsuo Yanagi[†]

*Graduate School of Biosphere Sciences, Hiroshima University, Higashihiroshima, Japan, [†]International EMECS Center, Kobe, Japan

1 INTRODUCTION

The most significant and serious change in the natural environment and socioeconomic conditions in Japan during the last 100 years occurred during the period of high national economic growth after World War II (WWII), corresponding approximately to the period from the mid-1950s to the mid-1970s. It was during this time that Japan's gross national product (GNP) reached the second highest level in the world (1968) and the nation's infrastructure, including highways, dams, ports, and harbors, were drastically developed and constructed all over the nation. At the same time, far-reaching socioeconomic changes occurred, such as urbanization, industrialization, and the "modernization of agriculture" to the extensive use of chemical fertilizers and pesticides. These extreme socioeconomic changes during this period were the primary drivers of a variety of environmental pressures such as an extreme increase in pollutant load, transformation of land use and coastlines, including huge areas of land reclamation along the coast.

These increased environmental pressures had strong adverse effects on water quality, sediment quality, living resources, habitat conditions, and the biological diversity of the coastal and marine environment. As a result, the coastal and marine environment and ecosystem, as well as ecosystem services, were seriously degraded. These processes of changes are the main reasons why the restoration of estuaries and bays is currently necessary in Japan. Although a variety of legal and technological countermeasures have been applied to this environmental deterioration, most of the once-deteriorated environment and ecosystem has not yet recovered.

In the years since the start of the 21st century, a new legal system on fisheries, the ocean, biological diversity, and water cycling has been established through the enactment of relevant basic laws to prepare for the improvement in the deteriorated environment, ecosystem, and material circulation. However, the restoration and improvement of the once-damaged estuaries and bays still remain a challenge. Although recent results of environmental monitoring generally indicate that pollutants load and water quality have significantly improved in many places compared with when they were at their worst stage, fisheries resources and habitat conditions such as seaweed beds and tidal flats are still seriously degraded. The recent decrease in total population of the nation and the development of an aging society, especially in rural and island areas, are very serious socioeconomic problems that are emerging in Japan because the number of hands, in particular the hands of young and active people who promote community-based environmental restoration, are becoming scarcer.

Environmental conservation and management policy in Japan after WWII first placed emphasis on water pollution control as a countermeasure against hazardous water pollution, including toxic substances such as heavy metals and PCBs. Following this, the major target of pollution control was to prevent red tides and eutrophication by using restrictive measures such as total pollution load control (TPLC) in terms of COD, total nitrogen (TN), and total phosphorus (TP). Environmental standards for seawater and effluent water from industries and domestic life were also established. The legislative TPLC system has been applied to Tokyo Bay, Ise Bay, and the Seto Inland Sea with their watershed areas (Fig. 1A). These three designated TPLC areas were the most eutrophic areas among the many enclosed coastal seas of Japan. However, this kind of passive conservation policy is gradually being shifted to active conservation such as *Satoumi*, which includes the restoration of biodiversity, biological productivity, and habitat as well as achieving a well-balanced nutrient cycle between land and sea (Berque and Matsuda, 2013). This is another reason why restoration of estuaries and bays is becoming a more significant issue (Fig. 1A and B).

Coasts and Estuaries. https://doi.org/10.1016/B978-0-12-814003-1.00023-X

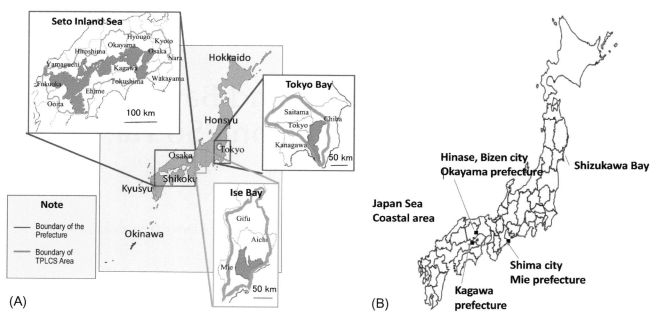

FIG. 1 (A) TPLC designated three areas in Japan including watershed. (B) Location of three case study sites on restoration and selected two research sites for a new research project, "Development of Coastal Management Methods to Achieve a Sustainable Coastal Sea."

Satoumi is defined as "A coastal sea with high biodiversity and productivity under suitable human interaction" (Yanagi, 2007, 2013). Holistic approaches such as ecosystem-based management (EBM) and integrated coastal management (ICM) are also being incorporated in a new management policy in Japan. As a typical example of the shift in management policy, in the newly revised governmental Basic Program for the Conservation of the Environment of the Inland Sea (Ministry of the Environment, 2015), the two major aims of the previous basic program, namely (1) conservation of water quality, and (2) conservation of natural landscape, have been reformed into following four new major aims (Fig. 2): (1) conservation and restoration of the coastal environment, (2) conservation and appropriate management of water quality, (3) conservation of natural and cultural landscapes, and (4) sustainable utilization of fish resources (Fig. 2).

From these changes in the legal system, it is evident that both conservation and restoration of the coastal environment are important nowadays. In the present study, three cases of restoration and related activities that have taken place in two

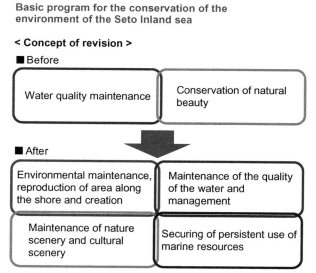

FIG. 2 Conceptual change of the national Basic Program for the Conservation of the Environment of the Seto Inland Sea.

cities and one prefecture are examined (Fig. 1B), with activities in Shima City, Mie Prefecture, in Bizen City, Okayama Prefecture, and in the entire area of Kagawa Prefecture. Bizen City and Kagawa Prefecture are located on the coast of the Seto Inland Sea, and therefore these areas are under the control of the Special Law for the Environmental Conservation of the Seto Inland Sea ("Seto Inland Sea Law"). All activities performed at the three sites are also closely related to and based on the concept of *Satoumi* (Secretariat of the Convention on Biological Diversity, 2011).

2   RESTORATION AND RELATED ACTIVITIES PERFORMED TO DATE

2.1   Shima City

Shima City (Fig. 1B) is located in the Ise-Shima National Park and implemented the first official municipal ICM Plan in Japan to be based on the basic plan established in 2011. The first PDCA (Plan-Do-Check-Action) cycle was completed in 2015, and the new second ICM Plan has been implemented since 2016. In the case of Shima City, the "Shima City ICM Basic Plan" is concurrently the "Shima City *Satoumi* Creation Basic Plan," which means that ICM and *Satoumi* cooperatively operate in the city.

During the drafting process for the new second ICM plan, the effects of the first basic plan were assessed. The characteristics and types of activity under this plan were examined and evaluated. During the period of the first plan, many activities were implemented according to the basic plan. Among these, there was particular focus on three activities: (1) restoration and revitalization of tidal wetlands, (2) documentation of local resources, and (3) establishment of *Satoumi* Academy (mainly for ecotourism).

Tidal flat restoration conducted at Ago Bay in Shima City clearly showed that the promotion of tidal exchange between the sea and once-closed wetlands on the inner side of artificial dikes improved both deteriorated sediment quality and macrobenthos conditions. This is similar to the improvement resulting from managed retreat policies in the United Kingdom (Boorman and Hazelden, 2017). This result suggests that biological diversity and biological productivity are improved by appropriate human interaction, which therefore also suggests that this kind of restoration activity matches the original concept of *Satoumi* (Matsuda, 2010; Matsuda and Kokubu, 2011). This type of restoration of tidal flats that were once isolated from the sea (Kokubu et al., 2008) is now going on at four sites in Ago Bay, Shima City, of which the case of tidal flat restoration in the Ishibuchi area is introduced in the following text.

Tidal flat restoration by opening the floodgates of dikes in the Ishibuchi area was conducted from April 2010 to February 2012. The location and an outline of the restoration site and sampling points are shown in Fig. 3. This approximately 20,000 m² site was reclaimed in the 1960s to cultivate rice fields. However, near the end of the 1990s, rice cultivation stopped because of changes in socioeconomic conditions, and the area became unused wetland inside the dike as a defense against the intrusion of seawater. In the tidal flat restoration here, a natural tidal exchange of water was introduced into the wetland area inside the dike by opening the floodgates. The tidal flat restoration area inside the dike was divided into two parts. The southern half of the restoration area was used for scientific investigation. Changes in sediment quality and macrobenthos were measured from April 2010 to February 2012, both at the site itself and also at the natural tidal flats outside the dike. Fig. 4 shows the seasonal changes in sediment quality in terms of AVS (Acid Volatile Sulfides), COD (Chemical Oxygen Demand), mud content, and TOC (Total Organic Carbon) at Stations 3, 5, 6, and 7 in the restoration site and the

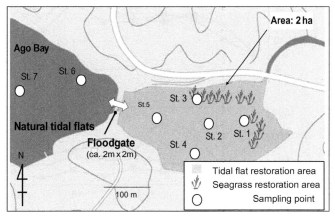

FIG. 3   The location and outline of the restoration site in Shima City.

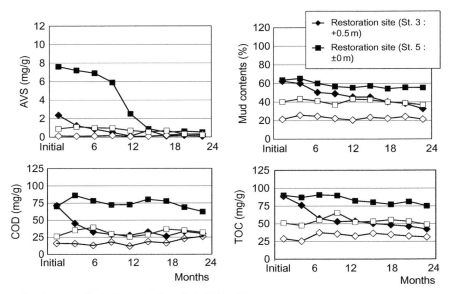

FIG. 4 Seasonal changes of sediment quality in the restoration site in Shima City.

natural tidal flats outside the dike. Before introducing seawater due to the tidal exchange into the inside of the dike, the sediment at the restoration site was muddy and anaerobic. AVS, COD, mud content, and TOC at the restoration site were all significantly higher than in the natural tidal flats outside the dike. Immediately after introducing of tidal exchanges, AVS at the restoration site decreased from 2.8 to 0.6 mg/g. Three months later, COD, mud content, and TOC also began to decrease gradually, especially in shallower areas (Station 3). In contrast, in the natural tidal flats outside the dike, the sediment quality did not change as conspicuously. Clearly the sediment at the restoration site inside the dike gradually changed from anaerobic to aerobic through tidal exchange, indicating that the organic compounds and reducing substances in the sediment of the restoration site inside the dike decreased (Figs. 3–5).

Fig. 5 shows the seasonal change in the number of species and the wet weight of macrobenthos (insects, crustaceans, gastropoda, bivalves, and polycesta) at Stations 3, 5, 6, and 7 in the restoration site and natural tidal flats outside the dike. Immediately after the introduction of tidal exchanges, the salinity at the site increased from 15 to 32. Before such an introduction, only six species living in brackish and eutrophic areas, such as *Capitella* sp. and *Chironomidae*, were found inside the dike. However, 3 months later, these insects disappeared, and polycesta, such as *Hediste* sp., increased. After 6 months of tidal exchange, mobile macrobenthos, such as *Batillaria cumingii*, *Cerithideopsilla cingulata*, and *Hemigrapsus penicillatus*,

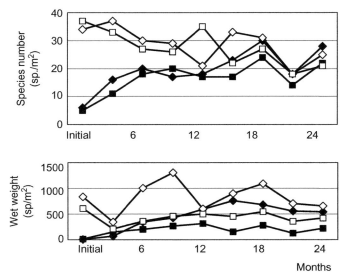

FIG. 5 Seasonal changes in the number of species and the wet weight of macrobenthos in the restoration site in Shima City. Symbols are the same as indicated in Fig. 4.

had clearly increased. Twenty species were found at Station 3 and eighteen species of macrobenthos were found at Stations 3 and 5, respectively. Twelve months later, fixed macrobenthos, such as *Musculus senhousia* and *Ruditapes philippinarum*, increased. The wet weight of the macrobenthos at the restoration site (Station 3) increased remarkably from 7.2 to $750 \, g/m^2$ over a 24-month period. Many juvenile fish, such as *Mugil cephalus*, *Lateolabrax japonicus*, and *Acanthogobius flavimanus*, were also found by visual observations at the restoration site after the introduction of seawater. Thus the macrofauna gradually increased after the introduction of tidal exchanges. It is expected that biological diversity and productivity in the once-abandoned coastal fallow fields will be further enhanced over time through continuous tidal exchange between the sea and the restoration site (Sohma et al., 2008).

2.2 Bizen City (Hinase Area)

Bizen City (Fig. 1B) in Okayama Prefecture established a Bizen City ICM Council in 2016. This initiative originated from the long-established seagrass bed restoration in *Satoumi* activities led by the fishermen's community from the Hinase area of Bizen City. Hinase is located on the southern coast of western Honshu Island and faces the central Seto Inland Sea, which is the largest enclosed coastal sea in Japan.

Hinase is one of the most commercially successful fishing areas along the coast of the Seto Inland Sea, and approximately 200 fishing families make their living from coastal fisheries with activities such as oyster farming, coastal pound netting, and direct selling of marine products at the Gomi-no-ichi Fisherman's Market run by the fishermen (*Gomi-no-ichi* in Japanese literally means five kinds of flavors).

All the fishing and related activities in the Hinase area are coordinated by the local Fisheries Cooperative Association (FCA), which also plays a key role in environmental restoration activities based on the *Satoumi* concept. The efforts to recover from the decline in coastal habitat and living resources by recreating the eelgrass (*Zostera marina*) beds have continued for more than 30 years since the mid-1980s.

The core members of the Hinase FCA initiated a project to restore the deteriorated eelgrass beds. Specialists from the Okayama Prefectural Fisheries Experimental Station supported the scientific and technical aspects of planting and propagating eelgrass. Members of the fishing community collected and prepared eelgrass seeds and undertook to plant them in the areas where they remembered the original eelgrass bed had been. These activities were aimed not only at recovering the stock of several commercial species but also at conserving the surrounding marine environment and ecosystem as a whole. Thus these activities contributed not only to fisheries but also to the ecosystem and also to the variety of ecosystem services around the area.

The original cooperation between the local fishing community and local scientists regarding the restoration of eelgrass beds in the Hinase area had been expanded, and it includes a wider variety of stakeholders such as citizens, cooperative consumers, students, NGOs, and NPOs with a certain degree of support from both central and local governments. The cooperation and collaboration among these components have been realized under a scheme similar to the *Satoumi* approach, which is closely related to the ecosystem-based management of fish stocks and the restoration of the natural habitat. The effect of this kind of cooperative approach was found to be very successful judging from the results of the 3-day Eelgrass Summit Meeting of Japan, held in Hinase in June 2016, with 2000 stakeholders attending from all over Japan.

The original area of eelgrass beds in Hinase was reported to have been approximately 600 ha in 1945, but it had decreased to just 12 ha in 1985, which was its worst condition (Fig. 6). As a result of long-lasting ongoing restoration, the eelgrass beds recovered to 80 ha in 2006, to almost 100 ha in 2009, and to approximately 200 ha in 2013 (Fig. 6).

Finally, it is expected that this successful experience of restoration in Hinase will be used to restore other areas around Bizen City, thanks to the leadership of the ICM Council under the Bizen City ICM Plan, and will then be expanded to cover wider areas. A lesson learned from this case study is that the degradation occurred in a very short time, whereas restoration needs enormous effort from many stakeholders. Therefore, precautions and a preventive approach are very important for the management of estuaries and coasts.

2.3 Kagawa Prefecture

The Kagawa Prefecture in Shikoku Island (Fig. 1B) published the "Kagawa's Vision for the Creation of *Satoumi*" in July 2014 on the occasion of the 80th anniversary of the establishment of the Seto Inland Sea National Park. As the primary objectives of the Vision, the aim of a sustainable living and society from a marine perspective is outlined, pointing out that

Satoumi describes an integrated understanding of the sea and land, as well as having an appropriate relationship with people, which keeps the sea healthy and supports a great diversity of life. This produces an abundant sea that provides many benefits, including not only marine resources, but also beautiful scenery, places for relaxation, food culture, and tourism. With the cooperation of a broad range of organization and individuals, we are working to identify a shared ideal for Satoumi that suits the unique characteristics of Kagawa.

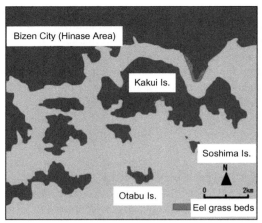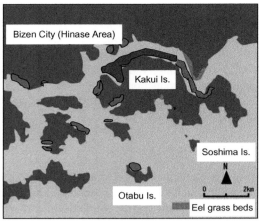

FIG. 6 Long-term change in eelgrass beds in Hinase area of Bizen City.

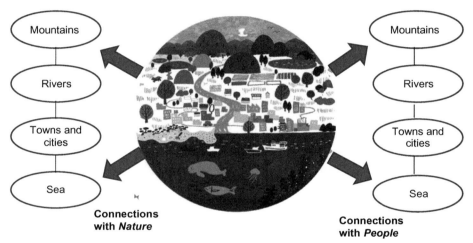

FIG. 7 Policy concept of Kagawa's *Satoumi*.

The brochure then listed five critical issues facing the Sea of Kagawa, namely (1) No signs of improvement in organic pollution, (2) Disruption of nutrient circulation balance, (3) Seaweed beds—increasing but still too few, (4) The urgent problem of sea litter, (5) Diminishing awareness of the relationship between people and the sea. It stressed that these five items must be the targets of restoration in a broad sense (Fig. 7).

As an approach to creating an ideal *Satoumi* for Kagawa, a "Scheme for the creation of new values by utilizing *Satoumi*," was proposed (Fig. 7). Kagawa Prefecture has now launched several practical projects, as well as research projects, to achieve a *Satoumi* policy. Promotions of marine litter management and ecotourism have been very active, and many citizens joined these activities. An integrated management system for marine litter by all the municipalities within the watershed area of individual river systems was also established, sharing a budget provided by all the local municipalities including upstream and downstream communities. Kagawa *Satoumi* College also opened to educate candidates to become nature guides.

3 FUTURE PERSPECTIVE

In 2014, the Ministry of the Environment, Japan, began a new research project, "Development of Coastal Management Methods to Achieve a Sustainable Coastal Sea" (PI: T. Yanagi), which will continue until the end of March, 2019 (http://www.emecs.or.jp/s-13/en/). The budget is approximately US $1.5 m/year. This project aims to provide a suitable ICM to achieve the goal of sustainable coastal communities. Three research sites (Fig.1A and B) have been selected, namely (1) the Seto Inland Sea, as a semienclosed coastal sea, (2) Shizukawa Bay along the Sanriku Coast, as an open coastal sea, and (3) the Japan Sea's coastal areas, where international management is necessary.

These locations have been chosen to clarify their natural characteristics from the viewpoint of physical, chemical, and biological oceanography. Social and human scientists are also included in this transdisciplinary study project to clarify the economic and cultural aspects of sustainable coastal communities. As part of the project, an integrated numerical model useful for policy decision-making in the coastal areas will be developed.

The target of this project is to achieve a "clean, productive, and prosperous coastal sea," which we call *Satoumi*. In other words, the aim of this project is to quantify *Satoumi*.

3.1   Clean, Productive, and Prosperous Coastal Sea

A Japanese proverb says, "Fish cannot live in a clean coastal sea." However, we say, "Fish cannot live in a coastal sea that is too clean, because the primary production is low because of low nutrient concentration and low phytoplankton density. But fish also cannot live in a dirty, coastal sea because red tides occur in the surface layer and hypoxia is generated in the bottom layer due to high nutrient concentration and high phytoplankton density." The highest primary production of the water column in the coastal sea will be possible under a midrange of transparency, where the sunlight can penetrate as far as the shallow sea bottom, and photosynthesis occurs not only thanks to floating phytoplankton but also because of phytoplankton attached to the sea floor, or on the leaves of flora and seaweed, or on the algae themselves.

Since 1973, the Ministry of the Environment, Japan, has continued environmental monitoring at approximately 200 sites (Fig. 8) four times a year (February, May, August, and November) in the Seto Inland Sea. Fig. 9 shows the year-to-year variation in (1) averaged transparency, (2) total fish catch, (3) averaged concentrations of TN, and (4) TP in the Seto Inland Sea. Rapid industrialization occurred during the 1960s, resulting in large amounts of nutrients and pollutants flowing into the Seto Inland Sea, and hence a decrease of transparency and an increase of fish catch occurred due to the eutrophication. The "Seto Inland Sea Law" was enacted in 1973, in which a decrease in COD load and the halting of land reclamation were decided on. However the number of red tides that occurred continued to increase, with the largest number of such occurrences (299) being recorded in 1976. To remedy that, TP load reduction was added to the Seto Inland Sea Law in 1979 and TN load reduction in 1986. Since introducing these policies of reducing both TP and TN loads, TP and TN concentrations have decreased in the Seto Inland Sea, as shown in Fig. 9. At the same time, transparency has recovered and the fish catch has decreased. A detailed history of such eutrophication and oligotrophication processes in the Seto Inland Sea is discussed in Yanagi (2015) (Figs. 8 and 9A–D).

Fig. 10 shows a significant correlation (1) between the 5-year running-mean transparency and the fish catch, and the TN concentration, and (2) between the fish catch and the average Trophic Level (TL) of fish caught in the Seto Inland

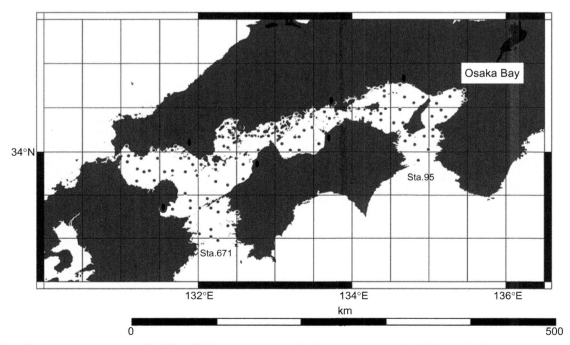

FIG. 8   Sampling sites for transparency, TP, TN, and DO concentrations in the Seto Inland Sea by the Ministry of the Environment, Japan, since the late 1980s.

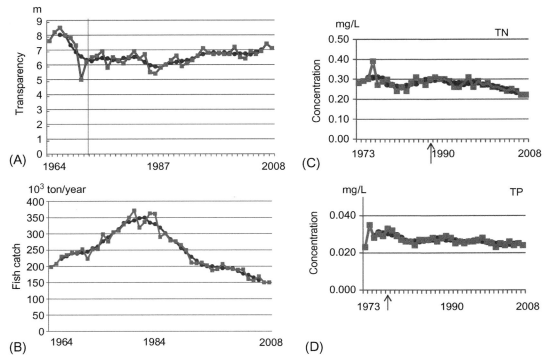

FIG. 9 (A) Variations in the average transparency, (B) fish catch, (C) surface TN, and (D) TP concentrations in the Seto Inland Sea. *Square*, yearly average. *Circles*, 5-year running mean. Arrows, year when the TN and TP loads control began. *(Reproduced from Transparency between 1964 and 1971 from Yanagi, T. (1988) Preserving the Inland Sea. Mar. Pollut. Bull., 19, 51–53, and transparency between 1972 (shown by vertical bar in (A)) and 2008 from the Ministry of the Environment, Japan (http://www.env.go.jp/water/heisa/heisa_net/setouchiNet/seto/index.html).)*

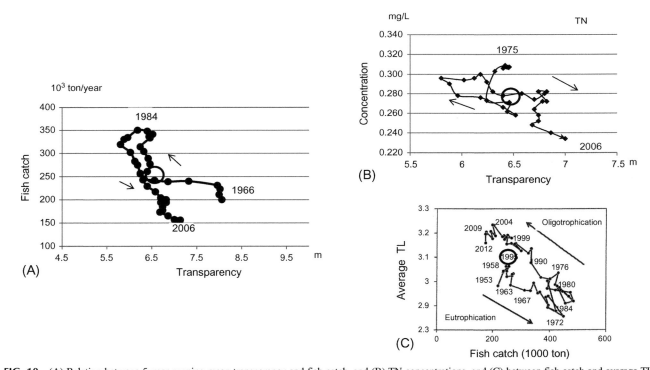

FIG. 10 (A) Relation between 5-year running mean transparency and fish catch, and (B) TN concentrations, and (C) between fish catch and average TL (Trophic Level) of fish caught in the Seto Inland Sea. Circle shows the proposed target transparency of 6.5 m in the Seto Inland Sea. *((C) Modified from Tanda, M., Yamashita, M., Harada, K., 2015. Trend of eutrophication and fish catch in the Seto Inland Sea. J. Environ. Conser. Eng., 44, 3, 122–127 (in Japanese).)*

Sea. Nitrogen is the limiting nutrient for photosynthesis in most areas of the Seto Inland Sea (Yanagi, 2015). There is a hysteresis, that is, the fish catch during eutrophication is higher than that during oligotrophication with the same transparency. TN concentrations during oligotrophication are higher than that during eutrophication. TP concentration has a similar hysteresis (not shown). Average TL during eutrophication is higher than that during oligotrophication. Such hysteresis is explained by the accumulation of organic matter in the bottom sediment during eutrophication, that is to say, the decrease in benthic biota due to hypoxia during eutrophication remains during oligotrophication and results in a decrease in fish catch. The release of accumulated TN from the bottom sediment results in a higher TN concentration during oligotrophication. The increase in plankton-feeder fish, such as the anchovy, with a lower trophic level during eutrophication, results in a lower average TL, and that of fish-feeder fish such as swordfish results in a higher average TL during oligotrophication (Fig. 10A–C).

It is proposed that the target transparency in the Seto Inland Sea should be 6.5 m, and the corresponding TN concentration and average TL 0.26–0.28 mg/L and 3.1, respectively, from Fig. 10, based on the aim to achieve high productivity and high biodiversity.

We did not propose a target transparency of about 6 m in the middle 1980s when the maximum fish catch was recorded (Fig. 9) because the biodiversity of fish caught was not very high at that time (<3.0), as shown in Fig.10C.

The target TN concentration when transparency was 6.5 m satisfied the Japanese Environmental standard of TN <0.3 mg/L in the coastal sea.

The hysteresis as shown in Fig. 10 disappeared approximately 30 years after the beginning of the total load reduction of TP and TN in the case of the Seto Inland Sea. That is to say, the biomass and diversity of the benthos has begun to increase since 2015 according to the field observations conducted by the Ministry of the Environment, Japan. The average species and individual numbers at 31 sampling points in Osaka Bay (Fig. 8) were 7 and $21/0.1\,m^2$ in 1993, 5 and $18/0.1\,m^2$ in 2003, but 12 and $95/0.1\,m^2$ in 2015.

Social and human scientists included in this project proposed an "integrated sustainability index," which consists of an economical index (population, fish catch, and so on), an environmental index (area of tidal flats, seagrass beds, and so on), and a social index (attendance to antipollution activities, fish food culture, and so on). We aim for the realization of a highly integrated sustainability index in the Seto Inland Sea, Shizukawa Bay along the Sanriku Coast, and the Japan Sea's coastal seas.

3.2 Management Method

We cannot directly manage water quality such as transparency, TP, TN and DO concentrations, and the number of red tides occurrences. However, we know the control processes of such water quality, for example: TP and TN loads from land, TP and TN release from the bottom, and the water exchange ratio between the coastal sea and the open ocean. We plan to take the actions to decrease or increase TP and TN loads from the land, to cover the seabeds with clean sand to decrease the bottom release, and/or to dredge the sea bottom to increase the water exchange ratio so as to change TP and TN concentrations and transparency in the coastal sea. Before such actions, we quantitatively check their effects using numerical models, and a cost-benefit calculation is also conducted. All other proposed actions, such as promoting the transfer efficiency from primary production to higher production, the establishment of MPAs (Marine Protected Areas), and economic and cultural methods to increase the sustainability of coastal communities are investigated using a similar methodology; they are all submitted to a committee whose members are fishermen, scientists, officers, and all stakeholders (Fig. 11).

In Shizukawa Bay along the Sanriku Coast (Fig. 7), a committee has been established to discuss methods of sustainable aquaculture in the area by including ourselves (that is, scientists), fishermen, local government officers, and NPOs. Based on their own experience and local knowledge of the environment, the fishermen decided to decrease the cultured oyster biomass to one-third the level of that before the complete destruction of culture facilities that resulted from the huge, 15-m high tsunami that hit the region on March 11, 2011. In the committee, they asked us (scientists), "How about the quantitative effect of this cultured oyster biomass decrease in Shizukawa Bay?" Our answer, based on our coupled (hydrological and ecological) numerical model, was that a one-third decrease in cultured oyster biomass is good from an economic and environmental viewpoint; in other words, the duration of the oyster culture for the harvest decreased from 2 to 1 year, and the hypoxia disappeared as a result of the combined effect of (1) decreasing the amount of oyster fecal and pseudo fecal pellets to one-third of previous levels, resulting in improvement of bottom sediment quality, and (2) changing the quality of oyster fecal and pseudo fecal pellets, that is, fecal pellets from older oysters (2 years) that includes organic matter that needs a significant amount of oxygen for decomposition compared to such matter from younger oysters (1 year).

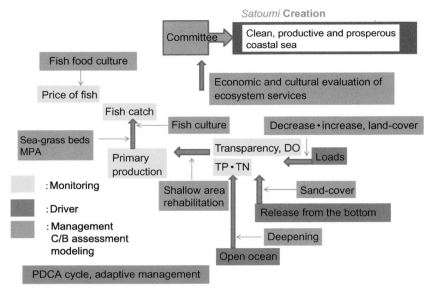

FIG. 11 Proposed *Satoumi* management method to achieve a clean, productive, and prosperous coastal sea.

We also monitored the growth rate of cultured oyster and the DO concentration in the bottom layer during the summer, and we found that the growth rate had roughly doubled from before and that hypoxia had disappeared. The first ASC (Aquaculture Stewardship Council) award in Japan was given to the cultured oyster in Shizukawa Bay in 2016 after the establishment of this successful and sustainable oyster culture system.

Certain decisions regarding actions needed to conserve the coastal environment will be decided in such committees. After consideration of such plans, actions will be performed and the effect of such actions will be checked. After they have been examined, further actions will be planned; this is adaptive management in practice using the PDCA (Plan-Do-Check-Action) cycle (Fig. 11).

In the case of the Japan Sea's coastal seas, we will draw up a proposal to be presented to NOWPAP (NOrth West Pacific Action Plan), which is part of UNEP (United Nations Environmental Programme), and for discussion with our colleagues from China, Korea and Russia, regarding suitable management methods; this will be based on our numerical model, which predicts water temperature and nutrients concentration in the Japan Sea.

ACKNOWLEDGMENTS

The transdisciplinary study project "Development of a Coastal Management Methods to Achieve Sustainable Coastal Seas" is performed under the umbrella of the Sustainable Initiative in the Marginal Seas of South and East Asia (SIMSEA) by the International Council for Science (ICSU) Regional Committee for Asia and the Pacific for "Future Earth" or "Future Coast."

This is a part of the Special Project by the Ministry of the Environment, Japan S-13 "Development of Coastal Management Methods to Achieve Sustainable Coastal Seas" (2014–18).

REFERENCES

Berque, J., Matsuda, O., 2013. Coastal biodiversity management in Japanese *Satoumi*. Mar. Policy 39, 191–200.

Boorman, L.A., Hazelden, J., 2017. Managed re-alignment; a salt marsh dilemma. Wetl. Ecol. Manag. 25, 387–403.

Kokubu, H., Okumira, H., Matsuda, O., 2008. Historical changes in the tidal flat and its effects on benthos and sediment quality in Ago Bay. J. Jpn. Soc. Water Environ. 31 (6), 305–311.

Matsuda, O., 2010. Recent attempts towards environmental restoration of enclosed coastal seas: Ago Bay restoration project based on the new concept of *Satoumi*. Bull. Fish. Res. Agency 29, 9–18.

Matsuda, O., Kokubu, H., 2011. Towards *Satoumi* in Ago Bay. CBD Technical Series No. 61 (Biological and Cultural Diversity in Coastal Communities, Exploring the Potential of *Satoumi* for Implementing the Ecosystem Approach in the Japanese Archipelago), pp.62–69.

Secretariat of the Convention on Biological Diversity, 2011. Biological and Cultural Diversity in Coastal Communities—Exploring the Potential of *Satoumi* for Implementing the Ecosystem Approach in the Japanese Archipelago. . CBD Technical Series No. 61, Montreal.

Sohma, A., Sekiguchi, Y., Kuwae, T., Nakamura, Y., 2008. A benthic-pelagic coupled ecosystem model to estimate the hypoxic estuary including tidal flats—model description and validation of seasonal/daily dynamics. Ecol. Model. 215, 10–39.

The Ministry of the Environment, Japan, 2015. 14 (in Japanese).

Yanagi, T., 2007. Sato-Umi: A New Concept for Coastal Sea Management. TERRAPUBU, Tokyo, p. 86.

Yanagi, T., 2013. Japanese Commons in the Coastal Seas: How the *Satoumi* Concept Harmonizes Human Activity in Coastal Seas with High Productivity and Diversity. Springer, Tokyo, p. 113.

Yanagi, T. (Ed.), 2015. Eutrophication and Oligotrophication in Japanese Estuaries. Springer, Dordrecht, p. 97.

Chapter 24

Challenges of Restoring Polluted Industrialized Muddy NW European Estuaries

R. Kirby
Ravensrodd Consultants Ltd., Liverpool, United Kingdom

1 INTRODUCTION

Estuaries have always been a focus of interest for man, who has used and misused them in a host of ways over the centuries. With population growth, urbanization, and industrialization, mainly in the last few hundred years, a number of quite gross scale examples of damage to natural systems have occurred. Some only died out arising from changing economics and industrial practices. In one notable, detrimental impact that of pollution by domestic and industrial discharges, the consequence for system health was so severe that initial practices soon had to be radically improved. Nevertheless many estuaries have remained in various degrees, polluted.

Estuaries have also been subject to major physical modification to suit the perceived needs of man. The myth had grown that our tidal rivers were infinitely adjustable to man-induced alterations. Only very recently, following rapid, gross-scale changes, has it begun to emerge that this is not so and a number of NW European estuaries have been permanently shifted into a new stable and undesirable phase. Now technological societies need to learn how to shift these back into a sustainable condition. This is an additional burden over and above costs of responding to sea-level rise.

In what follows earliest steps toward restoration are outlined in the context of emerging, improved technologies for managing fine cohesive sediment in harbors without gross-scale overdeepening and accompanied by new approaches of in situ bioremediation of sediment-adsorbed contaminants. It is already clear that reversing degradation set in train by excessive deepening will be much more challenging than the worst effects of pollution have proved.

2 ESTUARY MANAGEMENT

Perhaps "exploitation" would be a more pertinent term? This is on the grounds of the multiple ways in which muddy estuaries have been abused over the years. With steep rises in human communities along estuary shores in the course of the Industrial Revolution estuaries became subject to a broad range of exploitation. Intertidal margins were termed "wastelands" fit only for wildfowling and many were reclaimed. An invasive saltmarsh grass species, *Spartina*, was brought from the United States, hybridized into *Spartina anglica* and widely planted around European coasts as a cheap coast protection measure. It spreads widely and is subject to polarized views in respect of "its success." In some cases, for example coastal wetlands bordering the outer Thames, Blackwater, and Medway, a flourishing trade in the actual clay itself for brick and cement making was soon established, 1840–1920. Huge volumes were removed, finished products transported to satisfy building need in major cities such as London and even domestic and industrial waste brought back in the sailing barges to build causeways along which animals could be driven to graze on isolated salt marsh remnants. The water body faired no better, fisheries multiplied while industrial discharges from poorly regulated industries, generally combined with discharges of high volumes of untreated human domestic and animal sewage. Settlement growth was sufficiently rapid that by the early 1850s, during hot summers with low river flow, whole reaches of industrialized estuaries became anoxic and anaerobic. Intertidal oyster and mussel populations were suffocated and in many cases, even after installation of proper sewage treatment plants and better control of wastes later in the 1800s, have never recolonized their natural habitats. Clay extraction petered out gradually from the late 1800s but discharges of industrial and domestic wastes have continued.

Coasts and Estuaries. https://doi.org/10.1016/B978-0-12-814003-1.00024-1

In recent decades concern has risen in respect of foreshore losses due to sea-level rise. The phenomenon arises, called "coastal squeeze," where rising sea levels permanently cover the lower shore but high water mark is unable to roll inland owing to a sea defence. In western countries various pressures have led to this being addressed by moving defences inland. Such "set-back" is not a panacea. For example in the Severn Estuary, UK, broadening the shore encourages greater tidal flat erosion, the derived product being retained in the estuary and deposited in the immediate subtidal apron. The mean tidal flat erosion rate lies close to about 3.5 million tons/year and the redeposited mud beds blanket preexisting materials and are barren. What is perceived by some as an intertidal benefit becomes a major environmental dis-benefit in the immediate subtidal zone. Insufficient attention has been paid to prospects for regenerating the muddy foreshore and saltmarsh fringe. This is important as along many coastal stretches human investment features preclude set back.

A further "part-management solution" focused on the littoral margin is that of attempting to "feed" eroding foreshores by placing large quantities of poorly consolidated muddy dredge material in their immediately fronting subtidal aprons. This is in the hope that constructive waves will raise, carry up-shore, and allow deposition of a small component of the material. Such practices value shallow subtidal environments, where fauna is smothered, much less than intertidal mud flats. In any case on grounds of domination of the wave climate an erosional trend is unlikely to be reversed in such a crude manner. Instead, to be viable, much more attention would need to be given to wave attenuation.

Over the centuries estuaries have proved attractive sites not just for major cities but also for ports. Estuaries have experienced part management or exploitation to facilitate easy ship movement. In Victorian times (1842–1901) engineers considered it appropriate to apply "hard-engineered" solutions. Meandering navigation channels were often stabilized by constructing training walls. The Mersey and Ribble estuaries in the Irish Sea basin in United Kingdom, and the Weser estuary on the North Sea coast of Germany are notable examples. (Fig. 1). Experience showed that the shallow margins landward of such walls experienced rapid disposition of the ambient sediment. Walls were eventually overtopped by transported sediment driven into the system at an angle to the walls, so recreating a dredging need. The outer ends of paired walls became a new locus for deposition and engineers dealt with this challenge by further seawards extensions of such works. These days "soft-sediment engineering" solutions, notably capital dredging, are considered more appropriate.

To maintain a competitive edge ports have needed to keep abreast, wherever possible, with increase in ship size, especially draft. Demands for ever-larger vessels have soared, especially since the 1960s. Estuary specialists have, with prolonged experience, until now regarded estuaries as infinitely adaptable. Shallow intertidal margins have been reclaimed, braided sections and mid-channel banks eliminated, multiple channel systems dredged and regulated into a single channel, deepened by capital works, and kept stable by series of groynes perpendicular to the shore. These multiple alterations have only recently been shown by Winterwerp and Wang (2013a, b) and Winterwerp et al. (2017) to have set in motion a catastrophic deterioration of whole estuary systems, notably exemplified by the Loire (France) and Ems (Holland/Germany) but recognized in commercially exploited European estuaries elsewhere. Each change has led to a rise in tide height and velocity of the incoming flood tidal wave, accompanied by a reduction in bed friction. These estuaries have begun pumping and trapping mud, shifting their water-bodies into the hyper-turbid category. By analogy with the hyper-tidal, high turbidity Severn, UK, these systems exhibit severely impaired ecosystems, often verging on barren.

On grounds of this previously unknown disturbance out of regime, we have not initially known the tipping point? end point? or how to stop it? Our society has never before had to deal with the challenge of how to put such previously benign systems back into regime. The two, part-system management approaches to both foreshores and main channels in estuaries highlights the need for whole system and sustainable, resilient methods in future, a matter explored herein.

3 FUTURE SEA-LEVEL IMPACTS

Consequences of sea-level rise for estuaries are a big topic but one which must be touched on briefly if this issue is to be planned for. One recent and comprehensive modeling study by Idier et al. (2017) examines the entire western European continental shelf. It evaluates the 12 largest tidal components finding six of these to have demonstrable influence. It considers rises to 2100 AD including values of −0.25, 0, +0.25, +0.75, +1, +2, +3, +5, and +10 m above present day sea level. It finds that consequences at the coast differ depending on whether the high water mark is regarded as fixed or instead liable to flooding. It does not account for changes in nearshore sediment distribution which may be incurred from wave and current sorting. It reports that patterns of change in annual maximum water level are spatially similar over 70% of the area. However, notable increases in high tide levels are predicted, especially in the northern Irish Sea, southern part of the North Sea and the German Bight, with decreases forecast mainly in the western English Channel and Bay of Biscay.

This would imply the Seine and Gironde, if affected by the overdeepening, and Loire, known to be affected, would experience only modest exacerbation from this change, whereas major commercially important estuaries of the German Bight and southern North Sea including the Elbe, Weser, Ems, and Western Scheldt, all known to be affected by this man-induced

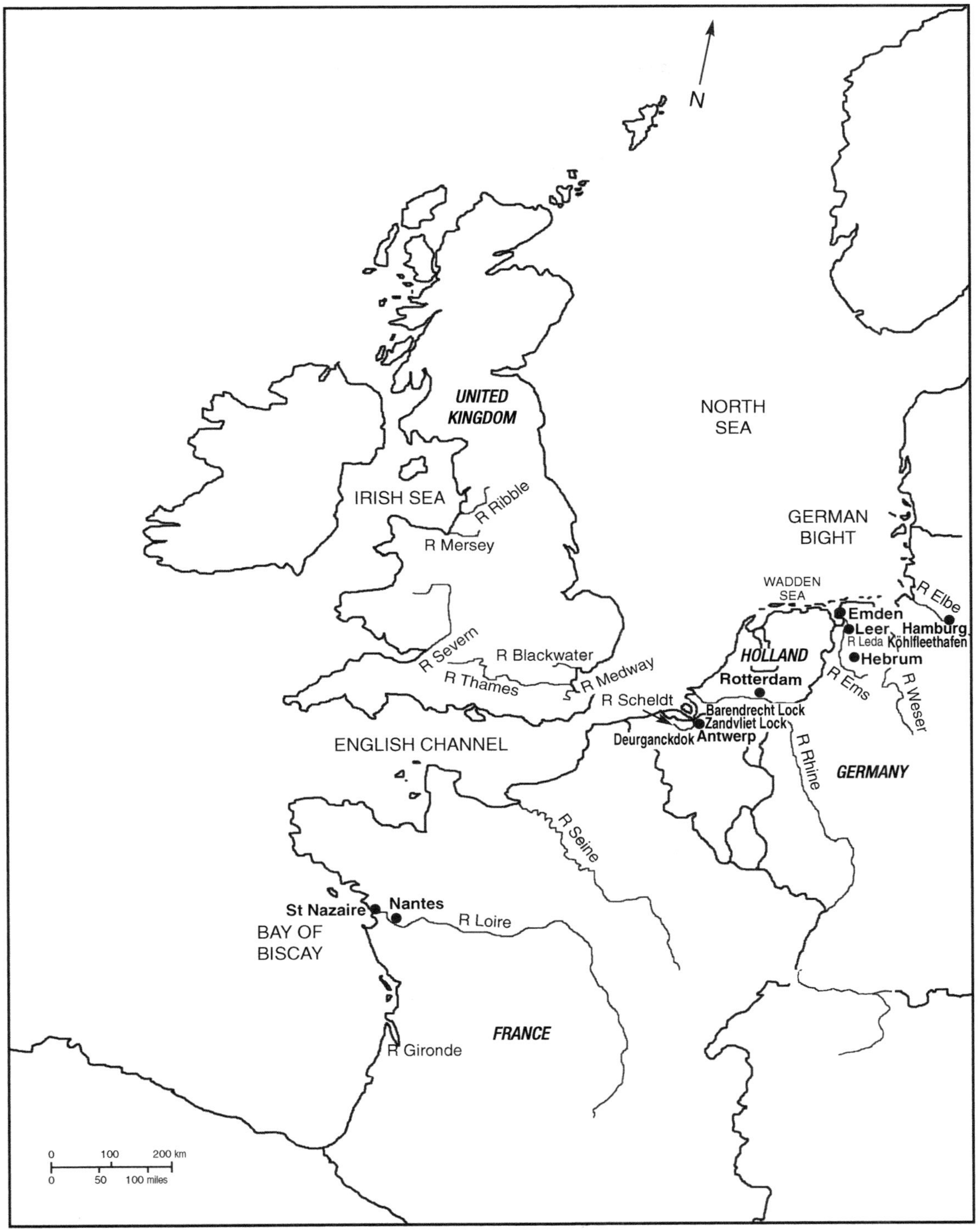

FIG. 1 Map of NW Europe showing localities mentioned in the text.

degradation, together with Dutch Wadden Sea, will all need to be managed to also cope with a more significant sea-level rise. Both mean sea level and, independently, the height of high water level will rise. The Lower Rhine is sufficiently canalized and overdeepened as to only show hyper-turbid conditions in the near-bed zone in its lower reaches, whereas British estuaries, being more numerous and not the site of such large ports may not be overdeepened to the same extreme degree. These forecasts will be important to implementation of the first coastal adaptation plans, which are a further component to the need to better manage commercial muddy estuaries, bringing them back into regime and adopting more sustainable methods to harness newly recognized fine cohesive sediment management attributes in future. These predictions by Idier et al. (2017) do not take account of the induced alterations in bed sediment distribution or transport that must be an important element in future management.

4 FUTURE COSTS

Hinkel et al. (2014) have provided estimates of the cost of protection and adaptation of world coasts under 21st century sea-level rise. These cost predictions took account of a wide range of uncertainties in socioeconomic development, sea-level rise, continental topography, population change, and adaptation strategies. In the absence of such measures and applying relative rises in the range 0.25–1.23 m, 0.2%–4.6% of global population will be flooded annually by 2100. In such circumstances global losses of GDP will range between 0.3% and 9.3%, with costs of adaptation and coastal defence reaching US $12–71 billion annually. These costs, naturally, take no account of need to restore muddy commercial estuaries degraded by overexploitation.

5 DEGRADED MAJOR ESTUARIES OF NW EUROPE AND THEIR RESTORATION

There are eight major estuaries providing the outlets to much of the land drainage of NW Europe. If the Rhine, Seine, and Gironde have been subject to stringent hydraulic analysis of their status this has yet to be widely publicized. Arising from strong evidence of man-induced regime shifts in four major estuaries, Elbe, Scheldt, Ems, and Loire, focused on interaction between effective hydraulic drag, fine sediment import and tidal amplification induced by engineering works, especially narrowing and deepening, detailed linear analytical modeling has been applied. A fifth estuary, the Weser, is believed to show comparable deterioration but seems, for now, not to have undergone such detailed study. The model solves the linearized shallow water equations in converging tidal rivers. By distinguishing reflecting and nonreflecting conditions, a nondimensional dispersion equation is derived which yields the real and imaginary wave numbers as a function of the estuarine convergence number and the effective hydraulic drag. This convergence number embraces the major geometrical features of an estuary intertidal area, as well as its convergence length and depth (Winterwerp and Wang, 2013a, b).

In the best documented example of this degradation, the Loire in Western France, reclamation of the intertidal zone has largely removed the so-called "accommodation space" for exchanging suspended fine sediment with the water body (Briere et al., 2012). This implies one step in the loss of "system resilience." In the last 50 years engineering work has focused on the multichannel estuarine system. These meandering reaches have been progressively realigned; axial banks and braided bed sections have been coalesced into a straightened single channel, which has steadily and regularly been deepened to meet the needs of deep-drafted ship traffic into port of Nantes. Each aspect of this channel re-alignment has reduced bed friction leading to a rising of tidal high water and an increase in flood tide velocities. It has been accompanied by import of fines from seaward, which can no longer be accommodated in the reclaimed intertidal zone. Suspended sediment concentrations have steadily risen such that the system has quite suddenly become hyper-turbid with extensive fluid mud zones. Channel deepening has been further exacerbated by placing of arrays of groins perpendicular to what is now this single main channel. The turbid near-bed layers further reduce hydraulic drag and promote enhanced flood tide velocities and dominance, trapping the fines.

The advecting fluid mud prevents daylight penetration and thus photosynthesis, while the dissolved oxygen (DO) levels remain negligible. In both respects, these intrinsic factors are hostile to faunal and floral colonization. On shore in the port city of Nantes the rise in high water height, accompanied by fall in low water elevation has altered ground water pressures and induced subsidence of adjacent buildings. It has been shown that bathymetric changes up-estuary of Nantes have become especially sensitive to these alterations. None of these shifts are desirable and the roots lie in the twinned coastal reclamation and gross-scale deepening.

As opposed to wise and proportionate management this exemplifies inadvertent but ultimately damaging exploitation. Yet until recently none of this was realized. Fig. 2 shows how the morphology of the Loire has been changed in the manner outlined above between 1887 and 2003. A similar degradation is in progress in the Ems. Fig. 3 shows how, following marginal reclamation and progressive deepening, during the short period from which records exist, the turbidity maximum zone has migrated up-estuary and the system has become steadily more turbid (de Jonge et al., 2014). Notwithstanding this plot of rather low values of mean surface concentration, hyper-turbid near-bed conditions can now be found along much of the Inner Ems fairway. Man-induced anaerobic fluid mud, as in the Loire and elsewhere, is similarly hostile to all oxygen-consuming photosynthesizing life forms. A complicating factor in the Ems is the weir across the estuary at Hebrum. Tidal amplification is affected by reflections from this, shortening further the already enhanced flood phase. Such degradation is also evident, though perhaps less so, in the Weser.

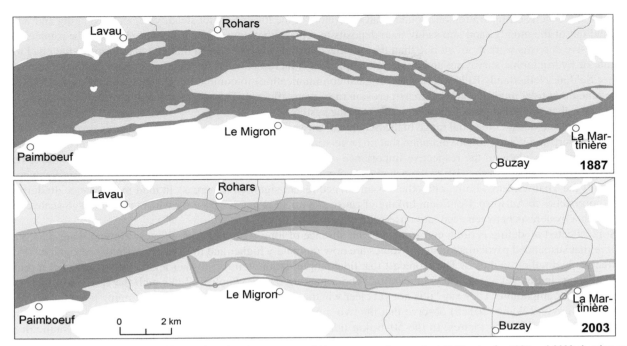

FIG. 2 Charts of Outer Loire Estuary, France, from just down-estuary of Nantes to near the mouth at St Nazaire for 1887 and 2003 showing extent of marginal reclamation, coupled with canalization over just less than 120 years. Channel alterations and overdeepening to favor deep-drafted vessels probably occurred in the 4 decades prior to 2003. *White*: high ground, *Green*: land, *Blue Grey*: former intertidal zone, *Blue*: water body. *Adapted from Winterwerp, J.C., Wang, Z.B., 2013b. Man-induced regime shifts in small estuaries-11: a comparison of rivers. Ocean Dyn. 63(11–12), 1293–1306.*

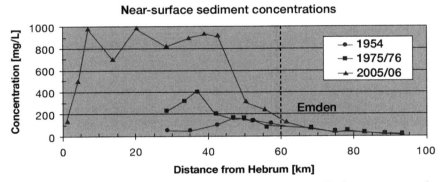

FIG. 3 Graphs showing enhancement of the turbidity maximum of the Ems estuary, Germany/Holland, as mean near-surface sediment concentrations versus down-estuary distance from the tidal limit at Hebrum, between 1954 and 2005/6. The system has become hyper-turbid with near-bed concentrations routinely reaching 40 g/L, loss of dissolved oxygen and absence of photosynthesis. *Simplified from de Jonge, V.N., Schuttelaars, H.M., van Beusekom, J.E.E., Talke, S.A. de Swart, H.E., 2014. The influence of channel deepening on estuarine turbidity levels and dynamics, as exemplified by the Ems estuary. Estuar. Coast. Shelf Sci. 139, 46–59. Elsevier.*

In the 1970s and 1980s mud maintenance dredging need in the inner Thames approaching the traditional impounded docks in the so-called Pool of London increased dramatically, threatening the commercial viability of port operations. Calculations showed that capital dredging works to deepen the approach to accommodate what were the deep-drafted vessels of the era had shifted the channel 0.3 m deeper than its equilibrium elevation. Permitting the bed to silt back up to this level reduced dredging need back to viable values, while the resulting shortening of port entry windows for ship operations had only a nominal effect on economic prospects: so this was a hydraulic issue.

In the 1980s and 1990s the Netherlands Government became increasingly concerned in respect of the severity of pollution in muds discharged into Dutch waters from the River Scheldt and its tributaries in Belgian and French territory. Eventually by international agreement watershed states were obliged to trap contaminated mud and carry it away into a series of confined disposal facilities (CDFs) along the waterway margins. An unanticipated consequence progressively arose for the Dutch administration on grounds that the waters of the lower Scheldt steadily began scavenging and winnowing mud out of the muddy sand and sandy mud deposits to replace that now trapped in inner reaches. The remaining now less cohesive deposits were more mobile than the stabilized mixed substrates, such that Dutch authorities then had to raise the rate of hydrographic surveying and charting, coupled with increased need to resite navigation channel marker buoys (Roger Seldon, Netherlands Rijkswaterstaat, personal communication) to combat more rapid channel meandering. Some complained that tidal mud flat and salt marsh erosion rates were affected. This, then, was a relatively large effect of withdrawing a rather small quantity of mud from an estuarine system.

In more recent decades the tidal range in the Scheldt at Antwerp has undergone an unanticipated increase of about 1.0 m. Yet Winterwerp and Wang (2013b) maintain that no large-scale engineering works have been carried out in the 20th century. However, they speculate on the respective importance of large-scale sand extraction (250 Mm^3), from the lower reaches coupled with up-estuary reaches being deepened by 1–2 m over considerable lengths. A parallel further consequence reminiscent of the Thames, Ems, Loire, and others is an alleged recent significant increase in mud maintenance dredging need in the approaches to Antwerp. Confident linking of cause and effect is not always easy when multiple large-scale changes, often with a hysteresis between each, are observed.

Sophisticated modeling is often constrained by an absence of long-term data sets. Winterwerp and Wang (2013b) also observe that suspended particulate material values are now regionally higher and this may explain the recognized decrease in effective hydraulic roughness. They forecast that tidal amplification will rise further and rapidly if regional turbidity in the Scheldt increases yet again. They are cautious about making overly simplistic links with the Ems and Loire, though patterns seem comparable. They say "it is not yet clear whether this estuary, too, is close to a tipping point."

Winterwerp and Wang (2013b) observe that the two-dimensional shape of the Elbe has been greatly changed and the fairway deepened by many metres. In the 50–60 km tidal river system down-estuary from Hamburg hydraulic drag has diminished dramatically since the 1970s. According to them no increase in regional suspended particulate matter has developed, as is elsewhere apparently linked to this. One possible explanation might be that Hamburg is much further inland than the other major ports discussed here. The German Bight to seaward and Lower Elbe is very largely sandy. Other sources Helmholtz Zentrum, Geesthact Center for Materials and Coastal Research report, in contrast, that hyper-turbid conditions are beginning to be encountered in the outer Elbe in the reaches around Cuxhaven. Furthermore, fines brought into the inner reaches from the Elbe River are in varying degrees trapped in the large number of semi-enclosed harbor basins opening off the tidal channel at Hamburg. Similarly, during the Cold War and for decades afterward highly contaminated mud from Eastern Bloc discharges into these inner estuary reaches were trapped, collected and processed for on-shore storage by a Confined Disposal Facility (CDF), the METHA 111 plant. These twin factors may contribute to the Elbe being comparatively mud starved. In view of the conflicting findings of Winterwerp and Wang (2013b) and the Helmholtz Zentrum perhaps onset of this hyper-turbid phase is simply emerging more slowly?

Winterwerp and Wang (2013b) search for reasons why decreased hydraulic drag has not been linked to turbidity increase in the Elbe? Without touching on the sediment supply issue they speculate that, instead, the Elbe is close to resonant conditions and that the tidal regime has become very sensitive to small changes in bathymetry?

In the context of these still relatively recent recognitions of gross-scale loss of bed friction, occasioned by the range of factors outlined above and exacerbating flood tide dominance, it may be instructive to reflect on a different muddy estuary, the Medway, Kent, UK (Fig. 4). This has yet to be subject to the rigorous mathematical model analysis of those above. In contrast to these others, as opposed to a decrease, this experienced a very large increase in its inter-tidal volume due to digging and removal of the salt marsh and tidal flat mud for the brick and cement industry mainly between 1840 and 1920. This industry has exacerbated widespread salt marsh cliff and tidal flat erosion attributed to sea-level rise and increased exposure to locally generated wind waves. Scrutiny of serial historic bathymetric surveys indicates that the estuary channels exhibit a slow net self-deepening trend. Unlike these perturbed systems above, the estuary is currently ebb-dominated. Fluorescent tracer tests confirm that the cohesive sediment derived from this mainly intertidal erosion is progressively flushed down-estuary and out into the Thames Estuary (Kirby, 2013). While able to over-winter large numbers of shore birds, this is a different example of a derelict industrial landscape severely shifted out of regime by man. Restoring this system would require technologies to raise low elevation tidal flats and grade them to a stable cross-sectional profile. Whereas the Medway contrasts with these mainland European commercial estuaries, its intrinsic characteristics further emphasize what large-scale disturbance by man can do (Kirby, 1990).

From the above broad understanding, a behavioral model has been drawn up (Fig. 5), which lays the responsibility for the recent degradations being observed in these muddy commercialized mainland European estuaries clearly at the door of the port industry and economic demands only arising in the last 40–50 years from need to accommodate deep-drafted vessels. Formulating a balanced policy for the future, we should be considering whether we might with economic and environmental benefit keep both ships and the sediment? One compromise might dictate moving major commercial ports to the

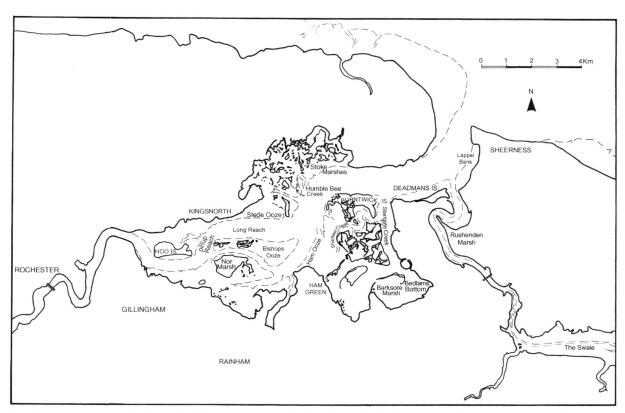

FIG. 4 Medway Estuary, UK, showing atypical fragmentary nature of salt marshes following intensive mud digging e.g., Stoke Marshes. What had formerly been Medway Saltings and Kingsferry Saltings is now a low, over-consolidated tidal flat, Ham Ooze. Bishops Ooze and Slede Ooze are, similarly, former salt marshes. These are but the most extensive of many other "Mud Holes."

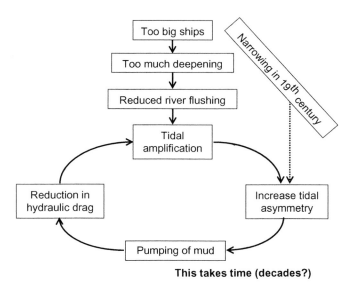

This takes time (decades?)

FIG. 5 Conceptual model showing the feed-back loop in deepened and canalized estuaries leading to very small tidal damping and hyper-concentrated (fluid mud) conditions. Loss of the multiple natural friction factors is called "Loss of Resilience." Technological societies need to learn how to reverse this degradation. *Adapted from Winterwerp, J.C., Wang, Z.B., 2013a. Man-induced regime shifts in small estuaries—1: theory. Ocean Dyn. 63(11–12), 1279–1292.*

mouths of estuaries, but even this offers only a part solution. This newly apparent recognition that estuaries are not, after all, infinitely adaptable to perturbations caused by man is, at least, disconcerting.

The next question to arise is: how can the degradation be reversed? This has never been attempted before. It is perhaps too negative to reflect on how to restore such an estuary if, in the meantime, a great city extending out to the main channel margin has been built. In reality cities have generally evolved over longer timescales than the deepening, though shorter than the reclamation.

All estuaries have their own idiosyncrasies indicating need for both general principles as well as bespoke tailoring. In what follows below, only the Ems and Loire are considered, though clearly others are involved. What is immediately apparent is that restoration is going to be time consuming and expensive. Authorities bordering the estuary have drawn up an "Ems Master-plan" extending forward until at least 2050 (Masterplan Ems 2050, 2018). One relatively modest revision to the way the hydrodynamic regime will be managed in future is to operate the tidal weir in the headwaters at Hebrum such that the flood tide is admitted through the weir in a manner aimed at reducing the new flood dominance and diminishing reflection back into the headwaters. A second endeavor common to all these newly hyper-turbid systems is how to remove or diminish the turbidity? In this case the authorities have begun or planned to pierce channel-side earth embankments and convey hyper-concentrated suspensions off into purpose-built settling basins. It is understood that early trials have been thwarted by heavy tracked vehicles becoming bogged down in the weak back-shore clay deposits.

In the case of the Loire mathematical modeling has made clear that the zone up-estuary of Nantes is acutely sensitive to the flood-enhancement induced by losses of bed friction. Fig. 6 shows block diagrams of "before" and "after" in which reclaimed marginal tidal channels can be re-excavated and allowed to flood and the main channel floor can be raised by placement of appropriate grade material. These various changes are aimed at raising bed friction and shifting back the resilience of these reaches. Modeling indicates that estuary reaches down-estuary of Nantes are far less sensitive, though still shifted permanently into the revised, flood-dominated hyper-turbid regime.

Various schemes to shift the system back from this new stable undesirable condition have been evaluated. One has been to use numbers of steep "thresholds" (cills), presumably gravel ridges, placed perpendicularly across the main channel floor. Another has been application of series of cross-channel air bubble curtains (ABCs) aimed at re-homogenizing stepped settling dense suspensions advecting up-estuary on the decelerating flood tide phase. In this case the object would have been to significantly diminish the extra component of up-estuary flow induced by settlement occurring during deceleration.

In reality we know from experience that dense advecting suspensions can flow up and over such hypothesized steps. Moreover we know from multiple measurements ahead of and immediately behind deep-drafted vessels trafficking through dense advecting near-bed suspensions that, once homogenized, these quickly discretize by settling into stepped suspensions again. ABCs require a shore-side compressor which needs to be more powerful and becomes more expensive as water depth increases. Unsuccessful prototype-scale experiments with ABCs in harbor basins in Antwerp, Rotterdam, and Hamburg are reported in PIANC Report 102 (2008). As far as is known no one has attempted to employ ABC's extending across an entire wide estuary channel or to employ multiple air lines stretched sequentially across such a waterway. French authorities currently accept that none of these prospects are economically viable and, as above, they probably would not be effective. All in all it is beginning to be recognized that putting such degraded systems back the way they were is going to take a lot of time and money. We are for the moment, merely on the nursery slopes.

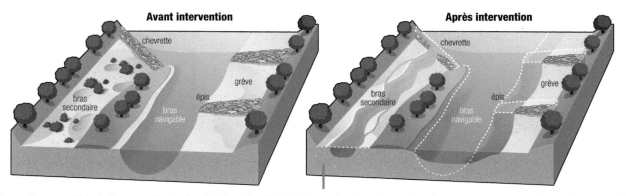

FIG. 6 Conceptual block diagrams, "Before" and "After" for 2012–2020 showing how the reaches of the Loire up-estuary of Port of Nantes will be restored. Previously reclaimed tributary branches are being re-opened and the main channel bed is being raised. Modeling indicates these reaches to be sensitive to such restoration, though reaches seaward of Nantes are much less so. *Adapted from Brière, C., Crebas, J., Becker, A., Winterwerp, J.C., 2012. (In French). Analysis of the Morphology of the Nantes Channel and Study of Its Restoration. Report to GIP (Groupement d'interpret Publique) 52 p.*

Maybe one route forward for the future resilience and sustainability of large estuary systems is to make more rapid progress in adopting modern generic sediment systems for waterside management of muddy ports.

6 GENERIC SEDIMENT MANAGEMENT SYSTEMS

Bearing in mind the exploitative nature of past estuarine use by man, not forgetting the four, part-system management procedures mentioned above, societies need to progress to favor more sustainable methods of managing fine cohesive sediment, without dredging and removing it from estuarine and near-shore sediment circulation cells, as well as finding ways of restoring estuarine resilience.

On grounds that demands created by ports have demonstrably proved capable of creating such serious deterioration to estuarine stability, we need more sympathetic, so-called generic methods. A generic method is one which, when tailored to local idiosyncrasies, applies to self-similar harbor configurations and problems wherever they occur. Such layouts include ship-locks, semi-enclosed basins, impounded docks, fairways, etc., all common harbor features around the world.

PIANC Report 102 (2008) reviews all successful and unsuccessful methods to Minimize Harbor Siltation around the world including not resorting to any conventional dredging. These are "anti-siltation technologies related to suspensions." It identified five methods, two of which are operated in unison.

6.1 Entrance Flow Optimization Structures

Entrance Flow Optimization Structures (EFOSs) including current deflecting walls, berms, and cills aimed at suppressing the eddy-induced shoaling otherwise caused by flow separation at the up-flow corners of entrances to semi-enclosed basins opening off a main estuary channel. Physical model testing coupled with results from full-scale devices at Köhlfleethafen in Hamburg and Deurganckdok on the Scheldt at Antwerp have reduced siltation by in the region 25%–40% and achieved reductions in siltation reaching 125,000 m^3/year. These are KSIS (keep sediment in the system) concepts and part solutions.

6.2 Trickle Auto-flushing Systems (TASs)

These address the especially challenging issue of "in-dock siltation" of impounded basins. They are conceptually simple and involve a settlement and collection system (gravity), though occasionally assisted, a bed-mounted pump, and a discharge pipeline back out into the main estuary. These discharge sediment at the same rate as that at which mud encroaches by density inflows through a ship lock. In other words the presence and operation of the dock has zero impact on the fine sediment circulation and budget of an estuary. An example at Leer on the River Leda, a tributary of the Ems, is a "total solution" having obviated any need for formerly damaging conventional mud maintenance dredging and disposal. There are other examples at each of Barendrecht and Zandvliet Locks and at Deurganckdok in Antwerp, though these only intercept and return part of the suspended sediment input back into the Scheldt.

6.3 Passive Nautical Depth (PND)

Many coastal zones and estuaries have long been known to exhibit the phenomenon called "fluid mud." Ports sited in susceptible areas have, perforce, had to learn how to deal with the issues raised by this phenomenon (navigating, surveying, dredging, and the inherent ubiquitous problems of severely adverse environmental impact, often coupled with the material providing a host for contaminants). As available port drafts have risen sharply in the last 60 years the number of fluid mud ports has increased. Fluid mud is a suspension with transitional properties between what is unambiguously navigable water and what is unnavigable settled bed material. These transitional zones often span a swathe extending to 3.0 m or more of the near-bed zone.

Vessels have sailed with their keels in fluid mud for more than 200 years (Stedman, 1806.) In the mid-20th century, with the switch from lead line to acoustic sounding records the latter often showed such dense suspensions as multiple near-bed reflectors. In these more demanding circumstances deciding in a rigorous manner what should be regarded as the bed became problematical. So this was a "technology-induced problem," a sounding lead sinking down to firm material, whereas a sonar record revealed these multiple, much shallower layers.

Accepting the shallower upper reflector on grounds of caution led to largely ineffective but also environmentally undesirable dredging of "black water." Natural fluid mud or this "man-induced" fluid mud is inevitably deoxygenated and at the same time, too fluid to support any macroscopic infauna. Its entrainment by dredgers, ships hulls, or propeller wash imposes a DO, burden on water bodies which may already exhibit low DO levels. From 1975 at the pioneer port, Rotterdam, then the world's biggest, shifting to in situ measurement of rheological properties led to the "nautical depth concept," a rigorous means to accurately survey deep within fluid mud layers a horizon still navigable by vessels, while maintaining an under-keel safety margin. Initially density, now more commonly a property even more intimately linked to navigability, such as shear strength or viscosity, is used. Millions of US $ have been earned and saved over the last 40 years this way, while appreciating that, in the limit this is no more than an alternative survey technique. Nevertheless, this is another important step in the movement toward adoption of more sustainable soft sediment management methods.

6.4 Active Nautical Depth

In a significantly more sophisticated advancement, together with "Conditioning," a coupled concept with much broader applicability in muddy estuaries of the world, this involves simultaneously manipulating the physical, chemical, and microbiological properties of mud in situ to create navigability and then having trading vessels sail through such material (Fig. 7). In this way deep-drafted ship operations can continue while muddy estuaries remain in regime. It permits commercial use for port operations to support an economy without degrading such systems as described above for a number of NW European waterways. Active nautical depth (AND) is accomplished by single, occasionally repeated manipulation by a conditioning vessel, in which developing interparticle bonds in a settling suspension are broken, the mud raised into a hopper and exposed to the atmosphere. Mud flowing from an inlet picks up oxygen, thus shifting its chemical and microbiological properties from an anaerobic-trending direction back into the aerobic sphere. It is then placed gently back onto the bed. The fluidization is termed "AND" on grounds of its being a deliberately engineered physical change, and the exposure of the slurry to air in an open hopper being termed "Conditioning" (by analogy with soil improvement, etc.). The process must be slow and gentle in order to avoid entraining large volumes of water.

This involves no more than "harnessing nature" on grounds that no additives of any kind are used and can be likened in science to creating a permanent hindered settling condition. It creates a suspension with properties much akin to non-drip paint. Oxygenating the slurry reactivates populations of aerobic bacteria, which accelerates the production of EPS (Exopolymeric Substances). So it is the voluminous EPS which blocks the pore spaces preventing dewatering, whereas the bacteria themselves fulfill a contrasted role (see later).

AND, coupled with Conditioning, provides a total solution and can be applied to ship lock entrances, impounded docks, semi-enclosed basins, anchorages and fairways. Maintenance dredging need at Emden has been shifted from 4 million tons/year to zero and costs reduced from Euro 12.5 m to Euro 4.0 m/year. It has been in use now for 28 years and at a number of NW European muddy ports. It is cheaper to apply and more environmentally acceptable than traditional maintenance or capital dredging. It is an inherently sustainable method which merits much wider adoption. So this might be termed man-induced "resource" fluid mud.

FIG. 7 Conditioning vessel "Meerval" in action in Emden entrance, Germany, superimposed on larger view of conditioned mud in the hopper. Slow and gentle manipulation of the physical, chemical and microbiological properties of the mud obviates need for any conventional maintenance dredging. "Man-induced resource fluid mud" is re-laid onto the bed permitting ships to sail through it. This KSIS technology is one route to more sustainable estuary management, while permitting current levels of economic activity. *Photographs kindly provided by N. Greiser.*

7 CLEANSING OF MUD IN CONTAMINATED INDUSTRIALIZED ESTUARIES

Fine cohesive sediment provides a host to a wide range of industrial pollutants which are adsorbed onto the poorly crystalline clay particles. There are more than 21,000 chemical contaminants in the subaqueous environment, a high proportion in estuaries. It is well-established that hyper-tidal and high macro-tidal muddy estuaries, with intrinsically variable tidal ranges, have the attribute of "natural self-cleansing" of contaminants from their mobile mud population. In such systems polluted mud is regularly cycled between anaerobic and aerobic chemical phases. Anaerobic bed deposits with stably adsorbed contaminants are entrained into oxygenated, turbulent water bodies, changing their chemical state, and inducing contaminant release. A well-documented example of this is the Severn Estuary, UK. Dissolved contaminants are progressively flushed seawards. Irrespective of whether this is considered desirable or not it is not preventable.

Another means of dealing with highly polluted estuarine mud developed and applied over the last few decades has been to dredge, collect, dewater, and isolate polluted mud from the environment by placing it in a CDF. Emphasis is almost always placed on minimizing the volume stored by fractionating off the inert sand fraction, squeezing out, treating, and disposing of the contaminated interstitial water and by densifying the residual mud fraction itself, converting it into "cake." Only the contaminated cake needs to be isolated. Such works are highly expensive and, in the limit, merely "relocate the problem."

An incidental unanticipated additional attribute of creating these long-lived aerobic "resource" fluid mud suspensions in polluted harbors has proved to be that the pernicious active biocide in a now-banned antifouling paint for ships hulls, tri-butyl tin, has been found to be destroyed in situ. (Table 1).The final product is inert tin oxide. Unlike the EPS (slime) which keeps the mud from dewatering and settling, in this case it is the bacteria themselves that do the work. Experimental testing is now in progress to learn how this works. Whether the mechanism can be speeded up and to what extent other contaminants can be dealt with in the same way is also now the topic of active research. Clearly this is a much more effective, cheaper, and better method that isolating it in a CDF. Unlike the fully mature generic sediment management system (SMS)'s these in situ bioremediation technologies are clearly in their infancy. They may, nevertheless, be another pointer to the future.

From the above, application of generic SMSs severely reduces or obviates need for mud maintenance dredging, and in the case of AND is intrinsically coupled with in situ destruction of a pernicious mud-bound contaminant. This, in turn, replaces need for mud removal to a C.D.F. Such advances possibly suggest a route to alternative and holistic sustainable estuary management? Experiments with application of AND in the inner Ems navigational fairway also proved successful. Possible shifting to favor all three of these in polluted, industrialized, and degraded muddy estuaries is a route to whole system, sustainable management? (Fig. 8).

8 CONCLUSIONS

The "part-system single-issue management" technologies we have been familiar with, set-back and dumping mud in the shallow subtidal zone, channel training walls, or overdeepening to accommodate deep-drafted commercial vessels, have each, at certain sites, through lack of a clear understanding, led to severe problems. Indeed, these can readily be regarded as single-beneficiary exploitation rather than management. Examples of digging away intertidal mud for the brick and cement industry or the recently recognized whole system destabilization due to overdeepening are but the most extreme examples. We begin to have ideas as to how climate change and sea-level rise will impact and change alluvial coasts and estuaries as

TABLE 1 Decline of Tributyl Tin (TBT) in Sediment in an Oxygenated Environment from the Test Site at Emden, Germany

Location	Date	No. of Samples	Top of Mud Mean Value (µg Sn/kg Dried Mud)	Mud Basal Zone (µg Sn/kg Dried Mud)
Emden entrance	Oct. 1992	9	283	–
	Feb. 1993	2	–	1560
Storage site	Dec. 1998	10	64	159
	Sep. 2000	10/8	16	29

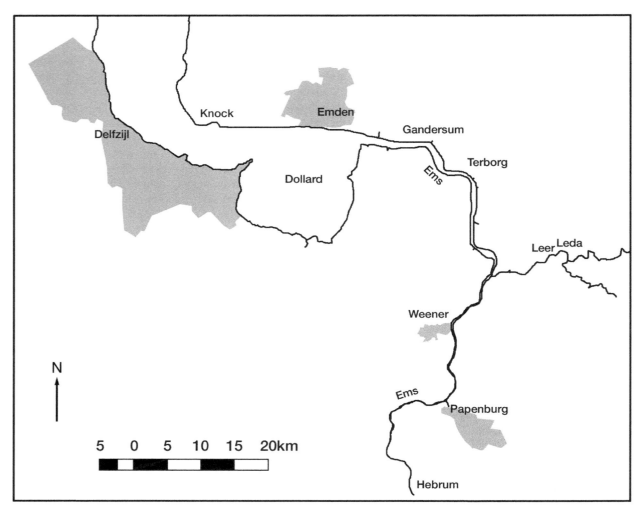

FIG. 8 Ems Estuary. Is this a blueprint for a resilient, sustainable muddy estuary? Adoption of active nautical depth at Emden and Delfzijl, together with the trickle auto-flushing systems at Leer and likely soon at Delfzijl imply no mud is dredged from these ports and approaches and is no longer removed from the estuary. In situ destruction of mud-adsorbed TBT by aerobic bacteria, an intrinsic by-product of active nautical depth at Emden and Delfzijl, also implies no mud is removed to confined disposal facilities ashore. Estuary restoration by allowing fairway mud deposits to rise to an equilibrium elevation but not settle would reestablish a natural regime, while still permitting ship passage. A viable maritime economy without gross-scale degradation would be instated. These are KSIS technologies. Attempts to restore former resilience via the Ems Masterplan remain embryonic but need to be supported. Modellers might advise on whether, where and by how much reclaimed marginal land might be returned to the water body, so facilitating sediment exchanges shown here to be necessary.

well as the costs of adapting to this. These are long term, slow changes we can anticipate and plan for. Where should we stand firm and where, instead, retreat?

It becomes axiomatic that "there can be no wise management without understanding." We need to be moving, for engineering, environmental, and economic reasons, to whole system, holistic sustainable management. Now that the severity of these multiple problems are starkly apparent, modern technological societies need to quickly find and implement solutions.

Costs of these will inevitably be high and will add to those arising from sea-level rise. No-one until now has tried to restore complex natural systems by "reverse-engineering" but early approaches to this are outlined here. Attempts at restoring the degraded Ems and Loire indicate how challenging, expensive, and time consuming this is likely to prove. Reversing an intertidal reclamation is one thing, making a wise compromise where, as often, some great city now extends to the waterway bank at an over-developed estuary is another. Similarly, reconfiguring deep-water ports in the inner reaches of major commercial estuaries is a particular challenge. Often it is easier to move the port to the estuary mouth. Enlightened management of sites such as the Medway would seek to re-categorize muddy dredge "arisings" as a resource and to devise technologies to return and retain it on severely eroded over-consolidated tidal mud flats. Even were this to be reached for, there currently seems no viable economic means at sites such as the Ribble, where a subtidal channel canalized by training walls is bordered by high elevation salt marshes such that a natural inter-tidal zone is virtually absent, to restore natural functioning of such a degraded system.

In the meantime, and perhaps fortuitously, we do now have access, in the case of muddy systems, to a suite of fully-evolved, mature, generic SMSs applicable to impounded docks, semi-enclosed basins and ship-locks, as well as to fairways and anchorages. These are cheaper, better, and infinitely sustainable as substitutes for old fashioned, damaging, capital, and maintenance dredging.

Whereas we recognize the efficiency in muddy port operations of modern trading vessels and of "landside management" in the form of container, bulk cargo, cruise liner terminals, etc., the vital "waterside management" has not evolved. These modern generic "SMSs" rely on the advances in this last generation of understanding of physics, chemistry, and microbiology of muds. This more fully evolved understanding means that the complexity and lack of understanding which previously held managers back can now be turned into an advantage. We are now looking to the ocean sciences rather than narrowly to traditional port engineering technologies. Increased understanding revises the number of options and a previously feared pernicious waste, fluid mud, has been converted into a resource with a beneficial use. "Dredging" involves digging up, carrying away, and dumping bed material. Now, instead, these are "anti-siltation technologies related to suspensions." As such they should not need a dredging or disposal licence. Now that these are fully evolved, mature technologies, the next step to work on is acceptance and adoption.

Furthermore, it has been unanticipated and fortuitous that AND serves, via its microbiology, not only to facilitate easy ship movement through lubricated mud via the intrinsic EPS supporting the medium, but in parallel, aerobic bacteria themselves destroy in situ, as opposed to "releasing," certain pernicious pollutants. This appears an important field of technical development for the future. As a corollary we need to recognize and categorize as desirable, or otherwise, "natural fluid-mud" which is always anaerobic and too weak to sustain a burrowing in-fauna, "man-induced accidental fluid mud," and similarly environmentally unwelcome, as is now to be found in the Rhine, Loire, Ems, etc., in recent years. Beds over which these lie or advect are always barren. These are entirely separate from "man-induced resource and aerobic fluid mud."

ACKNOWLEDGMENTS

The author wishes to thank the UK Natural Environment Research Council,latterly via its Institute of Oceanographic Sciences (now National Oceanography Centre) who supported shipboard & intertidal research on the muddy Medway & Severn Estuaries. Here,with colleagues, new instrumentation to document a number of previously unrecognised fine sediment phenomena were developed" Starting from 1965, etc.

REFERENCES

Brière C., Crebas J., Becker A. and Winterwerp J.C., 2012 (In French). Analysis of the Morphology of the Nantes Channel and Study of Its Restoration. Report to GIP (Groupement d'interpret Publique) 52 p.

de Jonge, V.N., Schuttelaars, H.M., van Beusekom, J.E.E., Talke, S.A., de Swart, H.E., 2014. The influence of channel deepening on estuarine turbidity levels and dynamics, as exemplified by the Ems estuary. Estuar. Coast. Shelf Sci. 139, 46–59.

Hinkel, J., Lincke, D., Vafeidis, A.T., Perrette, M., Nicholls, R.J., Tol, R.S.J., Marzeion, B., Fettweis, X., Ionescu, C., Levermann, A., 2014. Coastal flood damage and adaptation costs under 21st century sea-level rise. Proc. Natl. Acad. Sci. 111 (9), 3292–3297.

Idier, D., Paris, F., Le Cozannet, G., Boulahya, F., Dumas, F., 2017. Sea-level rise impacts on the tides of the European shelf. Cont. Shelf Res. 137, 56–71.

Kirby R., 1990. Historic and Modern Use of Brackish Water Clays by the Brick and Cement Industry in Europe. Ravensrodd Consultants Ltd., Unpublished Report 6 p+Figures and Tables.

Kirby, R., 2013. The long term sedimentary regime of the Outer Medway Estuary. Ocean Coast. Manag. 79, 20–33.

Masterplan Ems 2050, 2018. Available from: http://www.masterplan-ems.info/en/objectives.

PIANC, 2008. Minimising Harbour Siltation. Report 102 to MarCom, 75 p.

Stedman J.G., 1806. Narrative of a Five Years' Expedition. Against the Revolted Negroes of Surinam, in Guiana, on the Wild Coast of South America; from the Year 1772 to 1777: Elucidating the History of that Country, and Describing Its Productions, with an Account of the Indians of Guiana, & Negroes of Guinea. Google Books.

Winterwerp, J.C., Wang, Z.B., 2013a. Man-induced regime shifts in small estuaries—1: theory. Ocean Dyn. 63 (11–12), 1279–1292.

Winterwerp, J.C., Wang, Z.B., 2013b. Man-induced regime shifts in small estuaries-11: a comparison of rivers. Ocean Dyn. 63 (11–12), 1293–1306.

Winterwerp, J.C., Vroom, J., Wang, Z.B., Krebs, M., Hendriks, H.C.M., van Maren, D.S., Schrottke, K., Borgsmüller, C., Schöl, A., 2017. SPM response to tide and river flows in the hyper-turbid Ems River. Ocean Dyn. 67 (5), 559–583.

Chapter 25

Can Bivalve Habitat Restoration Improve Degraded Estuaries?

Ian Michael McLeod*, Philine S.E. zu Ermgassen[†], Chris L. Gillies[‡,#], Boze Hancock[§], Austin Humphries[¶,||]

*TropWATER, Centre for Tropical Water and Aquatic Ecosystem Research, James Cook University, Townsville, QLD, Australia., [†]School of GeoSciences, University of Edinburgh, Edinburgh, United Kingdom, [‡]The Nature Conservancy, Carlton, VIC, Australia, [#]James Cook University, Townsville, QLD, Australia, [§]The Nature Conservancy, Graduate School of Oceanography, University of Rhode Island, Narragansett, RI, United States, [¶]Department of Fisheries, Animal and Veterinary Science, University of Rhode Island, Kingston, RI, United States., [||]Graduate School of Oceanography, University of Rhode Island, Narragansett, RI, United States

1 INTRODUCTION: BIVALVES—THE FORGOTTEN HABITAT BUILDERS

Bivalve habitats have, until recent times, been generally overlooked as an important estuary habitat type. Historically, complex, three-dimensional habitats made up of dense aggregations of bivalves, their shells, associated species, and accumulated sediments were a dominant habitat type in temperate and subtropical estuaries around the world (Stenzel, 1971). These habitats were generally engineered by oyster (generally referred to as reefs) or mussel (generally referred to as beds) species. Until recent times these habitats were primarily managed as an important fisheries resource. Their historical extent and importance are difficult to estimate because bivalve habitats were often decimated before fisheries records were collected systematically, and there may be no remaining visible functioning bivalve habitats. Through the process of historical amnesia, or shifting baselines, successive generations of local people, and managers have grown accustomed to the new norm and have forgotten about the former abundant bivalve habitats.

Bivalve habitats are threatened globally. In a comprehensive review Beck et al. (2009, 2011) estimated that 85% of oyster reefs were lost globally and oyster reefs were functionally extinct (>99% loss) in 37% of estuaries. There are likely to be vast but largely unquantified losses of other habitat-forming bivalves. For example, formerly widespread green-lipped mussel (*Perna canaliculus*) beds in New Zealand, appear to occur at less than 1% of historical levels (McLeod, 2009; Paul, 2012). These losses are greater than those reported for other important estuary habitats including coral reefs, mangroves, and seagrasses (Grabowski et al., 2012). The loss of this fishery resource has had devastating effects on the coastal communities that relied on the harvest of bivalve habitats for employment and food.

Recently, benefits of bivalve habitats other than as a fishery resource have been recognized and bivalve restoration has expanded to focus on restoring reefs and beds to boost local fish and crustacean fisheries, improve water quality, and protect shorelines (zu Ermgassen et al., 2016a). The economic value of the full suite of ecosystem services derived from natural oyster reefs in North America was recently estimated to be as high as US$106,000 ha^{-1} year^{-1} (all values converted to 2017 $USD values by inflating in line with the annual average consumer price index; Grabowski et al., 2012), which is higher than estimates for other habitats such as mangroves ($82,000 ha^{-1} year^{-1}; Balmford et al., 2002), seagrass ($31,000 ha^{-1} year^{-1}; Grabowski et al., 2012), and permanent wetlands ($21,000 ha^{-1} year^{-1}; Sutton and Costanza, 2002).

Bivalve habitat restoration for ecosystem services has been scaling up in the United States and is increasingly being undertaken worldwide as a way to improve estuary condition and bring back imperiled ecosystems. However, estuaries have changed vastly through centuries of fishing, coastal development, habitat disturbance, sedimentation, and eutrophication. Bivalve habitats also face new challenges such as a changing climate, increasing ocean acidification and introduced predators, competitors, and diseases. This chapter discusses what scale of bivalve habitat restoration is possible worldwide, what positive influences restoration can make to degraded estuaries, and considers these within the context of a rapidly changing coastal environment.

Coasts and Estuaries. https://doi.org/10.1016/B978-0-12-814003-1.00025-3

2   WHAT ARE BIVALVE HABITATS?

There are numerous historical accounts of vast expanses of habitat-building bivalves throughout the historical literature. They describe highly complex structures of successive generations of bivalves forming expansive "barriers" or "banks" (Fig. 1; zu Ermgassen et al., 2016a), in some cases on the scale of "a mile in length" (Brooks et al., 1884). Defining bivalve reefs or beds is, however, something that has challenged observers since these early descriptions.

Bivalve reefs and beds are complex biogenic structures formed by successive generations of bivalves settling out and growing on top of one another. Within them, the habitat-building species are found at high density, while the dead shell material may dominate the structure. The term "bivalve reefs" generally applies to habitats with significant vertical relief (>0.5 m; Beck et al., 2009), whereas "bivalve beds" have a lower relief (sensu Coen and Grizzle, 2007). In both cases, the structures formed tend to accrete through time, as shell matter is deposited at rates greater than those lost to sedimentary dynamics (Mann et al., 2009). Most bivalve reefs today are less than 1 m high but there are massive, dead, biogenic reefs built by the European oyster, *Ostrea edulis* in the Bulgarian Black Sea up to 7 m above the seabed (Todorova et al., 2009) and subtidal shell reefs in Port Stephens, Australia up to 8 m above the seabed (Ogburn et al., 2007). Nevertheless, these habitats are far from permanent at individual locations within and along the dynamic estuaries and coasts where they are predominantly found. While habitat-forming bivalves are considered to be the basis of reef structures analogous to coral reefs but in temperate estuaries (Stenzel, 1971), the exact location and height of reefs within these systems has always changed over time, as

FIG. 1   Bivalve habitats. (A) Intertidal Sydney rock oyster, *Saccostrea glomerata*, growing on a mud bank in Port Stephens, New South Wales, Australia. (B) Subtidal green-lipped mussel, *Perna canaliculus*, bed growing on sand in an estuary channel in the Hauraki Gulf, New Zealand. (C) Subtidal eastern oyster, *Crassostrea virginica*, with a juvenile black sea bass, *Centropristis striata*, (located at the center of the image) Block Island, Rhode Island, United States. (D) Hooded oyster, *Saccostrea cucullata*, growing on a rocky shoreline, Hong Kong, China. (E) *S. glomerata* growing on wharf pilings in Port Stephens, Australia. (F) Leaf oysters, *Isognomon ephippium* growing on a mud bank in Hinchinbrook Channel, Queensland, Australia. (G) Liyashan Reef, made up of *Crassostrea sikamea* growing on mud flats, Jiangsu Province, China. (H) *S. glomerata*, growing on mangrove roots and pneumatophore in Port Stephens, New South Wales, Australia. (I) Subtidal flat oyster, *O. angasi*, reef in Tasmania, Australia. *(Photos from (A) I. McLeod. (B) I. McLeod. (C) S. Brown. (D) D. McAfee. (E) McLeod. (F) McLeod. (G) J. Cheng. (H) S. McOrrie (I) C. Gillies)*

vast historical fossilized oyster reefs in many Gulf of Mexico (United States) estuaries illustrate (May, 1971). Mussel beds are similarly known to be both transient and long lived in the estuaries in which they are found (Dankers et al., 2001). That said, at a large scale, these reefs and bed systems have been dominant in temperate waters for millennia (Stenzel, 1971).

It has been a challenge for ecologists and coastal managers to consistently define bivalve reefs for the purpose of mapping their extent and determining the extent and success of restoration efforts. Generally, definitions focus on the dominance of shell material in the structure of these bivalve habitats, and/or a minimum density of live bivalves (Baggett et al., 2014; OSPAR Commission, 2009), which serve as an indicator that the habitat is accreting, or at least persisting over time. Defining and mapping bivalve habitats are further challenged by the natural tendency for them to be patchy; often with reef "islands" separated by soft mud, or large reefs encompassing significant patches of soft bottom material (Fig. 1). These systems are highly fractal in their nature, with patchiness occurring at numerous spatial scales. It is therefore important to provide clear information about the scale and resolution at which mapping was undertaken. To this end, the United States National Oceanic and Atmospheric Administration (NOAA), The Nature Conservancy (TNC), and other partners have drafted guidelines for mapping and reporting oyster restoration efforts in the United States, which account for the challenges presented by varying scales and which provide a useful reference for other geographies at they move forward with their own restoration efforts (Baggett et al., 2014, 2015).

3 ECOSYSTEM SERVICES

The benefits that humans derive from nature are broadly referred to as ecosystem services. While all bivalve habitats are likely to provide some degree of ecosystem goods and services, the exact nature and quantification of these services is best studied in the eastern oyster, *Crassostrea virginica*. *C. virginica* forms extensive reefs in estuarine areas of the Atlantic and Gulf of Mexico coasts of the United States and have been documented to provide a suite of ecosystem services including, but not limited to improved water clarity and water quality, enhancing fish and invertebrate production, and reducing coastal erosion (Coen et al., 2007). The bivalve shell material forms complex three-dimensional habitats which can trap sediments and buffer wave energy and be used by sessile and mobile-associated species for attachment or protection. The bivalves produce feces and pseudofeces, which provide a rich material for detritivores and bacterial communities that remove nitrogen from the water column. These ecosystem goods and services are critical in supporting the livelihoods and social fabric of coastal communities.

Bivalves are filter feeders that improve water clarity by drawing down and filtering out particles from the water column. The edible particles are consumed and later deposited as feces, whereas the inedible particles are bound up in mucus and ejected as pseudofeces. In either case, the particles are drawn from the water column and deposited to the benthos, a process which both decreases turbidity in the water and which enriches the sediments with bioavailable carbon and nitrogen. The improved water clarity can both increase the amenity value of an area (Choe et al., 1996), and encourage the growth of seagrasses (Wall et al., 2008), which are themselves highly valuable habitats. Meanwhile, enriching the sediments with nutrient-rich compounds acts to stimulate the activity of denitrifying bacteria, which convert biologically active nitrogen to inert dinitrogen gas (Newell et al., 2002). Furthermore, the shell surface area and the additional structural complexity around the reef provides an ideal environment for this microbial action to take place, as it creates many sites where aerobic and anaerobic activity are in close proximity (Humphries et al., 2016). This process of enhanced denitrification alone has been valued at an average of $4050 \mathrm{ha}^{-1}\mathrm{year}^{-1}$ (Grabowski et al., 2012).

Oyster reefs are consistently found to support higher biodiversity and abundance of species than nearby unstructured habitats (e.g., Moebius, 1883; Shervette and Gelwick, 2008). The three-dimensional complex habitat provided by oyster reefs provides an important refuge from predation for many invertebrates and juvenile fish species (Tolley and Volety, 2005; Humphries et al., 2011), while the oysters themselves are prey for a number of larger fish species such as black drum, *Pogonias cromis* (Brown et al., 2008). A review of which species were consistently enhanced as juveniles by oyster reefs in the Atlantic coasts and Gulf of Mexico of the United States identified 12 and 19 species, respectively (zu Ermgassen et al., 2016b). This enhancement of large crustaceans and juvenile fishes is believed to contribute 2.8 and $5.3 \mathrm{t}^{-1}\mathrm{ha}^{-1}$ of oyster reef year^{-1}, respectively, to the system as a whole (zu Ermgassen et al., 2016b, updated tables available in zu Ermgassen et al., 2016a at http://oceanwealth.org/tools/oyster-calculator/). While quantitative evidence from other species is scant, green-lipped mussel, *Perna canaliculus*, beds in New Zealand have been shown to provide 3.5 times the productivity of invertebrates and host 13 times the density of small fishes than nearby soft sediments (McLeod et al., 2013). Furthermore, there are numerous qualitative accounts of bivalve species such as *Modiolus modiolus*, *Pinna* spp., *Atrinia* spp., *O. edulis*, and *Crassostrea rivularis* supporting enhanced biodiversity (Moebius, 1883; Barnes et al., 1973; Quan et al., 2012a; Ragnarsson and Burgos, 2012).

Oyster reefs can be robust structures with significant vertical relief that can have similar coastal defense properties as low-crested human-built structures such as break-waters, groynes, seawalls, dykes, or other rock-armored structures, through their effects on water circulation behavior and sediment transport. Therefore, they can be designed as effective coastal protection for erosion control and flood reduction (Reguero et al., 2018). There are reports oyster reefs as high as 3 m in the Yellow Sea (China), whereas in the United States, many extant oyster reefs are between 0.5 and 1 m in height. Where they are found in the shallow subtidal or intertidal zones, they have been documented to reduce coastal erosion of the shoreline, although typically only where wave energies are low (Piazza et al., 2005; Scyphers et al., 2011; La Peyre et al., 2015). This happens as a result of the oyster reefs absorbing waves in ways similar to a constructed breakwater and dissipating the energy. A generalized tool to help visualize the wave energy reduction from this "breakwater" effect of oyster reefs is available at https://vimeo.com/21810285. The effectiveness of this ecosystem service is dependent on the location and prevailing hydrodynamic conditions in each case (Piazza et al., 2005; Scyphers et al., 2011; La Peyre et al., 2015).

While the quantitative evidence for ecosystem service provision from other bivalve species is limited, there is good reason to suppose that all bivalve habitats provide at least some of these services. All habitat-building bivalve species are ecosystem engineers; creating structure from their successive generations of shell material, and producing biodeposits as a result of their feeding activity. The magnitude and degree of habitat building is dependent upon the population dynamics of bivalves, which are mediated by factors such as salinity, temperature, turbidity, substrate type, disease, and predation (Powell et al., 2003). It is the sustainable growth of bivalve habitats that facilitates ecosystem engineering properties and the basis of the ecosystem services they provide (Powell et al., 2006; Walles et al., 2015).

4 HISTORIC EXTENT AND FISHERIES

Like other marine resources, bivalve reefs and beds were once considered inexhaustible. This was a logical conclusion judging from their former vast abundance, and for thousands of years this was largely true. Habitat-forming bivalves have had high cultural value and served as the social backbone for indigenous populations around the world for thousands of years (Rick and Erlandson, 2009). They provided an easily accessible, protein-rich source of food and shells-were used as cutting and scraping tools, building materials, fish hooks, jewelry, and currency. The use of these resources is evidenced through the generation of historical bivalve middens (piles of discarded shells), which are found in most temperate coastal areas (Alleway and Connell, 2015). Some of these middens were massive. For example, one shell midden in New South Wales (Australia) was estimated to have a volume of 33,000 m^3, which contained 23,100 t of oyster shells (Bailey, 1975). These middens provide clues to the magnitude of indigenous harvests. Bailey (1975) estimated that the pre-European annual consumption was 17 t of oysters year^{-1}. This estimate is similar to the mean annual output of the local oyster fishery during their peak in the mid-20th century (Bailey, 1975). A recent study analyzing shells from middens around the Chesapeake Bay showed that harvests were sustained for 3000 years before European settlement (Rick et al., 2016). Other research from Florida (Sampson, 2015) and New York's Hudson River Estuary (Claassen and Whyte, 1995) also suggest limited pre-European impacts. However, research into indigenous harvest in the southeast United States (Dame, 2009) and Denmark (Milner, 2013) provided evidence of local depletion. Overall, it is likely that preindustrial populations mostly affected only local and shallow or intertidal populations, leaving subtidal populations as a source to replenish stocks.

It is difficult to comprehend the scale and importance of bivalve reefs and beds historically. Even in contemporary times comprehensive bivalve stock surveys are often only undertaken long after large-scale extraction had begun. Some examples illustrate the general trends. Oyster reefs were so extensive in estuaries on the Atlantic and Gulf Coasts of the United States that they were considered to be a navigation hazard (Coen and Grizzle, 2007). Whereas on the Pacific coast in Willapa Bay (Washington, United States) oysters may have dominated over a quarter of the bay bottom (Blake and Zu Ermgassen, 2015) with descriptions of "natural oyster-beds stretched over a distance of thirty miles in length and from four to seven in width" (Bancroft, 1890). The most extensive oyster grounds surveyed in North America included 25,500 ha in Tangier and Pocomoke Sounds (Chesapeake Bay, Virginia, United States) in 1878 and 16,500 ha in Matagorda Bay, Texas (United States) in 1907–75 (zu Ermgassen et al., 2012). Over a century ago, one-fifth of the Dutch part of the North Sea was covered with *O. edulis* beds (Gercken and Schmidt, 2014). Early explorers to Australia commonly described extensive oyster reefs. For example, the explorer Vancouver ran his vessel aground on a bank of oysters while attempting to leave a Western Australia estuary in 1791. Making light of the situation, Vancouver and his men feasted on the oysters and named the estuary Oyster Harbor (Gillies et al., 2015).

Commonly, intertidal bivalve populations in sheltered bays and coastal waters were the first to be harvested and overharvested because they were readily accessible. Increasing technology innovations such as improved boats, long-handled tongs, and small dredges allowed for more intensive harvesting of reefs and beds and access to deeper and more remote areas. Sailing cutters dragging small iron dredges were probably used as early as the 13th century (Seaman and Ruth, 1997). This early fishery was economically important, highlighted by often violent conflict between Danish and German fishers

(Gercken and Schmidt, 2014). Large-scale declines in bivalve habitats were recorded as early as 1695, with 10 oyster banks in the North Sea considered ruined by overfishing and cold winters (Gercken and Schmidt, 2014).

The scale of peak bivalve harvests worldwide is impressive. The mid to late 1800s marked peak harvest years in Europe, North America, and Australia and the rapid devastation of many reefs and beds. In 1864, 700 million *O. edulis* were consumed in London, employing up to 120,000 men in Britain to dredge oysters (MacKenzie et al., 1997). In France more than 100 million oysters were harvested annually during peak years in the 19th century (Yonge, 1960). In the Chesapeake Bay, peak production during the 1880s reached 20 million bushels of oysters annually (2 billion oysters or 900,000 t). These peak harvests took place during the development of machine-driven vessels that could deploy larger, heavier harvesting gear and allowed exploitation of deeper reefs, and access to more remote locations. Increasingly efficient transportation such as railways and greater use of preservation techniques such as using ice and canning, opened up new, inland markets for bivalve and shell products.

Bivalve habitats were not only harvested for their food value but also for their shells. These were used for landfill, road building, and construction, including large-scale burning for lime to create cement. Shells were also used for chemical production, soil conditioning, and fed to poultry. Between 1920 and 1944, 2.8 million t of shell products were produced from Chesapeake Bay oysters reefs (Hargis and Haven, 1999); of this over 1.5 million t were in the form of poultry grit and 1.2 million t of ground and/or burnt lime. In the gravel-poor coastal counties of Texas oyster shell was an important road building material, with nearly 30% of the more than 7.7 million cubic meters of oyster shell produced in Texas in 1955 alone going to road construction (Doran, 1965). In Australia, schooners supplying lime kilns simply berthed on oyster banks on low tide, then raked up live oyster and shell until the boat was full, a process referred to as "skinning" (Ogburn et al., 2007).

5 GLOBAL DECLINE OF BIVALVE HABITATS

Oysters and mussels fisheries have posed unique challenges for fisheries management because unlike fish and other mobile organisms, fisheries tend to simultaneously remove bivalves and their habitat. Larvae for many species preferentially settle on the shells of conspecifics so removing the habitat also reduces the available amount of suitable settlement substrate, thus limiting recruitment. The reef and bed structure is often bound together by the living bivalves. Once they are removed the physical structure of the reefs and beds are more vulnerable to being broken up by waves and currents. In addition, with the loss of vertical structure, remnant populations are more susceptible to smothering, predation, and disease.

Overfishing with destructive fishing gear is not the only driver of decline. There has been a long history of translocations and introductions of nonnative bivalves within and between bays and even countries in an attempt to revitalize struggling local fisheries (Beck et al., 2009; Gillies et al., 2015). All too often parasites, predators, and diseases were introduced with these relocations, or transported along with aquaculture gear. Severe disease and parasite outbreaks often followed these introductions driving native bivalve habitats to commercial and functional extinction in many coastal areas (Beck et al., 2009). While some native diseases were present their impacts were often exacerbated by a reduction in the vertical height of reefs and through increasing pollution, sediment, and nutrient loading in estuaries that could lead to decrease the resilience of shellfish to disease.

Peak harvest years were often coupled with rapid coastal land-use change and the clearing of local vegetation. This often led to large amounts of sediment entering estuaries and smothering bivalves. One example of this is evidenced in the wild oyster fishery of Rhode Island's Pt. Judith Pond (United States). In the late 1930s, development expanded in this area. As a result, local dredging became intensive and sediment transport from these activities served as a catalyst in burying oysters in one of the most productive oyster fisheries in the region (MacKenzie et al., 1997). In other cases, such as Yaquina Bay, Oregon, pollution from the associated industries drove the decline of existing oyster beds (Fasten, 1931).

Many bivalve habitats in intertidal and shallow subtidal areas have been eliminated by coastal development activities including filling ("land-reclamation") and dredging of shipping channels. In addition, reduced water quality and eutrophication have negatively affected many bivalve populations. Dam building and modification of water flow have also affected bivalve populations because many species have a relatively narrow range of salinity in which they thrive (Beck et al., 2009).

The catastrophic loss of bivalve reefs and beds is now well documented, along with the devastating consequences for local communities through loss of employment, food, and income security. Unfortunately, we are still managing the few remaining wild reefs without learning from past mistakes. Many oyster and mussel stocks are still being fished commercially using destructive methods despite being at less than 10% of their historical biomass (zu Ermgassen et al., 2012). Along with restoration the conservation of the last natural bivalve habitats should be a high priority for managers.

6 RESTORATION

Historically, restoration efforts have focused on fisheries enhancement with the goal of recovering lost or impaired bivalve fisheries. It is likely that people have translocated bivalves with them to establish easily accessible populations for millennia. There is a blurred line between fishing, local enhancement, and aquaculture. For instance, Native Americans in Rhode Island fished for oysters in deep waters of Rhode Island Sound, and then transplanted them to shallow areas in Pt. Judith Pond and Narragansett Bay for winter harvest.

Initial bivalve population recovery and restoration efforts included policies based around closed areas and closed seasons (Fig. 2). For areas where bivalve populations had collapsed, large-scale reintroductions were undertaken, sometimes between countries. For example, between 1894 and 1930, large amounts of spat (juvenile oysters) from the Netherlands, France, and Norway were distributed in the North Wadden Sea to restore beds for commercial fishing (Gercken and Schmidt, 2014). In the early 1880s, vast numbers of rock oysters, *Saccostrea glomerata* were transplanted from New Zealand to Australia to restore their stocks (Ogburn et al., 2007). Moving oysters from Scotland to replenish English and Dutch stocks was also a common strategy in Europe (Thurstan et al., 2013; Gercken and Schmidt, 2014). Although these translocations were sometimes successful in supporting fisheries over the short term, they often created new problems. Often exotic diseases, competitors and predators were introduced along with the bivalves (Wolff and Reise, 2002). Another strategy was the broad-scale placement of shell or shell fragments at high densities on the seafloor to create a new settlement surface. This led to a large scale, and reasonably successful "put and take" fishery in the United States, where shell is laid down on the seafloor to catch spat, then the oysters are dredged up once grown, and the cycle is repeated (Schulte, 2017).

FIG. 2 Bivalve habitat restoration. (A) Constructed oyster bank using oyster castles. Virginia, United States. (B) Granite rock being deployed as oyster settlement substrate in the Piankatank River, Virginia, United States. (C) Live adult green-lipped mussels, *Perna canaliculus*, being deployed in to form beds in the Hauraki Gulf, New Zealand. (D) Volunteers assisting with oyster castle deployment, United States. (E) Oyster gardeners with oyster basket, Queensland, Australia. (F) Volunteers moving bags of oyster shells for intertidal oyster restoration, United States. (G) Traditional Maori flax weaving being used to create mussel settlement substrate, Auckland, New Zealand (H) Living shoreline in North Carolina with bagged oyster shell and reef balls deployed to provide a settlement substrate for oysters and to protect the shoreline from erosion. (I) *Ostrea angasi* spat being grown out on scallop shells prior to deployments in Port Phillips Bay, Australia. *(Photos from (A) I. McLeod. (B) US Army/Patrick Bloodgood. (C) Shaun Lee. (D) Erika Norteman/ The Nature Conservancy. (E) Ian McLeod. (F) Erika Norteman/The Nature Conservancy. (G) Shawn Lee. (H) Jackeline M. Perez Rivera, US Marine Corps photo. (I) Ben Cleveland.)*

Since the 1990s, with a growing recognition of the ecosystem services provided by bivalve habitats, restoration efforts have started to focus on restoring reefs and beds for their structure and function (ecosystem services such as water quality improvements, shoreline protection and providing habitat, and food for harvested species) rather than just for their future harvest potential (Brumbaugh and Coen, 2009). Hundreds of bivalve restoration attempts have been made in the last three decades (Fig. 2; Kennedy et al., 2011). Restoration attempts have generally tried to overcome one or both of the two main limiting factors inhibiting natural recovery, substrate, and recruitment limitation.

Techniques used to restore populations limited by settlement substrate use shell or built three-dimensional reefs using rock or concrete. Attempts to restore populations with limited natural recruitment often included shells or other substrates seeded with juvenile oysters from hatcheries (Fig. 2). More recently there has been a focus on breeding disease-resistant bivalves for restoration efforts.

Unfortunately, most bivalve habitat restoration projects suffered from a lack of monitoring and poorly defined objectives. A review of available data on oyster restoration activities in the Chesapeake Bay in the period 1990–2007 found that few were monitored and the restoration project goals were often poorly defined (Kramer and Sellner, 2009; Kennedy et al., 2011). Many restoration projects were not protected from dredging leading to their failure (Schulte, 2017). Efforts to develop local and regional bivalve habitat restoration plans have been increasing in North America and Europe in recent times and guidelines have been provided such as that of Brumbaugh et al. (2006).

6.1 Large Scale Restoration Works

Recently successful bivalve restoration has been scaling up, particularly in the United States. This has been led top-down by large government initiatives, and bottom up by community groups. In 2004, the US Army Corps of Engineers constructed a 42 ha oyster reef by placing dredged and washed oyster shells in Great Wicomico River, Chesapeake Bay. Schulte et al. (2009) reported the success of the project with 180 million oysters present, making this the largest wild oyster population in the world. The success of this restoration was attributed to the absence of dredge fishing and to the high vertical relief of the reefs, which mimicked historical natural reefs. The largest current initiative is the Chesapeake Bay Executive Order, which requires the oyster populations of 20 Chesapeake Bay tributaries to be restored by 2025. One of the target tributaries is Harris Creek, where between 2012 and 2016, 142 ha of oyster reefs were successfully restored, at a cost of US$28 million (Box 1: Case Study 1). In areas outside of North America, bivalve restoration has also been scaling up. For example, there are plans to construct 20 ha of *O. angasi* beds in South Australia over the next few years. There is little information about the scale of restoration efforts in Japan and China as projects in these countries are not well covered in the western literature. However, some bivalve restoration initiatives in these countries have been substantial. For example, Box 2 (Case Study 2) describes a project in China where 100 km of oyster reefs were constructed.

BOX 1 Case Study 1: Large-scale Oyster Restoration in Harris Creek, Chesapeake Bay, United States

Historically, reefs of eastern oysters, *Crassostrea virginica*, supported extensive fisheries in Chesapeake Bay in eastern United States (Beck et al., 2011). Oyster restoration projects have been implemented in Chesapeake Bay for 30 years with the goals of restoring fisheries and ecosystem services. A recent state and federal policy called for a scaled-up approach, setting a goal of restoring oyster reefs in ten Chesapeake Bay tributaries by 2025. Resource managers and scientists collaboratively developed "Chesapeake Bay Oyster Metrics," criteria defining restoration success at both the reef and tributary levels. Harris Creek (Fig. B.1) was selected as the first tributary for restoration because it is an oyster sanctuary with no commercial fishing allowed, shows historical evidence of large oyster populations, and has a remnant oyster population. State and federal partners collected data on benthic habitat and oyster populations, along with scientific and public input, and developed a plan to restore the reefs. Areas with shell benthic habitat and >5 oysters per m^2 were treated with hatchery-produced seed oysters (spat-on-shell). Areas with no shell, or <5 oysters per m^2 (around 50% of areas), were treated by constructing a reef base (0.3 m high, from stone or shell), followed by seeding. Target planting density was 12.5 million seed per hectare. Between 2012 and 2016, 142 ha of reefs were restored, at a cost of US$28 million. Reefs are monitored 3 years after restoration; reefs seeded in 2012 and 2013 (78 ha) have been monitored. All but 1.2 ha exceeded the threshold oyster biomass (15 g dry tissue weight per m^2) and density (15 oysters per m^2). Forty seven hectares exceeded the higher, target biomass (50 g dry tissue weight per m^2) and density (15 oysters per m^2). Stone-base reefs averaged four times higher oyster densities than shell-base reefs and shell-base reefs showed higher densities than seed-only reefs. The model developed in Harris Creek is being used for oyster restoration for other Chesapeake Bay restoration projects and may be modified for oyster reef restoration projects in other locations and countries.

Author: Stephanie Reynolds Westby, NOAA

Continued

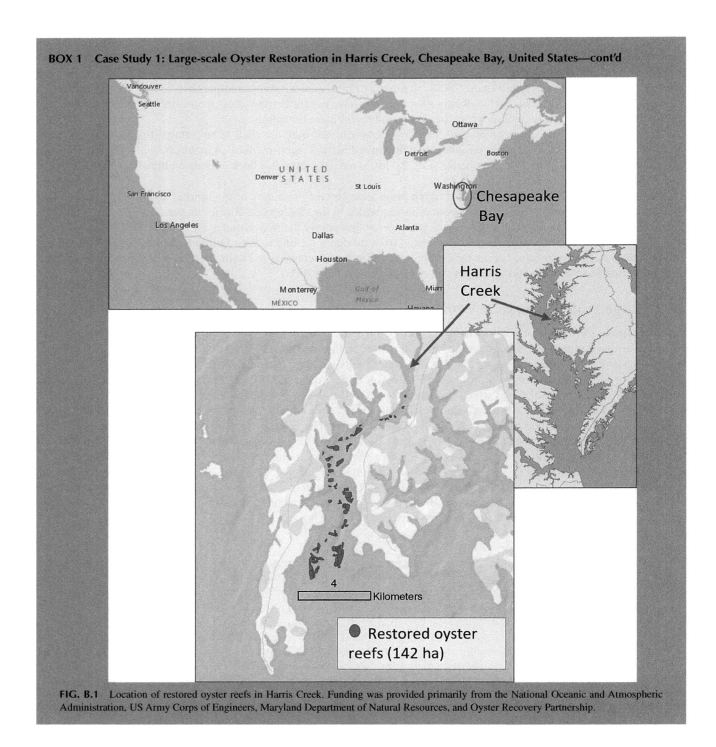

BOX 1   Case Study 1: Large-scale Oyster Restoration in Harris Creek, Chesapeake Bay, United States—cont'd

FIG. B.1   Location of restored oyster reefs in Harris Creek. Funding was provided primarily from the National Oceanic and Atmospheric Administration, US Army Corps of Engineers, Maryland Department of Natural Resources, and Oyster Recovery Partnership.

6.2   Community Restoration

Alongside large-scale often government-led restoration programs, community-led restoration is also scaling-up. Community-led restoration projects are usually relatively small scale and fall into three broad types, (1) oyster gardening of usually hatchery-produced oysters, (2) deployment of juvenile to adult bivalves within designated areas for stock enhancement, and (3) substrate enhancement using natural or recycled man-made materials, loose or in "bags" to enhance local settlement success (Brumbaugh and Coen, 2009). Such examples in Australia and New Zealand and in the United Kingdom are shown

BOX 2   Case Study 2: Oyster Restoration in China

Estuaries in China have an array of natural oyster (*Crassostrea ariakensis*) reefs as well as projects where active restoration methods are being tested, the most notable of which is at the mouth of the Yangtze River (Fig. B.2). In this urbanized setting where downtown Shanghai meets the coast, the Yangtze River was dredged for navigation purposes in 1997 and two ~50 km concrete dikes were constructed. In 2004, over 20t of hatchery-reared seed oysters from Xiangshan Bay were transplanted by the East China Sea Fisheries Research Institute and thus constituted one of the largest restoration projects in the world at that time. Since then, results from our studies indicate that the restoration project was a success: the oyster population is self-sustaining with normally distributed size classes (Quan et al., 2012a); species diversity and abundance of resident macrofauna are high (Quan et al., 2009, 2012a); reefs supported higher trophic organisms than adjacent salt marsh areas (Quan et al., 2012b). Continued monitoring of this restored oyster reef will be necessary and important for determining temporal trajectories of ecological change (La Peyre et al., 2014), as well as improve our understanding of oyster reef dynamics in multiuse urban environment.

Oyster reef restoration is gaining interest and momentum in China as local stewardship increases, however, little mechanistic information is available to guide best practices under local conditions. One exception is a recent study that found many different types of substrate may be used for successful oyster recruitment, from clam and oyster shell to limestone and clay brick (Quan et al., 2017). These types of studies addressing specific mechanisms that mediate restoration success will be necessary to progress the science and practice in China. Additionally, questions remain about how oyster aquaculture may influence reef restoration and be better designed contribute to larval supply for restoration efforts.

Authors: Weimin Quan (Chinese Academy of Fishery Sciences, Qingdao, China) and Austin Humphries

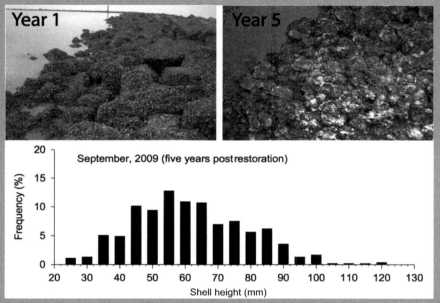

FIG. B.2   Restored oyster reef in the Yangtze River Estuary off the city shores of Shanghai, China. The concrete dike structure was built as part of a navigation channel project and then seeded with oysters (*Crassostrea ariakensis*) in 2004. The photo on the left dates from 2005, 1 year after the oyster restoration initiative, and the photo on the left was taken 5 years postrestoration. The graph below the photos shows the size distribution of oysters on the reef in 2009, indicating successful restoration with mature oysters. (*Modified from Quan, W., Humphries, A.T., Shen, X., Chen, Y., 2012a. Oyster and associated benthic macrofaunal development on a created intertidal oyster* (Crassostrea ariakensis) *reef in the Yangtze River estuary, China. J. Shellfish Res. 31(3), 599–610.*)

in Boxes 3 and 4, respectively. These initiatives are often led by citizen scientists (community, school, and Indigenous groups). Some of these programs are very large scale. In response to a lack of shell for restoration, shell recycling programs have started where shells are collected from restaurants and processing plants. One of the most ambitious projects is the Billion Oyster Project which aims to bring 1 billion oysters back into New York City waters. Along with providing much of the person power and resources for restoration projects these initiatives also empower local communities through education and training. To support these initiatives and ensure efforts are likely to be successful best practice guidelines have been developed such as that of Brumbaugh et al. (2006).

BOX 3   Case Study 3: Bivalve Restoration Down Under (Australia and New Zealand)

Two of Australia's most prevalent reef-building species, the rock oyster (*Saccostrea glomerata*) and the flat oyster (*Ostrea angasi*), are now largely expatriated from estuaries and coastal embayments with fewer than 10% of reefs remaining (Beck et al., 2011; Gillies et al., 2018). Since 2014, several projects to restore rock and flat oyster reefs have been established near most major capital cities (Gillies et al., 2018). These projects are largely based on successful methods employed in the United States and follow recommend guidelines for ecological and reef restoration including laying down limestone structures subtidally and reintroducing adult oysters (e.g., Brumbaugh et al., 2006). Bivalve habitat restoration in New Zealand has focused on the large green-lipped mussel *Perna canaliculus* (hereafter mussels). Vast beds of mussels grew in estuaries around northern New Zealand. Beds were fished from 1300 km^2 of the Hauraki Gulf, outside of Auckland by dredging. The dredging was unsustainable and the fishery collapsed in the late 1960s. Fifty years later, beds have not recovered with less 1 km^2 remaining. Large-scale restoration experiments using adult mussels have shown that they can survive across their former range, but recruitment remains low and this is the focus of ongoing research. A non-for-profit group "Revive our Gulf" is driving much of the mussel restoration efforts in New Zealand with plans for scale-up and trials in other locations. Critical to the success of these initial projects are the strong partnerships developed with the bivalve aquaculture industry which provide emerging restoration practitioners with technical guidance on animal husbandry, reproduction and practical deployment. Equally important has been the support of recreational fishing groups whose values align strongly with common objectives of bivalve restoration such as increasing estuary productivity, improving water quality, and reinstating habitat for fish and crayfish (Fig. B.3).

Authors: Chris Gillies and Ian McLeod

FIG. B.3   Juvenile crayfish with *Ostrea angasi* oysters, Tasmania, Australia. *(From Cayne Layton.)*

6.3   Restoration for Coastal Protection

Coastal erosion is a growing problem internationally because of sea-level rise driven by climate change, and increasing population and development in coastal areas. Living shorelines that include living elements (natural infrastructure), such as salt marsh and oyster reefs are increasingly being considered as an alternative to "hard" or "gray" coastal protection measures such as bulkheads, revetments, and concrete seawalls that may displace energy causing further erosion nearby. Living shorelines can include sand, plants, logs, oyster shell (often in bags), organic materials (e.g., biologs made out of jute), concrete, material filled structures, or other recycled or natural structural material to provide shoreline protection. These dissipate energy, trap sediments to encourage the growth of plants such as salt marsh or mangroves and provide a settlement structure for oysters (the living components of living shorelines). Living shorelines may have lower installation and maintenance costs compared to fully engineered alternatives and have other benefits and services.

There is growing interest in using oyster reefs and living shorelines for coastal protection in places especially vulnerable to sea-level rise such as Bangladesh, the southern states of the United States, and the Wadden Sea. There was concern that oyster reef vertical growth may not be able to keep pace with sea-level rise. However, recent direct measurements by Rodriguez et al. (2014) has shown that reef height accretion for intertidal *C. virginica* is up to 10 times faster than previously estimated and intertidal reefs studied in the mid-Atlantic US estuaries should be able to keep up with predicted sea-level rise.

BOX 4   Case Study 4: Essex Native Oyster Restoration Initiative

The River Blackwater has a long history fishing for the European native oyster, *Ostrea edulis*. Indeed, the *Colchester oyster* has been a sought-after delicacy for centuries (Sprat, 1669). In the 1800s *O. edulis* was overfished and impacted by poor water quality throughout its range, and since the 1980s from the introduced oyster disease, bonamia (Hudson and Hill, 1991). While the oyster fishery in the Blackwater persisted following the importation of broodstock, the fishery for the native oyster is now restricted to private grounds, with most oystermen focusing on harvesting the introduced Pacific oyster *Crassostrea gigas*. But the cultural heritage and fishermens' passion for the native oyster persists and it is the Blackwater Oysterman Association that has spearheaded the data gathering required to successfully designate the site a marine conservation zone for *O. edulis* habitat (beds) and the species, as well as the subsequent conservation efforts in partnership with conservation, industry and statutory bodies through the Essex Native Oyster Restoration Initiative (ENORI). Restoration work is being undertaken with three aims in mind: to recover the native oyster beds and species to self-sustaining levels as per the statutory conservation objectives, to restore the valuable ecosystem services provided by the native oyster, and to increase the oyster population to support the fishery of the species into the future. In 2015 25,000 adult oysters were purchased from the private fishery and re-laid in part of the public oyster grounds in the Blackwater Estuary to form a broodstock sanctuary. The relay site and surrounding area (totaling 200 ha) was voluntarily declared a no take zone within the estuary, which is more widely managed for the oyster fishery. This partnership and shared management approach is the first for this species in the United Kingdom. The laying of cultch and further broodstock enhancement is currently being planned in order to further restore the native oyster population (Fig. B.4).

Authors: Philine zu Ermgassen, Sarah Allison (Essex Wildlife Trust) and Alison Debney (Zoological Society of London)

FIG. B.4   *O. edulis* being deployed to restore populations in Blackwater Estuary. *(From Sarah Alison, Essex Wildlife Trust.)*

6.4   Should Nonnative Bivalve Species Be Used for Restoration?

Nonnative species of habitat-forming bivalve have been spread around the world. Many of these introductions have been deliberate in the attempt to restore depleted natural bivalve fisheries, or for aquaculture. Many other introductions have been accidental (transported on ships and equipment or in ballast water). These species can have positive or negative impacts on estuaries. Nonnative introductions have supported successful aquaculture industries in many countries. In the Wadden Sea the invasion of Pacific oyster, *Crassostrea gigas* has led to intertidal reef development that reduces coastal erosion and provides habitat for invertebrates (Herbert et al., 2016). In Australia, there have been no obvious negative consequences for the invertebrate communities as native *S. glomerata* reefs are replaced by *C. gigas* (Wilkie et al., 2012). However, the introduction of nonnative species has had devastating consequences for native bivalve species through the introduction of diseases, predators, competitors, and parasites (Wolff and Reise, 2002). Any future introductions should go through a rigorous risk assessment process that includes the risks to native bivalve species. https://link.springer.com/chapter/10.1007/978-94-015-9956-6_21.

7 THE FUTURE OF BIVALVE HABITAT RESTORATION

7.1 Social, Economic, and Environmental Benefits

The experiences from decades of restoration practice in the United States and elsewhere have demonstrated that bivalve habitat restoration can be a positive management action that results in multiple social, economic, and ecological benefits to estuaries and their surrounding communities. The restoration of bivalve habitats likely represents one of the few coastal habitats which have consistently achieved success at the "landscape-scale" across a variety of species, environmental conditions, and continents. One might ask why that has occurred for these systems and not for others? Perhaps part of the answer lies in the efforts of project proponents to emphasize the strong connection between community values and the social, economic, and environmental benefits derived from restoration such as job creation, livelihood benefits, and community economic benefits. Emphasis on communicating the social and economic benefits of bivalve habitat restoration enables communities to more easily align their own livelihood and wellbeing objectives with those of restoration proponents, thus enabling projects to receive support from a wider audience of diverse stakeholders rather than those focused solely on more narrow ecological (i.e., species recovery) benefits (Goldman and Tallis, 2009). Bivalve habitat restoration also provides communities the opportunity to collectively work on solutions to improve estuary health and coastal livelihoods rather than the more commonly divisive activities of problem definition and environmental regulation.

Key to the future growth of bivalve habitat restoration thus relies on the continued promotion of the natural and human benefits derived from the recovery of bivalve habitats in addition to the ecological benefits. Projects should consider the trade-off between setting purely ecological objectives (e.g., oyster specific metrics) against also including social and economic objectives such as job creation, community volunteering, and local business involvement. By documenting and communicating the social and economic benefits of projects in addition to ecological outcomes, economists, and estuary managers can more easily build the case for further investment in restoration activities and better assess trade-offs against investing in other social or environmental projects. Tools such as the oyster calculator produced by TNC (http://oceanwealth.org/tools/oyster-calculator) that allows managers to calculate how much oyster restoration is needed to reach water quality and fish productivity goals are likely to further build the case for sustained investments in bivalve habitat restoration. These initiatives also help to streamline objective setting, monitoring, and reporting across projects.

7.2 Global Expansion

Bivalve habitat restoration is on the cusp of expanding into new geographies such as Asia, Oceania, Africa, and Europe (where bivalve restoration is largely in the stages of early adoption [see case studies]). Partnerships between experienced organizations and early adopting organizations in new regions are an important step to scaling-up bivalve habitat restoration locally and globally. Such partnerships help transfer technical knowledge into new regions and expedite project development. They also provide opportunities for "knowledge donor" organizations to leverage their existing projects with new opportunities, resulting in a win-win scenario for both groups. Such partnerships also provide opportunities for global networks to strengthen and help build the foundation for global research, resourcing, and strategies. A global network of bivalve practitioners and researchers is likely to lead to stronger recognition of the role that restoration of bivalve (and other marine) habitats can play in meeting global development goals and international treaties. A global approach can also elevate the role that bivalve habitat restoration can play in helping to mitigate regional coastal threats such as pollution, coastal erosion, and ocean acidification, particularly around major coastal cities and communities.

7.3 Opportunities for Innovation

Like many habitat restoration projects, traditional project objectives have largely focused on recovering the primary habitat-forming species that was expatriated or degraded. Yet recent advancements in understanding ecosystem function, and thus ecosystem services, has helped expose a whole new range of possibilities for how habitat restoration can help address estuary threats and support livelihoods beyond just habitat loss. Estuary managers, communities, and coastal industries can now ask: what role can habitat restoration play in supporting commercial and recreational fisheries? How can bivalve habitat restoration support growth in the bivalve aquaculture industry? How can restoration help combat pollution and eutrophication or buffer shorelines from storm surges and sea-level rise?

With the advancement and application of bivalve habitat ecosystem services, innovative, and long-term financing mechanisms can be established to support restoration expansion as a method to help manage and mitigate broader estuary (and associated livelihood) threats. For instance, the denitrification and phosphorus removal benefits derived from bivalve habitats (Newell et al., 2002; Kellogg et al., 2013; Humphries et al., 2016) could provide a nutrient sink mechanism with funding for restoration activities derived from estuarine nutrient trading schemes, sewerage or pollution offsets. Such programs could operate in a similar way to freshwater protection funds which divert funding from downstream management interventions (e.g., desalination plants) to fund upper catchment restoration projects in order to secure clean water. The fisheries production benefits of bivalve habitats (zu Ermgassen et al., 2016b) could provide a model for ecosystem-based fisheries management, whereby restoration activities are funded through recreational fisheries license funds or commercial seafood levies. The shared costs associated with developing bivalve hatcheries or research and development in bivalve genetics, disease, and husbandry could be paid for in part by restoration projects, with industry cost savings returned back to bivalve habitat restoration projects (Box 5).

With the advancement in learning and application of habitat function and ecosystem services, stronger global networks, and an emphasis on incorporating social and economic (in addition to ecological) objectives, the future of bivalve habitat restoration is set to continue to expand in the future. Large-scale habitat restoration projects can be understood by estuary managers, communities, and industry as a viable and cost effective option for mitigating several environmental threats while also helping to boost coastal livelihoods and industries reliant on healthy estuaries.

BOX 5   Can Bivalve Aquaculture Replace the Lost Functioning of Bivalve Habitats?

Over the last five decades, aquaculture has become the fastest growing global food production sector (Diana, 2009; FAO, 2016). Bivalve aquaculture now dominates harvest from wild populations. For example, 95% of the world's demand for oysters is being met by aquaculture (Schulte, 2017). There are both synergies and conflicts between the goals of aquaculture and restoring lost bivalve habitats. Bivalve aquaculture requires high water quality and the bivalve aquaculture industry has helped improve water quality standards in many areas (e.g., through lobbying waste water treatment upgrades). The aquaculture industry also has vital bivalve husbandry knowledge and has developed disease-resistant stocks that are useful for restoration. Bivalve hatcheries that were primarily set up to produce spat (juvenile bivalves) for aquaculture are also needed to produce spat for large-scale bivalve restoration in recruitment limited systems.

Recently aquaculture has been considered as a potential tool to provide other ecosystem services. It has been suggested that bivalve aquaculture may provide a viable mechanism for reducing the eutrophication in estuaries as their role in improving water clarity and quality may be more powerful than natural reefs because of denser populations of bivalves. Lindahl et al. (2005) suggested that mussel aquaculture could provide a more cost-effective solution for reducing nitrogen in Swedish fiords than waste water treatment plants in some circumstances. Bivalve aquaculture provides habitat for many species of fish and invertebrates, and the benthic habitat below aquaculture facilities can be enhanced by the structured debris that falls from the aquaculture facility and the enriched biodeposits from bivalve filter feeding. However, at high densities bivalves can produce such a large quantity of nutrient-enriched biodeposits that the substrate below them becomes anoxic through bacterial decomposition rendering that habitat unsuitable for valued fish and invertebrate species.

Overall aquaculture can probably replace some, but not all the services provided by natural bivalve habitats, and the provision of services will be dependent on appropriate site selection. The biodiversity supported by aquaculture infrastructure and cultured bivalves is likely to be different from that of natural reefs, and aquaculture infrastructure is not likely to provide the same levels of coastal protection as natural reefs.

8 CONCLUSION

Bivalve habitats were once dominate dominant habitat types in temperate and subtropical estuaries and coastal waters, but are now greatly reduced over most of their former range. Given the important ecosystem services provided by bivalves this has contributed to the declining health of estuaries and the fisheries they support. In response to these declines, bivalve restoration is scaling up globally through large-scale government-led approaches and through small-scale local initiatives involving community groups and citizen scientists. Restoring bivalve habitats can improve the health of estuaries and coastal waters but is not a silver bullet. Restoration efforts will need to be coupled with improved management practices and the restoration of other habitat types such as seagrasses, salt marshes, and mangroves. Restoration of bivalves habitats is gaining support beyond conservation-focused groups with new projects focusing on food security, local employment, green engineering, shoreline protection, and nutrient trading. Estuaries will not be the same in the future with increasing coastal human populations and developments. Bivalve habitats of the future are unlikely to be restored to their historical extent and structure. However, we may able to recreate their historical functions and benefit from their ecosystem services through active restoration and by modifying aquaculture design, and incorporating bivalves habitats into coastal infrastructure planning.

REFERENCES

Alleway, H.K., Connell, S.D., 2015. Loss of an ecological baseline through the eradication of oyster reefs from coastal ecosystems and human memory. Conserv. Biol. 29, 795–804.

Baggett, L.P., Powers, S.P., Brumbaugh, R., Coen, L.D., DeAngelis, B., Green, J., Hancock, B., Morlock, S., 2014. Oyster Habitat Restoration Monitoring and Assessment Handbook. The Nature Conservancy, Arlington, VA. 96 pp.

Baggett, L.P., Powers, S.P., Brumbaugh, R.D., Coen, L.D., DeAngelis, B.M., Greene, J.K., Hancock, B.T., Morlock, S.M., Allen, B.L., Breitburg, D.L., Bushek, D., 2015. Guidelines for evaluating performance of oyster habitat restoration. Restor. Ecol. 23 (6), 737–745.

Bailey, G.N., 1975. The role of molluscs in coastal economies: the results of midden analysis in Australia. J. Archaeol. Sci. 2, 45–62.

Balmford, A., Bruner, A., Cooper, P., Costanza, R., Farber, S., Green, R.E., Jenkins, M., Jefferiss, P., Jessamy, V., Madden, J., Munro, K., 2002. Economic reasons for conserving wild nature. Science 297 (5583), 950–953.

Bancroft, H.H., 1890. History of Washington, Idaho and Montana 1845–1889. The History Company, San Francisco, CA. 836 pp.

Barnes, R.S.K., Coughlan, J., Holmes, N.J., 1973. A preliminary survey of the macroscopic bottom fauna of the solent, with particular reference to *Crepidula fornicata*, and *Ostrea edulis*. Proc. Malacol. Soc. Lond. 40, 253–275.

Beck, M.W., Airoldi, L., Carranza, A., Coen, L.D., Crawford, C.O., et al., 2009. Shellfish Reefs at Risk: A Global Analysis of Problems and Solutions. The Nature Conservancy, Arlington, VA. 56 pp.

Beck, M.W., Brumbaugh, R.D., Airoldi, L., Carranza, A., Coen, L.D., Crawford, C., Defeo, O., Edgar, G.J., Hancock, B., Kay, M.C., Lenihan, H.S., 2011. Oyster reefs at risk and recommendations for conservation, restoration, and management. Bioscience 61 (2), 107–116.

Blake, B., Zu Ermgassen, P.S.E., 2015. The history and decline of *Ostrea lurida* in Willapa Bay, Washington. J. Bivalve Res. 34, 273–280.

Brooks, W.K., Waddell, J.I., Legg, W.H., 1884. Report of the Oyster Commission of the State of Maryland, January 1884. 183 pp.

Brown, K.M., George, G.J., Peterson, G.W., Thompson, B.A., Cowan Jr., J.H., 2008. Oyster predation by black drum varies spatially and seasonally. Estuar. Coasts 31, 597–604.

Brumbaugh, R.D., Coen, L., 2009. Contemporary approaches for small-scale oyster reef restoration to address substrate vs recruitment limitatio: a review and comments relevant for the olympia oyster, *Ostrea lurida* Carpenter 1864. J. Shellfish Res. 28 (1), 147–161.

Brumbaugh, R.D., Beck, M.W., Coen, L.D., Craig, L., Hicks, P., 2006. Practitioners Guide to the Design and Monitoring of Shellfish Restoration Projects: An Ecosystem Services Approach. The Nature Conservancy, Arlington VA. MRD Education Report No. 22, 28 pp.

Choe, K.A., Whittington, D., Lauria, D.T., 1996. The economic benefits of surface water quality improvements in developing countries: a case study of Davao, Philippines. Land Econ., 519–537.

Claassen, C., Whyte, T., 1995. Biological remains at Dogan Point. In: Claassen, C. (Ed.), Dogan Point: A Shell Matrix Site in the Lower Hudson Valley. Occasional Publications Northeastern Anthropol, Bethlehem, PA, pp. 65–78.

Coen, L.D., Grizzle, R.E., 2007. The importance of habitat created by molluscan bivalve to managed species along the Atlantic Coast of the United States. Habitat Management Series No. 8, p. 115.

Coen, L.D., Brumbaugh, R.D., Bushek, D., Grizzle, R., Luckenbach, M.W., Posey, M.H., Powers, S.P., Tolley, S.G., 2007. Ecosystem services related to oyster restoration. Mar. Ecol. Prog. Ser. 341, 303–307.

Dame, R.F., 2009. Shifting through time: oysters and shell rings in past and present south-eastern estuaries. J. Shellfish Res. 28, 425–430.

Dankers, N., Brinkman, A.G., Meijboom, A., Dijkman, E., 2001. Recovery of intertidal mussel beds in the Wadden Sea: use of habitat maps in the management of the fishery. Hydrobiologia 465, 21–30.

Diana, J.S., 2009. Aquaculture and biodiversity conservation. Bioscience 59 (1), 27–38.

Doran Jr., E., 1965. Shell roads in Texas. Geogr. Rev. 55, 223–240.

FAO, 2016. The State of World Fisheries and Aquaculture 2016. Contributing to food security and nutrition for all, Rome. 200 pp.

Fasten, N., 1931. The Yaquina oyster beds of Oregon. Am. Nat. 65 (700), 434–468.

Gercken, J., Schmidt, A., 2014. Current status of the European Oyster (*Ostrea edulis*) and possibilities for restoration in the German North Sea, 2014. Bundesamt für Naturschutz Report.

Gillies, C.L., McLeod, I.M., Alleway, H.K., Cook, P., Crawford, C., et al., 2018. Australian shellfish ecosystems: past distribution, current status and future direction. PLOS ONE 13 (2), e0190914. https://doi.org/10.1371/journal.pone.0190914.

Shellfish reef habitats: a synopsis to underpin the repair and conservation of Australia's environmentally, socially and economically important bays and estuaries. In: Gillies, C.L., Creighton, C., McLeod, I.M. (Eds.), Report to the National Environmental Science Programme, Marine Biodiversity Hub. 2015. Centre for Tropical Water and Aquatic Ecosystem Research (TropWATER) Publication, James Cook University, Townsville. 68 pp.

Goldman, R.L., Tallis, H., 2009. A critical analysis of ecosystem services as a tool in conservation projects. Ann. N. Y. Acad. Sci. 1162, 63–78.

Grabowski, J.H., Brumbaugh, R.D., Conrad, R.F., Keeler, A.G., Opaluch, J.J., Peterson, C.H., Piehler, M.F., Powers, S.P., Smyth, A.R., 2012. Economic valuation of ecosystem services provided by oyster reefs. Bioscience 62 (10), 900–909.

Hargis Jr., W.J., Haven, D.S., 1999. Chesapeake oyster bars, their importance, destruction and guidelines for restoring them. In: Luckenbach, M.W., Mann, R., Wesson, J.A. (Eds.), Oyster Bar Habitat Restoration: A Synopsis and Synthesis of Approaches. Virginia Institute of Marine Science Press, Gloucester Point, VA, pp. 329–358.

Herbert, R.J., Humphreys, J., Davies, C.J., Roberts, C., Fletcher, S., Crowe, T.P., 2016. Ecological impacts of non-native Pacific oysters (*Crassostrea gigas*) and management measures for protected areas in Europe. Biodivers. Conserv. 25, 2835–2865.

Hudson, E.B., Hill, B.J., 1991. Impact and spread of bonamiasis in the UK. Aquaculture 93, 279–285.

Humphries, A.T., La Peyre, M.K., Kimball, M.E., Rozas, L.P., 2011. Testing the effect of habitat structure and complexity on nekton assemblages using experimental oyster reefs. J. Exp. Mar. Biol. Ecol. 409 (1), 172–179.

Humphries, A.T., Ayvazian, S.G., Carey, J.C., Hancock, B.T., Grabbert, S., Cobb, D., Strobel, C.J., Fulweiler, R.W., 2016. Directly measured denitrification reveals oyster aquaculture and restored oyster reefs remove nitrogen at comparable high rates. Front. Mar. Sci. 3, 74.

Kellogg, M.L., Cornwell, J.C., Owens, M.S., Paynter, K.T., 2013. Denitrification and nutrient assimilation on a restored oyster reef. Mar. Ecol. Prog. Ser. 480, 1–19.

Kennedy, V.S., Breitburg, D.L., Christman, M.C., Luckenbach, M.W., Paynter, K., Kramer, J., Sellner, K.G., Dew-Baxter, J., Keller, C., Mann, R., 2011. Lessons learned from efforts to restore oyster populations in Virginia and Maryland, 1990 to 2007. J. Shellfish Res. 30, 1–13.

Kramer, J.G., Sellner, K.G. (Eds.), 2009. ORET: metadata analysis of restoration and monitoring activity database, native oyster (*Crassostrea virginica*) restoration in Maryland and Virginia. An evaluation of lessons learned 1990–2007. Maryland Sea Grant Publication #UM-SG-TS-2009-02. CRC Publ. No. 09-168, College Park, MD, 40 pp.

La Peyre, M.K., Humphries, A.T., Casas, S.M., La Peyre, J.F., 2014. Temporal variation in development of ecosystem services from oyster reef restoration. Ecol. Eng. 63, 34–44.

La Peyre, M.K., Serra, K.T., Joyner, T.A., Humphries, A.T., 2015. Assessing shoreline exposure and oyster habitat suitability maximizes potential success for restored oyster reefs. Peer J. 3, e1317.

Lindahl, O., Hart, R., Hernroth, B., Kollberg, S., Loo, L.O., Olrog, L., Rehnstam-Holm, A.S., Svensson, J., Svensson, S., Syversen, U., 2005. Improving marine water quality by mussel farming: a profitable solution for Swedish society. Ambio 34 (2), 131–138.

MacKenzie Jr., C.L., Burrell Jr., V.G., Rosenfield, A., Hobart, W.L., 1997. The history, present condition, and future of the molluscan fisheries of North and Central America and Europe: volume 2, Pacific Coast and supplemental topics.

Mann, R., Harding, J.M., Southworth, M.J., 2009. Reconstructing pre-colonial oyster demographics in the Chesapeake Bay, USA. Estuar. Coast. Shelf Sci. 85, 217–222.

May, E.B., 1971. A survey of the oyster and oyster shell resources of Alabama. Ala. Mar. Res. Bull. 4, 1–53.

McLeod, I.M., 2009. Green-Lipped Mussels, *Perna canaliculus*, in Soft-Sediment Systems in Northeastern New Zealand. (Unpublished M.Sc. thesis) University of Auckland. 113 pp.

McLeod, I.M., Parsons, D.M., Morrison, M.A., Van Dijken, S.G., Taylor, R.B., 2013. Mussel reefs on soft sediments: a severely reduced but important habitat for macroinvertebrates and fishes in New Zealand. N. Z. J. Mar. Freshw. Res. 48, 48–59.

Milner, N., 2013. Human impacts on oyster resources at the Mesolithic-Neolithic transition in Denmark. In: Thompson, V.D., Waggoner Jr., J.C. (Eds.), The Archaeology and Historical Ecology of Small-Scale Economies. Univ Press of Florida, Gainesville, FL, pp. 17–40.

Moebius, K., 1883. The Oyster and Oyster-Culture. Report of Commissioner of Fish and Fisheries, pp. 683–747.

Newell, R.I.E., Cornwell, J.C., Owens, M.S., 2002. Influence of simulated bivalve biodeposition and microphytobenthos on sediment nitrogen dynamics: a laboratory study. Limnol. Oceanogr. 47, 1367–1379.

Ogburn, D.M., White, I., McPhee, D.M., 2007. The disappearance of oyster reefs from eastern Australian estuaries impact of colonial settlement or mud-worm invasion? Coast. Manag. 35, 271–287.

OSPAR Commission, 2009. Background document for *Ostrea edulis* and *Ostrea edulis* beds.

Paul, L.J., 2012. A history of the Firth of Thames dredge fishery for mussels: use and abuse of a coastal resource. New Zealand Aquatic Environment and Biodiversity Report No. 94, pp. 27.

Piazza, B.P., Banks, P.D., La Peyre, M.K., 2005. The potential for created oyster shell reefs as a sustainable shoreline protection strategy in Louisiana. Restor. Ecol. 13 (3), 499–506.

Powell, E.N., Klinck, J.M., Hofmann, E.E., McManus, M.A., 2003. Influence of water allocation and freshwater inflow on oyster production: a hydrodynamic–oyster population model for Galveston Bay, Texas, USA. Environ. Manag. 31 (1), 0100–0121.

Powell, E.N., Kraeuter, J.N., Ashton-Alcox, K.A., 2006. How long does oyster shell last on an oyster reef? Estuar. Coast. Shelf Sci. 69, 531–542.

Quan, W., Ni, Y., Shi, L., Chen, Y., 2009. Composition of fish communities in an intertidal salt marsh creek in the Changjiang River estuary. China. Chinese J. Oceanol. Limnol. 27 (4), 806–815.

Quan, W., Humphries, A.T., Shen, X., Chen, Y., 2012a. Oyster and associated benthic macrofaunal development on a created intertidal oyster (*Crassostrea ariakensis*) reef in the Yangtze River estuary, China. J. Shellfish Res. 31 (3), 599–610.

Quan, W.M., Humphries, A.T., Shi, L.Y., Chen, Y.Q., 2012b. Determination of trophic transfer at a created intertidal oyster (*Crassostrea ariakensis*) reef in the Yangtze River estuary using stable isotope analyses. Estuar. Coast. 35 (1), 109–120.

Quan, W., Fan, R., Wang, Y., Humphries, A.T., 2017. Long-term oyster recruitment and growth are not influenced by substrate type in China: implications for sustainable oyster reef restoration. J. Shellfish Res. 36 (1), 79–86.

Ragnarsson, S.A., Burgos, J.M., 2012. Separating the effects of a habitat modifier, *Modiolus modiolus* and substrate properties on the associated megafauna. J. Sea Res. 72, 55–63.

Reguero, B.G., Beck, M.W., Bresch, D.N., Calil, J., Meliane, I., 2018. Comparing the cost effectiveness of nature-based and coastal adaptation: a case study from the Gulf coast of the United States. PLoS One 13 (4), e0192132. https://doi.org/10.1371/journal.pone.0192132.

Rick, T.C., Erlandson, J.M., 2009. Coastal exploitation. Science 325, 952–953.

Rick, T.C., Reeder-Myers, L.A., Hofman, C.A., Breitburg, D., Lockwood, R., Henkes, G., Kellogg, L., Lowery, D., Luckenbach, M.W., Mann, R., Ogburn, M.B., 2016. Millennial-scale sustainability of the Chesapeake Bay native American oyster fishery. Proc. Natl. Acad. Sci. 113 (23), 6568–6573.

Rodriguez, A.B., Fodrie, F.J., Ridge, J.T., Lindquist, N.L., Theuerkauf, E.J., Coleman, S.E., Grabowski, J.H., Brodeur, M.C., Gittman, R.K., Keller, D.A., Kenworthy, M.D., 2014. Oyster reefs can outpace sea-level rise. Nat. Clim. Chang. 4 (6), 493.

Sampson, C.P., 2015. Oyster demographics and the creation of coastal monuments at Roberts Island Mound Complex, Florida. Southeast. Archaeol. 34 (1), 84–94.

Schulte, D.M., 2017. History of the Virginia Oyster Fishery, Chesapeake Bay, USA. Front. Mar. Sci. 4, 127.

Schulte, D.M., Burke, R.P., Lipcius, R.N., 2009. Unprecedented restoration of a native oyster metapopulation. Science 325 (5944), 1124–1128.

Scyphers, S.B., Powers, S.P., Heck, K.L., Byron, D., 2011. Oyster reefs as natural breakwaters mitigate shoreline loss and facilitate fisheries. PLoS ONE 6, e22396.

Seaman, M.N.L., Ruth, M., 1997. The molluscan fisheries of Germany. In: NOAA Technical Report NMFS 129: The History, Present Condition, and Future of the Molluscan Fisheries of North and Central America and Europe, Vol. 3: Europe, pp. 57–84.

Shervette, V.R., Gelwick, F., 2008. Seasonal and spatial variations in fish and macroinvertebrate communities of oyster and adjacent habitats in a Mississippi Estuary. Estuar. Coasts 31, 584–596.

Sprat, T., 1669. The history of the generation and ordering of green oysters, commonly called Colchester oysters. In: Sprat, T. (Ed.), The History of the Royal Society of London. third ed. Samuel Chapman, London, pp. 307–319.

Stenzel, H.B., 1971. Oysters. In: Moore, R.C. (Ed.), Treatise on Invertebrate Paleontology, Part N. University of Kansas Press, Kansas, pp. 953–1224.

Sutton, P.C., Costanza, R., 2002. Global estimates of market and non-market values derived from nighttime satellite imagery, land cover, and ecosystem service valuation. Ecol. Econ. 41, 509–527.

Thurstan, R.H., Hawkins, J.P., Raby, L., Roberts, C.M., 2013. Oyster (*Ostrea edulis*) extirpation and ecosystem transformation in the Firth of Forth, Scotland. J. Nat. Conserv. 21, 253–261.

Todorova, V., Micu, D., Klisurov, L., 2009. Unique oyster reefs discovered in the Bulgarian Black Sea. Dokl. Na Bolg. Akad. Na Nauk. 62, 871–874.

Tolley, S.G., Volety, A.K., 2005. The role of oysters in habitat use of oyster reefs by resident fishes and decapod crustaceans? J. Bivalve Res. 24, 1007–1012.

Wall, C.C., Peterson, B.J., Gobler, C.J., 2008. Facilitation of seagrass *Zostera marina* productivity by suspension-feeding bivalves. Mar. Ecol. Prog. Ser. 357, 165–174.

Walles, B., Mann, R., Ysebaert, T., Troost, K., Herman, P.M., Smaal, A.C., 2015. Demography of the ecosystem engineer *Crassostrea gigas*, related to vertical reef accretion and reef persistence. Estuar. Coast. Shelf Sci. 154, 224–233.

Wilkie, E.M., Bishop, M.J., O'Connor, W.A., 2012. Are native *Saccostrea glomerata* and invasive *Crassostrea gigas* oysters' habitat equivalents for epibenthic communities in south-eastern Australia? J. Exp. Mar. Biol. Ecol. 420–421, 16–25.

Wolff, W.J., Reise, K., 2002. Oyster imports as a vector for the introduction of alien species into northern and western European coastal waters. In: Gollasch, S., Leppakoski, E., Olenin, S. (Eds.), Invasive Aquatic Species of Europe—Distribution, Impacts and Management. Kluwer Academic Publishers, Dordrecht/Boston/London, pp. 193–205.

Yonge, C.M., 1960. Oysters. Collins, London. 209 pp.

zu Ermgassen, P.S.E., Spalding, M.D., Blake, B., Coen, L.D., Dumbauld, B., Geiger, S., Grabowski, J.H., Grizzle, R., Luckenbach, M., McGraw, K., Rodney, W., Ruesink, J.L., Powers, S.P., Brumbaugh, R.D., 2012. Historical ecology with real numbers: past and present extent and biomass of an imperilled estuarine ecosystem. Proc. R. Soc. B 279, 3393–3400.

zu Ermgassen, P., Hancock, B., DeAngelis, B., Greene, J., Schuster, E., Spalding, M., Brumbaugh, R., 2016a. Setting Objectives for Oyster Habitat Restoration Using Ecosystem Services: A Manager's Guide. The Nature Conservancy, Arlington, VA.

zu Ermgassen, P.S.E., Grabowski, J.H., Gair, J.R., Powers, S.P., 2016b. Quantifying fish and mobile invertebrate production from a threatened nursery habitat. J. Appl. Ecol. 53, 596–606.

Section F

Coral Reefs

Chapter 26

Successful Management of Coral Reef-Watershed Networks

Robert H. Richmond*, Yimnang Golbuu†, Austin J. Shelton III‡

*Kewalo Marine Laboratory, University of Hawaii at Manoa, Honolulu, HI, United States, †Palau International Coral Reef Center, Koror, Palau, ‡Center for Island Sustainability and Sea Grant Program, University of Guam, Mangilao, Guam

1   INTRODUCTION: IMPORTANCE OF LAND-SEA INTERACTIONS

The protection of coral reefs begins on land. This is true for reefs that are both adjacent to coasts as well as those over 100 km from shore. The problem lies in reductions of water and substratum quality from watershed discharges, and the effects on coral reef health and resilience. Such discharges can carry large volumes of freshwater, a variety of toxicants, sediment, nutrients, bacteria, and solid waste. Each of these elements, individually or in concert, interferes with key processes critical to the survival of coral reef ecosystems.

The two key processes responsible for the persistence of coral reefs over time, reproduction and recruitment, are chemically mediated. Subtle changes in water and bottom quality can have major effects on the success of these two linked elements of reef population replenishment. Reproduction is the process by which new individuals are produced from prior stock. In corals, reproduction can occur as a result of both sexual (fusion of sperm and egg) and asexual (parthenogenesis, fragmentation, tissue sloughing, polyp bailout) means. In sexual reproduction, fertilized eggs develop into larvae called planulae over a period ranging from about 18 h to 3 days, depending on the size of the egg and physical factors such as temperature and salinity. The resulting planula larvae are dispersed via currents, and while many coral reefs are "self-seeded" with larvae from the natal reef, others are repopulated by larvae that can traverse long distances with competency periods exceeding 100 days, creating connectivity among distant reef ecosystems. Asexual propagules, such as fragments and bits of ciliated tissue formed from a shedding process called "polyp bail out," have a limited dispersal range, and if they settle and survive, do so very close to the source. Successful sexual reproduction is critical to reef population maintenance, recovery, and resilience by adding an element of genetic variation not occurring in asexual processes, as well as producing seed material that can disperse over greater distances from a source site.

Water quality and substratum conditions are major factors affecting the success of reproduction and recruitment in corals. There are six critical stages in these linked processes, which are all chemically mediated (Fig. 1): gamete development prior to spawning, synchronization of gamete release among coral colonies of the same species, egg-sperm interactions that lead to fertilization of eggs, embryological development, substrate selection by competent larvae, and for most spawning species (as opposed to brooders), acquisition of symbiotic zooxanthellae, the single cell algae that provide much of a coral's daily energy needs. If any one of these key processes is affected, the natural replenishment of reef populations is prevented.

During mass coral spawning events, large quantities of eggs and sperm are released into the water column. Typically, these gametes are released in combined egg-sperm bundles. The eggs are rich in lipid, which makes them float to the surface of the ocean, carrying the sperm along with them. Once at the surface, these bundles burst apart, allowing the eggs and sperm to interact on the sea surface. Experiments have demonstrated that rates of self-fertilization are usually very low, on the order of 2% or less, while outcrossing of sperm and eggs from different colonies of the same species result in much higher fertilization rates, often over 90%. Hybridization can also occur among closely related species. It is clear that chemical cueing is important for successful egg-sperm interactions, and reductions in water quality, including the reduction in salinity or the presence of toxicants and sediment can result in reproductive failure.

Land-based sources of pollution (LBSP), directly caused by watershed discharges, have been identified by the US Coral Reef Task Force and other management-directed bodies as a primary cause of anthropogenic stress responsible for the decline of coral reefs worldwide. This is true for high islands and continental areas adjacent to coral reefs with well-defined

Coasts and Estuaries. https://doi.org/10.1016/B978-0-12-814003-1.00026-5

445

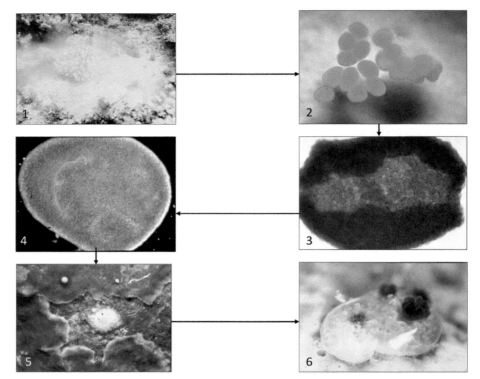

FIG. 1   A sketch of the six stages in the successful reproduction and recruitment of corals.

watersheds, as well as atolls and other low islands where watershed boundaries are less distinct. Coastal water quality can be compromised via both surface runoff and ground water discharges. Based on the size of the catchments, the extent of the effects of discharges can range from tens of meters to over 100 km.

2   MAJOR CONTRIBUTORS TO WATERSHED DISCHARGES

Urbanization, agricultural and industrial development, and housing and road construction are among the activities responsible for the alteration of watershed characteristics and hence, patterns of rain catchment and flow. The sediment and toxicant loads of discharges are tied to the specific activities that occur within watersheds, and the volume and velocity of flow in streams and rivers are directly related to the physical ground characteristics of permeability, rugosity, and slope. Reductions in the amounts of permeable surfaces through which rainwater can percolate are often a problem associated with housing, road, and industrial development. While agricultural development usually preserves permeable surfaces, alterations in plant ground cover and topography via plowing, terracing, and berm construction can enhance erosion and subsequent sedimentation. To prevent flooding, drainages within watersheds are often altered, with the increases in non-permeable surfaces adding substantial volume to runoff, and the straightening of streams and rivers, along with bottom hardening (lining channels with concrete) increases the velocity of water flow. Both of these activities result in greater impacts to coastal coral reefs and associated ecosystems, such as seagrass beds, than would occur under natural conditions.

3   CONTENTS OF WATERSHED DISCHARGES

3.1   Freshwater

Freshwater runoff alone can cause mortality in corals, and being more acidic than seawater, can result in the erosion of reef structures. The osmotic stress associated with freshwater runoff can cause cells to burst, and corals will often bleach and die following episodes of elevated freshwater exposure (Jokiel et al., 1993). Coral reef echinoderms are especially sensitive to drops in salinity, as they have no ability to osmoregulate.

As watershed discharges include freshwater, and freshwater floats on seawater, surface salinity is affected and even a 15% drop in surface water salinity from 34 to 28 ppt can prevent fertilization during spawning events (Richmond, 1994).

When sediment and/or toxicants are present, reproductive failure can occur with as little as a 5% drop in surface salinity tied to watershed runoff (Richmond, 1994).

3.2 Sediment

Sediment is one of the most damaging stressors affecting coastal coral reefs, and the effects can extend far offshore, with satellite images identifying plumes over 50 km from the source (Devlin and Brodie, 2005). Sediment includes a variety of particles that vary in grain size, composition, and physical attributes such as density and roughness. Watershed discharges can move materials including clay and colloidal-sized particles (<0.002 mm), silt (between 0.002 and 0.06 mm), sand (0.06–2 mm), gravel (2.0–63 mm), cobbles (63–200 mm), and boulders (>200 mm). Following large rain events, boulders over 1 m in diameter can be carried down channels and streams. The grain size and characteristics of sediment determine the types of impacts that will occur on affected reefs.

Sedimentation on coral reefs is measured three different ways: suspended sediment concentration (SSC; turbidity), which is expressed as weight per unit volume (mg/L), sedimentation rate is expressed as weight per area per unit of time (e.g., $mg\,cm^{-2}\,day^{-1}$), while sediment deposition is a linear measure of thickness (mm or cm).

These different parameters affect corals in different ways. The smallest particles, in the clay and colloidal size range, remain suspended the most in the water column and reduce the amount of sunlight reaching the corals. This turbidity cuts down the ability of the coral's algal symbionts, zooxanthellae, to photosynthesize and produce energy for the host coral. As light levels decrease, the algal symbionts shift from photosynthesis to photorespiration which may lead to oxidative stress in the host coral. Turbidity has a particularly negative metabolic effect on branching corals, as they have limited surface areas for light collection, and their colony morphology is more efficient at shedding particles of sediment. Plating and mounting corals are also affected by turbidity, but more so by the accumulation of particles including those in the sand size class that can be difficult to clear using mucus and cilia and may also be abrasive in the presence of wave action.

Fine particles are often a source of toxicant exposure to corals, as a variety of pollutants such as pesticides and hydrocarbons can adhere to these particles on land, with the chemicals being released once the particles enter the ocean or come into contact with corals. This is particularly true for non-water soluble compounds (lipophilic—"fat loving") that have a great affinity for living tissue with lipid content, such as cell membranes, eggs, and larvae present within coral colonies. Such contaminated particles can be ingested by coral polyps during heterotrophic feeding.

3.3 Toxicants

A variety of pollutants can be carried in watershed discharges, and these pollutants affect corals and other reef organisms. The impacts may include outright mortality or more subtle, sublethal effects including reduced growth rates, shifts in competitive ability against other organisms, reductions in fecundity, and, as previously mentioned, reproduction and recruitment failure. Water-soluble compounds can be carried long distances along discharge gradients, affecting corals km from shore. The nonsoluble compounds often adhere to organic material in runoff water, as well as to sediment particles, and are released once they come into contact with seawater.

Common classes of pollutants contained within watershed discharges include pesticides (both insecticides and herbicides), polycyclic aromatic hydrocarbons (PAHs; petroleum-based compounds including gasoline, diesel, and motor oil), personal care products (shampoos, cosmetics, hair care products, sunscreens), pharmaceuticals, and heavy metals.

A number of studies have been performed on insecticides and herbicides, and many are described in the accompanying safety information as being toxic to aquatic life. The organophosphate pesticide Chlorpyrifos, used on golf courses and in homes, was found to inhibit recruitment of planulae of the common Pacific coral *Pocillopora damicornis* (Peters et al., 1997). Experiments showed that at sublethal levels, exposed larvae didn't settle and metamorphose, and if "clean" larvae were presented with appropriate substrata that had been exposed to this chemical, recruitment was also affected. Herbicides can affect corals and zooxanthellate larvae by interfering with the ability of the algal symbionts to photosynthesize. Some insecticides are endocrine disrupting compounds, and it is no surprise that they can interfere with reproduction in corals, as these organisms possess the same proteins and metabolic pathways as found in higher organisms. The estrogenic compound estradiol 17β has been identified in waters surrounding corals during spawning events, and is also a pollutant associated with sewage effluent from un-metabolized birth control pills (Atkinson and Atkinson, 1992).

Petroleum products are very common in watershed discharges, and can be seen as a colored sheen in road drainages. Automobiles are a major source of these compounds and may also add heavy metals including cadmium from tire wear. Even when care is taken to educate consumers on the proper disposal of waste oil, small leaks multiplied by the number of cars operating within a watershed can contribute large amount of PAHs to coastal waters.

Personal care products are among the emerging chemicals of concern. Recent research on the UV absorbing compound oxybenzone found in many sunscreens and types of makeup, has demonstrated negative effects on coral reproduction, larval development, recruitment, and even colony viability (Downs et al., 2016). Parabens and a variety of compounds found in shampoos, soaps, and other personal hygiene products are being studied for the negative effects on a variety of marine organisms including corals. Chemicals used within watersheds will eventually make it to coastal waters via surface runoff, groundwater discharges, and sewage. Sewage outfalls are recognized as point sources of pollution, however, cesspools, septic tanks, leaky pipes, and sewage overflows following rain events are responsible for substantial quantities of chemicals and nutrients entering the ocean via streams, rivers, and groundwater.

3.4 Nutrients

Nutrients have been shown to negatively affect corals and coral reefs. They can alter the balance within the coral holobiont by increasing the division rate of zooxanthellae to the point where these algal cells become self-shading and the photosynthetic efficiency drops. At night, zooxanthellae respire oxygen and can cause oxidative stress in the coral host. Additionally, nutrients can alter another major component of the coral holobiont: bacteria. The important role of bacteria has been identified for providing recruitment cues for coral planula larvae, and for coral metabolism as has been determined in higher organisms. Nutrients also support the population growth of a variety of organisms that bio-erode corals and reef structures, such as sabellid and serpulid worms ("feather duster" and "Christmas tree" worms), boring sponges (clionids), and boring bivalves (e.g., Lithophaga). These organisms undermine the reef framework and accelerate the impacts of ocean acidification associated with climate change. Finally, there is a link between nutrients, plankton, and population outbreaks of the corallivorous Crown-of-Thorns starfish, *Acanthaster planci*, whose planktotrophic larvae benefit from increased food availability, which enhances their survival and recruitment rates.

3.5 Pharmaceuticals

Pharmaceuticals are another source of stress to corals, which originate within adjacent watersheds. The major categories of medications found in coastal areas and ground water include analgesics/anti-inflammatories, antidepressants, antibiotics, steroids, and those used to treat high cholesterol and blood pressure. While data are presently limited for the effective concentrations and potential synergisms among these compounds, recent studies of coral proteomics, genomics, metabolomics, and transcriptomics demonstrate the high level of concern warranted. The same problems expressed regarding the effects of nutrients on the bacterial abundance and diversity patterns on coral reefs holds for antibiotics. As more knowledge is gained about the important role of bacteria for corals and coral reefs, from mediating recruitment of larvae to the metabolic role of these microorganisms for the coral holobiont, it will be imperative that management efforts include plans for reducing exposure to such bioactive compounds. There are already programs in place to allow pharmacies and medical facilities to collect unused prescriptions for proper disposal.

3.6 The Sum of Stressors and Implications for Interventions on Land

With all elements of watershed discharges, it is important to separate the mean values of exposure to the extremes, and acute versus chronic. Annual rainfall values for any given watershed are not evenly distributed among days and months of the year, but rather, occur seasonally and in an episodic fashion. For example, a hurricane (typhoon or tropical cyclone) can dump over a meter of rainfall within several hours. These events can have drastic effects on coastal coral reefs and associated communities. From a biological perspective, these events can vastly overwhelm the organisms' ability to survive high levels of stress from salinity changes, toxicant levels, sediment loads, turbidity, and even temperature. Infrastructure has to be built, such as adequate water retention basins, which can address these larger storms to reduce the ecosystem damage associated with extreme events and enhance resilience of coastal marine communities.

4 THE KEY ROLE OF COASTAL OCEANOGRAPHY

Receiving water characteristics determine the spatial and temporal scales of watershed discharge impacts as well as their magnitude. Coral reefs within enclosed lagoons, with longer water residence times and lower flushing rates will respond differently than those along the open bays and coasts. Current patterns and wave action are also important parameters that affect coastal water quality. Waves can wash sediments away, but are also responsible for resuspension of accumulated particles. A study of the island of Molokai in Hawaii found 1 kg of sediment can have the impact of hundreds to thousands of kg of sediments due to resuspension. The size distribution of sediment particles will affect the amount of resuspension that occurs and the associated levels of turbidity. Sediment texture will affect the abrasive effects on corals that can occur during resuspension events.

Substratum characteristics are a major determinant of watershed discharge effects over longer time periods. Substantial fleshy and filamentous algal substratum cover serves as a reservoir of sediments that can be resuspended numerous times whenever there are waves. Once sediment becomes trapped on the bottom, flushing is inhibited and turbidity will have long-term consequences even in the absence of rainfall and the discharge of new sediments into the water column. The repeated vertical movement of accumulated sediments magnifies the negative effects of sedimentation. Bottom topography is another characteristic of coastal areas that effects sediment retention time and flushing rates.

5 CASE HISTORIES

5.1 Maunalua Bay, Hawaii

Maunalua Bay is located on the island of Oahu, Hawaii (Figs. 2 and 3). It is approximately 8 km long, with nine distinct sub-watersheds. It is surrounded by a high degree of development that began in the 1950s, with a mixture of residential communities, shopping malls, a marina, and a population of approximately 60,000 stakeholders. Flood control measures within the sub-watersheds to protect homes and businesses include channelization (straightening) of streams, with eight of the nine additionally having been hardened with concrete floors and walls. These engineering design approaches are effective at flood prevention, but do so at an extreme cost to coastal coral reef communities. The design allows for smaller rain events to fill the channels with contaminant-laden sediments that do not travel all the way to the ocean, but rather collect, dry due to evaporation and become higher in toxicant concentration with each smaller event. When a large rainfall occurs, these highly toxic muds enter into the coastal ocean, overwhelming the affected biota.

Following a heavy rainfall event in 2008, approximately 20 tons of sediment were deposited within the bay from the Kuli'ou'ou watershed due to erosion, transport, and settlement (Wolanski et al., 2009). Efforts to mitigate and reduce such excessive levels of coastal sedimentation include increased retention of rainwater on land through expansion and maintenance of ponding basins, the use of rain barrels for homes, and "softening" projects to allow runoff water from hardened surfaces to percolate back into the groundwater areas.

Maunalua Bay contains a large biomass of invasive algae, which trap sediments that are resuspended during wave events (Smith et al., 2002). In 2009, a project was undertaken to remove the invasive alga *Avrainvillea amadelpha* from the reef flat adjacent to the Paiko Lagoon to allow the flushing of accumulated sediments (http://www.malamamaunalua.org/habitat-restoration/the-great-huki/). Approximately 10 ha were cleared, removing 1.4 million kg of the mudweed. A study of the impacts of the algae removal found that instead of continuing to accumulate, the flushing rate was reduced to 4 years for the fine sediments and 6.6 years for the course particles (S. MacDuff, in prep.). While the value of the intervention was substantial, benefits will only be realized if such removal efforts are coupled with land-based remediation efforts to prevent additional sediment inputs from the adjacent watersheds.

5.2 La Sa Fu'a Watershed in Humåtak, Guam

Guam is a US territory located in the Western Pacific. The village of Humåtak (English: Umatac) is located in southwest Guam. It is a home to a community of 782 residents (U.S. Census Bureau, 2010), mainly from a handful of family clans who have lived in the village for generations. Traditional practices are perpetuated in the village through events such as fiesta celebrations and annual Guam Heritage and History Day festivities.

In 2002, fishermen in Humåtak noticed a decline in the size and quality of their catch. The Humåtak Mayor's Office reached out to University of Guam (UOG) Marine Laboratory researchers who identified accelerated land erosion, caused by poor land-use practices, as the likely cause of fish decline and environmental degradation in the area (Richmond et al., 2005). Coral reefs in the Humåtak area were severely impacted when the Agat-Umatac Road (Fig. 4) was reconstructed through the village from 1988 to 1990 at 183 m above sea level on steep sloping terrain (Richmond, 1993). The normal local government protocol to require environmental protection plans for this effort were politically waived and the resulting sedimentation caused by poor project management and severe degradation of the natural drainage system caused catastrophic coastal sedimentation and chemical (increased nutrient) pollution.

The Agat-Umatac Road was built without proper stormwater management and lacked engineering to slow, retain, and filter freshwater runoff before entering the ocean. A steep 1.67 km gutter bordering a section of the road drains almost directly into the La Sa Fu'a River causing high-velocity flash flooding with the power to rip 30-year-old coconut trees and

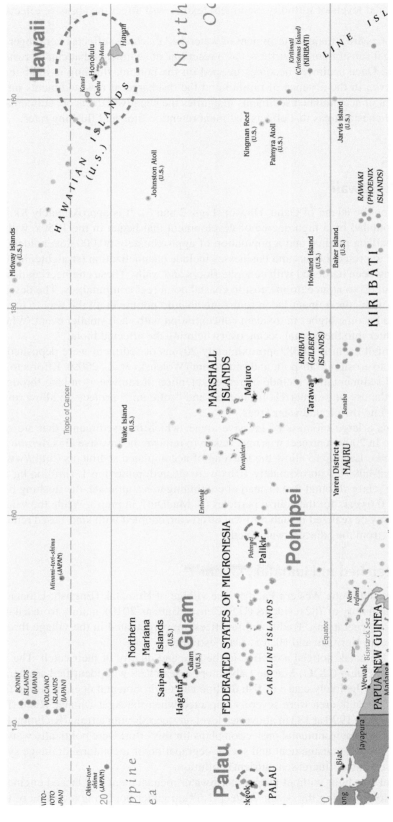

FIG. 2 A general location map of the Western Pacific showing Palau, Guam, Pohnpei, and Hawaii.

FIG. 3 Kuliouou watershed, showing poor land-use practices, ineffective sediment barriers, stream hardening and channelization, and resulting sediment/toxicant laden discharges.

FIG. 4 (*Top*) Fouha Bay during a typical sunny day. (*Bottom*) Fouha Bay after a large rain event. The sediment yield into Fouha Bay is 1714 tons per year (Rongo, 2004). *(Bottom photo by J. Lawrence, USDA NRCS Pacific Island Area.)*

FIG. 5 Mounds of soil left exposed during the 1988–1990 construction of the Agat-Umatac Road. This poor land-use practice resulted in the smothering and killing of corals after heavy rains washed the sediment into the ocean.

other vegetation from the riparian habitat. Large quantities of sediment were commonly washed into the ocean during heavy rains (Fig. 5). During low-wind conditions this allowed the sediment to be deposited on nearshore coral reefs without disturbance for weeks. Many corals were smothered and killed, and large areas of reef habitat shifted from a coral-dominated to an algal-dominated system, displacing populations of reef fish (Richmond, 1993; Wolanski et al., 2004).

Other anthropogenic impacts, such as fires set by deer poachers and the expanding population of feral ungulates, further remove groundcover and accelerate land erosion, preventing the recovery of the ecosystem without intervention (Wolanski et al., 2003).

6 REMEDIATION MEASURES

6.1 Humåtak Project

Concern for environmental degradation in the village gave rise to the Humåtak Project (www.humatakproject.org), a community-based environmental restoration effort. Restoration activities began in the La Sa Fu'a sub-watershed in Humåtak village. The sub-watershed catchment area is 3.2 km^2 and mostly undeveloped with the majority of Humåtak residents living in other areas of the village. The Humåtak Project pursues the following objectives:

1. Work with communities to address the environmentally damaging human behaviors and poor land-use practices responsible for the degradation of the coral reefs and associated fisheries of Fouha Bay.
2. Develop and test methods for reducing sediment inputs into the bay that are both economically feasible and culturally acceptable.
3. Evaluate the effectiveness of management-directed activities.

6.2 Community Engagement

Hundreds of volunteer hours have been contributed to educational outreach and watershed restoration activities since the start of the Humåtak Project. A six-part strategy was developed to form partnerships and engage community members in watershed restoration efforts: (1) establish trust with village leaders and residents, (2) educational outreach in classrooms, community events, and the field, (3) holding community meetings, (4) incorporating respect of culture, (5) promoting project awareness, and (6) maintaining interest through digital media.

Developed as a Humåtak Project signature outreach activity, Humåtak Watershed Adventures are interactive field trips that provide first-hand experiences for students and other members of the community to connect to their environment. Participants learn about natural resources, environmental threats, and the community actions needed to address these threats. Local terrestrial and marine scientists and managers deliver lectures to the participants at different sites throughout the adventure. Humåtak Watershed Adventures start in the upper watershed where causes and sources of erosion are showcased. In the mid-watershed, participants see where the Agat-Umatac Road and the La Sa Fu'a River intersect, and the importance of managing stormwater runoff is discussed. Participants conclude the adventure with a hike down to Fouha Bay to complete the comprehensive overview of watershed ecosystem connections.

6.3 Watershed Restoration Efforts and Its Effectiveness

As part of the Humåtak Project research efforts in Guam, the effectiveness of two watershed restoration tools was evaluated over a 21-month period between 2013 and 2015 (Shelton and Richmond, 2016). Tree seedlings and sediment filter socks (Fig. 6) were installed in eroding hillsides in the La Sa Fu'a Watershed by Humåtak Project volunteers. A soil probing method was developed to measure soil depth in order to determine how much sediment the watershed restoration tools trapped. An estimated 130 tree seedlings and 54 m of sediment filter socks trapped 112 tons of sediment on land in the main restoration study plot. In highly eroding badland sites where socks and trees were used in combination, the mean sediment trapping rate was $44 \, \text{kg} \, \text{m}^{-2} \text{year}^{-1}$.

In order to estimate the amount of restoration necessary to improve the coral reef health in Fouha Bay, Humåtak, the sediment delivery ratio (SDR) for the La Sa Fu'a Watershed was studied. The SDR is the proportion of gross eroded sediment in a catchment basin that reaches an outlet (SDR=sediment yield/gross erosion) (El-Swaify et al., 1982; Walling, 1983). Only a fraction of gross erosion reaches a watershed outlet because there are a series of sediment sources and sinks along the path of sediment transport.

Minton (2015) determined that a 75% reduction in sediment yield would be necessary to bring all of Fouha Bay below the severe-catastrophic sedimentation degree of impact ($>50 \, \text{mg} \, \text{cm}^{-2} \text{day}^{-1}$) proposed by Pastorok and Bilyard (1985). Assuming sedimentation rate correlates with sediment yield, a sediment-input reduction of 1379 tons would be required. Using a conservative SDR of 0.65, from the lower range of SDRs determined for other Guam watersheds (Hanson et al., 2007; NAVFAC, 2010), a 2121-ton sediment reduction would be required. The highest trapping efficiency of sediment filter socks and acacia trees (2 trees per $9 \, \text{m}^2$) observed by Shelton and Richmond, 2016 was $44 \, \text{kg} \, \text{m}^{-2} \text{year}^{-1}$. Therefore, the necessary sediment yield reduction could be achieved by treating a total area of $0.05 \, \text{km}^2$ with 19 km of 8 in diameter sediment filter socks and 11,000 acacia tree seedlings.

The decrease in sediment loading in turn will enable the recovery of corals in Fouha Bay within a stress gradient at a distance from the river mouth (Fig. 7).

Acacia tree seedlings and sediment filter socks are just two watershed restoration tools. Further studies that evaluate the sediment trapping efficiency of various plants and engineered tools will help grow the toolbox that natural resource managers use to restore watershed ecosystems.

7 CONTINUING EFFORTS

As a result of the growing awareness of Guam's environmental issues, multiple environmental initiatives have been undertaken in Humåtak. The Humåtak Project carries out efforts in partnership with the Humåtak Community Foundation, an organization which established the Umatac Coral Reef Ambassador Program to provide educational outreach on the

FIG. 6 Time series photographs of Humåtak Project restoration plot. An eroding hillside was chosen for treatment. Sediment filter socks were installed and *Acacia auriculiformis* tree seedlings were interspersed between the socks. The area was nearly completely revegetated after 11 months.

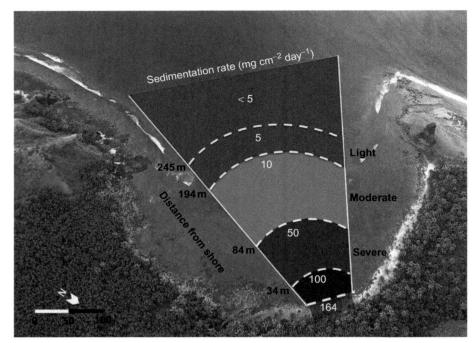

FIG. 7 The modeled sedimentation rates (Rongo, 2004; Minton, 2015) in Fouha Bay at increasing distance from shore and the corresponding degrees of sedimentation impact on coral reefs proposed by Pastorok and Bilyard (1985). *(Aerial image by John Jocson.)*

conservation and preservation of village resources. Originally coordinated under a National Oceanic and Atmospheric Grant to the Kewalo Marine Laboratory at the University of Hawai`i at Mānoa, the Humåtak Project now continues under the Guam Restoration of Watersheds (GROWs) Initiative led by the UOG Sea Grant Program and UOG Center for Island Sustainability.

7.1 Palau: Ngerikiil Bay

Palau is a sovereign nation in the Western Pacific that retains many cultural practices tied to natural resource sustainability. The largest island in the archipelago, Babeldaob, is volcanic in origin, with high relief and well-defined watersheds. The red soils are highly erodible following rain events. The population of Palau is about 20,000 with 14,000 Palauans and 6000 foreign workers. The coral reefs of Palau are world-renowned, making the nation an international dive destination. The near doubling of tourism from an annual 100,000 visitors per year to nearly 200,000 tourists has spurred hotel development, and a new road has opened once remote areas to a variety of agricultural and commercial activities.

The expanding tourism industry created a need for more housing near the main town of Koror, and due to extended family ownership of land, state owned areas, specifically intertidal Mangrove forests, were targeted for development. Mangroves trap about 30% of the sediment that runs off the land following rainstorms, and these were being removed in the village of Airai for the additional housing, resulting in extensive sedimentation of coastal reefs in Ngerikiil Bay (Golbuu et al., 2011a, b). Local fishers were among the first to notice the damage to reefs and the associated fisheries, creating a set of user conflicts.

Studies performed in Airai Bay, led by researchers from the Palau International Coral Reef Center, with active participation from local villagers, documented the amount of sediment deposition, resuspension, accumulation occurring following rain events, and the resulting effects on coral reefs and fisheries resources. A meeting was subsequently held with the stakeholders, including traditional and elected leaders, women's groups, and fishers. The outcome was a moratorium on the clearing of mangroves in Airai Bay, which eventually resulted in national legislation protecting mangroves in recognition of their value in protecting coastal coral reef and related resources of cultural, economic, and ecological value (Richmond et al., 2012). The success of this community-based effort led to further discussions on protecting coral reefs from land-based sources of stress, and the emergence of the Belau Watershed Alliance as a coordinating body for devising and implementing solutions.

FIG. 8 Upland taro field designed to retain runoff and slow down the outflow, and capture sediment in Palau. *(Photo courtesy of Faustina K. Rehuher-Marugg.)*

Taro fields were identified as a potential tool for further protecting coastal coral reefs from the effects of rainwater runoff and sedimentation. Taro is a plant that has been traditionally cultivated in Palau. The leaves are used in a local soup and its underground swollen stem is a starch that is a staple in the diet of Palauans and many other Pacific Islanders. As such, it is an important resource supporting local food security. Several varieties are cultivated and have different characteristics including optimal growth conditions (e.g., wet vs. dry). In addition to its value in the Palauan diet, activities associated with cultivation are a cornerstone of the culture, especially among women who tend the fields. Social and familial interactions are a major benefit that accrues from traditional taro cultivation.

Palau recently experienced storm and king tide events that flooded taro fields located in low-lying coastal areas, with the seawater intrusion resulting in loss of the crops. Such events are predicted to increase in frequency and magnitude as a result of sea level rise tied to global climate change. A study was performed on the use of taro fields as a buffer between land and sea, which included moving these higher up within watersheds. Measurements found that they trap between 60% and 90% of sediment in runoff, two to three times that of mangroves. The higher positioning within watershed reduced the volume and velocity of runoff and hence, the amount of sediment carried to the ocean. Two narrow channels constructed perpendicular to the fields trapped coarse and fine sediment, respectively, providing an easy means of harvesting this material for other purposes. By moving taro fields further up into watersheds, sedimentation impacts to coastal coral reefs have decreased, crops were protected from seawater inundation, and food security and cultural benefits have increased (Fig. 8) (Koshiba et al., 2014).

7.2 Enipein Watershed, Pohnpei

Pohnpei is one of four states in the Federated States of Micronesia. It is one of the more traditional islands in the region, with limited urbanization and tourism. It is a high, volcanic island, surrounded by a barrier and fringing reefs, with extensive coastal mangroves. Farming and fishing are staples of the local communities and economy, with sakau (kava), a kavalactone-containing pepper plant root, as the major cash crop. The root can be powdered or crushed in water to make a drink with narcotizing properties, such as muscle relaxation and numbness of the mouth and lips.

Pohnpei is a wet island, with an annual rainfall in the mountain rainforests of approximately 7600 mm. Sakau grows well in the rainforest, which has resulted in the clearing of these upland areas for cultivation. The result has been high levels of erosion of the red, volcanic soils, with resulting sedimentation impacts on coastal coral reefs. A major soil slumping event following heavy rains, resulted in the death of several children in a landslide.

A study of the impacts of high sedimentation rates found that both reef health and coastal fisheries were affected by the upland agricultural activities within the adjacent watersheds (Victor et al., 2006). Data collected from a locally managed marine protected area found no enhancement of herbivorous fish populations compared to adjacent areas that were regularly fished, which was attributed to the heavy sediment loads causing high levels of turbidity. This affects reef habitat quality, the ability of fish to clear their gills and to see approaching predators that use their lateral line systems that act as underwater radar, rather than sight to cue in on their prey.

A community-based approach was taken by a local NGO, the Conservation Society of Pohnpei, to move the sakau cultivation from the rainforest to low-lying areas ("grow low" program) in an effort to protect existing vegetation and support recovery of the forest. The approach garnered support from the farmers as well as the traditional and elected leaders (Richmond et al., 2007). This example served as a model for other communities in the region, with a focus on the need to address challenges through integration with local cultural practices.

8 A SYNTHESIS: SUCCESS AND FAILURES OF DIFFERENT APPROACHES

Approaches to protect coral reefs from watershed discharges include development and implementation of proactive policies, improvement of onsite retention of water and sediment on land, modified agricultural practices, and active intervention on reef sites to remove algae and facilitate flushing of accumulated sediments that are frequently resuspended. There were three main criteria applied to addressing the problems associated with watershed discharges on coral reefs: effectiveness, economic feasibility, and cultural acceptability.

The approach in Maunalua Bay, Hawaii, was the most challenging approach based on the scale of the problem. The algal removal effort was costly ($3.4 million US) and labor intensive, but successful in facilitating the flushing of sediment accumulated over decades. Bays act as bank accounts with regards to sediment, and the overall impacts over space and time are the result of the dynamics of deposits versus withdrawals. Thus, it is important to pair the responses to reduced inputs with those that facilitate flushing and reduce the problems associated with resuspension of accumulated materials. It is also important to reduce the magnitude of extreme events which can overwhelm the ability of corals to survive such pulses. The removal of benthic algae addressed the retention and resuspension issues, but not the flow of additional sediment into Maunalua Bay. More work needs to be performed within the adjacent watersheds to improve the flood control infrastructure, such as increased retention basins and softening of impermeable surfaces to allow percolation into the aquifer.

The projects in Guam demonstrated the value of revegetation and the use of sediment socks in reducing erosion and sedimentation into Fouha Bay. Much of the damage that occurred in the past could have been prevented by installing water catchment basins during and following road construction. Working with specific stakeholders, in this case hunters responsible for starting fires within the watershed, was also an effective approach. Finally, working with fishers to allow populations of grazing fish to recover and reduce benthic algal cover was found to be helpful. At all levels, community engagement was found to be a critical element for success.

The efforts in Palau and Pohnpei also included strong cultural and community elements that led to success. Bridging science and knowledge to policy is often a challenge, which in this case, was helped by the leadership of Palauan scientists who could present research findings in the Palauan language and in the appropriate cultural context. In addition to enhancing mangrove protections and using taro fields in a novel way, the Palauan government reviewed their existing regulatory framework and identified a problem with clearing and grading permits. Agricultural development was traditionally exempt from such permit requirements, which was appropriate when farms were small, at the family and community level. With the growing produce demands by the tourism sector, foreign-run commercial farms of hundreds of hectares or more sprang up, along with ineffective land-use practices.

9 MAJOR SOCIOECONOMIC-CULTURAL LESSONS LEARNED

The results of field studies in Guam, Hawaii, Pohnpei, and Palau supported several overarching conclusions. The first, and most obvious, is the need for integrated practices on land to reduce erosion of soils and prevent freshwater runoff, sediment, nutrients, and toxicants from reaching coastal and offshore coral reefs in the first place. These include onsite retention of water, improved agricultural practices, and protection of coastal mangroves and seagrass communities where they are part of the natural environment. Once sediment is deposited on coral reefs, resuspension causes a multiplier effect, whereas 1 kg of sediment can have the impact of tons. Benthic fleshy and filamentous algae become a reservoir for these sediments, reducing natural flushing rates, hence the value of protecting herbivores such as sea urchins and grazing species of fish. This requires cooperation of fishers, who need to understand the benefits associated with any sacrifice on their part. Voluntary compliance is a far more effective and efficient solution compared to using enforcement officers and the judicial system. Fishers, who understand the need for protecting populations of herbivorous reef fish, become part of the solution. Finally, the value of community engagement, outreach, and education is clear and essential for the success of watershed-based initiatives that support coral reef protection and recovery. Including and in some cases, revitalizing traditional cultural practices, such as the culture of taro, can be valuable approaches to improving watershed management, yielding multiple benefits to the participating communities.

While education and compliance are valuable tools, there is and always will be a need for sound legislation and regulation.

10 THE FUTURE: CLIMATE CHANGE ISSUES

Climate change is one of the main drivers of global coral reef losses through associated elevated seawater temperatures responsible for mass bleaching events and ocean acidification from elevated levels of atmospheric CO_2. However, this global driver also affects the impacts of local stressors, particularly by increasing the magnitude of watershed discharges. Predictions of increasing storm strength have come true, with an expectation that the trend will continue in the future. The result is more extreme events with larger amounts of rainfall, which in turn, increase the spatial and temporal scale of runoff impacts that exceed the ability of corals and reef communities to survive. Each extreme event that exceeds the adaptive and acclimation capacities of corals and other reef organisms results in mortality, the loss of genetic diversity, and reduction in ecosystem resilience in both the short and long term.

While local stressors operate at the level of tens to hundreds of km^2 climate change impacts have been affecting coral reefs at the regional level, at scales of thousands of km^2. Reefs far from watershed discharges and sources of coastal pollution have generally experienced better recovery rates than those nearer to shore, however, those surviving coral populations have lower genetic diversity. Additionally, reduced population size decreases that chance of successful sexual reproduction in corals as the distance among conspecific colonies increases following mass mortality events. Finally, surviving genotypes with resistance to elevated temperatures may not have a comparable resistance to diseases or other stressors, setting the stage for local and regional extinction events.

11 EVALUATION OF MITIGATION: METRICS OF SUCCESS

An important role for science in addressing the impacts of watershed discharges and other stressors on coral reefs is developing metrics for measuring the effectiveness of mitigation and conservation measures on reef vitality. In a world of limited financial, human, and institutional resources, it is critical that the allocation of these is performed in a manner that maximizes success. For example, if the target for sediment reduction is identified as needing to be 40% below existing peak levels, a 20% reduction shows progress, but may result in no measurable improvement. The goal then becomes finding ways to achieve the additional improvements.

Traditionally, coral reef monitoring programs have focused on percent coral cover or species abundance to detect changes for better or worse. Such measurements use mortality as the indicator of change, require longer times periods (years) for evaluation and miss subtleties such as genotypic diversity. New tools and technologies, specifically the "omics"— proteomics (protein expression), genomics (genetic diversity), and transcriptomics (gene expression) provide real time, quantitative data on molecular responses of corals and other organisms. Such tests are used in human medicine as diagnostic tools and can likewise be applied to corals (Downs et al., 2012).

12 CONCLUSIONS

Watershed discharges from land-based sources of pollution remain among the major causes of coral reef decline. Sediments, toxicants, nutrients, and freshwater all contribute to the loss of corals and other reef organisms. Effective solutions require an integration of approaches that are based on physical, biological, environmental, cultural, social, and economic considerations.

Our experience in the Pacific Islands has shown the value of community engagement in finding workable solutions that also require a degree of political will. While laws and regulations are important, a number of solutions can be implemented by communities and key stakeholder groups based on the "best available science," concurrently move with formal governmental actions. Public outreach and education of consumers can help reduce the amounts of toxicants such as pesticides, oil compounds, and personal care products that enter the coastal environment. Retention of rainwater at individual households and housing complexes can be achieved through the use of rain barrels and rain gardens. Hydromulching, use of sediment socks, revegetation, and improvements in agricultural practices can also be undertaken without government intervention. Large-scale infrastructure projects are often needed, which require substantial funding, and can be initiated through a public process.

New diagnostic tools for quantifying coral health, modeled after techniques used to assess human health, can provide key data on the effectiveness of practices and establishing target threshold levels. These data can also be used to support and guide and update development, implementation, and evaluation of regulations to better manage land-sea connections. Recent studies documenting the value of coral reefs and their ecosystem services demonstrate that investments in protecting coastal water quality are economically defensible, especially in a world of climate change-induced sea level rise. There is ample evidence from studies presented here and in the literature that coral reefs can and have recovered from the damage caused by a combination of local and global stressors, and it is important to balance expressions of urgency with the understanding that given the opportunity, a legacy of vital coral reefs is possible through appropriate actions.

REFERENCES

Atkinson, S., Atkinson, M.J., 1992. Detection of estradiol-17β during a mass coral spawn. Coral Reefs 11, 33–35.

Devlin, M.J., Brodie, J., 2005. Terrestrial discharge into the Great Barrier Reef Lagoon: nutrient behavior in coastal waters. Mar. Pollut. Bull. 51, 9–22.

Downs, C.A., Ostrander, G.K., Rougee, L., Rongo, T., Knutson, S., Williams, D.E., Mendioloa, W., Holbrook, J., Richmond, R.H., 2012. The use of cellular diagnostics for identifying sub-lethal stress in reef corals. Ecotoxicology 21, 768–782.

Downs, C.A., Kramarsky-Winter, E., Segal, R., Fauth, J., Knutson, S., Bronstein, O., Ciner, F., Jeger, R., Lichtenfeld, Y., Woodley, C., Pennington, P., Cadenas, K., Kushmara, A., Loya, Y., 2016. Toxicopathological effects of the sunscreen UV filter, Oxybenzone (Benzophenone-3), on coral planulae and cultured primary cells and its environmental contamination in Hawaii and the US Virgin Islands. Arch. Environ. Contam. Toxicol. 70 (2), 265–288.

El-Swaify, S.A., Dangler, E.W., Armstrong, C.L., 1982. Soil Erosion by Water in the Tropics: College of Tropical Agriculture and Human Resources. University of Hawaii, Hawaii. 173 p.

Golbuu, Y., Wolanski, E., Harrison, P., Richmond, R.H., Victor, S., Fabricius, K.E., 2011a. Effects of land use change on characteristics and dynamics of watershed discharges in Babeldaob, Palau, Micronesia. J. Marine Biol. 2011, 17.

Golbuu, Y., van Woesik, R., Richmond, R.H., Harrison, P., Fabricius, K.E., 2011b. River discharge reduces reef coral diversity in Palau. Mar. Pollut. Bull. 62, 824–831.

Hanson, K., Robotham, M., Pedone, P., Lawrence, J.H., Gavenda, R., 2007. Final Sasaatantano Watershed Resource Assessment: Focus on Sedimentation and Erosion Estimates for Uplands Draining into Apra Harbor. USDA-NRCS Pacific Islands Area. 109 p.

Jokiel, P.L., Hunter, C.L., Taguchi, S., Watarai, L., 1993. Ecological impact of a fresh-water "reef kill" in Kaneohe Bay, Oahu, Hawaii. Coral Reefs 12, 177–184.

Koshiba, S., Besebes, M., Soaladaob, K., Ngiraingas, M., Isechal, A.L., Victor, S., Golbuu, Y., 2014. 2000 years of sustainable use of watersheds and coral reefs in Pacific Islands: a review for Palau. Estuarine. Coasta Shelf Sci. 144 (2014), 19–26.

Minton, D., 2015. Changes in Coral Reef Structure along a Sediment Gradient inFouha Bay. Final Report for National Oceanic and Atmospheric AdministrationPacific Islands Regional Office. Contract: NFFT5000-12-04727JR. 192 pp.

NAVFAC, 2010. Final watershed assessment for potential coral mitigation projects; Umatac, Toguan, Geus and Ugum Watersheds, Guam. Department of the Navy, Naval Engineering Command, Pacific. June 28, 2010. 109 p.

Pastorok, R.A., Bilyard, G.R., 1985. Effects of sewage pollution on coral reef communities. Mar. Ecol. Prog. Ser. 21, 175–189.

Peters, E.C., Gassman, N.J., Firman, J.C., Richmond, R.H., Power, E.A., 1997. Ecotoxicology of tropical marine ecosystems. J. Environ. Toxicol. Chem. 16, 12–40.

Richmond, R.H., 1993. Coral reefs: present problems and future concerns resulting from anthropogenic disturbance. Am. Zool. 33, 524–536.

Richmond, R.H., 1994. Effects of coastal runoff on coral reproduction. In: Ginsburg, R.N. (Ed.), Proceedings of the Colloquium on Global Aspects of Coral Reefs: Health, Hazards and History, 1993. University of Miami, pp. 360–364.

Richmond, R.H., 2005. Recovering populations and restoring ecosystems: restoration of coral reefs and related marine communities. In: Norse, E.A., Crowder, L.B. (Eds.), Marine Conservation Biology, the Science of Maintaining the Sea's Biodiversity. Island Press, Washington, DC, pp. 393–409.

Richmond, R.H., Rongo, T., Golbuu, Y., Victor, S., Idechong, N., Davis, G., Kostka, W., Neth, L., Hamnett, M., Wolanski, E., 2007. Watersheds and coral reefs: conservation science, policy and implementation. Bioscience 57, 598–607.

Richmond, R.H., Golbuu, Y., Idechong, N., Wolanski, E., 2012. Integration of social and cultural aspects in designing ecohydrology and ecosystem restoration solutions. Treat. Estuar. Coast. Sci. 10, 71–80. Elsevier Science Publishers.

Rongo, T., 2004. Coral Community Change Along a Sediment Gradient in Fouha Bay, Guam. University of Guam, Master's Thesis, 83 p.

Shelton III, A.J., Richmond, R.H., 2016. Watershed restoration as a tool for improving coral reef resilience against climate change and other human impacts. Estuar. Coast. Shelf Sci. 183, 430–437.

Smith, J.E., Hunter, C.L., Smith, C.M., 2002. Distribution and reproductive characteristics of nonindigenous and invasive marine algae in the Hawaiian Islands. Pac. Sci. 56, 299–315.

U.S. Census Bureau, 2010. 2010 Census of Populations and Housing. U.S. Department of Commerce, Economics and Statistics Administration.

Victor, S., Neth, L., Golbuu, Y., Wolanski, E., Richmond, R., 2006. Sedimentation in mangroves and coral reefs in a wet tropical Island, Pohnpei, Micronesia. Estuar. Coast. Shelf Sci. 66, 409–416.

Walling, D.E., 1983. The sediment delivery problem. J. Hydrol. 65, 209e237.

Wolanski, E., Richmond, R.H., Davis, G., 2003. Water and fine sediment dynamics in transient river plumes in a small, reef-fringed bay. Guam. Estuar. Coast. Shelf Sci. 56, 1029–1040.

Wolanski, E., Richmond, R.H., McCook, L., 2004. A model of the effects of land-based human activities on the health of coral reefs in the Great Barrier Reef and in Fouha Bay, Guam, Micronesia. J. Mar. Syst. 46, 133–144.

Wolanski, E., Martinez, J., Richmond, R.H., 2009. Quantifying the impact of watershed urbanization on a coral reef: Maunalua Bay, Hawaii. Estuar. Coast. Shelf Sci. 84, 259–268.

Chapter 27

Challenges and Opportunities in the Management of Coral Islands of Lakshadweep, India

Purvaja, R., Yogeswari, S., Debasis, T., Hariharan, G., Raghuraman, R., Muruganandam, R., Ramesh Ramachandran

National Centre for Sustainable Coastal Management, Ministry of Environment, Forest and Climate Change, Government of India, Anna University Campus, Chennai, India

1 INTRODUCTION

Islands cover approximately 1.86% of the Earth's surface (Depraetere and Dahl, 2007), with 11% of the world's population (730 million) in 2017. National Oceanic Atmospheric Administration's (NOAA) satellite-originated Global Shoreline Database shows that there are 180,498 islands (all pieces of land >0.1 km^2) of which less than 5% of the islands are inhabited (Kakazu, 2011). These islands hold around two-thirds of the resources (e.g., fisheries, forests, minerals, etc.) of the planet (McCall, 1996). However, it should be noted that these islands have not been considered as prime subjects in planning research (Kakazu, 2011). Planning for small islands needs to accommodate the increasing pressures from changing environmental, social, and economic conditions (Coccossis, 1987; Van der Velde et al., 2007). The limited options available in small islands for development as shown by Hein (1990) have resulted in many island governments focussing development on existing socioeconomic activities rather than resource management (Boniface and Cooper, 2009; Kakazu, 2011).

1.1 India's Lakshadweep Islands

India has 1382 islands, which includes two major island territories—the Andaman and Nicobar group of Islands (836 islands) in the Bay of Bengal, and the Lakshadweep group of Islands (32 islands; Fig. 1) in the Arabian Sea.

The Lakshadweep archipelago comprises the most extensive coral atoll system in the Indian Ocean. The Lakshadweep Islands lies in the west coast of India, forming the northern segment of the Chagos-Maldive-Laccadive ocean ridge. A common feature of these islands is the presence of a large shallow lagoon, coral reef, and the sea. Lakshadweep is the smallest union territory of India remotely located at 440 km from mainland India. Out of 32 coralline islands, 10 are inhabited including Agatti, Kavaratti, Amini, Andrott, Kalpeni, Kadmat, Kiltan, Chetlat, Bitra, and Minicoy (Fig. 1). The islands of Lakshadweep are classified under three clusters as (i) Aminidivi group; (ii) Laccadive group, and (iii) Minicoy group in addition to submerged reefs and submerged banks (Table 1).

The Lakshadweep Islands cover a land area of 32 km^2, lagoon area of 4200 km^2, the territorial waters of 20,000 km^2, and an exclusive economic zone area covering 400,000 km^2. Average rainfall in the islands is 1600 mm. The total population of the 10 inhabited islands is 64,429 (33,106 males and 31,323 females) as per the 2011 Census of India, with a population density of 2013 persons per square kilometer. The mode of accessibility is mainly through shipping. An airport is established at Agatti, helicopter facilities are also provided but primarily for medical emergency purposes for the islanders. The Lakshadweep Islands receive the fuel, food, and other supplies from mainland India. Agriculture and fisheries are the main economic activities in the islands, coconut being the principal crop; pole and line fishing are the traditional fishing methods practiced. This paper addresses the key challenges prevalent in these small islands and suggests opportunities for sustainable development of the islands, commensurate with conservation of its pristine natural resources.

Coasts and Estuaries. https://doi.org/10.1016/B978-0-12-814003-1.00027-7

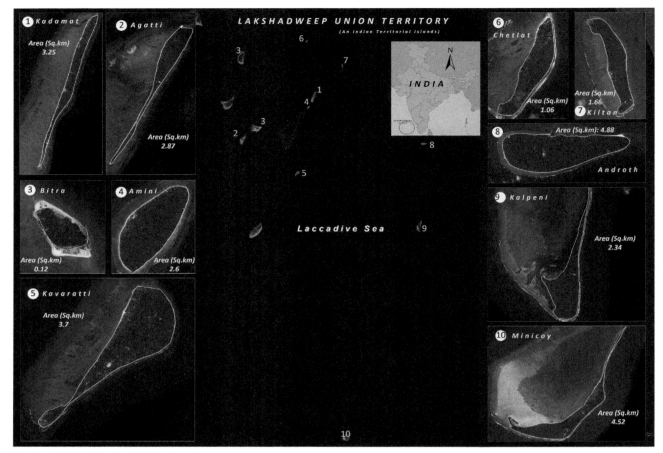

FIG. 1 Inhabited islands of Lakshadweep.

TABLE 1 Island Clusters and Names of Inhabited Islands of Lakshadweep

Name of the Cluster	Name of the Islands
Aminidivi group: 5 inhabited islands	1. Amini, 2. Kadmat, 3. Kiltan, 4. Chetlat, 5. Bitra
Laccadive group: 4 Inhabited & 12 Uninhabited islands	1. Androth, 2. Kavaratti, 3. Agatti, 4. Kalpeni, 5. Kalpitti, 6. Bangaram, 7. Tinakkara, 8. Parali, 9. Tilakkam, 10. Pitti, 11. Cheriyam, 12. Suheli, 13. Valiyakara, 14. PakshiPittiand, 15. Kodithala
Minicoy group: Two Islands	1. Minicoy, 2. Veiningili
Submerged Reefs	1. Beliapani (Chebeniani), 2. Cheriapani (Byramgore), 3. Perumul Par
Submerged Banks	1. Bassas de Pedro, 2. Sesostris Bank, 3. Cora Divh, 4. AminiPitti, 5. Kalpeni Bank

Data from Lakshadweep Action Plan on Climate Change, Department of Environment and Forestry, Union Territory of Lakshadweep, Supported by UNDP, 2012, pp. xiv–xix.

2 SWOT ANALYSIS

In order to understand the nature of challenges, a SWOT analysis (Table 2) focussing on the strengths, weaknesses, opportunities, and threats (SWOT), pertaining to the inhabited Islands of Lakshadweep, was made to determine appropriate strategies and interventions to propose an Integrated Island Management Plan (IIMP).

TABLE 2 SWOT Analysis of Lakshadweep Island and Its Resources

Strengths	Weakness
• Isolated location	• Small size of the Islands, with limited land availability
• Nature and climate	• Limited availability of water and energy
• Pristine environment	• Transport and communication
• Shallow lagoon with diverse coral reef and marine biodiversity	• Cultural and religious constraints
• Zero crime rate	• Accommodation facilities
• Polite, helpful and friendly local people	• Limited livelihood opportunities
	• Low carrying capacity
	• Shoreline management
	• Limited rights of the government on the land
	• Limited medical facilities
Opportunities	**Threats**
• Eco-friendly tourism	• Waste disposal (solid and waste water)
• Potential for water sports activities	• Oil spill in land and marine areas
• Potential for enhanced employment opportunities	• Salt water intrusion
• Scientific research	• Coastal erosion
• Preservation and conservation of unique marine species and biodiversity	• Planned, accidental or natural destruction of coral reef
• Potential for Blue flag beaches	• Climate change and sea level rise
	• Vulnerability to extreme weather conditions
	• Sociocultural issues
	• Vulnerable to epidemic diseases

The IIMP for individual islands is beyond the scope of this study. From our study of the inhabited islands, the nature of issues, challenges, and opportunities are identical in all aspects and hence the recommendations are provided for the 10 inhabited islands, rather than addressing them individually.

3   CHALLENGES

Despite the uniqueness and pristine environment of the islands, the smallness and its isolation from the mainland have resulted in problems associated with socioeconomic, environmental, and ecological concerns. Although environmentally and economically sustainable options of development can be effectively adopted, the sociocultural practices restrain from achieving it. Cultural restrictions are considered the biggest challenge in the expansion of tourism facilities in the inhabited islands. Some of the sector-wise key challenges in Lakshadweep Islands are given below:

(a) *Agriculture*: Small islands primarily depend on importing wide range of goods (capital and consumer goods), due to restricted resource base and production facilities, resulting in imbalanced trade economies. In the case of Lakshadweep, although it is primarily agrarian in nature, with coconut being the major crop of cultivation with high yield in terms of number, the size of the nuts is very small due to lack of nutrients in the soil. This coupled with inadequate irrigation facilities has limited the scope for agriculture and related employment.

(b) *Fisheries*: Fisheries are the secondary source of livelihood in Lakshadweep Islands of which, a large share of fishing is contributed by tuna fishery by conventional "pole line" fishing, using live bait fishes. The estimated fishery potential of Lakshadweep is 100,000 t, 50% of which comprises tunas. The current annual fish landings in Lakshadweep is about 15,000 t, which accounts for 15% of the estimated potential of the Islands. The fishing fleet comprises multi-day gillnet-cum-long liners (11), Maldivian fishing boats (11), Pablo boats (515), country crafts (752), and fiber-reinforced plastic (FRP) boats (83). The most popular gears in Lakshadweep include pole and line, hand line, gill nets, troll line, shore seines, and traps. The traditional pole and line fishery (from the Maldives and Minicoy) relies on an extensive range of coral reef and lagoon planktivores to use as live baits. These baitfish species range from a variety of fish families including clupeidae, caesionidae, apogonidae, pomentridae, atherinidae, and serranidae. Apart from Minicoy and Kavaratti, most other Lakshadweep Islands are solely relying on the clupeid bait fish Hondeli (*Spratelloides delicatulus*). *S. delicatulus* is often targeted at the time of spawning due to the ease of capture. The impact of this practice on their population is not clear, but declines in lagoon sprat and silverside populations have been observed. It is understood that there are no specific bait fisheries in the islands, and the individual fishers catch their own baits, leading to uncontrolled or overfishing of these live baits. Appropriate regulating measures need to be framed to reduce overfishing of the conventional bait fishes. Lack of infrastructure facilities have led to the reduction in fish catch in the territorial waters around these islands.

(c) *Island development and land availability*: Any kind of development in the islands is totally dependent on availability of land, fresh water, and energy (electricity) as also observed by Hall (2008) and Panakera et al. (2011). In Lakshadweep, there is limited availability of resources such as land, freshwater, energy (electricity) coexisting with the delicate ecosystem (coral reefs and lagoon) of the islands. Topography of the land, nature of the soil, drainage pattern along with the extent of available land area (0.16% of territorial water area) determines the nature of activities and development in these islands. This is further complicated by the presence of fragile ecosystems such as the coral reefs, seagrass, etc. that limit developmental activities. In addition, the islands are only 1–2 m above sea level and generally with a flat topography, hence are a major challenge for development.

(d) *Fresh water*: Freshwater resource is scarce despite high annual rainfall of 1600 mm. The key source of fresh water is groundwater, and rainwater harvested in ponds by individual household. However, the groundwater is highly contaminated due to the presence of septic tanks adjacent to the groundwater wells. The soils in atoll of Lakshadweep Islands are porous in nature, resulting in the lack of groundwater retention. The groundwater is available in the form of thin water lens varying from surface to 3.8 m below the ground level. The limited groundwater in these islands is also subjected to overextraction, resulting in sewage infiltration and saltwater intrusion (Planning Commission, 2007). Water management in small islands has been determined by various developmental issues along with environmental issues like disposal of solid waste and sewage. The quality of water plays a vital role in improving the environment of the island, along with its socioeconomic status (Van der Velde et al., 2007).

(e) *Sewage*: Considerable amount of sewage is being generated from the inhabited islands of Lakshadweep, which amounts to approximately 50,000–120,000 L/day. The households have conventional septic tanks for the collection of sewage. Discharge of waste water is through soak pits. But due to the porous nature of the soil, the wastewater seeps into the ground and contaminate the freshwater aquifer. As the freshwater extraction points such as dug well or bore well are located close to these septic tanks, the contaminated water seeps into the groundwater aquifer. They contain a high amount of nitrate along with harmful *Escherichia coli* bacteria, which are the primary cause for gastroenteric diseases prevalent in the inhabited islands of Lakshadweep.

(f) *Power*: The main source of power in the Lakshadweep Islands is through diesel generators. Diesel is transported to the islands and stored in containers on the islands. This is done well in advance for the energy requirement during the monsoon months, where transportation comes to a standstill due to harsh weather conditions. Transportation of diesel also contributes to the ecological damage, polluting both land and water in the fragile ecosystem. Further, improper storage of diesel results in the loss of fuel and ecological damage due to seepage of diesel through soil. The islands have also initiated harnessing power through renewable energy sources, primarily solar. The limited availability of land is a main constraint for establishing solar power stations.

(g) *Solid waste*: There is a considerable quantity of biodegradable and nondegradable solid waste generated; but data on the exact quantity of solid waste generated are unavailable. The biodegradable wastes primarily consist of waste from coconut retting, and nonbiodegradable waste consists of plastics, gunny bags, cement bags, glass, tin, metal electronic wastes, construction wastes, etc. (Fig. 2). The unused dried coconut biomass (dried husk, leaves or cadjans, shell, sawmill biomass, etc.) from the islands of Lakshadweep is a major environmental concern. Fishery waste is the second highest waste generated after coconut residue, which includes the heads and residues after removing the fillets. Usually these wastes are buried in pits dug near the seashore. Leaching from these pits enter the coastal waters leading to localized nutrient enrichment, oxygen depletion and increase in biological oxygen demand (BOD), and resultant localized eutrophication. Mass tourism results in large quantity of solid waste generation and the management of solid waste represents one of the most difficult challenges in the islands (Sealey and Smith, 2014; Rozelee et al., 2015).

FIG. 2   Types of degradable and nondegradable solid waste generated in the Lakshadweep Islands.

TABLE 3 Shoreline Change (in % of Total Length of Coast) Characteristics of the Inhabited Islands of Lakshadweep

Islands	Erosion	Artificial Coast*	Total Erosion	Accretion	Stable Coast
Agatti	38.32	24.48	62.80	36.82	0.38
Amini	18.92	59.35	78.27	21.23	0.50
Andrott	11.47	74.52	**85.99**	14.01	0.00
Bitra	55.19	16.80	71.99	28.02	0.00
Chetlat	32.54	20.32	52.86	47.14	0.00
Kadamat	13.08	57.39	70.47	29.38	0.15
Kalpeni	17.96	52.00	69.96	29.8	0.25
Kavaratti	4.47	78.98	**83.45**	16.55	0
Kiltan	23.02	36.89	59.91	38.9	1.18
Minicoy	29.26	19.02	48.28	51.32	0.39
Island Average	*24.42*	*43.98*	*68.40*	*31.32*	*0.29*

Artificial coasts indicate highly eroding sites that have been protected with tetrapods (seawalls). Highlighted numbers indicate highest erosion.

There have been a few initiatives taken up by the government in collection, segregation, and incineration of solid waste. However, there has been no record of the total quantity of waste generated and dumped on the ground. Leaching from the waste dump sites lead to ground water contamination, but the data are scant.

(h) *Coastal erosion*: Erosion is one of the key issues in all the islands of Lakshadweep. Almost all the islands are highly eroding at an average of 68% (ranging from 48% to 86%; Table 2). However, the intensity of coastal erosion differs for each island (Table 3; Fig. 3).

Factors attributing to coastal erosion in the islands of Lakshadweep are predominantly from human interventions (e.g., obtaining materials such as boulders, sand from the coast for construction activities, construction of jetties, etc.). In addition, increased wave action, particularly during monsoon, enhances erosion along the coasts of the islands. Apart from the above, unplanned and unscientific methods of shore protection, such as dumping of tetrapods (Fig. 4), hollow concrete blocks, and coir bags filled with pebbles/shingle along the coasts of the islands also make the islands more vulnerable to erosion.

(a) Sedimentation: Sedimentation in coral reefs is one of the major environmental concerns in the lagoon areas of the islands. Impact of motorized boats, unplanned anchoring, and mooring have caused churning and suspension and

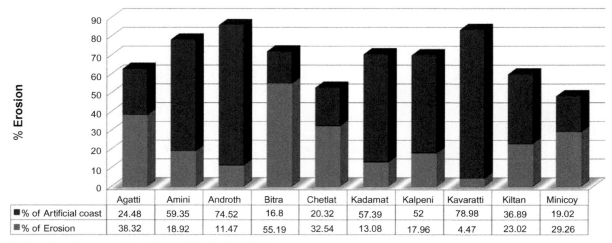

	Agatti	Amini	Androth	Bitra	Chetlat	Kadamat	Kalpeni	Kavaratti	Kiltan	Minicoy
■% of Artificial coast	24.48	59.35	74.52	16.8	20.32	57.39	52	78.98	36.89	19.02
■% of Erosion	38.32	18.92	11.47	55.19	32.54	13.08	17.96	4.47	23.02	29.26

FIG. 3 Coastal erosion status of the inhabited islands of Lakshadweep.

FIG. 4 Coastal erosion at Kavaratti Island, coast dominated by tetrapods as a coastal protection measure.

eventually settlement of sediments. Dredging to widen the lagoon is usually undertaken by the Administration, to maintain the navigation passage for the increasing numbers of ships. There are three main effects of sediment transport and deposition that has affected coral health in the lagoon. The cause-effect relationship indicates two primary causes: (i) turbidity of lagoon waters and (ii) benthic accumulation and entrainment of sediments, which result in reduced light penetration, smothering, and abrasion of coral and coral heads. The threats to the coral reef are shown in Fig. 5.

(b) Loss of biodiversity: Numerous threats like pollution, overexploitation of resources, land-use change still challenges the health of the corals and its diversity resulting in habitat loss of reef ecosystems (Newton et al., 2007; Díaz-Pérez et al., 2016; Graham et al., 2017). Sedimentation in coral reefs is one of the major concerns in the islands (Agatti, Kavaratti, Kadmat, etc.). Diving, snorkeling, reef fishing, motorized boats, unplanned anchoring, and mooring has caused churning, suspension, and eventually settlement of sediments on the coral reefs. Other activities such as lagoon fishing, waste disposal, tourism, and overharvesting of recourses are changing the ecological dynamics of reef ecosystems (Fig. 6).

Although tourism is responsible for reef damage (Kavaratti, Agatti), if carefully regulated, managed, and monitored it may actually help preserve these valuable habitats.

(c) Tourism: Tourism may be a good opportunity for the islanders; however, the probability of affecting the fragile environments and biodiversity is large. Tourism tends to provide economy to the islanders, but if this sector is dominated by the

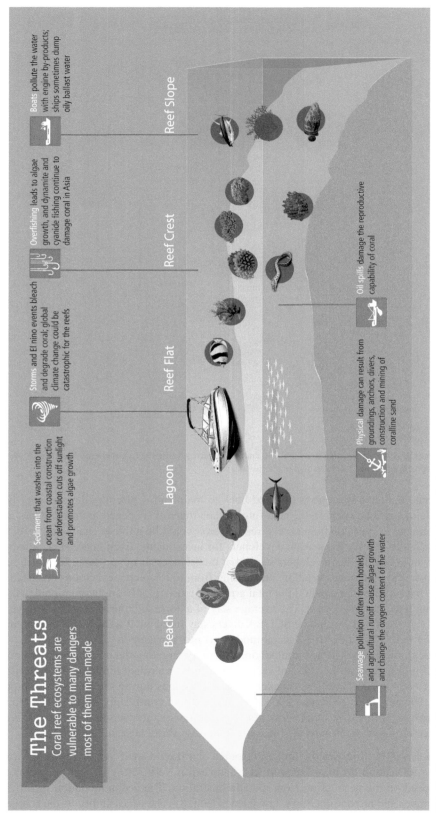

FIG. 5 Conceptual diagram of the threats to coral reefs in Lakshadweep Islands.

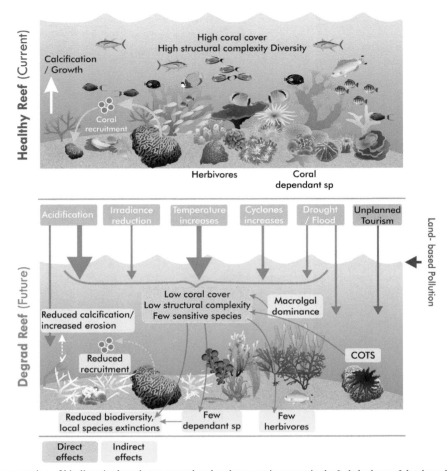

FIG. 6 Schematic representation of biodiversity loss due to natural and anthropogenic causes in the Lakshadweep Islands under (A) current and (B) in the future under ill-planned development scenarios. COTs: crown-of-thorns starfish.

mainlanders, the tourism-generated revenue is directed back to mainland. Currently, there is very limited source of livelihood for the people of Lakshadweep. In the near future when tourism is opened to the islanders, adequate finance for the necessary investments for establishing hotels, resorts, etc. would be a major challenge. With no other option, the islanders are compelled to involve the private developers for investments. Even under such situations, majority of the revenue generated would benefit the developers instead of the islanders if not properly administered. Immigration, water shortage, food insecurity, imported inflation, etc. arising from tourism creates various socioeconomic problems and uncertainties in the lives of the islanders. Of the total arrival of tourists, 14% are international visitors and 86% are from India and their length of stay also varies. The average stay of a domestic tourist is about 2–3 days while international tourists prefer to stay for 4–5 days and the SCUBA divers stay for about 2 weeks. The tourism inflow is growing steadily (Fig. 7), but the number of beds (i.e., accommodation) in the islands to support the tourism inflow has remained unchanged through the years. The accommodation is largely targeting the middle class and provides an exotic experience to tourists (Fig. 8). Tourism in Lakshadweep at present is controlled through a few specific tour packages offered as part of the cruise tourism. Tourists are taken to various islands during the day and are brought back to the ship the same evening. During 2006, 2007, and 2008, few international cruise ships were allowed to bring tourists to these islands, which had huge influx of visitors. However, as the infrastructure facilities and the resources were limited along sociocultural reservations in the islands the international cruise ships were not encouraged.

Despite the enormous potential for tourism the limited resource and fragile environment restrain it from achieving its full potential. Currently, the tourism in all the islands in Lakshadweep is very restrictive and is operated mostly as day tourism in tour packages based on the permission from the administration. The accommodation facilities (bed capacity) are extremely limited in number and are confined to a few islands (Fig. 8). The accessibility of the tourist to the island is primarily by ships and use of the only airport requires prior authorization from the Government of Lakshadweep.

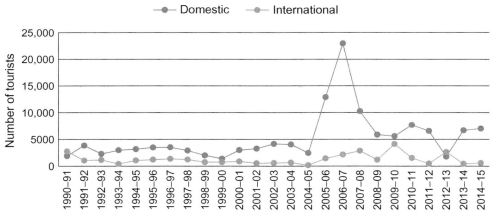

FIG. 7   Domestic and international tourist visit to the Lakshadweep Islands (1990–2015).

FIG. 8   Existing tourist accommodation in the islands.

As indicated by Douglas (2006) regulating the use of natural resources through sustainable development policies may affect the local people, whose livelihood is dependent on these resources. The major challenge lies in integrating sustainable use of natural resources while promoting economic development and environmental protection without altering/affecting the needs of the local people (McKee, 2013; Vogt et al., 2016).

4   INTERVENTIONS AND OPPORTUNITIES

Government of India has assigned top priority for the planned development of its island territories with special emphasis on sustainable growth of the economy and living standards of the islanders. With limited land-based resources, it is envisaged that the vast, comparatively shallow, practically calm, and protected lagoons of Lakshadweep, could provide excellent areas for various livelihood-related opportunities, especially in tourism. An integration of land and seascapes has been provided for the Lakshadweep Island, which includes inter-sectoral planning of the land area for the protection and conservation of the fragile coral lagoon. Management actions of the individual sectors would integrate or link with the activities of other sectors for an overall management of the Island. Each sector would incorporate these interlinking actions within its management plans, so that the combined action can be undertaken to achieve a specific objective of development and management of the Island. As an example, an integrated waste management plan that incorporates waste management with livelihood, tourism, and biodiversity conservation is provided.

5 INTEGRATED ISLAND MANAGEMENT PLAN

To address the challenges, in a sustainable manner, and to bring overall development in these islands, an integration of all the sectoral plans is recommended. Owing to the scarce resources and limited land area, the development in each island cannot be made in isolation. Thus, the development and management of these islands has to be carried out holistically, in clusters or in small groups. Each island has its own constraints. Certain infrastructure facilities which cannot be established in all islands are to be utilized by other islanders. For instance, due to limited land availability, it is not possible to establish airport in all the islands, therefore an airport in one strategic island can serve as common airport for a cluster of islands. Similarly, the management approach of the island has to take into account both the land and the marine area (lagoon and sea). The integration of sectoral plans and the output from each of the sectoral plans is shown in Fig. 9.

5.1 Freshwater Requirement

Recommendations/interventions to cater to the increased freshwater requirement for both residential and the tourist population include the following:

(a) obtaining water from the desalination plants already available in some of the inhabited islands, ground water replenishment through rainwater harvesting, and use of treated waste water (when it becomes available);
(b) use of treated waste water for noncontact and irrigation purposes;
(c) rainwater harvesting (storage and recharge); and
(d) reduce salinity ingress by using treated waste water to replenish the ground water table/aquifer.

5.2 Sewage

Availability of energy in Lakshadweep Islands is one of the factors that would determine the feasibility of setting up and operating a sewage treatment plant (STP). With establishment of an STP, it is possible to achieve zero-discharge as shown in Fig. 10. An integrated approach for managing the limited resources should be adopted for the sustainability of the island.

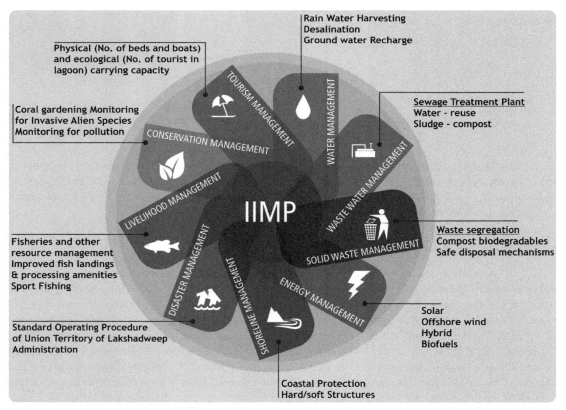

FIG. 9 Integration of management sub-plans.

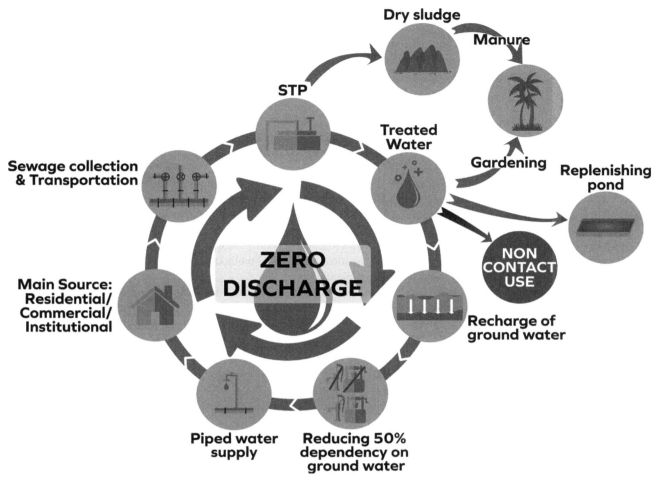

FIG. 10   Conceptual representation of zero discharge for Lakshadweep Islands.

For instance, water, sewage, and agriculture practices can be integrated as indicated in Fig. 10. Alternatively, considering energy as one of the major challenge, the Decentralized Waste Water Treatment System (DEWATS) for treatment of sewage is recommended. There are several advantages for preferring DEWATS for treatment of sewage over conventional STPs; however, feasibility studies need to be conducted prior to its establishment, operation, and maintenance.

5.3   Solid Waste Management

We have suggested the following steps to the Lakshadweep Administration to achieve a sustainable solid waste management system in the islands. The primary aim is to minimize the impact on ambient ecosystem by reducing, reusing, recycling/returning, through sustainable waste management. In addition, the following secondary measures were also suggested:

- formulation of a specific waste management plan for tourism;
- creating awareness on solid waste management among tourists;
- complete restriction on localized and unscientific dumping practices;
- strict implementation of zero tolerance for virgin or recycled plastic, less than 50 microns in thickness with penalization for violators;
- implementation of polluter pays principle (PPP) and penalization on violation of the norms.

Further, we have also suggested the following basic precautionary principles to the Government of Lakshadweep in order to combat the growing solid waste pollution in the islands.

a. *Segregation at the source:* One of the essential components of solid waste management is *segregation of waste at source* into biodegradable and nonbiodegradable wastes. This will not only reduce the cost of transportation for final disposal from the island to the mainland, but also provide segregated organic waste stock for possible generation of energy. Further, dumping/littering in open places (along the road or on beaches) shall not be permitted.

b. *Collection of wastes:* Separate waste bins shall be placed for the tourists to dispose biodegradable and nonbiodegradable wastes. Suitable measures are to be taken to ensure timely collection of the solid waste.

c. *Disposal of nonbiodegradable waste:* As the land area is limited, there exist fewer options for the management of nonbiodegradable solid wastes in the island. Best practices in waste management are Reduce, Reuse, and Recycle commonly known as 3Rs. It is often very difficult to recycle materials and goods because of the geographical isolation and uneconomically small size of the recycle market in the island. Similar on the lines of Fiji Island, a new R can be introduced for the interpretation of the 3R concept viz. "Reduce, Reuse, and Recycle/Return" by adding the word "Return (for recycle)" to the conventional definition of 3Rs. Options related to recycling of products need to be explored with self-help groups in the islands as well as the companies in mainland India.

d. *Disposal of biodegradable waste:* Practices of dumping of waste in the pit or backyard of shall be discouraged and discontinued if practiced. Additional measures and studies need to be conducted for the development of anaerobic digestion technologies, which would incorporate the microbial degradation of wastes. Composting has been suggested as alternate option which can be practiced as one of the most economical methods for producing manure from solid waste and the resulting compost shall be used for coconut plantations.

e. *Integrated waste management:* Alternatively, a group of sub-plans can also be integrated which would benefit the local community and conserve the fragile island environment. The schematic representation of integration and associated advantages affecting each of the sub-plans is shown in Fig. 11.

Coconut wastes are prevalent in all the islands of Lakshadweep. If properly collected, it could be a useful resource. The coconut husk and shells can be used as biomass fuel to generate energy. Activated charcoal can be manufactured from coconut shell, which can effectively remove impurities from waste water, which can be reused for irrigation and other noncontact purposes. The coconut tree trunk, the bark, the leaflets, and the coconut husk can be used as a building material for eco-friendly resorts to enhance tourism infrastructure. If trained, the local people can make various artistic handicrafts, which could be sold to the tourists, or national or international markets thereby enhancing the livelihood potentials of the islanders.

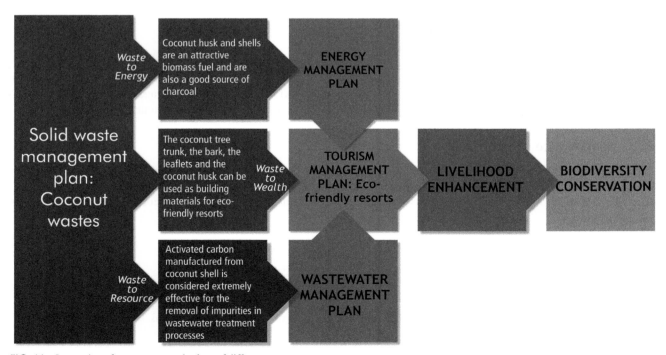

FIG. 11 Integration of management sub-plans of different sectors.

5.4 Island Shoreline Protection

Areas with high erosion have been protected by tetrapods, which provide a temporary protection from erosion. More scientific coastal protection methods such as establishment of geo-tubes or artificial reef structures are being evaluated to recommend the suitable protection strategy based on site-specific coastal processes.

5.5 Conservation of Corals

Considering the significance of the coral and associated biodiversity to the island ecosystem and for sustaining the livelihood activities of the islanders, the conservation perspective shall be integrated into every development activity in the island. Another important consideration is to assess the true economic value of reefs, including the tangible and intangible goods and services provided by the reefs to the island ecosystems, so that rational decisions can be made on the cost of management. From an ecological perspective, effective coral-reef conservation can be viewed as increasing or maintaining key ecosystem parameters such as fish biomass or coral cover, maintaining ecosystem processes and function, or increasingly, promoting resilience to disturbances and fluctuations. In order to sustain reef fisheries, the following suggestions were made to the Lakshadweep Administration:

- establish no-fishing zones/closed seasons and limitations on fishing gear;
- consider specific protection measures for algal grazers and corallivorous fishes;
- enforce legislation prohibiting destructive fishing practices;
- monitor the catch composition and size;
- develop alternative livelihoods for fishing communities as needed.

Given the ecological threats, environmental and resource constraints, and the opportunities, these islands have the capacity to adopt planning strategies from other islands that have overcome similar challenges or develop unique methods to achieve sustainable development. These practices should be implemented in a holistic manner, integrating various sectoral plans. Environment, eco-friendly, and socially acceptable intervention should be practiced to balance both development and nature. It is recognized that sustainable and eco-friendly development protects the delicate ecology and provide a greater platform for development. This would ensure protection of resources for future generations and would enhance the present ecology as well. Thus to enhance and utilize the existing opportunities, certain interventions are necessary as illustrated in Table 4.

TABLE 4 Key Issues, Challenges, Possible Interventions, and Opportunities for Lakshadweep Islands

Issues	Challenges	Recommended Interventions	Opportunities
Agriculture	Lack of nutrients in the soil	Use of sludge from STP as manure, post construction of STP in the islands	Enhanced production of coconut-based products
	Inadequate irrigation facilities	Use of recycled waste water for irrigation after construction of STP in the islands	
Fisheries	Overfishing of live bait fishes from the reef area	Closed season for bait fishing	(a) Conservation of live bait fishes for promote sustainable livelihood and fisheries development
		Avoiding use of certain species. (e.g., locals of Minicoy Island are against the use of certain bait fish varieties like Gumbala (clupeid) and Phitham (silver sides). They consider that these varieties are harmful to the tuna and tend to avoid using them). This must be followed by fishermen from other islands of Lakshadweep	(b) Enhancing harvest from the current 15%–40% of the fishery potential by 2025
		Restrictions on use of nets	
	Lack of storage and processing facilities	Establishment of adequate storage spaces and ensure immediate marketing of fish and fish products	Improved infrastructure facilities and livelihood of islanders
Land	Nonavailability of land area for any developmental activities	Expand development into uninhabited islands of Lakshadweep (e.g., Bangaram, Tinnakara Islands)	Balanced development of existing land and lagoon resources

Continued

TABLE 4 Key Issues, Challenges, Possible Interventions, and Opportunities for Lakshadweep Islands—cont'd

Issues	Challenges	Recommended Interventions	Opportunities
Fresh water	Limited availability of potable water	Rainwater harvesting for storage and recharge	Accessibility to clean drinking water for better quality of health
		Cleaning of existing ponds for storage of rainwater	
		Enhance current capacity of existing desalination plants	
		Floating desalinization plant as land area is scarce	
	Contamination due to sewage	Establishment of sewage treatment plant where feasible	
Sewage	No sewage treatment system	Follow construction and design standards and best practices (e.g., decentralized waste water treatment system and reuse- DEWATS)	Accessibility to clean drinking water. Sustainability of ecology and the environment.
	Seepage of sewage into groundwater aquifer	Establish Bio-toilets as an alternative to soak pits	
Solid waste	Disposal into coastal and marine areas	Reduce use of plastics in the islands by using local materials (e.g., coir bags)	Goal to become plastic-free islands by 2025 Improved high end tourism facilities and enhanced livelihoods
		Transport of various waste to the mainland for incineration or reuse (as in case of glass waste which is being transported to Mangalore for recycle and reuse)	
Energy	Use of diesel generators resulting in pollution of air, water and coastal ecosystems	Use of renewable energy (solar, wind)	Use of renewable green energy for maintaining clean and pristine islands with uninterrupted power supply
	Inadequate land area for establishing solar plants, wind turbines etc.	Floating solar panels can be used at appropriate locations in the lagoon taking care of the coral ecosystem Similarly, small-scale vertical axis wind turbines to be used instead of conventional offshore wind turbines for wind energy.	
Coastal Erosion	Dredging of reef areas for navigation	Protection of the reef areas by prohibiting dredging	Prevention of coastal land loss and promote access to beach areas under the scenario of tourism development
	Use of hard structures such as seawalls and tetrapods	Undertake coastal process studies to determine suitability of soft shore protection structures (e.g. artificial reefs/ geo-tubes etc.)	
	Loss of prime beach areas	Prevent further erosion by adopting site-specific shoreline management plan	
Sedimentation	Use of high speed boats causing churning of the sediments	Prevent entry of larger boats or ships within the lagoon area.	Conserve and protect the coral reef to improve health of the ecosystem
	Dredging of reefs for navigation	Dredging in live coral reef areas to be avoided	
Loss of biodiversity	Reduction in number of coral species and fisheries	Coral transplantation at selected locations	Improved health of coastal/ marine ecosystem resulting in improved (a) bait fishery resources, (b) enhanced tuna fisheries, and (c) enhanced well-being of island community
		Coral garden to be established to preserve native coral species	
	Decline in seagrass ecosystem	Seagrass beds to be restored by encouraging natural recolonization and by transplantation of mature seagrass plants taken from healthy donor beds	
Tourism	Low tourism carrying capacity	Establish infrastructure for potable water, sewage treatment system, and use of renewable energy. Encourage tourism in uninhabited islands of Lakshadweep	Achieve balance between conservation and island development for enhanced livelihood of the island community

Each developmental activity has its own impact on the ecology and environment, the same is applicable in the case of tourism development as well (Hall, 2008; Coccossis and Mexa, 2017). The need for establishing a mechanism which balances development, environment, and ensures ecological integrity arises and therefore the need for assessing carrying capacity becomes essential. In addition to this, most of the countries have their own legislation that provides for the conservation of the areas, which are ecologically sensitive and valuable (Tosun, 2001; Byrd, 2007; Coccossis and Mexa, 2017). High-end tourism, if managed and operated sustainably, can be an environmentally and economically sustainable option for economic development in Lakshadweep. The alternative option suggested is to encourage high-end tourism in uninhabited islands of Lakshadweep.

6 CONCLUSIONS

Several lessons emerged from the analysis of challenges and opportunities, which point to the need for integrated island management. These would in turn be linked as IIMP with spatial planning for the land and the marine areas (marine spatial planning) in the future. This also includes the preparation of sectoral management sub-plans that emphasizes protection of the island's vulnerable shoreline, conservation of fragile biodiversity, water management, sewage and solid waste management, disaster management, and pollution management. These spatial plans would conform to the local Development Control Regulations (DCRs). Considering the huge opportunity, the Lakshadweep Islands may promote sustainable tourism with measures to conserve its natural resources. Several suggestions for freshwater requirements, zero waste disposal, solid waste management, erosion control, improved fisheries, and conservation of the coral biodiversity have been addressed. Assessment of carrying capacities of tourism, ecology, and environment needs to be adopted and carried out periodically to balance development and conservation issues of the Lakshadweep Islands.

ACKNOWLEDGEMENTS

This study was undertaken as part of in-house research on "Assessment of Cumulative Coastal Environmental Impacts [ACCES]" (# IR 12003) of NCSCM. The authors acknowledge the financial support of the Union Territory of Lakshadweep (UTLA) and the Ministry of Environment, Forest and Climate Change, Government of India, under the World Bank assisted India ICZM Project. Conceptual diagrams in this chapter are made using the IAN Symbols, courtesy of the Integration and Application Network, University of Maryland Center for Environmental Science (ian.umces.edu/symbols/).

REFERENCES

Boniface, B., Cooper, C., 2009. Worldwide Destinations Casebook. The Geography of Travel and Tourism, second ed. Elsevier, New York, pp. 45–251.

Byrd, E.T., 2007. Stakeholders in sustainable tourism development and their roles: applying stakeholder theory to sustainable tourism development. Tour. Rev. 62 (2), 6–13.

Coccossis, H., Mexa, A., 2017. The Challenge of Tourism Carrying Capacity Assessment: Theory and Practice. Ashgate Publishing Ltd., Burlington, NJ2004.

Coccossis, H.N., 1987. Planning for islands. Ekistics 54, 84–87.

Díaz-Pérez, L., Rodríguez-Zaragoza, F.A., Ortiz, M., Cupul-Magaña, A.L., Carriquiry, J.D., Ríos-Jara, E., Rodríguez-Troncoso, A.P., María del Carmen García-Rivas, M.C., 2016. Coral Reef Health Indices versus the Biological, Ecological and Functional Diversity of Fish and Coral Assemblages in the Caribbean Sea. PLoS ONE 11 (8), e0161812.

Depraetere, C. Dahl, A.L., 2007. Island locations and classifications. In a world of islands: an island studies reader, edited by G. Baldacchino, 57-105. Luqa, Malta: Agenda Academic Publishers; Charlottetown, P.E.I.: University of Prince Edward Island, Institute of Island Studies.

Douglas, C.H., 2006. Editorial: small island states and territories: sustainable development issues and strategies—challenges for changing islands in a changing world. Sustain. Dev. 14, 75–80.

Graham, N.A., McClanahan, T.R., MacNeil, M.A., Wilson, S.K., Cinner, J.E., Huchery, C., Holmes, T.H., 2017. Human disruption of coral reef trophic structure. Curr. Biol. 27 (2), 231–236.

Hall, C.M., 2008. Tourism Planning: Policies, Processes and Relationships, second ed. Pearson Education Limited, Harlow.

Hein, P.L., 1990. Between Aldabra and Nauru. In: Beller, W., D'Ayala, P., Hein, P. (Eds.), Sustainable Development and Environmental Management of Small Islands. UNESCO, Paris.

Kakazu, H., 2011. Challenges and opportunities for Japan's remote islands. Eurasia Border Rev. 2 (1), 1–16.

McCall, G., 1996. Clearing confusion in a disembedded world: the case for nissology. Geogr. Z. 84 (2), 74–85.

McKee, T.L., 2013. Charting the course for sustainable small island tourist development. J. Environ. Prot. 4 (03), 258.

Newton, K., Cote, I.M., Pilling, G.M., Jennings, S., Dulvy, N.K., 2007. Current and future sustainability of island coral reef fisheries. Curr. Biol. 17 (7), 655–658.

Panakera, C., Willson, G., Ryan, C., Liu, G., 2011. Considerations for sustainable tourism development in developing countries: perspectives from the South Pacific. TOURISMOS: An Int. Multi. J. Tourism 6 (2), 241–262.

Planning Commission, 2007. Lakshadweep Development Report. Academic Foundation, Government of India, New Delhi, pp. 118–127.

Rozelee, S., Rahman, S., Omar, S.I., 2015. Tourists perceptions on environmental impact attributes of Mabul Island and their relationship with education factor. American-Eurasian J. Agric. Environ. Sci. 15, 146–152.

Sealey, K.S., Smith, J., 2014. Recycling for small island tourism developments: food waste composting at Sandals Emerald Bay, Exuma, Bahamas. Resour. Conserv. Recycl. 92, 25–37.

Tosun, C., 2001. Challenges of sustainable tourism development in the developing world: the case of Turkey. Tour. Manag. 22 (3), 289–303.

Van der Velde, M., Green, S.R., Vanclooster, M., Clothier, B.E., 2007. Sustainable development in small island developing states: agricultural intensification, economic development, and freshwater resources management on the coral atoll of Tongatapu. Ecol. Econ. 61 (2–3(1)), 456–468.

Vogt, C., Jordan, E., Grewe, N., Kruger, L., 2016. Collaborative tourism planning and subjective well-being in a small island destination. J. Destin. Mark. Manag. 5 (1), 36–43.

Chapter 28

The Future of the Great Barrier Reef: The Water Quality Imperative

Brodie, J. *, Grech, A. *, Pressey, B. *, Day, J. *, Dale, A.P. †, Morrison, T. *, Wenger, A. ‡

*ARC Centre of Excellence for Coral Reef Studies, James Cook University, Townsville, QLD, Australia, †The Cairns Institute, James Cook University, Cairns, QLD, Australia, ‡School of Earth and Environmental Sciences, University of Queensland, St. Lucia, QLD, Australia

1 INTRODUCTION—THE STATE OF THE GREAT BARRIER REEF

The Great Barrier Reef (GBR) is an extensive coral reef system off the northeast Australian coast (Fig. 1) comprising high value areas of coral reefs, seagrass meadows, and mangrove forests, and a range of iconic megafauna including whales, dugongs, turtles, sharks, dolphins, large fish, and valuable commercial, and recreational fisheries (Day and Dobbs, 2013). The area of the Great Barrier Reef World Heritage Area (GBRWHA) is 348,000 km² (Day and Dobbs, 2013) with 2300 km of coastline and an adjacent catchment area of 424,000 km² (Great Barrier Reef Marine Park Authority, 2014). The Australian and Queensland State Governments have intensively managed the GBR as a federal marine park since 1975 and a World Heritage Area (WHA) since 1981 (Day and Dobbs, 2013) focussing on ecosystem protection while allowing all reasonable uses.

Economists have estimated the direct and indirect economic contribution of the GBR at overall $6.4 billion in value added to the Australian economy in 2015–2016 (Deloitte Access Economics, 2017). Nearly 90% of this economic contribution (approximately $5.7 billion) was from tourism activities alone and the balance from recreation, commercial fishing, science, and research. In terms of employment, the GBR supported more than 64,000 full-time jobs in Australia (Deloitte Access Economics, 2017). Stoeckl et al. (2014) have estimated the value of ecosystem services provided by the GBR as between $15 and $20 billion per annum.

Key GBR ecosystems continue to be in poor (and declining into poorer) condition (Brodie and Waterhouse, 2012; Hughes et al., 2015), including drastically reduced coral cover (De'ath et al., 2012), seagrass health (Coles et al., 2015) and reduced populations of dugongs (Marsh et al., 2005), and turtles (Great Barrier Reef Marine Park Authority, 2014). Recent events have severely impacted GBR ecosystems. Prolonged periods of elevated sea surface temperatures associated with climate change led to coral bleaching and mortality (Hughes et al., 2017a,b). An unusually high frequency of category 4 and 5 tropical cyclones likely also exacerbated by climate change (Cheal et al., 2017) caused further damage, as did the progression of the fourth wave of crown-of-thorns starfish (CoTS) population outbreaks (Pratchett et al., 2017). Coral cover (the percentage live coral on a coral reef) on the GBR has declined from above 40% in the 1970s to 14% in 2012 (Hughes et al., 2011; De'ath et al., 2012).

Loss of ecosystem condition is also attributed to the impacts of terrestrially derived runoff associated with past and ongoing catchment and coastal development resulting in greatly increased loads of fine sediment, nutrients, and pesticides discharging to the GBR (Brodie et al., 2012; Waterhouse et al., 2017a). Increased nutrient discharges are associated with recurring population outbreaks of CoTS (Brodie et al., 2017a). Other causes of loss are a long history of commercial fishing (line, net, and trawl) and recreational fishing, including the previously carried out commercial harvest of dugong and turtle, coastal development causing the direct and indirect loss of habitats, shipping activities (oil and chemical spills), port development and maintenance (dredging, antifouling residues), and bather (shark) nets (associated with entangling dugong, turtle, dolphins, and other marine mammals) (Great Barrier Reef Marine Park Authority, 2014).

In addition, severe problems in the governance of the GBR have been recognized in recent analyses (Morrison, 2017). In particular, implementation failure (primarily at catchment scale) in water quality management (Queensland Audit Office, 2015; Dale et al., 2018) has been notable, and there are design concerns within 10 important governance subdomains that

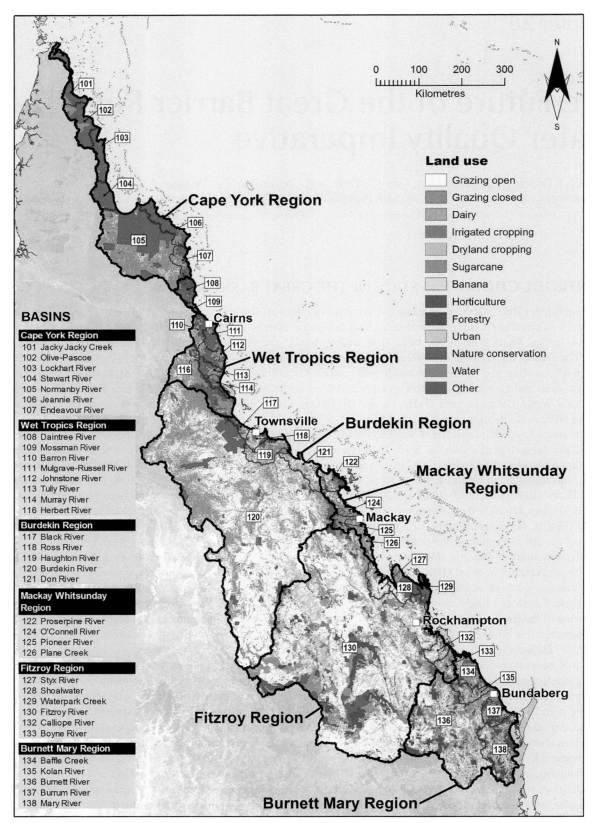

BASINS

Cape York Region
101 Jacky Jacky Creek
102 Olive-Pascoe
103 Lockhart River
104 Stewart River
105 Normanby River
106 Jeannie River
107 Endeavour River

Wet Tropics Region
108 Daintree River
109 Mossman River
110 Barron River
111 Mulgrave-Russell River
112 Johnstone River
113 Tully River
114 Murray River
116 Herbert River

Burdekin Region
117 Black River
118 Ross River
119 Haughton River
120 Burdekin River
121 Don River

Mackay Whitsunday Region
122 Proserpine River
124 O'Connell River
125 Pioneer River
126 Plane Creek

Fitzroy Region
127 Styx River
128 Shoalwater
129 Waterpark Creek
130 Fitzroy River
132 Calliope River
133 Boyne River

Burnett Mary Region
134 Baffle Creek
135 Kolan River
136 Burnett River
137 Burrum River
138 Mary River

Land use
Grazing open
Grazing closed
Dairy
Irrigated cropping
Dryland cropping
Sugarcane
Banana
Horticulture
Forestry
Urban
Nature conservation
Water
Other

FIG. 1 The GBR catchment, 35 major basins, and major land uses. *(Map produced by the Department of Natural Resources, Mines, and Energy, 2017).*

operate within GBR catchments (Dale et al., 2013, 2016, 2018). At a whole-of-GBR scale weak policy and poor integration of effort across competing governance subdomains has led to poor outcomes for water quality management. Changes in the overarching governance regimes have also caused management agencies to overestimate the success of policy options in the GBR's complex polycentric system as the GBR (Morrison, 2017). Management of port activities remains an issue with a weak environmental assessment regime (Grech et al., 2013) based on an inadequate use of available science (Brodie, 2014) and Environmental Impact Assessments undermined by inconsistent methods and a lack of independent evaluation, leading to inadequate scientific rigor (Sheaves et al., 2016). Management efforts to reduce the impacts of terrestrial pollutant discharge through the introduction of better management practices in agriculture (Thorburn et al., 2013; Eberhard et al., 2017a,b) have made some progress, but remain largely ineffectual with little chance of meeting the required targets in the specified timeframes (Brodie et al., 2017b; Waterhouse et al., 2017a). Analysts believe the causes of this failure lie in insufficient funding, where recent analysis estimate that $9 billion in expenditure is needed to reach water quality targets (Alluvium, 2016), and the lack of enforcement of existing legislation (Brodie and Pearson, 2016).

Action to reduce greenhouse gas emissions both at a national scale and globally have been assessed to be insufficient to keep temperature rises below the 2°C rise needed to preserve coral reefs (Heron et al., 2016). Fishing pressure on some species, for example, scallops, thought to be "sustainably managed" has resulted in the reduction of the stock to less than 6% of the original 1977 stock resulting in the fishery being severely restricted (Yang et al., 2016).

These conclusions were reinforced in the recently released 2017 Scientific Consensus Statement: Land use impacts on GBR water quality and ecosystem condition (Waterhouse et al., 2017a) commissioned by the Queensland and Australian governments to guide future GBR water quality policy. The conclusions (Waterhouse et al., 2017a) were as follows:

The overarching consensus is:
Key GBR ecosystems continue to be in poor condition. This is largely due to the collective impact of land runoff associated with past and ongoing catchment development, coastal development activities, extreme weather events, and climate change impacts such as the 2016 and 2017 coral bleaching events.
Current initiatives will not meet the water quality targets. To accelerate the change in on-ground management, improvements to governance, program design, delivery and evaluation systems are urgently needed. This will require greater incorporation of social and economic factors, better targeting and prioritisation, exploration of alternative management options and increased support and resources.

In this review, we assess the stressors affecting the GBR, focussing primarily on water quality impacts from catchment pollutants and those pollutants derived from coastal development, in the context of rapidly progressing climate change and continued exploitation of marine resources. Our focus is on the impacts and management of fine sediment and nutrients but with consideration also to the complete range of pollutants known to be affecting the GBR. These include pesticides, other potentially toxic organic chemicals [including endocrine disrupting substances, oil and polyaromatic hydrocarbons (PAHs), pharmaceuticals, and food additives], metals, and plastics.

2 TERRESTRIAL POLLUTION AND SOURCES

A total of 35 major basins (totally 424,000 km^2) drain into the GBR lagoon (Fig. 1). Agriculture is the major land use in the GBR catchment, covering more than 80% of the total area. Land uses in the GBR catchment include grazing (75%), nature conservation (13%), sugar cane (1%), and rain-fed summer and winter cropping (~3%) (Fig. 1; Waters et al., 2014). Relatively small areas of horticulture crops are grown in the high rainfall and coastal irrigation areas, with irrigated cotton mainly found in inland areas of the Fitzroy region. There are several large urban centers on the GBR coast such as Cairns, Townsville, Mackay, and Rockhampton (Fig. 2), and while the contributions from these areas may be relevant at a local scale, the overall contributions of pollutant loads are minor (<5% in most cases) compared to agricultural land uses (Waterhouse et al., 2012; Bartley et al., 2017).

Agricultural and urban development of the catchments since European settlement, sequentially from south to north since approximately 1830, has led to significant changes in the quantity and quality of water discharges into the GBR. Monitoring and modeling indicate that discharges of suspended sediments (SSs), nutrients, and pesticides to the GBR have increased greatly in this period (Waterhouse et al., 2017a; Kroon et al., 2012; Waters et al., 2014). Modeled estimates indicate that mean annual SS loads delivered to the GBR have increased by 5 times to 9900 kilotons/year, ranging between threefold and eightfold depending on the region; mean annual total nitrogen load has increased by 2.1 times to 55,000 tons/year, ranging between 1.2 and 4.7 times depending on the region; mean annual total phosphorus load has increased by 2.9 times to 8800 tons/year, ranging between 1.2 and 5.3 times depending on the region since European settlement (Bartley et al., 2017). At least 12,000 kg/year of herbicides are now discharged to the GBR compared to none in 1830.

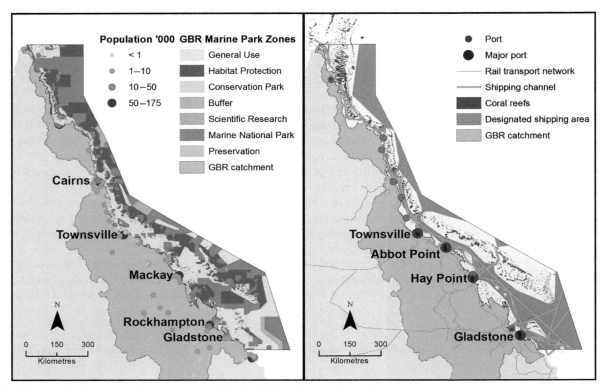

FIG. 2 Urban populations, ports, and transport networks in the GBR catchment and coastal regions.

3 STRESSORS AND THE IMPACTS

3.1 Fine Sediment

The effects of increased loads of SS on GBR marine ecosystems are as follows:

3.1.1 Light Reduction for Seagrass

Light is the primary limiting factor of seagrass production in the GBR (Collier and Waycott, 2009), and reductions in benthic light (particularly photosynthetically active radiation—PAR) have been directly linked to seagrass loss (Collier et al., 2011, 2012; Petus et al., 2014). Resuspended fine sediments strongly regulate light penetration in coastal waters less than about 15 m deep, which, unlike direct light loss during flood plume conditions, may occur year-round (Fabricius et al., 2013a, 2014, 2016) in windy conditions. Increased sediment loads cause decreases in seagrass-suitable habitat extent through changed clarity and hence benthic light (Saunders et al., 2017).

The seagrass growing season in the GBR occurs between August and December (Chartrand et al., 2016) when there is a peak in the formation of food reserves, sexual reproduction and the formation of seed banks. Seagrass is sensitive to low light associated with degraded water quality at this time as this effects the increase in aboveground biomass. Lower light during the wet season associated with increased SS (both from flood plumes and resuspension) can also drive rapid seagrass loss, and place seagrass beds in a poor state ahead of the following growing season (McKenzie et al., 2016). For example, substantial seagrass loss occurred in the Burdekin region between 2008 and 2012 (Petus et al., 2014). This was driven by large Burdekin River flows (Fabricius et al., 2014) and discharge of large loads of fine sediment. Consequential long-term resuspension of the benthic fine sediment and organic matter and unusual cloud cover in the dry season resulted in continuous low light (McKenzie et al., 2016). The capital dredging required for port developments can have negative impacts on seagrass through direct burial and/or physical removal, and indirectly from turbidity plumes and the associated reduction in available light (Chartrand et al., 2016). In the GBRWHA, recent studies have shown that these plumes can have a substantial impact on seagrass (York et al., 2015). Water quality is also reduced by the resuspension of maintenance dredge spoil which is regularly dumped adjacent to port developments.

3.1.2 Effects on Corals

High concentrations of fine SS (especially with a high organic content) can cause direct biological effects on coral (e.g., interfering with heterotrophic feeding), reduce the light quantity and quality (i.e., of PAR), and deposit on the corals' surface a fine layer of sediment and organic matter (Flores et al., 2012; Jones et al., 2015; Weber et al., 2006, 2012). Light reduction associated with increased turbidity reduces coral autotrophic feeding, increases coral bleaching and causes the death of zooxanthellae (Bessell-Browne et al., 2017a,b). Elevated SSs and sediment deposition negatively affect the reproductive cycle and early life histories of corals (Jones et al., 2015). Other negative effects of SSs are the entanglement and entrapment of coral sperm by sediment particles (Ricardo et al., 2015) and weighing down of the buoyant egg-sperm bundles such that they do rise to the top of the water column and suffer reduced dispersal (Ricardo et al., 2016a), both reducing fertilization success of corals. However, developing embryos and larvae can tolerate exposure to SSs through mechanisms to remove particles (Ricardo et al., 2016b). The subsequent coral life-history stage, that is, larval settlement, is reduced by a thin layer of fine, terrigenous sediment settlement (Perez et al., 2014; Jones et al., 2015).

In contrast turbid water can have a protective effect against coral bleaching, as has been observed in the Caribbean when turbid plumes from South American rivers reduced bleaching responses in Barbados (Oxenford and Vallès, 2016) and turbidity and upwelling reduced bleaching in Columbia (Bayraktarov et al., 2013). Turbid conditions in some locations were also attributed to reduced bleaching in Palau (Van Woesik et al., 2012). The protective effect of turbidity in terms of reducing irradiance, particularly UV radiation, is also the likely cause of reduced bleaching on the northern GBR in 2016 in those areas where waters were turbid (Hughes et al., 2017a), even though water temperatures were greatly above normal.

3.1.3 Effects on Fish

Increased benthic sediment loads, depositing into the epilithic algal matrix (EAM) found on coral reefs, can suppress fish herbivory (on the algae) and detritivory (on detritus within the EAM) (Bellwood and Fulton, 2008; Goatley and Bellwood, 2013; Gordon et al., 2016). This can change the EAM from palatable, short, productive algal turfs to unpalatable, long sediment-laden algal turfs. These changes can reduce the resilience of coral reefs to other stressors (Goatley et al., 2016). Finer sediments have greater effects on the suppression of herbivory (Tebbett et al., 2017a,b). Benthic sediment load appears to be a highly sensitive indicator and driver of ecosystem change on coral reefs. Changes in sediment loads may also indicate forthcoming changes in herbivory and other processes, for example, coral recruitment (Birrell et al., 2008) and may prove useful as a predictor for ecosystem transitions or as an early indication of a shift to an alternative stable state (Goatley et al., 2016).

Coral reef-associated damselfishes suffer adverse effects on their larval development, foraging success, and habitat use due to elevated concentrations of SS, at concentrations that have been observed at GBR inner-shelf reefs (Johansen and Jones, 2013; Wenger et al., 2012, 2014, 2016; Wenger and McCormick, 2013; Hess et al., 2017). Other fish such as juvenile bumphead parrotfish (*Bolbometopon muricatum*) depend on a highly specific microhabitat that is vulnerable to sedimentation from logging operations, as has been shown in the Solomon Islands (Hamilton et al., 2017).

3.2 Nutrients

Excess nutrient pollutant export from the rivers in the GBR has been associated with several ecosystem impacts (Brodie et al., 2011; Fabricius, 2011; Schaffelke et al., 2017). These include reef degradation and overall reduced coral biodiversity between Townsville and Cooktown, with a reduction in species richness of 40 species compared with the expected value in this region (DeVantier et al., 2006); enhanced vulnerability of reef corals to thermal bleaching stress (e.g., Wooldridge, 2016); increased presence of macroalgae on reefs, which can affect coral diversity and/or larval coral recruitment (De'ath and Fabricius, 2010); and reef damage from coral-eating CoTS (*Acanthaster planci*) outbreaks (Fabricius et al., 2010). Coastal runoff-induced phytoplankton blooms impose nutrient stress on coral reefs through a range of mechanisms (D'Angelo and Wiedenmann, 2014). Most detrimental effects are attributed to nitrogen excesses, but phosphorus may have an important role as well, if poorly understood (Furnas et al., 2005, 2011).

For all of these nutrient effects, either through direct nutrient effects or through enhanced phytoplankton biomass, nitrogen must be in an immediately or potentially bioavailable form; for example, while nitrate is immediately bioavailable, bacterial action can transform organic nitrogen to nitrate (known as mineralization), making it bioavailable. Dissolved inorganic nitrogen (DIN—nitrate plus ammonium) derived from agricultural fertilizer losses immediately becomes bioavailable. The particulate nitrogen (PN) derived from soil erosion is likely to become bioavailable through mineralization within the lagoon waters or in the sediment (Brodie et al., 2015).

3.2.1 CoTS (Crown-of-Thorns Starfish)

The CoTS are one of the major causes of coral mortality in the GBR (De'ath et al., 2012; Osborne et al., 2011). River nutrients can influence CoTS outbreak dynamics (Schaffelke et al., 2017) when wet season discharges from the Wet Tropics and the Burdekin rivers occur in the region between Ayr and Cooktown, in the period when phytoplankton-feeding CoTS larvae are present in the water column (November to March) (Brodie et al., 2005, 2017a; Fabricius et al., 2010).

A wave of outbreaks is initiated when these favorable conditions are reinforced by favorable hydrodynamic conditions in the area between Cairns and Lizard Island (Hock et al., 2014; Wooldridge and Brodie, 2015) and sufficient coral cover to sustain the outbreaks (Fabricius et al., 2010). After outbreaks are initiated in the Cairns to Lizard Island area, as they were in 1962, 1978, 1993, and 2009, they progress southward via larval transport, mainly on mid-shelf reefs, over a period of about 12 years to the areas approximately offshore from Mackay. It is generally assumed (Brodie, 1992) that numbers of outbreaks have increased in the period since predevelopment, and the frequency of CoTS waves has increased greatly from possibly as low as every 50–80 years to about every 15 years since preindustrial times (Fabricius et al., 2010).

The Chl-a concentrations at which the probability of larval survivorship change rapidly are not a single point but on a continuum (Fabricius et al., 2010). Recent studies show that the most favorable concentration for larval survival is estimated to be at 0.6–1 μg L^{-1} (Brodie et al., 2017a; Pratchett et al., 2017) with a high biomass of larger phytoplankton with algal cell numbers greater than 1000 cells mL^{-1} (Uthicke et al., 2015, 2018).

3.2.2 Macroalgae versus Coral Diversity

Enhanced nutrient availability can promote the growth of macroalgae and, in general, nutrient enrichment negatively affects coral physiology and ecosystem functioning (D'Angelo and Wiedenmann, 2014). Macroalgae are more abundant on reefs in waters with higher concentrations of water column Chl-a (above 0.45 μg L^{-1}), due to increased nutrient availability (De'ath and Fabricius, 2010). High macroalgal biomass has a number of adverse effects on corals through: space competition (McCook et al., 2001); altering the corals' microbial environment which affects coral metabolism (Hauri et al., 2010; Zaneveld et al., 2016) and larval survival (Morrow et al., 2017); reducing coral settlement (Birrell et al., 2008) and increasing the susceptibility to coral disease (Morrow et al., 2012; Vega Thurber et al., 2014).

3.2.3 Increased Coral Bleaching Susceptibility

The DIN availability is important in the functioning of the coral-algae symbiosis, and elevated DIN concentrations can cause changes that disrupt the ability of the coral host to maintain an optimal population of algal symbionts (Wooldridge, 2016; Wooldridge et al., 2017). Together with increased temperature, elevated DIN concentrations and changes in N:P ratios can increase the susceptibility of corals to bleaching (Wooldridge, 2016; Wooldridge et al., 2017; Vega Thurber et al., 2014; Wiedenmann et al., 2013; D'Angelo and Wiedenmann, 2014; Rosset et al., 2017; Fabricius et al., 2013b; Humanes et al., 2016).

3.2.4 Bioerosion

Bioerosion of living and dead corals occurs via a range of mechanisms involving many different organisms. These range from minute, primarily intra-skeletal organisms, the microborers (e.g., algae, fungi, bacteria) to larger and often externally visible macroboring invertebrates (e.g., sponges, polychaete worms, sipunculans, molluscs, crustaceans, echinoids), and fish (e.g., scarids, acanthurids) (Glynn and Manzello, 2015; Hutchings et al., 2005; Chazottes et al., 2017). Nutrient enrichment can increase the growth of both of these types of borers. For example, algal borers grow well with increased dissolved inorganic nutrient availability while filter-feeding sponges, worms, and bivalves grow well with increased phytoplankton (and zooplankton) biomass (LeGrand and Fabricius, 2011). Eutrophication of reefal waters by land-based sources of nutrient pollution can magnify the effects of ocean acidification through nutrient-driven bioerosion (Prouty et al., 2017). Thus increased bioerosion by these organisms can interact with reduced calcification due to ocean acidification to additively reduce reef net calcification (DeCarlo et al., 2015; Glynn and Manzello, 2015). Recent analysis on the GBR over the period 2004–2012 has shown that while bioeroding sponges such as *Cliona orientalis* vary in abundance with location no overall increase in percent cover has occurred during this period of declining coral cover (Ramsby et al., 2017).

3.2.5 Coral Diseases

Coral disease is a considerable contributor to coral cover declines on the GBR (Osborne et al., 2011) and is predicted to worsen with global pressures of increasing temperature and ocean acidification (Maynard et al., 2015; O'Brien et al., 2016a,b). Coral disease manifests as a general response to multiple stressors of corals, it has been positively correlated to sedimentation, elevated concentrations of nutrients and organic matter, and increased plastic pollution (Harvell et al., 2007; Haapkylä et al., 2011; D'Angelo and Wiedenmann, 2014; Vega Thurber et al., 2014; Thompson et al., 2014; Pollock et al., 2014, 2016; Lamb et al., 2016, 2018; Zaneveld et al., 2016).

3.2.6 Light Reduction

Algal blooms associated with flood plumes due to inputs of river-derived nutrients reduce water clarity. These phytoplankton blooms as well as non-algal suspended particulate matter (detritus, clay particles) in the plume reduce light availability for benthic plant communities including seagrass and coral (Collier et al., 2016a; Petus et al., 2014). In inner-shelf waters, the reduction of in situ light penetration due to resuspended sediment is usually a more dominant effect, but in deeper waters (>15 m) where resuspension does not normally occur (except in cyclonic conditions), the light reduction due to phytoplankton (and zooplankton) may be an important factor for communities such as deepwater seagrasses (Collier et al., 2016a,b) and mid-shelf coral reefs.

3.3 Pesticides

A large number of individual pesticides used mainly in agriculture are lost from farming operations and discharged into streams and the GBR (Lewis et al., 2009). The largest group present in relatively high concentrations are herbicides, with more than a dozen known to present some risk to GBR coastal, estuarine, and lagoonal habitats and species (Lewis et al., 2012, 2016). Pesticides are found throughout the GBR lagoon following river discharge events (Kennedy et al., 2012a,b) but the main areas of concern are in coastal wetlands and estuarine and coastal waters (Lewis et al., 2012; Davis et al., 2013) where pesticides are regularly found at above Australian trigger value concentrations (O'Brien et al., 2016a,b). Insecticides are also of some concern especially the neonicotinoid insecticides such as imidacloprid, which is used widely in sugarcane cultivation to control cane grubs and found commonly in streams draining cane lands (O'Brien et al., 2014, O'Brien et al., 2016a).

In some respects, pesticides are easier to manage than erosion/fine sediments and nutrients from fertilizer loss. Better spray targeting and hence use of greatly reduced applications reduce losses to waterways dramatically (Oliver et al., 2014; Davis and Pradolin, 2016). However, management of pesticides at an Australian national scale is poor (King et al., 2013). Well-meaning replacement of one pesticide by another "more environmentally friendly" one can lead to unintended consequences (Davis et al., 2014) with the replacement turning out to be, through inadequate risk assessment, more of a risk to aquatic ecosystems than the original banned pesticide.

3.4 Other Pollutants

These include potentially toxic organic chemicals, including endocrine disrupting substances, oil and PAHs, pharmaceuticals, and food additives, as well as metals and plastics. Berry et al. (2013) and Kroon et al. (2015) have reviewed the risk to the GBR from this diverse range of substance with some consensus that plastics (both macro and micro) likely presented the greatest risk at a GBR scale. A qualitative risk assessment of pollutants other than sediment, nutrients, and pesticides (Kroon et al., 2015) concluded that marine plastic pollution poses the highest relative risk to the GBR marine ecosystems, particularly in the Cape York NRM region due to exposure to oceanic and local shipping sources. Both macro and microplastic particles have been found in the gut of stranded dead green turtles from GBR waters (Caron et al., 2018).

The assessment also highlighted chronic contamination of water and sediments with antifouling paint components and exposure to certain personal care products in natural resource management regions south of Cape York. The relative risks of other pollutants are likely to be relatively low with some minor differences between natural resource management regions.

Certain metals, particularly cobalt, and a range of organic compounds have been found in unusual concentrations in the blood of green turtles (Villa et al., 2017; Heffernan et al., 2017; Dogruer et al., 2018) and are associated with reduced health indicators in the turtles. These substances may also be associated with documented mortality events in the central GBR, however, further research is needed to confirm this attribution.

3.5 Risk Summary

Recent analysis (Waterhouse et al., 2017b) of the risk profile of pollutants to GBR ecosystems shows that the primary pollutants of concern to GBR coastal and marine ecosystems, that is, sediments, nutrients, and pesticides, are all important at different scales and different locations. The greatest water quality risks to the GBR are from: (1) nutrient discharge, particularly associated with (a) CoTS outbreaks and their destructive effects on mid-shelf coral reefs between Townsville and Lizard Island; and (b) adverse effects on coral diversity in inner-shelf reefs associated with macroalgal growth; (2) fine sediment discharge which reduces the light available to seagrass ecosystems and inshore coral reefs; and (3) pesticides which pose a risk to freshwater and some inshore and coastal ecosystems (Waterhouse et al., 2017b).

4 THE CURRENT WATER QUALITY MANAGEMENT RESPONSE AND PROGRESS

4.1 Governance

Overall, societal governance systems have multiple influences on outcomes in the GBR, with economic, social development, and environmental governance themes all playing a role (Dale et al., 2018). Within this wider complexity, several fragmented water, coastal and reef focussed policy, and legal frameworks guide the management of the GBR at multiple decision-making levels (Day, 2016; Morrison, 2017). Mounting pressure on the GBR's outstanding universal value suggests that further improvements in the overall system of governance and GBR-focused management arrangements are required (Vella and Baresi, 2017). As the governance of the GBR sits within a much wider governance system with many different influences on key ecosystem health outcomes in the GBR, particularly water quality, societal-wide attention to governance reform is required. Morrison (2017) and Dale et al. (2018) have extensively benchmarked the health of this much wider and overarching system, identifying major systemic problems resulting in poor GBR outcomes (Dale et al., 2013, 2016, 2018; Morrison, 2017; Vella and Forester, 2018; Vella and Baresi, 2017). Positive developments include the Reef Water Protection Plan and, more recently, the Reef 2050 Plan. There are, however, still weaknesses in these embryonic approaches including in the design and empowerment of catchment scale delivery systems, and the lack of integration between Reef 2050 Plan and several governance subdomains (Morrison, 2017; Dale et al., 2018).

4.2 Ports and Shipping

In all 12 ports are located along the GBR coast, including the major ports of Gladstone, Hay Point, Abbot Point, and Townsville (Fig. 2). The four major ports have a throughput (imports and exports) of over 200 million tons annually, and almost 4000 individual ships call in at GBR ports every year (Great Barrier Reef Marine Park Authority, 2014). Major expansions are underway or proposed for the Ports of Hay Point, Abbot Point, Townsville, and Gladstone, and port capacity is expected to triple in the region by 2020 from a 2012 baseline (Bureau of Resources and Energy Economics [BREE], 2012), mainly driven by growth in demand from the mining sector. Analysts predict that the projected increase in ship numbers calling into ports to meet this demand will exceed 10,000 by 2032 (PGM Environment, 2012).

Ports and shipping reduce water quality in the GBR both within and adjacent to port limits and shipping lanes (Grech et al., 2013). Water quality stressors from ports include leaching of toxic antifoulants and contamination caused by the release of hydrocarbons, coal dust, heavy metals, and synthetic materials from storage facilities and during transfer. Dredging operations in ports and shipping lanes cause water turbulence, resulting in increased SSs and hence turbidity and loss of light, the siltation and smothering of bottom habitats, and the remobilization of synthetic compounds, hydrocarbons, and heavy metals (Erftemeijer et al., 2012; Wenger et al., 2017) and the consequent sea dumping of maintenance dredge spoil. Propeller and ship movement cause increased turbidity, siltation, and smothering of bottom habitats and the remobilization of synthetic compounds, hydrocarbons, and heavy metals. Oil is also discharged from ships during both normal operations and shipping accidents.

There are multiple regulatory and legislative mechanisms for managing the stressors exerted by ports and shipping in the GBR. Port activities are governed by a combination of local, state, and national requirements, given that ports cross-jurisdictional boundaries including State (Queensland) lands and waters, the Great Barrier Reef Marine Park (GBRMP), the GBRWHA, and Commonwealth (Australian) waters. The impacts of existing port operations are monitored by compulsory environmental compliance monitoring programs. Environmental Impact Assessments (EIAs)—which provide information on the potential biophysical, social, and economic effects of proposed actions—support decisions on new port developments. The goal of EIAs is to ensure that significant effects on biodiversity are avoided, or that appropriate mitigation measures are in place. The GBRMP Authority, Australian Maritime Safety Authority, and Marine Safety Queensland jointly manage shipping activities under international treaty law as well as other domestic sources of law. There are multiple strategies to avoid shipping impacts and accidents in the GBR, including designated shipping lanes and anchorage sites, ship inspections, and compulsory pilotage.

However, there are serious deficiencies in the current management arrangements of ports (and for other coastal developments) in the GBR. Many analysts believe the EIA process to be flawed, due to a failure to manage potential conflicts of interest on the part of Governments and engaged environmental consultants (Grech et al., 2013). The methodologies used to collect data that inform EIA are not consistent across the region, nor are they independently peer-reviewed or made available to the public (Sheaves et al., 2016). The EIAs rarely consider the Total Economic Value of all possible options, including comprehensive assessments of ecosystem services. Flawed data collection methodologies and lack of transparency and independent review also undermine compliance monitoring programs of existing port developments.

4.3 Pollutant Loads Reduction

Reef Report Cards, released annually, report progress toward the 2013 Reef Plan targets (target date is 2018). Report Card 2015 (Australian and Queensland governments, 2016) and Report Card 2016 (Australian and Queensland governments, 2017) reported limited success across indicators and regions. Some of the key indicators are summarized in Table 1 from the 2015 report (see also Eberhard et al., 2017a,b) (see also Figs. 3 and 4). The results show that a large proportion (in some cases, up to 77%) of agricultural land in the main industries (sugarcane cultivation and beef grazing) is managed using practices that are below best practice for water quality. This lack of progress toward better water quality demonstrates the challenges in implementing improved (lower water quality risk) land management practices, and highlights the limited progress toward achieving the management practice adoption targets over the first 8 years of Reef Plan (Eberhard et al., 2017a; Waterhouse et al., 2017a).

In addition, the load reduction targets (initially for 2018 and then new targets for 2025) are highly unlikely to be met by either 2018 or 2025. This is been shown in Fig. 3 for SS and Fig. 4 for DIN where projections are based on "business as usual." Continued similar funding per year and no use of Federal legislation means that levels will fall far short of the required targets. This likely failure to meet any of the targets has been noted by UNESCO (UNESCO, 2015, 2017) as a major concern in the assessment of the possibility of putting the GBRWHA on the "World Heritage in danger" status. The UNESCO report criticized Australia's lack of progress toward achieving its 2050 water quality targets and failure to pass land clearing legislation (discussed below).

Modeled pollutant load reductions provide estimates of the achievements in improved management practices (Waters et al., 2014; McCloskey et al., 2017b). Measurable reductions in nutrient and sediment loads lag behind practice change due to the time for practice to take effect such as the time needed for increased vegetation cover to prevent erosion, which will vary with catchment, position in catchment, stream power and flow events, and time taken for riparian trees and hillslope pasture to reestablish and grow (Bainbridge et al., 2009).

Analysis of the report cards from recent years shows that the rate of progress toward the targets is slowing (Eberhard et al., 2017a, b). Comparison of the rate of reduction in loads of sediment, nutrients, and pesticides between the periods 2009–2013 and 2013–2015 (Figs. 3 and 4) are as follows:

- Sediment: 10% reduction from 2009 to 2013, that is, 2% per year, but only another 2% between 2013 and 2015 at 1% per year.
- DIN: 17% reduction from 2009 to 2013, that is, 3.5% per year, but only another 1.5% from 2013 to 2015 at 0.75% per year.

TABLE 1 Progress Towards Targets and Assigned Scores in the 2015 Great Barrier Reef Report Card

Activity	2018 Target[1]	Progress to Target as at 2014–2015[2] (%)	2015 Report Card Score[2]
DIN load	50% reduction	18	Poor "E"
Sediment load	20% reduction	12	Moderate "C"
Sugarcane land—managed to best practice standard	90%	23	Poor "D"
Cattle grazing land—managed to best practice standard	90%	36	Poor "D"

[1] Load targets are the reduction in anthropogenic loads based on modeled estimation of anthropogenic loads.

[2] 2015 "ABCDE" scoring system.

(From Eberhard, R., Thorburn, P., Rolfe, J., Taylor, B., Ronan, M., Weber, T., Flint, N., Kroon, F., Brodie, J., Waterhouse, J., Silburn, M., Bartley, R., Davis, A., Wilkinson, S., Lewis, S., Star, M., Poggio, M., Windle, J., Marshall, N., Hill, R., Maclean, K., Lyons, P., Robinson, C., Adame, F., Selles, A., Griffiths, M., Gunn, J., McCosker, K., 2017a. Scientific consensus statement 2017: a synthesis of the science of land-based water quality impacts on the Great Barrier Reef, Chapter 4. Management options and their effectiveness. State of Queensland, 2017. Available from http://www.reefplan.qld.gov.au/about/assets/2017-scientific-consensus-statement-summary-chap04.pdf).

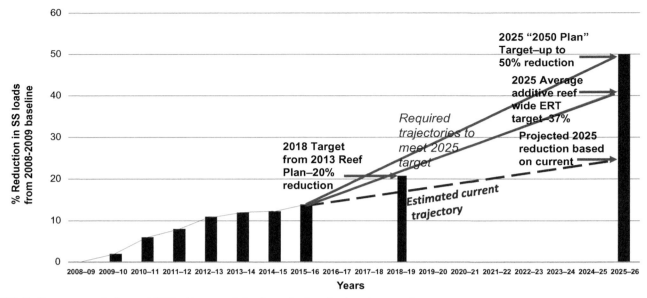

FIG. 3 Progress on reducing total GBR wide suspended sediment loads and trajectories toward targets.

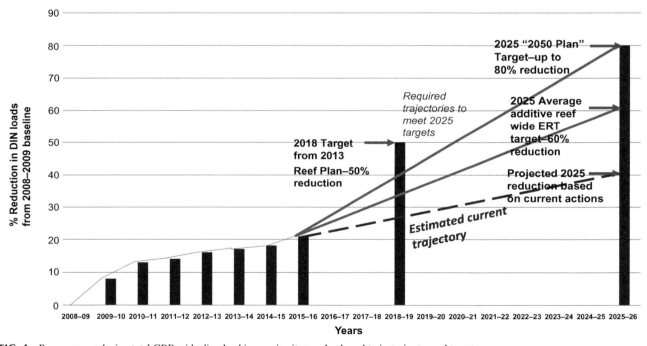

FIG. 4 Progress on reducing total GBR wide dissolved inorganic nitrogen loads and trajectories toward targets.

● Pesticides: 28% reduction from 2009 to 2013, that is, 5.5% per year, but only another 5% from 2013 to 2015 at 2.5% per year.

There was little further improvement between 2015 and 2016 as shown in the 2016 Report Card and as summarized in Figs. 3 and 4.

The overall condition of the inshore marine environment (water quality, seagrass, and coral) remains poor, and has not changed greatly since the 2011 Report Card. Marine water quality remained in generally "D" condition in 2015 but some areas have improved to "C" due to reduced rainfall and river discharge. It must be understood, however, that the methods used to assess water quality, reliant on remote sensing algorithms for chlorophyll and TSS, are highly inaccurate and biased (Waterhouse and Brodie, 2015) and thus our understanding of the true state of water quality is very limited, especially for inshore areas. In addition, results of the AIMS Long-Term Monitoring Program (mainly confined to mid- and outer-shelf reefs) and post-bleaching surveys of corals continue to show variations in coral cover often depending on climatic factors, but generally a continuing decline has the apparent frequency of events leading to coral mortality increases (e.g., Hughes et al., 2017b).

4.4 Crown-of-Thorns Starfish Management

The CoTS are a major cause of coral mortality in the GBR (De'ath et al., 2012) and a fourth wave of outbreaks is currently causing major coral mortality on mid-shelf reefs north of Townsville (Pratchett et al., 2017). Efforts to control CoTS numbers rely on two main management mechanisms: (1) culling CoTS at important tourism reef sites (Babcock et al., 2016) but with a recognition that control by culling at a whole of GBR scale is unlikely to be effective (Pratchett et al., 2014); and (2) reducing nutrient runoff from the Burdekin and Wet Tropics regions through better management practices, particularly in sugarcane cultivation (Brodie et al., 2012; Waterhouse et al., 2017b).

Culling CoTS at small-scale reef sites has been claimed by some to be locally successful if sufficient visits were made to high value reef areas. However, recent data cast doubt on the long-term effectiveness of culling even at small scales (Udo Englehardt, unpublished data). Other studies show that incomplete eradication at specific locations may simply prolong outbreaks and fail to actually protect local coral assemblages (Pratchett et al., 2017). Currently, more than $6 million per year is spent on CoTS culling with further funding recently committed by the Federal Government. This suggests that continuing improvement and refinement of these programs is required.

In Pratchett et al. (2017), a comprehensive review of our current knowledge on CoTS, it is noted that a major current research gap is: Can intensive culling contain or prevent, rather than eliminate, outbreaks? What are the detection limits and culling efficiencies for immature starfish? What are the long-term versus short-term benefits of direct control?

Control of CoTS populations through reducing nutrient discharge from fertilizer use (in sugarcane and banana cultivation) in the Wet Tropics and Burdekin regions is considered a potential management technique (Brodie et al., 2017a). However, reductions in loads of nutrients have never been sufficient (see Fig. 4) to assess whether this is working. Nutrient loads would have to be reduced far beyond what is being achieved at the moment for this to have any chance of success. Newly set targets for the rivers of the Wet Tropics (Brodie et al., 2017a) have required reductions of the order of 50%–70% of the anthropogenic load for DIN whereas actual reductions through management are still only around 20% (Australian and Queensland governments, 2017) and only predicted to reach an optimistic 40% by 2025 (Fig. 4). Recent studies (Walshe et al., 2017) have shown that substantially reducing Wet Tropics loads of nitrogen (via getting all sugarcane fertilizer use in sugarcane to "B" class practice) at a cost of $51.9 million could delay the start of the next wave of outbreaks by about 1.4 years. More drastic measures involving land retirement and greater costs could lengthen the return period further. In contrast, a greatly extended culling regime had no predicted effect on the timing of the next wave (expected to be in about 2024, 15 years after the initiation of the fourth wave in 2009). Even with sufficient resources to fund 20 surveillance boats and 40 cull boats (at an estimated cost of $3.5 billion), the mean return time under currently imperfect and difficult probabilities of detectability was less than 17 years, only a 1-year delay (Walshe et al., 2017). The assessment of surveillance and cull options relates only to the objective of arresting outbreaks in the initiation phase. The analyses did not extend to exploration of the effectiveness of options for protecting high-value reefs or reducing the rate of spread or intensity of downstream secondary impacts.

4.5 Tree Clearing

Land clearing threatens biodiversity and the functioning of terrestrial, freshwater, and marine ecosystems (Reside et al., 2016, 2017) and is known to increase erosion in rates in the GBR catchment and hence increase sediment loads from rivers (Hairsine, 2017). While many land managers have improved property management, tree clearing in environmentally sensitive GBR catchment areas continues to have the potential to impact Reef water quality and the ecological integrity of catchments, with high levels from 2014 (WWF, 2017). The 2014–2015 Statewide Land Cover and Tree Study (State of Queensland, 2015) showed the rate of clearing in Queensland had increased by 46% since 2011–2012. Across the state tree clearing is continuing at a level of almost 300,000 ha per year since the modification of the tree clearing laws in 2013 (Reside et al., 2017).

Continuation of the current rates of land clearing in Queensland will also undermine the recent State and Federal Government investments in improving water quality in the GBR (Reside et al., 2017), and will increase the cost of counteracting water quality decline estimated at $5–10 billion over the next 10 years (Alluvium, 2016; Brodie and Pearson, 2016).

5 THE FUTURE BASED ON CURRENT MANAGEMENT REGIME

Given the estimated investment of around $700 million from 2009 to 2015 (Brodie and Pearson, 2016), progress toward reaching the water quality targets appears to be slow, albeit against the massive scale of the GBR catchment. Modeling has also shown that even complete adoption of existing industry "best management practices" is not expected to achieve sediment and nutrient targets (Thorburn and Wilkinson, 2013; Alluvium, 2016)—see Figs. 3 and 4.

Multiscale drivers have confounded efforts to manage the GBR in recent times (Morrison, 2017). Between 2004 and 2011 an unprecedented boom in industrialization and urbanization in the wider GBR catchment caused a near doubling of shipping exports combined with high population growth. In 2008, the global financial crisis caused the AAA credit rating for Queensland and Australia to be downgraded, triggering public spending cuts. Both Queensland and Australia were increasingly dependent on the coal industry to avoid recession. In 2011, the boom ended altogether, accompanied by more public spending cuts. The coal industry successfully campaigned for environmental deregulation to maintain profit levels (Morrison, 2017). Morrison (2017) has comprehensively assessed the health of the overall GBR governance regime, showing that increasing complexity has masked these regressive changes to the regime. More recently, increasing global outcry and the 2016 and 2017 mass coral bleaching events have caused the Australian government to respond with additional funding for the GBR and a review of the governance of GBRMPA. While this is good news, the funding has been critiqued for being too lean and poorly directed, the review of GBRMPA for merely improving the role and composition of the board and executive management and not the ministerial council, and overall for failing to address climate mitigation (Morrison, 2017).

The Great Barrier Reef Water Science Taskforce Report (Great Barrier Reef Water Science Taskforce, 2016), highlighted "the poor outcome of continued business-as-usual as per current investment, and an indicative steep trajectory that will be needed to meet water quality targets." Tarte et al. (2017) also noted that "…as the GBR Water Science Taskforce Report and the 2015 Report Card assessment clearly show, progress with water quality load targets is not 'on-track' and it is highly likely that most 2018 targets will not be met. Consequently, if the 2018 targets are not met, it will be extremely challenging to meet the 2025 targets, particularly for DIN, which is the highest target to achieve (up to 80%), but has the worst performance to date." Similarly in the water quality assessment of the progress of the 2050 GBR Long Term Sustainability Plan (Section 2.3) (Roth et al., 2017) it was noted that no comprehensive or effective regulatory and compliance regime to improve water quality in the GBR has been enacted for diffuse sources of pollution (i.e., agricultural land uses), the main source of sediment, nutrients, and pesticides entering the reef (Kroon et al., 2016; Tarte et al., 2017).

Several studies have emphasized the lack of effective legislative and regulatory instruments governing agricultural land uses and management in catchments discharging into the GBR WHA (Jacobs Group (Australia) Pty Limited, 2014; McGrath, 2010; Wulf, 2004), including the lack of collaborative design underpinning such systems (Dale et al., 2018). For example, Wulf (2004) reiterated that the Australian Government has the power under two of the Acts listed in Reef 2050 Plan, the GBRMP Act 1975 and the Environment Protection and Biodiversity Conservation Act 1999, to better control land-based pollution into the GBR Marine Park and WHA, but has to date not applied this provision.

6 WHAT WOULD SUCCESS LOOK LIKE?

The management of terrestrial pollutant discharge into the GBR implicitly assumes that the impacts of the current loads of nutrients, sediments, and pesticides would be reversed if the loads were reduced. Thus there is an implicit aim that if loadings are reduced enough the continuing decline of species and ecosystems impacted by pollution can be reversed and hopefully system restoration may be achieved. Such restoration has been observed after, for example, nutrient management in Tampa, Florida, when seagrass meadows were restored to near their pre-pollution condition (Greening et al., 2014). In the GBR case, the restoration possibilities are complicated by the reality of multiple stressors, particularly those associated with climate change. Whether water quality management alone, even if successful, will be adequate to reverse the decline in coral cover on the GBR is probably unlikely given minimal or no action globally or nationally on climate change.

In nutrient-enriched conditions there are well-documented cases of eutrophied marine systems, dominated by algae, where reductions in nutrient loading have not returned the systems to their original ecological status (e.g., Duarte et al., 2009) or where recovery has been partial at best (Elliott et al., 2016; Borja et al., 2010; Gross and Hagy, 2017). This may be attributed to the range of other factors in the system that have dramatically changed during the period of increased nutrient loading, such as human population increases, increased carbon dioxide in the atmosphere, freshwater runoff changes, global temperature increases, and fish stock losses. Alternatively the management regime that enabled the nutrient loading reductions may have weakened or been repealed in the case of legal solutions (Gross and Hagy, 2017). In Moreton Bay, Queensland, nitrogen reductions have not reduced algal growth as the system is possibly phosphorus limited (Wulff et al., 2011) although increased growth of one species has been observed [nitrogen (N2)-*fixing Lyngbya majuscula*].

Successful examples in tropical seagrass/coral reef settings of management of terrestrial discharges such that ecosystem restoration occurred are as follows:

1. Tampa Bay, Florida (Greening et al., 2014) where reduction in wastewater nutrient loading of approximately 90% in the late 1970s lowered external total nitrogen loading by more than 50% within three years. After 1997–1998, water clarity increased significantly and seagrass is expanding significantly. Other contributing factors included a diversity of approaches that spanned:

   (a) active community involvement, including clear agreements about quantifiable restoration goals;
   (b) (b) regulatory, incentive-based, and voluntary reduction in nutrient loadings from point, atmospheric, and non-point sources;
   (c) long-term water quality and seagrass monitoring to assess the success of the program; and
   (d) continuing commitment from public and private sectors to work together to attain restoration goals.

2. Kaneohe Bay, Oahu, Hawaii (e.g., Stimson, 2015, 2018; Bahr et al., 2015). Sewage discharges into Kaneohe Bay, Hawaii increased from the end of the Second World War due to increasing population and urbanization to 20 ml/day in 1977. This chronic discharge into the lagoon introduced high levels of inorganic N and inorganic P, and southern lagoon waters become increasingly rich in phytoplankton. Reefs closest to the outfall become overgrown by filter-feeding organisms, such as sponges, tube worms, and barnacles. Reefs in the center of the bay further from the outfalls were overgrown by the indigenous green algae *Dictyosphaeria cavernosa*. After diversion of the outfalls into the ocean in 1978, in-water nutrient levels reduced, phyto- and zooplankton populations declined, and *D. cavernosa* abundance declined to 25% of the previous level. At the same time, increases in the abundance and distribution of coral species were reported, and the reefs slowly recovered. A drastic decline in previously dominant *D. cavernosa* occurred in 2006 is attributed to a gradual return to a coral-dominated state.

3. The temperate waters in Denmark provide an example of watershed management leading to some marine ecosystem recovery (Danish, 2012; Nimmo Smith et al., 2007; Riemann et al., 2015). Nutrient inputs from land were reduced by ~50% for nitrogen and 56% for phosphorus in the period from 1990 to 2015. These reductions resulted in significant and parallel declines in nutrient concentrations, and initiated a shift in the dominance of primary producers toward reduced phytoplankton biomass and increased cover of macroalgae in deeper waters. In the period 2009–2015 eelgrass (seagrass) meadows have also expanded toward deeper waters, in response to improving water clarity (Riemann et al., 2015). This success was highly dependent on a strong regulatory regime based on the European Framework Directive (Hering et al., 2010) which introduced a new legislative approach for managing and protecting water, based not on national or political boundaries but on natural geographical and hydrological formations: river basins. It also requires coordination of different EU policies, and sets out a precise timetable for action, with 2015 as the target date for getting all European waters into good condition.

The only management approaches that have resulted in measurable reductions in agricultural pollution to coastal ecosystems and an improvement of the marine ecosystem at risk have all taken integrated approaches inclusive of a strong regulatory component and have not relied on voluntary action alone (Greening et al., 2014; Stimson, 2015, 2018; Bahr et al., 2015; Riemann et al., 2015). Thus, it is likely that voluntary approaches alone may not be sufficient to achieve measurable improvements in GBR ecosystems via reducing pollutant loads to marine waters (Kroon et al., 2014, 2016). Hence for the targets to be achieved a strong regulatory component of the policy mix will be required (Brodie and Pearson, 2016; Kroon et al., 2014, 2016). There is also evidence, however, of the need for greater co-management approaches and increasingly adaptive and flexible cultures in the design and deployment of regulation (Dale et al., 2018).

Currently, ecologically relevant targets for pollutant load reductions have been set for 35 basins of the GBR catchment (Brodie et al., 2017b). These targets incorporate an ecological endpoint in the GBR (e.g., Brodie et al., 2016b) and, using modeling, a sufficient reduction is made such that the ecological endpoints are achieved. Thus the scientific underpinnings of these basin-scale targets are to achieve a restoration outcome for specific GBR ecosystems, and, in particular, in this case coral and seagrass status. This can be compared to the case studies referred to above where (a) seagrass was restored in Tampa Bay after an 80% reduction in total nitrogen loading; and (b) coral recovery occurred (albeit after a time lag of 30 years) in Kaneohe Bay after sewage effluent loading was largely eliminated by diversion to oceanic waters.

In essence it is assumed that reductions in pollutant loading to the GBR, to the extent of the new targets, will also achieve a recovery of coral (cover, diversity, and community structure) and seagrass (cover, biomass, spatial extent, community structure) to a significant degree. This restoration will then also benefit "downstream" species which are dependent on good coral or seagrass status, for example, dugongs. A complicating factor is, of course, that other stressors besides pollution are also impacting corals and seagrass of the GBR. The most prominent and important of these stressors is climate change. As climate change impacts accelerate (e.g., coral bleaching) even highly effective pollution management may not restore coral and seagrass to our projected restoration objectives.

However, local action to reduce sediment and nutrient loading is still likely important as it is known that nutrients are involved in subtle ways in the bleaching response (e.g., Wooldridge, 2017; Wooldridge et al., 2017). Recently released research shows that sub-bleaching temperatures and excess nitrogen promotes symbiont (zooxanthellae) parasitism in corals (Baker et al., 2018). These results and others show that reducing nutrient exposure is beneficial in reducing the susceptibility of corals to bleach.

7 WHAT COULD BE DONE TO IMPROVE GOVERNANCE AND MANAGEMENT?

The GBR Water Science Taskforce found that effective governance of water quality improvement in the GBR would require (Great Barrier Reef Water Science Taskforce, 2016):

- A mix of strategies and tools, including incentives, regulation, and innovation. Everyone including farmers, graziers, developers, the resources sector, community members, traditional owners, and tourism operators to be part of the solution.
- Recognition that achieving water quality targets in the timescale proposed would be well beyond the funds currently allocated by the Queensland and Australian governments.
- A move away from just voluntary compliance to mandatory and enforced regulations with sufficient penalties to ensure compliance.

Progress toward the Reef Plan targets has been slow and the present trajectory will not meet the current targets in 2018 or 2025 (Eberhard et al., 2017a,b). Greater combined investment in voluntary practice change programs and the use of regulatory tools and other policy mechanisms to accelerate practice change adoption are required. Gullies and streambank erosion are more important sources of sediment than previously thought, and management of these sources is a substantial challenge. There are new approaches to fertilizer management that, when fully developed and implemented, are likely to reduce nutrient losses. Policymakers can improve programs by making better use of social, economic, and institutional research, further collaboration with farmers and other stakeholders, and strong evaluation systems (Eberhard et al., 2017a).

As a result of the findings of the Taskforce (Great Barrier Reef Water Science Taskforce, 2016), the Queensland Government allocated the balance of its $100 million commitment to improving Reef water quality, establishing a range of targeted programs including: (i) the development of two major integrated programs (MIPs); (ii) extension; (iii) innovation; (iv) governance improvement; (v) monitoring and evaluation; and (vi) communication (Roth et al., 2017).

The MIPs have been funded and implemented (from 2017) in the Wet Tropics Basins of the Johnstone and the Tully in sugarcane cultivation management (https://terrain.org.au/wp-content/uploads/2017/08/Wet-Tropics-Major-Integrated-Project-Monday-final-5-July-2017.pdf), and in the Burdekin basin for erosion control focusing on gully remediation (http://www.nqdrytropics.com.au/tag/major-integrated-project/).

A significant project has for the first time sought to define and then benchmark the health of the over-arching system of governance affecting social, economic, and environmental outcomes relative to the GBR (Dale et al., 2016). This GBR-focussed application of the new Governance System Analysis (GSA) technique identifies those governance subdomains that present a high, medium, or low risk of failure to produce positive outcomes for GBR. This system-wide analysis tool system was developed specifically for use in analyzing the health of the system of governance affecting outcomes in the GBR. This approach importantly determined that three "whole of system" governance problems could undermine GBR outcomes.

- While stressing the integrative importance of the Reef 2050 Subdomain, the work considered that, due to its embryonic state, it faces several internal governance challenges.
- The identification of a major risk of implementation failure in the achievement of GBR water quality actions due to a lack of system-wide focus on building strong, stable delivery systems at catchment scale within 10 key governance subdomains (catchment delivery challenges).
- A finding that the Reef 2050 Subdomain currently has too limited a mandate/capacity to influence several other high-risk subdomains or competing policy challenges that must be more strongly aligned with Reef management such as the Greenhouse Gas Emission Management or Northern Development Subdomains.

8 A WAY FORWARD

Many studies examining management of water quality for the GBR in the face of climate change have compiled lists of well thought-out recommendations which the authors concluded would expedite successful management. These studies include the 2017 Scientific Consensus Statement (Waterhouse et al., 2017a; Eberhard et al., 2017a), the GBR water Science Taskforce (Great Barrier Reef Water Science Taskforce, 2016), and Brodie and Pearson (2016). Here we use these recommendations (sometimes modified), and others drawn from this study, to document our set of recommendations.

(1) Use a mix of policy instruments, including both voluntary and regulatory approaches (Brodie and Pearson, 2016; Gardner and Waschka, 2012; Waschka and Gardner, 2016; Great Barrier Reef Water Science Taskforce, 2016; Waterhouse et al., 2017a), initially with a greater focus on mandatory regulation with strong penalties for noncompliance. The revised Queensland Enhanced Regulations (open for comment in early 2018 in a Regulatory Impact Statement phase), may provide a much stronger regulatory regime for reducing nitrogen runoff to the GBR. It incorporates land retirement and change of land use from intensive industries (e.g., sugarcane) to less intensive (e.g., grazing). In addition stronger tree clearing legislation needs to be reinstated after the legislation was weakened by a previous government. Trees and grass both have a role in preventing erosion and the lack of effective vegetation management legislation is likely to increase erosion across the GBR catchment. A new "Vegetation Management" bill has now been introduced to the Queensland parliament (March 2018) to achieve this aim.

(2) Implement staged regulations for the Federal legislation to meet reef outcomes (e.g., Great Barrier Reef Water Science Taskforce, 2016; Kroon et al., 2014). Consideration should be given to the use of the GBRMP Act and the EPBC Act to regulate catchment activities that lead to damage to the Greater GBR (in conjunction with the relevant Queensland legislation as in Recommendation (1) (Brodie and Pearson, 2016; Eberhard et al., 2017b). More codesign is required in development and monitoring new approaches.

(3) Fully quantify the water quality and cost benefits of innovative management practice and system repair management options to complement existing options (e.g., Star et al., 2017).

(4) 4. Prioritize much more action within local regions (e.g., the MIPs approach). Strengthen management in the areas of the GBR with ecosystems with perceived higher resilience, for example, Cape York. Given existing limited funding, heavily reprioritize current spending to areas where a difference can still be made. For example, prioritize funding for sugarcane fertilizer usage reductions in the Wet Tropics where it may be effective in reducing and/or delaying CoTS outbreaks (Alluvium, 2016; Walshe et al., 2017; Brodie et al., 2017a). Similarly heavily prioritize erosion control management in the Mary and Burdekin Basins where reduction in SS discharge can make a difference to important seagrass meadows (Fabricius et al., 2014; Brodie et al., 2017a; Wooldridge, 2017). Both of these priorities largely fit the current MIPs approach. Commence more enhanced triage, protection, and restoration (Waterhouse et al., 2016).

(5) 5. GBR basin-specific pollutant load reduction targets (Brodie et al., 2017b) have been adopted but must be supported by a detailed, comprehensive, costed water quality management plan for the GBR (Waterhouse et al., 2017a; Great Barrier Reef Water Science Taskforce, 2016; Tarte et al., 2017; Brodie and Pearson, 2016; Eberhard et al., 2017a,b).

(6) 6. Increase funding to support catchment and coastal management delivery systems to the required levels identified to address the pollution issues for the GBR by 2025, that is, meet the water quality targets (Great Barrier Reef Water Science Taskforce, 2016; Tarte et al., 2017). This is likely to require expenditure of 1 billion dollars annually over the next 5–7 years in contrast to the current expenditure of about one-tenth of this (Brodie and Pearson, 2016; Alluvium, 2016).

(7) 7. Strengthened governance through improved independence (of core agencies from politics and exogenous pressure), improved implementation and delivery systems, more regular updating of governance arrangements, better horizontal and vertical policy integration, smarter use of available regulatory mechanisms, and better use of independent watchdogs (Dale et al., 2016; Morrison, 2017; Vella and Baresi, 2017; Roth et al., 2017).

(8) Explore alternative financing, including the establishment of structured ecosystem service markets, funding from private sources, and enhanced market delivery models.

(9) Strengthen the EIA process through the implementation of independent quality control and peer review of EIA as well as compliance monitoring programs (Grech et al., 2013; Sheaves et al., 2016).

(10) The need to modify our objectives in the light of climate change and the now limited capacity for recovery (e.g., Osborne et al., 2017). Hughes et al. (2017b) explained the need to adjust expectations (and see also Adams et al., 2017; Webster et al., 2017; Fraser et al., 2017; Roth et al., 2017).

Although many management approaches applied in the past in the GBR may have been both effective and appropriate, clearly today's pressures and cumulative impacts mean that these past approaches are not enough to retain the values for which the GBR has long been acclaimed (Day, 2016). The Australian Government's GBR Outlook Report 2014 (Great Barrier Reef Marine Park Authority, 2014, p. v), concluded:

> *Even with the recent management initiatives to reduce threats and improve resilience, the overall outlook for the GBR is poor, has worsened since 2009 and is expected to further deteriorate in the future. Greater reductions of threats at all levels, Reef-wide, regional, and local, are required to prevent the projected declines in the GBR and to improve its capacity to recover.*

Addressing these concerns is doable—but the longer the GBR is left without appropriate levels of restoration and effective management, the harder and more expensive it will become to address.

ACKNOWLEDGMENT

We would like thank Jane Waterhouse for proofreading the ms several times and helping draft the figures.

REFERENCES

Adams, V.M., Álvarez-Romero, J.G., Capon, S.J., Crowley, G.M., Dale, A.P., Kennard, M.J., Douglas, M.M., Pressey, R.L., 2017. Making time for space: the critical role of spatial planning in adapting natural resource management to climate change. Environ. Sci. Pol. 74, 57–67.

Alluvium, 2016. Costs of achieving the water quality targets for the GBR by Alluvium Consulting Australia for Department of Environment and Heritage Protection, Brisbane. Available from: www.openchannels.org/sites/default/files/literature/Costs%20of%20achieving%20the%20water%20quality%20targets%20for%20the%20Great%20Barrier%20Reef.pdf

Australian and Queensland Governments, 2016. Great Barrier Reef Report Card 2015 Reef Water Quality Protection Plan. Queensland Government, Brisbane, Australia.

Australian and Queensland Governments, 2017. Great Barrier Reef Report Card 2016 Reef Water Quality Protection Plan. Queensland Government, Brisbane, Australia.

Babcock, R.C., Dambacher, J.M., Morello, E.B., Plaganyi, E.E., Hayes, K.R., HPA, S., et al., 2016. Assessing different causes of crown-of-thorns starfish outbreaks and appropriate responses for management on the Great Barrier Reef. PLoS One 11 (12), e0169048. https://doi.org/10.1371/journal.pone.0169048.

Bahr, K.D., Jokiel, P.L., Toonen, R.J., 2015. The unnatural history of Kāneʻohe Bay: coral reef resilience in the face of centuries of anthropogenic impacts. Peer J 3, e950.

Bainbridge, Z.T., Brodie, J.E., Lewis, S.E., Waterhouse, J., Wilkinson, S.N., 2009. In: Amderssen, B., et al. (Eds.), Utilising catchment modelling as a tool for monitoring Reef rescue outcomes in the Great Barrier Reef catchment area. 18th IMACS World Congress—MODSIM International Congress on Modelling and Simulation, 13-17 July 2009, Cairns, Australia. ISBN: 978-0-9758400-7-8. Available from: http://mssanz.org.au/modsim09.

Baker, D.M., Freeman, C.J., Wong, J.C., Fogel, M.L., Knowlton, N., 2018. Climate change promotes parasitism in a coral symbiosis. The ISME Journal 1.

Bartley, R., Waters, D., Turner, R., Kroon, F., Wilkinson, S., Garzon-Garcia, A., Kuhnert, P., Lewis, S., Smith, R., Bainbridge, Z., Olley, J., Brooks, A., Burton, J., Brodie, J., Waterhouse, J., 2017. Scientific Consensus Statement 2017: A Synthesis of the Science of Land-based Water Quality Impacts on the Great Barrier Reef, Chapter 2: Sources of Sediment, Nutrients, Pesticides and Other Pollutants to the Great Barrier Reef. State of Queensland, 2017.

Bayraktarov, E., Pizarro, V., Eidens, C., Wilke, T., Wild, C., 2013. Bleaching susceptibility and recovery of Colombian Caribbean corals in response to water current exposure and seasonal upwelling. PLoS One 8, e80536https://doi.org/10.1371/journal.pone.0080536.

Bellwood, D.R., Fulton, C.J., 2008. Sediment-mediated suppression of herbivory on coral reefs: decreasing resilience to rising sea-levels and climate change? Limnol. Oceanogr. 53 (6), 2695–2701.

Berry, K.L.E., O'Brien, D., Burns, K.A., Brodie, J., 2013. Unrecognized Pollutant Risk to the Great Barrier Reef. Great Barrier Reef Marine Park Authority, Townsville, Australia.

Bessell-Browne, P., Negri, A.P., Fisher, R., Clode, P.L., Duckworth, A., Jones, R., 2017a. Impacts of turbidity on corals: the relative importance of light limitation and suspended sediments. Mar. Pollut. Bull. 117 (1-2), 161–170.

Bessell-Browne, P., Negri, A.P., Fisher, R., Clode, P.L., Jones, R., 2017b. Impacts of light limitation on corals and crustose coralline algae. Sci. Rep. 7 (1), 11553.

Birrell, C.L., McCook, L.J., Willis, B.L., Harrington, L., 2008. Chemical effects of macroalgae on larval settlement of the broadcast spawning coral Acropora millepora. Mar. Ecol. Prog. Ser. 362, 129–137.

Borja, Á., Dauer, D.M., Elliott, M., Simenstad, C.A., 2010. Medium-and long-term recovery of estuarine and coastal ecosystems: patterns, rates and restoration effectiveness. Estuar. Coasts 33 (6), 1249–1260.

Brodie, J.E., 1992. Enhancement of larval and juvenile survival and recruitment in *Acanthaster planci* from the effects of terrestrial runoff: a review. Mar. Freshw. Res. 43 (3), 539–553.

Brodie, J., 2014. Dredging the Great Barrier Reef: use and misuse of science. Estuar. Coast. Shelf Sci. 142, 1–3.

Brodie, J., Pearson, R.G., 2016. Ecosystem health of the Great Barrier Reef: time for effective management action based on evidence. Estuar. Coast. Shelf Sci. 183, 438–451.

Brodie, J., Waterhouse, J., 2012. A critical review of environmental management of the "not so Great" Barrier Reef. Estuar. Coast. Shelf Sci. 104-105, 1–22.

Brodie, J., Fabricius, K.E., De'ath, G., Okaji, K., 2005. Are increased nutrient inputs responsible for more outbreaks of crown-of-thorns starfish? An appraisal of the evidence. Mar. Pollut. Bull. 51, 266–278.

Brodie, J.E., Devlin, M.J., Haynes, D., Waterhouse, J., 2011. Assessment of the eutrophication status of the Great Barrier Reef lagoon (Australia). Biogeochemistry 106, 281–302. https://doi.org/10.1007/s10533-010-9542-2.

Brodie, J., Kroon, F., Schaffelke, B., Wolanski, E., Lewis, S., Devlin, M., Bainbridge, Z., Waterhouse, J., Davis, A., 2012. Terrestrial pollutant runoff to the Great Barrier Reef: current issues, priorities and management responses. Mar. Pollut. Bull. 65, 81–100.

Brodie, J., Burford, M., Davis, A., da Silva, E., Devlin, M., Furnas, M., Kroon, F., Lewis, S., Lønborg, C., O'Brien, D., Schaffelke, B., Bainbridge, Z., 2015. The Relative Risks to Water Quality from Particulate Nitrogen Discharged from Rivers to the Great Barrier Reef in Comparison to Other Forms of Nitrogen. Centre for Tropical Water & Aquatic Ecosystem Research, James Cook University, Townsville.

Brodie, J.E., Lewis, S.E., Collier, C.J., Wooldridge, S., Bainbridge, Z.T., Waterhouse, J., Rasheed, M.A., Honchin, C., Holmes, G., Fabricius, K., 2016b. Setting ecologically relevant targets for river pollutant loads to meet marine water quality requirements for the Great Barrier Reef, Australia: a preliminary methodology and analysis. Ocean Coast. Manag. 143, 136–147.

Brodie, J., Devlin, M., Lewis, S., 2017a. Potential enhanced survivorship of crown of thorns starfish larvae due to near-annual nutrient enrichment during secondary outbreaks on the central mid-shelf of the Great Barrier Reef, Australia. Diversity 9 (1), 17.

Brodie, J., Baird, M., Waterhouse, J., Mongin, M., Skerratt, J., Robillot, C., Smith, R., Mann, R., Warne, M., 2017b. In: Development of basin-specific ecologically relevant water quality targets for the Great Barrier Reef. TropWATER Report No. 17/38, James Cook University, Published by the State of Queensland, Brisbane, Australia. pp. 68.

Bureau of Resources and Energy Economics [BREE], 2012. Australian Bulk Commodity Export and Infrastructure: Outlook to 2025. Australian Department of Resources, Energy and Tourism, Canberra, Australia, p. 150.

Caron, A.G., Thomas, C.R., Berry, K.L., Motti, C.A., Ariel, E., Brodie, J.E., 2018. Ingestion of microplastic debris by green sea turtles (Chelonia mydas) in the Great Barrier Reef: validation of a sequential extraction protocol. Mar. Pollut. Bull. 127, 743–751.

Chartrand, K.M., Bryant, C.V., Carter, A.B., Ralph, P.J., Rasheed, M.A., 2016. Light thresholds to prevent dredging impacts on the Great Barrier Reef seagrass, *Zostera muelleri ssp. capricorni*. Front. Mar. Sci. 3, 106.

Chazottes, V., Hutchings, P., Osorno, A., 2017. Impact of an experimental eutrophication on the processes of bioerosion on the Reef: one tree island, Great Barrier Reef, Australia. Mar. Pollut. Bull. 118 (1–2), 125–130.

Cheal, A.J., MacNeil, M.A., Emslie, M.J., Sweatman, H., 2017. The threat to coral reefs from more intense cyclones under climate change. Glob. Chang. Biol. 23, 1511–1524.

Coles, R.G., Rasheed, M.A., McKenzie, L.J., Grech, A., York, P.H., Sheaves, M., McKenna, S., Bryant, C., 2015. The Great Barrier Reef world heritage area seagrasses: managing this iconic Australian ecosystem resource for the future. Estuar. Coast. Shelf Sci. 153, A1–A12.

Collier, C., Waycott, M., 2009. Drivers of change to seagrass distributions and communities on the Great Barrier Reef: literature review and gaps analysis. Reef and Rainforest Research Centre.

Collier, C.J., Uthicke, S., Waycott, M., 2011. Thermal tolerance of two seagrass species at contrasting light levels: implications for future distribution in the Great Barrier Reef. Limnol. Oceanogr. 56, 2200–2210.

Collier, C.J., Waycott, M., Giraldo Ospina, A., 2012. Responses of four Indo-West Pacific seagrass species to shading. Mar. Pollut. Bull. 65, 342–354.

Collier, C.J., Adams, M., Langlois, L., Waycott, M., O'Brien, K., Maxwell, P., McKenzie, L., 2016a. Thresholds for morphological response to light reduction for four tropical seagrass species. Ecol. Indic. 67, 358–366.

Collier, C.J., Chartrand, K., Honchin, C., Fletcher, A., Rasheed, M., 2016b. In: Light thresholds for seagrasses of the GBR: a synthesis and guiding document. Including knowledge gaps and future priorities. Report to the National Environmental Science Programme. Reef and Rainforest Research Centre Limited, Cairns. pp. 41.

D'Angelo, C., Wiedenmann, J., 2014. Impacts of nutrient enrichment on coral reefs: new perspectives and implications for coastal management and reef survival. Curr. Opin. Environ. Sustain. 7, 82–93.

Dale, A., Vella, K., Pressey, R.L., Brodie, J., Yorkston, H., Potts, R. 2013. A method for risk analysis across governance systems: a Great Barrier Reef case study Environ. Res. Lett., 8 (1), art. no. 015037.

Dale, A.P., Vella, K., Pressey, R.L., Brodie, J., Gooch, M., Potts, R., Eberhard, R., 2016. Risk analysis of the governance system affecting outcomes in the Great Barrier Reef. J. Environ. Manag. 183, 712–721.

Dale, A., Vella, K., Gooch, M., Potts, R., Pressey, R.L., Brodie, J., Eberhard, R., 2018. Avoiding implementation failure in catchment landscapes: a case study in Governance of the Great Barrier Reef. Environ. Manag. 62 (1), 70–81.

Danish EPA. 2012. Nitrate action programme 2008–2015. Danish Environmental Protection Agency. Available from: http://www.mst.dk/English/Agriculture/nitrates_directive/nitrate_action_programme/.

Davis, A.M., Pradolin, J., 2016. Precision herbicide application technologies to decrease herbicide losses in furrow irrigation outflows in a northeastern Australian cropping system. J. Agric. Food Chem. 64 (20), 4021–4028.

Davis, A.M., Thorburn, P.J., Lewis, S.E., Bainbridge, Z.T., Attard, S.J., Milla, R., Brodie, J.E., 2013. Environmental impacts of irrigated sugarcane production: herbicide run-off dynamics from farms and associated drainage systems. Agric. Ecosyst. Environ. 180, 123–135.

Davis, A.M., Lewis, S.E., Brodie, J.E., Benson, A., 2014. The potential benefits of herbicide regulation: a cautionary note for the Great Barrier Reef catchment area. Sci. Total Environ. 490, 81–92.

Day, J.C. 2016. The Great Barrier Reef marine park—the grandfather of modern MPAs. Chapter 5 (pp. 65-97) In: Big, Bold and Blue: Lessons from Australia's Marine Protected Areas, (Eds.), Fitzsimmons and Wescott. CSIRO Publishing, Canberra, Australia. ISBN: 9781486301942.

Day, J.C., Dobbs, K., 2013. Effective governance of a large and complex cross-jurisdictional marine protected area: Australia's Great Barrier Reef. Mar. Policy 41, 14–24.

De'ath, G., Fabricius, K., 2010. Water quality as a regional driver of coral biodiversity and macroalgae on the Great Barrier Reef. Ecol. Appl. 20, 840–850.

De'ath, G., Fabricius, K.E., Sweatman, H., Puotinen, M., 2012. The 27–year decline of coral cover on the Great Barrier Reef and its causes. Proc. Natl. Acad. Sci. U. S. A. 109 (44), 17734–17735.

DeCarlo, T.M., Cohen, A.L., Barkley, H.C., Cobban, Q., Young, C., Shamberger, K.E., Brainard, R.E., Golbuu, Y., 2015. Coral macrobioerosion is accelerated by ocean acidification and nutrients. Geology 43 (1), 7–10.

Deloitte Access Economics, 2017. At What Price? The Economic, Social and Icon Value of the Great Barrier Reef. Deloitte Access Economics, Brisbane. Available from: https://www2.deloitte.com/content/dam/Deloitte/au/Documents/Economics/deloitte-au-economics-great-barrier-reef-230617.pdf.

DeVantier, L.M., De'ath, G., Turak, E., Done, T.J., Fabricius, K.E., 2006. Species richness and community structure of reef-building corals on the nearshore Great Barrier Reef. Coral Reefs 25 (3), 329–340.

Dogruer, G., Weijs, L., Tang, J.Y.M., Hollert, H., Kock, M., Bell, I., Hof, C.A.M., Gaus, C., 2018. Effect-based approach for screening of chemical mixtures in whole blood of green turtles from the Great Barrier Reef. Sci. Total Environ. 612, 321–329.

Duarte, C.M., Conley, D.J., Carstensen, J., Sánchez-Camacho, M., 2009. Return to Neverland: shifting baselines affect eutrophication restoration targets. Estuar. Coasts 32 (1), 29–36.

Eberhard, R., Thorburn, P., Rolfe, J., Taylor, B., Ronan, M., Weber, T., Flint, N., Kroon, F., Brodie, J., Waterhouse, J., Silburn, M., Bartley, R., Davis, A., Wilkinson, S., Lewis, S., Star, M., Poggio, M., Windle, J., Marshall, N., Hill, R., Maclean, K., Lyons, P., Robinson, C., Adame, F., Selles, A., Griffiths, M., Gunn, J., McCosker, K., 2017a. Scientific consensus statement 2017: a synthesis of the science of land-based water quality impacts on the Great Barrier Reef, Chapter 4. In: Management options and their effectiveness. State of Queensland. Available from: http://www.reefplan.qld.gov.au/about/assets/2017-scientific-consensus-statement-summary-chap04.pdf.

Eberhard, R., Brodie, J., Waterhouse, J., 2017b. Managing water quality for the Great Barrier Reef. In: Hart, B., Doolan, J. (Eds.), Decision Making in Water Resources Policy and Management: An Australian Perspective. Elsevier, Collingwood, Australia, pp. 265–289.

Elliott, M., Mander, L., Mazik, K., Simenstad, C., Valesini, F., Whitfield, A., Wolanski, E., 2016. Ecoengineering with ecohydrology: successes and failures in estuarine restoration. Estuar. Coast. Shelf Sci. 176, 12–35.

Erftemeijer, P.L., Riegl, B., Hoeksema, B.W., Todd, P.A., 2012. Environmental impacts of dredging and other sediment disturbances on corals: a review. Mar. Pollut. Bull. 64 (9), 1737–1765.

Fabricius, K.E., 2011. Factors determining the resilience of coral reefs to eutrophication: a review and conceptual model. In: Dubinsky, Z., Stambler, N. (Eds.), Coral Reefs: An Ecosystem in Transition. Springer, Netherlands, pp. 493–505.

Fabricius, K., Okaji, K., De'ath, G., 2010. Three lines of evidence to link outbreaks of the crown-of-thorns seastar *Acanthaster planci*; to the release of larval food limitation. Coral Reefs 29, 593–605.

Fabricius, K., De'ath, G., Humphrey, C., Zagorskis, I., Schaffelke, B., 2013a. Intra-annual variation in turbidity in response to terrestrial runoff at nearshore coral reefs of the Great Barrier Reef. Estuar. Coast. Shelf Sci. 116, 57–65.

Fabricius, K.E., De'ath, G., Noonan, S., Uthicke, S., 2013b. Ecological effects of ocean acidification and habitat complexity on reef-associated macroinvertebrate communities. Proc. R. Soc. B Biol. Sci. 281 (1775).

Fabricius, K.E., Logan, M., Weeks, S., Brodie, J., 2014. The effects of river run-off on water clarity across the central Great Barrier Reef. Mar. Pollut. Bull. 84 (1-2), 191–200.

Fabricius, K.E., Logan, M., Weeks, S.J., Lewis, S.E., Brodie, J., 2016. Changes in water clarity related to river discharges on the Great Barrier Reef continental shelf: 2002–2013. Estuar. Coast. Shelf Sci. 173, A1–A5.

Flores, F., Hoogenboom, M.O., Smith, L.D., Cooper, T.F., Abrego, D., Negri, A.P., 2012. Chronic exposure of corals to fine sediments: lethal and sublethal impacts. PLoS One 7 (5), e37795.

Fraser, K.A., Adams, V.M., Pressey, R.L., Pandolfi, J.M., 2017. Purpose, policy, and practice: intent and reality for on-ground management and outcomes of the Great Barrier Reef Marine Park. Mar. Policy 81, 301–311.

Furnas, M., Mitchell, A., Skuza, M., Brodie, J., 2005. In the other 90%: phytoplankton responses to enhanced nutrient availability in the Great Barrier Reef Lagoon. Mar. Pollut. Bull. 51 (1–4), 253–265.

Furnas, M., Alongi, D., McKinnon, D., Trott, L., Skuza, M., 2011. Regional-scale nitrogen and phosphorus budgets for the northern (14 S) and central (17 S) Great Barrier Reef shelf ecosystem. Cont. Shelf Res. 31 (19-20), 1967–1990.

Gardner, A., Waschka, M., 2012. Using regulation to tackle the challenge of diffuse water pollution and its impact on the Great Barrier Reef: the challenge of managing diffuse source pollution from agriculture: GBR case study and Queensland's reef protection legislation: an analysis. Austl. J. Nat. Resour. Law Pol. 15 (2), 109–147.

Glynn, P.W., Manzello, D.P., 2015. Bioerosion and coral reef growth: a dynamic balance. In: Birkeland, C. (Ed.), Coral Reefs in the Anthropocene. Springer, Dordrecht, pp. 67–97.

Goatley, C.H., Bellwood, D.R., 2013. Ecological consequences of sediment on high-energy coral reefs. PLoS One 8 (10), e77737.

Goatley, C., Bonaldo, R., Fox, R., Bellwood, D., 2016. Sediments and herbivory as sensitive indicators of coral reef degradation. Ecol. Soc. 21 (1).

Gordon, S.E., Goatley, C.H., Bellwood, D.R., 2016. Low-quality sediments deter grazing by the parrotfish *Scarus rivulatus* on inner-shelf reefs. Coral Reefs 35 (1), 285–291.

Great Barrier Reef Marine Park Authority, 2014. Great Barrier Reef Outlook Report 2014. Great Barrier Reef Marine Park Authority, Townsville, Australia, p. 328. Available from: http://www.gbrmpa.gov.au/managing-the-reef/great-barrier-reef-outlook-report.

Great Barrier Reef Water Science Taskforce, 2016. Final Report. The Great Barrier Reef Water Science Taskforce, and the Office of the Great Barrier Reef. Department of Environment and Heritage Protection, Brisbane, Australia, p. 94. Available from: www.gbr.qld.gov.au/taskforce/final-report/.

Grech, A., Bos, M., Brodie, J., Coles, R., Dale, A., Gilbert, R., Hamann, M., Marsh, H., Neil, K., Pressey, R.L., Rasheed, M.A., Sheaves, M., Smith, A., 2013. Guiding principles for the improved governance of port and shipping impacts in the Great Barrier Reef. Mar. Pollut. Bull. 75 (1-2), 8–20.

Greening, H., Janicki, A., Sherwood, E.T., Pribble, R., Johansson, J.O.R., 2014. Ecosystem responses to long-term nutrient management in an urban estuary: Tampa Bay, Florida, USA. Estuar. Coast. Shelf Sci. 151, A1–A16.

Gross, C., Hagy III, J.D., 2017. Attributes of successful actions to restore lakes and estuaries degraded by nutrient pollution. J. Environ. Manag. 187, 122–136.

Haapkylä, J., Unsworth, R.K.F., Flavell, M., Bourne, D.G., Schaffelke, B., Willis, B.L., 2011. Seasonal rainfall and runoff promote coral disease on an inshore Reef. PLoS One 6 (2), e16893.

Hairsine, P.B., 2017. Sediment-related controls on the health of the Great Barrier Reef. Vadose Zone J. 16 (12).

Hamilton, R.J., Almany, G.R., Brown, C.J., Pita, J., Peterson, N.A., Choat, J.H., 2017. Logging degrades nursery habitat for an iconic coral reef fish. Biol. Conserv. 210, 273–280.

Harvell, D., Jordán-Dahlgren, E., Merkel, S., Rosenberg, E., Raymundo, L., Smith, G., Weil, E., Willis, B., 2007. Coral disease, environmental drivers, and the balance between coral and microbial associates. Oceanography 20, 172–195.

Hauri, C., Fabricius, K.E., Schaffelke, B., Humphrey, C., 2010. Chemical and physical environmental conditions underneath mat- and canopy-forming macroalgae, and their effects on understorey corals. PLoS One 5: e12685. doi:12610.11371/journal.pone.0012685.

Heffernan, A.L., Gómez-Ramos, M.M., Gaus, C., Vijayasarathy, S., Bell, I., Hof, C., Mueller, J.F., Gómez-Ramos, M.J., 2017. Non-targeted, high resolution mass spectrometry strategy for simultaneous monitoring of xenobiotics and endogenous compounds in green sea turtles on the Great Barrier Reef. Sci. Total Environ. 599, 1251–1262.

Hering, D., Borja, A., Carstensen, J., Carvalho, L., Elliott, M., Feld, C.K., Heiskanen, A.S., Johnson, R.K., Moe, J., Pont, D., Solheim, A.L., 2010. The European Water Framework Directive at the age of 10: a critical review of the achievements with recommendations for the future. Sci. Total Environ. 408 (19), 4007–4019.

Heron, S.F., Maynard, J.A., Van Hooidonk, R., Eakin, C.M., 2016. Warming trends and bleaching stress of the world's coral reefs 1985–2012. Sci. Rep. 6, 38402.

Hess, S., Prescott, L.J., Hoey, A.S., McMahon, S.A., Wenger, A.S., Rummer, J.L., 2017. Species-specific impacts of suspended sediments on gill structure and function in coral reef fishes. Proc. R. Soc. B 284, 20171279.

Hock, K., Wolff, N.H., Condie, S.A., Anthony, K.R.N., Mumby, P.J., 2014. Connectivity networks reveal the risks of crown-of-thorns starfish outbreaks on the Great Barrier Reef. J. Appl. Ecol. 51 (5), 1188–1196.

Hughes, T.P., Bellwood, D.R., Baird, A.H., Brodie, J., Bruno, J.F., Pandolfi, J.M., 2011. Shifting base-lines, declining coral cover, and the erosion of reef resilience: Comment on Sweatman et al. (2011). Coral Reefs 30 (3), 653–660.

Hughes, T.P., Day, J., Brodie, J., 2015. Securing the future of the Great Barrier Reef. Nat. Clim. Chang. 5, 508–511.

Hughes, T.P., Kerry, J.T., Álvarez-Noriega, M., Álvarez-Romero, J.G., Anderson, K.D., Baird, A.H., Babcock, R.C., Beger, M., Bellwood, D.R., Berkelmans, R., Bridge, T.C., 2017a. Global warming and recurrent mass bleaching of corals. Nature 543 (7645), 373–377.

Hughes, T.P., Barnes, M.L., Bellwood, D.R., Cinner, J.E., Cumming, G.S., Jackson, J.B., Kleypas, J., van de Leemput, I.A., Lough, J.M., Morrison, T.H., Palumbi, S.R., 2017b. Coral reefs in the anthropocene. Nature 546 (7656), 82–90.

Humanes, A., Noonan, S.H.C., Willis, B.L., Fabricius, K.E., Negri, A.P., 2016. Cumulative effects of nutrient enrichment and elevated temperature compromise the early life history stages of the coral *Acropora tenuis*. PLoS One 11 (8), e0161616.

Hutchings, P., Peyrot-Clausade, M., Osnorno, A., 2005. Influence of land runoff on rates and agents of bioerosion of coral substrates. Mar. Pollut. Bull. 51 (1), 438–447.

Johansen, J.L., Jones, G.P., 2013. Sediment-induced turbidity impairs foraging performance and prey choice of planktivorous coral reef fishes. Ecol. Appl. 23 (6), 1504–1517.

Jones, R., Ricardo, G.F., Negri, A.P., 2015. Effects of sediments on the reproductive cycle of corals. Mar. Pollut. Bull. 100 (1), 13–33.

Kennedy, K., Devlin, M., Bentley, C., Lee-Chue, K., Paxman, C., Carter, S., Lewis, S.E., Brodie, J., Guy, E., Vardy, S., Martin, K.C., Jones, A., Packett, R., Mueller, J.F. 2012a. The influence of a season of extreme wet weather events on exposure of the World Heritage Area Great Barrier Reef to Pesticides, Marine Pollution Bulletin 64 (7),1495-1507.

Kennedy, K., Schroeder, T., Shaw, M., Haynes, D., Lewis, S., Bentley, C., Paxman, C., Carter, S., Brando, V.E., Bartkow, M., Hearn, L., 2012b. Long term monitoring of photosystem II herbicides–Correlation with remotely sensed freshwater extent to monitor changes in the quality of water entering the Great Barrier Reef, Australia. Mar. Pollut. Bull. 65 (4-9), 292–305.

King, J., Alexander, F., Brodie, J., 2013. Regulation of pesticides in Australia: the Great Barrier Reef as a case study for evaluating effectiveness. Agric. Ecosyst. Environ. 180, 54–67.

Kroon, F.J., Kuhnert, P.M., Henderson, B.L., Wilkinson, S.N., Kinsey-Henderson, A., Abbott, B., Brodie, J.E., Turner, R.D.R., 2012. River loads of suspended solids, nitrogen, phosphorus and herbicides delivered to the Great Barrier Reef Lagoon. Mar. Pollut. Bull. 65 (4-9), 167–181.

Kroon, F.J., Schaffelke, B., Bartley, R., 2014. Informing policy to protect coastal coral reefs: insight from a global review of reducing agricultural pollution to coastal ecosystems. Mar. Pollut. Bull. 85 (1), 33–41.

Kroon, F.J., Berry, K.L.E., Brinkman, D.L., Davis, A., King, O., Kookana, R., Lewis, S., Leusch, F., Makarynskyy, O., Melvin, S., Müller, J., Neale, P., Negri, A., O'Brien, D., Puotinen, M., Smith, R., Tsang, J., van de Merwe, J., Warne, M., Williams, M., 2015. In: Identification, impacts, and prioritisation of emerging contaminants present in the GBR and Torres Strait marine environments. Report to the National Environmental Science Programme Reef and Rainforest Research Centre Limited, Cairns. pp. 138.

Kroon, F.J., Thorburn, P., Schaffelke, B., Whitten, S., 2016. Towards protecting the Great Barrier Reef from land-based pollution. Glob. Chang. Biol. 22 (6), 1985–2002.

Lamb, J.B., Wenger, A.S., Devlin, M.J., Ceccarelli, D.M., Williamson, D.H., Willis, B.L., 2016. Reserves as tools for alleviating impacts of marine disease. Philos. Trans. R Soc. Lond. B Biol. Sci. 371 (1689).

Lamb, J.B., Willis, B.L., Fiorenza, E.A., Couch, C.S., Howard, R., Rader, D.N., True, J.D., Kelly, L.A., Ahmad, A., Jompa, J., Harvell, C.D., 2018. Plastic waste associated with disease on coral reefs. Science 359 (6374), 460–462.

Lewis, S.E., Brodie, J.E., Bainbridge, Z.T., Rohde, K.W., Davis, A.M., Masters, B.L., Maughan, M., Devlin, M.J., Mueller, J.F., Schaffelke, B., 2009. Herbicides: a new threat to the Great Barrier Reef. Environ. Pollut. 157, 2470–2484. https://doi.org/10.1016/j.envpol.2009.03.006.

Lewis, S.E., Schaffelke, B., Shaw, M., Bainbridge, Z.T., Rohde, K.W., Kennedy, K.E., Davis, A.M., Masters, B.L., Devlin, M.J., Mueller, J.F., Brodie, J.E., 2012. Assessing the risks of PS-II herbicide exposure to the Great Barrier Reef. Mar. Pollut. Bull. 65, 280–291.

Lewis, S., Silburn, D.M., Shaw, M., Kookana, R., 2016. Pesticide behaviour, fate and effects in the tropics: a review of current knowledge. J. Agric. Food Chem. 64, 3917–3924.

Marsh, H., De'ath, G., Gribble, N., Lane, B., 2005. Historical marine population estimates: triggers or targets for conservation? The dugong case study. Ecol. Appl. 15 (2), 481–492.

Maynard, J., van Hooidonk, R., Eakin, C.M., Puotinen, M., Garren, M., Williams, G., Heron, S.F., Lamb, J., Weil, E., Willis, B., Harvell, C.D., 2015. Projections of climate conditions that increase coral disease susceptibility and pathogen abundance and virulence. Nat. Clim. Chang. 5 (7), 688–694.

McCloskey, G.L., Waters, D., Baheerathan, R., Darr, S., Dougall, C., Ellis, R., Fentie, B., Hateley, L., 2017b. In: Modelling reductions of pollutant loads due to improved management practices in the Great Barrier Reef catchments: updated methodology and results. Technical Report for Reef Report Card 2015, Queensland Department of Natural Resources and Mines, Brisbane, Queensland.

McCook, L.J., Jompa, J., Diaz-Pulido, G., 2001. Competition between corals and algae on coral reefs: a review of evidence and mechanisms. Coral Reefs 19 (4), 400–417.

McGrath, C., 2010. Does Environmental Law Work? How to Evaluate the Effectiveness of an Environmental Legal System. Lambert Academic Publishing, Köln, Germany.

McKenzie, L.J., Collier, C.J., Langlois, L.A., Yoshida, R.L., Smith, N., Waycott, M., 2016. In: Marine monitoring program—inshore seagrass. Annual Report for the sampling period 1st June 2014–31st May 2015. Centre for Tropical Water & Aquatic Ecosystem Research. James Cook University, Cairns.

Morrison, T.H., 2017. Evolving polycentric governance of the Great Barrier Reef. Proc. Natl. Acad. Sci. E3013–E3021. https://doi.org/10.1073/pnas.1620830114.

Morrow, K.M., Ritson-Williams, R., Ross, C., Liles, M.R., Paul, V.J., 2012. Macroalgal extracts induce bacterial assemblage shifts and sublethal tissue stress in Caribbean corals. PLoS One 7 (9), e44859.

Morrow, K.M., Bromhall, K., Motti, C.A., Munn, C.B., Bourne, D.G., 2017. Allelochemicals produced by brown macroalgae of the Lobophora genus are active against coral larvae and associated bacteria, supporting pathogenic shifts to Vibrio dominance. Appl. Environ. Microbiol. 83 (1). e02391-16.

Nimmo Smith, R.J., Glegg, G.A., Parkinson, R., Richards, J.P., 2007. Evaluating the implementation of the nitrates directive in Denmark and England using an actor oriented approach. Eur. Environ. 17 (2), 124–144. https://doi.org/10.1002/eet.440.

O'Brien, D., Davis, A., Nash, M., Di Bella, L., Brodie, J., 2014. In: Herbert water quality monitoring project: 2011–2013 results. Proceedings of the 36th Conference of the Australian Society of Sugar Cane Technologists. Gold Coast, Queensland, Australia, 29 April–1 May 2014.

O'Brien, D., Lewis, S., Davis, A., Gallen, C., Smith, R., Turner, R., Warne, M., Turner, S., Caswell, S., Mueller, J., Brodie, J., 2016a. Spatial and temporal variability in pesticide exposure downstream of a heavily irrigated cropping area: the application of different monitoring techniques. J. Agric. Food Chem. 64, 3975–3989.

O'Brien, P.A., Morrow, K.M., Willis, B.L., Bourne, D.G., 2016b. Implications of ocean acidification for marine microorganisms from the free-living to the host-associated. Front. Mar. Sci. 3, 47.

Oliver, D.P., Anderson, J.S., Davis, A., Lewis, S., Brodie, J., Kookana, R., 2014. Banded applications are highly effective in minimising herbicide migration from furrow-irrigated sugar cane. Sci. Total Environ. 466–467, 841–848.

Osborne, K., Dolman, A.M., Burgess, S.C., Johns, K.A., 2011. Disturbance and the dynamics of coral cover on the Great Barrier Reef (1995–2009). PLoS One 6 (3), e17516.

Osborne, K., Thompson, A.A., Cheal, A.J., Emslie, M.J., Johns, K.A., Jonker, M.J., Logan, M., Miller, I.R., Sweatman, H., 2017. Delayed coral recovery in a warming ocean. Glob. Chang. Biol. 23 (9), 3869–3881.

Oxenford, H.A., Vallès, H., 2016. Transient turbid water mass reduces temperature-induced coral bleaching and mortality in Barbados. Peer J. 4, e2118.

Perez III, K., Rodgers, K.S., Jokiel, P.L., Lager, C.V., Lager, D.J., 2014. Effects of terrigenous sediment on settlement and survival of the reef coral *Pocillopora damicornis*. Peer J. 2, e387.

Petus, C., Collier, C., Devlin, M., Rasheed, M., McKenna, S., 2014. Using MODIS data for understanding changes in seagrass meadow health: a case study in the Great Barrier Reef (Australia). Mar. Environ. Res. 98, 68–85.

PGM Environment, 2012. Available from: http://www.environment.gov.au/system/files/pages/884f8778-caa4-4bd9-b370-318518827db6/files/23qrc-doc3.pdf.

Pollock, F.J., Lamb, J.B., Field, S.N., Heron, S.F., Schaffelke, B., Shedrawi, G., Bourne, D.G., Willis, B.L., 2014. Sediment and turbidity associated with offshore dredging increase coral disease prevalence on nearby reefs. PLoS One 9 (7), e102498.

Pollock, F.J., Lamb, J.B., Field, S.N., Heron, S.F., Schaffelke, B., Shedrawi, G., Bourne, D.G., Willis, B.L., 2016. Correction: sediment and turbidity associated with offshore dredging increase coral disease prevalence on nearby reefs. PLoS One 11 (11), e0165541.

Pratchett, M.S., Caballes, C.F., Rivera-Posada, J.A., Sweatman, H.P.A., 2014. Limits to understanding and managing outbreaks of crown-of- thorns starfish (*Acanthaster* spp.). Oceanogr. Mar. Biol. 52, 133–200.

Pratchett, M.S., Caballes, C.F., Wilmes, J.C., Matthews, S., Mellin, C., Sweatman, H.P., Nadler, L.E., Brodie, J., Thompson, C.A., Hoey, J., Bos, A.R., 2017. 30 years of research on crown-of-thorns starfish (1986-2016): scientific advances and emerging opportunities. Diversity 9, 41. https://doi.org/10.3390/d9040041. http://www.mdpi.com/1424-2818/9/4/41.

Prouty, N.G., Cohen, A., Yates, K.K., Storlazzi, C.D., Swarzenski, P.W., White, D., 2017. Vulnerability of coral reefs to bioerosion from land-based sources of pollution. J. Geophys. Res. Oceans 122, 9319–9331.

Queensland Audit Office, 2015. Managing water quality in Great Barrier Reef catchments (Report 20: 2014–15). The State of Queensland, Brisbane.

Ramsby, B.D., Hoogenboom, M.O., Whalan, S., Webster, N.S., Thompson, A., 2017. A decadal analysis of bioeroding sponge cover on the inshore Great Barrier Reef. Sci. Rep. 7 (1), 2706.

Reside, A.E., Bridge, T.C., Rummer, J.L., 2016. Great Barrier Reef: clearing the way for reef destruction. Nature 537 (7620), 307.

Reside, A.E., Beher, J., Cosgrove, A.J., Evans, M.C., Seabrook, L., Silcock, J.L., Wenger, A.S., Maron, M., 2017. Ecological consequences of land clearing and policy reform in Queensland. Pac. Conserv. Biol. 23, 219–230.

Ricardo, G.F., Jones, R.J., Clode, P.L., Humanes, A., Negri, A.P., 2015. Suspended sediments limit coral sperm availability. Sci. Rep. 5, 18084.

Ricardo, G.F., Jones, R.J., Negri, A.P., Stocker, R., 2016a. That sinking feeling: suspended sediments can prevent the ascent of coral egg bundles. Sci. Rep. 6, 21567.

Ricardo, G.F., Jones, R.J., Clode, P.L., Negri, A.P., 2016b. Mucous secretion and Cilia beating defend developing coral larvae from suspended sediments. PLoS One 11 (9), e0162743.

Riemann, B., Carstensen, J., Dahl, K., Fossing, H., Hansen, J.W., Jakobsen, H.H., et al., 2015. Recovery of Danish coastal ecosystems after reductions in nutrient loading: a holistic ecosystem approach. Estuar. Coasts 39 (1), 82–97.

Rosset, S., Wiedenmann, J., Reed, A.J., D'Angelo, C., 2017. Phosphate deficiency promotes coral bleaching and is reflected by the ultrastructure of symbiotic dinoflagellates. Mar. Pollut. Bull. 118 (1), 180–187.

Roth, C.H., Addison, J., Anthony, K., Dale, A., Eberhard, R., Hobday, A., Horner, N.J., Jarvis, D., Kroon, K., Stone-Jovicich, S., Walshe, T., 2017. Reef 2050 Plan Review Options—Appendix to Final Report. Submitted to the Department of the Environment and Energy, Australia.

Saunders, M.I., Atkinson, S., Klein, C.J., Weber, T., Possingham, H.P., 2017. Increased sediment loads cause non-linear decreases in seagrass suitable habitat extent. PLoS One 12 (11), e0187284. https://doi.org/10.1371/journal.pone.0187284.

Schaffelke, B., Brodie, J., Collier, C., Kroon, F., Lough, J., McKenzie, L., Ronan, M., Uthicke, S., 2017. 2017 Scientific Consensus Statement: Land Use Impacts on Great Barrier Reef Water Quality and Ecosystem Condition. Chapter 1: The Condition of Coastal and Marine Ecosystems of the Great Barrier Reef and their Responses to Water Quality and Disturbances. The Reef Water Quality Protection Plan Secretariat, Brisbane, Australia, p. 80.

Sheaves, M., Coles, R., Dale, P., Grech, A., Pressey, R.L., Waltham, N.J., 2016. Enhancing the value and validity of EIA: serious science to protect Australia's Great Barrier Reef. Conserv. Lett. 9 (5), 377–383.

Star, M., Rolfe, J., Pringle, M., Shaw, M., McCosker, K., Ellis, R., Waters, D., Tindal, D., Turner, R., 2017. Targeting for pollutant reductions in the Great Barrier Reef management units. Central Queensland University, Rockhampton, Australia.

Stimson, J., 2015. Long-term record of nutrient concentrations in Kāne'ohe Bay, O'ahu, Hawai'i, and its relevance to onset and end of a phase shift involving an Indigenous alga, *Dictyosphaeria cavernosa*. Pac. Sci. 69 (3), 319–339.

Stimson, J., 2018. Recovery of coral cover in records spanning 44 yr for reefs in Kāne 'ohe Bay, Oa 'hu, Hawai 'i. Coral Reefs 37, 55–69.

Stoeckl, N., Farr, M., Larson, S., Adams, V.M., Kubiszewski, I., Esparon, M., Costanza, R., 2014. A new approach to the problem of overlapping values: a case study in Australia's Great Barrier Reef. Ecosys. Ser. 10, 61–78.

Tarte, D., Hart, B., Hughes, T., Hussey, K., 2017. Reef 2050 long-term sustainability plan progress on implementation. Review by Great Barrier Reef Independent Review Group. February 2017. Great Barrier Reef Independent Review Group, Australia. Available from: https://independent.academia.edu/DiTarte

Tebbett, S.B., Goatley, C.H., Bellwood, D.R., 2017a. Fine sediments suppress detritivory on coral reefs. Mar. Pollut. Bull. 114 (2), 934–940.

Tebbett, S.B., Goatley, C.H., Bellwood, D.R., 2017b. Algal turf sediments and sediment production by parrotfishes across the continental shelf of the northern Great Barrier Reef. PLoS One 12 (1), e0170854.

Thompson, A., Schroeder, T., Brando, V., Schaffelke, B., 2014. Coral community responses to declining water quality: Whitsunday Islands, Great Barrier Reef, Australia. Coral Reefs 33 (4), 923–938.

Thorburn, P.J., Wilkinson, S.N., 2013. Conceptual frameworks for estimating the water quality benefits of improved agricultural management practices in large catchments. Agric. Ecosyst. Environ. 180, 192–209.

Thorburn, P.J., Wilkinson, S.N., Silburn, D.M., 2013. Water quality in agricultural lands draining to the Great Barrier Reef: causes, management and priorities. Agric. Ecosyst. Environ. 180, 4–20.

UNESCO, 2015. World Heritage committee. Decision regarding the Great Barrier Reef. 39 COM 7B.7. Available from: http://whc.unesco.org/en/decisions/6216.

UNESCO, 2017. World heritage committee. Decision regarding the Great Barrier Reef. 41 COM 7B.24. Available from: http://whc.unesco.org/en/decisions/7027.

Uthicke, S., Logan, M., Liddy, M., Francis, D., Hardy, N., Lamare, M., 2015. Climate change as an unexpected co-factor promoting coral eating seastar (*Acanthaster planci*) outbreaks. Sci. Rep. 5, 8402.

Uthicke, S., Liddy, M., Patel, F., Logan, M., Johansson, C., Lamare, M., 2018. Effects of larvae density and food concentration on crown-of-thorns seastar (*Acanthaster cf. solaris*) development in an automated flow-through system. Sci. Rep. 8 (1), 642.

Van Woesik, R., Golbuu, Y., Houk, P., Isechal, A.L., Idechong, J.W., Victor, S., 2012. Climate change refugia in the sheltered bays of Palau: analogs of future reefs. Ecol. Evol. 2, 2474–2484. https://doi.org/10.1002/ece3.363.

Vega Thurber, R.L., Burkepile, D.E., Fuchs, C., Shantz, A.A., McMinds, R., Zaneveld, J.R., 2014. Chronic nutrient enrichment increases prevalence and severity of coral disease and bleaching. Glob. Chang. Biol. 20 (2), 544–554.

Vella, K., Baresi, U., 2017. Understanding how policy actors improvise and collaborate in the Great Barrier Reef. Coast. Manag. 45 (6), 487–504.

Vella, K., Forester, J., 2018. Creating spaces for action: lessons from front line planners in the Great Barrier Reef. In: Sipe, N., Vella, K. (Eds.), The Routledge Handbook of Australian Urban and Regional Planning. Routledge, New York, NY.

Villa, C.A., Flint, M., Bell, I., Hof, C., Limpus, C.J., Gaus, C., 2017. Trace element reference intervals in the blood of healthy green sea turtles to evaluate exposure of coastal populations. Environ. Pollut. 220, 1465–1476.

Walshe, T., Brodie, J., da Silva, E., Devlin, M., Ewels, C., Hock, K., MacNeil, A., Mumby, P., Park, G., Pratchett, M., Roberts, A., Waterhouse, J., Anthony, K., 2017. Options for early prevention of crown-of-thorns starfish outbreaks on the Great Barrier Reef. In: Accelerate Partnership Final Report submitted to the Queensland Department of Science, Information, Technology and Innovation. Australian Institute of Marine Science, Townsville, pp. 74.

Waschka, M., Gardner, A., 2016. Diffuse source pollution and water quality law for the Great Barrier Reef. In: Chapter 11: In Trans-jurisdictional Water Law and Governance. Routledge, London and New York, pp. 195–213.

Waterhouse, J., Brodie, J., 2015. In: Recent findings of an assessment of remote sensing data for water quality measurement in the Great Barrier Reef: supporting information for the GBR water quality improvement plans. Report to Cape York NRM, NQ Dry Tropics and Fitzroy Basin Association. TropWATER Report 2015, Townsville, Australia.

Waterhouse, J., Brodie, J., Lewis, S., Mitchell, A., 2012. Quantifying the sources of pollutants to the Great Barrier Reef. Mar. Pollut. Bull. 65, 394–406.

Waterhouse, J., Brodie, J., Audas, D., Lewis, S., 2016. Land-sea connectivity, ecohydrology and holistic management of the Great Barrier Reef and its catchments: time for a change. Ecohydrol. Hydrobiol. 16, 45–57.

Waterhouse, J., Brodie, J., Tracey, D., Smith, R., Vandergragt, M., Collier, C., Sutcliffe, T., Petus, C., Baird, M., Kroon, F., Mann, R., Adame, F., 2017a. 2017 Scientific Consensus Statement: Land Use Impacts on Great Barrier Reef Water Quality and Ecosystem Condition. Chapter 3: The Risk From Anthropogenic Pollutants to Great Barrier Reef Coastal and Marine Ecosystems. Reef Water Quality Protection Plan Secretariat, Brisbane, Australia, p. 184.

Waterhouse, J., Schaffelke, B., Bartley, R., Eberhard, R., Brodie, J., Star, M., Thorburn, P., Rolfe, J., Ronan, M., Taylor, B., Kroon, F., 2017b. 2017 Scientific Consensus Statement: Land Use Impacts on Great Barrier Reef Water Quality and Ecosystem Condition. Chapter 5: Overview of Key Findings, Management Implications and Knowledge Gaps. State of Queensland, 2017. Reef Water Quality Protection Plan Secretariat, Brisbane, Australia, p. 57.

Waters, D.K., Carroll, C., Ellis, R., Hateley, L., McCloskey, G.L., Packett, R., Dougall, C., Fentie, B., 2014. Modelling Reductions of Pollutant Loads Due to Improved Management Practices in the Great Barrier Reef Catchments—Whole of GBR. Queensland Department of Natural Resources and Mines, Toowoomba, Australia, p. 120.

Weber, M., Lott, C., Fabricius, K.E., 2006. Sedimentation stress in a scleractinian coral exposed to terrestrial and marine sediments with contrasting physical, organic and geochemical properties. J. Exp. Mar. Biol. Ecol. 336 (1), 18–32.

Weber, M., de Beer, D., Lott, C., Polerecky, L., Kohls, K., Abed, R.M., Ferdelman, T.G., Fabricius, K.E., 2012. Mechanisms of damage to corals exposed to sedimentation. Proc. Natl. Acad. Sci. 109 (24), E1558–E1567.

Webster, M.S., Colton, M.A., Darling, E.S., Armstrong, J., Pinsky, M.L., Knowlton, N., Schindler, D.E., 2017. Who should pick the winners of climate change? Trends Ecol. Evol.

Wenger, A.S., McCormick, M.I., 2013. Determining trigger values of suspended sediment for behavioral changes in a coral reef fish. Mar. Pollut. Bull. Available from: https://doi.org/10.1016/j.marpolbul.2013.02.014.

Wenger, A.S., Johansen, J.L., Jones, G.P., 2012. Increasing suspended sediment reduces foraging, growth and condition of a planktivorous damselfish. J. Exp. Mar. Biol. Ecol. 428, 43–48.

Wenger, A.S., McCormick, M.I., Endo, G.G.K., McLeod, I.M., Kroon, F.J., Jones, G.P., 2014. Suspended sediment prolongs larval development in a coral reef fish. J. Exp. Biol. 217 (7), 1122–1128.

Wenger, A.S., Williamson, D.H., da Silva, E.T., Ceccarelli, D.M., Browne, N.K., Petus, C., Devlin, M.J., 2016. Effects of reduced water quality on coral reefs in and out of no-take marine reserves. Conserv. Biol. 30 (1), 142–153.

Wenger, A.S., Harvey, E., Wilson, S., Rawson, C., Newman, S.J., Clarke, D., Saunders, B.J., Browne, N., Travers, M.J., Mcilwain, J.L., Erftemeijer, P.L., 2017. A critical analysis of the direct effects of dredging on fish. Fish Fish. 18 (5), 967–985.

Wiedenmann, J., D'Angelo, C., Smith, E.G., Hunt, A.N., Legiret, F.-E., Postle, A.D., Achterberg, E.P., 2013. Nutrient enrichment can increase the susceptibility of reef corals to bleaching. Nat. Clim. Chang. 3 (5), 160–164. Available from: http://www.nature.com/nclimate/journal/vaop/ncurrent/abs/nclimate1661.html#supplementary-information.

Wooldridge, S.A., 2016. Excess seawater nutrients, enlarged algal symbiont densities and bleaching sensitive reef locations: 1. Identifying thresholds of concern for the Great Barrier Reef, Australia. Marine Poll. Bull. 114 (1), 343–354.

Wooldridge, S.A., 2017. Preventable fine sediment export from the Burdekin River catchment reduces coastal seagrass abundance and increases dugong mortality within the Townsville region of the Great Barrier Reef, Australia. Mar. Pollut. Bull. 114 (2), 671–678.

Wooldridge, S.A., Brodie, J.E., 2015. Environmental triggers for primary outbreaks of crown-of-thorns starfish on the Great Barrier Reef, Australia. Mar. Pollut. Bull. 101 (2), 805–815.

Wooldridge, S.A., Heron, S.F., Brodie, J.E., Done, T.J., Masiri, I., Hinrichs, S., 2017. Excess seawater nutrients, enlarged algal symbiont densities and bleaching sensitive reef locations: 2. A regional-scale predictive model for the Great Barrier Reef, Australia. Marine Poll. Bull. 114 (1), 343–354.

Wulf, P., 2004. Diffuse land-based pollution and the Great Barrier Reef World Heritage Area: the Commonwealth's responsibilities and implications for the Queensland sugar industry. Environ. Plan. Law J. 21, 424–444.

Wulff, F., Eyre, B.D., Johnstone, R., 2011. Nitrogen versus phosphorus limitation in a subtropical coastal embayment (Moreton Bay; Australia): implications for management. Ecol. Model. 222, 120–130.

WWF, 2017. In: Accelerating bushland destruction in Queensland: clearing under self assessable codes takes major leap upward. World Wide Fund for Nature Australia. Available from: http://www.wwf.org.au/Article Documents/360/pub-accelerating-bushland-destruction-in-queensland- 21mar17. pdf.aspx.

Yang, W.-H., Wortmann, J., Robins, J.B., Courtney, A.J., O'Neill, M.F., Campbell, M.J., 2016. Quantit7ative assessment of the Queensland saucer scallop (*Amusium balloti*) fishery, 2016. State of Queensland 2016.

York, P.H., Carter, A.B., Chartrand, K., Sankey, T., Wells, L., Rasheed, M.A., 2015. Dynamics of a deep-water seagrass population on the Great Barrier Reef: annual occurrence and response to a major dredging program. Sci. Rep. 5, 13167. https://doi.org/10.1038/srep13167.

Zaneveld, J.R., Burkepile, D.E., Shantz, A.A., Pritchard, C.E., McMinds, R., Payet, J.P., Welsh, R., Correa, A.M., Lemoine, N.P., Rosales, S., Fuchs, C., Maynard, J.A., Vega Thurber, R., 2016. Overfishing and nutrient pollution interact with temperature to disrupt coral reefs down to microbial scales. Nat. Commun. 7, 11833.

Section G

Over-Arching Topics

Chapter 29

Estuarine Ecohydrology Modeling: What Works and Within What Limits?

Eric Wolanski

TropWATER and College of Marine & Environmental Sciences, James Cook University and Australian Institute of Marine Science, Townsville, QLD, Australia

1 INTRODUCTION: THE NEED FOR MODELS

The chapters in this book highlight a multitude of environmental problems with estuaries and coastal waters worldwide. In some cases practical solutions are also proposed and in many cases are being implemented. The ultimate aim is to restore these ecosystems to something that resembles the original state that existed before degradation from human activities. Humans have many tools at their disposition to achieve such goals (Elliott et al., 2016, this book). While in some cases the use of such tools has resulted in recovering parts of the ecosystem (Giani et al., 2012; Greening et al., 2014), the full ecosystem is seldom if ever recovered for a variety of ecological reasons as well as climate change and the human impact that remains (Duarte et al., 2009; Nixon et al., 2009; Tucker et al., 2014; Elliott et al., 2016). The lesson from comparing partial successes and failures in attempting restoration is that basically each estuary is unique (EPA (U.S. Environmental Protection Agency), 2010). What seems to determine why that is so is to what degree the management of human activities (e.g., land use and water management) throughout the catchment relies on ecohydrology science, that is, on understanding the biophysics of the ecosystem and managing human activities within these biophysical constraints so as to keep the system resilient. Estuarine management has often been done empirically by relying on experience, that is, adaptive management. Models are needed to facilitate this process and to go from the qualitative to the quantitative when assessing environmental impacts of development proposals and the likely effectiveness of restoration efforts (Wolanski and Elliott, 2015; Elliott et al., 2016). Numerical models are indeed increasingly used for that purpose but they often meet stumbling blocks (Barnes and Mazzotti, 2005; Jorgensen, 2011; Wolanski and Elliott, 2015; Ganju et al., 2016).

This chapter addresses the question "How reliable and how useful in practice are these models?" To have a chance to be useful in practice, the model needs to reproduce reliably the biophysics. Thus, firstly, the model must recognize that the physical system influences the biological system and vice versa. Also, the estuarine ecosystem does not stop at the tidal limit. Indeed, it comprises the whole river catchment including the river and its riverine wetlands, the estuarine wetlands, and the coastal waters with its own ecology including its intertidal areas, seagrass, as well as its rocky, limestone, or coral reefs (Fig. 1). Secondly, the first step in modeling an estuarine ecosystem is to develop a water circulation model and, for turbid estuaries, either a sediment dynamics model or bypassing the sediment model and instead collecting extensive field data on the suspended sediment concentration. This is an essential step because water moves suspended matter, and because sediments determine the fundamental niche for the fauna and flora on the bottom, they influence organisms in the water column by restricting the photic zone, and they act as a nutrient sink receiving nutrients from the catchment or being a source of nutrients to the adjacent coast (de Jonge and Elliott, 2001; Orive et al., 2002). In all cases additional data need to be collected about the partitioning of nutrients between the particulate and the dissolved phases. Thirdly, the water and sediment circulation models are used to drive an estuarine ecological model. The combined model is called an estuarine ecohydrology model.

If there are enough field data to verify these sub-models, then the ecohydrology model can be used, with quantifiable error bars, to predict how the ecosystem will vary for various development proposals or remediation efforts. In practice, the parameters commonly used in this assessment are either the sediment and/or the nutrients. For sediment, the common question is to assess how the patterns of sediment-induced turbidity and of sedimentation/erosion change with the human impacts. For nutrients, a common question is to assess the fate of nutrients in the estuary and, in particular, if eutrophication

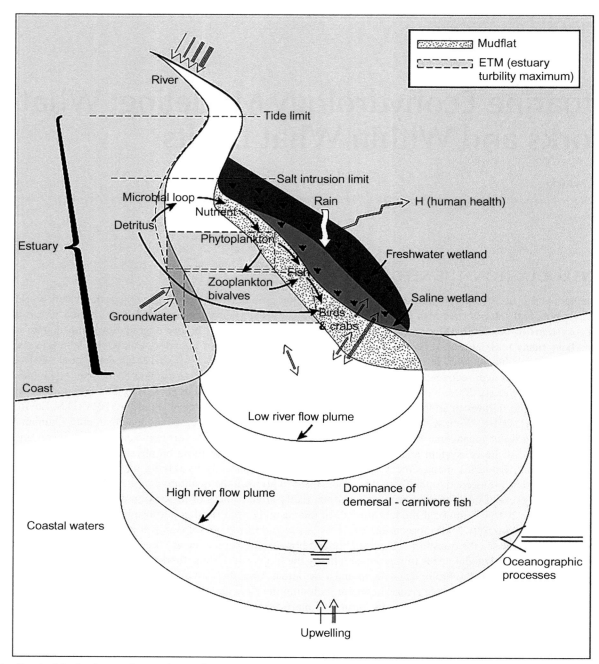

FIG. 1 Sketch of the dominant pathways of water, fine sediment, nutrients, and the food web in a moderately turbid estuary, together with the possible impact on human health. *(Modified from Wolanski, E., Elliott, M., 2015. Estuarine Ecohydrology. An Introduction. Amsterdam: Elsevier, 322 p.)*

and (at the worst case) toxic algae blooms may occur. More recently such models have been used to assess the impact of finfish fishing on populations of jellyfish (see below). Some models even add a human health sub-model (Fig. 1), principally for cholera and mosquito-borne diseases; these are still basic and they not reviewed here (see Wolanski and Elliott, 2015, for more details).

To describe such models, this chapter starts (Section 2) with a review of the models of physical processes, specifically (Section 2.1) water circulation models, and (Section 2.2) sediment dynamics models (describing separately sand dynamics, mud dynamics, and muddy sand dynamics). Section 3 reviews models of nutrient sequestration by fine sediment; this is an important process controlling the fate of nutrients in turbid estuaries, yet until recently this has been commonly ignored by modelers who considered only dissolved nutrients. Section 4 describes the most commonly used ecohydrology models; it starts with an introduction to explain the hypotheses and assumptions behind such modeling. Then it discusses the LOICZ biophysical model of nutrient budgets in an estuary, the NPZ estuarine ecosystem models, the UNESCO estuarine ecohydrology model, the Ecopath model, harmful algae blooms (HABs) models, and finally hypoxia models. Section 5 provides a synthesis.

2 MODELS OF PHYSICAL PROCESSES

2.1 Water Circulation Models

The water circulation in estuaries is controlled by physical forcing at the open boundaries, namely the river inflow, the groundwater inflow, the wind, and the oceanic forcings mainly in terms of tides, fluctuations of the mean sea level (e.g., storm surges and shelf waves), the waves, and the net oceanic currents. This circulation has been extensively studied and this knowledge has been used to develop numerical models of the water circulation; these models are generally reliable (Valle-Levinson, 2010; Uncles and Monismith, 2011; Ganju et al., 2016; Bruner de Miranda et al., 2017). These models are routinely used to study the distribution of passive waterborne particles that behave conservatively (i.e., that do not settle or erode from the bottom or change while in the water column). Salinity is such a conservative tracer; thus, these models can readily be used to study changes in salinity of an estuary due to dredging, changing river flows, or passing storms. However, these models still do not reproduce well estuarine fronts, that is, discontinuities in the salinity; yet, these fronts are important for fish larvae in their strategy to recruit (Kingsford and Suthers, 1994; Uncles, 2011; Teodosio et al., 2016). Thus there are strong limitations to the use of these models for modeling fish dynamics in a stratified estuary.

2.2 Sediment Dynamics Models

Modeling the dynamics of sediments in estuaries is more difficult than modeling the water circulation, because the sediment particles are not transported passively by the water currents; they settle and they are resuspended at a range of time scales, from millennia (the geomorphological time scales), to tides, storms, river flows, and particularly river floods and droughts (hours to months), to turbulence (sec to min). A key parameter determining the ability of sediment particles to be transported by water currents is the cohesion. Cohesive sediment (mud) has a median particle size d_{50} (also called d_s) $< 4\,\mu m$ (microns; $1\,mm = 1000\,\mu m$). Non-cohesive sediment (e.g., sand) has a $d_{50} > 64\,\mu m$. Silt has a d_{50} in between those of clay and sand. Mud is mainly carried is suspension within the water column while sand is mainly carried along the bottom and very close to the bottom. Mud is the dominant sediment in estuaries that are turbid. Thus, models for mud dynamics are different to those for sand dynamics and models for muddy sands are a further "mixed" model between the mud model and the sand model. A new type of sediment dynamics models, system behavior modeling, is emerging with simplified dynamics processes but needing extensive data assimilation; this is still work in progress (Horrillo-Caraballo et al., 2014).

2.2.1 Sand Dynamics

There are several engineering models to calculate the bed load; these include the Meyer-Peter, Einstein, and Ackers-White formulae (Raudkivi, 1967; Postma, 1967; Dyer, 1986, 1994; Camenen and Larson, 2005; Chanson, 1999). The models generally assume either that the sand moves as a creeping movement of layers along the bottom (Fig. 2A) or that the sand particles are rolled and bounced over the bottom in a process called saltation (Fig. 2B). They rely on laboratory-derived empirical formulae to calculate the bed-load transport as a function of the current-induced stress on the bottom, a threshold stress for bed load to be initiated, the density of the sediment, and the mean sand particle size d_{50}. These formulae yield bed-load predictions that can vary by a factor of 10 with each other and from field measurements; thus, the reliability of these equations is very limited in field applications (Borsje et al., 2013).

One reason for this difficulty is that field data for model calibration are very difficult to obtain. Indeed, there are no instruments to nonintrusively measure the bed-load transport in estuaries; thus, the net bed load must be determined by repeated bathymetric surveys to assess changes in sediment storage and by inference from the shape of sand shoals and asymmetries in bed forms (FitzGerald et al., 2005), or by measuring the bed-load surface velocity sediment ($[n-1]\Delta v$ in Fig. 2A) using a downward looking acoustic Doppler current meter (Wolanski et al., 2006a, b), but there are still no methods to measure the thickness of the bed-load layer, thus estimating the total bed-load flux remains elusive. All these field measurements come with large error bars because of the effects of bedforms that include increased bottom roughness length, change the flows and wave propagation in shallow waters, and the patchiness of sediment transport especially near crests of sand waves and ripples (Dyer, 1986; Ke et al., 1994; Le Hir et al., 2000; Dronkers, 2005). Thus, the value of several

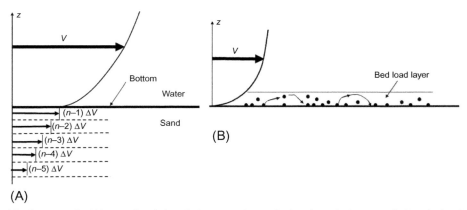

FIG. 2 Models of bed-load transport by (A) creep flow below the bottom and (B) saltation above the bottom. V is the velocity, Δv is the velocity increment. *(Modified from Wolanski, E., Elliott, M., 2015. Estuarine Ecohydrology. An Introduction. Amsterdam: Elsevier, 322 p.)*

empirical coefficients in the bed-load transport formulae needs to be assessed on a case by case basis for the formulae to fit the local field data.

These formulae are unreliable in estuaries for unsteady tidal flows. They are very unreliable in the presence of waves that resuspend the sediment. Thus, modeling the geomorphology of sandy estuaries is still an art based on local knowledge and experience.

2.2.2 Mud Dynamics

Cohesive sediment (mud) behaves differently than non-cohesive sediment (sand) as suspended mud particles forms flocs and these flocs depend not just on physics and chemistry of flocculation, but also on the biology, namely microorganisms and mucopolysaccharides produced by diatoms and larger organisms. Based on laboratory experiments with no biology, the erosion rate, E, and the settling rate, D, are parameterized as a function of the water velocity u (Partheniades, 1965)

$$E = \{0, \text{ if } u < u_c; \ A\,(u/u_c - 1)^n, \text{ if } u > u_c\} \tag{1}$$

$$D = \{\text{SSC } w\,(1 - u/u_d)^2, \text{ if } u < u_d; 0, \text{ if } u > u_d\} \tag{2}$$

where A is an erosion parameter, u_c and u_d are threshold velocities for, respectively, entrainment and deposition, n is a constant, SSC is the suspended solid concentration, and w is the settling velocity.

There have been attempts to ignore the biology to predict the value of w using physical flocculation models and fractal theory; the results are discouraging (Winterwerp, 2011). Thus, these formulae cannot be readily applied in the field without extensive field data for a number of reasons. Firstly, the erosion parameter A varies with the type of mud and the age of the mud (i.e., the time it had to consolidate in the bottom and whether or not laminae are present in the consolidated mud; Wartel, 2003); further in the presence of waves, the value of A is greatly increased by wave-induced pore pressure buildup (Maa and Mehta, 1987; Wolanski and Spagnol, 2003). Also the value of A depends strongly on the biology of the bed, particularly if there are borers digging burrows in the sediment and facilitating erosion (Mazik et al., 2008; Andersen and Pejrup, 2011; Bentley et al., 2014), if there are diatoms producing air bubbles that destabilize the sediment (Perillo, 2009), if there is an algal mat or a muddy biofilm that protects the mud from erosion (Andersen et al., 2005; Lumborg et al., 2006; Hubas et al., 2010; Passarelli et al., 2014), or if there are bivalves such as mussels armoring the bed (van Leeuwen, 2008). Secondly, the settling velocity w is not a constant; for $\text{SSC} < 1\,\text{g}\,\text{L}^{-1}$, w increases with increasing SSC values; for $\text{SSC} > 5\,\text{g}\,\text{L}^{-1}$, w decreases with increasing SSC values. In addition, the settling velocity w also varies with the density of the flocs, and this is primarily controlled by the biology. For instance, flocs formed by feeding bivalves are much denser and have a larger settling velocity than other biology-rich flocs (Andersen and Pejrup, 2011). Thus, the settling velocity, w, of individual mud flocs at a given site varies by a factor of 10 during one tidal cycle (Manning et al., 2007). Thirdly, it is unclear how to parameterize the water velocity u in Eqs. (1) and (2) when both tidal and wave-induced currents are present (Maa and Mehta, 1987; Aldridge and Rees, 1997; Lambrechts et al., 2010). Finally, the value of n is unknown a priori and it depends on the local conditions; typically, $n = 2–4$ in the laboratory and $n = 4–6$ in the field (Lambrechts et al., 2010; Wolanski and Elliott, 2015).

Thus, with all these caveats, the models generally can reproduce only qualitatively the pathways of fine sediment in shallow, muddy estuaries, which are sketched in Fig. 3. Just as for sandy estuaries, modeling muddy estuaries is still an art more than an exact science and it requires extensive field data.

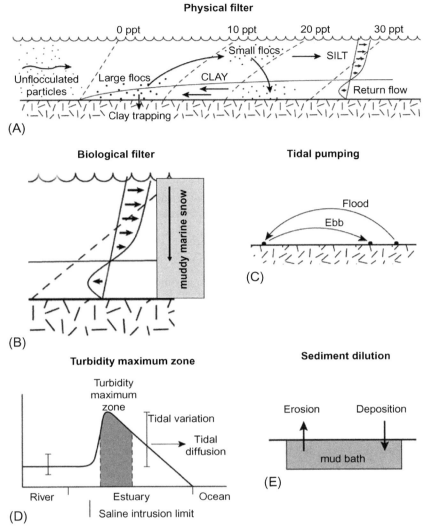

FIG. 3 (A) On reaching seawater—salinity is shown in ppt—the suspended riverine mud meets a physical filter. The mud particles flocculate and form large flocs near the salinity intrusion limit. The small flocs escape seaward, grow larger, settle to the bottom layer and are advected back landward to the turbidity maximum zone by the salinity-induced secondary circulation. Silt-based flocs are weaker, readily break up by turbulence, and are more readily exported. (B) On reaching clearer water downstream of the turbidity maximum zone, large muddy marine snow flocs form, and settle down to be brought back landward in the estuary by the salinity-driven secondary circulation. (C) Tidal pumping preferentially drives the sediment landward. (D) These processes result in forming a turbidity maximum zone. Some sediment is exported seaward by tidal diffusion. (E) The bulk of the fine sediment remains in the estuary, forming a "mud bath" where "new" riverine sediment is diluted with "old" sediment. *(Adapted from Syvitski, J.P.M., Harvey, N., Wolanski, E., Burnett, W.C., Perillo, G.M.E., Gornitz, V., 2005. Coastal Fluxes in the Anthropocene. Berlin: Springer-Verlag, pp. 39–94, 231 p.; Wolanski, E., Elliott, M., 2015. Estuarine Ecohydrology. An Introduction. Amsterdam: Elsevier, 322 p.)*

Further there are several additional feedback processes between mud, the biota and the water circulation that fine sediment models do not capture yet. Firstly, fluid mud lubricates the water flow over the bottom. Thus the apparent bottom friction coefficient in a muddy estuary can be much smaller than that in a sandy estuary; this affects the propagation of tides in an estuary (McAnnally Jr. and Hayter, 1990; Mehta and Srinivas, 1993; King and Wolanski, 1996). Secondly, the vegetation also affects the stability of the banks against erosion; the roots of plants and trees are particularly important (Feagin et al., 2009; D'Alpaos et al., 2009; Gedan et al., 2011; Chen et al., 2012). Thirdly, surface waves in coastal waters with a muddy bottom can fluidize the mud and form a fluid mud layer; energy is extracted from the surface waves to generate interfacial waves at the fluid mud interface and this reduces wave energy. As a result, the surface waves can be considerably attenuated without breaking by the time they reach the coast, and this protects coastal mudbanks (Wells and Coleman, 1981; Wells, 1983; Rodriguez and Mehta, 1998; Gensac et al., 2015).

2.2.3 Muddy Sand

Sand is often mixed with mud in estuaries. Its bed-load transport can be inhibited, or enhanced, by the presence of mud particles within the sand. This depends on whether the mud particles cement the sand particles (Wolanski et al., 2006a) or lubricate the sand by preventing sand particles to interlock (Barry et al., 2006). This modifies the bed load by a factor of up to 4; for modeling, this requires modifying the erosion constants in the transport equations, but the field data are still lacking. This sediment can so far only be modeled qualitatively at best.

3 MODELS OF NUTRIENT SEQUESTRATION BY FINE SEDIMENT

The fine sediment is important to the ecology because it determines the habitats and it increases turbidity and hence decreases light penetration and the photosynthetically active radiation (PAR). Fine sediment is also important to the ecology because it controls the fate of riverine nutrients in an estuary through the partition coefficient K_d, which is the fraction of the nutrients in particulate form. There is no generic theoretical formula for this process. From field data, K_d is found to depend on the SPM (suspended particulate matter), and empirical relationships like the following one can be obtained by curve fitting the field data:

$$K_d = \frac{\text{SPM}}{\text{SPM} + \delta} \tag{3}$$

where SPM is expressed in mg/L and δ is a constant that depends on the local sediment. Typically, with a large scatter in the field data, $\delta = 212$ in the Yangtze Estuary, China, and $\delta = 72$ in northern European muddy estuaries (Middelburg and Hermann, 2007; Xu et al., 2015). Thus, for European estuaries when SPM $= 10 \, \text{mg} \, \text{L}^{-1}$, 12% of the nutrients are in particulate form, but when SPM $= 1000 \, \text{mg} \, \text{L}^{-1}$, 93% of the nutrients are in particulate form. Such values of SPM are common as the riverine water transits through the turbidity maximum zone (Fig. 3) to reach clear coastal waters; during this transit, there are significant transfers of nutrients back and forth between the particulate and dissolved forms. As the models of fine sediment do not reproduce well the observed dynamics, they reproduce even less well the fate of nutrients. Nevertheless, with all these caveats, this is the state of knowledge and, in the absence of anything better, these models are all that we have to determine the key parameters of interest, principally the SSC (hence the light attenuation), the net movement of the sediment and the sequestration of nutrients by fine sediment. Without substantive field data, these models are at best qualitative when estimating the ecology of muddy estuaries; they must be used with great caution and the realization that there are large error bars in the predictions of the movement of mud banks and the fate of nutrients and pollutants in estuaries (Middelburg and Hermann, 2007; Xu et al., 2013, 2015; Millward and Liu, 2003; Fitzsimons et al., 2011; Du Laing, 2011; Covelli, 2012; Varma et al., 2013).

4 ESTUARINE ECOHYDROLOGY MODELS

4.1 Introduction

The earliest such models date from the 1950s and focused on anoxia and hypoxia and the resulting fish kills, resulting from the discharge of raw sewage and mill waste in rivers and estuaries. The models used the biological oxygen demand (BOD) of the waste to calculate the dissolved oxygen (DO) concentration in rivers and estuaries. The basic continuity equation for DO in 2D is

$$\partial \frac{\text{DO}}{\partial t} + \mathbf{u} . \nabla \text{DO} = -k_1 \text{BOD} + A \left(\text{DO}_0 - \text{DO} \right) \tag{4}$$

where $\mathbf{u}$ is the horizontal velocity vector, $\nabla = (\partial/\partial x + \partial/\partial y)$ and where x and y are the horizontal axes, k_1 is the rate of consumption of DO by the decay of the organic matter (BOD), A is the aeration rate and DO_0 is the saturation value of DO. This equation yields a DO sag curve; the DO values that initially decrease until a minimum DO value is reached, and there is a recovery period afterward that lasts much longer (typically 5 days) than the decay period. In practice, that simple model is over-simplified for several reasons. For instance, some of the organic material may not be carried away by the currents; instead it settles and decays in situ, thus locally decreasing the DO. Also, this model basically assumes that the water is so grossly polluted that only bacteria thrive; in practice, other aquatic biology processes occur that can assimilate the BOD; there may be day/night fluctuations in DO whereby algae and phytoplankton aerate the water column in daytime by photosynthesis and depress the DO at nighttime by respiration. Fish may also be important in improving the DO by feeding on the solid waste. Finally, there may be interaction with the benthos whereby some of the waste may be removed by filter feeders such as bivalves. Thus, the biology is important, but it is still usually neglected in these engineering applications.

The questions asked to models are much more advanced now. In view of the eutrophication of estuaries, the key question is "Where do the riverine nutrients go in an estuary?" To do that, sub-models describing the estuarine ecology (and if needed the chemistry) are linked with models of the water circulation; they are also linked, if needed, with models of the sediment dynamics. The water circulation model moves the nutrients around within the model estuary, which is represented by a number of cells. The sediment model moves the sediment around the estuary model between the cells. The nutrient model moves the nutrients within the food web in each cell of the model. However, this linking of models is not straightforward. This is because water circulation models generally have a high spatial resolution, so that hundreds if not thousands of cells are needed for an estuary. Taken together the combined physics-biology model can become impractical. To overcome that, several types of models have been proposed. The simplest such models describe spatially and temporally averaged ecosystems based on a few variables, while the most complex ones describe the spatial and temporal distribution of a large number of variables, including a limiting nutrient (Engqvist, 1996; Jorgensen et al., 1986; Jorgensen and Bendoricchio, 2001; Jorgensen, 2011). The latter models are still impractical to use because they have several hundreds of parameters, most of them are unknown (e.g., Fulton et al., 2004). All models require some level of simplicity to be practical.

To solve that problem, the engineering approach is to link high-resolution water circulation models with a simplified food web model that is commonly restricted at best to dissolved nutrients, bacteria, phytoplankton, and zooplankton. These models are hybrids, with a high spatial resolution of the water circulation and a coarse resolution of the ecology. As a result, these models basically cannot model higher components of the food web, such as fish; for that, simple regression analysis is often used relying on statistical summaries of field data, which are usually extremely noisy.

To solve that impasse, box models have been proposed that are a compromise between the above models. They divide the estuary in a small number (typically <50) of boxes; they transfer water (and the nutrients and the living organisms that they contain) from box to box at rates determined by the oceanography, and they model the dominant ecological processes within each box. The upstream box is usually the river inflow, and the downstream box is the coastal water.

4.2 The LOICZ Model

The simplest, yet realistic, such box model is the LOICZ model (Gordon et al., 1996; Swaney et al., 2011). It relies on field data about the river discharge, the salinity, the SPM and the dissolved inorganic nitrogen (DIN) and phosphorus (DIP) in the river, in the estuary and in coastal waters. Water and salinity is conserved, hence the total water and salt flux into the estuary is equal to the total outflow of water and salt from the estuary; based on that, a simple mass balance calculation yields the water fluxes Q in and out of the estuary, and thus also the residence time of water in the estuary (Fig. 4).

The nutrients (Y, which stands for DIN or DIP) enter the estuary at a rate IN and leave the estuary at a rate OUT. The net budget is:

$$\Delta Y = \left(\text{OUT} - \text{IN} \right) \tag{5}$$

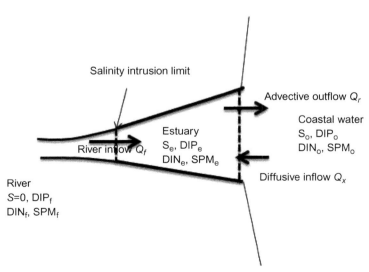

FIG. 4 Sketch of the key components of the LOICZ nutrient budget model for an estuary represented by one box and connected to the river upstream and the coastal water downstream. *(Adapted from Wolanski, E., Elliott, M., 2015. Estuarine Ecohydrology. An Introduction. Amsterdam: Elsevier, 322 p.)*

If the nutrients were conservative, then would simply transit through the estuary and it results $\Delta Y = 0$. But they are not conservative, that is, they are consumed or produced by the biology in the estuary; hence $\Delta Y \neq 0$. When Y is carbon, ΔY is the estuarine net ecosystem metabolism (NEM), also called $[p-r]$ (i.e., production minus respiration).

For DIP, the inflow is:

$$\text{IN} = \text{DIP inflow from the river} + \text{DIP inflow from the ocean} \tag{6}$$

$$= Q_f \text{DIP}_f + Q_x \left(\text{DIP}_o - \text{DIP}_e \right)$$

where Q_f is the river discharge, DIP_f, DIP_e, and DIP_o are the DIP in, respectively, the river, the estuary, and coastal waters, to which one can add the DIP inflow from rainfall and groundwater if relevant. The outflow of DIP from the estuary is:

$$\text{OUT} = Q_r \text{DIP}_r \tag{7}$$

where $\text{DIP}_r = \text{DIP}$ at the mouth the estuary $= 0.5 \, (\text{DIP}_e + \text{DIP}_o)$.

Having thus calculated IN and OUT for DIP from Eqs. (7–8), the net budget of DIP (ΔDIP) is calculated from Eq. (5). DIP is used as a proxy for C assuming stoichiometry (i.e., C:P = 106:1). Thus:

$$\text{NEM} = \left[p - r \right] = 106 \, \Delta \text{DIP} \left(\text{mol C} / \text{s} \right) \tag{8}$$

If the system is a net producer of organic matter ($[p-r] > 0$), then DIP is taken up ($\Delta \text{DIP} < 0$); if the system is a net consumer of organic matter, then $\Delta \text{DIP} > 0$. If NEM > 0 the system is autotrophic. If NEM < 0 it is heterotrophic. The net budget of DIN is similarly calculated assuming stoichiometry (i.e., $N:P = 16:1$), and the difference between expected and observed ΔDIN is nitrogen fixation minus denitrification [Nfix−denit]. If (Nfix−Denit) < 0, the estuary is net denitrifying system, that is, there is an additional sink of DIN. If (Nfix−Denit) > 0, the estuary is N fixing, that is, there is an additional source of DIN.

For turbid estuaries, a variant of the LOICZ model estimates the effect of the mud in sequestering nutrients by using the partition coefficient K_d Eq. (3). Nutrients are not transferred conservatively by the water from box to box; according to the difference in SPM between the boxes they are absorbed or released from particulate form following Eq. (3); in practice nutrient partitioning is found to be important when evaluating the nutrient budget of turbid estuaries (Xu et al., 2013, 2015).

The LOICZ model can be readily modified by adding additional, connected boxes one next to the other for branched estuaries, and boxes one below the other for stratified estuaries. The model is simple but robust and offers a reliable, order-of-magnitude estimate of the fate of nutrients provided field data are available; for that reason it has been used in >250 sites worldwide (Swaney et al., 2011). Nevertheless the LOICZ model does not consider higher components of the food web, such as fish, so its application is somewhat limited. Furthermore, the model ignores a most important source source of nutrients, namely organic detritus that can be transferred to the food chain through remineralization or feeding by detritivores. For detritus-rich estuaries the LOICZ model predictions are unrealistic (Bonthu et al., 2016; Kiwango et al., 2018). As the LOICZ model depends on field data from the river, the river and coastal waters, it cannot readily be used in a forecasting mode to assess the impact of development proposals or remediation measures.

4.3 NPZ Estuarine Ecosystem Models

This class of models is generally referred to as "NPZ" models (Nutrients, Phytoplankton, Zooplankton) and has a wide literature that includes Steele and Henderson (1981), Edwards and Brindley (1999), Swaney et al. (2008), Soetaert and Middelburg (2009), and Ganju et al. (2016).

To be practical, the models need to simplify the food web. An ecosystem model typically puts together in one "phytoplankton" category all the size ranges of phytoplankton, and all the size ranges of zooplankton into one "zooplankton" category. There remains the difficult question of how to represent mathematically the prey-predator relationship. Starting with phytoplankton and zooplankton, the conservation equation expresses the fact that mass is conserved; the equation is generally assumed to be of the Lotka-Volterra type although this form of equation itself is semiempirical (May, 1974; Flint and Kamp-Nielsen, 1997; Hilborn and Mangel, 1997; Kot, 2001; Jorgensen and Bendoricchio, 2001). Thus, a simple form of the equation for the growth rate of phytoplankton may be:

$$d\frac{\text{Phyto}}{dt} = k_1 \frac{N_{utrient}}{\left(K_1 + N_{utrient} \right) - G} \tag{9}$$

where Phyto is the phytoplankton concentration, k_1 and K_1 are in theory constants although k_1 may also vary with the PAR, hence also on SPM, $N_{utrient}$ is the concentration of the limiting nutrient, and G is the phytoplankton loss rate due to grazing by zooplankton. G is not a constant because it depends on both the phytoplankton biomass and the zooplankton biomass, both of which vary in time. $N_{utrient}$ is not constant either because it is depleted by phytoplankton or replenished by remineralization and external input. There are similar complex equations for $N_{utrient}$ and G. There are typically 12 parameters for each predator-prey relationship and the simplest food web may need over 50 parameters for which data are usually unavailable or insufficient (Flint and Kamp-Nielsen, 1997). According to the values of these parameters and the details of the equations, several mathematical solutions emerge. In one solution, the phytoplankton biomass reaches a steady state after an S-growth curve; in a second solution, a succession of population explosion and population crash occurs in time for both the prey and the predator; in a third solution, the populations of the predator and the prey fluctuate smoothly with time lags (May, 1974; Jorgensen and Bendoricchio, 2001). The robustness of the model, or its failure, is thus in the details of the mathematical formulation and the availability, or otherwise, of a sound data set to verify the model performance for a wide range of situations including some that the model may not have been designed for (Hilborn and Mangel, 1997). Often, such extensive data sets are unavailable and the models thus remain more curiosity driven than practical (Radach and Moll, 2006). Such models cannot reliably deal with estuarine ecosystems where the food web structure changes because the mathematical formulae expressing ecological phase shifts in an estuary are not known a priori.

4.4   The UNESCO Estuarine Ecohydrology Model

The estuary is divided into a series of connecting boxes (cells). As sketched in Fig. 5, in the physical sub-model these cells exchange water with each other by diffusion (tidal mixing) and by advection (the mean currents driven by the river runoff, the wind, and the oceanic inflow); these exchange rates are provided by a separate estuarine oceanography model (see Section 2.1). The upstream cell is at the tidal limit; it receives freshwater, fine sediment with its particulate nutrients, dissolved nutrients, organic detritus, and freshwater plankton. The observed values of salinity, detritus, SSC, and all of the elements in the model food web are imposed at the downstream cell in coastal water. All these require field data. The SSC in the estuary is imposed from field data; this can include an estuarine turbidity maximum zone (ETM). This is realistic for vertically fairly well-mixed estuaries.

This physical sub-model model is linked to an ecological sub-model. The mass transfer equations for the food web are simplified upfront by assuming that there are no blooms and no ecological phase shifts, that is, the basic structure of the

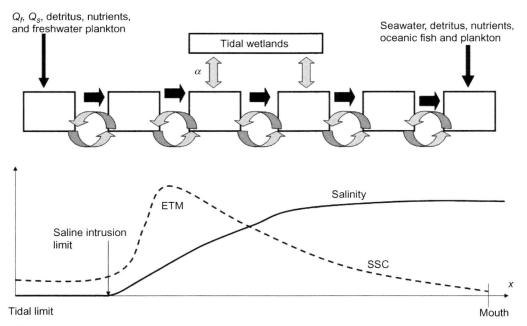

FIG. 5 Sketch of the estuarine ecohydrology physical sub-model. The top figure shows how the physical sub-model may divide the estuary into cells spread along the channel from the tidal limit to the mouth. The upstream cell receives the riverine discharge of water (Qf), sediment (Qs), detritus, freshwater plankton, and nutrients. The downstream cell receives oceanic water, detritus, nutrients, plankton, and fish. There is a net seaward flux (*black arrows*) as a result of the freshwater discharge, and a bidirectional tidal mixing between the cells (*curved arrows*). The model includes (*bottom figure*) a predetermined estuarine turbidity maximum (ETM) as well as a salinity intrusion limit determined by the model. In every cell, a food web exists (Fig. 6 discussed later). There may be also a lateral import, or an export, α, from/to tidal wetlands or discharges from human activities. (*Adapted from Wolanski, E., Elliott, M., 2015. Estuarine Ecohydrology. An Introduction. Amsterdam: Elsevier, 322 p.*)

food web does not change; S-shape population growth curves result based on a growth rate and minimum and a maximum values for each population—the data for this parameterization are generally available or can be inferred from long-term monitoring data (Shi et al., 2017). The prey-predator relationship is thus assumed to be:

$$dX/dt = \beta X \left(1 - X/X_o\right) H\left(Y, Y_{o1}\right) \tag{10}$$

where X is biomass of the predator, Y is the biomass of the prey, β is the predator biomass growth rate, X_o is the predator saturation biomass, Y_{o1} is the biomass of the prey when the predator stops hunting it, and H is the Heaviside function ($H=0$ if $Y < Y_{o1}$, $H=1$ otherwise). There is corresponding differential equation for Y, with an opposite sign, so that basically the mass consumed by the predator is the same as the mass lost by the prey. If there are several predators or several preys, Eq. (10) is used for each prey-predator relationship and these are added. Finally there is a natural death rate δ for every species. All dead material becomes detritus that is available in the food web. The experience with various systems (see Figs. 7–9 discussed later) suggests that the upper and lower limits are not of critical importance in the modeling (because blooms are not modeled) and thus the number of independent critical parameters in each prey-predator relationship is in practice decreased from 12 to 1, with the addition of one parameter for each species (its natural death rate δ). This greatly simplifies the model and allows more realistic food web models that do not stop at zooplankton. The disadvantage of that method is that biological processes may be over-simplified. Thus, within each cell of the model (Fig. 5), a food web is assumed with detritus playing a key role, and this food web varies from estuary to estuary. Fig. 6A shows the food web used in the Guadiana Estuary, Portugal, where the exchange of nutrients between the water and the substrate is negligible compared to that between the water and the wetlands, and where filtration by bivalves is important to the ecology (Wolanski et al., 2006b), as it is for instance in San Francisco Bay (Dugdale et al., 2016). The food web is more complex in the Chilika Lagoon, India, where fishing strongly affects the ecology (Fig. 6B; Bonthu et al., 2016), in the Wami Estuary, Tanzania, where hippos and mangroves are important in the ecology (Fig. 6C; Kiwango et al., 2018), and in Laizhou Bay, China (Fig. 6D) where jellyfish, resuspension by winter storms of nutrient-rich sediment, and overfishing are all important for the ecology (Y. Li, pers. comm.).

In all cases the model appears to yield qualitatively encouraging results when compared with observations (Figs. 7–9). In these relatively data-rich applications, one can probably test scenarios with reasonable error bars. For instance for Laizhou Bay the model was used to assess the impact of finfish fishing on jellyfish, and the prediction that jellyfish biomass has increased 10-fold as a result of finfish overfishing appears realistic in view of field observations over the last decade.

Nevertheless, at all these sites the data are still insufficient for both model calibration and verification. In most cases there are not enough field data to estimate the seasonal and interannual error bars; thus, the models have to be used with caution. Clearly the model may oversimplify prey-predator dynamics. However, on the positive side, it is mathematically stable because it has equal complexity at the top and bottom of the food web (Brauer and Castillo-Chavez, 2001), and it incorporates both bottom-up and top-down ecological controls; it considers the whole estuarine ecosystem including the riverine and oceanic influences; it can be readily modified to incorporate additional processes as new data become available. Any of these ecology sub-models can readily be made more complex by adding other interactions; for instance, one could add predation by minor carnivorous fish (Baker and Sheaves, 2009); however, this requires more parameters and more field data. The minimum complexity of the ecology sub-model that is needed for realism is a judgment call by the modeller working in close collaboration with the physical oceanographer and the ecologist familiar with the local conditions.

It is tempting to use these models top test scenarios, such as to evaluate the influence of river freshets from dams, of irrigation farming, of water extraction, of reclaiming tidal wetlands, of engineering activities at the river mouth, of dredging, and of fishing. The resulting predictions appear reasonable but there are not enough field data to estimate the error bars in the predictions.

Just like the NPZ models, of which it is a subset, this model cannot reliably deal with ecological phase shifts in an estuary. It is worth noting here that this ecohydrology model, after adjusting for different open boundaries and a different food web, has also been used for the central region of the Great Barrier Reef, where it was verified against time-series data of coral cover and the crown-of-thorns starfish population (Wolanski and De'ath, 2005), and for the Serengeti ecosystem, where it was verified against time-series data of the population of lions and wildebeest (Gereta et al., 2002).

4.5 The Ecopath Model

The Ecopath model has been widely used for describing ecosystems near steady state and the impact of fisheries (Heymans et al., 2011). The ecosystem is structured as a food web. Two equations are used, namely the conservation of mass (i.e., the production) and the conservation of energy (e.g., the consumption by group i is equal to the production by group i plus respiration by i and unassimilated food by i). The user needs to know the food web, the biomass, the production/biomass

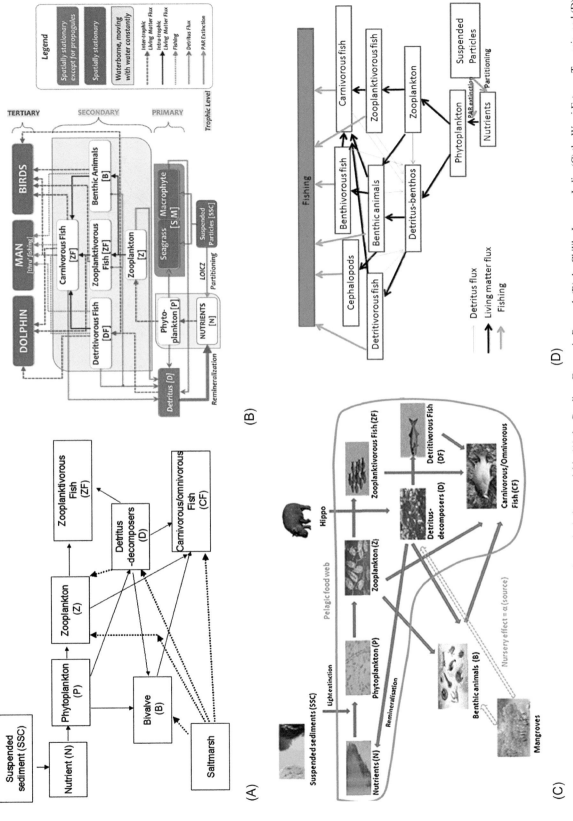

FIG. 6 The food web assumed in the UNESCO estuarine ecohydrology model in (A) the Guadiana Estuary in Portugal, (B) the Chilika Lagoon, India, (C) the Wami Estuary, Tanzania, and (D) Laizhou Bay, China. The microbial loop is the dominant process to remobilize the nutrients in the detritus. (*Adapted from Wolanski, E., Chicharo, L., Chicharo, M., Morais, P. 2006b. An ecohydrology model of the Guadiana Estuary (South Portugal). Estuar. Coast. Shelf Sci. 70, 132–143; Bonthu, S. Ganguly, D., Ramachandran, R., Pattnaik, A.K., Wolanski, E. 2016. Both riverine detritus and dissolved nutrients drive lagoon fisheries. Estuar. Coast. Shelf Sci. 183, 360–369; Kiwango, H., Njau, K.N., Wolanski, E. 2018. The application of nutrient budget models to determine the ecosystem health of the Wami Estuary, Tanzania. Ecohydrol. Hydrobiol. 18, 107–119.*)

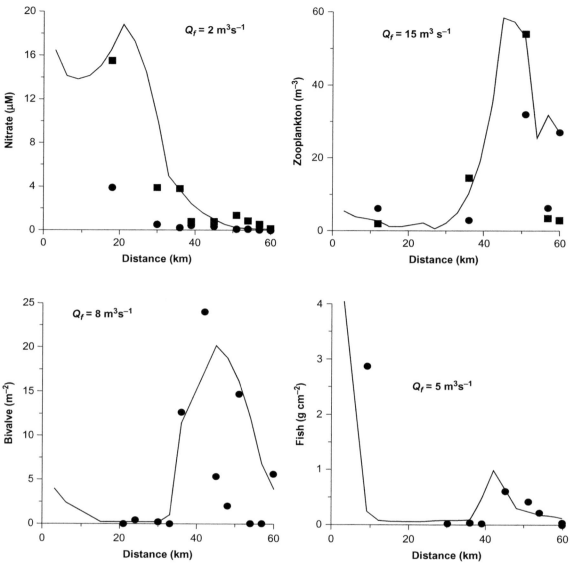

FIG. 7 Observed (● and ■ for different years) and predicted (line) along-channel distribution in the Guadiana Estuary, Portugal, of nitrate, zooplankton, bivalve and fish during low-flow condition for various freshwater flow rates Qf. The distance is measured from the tidal limit. The data were collected at different times corresponding to different river flow rates and there are insufficient data to estimate the error bars in the field data. *(Adapted from Wolanski, E., Chicharo, L., Chicharo, M., Morais, P., 2006b. An ecohydrology model of the Guadiana Estuary (South Portugal). Estuar. Coast. Shelf Sci. 70, 132–143.)*

ratio, the consumption/biomass ratio, and the unexplained mortality for each group. This model is data intensive and few estuarine ecosystems are sufficiently data rich to fully exploit this model.

Complex, spatially explicit, individual-based models of fish stocks exist and they are even more data intensive (Johanson et al., 2017). Models of maximum sustainable yield based on stock-recruitment relationships are based on similar concepts and are similarly very data intensive (Lemos, 2016). The Ecopath model has been expanded to the Ecospace model to include a spatial dimension, and this model is even more data intensive (Romagnoni et al., 2015); only probably in the data-rich North Sea are there sufficient data to apply this model but then the results are quite encouraging.

4.6 Harmful Algae Blooms Models

The modeling of HABs is still an art more than a formal science; there are many unknowns and large error bars because HABs have a rapidly time-varying history and a complex ecology (Walsh et al., 2011; Davidson et al., 2012). They are

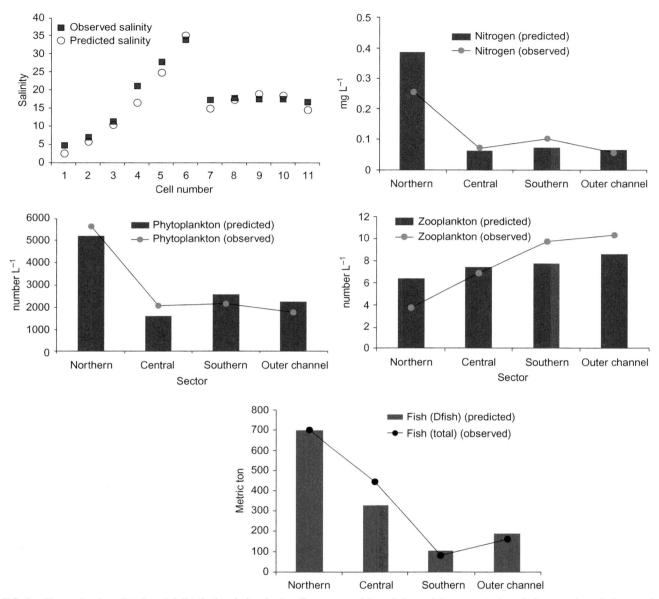

FIG. 8 Observed and predicted spatial distribution during the low-flow season of the salinity, and the concentration of nitrogen, phytoplankton, and zooplankton and the biomass of fish in Chilika Lagoon. There are insufficient data to estimate the error bars in the field data. *(Adapted from Bonthu, S., Ganguly, D., Ramachandran, P., Ramachandran, R., Pattnaik, A.K., Wolanski, E., 2016. Both riverine detritus and dissolved nutrients drive lagoon fisheries. Estuar. Coast. Shelf Sci. 183, 360–369.)*

generally based on the Silicon (Si) limitation hypothesis that is generally supported by field data on the nuisance flagellate *Phaeocystis*. However, the application of this hypothesis is not straightforward because high N:Si ratios do not consistently promote harmful dinoflagellates. Furthermore, this ratio does influence toxin production; the influence varies with different genus and species. No two HABs are similar in terms of the timing and abundance of the precursors (e.g., the cyanophyte *Trichodesmium erythraeum*), the dominant ichthyotoxic dinoflagellates (e.g., *Karenia brevis* and its toxic dinoflagellate competitors *Alexandrium monilatum* and *Pyrodinium bahamense*), and the presence of parasites and of benign species, as well as the fishing pressure that affects the ecology of the system. Thus, to model HABs, the modeler needs to know in advance which species will dominate in a developing HAB. The HAB model can be very complex with 27 state variables with multiple nutrient sources and phytoplankton groups (Walsh et al., 2011). It can also be very simple comprising just fish-fed and fish-starved red tides of *K. brevis*, or it can use statistical models for predicting bloom and toxin likelihoods

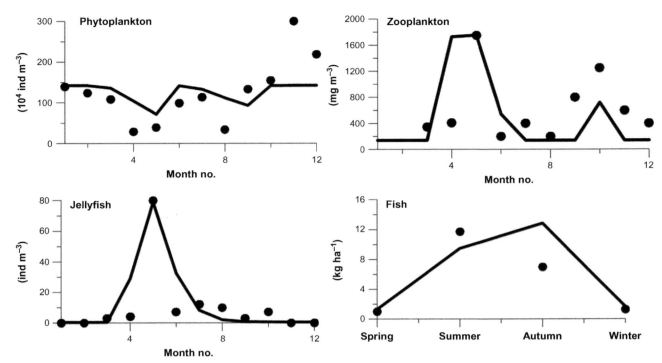

FIG. 9 Time-series plot of the spatially-averaged observed and predicted distribution of phytoplankton, zooplankton, jellyfish, and the fish biomass in Laizhou Bay, China. The field data were kindly provided by Yanfang Li from a search of the literature in Chinese. There are insufficient data to estimate the error bars in the field data.

once the physics is known (Anderson et al., 2016). Operational HAB models rely on assimilating satellite observations and daily ship-borne observations in ecological models whose complexity (e.g., which species to include) is dictated by the observations.

4.7 Hypoxia Models

Typically such models link a two-layer oceanographic model with an ecological model kept as simple as possible (Eldridge and Roelke, 2010; Beck et al., 2017). The model has an aerated surface mixed layer, a poorly aerated bottom layer, and an organic-rich sediment layer (Fig. 10). The ecological model includes SPM, DO, DIC, N, and P; organic matter can include both labile and refractory components. Light penetration is controlled by the SPM and a fraction of the SPM may be organic and a fraction of that may be labile and thus used in the food web. Such models are data-intensive and generally such data are unavailable for real-time modeling.

5 A SYNTHESIS

There is a need for ecohydrology models to test real scenarios, for example, to predict the likely impact of on the estuarine ecosystem of various developments such as a new dam changing the hydrology, land reclamation and dredging that change the water and sediment circulation, and receiving the effluent from a sewage treatment plant, agriculture, and aquaculture. This chapter summarizes the state of knowledge of such models. They encompass hydrodynamic modeling, sediment dynamic modeling, and estuarine ecological modeling. The combined model is called an estuarine ecohydrology model. Hydrodynamic-sediment-ecology models must be coupled to represent the dynamic feedbacks between the physics and the biology. The most informative models are process based, that is, they rely on equations that express the pathways of nutrients.

The water currents can be reliably modeled, except for the case of estuarine fronts—and this prevents the reliable modeling of the recruitment of fish larvae in a stratified estuary. The fine sediment dynamics also need to be modeled because fine sediment affects the turbidity, hence light penetration and photosynthesis occur; they also controls the habitats, hence

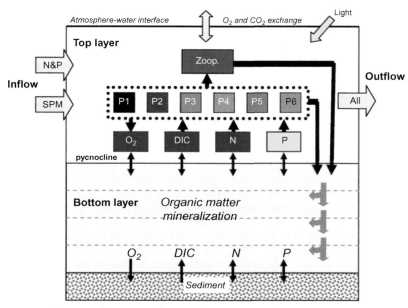

FIG. 10   The three-compartment (a surface aerated layer, a poorly aerated bottom layer and an organic-rich sediment layer) hypoxia model of Eldridge and Roelke (2010). There are six phytoplankton groups P1–P6. There is an inflow on one side and an outflow on the other side, determined by the oceanography.

the niches for the ecology. Fine sediment dynamics is also strongly influenced by the biology. Because the interaction between physics and biology controls the fine sediment dynamics, laboratory data on sediment dynamics that ignore the biology may not be useful for the model (Wolanski and Elliott, 2015). In addition, the suspended fine sediment also controls the fluxes of nutrients between the dissolved and particulate phases. Knowledge about all these processes is still rudimentary and thus extensive field data are needed to parameterize these processes for the model for each estuary because these various processes may have different importance from estuary to estuary as no two estuaries are the same, thus the models are site specific.

The estuarine ecohydrology modeler has to combine an ecology model with an estuarine water model and a sediment circulation model (or field data for the sediment). The food web model had to be simplified to a workable spatial and temporal complexity determined by the field data. The minimum complexity of the ecology sub-model that is needed for realism is a judgment call by the modeller working in close collaboration with the physical oceanographer and the ecologist familiar with the local conditions. Thus all sub-models should be simplified to avoid being drowned in complexity but they must not be over-simplified so that they still retain realism.

Ideally the ecology model and the water and sediment circulation models have similar levels of complexity, something that was envisioned by Nihoul (1975) over 40 years ago and largely ignored in practice. Indeed, because the physics of water movement are better known, there has been a tendency to develop highly complex, high-resolution (spatial and temporal) hydrodynamic models. However, such models are difficult to link with the ecology models, because estuarine ecology models are commonly not backed by sufficient field data and knowledge on the ecosystem functioning to justify high-resolution modeling (Jorgensen, 2011). Generally the practical solution is for the modeler to spatially smooth out the water, sediment and food web models to spatial scales that are coarse but till adequately represent the estuary, and also more often than not smooth out over time to arrive at the time scales of interest (e.g., averaging out the tides).

The usefulness of the model, its complexity and its transparency, and the type of questions that the models can answer and with what error bars, depends on the availability of field data. For most estuaries such complete data sets are unavailable and in such cases the model is only qualitative. Provided that suitable field data are available, estuarine ecohydrology modeling is realistic and practical for systems where the food web structure basically stays unchanged; in practice it is not possible to reliably model ecological phase shifts in an estuary. As no two estuaries are the same, by and large there is no universal estuarine ecohydrology model; for each estuary the modeler needs to work closely with the physical oceanographer and the ecologist. And when the manager and other stakeholders use a verified model as a decision-support tool, this should be done at least initially in close collaboration with the modeler so as to ensure that the model is not pushed beyond its boundaries.

REFERENCES

Aldridge, J.N., Rees, J.M., 1997. Interpreting observations of near-bed sediment concentration and estimation of "pick-up" function constants. In: Burt, N., Parker, R., Watts, J. (Eds.), Cohesive Sediments. John Wiley & Sons Limited, Chichester, pp. 289–303. 458 pp.

Andersen, T.J., Pejrup, M., 2011. Biological influences on sediment behavior and transport. In: Uncles, R.J., Monismith, S.G. (Eds.), Water and Fine sediment Circulation. Treatise on Estuarine and Coastal Science. vol. 2. Elsevier, Amsterdam, pp. 289–309.

Andersen, T.J., Lund-Hansen, L., Pejrup, M., Jensen, K.T., Mouritzen, K.N., 2005. Biologically induced differences in erodibility and aggregation of subtidal and intertidal sediments: a possible cause for seasonal changes in sediment deposition. J. Mar. Syst. 55, 123–138.

Anderson, C.R., Kudela, R.M., Kahru, M., Chao, Y., Bahr, F.L., Rosenfeld, L., Anderson, D.M., Norris, T., 2016. Initial skill assessment of the California Harmful Algae Risk Mapping (C-HARM) system. Harmful Algae 59, 1–18.

Baker, R., Sheaves, M., 2009. Overlooked small and juvenile piscivores dominate shallow-water estuarine "refuges" in tropical Australia. Estuar. Coast. Shelf Sci. 85, 618–626.

Barnes, T.K., Mazzotti, F.J., 2005. Using conceptual models to select ecological indicators for monitoring. Restoration and management of estuarine ecosystems. In: Bortone, S.A. (Ed.), Estuarine Indicators. CRC Press, Boca Raton, FL, pp. 493–502.

Barry, K.M., Thieke, R.J., Mehta, A.J., 2006. Quasi-hydrodynamic lubrication effect of clay particles on sand grain erosion. Estuar. Coast. Shelf Sci. 67, 161–169.

Beck, M.W., Lehrter, J.C., Lowe, L.L., Jarvis, B.M., 2017. Parameter sensitivity and identifiability for a biogeochemical model of hypoxia in the northern Gulf of Mexico. Ecol. Model. 363, 17–30.

Bentley, S.J., Swales, A., Pyenson, B., Daw, J., 2014. Sedimentation, bioturbation, and sedimentary fabric evolution on a modern mesotidal mudflat: a multi-tracer study of processes, rates, and scales. Estuar. Coast. Shelf Sci. 141, 58–68.

Bonthu, S., Ganguly, D., Ramachandran, P., Ramachandran, R., Pattnaik, A.K., Wolanski, E., 2016. Both riverine detritus and dissolved nutrients drive lagoon fisheries. Estuar. Coast. Shelf Sci. 183, 360–369.

Borsje, B.W., Roos, P.C., Kranenburg, W.W., Hulscher, S.J.M.H., 2013. Modeling tidal sand wave formation in a numerical shallow water model: the role of turbulence formulation. Cont. Shelf Res. 60, 17–27.

Brauer, F., Castillo-Chavez, C., 2001. Mathematical Models in Population Biology and Epidemiology. Springer, Berlin. 416 p.

Bruner de Miranda, L., Andutta, F.P., Kjerfve, B., de Castro Filho, B.M., 2017. Fundamentals of Estuarine Physical Oceanography. Springer. 480 p.

Camenen, B., Larson, M., 2005. A general formula for non-cohesive bed load sediment transport. Estuar. Coast. Shelf Sci. 63, 249–260.

Chanson, H., 1999. The Hydraulics of Open Channel Flows: An Introduction. Butterworth-Heinemann, Oxford. 512 p.

Chen, Y., Thompson, C.E.L., Collins, M.B., 2012. Saltmarsh creek bank stability: biostabilisation and consolidation with depth. Cont. Shelf Res. 35, 64–74.

Covelli, S., 2012. The MIRACLE project: an integrated approach to understanding biogeochemical cycling of mercury and its relationship with lagoon clam farming. Estuar. Coast. Shelf Sci. 113, 1–6.

D'Alpaos, A., Lanzoni, S., Rinaldo, A., Marani, M., 2009. Intertidal eco-geomorphological dynamics and hydrodynamic circulation. In: Perillo, G.M.E., Wolanski, E., Cahoon, D.R., Brinson, M.M. (Eds.), Coastal Wetlands: An Integrated Ecosystem Approach. Elsevier, Amsterdam, pp. 159–184.

Davidson, K., Gowen, R.J., Tett, P., Bresnan, E., Harrison, P.J., McKinney, A., Milligan, S., Mills, D.K., Silke, J., Crooks, A.M., 2012. Harmful algal blooms: how strong is the evidence that nutrient ratios and forms influence their occurrence? Estuar. Coast. Shelf Sci. 115, 399–413.

de Jonge, V.N., Elliott, M., 2001. Eutrophication. In: Steele, J., Thorpe, S., Turekian, K. (Eds.), Encyclopedia of Ocean Sciences. 2. Academic Press, London, pp. 852–870.

Dronkers, J., 2005. Dynamics of coastal sediments. In: Advanced Series of Oceanic Engineering. vol. 25. World Scientific, pp. 519.

Du Laing, G., 2011. Redox metal processes and controls in estuaries. In: Shimmield, G. (Ed.), Geochemistry of Estuaries and Coasts. Treatise on Estuarine and Coastal Science. vol. 4. Elsevier, Amsterdam, pp. 115–141.

Duarte, C.M., Conley, D.J., Carstensen, J., Sánchez-Camacho, M., 2009. Return to *Neverland*: shifting baselines affect eutrophication restoration targets. Estuar. Coasts 32, 29–36.

Dugdale, R.C., Wilkerson, F.P., Parker, A.E., 2016. The effect of clam grazing on phytoplankton spring blooms in the low-salinity zone of the San Francisco Estuary: a modelling approach. Ecol. Model. 340, 1–16.

Dyer, K.R., 1986. Coastal and Estuarine Sediment Dynamics. John Wiley & Sons. 342 p.

Dyer, K.R., 1994. Estuarine sediment transport and deposition. In: Pye, K. (Ed.), Sediment Transport and Depositional Processes. Blackwell Scientific Publications, Oxford, pp. 193–218.

Edwards, A.M., Brindley, J., 1999. Zooplankton mortality and the dynamical behaviour of plankton population models. Bull. Math. Biol. 61, 303–339.

Eldridge, P.M., Roelke, D.L., 2010. Origins and scales of hypoxia on the Louisiana shelf: importance of seasonal plankton dynamics and river nutrients and discharge. Ecol. Model. 221, 1028–1042.

Elliott, M., Mander, L., Mazik, K., Simenstad, C., Valesini, F., Whitfield, A., Wolanski, E., 2016. Ecoengineering with ecohydrology: successes and failures in estuarine restoration. Estuar. Coast. Shelf Sci. 176, 12–35.

Engqvist, A., 1996. Long-term nutrient balances in the eutrophication of the Himmerfjarden estuary. Estuar. Coast. Shelf Sci. 42, 483–507.

EPA (U.S. Environmental Protection Agency), 2010. Nutrients in Estuaries. A Summary Report of the National Estuarine Experts Workgroup 2005–2007. Available from: https://www.epa.gov/sites/production/files/documents/nutrients-in-estuaries-november-2010.pdf.

Feagin, R.A., Lozada-Bernard, S.M., Ravens, T.M., Möller, I., Yeager, K.M., Baird, A.H., 2009. Does vegetation prevent wave erosion of salt marsh edges? Proc. Natl. Acad. Sci. U. S. A. 106, 10109–10113.

FitzGerald, D.M., Buynevich, I.V., Fenster, M.S., Kelley, J.T., Belknap, D.F., 2005. Coarse-grained sediment transport in northern New England estuaries: a synthesis. In: FitzGerald, D.M., Knight, J. (Eds.), High Resolution Morphodynamics and Sedimentary Evolution of Estuaries. Springer, Dordrecht, pp. 195–213.

Fitzsimons, M.F., Lohan, M.C., Tappin, A.D., Millward, G.E., 2011. The role of suspended particles in estuarine and coastal biogeochemistry. In: Shimmield, G. (Ed.), Geochemistry of Estuaries and Coasts. Treatise on Estuarine and Coastal Sciencevol. 4. Elsevier, Amsterdam, pp. 71–114.

Flint, M.R., Kamp-Nielsen, L., 1997. Modeling an estuarine eutrophication gradient. Ecol. Model. 102, 143–153.

Fulton, E.A., Smith, A.D., Johnson, C.R., 2004. Biogeochemical marine ecosystem models I: IGBEM—a model of marine bay ecosystems. Ecol. Model. 174, 267–307.

Ganju, N.K., Brush, M.J., Rashleigh, B., Aretxabaleta, A.L., del Barrio, P., Grear, J.S., Harris, L.A., Lake, S.J., McCardell, G., O'Donnell, J., Ralston, D.K., Signell, R.P., Testa, J.M., Vaudrey, J.M.P., 2016. Progress and challenges in coupled hydrodynamic-ecological estuarine modeling. Estuar. Coasts 39, 311–332.

Gedan, K.B., Kirwan, M.L., Wolanski, E., Barbier, E., Silliman, B.R., 2011. The present and future role of coastal wetlands in protecting shorelines: answering recent challenges to the paradigm. Clim. Chang. 106, 7–29.

Gensac, E., Gardel, A., Lesourd, S., Brutier, L., 2015. Morphodynamic evolution of an intertidal mudflat under the influence of Amazon supply—Kourou mud bank, French Guiana, South America. Estuar. Coast. Shelf Sci. 158, 53–62.

Gereta, E., Wolanski, E., Borner, M., Serneels, S., 2002. Use of an ecohydrological model to predict the impact on the Serengeti ecosystem of deforestation, irrigation and the proposed Amala weir water diversion project in Kenya. Ecohydrol. Hydrobiol. 2, 127–134.

Giani, M., Djakovac, T., Degobbis, D., Cozzi, S., Solidoro, C., Umani, S.F., 2012. Recent changes in the marine ecosystems of the northern Adriatic Sea. Estuar. Coast. Shelf Sci. 115, 1–13.

Gordon, D.C., Boudreau, P.R., Mann, K.H., Ong, J.E., Silvert, W.L., Smith, S.V., Wattayakorn, G., Wulff, F., Yanagi, T., 1996. LOICZ Biogeochemical Modelling Guidelines. vol. 5. LOICZ Core Project, Netherlands Institute for Sea Research, Texel, Netherlands.

Greening, H., Janicki, A., Sherwood, E.T., Pribble, R., Johansson, J.O.R., 2014. Ecosystem responses to long-term nutrient management in an urban estuary: Tampa Bay, Florida, USA. Estuar. Coast. Shelf Sci. 151, A1–A16.

Heymans, J.J., Coll, M., Libralato, S., Christensen, V., 2011. Ecopath theory, modelling, and application to coastal ecosystems. In: Baird, D., Mehta, A. (Eds.), Estuarine and Coastal Ecosystems Modeling. Treatise on Estuarine and Coastal Science. vol. 9. Elsevier, Amsterdam, pp. 93–113.

Hilborn, R., Mangel, M., 1997. The Ecological Detective. Confronting Models with Data. Princeton University Press, Princeton. 315 p.

Horrillo-Caraballo, J.M., Karunarathna, H.U., Reeve, D.E., Pan, S., 2014. A hybrid-reduced physics modelling approach applied to the Deben Estuary, UK. Coast. Eng. Proc. 1 (34), 76. https://journals.tdl.org/icce/index.php/icce/article/view/7820.

Hubas, C., Sachidhanandam, C., Rybarczyk, H., Lubarsky, H.V., Rigaux, A., Moens, T., Paterson, D.M., 2010. Bacterivorous nematodes stimulate microbial growth and exopolymer production in marine microcosms. Mar. Ecol. Prog. Ser. 419, 85–94.

Johanson, A.N., Oschlies, A., Hasselbring, W., Worm, B., 2017. SPRAT: a spatially-explicit marine ecosystem model based on population balance equations. Ecol. Model. 349, 11–25.

Jorgensen, S.E., 2011. Handbook of Ecological Models Used in Ecosystem and Environmental Management. CRC Press, Boca Raton, FL. 636 p.

Jorgensen, S.E., Bendoricchio, G., 2001. Fundamentals of Ecological Modelling. Elsevier. 530 p.

Jorgensen, S.E., Kamp-Nielsen, L., Christensen, T., Windolf-Nielsen, J., Westergaard, B., 1986. Validation of a prognosis based upon an eutrophication model. Ecol. Model. 32, 165–182.

Ke, X., Collins, M.B., Poulos, S.E., 1994. Velocity structure and sea bed roughness associated with intertidal (sand and mud) flats and saltmarshes of the Wash, U.K. J. Coast. Res. 10, 702–715.

King, B., Wolanski, E., 1996. Bottom friction reduction in turbid estuaries. American Geophysical Union. Mixing in Estuaries and Coastal Seas. Coast. Estuar. Stud. 50, 325–337.

Kingsford, M.J., Suthers, I.M., 1994. Dynamic estuarine plumes and fronts: importance to small fish and plankton in coastal waters of NSW, Australia. Cont. Shelf Res. 14, 655–672.

Kiwango, H., Njau, K.N., Wolanski, E., 2018. The application of nutrient budget models to determine the ecosystem health of the Wami Estuary, Tanzania. Ecohydrol. Hydrobiol. 18, 107–119.

Kot, M., 2001. Elements of Mathematical Ecology. Cambridge University Press, Cambridge. 464 p.

Lambrechts, J., Humphrey, C., McKinna, L., Gourge, O., Fabricius, K., Mehta, A., Wolanski, E., 2010. The importance of wave-induced bed fluidisation in the fine sediment budget of Cleveland Bay, Great Barrier Reef. Estuar. Coast. Shelf Sci. 89, 154–162.

Le Hir, P., Roberts, W., Cazaillet, O., Christie, M., Bassoullet, P., Bacher, S., 2000. Characterization of intertidal flat hydrodynamics. Cont. Shelf Res. 20, 1433–1460.

Lemos, R.T., 2016. An alternative stock-recruitment function for age-structured models. Ecol. Model. 341, 14–26.

Lumborg, U., Andersen, T.J., Pejrup, M., 2006. The effect of *Hydrobia ulvae* and microphytobenthos on cohesive sediment dynamics on an intertidal mudflat described by means of numerical modeling. Estuar. Coast. Shelf Sci. 68, 208–220.

Maa, P.-Y., Mehta, A.J., 1987. Mud erosion by waves: a laboratory study. Cont. Shelf Res. 7, 1269–1284.

Manning, A.J., Martens, C., De Mulder, T., Vanlded, J., Winterwerp, J., Ganderton, P., Graham, G.W., 2007. Mud floc observations in the turbidity maximum zone of the Scheldt Estuary during neap tides. J. Coast. Res. SI50, 832–836.

May, R.M., 1974. Stability and Complexity in Model Ecosystems. Princeton University Press, Princeton. 265 p.

Mazik, K., Curtis, N., Fagan, M.J., Taft, S., Elliott, M., 2008. Accurate quantification of the influence of benthic macro- and meio-fauna on the geometric properties of estuarine muds by micro computer tomography. J. Exp. Mar. Biol. Ecol. 354, 192–201.

McAnnally Jr., W.H., Hayter, E.J., 1990. Estuarine boundary layers and sediment transport. In: Cheng, R.T. (Ed.), Residual Currents and Long-Term Transport. Springer-Verlag, New York, pp. 260–275.

Mehta, A.J., Srinivas, R., 1993. Observations on the entrainment of fluid mud by shear flow. In: Mehta, A.J. (Ed.), Estuarine Cohesive Sediment Transport. American Geophysical Union, Washington, DC., pp. 224–246.

Middelburg, J.J., Hermann, P.M.J., 2007. Organic matter processing in tidal estuaries. Mar. Chem. 106, 127–147.

Millward, G.E., Liu, Y.P., 2003. Modelling metal desorption kinetics in estuaries. Sci. Total Environ. 314–316, 613–623.

Nihoul, J.C.J., 1975. Modelling of marine systems. Elsevier Oceanography Series. vol. 10. Elsevier Scientific, pp. 271.

Nixon, S.W., Fulweiler, R.W., Buckley, B.A., Granger, S.L., Nowicki, B.L., Henry, K.M., 2009. The impact of changing climate on phenology, productivity, and benthic–pelagic coupling in Narragansett Bay. Estuar. Coast. Shelf Sci. 82, 1–18.

Orive, E., Elliott, M., de Jonge, V., 2002. Nutrients and eutrophication in estuaries and coastal waters. In: Developments in Hydrobiology. 526 p.

Partheniades, E., 1965. Erosion and deposition of cohesive soils. J. Hydraul. Div. Am. Soc. Civil Eng. 91, 105–138.

Passarelli, C., Olivier, F., Paterson, D.M., Meziane, T., Hubas, C., 2014. Organisms as cooperative ecosystem engineers in intertidal flats. J. Sea Res. 92, 92–101.

Perillo, G.M.E., 2009. Tidal courses: classification, origin and functionality. In: Perillo, G.M.E., Wolanski, E., Cahoon, D.R., Brinson, M.M. (Eds.), Coastal Wetlands: An Integrated Ecosystem Approach. Elsevier, Amsterdam, pp. 185–210. 941 pp.

Postma, H., 1967. Sediment transport and sedimentation in the estuarine environment. In: Lauff, G.H. (Ed.), Estuaries. American Association for the Advancement of Science, pp. 158–179. Publication No. 83.

Radach, G., Moll, A., 2006. Review of three-dimensional ecological modeling related to the North Sea shelf system. Part II: model validation and data needs. Oceanogr. Mar. Biol. Annu. Rev. 44, 1–60.

Raudkivi, A.J., 1967. Loose Boundary Hydraulics. Pergamon Press, Oxford. 331 p.

Rodriguez, H.N., Mehta, A.J., 1998. Considerations on wave-induced fluid mud streaming at open coasts. In: Black, K.S., Paterson, D.M., Cramp, A. (Eds.), Sedimentation Processes in the Intertidal Zone. Geological Society, London, pp. 177–186. Special Publication 139.

Romagnoni, G., Mackinson, S., Hong, J., Eikeset, A.M., 2015. The ecospace model applied to the North Sea: evaluating spatial predictions with fish biomass and fishing effort data. Ecol. Model. 300, 50–60.

Shi, P.-J., Fan, M.-L., Ratkowsky, A.A., Huang, J.-G., Wu, H.-I., Chen, L., Fang, S.-Y., Zhang, C.X., 2017. Comparison of two ontogenetic growth equations for animals and plants. Ecol. Model. 349, 1–10.

Soetaert, K., Middelburg, J.J., 2009. Modeling eutrophication and oligotrophication of shallow-water marine systems: the importance of sediments under stratified and well-mixed conditions. In: Eutrophication in Coastal Ecosystems. Springer, Netherlands, pp. 239–254.

Steele, J.H., Henderson, E.W., 1981. A simple plankton model. Am. Nat. 117, 676–691.

Swaney, D.P., Scavia, D., Howarth, R.W., Marino, R.M., 2008. Estuarine classification and response to nitrogen loading: insights from simple ecological models. Estuar. Coast. Shelf Sci. 77, 253–263.

Swaney, D.P., Smith, S.V., Wulff, F., 2011. The LOICZ biogeochemical modeling protocol and its application to estuarine ecosystems. Treatise Estuar. Coast. Sci. 9, 136–159.

Teodosio, M.A., Morais, P., Paris, C.B., Wolanski, E., 2016. Biophysical processes leading to the ingress of temperate fish larvae into estuarine nursery areas. Estuar. Coast. Shelf Sci. 183, 187–202.

Tucker, J., Giblin, A.E., Hopkinson, C.S., Kelsey, S.W., Howes, B.L., 2014. Response of benthic metabolism and nutrient cycling to reductions in wastewater loading to Boston Harbor, USA. Estuar. Coast. Shelf Sci. 151, 54–68.

Uncles, R.J., 2011. Small-scale surface fronts in estuaries. In: Treatise on Estuarine and Coastal Science. vol. 2. pp. 53–74.

Uncles, R.J., Monismith, S.G., 2011. Water and fine sediment circulation. In: Treatise on Estuarine and Coastal Science. vol. 2. Elsevier, Amsterdam.

Valle-Levinson, A., 2010. Contemporary Issues in Estuarine Physics. Cambridge University Press. 326 p.

van Leeuwen, B., 2008. Mussel Bed Influence on Fine Sediment Dynamics on a Wadden Sea Intertidal Flat. Civil Engineering and Management MSc, Deltares and University of Twente. 110 p.

Varma, R., Turner, A., Brown, M.T., Millward, G.E., 2013. Metal accumulation kinetics by the estuarine macroalga, *Fucus ceranoides*. Estuar. Coast. Shelf Sci. 128, 33–40.

Walsh, J.J., Lenes, J.M., Darrow, B.P., Chen, F.R., 2011. Forecasting and modelling of harmful algal blooms in the coastal zone: a prospectus. In: Baird, D., Mehta, A. (Eds.), Estuarine and Coastal Ecosystems Modeling. Treatise on Estuarine and Coastal Science, vol. 9. Elsevier, Amsterdam, pp. 217–330.

Wartel, S., 2003. Mud layer and Cyclic Sedimentation Patterns in the Estuary of the Schelde (Belgium—The Netherlands). http://www.naturalsciences.be/sedimento/publications/Abstract_1.

Wells, J.T., 1983. Dynamics of coastal fluid muds: low, moderate and high tide range environments. Can. J. Fish. Aquat. Sci. 40 (Suppl 1), 130–142.

Wells, J.T., Coleman, J.M., 1981. Physical processes and finegrained sediment dynamics, coast of Surinam, South America. J. Sediment. Petrol. 51, 1053–1068.

Winterwerp, J.C., 2011. The physical analyses of muddy sedimentation processes. In: Uncles, R.J., Monismith, S.G. (Eds.), Water and Fine Sediment Circulation. Treatise on Estuarine and Coastal Science, Vol. 2. Elsevier, Amsterdam, pp. 311–360.

Wolanski, E., De'ath, G., 2005. Predicting the present and future human impact on the Great Barrier Reef. Estuar. Coast. Shelf Sci. 64, 504–508.

Wolanski, E., Elliott, M., 2015. Estuarine Ecohydrology. An Introduction. Elsevier, Amsterdam. 322 p.

Wolanski, E., Spagnol, S., 2003. Dynamics of the turbidity maximum in King Sound, tropical Western Australia. Estuar. Coast. Shelf Sci. 56, 877–890.

Wolanski, E., Williams, D., Hanert, E., 2006a. The sediment trapping efficiency of the macro-tidal Daly Estuary, tropical Australia. Estuar. Coast. Shelf Sci. 69, 291–298.

Wolanski, E., Chicharo, L., Chicharo, M., Morais, P., 2006b. An ecohydrology model of the Guadiana Estuary (South Portugal). Estuar. Coast. Shelf Sci. 70, 132–143.

Xu, H., Wolanski, E., Chen, Z., 2013. Suspended particulate matter affects the nutrient budget of turbid estuaries: modification of the LOICZ model and application to the Yangtze Estuary. Estuar. Coast. Shelf Sci. 127, 59–62.

Xu, H., Newton, A., Wolanski, E., Chen, Z., 2015. The fate of Phosphorus in the Yangtze (Changjiang) Estuary, China, under multi-stressors: hindsight and forecast. Estuar. Coast. Shelf Sci. 163, 1–6.

Chapter 30

Hypersalinity: Global Distribution, Causes, and Present and Future Effects on the Biota of Estuaries and Lagoons

James R. Tweedley*,†, Sabine R. Dittmann‡,a, Alan K. Whitfield§,a, Kim Withers¶,a, Steeg D. Hoeksema*,‖, Ian C. Potter*,†

*Centre for Sustainable Aquatic Ecosystems, Harry Butler Institute, Murdoch University, Murdoch, Western Australia, Australia, †School of Veterinary and Life Sciences, Murdoch University, Murdoch, Western Australia, Australia, ‡College of Science & Engineering, Flinders University, Adelaide, SA, Australia, §South African Institute for Aquatic Biodiversity, Grahamstown, South Africa, ¶Department of Life Sciences, Texas A&M University, Corpus Christi, TX, United States, ‖Department of Biodiversity, Conservation and Attractions, Bentley Delivery Centre, Western Australia, Australia

1 INTRODUCTION

An estuary is sometimes simplistically regarded as the place where a river meets the sea (Lyell, 1833; Ketchum, 1951). The most widely cited scientific definition of an estuary is that of Pritchard (1967), who described an estuary as "*a semi-enclosed coastal body of water which has a free connection with the open sea and within which sea water is measurably diluted with fresh water from land drainage.*" While the salinity in those estuaries that remain permanently open to the sea does decline below that of full-strength seawater (35 salinity), this does not always occur when an estuary becomes closed from the sea by the formation of a bar at its mouth (Day, 1980; Potter et al., 2010). Thus the latter authors defined an estuary as "*a partially enclosed coastal body of water that is either permanently or periodically open to the sea and which receives at least periodic discharge from a river(s), and thus, while its salinity is typically less than that of natural sea water and varies temporally and along its length, it can become hypersaline in regions when evaporative water loss is high and freshwater and tidal inputs are negligible.*" In the following meta-analysis, it is recognized that estuaries sometimes possess prominent lagoonal-like or lacustrine-like areas. Estuaries with these characteristics have been termed separately as estuarine lagoons or estuarine lakes, depending on their size (Whitfield and Elliott, 2011).

The recognition that estuaries are characterized by the presence of a well-defined river or rivers distinguishes this type of system from both coastal and atoll (coral) lagoons. Following Phleger (1981), Kjerfve (1994) defined a coastal lagoon as "*a shallow coastal water body separated from the ocean by a barrier, connected at least intermittently to the ocean by one or more restricted inlets, and usually oriented shore-parallel.*" This widely used definition is adopted in the meta-analysis. Some authors, restrict, however, the term coastal lagoon to systems that are permanently isolated from the sea (Barnes, 1980; Whitfield and Elliott, 2011). As the term estuary has sometimes been applied to embayments that are not supplied by at least one well-defined river (Nunes and Lennon, 1986; Catalá et al., 2013), and thus do not meet the above criteria for a typical estuary, these systems are referred to here as coastal embayments.

The widely used Venice System for classifying marine waters according to their salinity (Anonymous, 1958) distinguished between those ranging from 30 to 40 (euhaline) from those that are >40 (hyperhaline). While this distinction is recognized, the term hypersaline is employed rather than hyperhaline as it is used more frequently (e.g., Ajemian et al., 2018; Rybak, 2018).

Estuaries, coastal lagoons, and embayments are highly productive ecosystems, which often contain valuable recreational and commercial fisheries (Schelske and Odum, 1961; Potter et al., 2015). Estuaries are dynamic, with their physicochemical characteristics changing temporally and spatially. For example, the salinity at any point in a permanently open

---

a These authors made an equal contribution.

Coasts and Estuaries. https://doi.org/10.1016/B978-0-12-814003-1.00030-7

estuary changes during each tidal cycle and over the course of a year, particularly in relation to the timing and magnitude of fluvial discharge. While such changes are typical of macrotidal systems (>4 m tidal range), a combination of isolation from the sea by a bar at the mouth of microtidal estuaries (<2 m tidal range) in environments with low rainfall and high rates of evaporation can lead to salinities that exceed by several times that of full-strength seawater (Chuwen et al., 2009). Salinities can also become highly elevated in certain lagoons and the shallow, inner parts of coastal embayments, when evaporation rates are high and, through limited tidal exchange, residence times are long (El-Kassas et al., 2016).

The marked hypersaline conditions that sometimes develop in estuaries, lagoons, and coastal embayments expose the resident biota to considerable osmotic stress (Kinne, 1964). The resultant changes in metabolism influence the activity, growth, spawning success, and development of species in the faunal community (Smyth and Elliott, 2016; Hallett et al., 2018). The extent to which various animal and plant species are adapted, however, to different levels of hypersalinity varies (Nordlie, 2009; Gonzalez, 2012; Rybak, 2018). Thus, as the magnitude of hypersalinity increases, the number of species decreases and, under extreme conditions, result in mass mortalities and even the local extinction of flora and fauna (Gunter, 1952; Whitfield et al., 1981; Molony and Parry, 2006; Kim et al., 2013).

This chapter first employs a meta-analysis of data derived from numerous studies to describe the geographical distributions and characteristics of those estuaries, lagoons, and coastal embayments that become hypersaline. The second component comprises case studies that describe the morphology and physicochemical characteristics of selected hypersaline systems. The natural and anthropogenic factors that led to hypersalinity and their effect on the biota, together with the changes likely to occur as a result of climate change and other anthropogenic activities are identified. The selected systems are the Laguna Madre in the United States and Mexico, Lake St Lucia in South Africa, the Coorong in South Australia, and the Stokes, Hamersley, and Culham inlets in Western Australia.

2 META-ANALYSIS OF HYPERSALINE ESTUARIES, LAGOONS AND COASTAL EMBAYMENTS

A total of 60 lagoons, 34 estuaries and 7 coastal embayments, in which hypersaline conditions (salinity ≥40) were recorded, were identified from documents in the Scopus database using the search terms "hypersaline" or "hypersalinity" and "estuary" or "lagoon" (Fig. 1). Data were then extracted for a suite of geographical and morphological characteristics (Fig. 2) in each system by using those documents and Google Earth, supplemented with maximum tidal ranges from www.tide-forecast.com. Note that data were not always available, however, for each of the characteristics of every system. Each system was also assigned to a Köppen-Geiger climate class according to the classification produced by Kottek et al. (2006).

Hypersaline systems are found on all continents except Antarctica (Fig. 1A), but are most numerous in North America (27) and Oceania (20; Fig. 2A). Hypersaline lagoons are found mainly in Baja California, the eastern coast of Mexico, the Mediterranean, Red Sea, and the Arabian Gulf, while hypersaline estuaries are relatively numerous in southern Africa and south-western and north-eastern Australia and many of the hypersaline coastal embayments are located in Australia (Fig. 1). Systems that become hypersaline occur in each of the five main climates, albeit the vast majority of each type of system is located in warm temperate (36), arid (33), and tropical (29) areas (Fig. 2B). Lagoons are most prevalent in tropical savannah with a dry winter (Aw) and, to a lesser extent, hot desert (BWh), and hot-summer Mediterranean (Csa) climates, although two systems were located in the tundra climate (Et; Harris et al., 2017). Estuaries were most common in Aw, together with humid subtropical (Cfa), warm-summer Mediterranean (Csb), and hot semi-arid (BSh) climates. Coastal embayments are typically found in arid climates. Thus, the climates in which hypersaline systems are found are typically characterized by warm/hot temperatures and low and/or seasonal rainfall.

The number of systems in each salinity category declined with increasing salinity from 42 at 40–59 to only 2 at ≥350 (Fig. 2C). Typically, coastal embayments tend to be the least hypersaline, with their salinities never exceeding 70. In contrast, 30% of estuaries and 40% of lagoons experience salinities >80. The greatest salinity recorded in an estuary was the 312 in Culham Inlet (Australia), and the greatest in a lagoon the 350 in both San Salvador Island (Bahamas) and San Jose lagoon (Mexico; Fig. 1B). These latter values are slightly greater than the 332 recorded in the Dead Sea (Rosenthal et al., 2006).

The areas of estuaries and lagoons range from <0.1 km^2 up to 2300 and 5666 km^2 in the Bahía Blanca Estuary (Argentina) and Indian River Lagoon (USA), respectively (Fig. 2D). Coastal embayments tend to be larger, with the areas of Shark Bay, the Gulf of St Vincent, and Spencer Gulf (Australia) all exceeding 5000 km^2. All systems were relatively shallow (typically <20 m), and particularly in the case of lagoons and estuaries in which the maximum depths in 90% and 70% of these systems, respectively, were <2 m (Fig. 2E). Coastal embayments are deeper, with the maximum depth of all but one exceeding 5 m. The majority of hypersaline systems (76%) are located in microtidal regions where the maximum tidal range is <2 m (Fig. 2F) and, with the exception of coastal embayments, the waters that become hypersaline in macrotidal systems are typically located in shallow, upstream areas, which are subjected to reduced tidal effects, or in closed lagoons.

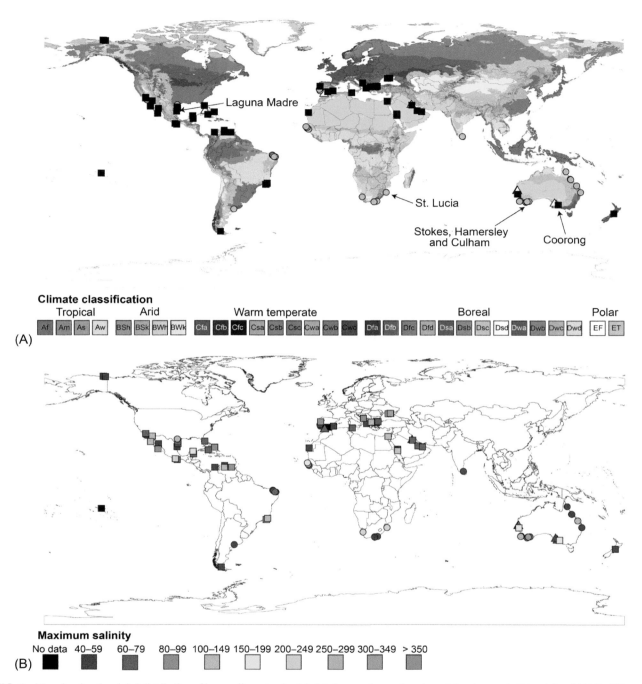

FIG. 1 Map showing the global distribution of hypersaline estuaries (circle), lagoons (square), and coastal embayments (triangle) and (A) the Köppen-Geiger climate classification (Kottek et al., 2006) and (B) the maximum salinity recorded. Climates are allocated to one of five main climate groups: A (tropical), B (arid), C (warm temperate), D (boreal), and E (polar), with the second and third letters indicating the seasonal precipitation type and the level of heat, respectively.

3 LAGUNA MADRE

3.1 Morphology and Physicochemical Environment

The Laguna Madre(s) in Texas and Tamaulipas (Fig. 1A) are variously considered separate systems, "sister" systems, or sometimes a single system bisected by the Rio Grande delta. Both lagoons are shallow (<1.5 m deep), elongate (~185 km long), and separated from the Gulf of Mexico by barrier islands (Tunnell, 2002c; Mendelssohn et al., 2017). They are each

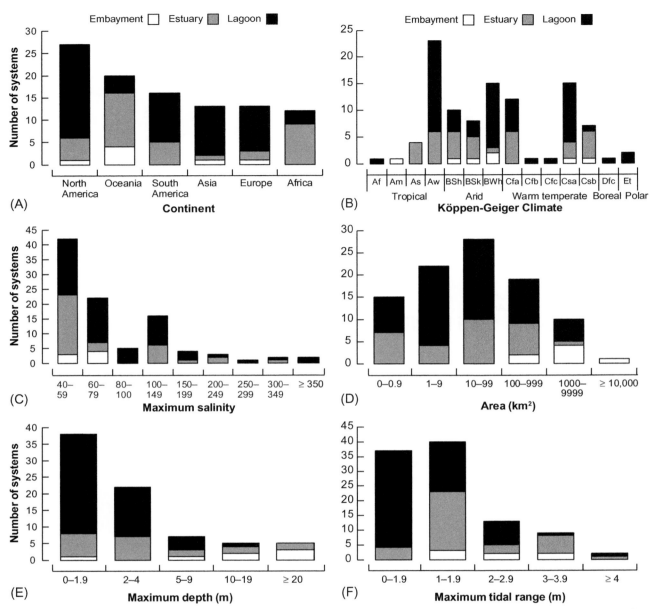

FIG. 2 Stacked bar graphs of the number of estuaries, lagoons, and coastal embayments that become hypersaline and their various geographical and morphological characteristics. (A) Continent in which system is located, (B) Köppen-Geiger climate classification, (C) maximum salinity recorded, (D) area of system, (E) maximum depth, and (F) tidal range in nearby coastal waters.

divided into upper (northern) and lower (southern) regions (Fig. 3A) by sandflats (the Saltillo Flats or "Land-Cut" in Texas; El Carrizal, in Tamaulipas) and both are relatively remote and difficult to access. The lagunas also share the following characteristics: a semi-arid climate, hypersalinity, limited freshwater inflow, clay dunes (lomas) along mainland shorelines and in the Rio Grande delta, a single secondary embayment (Baffin Bay and Bahia de Catán), numerous washover fans, extensive wind-tidal flats, and extensive seagrass meadows (Tunnell, 2002a). They also contain large populations of wintering and migrating shorebirds and productive fisheries.

Although no rivers flow into the Laguna Madre in Texas, several ephemeral streams and Petronila Creek discharge into Baffin Bay. The Arroyo Colorado, once a distributary of the Rio Grande, drains irrigation water from the Lower Rio Grande Valley into the lower Laguna Madre (Onuf, 2007). In Tamaulipas, the Río San Fernando flows into the northern laguna and is the origin of the sand of El Carrizal. Its flow, however, is constrained by dams (Hildebrand, 1980). The Río Soto la Marina flows past the terminus of the southern laguna and discharges into either the Gulf of Mexico or Laguna Morales.

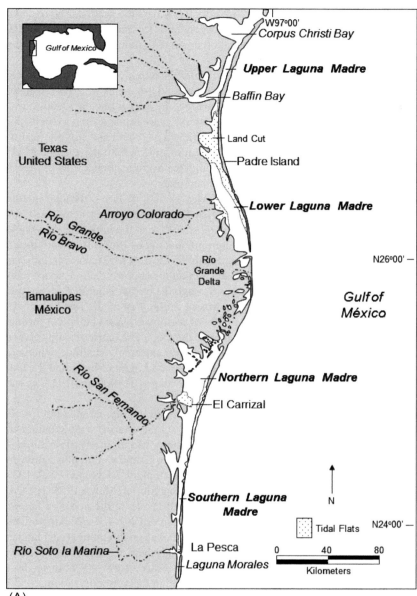

(A)

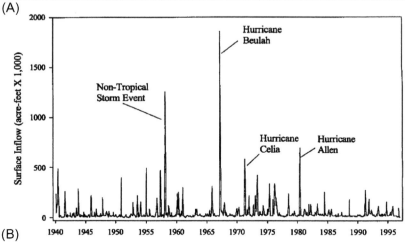

(B)

FIG. 3 (A) Map of the Laguna Madre in Texas and Tamaulipas and (B) Historical inflows into the Laguna Madre from 1940 to 2000, showing major tropical systems that contributed large volumes of surface inflow. 1 acre-foot is 1.23 megaliters. *(A) (Modified from Tunnell, Jr. J.W., 2002c. The Laguna Madre of Texas and Tamaulipas, Texas A&M University Press, College Station, TX, pp. 73–84.), (B) (From Tolan, J.M., Lee, W.Y., Chen, G., Buzan, D., 2004. Freshwater inflow recommendation for the Laguna Madre Estuary System. Texas Parks and Wildlife Department, Coastal Studies Program, Austin, TX, p. 114.)*

The shallowness of the southern Laguna Madre at La Pesca, as well as a roadway, effectively bars the northward movement of water from the Río Soto la Marina into the Laguna Madre system. The area between Tampico and the Rio Grande (Río Bravo), including the Laguna Madre de Tamaulipas, Laguna Morales, and several other small systems to the south, is often collectively referred to as "La Sistema."

The catchments in both Texas and Tamaulipas are very flat and near sea level, rainfall is extremely variable, but low in most years (i.e., 600–800 mm), and evaporation is 2–3 times greater than precipitation (Mendelssohn et al., 2017). Rainfall associated with storms, especially tropical systems, may exceed an entire year's average in only a few days and, due to the few sources of surface water, most inflow is attributable to precipitation (Fig. 3B). As evapotranspiration exceeds precipitation, the average inflow balance in the lower Laguna Madre is negative ($-451,882,931 \, \mathrm{m}^{-3} \mathrm{yr}^{-1}$) and slightly positive ($62,998,890 \, \mathrm{m}^{-3} \mathrm{yr}^{-1}$) in the upper Laguna Madre (Schoenbaechler and Guthrie, 2011a, b). Inflow into the Laguna Madre de Tamaulipas was estimated at $\sim 9 \times 10^6 \, \mathrm{m}^3$ in 1977–1978 (Hildebrand, 1980). Although the average annual flow of the Río San Fernando exceeds $665 \times 10^6 \, \mathrm{m}^3$, 60%–80% of the flow is diverted for irrigation and other uses.

"Fresher" (≤ 35) waters enter the Laguna Madre via exchange with the Gulf of Mexico and, in Texas, through interaction with Corpus Christi Bay. Historically, lower salinity water from Corpus Christi Bay was constrained by a sill, and mostly flowed within a relatively shallow, narrow natural channel along the mainland shoreline in the upper Laguna Madre (Tunnell, 2002a). A similar geological structure also constrains movement of water into and out of Baffin Bay, but this water tends to be as or more saline than that of the Laguna Madre proper. Exchange with the Gulf of Mexico is spatially and temporally limited. A few stabilized (i.e., jettied) inlets are found in both Texas and Tamaulipas but for the most part, interaction with the Gulf occurs after hurricanes, which cut numerous washover passes through the barrier islands. For example, in 1999, Hurricane Bret made landfall over Padre Island and the Texas Laguna Madre in the area of the Land-Cut, with $225 \, \mathrm{km} \, \mathrm{h}^{-1}$ winds, a 2–3 m storm surge, and torrential rainfall, which resulted in 10–12 washover passes being cut through Padre Island (Tunnell, 2002b) and allowing exchange of water between the lagoon and the Gulf. The barrier islands enclosing the Tamaulipan laguna are also often breached by hurricanes, but there are also several extremely shallow, ephemeral inlets along and between the barriers that allow intermittent and limited exchange between the Laguna Madre and the Gulf depending on water levels and other factors (Mendelssohn et al., 2017). The unnamed hurricanes of 1909 and 1933, as well as Hurricane Beulah in 1967, "completely flushed the system" (Tunnell, 2002b).

The primary causes of hypersalinity in the Laguna Madre are the high rates of evaporation, highly variable but usually small amounts of rainfall, and restricted freshwater inflow and limited exchange with lower salinity marine water. Natural climatic events, especially drought, and hurricanes, were historically the primary drivers of salinity variations: nearly fresh to 30 after wet tropical systems, euhaline (30–40) to hypersaline (40–80) during "normal" conditions (some wet months, some dry months), and then "brine" (>80) during extreme drought conditions (Tunnell, 2002c). For example, Hildebrand (1969, 1980) reported the following for Laguna Madre de Tamaulipas: (i) 1951–54: 39–48 in the southern lagoon, (ii) March 1955: 108–118 in the northern lagoon, (iii) summer 1960: 53–80 in the northern lagoon, (iv) summer 1961: 175 in the northern lagoon, (v) October 1961: after Hurricane Carla (dry) opened passes, 23–50 throughout lagoon, (vi) winter 1962–63: >92.5 in northern lagoon, (vii) December 1965: 295 "the waters of the laguna were saturated with salt" and salt was precipitating, (viii) January 1967: after storm surge and rain from Hurricane Inez partially refilled the lagoon; 117 in northern lagoon, (ix) September 1967: Hurricane Beulah, torrential rain, and washover passes opened; 0–6 throughout, (x) February 1969: 40–44 throughout, (xi) April 1969: 47–53 throughout, (xii) June 1969: 50–56 throughout, and (xiii) 1970: 47–55 in the northern lagoon.

Prior to channelization in 1949, the Texas Laguna Madre was hypersaline except after wet tropical systems, with salinities in the upper lagoon regularly exceeding 60–70 and salinities in Baffin Bay exceeding 100 (Hedgpeth, 1947). Massive fish kills were common once salinities exceeded 72, especially during summer when high salinities and very warm waters occurred with hypoxia. Salinities in the lower laguna reach, but rarely exceed, 60 (McKee, 2008).

3.2 Anthropogenic Influences and Hypersalinity

Historically, the Texas and Tamaulipan lagoons were characterized as having "boom-bust" years: wet years and/or tropical systems freshened the hypersaline waters, a few biologically productive years followed, then, after inlets and washover passes closed, salinities would increase to non-tolerable or lethal concentrations, causing productivity to decrease greatly. This is no longer the case as channelization and inlet stabilization have resulted in long-term moderation of salinities (Tunnell, 2002c). Dredging of the Gulf Intracoastal Waterway (GIWW, aka ICWW) in 1949 reconnected the upper and lower lagoons in Texas and set in motion hydrological processes that have reduced average salinities. Dredging and stabilization of four inlets (Pérez-Castaeda et al., 2010; Mendoza et al., 2011) after the 1970s has similarly reduced salinities in Tamaulipas. The reason salinities have decreased in both Texas and Tamaulipas is due to better exchange with adjacent, lower salinity waters and increased circulation. As a result, ecosystems have been altered.

After the GIWW was dredged, only 2.6%–2.8% of salinity measurements in the Laguna Madre exceeded 50 compared to 33%–38% prior to dredging (Quammen and Onuf, 1993). Since the mid-1970s, average annual salinity has ranged from 22 to 49 (Grand, 2014). In the most recent literature, salinities in Tamaulipas are reported as varying from 33 to 62 (Raz-Guzman and Huidobro, 2002; Pérez-Castaeda et al., 2010), with the lowest values near the tidal inlets. Isolated instances of salinities >60 may occur today, but the days of lethal, "briny" conditions (>80) are unlikely to recur, as long as maintenance dredging continues (Tunnell, 2002c).

3.3 Effects of Hypersalinity on the Biota

The Laguna Madres are enigmatic with regard to species diversity and biological productivity (Tunnell, 2002a). While depauperate biological communities are typically associated with hypersalinity this is not the case in the Texas Laguna Madre, especially in the 70 years since the GIWW was dredged. Prior to the 1950s, more than 50% of the commercial fishery landings on the Texas coast came from the Laguna Madre despite periodic massive fish kills resulting from high salinities and hot, shallow water. In Tamaulipas, commercial exploitation of finfish and shrimp was possible about 55% of years prior to when the inlets were dredged (Hildebrand, 1980). Many studies emphasize the significant biological productivity of the Laguna Madres and report higher than expected levels of species diversity and abundance in communities of algae, invertebrates, fish, and birds (see Tunnell et al., 2002). Thus, focus will be placed here on seagrasses in the Texas Laguna Madre and how salinity moderation has led to changes in their species composition, distribution, and extent.

In all, 81% of the seagrasses in Texas are found in Laguna Madre (Mendelssohn et al., 2017). Seagrasses thrive where average salinities are >20 with their upper limits of tolerance ranging from 50 to 70 depending on species (Withers, 2002b). Although they require significant amounts of light to penetrate the water this is possible in Laguna Madre because it is shallow with very little phytoplankton production (most of the time) due to high salinities and low freshwater inflow (Mendelssohn et al., 2017). Prior to dredging, the extent and species composition of seagrasses is unknown, but it was likely relatively sparse or episodic in the upper lagoon and confined to near Brazos Santiago Pass in the lower lagoon. By the mid-1960s, shoal grass (*Halodule wrightii*) dominated (~91%) seagrass cover throughout the Laguna Madre and was the only species in the upper laguna (Onuf, 2007). By 1988, *H. wrightii* cover had declined by 41% in the lower laguna as other species, for example, manatee grass (*Cymodocea filiformis*) and turtle grass (*Thalassia testudinum*) became more prominent. *Thalassia testudinum* continued to be the only species in the upper laguna and the cover it provided nearly doubled. In the lower laguna overall seagrass cover increased slightly by 1998, with continued decline of *H. wrightii* and increases of other species, especially *T. testudinum*.

The 1990s in the upper Laguna Madre were characterized by the "Texas brown tide." The synergism between drought and associated hypersalinity and a freeze and associated fish kill in 1989 resulted in a bloom of the pelagophyte *Aureoumbra lagunensis* that continued almost unabated for more than 8 years (Withers, 2002c). Thus, light penetration diminished, and with it biomass of seagrasses. *Halodule wrightii* cover declined by nearly 20 km^2 during the decade although overall cover increased by about the same amount due to expansion of *C. filiformis* (Onuf, 2007), which performs better than *H. wrightii* in low-light conditions. Remineralization of nutrients from dying seagrasses likely helped sustain the bloom. Hurricane Bret in 1999 helped end the bloom by reducing salinities and flushing the system. Onuf (2007) noted that there had been little recovery in areas of seagrass loss in 2002 with 63% seagrass cover overall.

3.4 The Future With Climate Change

Changes in temperature and precipitation, as well as the timing of precipitation and frequency of extreme events, are likely to affect the Laguna Madre in both Texas and Tamaulipas. Air temperatures in Texas are projected to increase by ~4 °C by 2059 (Nielsen-Gammon, 2011). The chance of a major hurricane hitting the Texas coast and the frequency and severity of droughts are predicted to increase over the next 50 years. Because the coastal plain is flat, low-lying, and subsiding, sea-level rise, which is estimated at 0.3–0.9 m on the mid- to-lower coast by the end of the century, will be more pronounced (Montagna et al., 2011). All of these factors have the potential to greatly alter hydrodynamics and ecosystem functioning in Laguna Madre. Water temperature is increasing and dissolved oxygen is decreasing at faster rates in the Laguna Madre than in most other systems along the Texas coast. In the absence of increased rainfall, hypoxia may increase as more briny conditions return. Alternatively, if rainfall increases and the barrier island is more frequently breached by hurricanes or sink due to subsidence and/or sea-level rise, salinities may decline even more, setting the stage for changes in seagrass coverage, and potentially loss, if average salinities decline.

Perhaps the most important change to habitat and ecosystem functioning is already well underway, that is, the loss or conversion of wind-tidal flat habitats. Wind-tidal flats are intertidal sand and mudflats that are irregularly inundated and exposed by wind seiching and are often covered by a cyanobacterial mat. These flats are some of the most important wintering shorebird habitats in North America (Withers, 2002a). In the upper Laguna Madre region, more than 50% of the spatial cover has already been lost (submerged) or converted to other habitats (marsh/mangrove, seagrass) since the 1950s, largely due to sea-level rise (Withers and Tunnell, 1998). Irregular inundation maintains cyanobacterial mats and prevents colonization by marsh species and non-mat-forming microalgae. However, inundation may be becoming more regular as mat species composition in some areas is shifting from almost entirely mat-forming cyanobacterial species to 50:50 mat-forming species and diatoms, with a small percentage of non-mat-forming species (Paul Zimba, pers. comm.). These changes could reduce the availability, abundance, and nutritional quality of food resources for birds. Increased inundation also reduces the amount of habitat available for birds since most prefer exposed substrates.

4 LAKE ST LUCIA

4.1 Morphology and Physicochemical Environment

Lake St Lucia, situated on the east coast of South Africa (Fig. 1A), is the oldest formally protected estuary in the world and was initially proclaimed as a game reserve in 1895. The system is a Ramsar Site of International Importance, as well as a UNESCO World Heritage Site. The water surface area is approximately 350 km^2 when the estuary is full, making the system one of the largest estuaries in Africa. A considerable amount of research has been conducted on the system, much of which is synthesized in a book focused on the ecology and conservation of St Lucia by Perissinotto et al. (2013). The system comprises three major compartments, False Bay, North Lake, and South Lake, which are linked to the sea via a 15 km long channel known as the Narrows (Fig. 4A). The lake compartments are large but shallow, with an average water depth of just over 1 m when full (Day et al., 1954). Three river systems enter False Bay, one enters North Lake, and the Mfolozi River joined the St Lucia Estuary in the mouth region prior to artificial separation in the 1950s.

Most rain falls during summer, declining from east to west across the catchment, and with the inland areas being dry in winter. River flow is intermittent, primarily due to the lack of precipitation during winter, but also due to increasing freshwater abstraction for agriculture, forestry, and human settlements. Between 1950 and 2010, little or no river flow from the Mfolozi River entered the St Lucia system due to the creation of a new river mouth for this system to the south (Fig. 4A).

In years of above average rainfall, the lake level rises above sea level and water then flows down the system via the Narrows to the Pacific Ocean. During prolonged droughts, some of which may last a decade, the lake level falls due to evaporation exceeding freshwater inputs, with seawater flowing up the Narrows and into the lake (Hutchison and Midgley, 1978). Maximum salinities (sometimes > 100) are often recorded in North Lake and False Bay under these conditions (Fig. 4B), thus creating a reverse salinity gradient for the overall system. Salinities in South Lake seldom exceed 50, but a connection with the Mfolozi River system is key to keeping the salinity within the euhaline range during droughts.

4.2 Anthropogenic Influences and Hypersalinity

Under natural conditions the St Lucia Estuary mouth would close during prolonged droughts and any run-off from the Mfolozi system would then flow northward into the lake system to replace freshwater lost by net evaporation. With the canalization and draining of the Mfolozi Swamps in the 1930s, the natural sediment filtering capacity of the swamps was lost and the large annual sediment load (0.68×10^6 tons) was then deposited in the St Lucia Estuary and lower Narrows (Taylor, 2013). During a prolonged drought in the 1950s the entire St Lucia mouth became silted up, completely blocking both the St Lucia and Mfolozi systems from the ocean. To prevent low-lying farms on the Mfolozi floodplain from becoming inundated due to the back-up of Mfolozi River water, a canal was dredged 1.5 km south of the St Lucia Estuary and maintained as an alternative Mfolozi Estuary mouth for the following 50 years (Whitfield and Taylor, 2009).

The absence of Mfolozi River water entering the St Lucia system, together with increasing freshwater abstractions from the other rivers flowing into the lake, has contributed to increasingly more severe salinity extremes in Lake St Lucia (Fig. 3B; Lawrie and Stretch, 2011). In the past two decades, desiccation of large areas of False Bay, North and South Lake has occurred, due primarily to natural estuary mouth closure in combination with prolonged droughts and unnaturally low freshwater inflows during the closed phase (Cyrus et al., 2010). These events pushed the system into an extreme state that had not been previously recorded and would not have occurred if Mfolozi River water had been available to the St Lucia system over this period. Forestry plantations in some of the catchment areas further exacerbated the freshwater supply situation, although removal of pine plantations on the Eastern and Western Shores of the lake has helped restore some groundwater flows to the system from these areas.

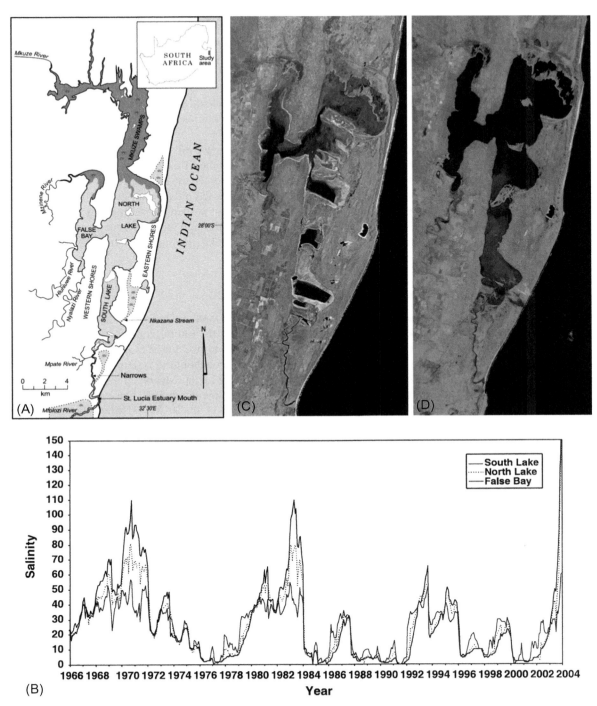

FIG. 4 (A) Map of Lake St Lucia showing the rivers, three lake compartments, and swamps associated with the system. (B) Salinity fluctuations between 1966 and 2004 in the three major lake compartments of the St Lucia system. Landsat image (C) toward the end of a prolonged drought (February 2016) when only 10% of Lake St Lucia was covered in water, and (D) following good seasonal rains and shows the entire system covered in water (May 2017). Images and information provided by the iSimangaliso Wetland Authority. *(A) (From Whitfield, A.K., Taylor, R.H., Fox, C., Cyrus, D.P., 2006. Fishes and salinities in the St Lucia estuarine system—a review. Rev. Fish Biol. Fish. 16,1–20.), (B) (After Whitfield, A.K., Taylor, R.H., Fox, C., Cyrus, D.P., 2006. Fishes and salinities in the St Lucia estuarine system—a review. Rev. Fish Biol. Fish. 16, 1–20.)*

During the 1970s and 1980s, several attempts were made to bring freshwater to St Lucia using various engineering interventions, mainly around the construction of a link canal that would transfer freshwater from the Mfolozi River during low flow periods into the St Lucia system. Pits were included in the canal design to trap sediments that would otherwise have entered the Narrows. Unfortunately, the almost completed link canal and intake works were destroyed by the floods caused by Cyclone Domoina in 1984 and never rebuilt (Forbes and Cyrus, 1992).

A prolonged drought in the region during the entire first decade of the 21st century highlighted the precarious ecological status of the St Lucia system. The loss of freshwater inputs from the Mfolozi River resulted in extreme hypersalinity in all three lake compartments, eventually leading to the almost complete evaporation of surface water within these compartments (Fig. 3C; Cyrus et al., 2011).

In 2010, a scientific workshop was convened at St Lucia Village to review available information and knowledge gaps associated with the need and potential implications of a full reconnection of the Mfolozi River to the St Lucia system. The Water Research Commission Report that arose from this workshop fully endorsed the re-linkage of the Mfolozi River to the St Lucia Estuary and the implementation of measures to minimize excessive sediment input. From 2008 onward measures were already being implemented to bring low winter flows (which contain less suspended sediment than the higher summer inputs) from the Mfolozi River into the St Lucia Estuary via back channels through the mangrove swamp (Whitfield et al., 2013). These flows reduced salinities in the Narrows and parts of South Lake, thus providing refuge areas for biota that could subsequently recolonize the rest of the St Lucia system.

4.3 Effects of Hypersalinity on the Biota

The salinity of Lake St Lucia has a major impact on the aquatic flora of the system. Under oligohaline conditions, the submerged macrophyte *Stuckenia pectinata* thrives and can spread throughout the littoral zones of all lake compartments if these low-salinity conditions persist for more than a year. As salinities begin to rise due to evaporation and seawater inflow, *Ruppia cirrhosa* begins to compete with *S. pectinata* and can also spread throughout the system under mesohaline, polyhaline, and euhaline conditions. Similarly *Zostera capensis* expands from the St Lucia Estuary, up the Narrows, and into South Lake where extensive beds have been recorded under polyhaline and euhaline conditions. Once salinities increase above 20, growth of *S. pectinata* ceases and these plants begin to senesce (Adams et al., 2013). Salinities above 50 appear to be lethal to both *R. cirrhosa* and *Z. capensis* and the lake then switches from a submerged macrophyte-dominated system into a phytoplankton-dominated system. Blooms of the dinoflagellate *Noctiluca* and cyanobacterium *Cyanothece* can occur under hypersaline conditions and may even result in fish kills at the lake (Grindley and Heydorn, 1970; Muir and Perissinotto, 2011).

Zooplankton species appear tolerant of hypersalinity, provided salinities remain below 60. According to Grindley (1976) the two dominant calanoid copepods *Pseudodiaptomus stuhlmanni* and *Acartiella natalensis* tolerate salinities up to 80 (Carrasco et al., 2013). The dominant mysid *Mesopodopsis africana* tolerates a salinity of 60 and even breeds in salinities above 50 (Grindley, 1982). Five species of penaeid prawn have been recorded in the lake, but their numbers decline significantly as salinities approach 60 (Champion, 1976). Although chironomid larvae and ostracods can survive salinities of 55, most of the zoobenthos appears to disappear from the lake as salinities rise above 50 (Boltt, 1975). Recolonization of the system following a hypersaline event takes place from South Lake northward, with those species having short life cycles and planktonic larvae being the first to colonize North Lake and False Bay.

Both species composition and abundance of fishes appear to respond to temporal salinity changes. Salinities between 10 and 120 are inversely related with numbers of fish species (Whitfield et al., 2006). Abundance is also affected, with gill net catch rates at salinities <20 being twice those recorded when salinities exceeded 50 (van der Elst et al., 1976). These results are supported by Mann et al. (2002) who found an inverse relationship between anglers CPUE and lake salinities. Although the lower gill net and angler catches during the hypersaline period could have been due to osmoregulatory stress forcing particular fish taxa out of high salinity areas, the disappearance of certain benthic food resources (Boltt, 1975) may also have played a role in reducing fish abundance, particularly zoobenthic feeders.

An interesting trend in gill net CPUE was recorded by Wallace (1975) who found numbers of the sciaenid *Argyrosomus japonicus* in South Lake increased as salinities in North Lake and False Bay went above 80, forcing this species into the lower salinity South Lake. According to Forbes and Cyrus (1993), piscivores, zooplanktivores, zoobenthivores, and detritivores are all represented in low and high salinities. However, the normally abundant haemulid *Pomadasys commersonnii* and sparid *Rhabdosargus sarba*, captured in areas where the salinity exceeded 70, were no longer feeding on normal molluscan and crustacean prey, but were consuming filamentous algae (Wallace, 1975). This suggests that hypersalinity decimates or eliminates most macrobenthic taxa which are not replaced by alternative invertebrates in such areas. In contrast, the planktonic components in the lake appear to proliferate during such conditions and flamingos foraging on this food chain become abundant during drought years and high salinities (Scharler and MacKay, 2013).

The freshwater fish group is most diverse and abundant under oligohaline lake conditions, although the cichlid *Oreochromis mossambicus* is common under all salinity regimes. Estuary-resident species also thrive under oligohaline conditions but, in contrast to freshwater taxa, are well represented in salinities up to 40. The marine group is most diverse and abundant within the salinity range 10–40, but a large number of species can also be found in salinities up to 70 (Whitfield et al., 2006). Very few fish species were able to tolerate salinities between 70 and 110, with only *O. mossambicus* surviving for extended periods in salinities above 110.

The food resources least affected by extreme hypersalinity are the microphytobenthos and detritus food chains, with detritivorous fishes dominant when the lake is in this state. Mass mortalities of fishes in Lake St Lucia have been recorded in both low (<5) and high salinity (>70) conditions (Wallace, 1975; Blaber and Whitfield, 1976). The fish kills are often triggered by exceptionally low or high water temperatures which affect the osmoregulatory abilities of these species. Hypersaline conditions and fish mortalities under extended closed estuary mouth conditions (2002–2010) and a lack of Mfolozi River water can result in almost the entire lake evaporating and causing loss of virtually all submerged aquatic macrophytes and mass mortalities of both aquatic invertebrates and fish (Whitfield and Taylor, 2009). However, such extreme conditions are artificial and would not have occurred if the Mfolozi River had remained linked to the St Lucia Estuary.

4.4 The Future With Climate Change

The scientists and environmental managers associated with Lake St Lucia have realized that reconnection of the Mfolozi River to the St Lucia Estuary is imperative for the short- and long-term survival of this World Heritage Site. Management will need to take cognizance of the high silt loads carried by the Mfolozi and the need to avoid large sediment inputs to the St Lucia system. This can be achieved by improved catchment rehabilitation and wise manipulation of the opening of the mouth following Mfolozi River flooding. This reconnection process has recently been facilitated by the removal of more than 1.3 million m^3 of dredge spoil between the two systems in May and June 2017, and will promote more natural hydrodynamic processes in the joint mouth opening.

Sugar cane farms that were created by the draining and cultivation on land previously comprising the Mfolozi Swamps are directly affected by back-flooding when the estuary mouth is closed. This has recently led to the affected farmers taking the iSimangaliso Wetland Park Authority to court in order to prematurely breach the now re-established joint Mfolozi and St Lucia mouth. However, in April 2017 the court found in favor of the Wetland Park Authority, with the Durban judge ruling that the "selfish and outdated" interests of a small group of farmers cannot be compared to the threat posed to the ecological health of Lake St Lucia by trying to force the much needed river flow into the sea.

Sea-level rise in association with climate change may also help to restore St Lucia, by temporarily reversing the natural shallowing of the system and increasing the water surface to volume ratio. Such a trend will also threaten the sugar cane farms in the lower Mfolozi River floodplain and may therefore accelerate the purchase of this land for rehabilitation into a functional swamp that can then be used to filter sediments from the Mfolozi River. In addition, climate change predictions for the area are that higher than current average rainfall will be recorded on an annual basis, thus further reducing the future hypersaline extremes that at present threaten the very existence of Lake St Lucia (Mather et al., 2013).

5 COORONG

5.1 Morphology and Physicochemical Environment

The Coorong is a choked coastal lagoon located at the terminus of the estuary of Australia's largest river system, the Murray-Darling, which has a catchment area of 1,060,000 km^2 (Fig. 1A; Kjerfve, 1986; Leblanc et al., 2012). This river system passes through several states and reaches the southern coast of Australia at Encounter Bay in South Australia (Fig. 5A). The river widens into the Lower Lakes (Lake Alexandrina and Lake Albert) before connecting to the sea through the Murray Mouth. The Coorong has a length of over 110 km and is sheltered from the Southern Ocean by the Younghusband Peninsula. A peninsula at Parnka Point narrows the water exchange between the North and South Lagoon, with the extent of hypersalinity increasing toward the southern end of the Coorong (Fig. 5A). The average width ranges from 1.5 to 2.5 km in the North and South Lagoon, with an average water depth of 1.2 and 1.4 m, respectively (Webster, 2011). Several small islands are scattered throughout the Coorong and the shorelines include sandy beaches, mudflats, sheltered bays, headlands, saltmarsh, and small rocky outcrops (Phillips and Muller, 2006).

A series of five barrages has divided the estuary since the late 1930s (Fig. 5A), to maintain the Lower Lakes as a freshwater reservoir. Flow through the barrages into the estuary and lagoon can only occur when water levels in the lakes exceed about 0.7 m AHD. The first fishways were constructed in 2004 and more are being installed, with their operation subject to

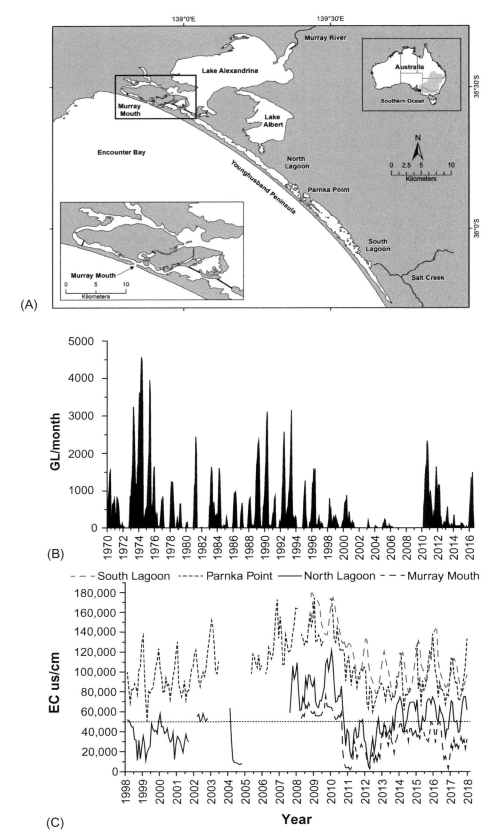

FIG. 5 (A) Map of the Coorong in South Australia, showing the lakes, estuary, and lagoon. (B) Monthly flow over the barrages into the Coorong since 1970, based on modeled monthly outflow data from the Murray Darling-Basin Authority. (C) Salinity in EC (μm/s) recorded from data loggers in the Murray Mouth, North Lagoon, Parnka Point, and the South Lagoon (https://www.waterconnect.sa.gov.au/Systems/SitePages/Surface%20Water%20Data. aspx, Accessed 1 February 2018). Full strength seawater salinity is ~50,000 μm/s.

water availability in the Lower Lakes (Bice et al., 2017). Hunters Creek on Hindmarsh Island provides the only unrestricted natural connectivity between the Lower Lakes and Coorong.

Flow over the barrages is the main freshwater supply for the Coorong. The only direct freshwater inflow to the South Lagoon occurs from the Upper South-East Drainage Scheme entering the South Lagoon at Salt Creek (Kingsford et al., 2011). Groundwater discharge occurs mainly into the South Lagoon, but is of unknown scale (Haese et al., 2008). The salinity and water level in the Coorong are driven by the inflow of freshwater and salt water from the ocean through the Murray Mouth. The Murray Mouth provides the only connection between the Coorong and the ocean, but the width of the mouth is highly dynamic, being subject to water flows over the barrages (Shuttleworth et al., 2005; Webster, 2011; Kämpf and Bell, 2014). Tides are microtidal and only affect the Murray Mouth area and parts of the North Lagoon when the mouth is open. Wind plays a major role in moving water inside the Coorong and episodically covering mudflats through wind seiching and low pressure over Encounter Bay pushing seawater into the system (Noye and Walsh, 1976; Webster, 2010). The Murray Mouth can close during droughts and under restricted water releases over the barrages, as occurred in 1981 and almost in 2003 (Webster, 2010, 2011; Kämpf and Bell, 2014).

The climate in South Australia is Mediterranean, with mostly winter rainfall and very hot summers (CSIRO, 2008). Rainfall in the catchment is affected by rainfall in the subtropics and over Australia's Great Divide and varies with the length and strength of the El Nino/La Nina cycle (CSIRO, 2008). In years with high rainfall and river flows, the highest water flow reaches the Lower Lakes in spring/early summer. Water volumes in the river and flow over the barrages into the Coorong are regulated by water management (see below).

The Ngarrindjeri are the traditional owners and custodians of the Coorong (Kämpf and Bell, 2014). The Lower Lakes, Coorong, and Murray Mouth form the Coorong National Park and became a Ramsar site in 1985 (Phillips and Muller, 2006). It is an Icon Site of Australia's largest river restoration scheme, The Living Murray Program (DEWNR, 2017). The Murray Mouth region is also protected (Habitat Protection Zone) under the Encounter Bay Marine Park. Research on the Coorong, Lower Lakes, and Murray Mouth has intensified over the last 15 years to investigate the major drivers for ecosystem processes, the response to water flow and monitoring the condition and effects of management interventions (Royal Society of South Australia, 2018).

5.2 Anthropogenic Influences and Hypersalinity

The Murray-Darling is a highly regulated river system with a series of locks. It supports intensive irrigation farming along its length, which is managed through water entitlements and trading by the Murray-Darling Basin Authority that also allocates environmental flow (Leblanc et al., 2012). Since construction of the barrages in the 1930s, water flow into the estuary has been regulated and subject to water extraction upstream, drought conditions, and the politics of water management (Kingsford et al., 2011; Leblanc et al., 2012). The average annual freshwater flow into the Coorong over recent decades has been reduced by about 70% and is highly variable, with no flow over the barrages for several consecutive years during the Millennium Drought (1997–2009) (Fig. 5B; Kingsford et al., 2011; Leblanc et al., 2012). Closure of the Murray Mouth during this drought was prevented through dredging operations which commenced in 2002 (Webster, 2010). Unregulated flows are occurring only after exceptional rainfall in the catchment and when major flood peaks reach the estuary, as in 2010/11 and 2016/17.

The reduced freshwater inflows have intensified the pronounced salinity gradient in the Coorong, with salinities increasing towards the South Lagoon (Fig. 5C). The Murray Mouth area can vary from fresh to slightly hypersaline. The greatest rate of change in salinity over time as well as over space occurs in the North Lagoon (Fig. 5C). The South Lagoon has been hypersaline for several decades, with salinities rarely below 60, and exceeding 180 during a prolonged drought. Recent flood events reduced salinities in the South Lagoon, but these were short-lived (Fig. 5C).

A reduced inflow of water also leads to prolonged periods of low water levels, which restricts the habitat value for the aquatic organisms in the Coorong. During the Millennium Drought, parts of the Coorong could be crossed on foot and sediment in the mudflats was too dried out from exposure to contain any organisms, or allow shorebirds to feed.

Palaeoenvironmental studies revealed that the increased salinity and a shift from a macrophyte to a microalgal and bacteria dominated system were unprecedented in the ~6000 year existence of the Coorong and occurred only during the last 50 years, coinciding with intensified water diversion upstream (Krull et al., 2009; Tulipani et al., 2014). Archaeological otoliths from shell middens and fossil evidence from the Coorong corroborate that hypersaline conditions and changes to invertebrates and algal assemblages have occurred only since European settlement (Dick et al., 2011; Disspain et al., 2011; Lower et al., 2013; Reeves et al., 2015). The condition of the Coorong as a hypersaline lagoon is thus a very recent phase in its Holocene history.

5.3 Effects of Hypersalinity on the Biota

The construction of the barrages and intensified dam storage upstream, exacerbated hypersaline conditions in the Coorong, affecting the entire ecosystem. Populations of the seagrass *Ruppia megacarpa* have collapsed and phytoplankton production increased since the 1950s (Krull et al., 2009; Reeves et al., 2015). *Ruppia tuberosa*, which is more salt-tolerant than *R. megacarpa*, had become the more abundant seagrass in the South Lagoon, but it has declined since the mid-1990s when extreme hypersalinity became more persistent (Dick et al., 2011). Hypersalinity reduces the reproductive capabilities of *Ruppia* by inhibiting flower and seed production and causing germination failures (Kim et al., 2013, 2015). Furthermore, prolific growth of filamentous algae has intensified in recent years, smothering *Ruppia* and thus preventing completion of its life cycle. The loss of *Ruppia* diminishes a further food source for birds (Kim et al., 2015).

Phytoplankton communities changed between drought and flow periods, and between the North and South Lagoon, due mostly to changes in salinity and also nutrients (Leterme et al., 2015; Hemraj et al., 2017a, b). Diatoms dominated under hypersaline conditions in the South Lagoon, and also dominated the phytoplankton during drought, together with dinoflagellates (Leterme et al., 2015). Under low flow, the planktonic food web was characterized by interactions between bacteria, viruses, and nano/picoplankton (Hemraj et al., 2017a). Hypersalinity during the Millennium Drought led to lower diversity and abundance of zooplankton (Geddes et al., 2016). Zooplankton communities were less distinct along the salinity gradient of the Coorong, but meroplanktonic larvae of polychaetes and gastropods were confined to lower salinities (Hemraj et al., 2017b), corresponding with the distribution of their later life stages (Dittmann et al., 2015).

The brine shrimp *Parartemia zeitziana* was very abundant and the only faunal species collected in salinities of 130 in the Coorong (Geddes et al., 2016). Benthic macroinvertebrate communities could be clearly separated by a salinity threshold of 64, although several species, particularly insect larvae, ostracods and some amphipods, were found under higher salinities in the southern reaches of the Coorong (Dittmann et al., 2015). Chironomid larvae characterize the community in the South Lagoon, where the presence of other macroinvertebrates is almost negligible (Dittmann et al., 2015). The diversity, abundance, and biomass of macroinvertebrates has been consistently highest in the Murray Mouth and northern sections of the North Lagoon, but during the Millennium Drought, the community became more similar to that in the hypersaline South Lagoon (Dittmann et al., 2015). The distribution ranges of the most common amphipods, polychaetes, and bivalves contracted during the drought, and recovery was species specific and subject to their life history. Some species, such as the micro-mollusc *Arthritica helmsi*, took several years to return after the drought.

The Coorong is a significant overwintering site for migratory shorebirds and also hosts a large number of water birds, including a significant population of pelicans (Phillips and Muller, 2006). Paton et al. (2009) found distinct bird communities across the regions of the Coorong. Compared to the 1980s, when the Coorong became a Ramsar site, the numbers of several bird species have declined by about 30%. Only banded stilts (*Cladorhynchus leucocephalus*) became more abundant during the drought, preying on the increased abundance of brine shrimp (Paton et al., 2009). Phenology comparisons indicate that some species of shorebirds did not undertake return migrations to their breeding grounds in the northern hemisphere in severe drought years (2007–2009), when benthic food availability was lowest (Dittmann et al., unpublished). Flow restoration in 2010 led to an increase in the abundance of some shorebird species, but many still occur in low numbers or continue to decline (O'Connor and Rogers, 2013).

Hypersalinity affects the diversity, abundance, distribution, and productivity of fish species, subject to their salinity tolerance and life history, with diadromous species most impacted by a loss of connectivity between river and the sea (Zampatti et al., 2010; Wedderburn et al., 2012; Ferguson et al., 2013; Brookes et al., 2015; Ye et al., 2016; Hossain et al., 2017). The salt-tolerant small-mouth hardyhead (*Atherinosoma microstoma*), which was one of the few fish species occurring in the hypersaline lagoon, was also affected by prolonged drought periods (Wedderburn et al., 2016; Ye et al., 2016). For most small-bodied fish, the high salinities occurring during drought conditions caused physiological stress (Hossain et al., 2016; Wedderburn et al., 2016) and increased dietary overlap between species (Hossain et al., 2017). For large-bodied native fish, including black bream (*Acanthopagrus butcheri*), drought can exacerbate the effects of overfishing (Ferguson et al., 2013; Earl et al., 2016). Under drought conditions, fish diets revealed that the food web in the Coorong was driven by autochthonous benthic and pelagic primary production (Lamontagne et al., 2016). Disconnection from freshwater flow and reduced connectivity to the ocean also affected the larval assemblage of fish (Bucater et al., 2013), as well as population characteristics and movement patterns of greenback flounder (*Rhombosolea tapirina*; Earl et al., 2014, 2017). Populations of several ecologically and economically important small- and large-bodied fish species in the Coorong have not fully recovered since flows resumed (Ye et al., 2017).

5.4 The Future With Climate Change

Estuaries will be affected by drought periods and extreme weather events induced by climate change (Wetz and Yoskowitz, 2013). Modeling has shown that under drought conditions, increased environmental flows are needed in the highly managed Murray-Darling river system to obtain environmental benefits (Kingsford et al., 2011). The projected future decline in rainfall under a medium emission scenario is similar to what occurred under the Millennium Drought (Leblanc et al., 2012). A repeat of the Millennium Drought, which exposed a very large extent of acid sulfate soils when water levels dropped by more than 1 m in the Lower Lakes and caused markedly hypersaline conditions throughout most of the length of the Coorong, would have far reaching effects on ecosystem functions and services, necessitating costly interventions (Kingsford et al., 2011; Banerjee et al., 2013). The frequency and duration of droughts matter for the ecosystem response, as Dittmann et al. (2015) revealed hysteresis for macroinvertebrate communities, with recovery taking longer when flows resumed after a prolonged drought than after a short drought. Reducing upstream extraction of water emerged as most effective to reduce the degradation of the hypersaline Coorong under future climate change scenarios (Lester et al., 2013). With rising sea levels, the purpose of the barrages may, however, change from retaining river water to barriers against intruding seawater.

6 STOKES, HAMERSLEY AND CULHAM INLETS

6.1 Morphology and Physicochemical Environment

The type of estuaries in south-western Australia form a gradient eastward from permanently open to intermittently and seasonally open to normally closed, paralleling a decline in rainfall and catchment size and thus fluvial discharge (Brearley, 2005). The Stokes, Hamersley, and Culham inlets are located along ~100 km of the south coast of Western Australia (Fig. 1) in a semi-arid region, in which median annual rainfall in their catchments range from 452 to 539 mm (Bureau of Meteorology, 2017). The estuaries and catchments are both small, ranging from 2 to 14 km^2 and 1268 to 5300 km^2, respectively, and typically shallow, with maximum depths <4 m (Hoeksema et al., 2018). As with many estuaries in microtidal regions, these systems comprise a wide basin area and the saline downstream reaches of rivers that discharge into the basin. Pronounced sand bars (up to 5 m above sea level) form at the mouth of each estuary through the wave-driven import of sediment from the marine environment, thus leading to discontinuity between the waters of the estuary and the ocean (Fig. 3A). As the mouth of each of these estuaries remains closed for several years, these systems are classified as normally closed (Hodgkin and Hesp, 1998).

The sand bars that form at the mouth of estuaries throughout the world become breached when, through rainfall and therefore fluvial discharge, water builds-up within the estuary to a point where large amounts of sediment from the bar are exported (Ranasinghe and Pattiaratchi, 1999; Rich and Keller, 2013; Slinger, 2017). The bar of Stokes Inlet generally breaches after exceptional winter rainfall, whereas that in Hamersley and Culham usually breaks following atypically high summer and autumn rainfall, often associated with cyclonic activity (Fig. 6A; Hoeksema et al., 2018). The frequency and timing of bar breaching of these three normally closed estuaries differ markedly. Thus, during the 45 years between 1972 and 2016, it occurred in 12 years in Stokes, ≥8 in Hamersley and only 3 in Culham. Moreover, the mouth of Culham Inlet was closed for the previous 52 years (Hodgkin, 1997). Differences in the frequency of breaching reflect variations in the amount of rainfall and thus discharge, the area and depth of the basin relative to fluvial discharge and the resilience of the bar (Hoeksema et al., 2018). For example, annual rainfall is ~100 mm greater in the catchment of Stokes than in those of Hamersley and Culham and the ratio of the area of basin to total annual discharge decreases progressively from Stokes to Hamersley to Culham, whereas the depth of their basins follows the reverse trend (i.e., 4, 1.3, and 1 m, respectively).

6.2 Anthropogenic Influences and Hypersalinity

The extent to which natural vegetation in the catchments of these three estuaries has been cleared for agriculture since the 1950s had, by 1995, reached 37% in Hamersley, 50% in Culham, and 68% in Stokes (Hoeksema et al., 2018). Clearing has increased the runoff from land in south-western Australia by between 2 and 4 times (Schofield, 1990; Ritson et al., 1995), accounting for the increase in the frequency with which the bars of the three estuaries become breached (Hodgkin and Clark, 1989). For example, the bar at the mouth of Culham broke 4 times in 24 years between 1993 and 2017, but only 4 times in the 144 years prior to 1993 (Hodgkin, 1997; Hoeksema et al., 2018).

The rivers of the Hamersley (Hamersley River) and Culham inlets (Phillips River) in their catchments above the upper limit of the estuary have been classified as brackish (salinity = 3–10) and saline (salinity > 10), respectively, and the

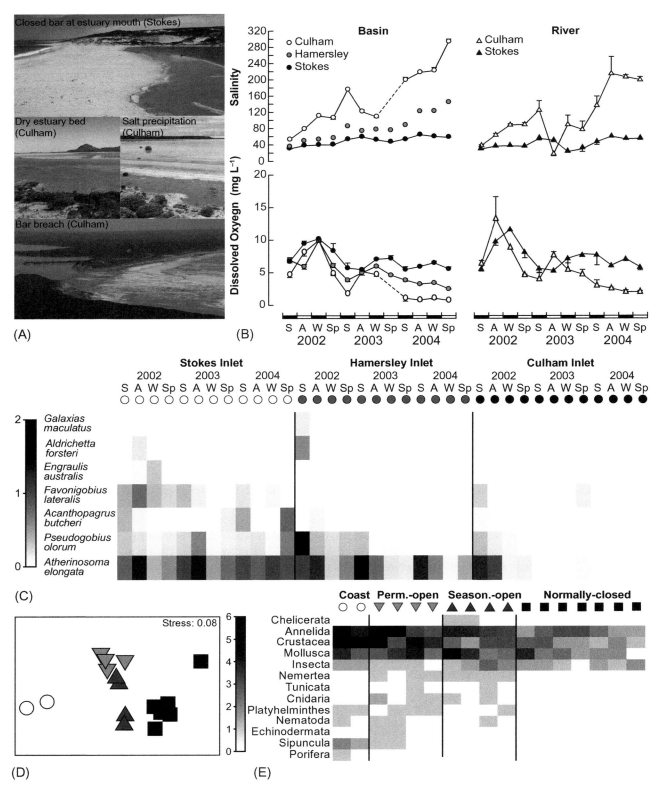

FIG. 6 (A) Photographs showing the topography of Stokes and Culham inlets. (B) Mean seasonal salinities and dissolved oxygen concentrations of the basin and rivers of Stokes, Hamersley, and Culham inlets. (C) Shade plot of the dispersion-weighted and square-root transformed densities $100\,m^{-2}$ of each fish species recorded in the nearshore waters of the basin of the three estuaries (Hoeksema et al., n.d.). (D) Nonmetric multidimensional scaling plot of the presence/absence of benthic macroinvertebrate species in coastal environments and permanently open, seasonally open, and normally closed estuaries in south-western Australia. (E) Shade plot showing the square-root transformed number of species belonging to higher groups of benthic macroinvertebrates. *((B) From Hoeksema, S.D., Chuwen, B.M., Tweedley, J.R., Potter, I.C., 2018. Factors influencing marked variations in the frequency and timing of bar breaching and salinity and oxygen regimes among normally-closed estuaries. Estuar. Coast. Shelf Sci. 208, 205–218.)*

corresponding regions of the Lort and Young rivers that flow into Stokes Inlet as saline and brackish/saline, respectively (Pen, 1999). The salinity in these rivers results from the drainage of salt lakes in the headwater catchments and/or from saline groundwater, augmented by secondary salinization due to clearing of deep-rooted vegetation (Clarke et al., 2002; Halse et al., 2003). Land clearing has also increased the sediment loads of rivers (Pen, 1999). For example, 50–60 cm of wet sediment (20–25 cm dry) accumulated in the basin of Stokes Inlet in the 30 years between 1957 and 1987, representing 10 times the estimated natural rate of sedimentation (Hodgkin, 1998). The depositing of this sediment will continue to effect the depth of this shallow basin, much of which is <1 m deep (Hodgkin and Clark, 1989).

Between summer 2002 and spring 2004, each of the three estuaries was closed to the ocean, following major natural breaches of their bars and the resultant influx of substantial volumes of oceanic water (Hoeksema et al., 2018). Lower average rainfall in 2002–2004, than over the 45 years between 1972 and 2016, i.e., by 28 mm in Stokes and by ~85 mm in Hamersley and Culham, allied with the closure of these estuaries and high evaporation, led to marked hypersalinity and reductions in oxygen concentration, the extents of which varied markedly among estuaries. Thus, mean salinities in the basins of Stokes, Hamersley and Culham inlets increasingly diverged from their respective minima of 30, 35, and 52 in summer 2002 to reach maxima of 64, 143, and 293 (Fig. 6B). The greatest individual salinity of 148 in Hamersley Inlet and the remarkable 312 in Culham Inlet are greater than those recorded in any other south-western Australian estuary, and the corresponding maximum of 65 in Stokes has only been exceeded by the 122 recorded in Wellstead, another normally closed estuary (Young and Potter, 2002), and by the 132 in the intermittently open Vasse-Wonnerup (Tweedley et al., 2014). As in the basins, mean salinities in the estuarine reaches of the rivers of Stokes, and particularly of Culham, increased markedly between 2002 and 2004, rising from 29 to 59 in Stokes and from 38 to 208 in Culham (Fig. 6B).

The trends exhibited by dissolved oxygen concentration in the basins of each of the three estuaries are inversely related to salinity, with values declining from maxima of ~10 mg L^{-1} in each estuary in winter 2002 to 5.9, 2.5, and 0.6 mg L^{-1}, in Stokes, Hamersley, and Culham inlets, respectively (Fig. 6B). The trend in the rivers of Stokes and Culham largely parallel those in the basin, with oxygen concentration declining to 5.9 in Stokes and 2.0 mg L^{-1} in Culham. The combination of very high salinities and low oxygen concentrations in these estuaries provides a challenging environment for their fauna, and particularly so in Hamersley and Culham inlets.

6.3 Effects of Hypersalinity on the Biota

The salinity in estuaries, which is related to fluvial discharge and evaporation rate, plays a major role in structuring the fish communities in these systems (Wedderburn et al., 2016). This applies particularly to normally closed estuaries, as few species are capable of surviving in salinities >50 (Gonzalez, 2012) and these systems frequently become markedly hypersaline. The exceptions include species of atherinid and mugilid, which are capable of osmoregulation in highly elevated salinities (Brauner et al., 2013).

As shown above, the extent of hypersalinity in Stokes, Hamersley, and Culham inlets is inextricably linked with the period over which the mouth of the estuary is closed. The ichthyofauna in each of these systems is highly depauperate and, in terms of abundance, is dominated by a very small number of species (Hoeksema et al., n.d.). Thus, seasonal seining in nearshore waters throughout each estuary between 2002 and 2004 yielded only five or six species, with *Atherinosoma elongata* by far the most abundant fish in both regions of all estuaries and numerically contributing >75%, with another atherinid and a gobiid the next most abundant (Hoeksema et al., n.d.). Corresponding gill netting in offshore waters yielded between one and eight species, among which the sparid *A. butcheri* overwhelmingly dominated the catches, with contributions ranging from 87% to 98% (Hoeksema et al., n.d.).

Each of the seven most numerous species in nearshore waters complete their life cycles within the estuaries and comprise virtually the whole catch in those waters. The two remaining species are represented by only seven individuals of the marine species *Aldrichetta forsteri* and nine of the freshwater species *Galaxias maculatus*. Similarly, estuarine residents accounted for >97% of the total catch in offshore waters of Stokes and Culham and 84% in Hamersley (Hoeksema et al., n.d.). The latter lower value is attributable to the presence of larger individuals of marine species, including those of the Mugilidae and Arripidae, which would have entered the estuary when it was previously open, but were subsequently unable to migrate back to marine waters to spawn.

The overwhelming domination of the ichthyofaunas by species whose individuals complete their life cycles within these systems reflects a strong selective advantage for this mode of life in normally closed estuaries. This is particularly true for small, short-lived species, such as certain atherinids and gobiids, which do not have to rely on a continuously or regularly open estuary mouth to provide a route to and from the ocean to complete their life cycle, as is the case with marine species. This accounts for estuarine residents making a large contribution to the total catch in estuaries that are intermittently or seasonally open and conversely for the far greater percentage contribution made by marine species to the abundance of the ichthyofaunas of permanently open estuaries (Tweedley et al., 2016).

During the 3 years in which the three estuaries were closed and their salinities increased, the mean number of species declined and ichthyofaunal composition changed, with their extents increasing markedly from Stokes to Hamersley to Culham, thus paralleling the trends by salinity (Fig. 6C). Thus, the rise in salinities to 175 in the basin of Culham Inlet over five seasons and then a decline to ~115 in the following three seasons was accompanied firstly by the "loss" of all fish and then the "return" of two species. The implication that salinity tolerances vary markedly among species is consistent with the absence of *G. maculatus* and *Leptatherina wallacei* in samples from any region of the three estuaries when salinities exceeded 40 and 60, respectively, whereas *Pseudogobius olorum* and *A. elongata* were recorded in salinities as high as 114 and 147, respectively. The latter is close to the highest in which the congeneric *A. microstoma* has been caught (Geddes, 1987) and which is apparently the highest recorded for a teleost.

A striking example of the way in which hypersalinity can lead to mass mortality is provided by *A. butcheri* in the river of Culham Inlet in 2001, with this species becoming stressed at a salinity of ~60 and ~1.3 million dying when salinity reached 83–85 (Hoeksema et al., 2006). The presence of a rock bar across the river had prevented these fish moving upstream to refugia, such as isolated pools in which salinities were always <30 and contained substantial numbers of *G. maculatus* and *L. wallacei*, which apparently have a "preference" for lower salinities.

Although the benthic macroinvertebrate faunas of normally closed hypersaline estuaries in south-western Australia have not been studied in detail, there are sufficient presence/absence data for their species in these estuaries (Hodgkin and Clark, 1987, 1988, 1989, 1990) to construct a nMDS and shade plot (Clarke et al., 2014) to facilitate comparisons with those of seasonally (Tweedley et al., 2012) and permanently open estuaries (Wildsmith et al., 2009, 2011) and coastal waters (Wildsmith et al., 2005). As with fishes, the invertebrate fauna in normally closed estuaries is depauperate, being restricted to a limited number of annelid, crustacean, mollusc, and insect species (Fig. 6D, E).

The diet of *A. butcheri*, a highly opportunistic species when sourcing its food (Sarre et al., 2000; Poh et al., 2018), is considerably more diverse in Stokes than in Hamersley and Culham, which are far more hypersaline (Chuwen et al., 2007). The diet of this sparid in the latter two estuaries contained a much greater proportion of larval chironomids, which are particularly abundant in hypersaline wetlands in south-western Australia (Davis and Christidis, 1997). In contrast, the diet in Stokes Inlet comprised larger volumes of polychaetes and amphipods, which are less tolerant of hypersaline conditions (Dittmann et al., 2015).

6.4 The Future With Climate Change

Climate change is predicted to influence the timing and magnitude of rainfall, air temperatures, and evaporation and sea-level rise, all of which, in turn, influence the environmental conditions of estuaries (Poloczanska et al., 2007; Hallett et al., 2018). While rainfall in the westward parts of south-western Australia decreased by 25% over the last 100 years, including 15%–20% in late autumn-winter rainfall since the 1970s (Hughes, 2003; Hope et al., 2015), no such decline was detected in the catchments of the Stokes, Hamersley, and Culham inlets (Hoeksema et al., 2018). Climate models, however, are predicting a decrease in winter rains in these catchments due to a southward shift in winter storm systems (Andrys et al., 2017; Hallett et al., 2018), which would be expected to lead to a decrease in the frequency that the bar at the mouth of Stokes Inlet is breached, as this system typically breaches in response to particularly heavy winter rainfall. These changes are less likely to directly influence breaching in Hamersley and Culham, as these systems typically breach following rainfall derived from cyclonic activity in summer and autumn. However, a decrease in rainfall and thus fluvial discharge could result in the maintenance of higher salinities. As some climate models predict an increase in summer rainfall and others a reduction (Andrys et al., 2017), it is difficult to hypothesize as to whether and how the estuarine hydrology of these systems might change in the future.

Decreases in rainfall and potential increases in evapotranspiration are projected to lead to a decrease in soil moisture and runoff (Hope et al., 2015), which would reduce the magnitude of fluvial discharge and potentially also reduce the frequency of bar opening, therefore leading to increased hypersalinity. While mean air temperatures and the number of days of hot (>35°C) and extreme (>40°C) temperatures are likely to increase, this may not lead to an increase in pan evaporation, as the magnitude of this type of evaporation has not increased since 1970 (Hope et al., 2015).

Sea level at Esperance increased by an average of $3.1\,mm\,yr^{-1}$ between 1993 and 2009 and is predicted to increase by 8–18 cm and 30–65 cm by 2030 and 2090, respectively, under the IPCC scenario RCP4.5 (Hope et al., 2015). An increase in sea level of 50 cm is predicted to be accompanied by a 100- to 1000-fold increase in the frequency of extreme sea-level events (Braganza et al., 2014; Hallett et al., 2018). These and/or an increase in storm surges may reduce the resilience of the bar to breaching, while overtopping could decrease the magnitude of any hypersalinity. However, if these events result in the transport of large amounts of sediment onshore, this could increase the height of the bar or cause the premature closure of any breach.

7 SUMMARY

Estuaries, lagoons, and coastal embayments that become hypersaline (salinity >40) are found worldwide, but are most common in tropical, arid, and warm temperate climates with low and/or highly seasonal rainfall. These systems, in which salinities can reach 312, 350, and 70, respectively, range markedly in area from <0.1 to >10,000 km^2, but are generally shallow (<2 m deep) and located in microtidal regions (tidal range <2 m). Hypersalinity is typically caused by (i) a reduced or closed connection with the ocean, (ii) high evaporation, and (iii) low freshwater input due to one or more of the following: arid climate and thus low rainfall, drought, water extraction, river diversion and, in the case of lagoons, also the absence of a river or rivers. The biota in hypersaline systems changes markedly in response to increased salinity, with very few species able to tolerate salinities over 100, which can result in mass mortalities of flora and fauna and simplification of the biotic community and food web. Some taxa, however, can survive in extreme salinities including brine shrimp, chironomid larvae, and certain fish species such as those in the Atherinidae and Cichlidae. Climate change will potentially influence the extent of hypersalinity in estuaries and lagoons, through altering, for example, the frequency and magnitude of rainfall and sea-level height. The predictions of the effects of climate change on an individual system vary with region and the type of that system.

ACKNOWLEDGMENTS

Gratitude is expressed to Clara Obregón Lafuente and Jessica Blakeway for assisting with the meta-analysis and Dr. Wes Tunnell for providing a critical review of components of the manuscript. Ryan Baring and Orlando Lam Gordillo are thanked for preparing the figure of the Coorong.

REFERENCES

Adams, J.B., Nondoda, S., Taylor, R.H., 2013. Macrophytes. In: Perissinotto, R., Stretch, D.D., Taylor, R.H. (Eds.), Ecology and Conservation of Estuarine Ecosystems: Lake St Lucia as a Global Model. Cambridge University Press, Cambridge, pp. 209–225.

Ajemian, M.J., Mendenhall, K.S., Pollack, J.B., Wetz, M.S., Stunz, G.W., 2018. Moving forward in a reverse estuary: habitat use and movement patterns of Black Drum (*Pogonias cromis*) under distinct hydrological regimes. Estuar. Coasts 41, 1410–1421.

Andrys, J., Kala, J., Lyons, T.J., 2017. Regional climate projections of mean and extreme climate for the southwest of Western Australia (1970–1999 compared to 2030–2059). Clim. Dyn. 48, 1723–1747.

Anonymous, 1958. The Venice system for the classification of marine waters according to salinity. Limnol. Oceanogr. 3, 346–347.

Banerjee, O., Bark, R., Connor, J., Crossman, N.D., 2013. An ecosystem services approach to estimating economic losses associated with drought. Ecol. Econ. 91, 19–27.

Barnes, R.S.K., 1980. Coastal Lagoons. Cambridge University Press, Cambridge.

Bice, C.M., Zampatti, B.P., Mallen-Cooper, M., 2017. Paired hydraulically distinct vertical-slot fishways provide complementary fish passage at an estuarine barrier. Ecol. Eng. 98, 246–256.

Blaber, S.J.M., Whitfield, A.K., 1976. Large scale mortality of fish at St Lucia. S. Afr. J. Sci. 72, 218.

Boltt, R.E., 1975. The benthos of some southern African lakes. Part 5: the recovery of the benthic fauna of St Lucia following a period of excessively high salinity. Trans. R. Soc. S. Afr. 41, 295–323.

Braganza, K., Hennessy, K., Alexander, L., Trewin, B., 2014. Changes in extreme weather. In: Christoff, P. (Ed.), Four Degrees of Global Warming: Australia in a Hot World. Routledge, London, pp. 33–59.

Brauner, C., Gonzalez, R.J., Wilson, J.M., 2013. Extreme environments: hypersaline, alkaline, and ion-poor waters. Fish Physiol. 32, 435–476.

Brearley, A., 2005. Ernest Hodgkin's Swanland, first ed. University of Western Australia Press, Crawley.

Brookes, J.D., Aldridge, K.T., Bice, C.M., Deegan, B., Ferguson, G.J., Paton, D.C., Sheaves, M., Ye, Q.F., Zampatti, B.P., 2015. Fish productivity in the lower lakes and Coorong, Australia, during severe drought. Trans. R. Soc. S. Aust. 139, 189–215.

Bucater, L.B., Livore, J.P., Noell, C.J., Ye, Q., 2013. Temporal variation of larval fish assemblages of the Murray Mouth in prolonged drought conditions. Mar. Freshw. Res. 64, 932–937.

Bureau of Meteorology, 2017. Australian Government, Bureau of Meteorology.

Carrasco, N.K., Perissinotto, R., Jerling, H.L., 2013. Zooplankton. In: Perissinotto, R., Stretch, D.D., Taylor, R.H. (Eds.), Ecology and Conservation of Estuarine Ecosystems: Lake St Lucia as a Global Model. Cambridge University Press, Cambridge, pp. 247–268.

Catalá, T.S., Mladenov, N., Echevarría, F., Reche, I., 2013. Positive trends between salinity and chromophoric and fluorescent dissolved organic matter in a seasonally inverse estuary. Estuar. Coast. Shelf Sci. 133, 206–216.

Champion, H.F.B., 1976. In: Heydorn, A.E.F. (Ed.), Recent prawn research at St Lucia with notes on the bait fishery. Proceedings of the St Lucia Scientific Advisory Council Workshop Meeting. Natal Parks, Game and Fish Preservation Board, Pietermaritzburg, Charters Creek, 15–17 February 1976. pp. 20.

Chuwen, B., Platell, M., Potter, I., 2007. Dietary compositions of the sparid *Acanthopagrus butcheri* in three normally closed and variably hypersaline estuaries differ markedly. Environ. Biol. Fish 80, 363–376.

Chuwen, B.M., Hoeksema, S.D., Potter, I.C., 2009. The divergent environmental characteristics of permanently-open, seasonally-open and normally-closed estuaries of south-western Australia. Estuar. Coast. Shelf Sci. 85, 12–21.

Clarke, C.J., George, R.J., Bell, R.W., Hatton, T.J., 2002. Dryland salinity in south-western Australia: its origins, remedies, and future research directions. Soil Res. 40, 93–113.

Clarke, K.R., Tweedley, J.R., Valesini, F.J., 2014. Simple shade plots aid better long-term choices of data pre-treatment in multivariate assemblage studies. J. Mar. Biol. Assoc. UK 94, 1–16.

CSIRO, 2008. Water Availability in the Murray-Darling Basin. In: A Report to the Australian Government from the CSIRO Murray-Darling Basin Sustainable Yields Project. CSIRO, Australia, pp. 67.

Cyrus, D., Jerling, H., MacKay, F., Vivier, L., 2011. Lake St Lucia, Africa's largest estuarine lake in crisis: combined effects of mouth closure, low levels and hypersalinity. S. Afr. J. Sci. 107.

Cyrus, D.P., Vivier, L., Jerling, H.L., 2010. Effect of hypersaline and low lake conditions on ecological functioning of St Lucia estuarine system, South Africa: an overview 2002–2008. Estuar. Coast. Shelf Sci. 86, 535–542.

Davis, J., Christidis, F., 1997. A Guide to Wetland Invertebrates of Southwestern Australia. Western Australian Museum, Perth.

Day, J.H., 1980. What is an estuary? S. Afr. J. Sci. 76, 198.

Day, J.H., Millard, N.A.H., Broekhuysen, G.J., 1954. The ecology of South African estuaries. Part 4: the St Lucia system. Trans. R. Soc. S. Afr. 34, 129–156.

DEWNR, 2017. Department of Environment Water and Natural Resources (2017) Condition Monitoring Plan (Revised) 2017. In: The Living Murray—Lower Lakes, Coorong and Murray Mouth Icon Site. DEWNR Technical Report 2016–17, Adelaide. pp. 106.

Dick, J., Haynes, D., Tibby, J., Garcia, A., Gell, P., 2011. A history of aquatic plants in the Coorong, a Ramsar-listed coastal wetland, South Australia. J. Paleolimnol. 46, 623–635.

Disspain, M., Wallis, L.A., Gillanders, B.M., 2011. Developing baseline data to understand environmental change: a geochemical study of archaeological otoliths from the Coorong, South Australia. J. Archaeol. Sci. 38, 1842–1857.

Dittmann, S., Baring, R., Baggalley, S., Cantin, A., Earl, J., Gannon, R., Keuning, J., Mayo, A., Navong, N., Nelson, M., Noble, W., Ramsdale, T., 2015. Drought and flood effects on macrobenthic communities in the estuary of Australia's largest river system. Estuar. Coast. Shelf Sci. 165, 36–51.

Earl, J., Fowler, A.J., Ye, Q., Dittmann, S., 2017. Complex movement patterns of greenback flounder (Rhombosolea tapirina) in the Murray River estuary and Coorong, Australia. J. Sea Res. 122, 1–10.

Earl, J., Fowler, A.J., Ye, Q.F., Dittmann, S., 2014. Age validation, growth and population characteristics of greenback flounder (Rhombosolea tapirina) in a large temperate estuary. N.Z. J. Mar. Freshw. Res. 48, 229–244.

Earl, J., Ward, T.M., Ye, Q., 2016. Black bream (Acanthopagrus butcheri) Stock Assessment Report 2014/15. In: Report to PIRSA Fisheries and Aquaculture. South Australian Research and Development Institute (Aquatic Sciences), Adelaide. SARDI Publication No. F2008/000810-2. SARDI Research Report Series No 885, 44 pp.

El-Kassas, H.Y., Nassar, M.Z.A., Gharib, S.M., 2016. Study of phytoplankton in a natural hypersaline lagoon in a desert area (Bardawil Lagoon in Northern Sinai, Egypt). Rend. Lincei 27, 483–493.

Ferguson, G.J., Ward, T.M., Ye, Q.F., Geddes, M.C., Gillanders, B.M., 2013. Impacts of drought, flow regime, and fishing on the fish assemblage in Southern Australia's largest temperate estuary. Estuar. Coasts 36, 737–753.

Forbes, A.T., Cyrus, D.P., 1992. Impact of a major cyclone on a southeast African estuarine lake system. Neth. J. Sea Res. 30, 265–272.

Forbes, A.T., Cyrus, D.P., 1993. Biological effects of salinity gradient reversals in a southeast African estuarine lake. Neth. J. Aquat. Ecol. 27, 483–488.

Geddes, M.C., 1987. Changes in salinity and in the distribution of macrophytes, macrobenthos and fish in the Coorong lagoons, South Australia, following a period of River Murray flow. Trans. R. Soc. S. Aust. 111, 173–181.

Geddes, M.C., Shiel, R.J., Francis, J., 2016. Zooplankton in the Murray estuary and Coorong during flow and no-flow periods. Trans. R. Soc. S. Aust. 140, 74–89.

Gonzalez, R.J., 2012. The physiology of hyper-salinity tolerance in teleost fish: a review. J. Comp. Physiol. B. 182, 321–329.

Grand, H., 2014. Dynamics of a salty regime. Texas Saltwater Fishing Magazine, 54–57. December 24.

Grindley, J.R., 1976. Zooplankton of St. Lucia. In: Heydorn, A.E.F. (Ed.), Proceedings of the St Lucia Scientific Advisory Council Workshop Meeting, Charters Creek, 15–17 February 1976. Natal Parks, Game and Fish Preservation Board, Pietermaritzburg.

Grindley, J.R., 1982. The role of zooplankton in the St Lucia estuary system. In: Taylor, R.H. (Ed.), St Lucia Research Review. Natal Parks Board, Pietermaritzburg, pp. 88–107.

Grindley, J.R., Heydorn, A.E.F., 1970. Red water and associated phenomena in St Lucia. S. Afr. J. Sci. 67, 210–213.

Gunter, G., 1952. The import of catastrophic mortalities for marine fisheries along the Texas coast. J. Wildl. Manag. 16, 63–69.

Haese, R.R., Gow, L., Wallace, L., Brodie, R.S., 2008. Identifying groundwater discharge in the Coorong (South Australia). Ausgeo News, 1–6.

Hallett, C.S., Hobday, A.J., Tweedley, J.R., Thompson, P.A., McMahon, K., Valesini, F.J., 2018. Observed and predicted impacts of climate change on the estuaries of south-western Australia, a Mediterranean climate region. Reg. Environ. Chang. 18 (5), 1357–1373.

Halse, S.A., Ruprecht, J.K., Pinder, A.M., 2003. Salinisation and prospects for biodiversity in rivers and wetlands of south-west Western Australia. Aust. J. Bot. 51, 673–688.

Harris, C.M., McClelland, J.W., Connelly, T.L., Crump, B.C., Dunton, K.H., 2017. Salinity and temperature regimes in Eastern Alaskan Beaufort Sea lagoons in relation to source water contributions. Estuar. Coasts 40, 50–62.

Hedgpeth, J.W., 1947. The Laguna Madre of Texas. N. Am. Wildlife Conf. 12, 364–380.

Hemraj, D.A., Hossain, A., Ye, Q.F., Qin, J.G., Leterme, S.C., 2017a. Anthropogenic shift of planktonic food web structure in a coastal lagoon by fresh-water flow regulation. Sci. Rep. 7, 44441.

Hemraj, D.A., Hossain, M.A., Ye, Q., Qin, J.G., Leterme, S.C., 2017b. Plankton bioindicators of environmental conditions in coastal lagoons. Estuar. Coast. Shelf Sci. 184, 102–114.

Hildebrand, H., 1980. The Laguna Madre de Tamaulipas: Its Hydrography and Shrimp Fishery. (Unpublished manuscript, submitted to National Marine Fisheries Service, on file at Texas A&M University-Corpus Christi). Center for Coastal Studies Library, Corpus Christi, TX, p. 42.

Hildebrand, H.H., 1969. Laguna Madre de Tamaulipas: observation son its hydrography and fisheries. In: Ayala Castañares, A., Phleger, F.B. (Eds.), Coastal Lagoons: A Symposium. Universidad Nacional Autónoma de México, Mexico, pp. 679–686.

Hodgkin, E.P., 1997. Culham Inlet: the history and management of a coastal salt lake in southwestern Australia. J. R. Soc. West. Aust. 80, 239–247.

Hodgkin, E.P., 1998. The future of the estuaries of south-western Australia. J. R. Soc. West. Aust. 81, 225–228.

Hodgkin, E.P., Clark, R., 1987. Wellstead Estuary, the Estuary of the Bremer River. Estuaries and Coastal Lagoons of South Western Australia. Environmental Protection Authority, Perth, Western Australia.

Hodgkin, E.P., Clark, R., 1988. Nornalup and Walpole Inlets, and the Estuaries of the deep and Frankland Rivers. Estuaries and Coastal Lagoons of South Western Australia. Environmental Protection Authority, Perth, Western Australia.

Hodgkin, E.P., Clark, R., 1989. Stokes Inlet and Other Estuaries of the shire of esperance. Estuaries and coastal lagoons of South Western Australia. Environmental Protection Authority, Perth, Western Australia.

Hodgkin, E.P., Clark, R., 1990. Estuaries of the Shire of Ravensthorpe and the Fitzgerald River National Park. Estuaries and Coastal Lagoons of South Western Australia. Environmental Protection Authority, Perth, Western Australia.

Hodgkin, E.P., Hesp, P., 1998. Estuaries to salt lakes: holocene transformation of the estuarine ecosystems of south-western Australia. Mar. Freshw. Res. 49, 183–201.

Hoeksema, S.D., Chuwen, B.M., Potter, I.C., 2006. Massive mortalities of black bream, *Acanthopagrus butcheri* (Sparidae) in two normally-closed estuaries, following extreme increases in salinity. J. Mar. Biol. Assoc. U.K. 86, 893–897.

Hoeksema, S.D., Chuwen, B.M., Tweedley, J.R., Potter, I.C., 2018. Factors influencing marked variations in the frequency and timing of bar breaching and salinity and oxygen regimes among normally-closed estuaries. Estuar. Coast. Shelf Sci. 208, 205–218.

Hoeksema, S.D., Chuwen, B.M., Tweedley, J.R., Potter, I.C., (n.d.) Ichthyofaunal characteristics of normally-closed estuaries: relationship with opening of estuary mouth and extents of change in salinity and oxygen. Estuar. Coast. Shelf Sci. (in preparation).

Hope, P., Abbs, D., Bhend, J., Chiew, F., Church, J., Ekström, M., Kirono, D., Lenton, A., Lucas, C., McInnes, K., Moise, A., Monselesan, D., Mpelasoka, F., Timbal, B., Webb, L., Whetton, P., 2015. In: Ekström, M., Whetton, P., Gerbing, C., Grose, M., Webb, L., Risbey, J. (Eds.), Southern and South-Western Flatlands Cluster Report: Climate Change in Australia Projections for Australia's NRM Regions. CSIRO, Australia, pp. 58.

Hossain, M.A., Aktar, S., Qin, J.G., 2016. Salinity stress response in estuarine fishes from the Murray Estuary and Coorong, South Australia. Fish Physiol. Biochem. 42, 1571–1580.

Hossain, M.A., Ye, Q., Leterme, S.C., Qin, J.G., 2017. Spatial and temporal changes of three prey-fish assemblage structure in a hypersaline lagoon: the Coorong, South Australia. Mar. Freshw. Res. 68, 282–292.

Hughes, L., 2003. Climate change and Australia: trends, projections and impacts. Austral. Ecol. 28, 423–443.

Hutchison, I.P.G., Midgley, D.C., 1978. Modelling the water and salt balance in a shallow lake. Ecol. Model. 4, 211–235.

Kämpf, J., Bell, D., 2014. The Murray/Coorong estuary: meeting of the waters? In: Wolanski, E. (Ed.), Estuaries of Australia in 2050 and Beyond. Springer, Dordrecht, pp. 31–47.

Ketchum, B.H., 1951. The exchange of fresh and salt water in estuaries. J. Mar. Res. 10, 18–38.

Kim, D., Aldridge, K.T., Ganf, G.G., Brookes, J.D., 2015. Physicochemical influences on Ruppia tuberosa abundance and distribution mediated through life cycle stages. Inland Waters 5, 451–460.

Kim, D.H., Aldridge, K.T., Brookes, J.D., Ganf, G.G., 2013. The effect of salinity on the germination of *Ruppia tuberosa* and *Ruppia megacarpa* and implications for the Coorong: a coastal lagoon of southern Australia. Aquat. Bot. 111, 81–88.

Kingsford, R.T., Walker, K.F., Lester, R.E., Young, W.J., Fairweather, P.G., Sammut, J., Geddes, M.C., 2011. A Ramsar wetland in crisis—the Coorong, Lower Lakes and Murray Mouth, Australia. Mar. Freshw. Res. 62, 255–265.

Kinne, O., 1964. The effects of temperature and salinity on marine and brakish water animals. II. Salinity and temperature-salinity combinations. Oceanogr. Mar. Biol. Annu. Rev. 2, 281–339.

Kjerfve, B., 1986. Comparative oceanography of coastal lagoons. In: Wolfe, D.A. (Ed.), Estuarine Variability. Academic Press, Orlando, FL, pp. 63–81.

Kjerfve, B., 1994. Coastal Lagoon Processes. Elsevier, Amsterdam.

Kottek, M., Grieser, J., Beck, C., Rudolf, B., Rubel, F., 2006. World Map of the Köppen-Geiger climate classification updated. Meteorol. Z. 15, 259–263.

Krull, E., Haynes, D., Lamontagne, S., Gell, P., McKirdy, D., Hancock, G., McGowan, J., Smernik, R., 2009. Changes in the chemistry of sedimentary organic matter within the Coorong over space and time. Biogeochemistry 92, 9–25.

Lamontagne, S., Deegan, B.M., Aldridge, K.T., Brookes, J.D., Geddes, M.C., 2016. Fish diets in a freshwater-deprived semiarid estuary (The Coorong, Australia) as inferred by stable isotope analysis. Estuar. Coast. Shelf Sci. 178, 1–11.

Lawrie, R.A., Stretch, D.D., 2011. Anthropogenic impacts on the water and salt budgets of St Lucia estuarine lake in South Africa. Estuar. Coast. Shelf Sci. 93, 58–67.

Leblanc, M., Tweed, S., Van Dijk, A., Timbal, B., 2012. A review of historic and future hydrological changes in the Murray-Darling Basin. Glob. Planet. Chang. 80–81, 226–246.

Lester, R.E., Fairweather, P.G., Webster, I.T., Quin, R.A., 2013. Scenarios involving future climate and water extraction: ecosystem states in the estuary of Australia's largest river. Ecol. Appl. 23, 984–998.

Leterme, S.C., Allais, L., Jendyk, J., Hemraj, D.A., Newton, K., Mitchell, J., Shanafield, M., 2015. Drought conditions and recovery in the Coorong wetland, south Australia in 1997–2013. Estuar. Coast. Shelf Sci. 163, 175–184.

Lower, C.S., Cann, J.H., Haynes, D., 2013. Microfossil evidence for salinity events in the Holocene Coorong Lagoon, South Australia. Aust. J. Earth Sci. 60, 573–587.

Lyell, C., 1833. Principles of Geology, third ed. John Murray, London.

Mann, B.Q., James, N.C., Beckley, L.E., 2002. An assessment of the recreational fishery in the St Lucia estuarine system, KwaZulu-Natal, South Africa. S. Afr. J. Mar. Sci. 24, 263–279.

Mather, A.A., Stretch, D.D., Maro, A.Z., 2013. Climate change impacts. In: Perissinotto, R., Stretch, D.D., Taylor, R.H. (Eds.), Ecology and Conservation of Estuarine Ecosystems: Lake St Lucia as a Global Model. Cambridge University Press, Cambridge, pp. 397–413.

McKee, D.A., 2008. Fishes of the Texas Laguna Madre. Texas A&M University Press, College Station, TX.

Mendelssohn, I.A., Byrnes, M.R., Kneib, R.T., Vittor, B.A., 2017. Coastal habitats of the Gulf of Mexico. In: Ward, C.H. (Ed.), Habitats and Biota of the Gulf of Mexico: Before the Deepwater Horizon Oil Spill. vol. 1. Springer, New York, pp. 359–640.

Mendoza, R., Arreaga, N., Hernández, J., Segovia, V., Jasso, I., Pérez, D., 2011. Aquatic invasive species in the Río Bravo/Laguna Madre ecological region. Background Paper 2011-02. Commission for Environmental Cooperation, Montreal, Canada, p. 146.

Molony, B.W., Parry, G.O., 2006. Predicting and managing the effects of hypersalinity on the fish community in solar salt fields in north-western Australia. J. Appl. Ichthyol. 22, 109–118.

Montagna, P.A., Brenner, J., Gibeaut, J., Morehead, S., 2011. Coastal impacts. In: Schmandt, J., North, G.R., Clarkson, J. (Eds.), The Impact of Global Warming on Texas. University of Texas Press, Austin, TX, pp. 96–123.

Muir, D.G., Perissinotto, R., 2011. Persistent phytoplankton bloom in Lake St. Lucia (iSimangaliso Wetland Park, South Africa) caused by a cyanobacterium closely associated with the genus *Cyanothece* (Synechococcaceae, chroococcales). Appl. Environ. Microbiol. 77, 5888–5896.

Nielsen-Gammon, J.W., 2011. The changing climate in Texas. In: Schmandt, J., North, G.R., Clarkson, J. (Eds.), The Impact of Global Warming on Texas. University of Texas Press, Austin, TX, pp. 39–68.

Nordlie, F.G., 2009. Environmental influences on regulation of blood plasma/serum components in teleost fishes: a review. Rev. Fish Biol. Fish. 19, 481–564.

Noye, B., Walsh, P., 1976. Wind-induced water level oscillations in shallow lagoons. Mar. Freshw. Res. 27, 417–430.

Nunes, R.A., Lennon, G.W., 1986. Physical property distributions and seasonal trends in spencer gulf, south Australia: an inverse estuary. Mar. Freshw. Res. 37, 39–53.

O'Connor, J.A., Rogers, D.J., 2013. Response of Waterbirds to Environmental Change in the Lower Lakes, Coorong and Murray Mouth Icon Site. South Australian Department of Environment, Water and Natural Resources, Adelaide, p. 52.

Onuf, C.P., 2007. Laguna Madre. In: Handley, L., Altsman, D., DeMay, R. (Eds.), Seagrass Status and Trends in the Northern Gulf of Mexico 1940–2002. pp. 29–40. USGS Scientific Investigations Report 2006-5287 and USEPA 855-R-04-003.

Paton, D.C., Rogers, D.J., Hill, B.M., Bailey, C.P., Ziembicki, M., 2009. Temporal changes to spatially stratified waterbird communities of the Coorong, South Australia: implications for the management of heterogenous wetlands. Anim. Conserv. 12, 408–417.

Pen, L.J., 1999. Managing Our Rivers. Waters and Rivers Commission (Western Australia), Perth.

Pérez-Castaeda, R., Blanco-Martnez, Z., Snchez-Martnez, J.G., Rbago-Castro, J.L., Aguirre-Guzmn, G., De La Luz Vzquez-Sauceda, M., 2010. Distribution of *Farfantepenaeus aztecus* and *F. duorarum* on submerged aquatic vegetation habitats along a subtropical coastal lagoon (Laguna Madre, Mexico). J. Mar. Biol. Assoc. U.K. 90, 445–452.

Perissinotto, R., Stretch, D.D., Taylor, R.H., 2013. Ecology and Conservation of Estuarine Ecosystems: Lake St Lucia as a Global Model. Cambridge University Press, Cambridge.

Phillips, B., Muller, K., 2006. Ecological Character of the Coorong, Lakes Alexandrina and Albert Wetland of International Importance. South Australian Department for Environment and Heritage. Adelaide Press, Adelaide, p. 238.

Phleger, F.B., 1981. A review of some general features of coastal lagoons. Coastal lagoon research, present and future. UNESCO Technical Papers in Marine Science 33, United Nations Educational, Scientific, and Cultural Organization, Paris, France, p. 7–14.

Poh, B., Tweedley, J.R., Chaplin, J.A., Trayler, K.M., Loneragan, N.R., 2018. Estimating predation rates of restocked individuals: the influence of timing-of-release on metapenaeid survival. Fish. Res. 198, 165–179.

Poloczanska, E.S., Babcock, R.C., Butler, A., Hobday, A.J., Hoegh-Guldberg, O., Kunz, T.J., Matear, R., Milton, D., Okey, T.A., Richardson, A.J., 2007. Climate change and Australian marine life. Oceanogr. Mar. Biol. Annu. Rev. 45, 409–480.

Potter, I.C., Chuwen, B.M., Hoeksema, S.D., Elliott, M., 2010. The concept of an estuary: a definition that incorporates systems which can become closed to the ocean and hypersaline. Estuar. Coast. Shelf Sci. 87, 497–500.

Potter, I.C., Warwick, R.M., Hall, N.G., Tweedley, J.R., 2015. The physico-chemical characteristics, biota and fisheries of estuaries. In: Craig, J. (Ed.), Freshwater Fisheries Ecology. Wiley-Blackwell, Chichester, pp. 48–79.

Pritchard, D.W., 1967. What is an estuary: a physical viewpoint. Am. Assoc. Advance. Sci. 83, 3–5.

Quammen, M.L., Onuf, C.P., 1993. Laguna Madre: seagrass changes continue decades after salinity reduction. Estuaries 16, 302–312.

Ranasinghe, R., Pattiaratchi, C., 1999. The seasonal closure of tidal inlets: Wilson Inlet—a case study. Coast. Eng. 37, 37–56.

Raz-Guzman, A., Huidobro, L., 2002. Fish communities in two environmentally different estuarine systems in Mexico. J. Fish Biol. 61, 182–195.

Reeves, J.M., Haynes, D., Garcia, A., Gell, P.A., 2015. Hydrological change in the Coorong Estuary, Australia, past and present: evidence from fossil invertebrate and algal assemblages. Estuar. Coasts 38, 2101–2116.

Rich, A., Keller, E.A., 2013. A hydrologic and geomorphic model of estuary breaching and closure. Geomorphology 191, 64–74.

Ritson, P., Boyd, D.W., Bari, M.A., 1995. Effects of forest clearing on streamflows and salinity at Wrights catchment, Western Australia. . Western Australian Water Authority Report No. WSIG5.

Rosenthal, E., Flexer, A., Möller, P., 2006. The paleoenvironment and the evolution of brines in the Jordan-Dead Sea transform and in adjoining areas. Int. J. Earth Sci. 95, 725–740.

Royal Society of South Australia, 2018. Natural History of the Coorong, Lower Lakes and Murray Mouth (Yarluwar-Ruwe). In: Mosley, L., Ye, Q., Shepherd, S., Hemming, S., Fitzpatrick, R. (Eds.), University of Adelaide Press, Adelaide.

Rybak, A.S., 2018. Species of *Ulva* (Ulvophyceae, Chlorophyta) as indicators of salinity. Ecol. Indic. 85, 253–261.

Sarre, G.A., Platell, M.E., Potter, I.C., 2000. Do the dietary compositions of *Acanthopagrus butcheri* in four estuaries and a coastal lake vary with body size and season and within and amongst these water bodies? J. Fish Biol. 56, 103–122.

Scharler, U.M., MacKay, C.F., 2013. Food webs and ecosystem functioning. In: Perissinotto, R., Stretch, D.D., Taylor, R.H. (Eds.), Ecology and Conservation of Estuarine Ecosystems: Lake St Lucia as a Global Model. Cambridge University Press, Cambridge, pp. 382–395.

Schelske, C.L., Odum, E.P., 1961. Mechanisms maintaining high productivity in Georgia estuaries. Proc. Gulf Caribbean Fish. Inst. 14, 75–80.

Schoenbaechler, C., Guthrie, C.G., 2011a. Coastal Hydrology for the Laguna Madre Estuary, with Emphasis on the Lower Laguna Madre. Texas Water Development Board, Austin, TX, p. 29.

Schoenbaechler, C., Guthrie, C.G., 2011b. Coastal Hydrology for the Laguna Madre Estuary, with Emphasis on the Upper Laguna Madre. Texas Water Development Board, Austin, TX, p. 28.

Schofield, N.J., 1990. Water interactions with land use and climate in south-west Australia. . Western Australian Water Authority Report No. WS60.

Shuttleworth, B., Woidt, A., Paparella, T., Herbig, S., Walker, D., 2005. The dynamic behaviour of a river-dominated tidal inlet, River Murray, Australia. Estuar. Coast. Shelf Sci. 64, 645–657.

Slinger, J.H., 2017. Hydro-morphological modelling of small, wave-dominated estuaries. Estuar. Coast. Shelf Sci. Part B 198, 583–596.

Smyth, K., Elliott, M., 2016. Effects of changing salinity on the ecology of the marine environment. In: Solan, M., Whiteley, N.M. (Eds.), Stressors in the Marine Environment. Oxford University Press, Oxford, pp. 161–174.

Taylor, R.H., 2013. Management history. In: Perissinotto, R., Stretch, D.D., Taylor, R.H. (Eds.), Ecology and Conservation of Estuarine Ecosystems: Lake St Lucia as a Global Model. Cambridge University Press, Cambridge, pp. 21–45.

Tulipani, S., Grice, K., Krull, E., Greenwood, P., Revill, A.T., 2014. Salinity variations in the northern Coorong Lagoon, South Australia: significant changes in the ecosystem following human alteration to the natural water regime. Org. Geochem. 75, 74–86.

Tunnell, J.W., Hilbun, N.L., Withers, K., 2002. Comprehensive Bibliography of the Laguna Madre of Texas and Tamaulipas. Center for Coastal Studies, Texas A&M University, Corpus Christi, TX, p. 109.

Tunnell Jr., J.W., 2002a. Introduction. In: Tunnell, J.W.J., Judd, F.W. (Eds.), The Laguna Madre of Texas and Tamaulipas. Texas A&M University Press, College Station, TX, pp. 3–6.

Tunnell Jr., J.W., 2002b. Geography, climate, hydrography. In: Tunnell, J.W.J., Judd, F.W. (Eds.), The Laguna Madre of Texas and Tamaulipas. Texas A&M University Press, College Station, TX, pp. 7–26.

Tunnell Jr., J.W., 2002c. The environment. In: Tunnell, J.W.J., Judd, F.W. (Eds.), The Laguna Madre of Texas and Tamaulipas. Texas A&M University Press, College Station, TX, pp. 73–84.

Tweedley, J.R., Keleher, J., Cottingham, A., Beatty, S.J., Lymbery, A.J., 2014. The Fish Fauna of the Vasse-Wonnerup and the Impact of a Substantial Fish Kill Event. Murdoch University, Perth, p. 113.

Tweedley, J.R., Warwick, R.M., Potter, I.C., 2016. The contrasting ecology of temperate macrotidal and microtidal estuaries. Oceanogr. Mar. Biol. Annu. Rev. 54, 73–171.

Tweedley, J.R., Warwick, R.M., Valesini, F.J., Platell, M.E., Potter, I.C., 2012. The use of benthic macroinvertebrates to establish a benchmark for evaluating the environmental quality of microtidal, temperate southern hemisphere estuaries. Mar. Pollut. Bull. 64, 1210–1221.

van der Elst, R.P., Blaber, S.J.M., Wallace, J.H., Whitfield, A.K., 1976. In: Heydorn, A.E.F. (Ed.), The fish fauna of Lake St Lucia under different salinity regimes. Proceedings of the St Lucia Scientific Advisory Council Workshop Meeting—Charters Creek, 15–17 February 1976. Natal Parks, Game and Fish Preservation Board, Pietermaritzburg. pp. 22.

Wallace, J.H., 1975. The estuarine fishes of the east coast of South Africa. Part 1. Species composition and length distribution in the estuarine and marine environments. Part 2. Seasonal abundance and migrations. Investig. Rep.—Oceanographic Res. Inst. 40, 1–72.

Webster, I.T., 2010. The hydrodynamics and salinity regime of a coastal lagoon—the Coorong, Australia—seasonal to multi-decadal timescales. Estuar. Coast. Shelf Sci. 90, 264–274.

Webster, I.T., 2011. Dynamic assessment of oceanic connectivity in a coastal lagoon-the Coorong, Australia. J. Coast. Res. 27, 131–139.

Wedderburn, S.D., Bailey, C.P., Delean, S., Paton, D.C., 2016. Population and osmoregulatory responses of a euryhaline fish to extreme salinity fluctuations in coastal lagoons of the Coorong, Australia. Estuar. Coast. Shelf Sci. 168, 50–57.

Wedderburn, S.D., Hammer, M.P., Bice, C.M., 2012. Shifts in small-bodied fish assemblages resulting from drought-induced water level recession in terminating lakes of the Murray-Darling Basin, Australia. Hydrobiologia 691, 35–46.

Wetz, M.S., Yoskowitz, D.W., 2013. An "extreme" future for estuaries? Effects of extreme climatic events on estuarine water quality and ecology. Mar. Pollut. Bull. 69, 7–18.

Whitfield, A., Elliott, M., 2011. Ecosystem and biotic classifications of estuaries and coasts. In: Wolanski, E., McLusky, D.S. (Eds.), Treatise on Estuarine and Coastal Science. Academic Press, Waltham, pp. 99–124.

Whitfield, A.K., Bate, G.C., Forbes, T., Taylor, R.H., 2013. Relinkage of the Mfolozi River to the St. Lucia estuarine system—urgent imperative for the long-term management of a Ramsar and World Heritage Site. Aquat. Ecosyst. Health Manage. 16, 104–110.

Whitfield, A.K., Blaber, S.J.M., Cyrus, D.P., 1981. Salinity ranges of some southern African fish species occurring in estuaries. Afr. Zool. 16, 151–155.

Whitfield, A.K., Taylor, R.H., 2009. A review of the importance of freshwater inflow to the future conservation of Lake St Lucia. Aquat. Conserv. Mar. Freshwat. Ecosyst. 19, 838–848.

Whitfield, A.K., Taylor, R.H., Fox, C., Cyrus, D.P., 2006. Fishes and salinities in the St Lucia estuarine system—a review. Rev. Fish Biol. Fish. 16, 1–20.

Wildsmith, M.D., Potter, I.C., Valesini, F.J., Platell, M.E., 2005. Do the assemblages of the benthic macroinvertebrates in nearshore waters of Western Australia vary among habitat types, zones and seasons? J. Mar. Biol. Assoc. U.K. 85, 217–232.

Wildsmith, M.D., Rose, T.H., Potter, I.C., Warwick, R.M., Clarke, K.R., 2011. Benthic macroinvertebrates as indicators of environmental deterioration in a large microtidal estuary. Mar. Pollut. Bull. 62, 525–538.

Wildsmith, M.D., Rose, T.H., Potter, I.C., Warwick, R.M., Clarke, K.R., Valesini, F.J., 2009. Changes in the benthic macroinvertebrate fauna of a large microtidal estuary following extreme modifications aimed at reducing eutrophication. Mar. Pollut. Bull. 58, 1250–1262.

Withers, K., 2002a. Wind-tidal flats. In: Tunnell, J.W.J., Judd, F.W. (Eds.), The Laguna Madre of Texas and Tamaulipas. Texas A&M University Press, College Station, TX, pp. 114–126.

Withers, K., 2002b. Seagrass meadows. In: Tunnell, J.W.J., Judd, F.W. (Eds.), The Laguna Madre of Texas and Tamaulipas. Texas A&M University Press, College Station, TX, pp. 85–101.

Withers, K., 2002c. Red and brown tides. In: Tunnell, J.W.J., Judd, F.W. (Eds.), The Laguna Madre of Texas and Tamaulipas. Texas A&M University Press, College Station, TX, pp. 255–258.

Withers, K., Tunnell, J.W., 1998. Identification of Tidal Flat Alterations and Determination of Effects on Biological Productivity of These Habitats Within the Coastal Bend. Corpus Christi Bay National Estuary Program, Corpus Christi, TX, p. 170.

Ye, Q., Bice, C.M., Bucater, L., Ferguson, G.L., Giatas, G.C., Wedderburn, S.D., Zampatti, B.P., 2016. Fish monitoring synthesis: understanding responses to drought and high flows in the Coorong, Lower Lakes and Murray Mouth. South Australian Research and Development Institute (Aquatic Sciences), Adelaide. SARDI Publication No. F2016/000348-1. SARDI Research Report Series No. 909, 39 pp.

Ye, Q., Bucater, L., Short, D., 2017. Coorong fish condition monitoring 2015/16: black bream (*Acanthopagrus butcheri*), greenback flounder (*Rhombosolea tapirina*) and smallmouth hardyhead (*Atherinosoma microstoma*) populations. South Australian Research and Development Institute (Aquatic Sciences), Adelaide. SARDI Publication No. F2011/000471-5. SARDI Research Report Series No. 943, 89 pp..

Young, G.C., Potter, I.C., 2002. Influence of exceptionally high salinities, marked variations in freshwater discharge and opening of estuary mouth on the characteristics of the ichthyofauna of a normally-closed estuary. Estuar. Coast. Shelf Sci. 55, 223–246.

Zampatti, B.P., Bice, C.M., Jennings, P.R., 2010. Temporal variability in fish assemblage structure and recruitment in a freshwater-deprived estuary: the Coorong, Australia. Mar. Freshw. Res. 61, 1298–1312.

Chapter 31

Alien Species Invasion: Case Study of the Black Sea

Nickolai Shalovenkov

The Centre for Ecological Studies, Russia

1 INTRODUCTION

The fauna and flora of the Black Sea are formed mainly by the Mediterranean and Pontian relic complexes of species. The invasion by the Mediterranean species of the Black Sea (or "Mediterranization"; Pusanow, 1967; Puzanov, 1967) is a natural process which started after the opening of the Bosphorus Strait (Istanbul) about 5000–7000 years ago (Arkhangelsky and Batalina, 1929). After this, the relic fauna remained in the brackish-water areas of the sea only (Morduchai-Boltovskoi, 1960; Zenkevich, 1963).

The Mediterranean euryhaline species that could withstand salinity of less than 19 ppt are established in the Black Sea. The Black Sea hosts altogether almost 800 phytoplankton species, 2000 invertebrates, and 200 fish species (Morduchai-Boltovskoi, 1972; Greze, 1979; Kisileva, 1979; Pitsyk, 1950, 1979; Koval, 1984; Zaitsev and Mamaev, 1997; Zaitsev and Alexandrov, 1998; Boltachev and Karpova, 2013b in WoRMS).

Based on the articles, review papers, and databases, 69 alien species of phytoplankton, 22 of zooplankton and 64 of benthos were observed during the last 50–100 years (Zaitsev and Öztürk, 2001; Aleksandrov et al., 2007; Alexandrov, 2017; Zaitsev, 2011; AquaNIS, n.d.; WoRMS, n.d.). In the ichthyofauna, 3 alien and 22 Mediterranean fish species were observed also during the same period (Boltachev et al., 2010a,b; Yankova et al., 2013; Boltachev and Karpova, 2014; Yankova, 2016). The species nomenclature of taxonomic groups was checked following the World Register of Marine Species.

In addition, Mediterranean and the Marmara Sea are the main transit corridors for the introduction of alien species in the Black Sea. However, the spread of alien species by currents is limited by hydrological features in the transformation area of the water masses of the Marmara Sea on the south-west shelf of the Black Sea, which create an ecological barrier (Jakubova, 1948). This ecological barrier results from temperature and salinity spatial gradients that limit the spread of species and create conditions for the relative isolation of habitats (Bogdanova, 1959, 1964; Zenkevich, 1963; Oğuz and Öztürk, 2011). Spatial gradients of temperature and salinity also prevail in other areas of the coastal shelf of the Black Sea that create six ecological areas (Varna and Burgas Bays, Danube, North-Western, Crimean, Caucasian and Anatolian) separated by conditional boundaries because of the spatiotemporal variability of temperature and salinity in the coastal shelf of the sea (Fig. 1). Each ecological area has distinctive features of species richness and composition of alien species.

2 ALIEN SPECIES INVASION OF THE BLACK SEA

2.1 Phytoplankton Alien Species

In the Black Sea, 69 alien species of phytoplankton have been recorded after 100 years of observations. Different ecological areas of the coastal shelf have a different number and composition of alien species of phytoplankton. The largest number of species (over 30) is recorded in the Varna and Burgas Bays, and also in the northwestern part of the sea while a smaller number of alien (24–27) occurs in n the Caucasian and Danube areas and the least number of species (8) is found near the Anatolian coast. In addition, a small (16) number of alien phytoplankton species occurs along the Crimean coast.

Only two nonnative species of the microalgae, the diatom *Pseudosolenia calcar-avis* (Schultze) B.G. Sundström, 1986 and the dinoflagellate *Scrippsiella trochoidea* (Stein) Loeblich III, 1976 are registered by researchers in all six ecoregions. The diatom *P. calcar-avis* was found for the first time at the Bulgarian coast and the northwest part of the Black Sea in 1926 (Usachev, 1928). This alien species occurs in the plankton all year round with maximum development indices at the end of

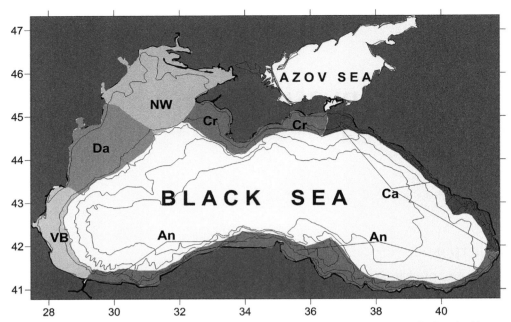

FIG. 1 The six ecological areas along the coastal shelf of the Black Sea: "VB"—Varna and Burgas Bays, "Da"—Danube (the area of direct Danube river-water influence on the shelf), "NW"—North-Western, "Cr"—Crimean, "Ca"—Caucasian, and "An"—Anatolian.

summer and in autumn, being most abundant in the North-West area (Morozova-Vodjanitckaja, 1957; Pitsyk, 1950,1979; Konsulova et al., 1991; Georgieva, 1993; Senicheva, 2002; Caraus, 2002; Selifonova and Yasakova, 2012; Terenko and Terenko, 2007; Yasakova, 2011; Karanda, 2014; Petrova and Gerdzhikov, 2015). The dinoflagellate *S. trochoidea* spread throughout the sea for 30 years after first found in the northwestern part of the Black Sea in 1960 (Ivanov, 1965; Caraus, 2002; Bodeanu, 2002; Çinar et al., 2005; Yasakova, 2011; Selifonova and Yasakova, 2012; Terenko and Terenko, 2007; Karanda, 2014; Petrova and Gerdzhikov, 2015).

Other two alien species of microalgae, the diatom *Lioloma pacificum* (Cupp) Hasle, 1996 and the dinoflagellate *Alexandrium tamarense* (Lebour, 1925) Balech, 1995, also have a wide distribution in the Black Sea, with the exception of some ecoregions. For instance, the diatom *L. pacificum* is not registered along the Anatolian coast, and the dinoflagellates *A. tamarense*—at the Crimean coast. This diatom does not form stable populations in its distribution areas (Terenko and Terenko, 2000; Senicheva, 2002; Yasakova, 2011). It should be particularly noted that the dinoflagellate *A. tamarense* is a toxic species that can form nuisance outbreaks or blooms (Yasakova, 2013).

In addition to these two microalgae, three more alien species are widespread in several ecological areas, but are not found near the Anatolian and Crimean coasts. These are the invaders *Prorocentrum cordatum* (Ostenfeld) J.D. Dodge, 1975, *Distephanus speculum f. octonarius* (Ehrenberg) S. Locker & E. Martini and *Phaeocystis pouchetii* (Hariot) Lagerheim, 1896. These nonnative species were noted by researchers in the phytoplankton only in the eastern and western parts of the Black Sea (Morozova-Vodjanitckaja, 1948; Petrova-Karadjova, 1990; Senichkina, 1983; Moncheva et al., 1995; Gvarishvili, 1998; Georgieva and Senichkina, 1996; Terenko and Terenko, 2000; Caraus, 2002; Melinte-Dobrinescu and Ion, 2013; Yasakova, 2013, 2014).

The alien species of phytoplankton which have a wide distribution in the eco-areas of the sea make up 10% only of the total number of species. Most (43%) invaders have been registered only in one of the six ecological areas of the sea (Table 1). Moreover, almost all these species invaded the Black Sea within the last 10–15 years. The exception is four species recorded in the phytoplankton near the Crimean coast in the 1990s that are still not found in other areas of the Black Sea (Table 1). The other half (47%) of nonnative species is found in two or three usually neighboring ecological areas, that is, they were able to spread beyond one ecological area after the initial invasion.

The composition of alien species in phytoplankton has distinct features in each ecological area. The level of similarity (or dissimilarity) of species composition of invaders between the ecological areas was calculated using multidimensional statistics. As followed from the results of the cluster analysis, the degree of similarity of the compositions of alien species between the four ecological areas (Anatolian, Caucasian, Crimean, and North-Western) was very low, since the linkage distances of their clustering (D) were high ($D=0.87$–0.93; Fig. 2A). The greatest similarity in species composition is recorded

TABLE 1 Alien Phytoplankton Species That Were Recorded in One From Six Areas of the Black Sea Only

Areas	Part From All Species in the Area (%)	Species	Year of the Find	Author
Anatolian	–	–		
Varna and Burgas Bays	24	*Azadinium spinosum* Elbrächter & Tillmann, 2009	2005	Mavrodieva (2012)
		Gymnodinium fuscum (Ehrenberg) Stein, 1878	1980	Vershinin (2008)
		Gymnodinium nanum J. Schiller, 1928	2008	Alexandrov et al. (2017)
		Gymnodinium pulchrum J. Schiller, 1928	1999	Alexandrov et al. (2017)
		Gyrodinium cochlea Lebour, 1925	2008	Alexandrov et al. (2017)
		Gyrodinium flagellare J. Schiller, 1928	2008	Alexandrov et al. (2017)
		Gyrodinium varians (Wulff) Schiller, 1933	2008	Alexandrov et al. (2017)
		Scrippsiella operosa (Deflandre) Montresor, 2003	2008	Rubino et al. (2010)
Danube shelf	–			
North-Western	37	*Alexandrium acatenella* (Whedon & Kofoid) Balech, 1985	2001	Alexandrov et al. (2004)
		Alexandrium affine (H. Inoue & Y. Fukuyo) Balech, 1995	2001	Alexandrov et al. (2004)
		Alexandrium pseudogonyaulax (Biecheler) Horiguchi ex Kita & Fukuyo, 1992	2002	Aleksandrov et al. (2007)
		Attheya decora T. West, 1860	2000	Terenko (2005)
		Dinophysis islandica Paulsen, 1949	2001	Terenko (2011)
		Dinophysis nasuta (Stein) Parke & Dixon, 1968	2008	Terenko (2011)
		Dinophysis recurva Kofoid & Skogsberg, 1928	2001	Terenko (2011)
		Dinophysis schilleri Sournia, 1973	2001	Nesterova et al. (2006)
		Levanderina fissa (Levander) Ø. Moestrup, P. Hakanen, G. Hansen, N. Daugbjerg & M. Ellegaard, 2014	2003	Aleksandrov et al. (2007)
		Pyramimonas longicauda L. Van Meel, 1969	2001	Aleksandrov et al. (2007)
		Spatulodinium pseudonoctiluca (Pouchet) J. Cachon & M. Cachon, 1968	2001	Alexandrov et al. (2004)
Crimean	25	*Oxytoxum gladiolus* Stein, 1883	1989	Senichkina et al. (2001)
		Pronoctiluca pelagic Fabre-Domergue, 1889	1983	Aleksandrov et al. (2007)
		Pseudonitzschia inflatula (Hasle) Hasle, 1993	1990	Senicheva (2002)
		Syracolithus dalmaticus (Kamptner) Leoblich Jr. & Tappan, 1966	1990	Bryantseva (2000) and Polikarpov et al. (2003)
Caucasian	26	*Alexandrium ostenfeldii* (Paulsen) Balech & Tangen, 1985	2004	Vershinin (2008) and Yasakova (2010)
		Chaetoceros throndsenii (Marino, Montresor, & Zingone) Marino, Montresor & Zingone, 1991	2005	Silkin et al. (2011)
		Gymnodinium stellatum Hulburt, 1957	2008	Gvarishvili et al. (2010) and Yasakova (2010)
		Oxytoxum variabile Schiller, 1937	2008	Yasakova (2010)
		Archaeperidinium minutum (Kofoid) Jørgensen, 1912	1999	Vershinin (2008)
		Protoperidinium parthenopes A. Zingone & M. Montresor, 1988	2001	Vershinin (2008)
		Pseudonitzschia pungens (Grunow in Cleve & Möller) Hasle, 1993	2000	Vershinin (2008)

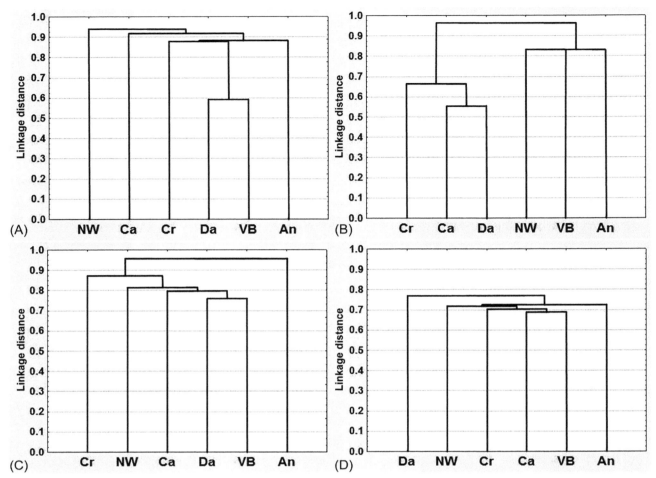

FIG. 2 Dendrograms of similarity (dissimilarity) between the six areas ("VB"—Varna and Burgas Bays, "Da"—Danube, "NW"—North-Western, "Cr"—Crimean, "Ca"—Caucasian and "An"—Anatolian) by the composition of alien species of (A) phytoplankton, (B) zooplankton, (C) zoobenthos, and (D) new Mediterranean fishes.

between the Danube region and Varna-Burgas Bays, which are in the zone of influence of the transformed waters of the Danube River. At the same time, the degree of similarity of the compositions of alien species was low for these two regions, because the distance of their clustering was $D=0.6$ only (Fig. 2A).

Thus, the high level of differences in the composition of alien phytoplankton species in six ecological areas of the Black Sea indicates a relative isolation of phytoplankton communities in these areas.

2.2 Zooplankton Alien Species

Twenty-six alien species were discovered in the zooplankton of the Black Sea from the mid-20th century. This is almost three times less than the number of alien species in phytoplankton. The largest number (18) of zooplankton nonnative species is found near the Crimean coast (Table 2). In other areas the number of invader species in zooplankton is smaller, varying from 5 to 10 species.

Only three species of invaders have been recorded in zooplankton in all areas of the sea. These are two types of macrozooplankton: the ctenophores *Mnemiopsis leidyi* A. Agassiz, 1865 and *Beroe ovata* Bruguière, 1789, and one species of microzooplankton: the tintinnida *Eutintinnus lusus-undae* (Entz, 1885).

The ctenophore *M. leidyi* was found in all areas of the Black Sea for 4 (1982–86) years (Pereladov, 1988). The appearance of the ctenophore has significantly impacted the zooplankton communities: a change in the species structure followed, the development indices decreased sharply and the number of once massive zooplankton species decreased (Pereladov, 1988; Zaitsev et al., 1988; Konsulov, 1989; Vinogradov et al., 1989, 1992; Vinogradov and Shushkina, 1992;

TABLE 2 Alien Zooplankton Species That Were Recorded in One of Six Areas of the Black Sea Only

Areas	Part From All Species in the Area (%)	Species	Year of the Find	Author
Anatolian	38	*Chrysaora hysoscella* (Linnaeus, 1767)	2009	Aleksandrov et al. (2017)
		Paraphyllina ransoni Russell, 1956	2011	Öztürk et al. (2011)
		Solmundella bitentaculata (Quoy & Gaimard, 1833)	2010	Isinbilir et al.,(2010)
Varna and Burgas Bays	11	*Euryte longicauda* Philippi, 1843	2000	Aleksandrov et al. (2017)
Danube shelf	–			
North-Western	20	*Protocystis (swerei) harstoni* (Murray, 1885)	1950	Pusanow (1967)
Crimean	28	*Eutintinnus apertus* Kofoid & Campbell, 1929	2002	Aleksandrov et al. (2007)
		Tintinnopsis mortensenii Schmidt, 1902	2009	Gavrilova (2017)
		Rhizodomus tagatzi Strelkow & Wirketis, 1950	2000	Gavrilova (2010)
		Favella brevis (Laackmann, 1909)	2001	Gavrilova (2005)
		Sarsamphiascus tenuiremis (Brady, 1880)	2000	Zagorodnyay and Kolesnikova (2003)
Caucasian	10	*Tintinnopsis tocantinensis* Kofoid & Campbell, 1929	2010	Selifonova (2011)

Kideys and Niermann, 1994; Gomoiu and Skolka, 1996; Shiganova, 1998; Mirzoyan, 2000; Gubanova et al., 2001). The situation in the planktonic communities of the Black Sea has radically changed after the invasion of another alien species, *B. ovata*, in 1996; this is a predator of the ctenophore *M. leidyi* (Konsulov and Kamburska, 1998; Shiganova et al., 2000; Martynyuk et al., 2001; Gubanova, 2003, Gubareva et al., 2004). As a result, the pressure exerted by the ctenophore *B. ovata* on the ctenophore *M. leidyi* began to manifest itself in the zooplankton communities dynamics as there resulted an increase in the species richness, abundance and biomass of zooplankton in the Black Sea (Shiganova et al., 2000; Martynyuk et al., 2001; Gubanova, 2003; Gubareva et al., 2004).

The microzooplankton representative, the tintinnida *E. lusus-undae*, was found in all areas of the Black Sea since 2001 (Gavrilova, 2001, 2005; Polikarpov et al., 2003; Aleksandrov et al., 2007). The tintinnida reaches a high abundance in zooplankton communities in certain years displacing native species of microzooplankton (Selifonova, 2011).

In addition to the above-mentioned alien species, two more representatives of mesoplankton also now have a wide distribution in the pelagic communities of the Black Sea. These are the copepods *Acartia tonsa* (*Acanthacartia*) Dana, 1849 and *Oithona davisae* Ferrari FD & Orsi, 1984 (Pereladov, 1988; Konsulov, 1989; Belmonte et al., 1994; Gomoiu and Skolka, 1996; Gubanova, 1997, 2000; Kamburska, 2004; Mihneva and Stefanova, 2013). They were not recorded earlier in some areas, although they are now abundant in the zooplankton of the Black Sea (Zaitsev et al., 1988; Aleksandrov et al., 2007). The copepod *A. tonsa* was not found in the Danube and North-West areas, and the copepod *O. davisae* was found in the Anatolian coast.

Half (13 species, i.e., 50%) of zooplankton invaders have been found in only one of the six ecological areas of the sea (Table 2). Almost all these nonnative species have been recorded in the plankton for the last 10–15 years. Alien species that have been recorded in two or three areas of the sea comprise 30% of the list of zooplankton invaders. Thus, alien zooplankton in different regions of the Black Sea has distinctive species structure.

Cluster analysis revealed some similarity in the species composition of zooplankton alien species for the Danube region and the Caucasian coast (Fig. 2B). Their linkage D is 0.56. It should be noted that these two regions are characterized by the greatest level of salinity reduction in the coastal shelf (Artamonov et al., 2012). The species composition of the alien zooplankton of the Crimean coast is very similar to that in these two regions ($D=0.68$, Fig. 2B). The lowest degree of similarity in the invader species compositions is for the zooplankton of the Anatolian coast, the Northwest region and the Varna-Burgas Bays ($D=0.83$–0.95).

The results of calculations indicate a high level of difference in the composition of alien species of zooplankton (as well as of phytoplankton) between six areas of the Black Sea.

2.3 Zoobenthos Alien Species

The first alien invasive species of zoobenthos are the shipworm *Teredo navalis* (Linne, 1758) and the barnacle *Amphibalanus improvisus* (Darwin, 1854) that arrived in the Black Sea even before the 20th century, namely 750–500 BC and in the middle of the 19th century, respectively (Gomoiu and Skolka, 1996; Skolka and Preda, 2010). Over the last century, 61 alien zoobenthic species have been recorded in bottom communities of the Black Sea. The main groups among alien species were the Polychaetes, the Crustaceans, and the Molluscs, that is, these were the same taxonomic groups of the benthic animals which prevailed in the bottom native fauna of the Black Sea. Only five species are found in all areas from the entire list of zoobenthic alien species for the Black Sea. These are crustaceans *A. improvisus* and *Rhithropanopeus harrisii* (Gould, 1841), molluscs *Anadara kagoshimensis* (Tokunaga, 1906), *Rapana venosa* (Valenciennes, 1846), and *T. navalis*. The others 58 nonnative species were recorded only in some areas of the Black Sea.

The *R. harrisii* mud crab and the *Rapana* sea snail, as predators, can have a negative impact on the native benthic communities. The average biomass and abundance for the crab *harrisii* has a low value on the Black Sea shelf, with the exception of the salty lakes of Varna and Beloslav, where the maximum values of the biomass are up to $640 \, g/m^2$ and the abundance is up to $260 \, ind/m^2$ (Todorova and Konsulova, 2008). Usually, the maximum density of the *Rapana* mollusc-predator is $10–12 \, ind/m^2$ in benthic communities (Gomoiu and Skolka, 2005; Todorova and Konsulova, 2008; Chikina, 2009; Alymov and Tikhonova, 2012; Zolotarev and Terentev, 2012; Snigerev, 2012; Shalovenkov, 2017).

A large population of the mollusc *A. kagoshimensis* is found on the coastal shelf of the Caucasus and in the area of influence of the transformed Danube River water where the biomass of the molluscs reached $450–2700 \, g/m^2$ and the abundance of the molluscs was $220–1160 \, ind/m^2$ (Chikina and Kucheruk, 2005; Abaza et al., 2006; Shurova and Zolotarev, 2007; Chikina, 2009).

The shellfish *Mya*, as successful invaders, have higher biomass and abundance on the north-western area of the Black Sea shelf, although it has settled in most parts of the Black Sea. The average biomass of the soft-clam, *Mya arenaria*, varied from 300 to $900 \, g/m^2$ in the 1980s, but now the clam biomass is not registered at more than $100 \, g/m^2$ in the Black Sea (Zolotarev et al., 1990; Makarov and Kostylyov, 2001; Gomoiu and Skolka, 2005; Abaza et al., 2006; Sinegub, 2006; Shurova and Zolotarev, 2007; Ivanov and Sinegub, 2008; Todorova and Konsulova, 2008; Stadnichenko and Zolotarev, 2009).

Alien species that were registered only in one of the six sea areas accounted for 41% (or 26 species) of the list of all zoobenthos invaders (Table 3). Half (32 species or 51%) of zoobenthos nonnative species have been registered only in two or three areas of the Black Sea. These data indicate the existence of differences in the species composition of alien species of benthic animals in various areas of the Black Sea.

The degree of similarity (or dissimilarity) of the species composition of the zoobenthos invasive species between areas was assessed using cluster analysis. From the calculation results, all six areas (Varna and Burgas Bays, Danube, North-Western, Crimean, Caucasian, and Anatolian) of the Black Sea have a high level of dissimilarity in the species composition of alien species. The clusters are formed at the last analysis stages and the linkage distance (D) of the ecological areas was in the range of 0.77–0.95 (Fig. 2C). The results of calculations indicate a high level of difference in the composition of alien species of zoobenthos (as well as of phytoplankton and zooplankton) between six areas of the Black Sea.

2.4 Fish Alien Species

Four alien and 25 new Mediterranean ("Mediterranization") species have been recorded in the ichthyofauna of the Black Sea from the middle of the last century. Alien species are the mullet haarder, *Liza haematocheila* (Temminck & Schlegel, 1845), the chameleon goby, *Tridentiger trigonocephalus* Gill, 1859, the barracuda, *Sphyraena pinguis* Gunther, 1874, and the pennant coralfish *Heniochus acuminatus* (Linnaeus, 1758).

The Far Eastern mullet haarder was naturalized as a species in the ichthyofauna of the Black Sea after its introduction in lagoons (the brackish Shabolat Lagoon) of the northwestern part of the sea for aquaculture for more than 40 years (Starushenko and Kazansky, 1996). In the late 1980s, the release in the sea of a small number of the mullet haarder led to the formation of a self-reproducing population in the Black Sea. The increase in the abundance of the mullet population has resulted in its occurrence in many lagoons, river mouths, and coastal areas of the Black Sea and then of the Mediterranean Sea. Since 1988, commercial fishing of the mullet haarder has been conducted regularly in the Black Sea (Zaitsev and Starushenko, 1997).

TABLE 3 Alien Zoobenthos Species That Were Recorded in One From Six Areas of the Black Sea Only

Areas	Part From All Species in the Area (%)	Species	Year of the Find	Author
Anatolian	28	*Asterias rubens* Linnaeus, 1758	2003	Karhan et al. (2008)
		Capitellethus dispar (Ehlers, 1907)	1959	Rullier (1963)
		Nephtys ciliata (Müller, 1788)	1965	Caspers (1968)
		Sirpus zariquieyi Gordon, 1953	1982	Zaitsev and Öztürk (2001)
		Sorites orbiculus (Forskål, 1775)	2010	Meriç et al. (2010)
Varna and Burgas Bays	9	*Streblospio shrubsolii* (Buchanan, 1890)	1957	Marinov (1957)
		Streptosyllis varians Webster & Benedict, 1887	1964	Kaneva-Abadjieva and Marinov (1966)
Danube shelf	13	*Arcuatula senhousia* (Benson, 1842)	2002	Micu (2004)
		Hemigrapsus sanguineus (De Haan, 1835)	2008	Micu et al. (2010)
		Molgula manhattensis (De Kay, 1843)	1971	Băcescu et al. (1971)
		Styela clava Herdman, 1881	2004	Micu and Micu (2004)
North-Western	27	*Corbicula fluminea* (O. F. Müller, 1774)	1995	Son (2007)
		Ercolania viridis (A. Costa, 1866)	2001	Zaitsev et al. (2004)
		Mytilopsis leucophaeata (Conrad, 1831)	2000	Therriault et al. (2004)
		Mytilus edulis Linnaeus, 1758	1990	Zaitsev et al. (2004)
		Mytilus trossulus Gould, 1850	2001	Zaitsev et al. (2004)
		Pachycordyle navis (Millard, 1959)	2001	Koshelev (2003)
		Penaeus japonicus Spence Bate, 1888	1960	Zaitsev and Öztürk (2001)
		Tubificoides benedii (d'Udekem, 1855)	1916	Shurova (2006)
Crimean	18	*Eudendrium capillare* Alder, 1856	1990	Shadrin (1999)
		Eudendrium vaginatum Allman, 1863	1990	Shadrin (1999)
		Hydroides dianthus (Verrill, 1873)	2009	Boltacheva et al. (2011)
		Perna viridis (Linnaeus, 1758)	2000	Mironov et al. (2002)
		Trinchesia perca (Marcus, 1958)	2007	Martynov et al. (2007)
Caucasian	7	*Penaeus semisulcatus* De Haan, 1844 [in De Haan, 1833–1850]	2005	Khvorov et al. (2006)

The Far East chameleon goby was first recorded in the Sevastopol Bay in 2006 after its introduction in 1980 (Boltachev and Karpova, 2014). In the Bay, the goby *T. trigonocephalus* has already formed a local population (Boltachev and Karpova, 2010). This species has not yet been found in the Black Sea outside of the Sevastopol Bay.

Two barracuda, *S. pinguis*, specimens were once caught in Balaklava Bay (Crimean coast) in 1999 (Boltachev and Yurakhno, 2002). This is the only find of this fish alien species in the Black Sea. It is believed that these two specimens migrated to the Crimean coast from the Mediterranean Sea (Boltachev and Yurakhno, 2002; Boltachev and Karpova, 2014) where this exotic species was already naturalized (Gücü et al., 1994; Taskavak and Bilecenoglu, 2001; Yaglioglu et al., 2015).

In Balaklava Bay, the pennant coralfish, *H. acuminatus*, was also caught only once in 2003. This exotic Indo-Pacific species was probably transferred to coastal waters of the Crimea through ship ballast waters (Boltachev and Astakhov, 2004; Boltachev et al., 2010a,b).

From 25 new fish species of the Mediterranean Sea, two fish species have since become established in some areas of the Black Sea (Boltachev and Karpova, 2014). These are the fish salema, *Sarpa salpa* (Linnaeus, 1758), and the gilt-head bream, *Sparus aurata* (Linnaeus, 1758).

The salema, *S. salpa*, and the gilthead bream, *S. aurata*, were caught for the first time in the Black Sea, the Batumi Bay, in the 1940s (Svetovidov, 1964). During this period, these two species of Mediterranean immigrants were noted in the southern Black Sea areas only. At the end of the 20th century, the salema and the gilthead bream began to be found in the northern areas including near the coast of the Crimea and in the northwestern part of the Black Sea (Svetovidov, 1964; Vasil'eva, 2007; Boltachev et al., 2010a,b; Tkachenko, 2012; Boltachev and Karpova, 2014).

The number of alien and Mediterranean fish species that were caught in only one of the six areas of the sea is 42% (Table 4). Most of these fish species have been registered near the Crimean coast during the last 10–15 years.

More than half (48% or 14 species) of new fish species are recorded in two or three usually neighboring ecological areas, that is, they were able to spread beyond one ecological area after the invasion. The widening of the range for most of them was not accompanied by an increase in their occurrence in habitats (Boltachev and Karpova, 2014).

A high degree of differentiation between the ecological areas (Varna and Burgas Bays, Danube, North-Western, Crimean, Caucasian, and Anatolian) in the species compositions of alien and new Mediterranean fishes was obtained from the results of cluster analysis. All six regions were combined into one cluster at the final stages of the calculation at the linkage distances of 0.64–0.77 (*D*) that is, demonstrated by their dendrogram (Fig. 2D).

TABLE 4 Alien and New Mediterranean Species of the Fishes That Were Recorded in One From Six Areas of the Black Sea Only

Areas	Part From All Species in the Area (%)	Species	Year of the Find	Author
Anatolian	8	*Serranus hepatus* (Linnaeus, 1758)	2012	Dalgiç et al. (2013)
Varna and Burgas Bays	9	*Pomatoschistus marmoratus* Risso, 1810	2010	Apostolou et al. (2011)
Danube shelf	38	*Centracanthus cirrus* Rafinesque, 1810	2004	Abaza et al. (2006)
		Prionace glauca (Linnaeus, 1758)	1940s	Cărăuşu (1952)
		Sphyrna zygaena (Linnaeus, 1758)	1950s	Zaitsev (2011)
North-Western	–			
Crimean	37	*Chromogobius zebratus* Kolombatovic, 1891	2013	Boltachev and Karpova (2013a,b)
		Epinephelus caninus (Valenciennes, 1843)	2012	Kovtun (2012)
		Gammogobius steinitzi Bath, 1971	2003	Boltachev and Astakhov (2004)
		Heniochus acuminatus (Linnaeus, 1758)	1999	Boltachev et al. (1999)
		Micromesistius poutassou (Risso, 1827)	2009	Boltachev et al. (2010a)
		Millerigobius macrocephalus (Kolombatovic, 1891)	2006	Boltachev et al. (2010b)
		Tridentiger trigonocephalus Gill, 1859	2006	Boltachev et al. (2007)
Caucasian	–			

3 GRADIENTS OF TEMPERATURE AND SALINITY AS ECOLOGICAL BARRIERS

In the Black Sea, the water temperature of the surface layer varies in a wide range: from about 0°C in winter to 28–29°C in summer (Ivanov and Belokopytov, 2011). The sea surface temperature field has a permanent gradient from the north-western to the southeast of the Black Sea. Spatial temperature gradients are most pronounced in winter, and in the spring-summer period they are significantly smoothed. In February, the surface water temperature in the northwest and the southeast is typically 0°C and 9°C, respectively (Blatov et al., 1984). In May, this temperature difference is about 5.5°C (11°C in the northwest and 16.5°C in the southeast), and in August it is about 4°C (21°C in the northwest and 25°C in the southeast). In November, the structure of the surface temperature field is similar to that in May (10.5–15.5°C).

The water salinity in the coastal areas of the Black Sea varies widely between 0 and 37 psu. The minimum salinity is observed near the mouths of rivers during periods of floods. This lower salinity value is typical for the northwestern and southeastern parts of the Black Sea, and also to some parts of the Anatolian coast. High salinity (up to 34–37 psu) is found in the near-bottom layer of the Bosphorus Strait and in the adjoining parts of the Black Sea where the Mediterranean waters penetrate. In February, the salinity of surface waters is the highest in all areas of the sea. Waters with salinity of less than 17 psu occupy less than 5% of the Black Sea area. In May, the surface waters under the influence of river spring flood in the southeastern part of the sea are reduced to 15.5 psu, in the northwest part—to 10 psu and even less (Blatov et al., 1984). A gradual expansion of the low salinity zone is observed in the Black Sea in August and is due to diffusion of water from the melting of mountain glaciers throughout summer. In November the surface salinity of the Black Sea approximates that in winter. The salinity field of the surface waters of the Black Sea differs from the temperature field by a having a more stable distribution during the year in general.

Thus, the spatial heterogeneity of the salinity and temperature fields in the surface layer determines the differences in habitats for hydrobionts in areas of the Black Sea. Thus, six areas have differences in thermohaline characteristics (T, S) of coastal waters. Moreover, each region had distinctive values of T-S characteristics during all seasons (Fig. 3). However two areas of the Black Sea, Varna-Burgas Bays and the Caucasus coast, are an exception. Close values of thermohaline characteristics of water are observed between these two areas, although they are a considerable distance from each other (about 800 km). Also, these two areas had the greatest similarity degree among six areas in the species composition of alien and new Mediterranean fish (Fig. 2D). Fishes, unlike other hydrobionts, are capable of active swimming in search of an optimal habitat. This may explain the relative similarity in the compositions of new fish species for these two areas that have close values for temperature and salinity throughout the year.

However, the differences in thermohaline characteristics do not satisfactorily explain the differences between the six regions in the composition of alien species for phytoplankton, zooplankton, zoobenthos, and fishes. Most likely, the features of the compositions of alien species in six regions are determined not only by their water temperature and salinity as but also by the gradients of temperature and salinity. Temporal temperature and salinity gradients forming hydrological fronts are barriers that restrict the exchange of alien species between sea areas. Therefore, practically half of nonnative species

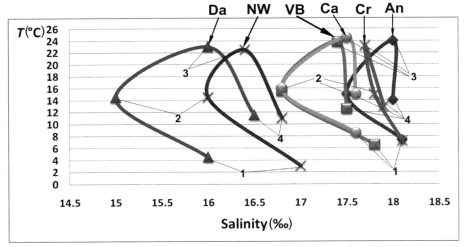

FIG. 3 Seasonal variability of thermohaline (T, S) characteristics in the regions: "VB"—Varna and Burgas Bays, "Da"—Danube, "NW"—North-Western, "Cr"—Crimean, "Ca"—Caucasian, and "An"—Anatolian; 1—February, 2—May, 3—August, 4—November. *(Based on data from Ivanov, V.A., Belokopytov, V.N., 2011. Oceanography of the Black Sea. Marine Hydrophysical Institute, EKOSI-Gidrofzika, Sevastopol, pp. 1–209 (in Russian).)*

have been registered only in one of the six ecological areas of the sea (Tables 1–4). Almost all these alien species have been recorded during the last 10–15 years and they did not expand their areas to neighboring areas during this period. As an example, the absence of alien mollusca on the shelf of the Crimea for many years after their settlement in the Black Sea is an indication of the relative "isolation" of regional benthic communities (Shalovenkov, 2017). At the same time, the water masses of the coastal shelf of the Crimea have regional peculiarities of hydrophysical fields and are characterized by relatively stable hydrological fronts on its borders (Zats et al., 1966; Blatov et al., 1984; Blatov and Ivanov, 1992; Artamonov et al., 2012).

Two main thermal frontal zones occur on the north-western shelf and in the eastern part of the sea (Fig. 4A). These frontal zones reach the highest intensity at the beginning of winter. Another thermal frontal, located along the Anatolian

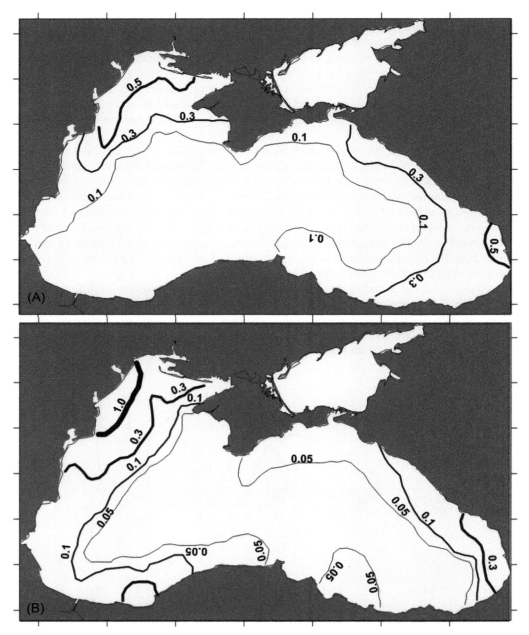

FIG. 4 The spatial distribution of the mean monthly values of the horizontal gradients of surface water (A) temperature (°C per 10 km), and (B) salinity (psu per 10 km). *(Modified from Ivanov, V.A., Belokopytov, V.N., 2011. Oceanography of the Black Sea. Marine Hydrophysical Institute, EKOSI-Gidrofzika, Sevastopol, pp. 1–209 (in Russian).)*

and Caucasian coast, near the Bosphorus Strait, is created by the strong upwelling in summer season (Oğuz et al., 1992; Korotaev et al., 2003; Oğuz and Öztürk, 2011). The frontal zones of salinity are clearly evidenced in the surface layer along the western and eastern coasts of the Black Sea (Fig. 4B). They reach their greatest intensity in the spring as a response to the greatest river inflow (Blatov et al., 1984; Blatov and Ivanov, 1992; Artamonov et al., 2012).

The ecological barriers formed by the frontal zones create obstacles in the exchange of alien species between the regions of the Black Sea. However, the most common way in which invasive species are introduced in the Black Sea is via ships. Alien species usually travel by ships either as fouling of the hulls or in ballast water (Gomoiu, 2001; Alexandrov and Berlinsky, 2005; Aleksandrov et al., 2007, Aleksandrov, 2015). Therefore, the location of shipping companies' routes helps determine the features of the invasive species composition in the six regions of the Black Sea.

4 LARGE-SCALE CURRENTS AND ALIEN SPECIES

The cyclonic system of surface water currents affects the heterogeneity in the mesoscale distribution of biological indicators in the Black Sea (Oğuz et al., 2002). The large-scale structure of the sea currents (Fig. 5) is composed of two large-scale cyclonic basin-wide gyres in the eastern and western parts of the Black Sea, the Rim Current circulation (the Main current of the Black Sea) in the zone of the continental slope, and anticyclonic eddies in the coastal zone (Blatov et al., 1984; Blatov and Ivanov, 1992; Oğuz et al., 1992; Korotaev et al., 2003). The cyclonic Rim Current has a seasonal cycle, with its characteristic features repeating yearly, although with some interannual variability (Korotaev et al., 2003). This current forms a frontal zone between the coastal waters and the open sea (Ovchinnikov et al., 1993; Artamonov et al., 2012). This frontal zone is clearly detected through its *T-S* characteristics; during the spring river floods, the frontal zone is evidenced by the horizontal salinity gradients, and in winter by the horizontal temperature gradients (Ivanov and Belokopytov, 2011). Two large-scale cyclonic basin-wide gyres, in turn, divide the sea into two regions (eastern and western) enabling the transport of water between coastal areas and the open sea.

After passaging the Bosphorus Strait, the ships carry out the replacement of ballast water in the Black Sea, usually at a distance of 100 km from the strait (Aleksandrov, 2015). The ballast replacement process continues along the vessel's route

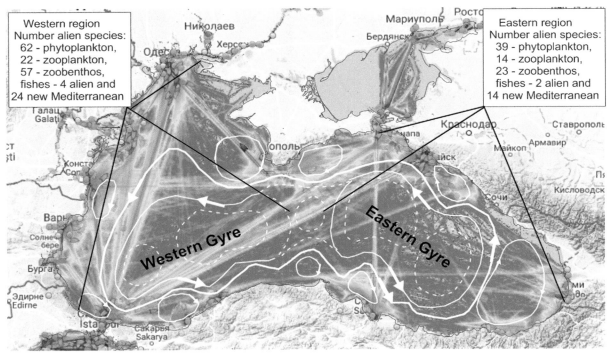

FIG. 5 The number of alien species registered in the western and eastern regions of the Black Sea, the maritime annual traffic density for 2016 (*color:* increase from *green* to *yellow* to *red*, and the large-scale structure of the sea currents—*white solid* and *dotted lines*). (*Modified from www.marinetraffic. com and Korotaev, G., Oguz, T., Nikiforov, A., Koblinsky, C., 2003. Seasonal, interannual, and mesoscale variability of the Black Sea upper layer circulation derived from altimeter data. J. Geophys. Res. 108 (C4), 3122.)*

to the port of call where the final stage of ballast water replacement is made. The disposal of ships' ballast water near the Bosphorus Strait promotes the spread of alien species that are dispersed by the currents and ultimately to the coastal zone where favorable abiotic conditions occur.

Indirect indicator of the level of anthropogenic load in the Black Sea areas by navigation is the high maritime annual traffic density both in the ports area and in the open sea (Fig. 5 for 2016). The greatest density of sea traffic is observed in the western region of the Black Sea, that is, in the zone of influence of the western cyclonic circulation and the western section of the Black Sea Main current. The largest number of alien marine species is recorded for this region of the Black Sea as well.

Nonnative species were recorded in two regions of the Black Sea in the following quantities: in the western region—62 for phytoplankton, 22 for zooplankton, and 57 for zoobenthos; in the eastern—30 for phytoplankton, 14 for zooplankton, and 23 for zoobenthos (Fig. 5). For fishes, the number of nonnative species was as follows: in the western—4 alien and 24 new Mediterranean; in the eastern—2 alien and 14 new Mediterranean. Thus, the number of alien and new Mediterranean (fishes) species in the western region exceeded by a factor of about two, their number in the eastern region. These indicators clearly correspond to differences in the levels of anthropogenic load from the density of sea traffic in the western and eastern regions of the Black Sea.

5 TRENDS OF INVASION OF ALIEN SPECIES

A small number of the alien species was recorded at the beginning of the 20th century (Fig. 6A–D). Temporal trends in invasive species were characterized by an increase in the number of alien species for all groups of the hydrobionts in the Black Sea in the second half of the 20th century. The deteriorating environmental conditions in the 1980s when large-scale eutrophication and hypoxia occurred in the Black Sea, prevented the introduction of nonnative species, except for some

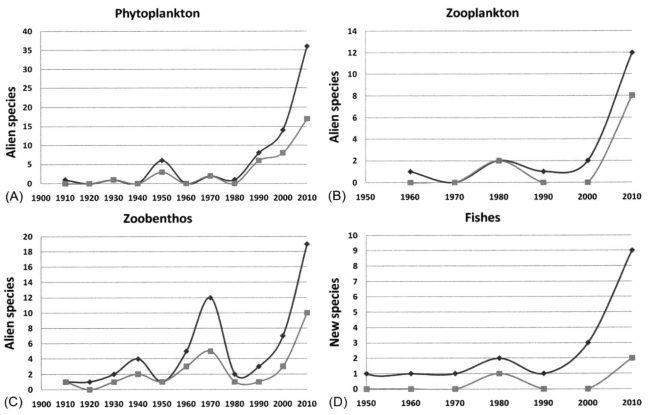

FIG. 6 Interannual variability of introduction of the alien species in the Black Sea for the last 100 years: (A) phytoplankton, (B) zooplankton, (C) zoobenthos, (D) fishes (alien and Mediterranean); *black line* (A–D) the total number of alien species, *gray line* (A–C) the number of alien species from the Indo-Pacific areas, *gray line* (D) alien species.

phytoplankton species, during that period. In 1990–2010, the number of alien species increased by 58 for phytoplankton, 15 for zooplankton, 29 for zoobenthos, and by 13 new Mediterranean fishes that made almost 50% of the quantity of the new species installed over the last century in the Black Sea.

Since the end of the 20th century, an increase of the number of alien species has been observed simultaneously with the increase in the ship traffic to the Black Sea and, as a consequence, with the increase in the volume of ballast water discharged into the sea. Thus, the number of ships passing through the Bosphorus Strait for the year has increased five-fold over 50 years: 1950—about 10,000 ships, and in 2000—about 50,000 ships (Öztürk, 2002). After 2000, the number of ships passing through the Bosphorus varied between 42,000 and 55,000 per year (Birpinar et al., 2005, 2009). At the same time, the number of alien species increased by two to six times (for different groups of hydrobionts) in the Black Sea (Fig. 6).

This latter increase in the number of alien species may be due to climate changes. Indeed, the increase of the thermal background and the weakening of the dynamics of the near-water atmosphere were accompanied by the warming of sea coast water and in the open sea from 0.05°C to 2.00°C for every 10 years in the last two decades (Ilyin, 2009, 2010, 2012; Belokopytov, 2013, 2014). The climatic changes in the Black Sea region are adequately described by the fluctuations of the Atlantic Multidecadal Oscillation (AMO) index at interdecadal to century time scales (Oğuz et al., 2006; Ilyin, 2010; Polonsky et al., 2013). The changes in the number of alien species of the some groups of hydrobionts in the Black Sea are significantly correlated with the AMO Index over the past 50 years (Fig. 7A and B); indeed the correlation coefficient between the number of alien species of phytoplankton, zooplankton, zoobenthos, fish, and the AMO Index varied from 0.48 to 0.68 at decadal time scale. It follows that the changing climate probably facilitates the process of invasion of the Black Sea

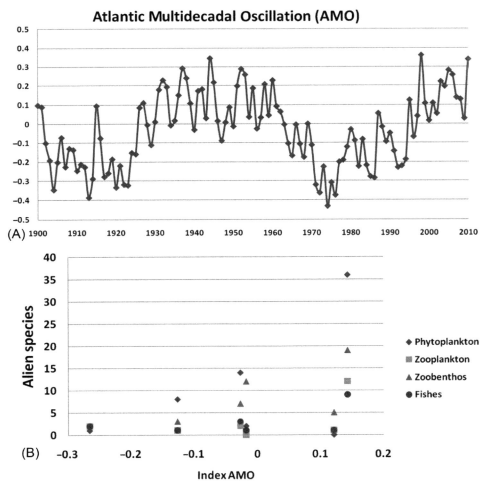

FIG. 7   (A) Changes in the Atlantic Multidecadal Oscillation (AMO) index over the last 100 years and (B) the ratio between the number of alien species (phytoplankton, zooplankton, zoobenthos, and fishes) and the AMO index for the last 50 years.

by alien and Mediterranean species. Further the increase in the mean annual temperature explains also the increase in the proportion of alien species of zooplankton and zoobenthos from warmer seas, that is, from the Indo-Pacific region (Fig. 6B and C). Thus, increasing sea surface temperatures have probably affected the habitat of hydrobionts in the Black Sea and facilitated the spread of alien species in the coastal shelf in the 10 years.

6 INVASIVE CORRIDORS OF THE BLACK SEA BASIN

6.1 The Atlantic and Indo-Pacific Corridors

Two invasive corridors exist for alien species from different geographic regions for penetration into the Black Sea (Fig. 8). The Atlantic corridor passes through the Strait of Gibraltar and the Mediterranean Sea. The Indo-Pacific Corridor is a more complex transcontinental route for long-distance extensions of the alien species areal. Two routes are from the Indo-Pacific region: one through the Suez Canal and the Mediterranean, the second—the Panama Canal, the Strait of Gibraltar, and the Mediterranean Sea.

As noted above, the main vector of invasions by alien species in the Black Sea is ballast water and fouling of ships' hulls (Gomoiu, 2001; Alexandrov and Berlinsky, 2005; Aleksandrov et al., 2007; Aleksandrov, 2015). Replacement of ballast waters is carried out at the entrance of ships to the Mediterranean Sea and the ballast exchange process continues further on route. The oceanographic transport of alien species from the Mediterranean to the Black Sea is inhibited by ecological barriers formed by strong physical gradients along the Turkish Straits System (Oğuz and Öztürk, 2011). For this reason, the number of alien species in the Mediterranean Sea is almost four times greater than in the Black Sea (Zenetos et al., 2017). Therefore, the Mediterranean Sea can be considered a kind of "storage" for alien species that can then be transported with ballast water to the Black Sea.

The ratio of nonnative Black Sea species by geographic origin is almost equal for the Atlantic and Indo-Pacific regions. Thus, alien species from the Indo-Pacific region account for 53% in phytoplankton, 56% in zooplankton, and 49%

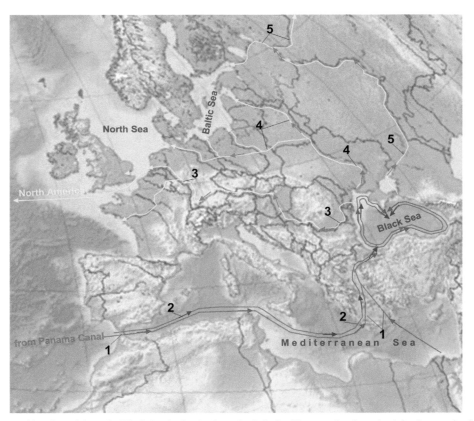

FIG. 8 The invasion corridors in and from the Black Sea basin: 1—from the Indo-Pacific areas, 2—from the Atlantic area, 3—South Ponto-Caspian, 4—Central Ponto-Caspian, 5—North Ponto-Caspian.

in zoobenthos of the Black Sea. As noted above, significant increase in the proportion of alien species in zooplankton and zoobenthos from the Indo-Pacific region was observed recently together with an increase in the mean annual surface temperature of the Black Sea (Fig. 6B and C).

6.2 Ponto-Caspian Corridor

The endemic species from estuaries of the Black Sea were the source of the alien species in the seas and rivers of Europe (Gomoiu et al., 2002; Semenchenko et al., 2009, 2013, 2014; Slynko et al., 2010). The Danube, Dnieper, and Volga Rivers are three main corridors of biological invasion for the Ponto-Caspian species (Ketelaars et al., 1999; Bij de Vaate et al., 2002; Devin et al., 2005; Karataev et al., 2007; Leuven et al., 2009; Slynko et al., 2010). The Southern corridor, from the Black Sea basin to the North Sea region, is the Danube-Main-Rhine waterway (Fig. 8). The Central corridor, from the Black Sea basin to the Baltic Sea region, is the Dnieper and Bug-Pripyat Canal via a network of waterways. The Northern corridor, from the Black, Azov, and Caspian Seas to the Baltic and White Seas regions, is the Volga River, the Volga-Don Canal, the Volga-Baltic Canal, and White Sea-Baltic Sea Canal waterway.

The large rivers of Europe underwent regulation of the flow in the last century and were turned into a chain of water-storage basins, mainly lake-channel type (Slynko et al., 2010). The network of channels has connected the river basins of the Black Sea with the regions of Northern Europe (Bij de Vaate et al., 2002). Also, a large number of deliberate introductions of new species of fish and invertebrates were conducted in rivers, lakes, and water-storage basins of Europe during this period (Karpova et al., 1996; Gherardi et al., 2009; Slynko et al., 2010). The habitat of hydrobionts has strongly changed: the speeds of a current in water basins have considerably decreased, the heat capacity of water masses has increased, and the mineralization and heterogeneity of environment has considerably increased also (Denisova, 1979; Semenchenko et al., 2009, 2013, 2014; Slynko et al., 2010). Along with these changes, the intensity of navigation and the expansion of the operation of river-sea ships cruising in both rivers and seas undoubtedly exacerbated the invasion of the Ponto-Caspian species in Northern Europe.

Having overcome the North Sea and the Atlantic Ocean, the Ponto-Caspian fauna penetrated into the water bodies of Great Britain and the North America Great Lakes. Currently, 23 Ponto-Caspian species have already been recorded in the freshwater ecosystems of Great Britain and 8 Ponto-Caspian species have successfully established in the Laurentian Great Lakes (Ricciardi and MacIsaac, 2000; Vanderploeg et al., 2002; Grigorovich et al., 2003; Gallardo and Aldridge, 2013, 2015). Global climate change may also have favored this biological invasion in Europe and America. Thus, the spatial scale of the impact of alien species on biodiversity has a complex structure of interactions between the basins of the seas of Europe.

7 INVASIONS OF ALIEN SPECIES IN THE BLACK SEA—THE FUTURE

Alien species are introduced in the Black Sea mainly by ships through ballast water and fouling of ship hulls. Therefore, the number of new nonnative species strongly depends on the expansion of seaborne trade and the volume of shipment. Measures are clearly necessary to decrease the anthropogenic factor enhancing the invasion of the Black Sea by nonnative species. The International Convention on the Control and Management of Ships' Ballast Water and Sediments, BWM Convention—2004, entered into force in September 2017; 59 states have accepted the Convention. In accordance with the Convention, all commercial vessels are obliged to control the discharge and exchange of ballast water. Therefore, ships will be equipped with systems for cleaning and disinfection of ballast water—the Ballast Water Management System.

However, in addition to the anthropogenic factor, climatic fluctuations will continue to affect the level of invasions of nonnative species. The Mediterranean Sea is a kind of "storehouse" for alien species, in which their number exceeds 800 species (Zenetos et al., 2017). With the rise in temperature, the invasion of the Black Sea by new Mediterranean species of fauna and flora ("Mediterranization"), and alien species that have already colonized coastal and estuarine regions of the Mediterranean Sea will continue. It is also likely that the proportion of alien species from the Indo-Pacific region will further increase with the warming of the waters.

The scientific community actively promotes the integration of scientific research in the development of international legislative directives for the protection of marine ecosystems from bio-invasions (Ojaveer et al., 2014a,b). This brings some hope that humanity can solve this important environmental problem of minimizing the impact of alien species on biodiversity and promoting the stability of local marine ecosystems in the Black Sea. In turn this will help limit the invasion of alien species in Northern Europe through the Black Sea.

REFERENCES

Abaza, V., Boicenco, L., Bologa, A.S., Dumitrache, C., Moldoveanu, M., Sburlea, A., et al., 2006. Biodiversity structure from the Romanian marine area. Cercetari Marine—Recherches Marines, vol. 36. INCDM, Constanta, pp. 15–29.

Aleksandrov, B.A., 2015. Regularities of new species invasions into the Black Sea and some approaches to their study. Scientific notes of the Ternopol national pedagogical university. Ser. Biol. 64 (3–4), 29–32 (in Russian).

Aleksandrov, B., Boltachev, A., Kharchenko, T., Liashenko, A., Son, M., Tsarenko, P., et al., 2007. Trends of aquatic alien species invasion in Ukraine. Aquat. Invasions Eur. J. Appl. Res. Biol. Invasion Aquat. Ecosyst. 2 (3), 215–242.

Alexandrov, B., Bashtannyy, R., Clarke, C., Hayes, T., Hilliard, R., Polglaze, J., et al., 2004. Ballast Water Risk Assessment, Port of Odessa, Ukraine, October 2003: Final Report, Published in March 2004 by the Programme Coordination Unit Global Ballast Water, 2004. Globallast monograph series. vol. 10. 1–135.

Alexandrov, B., Moncheva, S., Stefanova, K., Raykov, V., Dencheva, K., Gvarishvili, T., et al. (Eds.), 2017. Black Sea Non-Indigeneous Species. Black Sea Commission Publication, pp. 1–40. https://www.cbd.int/doc/meetings/mar/ebsaws-2017-01/other/ebsaws-2017-01-bsc-submission-03-en.pdf.

Alexandrov, B., Berlinsky, N., 2005. Introduced species in the Black Sea: the role of ballast water at Odessa Port, Ukraine. NEAR Curriculum in Natural Environmental Science, vol. 50. Terre et Environnement, pp. 141–154.

Alymov, S., Tikhonova, E., 2012. The Kerch strait sediment pollution indices and malacofauna characteristics. Scientific notes of the Ternopol national pedagogical university. Ser. Biol. 2 (51), 13–17 (in Russian).

Apostolou, A., Ivanova, P., Velkov, B., Vassilev, M., Dobrev, D., Dobrovolov, I., 2011. Pomatoschistus marmoratus (RISSO 1810), is it really a "New" species for Bulgarian Ichthyofauna? Acta Zool. Bulg. 63 (3), 289–294.

AquaNIS, n.d. AquaNIS: the information system on aquatic non-indigenous and cryptogenic species. http://www.corpi.ku.lt/databases/index.php/aquanis/.

Arkhangelsky, A., Batalina, M., 1929. In: To the knowledge of the history and development of the Black Sea. Proceedings of the USSR Academy of Sciences, VII Series. vol. 8. Department of Physical and Mathematical Sciences, pp. 691–706 (in Russian).

Artamonov, J.V., Belokopitov, V.N., Skripaleva, E.A., 2012. The features of variability of hydrological and biooptical characteristics on the Black Sea surface according to satellite and contact measurements. In: Eremeev, V., Konovalov, S. (Eds.), Stability and Evolution of Oceanologic Characteristics of Black Sea Ecosystem. Ecosi-Gidrofizika, Sevastopol, pp. 88–115 (in Russian).

Băcescu, M., Müller, G., Gomoiu, M.-T., 1971. Cercetări de ecologie bentală în Black Sea, Analiza cantitativă, calitativa şi comparată a faunei bentale pontice. Ecologie marină. 4. 1–357.

Belmonte, G., Mazzocchi, M.G., Prusova, I.Y., Shadrin, N.V., 1994. Acartia tonsa: a species new for the Black Sea fauna. Hydrobiologia 292 (1), 9–15.

Belokopytov, V.N., 2013. About the climatic variability of thermohaline structure of the Black Sea. Ecologicyal safety of coastal and shelf zones and complex resources use of the shelf. Proc. MHI Sevastopol 27, 226–230 (in Russian).

Belokopytov, V.N., 2014. Climatic variability of the Black Sea density structure. Ukr. Hydrometeorol. J. 14, 225–227 (in Russian).

Bij de Vaate, A., Jazdzewski, K., Ketelaars, H.A.M., Gollasch, S., Van der Velde, G., 2002. Geographical patterns in range extension of Ponto-Caspian macroinvertebrate species in Europe. Can. J. Fish. Aquat. Sci. 59, 1159–1174.

Birpinar, M.E., Talu, G.F., Su, G., Gulbey, M., 2005. The Effect of Dense Maritime Traffic on the Bosphorus Strait and Marmara Sea Pollution. Ministry of Environment and Forestry, The Regional Directorate of Istanbul, Istanbul. http://www.teknolojikkazalar.org/get_file?id=480e1a08740b4.

Birpınar, M.E., Talu, G.F., Gönençgil, B., 2009. Environmental effects of maritime traffic on the İstanbul strait. Environ. Monit. Assess. 152, 13. https://doi.org/10.1007/s10661-008-0292-8.

Blatov, A., Ivanov, V., 1992. Hydrology and Hydrodynamics of the Black Sea Shelf Zone (for Example, the Southern Coast of the Crimea). Naukova Dumka Publishers, Kiev, pp. 1–242 (in Russian).

Blatov, A., Bulgakov, N., Ivanov, V., Kosarev, A., Tuzhilkin, V., 1984. Variability of Hydrophysical Fields of the Black Sea. Gidrometeoizdat Publishers, Leningrad, pp. 1–231 (in Russian).

Bodeanu, N., 2002. Algal blooms in Romanian Black Sea waters in the last two decades of the 20th century. Cercetari Marine, vol. 34. INCDM, pp. 7–22.

Bogdanova, A.K., 1959. In: The water exchange through the Bosporus and its role in mixing the waters of the Black Sea. Proceeding Sevastopol Biological Station. vol. 12, pp. 401–420 (in Russian).

Bogdanova, A.K., 1964. In: Role of the negative-positive circulation in water exchange through the Bosporus. Proceeding Sevastopol Biological Station. vol. 15, pp. 534–549 (in Russian).

Boltachev, A.R., Astakhov, D.A., 2004. An unusual finding of pennant coralfish Heniochus acuminatus (Chaetodontidae) in Balaklava Bay Sevastopol Southwestern Crimea. J. Ichthyol. 44 (6), 853–854.

Boltachev, A.R., Karpova, E.P., 2010. Naturalization of the Pacific Chameleon Goby Tridentiger trigonocephalus (Perciformes, Gobiidae) in the Black Sea (Crimea, Sevastopol Gulf). J. Ichthyol. 50 (2), 231–239.

Boltachev, A.R., Karpova, E.P., 2013a. First record of dogtooth grouper Epinephelus caninus (Valenciennes, 1834), Perciformes, Serranidae, in the Black Sea. BioInvasions Rec. 2 (3), 257–261.

Boltachev, A.R., Karpova, E.P., 2013b. Experts in the world register of marine species "WoRMS". http://www.marinespecies.org/aphia.php?p=sourcedetails&id=149025 (Accessed 1 December 2013).

Boltachev, A.R., Karpova, E.P., 2014. Faunistic revision of alien fish species in the Black Sea. Russ. J. Biol. Invasions 5 (4), 225–241.

Boltachev, A.R., Yurakhno, V.M., 2002. New evidence of ongoing Mediterranization of the Black Sea's ichthyofauna. J. Ichthyol. 42 (9), 713–719.

Boltachev, A.R., Gaevskaya, A.V., Zuev, G.V., Yurakhno, V.M., 1999. The blue whiting (Micromesistius poutasssou Riss, 1826) (Piscel: Gadidae) is the new species for the Black Sea ichthiofauna. Mar. Ecol. Sevastopol 48, 79–82 (in Russian).

Boltachev, A.R., Vasil'eva, E.D., Danilyuk, O.N., 2007. The first finding of the striped tripletooth goby *Tridentiger trigonocephalus* (Perciformes, Gobiidae) in the Black Sea (the estuary of the Black River, Sevastopol Bay). J. Ichthiol. 47 (9), 802–805.

Boltachev, A.R., Karpova, E.P., Machkevskiy, V.K., 2010a. Naturalization of Miller's bull had *Millerigobius macrocephalus* (Perciformes, Gobiidae) in Sevastopol Bay. Mar. Ecol. J. Sevastopol 9 (1), 32 (in Russian).

Boltachev, A.R., Karpova, E.P., Klimova, T.N., Chesalin, M.V., Chesalina, T.L., 2010b. Fishes (pisces). In: Matishov, G.G., Boltachev, A.R. (Eds.), The Introducers in the Biodiversity and Productivity of the Sea of Azov and the Black Sea. Southern Scientific Center of the Russian Academy of Sciences Publications, Rostov-on-Don, pp. 76–113 (in Russian).

Boltacheva, N.A., Lisizkaja, E.V., Lebedovskja, M.V., 2011. New for the Black Sea species of the polychaetes *Hydroides dianthus* (Verrill, 1873) (Polychaeta: Serpulidae) from the coastal waters of the Crimea. Mar. Ecol. J, Sevastopol 10 (2), 34–38 (in Russian).

Bryantseva, Y., 2000. Variability of the Black Sea Phytoplankton Structural Characteristics (Ph.D. thesis). Institute of Biology of the Southern Seas, NASU, Sevastopol, pp. 1–19 (in Russian).

Caraus, I., 2002. The Algae of Romania. vol. 7. Studii si Cercetari, Universitatea Bacau, Biologie1–694.

Cărăuşu, S., 1952. Tratat de Ichtiologie. Bucuresti: Edit. Acad. RPR, 1–802.

Caspers, H., 1968. La macrofaune benthique du Bosphore et les probiemes de infiltration des elements mediterraneans dans la mer Noire. Rapp. Comm. Int. Mer. Medit. 19, 107–115.

Chikina, M., 2009. The macrozoobenthos of soft bottom in the North Caucasian coast of the Black Sea: the spatial structure and long-term dynamics (Ph.D. thesis). Institute of Oceanology of the Russian Academy of Sciences, Moscow, pp. 1–25 (in Russian).

Chikina, M., Kucheruk, N., 2005. Long-term changes in the structure of benthic communities in the northeasten part of the Black Sea. Influence of alien species. Oceanology 45 (1), 176–182 (in Russian).

Çinar, M.E., Bilecenoğlu, M., Öztürk, B., Katagan, T., Aysel, V., 2005. Alien species on the coasts of Turkey. Mediterr. Mar. Sci. 6 (2), 119–146.

Dalgiç, G., Gümüş, A., Zengin, M., 2013. First record of brown comber *Serranus hepatus* (Linnaeus, 1758) for the Black Sea. Turk J Zool. 37, 523–524.

Denisova, A.I., 1979. The Formation of the Hydrochemical Regime of the Dnieper Reservoirs and Methods of Its Prediction. Naukova Dumka Press, Kiev, pp. 1–292.

Devin, S., Bollache, L., Noël, P.-Y., Beisel, J.-N., 2005. Patterns of biological invasions in French freshwater systems by non-indigenous macroinvertebrates. Hydrobiologia 551, 137–146.

Gallardo, B., Aldridge, D.C., 2013. Review of the ecological impact and invasion potential of Ponto Caspian invaders in Great Britain. Cambridge Environmental Consulting. 1–130.

Gallardo, B., Aldridge, D.C., 2015. Is Great Britain heading for a Ponto–Caspian invasional meltdown? J. App. Ecol. 52 (1), 41–49.

Gavrilova, N.A., 2001. Eutintinnus (Ciliophora: Oligotrichida: Tintinnidae) a new genus of Tintinnids for the Black Sea fauna. Mar. Ecol. Sevastopol 58, 29–31 (in Russian).

Gavrilova, N.A., 2005. New for the Black Sea Tintinnids species. Mar. Ecol. Sevastopol 69, 5–11 (in Russian).

Gavrilova, N.A., 2010. Microzooplankton: Tintinnidae. The Black Sea. In: Matishov, G., Boltachyov, A. (Eds.), The Introducers in the Biodiversity and Productivity of the Sea of Azov and the Black Sea. SSC Ras Publishing, Rostov-on-Don, pp. 63–69 (Chapter 2).

Gavrilova, N.A., 2017. About first recordsof tintinnid *Tintinnopsis mortensenii* Schmidt, 1902 (Spirotrichea, Choreotrichia, Tintinnida, Codonellidae) in the Sevastopol Bay plankton (Black Sea). Mar. Biol. J. 2 (1), 86–87 (in Russian).

Georgieva, L.V., 1993. Phytoplankton. Species composition and dynamics of phytoplankton. In: Plankton of the Black Sea. Naukova Dumka, Kiev, pp. 31–55 (in Russian).

Georgieva, L.V., Senichkina, L.G., 1996. Phytoplankton of the Black Sea: current situation and the research prospects. Mar. Ecol. Sevastopol 45, 6–13 (in Russian).

Gherardi, F., Gollasch, S., Minchin, D., Olenin, S., Panov, V.E., 2009. Alien invertebrates and fish in European inland waters. In: Handbook of Alien Species in Europe. Springer, Dordrecht, pp. 81–92 (Chapter 6).

Gomoiu, M.-T., 2001. Impacts of naval transport development on marine ecosystems and invasive species problems. J. Environ. Prot. Ecol. 2 (2), 475–481.

Gomoiu, M.T., Skolka, M., 1996. Changements récents dans la biodiversité de la Mer Noire dus aux immigrants. Geo-Eco-Marina 1, 49–66.

Gomoiu, M.T., Skolka, M., 2005. Invasive Species in the Black Sea. Ovidius University Press Publishers, Constanta, pp. 1–150.

Gomoiu, M.T., Alexandrov, B., Shadrin, N., Zaitsev, Y., 2002. The Black Sea—a recipient, donor and transit area for alien species. In: Leppäkoski, E., Gollasch, S., Olenin, S. (Eds.), Invasive Aquatic Species of Europe, Distribution, Impacts and Management. Springer, Dordrecht, pp. 341–350.

Greze, V.N., 1979. Zooplankton. In: Greze, V.N. (Ed.), The Foundations of Biological Productivity of the Black Sea. Naukova Dumka, Kiev, pp. 143–168 (in Russian).

Grigorovich, I.A., Colautti, R.I., Mills, E.L., Holeck, K., Ballert, A.G., MacIsaac, H.J., 2003. Ballast-mediated animal introductions in the Laurentian Great Lakes: retrospective and prospective analyses. Can. J. Fish. Aquat. Sci. 60, 740–756.

Gubanova, A.D., 1997. In: To a question on occurrence of *Acartia tonsa* Dana in the Black Sea. Materials of the Second Congress of the Hydrobiological Society of Ukraine (27–31 October, Kiev, Ukraine), Kiev, pp. 24–25 (in Russian).

Gubanova, A.D., 2000. Occurrence of *Acartia tonsa* Dana in the Black Sea. Was it introduced from the Mediterranean? Mediterr. Mar. Sci. 1 (1), 105–109.

Gubanova, A.D., 2003. Changes in mesozooplankton community in 2002, in comparison with 1990s. In: Eremeev, V.N., Gaevskaya, A.V. (Eds.), Modern Condition of Biological Diversity in Near–Shore Zone of Crimea (the Black Sea Sector). Ecosi-Gidrophizika, Sevastopol, pp. 84–94 (in Russian).

Gubanova, A.D., Prusova, I.Y., Niermann, U., Shadrin, N.V., Polikarpov, I.G., 2001. Dramatic change in the copepod community in Sevastopol Bay (Black Sea) during two decades (1976–1996). Senckenberg Marit. 31, 17–27.

Gubareva, E.S., Svetlichny, L.S., Romanova, Z.A., Abolmasova, G.I., Aninski, B.E., Finenko, G.A., et al., 2004. Zooplankton community state in Sevastopol Bay after the invasion of ctenophore *Beroe ovata* into the Black Sea. Mar. Ecol. J. Sevastopol 3 (1), 39–46 (in Russian).

Gücü, A.C., Bingel, F., Avsar, D., Uysal, N., 1994. Distribution and occurrence of Red Sea fish at the Turkish Mediterranean coast—northern Cilician basin. Acta Adriat. 34 (1/2), 103–113.

Gvarishvili, T., 1998. Species composition and biodiversity of Georgian Black Sea phytoplankton. In: Kotlyakov, V., Uppenbrink, M., Metreveli, V. (Eds.), Conservation of the Biological Diversity as a Prerequisite for Sustainable Development in the Black Sea Region. NATO ASI Series, Kluwer Academic Publishers, Dordrecht, pp. 95–100.

Gvarishvili, T., Mikashavidze, E., Mgeladze, M., Diasamidze, R., Zhgenti, D., Janelidze, N., et al., 2010. Seasonal dynamics of the phyto- and zooplankton in the Georgian coastal zone of the Black Sea. Ga. J. Sci. 8 (1–2), 47–58.

Ilyin, Y.P., 2009. Observed long-term changes in the Black Sea physical system and their possible environmental impacts. In: Climate Forcing and Its Impacts on the Black Sea Marine Biota. CIESM Monograph, 39, pp. 35–43.

Ilyin, Y.P., 2010. Climatic variability of salinity features on the Bosporus and North-Western shelves revealed from observational data. J. Environ. Prot. Ecol. 11 (3), 993–1000.

Ilyin, Y.P., 2012. The contribution of regional and global factors in interannual variability of meteorological conditions of the coastal zone of the Black Sea. Ecological safety of coastal and shelf zones and complex resources use of the shelf. Proc. MHI Sevastopol 26 (1), 117–122 (in Russian).

Isinbilir, M., Yilmaz, I.N., Pirano, S., 2010. New contributions to the jellyfish fauna in the Marmara Sea. Ital. J. Zool. 77 (2), 179–185.

Ivanov, A.I., 1965. Characteristics of the qualitative composition of the phytoplankton of the Black Sea. In: Vodjanitcky, V.A. (Ed.), Investigations of Plankton of the Black Sea and Sea of Azov. Naukova dumka, Kiev, pp. 17–34 (in Russian).

Ivanov, V.A., Belokopytov, V.N., 2011. Oceanography of the Black Sea. Marine Hydrophysical Institute, EKOSI-Gidrofzika, Sevastopol, pp. 1–209 (in Russian).

Ivanov, D.A., Sinegub, I.A., 2008. In: Transformation of bottom biocenosis of the Kerch Strait after the invasion of the predatory snail *Rapana thomasian* and bivalves Mya arenaria and *Cunearca cornea*. Modern problems of the Azov-Black Sea region, Proceedings of the III-rd International Conference. Kerch 10–11 November 2007. YugNIRO, Kerch, pp. 45–51 (in Russian).

Jakubova, L.I., 1948. In: Features of biology near the Bosporus site of the Black Sea. Proceeding Sevastopol Biological Station. vol. 6, pp. 274–285 (in Russian).

Kamburska, L., 2004. Effects of Beroe cf ovata on gelatinous and other zooplankton along the Bulgarian Black Sea Coast. In: Dumont, H.J., Shiganova, T.A., Niermann, U. (Eds.), Aquatic Invasions in the Black, Caspian, and Mediterranean Seas. Part of the Nato Science Series: IV: Earth and Environmental Sciences Book Series, vol. 35. Kluwer Academic, Dordrecht, pp. 137–154.

Kaneva-Abadjieva, V., Marinov, T., 1966. Distribution of sandy macrozoobenthos community along Bulgarian Black Sea coast. Proc. NIRSO 7, 69–96.

Karanda, O.B., 2014. Seasonal dynamics of the phytoplankton of the coastal zone of the Odessa Bay. In: Collection of Scientific Articles of the Fifth All-Ukrainian Scientific and Practical Conference of Young Scientists and Students. Zhytomyr Ivan Franko State University, Zhitomir, pp. 59–61 (in Russian).

Karataev, A.Y., Mastitsky, S.E., Burlakova, L.E., Olenin, S., 2007. Past, current, and future of the central European corridor for aquatic invasions in Belarus. Biol. Invasions 10, 215–232.

Karhan, S.Ü., Kalkan, E., Yokeş, M.B., 2008. First record of the Atlantic starfish, *Asterias rubens* (Echinodermata: Asteroidea) from the Black Sea. JMBA Mar. Biodivers. Rec. e63. https://doi.org/10.1017/S175526720700663X.

Karpova, E.I., Petr, T., Isaev, A.I., 1996. Reservoir fisheries in the countries of the commonwealth of independent states. FAO Fish. Circ. 915, 1–132.

Ketelaars, H.A.M., Lambreqts-van de Clundert, F.E., Carpentier, C.J., Waqenvoort, A.J., Hooqenboezem, W., 1999. Ecological effects of the mass occurrence of the Ponto-Caspian invader, *Hemimysis anomala* G. O. Sars, 1907 (Crustacea: Mysidacea), in a freshwater storage reservoir in the Netherlands, with notes on its autecology and new records. Hydrobiologia 394, 233–248.

Khvorov, S.A., Boltachov, A.R., Reshetnikov, S.I., Pashkov, A.N., 2006. First record of the green tiger prawn *Penaeus semisulcatus* (Penaeidae, Decapoda) in the Black Sea. Mar. Biol. 72, 65–69 (in Russian).

Kideys, A.E., Niermann, U., 1994. Occurrence of Mnemiopsis along the Turkish coast. J. Mar. Sci. 51, 423–427.

Kisileva, M.I., 1979. Zoobenthos. In: Greze, V.N. (Ed.), The Foundations of Biological Productivity of the Black Sea. Naukova Dumka, Kiev, pp. 208–241 (in Russian).

Konsulov, A., 1989. New invider in the Black Sea—is it dangerous or harmless for the ecology. Mar. World J. 2, 1–8.

Konsulov, A.S., Kamburska, L.T., 1998. In: Ecological determination of the new Ctenophora? *Beroe ovata* invasion in the Black Sea. Proceedings of the Institute of Oceanology, Varna. vol. 2, pp. 195–198.

Konsulova, T., Konsulov, A., Moncheva, S., 1991. Ecological characteristic of Varna Bay (Black Sea) coastal ecosystem under summer "bloom" conditions. C. R. Acad. Bulg. Sci. 44 (8), 115–117.

Korotaev, G., Oguz, T., Nikiforov, A., Koblinsky, C., 2003. Seasonal, interannual, and mesoscale variability of the Black Sea upper layer circulation derived from altimeter data. J. Geophys. Res. 108 (C4), 3122. http://www.ims.metu.edu.tr/cv/oguz/PDFs/2002JC001508.pdf.

Koshelev, A.V., 2003. In: Cordylophora inkermanica (Hydrozoa, Clavidae)—new species of periphyton fauna in limans in the North-West part of the Black Sea. Proceedings of Zoology. vol. 37 (4). Schmalhausen Institute of Zoology NASU, Kiev, pp. 84. (in Russian).

Koval, L.G., 1984. Zoo- and Necro-Zooplankton of the Black Sea. Nauka Dumka, Kiev1–127. (in Russian).

Kovtun, O.A., 2012. The first find of bull had *Gammogobius steinitzi* Bath, 1971 (Actinopterygii, Perciformes, Gobiidae) in the submarine caves of western Crimea (draft report). Mar. Ecol. J. Sevastopol 11 (3), 56. (in Russian).

Leuven, R.S.E.W., Van der Velde, G., Baijens, I., Snijders, J., Van der Zwart, C., Lenders, H.J.R., et al., 2009. The river Rhine: a global highway for dispersal of aquatic invasive species. Biol. Invasions 11, 1989–2008.

Makarov, Y.N., Kostylyov, E.F., 2001. Molluscs in eutrophic areas of the Ukrainian shelf of the Black Sea (observations 1997-1998). In: Scientific Notes of the Zhytomyr Ivan Franko State University. vol. 10, pp. 120–122. (in Russian).

Marinov, T., 1957. Beitrag zur Kenntnis unserer Schwarzmeer Polychaeten fauna. Proc. Mar. Biol. 19, 105–119.

Martynov, A.V., Korshunova, T.A., Grintsov, V.A., 2007. Opisthobranch molluscs of the Northern Black Sea. I. Short history of studies and the first record of a non-indigenous nudibranch species *Trinchesia perca* Nudibranchia: Tergipedidae. RUTHENICA: Russ. Malacol. J. 17 (1), 43–54.

Martynyuk, M.L., Mirzoyan, Z.A., Studenikina, E.I., 2001. In: Structural and functional changes in zooplankton of the North-Eastern part of the Black Sea in connection with the appearance of *Beroe ovate*. Problems of Conservation of Ecosystems and Rational Use of Bioresources of the Azov-Black Sea Basin: Materials of the International Scientific Conference, Rostov-on-Don, pp. 134–136 (in Russian).

Mavrodieva, R., 2012. Mechanisms of interactions between phytoplankton and marine environmental factors (biotic and abiotic) in the Bulgarian Black Sea coastal waters (Sozopol Bay) (Ph.D. thesis). Institute of Biodiveristy and Ecosystem Research (Bulgarian Academy of Science), Sofia, pp. 1–33.

Melinte-Dobrinescu, M., Ion, G., 2013. Emiliania huxleyi fluctuation and associated microalgae in superficial sediments of the Romanian Black Sea Coast. Geo-Eco-Marina 19, 129–135.

Meriç, E., Yokeş, B., Avşar, N., Dinçer, F., 2010. In: Indo–Pasifik kökenli Göçmen Foraminiferler Karadeniz'e Ulaçyorlar mi. 45. Yil Jeoloji Sempozyumu, Bildiri Özleri Kitabi, 13–16 Ekim 2010, Trabzon, pp. 149–151 (in Turkish).

Micu, D., 2004. In: First record of *Musculista senhousia* (Brenson in Cantor, 1842) from the Black Sea. Abstracts of the International Symposium of Malacology (Romania, Sibiu, 2004), Sibiu, pp. 47.

Micu, D., Micu, S., 2004. In: A new type of macrozoobenthic community from the rocky bottoms of the Black Sea. International Workshop on Black Sea Benthos, 18–23 April 2004, İstanbul, Turkey, pp. 70–83.

Micu, D., Niţă, V., Todorova, V., 2010. First record of the Japanese shore crab *Hemigrapsus sanguineus* (de Haan, 1835) (Brachyura: Grapsoidea: Varunidae) from the Black Sea. Aquat. Invasions 5 (2), 4.

Mihneva, V., Stefanova, K., 2013. The non-native copepod *Oithona davisae* (Ferrari F.D. and Orsi, 1984) in the Western Black Sea: seasonal and annual patterns of abundance. BioInvasions Rec. 2 (2), 119–124.

Mironov, S., Shadrin, N.V., Grinzov, V.A., 2002. New species of mollusks in the marine and continental waters of the Crimea. Mar. Ecol. Sevastopol 61, 43 (in Russian).

Mirzoyan, Z.A., 2000. Changes in structure and productivity of zooplankton community of the Sea of Azov after invasion of the ctenophore. In: Volovik, S.P. (Ed.), Ctenophore *Mnemiopsis leidyi* (A. Agassiz) in the Azov and Black Seas: Biology and the Consequences of the Invasion, Rostov-on-Don, pp. 189–207 (in Russian).

Moncheva, S., Petrova-Karadjova, V., Palasov, A., 1995. Harmful algal blooms along the Bugarian Black Sea Coast and Possible patterns of Fish and Zoobenthic Mortalities. In: Lassus, P., Arzul, G., Denn, E., Gentien, P. (Eds.), Harmful Marine Algal Blooms. Lavoisier Publ. Inc., Paris, pp. 193–198.

Morduchai-Boltovskoi, F.D., 1960. Caspian Fauna in the Azov-Black Sea Basin. Academy of Sciences of the USSR, M-L1–288 (in Russian).

Morduchai-Boltovskoi, F.D., 1972. General characteristics of the fauna of the Black and Azov Seas. In: Vodjanitcky, V.A. (Ed.), Identification Manual of the Fauna of the Black and Azov Seas. vol. 3. Naukova Dumka, Kiev, pp. 316–324 (in Russian).

Morozova-Vodjanitckaja, N.V., 1948. In: Phytoplankton of the Black Sea. Part 1. Proceeding of Sevastopol Biological Station. vol. 7, pp. 39–172 (in Russian).

Morozova-Vodjanitckaja, N.V., 1957. In: Phytoplankton in the Black Sea and its quantitative development. Proceeding of Sevastopol Biological Station. vol. 9, pp. 3–13 (in Russian).

Nesterova, D.A., Terenko, L.M., Terenko, G.V., 2006. List of phytoplankton species, North–Western part of the Black Sea: biology and ecology. Kiev: Naukova dumka. 557–576 (in Russian).

Oğuz, T., Öztürk, B., 2011. Mechanisms impeding natural Mediterranization process of Black Sea fauna. J. Black Sea/Med. Environ. 17 (3), 234–253.

Oğuz, T., La Violette, P.E., Ünlüata, Ü., 1992. The upper layer circulation of the Black Sea: its variability as inferred from hydrographie and satellite observations. J. Geophys. Res. 97, 12569–12584.

Oğuz, T., Ashwini, D., Malanotte-Rizzoli, P., 2002. On the role of mesoscale processes controlling biological variability in the Black Sea: inferences from SeaWiFS-derived surface chlorophyll field. Cont. Shelf Res. 22, 1477–1492.

Oğuz, T., Dippner, J.W., Kaymaz, Z., 2006. Climatic regulation of the Black Sea hydro-meteorological and ecological properties at interannual-to-decadal time scales. J. Mar. Syst. 60 (3–4), 235–254.

Ojaveer, H., Galil, B.S., Gollasch, S., Marchini, A., Minchin, D., Occhipinti-Ambrogi, A., et al., 2014a. Identifying the top issues of marine invasive alien species in Europe. Manag. Biol. Invasion 5 (2), 81–84.

Ojaveer, H., Galil, B.S., Michin, D., Olenin, S., Amorim, A., Canning-Clode, J., et al., 2014b. Ten recommendations for advancing the assessment and management of non-indigenous species in marine ecosystems. Mar. Policy 44, 160–165.

Ovchinnikov, I.M., Titov, V.B., Krivosheja, V.G., Popov, Y.I., 1993. Basic hydrophysical processes and their role in the ecology of the Black Sea water. Oceanology 33 (6), 801–807 (in Russian).

Öztürk, B., 2002. The Ponto-Caspian region: predicting the identity of potential invaders. In: Spread of Alien Species by Ships. CIESM Monograph, 20, pp. 75–78 (Chapter 10).

Öztürk, B., Mihneva, V., Shiganova, T., 2011. Fist records of *Bolinopsis vitrea* (L. Agassiz, 1860) (Ctenophora: Lobata) in the Black Sea. Aquat. Invasions 6 (3), 455–460.

Pereladov, M.V., 1988. In: Some observations for biota of Sudak Bay of the Black Sea. The Third All-Union Conference on Marine Biology. Naukova Dumka, Kiev, pp. 237–238. (in Russian).

Petrova, D., Gerdzhikov, D., 2015. Phytoplankton taxonomy in the Bulgarian coastal (2008–2010). Bulg. J. Agric. Sci. 21 (1), 90–99.

Petrova-Karadjova, V.J., 1990. Monitoring the blooms along the Bulgarian Black Sea Coast. Rapp. Comm. Int. Mer. Medit. 32 (1), 209.

Pitsyk, G.K., 1950. On the abundance, composition, and distribution of phytoplankton in Black Sea. Proc. All-Union Inst. Fish. Oceanogr. Moscow 14, 215–245 (in Russian).

Pitsyk, G.K., 1979. The systematic composition of phytoplankton. In: Greze, V.N. (Ed.), The Foundations of Biological Productivity of the Black Sea. Naukova Dumka, Kiev, pp. 63–70 (in Russian).

Polikarpov, I.G., Saburova, M.A., Manzhos, L.A., Pavlovskaya, T.V., Gavrilova, N.A., 2003. Biological diversity of the Black Sea coastal microplankton nearshore Sevastopol (2001-2003). Planktonic ciliates. In: Eremeev, V.N., Gaevskaya, A.V. (Eds.), Modern Condition of Biological Diversity in Near–Shore Zone of Crimea (The Black Sea Sector). Ecosi-Gidrophizika, Sevastopol, pp. 37–51 (in Russian).

Polonsky, A.B., Shokurova, I.G., Belokopytov, V.N., 2013. Decadal variability of temperature and salinity in the Black Sea. Mar. Hydrophys. J. 6, 27–41. (in Russian).

Pusanow, I.I., 1967. Über die sukzessiven Stadien der Mediterranisation des Schwarzen Meeres. Intern. Rev. Gesammten Hydobiol. 52 (2), 219–236.

Puzanov, I.I., 1967. Mediterranization of the Black Sea and the prospects for its intensification. Eur. Zool. J. 46 (9), 1287–1297 (in Russian).

Ricciardi, A., MacIsaac, H.J., 2000. Recent mass invasion of the North American Great Lake by Ponto-Caspian species. Trends Ecol. Evol. 15 (2), 62–65.

Rubino, F., Belmonte, M., Moncheva, S., Slabakova, N., Kamburska, L., 2010. Resting stages produced by plankton in the Black Sea—biodiversity and ecological perspective. Rapp. Comm. Int. Mer. Medit. 39, 399.

Rullier, F., 1963. Les annélides polychètes du Bosphore, de la Mer de Marmara et de la Mer Noire, en relation avec celles de la Méditerranée. Rapp. Comm. Int. Mer. Medit. 17 (2), 161–260.

Selifonova, Z.P., 2011. *Amphorellopsis acuta* (Ciliophora: Spirotrichea: Tintinnida)—new spcies of tintinnid in the Black Sea. Mar. Ecol. J. Sevastopol 10 (1), 85.

Selifonova, Z.P., Yasakova, O.N., 2012. Phytoplankton of areas of the seaports of the northeastern the Black Sea. Mar. Ecol. J. Sevastopol 11 (4), 67–77 (in Russian).

Semenchenko, V.P., Rizevsky, V.K., Mastitsky, S.E., Vezhnovets, V.V., Pluta, M.V., Razlutsky, V.I., et al., 2009. Checklist of aquatic alien species established in large river basins of Belarus. Aquat. Invasions 4 (2), 337–347.

Semenchenko, V.P., Vezhnovets, V.V., Lipinskaya, T.P., 2013. Alien species of Ponto-Caspian amphipods (Crustacea, Amphipoda) in the Dnieper River basin (Belarus). Russ. J. Biol. Invasions 4, 269–275.

Semenchenko, V.P., Son, M.O., Novitsky, R.A., Kvatch, Y.V., Panov, V.E., 2014. Alien macroinvertebrates and fish in the Dnieper river basin. Russ. J. Biol. Invasions 4, 76–95.

Senicheva, M.I., 2002. New and rare species of diatom and dinophyte algae for the Black Sea. Mar. Ecol. Sevastopol 62, 25–29 (in Russian).

Senichkina, L.G., 1983. Phytoplankton of the north–western part of the Black Sea in winter period. In: Seasonal Changes in the Black Sea plankton. Nauka, Moscow, pp. 55–65 (in Russian).

Senichkina, L.G., Altukhov, D.A., Kuzmenko, L.V., Georgieva, L.V., Kovaleva, T.M., Senicheva, M.I., 2001. Species diversity of Black Sea phytoplankton in the southeastern coast of Crimea. In: Karadag: History, Biology, Archaeology. Collection of Papers Dedicated to 85th Anniversary of Karadag Scientific Station. Sonat Press, Simferopol, pp. 119–125.

Shadrin, N.V., 1999. Ecosystem functioning and economics: interrelations on global and local scales. In: Sevastopol Aquatory and Coast: Ecosystem Processes and Services for Human Society. Aquavita, Sevastopol, pp. 10–24. (in Russian).

Shalovenkov, N., 2017. Non-native zoobenthic species at the Crimean Black Sea Coast. Mediterr. Mar. Sci. 18 (2), 260–270.

Shiganova, T.A., 1998. Invasion of the Black Sea by the ctenophore *Mnemiopsis leidyi* and recent changes in pelagic community structure. Fish. Oceanogr. 7 (3–4), 305–310.

Shiganova, T.A., Bulgakova, Y.V., Volovik, S.P., Mirzoyan, Z.A., Dudkin, S.I., 2000. In: New invasive specie *Beroe ovata* and its impact on the ecosystem of the Azov-Black Sea basin in August-September 1999. Problems of Conservation of Ecosystems and Rational Use of Bioresources of the Azov-Black Sea basin: Materials of the International Scientific Conference, Rostov-on-Don, pp. 432–449 (in Russian).

Shurova, N.M., 2006. The appearance in the Black Sea of Atlantic Oligochaeta *Tubificoides benedii* (Annelida, Oligochaeta) and peculiarities of its distribution on the north–western Black Sea shelf. In: Proceedings of Zoology. vol. 40 (5). Schmalhausen Institute of Zoology NASU, Kiev, pp. 453–455 (in Russian).

Shurova, N., Zolotarev, V., 2007. Structure of populations of marine bivalve mollusks in the Danube delta area. Ecological safety of coastal and shelf zones and complex resources use of the shelf. Proc. MHI Sevastopol 15, 556–566 (in Russian).

Silkin, V.A., Abakumov, A.I., Pautova, L.A., Mikaelyan, A.S., Chasovnikov, V.K., Lukashova, T.A., 2011. Co-existence of non-native and the Black Sea phytoplankton species: invasion hypotheses discussion. Russ. J. Biol. Invasions 3, 24–35.

Sinegub, I.A., 2006. Macrozoobenthos. Bottom communities. 1984-2002. In: Northwestern Part of the Black Sea: Biology and Ecology. Naukova dumka, Kiev, pp. 276–286 (in Russian).

Skolka, M., Preda, C., 2010. Alien invasive species at the Romanian Black Sea Coast—present and perspective. Trav. Mus. Natl. Hist. Nat. Grigore Antipa 53, 443–467.

Slynko, Y.V., Dgebuadze, Y.Y., Novitskiy, R.A., Kchristov, O.A., 2010. Invasions of alien fishes in the basins of the largest rivers of the Ponto-Caspian basin: composition, vectors, invasion routes, and rates. Russ. J. Biol. Invasions 4, 74–89.

Snigerev, S., 2012. In: Current status of the Rapana venous *Rapana thomasiana thomasiana* (Crosse, 1861) in coastal waters of the Zmeini Island. Proceedings of the VII International Conference "Modern Fisheries and Environmental Problems of the Azov-Black Sea", Kerch, 20–23 June 2012, pp. 137–139 (in Russian).

Son, M.O., 2007. Invasive Molluscs in Fresh and Brackish Waters of the Northern Black Sea Region. Druk LTD, Odessa, pp. 1–132 (in Russian).

Stadnichenko, S.V., Zolotarev, V.N., 2009. Population structure of marine bivalve mollusks in the Danube Delta region in 2007–2008. Ecological safety of coastal and shelf zones and complex resources use of the shelf. Proc. MHI Sevastopol 20, 248–261 (in Russian).

Starushenko, L.I., Kazansky, A.B., 1996. Introduction of mullet harder (Mugil soiuy Basilewsky) into the Black Sea and the Sea of Azov. Stud. Rev. Gen. Fish. Council Med. 67, 29.

Svetovidov, A.N., 1964. The Fishes of the Black Sea. Russian Academy of Sciences, Institute of Zoology, Nauka, Moscow-Leningrad, pp. 1–546 (in Russian).

Taskavak, E., Bilecenoglu, M., 2001. Length–weight relationships for 18 Lessepsian (Red Sea) immigrant fish species from the eastern Mediterranean coast of Turkey. J. Mar. Biol. Assoc. UK 81 (5), 895–896.

Terenko, L., 2005. New dinoflagellate (Dinoflagellate) species from the Odessa Bay of the Black Sea. Algology 15 (2), 236–244 (in Russian).

Terenko, L.M., 2011. Genus Dinophysis Ehrenb. (Dinophyta) in Ukrainian coastal waters of the Black Sea: species composition, distribution and dynamic. Algology 21 (3), 346–356 (in Russian).

Terenko, L.M., Terenko, G.V., 2000. Species diversity of the plankton phytocenosis in the Odessa Bay of the Black Sea. Mar. Ecol. Sevastopol 52, 56–59 (in Russian).

Terenko, L.M., Terenko, G.V., 2007. In: Dynamics of *Scrippsiella trochoidea* (Dinophyceae) blooms in Odessa Bay of the Black Sea (Ukraine). XXVI Intern. Phycol. Conf. (17-20 May 2007, Lublin/Nałęczyw, Poland), pp. 91–92.

Therriault, T.W., Docker, M.F., Orlova, M.I., Heath, D.D., MacIsaac, H.J., 2004. Molecular resolution of the family Dreissenidae(Mollusca:Bivalvia) with emphasis on Ponto-Caspian species, including first report of Mytilopsis leucophaeata in the Black Sea basin. Mol. Phylogenet. Evol. 30, 479–489.

Tkachenko, P.V., 2012. Fishes of Tendrovskoe and Yagorlytskoe Bays, and adjacent waters of the Black Sea. In: Natural Almanac. Biological Sciences. 18. Mis'ka Drukarnya Press, Kherson, pp. 181–193 (in Russian).

Todorova, V., Konsulova, T., 2008. In: Ecological state assessment of zoobenthic communities on the Northwestern Black Sea shelf—the performance of multivariate and univariate approaches. Proceedings of the 1st Biannual Scientific Conference: Black Sea Ecosystem 2005 and Beyond, Biodiversity, Ecophysiology, Istanbul, 8-10 May 2006. vol. 1, pp. 726–742.

Usachev, P.I., 1928. On the phytoplankton of the northwestern part of the Black Sea. In: Diary of the All-Union Congress of Botany. Leningrad, pp. 163–164 (in Russian).

Vanderploeg, H.A., Nalepa, T.F., Jude, D.J., Mills, E.L., Holeck, K.T., Liebig, J.R., et al., 2002. Dispersal and emerging ecological impacts of Ponto-Caspian species in the Laurentian Great Lakes. Can. J. Fish. Aquat. Sci. 59, 1209–1228.

Vasil'eva, E.D., 2007. Fishes of the Black Sea. Identification Manual of the Marine, Brakish-Water, Eurihaline and Anadromous Species With Color Illustrations Collected by Bogorodsky SV. VNIRO Publishers, Moscow, pp. 1–238 (in Russian).

Vershinin, A., 2008. In: Phytoplankton of north–eastern Black Sea: composition, annual changes in community structure, and the problem of invasive species. Materials In. Sci. Conf. and VII Marine Biology School "Modern Problems of Algology" (Rostov-on-Don, 9–13 June 2008), Rostov-on-Don, pp. 403–415 (in Russian).

Vinogradov, M.E., Shushkina, E.A., 1992. Temporary changes of the zoocenosis in the open parts of the Black Sea. Oceanology 32 (4), 709–717 (in Russian).

Vinogradov, M.E., Shushkina, E.A., Musaeva, E.I., Sorokin, P.Y., 1989. Ctenophore *Mnemiopsis leidyi* (A. Agassiz) (Ctenophora: Lobata)—new settlers in the Black Sea. Oceanology 29, 293–299 (in Russian).

Vinogradov, M.E., Sapozhnikov, V.V., Shushkina, E.A., 1992. The Black Sea Ecosystem. Nauka, Moscow, pp. 1–112 (in Russian).

WoRMS, n.d. WoRMS—the world register of marine species. http://www.marinespecies.org.

Yaglioglu, D., Turan, C., Gurlek, M., 2015. Population structure analysis of Red Barracuda, *Sphyraena pinguis* (Perciformes, Sphyraenidae) in the North-Eastern Mediterranean sea using morphometric and meristic characters. Fresenius Environ. Bull. 24 (1b), 263–268.

Yankova, M., 2016. Alien invasive fish species in Bulgarian waters: an overview. Int. J. Fish. Aquat. Stud. 4 (2), 282–290.

Yankova, M., Pavlov, D., Ivanova, P., Karpova, M., Boltachev, A., Bat, L., et al., 2013. Annotated check list of the non-native fish species (Pisces) of the Black Sea. J. Black Sea/Medit. Environ. 19 (2), 247–255.

Yasakova, O.N., 2010. The new species of phytoplankton in the north–eastern part of the Black Sea. Russ. J. Biol. Invasions 4, 90–97.

Yasakova, O.N., 2011. New species of phytoplankton in the northeastern part of the Black Sea. Russ. J. Biol. Invasions 2 (1), 63–67.

Yasakova, O.N., 2013. The seasonal dynamics of potentially toxic and harmful phytoplankton species in Novorossiysk Bay (Black Sea). Russ. J. Mar. Biol. 39 (2), 107–115 (in Russian).

Yasakova, O.N., 2014. The current state of phytoplankton in the bays of the cities of Anapa and Gelendzhik, the Black Sea. Bull. South. Sci. Center Russ. Acad. Sci. 10 (1), 35–48 (in Russian).

Zagorodnyay, Y.A., Kolesnikova, E.A., 2003. In: To the problem of the alien copepod species introduction into the Black Sea. Evolution of marine ecosystems impacted by introducers and artificial mortality of fauna. Abstracts of International Conferences in Azov, June 15-18, Rostov-on-Don, pp. 80–82.

Zaitsev, Y.P., 2011. New peculiarities of diversicology of Ponto–Azov. Ecological safety coastal and shelf areas and the complex use of shelf's resources. Proc. MHI Sevastopol 25 (1), 274–285 (in Russian).

Zaitsev, Y.P., Alexandrov, B.G., 1998. Black Sea Biological Diversity. United Nations Publications, Ukraine/New York, pp. 1–351.

Zaitsev, Y., Mamaev, V., 1997. Marine Biological Diversity in the Black Sea. A Study of Change and Decline. vol. XV. United Nations Publications, New York, pp. 1–208.

Zaitsev, Y.P., Öztürk, B. (Eds.), 2001. Exotic Species in the Aegean, Marmara, Black, Azov and Caspian Seas. vol. 8. Turkish Marine Research Foundation, Istanbul, pp. 1–267.

Zaitsev, Y.P., Starushenko, L.I., 1997. Haarder (Mugil soiuy Bas.)—a new commercial fish in the Black and Azov Seas. Hydrobiol. J. Kiev 33, 29–37.

Zaitsev, Y.P., Vorobyova, L.V., Aleksandrov, B.G., 1988. A northern source of replenishment of the Black Sea fauna. In: Reports of the Academy of Sciences of the Ukrainian SSR, 11, pp. 61–63 (in Russian).

Zaitsev, Y., Alexandrov, B., Berlinsky, N., Bogatova, Y., Bolshakov, V., Bushuev, S., et al., 2004. Base biological investigations of Odessa sea port: August–December, 2001. Final report. In: Monographs Series of Odessa Site of GloBallast Project, vol. 7, pp. 171 (in Russian).

Zats, V., Lukjanenko, O., Jazevich, G., 1966. Hydrometeorological Regime of the Southern Coast of the Crimea. Gidrometeoizdat Publishers, Leningrad, pp.1–120 (in Russian).

Zenetos, A., Çinar, M.E., Crocetta, F., Golani, D., Rosso, A., Servello, G., et al., 2017. Uncertainties and validation of alien species catalogues: the Mediterranean as an example. Estuar. Coast. Shelf Sci. 191, 171–187.

Zenkevich, L.A., 1963. Biology of the seas of URSS. Pub. Acad. Sci. USSR Moskow, 1–739 (in Russian).

Zolotarev, P., Terentev, A., 2012. Changes in the macrobenthic communities Gudauta oyster banks. Oceanology 52 (2), 251–257 (in Russian).

Zolotarev, P.N., Rubinshtein, I.T., Larchenko, N.A., Povchun, A.S., 1990. Condition of benthos of the Karkinitsky Bay of the Black Sea in the 80s. In: Deposited Manuscript in VINITI № 5447. IBSS NASU, Sevastopol, pp. 1–34 (in Russian).

Chapter 32

Coastal Fisheries: The Past, Present, and Possible Futures

Maria-Lourdes D. Palomares, Daniel Pauly

Institute for the Oceans and Fisheries, University of British Columbia, Vancouver, BC, Canada

1   INTRODUCTION

Fisheries contribute greatly to the food and livelihoods of people throughout the world. Coastal fisheries contribute much, if not most, of global catches, but quantitative estimates of the extent of their contribution relative, for example, to high seas fisheries, or to inland fisheries depend on how coastal fisheries are defined. In the absence of a consensus definition of "coastal" by an authoritative international agency, we will be using here the definition initially proposed by Chuenpagdee et al. (2006) for the area accessible to small-scale fisheries, that is, the areas at most 50 km from inhabited coastlines or down to a depth of up to 200 m, whichever comes first. This definition describes the "inshore fishing area" (IFA; about 3% of the ocean surface), to which all the catches of small scale, that is, artisanal, subsistence, and recreational fisheries are assumed to have been taken (Pauly and Zeller, 2016a, b, Zeller et al., 2016). Note that this definition excludes the coastlines of uninhabited regions (e.g., much of the Arctic; Zeller et al., 2011) or islands (e.g., Kerguelen), the latter frequently having industrial fisheries fishing around them, but not "coastal" fisheries (see, e.g., Palomares and Pauly, 2011).

In addition to small-scale fisheries, large-scale industrial fisheries also frequently operate within the IFA of various countries, especially in the tropics, for reasons to be described below. In fact, this is the major cause of the frequent conflicts between industrial fisheries and small-scale, notably, artisanal fisheries. Also, because the key features of industrial and artisanal fisheries, and the relationships between them strongly varies between different parts of the world, notably, as a function of their ecology, economics, and governance, we have grouped the world's maritime countries into nine aggregates, here called "socio-ecological areas" (SEAs; Fig. 1).

This chapter reviews the major features of coastal fisheries and their catches since 1950, the first year that annual global fisheries statistics were published by the newly created Food and Agriculture Organization of the United Nations (FAO).

Important here is that the catch data we present and use for inferring ecosystem state (see Section 6) and the effects of global warming (see Section 7), are "reconstructed," that is, the FAO catch data submitted by its member countries were complemented with the catches that typically are omitted in official statistics (most artisanal catches, virtually all subsistence catches, all catches by fleets operating illegally, and all discarded fish). Adding estimates of these neglected catches added about 50% to the global marine catch (Pauly and Zeller, 2016a, b), most of it to coastal fisheries.

Note also that the reconstructed catches are assigned to ½° latitude/longitude cells (Palomares et al., 2016; Lam et al., 2016) of which there are about 180,000 in the world oceans, and of which about 20,000 are coastal (i.e., also cover some land area). The time series of catches in coastal areas presented here are assembled by pooling the catches in coastal and adjacent spatial cells for the different seas defined in Fig. 1, such as to meet the definition of IFAs. The inference of fishing down and global warming effects is based on the trophic levels and to the preferred temperatures, respectively, of the taxa whose catches are distributed among spatial cells (see www.seaaroundus.org for details).

2   COASTAL FISHERIES AS A KEY COMPONENT OF GLOBAL FISHERIES

Fig. 2 presents time series of global reconstructed marine catches for the period from 1950 to 2014. Fig. 2A distinguishes between the catch of coastal fisheries as defined above (i.e., with the IFAs of all maritime countries and their territories),

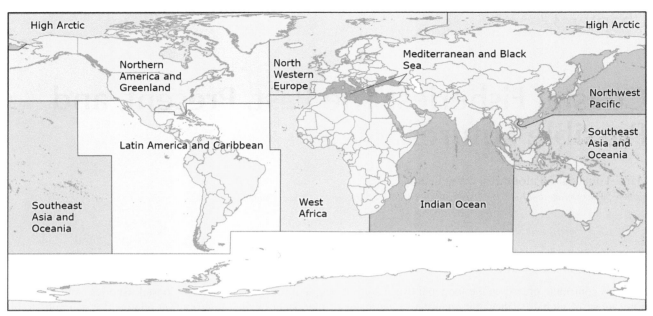

FIG. 1 The nine socio-ecological areas (SEAs) used here to differentiate between the coastal fisheries of different parts of the world. The numbers refer to the mean annual coastal and total catch in each SEA in 2000–14.

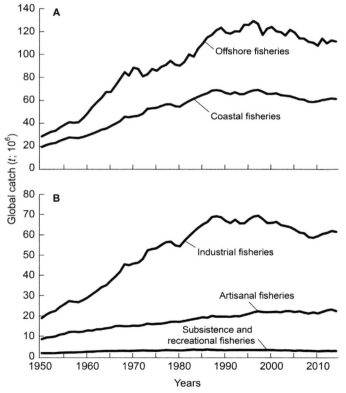

FIG. 2 Catches from 1950 to 2014 of (A) the world's marine fisheries and (B) coastal fisheries as defined here (see text).

the catches of industrial fisheries in the offshore (non-IFA) parts of countries' 200 mile exclusive economic zones (EEZs), and high sea (non-EEZ) catches. Fig. 2B separates the coastal catches into industrial, artisanal, subsistence, and recreational fisheries, with the industrial catch separated into landings and discards (small-scale fisheries have negligible discards).

As shown in Fig. 2A, coastal fisheries as defined here, made up an average of 55% of global marine fisheries in the 5-year period from 2010 to 2014, while Fig. 2B shows that small-scale fisheries (i.e., artisanal, subsistence, and recreational) in the same period, contributed 36% of the marine catches consumed directly by people, that is, excluding fish reduced to fish meal or discarded.

3 REGIONAL AND TEMPORAL DIFFERENCE IN COASTAL FISHERIES

The nine panels in Fig. 3 document the different components of different SEAs of the world in terms of the catches. Fig. 3 demonstrates that small-scale fisheries generate, in all SEAs, a substantial part of the catches, even exceeding industrial catches in the Arctic and Indian Oceans.

An important feature of small-scale fisheries is that they do not fluctuate as strongly as industrial fisheries, for two basic reasons: (1) the fishing power of artisanal fisheries is rarely sufficient to annihilate marine fish stocks, something which industrial fisheries are not only capable of, but also frequently do and (2) when a species targeted by a small-scale fishery declines, other species are exploited, so that overall catches are maintained.

However, another important reason for the relative stability of small-scale fisheries in Fig. 3 is that their catches were often estimated indirectly, from the demography of fisher communities and slowly changing individual catch rates, or from fish consumption data in places where industrially caught fish was not available. Still, we believe that the relatively stable or slowly increasing artisanal catches in Fig. 3 are realistic. Moreover, they would be higher if it were not for competition from industrial fishing, our next topic.

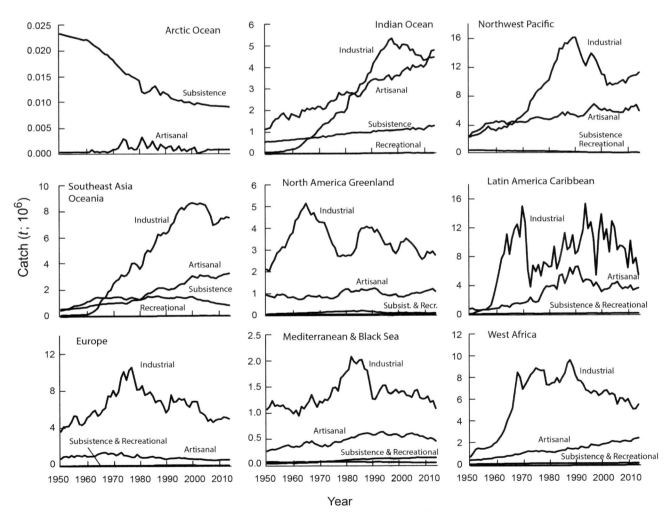

FIG. 3 Catches of industrial, artisanal, subsistence, and recreational fisheries by SEAs, 1950–2014.

4   LARGE-SCALE INDUSTRIAL VERSUS SMALL-SCALE ARTISANAL AND RECREATIONAL FISHERIES

The co-occurrence in the IFA of industrial and small-scale, particularly artisanal fishing, is bound to generate conflicts between these two sectors. This is particularly true in the tropics, where marine primary and secondary production occur closer inshore than that in higher-latitude areas. The reason for this is that in the tropics, the phytoplankton production is almost entirely consumed in the surface waters, with very little marine snow (dead phytoplankton debris, mostly from the fecal pellets of herbivorous zooplankton) reaching deeper waters, due to the high temperatures accelerating microbial degradation in the water column (Longhurst and Pauly, 1987).

Conversely, in high latitudes, the marine snow resulting from autumnal phytoplankton blooms tends to be accumulated on the sea floor of shelf and slope areas, where it then sustains deep benthic communities through the winter (Ursin, 1984). These benthic communities, in turn sustain benthic fish populations, which themselves sustain trawl fisheries which may be operating far from coastlines. In the tropics, in contrast, the coupling of pelagic with benthic food webs occurs only inshore, mostly in shallow waters, down to a depth of about 50 m. Hence, trawlers must operate close inshore to maintain reasonable catches, especially given that they target exportable penaeid shrimps. This targeting, incidentally, is the reason why coastal industrial fisheries generate so much as discards of edible fish (Zeller et al., 2017). Artisanal fisheries, on the other hand, target mainly fish for local consumption, in many cases the very fish that shrimp trawlers tend to discard as unwanted bycatch. This is, at its most simple, at the heart of the perennial conflict between artisanal and industrial fishers, especially along tropical coastlines (Pauly, 2006).

The response of governments the world over has been attempts to separate the protagonists, and thus many countries restrict industrial fishing, especially trawling in coastal areas. These restrictions are often not respected, however, which then leads to questions about the seriousness of various governments in protecting small-scale fisheries, and the livelihood they provide. This also leads to questions about equity, as the industrial vessels frequently operate (legally or not) in the coastal areas of developing countries (Belhabib et al., 2015). There, they compete with artisanal fishers for fish and invertebrate resources, which then either ends by supplying local markets or the markets of European countries, the United States, and other advanced economies (Swartz et al., 2010; Pauly et al., 2014).

In many, predominantly higher latitude countries, the conflicts alluded to above often take the form of a competition between "commercial" fisheries (an ambiguous term, covering both industrial and artisanal fisheries) and recreational fisheries. Here, however, the issue is not one of mitigating conflicts, but one of making the best of fish resources. Thus, a fish caught in West Africa by a recreational fisher (yes, there are recreational fisheries along the coast of West Africa) will generate seven times as much revenue to the country where it is caught than the same fish caught commercially (Belhabib et al., 2016); similar ratios have been documented from other parts of the world.

5   A NEGLECTED SECTOR: SUBSISTENCE FISHERIES

Subsistence fisheries exist throughout the world and essentially consist of coastal people catching fish or gathering invertebrates in shallow waters or intertidal areas for their own and their family's consumption, or for bartering against other goods. Only a few countries or territories monitor and report on their subsistence fisheries, Alaska being one of the positive exceptions.[1] However, despite near universal official neglect, subsistence fisheries are extremely important to the food security of numerous countries, particularly small island developing states (SIDS) in the Pacific and Indian Ocean (Chapman, 1987; Hauzer et al., 2013).

The *Sea Around Us* catch reconstructions (Pauly and Zeller, 2016a, b) involved quantifying the subsistence catch of all maritime countries and their aggregate catch; in the period 2010–14, this was 3.6 million tons per year. As such, it represents about 3% of the world's marine fisheries catch, but at the same time it represents almost 5% of that part of the marine catch that is used for human consumption. Also, because much of the subsistence catch in SIDS is caught by women (Chapman, 1987; Harper et al., 2013), it contributes directly to the well-being of coastal families, which is not necessarily the case for the catches made by men, whose income is often diverted away from their families. Finally, we note that the global recreational catch, which is also taken overwhelmingly in the IFA as defined here, is approximately 1 million tons per year, and that some of this may also be considered subsistence catches.

---

[1]   www.adfg.alaska.gov/index.cfm?ADFG=fishingSubsistence.main.

6 "FISHING DOWN" AND OTHER ECOSYSTEM IMPACTS OF COASTAL FISHERIES

Pauly et al. (1998) demonstrated the occurrence in many exploited aquatic ecosystems of a process now widely known as "fishing down marine food webs" (FD). The process, due to the greater decline under exploitation of larger high trophic level fish in ecosystems relative to the smaller, low trophic level fish and invertebrates, can be demonstrated using time series on the mean trophic level (mTL) of fisheries catches which tend to decline overtime. This assumes that the different fisheries jointly exploiting an ecosystem can access all of its fishable species, which generally occurs once that ecosystem is exploited industrially.

In the two decades since it was first described, the ubiquity of FD has been widely confirmed, for example, by Bhathal and Pauly (2008) for India, Gascuel et al. (2016) for Northwestern Europe, Pauly et al. (2001) for Canadian waters, and Liang and Pauly (2017) for Chinese waters (see also www.fishingdown.org). However, FD can easily be masked by fishing operations moving gradually offshore when their catches and the mTL of their catch decline (Bhathal and Pauly, 2008; Liang and Pauly, 2017), and thus again catch large, high mTL fish (Fig. 4). Thus, Kleisner et al. (2014), based on the logic of Fig. 4, and with considerations in Bhathal and Pauly (2008), devised an approach and software to reexpress a single aggregate series of mTL data into distinct time series of mTL pertaining to what can be conceived as a strip more or less parallel to a coastline.

Fig. 5 shows the result of an application of this approach to the global catch of marine fisheries. Therein, the longest time series of mTL, which largely represents coastal fisheries, declined from 3.56 in the early 1950s to 3.06 in the early 2010s, that is, by 0.5 TL in 65 years, or 0.077 TL per decade. In the late 1950s, however, industrial fishing moved further offshore, and started to catch fish with an mTL of about 4.0, which declined to about 3.6 in the late 2010s, that is, 0.067 per

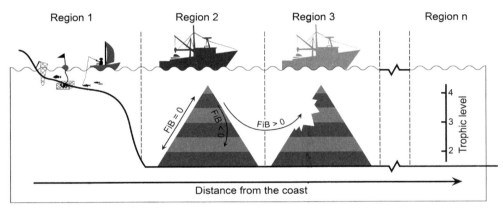

FIG. 4 Schematic representation of the offshore expansion of fisheries, which, following the initial addition of an industrial component to regions 1–2 will then expand to region 3, when costal stocks are depleted and the mTL of the catch has declined, etc. The catch in newly accessed regions will have a higher trophic level. *(Modified from Kleisner, K., H. Mansour and D. Pauly. 2014. Region-based MTI: resolving geographic expansion in the marine trophic index. Mar. Ecol. Prog. Ser., 512: 185–199; see text.)*

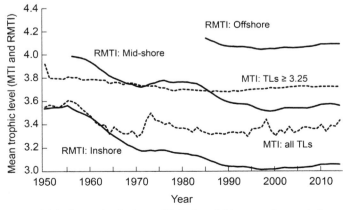

FIG. 5 Decline of the mean trophic level of fisheries catches in the world's inshore fishing areas (see text), demonstrating a "fishing down" effect. The dotted lines represent the marine trophic index (MTI) trends calculated directly from the taxa caught and their trophic levels, while the solid line is the output of the regional marine trophic index (RMTI) method of Kleisner et al. (2014).

decade. Finally, the increase of catches in the mid-1980s was based on catches further offshore, starting with an mTL above 4.1 and declining very slowly that is, remaining at TLs typical of the various species of tuna (see www.fishbase.org, from which most of the TL values used for this analysis originate).

Note that these mTL declines have occurred, since 1996, in conjunction with declining catches. Thus, the mTL declines are not due to new species of low-TL trophic levels being caught, while the high-TL species maintain their catch. Hence, we do not have here a case of fishing "through" the food web, which would imply increasing catches (Essington et al., 2006). Rather, Fig. 5 demonstrates that the world marine fisheries, and particularly coastal fisheries, suffer from fishing down.

7   COASTAL FISHERIES AND CLIMATE CHANGE

Coastal fisheries are beginning to be affected by climate change (Fig. 6) and this impact is only going to increase. Along subtropical and temperate coastlines, climate change will induce a change in the fish communities available for exploitation, as already noticeable in many countries. This is due to the poleward shift of fish and marine invertebrates attempting to maintain themselves in waters with suitable temperatures (Cheung et al., 2009, 2010). In the tropics, however, no warmer-adapted species will replace the fish and invertebrate species that are displaced by the high temperature, and hence there can only be a net loss as species relocate (Cheung et al., 2013). Thus, it is in the tropics that coastal fisheries will be affected most by global warming, that is, the regions that have least contributed to the greenhouse gas emissions that are the cause of the warming.

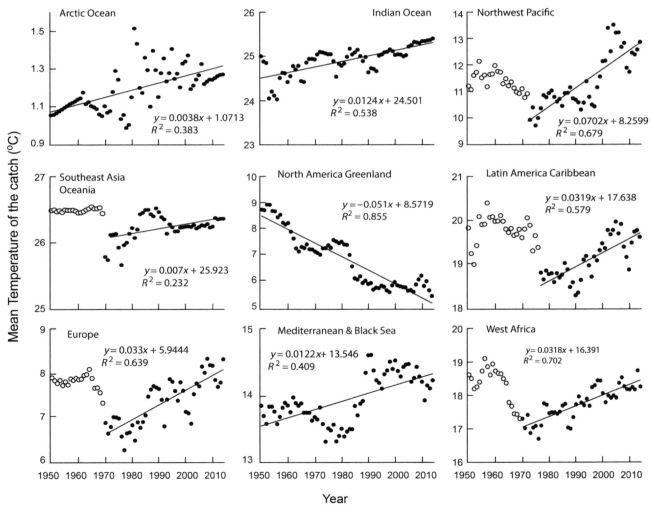

FIG. 6 Increase in the mean temperature of the catch from inshore areas (as defined in Cheung et al., 2013), as expected given global warming. North America and Greenland are an exception that still requires an explanation.

8 THE GOVERNANCE OF COASTAL FISHERIES

Jentoft and Chuenpagdee (2009)) suggested that the governance of coastal fisheries was a "wicked" problem, that is, extremely difficult to solve, and even to define. Yet, some of their issues are easily resolvable, once the major features are understood (Fig. 7). Thus, it is obvious that industrial fisheries, given their enormous catching capacity and their propensity for using habitat-destroying gear, cannot coexist with artisanal fisheries (Pauly, 2017). Thus, if a government has as an explicit policy (as many do) to encourage artisanal fisheries, it must also ensure industrial vessels do not operate inshore, both legally and in practice. At the same time—and this contributes to the "wickedness" alluded to above—uncontrolled small-scale fisheries can also deplete coastal stocks (Lam and Pauly, 2010), as demonstrated at numerous locations, for example, the Philippines. Thus, their development, and especially the increase of artisanal fishers' populations will have to be countered (Pauly, 2006). This is probably where the greatest challenges lie.

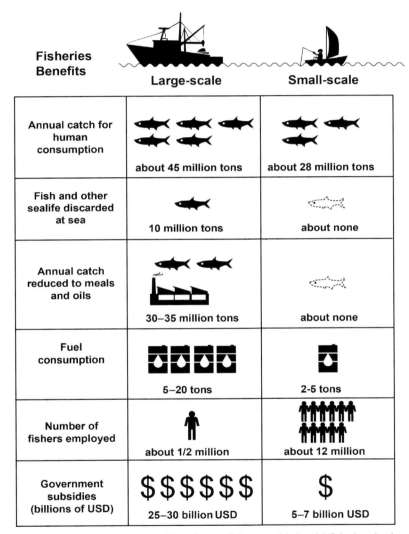

FIG. 7 Schematic illustration of the difference between (coastal) small-scale fisheries and industrial fisheries, that is, coastal and offshore. *(Updated from Thompson, D., 1988. The world's two marine fishing industries – how they compare. Naga, ICLARM Q. 11 (3), 17 and Pauly, D. 2006. Major trends in small-scale marine fisheries, with emphasis on developing countries, and some implications for the social sciences. Maritime Stud. (MAST) 4(2), 7–22).*

ACKNOWLEDGMENTS

This is a contribution of the *Sea Around Us*, a research initiative at the UBC's Institute for the Oceans and Fisheries funded by various philanthropic foundations, notably the Oak, Marisla, Paul M. Angell, and MAVA Foundations.

REFERENCES

Belhabib, D., Campredon, P., Lazar, N., Sumaila, U.R., Cheikh Baye, B., Abou Kane, E., Pauly, D., 2016. Best for pleasure, not for business: evaluating recreational marine fisheries in West Africa using unconventional sources of data. Palgrave Commun. 2, 15050. https://doi.org/10.1057/palcomms.2015.50.

Belhabib, D., Sumaila, U.R., Lam, V.W., Zeller, D., Le Billon, P., Kane, E.A., Pauly, D., 2015. Euros vs. Yuan: comparing European and Chinese fishing access in West Africa. PLoS One 10 (3), e0118351.

Bhathal, B., Pauly, D., 2008. "Fishing down marine food webs" and spatial expansion of coastal fisheries in India, 1950–2000. Fish. Res. 91, 26–34.

Chapman, M.D., 1987. Women fishing in Oceana. Hum. Ecol. 15, 267–288.

Cheung, W.W.L., Lam, V.W.Y., Sarmiento, J.L., Kearney, K., Watson, R., Pauly, D., 2009. Projecting global marine biodiversity impacts under climate change scenarios. Fish Fish. 10, 235–251.

Cheung, W.W.L., Lam, V.W.Y., Sarmiento, J.L., Kearney, K., Watson, R., Zeller, D., Pauly, D., 2010. Large-scale redistribution of maximum fisheries catch potential in the global ocean under climate change. Glob. Chang. Biol. 16, 24–35.

Cheung, W.W.L., Watson, R., Pauly, D., 2013. Signature of ocean warming in global fisheries catch. Nature 497, 365–368.

Chuenpagdee, R., Liguori, L., Palomares, M.L.D., Pauly, D., 2006. Bottom-Up, Global Estimates of Small-Scale Marine Fisheries Catches. Fisheries Centre Research Reports 14(8) University of British Columbia, Vancouver. 112 p.

Essington, T.E., Beaudreau, A.H., Wiedenmann, J., 2006. Fishing through marine food webs. Proc. Natl. Acad. Sci. U. S. A. 103 (9), 3171–3175.

Gascuel, D., Coll, M., Fox, C., Guénette, S., Guitton, J., Kenny, A., Knittweis, L., Nielsen, J.R., Piet, G., Raid, T., Travers-Trolet, M., 2016. Fishing impact and environmental status in European seas: a diagnosis from stock assessments and ecosystem indicators. Fish Fish. 17, 31–55.

Harper, S., Zeller, D., Hauzer, M., Pauly, D., Sumaila, U.R., 2013. Women and fisheries: contributions to food security and local economies. Mar. Policy 39, 56–63.

Hauzer, M., Dearden, P., Murray, G., 2013. The fisherwomen of Ngazidja island, Comoros: fisheries livelihoods, impacts, and implications for management. Fish. Res. 140, 28–35.

Jentoft, S., Chuenpagdee, R., 2009. Fisheries and coastal governance as a wicked problem. Mar. Policy 33, 553–560.

Kleisner, K., Mansour, H., Pauly, D., 2014. Region-based MTI: resolving geographic expansion in the marine trophic index. Mar. Ecol. Prog. Ser. 512, 185–199.

Lam, M., Pauly, D., 2010. Who's right to fish? Evolving a social contract for ethical fisheries. Ecol. Soc. 15 (3), 16. Available from: www.ecologyandsociety.org/vol15/iss3/art16/.

Lam, V.W.Y., Tavakolie, A., Palomares, M.L.D., Pauly, D., Zeller, D., 2016. The *Sea Around Us* catch database and its spatial expression. In: Pauly, D., Zeller, D. (Eds.), Global Atlas of Marine Fisheries: A Critical Appraisal of Catches and Ecosystem Impacts. Island Press, Washington, DC, pp. 59–67.

Liang, C., Pauly, D., 2017. Fisheries impacts on China's coastal ecosystems: unmasking a pervasive "Fishing Down" effect. PLoS One 12 (3), e0173296. https://doi.org/10.1371/journal.pone.0173296.

Longhurst, A., Pauly, D., 1987. Ecology of Tropical Oceans. Academic Press, San Diego. 407 p.

Palomares, M.L.D., Pauly, D., 2011. A brief history of fishing in the Kerguelen Island, France. In: Harper, S., Zeller, D. (Eds.), Fisheries Catch Reconstruction: Islands, Part II. pp. 15–20. Fisheries Centre Research Reports 19(4).

Palomares, M.L.D., Cheung, W.W.L., Lam, V.W.Y., Pauly, D., 2016. The distribution of exploited marine biodiversity. In: Pauly, D., Zeller, D. (Eds.), Global Atlas of Marine Fisheries: A Critical Appraisal of Catches and Ecosystem Impacts. Island Press, Washington, DC, pp. 46–58.

Pauly, D., 2006. Major trends in small-scale marine fisheries, with emphasis on developing countries, and some implications for the social sciences. Maritime Stud. (MAST) 4 (2), 7–22.

Pauly, D., 2017. A vision for marine fisheries in a global blue economy. Mar. Policy https://doi.org/10.1016/j.marpol.2017.11.0.10.

Pauly, D., Zeller, D., 2016a. Global Atlas of Marine Fisheries: A Critical Appraisal of Catches and Ecosystem Impacts. Island Press, Washington, DC. xii+497 p.

Pauly, D., Zeller, D., 2016b. Catch reconstructions reveal that global marine fisheries catches are higher than reported and declining. Nat. Commun. 9. https://doi.org/10.1038/ncomms10244.

Pauly, D., Christensen, V., Dalsgaard, J., Froese, R., Torres, F.C., 1998. Fishing down marine food webs. Science 279, 860–863.

Pauly, D., Palomares, M.L.D., Froese, R., Sa-a, P., Vakily, M., Preikshot, D., Wallace, S., 2001. Fishing down Canadian aquatic food webs. Can. J. Fish. Aquat. Sci. 58, 51–62.

Pauly, D., Belhabib, D., Blomeyer, R., Cheung, W.W.L., Cisneros-Montemayor, A., Copeland, D., Harper, S., Lam, V.W.Y., Mai, Y., Le Manach, F., Österblom, H., Mok, K.M., van der Meer, L., Sanz, A., Antonio, S., Shon, U.R., Sumaila, W., Swartz, R., Watson, Y.Z., Zeller, D., 2014. China's distant-water fisheries in the 21st century. Fish Fish. 15 (3), 474–488.

Swartz, W., Sumaila, U.R., Watson, R., Pauly, D., 2010. Sourcing seafood for the three major markets: the EU, Japan and the USA. Mar. Policy 34 (6), 1366–1373.

Ursin, E., 1984. The tropical, the temperate and the arctic seas as media for fish production. Dana 3, 43–60.

Zeller, D., Booth, S., Pakhomov, E., Swartz, W., Pauly, D., 2011. Arctic fisheries catches in Russia, USA and Canada: baselines for neglected ecosystems. Polar Biol. 34 (7), 955–973.

Zeller, D., Palomares, M.L.D., Tavakolie, A., Ang, M., Belhabib, D., Cheung, W.W.L., Lam, V.W.Y., Sy, E., Tsui, G., Zylich, K., Pauly, D., 2016. Still catching attention: *Sea Around Us* reconstructed catch data, their spatial expression and public accessibility. Mar. Policy 70, 145–152.

Zeller, D., Cashion, T., Palomares, M.L.D., Pauly, D., 2017. Global marine fisheries discards: a synthesis of reconstructed data. Fish Fish. https://doi.org/10.1111/faf.123.

Chapter 33

Temperate Estuaries: Their Ecology Under Future Environmental Changes

Ducrotoy J.-P.[*], Michael Elliott[*], Cutts N.D.[*], Franco A.[*], Little S.[†], Mazik K.[*], Wilkinson M.[‡],

[*]Institute of Estuarine and Coastal Studies, University of Hull, Hull, United Kingdom, [†]School of Animal, Rural and Environmental Sciences, Nottingham Trent University, Nottinghamshire, United Kingdom, [‡]Institute of Life and Earth Sciences, Heriot-Watt University, Edinburgh, United Kingdom

1 INTRODUCTION

Temperate estuaries are highly variable, contain many habitat types, and are the sites of extensive urban and industrial activities. Regional and local differences in trends of climate and human activities ensure highly variable long-term trajectories. They are highly dynamic environments with gradients of seawater and freshwater flows and with the tidal cycle, and are critical transition zones linking land, freshwater, and marine habitats (Elliott and Whitfield, 2011). They have particular habitats and biogeochemical cycling and produce ecosystem services which deliver societal goods and benefits. In temperate estuaries, C, N, and P recycling is high in these detritus-based systems and climate change affect these cycles and estuarine biodiversity through an increased variability due to the impact of human activities on ecosystems.

Given the above characteristics, there is a need to forecast how they could be influenced by the current increasing climate variability especially under site-specific conditions, thus requiring site-based management. It is especially important to understand how estuaries respond to climate change in order to adapt measures to unexpected evolutionary trajectories (Little et al., 2017). It is valuable to clarify and identify the elements which respond preferentially to new climatic conditions and the changes to the dynamic physical characteristics of the ecosystems considered.

This review focuses on the ecological changes compared to the physical responses although ecological understanding requires knowledge of rising sea levels and its influence on the hydrodynamics regime. Changes in climate (e.g., temperature rise, sea level rise, increased risks of floods, and droughts) may increase the risk of abrupt and nonlinear changes in many ecosystems, which would affect their composition, function, biodiversity, and productivity (Fig. 1). Changes in these and in the frequency of extreme events may disrupt ecosystems and alter societal goods and services provided (for more details, see Ducrotoy, 2011).

Accelerated sea level rise will deepen nearshore waters and increase the volume of tidal water entering the estuary channel, increasing wave height and frequency in outer estuarine channels (where tidal waves propagate). This will increase tidal intrusion, amplitude, and current speeds in inner and upper estuarine channels as a result of increased depth and decreased friction drag of the tidal wave (Holleman and Stacey, 2014).

Precipitation and river flow will be affected in terms of number of days of flood and low water in the estuary. The proportion of haline areas (oligo-, meso-, and eury-) will change together with depth and river flow into the estuary (Ysebaert et al., 2005). The surface of intertidal areas relates to the marine intrusion and bathymetry. Connectivity of estuarine and peri-estuarine zones depends on the velocity of axial currents. Turbidity, and its effect on oxygen content, is important in terms of reactive surface area in relation to local anthropogenic activities and movements of the turbidity maximum. In addition, the synergistic effect of turbidity and increased temperature will also increase water column oxygen demand.

Chemical and organic exposure, pathways, foodweb uptake, and pollution *per se* will change with climate change thus affecting toxic effects. Hydrological changes resulting from rainfall, runoff, severe weather events, and erosion will affect transport, dilution, and fate processes of harmful chemicals including volatilization, adsorption, hydrolysis, biodegradation, photo degradation, photo-enhanced toxicity, uptake, and metabolism. While some of these rates increase with increasing temperature and pH, quantitative predictions and assessments of interactions are complex. Changes in environmental conditions (temperature, pH, and oxygenation), expected to be associated with climate change and ocean acidification, will

Coasts and Estuaries. https://doi.org/10.1016/B978-0-12-814003-1.00033-2

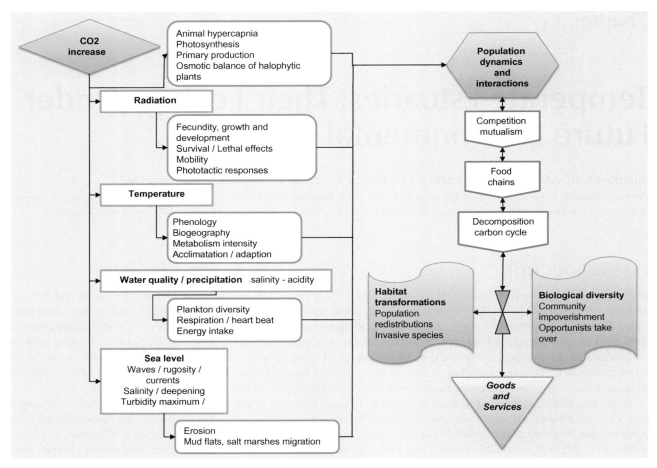

FIG. 1 Possible biological effects of climate change on estuarine ecosystems.

interact with the responses of organisms to contaminants, especially fish (Solan and Whiteley, 2016). The use of agrochemicals and other human activities such as the amount of urbanization and industrialization are also changing in response to climate change, including the increasing use of trace metals and broad-spectrum antibiotic drugs.

2 THE RESPONSE OF ESTUARINE ECOLOGICAL COMPONENTS TO CLIMATE CHANGE

2.1 Phytoplankton Primary Production

Globally, climate change (in the form of increased temperature) has altered precipitation patterns leading to extreme rainfall events and drought, and these patterns are expected to intensify over the next century (Voynova and Sharp, 2012; Wetz et al., 2011) and vary between tropical and temperate systems. These changes and extreme events will strongly influence freshwater flow and the supply of organic matter and nutrients to estuaries (Wetz and Yoskowitz, 2013). Any changes to freshwater flow, runoff and residence or flushing time (Maier et al., 2012), temperature (Hall et al., 2013), and stratification (Wikner and Andersson, 2012) will affect estuarine productivity as will osmotic stress, grazing pressure, and stage of the spring-neap tidal cycle (Stratham, 2012). Natural and anthropogenic variations in these factors, such as those associated with periodic climatic events such as the NAO and El Niño, together with non-climate-related variation may obscure any patterns linked to climate change. These patterns may also be further obscured by anthropogenic influences (e.g., urbanization and industrialization) and their effects on estuarine morphology, hydrodynamics, and nutrient status.

Increasing temperature has changed phytoplankton biomass and phenology in marine systems but this is less well observed in estuaries due to the difficulty in separating the effects of climate change from those associated with land use. For example, widespread eutrophication coupled with reduced nutrient and organic matter inputs to estuaries have led to long-term changes in phytoplankton biomass and species composition which are not related to climate change. Despite this, changes in phytoplankton biomass and phenology may be linked to those environmental factors varying with climate change. In a long-term (1978–2006) study phytoplankton dynamics in the Western Scheldt estuary, Kromkamp and Van Engeland (2010) identified a 7-year cycle in phytoplankton biomass linked to concurrent periodicity of rainfall and river discharge. Additionally, over 30 years, they found an earlier onset of the spring bloom by 1–2 days/year and, in the middle and outer estuary, the end of the bloom advanced by 1.7–1.9 days/year, giving no overall change in bloom duration. In contrast, in the inner estuary, the onset of the bloom advanced but its termination did not change thus extending the bloom by 1.4 days/year. In the brackish Baltic Sea, the onset of the summer bloom has advanced by 20 days over a 35-year period (Cloern et al., 2016). However, while the patterns identified by Kromkamp and Van Engeland (2010) could be linked to global warming, they provided no direct evidence of cause. This and other studies (Cloern and Jassby, 2010; Wetz and Yoskowitz, 2013) highlight the complexity of estuarine systems, emphasizing that the response of phytoplankton communities to climate change will be system-specific and may even vary throughout an estuary.

In addition to the timing of algal blooms, increased temperature can decrease the overall size of individual phytoplankton, and therefore community biomass, in both estuarine and marine systems (Guinder et al., 2013; Moreau et al., 2014). Increased temperature has also been linked to toxic algal blooms (Wetz and Yoskowitz, 2013) which, in nutrient-rich estuaries or enclosed seas (e.g., the Baltic), may confound the effect of temperature on cell size.

Changes in salinity, pH, CO_2, and exposure to UVB radiation are all associated with climate change. Lionard et al. (2005) highlighted the influence of salinity and light intensity on phytoplankton growth, although salinity in brackish waters appeared less important than other sources of climate change-related variation. The response to light, particularly with increased nutrient concentrations, is species-specific (Domingues et al., 2017) and despite light strongly influencing growth, there are no major effects of increased exposure to UVB (Moreau et al., 2014). In the Derwent River estuary (Australia) phytoplankton community, Nielsen et al. (2012) experimentally suggested that, as with marine systems, the predicted changes in pH and CO_2 over the next century would not significantly influence phytoplankton. In contrast, Huang et al. (2018) found CO_2 enrichment of 1000 µatm enhanced primary production.

Temperature increase is occurring with an increased frequency of extreme weather events, leading to abnormal physical conditions that can alter phytoplankton community dynamics with implications for foodweb function. However, due to the unpredictable nature of storms and the difficulties of strategic and robust monitoring of them, there is limited understanding of their immediate and long-term impacts on water quality and phytoplankton population and community dynamics (Filippino et al., 2017). Voynova and Sharp (2012) found that an anomalous high discharge event in the Delaware estuary (normally a well-mixed estuary) led to increased nutrients and abnormal stratification that abruptly increased phytoplankton biomass, coupled with bottom water oxygen depletion following phytoplankton decay. A climate change-induced shift in the Delaware estuary may become more frequent as about 50% of the extreme flood events recorded over the last century have occurred during the last decade. Howley et al. (2018) also reported an increase in phytoplankton biomass following a flood event. However, Zingone et al. (2010) and Wetz and Yoskowitz (2013) found that increased turbidity and rapid flushing may follow flood events, creating unfavorable conditions for phytoplankton growth and reductions in their productivity following floods and tropical storms (Abbate et al., 2017). Despite this, as conditions recover, phytoplankton blooms may occur weeks or months after the event (Wetz and Yoskowitz, 2013). Furthermore, the influence of storm-induced water quality changes on phytoplankton community composition and biomass varies along the estuarine salinity gradient (Filippino et al., 2017).

Maier et al. (2012) found an inverse relationship between flow and phytoplankton biomass in the Taw Estuary in SW England, where low flow conditions created the highest chlorophyll-a concentrations in the inner estuary. Spring tides and/or an increase in flow dispersed the phytoplankton, shifting the maximum concentrations to the outer estuary with high flows in combination with the spring tides; this reduced the phytoplankton biomass. Wikner and Andersson (2012) observed a similar relationship in the Baltic Sea and identified comparatively low phosphorous concentrations as the cause. Drought may reduce inputs and organic matter to estuaries (Wetz et al., 2011), thus reducing phytoplankton bloom activity and so being beneficial, although this may depend on anthropogenic nutrients behaving differently (particularly point source inputs) and leading to different responses to riverine inputs (Wetz and Yoskowitz, 2013). Similarly, Wetz et al. (2011) reported a 46%–68% reduction in dissolved inorganic nitrogen during a drought in the Neuse River estuary, North Carolina, coupled with a reduction in phytoplankton biomass. The effects of this at higher trophic levels (zooplankton and fish) were noticeable during the following years indicating that while phytoplankton communities can adjust to changing conditions relatively rapidly, the knock-on effects may be prolonged.

The relationship between phytoplankton dynamics and climate change is likely to be spatially variable within an estuary and is likely to be both estuary and event specific. Similarly, within a system, there are likely to be optimum conditions for increases in phytoplankton biomass (Hall et al., 2013) with the relationship between flow and biomass being described as parabolic, for example, as in the Douro estuary in Northern Portugal (Azevedo et al., 2014). Therefore, long-term predictions in phytoplankton dynamics, either generally or locally, according to climate change will be difficult. Phytoplankton respond rapidly to changes in flow, tidal flushing, nutrient availability, turbidity, temperature, hypoxia, and stratification and the effects may be short lived compared to those which have been observed in zooplankton, macrofauna, and fish (see below). Nevertheless, Maier et al. (2012) highlighted the importance of the effects of climate change, suggesting that it may alter the system trophic status, disrupt trophic interactions, and, possibly, favor the proliferation of nuisance species.

2.2 Zooplankton

Phytoplankton dynamics determine the structure of the zooplankton community followed by linkages between climate variability, contrasting hydrological regimes, and the development of dominant copepods, as well as the impact on the food web and fish recruitment. In the Seine estuary (France), reduced summer river discharge reduced phytoplankton biomass and estuarine zooplankton productivity with knock-on effects at higher trophic levels (Jakubavičiute et al., 2017).

The abundance and diversity of zooplankton increased during drought conditions in the Mondego estuary, Portugal, due to the reduced freshwater flow and hence salinity increased (Marques et al., 2007). The latter study also found zooplankton abundance correlated with phytoplankton biomass but there was insufficient evidence to link this to the drought. Indeed, Maier et al. (2012) and Wetz and Yoskowitz (2013) suggest that nutrient inputs (and therefore phytoplankton biomass) may be reduced during drought conditions, indicating that the changes in observed zooplankton community structure may not have been linked directly to phytoplankton biomass.

Increased temperature negatively affects both survival (Hammock et al., 2016) and mean size of zooplankton species. For example, for copepods in Long Island Sound, Rice et al. (2015) found a decrease in the mean size of *Acartia tonsa* and *Acartia hudsonica*, an increase in the proportion of the small *Oithona* sp. and the disappearance of large bodied genera since the 1950s. With decreased algal concentration (food availability), the upper salinity tolerance of the copepod *Eurytemora affinis* was reduced (Hammock et al., 2016). This species was also sensitive to temperature increase, although temperature did not act synergistically with food availability in the same way as salinity. Both studies suggested that, even where food is not limiting, climate change is likely to have impact on copepod communities, with implications for estuarine foodweb dynamics. Indeed, Taylor (1993) showed the differing salinity and temperature tolerances of the dominant estuarine holozooplankton species such that any changes to the estuarine tidal and freshwater dynamics will greatly influence the distribution of the species.

2.3 Macroalgae and Microphytobenthos

The macroscopic and microscopic estuarine algae living on hard surfaces in the intertidal and within the photic zone of the sublittoral, occasionally free-floating, or lying loose on sediment surfaces in sheltered locations, will be subject to the climate-related changes to estuarine dynamics, as well as the microphytobenthos, the predominantly surface-dwelling single-celled diatoms, euglenoids, and cyanobacteria although little is known about the effects of climate change on the structure and productivity of microphytobenthos communities. Their vertical distribution is determined by the opposite effects of their light-active vertical migration, which tends to concentrate them near the surface, and the overlying water mixing by bottom currents. Uncertainties in the vertical distribution are increased by their variable horizontal distribution. On the surface of mudflats, epipelic diatoms play an essential role in the stability of superficial sediments and in the food of zoobenthos (Checon et al., 2016) and depend closely on the nutrient content of the medium. Estuaries rich in dissolved matter are dominated by small algae of the genus *Navicula* and *Nitzschia*, while larger species populate the poorer estuaries. As emphasized by Cartaxana et al. (2015) elevated temperature will have a detrimental effect on their biomass and photosynthetic performance under elevated CO_2 concentrations. Furthermore, elevated temperature may increase the cyanobacteria and change the relative abundance of major benthic diatom species.

The red (Rhodophyta), brown (Heterokontophyta, Phaeophyceae), and green (Chlorophyta) seaweeds are most numerous on the open coast and the estuarine ones are mainly a small subset of species drawn from the open coastal flora. The estuarine flora is complex as it continuously varies along the salinity gradient of an estuary giving a widespread typical distribution pattern for many temperate areas in the Atlantic and some in the Pacific (Wilkinson et al., 1995). There is a reduced estuarine penetration from the mouth in species presence, ranging from a sheltered open coast shore and fucoid dominated with perhaps the same species richness as on the open coast—in Britain between 50 and 100 species. The sublittoral flora may be dominated by kelps, possibly with limited depth owing to the attenuation of light caused by turbidity from suspended estuarine sediment and the understory of the intertidal zone will have a mixture of red green and brown species with additional species in rock pools. This distribution changes rapidly with progression into the estuary with a selective attenuation firstly of red species, then of brown ones, so that green ones are the only ones that penetrate fully to the tidal limit.

The dominant perennial algae such as fucoids give way to short-lived opportunistic, mat-forming species which are joined in the upper reaches by conspicuous mats of other groups which are accepted as part of the attached "algal" flora of estuaries: cyanobacteria, the xanthophyte *Vaucheria*, and filamentous diatoms. The high turbidity and sediment accretion restricts the upstream penetration of sublittoral vegetation. Wilkinson et al. (1995) described this as three zones: zone A is the sheltered open coast at the mouth with fucoid domination and with many species forming a diverse understory; zone B is the transition in the lower part of the estuary where species richness declines and red species disappear; zone C is the upper estuary characterized by mat-forming, short-lived species of green algae with cyanobacteria, *Vaucheria* and diatoms. Algae may be absent in zone C although most estuaries have up to 10 species here, occasionally more, but all drawn from a pool of only 20–30 species, living on mud as well as on hard substrata. This common pool occurs both in NW Europe and many other similar areas globally.

The boundary between zones B and C usually delimits the upper limit of fucoids and in British estuaries only three *Fucus* species penetrate far into zone B: *Fucus vesiculosus*, *Fucus spiralis*, and *Fucus ceranoides* and usually only one reaches the inner part of zone B. *F. ceranoides* is regarded by many as the classic brackish-water macroalga but it is absent from approx. 35% of British estuaries, in most of which *F. vesiculosus* is the dominant. The relatively small and stable attached algae community in zone C and the inner part of zone B are very hardy, tolerant to low and variable salinities, to scouring by waterborne sediment, and to high water column turbidity. They include the species which are likely to be found in polluted and disturbed areas on the open coast and hence may be resilient to environmental changes caused by climate change effects. Using a multivariate analysis of 200 upper estuarine British sites, Wilkinson et al. (2007) did not find assemblages representing different water qualities, a situation different from those on the open coast.

Given the macroalgal salinity tolerances, climate change effects affecting the freshwater discharge, the stratification, and the tidal penetration will all affect the macroalgal penetration into estuaries. At the most simple, rising sea level will result in greater tidal penetration and isohalines moving upstream allowing marine species to migrate. However, this is further complicated as the salinity experienced by an alga fixed at an estuarine site is not characterized by the site average salinity but by the tidal fluctuation in salinity experienced by the organism which in turn is affected by the mixing/stratification pattern. In a highly stratified estuary, a sessile alga could experience all conditions from freshwater to full salinity and back again during a tidal cycle whereas in a well-mixed estuary there may only be a small salinity fluctuation at a fixed point over the tidal cycle.

Climatic changes will affect the mixing pattern of an estuary based on freshwater flow and tidal range with the former increased or decreased by rainfall patterns. Tidal range may only change if rising sea level causes greater upstream penetration of tides into narrower upstream reaches, possibly constrained in width by man-made walls or defenses, which could artificially raise the water level or alter circulation. Coastal squeeze caused by rising sea levels in areas with a constrained upper shore will limit the high tidal distribution of exposure-tolerant species.

Compared to the open coast, some important environmental changes on the macroalgae may be less important in estuaries. Firstly, geographical species range may be less affected given the relatively universal nature of the species—poor, hardy, upper estuarine species assemblage which comprises species with very wide geographic distribution. Despite this, there might be changes within estuaries due to increased saline penetration. In the estuarine Hardangerfjord in Norway, Husa et al. (2014) compared the 1950s to the present and concluded that although the large brown algae were resilient, there were more species in the estuary. There was an immigration of warmer water species and red algae had penetrated further upstream together with the change in salinity but interpretation of the patterns was confounded by anthropogenic factors.

It is of value to consider comparative changes to the higher latitude areas which may show even greater responses to climate change than temperate areas. An example of the interaction of temperature and salinity tolerance was a greater penetration of brown algae with taxa only on the open coast south of Iceland such as *Petalonia*, *Chordaria*, and *Dictyosiphon* penetrating much further up the estuaries (Munda, 1992). Therefore, rising temperatures in Icelandic estuaries may change them to the pattern seen further south in the British Isles with less brown algal penetration. Apart from this, upper estuarine algae are already accustomed to wider temperature fluctuations than those on the open coast so a slight rise in temperature may have little effect, although this was not the case in the Hardangerfjord example above where there may have also been anthropogenic effects. Duarte et al. (2013), considering range shifts of fucoids on the open coast in Spain, suggest that relicts of their populations remained unchanged in estuaries.

Although climate change may increase storminess, storms may have less effect on the sheltered estuarine environment and the effects of ocean acidification may be dampened and potentially buffering conditions in the highly variable environment of many estuaries; indeed some estuaries may already be acidified in receiving runoff from peat areas. The multivariate analysis of the species assemblages in zone C of many British estuaries mentioned earlier included some sites at the heads of estuaries fed directly by acidic upland waters. This is similar to previous concerns decades ago because of acid rain, which was shown to create no floral difference in zone C from other estuaries.

2.4 Angiosperms

Tidal wetlands actively sequester atmospheric carbon by burying sediments rich in organic matter and these represent an important part of the terrestrial biological carbon pool. Howes et al. (2010) consider that <10% of the carbon fixed through photosynthesis in saltmarshes is buried but since the rate of primary production is exceptionally high, these systems contribute significantly to carbon sequestration compared to other ecosystems This biophysical control over carbon assimilation, sequestration, and release strongly influences the dynamics of global climate systems.

Saltmarsh extent can be affected by alterations to the extent or frequency of tidal inundation of estuarine upper reaches. Tidal marshes along the margins are characterized by both horizontal (salinity) and vertical gradients. Changes in the salinity gradient will affect the distribution of different marsh types and freshwater reduction together with sea level rise will increase saline penetration and so affect very rare freshwater tidal marshes; a change able to be reversed with increased freshwater discharge. Even more important is the likely impact in the vertical dimension. Tidal marshes are typically characterized by pioneer vegetation below mean high water level followed by young and older marsh in elevated areas where vegetation characteristics change according to decreasing inundation frequency and sediment characteristics (Carus et al., 2017).

Creeks within marshes show a cyclic pattern of erosion and sedimentation which may increase the effects of sea level rise. Lower suspended solid levels in the water may hamper sedimentation and prevent the marsh plateau keeping pace with rising sea levels and so increase flooding. In contrast, the landward margin constrained by dykes or other hard structures will lead to coastal squeeze such that the surface available for marsh development is reduced as the marsh cannot migrate landwards. Increased hydrodynamic energy on the other hand might increase erosion and hence also lead to a loss of marshes. Only if there is sufficient sediment at marsh sites will the vegetation characteristics determine the extent to which sediment particles can be captured and accrete on the tidal marsh surface. Deflocculation/flocculation processes may lead to higher suspended particulate matter (SPM) values so tidal marsh will evolve more quickly to a climax vegetation state. This could result in limited plant diversity but enables tidal marshes to keep pace with the increase in mean high water level (MHWL) which can be considered as favorable for coastal protection (Bernárdez et al., 2017).

Increasing atmospheric CO_2 will enhance photosynthesis and therefore plant productivity and in saltmarsh ecosystems, carbon balance depends on the equilibrium between photosynthesis and respiration. Saltmarsh phanerogams can store carbon in sediments through photosynthesis and root exudation into the sediment and root decomposition is another important pathway. Duarte et al. (2014) found that the effect of photosynthesis as a sediment respiration modulator was not only seasonal, but also important diurnally and so C inputs to sediments increased significantly. Higher atmospheric CO_2 concentrations caused a decrease in CO_2 efflux from sediments, indicating other factors controlling sediment respiration. These authors hypothesized that a decreased root decomposition rate or a change in soil physical conditions could counteract the effect of increased C inputs from root biomass to the sediments; this questions the main factors driving respiration in the sediments.

Sediment microflora plays an important role in carbon cycling, working as a sink or source of carbon, and so sediment respiration is crucial and has gained importance due to the need to counteract climate change. Moreover, in the sediments, temperature has an opposite effect in decreasing CO_2 efflux by respiration. The high productivity of saltmarshes and their central role in organic matter decomposition result in these ecosystems having an intrinsic ability to adapt, counteracting the greenhouse effect and temperature increase by becoming more efficient carbon sinks, reducing the CO_2 emissions and therefore reducing climate change impacts in coastal areas. The salinity of sediment pore water is an important environmental factor influencing plant growth and so species composition in estuarine salt marshes. As with drought stress, increasing soil salinity decreases soil water potential and CO_2 fixation rates in marsh plants appear sensitive to increasing salinity (Maricle et al., 2007).

Many of the effects of salt on plants result from water stress and the tolerance to saline soils partly relates to specific biophysical, morphological, and biochemical adaptations in plants. The impacts of increasing temperature and changing nutrient concentrations on marsh vegetation are poorly understood. When estuarine plants suffer from large-scale changes in their chemical environment, their sensitivity to solar UV radiations (UVRs) may be influenced by environmental conditions. Solar UVR might cause inhibition of carbon fixation, damage DNA, destroy PSII photosystem, bleach pigments, increase membrane permeability, reduce nitrate or phosphorus uptake, and ultimately decrease the primary production. It can also alter the species composition and influence the food chain.

In high turbidity, higher availability of CO_2 could enhance the photosynthetic performance. Moderate levels of longer UV wavelengths (mostly UV-A) can result in positive effects, such as enhancing photosynthetic carbon fixation by plants, photorepairing of the UV-B-damaged DNA, and stimulation of the synthesis of mycosporine-like amino acids (MAAs) that play protective roles against UV-R (Rastogi and Sinha, 2011).

2.5 Benthic Invertebrates

The benthos has been less well studied than flora in the context of global changes and reliable reference conditions are more difficult to identify for estuaries because of the high level of natural variability in environmental conditions, especially in salinity (Elliott and Quintino, 2018). The extension of favorable salinity conditions will extend the landward range of marine and brackish-water species, while saline incursion into tidal freshwater areas may result in the local loss and downstream range retraction of some freshwater species (Little et al., 2016; Solan and Whiteley, 2016). Where the faunal response to saline incursion events (primarily driven by drought-induced low river flows) have been recorded, mobile marine-derived species have been observed to exhibit population changes and extend upstream (e.g., the shore crab *Carcinus maenas,* the common shrimp *Crangon crangon,* and the amphipod *Gammarus zaddachi* in the River Thames estuary; Attrill and Power, 2002). In tidal freshwater reaches, small increases in salinity can produce large changes in the upper estuarine community composition, with the fauna reduced to those species able to tolerate saline conditions. Altered species distributions are likely to result in changes in community structure, with a landward shift in patterns of estuarine benthic faunal composition based on salinity tolerance. This decreases the richness and abundance of freshwater and marine-derived species from the river and sea, respectively, toward the mid-estuary, where the community consists of a small number, but high abundance of estuarine specialists (Whitfield et al., 2012). Little et al. (2016), however, questions the widely held assumption that the salinity tolerance of benthic macrofauna alone will determine range expansion or retraction in estuaries in response to increased saline incursion. While salinity appears to be the primary environmental driver of the distribution and composition of benthic fauna in estuaries, it may act hierarchically with other abiotic factors (i.e., substratum type); superimposed upon these factors are biotic relationships of competition and predator-prey interactions (Gray and Elliott, 2009). As such, the assumption that increasing salinities will be directly reflected in a long-term, upstream shift in species distributions and patterns of community composition may be oversimplistic and not represent a complex and highly variable system and fauna (Little et al., 2017).

Freshwater-derived species inhabiting upper estuarine zones are perceived as being most at risk from increasing salinity (Little, 2012). However, the response of these species may be more complex and occur over longer time periods than is currently assumed, particularly as the physiological mechanisms that allow species and populations to survive saltwater incursion are not fully understood and neither are the timescales on which different physiological and evolutionary mechanisms could operate (Tills et al., 2010). For example, some freshwater macroinvertebrate species have considerable tolerance to salinity, particularly aquatic insects (traditionally viewed as the most sensitive taxa to salinity increases; Williams, 2009), with some species able to function normally in brackish estuarine environments. In the absence of extreme salinity increases (e.g., driven by droughts and storm surges) or "squeezing" of the tidal freshwater zone (i.e., against in-stream engineering structures such as weirs/dams/sluices), some of these species may persist in tidal conditions where salinity increases are gradual, particularly in estuaries where abiotic factors (such as substratum composition) provide suitable habitat and refuge (Williams and Hamm, 2002).

The role of tidal freshwater areas (and resident freshwater-derived fauna) on the structure and functioning of estuaries has not been quantified; however, any loss or change in these areas may be detrimental to estuarine food-web function. For example, the Aber estuary UK, recorded a >100 times increase in the mean weight of flounder (*Platichthys flesus*) (from 5 to 540 mg) between March and September, based on a diet that consisted largely of freshwater chironomid larvae (Williams and Williams, 1998).

Seasonal differences in saline incursion extents (driven by changes in river flow) may make the distributions of mobile estuarine epifauna more variable and unable to progress beyond early benthic community succession as the extension of favorable conditions extends and contracts (Ysebaert et al., 2005). The effect of rainfall in intertidal areas over a 3-year period showed the reduced rainy phases and the response of invertebrates (Ford et al., 2007). Increasing rainfall intensity, particularly associated with storm events, affect the quantity, timing, and quality of freshwater inflow into estuaries, rapidly reduce salinity, and increase nutrient loads and sediment inputs. Levinton et al. (2011) predicted that increases in precipitation in the Northeastern US may lower salinity to below the threshold required for oyster survival in estuarine regions, potentially disrupting population dynamics and impeding oyster restoration. Reduced salinity has reduced the diversity and abundance of benthic macrofauna, particularly where species are relatively immobile and reductions are pulsed. For example, all stenohaline marine sessile and infaunal molluscs died following heavy rain discharge which reduced salinities from up to 30 in 2 months in Mesquite Bay, Central Texas (Hoese, 1960).

Temperature is a major environmental variable affecting all ecological components, including the microbenthic fauna, especially in shallow waters (Solan and Whiteley, 2016). While all estuarine organisms accommodate temperature changes on daily, weekend, tidal, lunar, and seasonal cycles, their developing stages, larvae and postlarvae may be the most susceptible. Many benthic organisms have temperature thresholds for initiating breeding and settlement (Rasmussen, 1973), and so climate warming changes may bring forward the breeding period of summer spawners and delay that of winter spawners; this is analogous to changes seen near thermal effluents (McLusky and Elliott, 2004). The detailed analysis of Rasmussen (1973) presents a good example of the data required to predict thermal changes in the benthos.

Changes in mixing characteristics and residence times caused by alterations in river flow in addition to changes in water temperature and sea level may influence the vulnerability of estuaries to eutrophication (Scavia et al., 2002; Bishop et al., 2006). An increase in the frequency and duration of hypoxic events will modify the benthic fauna-dependent species tolerance to low oxygen conditions. Pulsed and short-duration hypoxic events can cause the high mortality of many species of amphipods and harpacticoid copepods, the emergence and mortality of infauna and a decline in abundance of surface-dwelling nematodes, oligochaetes and turbellarians. In contrast, some species, such as *Macoma (Limecola) balthica* can survive week-long hypoxia. The local loss of benthic invertebrate fauna will have an adverse effect at higher trophic levels. For example, degradation of *Macoma* spp. populations in the Neuse River estuary, North Carolina, due to seasonal hypoxia events appeared to reduce productivity of their commercial fish species predators (Powers et al., 2005).

Although proposed as a solution to high atmospheric CO_2, a main risk associated with carbon capture and storage (CCS) is the leakage of the stored CO_2, which can produce ecosystem effects. Prior to the 1970s, there were a few studies devoted to ocean acidification and benthic organisms. Although more studies have been carried out recently, few have considered the estuarine benthic fauna, with most research being carried out on calcareous plankton and the dependence of coral calcification rates on seawater pH and acidification (Solan and Whiteley, 2016). Changes to the seawater carbonate chemistry will affect benthic processes, organisms, communities, and ecosystems, particularly in estuaries where complex biogeochemical processes take place. However, the estuarine dynamics of pH are poorly understood as it is the ability of organisms to adapt to gradual changes in pH.

The thecosome pteropod *Limacina helicina* from Puget Sound, an urbanized estuary in the northwest continental US, experiences shell dissolution and altered mortality rates when exposed to high CO_2, low aragonite saturation state [Omega(a)] conditions. Under starvation conditions, pteropod survival may not be greatly affected by current and expected near-future aragonite saturation state but shell dissolution may be affected (Gobler and Baumann, 2016). However, the synergistic consequences of low oxygen and acidification for early life stage bivalves, and probably other marine organisms, are more severe than would be predicted hence with repercussions under future climate change scenarios.

With climate change, ultraviolet radiation (UVR) will become of increasing concern and has been shown to have different effects on different ecological components, areas, habitats, and latitudes (Solan and Whiteley, 2016). In coastal waters, due to high concentrations of suspended matter and yellow substance, the transmission of UV-B will be limited to surface layers <0.5–1 m although on tidal flats, effects on benthic organisms due to shallow water depth and emersion effects will be more direct. The UVR may directly damage cellular molecules such as DNA, proteins, and/or membranes and UV-A (315–400 nm) may also produce reactive oxygen species with its concomitant negative effects. In general terms, UVR may produce a direct and immediate decrease in survival (i.e., "lethal exposure").

Indirect, more subtle effects of UV-R that do not include immediate mortality (i.e., "sublethal exposure") are more varied and difficult to detect against background noise. The ecological implications of sublethal exposure by UV-R on natural estuarine populations are poorly known but studies have been carried out in coastal sheltered environments and allow prediction of various types of biological effects (Table 1). Eggs and larvae are usually regarded as being more vulnerable to solar UV-R radiation than older stages (Häder et al., 2011). Although planktonic organisms such as the larvae of the cockle (*Cerastoderma edule*) might not have mechanisms to avoid or to minimize UV-R-induced damage, benthic larvae [such as from the ragworm *Nereis (Hediste) diversicolor*] could be more adaptable (behavioral avoidance, bioaccumulation of UVR-absorbing compounds, efficient DNA repair systems, etc.); however, they may still be affected by solar radiation even when there is no immediate or evident effects on survival.

A changing climate may increase the likelihood of alien and invasive species colonizing estuaries. Drought-induced low freshwater flow conditions might provide ideal microhabitats for the propagation of invasive species, in particular, crustaceans, with increased larval retention due to low flows (preventing them from being washed out into the sea) and increased salinities, which are critical for larval development. The increased presence of nonindigenous species will occur due to an increased number of vectors, including opened navigation routes (such as through polar regions) and environmental conditions preferred by those species (Herborg et al., 2005).

TABLE 1 Ecological Implications of Estuarine Intertidal Benthos to "sublethal exposure" (see text)

- Decreased fecundity
- Decreased growth and development: photoperiod/solar radiation
- Adequate timing (e.g., appearance of juveniles)
- Tolerance vs age, development stage
- Decreased survival
- Decreased mobility
- Inhibition of phototactic and photophobic responses
- Decreased photorepair ability
- Change in species composition: temperature
- Growth and survival: increase in tolerance in genetic groups

Parasites can threaten populations (especially in the first year) and may increase susceptibility to environmental disturbances (such as climate or pollution) or decrease the growth rate. Hence, climate change, pollution, and parasitism can have synergistic effects. For example, weakened cockles (*C. edule*), with immune deficiency, will be more easily attacked by parasites, such as the tapeworm genus *Bucephalopsis*. Furthermore, the prevalence and intensity of parasitic infection can vary relative to the presence of predatory birds at key periods in bivalve development. While parasite infestation may not be modulated by the presence of metal or bacteria in the environment, metal bioaccumulation in turn was strongly influenced by the presence of one or several pathogens (Ibhadon and Ducrotoy, 2004). Beyond disrupting the accumulation of pollutants, the pathogens interfere with the cellular detoxification mechanisms including impairment of metallothionein synthesis.

Changes in species distribution may create new or modified community assemblages and so alter estuarine food webs through changing trophodynamics, predator-prey interactions, and intra- and interspecific competition. For example, increases in the landward extent of *C. maenas* associated with drought-driven increases in salinity, as in the Mondego estuary, Portugal (Bessa et al., 2010), could have a significant impact on prey populations of juvenile fish species (Taylor, 2005). However, replacing one species with a given set of traits with another species having similar traits will maintain the community functioning even if the species identities are changed.

2.6 Fish

Fish experience climate through temperature, winds, currents, and precipitation (Ottersen et al., 2010). In estuaries, additional climate effects on fish populations arise from alterations of habitat availability and suitability consequent to changes in the hydrological regime, saline intrusion, and water quality (Scavia et al., 2002; Roessig et al., 2004; Graham and Harrod, 2009; Gillanders et al., 2011; James et al., 2013; Robins et al., 2016). Many estuaries have shown long-term changes to fish habitat and climate change is expected to increase these changes (Amorim et al., 2017).

Most climate-related research on fish has concentrated on temperature as this has a major and direct influence on their ecology and physiology (Graham and Harrod, 2009). This affects key life history processes such as maturation, reproductive and recruitment success, egg and larval development, growth and somatic production, oxygen demand, swimming performance, immune function, etc. (Graham and Harrod, 2009 and references therein). These effects may vary with the fish life stage, reflecting ontogenetic variability in thermal tolerance ranges. The differential sensitivity to temperature may have a cascade effect on the overall viability and productivity of estuarine fish populations, particularly when affecting early life stages. Fish eggs and larvae may be particularly sensitive to temperature changes (e.g., plaice *Pleuronectes platessa*), while juveniles often have wider thermal tolerance ranges.

Other effects add to the complexity of the interaction between fish and temperature changes, for example, the impairment of antipredatory behavioral performance at higher temperature as observed for juveniles of European sea bass *Dicentrarchus labrax* (Malavasi et al., 2013). Similarly there may be the combined effect of ontogenesis on both physiological sensitivity and access to food resources (e.g., diet changes), as suggested by the faster growth of younger estuary perch *Percalates colonorum* in warmer years (Morrongiello et al., 2014). Hydrographic-driven changes in the connectivity between marine spawning sites and estuarine nursery areas may also affect recruitment success, in turn fluctuating fish stock productivity.

The response of fish to changes in the thermal regime may lead to shifts in their population extent, as observed for several marine species (Engelhard et al., 2011). Warmer water species will migrate toward temperate latitudes (Cheung et al., 2013) together with the shift in the distribution of colder water species toward northern latitudes or deeper areas (Dulvy et al., 2008; Engelhard et al., 2014). Northward migration of estuarine fish in European tidal estuaries has also been ascribed to global warming effects (Nicolas et al., 2011).

These effects combined may alter the available pool of species comprising the estuarine fish assemblage, thus potentially affecting the taxonomic composition and relative abundance of species. The direction and magnitude of such changes would be site-specific, due to the geographical constraints. For example, climatic variability influences long-term trends in fish species densities, as shown along the Dutch coast (Tulp et al., 2008); it also has a major influence on the structure of the estuarine fish assemblage, the growth of many resident juveniles, and the abundance of many of the dominant fish species using the estuary as a nursery area, as shown for the Thames Estuary (Attrill and Power, 2002).

Changes in water temperature may enhance phenological variability of fish reproduction (e.g., reproductive investment of the grass goby *Zosterisessor ophiocephalus* in the Venice lagoon; Zucchetta et al., 2010), and severely affect the reproductive and recruitment success of the species through match-mismatch dynamics (e.g., in maturation between sexes or between hatching and larval food availability; Zucchetta et al., 2012). The effects of temperature on recruitment success particularly affect the large-scale geographical distribution and productivity of fish stocks (Graham and Harrod, 2009).

Changes in the abundance and frequency of precipitation under future climate conditions may affect estuarine fish assemblages by altering the hydrological regime and the mixing of marine and fresh waters in these systems. Past climate-induced shifts in estuarine hydrological and salinity conditions, together with the warming of estuarine waters, have affected fish assemblages in the Gironde estuary (Chaalali et al., 2013). An increase in small marine pelagic fish (e.g., sprats, anchovies) and a decrease in flounder and catadromous species (e.g., smelt) was associated with periods of decreased river discharge and runoff and the consequent increased intrusion of marine waters (Pasquaud et al., 2012). Increased water salinity may reduce the species richness and diversity of estuarine fish assemblages, by reducing the contribution of freshwater and diadromous species [e.g., in Australian estuaries (Zampatti et al., 2010)].

Estuarine fish migrations may also be affected by climate-induced alteration of hydrological regimes; for example, for salmonid migrations in the southeast of England, with reduced runoff and water flow, higher water temperatures and lower water oxygenation decreasing the suitability of upstream (riverine) habitats (Solomon and Sambrook, 2004; Graham and Harrod, 2009, and references therein). The migratory behavior of flatfish larvae may also be influenced by the estuarine salinity gradient which acts as an external cue directing the colonization of estuarine nurseries from marine spawning areas (Bos and Thiel, 2006; Zucchetta et al., 2010); for example, selective tidal stream transport of larvae relies on interplay of salinity and flow conditions, both of which can be affected by climate change (Amorim et al., 2016). High freshwater discharges may also act as spawning cues for adults and provide favorable conditions for larvae, thus having an effect on the recruitment success of estuarine-dependent fish (e.g., estuary perch; Morrongiello et al., 2014).

The increase in the frequency of extreme drought events in future climate scenarios may alter fish assemblage structure and production in estuaries, particularly at lower latitudes (Dolbeth et al., 2008; Nyitrai et al., 2013; Boucek and Rehage, 2014). In the past, extreme drought events have depleted freshwater species, decreased estuarine resident fish, and increased marine stragglers in southern estuaries (Nyitrai et al., 2013). The significant reduction in the estuarine production in the driest years may also have affected the marine stocks via a reduced export (Dolbeth et al., 2008).

The loss of intertidal habitats (including tidal flats and saltmarshes) is an extreme pressure in estuarine areas, because of the cumulative impacts of several anthropogenic pressures, not least increasing land claim and habitat removal due to urbanization, agricultural occupation, and industrialization (Colclough et al., 2010). Through increasing wave height, surges and sea level rise, and the associated shifts in sediment supply and transport dynamics, climate change has the potential to modify the topography of estuarine intertidal habitats with possible effects on the overall estuarine production (Robins et al., 2016). Changes in the availability, configuration, and location of habitats are expected to be most notable in modified estuaries where the habitat high water mark is constrained by hard defenses producing coastal squeeze (Pontee, 2013).

The loss of estuarine intertidal habitat and associated food resources (e.g., decrease in benthic biomass; Yamanaka et al., 2013) may have important implications for the structure and function of fish assemblages and carrying capacity of important habitats such as nursery and feeding grounds (Laffaille et al., 2000; Elliott and Hemingway, 2002; Franco et al., 2006). This may result in a bottleneck effect for population size and productivity, particularly for those broadcast marine spawning species where the size of populations is determined by the size and availability of spawning and nursery habitats.

Climate-induced loss of estuarine marsh habitats is considered critical to the long-term survival potential of salmonids using these habitats for foraging and refuge (Koski, 2009). Similarly, the vulnerability of other highly productive estuarine nursery habitats to climate change will affect the survival and recruitment of young fish in transitional waters, and in turn the viability of fish populations there and at sea (Jones, 2014). Simplifying the estuarine habitat mosaic following habitat losses is likely to reduce the capacity of estuaries to support life-history diversity (Flitcroft et al., 2013), and to reduce the attractiveness of these systems to fish communities (Amorim et al., 2017). However, habitat fragmentation due to sea level rise may even positively affect fish nursery production (Fulford et al., 2014).

For both freshwater and marine fishes, water acidification may affect habitat selection, predator detection and avoidance behavior, as well as spawning and migration behavior (e.g., in salmonids), or mate choice and reproductive behavior (e.g., in common estuarine species, such as pipefish and sticklebacks), even when pH changes are relatively small (Williamson et al., 2013). Estuarine environments may be highly susceptible to reduced pH, due to their shallowness, lower salinity, and lower alkalinity compared to marine waters, and because of increased CO_2 via freshwater input (Miller et al., 2009). Conversely, areas experiencing naturally lowered pH my already be adapted to further acidification although further studies are required. Land use will also affect pH, for example, in estuarine catchments including peatlands. Prolonged exposure to lower pH will activate energetic trade-offs, with the additional energy needed for internal pH regulation being diverted from growth, maintenance, or reproduction (Williamson et al., 2013). Although the active high metabolism of teleost fish gives them an advantage, for example, compared to bivalves and echinoderms, in adapting to ocean acidification, a higher sensitivity has been observed for larval stages compared to adults, giving possible consequences for population success (Williamson et al., 2013; Solan and Whiteley, 2016). Ocean acidification may also influence fishery species indirectly, through negative impacts on trophically connected calcifying organisms (Le Quesne and Pinnegar, 2012).

Increased water temperature might also degrade the use of estuaries as pathways of migration for diadromous species by exacerbating deleterious effects of other environmental conditions (e.g., eutrophication, reduced levels of dissolved oxygen), particularly on those species that are sensitive to such conditions (e.g., salmonids, shads). Poor water quality with low oxygen concentration operates as a barrier to fish migration. It will become more prevalent with global warming, reducing the water-carrying capacity for oxygen (Elliott and Hemingway, 2002).

Dissolved oxygen decrease in warmer waters, coupled with the higher oxygen demand (due to higher metabolic rates), will influence many aspects of fish ecology, including habitat use, reproductive success, growth, and predation risk, and so negatively affect the carrying capacity of aquatic systems (Graham and Harrod, 2009). Effects on wind intensity and patterns, and ocean circulation, may affect the connectivity between estuarine and marine habitats, for example, by affecting the fish larval transport from spawning grounds to estuarine nurseries, with consequences there on the fish use and species composition. In addition, increased residence times in transitional waters will reduce the dilution of dissolved nutrients and pollutants and increase the flushing rate (Struyf et al., 2004). This may increase the risk and frequency of algal blooms, subsequent low oxygen conditions and greater exposure to pollutants, which may impact the estuarine fauna (Graham and Harrod, 2009).

2.7 Birds

Research into the direct and indirect effects of climate change on estuarine waterbird species has only recently been developed, reflecting the complex, multifactor influences acting on a highly mobile group. This includes direct anthropogenic effects, for example, wildfowling (hunting) effort and farming practices superimposed on climate change. For example, Eglington and Pearce-Higgins (2012) quantified the relative importance of land use and weather on bird populations and concluded both factors to influence annual population growth at a similar explanatory power.

In assessing the effects of climate change for estuarine birds, Pearce-Higgins et al. (2017) showed the globally declining status of Numeniini (curlews and godwits) populations, whereas at a species level climate change was included in a multiple parameter causal factor approach (Franks et al., 2017). This included investigating population declines in Eurasian Curlew (*Numenius arquata*) in Britain and changes in migration phenology of Jack Snipe (*Lymnocryptes minimus*) at a wetland site in Dublin Bay, Ireland in relation to recent warming (Cooney, 2017).

As a complicating factor, the highly mobile northwest European waterbirds associated with estuarine Important Bird Areas (e.g., species using the East Atlantic Flyway) for long distance migration and northwest European estuaries as staging or wintering sites along the route, means that the effects of climate change are confounded by other parameters. Phenological changes have occurred in recent decades (Cooney, 2017) and citizen science shows the potential to increase the detail of cover and the analytical power of the data (e.g., Newson et al., 2016).

All estuaries are showing changes in waterbird assemblage composition, resource utilization, and phenology, and as with the wider bird assemblage, a northward shift in distribution. Some wetland areas have species at the edges of their ranges more frequently and in greater abundance and also there are new species. However, there is also a reduction in the status of other species, and with species potentially moving along separate range trajectories depending on their specific ecological needs (e.g., Gillings et al., 2015).

The management of estuaries needs to be able to address these changes and incorporate an adaptive strategy into future plans. Using bioclimatic models to assess the ongoing efficacy of protected areas to support waterbird communities, Johnston et al. (2013) concluded that while some species are being affected by climate change, protected areas (often transitional water bodies) will continue to have a core importance for the conservation of waterbirds and the maintenance of flyway integrity.

Some studies using citizen science data, for example breeding birds records, investigated the potential for climate change to affect British birds and concluded that while some species may benefit from such change, those detrimentally affected include many of the species already listed as being of conservation concern. However, most studies, such as the one by Martay et al. (2017) focussed on terrestrial populations. This illustrates the difficulties in establishing robust conclusions of climate effects on mobile estuarine species and, in particular, water birds.

Temperature increase has a range of effects on the pathways for the alteration of breeding waterbird patterns, both in terms of triggers for migration and breeding commencement and for breeding season duration. Temperature change will also, however, alter prey availability and/or timing of prey recruitment, and the timing of fledging events when there is maximum prey abundance on breeding grounds may be detrimentally effected.

The potential for intertidal habitat loss on estuaries through coastal squeeze and through increasing urbanization and agriculture land gain, will affect habitat extent, assemblage, and associated functions for estuarine waterbirds, and have the greatest impact on wintering and passage birds. These affects will vary across an estuarine waterbird assemblage, but will be largely negative as there would be an expected net loss of habitat offering exploitable functions. However, it is also important to note that other groups of birds on estuaries will be affected by climate change. For instance, Musseau et al. (2018) showed a rapid loss of intertidal saltmarsh on the Gironde estuary, resulting from changes to physical process and relative sea level rise with an impact on suitable breeding habitat for marshland-dependent passerines. However, such climate-related modifications to physicochemical processes in estuaries can lead to deposition of sediment in some locations, and the development of marsh and reedbed habitats, with an associated increase in suitable breeding habitat for some species, for example, passerines, but a loss of foraging habitat for others, for example, waders (Cutts et al., 2012). The implications of climate change on waterbird population status are therefore complex.

Coastal squeeze is at least partially linked to climate change and the resulting loss of intertidal area will reduce carrying capacity through a direct loss of foraging and roost potential. However, this will also be influenced by a series of habitat quality parameters including physical criteria such as mudflat slope, elevation/exposure to air and scour, as well as effects on biological components such as prey composition and size. These multiple, complex factors will affect parameters such as "giving up densities" (GUDs, the prey remaining in an area after being exploited by predators) and linked to daily energy expenditure (DEE) requirements (Hagy and Kaminski, 2015). Coastal squeeze also may reduce the width of the intertidal zone, bringing birds closer to potential anthropogenic disturbance sources with implications for site take-up and DEE issues.

Furthermore, increased extreme spring/summer rainfall patterns may affect breeding success in several estuarine species, for example, avocet (*Recurvirostra avosetta*), and these can similarly be affected by tidal surge events. Impacts can occur directly through nest flooding and the drowning of young, as well as indirectly through decreased foraging potential and starvation. Storm surges may modify the morphology of the intertidal zone as well as directly affect prey availability, leading to short-term changes in the location and foraging potential of key intertidal areas. This in turn can affect the DEE for some species and may influence the suitability of an estuary for species with specific niche foraging and roost requirements. However, as noted above, climatic effects such as increased precipitation will vary spatially and some wetland areas, particularly in lower latitudes, may see a different combination of climatic variables effecting habitat and function both negatively and positively.

Although management actions may be able to address some of the alterations to community status resulting from climate change through building resistance and resilience, accommodative actions will also be required. These will include developing management actions for colonizing species, but will also need to acknowledge that where changing conditions are such that one or more species can no longer be supported within the system, then this has to be considered an acceptable loss. While some species may be detrimentally affected by climate change, some may benefit, and so a wider context of spatial distributional changes needs to be considered.

Changes in wintering population ranges reflect prey availability and environmental parameters. Warmer winters reduce the influx of continental birds, for example, mallard (*Anas platyrhynchos*), into adjacent islands such as the British Isles. Waders may undertake shorter movements or passage size and timing may alter. Breeding ranges may be altered but will vary with species and location with benefits and disadvantages for some species and linked largely to prey availability (e.g., location, abundance, and timing). As such, some species will increase their breeding range, while others will see a reduction. The net effect on breeding grounds is expected to be negative, while the net effect in wintering sites (and for passage movements) is expected to be positive. As such, an adaptive management process is required, not only to attempt to address detrimental impacts of climate change on waterbird populations, but also potentially to look toward new habitats and functions for a dynamic assemblage of currently non-native birds (Border et al., 2018). Furthermore, it might be expected that functional delivery in estuaries for the general waterbird assemblage may be at least partially maintained over time against a background of climate change, but with a change in the naturally dynamic assemblage in response to altering conditions (Johnston et al., 2013).

Finally, while the implications to waterbirds from direct climate change effects have predominantly been considered here, indirect impacts resulting from anthropogenic parameter shifts will also need to be considered. These include new farming practices, which may remove suitable roosting areas, and changes in recreational activity and Government policy such as the development of the low carbon economy and switches to renewable resources including tidal power.

3 FINAL DISCUSSION AND CONCLUSIONS

As shown here for all the ecological components, an understanding of the physics and chemistry of estuaries and the way in which climate change will change these ecosystems will be essential in showing how estuaries will adapt (or need to be adapted by human actions) to the influence of climate change. Their importance as critical transition zones linking land, freshwater, and marine habitats needs to be protected but this also may allow them to adapt. The influence of physics and chemistry on the biology with climate change superimposed over these ecosystems is crucial in understanding how ecosystems are adapting to current changing conditions. Hydromorphology is essential in structuring estuaries and in imposing essential ecological conditions to communities (Wolanski and Elliott, 2015). Assessing the effects of changes of physical conditions on communities relies on understanding the links between biology and hydromorphosedimentary factors. The challenge is now selecting for study and management the most important structural parameters in a given estuarine system.

In addition to global change, local conditions also might change markedly. Biological components have also been subject to human influences through commercial harvesting (e.g., fishing and aquaculture) of certain species as well as the introduction of alien species (either species which compete directly for resources or through the introduction of parasites and disease organisms). These may be regarded as endogenic managed pressures, that is, those pressures acting within a management area and for which both the causes and consequences of change can be addressed (Borja et al., 2011). It is the exogenic unmanaged pressures which also need to be tackled—those which emanate from outside the management area and in which management has to address the consequences whereas the causes can only be addressed externally (global action)—climate change belongs to this group of pressures.

In estuaries and lagoons, anthropogenic activities such as dredging, land claim, harbor and industrial development, urbanization and agriculture land gain, and recreational and tourism development have produced hydromorphological modifications. Furthermore, the water quality of these environments is also affected by complex discharges of pollutants such as domestic and industrial effluents. Therefore, aquatic systems are affected by multiple activities and human pressures and although there is a good understanding of individual activities and pressures, cumulative and in-combination effects are poorly understood (Elliott et al., 2018). These complex effects are then compounded by the repercussions of climate change, for example, the fate and effects of contaminants then being affected by warming, the waters, or changes to the salinity balance.

Estuaries differ by orders of magnitude in terms of size, yet they all have common properties and processes such as their connectivity, complexity, and variability, and their responses to changes in external influences are unlikely to be the same as other coastal ecosystems. For example, their variability may confer a resilience and resistance to climate change repercussions, defined as environmental homeostasis and the estuarine quality paradox (Elliott and Quintino, 2018), not shown in marine or freshwater environments. Furthermore, and perhaps again differing from other habitats, estuaries have always been modified throughout their geological history and even are regarded as ephemeral such that climate change may be regarded as yet another forcing factor to which they will adapt. Despite this, as shown here, for each ecological component there is a good conceptual knowledge and some good case studies with quantitative data, but it is difficult to predict precisely the amount of change in any component let alone the interactions between components. For example, while the benthic macrofauna may change their distributions and abundance with climate-related changes in salinity, the knock-on effect on their bird and fish predators is more difficult to predict.

Consequently, generalizations about long-term trends in estuaries are not possible without considering regional to local differences in forcing variables and responses. Hence climate change will affect the practice and future of the management of estuarine habitats. For instance, the impact of abstraction and flow regulation pressures, such as from dams, will be related to the reduction in freshwater flow into the transitional water body but climate change will affect the quantity and frequency of freshwater delivery to the estuary. In turn, the salinity balance in the estuary will then be influenced by this and the tidal pressure, in itself affected by sea level rise, the presence of storm surges, etc. As such, applying thresholds for high/medium/low sensitivity should primarily relate to the sensitivity of the ecology of the water body (under reference conditions) to changes in salinity, CO_2, or radiation.

Despite the above comments, on a time scale relevant to climate change, many other factors influencing estuaries will also change. As shown above for each component, these include the human population pressure, water abstraction for irrigation and cooling water, interference of freshwater discharge, discharge of sediments and substances, harvesting, and species introductions. Changes in climate and these more direct anthropogenic drivers affect estuaries mainly through their joint influences on freshwater flow, sediment budgets, alteration of vegetation, eutrophication, and shifts in species composition or distribution. The macrotidal, geologically young, flooded coastal plain estuaries with energy coming from the freshwater input and waves/tides from the sea are dominated by a series of ecotones, each of which will be affected by climate change (Basset et al., 2013). The position of these ecotones will change with climate and other anthropogenic changes. Some of the physical and biological parameters may be close to critical thresholds but these are not known and, as emphasized here, estuarine organisms with greater environmental tolerances may be more resistant and resilient to change (Duarte et al., 2015).

The impacts of human activities on biological communities (living resources) and habitats (mapping, evolutionary potential) in future will depend on the response of biogeochemical cycles and biotic communities to complex forcing due to climate change (Ducrotoy and Furukawa, 2016). As indicated above, there is the need to build scenarios of environmental impacts and improve monitoring and management. In order to gain an ecological vision of an estuary, research needs to focus on ecosystem functioning, the production of ecosystem services, and the delivery of societal goods and benefits (Turner and Schaafsma, 2015) rather than just on ecological structure given the dynamics of these systems. Hence monitoring needs to encompass sediment and water dynamics, including erosion and deposition cycles, water circulation along longitudinal and transverse directions, geomorphology, sedimentology, and biogeochemistry, the effects of these on all components of the biota and the ability of ecoengineering to accommodate the natural and anthropogenic change, including climate change (Elliott et al., 2016).

Finally, it is emphasized that better integrated monitoring programs that embrace physical, chemical, and biological characteristics are needed to better understand how climate change will affect ecological adjustments at local level. It is suggested here that whereas the ecological features of estuaries may adjust to climate change, it is the human features that will be more adversely affected and will require societal adaptation. Similarly, it is emphasized that in the management of such change in estuaries, we are managing the consequences of climate change rather than the causes—the latter is left to global agreements such as Paris COP 2016.

REFERENCES

Abbate, M.C.L., Molinero, J.C., Guinder, V.A., Perillo, G.M.E., et al., 2017. Time-varying environmental control of phytoplankton in a changing estuarine system. Sci. Total Environ. 609, 1390–1400.

Amorim, P., Perán, A.D., Pham, C.K., Juliano, M., Cardigos, F., Tempera, F., Morato, T., 2017. Overview of the ocean climatology and its variability in the Azores region of the North Atlantic including environmental characteristics at the seabed. Front. Mar. Sci. 4 (56), 1–16. https://doi.org/10.3389/fmars.2017.00056.

Amorim, E., Ramos, S., Elliott, M., Bordalo, A.A., 2016. Immigration and early life stages recruitment of the European flounder (Platichthys flesus) to an estuarine nursery: the influence of environmental factors. J. Sea Res. 107 (1), 56–66.

Attrill, M.J., Power, M., 2002. Climatic influence on a marine fish assemblage. Nature 417, 275–278.

Azevedo, I.C., Bordalo, A.A., Duarte, P., 2014. Influence of freshwater inflow variability on the Douro estuary primary productivity: a modelling study. Ecol. Model. 272, 1–15.

Basset, A., Barbone, E., Borja, A., Elliott, M., et al., 2013. Natural variability and reference conditions: setting type-specific classification boundaries for lagoon macroinvertebrates in the Mediterranean and Black Seas. Hydrobiologia 704 (1), 325–345.

Bernárdez, P., Prego, R., Filgueiras, A.V., et al., 2017. Lithogenic sources, composition and intra-annual variability of suspended particulate matter supplied from rivers to the Northern Galician Rias (Bay of Biscay). J. Sea Res. 130, 73–84.

Bessa, F., Baeta, A., Martinho, F., Marques, S., Pardal, M.A., 2010. Seasonal and temporal variations in population dynamics of the Carcinus maenas (L.): the effect of an extreme drought event in a southern European estuary. J. Mar. Biol. Assoc. U. K. 90, 867–876.

Bishop, M.J., Powers, S.P., Porter, H.J., Peterson, C.H., 2006. Benthic biological effects of seasonal hypoxia in a eutrophic estuary predate rapid coastal development. Estuar. Coast. Shelf Sci. 70, 415–422.

Border, J.A., Johnston, A., Gillings, S., 2018. Can climate change matching predict the current and future climatic suitability of the UK for the establishment of non-native birds. Bird Study 65, 72–83.

Borja, A., Barbone, E., Basset, A., Borgersen, G., et al., 2011. Response of single benthic metrics and multi-metric methods to anthropogenic pressure gradients, in five distinct European coastal and transitional ecosystems. Mar. Pollut. Bull. 62, 499–513.

Bos, A.R., Thiel, R., 2006. Influence of salinity on the migration of postlarval and juvenile flounder Pleuronectes flesus L. in a gradient experiment. J. Fish Biol. 68, 1411–1420.

Boucek, R.E., Rehage, J.S., 2014. Climate extremes drive changes in functional community structure. Glob. Chang. Biol. 20, 1821–1831.

Cartaxana, P., Vieira, S., Ribeiro, L., Calado, R., Da Silva, J.M., 2015. Effects of elevated temperature and CO2 on intertidal microphytobenthos. BMC Ecol. 15 (1). article number10.

Carus, J., Heuner, M., Paul, M., Schröder, B., 2017. Plant distribution and stand characteristics in brackish marshes: unravelling the roles of abiotic factors and interspecific competition. Estuar. Coast. Shelf Sci. 196, 237–247.

Chaalali, A., Beaugrand, G., Boët, P., Sautour, B., 2013. Climate-caused abrupt shifts in a European macrotidal estuary. Estuar. Coasts 36, 1193–1205.

Checon, H.H., Pardo, E.V., Zacagnini Amaral, A., 2016. Breadth and composition of polychaete diets and the importance of diatoms to species and trophic guilds. Helgol. Mar. Res., 70 Article Number 19.

Cheung, W.W., Watson, R., Pauly, D., 2013. Signature of ocean warming in global fisheries catch. Nature 497, 365–368.

Cloern, J.E., Jassby, A.D., 2010. Patterns and scales of phytoplankton variability in estuarine-coastal ecosystems. Estuar. Coasts 33 (2), 230–241.

Cloern, J.E., Abreu, P.C., Carstensen, J., Chauvaud, L., et al., 2016. Human activities and climate variability drive fast-paced change across the world's estuarine-coastal ecosystems. Glob. Chang. Biol. 22, 513–529.

Colclough, S., Fonseca, L., Watts, W., Dixon, M., 2010. In: High tidal flats, salt marshes and managed realignments as habitats for fish, science for nature conservation and management: the Wadden Sea Ecosystem and EU Directives. 12th International Scientific Wadden Sea Symposium 30 March–3 April 2009. Wadden Sea Secretariat, Wilhelmshaven, Germany, Wilhelmshaven, Germany. pp. 115–120.

Cooney, T., 2017. Migration phenology of Jack Snipe (Lymnacryptes minimus) at an Irish coastal wetland. Irish Birds 10, 463–468.

Cutts, N.D., Thompson, S.M., Franco, A., Hemingway, K.L., 2012. Temporal and spatial changes in habitat and waterfowl assemblage in the upper Humber. Research report to Natural England.

Dolbeth, M., Martinho, F., Viegas, I., Cabral, H., Pardal, M.A., 2008. Estuarine production of resident and nursery fish species: conditioning by drought events? Estuar. Coast. Shelf Sci. 78, 51–60.

Domingues, R.B., Guerra, C.G., Barbosa, A.B., Galvão, H.M., 2017. Will nutrient and light limitation prevent eutrophication in an anthropogenically-impacted coastal lagoon. Cont. Shelf Res. 141, 11–25.

Duarte, L., Viejo, R.M., Martinez, B., deCastro, M., Gomez-Gesteira, M., Gallardo, T., 2013. Recent and historical range shifts of two canopy-forming seaweeds in North Spain and the link with trends in sea surface temperatures. Acta Oecol. 51, 1–10.

Duarte, B., Santos, D., Silva, H., Marques, J.C., Caçador, I., Sleimi, N., 2014. Light-dark O_2 dynamics in submerged leaves of C3 and C4 halophytes under increased dissolved CO_2: clues for saltmarsh response to climate change. AoB Plants 6. http://dx.doi.org/10.1093%2Faobpla%2Fplu067.

Duarte, C.M., Borja, A., Carstensen, J., Elliott, M., Krause-Jensen, D., Marbà, N., 2015. Paradigms in the recovery of estuarine and coastal ecosystems. Estuar. Coasts 38 (4), 1202–1212.

Ducrotoy, J.-P., 2011. Ecological restoration of tidal estuaries in North Western Europe: an adaptive strategy to multi-scale changes. Plankton Benthos Res. 5 (Suppl) 4–14.

Ducrotoy, J.-P., Furukawa, K., 2016. Integrated coastal management: lessons learned to address new challenges. Mar. Pollut. Bull. 102 (2), 241–242.

Dulvy, N.K., Rogers, S.I., Jennings, S., Stelzenmüller, V., Dye, S.R., Skjoldal, H.R., 2008. Climate change and deepening of the North Sea fish assemblage: a biotic indicator of warming seas. J. Appl. Ecol. 45, 1029–1039.

Eglington, S.M., Pearce-Higgins, J.W., 2012. Disentangling the relative importance of changes in climate and land-use intensity in driving recent bird population trends. PLoS One 7 (3). https://doi.org/10.1371/journal.pone.0030407.

Elliott, M., Hemingway, K.L. (Eds.), 2002. Fishes in Estuaries. Blackwell Science, Oxford, p. 636.

Elliott, M., Quintino, V.M., 2018. The estuarine quality paradox concept. In: Encyclopaedia of Ecology, Second ed. (Editor-in-Chief) B. Fath. Elsevier, Amsterdam, ISBN: 9780444637680.

Elliott, M., Whitfield, A., 2011. Challenging paradigms in estuarine ecology and management. Estuar. Coast. Shelf Sci. 94, 306–314.

Elliott, M., Mander, L., Mazik, K., Simenstad, C., et al.Valesini, F., Whitfield, A., Wolanski, E., 2016. Ecoengineering with ecohydrology: successes and failures in estuarine restoration. Estuar. Coast. Shelf Sci. 176, 12–35.

Elliott, M., Boyes, S.J., Barnard, S., Borja, Á., 2018. Using best expert judgement to harmonise marine environmental status assessment and maritime spatial planning. Mar. Pollut. Bull. 133, 367–377.

Engelhard, G.H., Pinnegar, J.K., Kell, L.T., Rijnsdorp, A.D., 2011. Nine decades of North Sea sole and plaice distribution. ICES J. Mar. Sci. 68, 1090–1104.

Engelhard, G.H., Righton, D.A., Pinnegar, J.K., 2014. Climate change and fishing: a century of shifting distribution in North Sea cod. Glob. Chang. Biol. 20, 2473–2483.

Filippino, K.C., Egerton, T., Hunley, W.S., Mulholland, M.R., 2017. The influence of storms on water quality and phytoplankton dynamics in the Tidal James River. Estuar. Coasts 40, 80–94.

Flitcroft, R., Burnett, K., Christiansen, K., 2013. A simple model that identifies potential effects of sea-level rise on estuarine and estuary-ecotone habitat locations for Salmonids in Oregon, USA. Environ. Manag. 52, 196–208.

Ford, R.B., Anderson, M.J., Kelly, S., 2007. Subtle and negligible effects of rainfall on estuarine infauna: evidence from three years of event-driven sampling. Marine Ecol.-Prog. Ser. 340, 17–27.

Franco, A., Franzoi, P., Malavasi, S., Riccato, F., Torricelli, P., 2006. Fish assemblages in different shallow water habitats of the Venice Lagoon. Hydrobiologia 555, 159–174.

Franks, S.E., Douglas, D.J.T., Gillings, S., Pearce-Higgins, J.W., 2017. Environmental correlates of breeding abundance and population change of Eurasian Curlew Numenius arquata in Britain. Bird Study 64 (3), 393–409.

Fulford, R.S., Peterson, M.S., Wu, W., Grammer, P.O., 2014. An ecological model of the habitat mosaic in estuarine nursery areas: Part II—projecting effects of sea level rise on fish production. Ecol. Model. 273, 96–108.

Gillanders, B.M., Elsdon, T.S., Halliday, I.A., Jenkins, G.P., Robins, J.B., Valesini, F.J., 2011. Potential effects of climate change on Australian estuaries and fish utilising estuaries: a review. Mar. Freshw. Res. 62, 1115–1131.

Gillings, S., Balmer, D.E., Fuller, R.J., 2015. Directionality of recent bird distribution shifts and climate change in Great Britain. Glob. Chang. Biol. 21, 2155–2168.

Gobler, C.J., Baumann, H., 2016. Hypoxia and acidification in ocean ecosystems: coupled dynamics and effects on marine life. Biol. Lett. 12 (5). http://dx.doi.org/10.1098/rsbl.2015.0976.

Graham, C.T., Harrod, C., 2009. Implications of climate change for the fishes of the British Isles. J. Fish Biol. 74, 1143–1205.

Gray, J.S., Elliott, M., 2009. Ecology of Marine Sediments: Science to Management. Oxford University Press, Oxford260.

Guinder, V.A., Popovich, C.A., Molinero, J.C., Marcoveccio, J., 2013. Phytoplankton summer bloom dynamics in the Bahía Blanca Estuary in relation to changing environmental conditions. Cont. Shelf Res. 52, 150–158.

Häder, D.-P., Helbling, E.W., Williamson, C.E., Worrest, R.C., 2011. Effects of UV radiation on aquatic ecosystems and interactions with climate change. Photochem. Photobiol. Sci. 10, 242–260.

Hagy, H.M., Kaminski, R.M., 2015. Determination of foraging thresholds and effects of application on energetic carrying capacity for waterfowl. PLoS One 10 (3), e0118349. https://doi.org/10.1371/journal.pone.0118349.

Hall, N.S., Paerl, H.W., Peierls, B.L., Whipple, A.C., Rossignol, K.L., 2013. Effects of climatic variability on phytoplankton community structure and bloom development in the eutrophic, microtidal, New River Estuary, North Carolina, USA. Estuar. Coast. Shelf Sci. 117, 70–82.

Hammock, B.G., Lesmeister, S., Flores, I., Bradhurd, G.S., et al., 2016. Low food availabilty narrows the tolerance of the copepod Eurytemora affinis to salinity, but not to temperature. Estuar. Coasts 39, 189–200.

Herborg, L.M., Rushton, S.P., Clare, A.S., Bentley, M.G., 2005. The invasion of the Chinese mitten crab (*Eriocheir sinensis*) in the United Kingdom and its comparison to continental Europe. Biol. Invasions 7, 959–968.

Hoese, H.D., 1960. Biotic changes in a bay associated with the end of a drought. Limnol. Oceanogr. 5, 326–336.

Holleman, R.C., Stacey, M.T., 2014. Coupling of sea level rise, tidal amplification, and inundation. J. Phys. Oceanogr. 44, 1439–1455.

Howes, N.C., FitzGerald, D.M., Hughes, Z.J., Georgiou, I.Y., et al., 2010. Hurricane-induced failure of low salinity wetlands. Proc. Natl. Acad. Sci. 107 (32), 14014–14019.

Howley, C., Devlin, M., Burford, M., 2018. Assessment of water quality from the Normanby River catchment to coastal flood plumes on the northern Great Barrier Reef, Australia. Mar. Freshw. Res. 69 (6), 859–873.

Huang, Y., Liu, X., Laws, E.A., Chen, B., et al., 2018. Effects of increasing CO2 on the marine phytoplankton and bacterial metabolism during a bloom: a coastal mesocosm study. Sci. Total Environ. 633, 618–629.

Husa, V., Steen, H., Sjotun, K., 2014. Historical changes in macroalgal communities in Hardangerfjord (Norway). Mar. Biol. Res. 10, 226–240.

Ibhadon, A.O., Ducrotoy, J.-P., 2004. Ecological interpretation of metal contents and contaminant source characterisation of sediments from a megatidal estuary. J. Environ. Monit. 6 (8), 684–688.

Jakubavičiute, E., Casini, M., Ložys, L., Olsson, J., 2017. Seasonal dynamics in the diet of pelagic fish species in the southwest Baltic Proper. ICES J. Mar. Sci. 74 (3), 750–758.

James, N.C., van Niekerk, L., Whitfield, A.K., Potts, W.M., et al., 2013. Effects of climate change on South African estuaries and associated fish species. Clim. Res. 57, 233–248.

Johnston, A., Ausden, M., Dodd, A.M., Bradbury, R.B., et al., 2013. Observed and predicted effects of climate change on species abundance in protected areas. Nat. Clim. Chang. 3, 1055–1061.

Jones, C.M., 2014. Can we predict the future: juvenile finfish and their seagrass nurseries in the Chesapeake Bay. ICES J. Marine Sci.: J. Conseil 71, 681–688.

Koski, K., 2009. The fate of coho salmon nomads: the story of an estuarine-rearing strategy promoting resilience. Ecol. Soc. 14, 4.

Kromkamp, J.C., Van Engeland, T., 2010. Changes in phytoplankton biomass in the Western Scheldt estuary durin the period 1978–2006. Estuar. Coasts 33, 270–285.

Laffaille, P., Feunteun, E., Lefeuvre, J.C., 2000. Composition of fish communities in a European macrotidal salt marsh (the Mont Saint-Michel Bay, France). Estuar. Coast. Shelf Sci. 51, 429–438.

Le Quesne, W.J.F., Pinnegar, J.K., 2012. The potential impacts of ocean acidification: scaling from physiology to fisheries. Fish Fish. 13, 333–344.

Levinton, J., Doall, M., Ralston, D., Starke, A., Allam, B., 2011. Climate change, precipitation and impacts on an estuarine refuge from disease. PLoS One 6, e18849.

Lionard, M., Muylaert, K., Van Gansbeke, D., Vyverman, W., 2005. Influence of changes in salinity and light intensity on growth of phytoplankton communities from the Schelde river and estuary (Belgium/the Netherlands). Hydrobiologia 540, 105–115.

Little, S., 2012. The impact of increasing saline penetration upon estuarine and riverine benthic macroinvertebrates. Unpublished PhD Thesis, Loughborough University.

Little, S., Wood, P.J., Elliott, M., 2016. Quantifying salinity-induced changes on estuarine benthic fauna: the potential implications of climate change. Estuar. Coast. Shelf Sci. 198 (B), 610–625.

Little, S., Spencer, K.L., Schuttelaars, H.M., Millward, G.E., Elliott, M., 2017. Unbounded boundaries and shifting baselines: estuaries and coastal seas in a rapidly changing world. Estuar. Coast. Shelf Sci. 198 (B), 311–319.

Maier, G., Glegg, G., Tappin, A.D., Worsfold, P.J., 2012. A high resolution temporal study of phytoplankton bloom dynamics in the eutrophic Taw estuary (SW England). Sci. Total Environ. 434, 228–239.

Malavasi, S., Cipolato, G., Cioni, C., Torricelli, P., et al., 2013. Effects of temperature on the antipredator behaviour and on the cholinergic expression in the European sea bass (Dicentrarchus labrax L.) juveniles. Ethology 119, 592–604.

Maricle, B.R., Cobos, D.R., Campbell, C.S., 2007. Biophysical and morphological leaf adaptations to drought and salinity in salt marsh grasses. Environ. Exp. Bot. 60 (3), 458–467.

Marques, S.C., Azeiteiro, U.M., Martinho, F., Pardal, M.A., 2007. Climate variability and planktonic communities: the effect of an extreme event (severe drought) in a southern European estuary. Estuar. Coast. Shelf Sci. 73, 725–734.

Martay, B., Brewer, M.J., Elston, D.A., Bell, et al., 2017. Impacts of climate change on national biodiversity population trends. Ecography 40, 1139–1151.

McLusky, D.S., Elliott, M., 2004. The Estuarine Ecosystem; Ecology, Threats and Management, third ed. Oxford University Press, Oxford, p. 216.

Miller, A.W., Reynolds, A.C., Sobrino, C., Riedel, G.F., 2009. Shellfish face uncertain future in high CO2 world: influence of acidification on oyster larvae calcification and growth in estuaries. PLoS One 4, e5661.

Moreau, S., Mostajir, B., Almandoz, G.O., Demers, S., et al., 2014. Effects of enhanced temperature and ultraviolet B radiation on a natural plankton community of the Beagle Channel (Southern Argentina): a mesocosm study. Aquat. Microb. Ecol. 72, 155–173.

Morrongiello, J.R., Walsh, C.T., Gray, C.A., Stocks, J.R., Crook, D.A., 2014. Environmental change drives long-term recruitment and growth variation in an estuarine fish. Glob. Chang. Biol. 20, 1844–1860.

Munda, I.M., 1992. Gradient in seaweed vegetation patterns along the North Icelandic coast, related to hydrographic conditions. Hydrobiologia 242 (3), 133–147.

Musseau, R., Boutault, L., Beslic, S., 2018. Rapid losses of intertidal salt marshes due to global change in the Gironde estuary (France) and conservation implications for marshland passerines. J. Coast. Conserv. 22, 443–451.

Newson, S.E., Moran, N.J., Musgrove, A.J., Pearce-Higgins, J.W., et al., 2016. Long-term changes in the migration phenology of UK breeding birds detected by large-scale citizen science recording schemes. Ibis 158 (3), 481–495.

Nicolas, D., Chaalali, A., Drouineau, H., Lobry, J., et al., 2011. Impact of global warming on European tidal estuaries: some evidence of northward migration of estuarine fish species. Reg. Environ. Chang. 11, 639–649.

Nielsen, L.T., Hallegraeff, G.M., Wright, S.W., Hansen, P.J., 2012. Effects of experimental seawater acidification on an estuarine plankton community. Aquat. Microb. Ecol. 65, 271–285.

Nyitrai, D., Martinho, F., Dolbeth, M., Rito, J., Pardal, M.A., 2013. Effects of local and large-scale climate patterns on estuarine resident fishes: the example of Pomatoschistus microps and Pomatoschistus minutus. Estuar. Coast. Shelf Sci. 135, 260–268.

Ottersen, G., Kim, S., Huse, G., Polovina, J.J., Stenseth, N.C., 2010. Major pathways by which climate may force marine fish populations. J. Mar. Syst. 79, 343–360.

Pasquaud, S., Béguer, M., Larsen, M.H., Chaalali, A., Cabral, H., Lobry, J., 2012. Increase of marine juvenile fish abundances in the middle Gironde estuary related to warmer and more saline waters, due to global changes. Estuar. Coast. Shelf Sci. 104–105, 46–53.

Pearce-Higgins, J.W., Beale, C.M., Oliver, T.H., August, T.A., Carroll, M., Massimino, D., Ockendon, N., Savage, J., Wheatley, C.J., Ausden, M.A., Bradbury, R.B., Duffield, S.J., Macgregor, N.A., McClean, C., Morecroft, M.D., Thomas, C.D., Watts, O., Beckmann, B.C., Fox, R., Sutton, P.G., Crick, H.Q.P., 2017. A national-scale assessment of climate change impacts on species: assessing the balance of risks and opportunities for multiple taxa. Biol. Cons. 213, 124–134.

Pontee, N., 2013. Defining coastal squeeze: a discussion. Ocean Coast. Manag. 84, 204–207.

Powers, S.P., Peterson, C.H., Christian, R.R., Sullivan, E., Powers, M.J., Bishop, M.J., Buzzelli, C.P., 2005. Effects of eutrophication on bottom habitat and prey resources of demersal fishes. Mar. Ecol. Prog. Ser. 302, 233–243.

Rasmussen, E., 1973. Systematics and ecology of the Isefjord marine fauna (Denmark). Ophelia 11, 1–507.

Rastogi, R.P., Sinha, R.P., 2011. Solar ultraviolet radiation-induced DNA damage and protection/repair strategies in cyanobacteria. Int. J. Pharm. Biosci. 2, 271–288.

Rice, E., Dam, H.G., Stewart, G., 2015. Impact of climate change on estuarine zooplankton: surface water warming in Long Island Sound is associated with changes in copepod size and community structure. Estuar. Coasts 38, 13–23.

Robins, P.E., Skov, M.W., Lewis, M.J., et al., 2016. Impact of climate change on UK estuaries: a review of past trends and potential projections. Estuar. Coast. Shelf Sci. 169, 119–135.

Roessig, J.M., Woodley, C.M., Cech, J.J.J., Hansen, L.J., 2004. Effects of global climate change on marine and estuarine fishes and fisheries. Rev. Fish Biol. Fish. 14, 251–275.

Scavia, D., Field, J.C., Boesch, D.F., Buddemeier, R.W., et al., 2002. Climate change impacts on U.S. coastal and marine ecosystems. Estuaries 25, 149–164.

Solan, M., Whiteley, N. (Eds.), 2016. Stressors in the Marine Environment: Physiological and Ecological Responses: Societal Implications. Oxford University Press, Oxford.

Solomon, D.J., Sambrook, H.T., 2004. Effects of hot dry summers on the loss of Atlantic salmon, Salmo salar, from estuaries in South West England. Fish. Manag. Ecol. 11, 353–363.

Stratham, P.J., 2012. Nutrients in estuaries—an overview and the potential impacts of climate change. Sci. Total Environ. 434, 213–227.

Struyf, E., Van Damme, S., Meire, P., 2004. Possible effects of climate change on estuarine nutrient fluxes: a case study in the highly nutrified Schelde estuary (Belgium, The Netherlands). Estuar. Coast. Shelf Sci. 60, 649–661.

Taylor, C.J.L., 1993. The zooplankton of the Forth Estuary. Neth. J. Aquat. Ecol. 27 (2–4), 87–99.

Taylor, D.L., 2005. Predatory impact of the green crab (Carcinus maenas Linnaeus) on post-settlement winter flounder (Pseudopleuronectes americanus Walbaum) as revealed by immunological dietary analysis. J. Exp. Mar. Biol. Ecol. 324, 112–126.

Tills, O., Spicer, J.I., Rundle, S.D., 2010. Salinity-induced heterokairy in an upper-estuarine population of the snail Radix balthica (Mollusca: Pulmonata). Aquat. Biol. 9, 95–105.

Tulp, I., Bolle, L.J., Rijnsdorp, A.D., 2008. Signals from the shallows: in search of common patterns in long-term trends in Dutch estuarine and coastal fish. J. Sea Res. 60 (1-2), 54–73.

Turner, R.K., Schaafsma, M. (Eds.), 2015. Coastal Zones Ecosystem Services: From Science to Values and Decision Making. Springer Ecological Economic Series, Springer International Publications, Switzerland, ISBN: 978-3-319-17213-2.

Voynova, Y.G., Sharp, J.H., 2012. Anomalous biogeochemical response to a flooding event in the Delaware estuary: a possible typology shift die to climate change. Estuar. Coasts 35, 943–958.

Wetz, M.S., Yoskowitz, D.W., 2013. An "extreme" future for estuaries? Effects of extreme climatic events on estuarine water quality and ecology. Marine Poll. Bull. 69, 7–18.

Wetz, M.S., Hutchinson, E.A., Luneta, R.S., Paerl, H.W., Taylor, C.J., 2011. Sefere droughts reduce estuarine productivity with cascading effects on higher trophic levels. Limnol. Ocanogr. 56 (2), 627–638.

Whitfield, A.K., Elliott, M., Basset, A., Blaber, S.J.M., West, R.J., 2012. Paradigms in estuarine ecology—the Remane diagram with a suggested revised model for estuaries: a review. Estuar. Coast. Shelf Sci. 97, 78–90.

Wikner, J., Andersson, A., 2012. Increased freshwater discharge shifts the trophic balance in the coastal zone of the northern Baltic Sea. Glob. Chang. Biol. 18, 2509–2519.

Wilkinson, M., Telfer, T., Grundy, S., 1995. Geographical variations in the distributions of macroalgae in estuaries. Neth. J. Aquat. Ecol. 29, 359–368.

Wilkinson, M., Wood, P., Wells, E., Scanlan, C., 2007. Using attached macroalgae to assess ecological status of British estuaries for the European water framework directive. Mar. Pollut. Bull. 55, 136–150.

Williams, D., 2009. Coping with saltwater: the conditions of aquatic insects in estuaries as determined by gut content analysis. Open Marine Bio. J. 3, 21–27.

Williams, D.D., Hamm, T., 2002. Insect community organisation in estuaries: the role of the physical environment. Ecography 2, 372–384.

Williams, D.D., Williams, N.E., 1998. Seasonal variation, export dynamics and consumption of freshwater invertebrates in an estuarine environment. Estuar. Coast. Shelf Sci. 46, 393–410.

Williamson, P., Turley, C., Brownlee, C., Findlay, H.S., et al., 2013. Impacts of ocean acidification. MCCIP Sci. Rev. 34–48.

Wolanski, E., Elliott, M., 2015. Estuarine Ecohydrology: An Introduction. Elsevier, Amsterdam. ISBN: 978-0-444-63398-9, p. 322.

Yamanaka, T., Raffaelli, D., White, P.C., 2013. Non-linear interactions determine the impact of sea-level rise on estuarine benthic biodiversity and eco-system processes. PLoS One 8, e68160.

Ysebaert, T., Fettweis, M., Meire, P., Sas, M., 2005. Benthic variability in intertidal soft-sediments in the mesohaline part of the Schelde estuary. Hydrobiologia 540, 197–216.

Zampatti, B.P., Bice, C.M., Jennings, P.R., 2010. Temporal variability in fish assemblage structure and recruitment in a freshwater-deprived estuary: the Coorong, Australia. Mar. Freshw. Res. 61, 1298–1312.

Zingone, A., Philips, E.J., Harrison, P.J., 2010. Multiscale variability of twenty-two coastal phytoplankton time series: a global scale comparison. Estuar. Coasts 33, 224–229.

Zucchetta, M., Franco, A., Torricelli, P., Franzoi, P., 2010. Habitat distribution model for European flounder juveniles in the Venice lagoon. J. Sea Res. 64, 133–144.

Zucchetta, M., Cipolato, G., Pranovi, F., Antonetti, P., Torricelli, P., Franzoi, P., Malavasi, S., 2012. The relationships between temperature changes and reproductive investment in a Mediterranean goby: insights for the assessment of climate change effects. Estuar. Coast. Shelf. S. 101, 15–23.

Chapter 34

Plastic Pollution in the Coastal Environment: Current Challenges and Future Solutions

K. Critchell[*,†], A. Bauer-Civiello[*,a], C. Benham[*], K. Berry[*], L. Eagle[‡], M. Hamann[*], K. Hussey[§], T. Ridgway[¶]

[*]College of Science and Engineering, James Cook University, Townsville, QLD, Australia, [†]Marine Spatial Ecology Lab, University of Queensland, Brisbane, QLD, Australia, [‡]College of Business, Law and Governance, James Cook University, Townsville, QLD, Australia, [§]Centre for Policy Futures, Faculty of Humanities and Social Sciences, The University of Queensland, St Lucia, QLD, Australia, [¶]Global Change Institute, The University of Queensland, St Lucia, QLD, Australia

1 PLASTIC POLLUTION IN THE MARINE ENVIRONMENT: AN EMERGING CONTAMINANT OF GLOBAL CONCERN

Over the past century, plastic has become an increasingly common and convenient manufacturing material, replacing more traditional materials such as glass, aluminum, and natural fibers (Andrady and Neal, 2009). Across the globe, the use and production of polymers has grown exponentially (Gross, 2017) and consequently the proportion of plastic-based products entering the waste steam has increased (Jambeck et al., 2015). Plastic items now comprise up to 80% of marine litter (Barnes et al., 2009; Boteler et al., 2015; Watkins et al., 2015; Veiga et al., 2016), and estimates of abundance in global oceans range from 5.2 trillion pieces (Eriksen et al., 2014) to 15–51 trillion pieces (Van Sebille et al., 2015). The plastic load entering the oceans from land is large and increasing, with estimates between 4 and 12 million tonnes per year (Jambeck et al., 2015). Due to the nature of manufacturing, and the longevity of the product, plastics remain in the environment for long periods of time. This makes plastic pollution one of the most ubiquitous and pressing contemporary threats to the world's coastal and marine environments (Thompson et al., 2009; Worm et al., 2017).

Microplastics (plastic particles <5 mm in size) are now recognized as a globally ubiquitous contaminant, and have received an increasing amount of research and policy attention over the last decade. Microplastics represent a diverse mix of polymer types, virgin plastics, and secondary fragments arising from degraded macroplastics (Browne et al., 2007). Microplastic pollution was first identified in marine and coastal environments in the 1960s; by the early 1970s the first accounts of microplastic ingestion by birds were identified; by the 1990s it was clear that the ingestion of microplastics had a toxic effect on some species; and by the 2000s implications for human health emerged (see Vegter et al., 2014 and Chae and An, 2017 for reviews). Although microplastics were recognized as a pollutant of concern in the 1990s, most of the early research on plastics in marine and coastal environments focused on macroplastics and the risk they pose to animal entanglement (Vegter et al., 2014). Despite the increasing knowledge regarding the potential harmful effects of microplastics, implementation of policy and management of microplastics remains in its infancy.

There is an increasing body of literature indicating that microplastic pollution is accumulating in coastal and marine environments (Derraik, 2002; Jambeck et al., 2015), and is having a detrimental impact to species and habitats (Laist, 1997; Henderson, 2001; Boren et al., 2006; Bravo Rebolledo et al., 2013), including humans (Chae and An, 2017). Ingestion of microplastic particles is known to occur at multiple trophic levels, from nonspecific feeders, such as plankton (Cole et al., 2013, 2015), coral (Hall et al., 2015), mussels (Castro et al., 2016), and lug worms (Van Cauwenberghe et al., 2015) through to larger animals such as marine turtles (Schuyler et al., 2012) and cetaceans (Baulch and Perry, 2014). Some of the best described interactions occur in groups of surface feeding sea birds (Ryan, 1988; Moser and Lee, 1992; Acampora et al., 2014;

---

a The order of the authors is alphabetical after the first author.

Verlis et al., 2014). Although microplastic research is growing, most studies have been focused on effects on the individual and species-level effects such as growth, fecundity, and health associated with microplastic consumption (Cole et al., 2015; Watts et al., 2015), and fewer have examined tropic transfer, population-level effects (including humans), or the long-term effects of ingestion (Vegter et al., 2014; Nelms et al., 2016).

It is clear that microplastic pollution arises from a diffuse array of land or catchment-based sources, for example, urban and industry-generated runoff or discharges, agricultural runoff, and wastewater treatment discharge (Batista et al., 2014). From these sources, plastics are often transported into freshwater or coastal habitats and then distributed and concentrated along the coasts or offshore by oceanic and atmospheric conditions. While microplastic pollutants are being passively distributed throughout the world's coastal and marine habitats, the recognized and potential solutions require innovative thinking and improved understanding of the social, economic, and political aspects of use and disposal.

In this chapter, we explore the issue of plastic pollution in coastal and marine environments. We discuss the sources and pathways for movement of plastics, challenges associated with addressing this issue of emerging global concern, and current initiatives to reduce the current load and future input of plastic pollution to marine and coastal environments. In particular, we examine the potential for economic, social, and behavioral change to provide lasting and systemic solutions to plastic pollution.

2 SOURCES AND METHODS OF DISPERSAL OF MICROPLASTIC POLLUTION IN THE COASTAL AND MARINE ENVIRONMENT

Microplastic particles generally enter the coastal and marine environment through accidental loss during manufacture and transport, urban-based litter flowing into storm drains rivers and aquatic systems, and as part of treated sewage outfall. However, the relative contribution of these source types to coastal and marine microplastic loads is difficult to quantify and is likely to vary on temporal and spatial scales. In this section, we explore current knowledge on the three main sources of microplastic pollution and the dispersal mechanisms into and within coastal and marine systems.

3 LOSS OF VIRGIN MICROPLASTICS DURING MANUFACTURE OR TRANSPORT

Virgin pellets, or pellets manufactured from recycled plastics, are raw forms of plastic and they are often transported from the point of manufacture to a secondary location for processing into new materials. These virgin pellets are shaped like small beads to act like a fluid, which can be poured and leveled easily, making transport and processing more cost effective. They are lost into the environment from accidental spills or through careless on-site transfer, cleanup, or processing systems at processing plants (Mato et al., 2001; Turner and Holmes, 2011). It is difficult to quantify the amount of these virgin microplastic pellets coming from different sources. However, virgin pellets have been found in vast quantities in areas in close proximity to plastic processing industries, and this inspired the industry lead, voluntary initiative: "Operation Clean Sweep" (American Chemistry Council, 2012), which aims to reduce loss of pellets to zero.

4 MICROPLASTICS FROM HOUSEHOLDS—FIBERS AND MICROBEADS

Another component of virgin plastic is the microplastic fibers and microbeads which are used in fabric and health care products, respectively. Synthetic fabrics, which lead to microplastic fibers, are becoming one of the most commonly used materials in the manufacture of clothing, towels, upholstery, and linen, especially in the era of fast fashion. Plastic microbeads are a common ingredient in personal care products like face wash, skin cleansers and industrial, domestic, and household cleaning products (Chang, 2015). Given their very small size (<50 μm), they tend not to be captured by filtration devices or screens in the appliance (e.g., washing machines) or at sewage treatment facilities. Few estimates of the volume of microbeads in marine environments exist, however, Hartline et al. (2016) estimate that up to 0.3% of a garments mass can be released as microfibers in a conventional washing machine (Hartline et al., 2016). While knowledge of their contribution to microplastic pollution is increasingly being recognized, their contribution remains largely unquantified and the ecological affects are not well studied. Microplastics in the form of fibers and microbeads from personal care products are primarily generated at household scales and make their way to marine and coastal habitats via sewage or gray water systems.

5 BREAKDOWN OF LARGE PLASTICS

A large, but mostly unquantified, source of microplastic in the coastal and marine environment is a result of fragmented macroplastics or whole plastic items. Many of the plastic objects used by people every day are considered single use, convenient, and disposable. The larger objects can enter the environment from many sources, generally split into oceanic or land based. Oceanic sources include; fishing, boating, and shipping—including gradual breakdown of rope and polymer-based paints. Land-based sources include, primary industry, litter, sewage, and storm water (Cunningham and Wilson, 2003; Sheavly and Register, 2007; Kuo and Huang, 2014; Verlis et al., 2014). Once macroplastics are in the environment, particularly on beaches, they eventually degrade under high exposure to ultraviolet (UV) light, coupled with physical abrasion due to wave action and sediment movements. Exposure eventually weakens the polymer bonds of plastics and they fragment, which leads to secondary microplastics, adding to the microplastic load (Welden and Cowie, 2017). When degradation occurs outside of a suitable disposal area, the secondary plastics become a component of environmental microplastic pollution (Fig. 1).

FIG. 1 Examples of types of plastic pollution. (A) Plastic debris accumulated on the coast of Orpheus Island, Australia, (B) plastic fragments resulting from the degradation of larger items, (C) virgin resin pellets found on a beach in the remote Keppel Islands, Australia, (D) very small fragments and fibers from environmental samples, and (E) microplastics as a result of degradation of larger items. *(Photo credit for all images A. Bauer-Civiello.)*

6 MICROPLASTIC POLLUTION IN THE COASTAL AND MARINE ENVIRONMENT

In many of the world's coastal cities and urban areas, storm water and sewage discharge is directed into freshwater or coastal systems. Although freshwater and sewage discharge is regarded as one of the major inputs of plastics into the coastal and marine environment, surprisingly few data exist to quantify rates of input, either into the freshwater system or from the freshwater into the coastal system (Eerkes-Medrano et al., 2015). For example, microplastic pollution has been detected in the surface water of the Great Lake system on the North American continent and within the sediments of the St. Lawrence River, which connects the Great Lake system to the ocean (Eriksen et al., 2013; Castañeda et al., 2014). This indicates pathways of plastic pollution from inland states of the United States and Canada into freshwater systems, which then facilitates transports of microplastic particles to the ocean. Indeed, Lebreton et al.(2017) showed that large amount for mismanaged plastic waste enters the ocean every year, with 67% of the global total coming from just 20 rivers (Fig. 2).

The polymer type and size of the microplastic particle, as well as the salinity of the aquatic system underpin the ability of microplastic particles to be moved from freshwater to salt water environments. Plastic polymers vary in density, from low-density polymers, for example, polypropylene (PP), which has a density of approximately 0.9 g/mL to high-density polytetrafluoroethylene (PTFE, or Teflon) which is approximately 2.2 g/mL. This range in polymer chemistry and density results in variation in buoyancy once the plastic particles enter a waterbody, and influences the polymer types that tend to sink in the freshwater system or move with water flow into the coastal, estuarine zones. Most plastics are denser than freshwater (density ~1.0 g/mL), and will therefore accumulate on the riverbed—hence freshwater systems can be an effective sink for plastic pollution. However, in the presence of vertical or horizontal water flow and turbulence, these currents can move the more buoyant plastics from the freshwater system into the coastal zone, especially in a strong flood.

Many commonly used polymers remain negatively buoyant in oceanic waters, such as polyvinyl chloride (PVC), and it is theorized that the majority of plastics accumulate in benthic habitats (Backhurst and Cole, 2000; Angiolillo et al., 2015). The plastic particles which settle at, or close to, the bottom of a river or in the benthic habitats of the coastal zone are still moved by water circulation, albeit slowly, and it has been suggested that there could be a plastic ring around continental shelves (Chubarenko and Stepanova, 2017). However, this is largely unquantified because benthic habitats, especially those located offshore of continental shelves, are challenging habitats to survey, and thus remain some of the least explored marine systems on the planet. Understanding benthic ecosystems and the degree to which they are impacted by microplastic pollution remains an area with considerable scope for research.

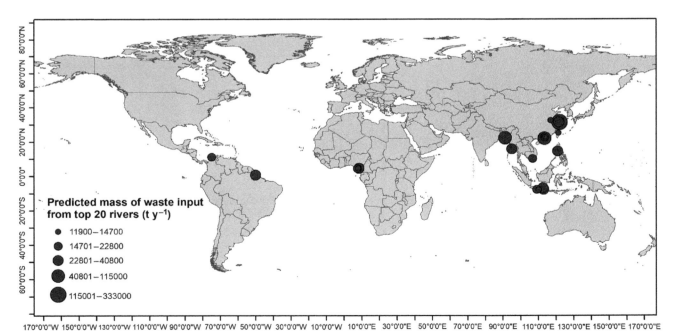

FIG. 2 Map showing the input from the top 20 waste contributors, as predicted by Lebreton et al. (2017). The size of the circle represents the size of the relative contribution of the river system. *(Adapted from Lebreton, L.C.M., van der Zwet, J., Damsteeg, J.-W., Slat, B., Andrady, A., Reisser, J., 2017. River plastic emissions to the world's oceans. Nat. Commun. 8, 15611. https://doi.org/10.1038/ncomms15611.)*

Once microplastics particles are in coastal and marine systems, they are transported by oceanic and atmospheric processes. Larger, more buoyant items are strongly influenced by wind and currents and the more buoyant items often accumulate on beaches facing dominant wind directions or accumulate within oceanic gyres. Smaller or less buoyant items that do not tend to break the surface have lower surface areas exposed to wind. Hence, they will not be as directly influenced by the wind speed and direction and are more likely to move with the water currents only (Daniel et al., 2002; Critchell et al., 2015), aggregating in areas of water convergence zones. This difference in dispersal mechanisms can result in a different pattern of coastal accumulation for microplastics than for macroplastics (Isobe et al., 2014) and for microplastics of difference buoyancies (Kataoka et al., 2015), therefore, monitoring programs need to be tailored for the types of plastic they aim to quantify.

Some of the most recognizable impacts of plastic pollution are observed on beaches. Many beaches around the world accumulate macroplastic and these can act as sources of microplastics, as larger debris items breakup, increasing the microplastic load on the coastline. An unknown proportion of these microplastic particles is then transported by wave and tide energy back into, and sometimes beyond, the nearshore waters. In addition, many of the accounts of ecological impact came from ecologists reporting high rates of plastic ingestion in deceased seabirds within their breeding rookeries (Moser and Lee, 1992; Ryan, 2008).

6.1 Dispersal and Accumulation Patterns

Advances in hydrodynamic modeling and access to remotely sensed data have meant that it is increasingly possible to predict the dispersal and movements of microplastic particles with mathematical modeling. Most of the original work focused on large-scale models, for example, Lebreton et al. (2012) modeled the accumulation of microplastic in the oceanic gyres, and Kako et al. (2011) developed a hind cast/forecast model for plastic pollution originating in Japan. These models highlight the need for international collaboration, however, they are too course in spatial resolution to answer local questions about plastic pollution accumulation. To answer the questions about dispersal at a local scale, fine-scale spatial resolution is required (e.g., Critchell et al., 2015). At this scale, the sources of the plastics (macro- or micro-) must be known, or well understood as this is the most important parameter in understanding local dispersal and accumulation (Critchell and Lambrechts, 2016).

Modeling-based research and field-based surveys have meant that the accumulation of buoyant plastics at the ocean surface has been well studied in some areas of the world, for example, in the North Pacific Ocean gyre. Indeed, the existence of plastics accumulating in the now infamous North Pacific gyre was first described in the 1980s (Day et al., 1989) and other accumulation areas have since been described in many of the world's large oceanic gyres. However, accumulation patterns and the dispersal of microplastic pollution in coastal waters have received far less research attention. A growing number of studies are using plankton trawl techniques to understand local-scale patterns in the presence and abundance of microplastic pollution but there are few demonstrating convincing patterns to suggest consistent accumulation areas (Reisser et al., 2013; Pedrotti et al., 2016). This avenue of research will undoubtedly benefit from the development and refinement of sampling and laboratory techniques to improve the accuracy of identification and quantification of very small ($<100\,\mu m$) particles. The published literature does demonstrate the high degree of spatial and temporal variability in microplastic accumulation. This variability is one of the many challenges in researching a subsequently managing microplastic pollution.

7 GOVERNANCE CHALLENGES AND CURRENT APPROACHES

Marine plastic pollution is a diffuse and evolving problem requiring innovative, multi-sectoral solutions. The scale and pervasiveness of marine micro- and macroplastic pollution, including the diversity and number of stakeholders with responsibility for aspects of the waste stream; the large number of sources and entry points; and the movement of plastics particles throughout the coastal and marine habitats demands a globally coordinated approach in union with national and local action. While we now have global commitments to reduce marine waste, these have made limited advances to date (United Nations Environmental Programme (UNEP), 2017). Marine plastic pollution is a highly challenging governance problem in part because the composition, distribution, and quantity of plastic pollution in the marine environment vary both spatially and seasonally—that is, the problem is not static (Hardesty and Wilcox, 2011; Reisser et al., 2013). Furthermore, the nature of the marine plastic pollution problem evolves overtime as new sources of plastic are introduced or removed from the waste stream, urban populations grow, and use patterns for plastics change (Gregory and Andrady, 2003; Cole et al., 2011). Microplastic pollution associated with personal care products is a key example of a new form of plastic pollution creating a "moving target" for policymakers.

Another challenging characteristic of marine plastic pollution is the lack of clear ownership over the problem. The ocean is a global "commons," with responsibility shared between a myriad of stakeholders and a range of government authorities operating at a variety of scales (local to global). The diffuse (nonpoint source) nature of microplastic pollution is a critical barrier to its successful mitigation as is the fact that the full costs of plastic pollution are difficult to assess and to attribute to the litterer or producer of the waste. A number of global agreements exist to address the problem of macroplastic pollution, including the International Convention for the Prevention of Pollution from Ships (MARPOL) Annex V 1988, the Convention on the Prevention of Marine Pollution by Dumping of Wastes and Other Matter 1972 (the London Convention), and the Manila Declaration on Furthering the Implementation of the Global Program of Action for the Protection of the Marine Environment from Land-based Activities 2–12 (the Manila Declaration), which led to the Global Partnership on Marine Litter (GPML). However, despite significant successes in some areas, these measures have largely been unable to effectively curb marine plastic pollution, and in particular microplastic pollution. This failure has been attributed to the inability of global partnership approaches to comprehensively manage sources and entry points for plastic pollution, as well as to develop consistent policy approaches and to monitor and enforce regulatory compliance across national borders (Gold et al., 2014; United Nations Environmental Programme (UNEP), 2017).

Lastly, although it is critical to continue removing and reducing the flow of plastic pollution into the world's oceans, as a society the circular economy approach suggests that we must also fundamentally reduce our dependence on plastic products, including by developing alternative materials and changing our lifestyles to reduce use of virgin single-use plastics. This is perhaps the most difficult challenge of all given the extent to which petroleum-based plastic products are embedded in the production and transport of consumer goods, their vast medical, industrial and agricultural applications, and their low cost relative to biodegradable and traditional (e.g., glass and aluminum) alternatives (Sheavly and Register, 2007; Hammer et al., 2012). Moreover, the increasingly fast-paced nature of many global societies and economies demands convenience, often wrapped in plastic. Thus, the changes required for a circular economy are likely to extend well beyond plastic pollution into the very heart of modern work and life.

8 A CIRCULAR ECONOMY APPROACH FOR MARINE PLASTIC POLLUTION

Despite the challenges associated with tackling microplastic pollution, there are reasons for optimism (see Sheavly and Register, 2007). The emergence of strong pro-environmental behaviors among younger generations represents an important opportunity and a driver of support for governments seeking to act to protect the natural environment (Benckendorff et al., 2012). Although initiatives that deal specifically with microplastic pollution are embryonic compared with those aimed at larger plastic items, governments worldwide are increasingly recognizing the threat associated with microplastics, and are attempting to curb the use of such products through legislation or voluntary schemes (see Table 1). It is now widely recognized that a circular economy approach that brings about fundamental changes to the production and consumption of plastic goods will ultimately be required if we are to effectively stem the tide of plastic pollution (World Economic Form (WEF), 2016; United Nations Environmental Programme (UNEP), 2017). Here, we outline how a circular economy approach that emphasizes the phasing out of virgin plastic products at the top of the supply chain (a hierarchical approach to waste management) can provide a framework for long-term, sustainable reduction of microplastic pollution in the marine environment. We use case study examples from the United States, Australia, and the European Union (EU) to illustrate a range of innovative governance approaches to address microplastic waste.

8.1 Waste Management and Marine Plastic

Marine plastic pollution and waste management are inextricably linked (Thevenon et al., 2014; Jambeck et al., 2015). Wastewater treatment is a particularly important contributor to microplastic pollution because sewage is a key source of marine litter. Wastewater contains items such as sanitary products, cotton buds, and microplastics in the form of fibers from laundered synthetic fabrics and microbeads from personal care products such as face washes and toothpastes (Fendall and Sewell, 2009; Thevenon et al., 2014; Napper and Thompson, 2016). Integrated Solid Waste Management (ISWM) is therefore essential for effective management of the waste stream, encompassing waste prevention, mitigation through recycling and disposal programs, and remediation efforts aimed at removing debris from waterways, beaches, and the ocean (see Modak, 2010).

There is international recognition of the need for integrated, precautionary approaches to waste management. The United Nations Environment Program (UNEP) Governing Council Decision 25/8 on waste management highlighted the need for integrated and holistic efforts on waste management, and a need for governments to further develop national

TABLE 1 Example Initiatives to Reduce Primary Microplastic Pollution

Abatement Measure	Implementing Sector	Instrument Type	Purpose of Abatement Measure	Intervention Point (Product-to-Waste Chain)	Scale of Implementation
Develop, implement, monitor, and review a coordinated marine waste reduction strategy including clear, measurable and achievable actions and targets to reduce microbead inputs into the marine and freshwater environment (all rivers and drains lead to the sea). Increase targets over time	Government	Strategic planning[a]	All	All	Global, national
Ensure that waste management policies reflect and support other policies (e.g., consider littering in local waste management plans and in river management, as well as for other policies that are not related to the environment)	Government	Strategic planning[a]	All	All	National, local
Ban microbeads in household cleaning products and personal care products	Government	Regulatory instrument	Prevention	Design and production	Global, National
Close "loopholes" in the definitions for microbeads in the cosmetic/cleaning industries	Government	Regulatory instrument	Prevention	Design and production	National
Legislate for disclosure of microplastic components in all products	Government	Regulatory instrument	Mitigation	Design and production	National
Legislate for consistent labeling for microbead ingredients on packaging	Government/industry	Regulatory instrument	Mitigation	Design and production	National
Foster effective enforcement of extended producer responsibility schemes	Government	Economic instrument	Mitigation	Design and production	Global, National
Economic incentives for industry to reduce use of microbead ingredients	Government	Economic instrument	Prevention	Design and production	National
Mandatory installation of microbead filter devices on washing machines	Government/Industry	Technological innovation and infrastructure	Mitigation	Collection and waste transfer	National, Local
Voluntary retailer bans on sale of products containing microbeads	Industry	Comanagement and voluntary initiatives	Mitigation	Distribution	National
Avoid purchasing products containing microbeads	Household/consumer	Comanagement and voluntary initiatives	Mitigation	Use and consumption	Global
Voluntary substitution of organic materials for plastic microbeads in personal care and cleaning products	Industry	Comanagement and voluntary initiatives	Prevention	Design and production	Global, national
Model good waste management practices in government operations (e.g., avoid plastic packaging in public procurements or at events)	Government	Co-management and voluntary initiatives	Prevention	Use and consumption	National, local

(Continued)

TABLE 1 Example Initiatives to Reduce Primary Microplastic Pollution—cont'd

Abatement Measure	Implementing Sector	Instrument Type	Purpose of Abatement Measure	Intervention Point (Product-to-Waste Chain)	Scale of Implementation
Share information with local authorities, NGOs and private industry stakeholders on marine litter, including supporting the use of public databases and common recording templates (Cheshire et al., 2009)	Government	Comanagement and voluntary initiatives		All	National, local
Support international activities to combat marine litter such as the Rio+20 commitment to act to achieve significant reductions in marine litter.	Government	Comanagement and voluntary initiatives		All	National
Deliver consumer education and awareness campaigns to assist consumers to interpret ingredient lists/recognize microbead ingredients in retail products; and raise awareness of the environmental impacts of microbeads, sustainable consumption and separation of recyclables.	Government/ NGO	Comanagement and voluntary initiatives	Mitigation	Design and production; use and consumption	National; local

[a] Strategic planning and research activities have not previously been included in typologies of waste management initiatives, but they nevertheless play a critical role and are also qualitatively different from other interventions, thus are included here as separate categories.

policies that promote ISWM. This approach should include the waste hierarchy six R's (reduce-redesign-remove-reuse-recycle-recover), which builds on the traditional three R's (reduce-reuse-recycle). Given the limited resources available to tackle plastic pollution, it is likely to be necessary to prioritize measures that can deliver the best outcome for the least cost. The concept of a hierarchy for waste management (see Fig. 3A), can inform decisions about abatement measures. The aim of hierarchical approaches is to prioritize initiatives that prevent the generation of waste and the transport of litter into the marine environment (Watkins et al., 2015, p. 10). Measures that improve waste management on land can help to capture waste that may otherwise become marine litter. Approaches that focus on cleaningup the ocean are lowest priority. Once in the environment, plastic items can realize their potential to cause harm and they are more difficult and costly to remediate; prevention is key. The circular economy concept also emphasizes the importance of prevention and mitigation of marine plastics impacts (Veiga et al., 2016; see Fig. 3B).

Transitioning effectively to a circular economy is likely to require a fundamental shift in the production and consumption of plastic products, recognizing that the present per capita use of energy and resources is both unsustainable and inequitable at a global scale (United Nations Environmental Programme (UNEP), 2017). This shift will likely require a staged approach involving the development of necessary infrastructure and investment incentives in addition to behavioral changes (World Economic Form (WEF), 2016; United Nations Environmental Programme (UNEP), 2017). However, a circular economy can also generate substantial economic benefits, allowing currently lost assets and costs to be recaptured (WEF, 2016). For instance, at present, 95% of the value of plastic packaging material, or USD$80–120 billion annually is lost to the economy after first use (World Economic Form (WEF), 2016). Under current arrangements, plastic waste in the marine environment is an external cost to society that is not adequately borne by the producer of the waste, and this needs to be addressed through initiatives like improved product design and extended producer responsibility (EPR) (Watkins et al., 2015).

There is now a recognition that new international legal mechanisms, operating in tandem with regional, national, and subnational programs are needed to prompt reductions in plastic marine litter, complemented by a strengthening of new and existing regional seas programs, and research programs to support and inform the development of more effective national policies (Gold et al., 2014). Single-use plastics are an example of a product that has been designed with a lack of guidelines or legal requirements for recyclability, with little value attributed to its end of life so these items are treated as disposable (Watkins et al., 2015). In addition, those items that can be recycled can only be recycled a small number of times, incurring a loss of materials and energy. A major study examining the socioeconomic costs of marine litter suggests that the costs of inaction are generally higher than the cost of action, with environmental damage to marine ecosystems alone estimated at USD$13 billion per year, including economic losses to fisheries, tourism, and time spent on beach cleaning (Watkins et al., 2015).

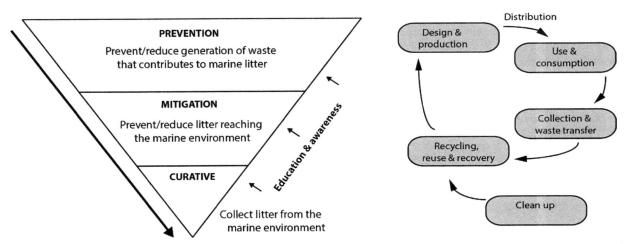

FIG. 3 (A) A hierarchy of marine litter initiatives. (B) A circular economy approach. *(Part figure (A) was adapted from Watkins E, Ten Brink P, Withana S, Mutafoglu K, Scheitzer J-P, Russi D, Kettunen M: Marine litter: socio-economic study, London, Brussels, 2015, Scoping Report. IEEP. Part figure (B) was adapted from Veiga, J.M., Vlachogianni, T., Pahl, S., Thompson, R.C., Kopke, K., Doyle, T.K., Hartley, B.L., Maes, T., Orthodoxou, D.L., Loizidou, X.I., Alampei, I., 32016. Enhancing public awareness and promoting co-responsibility for marine litter in Europe: the challenge of MARLISCO. Mar. Pollut. Bull. 102, 309–315.)*

8.2 Reducing Plastic Pollution in the Oceans

Successful reduction of plastic pollution in the oceans will require broad-based behavioral change at all institutional scales and in every sector, including households, industry, agriculture, and fisheries (Sheavly and Register, 2007). This change must be supported by a change in policies designed to promote sustainable waste management and prevent the influx of plastics into the marine environment. Currently, efforts to combat marine plastic pollution focus largely on reducing the volume of plastic that reaches the ocean, or on cleanup (Ocean Conservancy, 2015). In the past, such initiatives have focused largely on macro-sized plastic items, and secondary microplastic fragments. Measures aimed at stemming the flow of large plastic items can ultimately reduce the risk that such products will breakdown to form secondary microplastics in the marine environment. However, the widespread use of primary microplastics in personal care and cleaning products is likely also to require a suite of targeted and innovative policy responses aimed at multiple points of intervention in the production chain (see Table 1). These initiatives may include:

- Regulatory and policy instruments (e.g., legislation)
- Economic instruments (e.g., incentives, markets for plastics recycling)
- Comanagement and voluntary initiatives (including education and awareness campaigns)
- Technological innovation and improvement of infrastructure

Plastic pollution reduction policies should be developed within the broader landscape of environmental, economic, and social policy at the local, national, and international scales. The European Commission Strategy on Plastics in a Circular Economy provides an example of a strategic planning process designed to implement circular economy principles to reduce plastic waste at multiple points in the supply chain, while also integrating with other relevant policy areas. The strategy includes initiatives to decouple virgin plastic production from fossil fuel feedstocks, facilitate increased plastic recycling and reuse, and reduce plastic inputs into the environment. Importantly, however, it also seeks to align plastics policies with existing EU climate and energy goals (European Commission, 2017). Ongoing research into alternative packaging materials; more efficient waste management, recycling and energy recovery techniques; and behavioral change will also be critical to support a transition toward a circular economy.

While national governments are responsible for participating in international initiatives as well as developing and implementing some national-level policy and legislation, other, important initiatives to reduce marine plastic—such as recycling and the management of wastewater and solid wastes—often fall under subnational jurisdiction, for example, municipal, provincial, village, or community. Voluntary initiatives also form an important part of the marine waste abatement "continuum." The diversity of measures available to combat marine waste reflects the multi-stakeholder engagement necessary in bringing in abatement measures, and highlights the need for comanagement approaches involving a range of stakeholders across multiple sectors.

9 REDUCING MARINE PLASTIC POLLUTION: CASE STUDIES

In many countries, governments, industries, and communities are taking action against marine plastic pollution. Here, we provide examples of current measures for reducing microplastic waste. Many of these initiatives are still in the formative stages but they nevertheless provide useful illustrations of the strengths and pitfalls of different approaches to reduce the future input of microplastic pollution into the coastal and marine environments.

10 CASE STUDY 1: BANNING MICROBEADS IN PERSONAL CARE AND CLEANING PRODUCTS

A number of countries, including the United States, the United Kingdom, and Canada have acted to address microplastic pollution by banning the use of microbead exfoliants in personal care products. Others, such as Australia, have introduced voluntary phase-out schemes, with an undertaking to legislate if these schemes are not effective. A range of options exist to ban or phase-out microbeads through legislation. In Australia, for example, an immediate ban on microbeads could be achieved by:

"amending the Cosmetics Standard 2007, made under the Industrial Chemicals (Notification and Assessment) Act 1989 (Cth). The Standard currently sets out requirements for products like sunscreens and could be amended to specify that certain products, such as facial cleansers and toothpaste, must not contain microbeads. It is an offence to import or manufacture a product that does not meet the Standard. Since most products containing microbeads are likely to be imported from overseas, they could also

be listed under the Customs (Prohibited Imports) Regulations 1956. These regulations prohibit the importation of a wide range of goods and chemicals into Australia, but would not prevent Australian manufacture of products containing microbeads" (Power and Hanna, 2016).

Although there are numerous legislative avenues available, initiatives to reduce primary microbead pollution through voluntary or mandatory phaseouts have had mixed success to date. For example, the United States has initiated a ban on microbeads in personal care products, but this does not apply to other products, for example, commercial cleaning products. Thus, it is critical that any legislative measures are carefully designed to limit such loopholes.

11 CASE STUDY 2: EPR

As the name suggests, EPR is a regulatory intervention that extends a producer's responsibility for a product to the postconsumer stage of the product's life cycle. The Organization for Economic Cooperation and Development (OECD) definition points to two related features of the policy: (i) shifting responsibility (physically and/or economically; fully or partially) "upstream" toward the producer and away from municipalities for the management of waste created by a product, and consequently (ii) providing incentives for producers to incorporate environmental considerations into the design of their products (OECD, 2001). The concept gained considerable prominence when the EU adopted the Directive on Waste Electrical and Electronic Equipment (WEEE) in January 2003, a policy that was expressly aimed at enhancing resource efficiency through improved collection, treatment, and recycling of electronics at the end of their life.

Since that time, most OECD countries and many emerging economies have implemented EPR policies in key sectors such as packaging, electronics, batteries, and vehicles (OECD, 2014, p. 03). The concept's impact has largely been limited to expanding postconsumer recycling (Lifset et al., 2013, p. 162) but there have been significant efforts made in recent years to encompass a broader mix of policy instruments to increase its impact. The four types of intervention currently employed by the OECD are (from OECD, 2014, p. 6):

(1) *Product take-back requirements.* These require producers or retailers to collect plastic products at the postconsumer stage, including provision of incentives to encourage consumers to bring the used product back to the selling point.
(2) *Economic and market-based instruments*, such as deposit-refund schemes and upstream combination tax/subsidy (UCTS) that imposes a tax on a producer which is subsequently used to pay for collection and/or recycling efforts.
(3) *Regulations and performance standards* such as minimum recycled content. Standards can be mandatory or applied by industries themselves through voluntary programs. In the EU, there is a restriction on the use of certain hazardous substances in electrical and electronic equipment, and we are now seeing similar restrictions on certain plastics (see Case Study 1, above).
(4) *Information-based instruments*, which aim to indirectly support EPR programs by raising public awareness of the impacts of certain products and packaging. Such measures can include imposing labeling requirements on producers and informing recyclers about the materials used in products.

The chosen mix of instruments will depend on a complex mix of variables, including the product or range of products to be targeted; whether the instruments are mandatory or voluntary, and the level of government intervention in collection, treatment, and recycling of plastic products. Clearly, each of these considerations varies from one country and jurisdiction to another, but the underlying philosophy is consistent: that producers have a responsibility for their products long after the consumer has used them.

Overall, there is evidence to suggest that EPR has been successful in decreasing the volume of waste generated (see Case Study Box 1), increasing recycling rates, and provoking innovative approaches to packaging and product design and source materials (see Herrmann et al., 2014). Interestingly, EPR is also thought to have generated a range of other benefits, including innovation in both technology and organizational design, and improved organization of supply chains thanks to the emergence of international operators in the recycling sector (OECD, 2014, p. 09).

12 BEHAVIORAL CHANGE—LITTERING AND PLASTIC POLLUTION

Although rarely quantified, littering by individuals or groups is well recognized as a significant input of macro- and microplastic plastic waste to the marine and coastal environments. There are two major weaknesses in strategies

Box 1   Benchmark Criteria and Key Components

1. *Behavior*: Focus on influencing specific behaviors not just knowledge, attitudes, and beliefs.
2. *Customer orientation*: Use a mix of data sources and research methods to understand targets' lives and behaviors.
3. *Behavioral theories*: Use of individual theories or combinations to understand behavior and inform the intervention, including testing of theoretical assumptions
4. *Insight*: Understanding of what moves and motivates the target
5. *Exchange*: Maximize benefits and minimizing costs of adopting and maintaining desired behaviors
6. *Competition*: Understanding of what competes for time, attention, and inclination to behave in a particular way
7. *Segmentation*: Avoids one size fits all approach
8. *Methods mix*: Uses a mix of methods to bring about behavior change. Does not rely solely on raising awareness

aimed at changing the behavior of individuals and groups. The first is the assumption that problematic behavior such as littering occurs as a result of a lack of knowledge of the impact of behaviors (Toomey et al., 2016) and the second is that this deficiency can be rectified by information provision (Bates, 2010; Miller et al., 2010). However, attempts to provide top-down information provision alone have been proven to be insufficient to change behaviors (Catney et al., 2013). As a result, there is now a greater focus on the use of social marketing principles to "deliver social change programs that are effective, efficient, equitable, and sustainable" (iSMA, ESMA, and AASM, 2013). Effective campaigns have been identified as including the following benchmarks and key components (National Social Marketing Centre, n.d.):

Benchmarks 1, 2, 4–6 illustrate the need to understand the drivers of current behaviors, barriers to them, and potential enablers of behavior change, including influence of social groups and perceived norms on both individual and group behaviors. The use of theory (benchmark 3) to guide activity aimed at achieving long-term behavioral change has been proven to result in greater effectiveness than activity that is not at least informed by theory (Truong, 2014; Almosa et al., 2017). Benchmark 7 recognizes that different population segments will have different attitudes, motivations, and abilities regarding behavior change and thus different approaches may be necessary across segments (Dietrich et al., 2016). Benchmark 8 indicates both the need for different forms of communication but also the need to combine information, education, persuasion, infrastructure (such as increased litter disposal facilities), and legislation. While the first four strategies may be effective for the majority of the population, legislation may be needed for recalcitrant sectors (Rothschild, 1999).

The use of behavioral change strategies is common in the health and education sectors (Eagle et al., 2013). Two strategies which have been shown to be effective in generating pro-environmental behavior are two well-known anti-littering campaigns, the long-running American "Don't Mess with Texas" campaign (Miller and Prentice, 2013) and the British "Keep Britain Tidy" campaign (Campbell, 2007), both of which used combinations of strategies to achieve their goals. However, there is a growing need to expand the application of social marketing strategies to generate pro-environmental behaviors (see Kamrowski et al., 2014; Eagle et al., 2016; Mellish et al., 2016) related to littering and management of plastic pollution. We see substantial scope for a partnership approach between local and state governments with other sectors invested in the coastal and marine zones such as tourism, recreational and commercial fisheries, and community groups to develop and deliver coordinated strategies to improve or maintain the condition of natural coastal values (Eagle et al., 2016; Mellish et al., 2016).

13   CONCLUSION

The study and management of plastic pollution in the oceans is among the greatest environmental challenges of the 21st century. Plastic in the marine environment is a ubiquitous and highly diffuse pollutant, and its concentration and characteristics vary across space and time, creating a "moving target" for managers and policymakers. Moreover, plastics are particularly difficult to manage because they have become integral to modern economies and societies. However, strategies such as the circular economy approach, that emphasize the reduction of plastic use at the source in tandem with cleanup, resource recovery, and recycling of plastic products, provide a glimpse of a future in which plastic pollution and its associated risks can be mitigated. A number of countries are already acting to recognize and address marine plastic pollution, including microplastics. Successfully mitigating plastic pollution requires a commitment to action at all societal scales, from the household, through to the community, national, and global levels. It is only through such collaboration and commitment that the threat of plastic pollution to people and the environment can be successfully managed.

REFERENCES

Acampora, H., Schuyler, Q.A., Townsend, K.A., Hardesty, B.D., 2014. Comparing plastic ingestion in Juvenile and adult stranded short-tailed shearwaters (Puffinus Tenuirostris) in Eastern Australia. Mar. Pollut. Bull. 78, 63–68.

Almosa, Y., Parkinson, J., Rundle-Thiele, S., 2017. Littering reduction: a systematic review of research 1995–2015. Soc. Mark. Q. 1524500417697654, 203–222.

American Chemistry Council, 2012. Operation clean sweep [Online]. Available from: https://opcleansweep.org. (Accessed November 2017).

Andrady, A.L., Neal, M.A., 2009. Applications and societal benefits of plastics. Philos. Trans. R Soc. Lond. B Biol. Sci. 364 (1526), 1977–1984.

Angiolillo, M., Lorenzo, B.D., Farcomeni, A., Bo, M., Bavestrello, G., Santangelo, G., Cau, A., Mastascusa, V., Cau, A., Sacco, F., Canese, S., 2015. Distribution and assessment of marine debris in the deep Tyrrhenian Sea (Nw Mediterranean Sea, Italy). Mar. Pollut. Bull. 92, 149–159.

Backhurst, M.K., Cole, R.G., 2000. Subtidal benthic marine litter at Kawau Island, North-Eastern New Zealand. J. Environ. Manag. 60, 227–237.

Barnes, D.K.A., Galgani, F., Thompson, R.C., Barlaz, M., 2009. Accumulation and fragmentation of plastic debris in global environments. Philos. Trans. R Soc. Lond. B Biol. Sci. 364, 1985–1998.

Bates, C.H., 2010. Use of social marketing concepts to evaluate ocean sustainability campaigns. Soc. Mark. Q. 16 (1), 71–96.

Batista, M.I., Henriques, S., Pais, M.P., Cabral, H.N., 2014. Assessment of cumulative human pressures on a coastal area: integrating information for MPA planning and management. Ocean Coast. Manag. 102, 248–257.

Baulch, S., Perry, C., 2014. Evaluating the impacts of marine debris on cetaceans. Mar. Pollut. Bull. 80, 210–221.

Benckendorff, P., Moscardo, G., Murphy, L., 2012. Environmental attitudes of generation Y students: foundations for sustainability education in tourism. J. Teach. Travel Tourism 12, 44–69.

Boren, L.J., Morrissey, M., Muller, C.G., Gemmell, N.J., 2006. Entanglement of New Zealand fur seals in man-made debris at Kaikoura, New Zealand. Mar. Pollut. Bull. 52, 442–446.

Boteler, B., Abhold, K., Oosterhuis, F., Fernandez, P., Hadzhiyska, D., Pavlova, D., Mira Veiga, J., 2015. Best Practice Examples of Existing Economic Policy Instruments and Potential New Economic Policy Instruments to Reduce Marine Litter and Eliminate Barriers to Ges. CleanSea, Amsterdam.

Bravo Rebolledo, E.L., Van Franeker, J.A., Jansen, O.E., Brasseur, S.M.J.M., 2013. Plastic ingestion by harbour seals (Phoca Vitulina) in the Netherlands. Mar. Pollut. Bull. 67, 200–202.

Browne, M.A., Galloway, T., Thompson, R., 2007. Microplastic—an emerging contaminant of potential concern? Integr. Environ. Assess. Manag. 3, 559–561.

Campbell, F., 2007. People Who Litter. Environmental Campaigns (ENCAMS), Wigan.

Castañeda, R.A., Avlijas, S., Simard, M.A., Ricciardi, A., 2014. Microplastic pollution in St Lawrence river sediments. Can. J. Fish. Aquat. Sci. 71, 1767–1771.

Castro, R.O., Silva, M.L., Marques, M.R.C., De Araújo, F.V., 2016. Evaluation of microplastics in Jurujuba Cove, Niterói, Rj, Brazil, an area of mussels farming. Mar. Pollut. Bull. 110, 555–558.

Catney, P., Dobson, A., Hall, S.M., Hards, S., Macgregor, S., Robinson, Z., Ormerod, M., Ross, S., 2013. Community knowledge networks: an action-orientated approach to energy research. Local Environ. 18, 506–520.

Chae, Y., An, Y.-J., 2017. Effects of micro- and nanoplastics on aquatic ecosystems: current research trends and perspectives. Mar. Pollut. Bull. 124, 624–632.

Chang, M., 2015. Reducing microplastics from facial exfoliating cleansers in wastewater through treatment versus consumer product decisions. Mar. Pollut. Bull. 101, 330–333.

Cheshire, A.C., Adler, E., Barbière, J., Cohen, Y., Evans, S., Jarayabhand, S., Jeftic, L., Jung, R.T., Kinsey, S., Kusui, E.T., Lavine, I., Manyara, P., Oosterbaan, L., Pereira, M.A., Sheavly, S., Tkalin, A., Varadarajan, S., Wenneker, B., Westphalen, G., 2009. UNEP/IOC Guidelines on Survey and Monitoring of Marine Litter. UNEP Regional Seas Reports and Studies, No. 186; IOC Technical Series No. 83: xii + 120 pp.

Chubarenko, I., Stepanova, N., 2017. Microplastics in sea coastal zone: lessons learned from the Baltic Amber. Environ. Pollut. 224, 243–254.

Cole, M., Lindeque, P., Halsband, C., Galloway, T.S., 2011. Microplastics as contaminants in the marine environment: a review. Mar. Pollut. Bull. 62, 2588–2597.

Cole, M., Lindeque, P., Fileman, E., Halsband, C., Goodhead, R., Moger, J., Galloway, T.S., 2013. Microplastic Ingestion by Zooplankton. Environ. Sci. Technol. 47, 6646–6655.

Cole, M., Lindeque, P., Fileman, E., Halsband, C., Galloway, T.S., 2015. The impact of polystyrene microplastics on feeding, function and fecundity in the Marine Copepod Calanus Helgolandicus. Environ. Sci. Technol. 49, 1130–1137.

Critchell, K., Lambrechts, J., 2016. Modelling accumulation of marine plastics in the coastal zone; what are the dominant physical processes? Estuar. Coast. Shelf Sci. 171, 111–122.

Critchell, K., Grech, A., Schlaefer, J., Andutta, F.P., Lambrechts, J., Wolanski, E., Hamann, M., 2015. Modelling the fate of marine debris along a complex shoreline: lessons from the Great Barrier Reef. Estuar. Coast. Shelf Sci. 167, 414–426.

Cunningham, D., Wilson, S.P., 2003. Marine debris on beaches of the greater Sydney region. J. Coast. Res. 421–430.

Daniel, P., Jan, G., Cabioc'h, F., Landau, Y., Loiseau, E., 2002. Drift modeling of cargo containers. Spill Sci. Technol. Bull. 7, 279–288.

Day, R. H., Shaw, D. G. & Ignell, S. E. The quantitative distribution and characteristics of neuston plastic in the North Pacific Ocean, 1985–88. *In:* Shomura, R. S. & Godfrey, M. L., eds. The Second International Conference on Marine Debris, 1989 Honolulu, Hawaii. 247-266.

Derraik, J.G.B., 2002. The pollution of the marine environment by plastic debris: a review. Mar. Pollut. Bull. 44, 842–852.

Dietrich, T., Rundle-Thiele, S., Kubacki, K., 2016. Segmentation in Social Marketing: Process, Methods and Application. Springer, Singapore.

Eagle, L., Dahl, S., Hill, S., Bird, S., Spotswood, F., Tapp, A., 2013. Social Marketing. Pearson Education, London, p. 341.

Eagle, L., Hamann, M., Low, D.R., 2016. The role of social marketing, marine turtles and sustainable tourism in reducing plastic pollution. Mar. Pollut. Bull. 107, 324–332.

Eerkes-Medrano, D., Thompson, R.C., Aldridge, D.C., 2015. Microplastics in freshwater systems: a review of the emerging threats, identification of knowledge gaps and prioritisation of research needs. Water Res. 75, 63–82.

Eriksen, M., Maximenko, N., Thiel, M., Cummins, A., Lattin, G., Wilson, S., Hafner, J., Zellers, A., Rifman, S., 2013. Plastic pollution in the South Pacific subtropical gyre. Mar. Pollut. Bull..

Eriksen, M., Lebreton, L.C., Carson, H.S., Thiel, M., Moore, C.J., Borerro, J.C., et al., 2014. Plastic pollution in the world's oceans: more than 5 trillion plastic pieces weighing over 250,000 tons afloat at sea. PloS One 9 (12), e111913.

European Commission, 2017. Strategy on Plastics in a Circular Economy Roadmap. European Commission, Brussels, pp. 24.

Fendall, L.S., Sewell, M.A., 2009. Contributing to marine pollution by washing your face: microplastics in facial cleansers. Mar. Pollut. Bull. 58, 1225–1228.

Gold, M., Mika, K., Horowitz, C., Herzog, M., Leitner, L., 2014. Stemming the tide of plastic marine litter: a global action agenda. Tul. Environ. Law J. 27, 165–203.

Gregory, M.R., Andrady, A.L., 2003. Plastics in the marine environment. Plast. Environ. 379, 389–390.

Gross, M., 2017. Our planet wrapped in plastic. Curr. Biol. 27 (6), R785–R788.

Hall, N.M., Berry, K.L.E., Rintoul, L., Hoogenboom, M.O., 2015. Microplastic ingestion by Scleractinian Corals. Mar. Biol. 162, 725–732.

Hammer, J., Kraak, M.H.S., Parsons, J.R., 2012. Plastics in the marine environment: the dark side of a modern gift. In: WHITACRE, D.M. (Ed.), Reviews of Environmental Contamination and Toxicology. Springer, New York.

Hardesty, B.D., Wilcox, C., 2011. Understanding the Types, Sources and at-Sea Distribution of Marine Debris in Australian Waters. CSIRO, Hobart.

Hartline, N.L., Bruce, N.J., Karba, S.N., Ruff, E.O., Sonar, S.U., Holden, P.A., 2016. Microfiber masses recovered from conventional machine washing of new or aged garments. Environ. Sci. Technol. 50, 11532–11538.

Henderson, J.R., 2001. A pre-and post-MARPOL Annex v summary of Hawaiian monk seal entanglements and marine debris accumulation in the Northwestern Hawaiian Islands, 1982–1998. Mar. Pollut. Bull. 42, 584–589.

Herrmann, C., Schmidt, C., Kurle, D., Blume, S., Thiede, S., 2014. Sustainability in manufacturing and factories of the future. Int. J. Precis. Eng. Man.-GT. 1, 283–292.

Isobe, A., Kubo, K., Tamura, Y., Kako, S., Nakashima, E., Fujii, N., 2014. Selective transport of microplastics and mesoplastics by drifting in coastal waters. Mar. Pollut. Bull. 89, 324–330.

Jambeck, J.R., Geyer, R., Wilcox, C., Siegler, T.R., Perryman, M., Andrady, A., Narayan, R., Law, K.L., 2015. Plastic waste inputs from land into the ocean. Science 347, 768–771.

Kako, S.I., Isobe, A., Magome, S., Hinata, H., Seino, S., Kojima, A., 2011. Establishment of numerical beach-litter hindcast/forecast models: an application to Goto Islands, Japan. Mar. Pollut. Bull. 62, 293–302.

Kamrowski, R.L., Sutton, S.G., Tobin, R.C., Hamann, M., 2014. Potential applicability of persuasive communication to light-glow reduction efforts: a case study of marine turtle conservation. Environ. Manag. 54, 583–595.

Kataoka, T., Hinata, H., Kato, S., 2015. Backwash process of marine macroplastics from a beach by nearshore currents around a submerged breakwater. Mar. Pollut. Bull. 101, 539–548.

Kuo, F.-J., Huang, H.-W., 2014. Strategy for mitigation of marine debris: analysis of sources and composition of marine debris in Northern Taiwan. Mar. Pollut. Bull. 83, 70–78.

Laist, D.W., 1997. Impacts of marine debris: entanglement of marine life in marine debris including a comprehensive list of species with entanglement and ingestion records. In: Coe, J., Rogers, D. (Eds.), Marine Debris. Springer, New York.

Lebreton, L., Greer, S., Borrero, J., 2012. Numerical modelling of floating debris in the world's oceans. Mar. Pollut. Bull. 64, 653–661.

Lebreton, L.C.M., van der Zwet, J., Damsteeg, J.-W., Slat, B., Andrady, A., Reisser, J., 2017. River plastic emissions to the world's oceans. Nat. Commun. 8, 15611. https://doi.org/10.1038/ncomms15611.

Lifset, R., Atasu, A., Tojo, N., 2013. Extended producer responsibility: national, international, and practical perspectives. J. Indus. Ecol. 17 (2), 162–166.

Mato, Y., Isobe, T., Takada, H., Kanehiro, H., Ohtake, C., Kaminuma, T., 2001. Plastic resin pellets as a transport medium for toxic chemicals in the marine environment. Environ. Sci. Technol. 35, 318–324.

Mellish, S., Pearson, E.L., Sanders, B., Litchfield, C.A., 2016. Marine wildlife entanglement and the seal the loop initiative: a comparison of two free-choice learning approaches on visitor knowledge, attitudes and conservation behaviour. Int. Zoo Yearb. 50, 129–154.

Miller, D.T., Prentice, D., 2013. Psychological levers of behavior change. Behav. Found. Public Pol. 301–309.

Miller, G., Rathouse, K., Scarles, C., Holmes, K., Tribe, J., 2010. Public understanding of sustainable tourism. Ann. Tour. Res. 37 (3), 627–645.

Modak, P., 2010. Municipal solid waste management: turning waste into resources. Chapter 5. In: United Nations Department of Economic and Social Affairs (UNDESA 2010). Shanghai Manual – A Guide for sustainable urban development in the 21st Century. Available from: http://www.un.org/esa/dsd/susdevtopics/sdt_pdfs/shanghaimanual/Chapter%205%20-%20Waste_management.pdf. Accessed 14 November 2016

Moser, M.L., Lee, D.S., 1992. A fourteen-year survey of plastic ingestion by Western North Atlantic Seabirds. Colonial Waterbirds 15, 83–94.

Napper, I.E., Thompson, R.C., 2016. Release of synthetic microplastic plastic fibres from domestic washing machines: effects of fabric type and washing conditions. Mar. Pollut. Bull. 112, 39–45.

National Social Marketing Centre n.d.. Social Marketing Benchmark Criteria. London: National Social Marketing Centre.

Nelms, S.E., Duncan, E.M., Broderick, A.C., Galloway, T.S., Godfrey, M.H., Hamann, M., Lindeque, P.K., Godley, B.J., 2016. Plastic and marine turtles: a review and call for research. ICES J. Mar. Sci. 73, 165–181.

Ocean Conservancy, 2015. Stemming the tide: Land-based strategies for a plastic-free ocean. Ocean Conservancy and McKinsey Center for Business and Environment. 48pp.

OECD, 2001. Extended Producer Responsibility: A Guidance Manual for Governments. OECD, Paris, France, pp. 159.

OECD, 2014. The State of Play on Extended Producer Responsibility (EPR): Opportunities and Challenges. OECD, Paris, France, pp. 17.

Pedrotti, M.L., Petit, S., Elineau, A., Bruzaud, S., Crebassa, J.-C., Dumontet, B., Martí, E., Gorsky, G., Cózar, A., 2016. Changes in the floating plastic pollution of the mediterranean sea in relation to the distance to land. PLoS One 11, e0161581.

Power, S., Hanna, E., 2016. Tiny plastics causing big problems in our oceans. Parliament of Australia. Available from: https://www.aph.gov.au/About_Parliament/Parliamentary_Departments/Parliamentary_Library/FlagPost/2016/June/Marine_microplastics. (Accessed October 12, 2016).

Reisser, J., Shaw, J., Wilcox, C., Hardesty, B.D., Proietti, M., Thums, M., Pattiaratchi, C., 2013. Marine plastic pollution in waters around Australia: characteristics, concentrations, and pathways. PLoS One 8, e80466.

Rothschild, M., 1999. Carrots, sticks, and promises: a conceptual framework for the management of public health and social issue behaviors. J. Mark. 63, 24–37.

Ryan, P., 1988. Effects of ingested plastic on seabird feeding: evidence from chickens. Mar. Pollut. Bull. 19, 125–128.

Ryan, P.G., 2008. Seabirds indicate changes in the composition of plastic litter in the Atlantic and South-Western Indian Oceans. Mar. Pollut. Bull. 56, 1406–1409.

Schuyler, Q., Hardesty, B.D., Wilcox, C., Townsend, K., 2012. To eat or not to eat? Debris selectivity by marine turtles. PLoS One 7, e40884.

Sheavly, S.B., Register, K.M., 2007. Marine debris & plastics: environmental concerns, sources, impacts and solutions. J. Polym. Environ. 15, 301–305.

Thevenon, F., Carroll, C., Sousa, J. (Eds.), 2014. Plastic Debris in the Ocean: The Characterisation of Marine Plastics and Their Environmental Impacts, Situation Analysis Report. IUCN, Gland, Switzerland, pp. 52.

Thompson, R.C., Moore, C.J., Von Saal, F.S., Swan, S.H., 2009. Plastics, the environment and human health: current consensus and future trends. Philos. Trans. R. Soc. Lond. B Biol. Sci. 1, 1–14.

Toomey, A.H., Knight, A.T., Barlow, J., 2016. Navigating the Space between Research and Implementation in Conservation. Conserv. Lett. 10 (5), 619–625. https://doi.org/10.1111/conl.12315.

Truong, V.D., 2014. Social marketing a systematic review of research 1998–2012. Soc. Mark. Q. 20, 15–34.

Turner, A., Holmes, L., 2011. Occurrence, distribution and characteristics of beached plastic production pellets on the Island of Malta (Central Mediterranean). Mar. Pollut. Bull. 62, 377–381.

United Nations Environmental Programme (UNEP), 2017. Marine Plastic Debris and Microplastics: Global Lessons and Research to Inspire Action and Guide Policy Change. UNEP, Nairobi.

Van Cauwenberghe, L., Claessens, M., Vandegehuchte, M.B., Janssen, C.R., 2015. Microplastics are taken up by mussels (Mytilus edulis) and lugworms (Arenicola marina) living in natural habitats. Environ. Pollut. 199, 10–17.

Van Sebille, E., Wilcox, C., Lebreton, L., Maximenko, N., Hardesty, B.D., Van Franeker, J.A., et al., 2015. A global inventory of small floating plastic debris. Environ. Res. Lett. 10 (12), 124006.

Vegter, A., Barletta, M., Beck, C., Borrero, J., Burton, H., Campbell, M., Eriksen, M., Eriksson, C., Estrades, A., Gilardi, K., Hardesty, B.D., Sul, J.a.I.D., Lavers, J.L., Lazar, B., Lebreton, L., Nichols, W.J., Ribic, C., Ryan, P., Schuyler, Q.A., Smith, S.D.A., Takada, H., Townsend, K., Wabnitz, C.C.C., Wilcox, C., Young, L., Hamann, M., 2014. Global research priorities to mitigate plastic pollution impacts on marine wildlife. Endanger. Species Res. 25, 225–247.

Veiga, J.M., Vlachogianni, T., Pahl, S., Thompson, R.C., Kopke, K., Doyle, T.K., Hartley, B.L., Maes, T., Orthodoxou, D.L., Loizidou, X.I., Alampei, I., 2016. Enhancing public awareness and promoting co-responsibility for marine litter in Europe: the challenge of MARLISCO. Mar. Pollut. Bull. 102, 309–315.

Verlis, K.M., Campbell, M.L., Wilson, S.P., 2014. Marine debris is selected as nesting material by the brown booby (Sula leucogaster) within the Swain Reefs, Great Barrier Reef, Australia. Mar. Pollut. Bull. 87, 180–190.

Watkins, E., Ten Brink, P., Withana, S., Mutafoglu, K., Scheitzer, J.-P., Russi, D., Kettunen, M., 2015. Marine Litter: Socio-Economic Study. Scoping Report. IEEP, London, Brussels.

Watts, A.J.R., Urbina, M.A., Corr, S., Lewis, C., Galloway, T.S., 2015. Ingestion of plastic microfibers by the Crab Carcinus maenas and its effect on food consumption and energy balance. Environ. Sci. Technol. 49, 14597–14604.

Welden, N.A., Cowie, P.R., 2017. Degradation of common polymer ropes in a sublittoral marine environment. Mar. Pollut. Bull. 118, 248–253.

World Economic Forum (WEF), 2016. The New Plastics Economy. Rethinking the Future of Plastics. WEF, Geneva.

Worm, B., Lotze, H.K., Jubinville, I., Wilcox, C., Jambeck, J., 2017. Plastic as a persistent marine pollutant. Annu. Rev. Environ. Resour. 42, 1–26.

Chapter 35

Changing Hydrology: A UK Perspective

Peter E. Robins, Matt J. Lewis

School of Ocean Sciences, Marine Centre Wales, Bangor University, Menai Bridge, United Kingdom

1 INTRODUCTION

Estuaries define the confluence linking terrestrial and marine processes, making them critical regions for our holistic understanding of hydrological and biogeochemical cycles. Estuaries are where we predominantly live and where many of our major cities are located (e.g., New Orleans, New York, New Delhi, and London). Consequently, waterborne effluents and wastes derived from industries, urban areas, and agriculture are transported through rivers and estuaries to the coast. Yet, there are fundamental physical processes through this continuum that we poorly understand and models are often applied with high levels of uncertainty so that we cannot predict future impact with sufficient accuracy. As such, we urgently need a multidisciplinary approach to improve our knowledge of how estuaries and the life within them may change faced with current climate change projections.

Amongst all natural environments, estuaries are some of the most biologically productive, but they are also vulnerable to stress and to change because of their sensitivity to large and variable physical drivers. Future sea states (sea levels, tidal characteristics, storm surges, and waves) and future river flows are projected to be impacted by climate change (Robins et al., 2016). Together, these drivers form a poorly understood combination hazard risk, where we do not know the impact of different combinations of events nor how likely they are to happen. People who use and manage estuaries and coasts are therefore concerned about the many potential consequences of climate change. Changing fluxes of nutrients, litter (e.g., plastics), and microbial pathogens from terrestrial anthropogenic sources may lead to increased degradation of coastal ecosystems and human health risk through poor water quality and a risk to food security (Robins et al., 2016 and references therein). Estuarine ecosystems and cycles do tolerate change due to natural variability and extreme events; however, our ability to understand and predict change is often necessarily limited in both observations and in model predictions, especially as we tend to quantify fluxes through the river-estuary-coast continuum at national scale (e.g., Greene et al., 2015). However, downscaling from global to national to catchment scales requires considerable observational and computational resources. Because estuaries hold high socioeconomical value, they require operational management and "future proofing"; primarily for flood risk, water quality, ecosystem health, and food security. Therefore, there is a growing need for improved modeling methods, improved model parameterizations, and reduced model uncertainties to understand how these systems may respond to change.

Across a range of estuary types, it is difficult to characterize common patterns of impact, due to the complexity of interacting drivers in different environments. For example, a storm passing through the United Kingdom will lead to a range of river hydrograph types depending on catchment characteristics, and therefore carry different loads (e.g., sediments, nutrients, pathogens, and microplastics) into different estuary types. Moreover, the phased occurrence of such loads with marine drivers will differ between systems. We therefore rely on bespoke models per system, which simulate the hydrodynamic-ecosystem processes, to predict the probabilistic hazard; however, current methodologies and techniques to predict change raise uncertainty. Faced with a range of uncertainties in knowledge and in model parameterization, there is a real question as to whether model predictions of land-to-sea processes are useful in informing management decisions about flood risk, water quality, and ecological status (Robins et al., 2018). One important unknown is how sensitive estuarine models are to boundary forcing such as river flow, and what steps might be needed to improve their usefulness as a management tool for impact. In this chapter, we focus on the United Kingdom, where the future impact on estuaries due to changes in the hydrology may be as important, if not more so, than due to other climate drivers such as sea-level change or morphological change.

Coasts and Estuaries. https://doi.org/10.1016/B978-0-12-814003-1.00035-6

2 SENSITIVITY OF UK ESTUARIES TO RIVER FLOWS

In the United Kingdom, the majority of estuaries[1] are relatively shallow with large tidal forcing (e.g., having a tidal range greater than 2 m) and variable freshwater inputs. A lot of the smaller estuaries—typically less than 30 km in length, of which there are many in the United Kingdom—drain almost completely during the ebbing tide exposing substantial intertidal regions comprising salt marshes and tidal flats. From a hydrodynamic perspective, this means that the local water cycle is strongly influenced by advective tidal dynamics, meaning that circulation is controlled by the back-and-forth nature of the flooding and ebbing tide, but also by river flow variability. On the other hand, circulation in larger and deeper estuaries like the Humber, Severn, Forth, and Thames will be less influenced by hour-by-hour tidal oscillations and river flows, but instead respond to net circulation patterns that act on longer timescales and are caused by differences in salinity structure. Typical estuarine net circulation results in near-surface freshwater outflow to the sea and inflow of coastal saltwater at depth. Depending on the strength of the tidal flux relative to the river flux, and to the estuary shape itself, an estuary can experience stratification and hence strong estuarine circulation. These stratified estuarine systems, termed "salt wedge" or "partially mixed," are usually the result of strong river flows relative to the tide. Alternatively, strong tides relative to the river flow lead to well-mixed waters which tend to experience weaker estuarine circulation. The estuarine circulation is subject to augmentation or breakdown due to a range of factors, most notably tidal state (e.g., flood vs. ebb and spring vs. neap) and the passing of storm tracks that typically last 1–3 days for UK catchments. In turn, the strength of the tide vs. river dynamics in small estuaries, or the residual circulation in larger estuaries, will influence the residence time of water and biogeochemical cycles and, hence, the overall estuary health. Both strongly stratified and well-mixed estuaries respond rapidly to changes in river flow and tidal mixing, while partially mixed estuaries respond more slowly, although have the greatest sensitivity to change (MacCready, 1999).

It is therefore clear that of high importance for predicting future impacts to estuaries are simulating variabilities in river flows—particularly due to the nature and frequency of high flow events caused by intense rainfall. Rivers in steep catchments, for example, mountainous regions of the United Kingdom's west coast, tend to rise very quickly after rainfall—over a period of hours rather than days, especially where catchments are small and/or impermeable. As such, the connecting estuaries are often in a state of nonequilibrium in terms of salt balance and biogeochemical cycles, since steady-state recovery can take several weeks. As an example, Robins et al. (2018) analyzed a 35-year record of river flows (15-min sampling) for the River Conwy, western United Kingdom, which were used to characterize variabilities in storm hydrograph type and simulate the estuarine response in terms of the salt balance and also nutrient transport. For this fast-responding catchment-estuary system, the natural variability in storm type ranged from several hours to a few days in storm duration and up to 250 m^3/s in flow magnitude. This generated large variability in the downstream estuarine circulation and mixing, meaning that the estuarine salt content and nutrient concentrations varied considerably in response to the river flow. Crucially for future impact modeling studies, this pattern was not captured by simulations forced with daily-averaged river flows (because the storm type was not represented) but was captured using 15-min river forcing.

In contrast, a model of a slower-responding system (the Humber, eastern United kingdom, which has a catchment that covers a third of England) with much longer-lasting hydrographs (2–6 days) showed less variability in estuarine salt and nutrient budgets due to different storm types. Moreover, the model showed good approximation when forced with daily-averaged flows (Robins et al., 2018). Because the river response to rainfall was relatively gradual and the estuary is relatively large, changes in river flows due to individual storms did not strongly influence the total estuarine salt and nutrient content. Instead, the estuary responded over atmospheric frontal time scales of several days. Moreover, these systems will be sensitive to projected seasonal shifts in our climate of "wetter winters and drier summers" (Christierson et al., 2012). These results have implications for entire system impact modeling; when we determine future changes to estuaries, some systems will need high-resolution future river flow estimates (i.e., sub-hourly), whereas other systems may be well approximated by daily mean flow estimates as is the *status quo*. One improvement that could be made for catchment-to-coast modeling, therefore, is the downscaling of climate models to predict sub-daily rainfall. Projected changes in future weather patterns can then be linked directly to changes in hydrograph type and biogeochemical flux and in turn changes to estuarine impact. One must, however, take into account of other sources of model uncertainty, such as model sensitivity to changes in sediment dynamics and morphology. Indeed, the importance of sedimentation as a driver for estuary impact is unclear in relation to river flow and sea-level rise drivers.

---

1. We are not including fjords found in Scotland in this definition.

3 PAST TRENDS AND FUTURE PROJECTIONS FOR HYDROLOGY

River flows represent the integrated response of all hydrometeorological processes acting upon a catchment, and are therefore a key indicator of potential impacts of climate change to UK estuaries (Hannaford, 2015). Atmospheric warming has been shown by Watts et al. (2015) to intensify the hydrological cycle; although there is a complex, nonlinear process leading from increasing temperatures, through changes in precipitation to river flow response. Past trends show that, while annual-mean rainfall in the United Kingdom has not changed over the past 50 years, winter rainfall has intensified and increasingly occurred in clustered events (Watts et al., 2015). For the same period, river flows have increased in winter also, while there was no consistent change in the summer (Hannaford, 2015; Watts et al., 2015). For past extreme events, data from extensive UK river-gauge networks show that the relationship between precipitation and river flow is spatially variable and dependent on catchment characteristics (Keef et al., 2009).

Several studies have investigated potential changes to future UK river flows (e.g., Prudhomme and Davies, 2009a,b), including impacts to groundwater (e.g., Jackson et al., 2011), and analysis of the latest UKCP09 future river flow data series (1 km-resolution and catchment-bias-corrected; Prudhomme et al., 2012). The scientific consensus on projected future UK precipitation patterns is the "wetter winters and drier summers" signal (Christierson et al., 2012), but with some spatial variability. Indeed, the UKCP09 Report explains that little change is expected in the median precipitation amount by 2100, but significant changes in the trends of winter and summer precipitation are expected. The report generally found a slight increase in the winter mean precipitation and a decrease in summer mean precipitation for the western United Kingdom. Recent work that includes high-resolution and convective systems modeling suggests future heavier rainfall events during summer (Kendon et al., 2014). River catchment models indicate river flows are projected to reduce in summer (by 40%–80%) and increase in winter (by up to 25%), again particularly in mountainous regions of the western United Kingdom (Prudhomme et al., 2012). These projections were biased through inability to capture monthly precipitation climates (Smith et al., 2013). Furthermore, when future precipitation patterns are projected through to future river flow rates there is considerable uncertainty in general circulation model (GCM) projections (Prudhomme and Davies, 2009a), as well as downscaling and catchment model uncertainties (Prudhomme and Davies, 2009b).

Another important driver in river flow climatology is land use, which can alter river hydrographs; for example, hydroelectric and water-resource dams, and flood mitigation measures—however, this will not be considered here. Therefore, it is clear that there is high uncertainty in predicting future changes to UK river flows, due to the spatial variability associated with projected changes to rainfall patterns. Furthermore, considering climate model uncertainties, downscaling uncertainties, hydrological model uncertainties and land use changes, as well as variability of the climate system, prediction of future estuary changes requires further research to determine confidence in future river flow rates.

An ongoing analysis by researchers at the Bristol and Bangor Universities (unpublished), using measured flows from over 100 UK rivers covering several decades, has enabled storm hydrograph statistics to be characterized. How different hydrograph types impact upon estuaries can then be determined, and how these hydrograph types may change in the future can be predicted based on projected rainfall patterns. The analysis has revealed that storm hydrographs typically last several hours in small or steep catchments and up to several days in larger catchments. This information is important because it can be used to help to improve model accuracy for predicting future impact to UK estuaries and coasts, for example, for predicting coastal flood risk. Winter hydrograph types were characteristically more "flashy" (i.e., greater flux per unit time) than those during summer, presumably because of increased winter groundwater saturation. Other than this, trends in storm type and frequency were not significant at a national scale, although there was generally a relationship between storm type and the North Atlantic Oscillation (NAO)—a time-varying index for broad-scale atmospheric conditions influencing the United Kingdom and therefore a proxy index for climate change. As an example, Fig. 1 shows the typical hydrograph types expected for the River Conwy (the United Kingdom) and how they vary (natural variability) over different winters; some winters being more flashy than other winters, on average. The figure also shows the correlation between these hydrograph types and the NAO index for each winter season. The clear negative relationship shown in Fig. 1B implies that UK river patterns are sensitive to the NAO index, and therefore will change in response to climate change. Several climate models suggest higher (more positive) average values of NAO index during the next 100 years (Woodworth et al., 2007). Furthermore, the many small estuaries around the United Kingdom that are sensitive to river flow variability will in turn experience change. An obvious question then arises for future research: how important are the many small estuaries that are susceptible to changes in weather patterns when compared with the fewer larger ones that are more resilient to these changes? Changing hydrology is therefore not solely caused by land use change, such as the effects of deforestation or

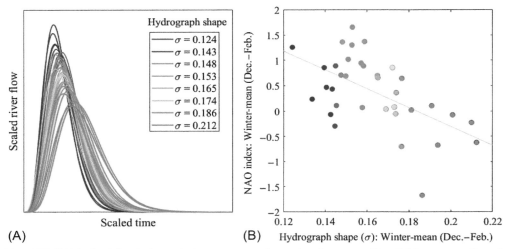

FIG. 1 Conwy River, UK: (A) Idealized river hydrograph shapes corresponding to the winter-mean (Dec.–Feb.) storm type for each year 1980–2015. The hydrographs are sorted from most flashy (low σ, colored blue) to least flashy (high σ, colored red); see Robins et al. (2018) for the methodology. (B) Correlation between winter-mean hydrograph shape (σ) and winter-mean North Atlantic Oscillation (NAO) index. The linear line of best fit is plotted to show the trend, with a Pearson's Rank correlation of 0.0005, based on 35 samples (years), and a coefficient of determination $r^2 = 0.25$. See Neill and Hashemi (2013) for justification of using winter-mean (Dec.–Feb.) NAO values. *(Modified from Robins PE, Lewis MJ, Freer J, Cooper DM, Skinner CJ, Coulthard TJ: Improving estuary models by reducing uncertainties associated with river flows, Estuar. Coast. Shelf Sci., 2018).*

urbanization, which are not discussed further here other than to say that these effects left unmanaged are likely to increase direct runoff and lead to more flashy hydrographs. Note that the all rivers and estuaries are susceptible to change through broader-scale (e.g., seasonal) climate change, such as projected wetter winters and drier summers for the United Kingdom, and it would be interesting to investigate how estuaries will respond to climate change on different temporal scales in relation to changes in sedimentation and sea-level rise.

4 POTENTIAL IMPACTS TO ESTUARIES FROM CHANGING HYDROLOGY

4.1 Flooding and Inundation

Increased estuarine flood risk and inundation occurrence are caused primarily by a combination of increased sea level (sea-level rise and changes to the storm surge and wave climates) and changes to incidences of high river flow events (Robins et al., 2016). Further, there is evidence for a positive relation in some UK estuaries between storm surge and peak river flows (i.e., high surge occurs simultaneously with peak river flow), suggesting that surges should to be considered for flood risk estimation (Svensson and Jones, 2004). High correlation generally occurred for south and west UK catchments that were hilly (i.e., fast responding) and south or west facing (i.e., facing into the predominant storm tracks). Although such analyses have so far been conducted at daily resolution, and it is unclear how phased occurrence of surge and peak river flow at sub-daily scales will affect flood risk. Indeed, the co-occurrence (and how this may change in the future) of tides, surges, waves, and river flows—and their combined impact in terms of estuarine flood risk—is poorly understood at present. Further, determining the pollution risk from one source within the context of a combination hazard is unknown.

Both sea-level and river flow climates in the United Kingdom appear to be changing due to climate change, potentially resulting in increased flood risk to estuaries (Robins et al., 2016). Although projected increases in UK average temperatures will increase evapotranspiration, altering vegetation types and reducing soil moisture—all of which could reduce runoff and potentially dampen flood risk slightly during summer (Robins et al., 2016). Potential negative impacts of increased flooding to ecosystem goods and services, in the absence of mitigation, include: damage to infrastructure and population displacement, increased erosion and loss of land to the sea, pollution and increased human health risk, and significant loss of wetland habitats and ecosystems. Evidently, it is important for modeling approaches to simulate the nonlinear coupling of tide-surge-wave-river events at appropriate spatial and temporal scales for each unique system, since static (sea-level only) events will underestimate flood risk (King, 2004; Quinn et al., 2013).

4.2 Water Quality

By the end of this century, projected sea-level rise and increased temperatures together with changes in river flow patterns will ultimately have impacts upon UK estuarine water quality. Estuaries play a crucial role in biogeochemical cycles, transporting, and processing macronutrients with spatial discontinuities in retention and storage, which in turn influence the growth and resilience of plants and animals. Projections suggest a possible rise of macronutrient concentrations (such as nitrogen and phosphorus) under climate change conditions, with consequent impacts of augmented eutrophication, hypoxia, and harmful algal blooms (Statham, 2012). Projected sea-level rise, combined with reduced summer river flows, could heighten the likelihood of estuarine nutrient trapping, in between river flushing events. Nutrification in summer is more prone to cause eutrophication than in winter: only summer production is nutrient limited; winter production is light limited. Thus, the dry summer effect has the greatest potential for augmenting eutrophication.

Projected increased precipitation and flooding may heighten the risk of waterborne and foodborne diseases caused by microbial pathogens (e.g., *Cryptosporidium*, Norovirus) present in untreated sewage discharge and agricultural runoff (McMichael et al., 2006). In addition, toxic pollutants, ammonia, organic solids, and oxygen-demanding substances are also discharged into the water system (Stachel et al., 2004) and may increase concurrently with high precipitation events. It is well established that increased rainfall may influence the concentrations of toxic metals, organic chemicals, algal toxins, and human pathogen contaminants in seafood (Marques et al., 2010). Conversely, drought conditions will result in decreased river flow and lower concentrations of microbial contaminants (Rose et al., 2001). The mobilization of historic, consolidated, and heavy metal contaminants in semi-industrialized UK estuaries (e.g., from mudflats and saltmarshes in the Humber and Mersey systems, UK) poses a serious risk to water quality and human health (Lee and Cundy, 2001). On the other hand, a potential positive outcome is that this re-mobilization, which is often a result of extreme stormy weather, accelerates sediment exchange with the open sea (e.g., the exchange of mercury-rich fine sediments from the Mersey Estuary into Liverpool Bay).

4.3 Habitats

Estuarine intertidal habitats and biogenic reefs—often priority habitats for protection in the United Kingdom—are intense production areas of invertebrates that feed higher trophic levels, including birds (McLusky and Elliott, 2004). These habitats are already undergoing geomorphological change from climatic drivers (wave height, surges, and sea-level rise); coastal squeeze and temperature warming being primarily responsible for reduced species diversity (Robins et al., 2016). These environments are also likely to respond to changes in river flow, although there are no detailed studies to support this prediction in the United Kingdom. Eventually, climate change may have implications on the conservation status of UK estuaries.

5. SUMMARY

We summarize the potential impacts of changing hydrology on estuaries. Climatologists predict seasonal hydrological cycles are to change over the coming century, based on changing seasonal rainfall and evaporation projections, although there is high uncertainty when predicting changes to storm tracks and consequently changes to hydrograph shapes. Nevertheless, we demonstrate, through published and ongoing research, that hydrograph shape is sometimes important for estuarine response, and that current modeling approaches may need to be refined if we are to produce meaningful simulations for management. For example, the downscaling of climate modeled rainfall to hourly resolution in order to predict hydrograph shape. Then research can address the importance of changing hydrology on estuaries relative to other important drivers such as sea-level rise, changing land use, and changing morphology.

Fig. 2 summarizes the likely, but not all, pathways from climatic drivers to processes to impacts on UK estuaries (taken from Robins et al., 2016). There is also no indication from the figure of the scale of impact; for example, the consequences of flooding may be more significant than shifts in frontal positions, although this is difficult to quantify. In addition, the figure does not consider changes in storm hydrograph shape, only seasonal changes such as increased winter flows. Although the likely network of processes and impacts are clearly laid out, indicating a general negative future picture with only few positive impacts.

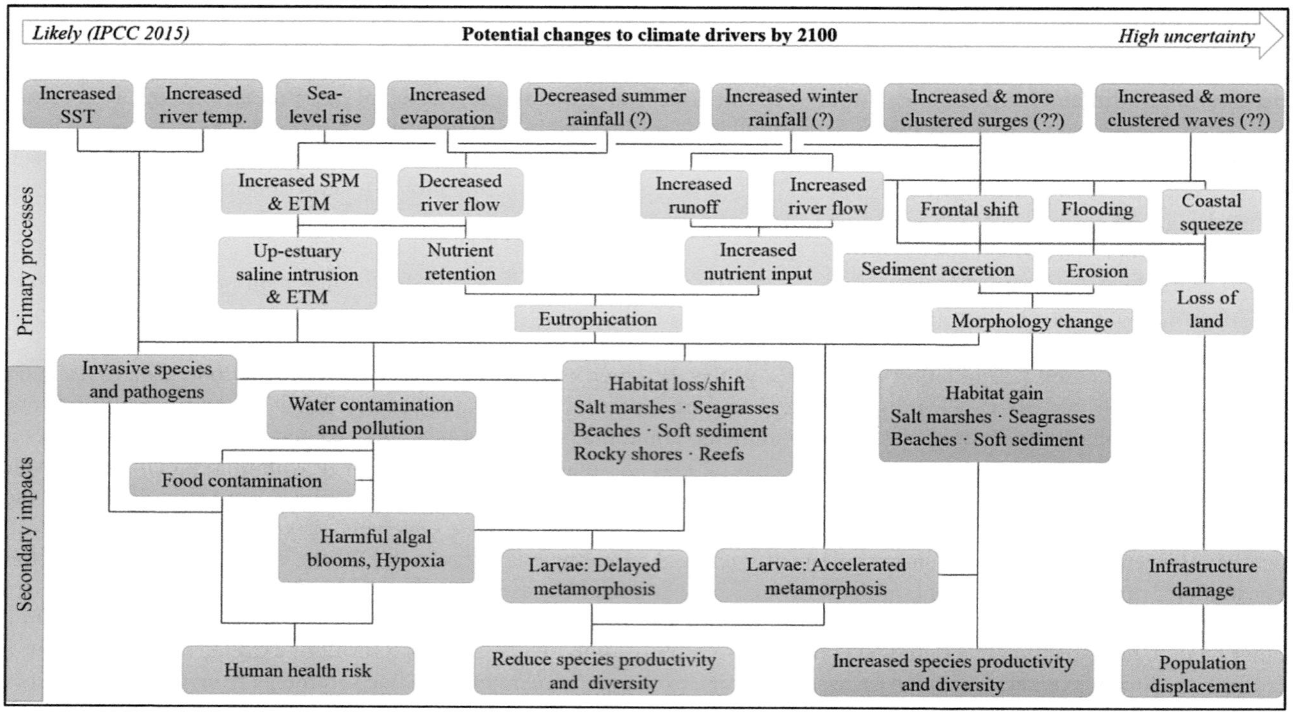

FIG. 2 Flow diagram showing the main climatic drivers, and primary and secondary impacts of climate change to UK estuaries. Pathways (black lines) are directed *from* the bottom/side *to* the top of a "driver/impact box." Predominantly positive impacts are colored green. SPM refers to suspended particulate matter and ETM refers to estuarine turbidity maximum. *(From Robins, P.E., Skov, M.W., Lewis, M.J., Giménez, L., Davies, A.G., Malham, S.K., Neill, S.P., McDonald, J.E., Whitton, T.A., Jackson, S.E., Jago, C.F., 2016. Impact of climate change on UK estuaries: a review of past trends and potential projections. Estuar. Coast. Shelf Sci. 169, 119–135.)*

REFERENCES

Christierson, B.V., Vidal, J.P., Wade, S.D., 2012. Using UKCP09 probabilistic climate information for UK water resource planning. J. Hydrol. 424, 48–67.

Greene, S., Johnes, P.J., Bloomfield, J.P., Reaney, S.M., Lawley, R., Elkhatib, Y., Freer, J., Odoni, N., Macleod, C.J.A., Percy, B., 2015. A geospatial framework to support integrated biogeochemical modelling in the United Kingdom. Environ. Model. Software 68, 219–232.

Hannaford, J., 2015. Climate-driven changes in UK river flows: a review of the evidence. Prog. Phys. Geogr. 39 (1), 29–48.

Jackson, C.R., Meister, R., Prudhomme, C., 2011. Modelling the effects of climate change and its uncertainty on UK Chalk groundwater resources from an ensemble of global climate model projections. J. Hydrol. 399 (1-2), 12–28.

Keef, C., Svensson, C., Tawn, J.A., 2009. Spatial dependence in extreme river flows and precipitation for Great Britain. J. Hydrol. 378 (3-4), 240–252.

Kendon, E.J., Roberts, N.M., Fowler, H.J., Roberts, M.J., Chan, S.C., Senior, C.A., 2014. Heavier summer downpours with climate change revealed by weather forecast resolution model. Nat. Clim. Chang. 4 (7), 570.

King, D.A., 2004. Climate change science: adapt, mitigate, or ignore? Science 303 (5655), 176–177.

Lee, S.V., Cundy, A.B., 2001. Heavy metal contamination and mixing processes in sediments from the Humber Estuary, Eastern England. Estuar. Coast. Shelf Sci. 53 (5), 619–636.

MacCready, P., 1999. Estuarine adjustment to changes in river flow and tidal mixing. J. Phys. Oceanogr. 29 (4), 708–726.

Marques, A., Nunes, M.L., Moore, S.K., Strom, M.S., 2010. Climate change and seafood safety: human health implications. Food Res. Int. 43 (7), 1766–1779.

McLusky, D.S., Elliott, M., 2004. The Estuarine Ecosystem: Ecology, Threats and Management. Oxford University Press on Demand.

McMichael, A.J., Woodruff, R.E., Hales, S., 2006. Climate change and human health: present and future risks. Lancet 367 (9513), 859–869.

Neill, S.P., Hashemi, M.R., 2013. Wave power variability over the northwest European shelf seas. Appl. Energy 106, 31–46.

Prudhomme, C., Davies, H., 2009a. Assessing uncertainties in climate change impact analyses on the river flow regimes in the UK. Part 1: baseline climate. Clim. Chang. 93 (1-2), 177–195.

Prudhomme, C., Davies, H., 2009b. Assessing uncertainties in climate change impact analyses on the river flow regimes in the UK. Part 2: future climate. Clim. Chang. 93 (1-2), 197–222.

Prudhomme, C., Dadson, S., Morris, D., Williamson, J., Goodsell, G., Crooks, S., Boelee, L., Davies, H., Buys, G., Lafon, T., Watts, G., 2012. Future flows climate: an ensemble of 1-km climate change projections for hydrological application in Great Britain. Earth Sys. Sci. Data 4 (1), 143–148.

Quinn, N., Bates, P.D., Siddall, M., 2013. The contribution to future flood risk in the Severn Estuary from extreme sea level rise due to ice sheet mass loss. J. Geophys. Res. Oceans 118 (11), 5887–5898.

Robins, P.E., Skov, M.W., Lewis, M.J., Giménez, L., Davies, A.G., Malham, S.K., Neill, S.P., McDonald, J.E., Whitton, T.A., Jackson, S.E., Jago, C.F., 2016. Impact of climate change on UK estuaries: a review of past trends and potential projections. Estuar. Coast. Shelf Sci. 169, 119–135.

Robins, P.E., Lewis, M.J., Freer, J., Cooper, D.M., Skinner, C.J., Coulthard, T.J., 2018. Improving estuary models by reducing uncertainties associated with river flows. Estuar. Coast. Shelf Sci.

Rose, J.B., Epstein, P.R., Lipp, E.K., Sherman, B.H., Bernard, S.M., Patz, J.A., 2001. Climate variability and change in the United States: potential impacts on water-and foodborne diseases caused by microbiologic agents. Environ. Health Perspect. *109* (Suppl 2), 211.

Smith, A., Bates, P., Freer, J., Wetterhall, F., 2013. Future flood projection: investigating the application of climate models across the UK. Hydrol. Process. *28* (2810), e2823.

Stachel, B., Götz, R., Herrmann, T., Krüger, F., Knoth, W., Päpke, O., Rauhut, U., Reincke, H., Schwartz, R., Steeg, E., Uhlig, S., 2004. The Elbe flood in August 2002-occurrence of polychlorinated dibenzo-p-dioxins, polychlorinated dibenzofurans (PCDD/F) and dioxin-like PCB in suspended particulate matter (SPM), sediment and fish. Water Sci. Technol. 50 (5), 309–316.

Statham, P.J., 2012. Nutrients in estuaries—an overview and the potential impacts of climate change. Sci. Total Environ. 434, 213–227.

Svensson, C., Jones, D.A., 2004. Dependence between sea surge, river flow and precipitation in south and west Britain. Hydrol. Earth Syst. Sci. 8 (5), 973–992.

Watts, G., Battarbee, R.W., Bloomfield, J.P., Crossman, J., Daccache, A., Durance, I., Elliott, J.A., Garner, G., Hannaford, J., Hannah, D.M., Hess, T., 2015. Climate change and water in the UK–past changes and future prospects. Prog. Phys. Geogr. 39 (1), 6–28.

Woodworth, P.L., Flather, R.A., Williams, J.A., Wakelin, S.L., Jevrejeva, S., 2007. The dependence of UK extreme sea levels and storm surges on the North Atlantic Oscillation. Cont. Shelf Res. 27 (7), 935–946.

Section H

Management of Change

Chapter 36

Global Change Impacts on the Future of Coastal Systems: Perverse Interactions Among Climate Change, Ecosystem Degradation, Energy Scarcity, and Population

John W. Day[*], John M. Rybczyk[†]

[*]*Department of Oceanography and Coastal Sciences, College of the Coast and Environment, Louisiana State University, Baton Rouge, LA, United States, [†]Department of Environmental Science, Western Washington University, Bellingham, WA, United States*

1 INTRODUCTION

Global climatic change is projected to have a number of serious impacts including increases in temperature, accelerated sea-level rise (SLR), an increase in extreme weather events, changes in freshwater discharge, and changes in the frequency and intensity of tropical storms (Day and Templet, 1989; Twilley et al., 2001; Poff et al., 2002; Scavia et al., 2002; Day et al., 2005, 2008; Yáñez-Arancibia and Day, 2005; Hoyos et al., 2006; Vermeer and Rahmstorf, 2009; McCarthy, 2009). These changes will have tremendous human, economic, and ecological impacts.

Climate change impacts on coastal systems include high rates of SLR due to both subsidence as well as accelerated eustatic sea-level rise (ESLR) (FitzGerald et al., 2008; Pfeffer et al., 2008; Vermeer and Rahmstorf, 2009; IPCC, 2007, 2013; Koop et al., 2016), more severe tropical cyclones (Emanuel, 2005; Goldenberg et al., 2001; Hoyos et al., 2006; Kaufmann et al., 2011; Webster et al., 2005; Mei et al., 2015), drought (IPCC, 2007), and more erratic and intense weather (Min et al., 2011; Pall et al., 2011). These climate impacts are affecting coastal systems directly, at the scale of the drainage basin, and globally. Basin-level impacts include both increases (e.g., Mississippi) and decreases (e.g., Colorado) in projected precipitation and river discharge.

Global climate change impacts on coastal systems will have highly significant impacts on coastal ecosystems (FitzGerald et al., 2008; Day et al., 2008, 2011c; Rybczyk et al., 2013). Coastal zones with large areas of intertidal wetlands, especially deltas, will be strongly impacted by SLR (Day et al., 2016b). Rainfall also plays an important role in the response of coastal systems to climate change. In areas that already have low rainfall, further reductions in rainfall will likely have serious impacts on coastal vegetation (Day and Templet, 1989; IPCC, 2007).

Thus, there is strong evidence that climate change will strongly impact coastal systems in the 21st century but what is less understood is that climate change will interact with other global change forcings to threaten both natural and human systems in the coastal zone. These forcings include ecosystem degradation, resource scarcity especially energy, population growth, and a flawed economic system (Hall and Klitgaard, 2018; Day et al., 2018). Our objective in this paper is to present the evidence for climate change and describe its physical, hydromorphological, ecological, and human impacts on coastal systems. We then discuss how climate change interacts with other global change forcings often in a way that makes sustainable restoration more difficult and challenging.

Coasts and Estuaries. https://doi.org/10.1016/B978-0-12-814003-1.00036-8

2 GLOBAL CLIMATE CHANGE: PAST TRENDS, FUTURE PREDICTIONS, AND SYSTEM IMPACTS

There has been considerable study of global climate change and its impacts. The most well-known reports on climate change are those of the Intergovernmental Panel on Climate Change (IPCC), which has issued assessment reports in 1990, 1996, 2001, 2007, 2013, and 2017 as well a number of other synthesis articles (e.g., Poff et al., 2002; Day et al., 2008; McCarthy, 2009; Rybczyk et al., 2013).

2.1 Temperature

Global temperatures will likely rise 1–5°C during the 21st century (IPCC, 2007, 2013) compared to an increase in temperature of approximately 0.85°C during the 20th century. Over the same period, CO_2 levels increased from preindustrial levels of about 280 to over 400 ppm between 1880 and 2010, IPCC (2013). Increasing temperatures will affect precipitation patterns, SLR, the intensity and frequency of tropical cyclones and other storms, the frequency of extreme weather events, and biological processes.

The IPCC publishes predictions of global temperature increase for various scenarios of greenhouse gas (GHG) emissions called representative concentration pathways (RCP), and are based on different socioeconomic assumptions such as economic activity, fossil fuel use, and population growth (Moss et al., 2008). Predictions range from as low as 0.3°C of warming by 2100 for RCP2.6, and as high as 4.8°C for RCP8.5. Friedlingstein et al. (2014) reported that CO_2 levels in the atmosphere are increasing at rates consistent with the highest IPCC scenarios, indicating that temperature projections will likely be on the high end.

Many coastal species, such as mangroves, inhabit restricted ranges in the latitudinal temperature gradient that exists from the equator to the poles. Climate-induced changes in regional temperature regimes will induce a wide range of ecological responses such as local extinctions, shifts in geographic locations, changes in biodiversity, and changes in rates of primary production and decomposition. Many species will shift their geographic ranges to higher latitudes as is happening to mangroves worldwide (e.g., Yáñez-Arancibia et al., 2010; Twilley and Day, 2013; Osland et al., 2016; Dangremond and Feller, 2016; Doughty et al., 2016; Saintilan et al., 2014) and many areas that are now subtropical or subtemperate will become tropical in a process that is called *tropicalization* (Day et al., 2010c).

2.2 Sea-Level Rise

Accelerated ESLR during the 21st century will impact coastal ecosystems globally. During the last glacial maximum, sea level was over 100 m below current levels but rose at the end of the last glacial epoch as land-based ice masses melted and ocean water expanded. Currently, about half of SLR is due to this thermal expansion but it will proportionally decrease with continued warming as more land-based ice masses melt. There is enough water locked in land-based ice to raise sea levels by 80 m (Emery and Aubrey, 1991).

SLR for most of the 20th century averaged 1.7 mm/year (FitzGerald et al., 2008; Hansen et al., 2015, Fig. 1) and was 3–4 mm/year from 1993 to 2010 (IPCC WG1 2013; Nerem et al., 2010). Hansen et al. (2015) reported that global SLR was 0.6 mm/year between 1900 and 1939 and 3.3 mm/year between 1993 and 2015 (Fig. 1). Thus, the rate for the past two decades was nearly twice the average of the 20th century as a whole and over five times higher than the beginning of the 20th century.

The IPCC (2013) projects SLR between 0.3 and 1.0 m by 2100. Recent evidence suggests that this may be conservative. For example, from 1993 to 2003 observed ESLR was 3.1 mm/year, exceeding the lower limit predicted by the IPCC, 2007 assessment. Semiempirical models project sea level may rise nearly 2 m by 2100 (Rahmstorf, 2007; Vermeer and Rahmstorf, 2009; Horton et al., 2014, Fig. 2). Koop et al. (2016) project a rise of 52–131 cm by 2100 for the RCP8.5 scenario. The NOAA National Climate Assessment (Parris et al., 2012) assigns plausible lower and upper limits of 0.2 and 2.0 m, where the upper limit considers the maximum possible glacier and ice sheet loss (Pfeffer et al., 2008). Currently, globally averaged SLR likely exceeds 4.0 mm/year (McCarthy, 2009). Recently, there has been increasing concern over the vulnerability of the West Antarctica and Greenland ice sheets to global warming (Rignot et al., 2011, 2014; Tollefson, 2017), which suggests future rates of ESLR closer to the upper end of projections. DeConto and Pollard (2016) report that if CO_2 emissions continue unabated, Antarctic contribution to SLR may add an additional meter resulting in a total SLR of more than 2 m or more by 2100 (Hoyos et al., 2006; Rahmstorf, 2007; Pfeffer et al., 2008; Rohling et al., 2008; Mitrovica et al., 2009; Vermeer and Rahmstorf, 2009, Fig. 2). Sea level will continue rising over the next couple of centuries after 2100 by 2–3 m or higher depending on different climate scenarios. It is worth noting that during the last interglacial period,

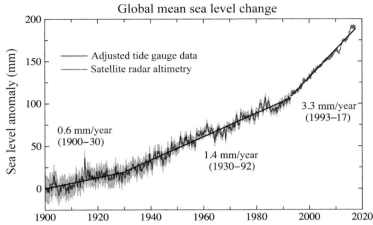

FIG. 1 Rates of sea-level rise from 1900 to 2015. *(From Hansen, J., Sato, M., Hearty, P., Ruedy, R., Kelley, M., Masson-Delmotte, V., Russell, G., Tselioudis, G., Cao, J., Rignot, E., Velicogna, I., 2005. Ice melt, sea level rise and superstorms: evidence from paleoclimate data, climate modeling, and modern observations that 2 C global warming is highly dangerous. Atmos. Chem. Phys. 16(6), 3761–3812.)*

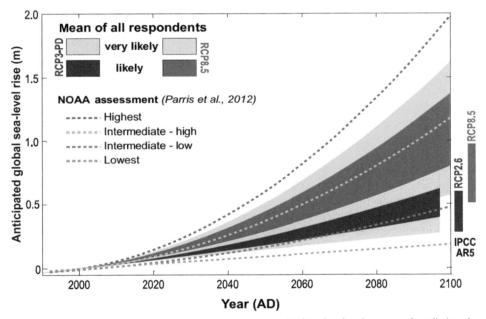

FIG. 2 Range of sea-level rise projections based on expert assessment. These projections bracket the range of predictions based on modeling and semiempirical projections. *(From Horton, B.P., Rahmstorf, S., Engelhart, S., Kemp, A., 2014. Expert assessment of sea-level rise by AD 2100 and AD 2300. Quat. Sci. Rev. 84, 1–6.)*

about 130,000–115,000 years ago when global temperatures were only slightly warmer than today, the sea-level rose 3–5 m. This has been attributed to subsurface ocean warming and rapid ice loss in Antarctica (Hansen et al., 2015).

Relative sea-level rise (RSLR) is the combination of ESLR and subsidence. For example, high rates of subsidence commonly occur in deltas due to compaction, consolidation, and dewatering of sediments (Syvitski et al., 2009; Day et al., 2016b). RSLR for much of the Mississippi delta is in excess of 10 mm/year (Baumann et al., 1984; Day et al., 2007). RSLR in the Nile delta region is as high as 5 mm/year (Stanley, 1988; Stanley and Warner, 1993; Milliman et al., 1989) and is between 2 and 6 mm/year for the Rhone and Ebro deltas (Sestini, 1992, 1996; Ibáñez et al., 1996; Pont et al., 2002). Because of the high rates of RSLR, deltas will be especially impacted by accelerated ESLR (Syvitski et al., 2009; Giosan et al., 2014). Humans have also accelerated RSLR by drainage leading to oxidation of organic soils and withdrawal of subsurface fluids such as groundwater or shallow deposits of oil and natural gas (Sestini, 1992; Morton et al., 2002; Ko and Day, 2004). An understanding of vegetation response in areas with high RSLR can provide insights into the effects of accelerated ESLR in the future (Stanley, 1988; Day and Templet, 1989; Stevenson et al., 1985; Day et al., 1997, 2005, 2007; Pont et al., 2002). Because of this, deltas serve as models of the response of coastal systems generally to ESLR (Day and Templet, 1989; Day et al., 1997, 2016b; Pont et al., 2002). RSLR varies globally due to a number of factors such as the rate of subsidence and isotatic rebound at high latitudes (Pirazzoli, 1996; Lambeck et al., 2010).

3 COASTAL WETLAND RESPONSE TO TEMPERATURE AND ACCELERATED SLR

One important likely result of increasing temperature will be reduction in the frequency of extreme freezes at the tropical-temperate interface and a northward migration of the tropical zone. Mangroves will move toward higher latitudes. For example, mangroves are replacing salt marshes along the northern Gulf of Mexico (Bianchi et al., 2013; Osland et al., 2016). The black mangrove [*Avicennia germinans* (L.)] is more tolerant of freezes than other mangrove species (McMillan and Sherrod, 1986) and its range extends farther north in the Gulf of Mexico than other mangrove species. This is happening to mangroves worldwide (e.g., Yáñez-Arancibia et al., 2010; Twilley and Day, 2013; Osland et al., 2016; Dangremond and Feller, 2016; Doughty et al., 2016; Saintilan et al., 2014).

As sea levels rise, coastal wetlands can respond in one of three ways (Wolanski et al., 2009; Day et al., 2011a, b; Rybczyk et al., 2013). *First,* they can accrete inorganic and organic sediments at a rate equal to or greater than the rate of SLR, resulting in no net change in habitat area or type. Estuarine wetland elevation, relative to sea level, is a function of numerous processes including mineral and organic matter accretion, sediment compaction, deep subsidence, and ESLR, all operating at different time scales.

Coastal wetlands can persist if they accrete vertically at a rate at least equal to water level rise (Cahoon et al., 1995) and studies have shown that coastal marshes are survive able to survive historical SLR 1–2 mm and persist for hundreds to thousands of years (Redfield, 1972; McCaffey and Thompson, 1980; Morris et al., 2002; Rybczyk and Cahoon, 2002). Accelerated SLR means that surface elevation of coastal wetlands will have to significantly increase in order to survive. Some coastal wetlands, in the Mississippi delta, can survive high rates of SLR in excess of 10 mm/year (Day et al., 2000, Day et al., 2010a). Wetland plants can modify their environment and enhance their ability to survive rising sea levels by trapping sediments and increasing organic soil formation. Stems slow the flow of sediment-laden water and thus enhance deposition. Roots and rhizomes further sequester and hold the sediment in place in addition to organic matter production.

To cope in the intertidal environment characterized by alternate flooding and draining, waterlogged soils, depletion of oxygen, and the production of natural toxins such as sulfides that inhibit plant growth, coastal plants have a number of adaptations, including the production of "aerial" roots and aerenchyma tissue and physiological changes that allow the plant to deal with stressful conditions of excessive inundation, anoxia, and high sulfide concentrations (Mendelssohn and Morris, 2000). Because plants become progressively more stressed and ultimately die if they are inundated for too long (Mendelssohn and McKee, 1988; McKee and Patrick, 1988), an increase in water levels due to SLR can severely stress the integrity of coastal wetland ecosystems. Riverine sediment input can enhance the ability of coastal wetland to survive rising water levels by input of sediments (Day et al., 2011a, b; Roberts et al., 2015; Delaune et al., 2016; Twilley et al., 2016). Rivers also supply freshwater that buffers saltwater intrusion and iron that precipitates toxic sulfides (Delaune and Pezeshki, 2003; DeLaune et al., 2003).

Second, estuarine habitat can migrate upslope as sea levels rise. At the wetland-upland transition, upland areas can transition to wetlands as wetlands "migrate" inland (Brinson et al., 1995). However, in many locations, coastal wetlands are bordered by development, diked farmland, or steep habitats, such that migration would not be possible (Titus et al., 1991), in which case remaining habitat would be slowly "squeezed" out of existence.

Third, if wetlands cannot keep pace with rates of SLR, an increase in inundation frequency and duration leads to a shift in the distribution of vegetated habitats across a wetland (Warren and Niering, 1993; Kirwan and Murray, 2008) (e.g., shifts from areas dominated by *Spartina patens* to *Spartina alterniflora*). Over longer time periods, these types of changes lead to conversion of more and more wetland area to unvegetated mudflats or even subtidal open water (Fig. 3).

SLR has already led to significant geomorphological changes of coastal systems, salinity intrusion in estuaries, and loss of associated wetlands around the world, including Chesapeake Bay (Stevenson et al., 1985) and other mid-Atlantic estuaries (Kana et al., 1986; Hackney and Cleary, 1987), Long Island Sound (Clark, 1986), the Mississippi delta (Conner et al., 2014, 1989; Day et al., 2000, 2007), Rhone (Pont et al., 2002), Nile and Ganges (Stanley, 1988; Milliman et al., 1989), Indus (Snedaker, 1984) and Ebro (Ibáñez et al., 1996, 1997) deltas, and Venice Lagoon (Pirazzoli, 1987; Sestini, 1992, 1996; Day et al., 1999, 2011b).

FIG. 3 Past and projected wetland loss in the Mississippi delta. (US Geological Survey).

4 THE IMPACTS OF CHANGES IN FRESHWATER INPUT ON COASTAL ECOSYSTEMS

The IPCC (2013) predicts that globally there will be both increases and decreases in precipitation in the 21st century, with changes both temporally and spatially. A general prediction of global climate change is that the outer tropics and subtemperate zone will be drier, the inner tropics will be wetter, and high latitudes will become wetter as warmer temperatures lead to the movement of greater amounts of water vapor away from tropical and subtemperate regions. Areas that are currently dry will likely become drier. In general, this indicates that freshwater runoff to coastal areas will decrease in mid latitudes and increase around the equator and at higher latitudes (Fig. 4).

Increased freshwater input can have both beneficial and detrimental impacts on coastal systems. The benefits of freshwater input stimulating vertical soil accretion have already been highlighted. An additional benefit of increased freshwater inflow is an increase in fisheries production due to increased primary production (Nixon, 1988). In the coastal marshes of the southern Everglades and in the Mississippi delta, there are current management plans to increase freshwater flow to coastal areas mainly for habitat restoration (Day et al., 2007, 2018, 2019; Sklar et al., 2005, 2019). Similar plans have been suggested for the Ebro and Rhone deltas (Ibáñez et al., 1997; Day et al., 2019; Pont et al., 2002, 2017). If freshwater input decreases, it will likely lead to less accretion, lowered productivity, and saltwater intrusion (Herrera-Silveira et al., 2019; Kemp et al., 2016). Coastal systems with decreased freshwater input will become less sustainable (Day et al., 2016b). Reduced freshwater input to the East China Sea from the Yangtze River due to the Three Gorges Dam reduced diatom populations in the coastal ocean (Gong et al., 2006) but in the Nile coastal waters fish production has increased due to inflow of sewage water (Chen, 2019).

Increased freshwater runoff can lead to eutrophication in coastal waters. Agricultural runoff and wastewater in tributary watersheds is degrading many coastal ecosystems. Nuisance algal blooms and low oxygen in bottom has been documented in Chesapeake Bay (Harding and Perry, 1997; Kemp et al., 1992) and off the coast of Louisiana (Rabalais et al., 1996), China (Liu et al., 2013), and in many other areas worldwide (Diaz and Rosenberg, 2008; Wolanski and Elliott, 2015). Liu et al. (2013) described the world's largest macroalgal bloom in the Yellow Sea. A number of management suggestions have

Projected patterns of precipitation changes

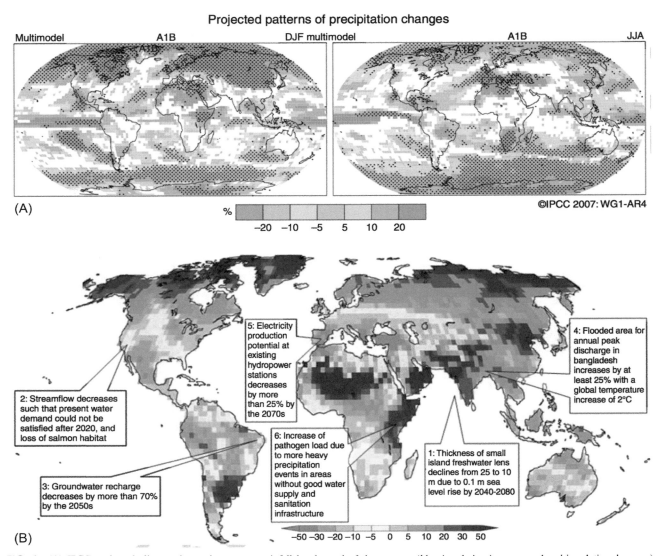

FIG. 4 (A) IPCC projected climate change impacts on rainfall by the end of the century (*blue* is relative increase and *red* is relative decrease). (B) Projected impacts on freshwater runoff by 2100 (IPCC, 2007). Changes in freshwater input can vary significantly depending on human changes in the catchment. For example, upstream dams in the Indus, Mekong, and Nile basins have led to reductions in both freshwater and sediment input (Kidwai et al., 2019; Nhan and Cao, 2019).

been made to reduce high nutrient loading to streams and coastal waters. For the Mississippi River basin, for example, these recommendations include changes in farming practices, buffer strips along streams, use of wetlands to improve water quality, and reduction of nitrate in river water by diversions into riparian ecosystems and the Mississippi delta (Mitsch et al., 2001). The increased runoff may also lead to problems with toxic pollutants (e.g., heavy metals and organic chemicals) if there are high levels of these chemicals in the runoff.

Tao et al. (2014) reported that peak discharge of the Mississippi River will likely increase by 10%–60% during this century due to interactive effects of climate change and land use even given the presence of many dams, mainly on the Missouri River. The Ohio River, the primary source of water remains largely undammed. Such large increases may compromise the flood control system on the Mississippi River in times of increasing energy scarcity (Kemp et al., 2014; Day et al., 2018, 2019).

5 TROPICAL CYCLONES

Increasing warming of ocean surface may lead to an increase in the frequency of tropical cyclones but there has been much uncertainty in such predictions (IPCC, 2013) but tropical cyclones have become more frequent in the North Atlantic and globally (Webster et al., 2005). Recent reports, however, have drawn stronger conclusions concerning tropical cyclone. Emanuel (2005) reported that a 1°C increase in sea surface temperature in the tropics over the past half century and during this same period, total hurricane intensity or power increased by about 80%. Webster et al. (2005) reported an increase in the number of category 4 and 5 storms over the past several decades. It has been argued that these changes were not linked to climate change but are due to decadal cycles in tropical storm activity. Hoyos et al. (2006), however, analyzed factors contributing to hurricane intensity and concluded that the increasing numbers of categories 4 and 5 for the period 1970–2004 were directly linked to the increase in sea surface temperatures. Elsner et al. (2008) also concluded that there was an increasing intensity of the strongest tropical storms. Bender et al. (2010) used a hurricane prediction model that projected a nearly doubling of the frequency of category 4 and 5 storms by the end of the 21st century. Knutson et al. (2010) reported that the frequency of tropical cyclones will decrease or remain unchanged, but the frequency of the most intense hurricanes will increase. Mei et al. (2015) reported that typhoon intensity has increased in the Pacific region and predicted climate change will increase average typhoon intensity in the Pacific area by 14% by 2100. NOAA (2018) stated that it is premature to conclude that GHG emissions have already had a detectable impact on Atlantic hurricane or global tropical cyclone activity but they state that anthropogenic warming will lead to an increase in the occurrence of very intense tropical cyclones and more intense rainfall. Regardless of whether the intensification of hurricanes is due to climate change or is part of a decade's long cycle, it is likely that there will be more and stronger hurricanes in the coming decades and this will interact with other climate forcings to impact estuarine ecosystems. Changes in the intensity and frequency of storms can have a variety of impacts, especially on coastal wetlands, as outlined below.

In general, long-term changes in the frequency and intensity of strong storms will most likely alter the species composition and biodiversity of coastal wetlands, as well as important ecosystem processes as nutrient cycling and primary and secondary productivity (Michener et al., 1997) leading to both positive and negative effects. For example, hurricanes greatly increase the rate of soil accretion in marshes helping to offset accelerated SLR (Cahoon et al., 1995). Runoff generated by hurricanes introduces freshwater and nutrients which can enhance coastal wetland productivity (Conner et al., 1989). In the arid areas, freshwater input can also have a stimulatory impact by reducing salinity stress (Conner et al., 1989).

On the negative side, tropical storms can reduce the structural complexity of forested wetlands such as mangroves and tidal freshwater forested wetlands (Doyle et al., 1995; Rybczyk et al., 1995). Hurricanes can also cause tree loss. During Hurricane Andrew in 1992, an extensive swath of mangrove trees was downed and defoliated across southern Florida (Doyle et al., 1995) and nearly 10% of trees in a freshwater forested wetland in Louisiana were blown down in this single storm (Rybczyk et al., 1995). The impact from the interaction between rising water levels and hurricanes may be amplified with an increase in hurricanes and hasten the loss of coastal forested wetlands. High runoff from hurricanes can also lead to excessive nutrient loading and eutrophication problems. For example, record runoff from Hurricane Floyd into the Pamlico Sound estuary, North Carolina, led to water quality problems (Paerl et al., 2000).

6 EXTREME WEATHER EVENTS

Evidence indicates that the intensification of extreme weather events will continue in a warming climate (Pachauri et al., 2014). Most intuitively, a shift toward warmer weather conditions will increase evaporation and surface drying, therefore increasing drought events (Dai, 2011). Warmer air also can hold more moisture if moisture is available. But in arid areas, moisture is much less available. Heavy precipitation events, and consequently flooding, are expected to increase in intensity with climate change (Groisman et al., 2005; Min et al., 2011; Pall et al., 2011; Prein et al., 2016). Several recent examples are available for South Louisiana. In 2000–01, an extreme drought raised salinities in western Lake Pontchartrain and Lake Maurepas from an average of 2–3 to 10–12 psu and led to mortality of cypress over a wide area (Day et al., 2012; Shaffer et al., 2016). Such droughts, in combination with increasing sea level and strong hurricanes, will lead to salinity stress on broad areas of freshwater wetlands. In August 2016, nearly a meter of rain fell in 3 days during a stalled front that was not associated with a hurricane led to extensive flooding east of the Mississippi River. Hurricane Harvey dumped over a meter of rain on the Houston in 2017 areas in a few days when its forward movement was stalled by a frontal system.

7 ENERGY SCARCITY AND COASTAL ADAPTATION AND RESTORATION

Decreasing energy availability and higher energy prices will limit options for coastal restoration and adaptation and complicate human response to climate change. Fossil fuels dominate world energy use and are expected to do so for decades to come (Day et al., 2007, 2018; Tessler et al., 2015; Wiegman et al., 2017, Fig. 5). The implication of a future where fossil

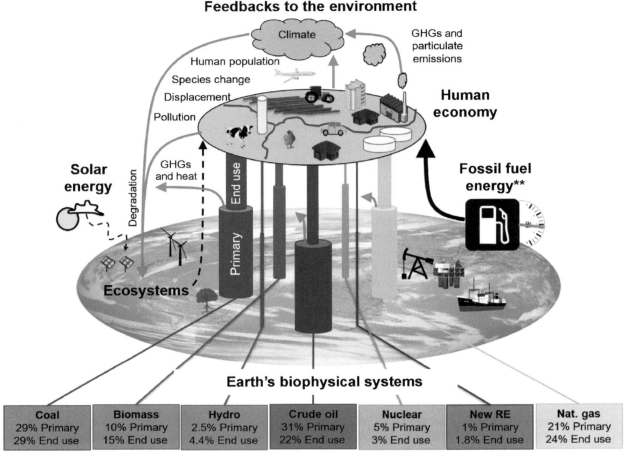

FIG. 5 Energy pillars diagram showing relative importance of different energy sources. This visualization shows how energy drawn from earth systems *(black lines)* supports the foundation of the economy, and how human activities *(red lines)* impact the environment. The size of the energy pillars beneath the economy is proportional to their share in the global primary energy (base of the pillar) and energy end use (top of the pillar, due to thermodynamic losses). The fuel gauge represents remaining cheap conventional oil; by 2013 society had burned ~1100 GB of about 2200 GB of recoverable conventional oil resources. *(From Wiegman, R., Day, J., D'Elia, C., Rutherford, J., Morris, J., Roy, E., Lane, R., Dismukes, D., Snyder, B., 2017. Modeling impacts of sea-level rise, oil price, and management strategy on the costs of sustaining Mississippi delta marshes with hydraulic dredging. Sci. Total Environ. https:// doi.org/10.1016/j.scitotenv.2017.09.314.)*

fuels reach a peak and then decline is that the cost of energy as well as other resources such as metals will be higher in coming decades (Campbell and Laherrere, 1998; Day et al., 2018; Wiegman et al., 2017) resulting in significant increases in the cost of energy-intensive activities. This is because fossil fuels account for more than 80% of world energy use and are especially important for energy-intensive coastal restoration (Fig. 7). Much restoration and management in coastal systems is based on large-scale energy intensive activities such as dredging, maintenance of navigation channels, building and maintaining dikes, transporting dredged sediments in pipelines, building and maintaining large water control structures (i.e., river diversions in the Mississippi delta), the massive flood protection system in the Netherlands, large fossil fuel-powered pumping works of Shanghai metropolitan area, and structures such as the Thames barrier and the storm barrier (the MOSE) in Venice, and maintaining flood control systems for areas near or below sea level (e.g., New Orleans, Venice Lagoon, Shanghai, the Po, and Ebro deltas). The situation in the Netherlands, with extensive system of dikes, water control structure, pumps, and large areas below sea level is perhaps the ultimate example of this type of management. As energy prices escalate in coming decades, the cost of maintaining such systems will likely become prohibitively expensive, especially with respect to climate projections.

Tessler et al. (2015) reported that the combination of land use changes, regional water management, subsidence, climate change, and increasing energy costs will significantly increase the risks of nonsustainable outcomes for deltas (Fig. 6). Although both developed and developing countries are exposed to risks, first-world countries presently limit their risk in the coastal zone by expensive and energy-intensive infrastructure and coastal defenses. In a future

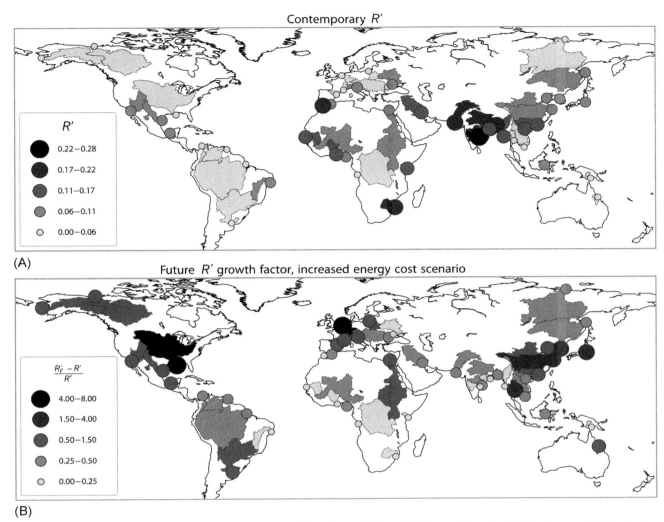

FIG. 6 Contemporary and 50-year risk of unsustainable outcomes for major world deltas. The 50-year risk trend relative to the current risk trend shows that the Rhine and Mississippi deltas have the highest potential for unsustainable outcomes. R' is risk due to current projected trends in relative sea-level rise. $R'f$ is the future risk trend with changing conditions especially in energy availability. *(From Tessler, Z.D., Vörösmarty, C., Grossberg, M., Gladkova, I., Aizenman, H., Syvitski, J., Foufoula-Georgiou, E., 2015. Profiling risk and sustainability in coastal deltas of the world. Science 349, 638–643.)*

characterized by scarce and expensive energy, maintaining such expensive infrastructure will likely become increasingly untenable. For example, expansion of dams in the Mekong, Nile, and Indus basins to provide hydroelectricity and irrigation water is reducing sediment discharge into the deltas, and causing severe erosion (Nhan and Cao, 2019; Kidwai et al., 2019). Tessler et al. (2015) project that the risks for unsustainable outcomes will rise by four to eight times in the Mississippi and Rhine deltas. The current emphasis on short-term solutions for the world's deltas will greatly constrain options for designing sustainable solutions in the long term. Wiegman et al. (2017) modeled the interactive impacts of SLR and increasing energy price on the cost of creating marshes in the Mississippi delta with dredged sediments transported in pipelines. They reported that the cost of such marsh creation would rise considerably for high-end scenarios of SLR and energy cost increases (Fig. 7). We believe that this is a general principle where climate change and resource scarcity occur at the same time and interact. This has implications, for example, for coastal megacities in developed and developing countries. In developing countries where very expensive flood protection systems are not affordable, there will likely have to be continuing adjustment to factors such as SLR and storms by inland migration and raising and storm proofing buildings. In developed countries, there will likely be continued reliance on highly engineered and expensive flood control systems. Increasing cost of energy may lead to failure to maintain these systems and potentially to catastrophic failure.

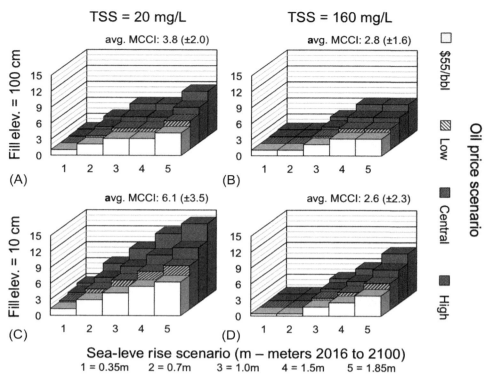

FIG. 7 The impact of oil price and sea-level rise the cost of sustaining coastal marsh with hydraulic dredging from 2016 to 2100. A marsh creation cost index (MCCI) of 1 equals $128,000/ha. MCCI is reported for fill height of 100 cm (A and B) and 10 cm (C and D) with total suspended sediment (TSS) of 20 mg/L (A and C) and 160 mg/L (B and D). *(From Wiegman, R., Day, J., D'Elia, C., Rutherford, J., Morris, J., Roy, E., Lane, R., Dismukes, D., Snyder, B., 2017. Modeling impacts of sea-level rise, oil price, and management strategy on the costs of sustaining Mississippi delta marshes with hydraulic dredging. Sci. Total Environ. https://doi.org/10.1016/j.scitotenv.2017.09.314.)*

8 COASTAL ENVIRONMENTAL DEGRADATION AS A SOCIETAL ENERGY SINK

As scientists studying coastal areas know, these are among the most productive ecosystems on earth. This is expressed in terms of primary production and coastal fisheries. Rich coastal resources and water transportation avenues are reasons that humans have always settled in coastal areas and coastal areas are where most of the initial state formation occurred (Day et al., 2012).

But human activity has caused severe environmental degradation in coastal areas reducing the provisioning of ecosystem goods and services (Rockström et al., 2009; Sverdrup et al., 2015). These pressures act as destabilizing energy sinks to the economy that have the potential to disrupt and reduce the functioning of productive coastal systems. Examples of environmental degradation in the coastal zone include eutrophication, destruction of coastal habitats especially wetlands, and at the global scale accelerated SLR. Environmental stability is a crucial factor influencing society. The onset of the stable Holocene epoch corresponds with the period that the first complex human civilizations emerged (Day et al., 2012). The current scale of coastal environmental degradation is occurring as the world transitions into the Anthropocene—a new geological epoch that will likely be less stable than the Holocene (Rockström et al., 2009; Steffen et al., 2015).

The manifestations of environmental degradation in the coastal zone are increasing with greater human population levels and resource use. In effect, this acts as an energy tax on society because as coastal ecosystems are degraded, society has to pay more the goods and services it needs (Day et al., 2018).

8.1 Population

Over half of the world's population currently lives in cities, and the proportion is projected to increase in the coming decades. Blackburn et al. (2019) discuss coastal megacities in the 21st century. The world's population has risen from one billion at the beginning of the 19th century to over 7.4 billion in 2017 and it is projected to reach between 9 and 11 billion by 2100. The growth of large cities has been even more dramatic; demographers call this the urban transition. In 1800, it is estimated that only one city had a population of more than one million; Beijing at 1.1 million. By 2017, there were over

500 cities with a million or more people, overwhelmingly in the developing world. So as world population grew by a little more than seven times from 1800 to 2017, the growth of cities of one million or more grew by more than 500 times (Day et al., 2016a).

Coastal areas especially have had high levels of urbanization (Nicholls et al., 2011). There are 72–187 million people in the coastal zone that would be displaced by 0.5–2.0 m of SLR (Nicholls et al., 2011). About 10% of the global population and almost two-thirds of cities with more than 5 million people are in coastal areas less than 10 m above sea level that are at high risk from stronger storms and other hazards caused by climate change (McGranahan et al., 2007; Day et al., 2016b). As the severe 2017 Gulf-Atlantic hurricane season indicates coastal societies will be forced to build and maintain large, expensive coastal protection systems and to use significant energy resources to emergency evacuations and rebuilding. Places that cannot afford such expensive systems will face recurring impacts from rising sea levels and coastal storms that will force large numbers of people to migrate inland or succumb to natural disasters.

While migration toward the coast during the 20th century created many large coastal megacities, migration away from delta megacities is likely for the mid to later 21st century due to SLR and stronger storms (Min et al., 2011; Horton et al., 2014; Koop et al., 2016). More than 10 million people are displaced from coastal areas each year due to storm surges and river flooding (20 million in 2013, Asian Development Bank, 2013; Norwegian Refugee Council, 2014). In coming decades, trillions of dollars of infrastructure are at risk in coastal megacities from the destructive force of extreme weather events (e.g., Hinkel et al., 2013). In essence, many urbanized coastal areas now function under the influence of human impacts. Little of the natural system remains for many coastal areas. Unless coastal cultures and inhabitants can develop approaches and infrastructure to survive future extreme weather, living in coastal areas will become disadvantageous.

Many large coastal urban areas are threatened by global change in the Anthropocene. These include Bangkok, Jakarta, Rangoon, Dhaka, Alexandria and Cairo, Seoul, Tokyo, Shanghai, Guangzhou, Tianjin, New Orleans, and arguably parts of Mumbai, Kolkata, Karachi, Lagos, Ho Chi Minh City, Rotterdam, and New York. Large coastal urban areas (and small ones as well) are impacted by climate change in the form of accelerated SLR, increasing frequency of powerful storms (Hurricane Katrina and superstorm Sandy), and extreme weather events (extensive flooding in Houston and Mumbai in 2017). Many large coastal urban areas have spread into low-lying areas that make them more susceptible global change impacts. This is exacerbated the destruction of productive coastal ecosystems that provided high levels of ecosystems services such as storm protection, food production, and water quality improvement. All of this reduces the resilience and sustainability of coastal urban areas, especially large ones.

8.2   Ranking Coastal Sustainability

Day et al. (2016b) ranked the sustainability of deltas based on a variety of environmental and socioeconomic forcings. Here, we apply these ideas to coastal systems in general. The rankings are based on the current status of individual coastal systems, their degree of human impact, and the ability to deal with limitations that will be imposed by global change in the 21st century including climate change, growing resource scarcity, coastal ecosystem degradation, and population growth in the coastal zone and globally.

8.3   Sea-Level Rise

SLR will impact all coastal systems, most importantly coastal wetlands and human development. The ability of coastal systems to cope with SLR will be much higher in some systems than others. For example, low-relief coastal areas will be more threatened. Coastal systems with high freshwater input and low human impact will best be able to cope with SLR. Systems with high subsidence rates such as occur in large deltas or due to human activities like subsurface fluid withdrawal will be less able to cope. Human development in low-lying areas, especially large urban areas are at risk of nonsustainable outcome due to destruction of productive coastal ecosystems, increasing cost of coastal defenses, and increased vulnerability to climate change.

8.4   Changes in Freshwater Input

Changes in freshwater input will impact many coastal systems. The inner tropics will experience increases in precipitation and generally higher discharge to tropical coasts. In areas where human activities have not affected freshwater input from the basin, inner tropical coasts will have higher potential for sustainability. However, basins that have significant reductions in freshwater and sediment input will be less sustainable. For example, in recent decades, many large dams have been constructed on the Mekong River and distributaries for hydroelectric production (see Nhan and Cao, 2019). The reduction of sediment input to the Mekong delta will reduce its sustainability. Major problems that the Mekong River Delta is facing are SLR, increasing frequency and magnitude of tropical storms due to climate change, saltwater intrusion, severe acid sulfate and saline soils, and upstream deforestation (Leinenkugel et al., 2014). The Mekong delta supports extensive rice culture and thus is sensitive to increases in salinity. Giosan et al. (2014) reported that the Mekong delta has a significant sediment deficit with respect to projected SLR in this century. The massive dam building program on the river will further impact both freshwater and sediment delivery to the delta thus compromising its sustainability. A similar scenario is predicted for the Indus and Nile deltas (Kidwai et al., 2019). In other regions with fewer large dams and more humid climates, river discharge is not projected to decrease. For example, the peak discharge of the Mississippi River is projected to increase by up to 60% due to both climate change and human impacts in the basin (Tao et al., 2014; Min et al., 2011; Pall et al., 2011; Prein et al., 2016).

8.5 Tropical Cyclones

It is projected that the frequency of category 4 and 5 hurricanes will increase. Coastal systems that will be most affected by tropical cyclones with large areas of near or below sea-level land and degraded coastal wetlands are most at risk of severe cyclone damage. The Mississippi delta is perhaps the prime example of this (Day et al., 2018, 2019) but deltas in on the western sides of the oceans in the subtropics to the low and mid-temperate zone are also threatened.

8.6 Below Sea-Level Coastal Areas

Below sea-level coastal areas will become increasingly untenable and unsustainable in this century due to growing climate impacts and the increasing costs of maintaining coastal defense systems (Tessler et al., 2015; Day et al., 2016b; Wiegman et al., 2017). A number of coastal systems have large areas below sea level. The majority the Po Delta is embanked and lies several meters below sea level (Sestini, 1992). Only the outer fringes of the delta are somewhat natural with high accretion rates at the river mouths (Day et al., 2011a, b). Most of the Rhine Delta is below sea level, is highly developed, and has been for centuries. The landscape below sea level developed during a time of relatively stable climate. The land was first drained with windmills beginning in the 16th century and then with fossil fuel-powered pumps. Most of the delta since then has been converted to fast lands for agriculture, industry, and urban development, and is no longer a functioning delta (Knights, 1979). During the 20th century, there was a massive increase in energy-intensive infrastructure with high maintenance costs (Meire, P. pers. comm.). The Mississippi delta has large areas below sea level especially in the New Orleans metropolitan area (Day et al., 2019). As the century progresses and climate impacts grow more severe and energy costs escalate, the risk of not maintaining the current system will grow (Tessler et al., 2015). Other below sea-level areas are near Shanghai, Bangkok, the Vistula delta, and in parts of the Rhone and Ebro deltas.

8.7 Arid and Semiarid Areas

Coastal systems with low rainfall both at the coast and in the drainage basin will have reduced sustainability. These areas are threatened by hyper salinity and loss of wetlands. A number of deltas are in arid areas. The Nile delta has been largely converted to agriculture and is dropping below sea level due to subsidence and is experiencing salt water intrusion (Stanley, 1988; Stanley and Warner, 1993; Chen, 2019). Construction of a large dam in Ethiopia will further reduce river discharge. Other threatened deltas include the Indus, Tigris-Euphrates, and Colorado. Arid and semiarid coasts include much of the Mediterranean and Mid-East, parts of west and East Africa (Niang et al., 2019), subtropical and subtemperate coasts of the Pacific coast of the Americas, southern Africa, and parts of Australia (Tweedly et al., 2019). It will be difficult to restore or maintain these areas in a healthy sustainable state, especially given climate impacts of SLR, decreasing rainfall, increasing human demand for water, and the increasing cost of energy to maintain defenses. In this volume, several papers address arid and semiarid coastal systems (Peel-Harvey Estuary: Valesini et al., 2019; Indus: Kidwai et al., 2019; Tigris-Euphates: Sharifinia et al., 2019; hypersaline systems, Tweedly et al., 2019).

8.8 Arctic Coastal Systems

These coastal systems face a variety of problems associated with climate change and human settlements in the Arctic. The Mackenzie and Lena discharge into the Arctic Ocean. Melting sea ice in the Arctic Ocean is exposing coastal systems there to increased wave attack as expanding open water increases fetch threatening both natural systems and human settlements.

This is especially true for deltas such as the Yukon, Mackenzie, and Lena (Forbes, 2019). Lovecraft and Meek (2019) explain the links between the ecosystems and humans in the Arctic in this changing climate.

9 ECOLOGICAL ENGINEERING AND ECOHYDROLOGY—SYSTEM FUNCTIONING AS A BASIS FOR SUSTAINABLE MANAGEMENT OF COASTAL SYSTEMS

Sustainable management of coastal systems during this century with accelerating climate change impacts, growing resource scarcity, growing coastal populations, increasing cost of energy, and increasing environmental degradation will necessitate working with natural systems to enhance their ability to survive these forcings. A fundamental concept to accomplish this is that sustainable management should be based on working with natural system functioning, this is addressed by the fields of ecological engineering and ecohydrology (Day et al., 1997, 2000; Wolanski and Elliott, 2015). In this way, the natural energetic pulses that built and sustained coastal systems can be reintegrated into management. If management is directed at maximizing natural processes toward sustaining coastal systems, then it will be much less expensive and more effective and sustainable. One approach for such endeavors is called "ecological engineering" and its principles and procedures point a way toward working with nature in achieving sustainability (Mitsch and Jørgensen, 2004). Ecological engineering is defined as "the design of sustainable ecosystems that integrate human society with its natural environment for the benefit of both." This approach combines basic and applied science for the restoration, design, and construction of ecosystems. Ecological engineering relies primarily on the energies of nature, with human energy used in design and control of key processes. Ecohydrology works with improving system resilience by improving and restoring system functioning and as much as possible reducing human impacts (Elliott et al., 2016).

9.1 Ocean Acidification

Because of increased atmospheric CO_2, ocean acidification is occurring and the pH levels will decrease from 8.1 to 7.8 by 2100 if atmospheric CO_2 increases from 390 to 750 ppm under the business-as-usual CO_2 emission scenario (IPCC, 2013). Ocean acidification reduces the ability of many marine organisms, including corals, to secrete their calcium carbonate skeleton; indeed, field observations around CO_2 vents in Papua New Guinea, where pH declines from 8.1 to less than 7.7 showed reductions in coral genetic diversity and biodiversity, recruitment, and colony density along the pH gradient (Fabricius et al., 2011). Recent studies suggest that reef sediments globally will transition from net precipitation to net dissolution probably by 2050 under the business as usual CO_2 emission scenario (Eyre et al., 2018). Similarly, lobsters, mussels, shrimps, and oysters will not be able to form strong enough shells in acidified waters. The ecological effects of ocean acidification in estuaries and coastal waters may thus be profound. Estuaries and bays with exceptionally slow flushing rates (residence time $\gg$ 30 days) may best adapt to ocean acidification (Golbuu et al., 2016).

9.2 Human Activity and Coastal Management

Increasing temperature, acceleration of SLR, changes in freshwater runoff, and increasing hurricane intensity are among the climate forcings that will affect coastal ecosystems. These forcings interact with each other and with human impacts to exacerbate the impact of climate change (Day et al., 2008, 2016). For example, SLR and lower freshwater inflow lead to both increased salinity and longer flooding duration, resulting in multiple stresses on wetland plants. This is especially important for large expanses of coastal wetlands such as exist in deltas (Syvitski et al., 2009; Day et al., 2016) and other low relief coastal systems such as the Everglades (see Sklar et al., 2019). Yáñez-Arancibia et al. (2009a, b) pointed out that sustainability of social and economic development must be based on the functioning of natural ecosystems, and climatic change effects must be incorporated into environmental planning because coastal erosion, salt intrusion, SLR, and rainfall variability will affect development initiatives. Yáñez-Arancibia et al. (2009a, b) suggested that economic development in the coastal zone should be based on ecological integrity. These resources represent "natural capital" that supports the economic health of society. The goods and services provided by natural capital represent the "interest" generated by human investment in natural ecosystems. The perspective of sustainable development should be linked to the ecological integrity of coastal ecosystems with healthy and resilient conditions. This is the reason that healthy ecosystems support a healthy economy. Climate change will impact the potential for sustainable management.

Anthropogenic impacts interact with other 21st century mega trends to cause impacts that are more severe than impact acting singly. Isolation of rivers from coastal systems and pervasive hydrologic alteration has caused a high rate of coastal wetland loss in many coastal ecosystems, especially deltas (Syvitski et al., 2009; Day et al., 2007; Pont et al., 2002). These changes have also made coastal systems more vulnerable to climate change, human development, and energy scarcity.

10 CONCLUSIONS: ECOSYSTEM GOODS AND SERVICES AND COST OF ENERGY

In this chapter, we summarized projected trends for climate change, energy scarcity, population, and environmental degradation and discussed how they threaten the sustainability of coastal systems. Projections indicate that if aggressive action is not taken very soon many coastal wetlands will disappear by 2100 (e.g., Blum and Roberts, 2009). However, some coastal wetlands in the Arctic will not disappear as they are forming as the land rises (Forbes, 2019). Energy scarcity threatens the economic sustainability of many coastal systems, both natural and human. Evidence suggests that fossil fuel production will peak before mid-century (e.g., Day et al., 2018), with oil peaking first, followed by natural gas and coal. Renewables have the potential to produce large amounts of energy for society, but at a much higher cost compared with conventional fossil fuels. Declining Energy Return on Investment (EROI) of the energy supply overtime means that energetic and monetary costs will increase. This will eventually make highly energy intensive flood control, natural resource management, and human systems extremely expensive.

Based on climate and energy trends, it will be necessary to implement management plans that allow coastal systems to become more resilient to climate change impacts, but also at the lowest operating energy cost. Day et al. (1997) suggest that only management that is based on the functioning of deltas is most sustainable in the long run. The dynamic pulsing nature of most natural coastal ecosystems has been reduced. Current energy intensive management of many coastal systems is becoming unsustainable because it has suppressed the natural pulsing of these systems. Tessler et al. (2015), Day et al. (2016b), and Wiegman et al. (2017) report the severity of climate change and energy scarcity will make coastal restoration and management even more challenging and expensive during this century, especially for first world systems where the most energy intensive management activities are used. Climate change and declining availability and accessibility of energy necessitate the need for an aggressive new approach to management of coastal ecosystems. Ecohydrology is a management framework that offers a more sustainable approach for coastal management and restoration based on system functioning.

Only management that is based on the functioning of deltas is sustainable in the long run (Day et al., 1997); for instance, the dynamic pulsing nature of most natural coastal ecosystems must be restored. Current energy intensive management of many coastal systems is becoming unsustainable because it has suppressed the natural pulsing of these systems (Tessler et al., 2015; Day et al., 2016d; Wiegman et al., 2017).

REFERENCES

Asian Development Bank, 2013. Climate Change and Migration in Asia and the Pacific. Asian Development Bank, Philippines, p. 77. Available from: www.adb.org.

Baumann, R., Day, J., Miller, C., 1984. Mississippi deltaic wetland survival: sedimentation vs. coastal submergence. Science 224, 1093.

Bender, M., Knutson, T., Tuleya, R., Sirutis, J., Vecchi, G., Garner, S., Held, I., 2010. Modeled impact of anthropogenic warming on the frequency of intense Atlantic hurricanes. Science 327, 454–458.

Bianchi, T.S., Allison, M.A., Zhao, J., Li, X., Comeaux, R.S., Feagin, R.A., Kulawardhana, R.W., 2013. Historical reconstruction of mangrove expansion in the Gulf of Mexico: linking climate change with carbon sequestration in coastal wetlands. Estuar. Coast. Shelf Sci. 119, 7–16.

Blackburn, S., Pelling, M., Marques, C., 2019. Megacities and the coast: global context and scope for transformation. In: Wolanski, E., Day, J., Elliot, M., Ramachandran, R. (Eds.), Coasts and Estuaries—The Future. Elsevier, Amsterdam.

Blum, M., Roberts, H., 2009. Drowning of the Mississippi delta due to insufficient sediment supply and global sea-level rise. Nat. Geosci. 2, 488–491.

Brinson, M.M., Christian, R.R., Blum, L.K., 1995. Multiple states in the sea-level induced transition from terrestrial forest to estuary. Estuaries 18, 648–659.

Cahoon, D.R., Reed, D.J., Day Jr., J.W., 1995. Estimating shallow subsidence in microtidal salt marshes of the Southeastern United States: Kaye and Barghoorn revisited. Mar. Geol. 128, 1–9.

Campbell, C.J., Laherrere, J.H., 1998. The end of cheap oil. Sci. Am. 278, 78–83.

Chen, Z., 2019. A brief overview of ecological degradation of the Nile Delta: what can we learn? In: Wolanski, E., Day, J., Elliot, M., Ramachandran, R. (Eds.), Coasts and Estuaries—The Future. Elsevier, Amsterdam.

Clark, J., 1986. Coastal forest tree populations in a changing environment, Southeastern Long Island, New York. Ecol. Monogr. 56, 259–277.

Conner, W.H., Day Jr., J.W., Baumann, R.H., Randall, J., 1989. Influence of hurricanes on coastal ecosystems along the northern Gulf coast. Wetl. Ecol. Manag. 1 (1), 45–56.

Conner, W., Duberstein, J., Day, J., Hutchinson, S., 2014. Impacts of changing hydrology and hurricanes on forest structure and growth along a flooding/elevation gradient in a south Louisiana forested wetland from 1986 to 2009. Wetlands 34, 803–814.

Dai, A., 2011. Drought under global warming: a review. Wiley Interdiscip. Rev. Clim. Chang. 2 (1), 45–65.

Dangremond, E.M., Feller, I.C., 2016. Precocious reproduction increases at the leading edge of a mangrove range expansion. Ecol. Evo. 6, 5087–5092.

Day Jr., J.W., Shaffer, G.P., Britsch, L.D., Reed, D.J., Hawes, S.R., Cahoon, D., 2000. Pattern and process of land loss in the Mississippi Delta: a spatial and temporal analysis of wetland habitat change. Estuaries 23, 425–438.

Day, J., Templet, P., 1989. Consequences of sea-level rise: implications from the Mississippi Delta. Coast. Manag. 17, 241–257.

Day, J., Martin, J., Cardoch, L., Templet, P., 1997. System functioning as a basis for sustainable management of deltaic ecosystems. Coast. Manag. 25, 115–154.

Day, J., Rybczyk, J., Scarton, F., Rismondo, A., Are, D., Cecconi, G., 1999. Soil accretionary dynamics, sea-level rise and the survival of wetlands in Venice Lagoon: a field and modeling approach. Estuar. Coast. Shelf Sci. 49, 607–628.

Day, J., Barras, J., Clairains, E., Johnston, J., Justic, D., Kemp, G., Ko, J., Lane, R., Mitsch, W., Steyer, G., Templet, P., Yáñez-Arancibia, A., 2005. Implications of global climatic change and energy cost and availability for the restoration of the Mississippi Delta. Ecol. Eng. 24, 253–265.

Day, J., Boesch, D., Clairain, E., Kemp, P., Laska, S., Mitsch, W., Orth, K., Mashriqui, H., Reed, D., Shabman, L., Simenstad, C., Streever, B., Twilley, R., Watson, C., Wells, J., Whigham, D., 2007. Restoration of the Mississippi Delta: lessons from Hurricanes Katrina and Rita. Science 315, 1679–1684.

Day, J., Christian, R., Boesch, D., Yáñez-Arancibia, A., Morris, J., Twilley, R., Naylor, L., Schaffner, L., Stevenson, C., 2008. Consequences of climate change on the ecogeomorphology of coastal wetlands. Estuar. Coasts 37, 477–491.

Day, J., Barras, J., Davis, D.W., Paul Kemp, G., Lane, R., Mitsch, W.J., Templet, P.H., 2010a. Integrated coastal management in the Mississippi delta: system functioning as the basis of sustainable management. In: Day, J., Yáñez-Arancibia, A. (Eds.), Ecosystem-Based Management. Vol. 5. The Gulf of Mexico: Its Origins, Waters, Biota. Texas A&M Univeristy Press, College Station, TX.

Day, J., Yáñez-Arancibia, A., Cowan, J., Day, R., Twilley, R., Rybczyk, J., 2010c. Global climate change impacts on coastal ecosystems in the Gulf of Mexico: considerations for integrated coastal management. In: Day, J., Yáñez-Arancibia, A. (Eds.), Ecosystem-Based Management. The Gulf of Mexico: Its Origins, Waters, Biota & Human Impacts Series. Texas A&M Univeristy Press, College Station, TX.

Day, J., Kemp, P., Reed, D., Cahoon, D., Boumans, R., Suhayda, J., Gambrell, R., 2011a. Vegetation death and rapid loss of surface elevation in two contrasting Mississippi delta salt marshes: the role of sedimentation, autocompaction and sea-level rise. Ecol. Eng. 37, 229–240. https://doi.org/10.1016/j.ecoleng.2010.11.021.

Day, J.W., Ibanez, C., Scarton, F., Pont, D., Hensel, P., Day, J.N., Lane, R., 2011b. Sustainability of Mediterranean deltaic and lagoon wetlands with sea-level rise: the importance of river input. Estuar. Coasts 34, 483–493. https://doi.org/10.1007/s12237-011-9390-x.

Day, J., Yáñez-Arancibia, A., Rybczyk, J.M., 2011c. Climate change: effects, causes, consequences: physical, hydrogeomorphological, ecophysiogical, and biogeographical changes (Chapter 8.15). In: Wolanski, E., McLusky, D. (Eds.), Treatise of Estuarine and Coastal Science. Elsevier, Oxford.

Day, J.W., Hunter, R., Keim, R., DeLaune, R., Shaffer, G., Evers, E., Reed, D., Brantely, C., Kemp, P., Day, J.N., Hunter, M., 2012. Ecological response of forested wetlands with and without large-scale Mississippi River input: Implications for management. Ecol. Eng. 42, 57–67.

Day, J., Hall, C., Roy, E., Moerschbaecher, M., D'Elia, C., Pimentel, D., Yanez, A., 2016a. America's Most Sustainable Cities and Regions - Surviving the 21st Century Megatrends. Springer, New York, p. 348.

Day, J., Agboola, J., Chen, Z., D'Elia, C., Forbes, D.L., Giosan, L., Kemp, P., Kuenzer, C., Lane, R.R., Ramachandran, R., Syvitski, J., Yanez, A., 2016b. Approaches to defining deltaic sustainability in the 21st century. Estuar. Coast. Shelf Sci. 183, 275–291. https://doi.org/10.1016/j.ecss.2016.06.018.

Day, J., D'Elia, C., Wiegman, A., Rutherford, J., Hall, C., Lane, R., Dismukes, D., 2018. The energy pillars of society: Perverse interactions of human resource use, the economy, and environmental degradation. Biophys. Eco. Res. Qual. 3, 2. https://doi.org/10.1007/s41247-018-0035-65.

Day, J.W., Colten, C., Kemp, G.P., 2019. Mississippi delta restoration and protection: shifting baselines, diminishing resilience, and growing nonsustainability. In: Wolanski, E., Day, J., Elliot, M., Ramachandran, R. (Eds.), Coasts and Estuaries—The Future. Elsevier, Amsterdam.

DeConto, R.M., Pollard, D., 2016. Contribution of Antarctica to past and future sea-level rise. Nature 531 (7596), 591–597.

Delaune, R.D., Pezeshki, S.R., 2003. The role of soil organic carbon in maintaining surface elevation in rapidly subsiding U.S. Gulf of Mexico coastal marshes. Water Air Soil Pollut. 3, 167–179.

DeLaune, R.D., Jugsujinda, A., Peterson, G., Patrick, W., 2003. Impact of Mississippi River freshwater reintroduction on enhancing marsh accretionary processes in a Louisiana estuary. Estuar. Coast. Shelf Sci. 58, 653–662.

DeLaune, R.D., Sasser, C.E., Evers-Hebert, E., White, J.R., Roberts, H.H., 2016. Influence of the Wax Lake Delta sediment diversion on aboveground plant productivity and carbon storage in deltaic island and mainland coastal marshes. Estuar. Coast. Shelf Sci. 177, 83–89.

Diaz, R., Rosenberg, R., 2008. Spreading dead zones and consequences for marine ecosystems. Science 321, 926–929.

Doughty, C.L., Langley, J.A., Walker, W.S., Feller, I.C., Schaub, R., Chapman, S.K., 2016. Mangrove range expansion rapidly increases coastal wetland carbon storage. Estuar. Coasts 39, 385–396.

Doyle, T.W., Smith, T.J., Robblee, M.B., 1995. Wind damage effects of hurricane Andrew on mangrove communities along the southwest coasts of Florida, USA. J. Coast. Res. Spec. Issue 21, 159–168.

Elliott, M., Mander, L., Mazik, K., Simenstad, C., Valesini, F., Whitfield, A., Wolanski, E., 2016. Ecoengineering with ecohydrology: successes and failures in estuarine restoration. Estuar. Coast. Shelf Sci. 176, 12–35.

Elsner, J., Dossin, J., Jagger, T., 2008. The increasing intensity of the strongest tropical cyclones. Nature 455, 92–95.

Emanuel, K., 2005. Increasing destructiveness of tropical cyclones over the last 30 years. Nature 436, 686.

Emery, K.O., Aubrey, D., 1991. Sea Levels, Land Levels, and Tide Gauges. Springer-Verlag, New York.

Eyre, B., Cyronak, T., Drupp, P., De Carlo, H., Sachs, J.P., Andersson, A.J., 2018. Coral reefs will transition to net dissolving before end of century. Science 359, 908–911.

Fabricius, K.E., Langdon, C., Uthicke, S., Humphrey, C., Noonan, S., De'ath, G., Okazaki, R., Muehllehner, N., Glas, M.S., Lough, J.M., 2011. Losers and winners in coral reefs acclimatized to elevated carbon dioxide concentrations. Nat. Clim. Chang. 1, 165–169.

FitzGerald, D., Fenster, M., Argow, B., Buynevich, I., 2008. Coastal impacts due to sea-level rise. Annu. Rev. Earth Planet. Sci. 36, 601–647.

Forbes, D.L., 2019. Arctic deltas and estuaries: a Canadian perspective. In: Wolanski, E., Day, J., Elliot, M., Ramachandran, R. (Eds.), Coasts and Estuaries – the Future. Elsevier, Amsterdam.

Friedlingstein, P., Meinshausen, M., Arora, V., Jones, C., Anav, A., Liddicoat, S., Knutti, R., 2014. Uncertainties in CMIP5 climate projections due to carbon cycle feedbacks. J. Clim. 27 (2), 511–526.

Giosan, L., Syvitski, J., Constantinescu, S.D., Day, J., 2014. Protect the world's deltas. Nature 516, 31–33.

Golbuu, Y., Gouezo, M., Kurihara, H., Rehm, L., Wolanski, E., 2016. Long-term isolation and local adaptation in Palau's Nikko Bay help corals thrive in acidic waters. Coral Reefs 35, 909.

Goldenberg, S., Landsea, C., Mestas-Nunez, A., Gray, W., 2001. The recent increase in Atlantic hurricane activity: causes and implications. Science 293, 474–479.

Gong, G., Chang, K., Chaing, T., Hsiung, C., Hung, S.D., Codispoti, L., 2006. Reduction of primary production and changing nutrient ratio in the East China Sea: effect of the Three Gorges Dam. Geophys. Res. Lett. 33. https://doi.org/10.1029/2006GL025800.

Groisman, P.Y., Knight, R.W., Easterling, D.R., Karl, T.R., Hegerl, G.C., Razuvaev, V.N., 2005. Trends in intense precipitation in the climate record. J. Clim. 18 (9), 1326–1350.

Hackney, C., Cleary, W., 1987. Saltmarsh loss in southeastern North Carolina lagoons: importance of sea-level rise and inlet dredging. J. Coast. Res. 3, 93–97.

Hall, C., Klitgaard, K., 2018. Energy and the Wealth of Nations, second ed. Springer, New York, p. 511.

Hansen, J., Sato, M., Hearty, P., Ruedy, R., Kelley, M., Masson-Delmotte, V., Russell, G., Tselioudis, G., Cao, J., Rignot, E., Velicogna, I., 2015. Ice melt, sea level rise and superstorms: evidence from paleoclimate data, climate modeling, and modern observations that 2 C global warming is highly dangerous. Atmos. Chem. Phys. 16 (6), 3761–3812.

Harding, L.W., Perry, E.S., 1997. Long-term increase of Phytoplankton Biomass in Chesapeake Bay, 1950–1994. Mar. Ecol. Prog. Ser. 157, 39–52.

Herrera-Silveira, J.A., Lara-Dominguez, A.L., Day, J.W., Yanez-Aranacibia, A., Ojeda, S.M., Hernandez, C.T., Kemp, G.P., 2019. Ecosystem functioning and sustainable management in coastal systems with high freshwater input in the Southern Gulf of Mexico and Yucatan peninsula. In: Wolanski, E., Day, J., Elliot, M., Ramachandran, R. (Eds.), Coasts and Estuaries–The Future. Elsevier, Amsterdam.

Hinkel, J., Lincke, D., Vafeidls, A., Parrette, M., Nicholls, R., Tol, R., Marzeion, B., Fettweis, X., Ionescu, C., Levermann, A., 2013. Coastal flood damage and adaptation cost under 21st sea-level rise. Proc. Natl. Acad. Sci. https://doi.org/10.1073/pnas.1222469111.

Horton, B.P., Rahmstorf, S., Engelhart, S., Kemp, A., 2014. Expert assessment of sea-level rise by AD 2100 and AD 2300. Quat. Sci. Rev. 84, 1–6.

Hoyos, C., Agudelo, P., Webster, P., Curry, J., 2006. Deconvolution of the factors contributing to the increase in global hurricane intensity. Science 312, 94–97.

Ibáñez, C., Prat, N., Canicio, A., 1996. Changes in the hydrology and sediment transport produced by large dams on the lower Ebro River and its estuary. Reg. River. 12, 51–62.

Ibáñez, C., Canicio, A., Day, J.W., Curco, A., 1997. Morphologic evolution, relative sea-level rise and sustainable management of water and sediment in the Ebre Delta. J. Coast. Conserv. 3, 191–202.

IPCC, 2007. Climate Change 2007: The Scientific Basis, Contribution of Working Group 1 to the Third Assessment Report. Cambridge University Press, Cambridge.

IPCC, 2013. Climate change 2013: The physical science basis. In: Contribution of Working Group 1 to the Fifth Assessment Report of the Intergovernmental Panel on Climate Change Cambridge, UK, p. 1535.

Kana, T.W, B.J. Baca, M.L. Williams. 1986. Potential impacts of Sea-level rise on Wetlands around Charleston, South Carolina. USEPA, 230-10-85-014, 65 pp.

Kaufmann, R.F., Kauppi, H., Mann, M., Stock, J., 2011. Reconciling anthropogenic climate change with observed temperature 1998–2008. Proc. Natl. Acad. Sci. 108 (29), 11790–11793. https://doi.org/10.1073/pnas.1102467108.

Kemp, W., Sampou, P., Garber, J., Tuttle, J., Boynton, W., 1992. Seasonal depletion of oxygen from bottom waters of Chesapeake Bay: roles of benthic and planktonic respiration and physical exchange processes. Mar. Ecol. Prog. Ser. 85, 137–152.

Kemp, P., Willson, C., Rogers, D., Westphal, K., Binselam, A., 2014. Adapting to change in the lowermost Mississippi River: implications for navigation, flood control, and restoration of the delta ecosystem. In: Day, J., et al. (Eds.), Perspectives on the Restoration of the Mississippi Delta. Springer, New York, pp. 51–85.

Kemp, P., Day, J., Rogers, D., Giosan, L., Peyronnin, N., 2016. Enhancing mud supply to the Mississippi River delta: dam bypassing and coastal restoration. Estuar. Coast. Shelf Sci. https://doi.org/10.1016/j.ecss.2016.07.008.

Kidwai, S., Ahmed, W., Tabrez, S.M., Zhang, J., Giosan, L., Clift, P., Inam, A., 2019. The Indus delta—catchment, river, coast and people. In: Wolanski, E., Day, J., Elliot, M., Ramachandran, R. (Eds.), Coasts and Estuaries—The Future. Elsevier, Amsterdam.

Kirwan, M.L., Murray, A.B., 2008. Ecological and morphological response of brackish tidal marshland to the next century of sea level rise: Westham Island, British Columbia. Glob. Planet. Chang. 60, 471–486.

Knights, B., 1979. Reclamation in the Netherlands. In: Knights, B., Philipps, A. (Eds.), Estuarine and Coastal Land Reclamation and Water Storage. Saxon House, London.

Knutson, T., McBride, J., Chan, J., Emanuel, K., Holland, G., Lansea, C., Held, I., Kossin, J., Srivastava, A., Masato, S., 2010. Tropical Cyclones and climate change. Nat. Geoscience. https://doi.org/10.1038/NGE0779.

Ko, J., Day, J., 2004. A review of ecological impacts of oil and gas development on coastal ecosystems in the Mississippi delta. Ocean Coast. Manag. 47, 671–691.

Koop, R., Kemp, A., Bitterman, K., Horton, B., Donnelly, J., Gehrels, W., Hay, C., Mitrovica, J., Morrow, E., Rahmstorf, S., 2016. Temperature-driven global sea-level variability in the common era. Proc. Natl. Acad. Sci. https://doi.org/10.1073/pnas.1517056113.

Lambeck, K., Woodroffe, C.D., Antonioli, F., Anzidei, M., Gehrels, W.R., Laborel, J., Wright, A.J., 2010. Paleoenvironmental records, geophysical modelling, and reconstruction of sea level trends and variability on centennial and longer timescales. In: Church, J.A., Woodworth, P.L., Aarup, T., Wilson, W.S. (Eds.), Understanding Sea Level Rise and Variability. Wiley-Blackwell, Chichester, pp. 61–121.

Leinenkugel, P., Wolters, M., Oppelt, N., Kuenzer, C., 2014. Tree cover and forest cover dynamics in the Mekong Basin from 2001 to 2011. Remote Sens. Environ. 158, 376–392.

Liu, D., Keesing, J., He, P., Wang, Z., Shi, Y., Wang, Y., 2013. The world's largest macroalgal bloom in the Yellow Sea, China: formation and implications. Estuar. Coast. Shelf Sci. 129, 2–10.

Lovecraft, A.L., Meek, C.L., 2019. Arctic coastal systems: evaluating the DAPSI(W)R(M) framework. In: Wolanski, E., Day, J., Elliot, M., Ramachandran, R. (Eds.), Coasts and Estuaries—The Future. Elsevier, Amsterdam.

McCaffey, R., Thompson, J., 1980. A record of the accumulation of sediment and trace metals in a connecticut salt marsh. In: Estuarine Physics and Chemistry: Studies in Long Island Sound. Academic Press, New York, pp. 165–236.

McCarthy, J., 2009. Reflections on: our planet and its life, origins, and futures. Science 326, 1646–1655.

McGranahan, G., Balk, D., Anderson, B., 2007. The rising tide: assessing the risks of climate change and human settlements in low elevation coastal zones. Environ. Urban. 19 (1), 17–37.

McKee, K., Patrick, W., 1988. The relationship of smooth cordgrass (Spartina alterniflora) to tidal datums – a review. Estuaries 11, 143–151.

McMillan, C., Sherrod, C.L., 1986. The chilling tolerance of black mangrove, *Avicennia germinans*, from the Gulf of Mexico coast of Texas, Louisiana, and Florida. Contrib. Mar. Sci. 29, 9–16.

Mei, W., Xie, S., Primeau, F., McWilliams, J., Pasquero, C., 2015. Northwestern Pacific typhoon intensity controlled by changes in ocean temperatures. Sci. Adv. https://doi.org/10.1126/sciadv.1500014.

Mendelsshohn, I., McKee, K., 1988. Spartina alterniflora die-back in Louisiana: time-course investigation of soil water logging effects. J. Ecol. 76, 509–521.

Mendelssohn, I.A., Morris, J.T., 2000. Eco-physiological controls on the productivity of Spartina alterniflora loisel. In: Weinstein, M.P., Kreeger, D.A. (Eds.), Concepts and Controversies in Tidal Marsh Ecology. Kluwer Academic, Boston, MA, pp. 59–80.

Michener, W.K., Blood, E.R., Bildstein, K.L., Brinson, M.M., Gardner, L.R., 1997. Climate change, hurricanes and tropical storms, and rising sea level in coastal wetlands. Ecol. Appl. 7, 770–801.

Milliman, J.D., Quraishee, G.S., Beg, M.A., 1989. Sediment discharge from the Indus River to the ocean; past, present and future. In: Haq, B.H., Milliman, J.D. (Eds.), Marine Geology and Oceanography of Arabian Sea and Coastal Pakistan. Van Nostrand Reinhold, New York, pp. 265–270.

Min, S.-K., Zhang, X., Zwiers, F.W., Hegeri, G.C., 2011. Human contribution to more-ntense precipitation extremes. Nature 470, 378–381.

Mitrovica, J., Gomez, J., Clark, P., 2009. The sea-level fingerprint of west Antarctic collapse. Science 323, 753.

Mitsch, W.J., Jørgensen, S.E., 2004. Ecological Engineering and Ecosystem Restoration. John Wiley, New York, p. 411.

Mitsch, W., Day, J., Gilliam, J., Groffman, P., Hey, D., Randall, G., Wang, N., 2001. Reducing nitrogen loading to the Gulf of Mexico from the Mississippi River basin: strategies to counter a persistent problem. Bioscience 51, 373–388.

Morris, J.T., Sundareshwar, P.V., Nietch, C.T., Kjerfve, B., Cahoon, D.R., 2002. Responses of coastal wetlands to rising sea level. Ecology 83, 2869–2877.

Morton, R.A., Buster, N., Krohn, M.D., 2002. Subsurface controls on historical subsidence rates and associated wetland loss in southcentral Louisiana. Gulf Coast Asso. Geo. Soc. 52, 767–778.

Moss, R., Babiker, M., Brinkman, S., et al., 2008. Towards New Scenarios for Analysis of Emissions, Climate Change, Impacts, and Response Strategies Technical Summary. Intergovernmental Panel on Climate Change, Geneva.

Nerem, R.S., Chambers, D.P., Choe, C., Mitchum, G.T., 2010. Estimating mean sea level change from the TOPEX and Jason altimeter missions. Mar. Geod. 33 (S1), 435–446.

Nhan, N.H., Cao, N.B., 2019. Damming the Mekong: impacts in Vietnam and solutions. In: Wolanski, E., Day, J., Elliot, M., Ramachandran, R. (Eds.), Coasts and Estuaries—The Future. Elsevier, Amsterdam.

Niang, A., Scheren, P., Diop, S., Kane, C., Koulibaly, C.T., 2019. The Senegal and Pangani rivers: examples of over-used river systems within water-stressed environments in Africa. In: Wolanski, E., Day, J., Elliot, M., Ramachandran, R. (Eds.), Coasts and Estuaries—The Future. Elsevier, Amsterdam.

Nicholls, R.J., Marinova, N., Lowe, J.A., Brown, S., Vellinga, P., De Gusmao, D., Hinkel, J., Tol, R.S., 2011. Sea-level rise and its possible impacts given a "beyond 4 C world" in the twenty first century. Philos. Trans. A Math. Phys. Eng. Sci. 369 (1934), 161–181.

Nixon, S., 1988. Physical energy inputs and the comparative ecology of lake and marine ecosystems. Limnol. Oceanogr. 33, 1005–1025.

NOAA, 2018. Global Warming and Hurricanes—An Overview of Current Research Results. Available from: https://www.gfdl.noaa.gov/global-warming-and-hurricanes/.

Norwegian Refugee Council, 2014. Global Estimates 2014 e People Displaced by Disasters. Internal Displacement Monitoring Center, Geneva, p. 64. Available from: www.internal-displacement.org.

Osland, M., Day, R., Hall, C., Brumfield, M., Dugas, J., Jones, W., 2016. Mangrove expansion and contraction at a poleward limit: climate extremes and land-ocean temperature gradients. Ecology 98, 125–137. https://doi.org/10.1002/ecy.1625.

Pachauri, R.K., Allen, M.R., Barros, V.R., Broome, J., Cramer, W., Christ, R., Church, J.A., Clarke, L., Dahe, Q., Dasgupta, P., Dubash, N.K., 2014. Climate change 2014: synthesis report. Contribution of working groups I, II and III to the fifth assessment report of the intergovernmental panel on climate change. IPCC, Geneva.

Paerl, H.W., Bales, J.D., Ausley, L.W., Buzzelli, C.P., Crowder, L.B., Eby, L.A., Go, M., Peierls, B.L., Richardson, T.L., Ramus, J.S., 2000. Hurricanes' hydrological, ecological effects linger in major U.S. estuary. Eos 81, 457.

Pall, P., Aina, T., Stone, D.A., Stott, P.A., Nozawa, T., Hilberts, A.G.J., Lohmann, D., Allen, M.R., 2011. Anthropogenic greenhouse gas contribution to flood risk in England and Wales in autumn 2000. Nature 470, 382–385.

Parris, A., Bromirski, P., Burkett, V., Cayan, D., Culver, M., Hall, J., Horton, R., Knuuti, K., Moss, R., Obeysekera, J., Sallenger, A., Weiss, J., 2012. Global sea level rise scenarios for the US National Climate Assessment. NOAA Tech Memo OAR CPO-1.

Pfeffer, W., Harper, J., O'Neel, S., 2008. Kinematic constraints on glacier contributions to 21st-century sea-level rise. Science 321, 1340–1343.

Pirazzoli, P., 1987. Recent sea-level changes and related engineering problems in the Lagoon of Venice, Italy. Prog. Oceanogr. 18, 323–346.

Pirazzoli, P.A., 1996. Sea Level Changes: The Last 20 000 years. J Wiley & Sons, Chichester.

Poff, L., Brinson, M., Day, J., 2002. Aquatic Ecosystems & Global Climate Change: Potential Impacts on Inland Freshwater and Coastal Wetlands Ecosystems in the United States. Pew Center on Global Climate Change, Arlington, VA, p. 44.

Pont, D., Day, J., Hensel, P., Franquet, E., Torre, F., Rioual, P., Ibáñez, C., Coulet, E., 2002. Response scenarios for the deltaic plain of the Rhône in the face of an acceleration in the rate of sea-level rise, with a special attention for Salicornia-type environments. Estuaries 25, 337–358.

Pont, D., Day, J., Ibáñez, C., 2017. The impact of two large floods (1993–1994) on sediment deposition in the Rhone delta: implications for sustainable management. Sci. Total Environ. doi.org/10.1016/j.scitotenv.2017.07.155.

Prein, A.F., Rasmussen, R.M., Ikeda, K., Liu, C., Clark, M.P., Holland, G.J., 2016. The future intensification of hourly precipitation extreme. Nat. Clim. Chang. 7 (1), 48–52.

Rabalais, N.N., Turner, R.E., Justiæ, D., Dortch, Q., Wiseman Jr., W.J., Sen Gupta, B.K., 1996. Nutrient changes in the Mississippi River and system responses on the adjacent continental shelf. Estuaries 19, 386–407.

Rahmstorf, S., 2007. A semi-empirical approach to predicting sea-level rise. Science 315, 368–370.

Redfield, A.C., 1972. Development of a New England salt marsh. Ecol. Monogr. 42, 201–237.

Rignot, E., Velicogna, I., van den Broeke, M.R., Monaghan, A., Lenaerts, J.T., 2011. Acceleration of the contribution of the Greenland and Antarctic ice sheets to sea level rise. Geophys. Res. Lett. 38 (5), L18601.

Rignot, E., Mouginot, J., Morlighem, M., Seroussi, H., Scheuchl, B., 2014. Widespread, rapid grounding line retreat of Pine Island, Thwaites, smith, and Kohler glaciers, West Antarctica, from 1992 to 2011. Geophys. Res. Lett. 41 (10), 3502–3509.

Roberts, H.H., DeLaune, R.D., White, J.R., Li, C., Sasser, C.E., Braud, D., Khalil, S., 2015. Floods and cold front passages: impacts on coastal marshes in a river diversion setting (Wax Lake Delta Area, Louisiana). J. Coast. Res. 31, 1057–1068.

Rockström, J., Steffen, W., Noone, K., Persson, Å., Chapin III, F.S., Lambin, E.F., Lenton, T.M., Scheer, M., Folke, C., Schellnhuber, H.J., Nykvist, B., et al., 2009. A safe operating space for humanity. Nature 461 (7263), 472–475.

Rohling, E., Grant, K., Hemleben, C., Siddall, M., Hoogakker, B., Bolshaw, M., Jucera, M., 2008. High rates of sea-level rise during the last interglacial period. Nat. Geosci. 1, 38–42.

Rybczyk, J., Cahoon, D., 2002. Estimating the potential for submergence for two wetlands in the Mississippi River delta. Estuaries 25, 985–998.

Rybczyk, J., Zhang, X., Day, J., Hesse, I., Feagley, S., 1995. The impact of hurricane Andrew on tree mortality, litterfall and water quality in a Louisiana coastal swamp forest. J. Coast. Res. 21, 340–353.

Rybczyk, J., Day, J., Yanez, A., Cowan, J., 2013. Global climate change and estuarine systems. In: Day, J., et al. (Eds.), Estuarine Ecology. Wiley-Blaclwell, New Jersey, pp. 497–519.

Saintilan, N., Wilson, N.C., Rogers, K., Rajkaran, A., Krauss, K.W., 2014. Mangrove expansion and salt marsh decline at mangrove poleward limits. Glob. Chang. Biol. 20, 147–157.

Scavia, D., Field, J., Boesch, D., Buddemeier, R., Burkett, V., Cayan, D., Fogarty, M., Harwell, M.A., Howarth, R.W., Mason, C., Reed, D.J., Royer, R.C., Sallenger, A.H., Titus, J.G., 2002. Climate change impacts on U.S. coastal and marine ecosystems. Estuaries 25, 149–164.

Sestini, G., 1992. Implications of climatic changes for the Po delta and Venice lagoon. In: Jeftic, L., Milliman, J., Sestini, G. (Eds.), Climatic Change and the Mediterranean. Edward Arnold, London.

Sestini, G., 1996. Land subsidence and sea-level rise: the case of the Po delta region, Italy. In: Milliman, J., Haq, B. (Eds.), Sea-Level Rise and Coastal Subsidence. Kluwer Academic, Dordrecth, pp. 235–248.

Shaffer, G., Day, J., Kandalepas, D., Wood, W., Hunter, R., Lane, R., Hillmann, E., 2016. Decline of the Maurepas swamp, Pontchartrain Basin, Louisiana and approaches to restoration. Water 7, https://doi.org/10.3390/w70x000x.

Sharifinia, M., Daliri, M., Kamrani, E., 2019. Estuaries and coastal zones in the Northern Persian Gulf (Iran). In: Wolanski, E., Day, J., Elliot, M., Ramachandran, R. (Eds.), Coasts and Estuaries—The Future. Elsevier, Amsterdam.

Sklar, F., Chimney, M., Newman, S., McCormick, P., Gawlik, D., Miao, S., McVoy, C., Said, W., Newman, J., Coronado, C., Crozier, G., Korvela, M., Rutchey, K., 2005. The ecological-societal underpinnings of Everglades restoration. Front. Ecol. Environ. 3 (3), 161–169.

Sklar, F.H., Meeder, J.F., Troxler, T.G., Drschel, T., Davis, S.E., Ruiz, P., 2019. Everglades. In: Wolanski, E., Day, J., Elliot, M., Ramachandran, R. (Eds.), Coasts and Estuaries—The Future. Elsevier, Amsterdam.

Snedaker, S., 1984. Mangroves: a summary of knowledge with emphasis on Pakistan. In: Haq, B.H., Milliman, J.D. (Eds.), Marine Geology and Oceanography of Arabian Sea and Coastal Pakistan. Van Nostrand Reinhold, New York, pp. 255–262.

Stanley, D., 1988. Subsidence in the northeastern Nile delta: rapid rates, possible causes, and consequences. Science 240, 497–500.

Stanley, D.J., Warner, A.G., 1993. Nile delta: geological evolution and human impact. Science 260, 628–634.

Steffen, W., Broadgate, W., Deutsch, L., Gaffney, O., Ludwig, C., 2015. The trajectory of the Anthropocene: the great acceleration. Anthr. Rev. 2 (1), 81–98.

Stevenson, J., Kearney, M., Pendleton, E., 1985. Sedimentation and erosion in a Chesapeake Bay brackish marsh system. Mar. Geol. 67, 213–235.

Sverdrup, H.U., Ragnarsdottir, K.V., Koca, D., 2015. An assessment of metal supply sustainability as an input to policy: security of supply extraction rates, stocks-in-use, recycling, and risk of scarcity. J. Clean Prod. 140, 359–372.

Syvitski, J., Kettner, A., Overeem, I., Hutton, E., Hannon, M., Brakenridge, R., Day, J., Vorosmarty, C., Saito, Y., Giosan, L., Nichols, R., 2009. Sinking deltas due to human activities. Nat. Geosci. https://doi.org/10.1038/NGE0629.

Tao, B., Tian, H., Ren, W., Yang, J., Yang, Q., He, R., Cai, W., Lohrenz, S., 2014. Increasing Mississippi River discharge throughout the 21st century influenced by changes in climate, land use, and atmospheric CO^2. Geophys. Res. Lett. 41, 4978–4986. https://doi.org/10.1002/2014GL060361.

Tessler, Z.D., Vörösmarty, C., Grossberg, M., Gladkova, I., Aizenman, H., Syvitski, J., Foufoula-Georgiou, E., 2015. Profiling risk and sustainability in coastal deltas of the world. Science 349, 638–643.

Titus, J., Park, R., Leatherman, S., Weggel, J., Green, M., Mausel, P., Brown, S., Gaunt, C., Trehan, M., Yohe, G., 1991. Greenhouse effect and sea-level rise: potential loss of land and the cost of holding back the sea. Coast. Manag. 19, 171–204.

Tollefson, J., 2017. Larsen C's big divide. Nature 543 (7643), 402–403.

Tweedly, J.R., Dittmann, S.R., Whitfield, A.K., Withers, K., Hoeksema, S.D., Potter, I.C., 2019. Australian systems and Colorado River. In: Wolanski, E., Day, J., Elliot, M., Ramachandran, R. (Eds.), Coasts and Estuaries—The Future. Elsevier, Amsterdam.

Twilley, R., Day, J., 2013. Mangrove wetlands. In: Day, J., et al. (Eds.), Estuarine Ecology. Wiley-Blaclwell, New Jersey, pp. 165–203.

Twilley, R., Barron, E., Gholz, H., Harwell, M., Miller, R., Reed, D., Rose, J., Siemann, E., Wetzel, R., Zimmerman, R., 2001. Confronting Climate Change in the Gulf Coast Region: Prospects for Sustaining Our Ecological Heritage. Union of Concerned Scientist, Cambridge, MA, and Ecological Society of America, Washington DC, p. 82.

Twilley, R.R., Bentley, S.J., Chen, Q., Edmonds, D.A., Hagen, S.C., Lam, N.S.-N., Willson, C.S., Xu, K., Braud, D., Peele, R.H., McCall, A., 2016. Co-evolution of wetland landscapes, flooding, and human settlement in the Mississippi River Delta Plain. Sustain. Sci. 11, 711–731.

Valesini, F.J., Hallet, C.S., Hipsey, M.R., Kilminster, K.L., Huang, P., Hennig, K., 2019. Peel-Harvey estuary, Australia. In: Wolanski, E., Day, J., Elliot, M., Ramachandran, R. (Eds.), Coasts and Estuaries—the Future. Elsevier, Amsterdam.

Vermeer, M., Rahmstorf, S., 2009. Global sea level linked to global temperature. Proc. Natl. Acad. Sci. 106, 21527–21532.

Warren, R.W., Niering, W.A., 1993. Vegetation change on a northeast tidal marsh: interaction of sea-level rise and marsh accretion. Ecology 74, 96–113.

Webster, J., Holland, G., Curry, J., Chang, H., 2005. Changes in tropical cyclone number, duration, and intensity in a warming environment. Science 309, 1844.

Wiegman, R., Day, J., D'Elia, C., Rutherford, J., Morris, J., Roy, E., Lane, R., Dismukes, D., Snyder, B., 2017. Modeling impacts of sea-level rise, oil price, and management strategy on the costs of sustaining Mississippi delta marshes with hydraulic dredging. Sci. Total Environ. https://doi.org/10.1016/j.scitotenv.2017.09.314.

Wolanski, E., Elliott, M., 2015. Estuarine. Ecohydrology an Introduction. Elsevier, Amsterdam, p. 322.

Wolanski, E., Brinson, M., Cahoon, D., Perillo, G., 2009. Coastal wetlands. A synthesis. In: Perillo, G., Wolanski, E., Cahoon, D., Brinson, M. (Eds.), Coastal Wetlands. An Integrated Ecosystem Approach. Elsevier, Amsterdam, pp. 1–62.

Yáñez-Arancibia, A., Day, J.W., 2005. Ecosistemas vulnerables, riesgo ecológico y el record 2005 de huracanes en el Golfo de México y Mar. Available from: http://www.ine.gob.mx/download/huracanes2005.pdf.

Yáñez-Arancibia, A., Ramírez Gordillo, J.J., Day, J.W., Yoskowitz, D., 2009a. Environmental sustainability of economic trends in the Gulf of Mexico: what is the limit for the Mexican coastal development? In: Cato, J.A. (Ed.), Ocean and Coastal Economy, vol. 2, The Gulf of Mexico: Its Origins, Waters, Biota & Human Impacts Series. Texas A&M Univeristy Press, College Station, TX.

Yáñez-Arancibia, A., Day, J.W., Currie-Alder, B., 2009b. Functioning of the Grijalva-Usumacinta river delta, Mexico: challenges for coastal management. Ocean Yearb. 23, 473–501.

Yáñez-Arancibia, A., J. W. Day, R. R. Twilley, and R. H. Day. 2010. Mangroves and climate change: global tropicalization of the Gulf of México? pp. 91-126. In: A. Yáñez-Arancibia (Ed.), Impacts of Climate Change in the Coastal Zone. INE-SEMARNAT, Mexico DF, 180 pp.

Chapter 37

Human-Nature Relations in Flux: Two Decades of Research in Coastal and Ocean Management

Bernhard Glaeser

Freie Universität, Berlin, Germany, German Society for Human Ecology (DGH), Berlin, Germany

1 PREAMBLE: AIM AND OVERVIEW

The aim of the paper is to highlight the observed paradigm shifts in the science-policy link within the larger context of coastal and ocean management (COM). It is the agenda setting that is focused upon rather than individual case study outcomes. The case studies reviewed were designed and developed in different national and political contexts as in different geographical and climate settings and followed different objectives. They are hardly comparable. The methodology used is historically descriptive. The analytical tool applied is the ex post judgment by the author to have discovered an unplanned development of an academic field, following changing societal needs and challenges. The outcome is an attempt to present evidence of social change in correspondence with natural changes on a *pars pro toto* scale.

The paper is organized as follows. In an introductory Section 2, the notion and necessity of COM are explained *vis-à-vis* the age-old issue of human-nature relations, as taken up by a novel interdisciplinary approach, linking natural and social sciences. The first major inter- and transdisciplinary (including different sciences and stakeholders) national coastal program in Europe (Section 3) was the Swedish SUCOZOMA research to identify coastal stakeholders and to identify and resolve coastal conflicts. The major (social) stakeholder group in consideration was fisheries; the major (natural) problem was eutrophication. Germany followed suit (Section 4), set up a comparable coastal research program for the North Sea and the Baltic that culminated in designing a national coastal and ocean strategy within a European framework. An unprecedented event changed the course of coastal management entirely: a new coastal and ocean research focus on natural calamities and coastal hazards emerged in the wake of the tsunami 2004 in South-East Asia and the 2005 hurricane Katrina in New Orleans (Section 5). Climate change became a major threat to the well-being of Indonesian fishermen (Section 6). It became apparent that global and local changes, socially as well as naturally, were interconnected. Finally, coastal and ocean typologies came to be built up as an analytical instrument and planning tool (Section 7), a comparative instrument to identify coastal issues and solutions and to implement it as to support various stakeholder groups around the globe in perhaps similar settings.

It seems obvious that the case studies and chapters feature different issues, involve different complexities and follow very different designs—resulting in different narratives and styles of reporting and writing. However, there is a common thread: human-nature interactions on coasts and oceans with respect to paradigm shifts in the research and policy linkages.

2 HUMAN-NATURE RELATIONS: WHY COM?

At least two-thirds of the human population live or work on coasts. They include not only fisher families whose livelihoods are shrinking but the pensioners, people related to tourism and an increasing population in megacities around the globe but also the shipping, harbor, and other industries. Coasts and oceans provide goods and ecosystem services. Major resources are food and minerals for an increasing human population. Oceans are the cradles of life on earth. They are known as the "unknown planet," largely unexplored, yet containing a maximum of biodiversity. A "census" of marine life has been initiated (Knowlton, 2010). For all these reasons, it is not amazing that vested interests are competing for space and resources, represented by political, economic, and scientific stakeholders. Conflicts arise among them, and it has been a prominent goal by ocean and coastal management to aim at resolving those conflicts.

Coasts and Estuaries. https://doi.org/10.1016/B978-0-12-814003-1.00037-X

The point of departure is the relation between humans and nature with the goal to present changing approaches to sustainable COM. Whereas it should have been theoretically self-evident and politically imperative not to separate social and ecological issues, it took a long time before research and politics recognized the need for integration. Concerning coastal areas, the first legislative effort globally to establish this connection was the US Coastal Zone Management Act (CZMA) of 1972 as a national policy to preserve, protect, develop, or restore coastal resources for present and future generations (https://coast.noaa.gov/czm/act/).

The challenges for science (including social science), policy, and civil society are vast and the complex issues framed pathways for a new paradigm. Novel fields of research emerge when society faces complex problems and needs to generate new knowledge in order to understand and regulate the real-world issues. This can be achieved by employing innovative, theory-driven empirical approaches. One such field is an integrated and sustainable COM. The predominant difference to most university disciplines is the approach of integration, which implies not only the need to link social and ecological systems but also emphasizes the integration of three geographical dimensions namely water, land, and air. It also tries to amalgamate three scientific dimensions, the natural and social sciences plus the humanities as they all represents the basic aspects of coastal life. Finally, COM links science to policy. The common denominator has always been sustainability, in theory and practice.

This paper reviews two decades of research in COM including the author's own experience in this emerging field. This study focused mainly on the multi- and interdisciplinary researches between 1996 and 2016 and discovered an amazing change of research focuses over these years. Generally, focal themes of research were suiting to the changing societal needs and scientific outlooks. Beginning with competition for coastal space and resources, ensuing conflicts and conflict resolution the resource governance gained prominence. Later on, disaster management and climate change emerged as new research topics. More recently, typologies of coastal and marine (CM) social-ecological systems (SESs) have focused on theory-driven issues.

Theoretical aspects have been embedded and grounded in empirical case studies. The case studies presented were taken from economically developed countries in the temperate zone defined by a high per capita gross domestic product (GDP) level, namely Sweden and Germany in Europe; and from an economically less-developed country in the tropical zone, defined by a lower per capita GDP level and a significant incidence of poverty, Indonesia in South-East Asia. The sections below represent individual cases:

- Stakeholder identification and conflict resolution: the example of Sweden
- National coastal and ocean strategies: the example of Germany
- Natural calamities and coastal hazards: the 2004 tsunami in South-East Asia and the 2005 hurricane Katrina in New Orleans
- Climate change: the example of Indonesia
- Coastal and ocean typologies: an analytical instrument and planning tool

3 STAKEHOLDER IDENTIFICATION AND CONFLICT RESOLUTION: THE EXAMPLE OF SWEDEN

The Swedish research program "Sustainable Coastal Zone Management" (SUCOZOMA) was initiated in 1997 and completed in 2004. SUCOZOMA's environmental research was funded by MISTRA, the Swedish Foundation for Strategic Environmental Research, as a first national coastal program in Europe, covered North Sea and Baltic sites and aspects, representing areas and institutions in West and East Sweden. The program was managed by social sciences, a novelty in those days, namely by the Human Ecology Section of Göteborg (Gothenburg) University (Glaeser, 1999).

Main goal of SUCOZOMA was to understand and compare different stakeholders' needs and interests and their stand in relation to the sustainable use of coastal ecosystems. Articulating different coastal activities and varied interests of different actors in an integrated manner in the planning and managment process especially while aiming to ensure environmentally sustainable, economically feasible, and equitable solutions (French, 1997: specifically Chapter 7) was truly challenging for the policy makers, planners, and managers.

Fig. 1 explains the program structure guided by a vision, represented by four goals which are summarized outside the circle viz. integrated management, sustainable fishery, enhanced recreational value, and restored biodiversity. The science-policy fields (ecology-economy, resource management, management institutions, competing interests) constitute the hub around which the project areas revolve: fishing technology, mussel farming, managing nutrient discharges, and stock enhancement.

The way to manage the program and to achieve its goals was an interdisciplinary research that interlinks diverse disciplines, such as ecology and fisheries science, economics and law, as well as human ecology to study the human-environment

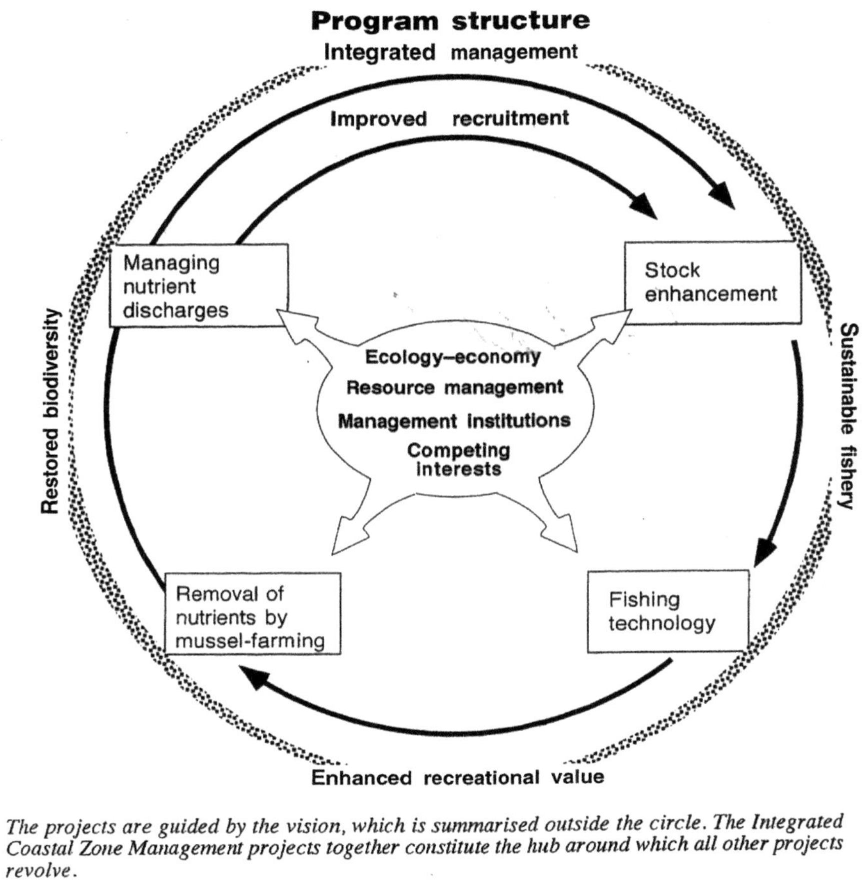

FIG. 1 SUCOZOMA program structure.

relations. The research projects were led by the Universities of Gothenburg and Stockholm as well as the National Board of Fisheries (Fiskeriverket). The program's primary concern was on sustainable management of water and water-based resources with greater emphasis on coastal fishery, mussel farming, and water quality management. Significant importance was given to conflict resolution between coastal stakeholders viz. agriculture vs coastal fisheries. The science-policy link included the National Board of Fisheres and West Coast communities.

In order to understand and compare different stakeholders' needs and interests, a stakeholder analysis with respect to mussel culture was done using an evaluation matrix to identify stakeholders' capability to influence mussel culture and or to be influenced by it in Sweden. The groups considered were regulating bodies, resource users, industries, local inhabitants, and political actors. A combined ranking list showed the provincial government (länsstyrelsen), land/water right owners, part-time residents, and neighborhood associations with the mussel culture farmers on top while the Nordic Council and the International Council for the Exploration of the Sea (ICES) at the bottom of the list (Ellegard, 1998a).

In addition to stakeholder analysis, media survey method was used to explore the press coverages related to various aspects of mussel culture. In the view on mussels, it was observed in the selected Swedish print media that the "food connection" was by far the strongest representing 52% of the articles analyzed. As a comparison, environmental issues were mentioned in 19% of the cases, poetry/archeology in 12%, and tourism in 5%. The neutral view on mussels was overwhelming in relation to negative connotations. The environmental message was found to be mixed: close to 50% of the texts dealt with environmental degradation, warnings, and alerts (Ellegard, 1998b).

All aquaculture in Sweden, independent of size, required a permit under the Fisheries Act of country's administrative board, that governs outbreak of diseases to fish and transfer of fish and shellfish. If the operational plan of aquaculture is to produce more than 10 tonnes of fishery resources, the farmer has to submit an application to the county administrative board which will direct the application to municipality (Ackefors and Grip, 1995). The process is to satisfy the regulation under the Natural Resources Management Act and to seek permit under the Fisheries Act, the Environment Protection Act, and the Nature Conservation Act. A permit from the county administrative board would include detailed conditions regarding the aquaculture farm development. A decision under the Environment Protection Act must be compatible with existing local plans.

The projects were originally assembled as scientifically promising and socially relevant research fields, represented by individually outstanding researchers, with coastal fisheries and eutrophication as the core. The social science, human ecology, and management-oriented projects were included to provide the program with an integrative principle. A core focus was to identify coastal stakeholders, their conflicts and to use inter- and transdisciplinary science in cooperation with coastal communities to approach solutions. It was agreed and it was necessary that all projects worked toward a common goal, a socially and environmentally sustainable coast (Glaeser and Grahm, 1998). SUCOZOMA was the first national coastal management program in Europe that integrated different disciplines, established a science-policy link, and aimed at sustainable coastal development. It set the standard for European coastal research in the decades to come (Glaeser, 1999).

4 NATIONAL COASTAL AND OCEAN STRATEGIES: THE EXAMPLE OF GERMANY

The German coastal management community learned from the Swedish experiences and set up two national coastal research programs, viz. "Coastal Futures" for the North Sea and, in close cooperation, "ICZM Oder" for the Baltic. Both were funded from 2004 until 2010 by the Federal Ministry of Education and Research (BMBF). Coastal stakeholders of concern were actors in tourism, harbors and the shipping industry, as well as the newly emerging offshore wind farming actors, in combination with fisheries and mussel farming operations around the windfarm premises (Glaeser, 2000, 2002, 2004).

A need was felt in Germany to embark on a coastal and ocean strategy as in the European Union (EU) and other EU countries (Europäische Union, 2002). On top of primary research tasks, the groundwork was laid to develop a strategic framework for Germany, in cooperation and as part of the EU efforts. The basic procedures for an Integrated Coastal Zone Management (ICZM) and an ocean strategy within Europe were defined by the following steps: the EU develops a framework and the national-level actors are responsible for implementing it. It was thereby required to focus on public participation and bottom-up approaches, a result of the previous EU demonstration programs (Europäische Kommission, 1999) in integrated and sustainable coastal management. A particular focus in Germany was on renewable energy in offshore wind farming in the exclusive economic zones (EEZs) (Fig. 2; BMU, 2002a,b,c; Glaeser et al., 2009).

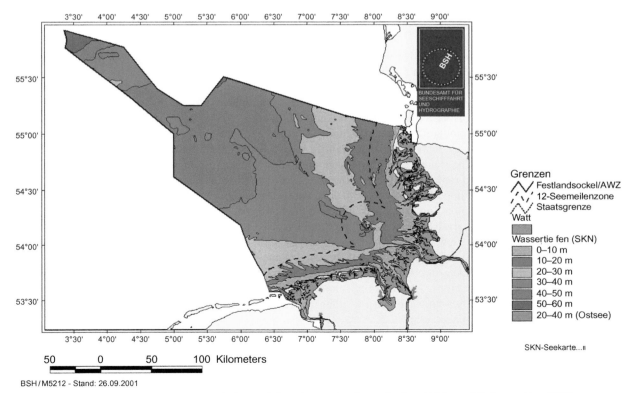

FIG. 2 Germany's exclusive economic zone (EEZ) in the North Sea. *(Source: Bundesamt für Seeschiffahrt und Hydrographie—BSH.)*

The European Council issued recommendations to the member countries as to developing a national ICZM strategy that included the following steps (Gee et al., 2004):

- perform an environmental stock-take;
- clarify the division of labor between different administrative levels;
- develop a combination of instruments for implementation, with special focus on bottom-up initiatives and public participation;
- implement an appropriate legal framework; and
- establish appropriate monitoring systems for the coastal zone.

The ministry in Germany, eventually responsible for a national ocean strategy, became the Federal Ministry for the Environment (BMU). The BMU announced assessment and steps toward a National ICZM Strategy (BMU, 2006):

1. ICZM promotes sustainable development of coastal zones with their specific ecological, economic, and social features.
2. ICZM guides political and social action at all levels and coordinates CZ development through integration of all concerns.
3. ICZM incorporates all relevant policy areas, economic and scientific actors, social groups, and levels of administration into a participatory process to identify consensual solutions and improve conflict management.
4. ICZM combines planning, implementation and evaluation of coastal changes, and transfer of experience.

The research project "Integrated Coastal Zone Management (ICZM): Strategies for coastal and marine spatial planning," was funded by the German Federal Ministry of Transport, Building and Urban Affairs (BMVBS) and the Federal Office for Building and Spatial Planning (BBR) from 2002 to 2006. Conceived to support the development and implementation of a national ICZM strategy, the project took stock of development trends, opportunities, and challenges faced by Germany's CM areas and made practical suggestions for implementing long-term ICZM processes from the special perspective of spatial planning. Project results were incorporated into the publication "Integrated Coastal Zone Management: Assessment and steps toward a national ICZM strategy," which was drafted by the Federal Ministry of Environment and adopted through cabinet decision on March 22, 2006 (http://www.ikzm-strategie.de/) (Gee et al., 2006). One could say that the project describes a combination of top-down (EU framework) and bottom-up (national strategy) approaches.

Germany has a longstanding tradition of spatial planning vis-à-vis the terrestrial environment, but only after 2002 did the country begin to consider similar approaches with respect to the marine environment. In order to become a successful instrument of Integrated Coastal Management (ICM), spatial planning must take into account the uniqueness and complexity of the coastal zone, including the multifaceted patterns of use and the growing potential for conflict (Glaeser et al., 2005).

The cornerstones of German national ICM policy were:

- developments in global trade, including maritime traffic and ports;
- developments in energy and climate policy, including large-scale offshore wind farms and cable connections to the mainland;
- developments of aqua and mariculture;
- replenishment of marine resources;
- meeting the targets and requirements of European environmental directives, including the designation of marine protected areas (MPAs); and
- meeting the targets and requirements of coastal protection, with particular respect to climate change (Gee et al., 2006).

A comprehensive stock-take identified significant trends, including offshore wind development, and made practical suggestions for implementing long-term ICZM processes from the special perspective of spatial planning on the coastal zone and also pointed out a range of structural needs that a national strategy was to address. Simultaneously, parallel conceptual developments took place at the EU level within the frame of international maritime policy. A more comprehensive view of spatial planning corresponded to the polycultural ICM approach and supported regional CM development by working as an additional tool to implement integrative use concepts.

Whilst it is clearly important to adapt spatial planning instruments to implement ICZM at a national level, it is equally important to take a more comprehensive and long-term view. In order to effectively deal with the expected economic, ecological, and social impacts, it is imperative to understand the future trends. Two large-scale research programs ("Coastal Futures" for the North Sea and "ICZM Oder" for the Baltic: see above) had been launched to provide this much-needed information to supplement the development of the national strategy (Gee et al., 2004; Glaeser et al., 2005). Spatial planning and coastal and ocean strategies at the national and EU level are another example of human-nature interactions with respect to paradigm shift in the research and policy linkages—the common thread of this paper.

5   NATURAL CALAMITIES AND COASTAL HAZARDS

A new coastal and ocean research focus on natural calamities and coastal hazards had been a product of an apparent increase of catastrophic coastal events, sparked off by two "marker stone" incidents. The tsunami in December 2004 in South-East Asia, including Indonesia, destroyed hundreds of thousands of lives, homes, and other livelihood assets. Hurricane Katrina in August 2005 devastated much of the low-lying areas of New Orleans, USA (Glaeser, 2008). This section differs from the two previous ones in that it does not report on a complex research program. It reviews documents in light of personal experience in the sense of participatory observation as the author was close to the disasters at stake and struck by the apparent vulnerability of the coastal environment and its population. Personal concern is reflected by a less formal style of writing. The common thread, however, is maintained. Natural calamities and coastal hazards changed coastal focuses and research tremendously. They are another example of human-nature interactions with respect to paradigm shift in the research and policy linkages.

5.1   The Example of the 2004 Tsunami in South-East Asia

On December 26, 2004, the tsunami disaster which affected a number of Indian Ocean nations impacted particularly hard on the northern parts of Sumatra (Indonesia), causing the deaths of well beyond 100,000 people and bringing further heavy destruction to coastal ecosystems and adjacent agricultural lands and infrastructure (Glaeser, 2008). A global natural hazard map is presented in Fig. 3 (National Geographic 2005). The peaks represent the respective hazards. The highest peak arises on the west coast of Sumatra. Fig. 4 highlights the depressing situation in Aceh province. In the aftermath, it became evident that an early warning system linked to improved disaster management and related to more prudent coastal planning and coordination would be required.

The city of Padang is the capital of the province of West Sumatra in Indonesia. Containing a million inhabitants, it is the largest city on the western coast of Sumatra. Padang is said to be the riskiest spots on earth when it comes to natural

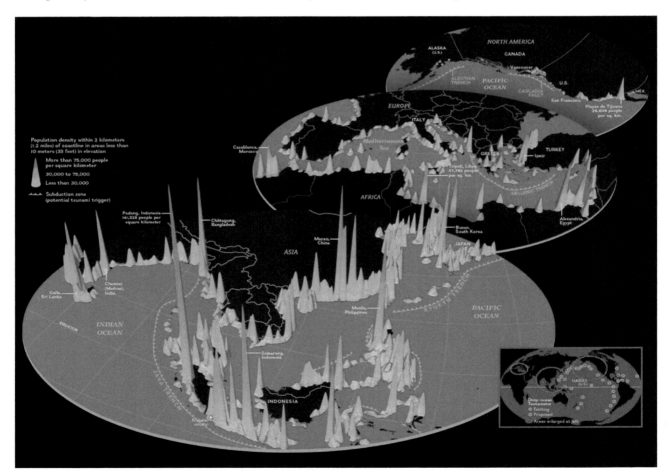

FIG. 3   Natural hazard map in the world. *(Source: National Geographic, April 2005.)*

FIG. 4 Forlorn after the tsunami. *(Source: Febrin Anas Ismail, Padang.)*

hazards (National Geographic, April 2005). Shortly after the tsunami, the Padang community, including the Mayor of Padang, Padang University, and civil society representatives especially the local NGO, KOGAMI (KOmmunitas SiaGA TsunaMI = Tsunami Ready Community) were preparing for the next natural disaster. KOGAMI, a group of local activists and professionals, determined to improve disaster prevention in Padang, has urged strongly that disaster preparedness for the community be accelerated. By early 2006, preparations were already well under way; evacuation exercises had been carried out (Fig. 5), and various stakeholder groups—that is, Indonesian government representatives, nongovernmental civil society organizations, and technical specialists and other academic experts—had begun working together (Communication with Febrin Anas Ismail, Padang, orally and by email). Joint efforts combined forces at different administrative levels as well as a mutual top-down and bottom-up approach.

A German team, representing part of the international community, was asked to join and to contribute to a series of planned disaster preparedness and management workshops that began in November 2006 in Bali and Padang. UNESCO supported KOGAMI in designing a standard operating procedure (SOP) of disaster management for Padang City. The competent Mayor of Padang presented this draft SOP at a national conference in August 2006, designed to motivate all Indonesian province leaders to begin with community preparedness in coastal areas. KOGAMI went one step further by requesting that this informal draft SOP be formalized into legal regulation (Kogami Indonesia, 2006). It was expected that this cooperation,

FIG. 5 Evacuation exercise at G. Pangilun. *(Source: F.A. Ismail, Padang.)*

the joining governance forces at the local, national, and international level, would prove highly useful for local communities and the Indonesian coastal regions. The program was finally launched in October 2008 (Kogami Indonesia, 2008).

The 2004 Indian Ocean tsunami, with a death toll of more than 228,000 people, left more than 2 million people homeless and cost US$8.71 billion in all affected countries: Burma, India, Indonesia, Kenya, Malaysia, Maldives, Seychelles, Somalia, Sri Lanka, and Thailand. Indonesia was hit the hardest, with 130,000 people killed and a damage of US $4.40 billion. The aid response was unprecedented for a natural disaster, with US$6.25 billion donated to a central UN relief fund (The Guardian, 2014).

The tasks ahead would have to be to include disaster preparedness as part of ICM and enforce Europe-Asia cooperation, one continent learning from the experience of the other in an effort to combat natural calamities and to cope with coastal hazards.

5.2 The Example of the 2005 Hurricane Katrina in New Orleans, USA

The catastrophe caused by hurricane Katrina came as no surprise—at least not to scientists and local administrators in the State of Louisiana and the city of New Orleans. Its horrendous impacts were predicted precisely; the exact timing alone remained uncertain. The precursors of hurricane Katrina, Cindy, and Dennis, which had occurred early in July 2005, were the ominous and stern warnings of what was to come. By the end of August 2005, Katrina had wiped the slate clean. This top "category 5" storm had barely slowed to "category 4" when it surprisingly veered to the right, the eye initially bypassing New Orleans, but with sufficient force to cause the levees, protecting that city, to break, allowing the below-sea-level basin to "fill up like a bathtub."

In 2005, the biannual US-based international Coastal Zone Conference was organized in New Orleans (CZ '05: designated "Balancing on the Edge") was held from July 17–21, 2005, little more than a month before Katrina. The participants learned that the French Quarter of New Orleans (Fig. 6) was like an island in a bowl. Once a storm destroys or opens a levy the water pours in and fills the bowl, namely the city of New Orleans. Any storm could do that, there was full risk from category 1 through category 5 storms. All of the local organizers, scientists as well as state and city officials knew that any next storm approaching New Orleans would do the job. This was an accident waiting to happen. The only information not known yet was the date and the name of the storm. This puzzle was solved 5 weeks later. The name was Katrina, a category 5 hurricane. It caused an exodus and took the lives of those who could not escape.

Walter Maestri, director of the Louisiana Emergency Management, on the occasion of the Coastal Zone '05 Conference, described the precarious situation of New Orleans thus:

New Orleans is a bowl, surrounded by levees and water, 6–8 feet below sea level. The levees are 14 feet high. Hurricanes can top 20-foot levees and fill the bowl. In that case, the water has to be pumped out; there is no gravitational drainage. The New Orleans population of 1.2 million people will have to be evacuated, which means that they may have to leave by car. (CZ '05 Local Plenary: Living on the edge, how do we keep the balance? Panel on July 20, 2005, pre-Katrina)

FIG. 6 Coastal zone 2005 in New Orleans. *(Photo: Bernhard Glaeser.)*

One month later, this is what happened: the levees were topped, the bowl flooded, and 1.2 million persons were left to fend for themselves. The levees in New Orleans were built to withstand category two to three storms, but even a category 1 storm could already cause the kind of a problem described. The reason is that there was nothing out there in the Gulf of Mexico to halt the progress of a storm. New Orleans had to take the first blow. Unlike the situation 30 years before, some of the barrier islands that protected the Gulf of Mexico and the coastline of the Gulf States had been lost. Ship Island, for example, had been cut into two by a storm. The coastline had changed. Finally, subsidence occurred; wetlands were inundated with salt water. The French Quarter, today in the heart of New Orleans, was the original settlement and the only high ground in the city.

According to Shea Penland (Pontchartrain Research Institute, University of New Orleans), flood control levees altered the water and sediment distribution. Mississippi River diversion control halted delta switching (repeated shifting of the locus of deltaic deposition), and subsidence destabilized the wetland surface. Out of a total land loss of 691,000 square miles, 36% (the top rank) were due to oil and gas exploration, 26% were destroyed by waves, and 22% were co-opted for multiple hydrological alternatives (CZ '05 Local Plenary: Living on the edge, how do we keep the balance? Panel on July 20, 2005, pre-Katrina).

Carlton Dufrechou (Executive Director, Lake Pontchartrain Basin Foundation) added the following, supplementing Penland's argumentation:

> *Levees are good for economics but destroy Louisiana's foundation. The Mississippi sediments had built and stabilized the coast. Now the coast subsides. Coastal wetlands have been sinking at a rate of 2 inches per annum. If a category 3–4 storm hits New Orleans the immediate problem is how to get people out. The next problem will be diseases. In the medium and long run, New Orleans' colorful past will be lost, and many of its cultural treasures will not be recoverable. The country's infrastructure, its economy will face a tremendous hit. (CZ '05 Local Plenary: Living on the edge, how do we keep the balance? Panel on July 20, 2005, pre-Katrina)*

People in hospitals died. The insurance damage was US $30 billion. The result was total collapse, including the coastal ecosystem, the culture of the city, and the local economy. The final question remained: Will "Big Easy" ever be the same again (Fig. 7)?

The lessons were learned on Cape Cod, a celebrity peninsula near Boston, some 1500 miles north, on the Atlantic coast. Experts claimed that a hurricane could hit any time and predicted a devastating season. Insurance companies tripled their premiums. The Cape Cod Commission knew that evacuation of the population was hardly possible because there were only two bridges leading to the mainland. Shelters were designated and storage units located along highways. They offered cots, blankets, water bottles, and heater meals. An Emergency Preparedness Handbook was composed which could be downloaded from the internet.

FIG. 7   Will "Big Easy" ever be the same again? *(Photo: Bernhard Glaeser.)*

One of the most serious and regrettable outcomes of hurricane Katrina in New Orleans was that hurricane victims had to wait for an unreasonable amount of time for food and water, while relief organizations were stuck on the periphery of the disaster area. What the American Red Cross learned from such badly managed incidents, was to avoid similar situations on Cape Cod by locating trailers and storage containers like portable on demand storage (PODS) system along the major highway. There should be a sufficient number of PODS units, containing cots, blankets, and water bottles, to take care of 8000 people for 1 day. Further, there should be a sufficient number of them to provide "heater meals" (self-heating food packets, each with its own biodegradable heat source) to up to 5000 people for 3 days. An online, downloadable "Emergency Preparedness Handbook" for Cape Cod residents in SLOSH zones could be found in the internet under: http://www.capecodcommission.org (Cape Cod Commission, 2007: accessed April 2007). Earlier in 2006, the American Red Cross Cape Cod Chapter embarked on a $400,000 fundraising campaign to purchase four trailers, five PODS units, and additional supplies (Cape Cod Times, 2006, July 14: A10).

The hurricane warning system for Cape Cod consisted of forecast advisory, hurricane watch, and hurricane warning. Advisory messages were issued at 6-h intervals by the National Hurricane Center for all tropical depressions (wind blows at speeds less than 39 mph), tropical storms (speeds range from 39 to 73 mph), and hurricanes (sustained winds 74 mph or more). The intervals were shortened when landfall was expected. Advisories specified the storm's location, intensity, direction, and speed. Hurricane watch indicated that a hurricane was near. Everybody should be ready to take protective measures if a hurricane warning was issued. This happened when a hurricane was expected within 24 h. All precautions should be taken immediately. In case of an emergency, the Massachusetts Emergency Management Agency, the state police, the governor, or the National Weather Service could activate the Emergency Alert System (EAS). Cape Cod radio stations would monitor the emergency broadcasts and then rebroadcast the information. A handbook compiled by local groups advised on precautions if one decided to stay at home. Hospitals were tied into the weather alert system. It enabled them to get prepared for a hurricane 72 h in advance. Hospital workers had been trained in emergencies. The lessons of hurricane Katrina were thus not lost to other hospitals, according to the Cape Cod Healthcare (Cape Cod Times, 2006, July 15: A12–A13).

Were the lessons really learned? A hurricane survey in April 2005 among 1100 adults in the Atlantic and Gulf coastal states of the United States (interviewed by telephone April 20–26, 2005) revealed that people who live in hurricane prone states—the focus was on Florida—were not ready (see http://www.myhurricanecenter.com/2005survey.php, accessed April 2007). Moreover, as in the Indonesian tsunami case, warning systems may alert and prepare people, but they cannot prevent storms. Only a few months after the storm disaster in Louisiana, an unpredicted freak winter gale hit Cape Cod on December 9, 2005, which deroofed houses, devastated gardens, and smashed stone walls (Cape Cod Times, 2006, July 15: A13, and personal communications with local inhabitants).

In 2007, a few Cape Cod towns held low-key simulation drills concerning emergency preparedness. Barnstable County's Regional Emergency Planning Committee had been working to create and implement a viable plan for response to emergency situations on the Cape and Islands. A plan was prepared to be instituted in 2008 (Barnstable Emergency Plan, 2007).

It is concluded that lessons from experience are to be learned to improve risk management. When it comes to disasters and disaster management, scales play a role. Appropriate governance links the local to the national and the supranational level. This includes good networking, open lines of communication, competent well-informed authorities with clearly assigned tasks, and decision-making responsibilities, in the right place at the right time to manage and coordinate activities efficiently. While climate change, sea-level rise, and weather monitoring need to take account of the global scale, its application and the measures to be taken are local, coordinated nationally and funded by national and sometimes international aid. Hence, it is important to develop national and international strategies to manage coasts. Disaster preparedness needs to be integrated into coastal management. The coastlines are the first to be affected by sea-level rise, a global phenomenon, or by storm floods and torrential rains, which happen repeatedly in Europe as elsewhere. The Cape Cod story is to be seen as an appendix only to the New Orleans story and provides some lessons learned far away from New Orleans.

6   CLIMATE CHANGE: THE EXAMPLE OF INDONESIA

Sea-level rise, larger storm surges, or torrential rains may be caused by climate change.

Climate change is followed by environmental, economic, and social change. Ecosystems lose their provisioning capacity resulting in less fishing and other sources of income. Fisher families may disintegrate because the fishermen as main bread winners may be replaced by their kids who work in tourism. Climate impacts are often poverty, social injustice, and uneven distributions of wealth and interests. Benefits and costs are not evenly divided either.

A climate change example with some of its repercussions is taken from Indonesia (Glaeser and Glaser, 2010). The Spermonde Archipelago is located in south-west Sulawesi, off the old merchant and harbor city of Makassar. Round about 100 mostly small and low-lying islands, some without fresh water supply constitute the archipelago (Glaeser et al. 2018). Their existence may be threatened by climate change. Increased storm surges are observed, which makes fishing ever more hazardous. Island flooding destroys streets and houses (Fig. 8). Villagers on one of the islands witnessed coastal threats and painted a gloomy picture: The whole island was hip-deep under water. Increased beach and coastal erosion began to change the size and shape of the island, resulting in loss of island surface areas and shoreline buildings (Fig. 9). More dangerous weather situations, destructive storms, and higher waves, made fishing more difficult, arduous, and less productive.

The author cooperated with Science for the Protection of Indonesian Coastal Ecosystems (SPICE), an Indonesian-German research program, funded by the German Ministry for Education and Research (BMBF 2007–2010, Grant No. 03F0474A). Two partners were the Leibniz Center for Tropical Marine Research (ZMT) in Bremen and the University of Makassar "UNHAS," the Coral Institute Marine Coast and Small Islands (MaCSI) in particular. SPICE research addressed

FIG. 8   Public toilets washed away. Bone Tambung Island. *(Photo: Bernhard Glaeser.)*

FIG. 9   Beach erosion. Lanyukan Island. *(Photo: Bernhard Glaeser.)*

social and ecological issues related to the management of the Indonesian coastal ecosystems and their resources. It included several field trips and involved Indonesian and German researchers and students. Social science methods applied were, among others, a quantitative household survey, participant observation, questionnaire development for local stakeholders, and qualitative in-depth interviews.

The overall picture was slightly depressing. Fish resources have been depleted, apparently not a recent problem (Erdmann, 1995). A 65-year-old Saugi (a Spermonde islander) fisherman phrased it like this: "Crabs are much smaller now. Formerly, four to five crabs weighed 1 kg, now it takes 12 crabs to reach that weight. Whereas the small crabs used to be thrown back, now everything is taken. During earlier times, fishers went for fishing only when the weather was good. Now, they go every day, even when it storms. In those days when storms hit fishers would throw the anchor and stay. Nowadays, the boat would sink. This proves that the waves are higher now." Climate change effects explained in simple words (Glaeser and Glaser, 2010).

In general, none of the interviewees had heard of climate change or sea-level rise. When asked, "What would you do if the waves eat your island?," they responded that their solution to such calamity would be to move to the mainland, live with relatives, or call for help from the local government. On some islands, wave breakers have been constructed. Many informants held grave concerns about erosion and flooding. Population pressure has increased tremendously. When the local midwife had arrived on one of the island in the 1940s as a child there existed "not more than seven houses and a jungle of trees, no floods and no erosion." Many local inhabitants acknowledged that if continue to adopt the current fishing practices would result in the degradation if not destruction of the reefs.

Among the interviewed, a strong perception prevailed that flooding events are becoming more common and severe. However, informants did not identify climate change as the cause of either erosion or of increased flooding. In fact, sea-level rise is threatening the islands. Houses have been destroyed, already. People moved to relatives or left their island. There are few sources for income offered or found. A vision or concept for future development, which would take into account the threats arising due to climate change seem to be nonexistent. The respondents simply hoped for more fish in the future. The conviction of a God-given destiny is ever present.

At the regional (subnational) level, coastal fishermen in small boats in different areas of the globe, including Indonesia, are risking their lives in climate change-related worsening storms. Internationally, the distribution of the relative costs and benefits of climate change between richer and poorer continents and nations is sufficiently known. A climate divide is apparent and growing between nations, between central and peripheral regions and between wealthier and impoverished marginal communities and households at the local level (Glaeser and Glaser, 2010).

In this Indonesian Spermonde study, the costs and sufferings at the local household level incurred in the course of climate change enhancing activities from the higher level of the climate divide hierarchy accrue at the lower levels of the hierarchy: to the residents on the poorer islands in Spermonde who were least likely to access alternatives in the form of eroding their shores and houses; as to the riverside residents in Riau, Sumatra, whose daily lives were hampered by polluted domestic water sources and dying food fish produced through interlinked deforestation and industrialization.

A former Indonesian Fisheries Minister, Freddy Numberi, advocated a policy shift to focus the country's development on the oceans: "An ocean policy, made with the cooperation of all stakeholders ..., will set us on the right course to realize the full potential of our marine resources" (Resource Focus, 2009, 4). It is not clear, however, where Indonesia is heading. On the one hand, if the ocean is a sink there is money to be made by means of international climate change agreement, which is a great attraction of course. On the other hand, political leaders call for care of the environment and decry the social consequences of climate change. Distribution and equity need to be linked to livelihoods and adaptation to climate change. Moreover, local knowledge and local culture are needed to protect coastal ecosystems and their resources. Both are endangered by climate change including sea-level rise and by other global developments, such as increasingly mobile international fisheries operations, including live reef food fisheries, ornamental fishing, and trawling (Glaeser and Glaser, 2009).

The section presents an attempt in multilevel social-ecological research. It links global and local issues, using case studies from Indonesia as a focus for a discussion of national policy and governance approaches and to illustrate how these relate to livelihoods and to CM resource management. Climate change is a major aspect of global including environmental change. Both are linked to the economic, social, and cultural dimension of change. Observations in Indonesia (Sumatra and Sulawesi) show that globalization and climate change produce repercussions on local coastal developments and livelihoods. The Indonesian government has set the stage for linking ocean developments and coastal threats to climate change. It also seems that the contemporary "climate divide" represents a new version of the old conflict between the developed and the developing world on environmental issues, globally as well as locally. The Indonesian study provides an example of the human-nature interactions at the local level with respect to paradigm shift in the research and policy linkages at the regional and national level, all influenced by driving forces at the global level. Multilevel, interdisciplinary social-ecological research is advocated in order to explore feedbacks between global change and local livelihood dynamics (Glaeser and Glaser, 2010).

7 COASTAL AND OCEAN TYPOLOGIES: AN ANALYTICAL INSTRUMENT AND PLANNING TOOL

More recently, multilevel social-ecological research has been taken from its empirical grounding to a more theoretical level to comparative and structural approach.

A consolidation can be observed in international research: an attempt to filter, to differentiate, and to unify the accumulated knowledge relating to coasts and oceans as an integrated whole. Coastal and ocean typologies, combining social and ecological factors, have gained importance. These typologies serve two purposes: they are developed as an analytical instrument and they serve as a governance tool in coastal and ocean planning and policy.

As the Anthropocene proceeds, regional and local sustainability problems are ever more likely to originate at multiple levels and scales of the earth system, including the global. The rate of global environmental change is now vastly outpacing our policy response. SESs analysis needs to pay attention to these changes in order to support environmental governance appropriately. This governance approach initiated the development of a coastal social-ecological typology and applied it in an exemplary fashion to nine CM case studies. A problem or issue-specific approach to the delineation of social-ecological units is the point of departure. Multiple levels and scales in CM SESs are reviewed. A current major challenge to SESs analysis is the issue specific identification of those key cross-level and cross-scale interactions and connectivities which play major roles in shaping CM social-ecological dynamics, patterns, and outcomes (Glaser and Glaeser, 2014).

An attempt is made to develop a CM social-ecological typology. An explicitly regional (subnational) focus is taken to explore how a regionally grounded, multi-scale analysis may support multilevel local to global sustainability efforts. The Indonesian case study (previous section) exemplifies this approach. Social-ecological sustainability problems, caused by drivers at different earth system levels, lead the way into the proposed typology. A SES consists of a biogeophysical territory, an identified issue and the associated social agents. It can extend across disciplines as well as across institutional levels and scales. A global sustainability research matrix, which is based on ecozones and problem types, can thus be constructed and serves as a research driven multilevel typology. The regional application links directly to stakeholder agendas at the problem level. It is argued that some of the central functions of CM SESs are resource provision, livelihood access, and storm and erosion protection, which need special attention in a CM social-ecological typology, as exemplified in the Indonesian case study used (Glaeser, 2016).

A two-dimensional (2D) hierarchical typology based on ecozones and problem types—with the Indonesian case taken as a specimen that links climate zones and ecozones (from tropical to polar) to several hierarchically nested levels of action (from global to local). The various cells contain independent and dependent social-ecological variables. The model starts from global change, including the economic and sociocultural drivers of change and the climatic pressure, which is exerted upon the climatically differentiated ecosystems. At the national level, a political and administrative response to the felt pressure may link the global and the local level. It is, finally, at the regional and the local level where the ecological and social impacts of global change are encountered. This is the place for political and administrative measures or for responses by the affected population itself. The responses to hazardous or threatening situations reach from passive adaptive strategies, such as migrating, to proactive responses, such as coastal protection, change of professions, or educational efforts.

The 2D multilevel global typology can be used to analyze CM SESs in a comparative way. It can classify the solicited case studies with respect to place-specific features under the conditions of climate change, employing natural and social science knowledge systems (Table 1; Glaeser, 2012).

Exploratory research is used to propose steps toward a multilevel typology. It is extended to the social-ecological subsystems such as natural, social, governance and applied to additional cases. The 2D hierarchical typology is proposed as a tool to analyze, compare, and classify CM systems. A subsystems appraisal typology is meant to evaluate action results.

The typology approach may be seen as a stepping stone in theory development. The CM typology is 2D, horizontally combining social and ecological parameters. It also contains a vertical component by stretching into different levels: tts nature is multilevel hierarchical. Empirical case studies are used as examples. This means: while the development of the typology is conceptually oriented, it is research driven at the same time. The next step is to subdivide the social-ecological hierarchical typology. The subsystems are the natural, the social, and governance. Governance consists of three elements: government, the markets, and civil society, including communities of all levels and nongovernmental organizations (Table 2).

An appraisal typology (Table 3) evaluates responses and achievements of measures taken by the governance system, first by local and regional-level civil society measures at the village, town, or district level. Government actions at the provincial and national level would then also have to be included. Markets for fish and other resources at different levels will certainly play a role and influence outputs and outcomes. In all three governance segments (government, market, and civil society), appropriateness, effectiveness, and acceptability of outputs and outcomes are appraised and valued.

TABLE 1 Two-Dimensional Research-Based Hierarchical Typology: An Example of Indonesia

Social-ecological multi-level typology for coastal and marine systems: With respect to-specific features under the condition of climate change Indonesia (SPICE case study) as a tropical zone example

Spatial Scale/ Level	Ecozones/Climate Zones			
	Temperate	Tropical/Subtropical		Polar
	Social Ecological	Social Ecological		Social Ecological
Global		Global climate and environmental change, linked to the economic, social, and cultural dimension of change		
National		National (e.g., Indonesian) coastal, ocean and environmental policies, linking ocean developments and coastal threats to climate change		
Regional and local		Local (e.g., Sulawesi) coastal developments and livelihoods. Social: Earning opportunities for women, including aquaculture (sea cucumbers, fish cages, algae cultivation). Ecological: Beach protection, mangrove plantation, reef management, and rehabilitation to reduce beach erosion.		

Source: Adapted from Glaeser B, 2012. Klimawandel und Küsten. In: Weixlbaumer N. (Ed.), Anthologie zur Sozialgeographie. Abhandlungen zur Geographie und Regionalforschung (Vienna), 16, 68-82 (in German): 78.

TABLE 2 Comparative Subsystems Typology: The Example of Indonesia (Design: Bernhard Glaeser)

Systems Features	Natural System	Social System	Governance System
Objectives	Stock preservation	Decent livelihood	Equitable regulations
System specific issues	Climate change	Overfishing	Regulations not implemented
Scale: system boundary	Ecological boundaries	Kinship, trade, market boundaries	Administrative boundary
Scale overlaps	Fishing	Markets	Rules/regulations
Level	Local to international	Local to global	Regional to national
Change major changes/impacts	Storm surges and beach erosion	A patron-client relation evolved among traders and fishermen	Change of administration, markets, civil society
Responses: governance subsystem (government, market, civil society)	Government regional to national, markets no, fisher community partly	Government little effort, Markets local to global, Fisher community and patron-client relation	Government national, Markets no, NGO's national and international
Appraisal: outputs, outcomes, ensuing goals and visions	Enabling conditions, changed behavior, achievements, sustain development	Enabling conditions, changed behavior, achievements, sustain development	Enabling conditions, changed behavior, achievements, sustain development

For purposes of generalization, it is necessary to look beyond the Indonesian case, which has helped set up 2D research-based typologies. More case studies are needed to check and control the viability of the typologies suggested as comparative tools. This has been done in a trial mode by using nine cases selected by IMBER Human Dimensions group members and one in this volume. Table 4 shows a 2D research-based typology representing types of SESs (coastal/islands, shelf/ open sea, river mouth, urban) and climate zones (temperate, tropical/subtropical, and polar). An additional, specifying level is inserted to identify the major problem foci (e.g., mass mortality, ocean temperature, and settlement density). In all, 10 researched cases include 6 IMBER affiliated Human Dimensions Working Group (HDWG) case studies (template for the description of IMBER ADApT case studies, completed by Cooley, 2013, Defeo, 2013, Glaeser et al., 2013, Guillotreau,

TABLE 3 Subsystems Appraisal Typology: An Example of Indonesia

Climate Zone, Ecoregion, SES			Outputs & Outcomes at the Case Study Level: Across the Natural, Social & Governance System				
Climate zone	Eco-system type	SES type	Outputs (Immediate results of responses)	Outcome: First order (enabling conditions)	Outcome: Second order (behavior change)	Outcome: Third order (vision achieved)	Outcome: Fourth order (sustainable development)
Polar	Shelf	Estuary					
Tropical	Coastal	Island	Local people construct wave breakers, use coral stone as building material which, led to a stop of government support. Governmental measures: Dams have been constructed, to reduce waves, to prevent coastal erosion and to establish harbors for fishing boats.	Fishers develop intricate patron-client links and informal rules on resource use. The patron-client system provides fishers with an alternative type of social security, allowing them to cope with short-term stressors (storms, loss of gear, localized overfishing). However, capacity to cope with more extensive stressors (complete loss of ecosystem services: fisheries resources, protection from erosion) appears limited Few alternative livelihood options. low level of education (Glaser et al., 2010)	Locally enforced rules for a specified but small sea territory surrounding the respective islands in the whole Archipelago. These regulations provide rules-in-use for marine resource use, creating a polycentric structure for area-based marine governance. However, their formation & enforcement depend on local leaders, which—in conjunction with the missing horizontal connectivity creates vulnerability	Informal governance system: NGOs are advocating for no-take zones, introduce 1. new technologies (coral management, mariculture cultivation of sea weeds), 2. credits to empower self-employment in mobile restaurants (soups, snacks, drinks) or shops (drinks, snacks, tools). Introduction of COREMAP (Coral Reef Rehabilitation and Management) program) to improve environmental awareness, alternative livelihood options, establish community-based MPAs (Ferse et al., 2014)	Not yet achieved
Tem- perate	Open Sea	Urban					

SES=Social-ecological systems.

Design: Bernhard Glaeser. Sources: Glaeser, B., Ferse, S., Gorris, P., 2018. Fisheries in Indonesia between livelihoods and environmental degradation: coping strategies in the Spermonde Archipelago, Sulawesi. In: Guillotreau, P., Bundy, A., Perry, R.I. (Eds.); Global Change in Marine Systems. Integrating Natural, Social and Governing Responses. Routledge, Studies in Environment, Culture, and Society: London and New York. Ferse, S.C.A., Glaeser, M., Neil, M., Schwerdtner Máñez, K., 2014. To cope or to sustain? Eroding long-term sustainability in an Indonesian coral reef fishery. In: Glaser, M., Glaeser, B. (Eds.): Linking Regional Dynamics in Coastal and Marine Social-Ecological Systems to Global Sustainability. Reg. Environ. Change 14(6), 2127–2138 (special issue). Glaser, M., Baitoningsih, W., Ferse, S.C.A., Neil, M., Deswandi, R., 2010. Whose sustainability? Top-down participation and emergent rules in marine protected area management in Indonesia. Mar. Policy 34, 2053–2066.

TABLE 4 Two-Dimensional Research Based Problem Focus Typology: Ten Case Studies

Case Study Typology (IMBER-HDWG): For the analysis of CM-SES change			
SES Types	Climate Zones/Ecoregions: Problem Foci		
	Temperate	Tropical/Subtropical	Polar
	Social Natural Governance	Social Natural Governance	Social Natural Govern
Coastal, Small Islands	S Cooley: USA, NW Pacific oyster – Mass mortality – Ocean acidification – Unemployment – Cultural identity P Guillotreau: France, W Oyster (*Crassostrea gigas*) – Mass mortality – Increasing sea temperature O Defeo: Uruguay, Atlantic Sandy beach yellow clams – Mass mortality – Increasing sea temperature – Unemployment	B Glaeser: Indonesia, Sulawesi Island development and livelihoods – Overfishing – Reef destruction – Coastal erosion – Less income H Tran: Vietnam, Tran Van Thoi Paddy/agriculture aquaculture – Climate change – Population pressure – Settlement density – Mangrove decrease – Past war	AL Lovecraft, CL Meek: pan-Arctic coasts Environmental management of social-ecological systems – Warming cryosphere (permafrost degradation) – Coastline collapse – Thermal erosion – Reduced food sovereignty – Cultural values devalued – Loss of fate control for nomadic groups – Arctic: not a governance entity, but loose system of eight nation-states
Shelf, Open sea	M Isaacs: S. Africa, W Small pelagic fishery (sardine, anchovy) – Shift in catch abundance – Unemployment		
River mouth, estuary, lagoon	M Canu et al: Italy, Venice Lagoon Clam (*Tapes philippinarum*) – Pollution – Social conflict – Illegal fishing	J Amparo et al: Philip-pines, MMO River System Fishery – Water/fish quality – Heavy metal contamination – Health risk	
Urban	M Makino: Japan, Tokyo Bay Shrimp fishery – Heavy industry development – Marine pollution – Water quality – Fish depletion		

IMBER-HDWG=Integrated Marine Biogeochemistry and Ecosystem Research—Working Group on Human Dimensions.
CM-SES=Coastal and Marine Social-Ecological Systems.
Design: Bernhard Glaeser

2013, Isaacs, 2013, Makino, 2013; for details see Bundy et al., 2016). Four external contributions are added (Amparo et al., 2014, Canu and Solidoro, 2014, Tran, 2014, Lovecraft and Meek, 2018). All 10 cases follow identical criteria and goals: to classify coast- and marine-related changes, to identify SES responses, and to assess long-term outcomes. They are applied to three social-ecological subsystems: the natural, the social, and the governance subsystem (Table 4; Bundy et al., 2016).

The case studies selected originate in three climate zones, temperate, tropical, and polar. As for the SESs involved, all types are represented, including an urban case. The focus is predominantly on mariculture and different fisheries, but includes also environmental governance, island development, and paddy rice. The problems encountered feature quite a variety of issues and deal—among others—with ocean acidification, pollution, water quality, health hazards, and gendered livelihood problems. The conclusion is to widen the geographical scope, include polar studies and generally invite more case studies to provide a more complete picture.

Why is it important or useful to develop and apply typologies? The scientific reasoning is to have at hand a comparative tool to analyze CM types, beginning with geological formations including ecological variations and ending with social specifics. From the administrative point of view, social-ecological typologies are to be employed as decision support tools to generate responses to real-world problems in coastal areas. From the methodological side, it ought to be added that typologies need a large number of cases, which means, a whole variety of empirical details. At the same time, the number of types, clusters, or cells should be limited in order to maintain the applicability of a typology to real-world issues and to find comparable cases and specific differences. *In toto*, the common thread of the paper, human-nature interactions with respect to paradigm shift in the research and policy linkages, culminated in a self-reflexive review of the field to strengthen its theoretical and methodological basis.

8 SUMMARY AND CONCLUSION

This contribution featured two decades of COM research and introduced ICM as a science and management approach which emerged in the 1970s. The initial goal was to identify coastal stakeholders to resolve their conflicts and to achieve sustainable coastal development. Later on, the issue of governance gained importance and in this regard, national and international coastal and ocean strategies were devised. In the meantime, however, the coastal hazards theme has come to dominate much of the coastal management in certain world regions. This shift in attention has surely come about because of an apparent increase in the frequency of catastrophic coast-related events, such as the December 2004 tsunami that devastated large parts of Southeast Asia, or the August 2005 hurricane Katrina, which destroyed or made uninhabitable much of the city of New Orleans. Disaster research and risk management have begun to receive greater consideration and sharpened focus, leading, for example, to increased collaboration efforts between Asia and Europe.

Climate change and sea-level rise have repercussions on income and livelihoods of coastal populations. The worldwide near collapse of coastal fisheries affects poor and impoverished communities at the local level, but at a global scale. It includes livelihoods in the North as in the South and serves as an indicator for global environmental problems. The conclusion is to call for sustainable COM. Coastal and ocean typologies promise to be of use in this respect. They are developed as a comparative analytical instrument on different administrative levels, extending from the local to the global and back. Such typologies may be applied and serve as a political planning tool, approaching social and ecological issues to solve problems in different parts of the world, independent of climate zones or political systems. Empirical case studies are growing, eventually, into theory building exercises, accompanied by a connected planning approach. The common thread, human-nature interactions with respect to paradigm shift in the research and policy linkages, has been taken to different fields and sites. Reviewing 20 years of coast and ocean research witnesses the maturation of a field.

REFERENCES

Ackefors, H., Grip, K., 1995. The Swedish model for coastal zone management. In: Report 4455. Swedish Environmental Protection Agency, Stockholm.

Amparo, J.M.S., Geges, D.B., Jimena, C.E.G., Malenab, M.C.T., Mendoza, M.E.T., Saguigui, S.L.C., Visco, E.S., Mendoza, M.D., Ibanez Ibanez, L.O., 2014. Case Study Philippines, Meycauayan—Marilao—Obando River System: pollution. Template for the description of ADApT_A case studies. IMBER HD-WG (unpublished).

Barnstable Emergency Plan, 2007. Available from: http://www.barnstablecounty.org/BulletinBoard/bcrepc@barnstablecounty.org. (Accessed May 2007).

BMU, 2002a. Strategie der Bundesregierung zur Windenergienutzung auf See im Rahmen der Nachhaltigkeitsstrategie der Bundesregierung, Januar 2002 (in German). Available from: http://www.bmu.de/files/windenergie_strategie_br_020100.pdf (Accessed 20 March 2006).

BMU, 2002b. Weiterer Ausbau der Windenergienutzung im Hinblick auf den Klimaschutz—Teil 1. Erarbeitet von Deutsches Windenergie-Institut Wilhelmshaven, November 2002 (in German). Available from: http://www.bmu.de/de/800/js/download/b_offshore02/ (Accessed 13 February 2006).

BMU, 2002c. Internationale Aktivitäten und Erfahrungen im Bereich der Offshore-Windenergienutzung. Deutsche WinGuard GmbH, Varel, Januar 2001 (in German). Available from: http://www.bmu.de/de/800/js/download/b_offshore02/ (Accessed 13 February 2006).

BMU, 2006. Integriertes Küstenzonenmanagement in Deutschland: Entwurf für eine nationale Strategie für ein Integriertes Küstenzonenmanagement (13 February 2006) (in German).

Bundy, A., Chuenpagdee, R., Cooley, S.R., Defeo, O., Glaeser, B., Guillotreau, P., Isaacs, M., Mitsutaku, M., Perry, R.I., 2016. A decision support tool for response to global change in marine systems: the IMBER-ADApT framework. Fish Fish. 17 (4), 1183–1193. https://doi.org/10.1111/faf.12110.

Canu, D.M., Solidoro, C., 2014. Case study Italy, Venice Lagoon: pollution. Template for the description of ADApT_A case studies. IMBER HD-WG (unpublished).

Cape Cod Commission, 2007. Emergency preparedness handbook. Available from: http://www.capecodcommission.org (Accessed April 2007).

Cape Cod Times 2006, July 14 and 15.

Cooley, S., 2013. Case study USA. North-west coast: Pacific oyster. Template for the description of ADApT_A case studies. IMBER HD-WG (unpublished).

Defeo, O., 2013. Case study Uruguay. Atlantic coast: Sandy beach yellow clams. Template for the description of ADApT_A case studies. IMBER HD-WG (unpublished).

Ellegard, A., 1998a. Mussel culture at stake: identifying the holders. SUCOZOMA 1.1, working paper, Göteborg University, Human Ecology Section, Göteborg, Sweden.

Ellegard, A., 1998b. Mussels in some Swedish media. SUCOZOMA 1.1, working paper, Göteborg University, Human Ecology Section, Göteborg, Sweden.

Erdmann, M., 1995. An ABC Guide to Coral Reef Fisheries in Southwest Sulawesi, Indonesia. NAGA, The ICLAM Quarterly (April)4–6.

Europäische Kommission, 1999. Schlussfolgerungen aus dem Demonstrationsprogramm der Europäischen Kommission zum integrierten Küstenzonenmanagement (IKZM). Amt für Veröffentlichungen der Europäischen Gemeinschaften, Luxemburg. (in German).

Europäische Union, 2002. Empfehlung des Europäischen Parlamentes und des Rates vom 30. Mai 2002 zur Umsetzung einer Strategie für ein integriertes Management der Küstengebiete in Europa (2002/413/EG). Amtsblatt der Europäischen Gemeinschaften L 148/24 vom 6. Juni 2002 (in German).

Ferse, S.C.A., Glaser, M., Neil, M., Schwerdtner Mánez, K., 2014. To cope or to sustain? Eroding long-term sustrainability in an Indonesian coral reef fishery. In: Glaser, M., Glaeser, B. (Eds.), Linking Regional Dynamics in Coastal and Marine Social-Ecological Systems to Global Sustainability. Reg Environ Change 14 (6), 2127-2138 (special issue).

Resource Focus, 2009. RI (Republic of Indonesia) still underestimates marine potential. The Jakarta Post February 26, 4.

French, P.W., 1997. Coastal and Estuarine Management. Routledge, London and New York.

Gee, K., Kannen, A., Glaeser, B., Sterr, H., 2004. National ICZM strategies in Germany: a spatial planning approach. In: Schernewski, G., Löser, N. (Eds.), BaltCoast 2004—Managing the Baltic Sea. Coastline Reports, 2004-2. pp. 23–33.

Gee, K., Kannen, A., Licht, K., Glaeser, B., Sterr, H., 2006. Integrated coastal zone management (ICZM): strategies for coastal and marine spatial planning (Az: Z6-4.4-02.119). In: Final report: The role of spatial planning and ICZM in the sustainable development of coasts and seas. Federal Ministry of Transport, Building and Urban Affairs (BMVBS) and Federal Office for Building and Spatial Planning (BBR), Berlin.

Glaeser, B., 1999. Integrated coastal zone management in Sweden: assessing conflicts to attain sustainability. In: Salomons, W., Turner, R.K., de Lacerda, L.D., Ramachandran, S. (Eds.), Perspectives on Integrated Coastal Zone Management. Springer, Berlin, pp. 355–375.

Glaeser, B., 2000. Coastal management and sustainability in Baltic East Germany: learning from Scandinavia? In: Pacyna, J.M., Kremer, H., Pirrone, N., Barthel, K.-G. (Eds.), Socioeconomic Aspects of Fluxes of Chemicals into the Marine Environment. European Commission, Community Research, Project Report. Series: Energy, Environment and Sustainable Development, Office for Offical Publications of the European Communities, Luxembourg, pp. 113–126.

Glaeser, B., 2002. Linking partners in joint coastal management research: Strategies toward sustainability. In: Schernewski, G., Schiewer, U. (Eds.), Baltic Coastal Ecosystems: Structure, Function and Coastal Zone Management. Series: Central and Eastern European Development Studies (CEEDES), Springer, Berlin, pp. 353–362.

Glaeser, B., 2004. The social science responses to new challenges for the coast. In: Schernewski, G., Dolch, T. (Eds.), Coastline Reports 2004-1. EUCC-The Coastal Union, Warnemünde, pp. 201–211.

Glaeser, B., 2008. Integrated coastal management (ICM) between hazards and development. In: Krishnamurthy, R., Kannen, A., Ramanathan, A., Tinti, S., Glavovic, B., Green, D., Han, Z., Agardy, T. (Eds.), Integrated Coastal Zone Management: The Global Challenge. Research Publishing, Singapore and Chennai, pp. xiii–xxi.

Glaeser, B., 2012. Klimawandel und Küsten. In: Weixlbaumer, N. (Ed.), Anthologie zur Sozialgeographie. Abhandlungen zur Geographie und Regionalforschung (Vienna), 16, 68-82 (in German).

Glaeser, B., 2016. From global sustainability research matrix to typology: a tool to analyze coastal and marine social-ecological systems. Reg. Environ. Chang. 16 (2), 367–383. https://doi.org/10.1007/s10113- 015-0817-y.

Glaeser, B., Ferse, S., Gorris, P., 2013. Case study Indonesia. Spermonde Archipelago: Island development & livelihoods. Template for the description of ADApT_A case studies. IMBER HDWG: Version 2013-05-17 (unpublished).

Glaeser, B., Ferse, S., Gorris, P., 2018. Fisheries in Indonesia between livelihoods and environmental degradation: coping strategies in the Spermonde Archipelago, Sulawesi. In: Guillotreau, P., Bundy, A., Perry, R.I. (Eds.), Global Change in Marine Systems. Integrating Natural, Social and Governing Responses. Routledge: Studies in Environment, Culture, and Society, London and New York.

Glaeser, B., Gee, K., Kannen, A., 2005. Germany going coastal: the national ICZM strategy. In: U.S. National Oceanic and Atmospheric Administration (NOAA). Coastal Service Center. 2005. Coastal Zone 05. In: Proceedings of the 14th Biennial Coastal Zone Conference, New Orleans, Louisiana, 17–21 July 2005. NOAA/CSC/20518-CD. CD-ROM. NOAA Coastal Services Center, Charleston, SC.

Glaeser, B., Glaser, M., 2009. Trip to Indonesia in February/March 2009. . LOICZ Reports INPRINT 1/09, 30-33.

Glaeser, B., Glaser, M., 2010. Global change and coastal threats: the Indonesian case. Human Ecol. Rev. 17 (2), 135–147.

Glaeser, B., Grahm, J. (Eds.), 1998. On Northern Shores and Islands. Human Well-Being and Environmental Change. Humanekologiska Skrifter 16 (Göteborg University, Human Ecology Section). Kompendiet, Göteborg.

Glaeser, B., Kannen, A., Kremer, H., 2009. The future of coastal areas. Challenges for planning practice and research. Gaia 18 (2), 145–149.

Glaser, M., Baitoningsih, W., Ferse, S.C.A., Neil, M., Deswandi, R., 2010. Whose sustainability? Top-down participation and emergent rules in marine protected area management in Indonesia. Mar. Policy 34, 2053–2066.

Glaser, M., Glaeser, B., 2014. Towards a framework for cross-scale and multi-level analysis of coastal and marine social-ecological systems dynamics. In: Glaser, M., Glaeser, B. (Eds.), Linking Regional Dynamics in Coastal and Marine Social-Ecological Systems to Global Sustainability. Reg. Environ. Change 14 (6), 2039-2052 (special issue). doi:https://doi.org/10.1007/s10113-014-0637-5

Guillotreau, P., 2013. Case study France. Atlantic coast: Oyster (*Crassostrea gigas*). Template for the description of ADApT_A case studies. IMBER HD-WG (unpublished).

Isaacs, M., 2013. Case study South Africa. West coast: small pelagic fishery. Template for the description of ADApT_A case studies. IMBER HD-WG (unpublished).

Knowlton, N., 2010. Citizens of the Sea. Wondrous Creatures from the Census of Marine Life. Washington, DC, National Geographic.

Kogami Indonesia, 2006. Meeting with the mayor of Padang. Email to bglaeser@wz-berlin.de from multiply@multiply.com on August 2, 2006.

Kogami Indonesia, 2008. Program launching of KOGAMI. Email to bglaeser@wzb.eu from multiply@multiply.com on October 21, 2008.

Lovecraft, A.L., Meek, C.L., 2018. Arctic coastal systems: evaluating the DAPSI(W)R(M) framework. . This book.

Makino, M., 2013. Case study Japan. Tokyo Bay: Shrimp fishery. Template for the description of ADApT_A case studies. IMBER HD-WG (unpublished).

National Geographic, 2005. Geographica: Tsunamis—Where Next? Vol. 207/4. US edition, no page number. German edition, 18-19. Source: Landscan global population database, Oak Ridge National Laboratory, NG maps.

The Guardian, 2014. Available from: https://www.theguardian.com/global-development/ng-interactive/2014/dec/25/human-financial-cost-indian-ocean-tsunami-interactive. (Accessed June 2018).

Tran, H. H., 2014. Case study Vietnam. Tran Van Thoi district: climate change. Template for the description of ADApT_A case studies. IMBER HDWG (unpublished).

Chapter 38

Megacities and the Coast: Global Context and Scope for Transformation

Sophie Blackburn*, Mark Pelling*, César Marques[†]

*Department of Geography, King's College London, London, United Kingdom, [†]National School of Statistical Science—Brazilian Institute of Geography and Statistics (ENCE/IBGE), Rio de Janeiro, Brazil

1 INTRODUCTION

This chapter explores the dynamic interactions between mega-scale urbanization and the coastal environment. The coastal zone is characterized by high biological diversity and highly dynamic geomorphology. It also offers a range of opportunities for industry, including access to valuable geological resources (particularly oil and gas reserves), national and international trade routes, and opportunities for tourism. As a result of these wide-ranging economic, ecological, aesthetic, and scientific values, human settlement in the coastal zone has long been intensive. Recent estimates suggest that in more than half of the world's coastal countries, at least 80% of the national population currently lives within 100 km of the coastline (Martínez et al., 2007). Global population projections suggest that, even under low-growth assumptions, the population living in low-elevation coastal zones could rise to 625 million in 2030 (Neumann et al., 2015).

Urbanization in the coastal zone has a wide range of intersecting environmental and human impacts. Whilst the manner of urban development and intensification is a critical driver of environmental transformations (e.g., coastal erosion, habitat destruction, and ecosystem destabilization), the direct and indirect outcomes of those environmental shifts in turn have dramatic ramifications for human life and industry (e.g., depletion of ecosystem services, impacts on environmental well-being). Urbanization on the coast also leads to increased exposure of human life and assets to the wide range of coastal hazards, including storm surge, cyclones and tsunami, as well as environmental health concerns. In the context of global environmental change, this interaction between human and geo-ecological spheres is facing added pressures—including resource depletion, environmental stress, sea-level rise, and shifts in hazard distribution, timing, and magnitude.

These cumulative and compounding risks are felt most keenly in the world's coastal megacities, defined here as urban agglomerations with more than 10 million inhabitants, located within 50 m elevation and 100 km distance of mean high water. The intense and dynamic pressures felt in those cities make social transformations toward sustainable development increasingly urgent, yet also highly challenging owing to the complexity of megacity governance. This chapter discusses the global geographic location and characteristics of mega-urban environments, before exploring in more detail the risks and opportunities for resilience and transformation that are found there. The chapter adopts an explicitly interdisciplinary approach, building on a global synthesis of environmental and social scientific literature (see Pelling and Blackburn, 2013).

2 LOCATING COASTAL MEGACITIES

Whilst more than half the world's population is now urban (UNFPA, 2007), only 2.8% of global land cover is urbanized (McGranahan et al., 2006). This intense concentration of population in urban spaces is most acute on the coast: two-thirds of cities with a population exceeding 5 million are located at least partially within 10 m above sea level (McGranahan et al., 2007), 15 of which are coastal megacities.

This chapter adopts the UN-DESA (2012a) definition of megacities as those with more than 10 million inhabitants; but expands it to also include "urban regions," defined as city-regions of comparable magnitude (i.e., cumulative population of at least 10 million) where some parts may lie outside of administratively designated urban zones. This includes areas with multiple overlapping jurisdictions (i.e., municipal, city, or local governments) and areas dominated by a high number of small, intervening settlements linked by semirural or peri-urban areas that, individually, would escape the megacity characterization. Our inclusion of urban regions seeks to take account of the economic and infrastructural significance of these

Coasts and Estuaries. https://doi.org/10.1016/B978-0-12-814003-1.00038-1

inter-connected areas. We define the coast geographically as the area within 50 m elevation and 100 km distance of mean high water, as discussed by UNEP (2005).

Table 1 identifies 15 coastal megacities (from a total of 23 megacities worldwide), all of which have significant populations below 20° above mean sea level. Of these, nine coastal megacities are in Asia; two each in Latin America and North America; and one each in Africa and Europe. These cities had a combined population of approximately 232 million in 2011 (indicating that out of a total global population of 7 billion, at least 3.3% of the world's population lives in coastal megacities). It must be noted that the list below is subject to various uncertainties and is open to debate as well as rapid change; for example, it includes an additional two Chinese megacities to von Glasow et al. (2012), despite drawing from data just 1 year later (von Glasow et al. draw on UN data from 2010, Table 1 from 2011).

Table 1 indicates that coastal megacities have emerged in a diverse range of environmental contexts in terms of physical form, geology (geomorphology), and climate. Each city's specific context determines, amongst other things, the range of biophysical hazards, the city experiences, the relative ease or difficulty of infrastructural development, and the proximate availability of ecosystem services and natural resources. Large urban areas have tended to be constructed near the mouths of major rivers, where locational and transportation advantages emerge through the linking of interior hinterlands with global trade through shipping (de Sherbinin et al., 2007). The intensity of urban development along the coastal strip is also a product of openness to international trading opportunities (Henderson and Wang, 2007), and has accelerated markedly in low- and middle-income countries over the last half-century (UN-HABITAT, 2008).

Of the cities outlined in Table 1, several are found in delta areas, which have significant land areas under 20 m above mean sea level. Being low lying, these are particularly prone to subsidence owing to sediment compaction (Vörösmarty et al., 2009). Other coastal urban types identified in Table 1 include those situated on steeply rising inland topography or even mountainous zones, such as characterizes many cities along the Mediterranean (Istanbul) and along the Brazilian coast (Rio de Janeiro). For some of these cities, there is a narrow coastal shelf which is densely populated, and in others the elevation rises abruptly from the coastline.

TABLE 1 Coastal Megacities

	City	Population of Urban Agglomerations (Millions)	Coastal Type
1	Tokyo	37.2	Coastal plain, some hills
2	New York-Newark	20.4	Island and coastal plain
3	Shanghai	20.2	Delta
4	Mumbai (Bombay)	19.7	Islands, delta, coastal plain
5	Kolkata (Calcutta)	14.4	Delta
6	Karachi	13.9	Coastal plain, delta
7	Buenos Aires	13.5	Coastal plain, part delta
8	Los Angeles-Long Beach-Santa Ana	13.4	Coastal plain
9	Rio de Janeiro	12.0	Mountainous, narrow coastal plain
10	Manila	11.9	Coastal plain
11	Osaka-Kobe	11.5	Coastal plain, mountainous
12	Lagos	11.2	Low lying coastal plain
13	Istanbul	11.3	Mountainous
14	Guangzhou	10.9	Delta
15	Shenzhen	10.6	Delta, coastal plain, islands

Based on Blackburn, S., Marques, C., 2013. Mega-urbanisation on the coast: global context and key trends in the twenty-first century. In: Pelling, M., Blackburn, S. (Eds.), Megacities and the Coast: Risk, Resilience and Transformation, 1–21, 2013, in which population statistics are sourced from UN-DESA, 2012b. World urbanization prospects: the 2011 revision. United Nations, Department of Economic and Social Affairs (UN-DESA), Population Division, CD-ROM Edition and location typologies from UN-DESA, 2009. World urbanization prospects: the 2009 revision. United Nations, Department of Economic and Social Affairs, Population Division, New York: United Nations.

3 CHALLENGES IN DEFINING COASTAL MEGACITIES

As noted above, classifying coastal megacities is a contested process. This is for several reasons. As Satterthwaite (2010) highlights, the timing, methodology, and quality of data collection varies widely between cities, making direct comparison between them problematic. For example, whilst many data sources (including the UN-DESA data) distinguish between "urban proper" (i.e., city center) and "urban agglomeration" (i.e., the wider or "greater" city area), others do not.

Connected to this is the issue of accurately and meaningfully delineating a city's boundaries, and thereby setting the limits within which the population data are included. Official city-level or national urban data tend to be based on administrative boundaries and may exclude a significant number of people—including those migrating to the city for work from adjacent or suburban areas, and populations living in informal settlements. Whilst making substantial contributions to the city economy and adding pressure to city infrastructures and resources, these groups may be invisible to official datasets (such as the census or tax records). Depending on methodology, these hidden populations can lead to significant overestimates or underestimates of megacity populations (Cohen, 2006; Satterthwaite, 2010).

A third issue which renders the classification of coastal megacities problematic (or, indeed, any city), is lack of clarity over the significance of such classification—that is, what does the designation of a region as a city, and hence as an urban space, communicate about its quality and form? And what specific indicators denote the emergence or growth of those spaces? The term urbanization can be used to describe (at least) three different processes, which do not necessarily coincide, and indeed are becoming increasingly distinct. Three alternative definitions of urbanization are summarized below:

(1) Urbanization as urban population growth
This is the process through which a population comes increasingly to reside in a space designated as urban, and an overall pattern of the spatial concentration of population. It is indicated by an increasing proportion of the population living in urban areas, accompanied by a relative decline in rural populations and enterprise. Urban population growth is the sum of natural increase plus the rate of in-migration. This is the most common definition of urbanization.

(2) Urbanization as urban physical expansion
Here, urbanization is equated to the spatial extent and expansion of built-up areas, and/or of the land administratively designated as urban, and the commensurate loss of rural land. This can be useful in evaluating cities' environmental impacts. However, current trends of declining urban density and fragmentation mean urban expansion often now takes place without urban population growth, and conversely involves of process of declining settlement density. In these cases, it is misleading to refer to this declining settlement density as urbanization, since increasing settlement density has historically been seen as one of the key features of urbanization.

(3) Urbanization as a process of socioeconomic transition
This definition refers to a transition between social, economic, cultural, and political systems, from rural areas to those associated with urban centers. This definition is now contested owing to well-documented flaws in a simplified rural-urban dichotomy, and the challenges associated with defining both "rural" and "urban" (see e.g., Uzzell, 1979; Pumain and Robic, 1996; Beguin, 1996). Lefebvre (2014) challenged the city-countryside dichotomy in his thesis on "urban society," arguing that in the current stage of capitalism the urban fabric (as a particular socio-spatial form) extends virtually beyond cities, reaching into immediate countryside and distant regional spaces (Monte-More, 2005). Rejection of the assumption that urban and rural societies have essential cultural differences stems from a recognition of the complexity of rural capital accumulation systems, and that defining rural simply as "not urban" denies its agency and uniqueness (Uzzell, 1979; Halfacree, 1993). As a result, treatment of "rural" and "urban" as two extremes of a continuum (rather than discrete categories) is now widely preferred (Champion and Hugo, 2004). Application of the continuum concept in land-use classifications in practice varies in complexity, but over time more elaborate conceptualizations of interconnections of urban-rural interconnections have emerged. Settlement classifications are not often based on relative, measurable parameters including population size and density, diversity of economic activity, level of "built-up-ness" (related either density of construction or average building height), and level of capital generation.

A number of trends are additionally challenging traditional conceptions of what it means and looks like to be urban, most pertinently: (i) the most rapid urban growth now occurring in low- and middle-income countries, rather than the global north [although Satterthwaite, 2010 and others are quick to dispel popular myths of a future led by "explosive" urbanization]; (ii) declining urban densities, partly due to the rapidity of suburban expansion and transitions within the "peri-urban interface" (see e.g., Allen, 2003; Cohen, 2006); and (iii) the growth of poly-centric megaregions, urban corridors, and city-regions, which form through the overlapping sprawl of multiple urban centers (Aguilar et al., 2003). The shifting urban form means that whereas once the urbanization of population, space, and culture might be treated as different aspects of the same transformation, this is now no longer the case. As Champion and Hugo (2004) argue, the key in any study of urban phenomena is to be clear, and to avoid definitive statistics in the absence of clear qualification as to their meaning.

As with urban areas, there is no single definition of the coastal zone, and use of this term varies significantly across disciplines and between locations. Kay and Alder (2005) identify four typologies of coastal definitions, used in the policy realm: fixed distance definitions, variable distance definitions, definition according to the use, and hybrids of these (pp. 4). Others still have defined the coast in terms of elevation relative to the sea or ocean (McGranahan et al., 2007), or a combination of distance and elevation criteria (Klein et al., 2003; UNEP, 2005).

These measures of physical proximity, while useful, fail to capture the three-dimensional (3D) dynamics such as atmospheric flows and exchanges characteristic of the coastal zone, which have fundamental implications at ground level (e.g., controlling the hydrological cycle and exposure to tropical cyclones). Furthermore, strict use of elevation-based definitions excludes some cities from the "coastal megacity" bracket which might intuitively be included—for example, São Paulo Metropolitan area (Brazil), which lies only 50 km from the coast but 800 m above sea level (Nicholls, 1995). von Glasow et al., 2012 who include São Paulo in their classification, acknowledge, and address the difficulty of delineating between "coastal" and "not coastal" by identifying both "coastal agglomerations" and "those with coastal influence" (the latter including Dhaka, for example). Such an approach could also arguably include Greater London (although von Glasow et al. do not), since the Thames—which historically and currently has been a hugely significant source of both economic opportunity and environmental hazard in London—is highly tidal, indicated by the need for large-scale flood defence structures such as the Thames Barrier. These arguments also support the inclusion of those parts of urban regions which lie beyond the coastal plain (in places such as Los Angeles, Seoul, and Istanbul) in accounts of coastal urbanization.

4 RISK, VULNERABILITY, AND RESILIENCE IN COASTAL MEGACITIES

The growth in the number and size of coastal megacities worldwide, particularly those in low- and middle-income countries, is increasing pressures on the natural environment. It is also changing the way that environmental risk is produced and experienced due to land reclamation and the development of flood-prone coastal zones. These goalposts are shifting further as a result of global environmental change, which is anticipated to bring changes in the behavior and frequency of extreme climate events and cause additional stress to human and environmental systems, through raised sea levels and potential for increased wind storms and onshore wave height. These local and global systems are best viewed as being coupled, because disturbances in one can cause and reinforce disturbances in another through complex feedback mechanisms. For example, urban expansion along the coast often results in concrete reinforcement of the coastline, which interrupts natural sediment transport, which can increase flood risk; also urban expansion leads to pollution and exploitation of resources which harms ecosystems, and ultimately reduces the availability and quality of ecosystem services and resources on which the urban population depends—with negative implications for human health, economies, and well-being. In both cases, greater attention to finding sustainable forms of urban growth, in which human and environmental systems can coexist in a balanced and symbiotic way, is necessary.

Vulnerability to hazards is defined as human sensitivity to hazard impacts and is the product of a wide range of factors (Wisner et al., 2004). Resilience, simply put, is the ability to withstand hazard impact, such that the system affected is able to continue to function, as well as learn from and adapt to new and unanticipated sources of hazard (Cutter et al., 2008). Whilst related, urban resilience and vulnerability should not be viewed simply as opposites (Pelling, 2011b). The factors affecting vulnerabilities and resilience in coastal megacities are addressed in this section.

4.1 Drivers of Risk in Coastal Megacities

The hazards to which a city is exposed depend upon the geographical location and physical characteristics of the city in question. Flood risk, for example, is greatest in cities that are either in close proximity to major rivers or on low-lying coasts exposed to storm surges (Prasad et al., 2009; Dasgupta et al., 2009). Elevation is another strong predicator of coastal flood risk, and coastal plains may or may not be sufficiently raised above sea level to mitigate exposure to coastal flooding as well as near- or medium-term projections for sea-level rise due to climate change. Deltaic settings tend to be particularly vulnerable for this reason, although Manila and Mumbai are examples of non-deltaic but very low-lying exposed cities. Some key factors which contribute to heightened exposure to current and future risks in coastal megacities are outlined in Box 1.

The construction of risk, however, is more than simply exposure to hazard; risk is also the product of vulnerability (Wisner et al., 2004). Pelling (2011a) observes seven features of coastal megacities which influence their vulnerability to natural hazards, relative to other types of human settlement in other locations:

(1) High concentration of physical assets, industries, energy installations, and exposed populations;
(2) A significant migrant population and cultural and socioeconomic diversity;

BOX 1 Geographical Characteristics Increasing Hazard Exposure in Coastal Megacities

Low elevation deltas—Coastal cities tend to be highly exposed to coastal flooding, due to changes in sea level, tidal waves, or the effects of cyclones or frontal systems. Many coastal cities are also located by rivers. Given how floods propagate downstream through catchments, this also makes coastal cities vulnerable to flooding from upstream (e.g., the flooding of the Mississippi which affected New Orleans in May 2011).

Topography: Coastal cities may be surrounded by mountains or topographic barriers which enhance local precipitation as onshore, saturated winds are forced to rise. Weather systems flowing on-land from features such as persistent trade winds, interact with these physical barriers resulting in heavy rainfall rates and runoff. Even in the absence of significant topography, changes in surface roughness affect the movement of onshore storm systems.

Land use: Many large coastal cities are also ports and industrial processing areas, for example, Shanghai is amongst the world's busiest ports. Some associated industrial activities, for example, oil refineries, result in huge emissions with significant implications for local and regional air quality. The ratio of green space vs impermeable surface in a coastal megacity also has significant implications for flood risk, as it affects the infiltration rate of the land area.

Sea/land breezes: The differential heating of water and land generates daytime sea breezes (onshore flows of air) and land breezes (offshore flows of air). In summer, sea breezes are important in coastal cities in mitigating heat stress, with implications for thermal stress and air quality. Tokyo, for instance, is considering removing buildings to allow the sea breeze to penetrate and aerate the city. Other cities are also considering the orientation of the buildings in order to affect the inflow of sea air (e.g., Hong Kong and Singapore). This needs to take into account both wind flow and solar gain (e.g., Ng et al., 2011). Sea/land breezes also serve to concentrate and recirculate pollutants across coastal cities with important implications, particularly at night when the urban boundary layer (and thus atmospheric mixing) is diminished.

Population density and heat island effect: Coastal megacities tend to be denser than megacities in other geographical settings, in part because of competing land uses and topography, and as a result some 13 of the 20 densest cities worldwide are in the coastal zone (McGranahan et al., 2007). Thus, coastal cities tend to have a high proportion of impervious cover, which is a key determinant of energy flux partitioning—latent heat fluxes (evapotranspiration rates) are suppressed, while sensible and storage heat fluxes are enhanced. This results in greater heating of the air and substrate, culminating in the "heat island effect." High population densities (which often coincide with high poverty levels) experience heightened exposure to concentrated pollution, which exacerbates environmental health problems and enhances greenhouse gas emissions—contributing further to urban warming effects.

(Source: Grimmond CSB. Urban climate in the coastal zone: resilience to weather extremes. In: Pelling M. Megacities and the Coast: Transformation for Resilience. A Preliminary Review of Knowledge, Practice and Future Research. June, LOICZ, 2011.)

---

(3) Extensive networks of infrastructure both coastal and interior;

(4) Exposure to multiple hazards: subsidence, salinization, liquefaction, sea-level rise, etc., which makes risk management more complex and raises the likelihood of risk reduction policies with regards one risk having adverse effects on vulnerability to another;

(5) A capacity to trigger economic contagion at different scales through their strategic importance as finance and trade centers;

(6) The possibilities of hotspot growth for new ecological assemblages as a result of degradation and interruption of ecosystems (with feedback implications for hazard creation and exposure); and

(7) An extensive source of intervention in biophysical systems through extractive industry, water and air pollution, and others, implicated in the creation and reinforcement of urban hazards by reducing the sustainability and environmental quality of the city.

These features highlight that it is the unique juxtaposition between large-scale settlement and environmental processes that exists in coastal megacities which culminates in heightened risk in coastal megacities. It is where these systems interact that the potential for hazard arises. However, it is also in this space of overlap that the potential for adaptation and resilience arises, depending on cities' responses to the source of environmental pressure.

4.2 Possibilities for Resilience and Transformation

Megacities and coastal zones are arguably the two most dynamic human and natural environments on the planet. It is little wonder that both have attracted so much study. It is surprising though that little work has prioritized the integration of knowledge and experiences between these two realms. Moving beyond a disciplinary orientation to a systems viewpoint that can draw out the feedbacks, trade-offs, and compounding processes in the urban-coast space is long overdue.

The task of bringing together the urban planning and coastal management communities is complicated by the multiple levels at which they interact. At the forefront of a good deal of contact are marginalized populations—such as low-income housing settlements on the coastal fringes or artisanal and informal sector fishers. In New York, the fringe communities of Jamaica Bay which flooded during Hurricane Sandy included a disproportionate number of low-income and rental households and residential homes for the elderly. These are not easy populations to integrate into planning processes, nor have planners and urban decision-making institutions more broadly often proven their legitimacy to these groups. More often, planning has its de facto priority economic growth, and this may not coincide with the interests of the urban poor. Redevelopment of low-income settlements post-disaster or as part of disaster risk reduction (DRR) or climate change adaptation (CCA) that does not provide enhanced access to basic needs and working opportunities is all too common. The dynamic processes that lie behind the coevolution of political, economic, and physical systems are also ill-captured by standard static development data.

The existing literature and underlying practice on climate change and large-scale urbanization is unhelpful in its polarization between rich and poor contexts (Romero-Lankao and Dodman, 2011). There are many examples of climate change mitigation projects in richer cities (normally driven by technological innovation), and of vulnerability reduction in poorer cities (driven by social or economic or land-use policy). However, this bifurcation in policy and vision is unhelpful in so far as it makes transfer of knowledge and expertise difficult. Vulnerability reduction is tainted as part of the policy discourse of a city at risk (with all the implications this has for poor governance, widespread poverty, etc.). The result imaginative opportunities for learning from the global South are lost. A review of how cities worldwide (low, middle, and high income) is responding to the UNISDR's Making Cities Resilient campaign revealed many to be making calculated adjustments to their institutional structures, funding streams, building codes, development priorities, and monitoring frameworks (UNISDR, 2012; Johnson and Blackburn, 2014). Greater capture of spontaneous and strategic adaptation in low-income cities is needed—and should include the activities of individuals and community groups as well as governments. Local governments are the central actors in urban risk management and should be central for any revisioning of the city and for shared learning going forward (Pelling, 2003). Global and regional networks such as Local Governments for Sustainability (ICLEI) offer some scope for moving beyond this, and a focus on coastal cities as a "hotspot" as adopted by the international scientific collaborative Future Earth Coasts can facilitate better flow of ideas and experience globally.

Similarly, the high technology demonstrated through climate change mitigation schemes, and large-scale local hazard mitigation (e.g., through flood defences, drainage, etc.) may not be possible for poorer cities. Megacities have an advantage being centers of global wealth creation, even in poorer countries, so that opportunities for taking up high-tech and large-scale solutions can be considered. Yet, a tension lies in transferring technology-based solutions into contexts where high levels of poverty make the poor vulnerable to development; the poor have a small voice in strategic development of this scale, even when it is their livelihoods, homes, and communities that might be the immediate "losers." Technology needs to be embedded in a social solution in such contexts: mitigation not only in industry but also in vulnerability reduction and livelihood enhancement. Integration, again, is the watchword.

So how might we reframe megacities and large urban regions on the coast as integrated systems? A starting point is to consider the multiple and overlapping boundary spaces between the coast and interior—boundaries set by wind, temperature, water, and chemical flows. These will have atmospheric, fluvial, surface, and underground expressions, each formed by the interaction of physical process and built form. Social-ecological theory presents a number of metaphors which could usefully push an integrated research agenda on coastal cities: for example, thresholds at which systems flip from one stable state to another (e.g., the point at which urban infilling increases populations in exposed areas beyond a capacity to respond to flooding), and the extent to which changes in systems at one scale might infect or be suppressed, even provoked, by other scaled subsystems (e.g., successful community mobilization to lobby for change in insurance or construction regulation can have implications for policy at the city and national scale). Finally, it is necessary to consider the relative merits of resilience being understood as stability-seeking or transformation. Stability-seeking builds resilience through protecting existing assets and processes, for example, a seawall to protect physical assets (i.e., buildings). Transformative work sees development processes as the problem to be solved, such as regulating land speculation to allow more affordable housing to be build and to reduce the number of poor quality dwellings that lead to risk-exposed residents. This question around resilience vs transformation is perhaps the most profound and offers a mechanism for linking two great challenges faced by cities around the world in an era of climate change: mitigation and adaptation. Can there be pathways for transforming development so that it better meets social needs, through risk reduction at the local and global scales?

Important to improving synergies are addressing information and data deficiencies, administrative and governance constraints, funding limitations, as well as a need to define basic metrics of progress—for example, measures of increased resilience and adaptive capacity and decrease vulnerability (Tol and Yohe, 2007; da Silva et al., 2012). Building risk management, CCA, and mitigation into urban planning for the future can draw on many existing experiences, from Integrated Coastal Zone Management to multi-hazard risk analysis and the multi-agency approaches employed for water and transport systems management in many cities today. The required step change in political will and policy focus, and the necessity for broader, inclusive approaches that involve the most vulnerable urban communities, suggests the following priorities for furthering risk management in coastal megacities:

- High-level, proactive leadership to initiate and coordinate adaptive planning for the critical infrastructure of cities and surroundings.
- The incorporation of climate change risks into stakeholder agency and organization operations, management, and planning. This in turn requires long-term planning horizons and decision-making procedures that can expose and tolerate uncertainty. Most immediately this supports the identification of low-regrets policies.
- To promote a dynamic planning process among city government, public and private stakeholders, and experts to support a risk-management approach to incorporate climate change and to begin to implement flexible adaptation pathways.
- To develop linkages (including organizational forms or joint working groups on legislation) to enable cross-agency and cross-policy coordination in adaptation plans to cope with extreme and compound hazard events, ecological stresses, and human impacts in the coastal zone.

In tandem with an expanding knowledge base, there is likely to be a multiplicity of competing sources and formats of information on hazards, including research, monitoring, forecasts, and early warning systems. Therefore, actors and organizations with humanitarian responsibilities for urban regions on the coast need to ensure their information systems and disaster management procedures are effectively coordinated and codified in order to facilitate effective decision making. Early warning systems in places exposed to more than one hazard (e.g., tsunami, storm surge, and pluvial flooding) should aim at common communication strategies. Lessons could be learnt here from the development of the Integrated Food Security Phase Classification, a standardized scale (integrating food security, nutrition, and livelihood data) developed by a group of leading international humanitarian agencies in 2004 to help to coordinate response in food security emergencies worldwide. The innovative use of social media and crowdsourcing technologies can also provide a means to streamline information management systems in emergency response. For example, following the Haiti earthquake in 2010, users of openstreetmap. org assisted rescue workers and humanitarian agencies by providing missing information on the layout of Port-au-Prince; following the Chennai floods in 2015, residents and relief workers used social media to coordinate and prioritize rescue efforts and to disseminate real-time updates (Sen, 2018). The technology is now also being used to enhance preparedness efforts in coastal cities such as Padang, Indonesia, as well as in many other crisis-prone situations across the world.

Projecting from our current situation, it is possible we may see increasing resistance from host country governments to disaster response led by the traditional actors of the international humanitarian system—United Nations agencies, the Red Cross and Red Crescent movement, and international non-governmental organizations (NGOs). Such reticence was evident, for example, in Myanmar in 2008 when Cyclone Nargis devastated coastal areas of the country. This ambivalence also relates to the increasing political centrality of crises and resulting moves toward the reassertion of sovereignty on behalf of crisis-affected states, particularly in regions such as South-East Asia. In this context, regional disaster management frameworks are becoming increasingly important as a counter pressure, especially for urban regions on the coast, as seen in the case of the Association of South-East Asian Nations' (ASEAN) Agreement on Disaster Management and Emergency Response, signed in 2009 (ASEAN, 2009). The development of such frameworks will be similarly important for other regional organizations, and some are already responding to this need. For example, the Economic Community of West African States (ECOWAS) is currently implementing a program of action running to 2014 to operationalize its own DRR policy (ECOWAS, 2006).

Given the evidence and arguments developed in this chapter, the pressure on policy makers with risk management responsibilities to meet the needs of vulnerable urban populations on the coast looks likely only to increase in the coming decades. Dealing with this challenge through an integrated, collaborative, and strategic approach that can consider short- and long-term needs is likely to prove essential to meeting the challenge of delivering safer and more resilient coastal megacities for their future inhabitants.

5   CONCLUSIONS: URBAN TRANSITIONS, URBAN FUTURES

At a global scale, the urban transition—a shift from a predominantly rural population to a predominantly urban population—has been accompanied by a demographic transition—a shift from high birth and death rates, to low birth and death rates (de Sherbinin and Martine, 2007). However, the impact of the urban transition on the environment varies according to many factors other than population changes (Bruns et al., 2013). Factors affecting the sustainability potential of megacities include the spatial extent, type of land, form of occupancy, and ecological condition of urban residence (de Sherbinin and Martine, 2007). Some cities, for example Lagos, are aggressively expanding into coastal areas whilst simultaneously failing to provide basic services in low-income settlements exposed to coastal hazards. As well as environmental impacts, the damage to ecosystem services including clean water, inadequate access to or disproportionate demand for available resources, are drivers of environmental health concerns in coastal megacities (McGranahan et al., 2006). Literature on the urban environmental transition states that the majority of urban environmental health problems are actually associated with large-scale poverty rather than urbanization per se, and that wealth accumulation is associated with negative environmental impacts being located increasingly remotely from the point of consumption (McGranahan and Songsore, 1994). Poverty and the distribution of environmental impacts are an outcome of economic opportunity and social policy: this places the ambition for sustainable coastal urbanization within the grasp of urban populations and their leaders (Martine et al., 2008).

ACKNOWLEDGMENTS

This chapter draws from Pelling and Blackburn's edited assessment review *Megacities and the Coast: Risk, Resilience and Transformation,* undertaken with support from the International Geosphere-Biosphere Program and Land-Ocean Interaction on the Coastal Zone (now Future Earth Coasts). In particular, it draws on Blackburn and Marques (2013) and Birkmann et al. (2013).

REFERENCES

Aguilar, A., Ward, P., Smith Sr., C., 2003. Globalization, regional development, and mega-city expansion in Latin America: analyzing Mexico City's peri-urban hinterland. Cities 20 (1), 3–21.

Allen, A., 2003. Environmental planning and management of the peri-urban interface: perspectives on an emerging field. Environ. Urban. 15 (1), 135–148.

ASEAN (2009) Regional disaster management agreement enters into force. [online]. Available from http://www.asean.org/24136.htm (Accessed 9 March 2012)

Beguin, H., 1996. Faut-il définir la ville? In: Derycke, P., Huriot, J., Pumain, D. (Eds.), Penser la ville: Théories et modèles. Anthropos, Collection Villes, Paris, pp. 301–320.

Birkmann, J., Garschagen, M., Lopez, A., Pelling, M., Qaiem Maqami, N., Yu, Q., 2013. Urban Development, climate change and disaster risk reduction: interaction and integration. In: Pelling, M., Blackburn, S. (Eds.), Megacities and the Coast: Risk, Resilience and Transformation. pp. 172–199.

Blackburn, S., Marques, C., 2013. Mega-urbanisation on the coast: global context and key trends in the twenty-first century. In: Pelling, M., Blackburn, S. (Eds.), Megacities and the Coast: Risk, Resilience and Transformation, pp. 1–21.

Bruns, A., et al., 2013. The environmental impacts of megacities on the coast. In: Pelling, M., Blackburn, S. (Eds.), Megacities and the Coast: Risk, Resilience and Transformation, pp. 22–69.

Champion, T., Hugo, G., 2004. Introduction: moving beyond the urban-rural dichotomy. In: Champion, T., Hugo, G. (Eds.), New Forms of Urbanisation: Beyond the Urban-Rural Dichotomy. Ashgate, Aldershot, Hants, pp. 3–24.

Cohen, B., 2006. Urbanization in developing countries: current trends, future projections, and key challenges for sustainability. Technol. Soc. 28, 63–80.

Cutter, S., Barnes, L., Berry, B., Burton, C., Evans, E., Tate, E., Webb, J., 2008. A place-based model for understanding community resilience to natural disasters. Glob. Environ. Chang. 18 (4), 598–606.

da Silva, J., Kernaghan, S., Luque, A., 2012. A systems approach to meeting the challenges of urban climate change. Int. J. Urban Sustain. Develop. 1, 1–21.

Dasgupta, S., B. Laplante, S. Murray, and D. Wheeler. (2009) Climate change and the future impacts of storm-surge disasters in developing countries. Centre for Global Development Working Paper 182.

de Sherbinin, A., Martine, G., 2007. Urban Population, Development and Environment Dynamics: Situating PRIPODE. Committee for International Cooperation in National Research in Demography (CICRED), Paris.

de Sherbinin, A., Schiller, A., Pulsipher, A., 2007. The vulnerability of global cities to climate hazards. Environ. Urban. 19 (1), 39–64.

ECOWAS, 2006. ECOWAS policy for disaster risk reduction. Humanitarian Affairs Department (DHA) (August 2006).

Halfacree, K., 1993. Locality and social representation: space, discourse and alternative definitions of the rural. J. Rural. Stud. 9 (1), 23–37.

Henderson, J., Wang, H., 2007. Urbanization and city growth: the role of institutions. Reg. Sci. Urban Econ. 37 (3), 283–313.

Johnson, C., Blackburn, S., 2014. Advocacy for urban resilience: UNISDR's making cities resilient campaign. Environ. Develop. 26, 29–52.

Kay, R., Alder, J., 2005. Coastal Planning and Management, second ed. Taylor and Francis, Abingdon, OX.

Klein, R.J.T., Nicholls, R.J., Thomalla, F., 2003. The resilience of coastal megacities to weather-related hazards. In: Kreimer, A., Arnold, M., Carlin, A. (Eds.), Building Safer Cities: The Future of Disaster Risk. Disaster Risk Management Series No3. World Bank Publications, Washington, DC, pp. 101–120.

Lefebvre, H., 2014. The Urban Revolution. University of Minnesota Press, Minneapolis.

Martine, G., McGranahan, G., Montgomery, M., Fernandez-Castilla, R., 2008. Introduction: the new global frontier: cities, poverty and environment in the 21st century. In: Martine, G., McGranahan, G., Montgomery, M., Fernandez-Castilla, R. (Eds.), The New Global Frontier: Urbanization, Poverty and Environment in the 21st Century. Earthscan, London, pp. 1–13.

Martínez, M., Intralawan, A., Vázquez, G., Pérez-Maqueo, O., Sutton, P., Landgrave, R., 2007. The coasts of our world: ecological, economic and social importance. Ecol. Econ. 63, 254–272.

McGranahan, G., Songsore, J., 1994. Wealth, health and the urban household. Environment 36 (6), 4–45.

McGranahan, G., Marcotullio, P., Bai, X., Balk, D., Braga, T., Douglas, I., Elmqvist, T., Rees, W., Satterthwaite, D., Songsore, J., Zlotnik, H., Eades, J., Ezcurra, E., 2006. Chapter 27: urban systems. In: Ecosystems and Human Well-Being: Current State and Trends, Millennium Ecosystem Assessment. Island Press, Washington, DC, pp. 795–825.

McGranahan, G., Balk, D., Anderson, B., 2007. The rising tide: assessing the risks of climate change and human settlements in low-elevation coastal zones. Environ. Urban. 19 (1), 17–37.

Monte-More, R.L. (2005) What is the urban in the contemporary world?. Cad. Saúde Pública [online], 21(3), pp.942-948. doi:dx.doi.org/10.1590/S0102-311X2005000300030.

Neumann, B., Vafeidis, A., Zimmermann, J., Nicholls, R., 2015. Future coastal population growth and exposure to sea-level rise and coastal flooding—a global assessment. PLoS One 10 (6), e0118571. https://doi.org/10.1371/journal.pone.0118571.

Ng, E., Yuan, C., Chen, L., Ren, C., Fung, J.C.H., 2011. Improving the wind environment in high density cities by understanding urban morphology and surface roughness: A study in Hong Kong. Landsc. Urban Plan. 101, 59–74.

Nicholls, R.J., 1995. Coastal megacities and climate change. GeoJournal 37 (3), 369–379.

Pelling, M., 2003. The Vulnerability of Cities: Social Resilience and Natural Disaster. Earthscan, London.

Pelling, M. (2011a). Introduction. In: Pelling, M. Megacities and the Coast: Transformation for resilience. A Preliminary Review of Knowledge, Practice and Future Research. June, LOICZ. Available from: http://www.loicz.org/science/hotspot/Urbanization/News_Reports/index.html.en (Accessed 30 October 2012)

Pelling, M., 2011b. Adaptation to Climate Change: From Resilience to Transformation. Routledge, London.

Pelling, M., Blackburn, S. (Eds.), 2013. Megacities and the Coast: Risk, Resilience and Transformation. Routledge, London.

Prasad, N., Ranghieri, F., Shah, F., Trohanis, Z., Kessler, E., Sinha, R., 2009. Climate Resilient Cities: A Primer on Reducing Vulnerability to Disasters. World Bank, Washington, DC.

Pumain, D. and Robic, M. (1996). Théoriser la ville. In: Derycke, P.; Huriot, J.; Pumain, D. (coord.), Penser la ville, Théories et modèles, Paris: Anthropos, Collection Villes, pp. 107-161.

Romero-Lankao, P., Dodman, D., 2011. Cities in transition: transforming urban centres from hotbeds of GHG emissions and vulnerability to seedbeds of sustainability and resilience. Curr. Opin. Environ. Sustain. 3 (3), 113–120.

Satterthwaite, D. (2010) Urban myths and the Mis-use of data that underpin them, UNU-WIDER Working Paper No.2010/28.

Sen, S. (2018) #Chennaifloods: use of social media and collective intelligence in disasters. Contested Development Working Paper Series, 79, King's College London. Available from: www.kcl.ac.uk/sspp/departments/geography/research/Research-Domains/Contested-Development/workingpapers.aspx.

Tol, R.S.J., Yohe, G.W., 2007. The weakest link hypothesis for adaptive capacity: an empirical test. Glob. Environ. Chang. 17 (2), 218–227.

UN-DESA, 2012a. World urbanization prospects: the 2011 revision—highlights. In: United Nations Department of Economic and Social Affairs, Population Division. United Nations, New York.

UNEP (2005) Assessing coastal vulnerability: developing a global index for measuring risk, report prepared by A. Singh, S. Pathirana & H. Shi, Division of Early Warning and Assessment, UNEP, Nairobi, Kenya.

UNFPA, 2007. State of the World Population 2007: Unleashing the Potential of Urban Growth. United Nations Population Fund, New York.

UN-HABITAT, 2008. State of the World's Cities 2008/2009: Harmonious Cities, United Nations Human Settlements Programme. Earthscan, London.

UNISDR (2012), Making cities resilient report 2012: a snapshot of how local governments reduce disaster risk. 116 pp., Available from: www.unisdr.org/we/inform/publications/28240.

Uzzell, D., 1979. Conceptual fallacies in the rural-urban dichotomy. Urban Anthropol. 8 (3/4), 333–350.

von Glasow, R., Jickells, T., Baklanov, A., Carmichael, G., Church, T., Gallardo, L., Hughes, C., Kanakidou, M., Liss, P., Mee, L., Raine, R., Ramachandran, P., Ramesh, R., Sundseth, K., Tsunogai, U., Uematsu, M., Zhu, T., 2012. Megacities and large urban agglomerations in the coastal zone: interactions between atmosphere, land, and marine ecosystems. Ambio 42 (1), 13–28. https://doi.org/10.1007/s13280-012-0343-9.

Vörösmarty, C., Syvitski, J., Day, J., de Sherbinin, A., Giosan, L., Paola, C., 2009. Battling to save the world's river deltas. Bull. At. Sci. 65 (2), 31–43.

Wisner, B., Blaikie, P., Cannon, T., Davis, I., 2004. At Risk: Natural Hazards, People's Vulnerability and Disasters (second ed.). Routledge, London.

Chapter 39

Arctic Coastal Systems: Evaluating the DAPSI(W)R(M) Framework

Amy Lauren Lovecraft*, Chanda L. Meek[†]

*Center for Arctic Policy Studies, University of Alaska Fairbanks, Fairbanks, AK, United States, [†]Department of Political Science, University of Alaska Fairbanks, Fairbanks, AK, United States

1 INTRODUCTION

This chapter provides an overview of Arctic coasts in relation to the DAPSI(W)R(M) framework for coastal environmental management with the aim of informing sustainable stewardship for human-environment coupled systems (Chapin et al., 2006, Chapin et al., 2009). It is impossible within the scope of this chapter to cover every dimension of pan-Arctic coastal systems in relation to their physical natures and social qualities. As such, this paper develops the context for future coastal research with an interdisciplinary overview of the current state of the Arctic's coasts from a social-ecological perspective (Fischer et al., 2015). It then applies the DAPSI(W)R(M) framework to the entirety of the pan-Arctic coastal system but with local-scale examples to illustrate key aspects of this analytic tool.

It is helpful to note that the word "Arctic" can have several meanings related to the high North. The Arctic Circle is bounded by the latitude of 66.30° North and within this region the sun is both above the horizon and below the horizon for a period of 24 h depending on the season. However, others have argued for a more flexible definition based on the extent of permafrost, sea ice coverage, temperature ranges, or northern tree line. For example, Canada's internal delineation is north of 60°N. Depending on one's disciplinary background, or rhetorical needs, the Arctic can be tied to physical, symbolic, or political boundaries. For the purpose of considering the management of Arctic coastal systems, the Arctic Monitoring and Assessment Program (AMAP) working group of the Arctic Council nicely balances these dimensions of "Arcticness" in its geographical coverage (http://www.amap.no/). This paper considers the Arctic as the AMAP area (Fig. 1).

1.1 The Rapidly Changing Arctic

While the earth is generally warming, the rates at the poles are much faster than mid-latitudes and Arctic scientists often prefer to discuss "climate change" as a suite of factors rather than unidirectional rise in temperature alone. This is because changing temperatures are only one part of the earth system flux occurring due to significant greenhouse gas additions to the earth's atmosphere over the last century (Solomon et al., 2009). For the Arctic, it is not only ocean warming and acidification, but also the changing nature of the earth's cryosphere—a word stemming from the Greek word "kryos" referring to cold or frost—that will bring the most monumental changes, in particular to coastal regions. In modern usage, the cryosphere refers to all locations on the planet where water is in its solid form either above ground as freshwater ice or sea ice, glaciers, and snow, or below ground as permafrost. In addition, one cannot ignore the impacts on lower latitudes. For example, changes in the northern circumpolar jetstream due to Arctic warming are linked to increased frequency of extreme winter weather in the United States (Cohen et al., 2018) and the mid-latitudinal changes tied to seasonal cycles of, for example, precipitation (Francis et al., 2017).

In brief, the most recent NOAA Arctic Report Card (2017) indicates that the average surface air temperature for the year ending September 2017 is the second warmest, with 2016 being the warmest, since 1900. In early March 2017, satellites observed the lowest winter maximum in sea ice on record (1979–present). The March maximum was 8% lower and the September minimum was 25% lower than the 1981–2010 average sea ice extent. The sea ice cover continues to be relatively young and thin with multiyear ice (more than 1 year old) comprising only 21% of ice cover in 2017 compared to 45% in 1985. The August 2017, sea surface temperatures in the Barents and Chukchi seas were up to 4°C warmer

Coasts and Estuaries. https://doi.org/10.1016/B978-0-12-814003-1.00039-3

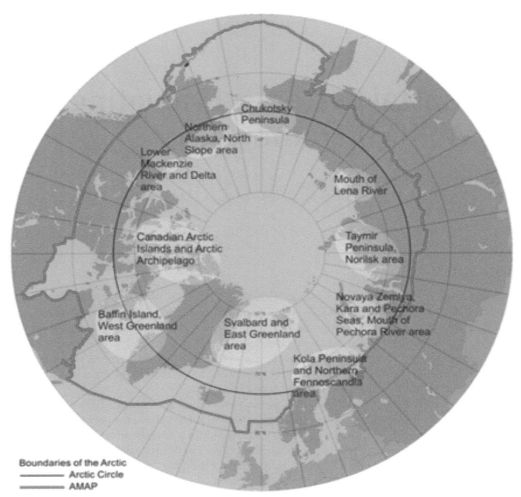

FIG. 1 The Arctic boundaries and 10 coastal areas targeted for pollution studies. *(Adapted from https://www.amap.no/about/geographical-coverage.)*

than average, contributing to a delay in the autumn freeze-up in these regions. Recent data on sea ice extent indicate that 50% of the summer sea ice extent and 60% of its volume have disappeared in the last several decades (Meier et al., 2014). Inside the earth, permafrost is thawing with new record high mean annual ground temperatures observed at many permafrost observation stations across the Arctic and near the surface the "active layer," where permafrost thaws and freezes seasonally, is freezing up 2 months later than usual (Romanovsky et al., 2017). Permafrost temperatures in 2016 at many observation sites around the Arctic were among the highest on record (as long as 1978–present, but duration of records varies). Increases in permafrost temperature, since 2000, have been greatest in permafrost of the Alaskan Arctic, Canadian high Arctic, and Svalbard. Land ice is also in retreat across the Arctic with measurement of retreat rates for marine-terminating outlet glaciers in the Atlantic Arctic seeing a threefold increase (Carr et al., 2017). Including other ice sheet data such as the expanding surface melt area in Greenland (Tadesco et al., 2015) further indicates that land ice loss has significant consequences for coastal planners, largely related to sea-level rise (Larour et al., 2017), but also coastal erosion. These trends are not expected to slow.

1.2 Arctic Ecosystem Services

The relatively recent International Polar Year, 2007, focused attention on the changes in the Arctic and, unlike earlier polar years, it included the "human dimension." In the last 60 years, humans have changed planetary ecosystems more quickly and extensively than at any comparable period of human history and we are just beginning to understand the implications (Chapin et al., 2006; MEA, 2005, Post et al., 2009). Important advances in understanding how societies perceive, use, and manage their environment have been made by focusing on processes that link ecological and social subsystems through their effects on human actors (Young, 2002; Dietz et al. 2003). Most simply, these linkages are the feedback loops created by the cycle of societal outputs that affect ecological processes and the inputs, generally benefits, which humans derive from their ecosystems.

Addressing the latter first, the Millennium Ecosystem Assessment completed a multiyear synthesis project to explain global conditions and assess the relationships between ecosystems and human well-being. Ecosystem services are generally classified as four types (Millennium Ecosystem Assessment, 2005). *Provisioning* services are the tangible goods people obtain from their natural environment such as food, fiber, fuel, and fresh water. Second, ecosystems provide societies with *cultural* services through spiritual enrichment, knowledge systems, social relationships, aesthetic beauty, and other nonmaterial benefits. Third, *regulating* services refer to the benefits derived from the regulation of biochemical processes by the ecosystem itself such as climate regulation from the land cover, pollination through plant and animal interactions, and water quality patterns affected by soil erosion, coastal filtration, and marine ecosystem uptake. Lastly, the ecological foundations of societies rely on *supporting* services that are necessary for the production of all the services noted above such as soil formation, photosynthesis, nutrient cycling, and other fundamental processes key to life in ecosystems (Fig. 2).

The Millennium Ecosystem Assessment (MEA) explains human well-being as having five components derived from these services: security, basic material for good life, health, and good social relations. These four support the fifth—freedom of choice and action, meaning "opportunity to be able to achieve what an individual values doing and being" (Millennium Ecosystem Assessment, 2005, vi). It is this fifth component that requires careful examination because when one's freedom to choose and act according to one's values depends on a particular set of ecosystem traits that ecosystem is fundamental to well-being. And yet, no arrows point back to the ecosystem from human well-being—but we know freedom of choice and action in terms of ecosystem service management will directly affect the constituents of well-being. Given this, what are the appropriate management goals and mechanisms to ensure human well-being in relation to Arctic resources? Across

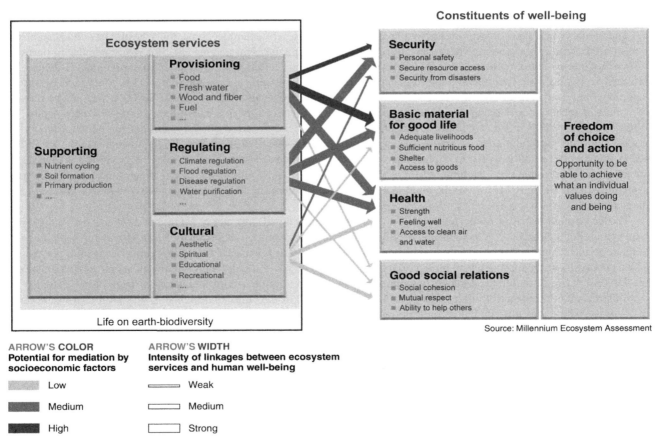

FIG. 2 Ecosystem services and constituents of well-being. *(Source: Millennium Ecosystem Assessment: Ecosystems and Human Well-being: Synthesis. Island Press, Washington, DC, 2005).*

the Arctic's systems of coastlines, there is a mixture of highly localized natural resource-dependent communities alongside, although often not spatially near, other communities whose food security, jobs, and social lives are not as tightly tied to the region's ecology.

2 THE ARCTIC COASTAL MARGIN AND ITS SOCIAL-ECOLOGICAL SYSTEM

In this chapter, we treat the Arctic as a whole, and its coastal systems, from the perspective of social-ecological (or social-environmental) systems (SES) thinking. The coastal inhabitants across the Arctic live in communities that are dependent upon terrestrial and marine systems for their ecosystem services, not only in terms of provisioning, but also for services such as transportation and cultural continuity (ACIA, 2004; Millennium Ecosystem Assessment, 2005; Carson and Peterson, 2016). In this and other ways, the social systems and the ecological systems are coupled, so that feedbacks form across the subsystems (Chapin et al., 2006). Some of the most dramatic fluctuations occurring in the Arctic, many of which have garnered international media attention as symbols of the rapidly warming Arctic, happen on the *coastal margin*. We use this phrase intentionally. The coastlines, deltas, estuaries, and other features of the Arctic Ocean system are marginal if we understand the word "margin" to mean "situated on a border or edge."

While we are mindful that the products of cryospheric and coastal services are subject to becoming private or public goods based on national jurisdiction, we argue that the seasonal marine cycle, and its effects on coastal areas, creates an inherently interdependent set of common pool resources across the Arctic. System users are *socioecologically* interdependent (Sarker et al., 2008) because, externalities, positive or negative, in one area of management (e.g., ship traffic, oil exploration) will likely impact another (e.g., search and rescue, marine mammal harvests). The interdependencies currently "transcend the space and levels of management" of the resource system (Brondizio et al., 2009), but this does not imply that the policy goals of all interested parties are the same. Using sea ice as an example, a review of sea ice system services (Lovecraft and Eicken, 2009; Eicken et al., 2009) demonstrates that Arctic sea ice has been a part of human social practices as far south as Japan and as long as 10,000 years or more. Myriad examples include the ice as a place of enculturation, a platform for industry, a habitat for animals, a hazard for traffic, travelways for hunters and explorers, a buffer to coastal communities, and a part of spiritual identity.

The natures of coastal communities across the Arctic are diverse, for example, some are large and have had established ports on the Arctic Ocean since the medieval era such as Arkhangelsk, Russia (Fig. 3). But many people, in particular in North America and Greenland, live in small coastal settlements, often with a majority of indigenous residents, and primarily rely on subsistence fishing, hunting, and gathering for food and cultural continuity. This contrast in livelihoods and place-based experiences for different coastal locations means any evaluations with DAPSI(W)R(M) must be sensitive to scale and local conditions when evaluating the connectivity between marine ecosystems, terrestrial ecosystems, and further out at sea. This inherently includes indigenous perspectives but is not limited to them nor intended to rank or otherwise value any particular approach more than another. The SES mode of analysis has similar attributes to the DAPSI(W)R(M) framework in that it works to capture the iterative and complex interactions between the governed system and its societies. In this chapter, one can think of SES as the theoretical structure guiding the explanation of the pan-Arctic coastal system and DAPSI(W)R(M) as a toolkit that can be applied to determine key variables in the system for environmental management.

The ecosystem services model employed in the MEA has shaped how many scientists describe the relationships between ecological and social systems especially in the previously hidden or taken-for-granted ways in which ecosystems coproduce human livelihoods and well-being. Since the initial release of the Millennium Ecosystem Assessment (2005) many different approaches to model relationships between societies and their environments have been developed, the DAPSI(W)R(M) included. A recent version that has gained traction among managers is Díaz et al. (2015) which demonstrates that ecosystem thinking has gone beyond inclusion of the "human dimension" solely in terms of economics or governance and now includes multiple perspectives toward human-environmental relationships and knowledge systems. For instance, in the Intergovernmental Platform on Biodiversity and Ecosystem Services (IPBES) framework (a science-policy bridging organization that advises parties to the Convention on Biological Diversity, among other treaties), ecosystem services are also defined by some stakeholders as "nature's gifts to people"—a concept that aids the advice body in more correctly identifying values defining how society and the environment are related as well as issues to be aware of in promoting governance advice. IPBES also considers indigenous knowledge and local knowledges as critical inputs into how the body "knows" what dynamics it is considering. We mention the work of the IPBES because the Arctic cannot adequately be described in terms of instrumental uses, such as services that may have substitutions available, or knowledge based solely on scientific studies. To do so would be to ignore the vast differences in cosmology and observational experiences that natural resource agencies and indigenous peoples in the Arctic bring to discussions

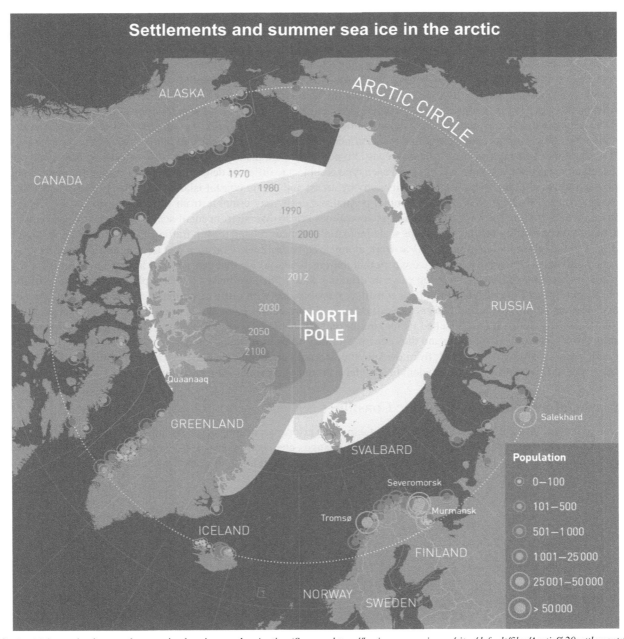

FIG. 3   Major and minor settlements in the circumpolar Arctic. *(Source: https://horizon-magazine.eu/sites/default/files/Arctic%20settlements%20 graphic.jpg.)*

related to lands, waters, living resources, and the often shared governance thereof where knowledge is often "coproduced" (Daniel and Behe, 2017).

2.1   Arctic Coastal Ecological Systems

The coastal zones of the Arctic are diverse in their physical nature and extensive. The Arctic Ocean occupies a roughly circular basin and covers an area of about 14,056,000 km² (5,427,000 sq mi), almost the size of Antarctica. The coastline of the Arctic Ocean is approximately 45,390 km (28,200 mi) long and the Arctic Ocean, while the smallest of the world's five oceans, is over 14 million km² (The World Factbook, 2018). For Russia and Canada, their Arctic coastlines constitute a majority of their coastal possessions. These areas are almost entirely without trees. In some places, such as Greenland

and Norway, the coastlines are mountainous terrain; in other places, there are extensive wetlands and lakes that end in high bluffs such as in Russia. These different ecological features mean the fauna, while all similar in terms of adaptation to Arctic conditions (e.g., animals with blubber layers, migration strategies), and flora are also varied. The defining physical feature for Arctic coasts is ice - both marine and terrestrial. Sea ice, in particular the shore-fast sea ice in winter, calms wave activity and creates a "… deceptively static counterpart to southern coasts where the activity of the ocean conveys an impression of ceaseless dynamism. Sea ice, though, can move rapidly and with great power under the influence of wind and currents" (Atkinson, 2011 p. 219). The importance of sea ice to the pan-Arctic SES is reflected in its study as a system in which sea ice provides services that are regulating, provisioning, cultural, and supporting (Eicken et al., 2009). The other elements of the cryosphere—snow, land ice, and permafrost—also affect coastal zones and are shifting in their parameters as noted in 1.1.

The circum-Arctic region has a variety of coastal types. These are discussed below with material primarily drawn from the excellent overview by Atkinson (2011) and each type is subject to different deterioration in response to climate changes— deltaic and lowland plain, permafrost coastlines, rocky areas, and Arctic coastal islands. All of these types are vulnerable to changes due to thermal erosion, for example, permafrost cliffs may crumble from increased ground temperatures and without a sea ice buffer. The human relationship to coastlines is also diverse from small-scale subsistence, often regulated by millennia-old community practices, to large-scale industrial practices related to ship traffic and governed by the Polar Code.

Fig. 4 details the ecological processes at work along Arctic coastlines. Consider that the image, however, starts on the left-hand side with multiyear ice, or the concept that there is thick old ice that can serve as a platform. Multiyear ice contains air pockets and so little brine it can serve as drinking water; this also makes it a firm platform and more difficult for icebreakers to pass. Further left would be a coastline, possibly one with cliffs of permafrost or maybe a marine terminal glacier. Sea ice calms waves and buffers shorelines not only protecting lands but also providing an extension of land for people. This year Arctic sea ice reached its maximum on March 17 at 14.48 million km^2. This is the second lowest maximum in the 39 years of satellite data and the four lowest maximums have occurred in the last 4 years. As sea ice diminishes and dark water absorbs more warmth from the sun, the warming effects pervade the cyrosphere of the coastal regions and destabilize a SES that has grown dependent on regular patterns of seasonal ice formation.

2.2 Social Systems Among Arctic Coastlines

People living on Arctic coastlines are generally facing precarious conditions of geographical isolation, limited transportation options, shifting seasonality and unidirectional changes characterized by erosion and species availability, and the effects of ice and cold on equipment and infrastructure. As Fig. 3 demonstrates there are some large settlements on Arctic

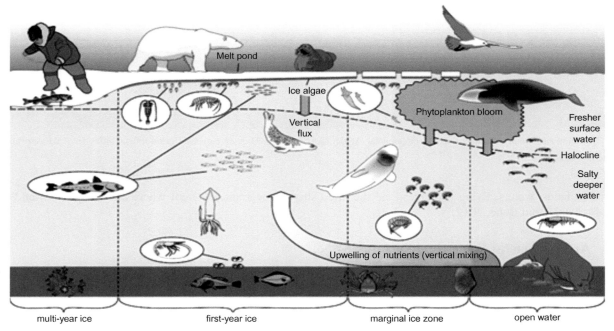

FIG. 4 The ecological processes at work along Arctic coastlines. *(Source: CAMAP.)*

coasts, primarily in Russia that have longstanding and significant infrastructure, but along most coastlines populations are sparse. In terms of human settlement, it is important to note that prior to the 20th century the norm was annual cycles of migration around Arctic coastlines with nomadic and seminomadic groups adapting seasonally for different activities. The modern era demonstrates a repeated pattern of forced relocation as different jurisdictional hegemons dominated coastal regions, expanding ports and following markets for whale, fur, or other circumpolar commodities. The persistent effects of this imperialism and colonialism have resulted in numerous pan-Arctic state-sponsored projects to settle indigenous communities into sedentary patterns surrounding western "infrastructural nodes such as churches, schools, and stores (many of these nodes were along the coast due to colonial transportation networks and technologies)" (Schweitzer, 2011, p255). As a consequence, there are communities throughout the Arctic—with the possible exception of Fennoscandia coasts—located on state-selected coastlines that are now literally crumbling due to increasing storm activities, warming, and coastline collapse (*ibid*). In short, as one considers management strategies of a coastal system from a perspective that considers both human and environmental assets, it is worth noting that the vulnerability of many coastal communities in the Arctic has been socially constructed (Marino, 2012). It is far less a product of the communities themselves—their people and management of natural resources—and far more a result of colonial and settler government decisions.

2.2.1 Governance of the Arctic System

In terms of international governance, the Arctic is made up of eight sovereign nations often divided up as the Nordic countries—Norway, Sweden, Finland, Denmark (Greenland), and Iceland; North America—Canada and the United States; and Russia. This list includes what is sometimes discussed as a ninth polity, the Inuit—via the Inuit Circumpolar Council (ICC). Of the approximately 4 million people living in the Arctic nearly half a million are indigenous. The eight nations developed the Arctic Council in 1996 and have included six indigenous peoples' organizations as permanent participants. These include: include: the Aleut International Association, the Arctic Athabaskan Council, Gwich'in Council International, the ICC, the Russian Association of Indigenous Peoples of the North, and the Saami Council. The permanent participants have "full consultation rights." Occasionally, there is a further split in terms of policy by discussing the five coastal (littoral) nations—Russia, Canada, Norway, Denmark, and the United States—which are also in possession of the majority of the land and sea in the Arctic, thus the coastal systems. In addition, the Arctic Council provides observer status for non-Arctic states, such as the United Kingdom and the People's Republic of China, along with other entities (e.g., Nordic Council of Ministers, International Federation of Red Cross and Red Crescent Societies, Association of World Reindeer Herders, University of the Arctic).

The Arctic Council, while the most significant of the pan-Arctic cooperative organizations, is not actually a governing body. It grew out of an initial 1991 environmental cooperative agreement, the Arctic Environmental Protection Strategy, that became the intergovernmental forum of today. Primarily, the Arctic Council operates through six working groups designed around monitoring, conservation, emergency response, and sustainability development. These groups produce assessments and recommendations related to a variety of Arctic social and environmental conditions that tend to drive policy making and agenda setting in the region. The Arctic Council also serves as a forum for legally binding instruments across nations such as the cooperative Arctic Search and Rescue agreement in 2011 and more recently the Agreement on Enhancing International Arctic Scientific Cooperation in 2017. The Chairmanship rotates every 2 years among Arctic states with the United States, having just passed this position to Finland in 2017.

Despite their size variations, the Arctic coastal communities of these Council nations are marginal in the sense that their sociocultural and political-economic identities are tied to the edge, the Arctic Ocean, with a majority located on the periphery of national governance and major population centers. This distance is most pronounced in Canada, the United States, Greenland (which is still technically a part of Denmark), and parts of Russia but less so for the smaller Nordic countries. In addition, the federal nature of the governments in North America and Russia as well as the 2009 Act on Greenland Self-Government, that further extends powers enacted in the 1979 Home Rule Act, create subgovernments that may have significant differences of opinion and priorities from national legislatures, agencies, and institutions. For example, the United States is an Arctic nation due to a single state, Alaska. But, the state of Alaska has priorities and approaches toward the Arctic that may not always correspond with national concerns. For example, in North America low cost of oil on international markets generally benefits consumers at a national level but it can deeply hurt communities in Alaska largely tied to oil revenues.

2.3 The Significance of Arctic Coasts for Subsistence Livelihoods

One way to clearly see the linkages between the physical nature of Arctic coastal systems and social systems is through food, or subsistence practices. These concepts can mean different things in the North. The indigenous population surrounding the Arctic Ocean is primarily Inuit. The ICC is a nongovernment entity that represents more than 160,000 Inuit people in Canada, the United States, Greenland, and Russia. The Inuit describe themselves *"We Eskimo are an international community sharing common language, culture, and a common land along the Arctic coast of Siberia, Alaska, Canada and Greenland. Although not a nation-state, as a people, we do constitute a nation"* (ICC—Canada). As noted above, they are only one of six permanent participants, but the Inuit are the primary indigenous group along Arctic coasts except for Fenno-Scandia areas and the majority of the Russian coast. In these areas reindeer herding cultures dominate. In Northern Europe the Sami people's territory, or Sapmi as the indigenous people call what others, in English, have labeled "Lapland" covers approximately 400,000 km2 and has a population of nearly 2 million people, though not all Indigenous. In their traditional areas of Northern Scandinavia, there has been a long discussion in relation to Mountain Sami rights vs Coastal Sami rights. Currently, Sami rights to marine resources are still contested along the Northern Norwegian coast (Brattland, 2010). The Sami are permanent participants but hold no fishing or marine hunting/harvest rights in their nation-states (Latsch, 2012). In Northwest Arctic Russia along the Kara Sea, researchers have documented Nenets reindeer herding as it is affected by coastal oil fields development (Kumpula et al., 2011). Other reindeer-herding societies along the Russian coasts also use the coastal plains for grazing.

When facing potential pressures from industrial development and other anthropogenic drivers of change, Inuit across the Arctic reaffirm their need for food security. For the Alaskan branch of the ICC, food security is "…characterized by environmental health and is made up of six interconnecting dimensions: (1) Availability, (2) Inuit Culture, (3) Decision-Making Power and Management, (4) Health and Wellness, (5) Stability, and (6) Accessibility" (Inuit Circumpolar Council - Alaska, 2015: 31). The word subsistence across northern indigenous cultures is often tied to food, but it is often expressed as a way of life, the ability of one to be out on the lands and waters, gathering foods, processing them, sharing them, and nourishing communities, cultures and traditions (Natcher, 2009). As traditional foods and their availabilities change with unexpected seasonal variations and economic development, changes in community structure and function have already been noticed. Natcher (2009) writes that,

> Owing to the incompatibility of subsistence and "modern" wage economies, Inuit family structure, values, and expectations have been altered to the point where traditional forms of socialization are being devalued (Hund, 2004: 1). As a result, "the functioning of social networks have been affected by a decrease in importance of the extended family unit and the emergence of intergenerational segregation, a decline in the practice of traditional cultural values, a concentration of resources in fewer hands, and the emergence of social conflict". (Ford et al., 2008: 54)

In relation to the DAPSIWRM framework, capturing the linkages of "living resources" to both state and driver variables means that decision makers must manage for multiple objectives. For example, the human-walrus relationship that functions as the means of experience-as-cultural existence for Inuit in the Bering Strait region (detailed in Metcalf and Robards, 2008) is directly affected by drivers such as subsistence food and industrial fisheries markets as well as feedbacks from the measures taken by resource managers in relation to state and driver variables. The concept of food sovereignty, in which indigenous peoples are the ultimate stewards of their food sources (Fig. 5), is both a goal of the ICC as well as other indigenous peoples and recognition of this standpoint would perhaps turn upside down the current "balance" between dominant governmental acceptance of industrial uses of the coast and the living resources of the people living there.

3 THE COMPLEXITY OF ECOSYSTEM MANAGEMENT

The ecosystem management literature discusses the shift from an original framework of drivers-pressures-state-impact-response (DPSIR) to one that has clarified several aspects of these variables and is now called DAPSI(W)R(M) (Elliott et al., 2017; Scharin et al., 2016). DAPSI(W)R(M) is also intentionally designed to assist environmental managers in "capturing" the system to be governed (Scharin et al., 2016). In doing so, it is similar to models of ecosystem services provisions in SESs but with a more extensive focus on how, as societies acquire the constituents of well-being, this creates drivers and eventually impacts environmental, in this case coastal Arctic, systems thus requiring measures. However, because the Arctic has diverse peoples with often distinctly different relationships to what has been labeled in the framework as drivers it is worth a brief review to understand the framework's underlying assumptions. Elliot et al. (2017) use Maslow's hierarchy, the original and updated versions (1943, 1970) to explain drivers as basic human needs through self-actualization. While this is a solid place to start when one begins to consider seriously the governance of Arctic coastlines from pan-Arctic to local perspectives, several critiques of Maslow's work become important. First, his work using biographical analysis excluded people from oral traditions, as well as having a very small sample of women. In addition, lower and higher needs are not necessarily causally related (with a person having to fill lower before higher) as there are numerous examples of places where there is persistent and dramatic poverty and yet people are still able to experience love and a sense of belonging. Consider the need to eat. Basic human sustenance is a driver. But how one fulfills this, the activities that create food for

FIG. 5   The Inuit (Alaska) perspective related to food security as interlinking puzzle pieces that all demonstrate an interlinking between social and natural phenomena. *(Source: Inuit Circumpolar Council—Alaska. 2015. Alaskan Inuit Food Security Conceptual Framework: How to Assess the Arctic from an Inuit perspective. Technical Report. Anchorage: ICC-Alaska. Available from www.iccalaska.org.)*

people, can be dramatically different. Tay and Diener (2011) tested Maslow's theory comparatively using 60,865 participants from 123 countries between 2005 and 2010. They found that there are universal human needs existing regardless of cultural differences, but how these are ordered within the low-to-high hierarchy proved false. In short, one need not have food to have fulfillment of friendships. Elliott et al. (2017) discuss this in relation to the satisfaction of human wants through modern market mechanisms, noting, "higher order needs are not satisfied by ever increasing want satisfaction" (p. 31). But for some cultures specific foods procured in particular ways are vital to sustaining knowledge systems and the capacity for adaptation and self-determination.

Nation-states promised to help to sustain indigenous cultural pathways and knowledge systems when they signed the United Nations Declaration on the Rights of Indigenous Peoples. All Arctic States with the exception of Russia have expressed support for the declaration and are at various stages in meeting its goals. The Arctic Council reports that roughly half a million indigenous people live throughout the Arctic, speaking over 40 languages. Such metrics are complicated by the fact that countries define and count "indigenous" differently. Nonetheless these peoples, in particular the Inuit, have worldviews rooted in indigenous knowledge and lived experiences that often create profoundly different problem definitions (Rochefort and Cobb, 1993; Stone, 2012) for environmental management. Thus, management expectations, data gathering strategies, goals for resources and monitoring, and cultural relationships with ecological systems often differ from Western realities using scientific approaches (Harding, 2004).

3.1   Applying the DAPSI(W)R(M) Framework to Arctic Coasts

Multiple authors in this edited volume have explained the DAPSI(W)R(M) framework. We do not think it needs repeating. However, we do follow the line of thought that any one Arctic coast's context will have its own interdependencies and cycles within the framework. Similar to the overview of Baltic Sea by Scharin et al. (2016), the Arctic coastal regions have "interdependencies between environmental state changes impacts and policy responses" (p. 56) in addition one could apply the DAPSI(W)R(M) cycle at different scales or along different research trajectories such as a particular environmental problem (e.g., marine pollution), or a location (e.g., Baffin Bay), jurisdictional unit (e.g., Canada), or interest (e.g., shipping) meaning that "these cycles can be linked and nested within a system to provide a more holistic view of the complexity of the state of the marine environment" *(Ibid)*. Again, after Scharin et al., we consider that the "nested cycles for different drivers (the need for goods, transport, living space, etc.) constitute endogenic managed pressures (EnMP) onto which are superimposed the effects of Exogenic Unmanaged Pressures (ExUP)" (2016, p58). The Arctic has been subject to numerous recent studies related to resource management pressures (e.g., ACIA, 2004, Carson and Peterson, 2016; Larsen and Fondahl, 2015; Rasmussen, 2011) but our chapter has a more modest goal, to inform users of the DAPSI(W)R(M) decision support tool of pertinent information related to environmental management on Arctic coasts (Table 1).

Drivers are meant to be the basic human needs often met through social processes that directly or indirectly drive human systems such as marketplaces, identity formation, governance, spirituality, and food production. We argue that all drivers, and related metrics must be considered within the umbrella of the Anthropocene, the distinct geological epoch defined by significant alteration of the earth by humans that would not be otherwise present (Crutzen, 2002, Waters et al., 2016). Elliott et al. place aesthetic and transcendence needs into drivers (Elliott et al., 2017). We would not disagree but argue that the human need to satisfy wants includes exchanges of goods and services and the jurisdictional attributes of the world. This need is, for the purposes of using this framework for current and near future management, fairly fixed by the socio-environmental features in the Anthropocene (e.g., climate changes, landscape fragmentation, and ocean pollution) in world dominated by nation-states and liberal economics. In fact, resistance against any of these drivers demonstrates their impact on the nature of human understanding; especially if we are discussing coastal zone management. We place any of these generally long-term attributes of a globalized world in the drivers category rather than activities. In the globalized world of commodity and financial flows alongside near instantaneous communication pathways, it is very difficult to disentangle a human want, or need, from the systems of international trade and communication that both causes and provides for the want. Do we desire self-governance because of or in spite of global governing regimes and the hegemonic wielding of power by market actors and nation-states? To avoid a much more lengthy parsing problem, we separate drivers and activities by permit requirement. If you need a permit or any form of government (at any scale, including informal permission from cultural groups) permission to act you fall in the activity category. Consequently, if we think of market forces placing pressing into the Arctic via marine traffic, the press of market forces is a driver, but the entrance of marine traffic into the Arctic Ocean requires following permitting and regulations so is an activity.

Some activities tied to Arctic coasts are similar to many other coastal regions. Extraction of living and nonliving resources, aquaculture, transport and shipping, and navigational and infrastructure efforts to support them, energy production (renewable and nonrenewable), tourism and leisure, industry on shore including agriculture, military, search and rescue, scientific research, and education would encompass the major activities (Smith et al., 2016; Scharin et al., 2016). We would add to this list a phrase of "home" or simply "living life" to try to capture a sense of the Inuit perspective of the Arctic Ocean—it is not a place of discrete activities but of living one's life.

Pressures are the result of activities that stem from the world of human wants and the systems that create wants. These can either be endogenous to the system—in this case the pan-Arctic coastal system surrounding the Arctic Ocean—or exogenous from it. The boundary between these is tricky. When we think of the Arctic people generally talk about climate changes as key factors that place pressure on SESs. Climate forcers can be short term such as aerosols and black carbon and long term such as greenhouse gases. While the Arctic does produce greenhouse gases that are warming the ocean, air, and permafrost it is not the major producer of them (one could consider them exogenous) and yet local Arctic demand for heating one's house is an endogenous feature contributing, on a much smaller scale, to global warming. For example, the reliance on global production systems of oil and gas for home heating crosses the exogenous/endogenous Arctic boundary. This creates short-term forcers (e.g., black carbon) that are commonly produced related to heating homes with biomass or diesel.

Pressures are, or stem from, the externalities of activities. They are what create state changes (environmental impacts) and impacts (impacts on human well-being). The ecological pressures have been noted earlier in the chapter, the human welfare question depends on one's standpoint. If one is an employee of a multinational shipping company in the Arctic, the pressures of climatic change such as diminishing sea ice creates opportunities for advancing a market-based approach to accumulation of material goods. If one hunts seal from or travels distances to visit family on sea ice its decline creates danger, diminishes family and cultural ties of sharing, and reduces healthy caloric intake. It is in this way that the DAPSI(W)R(M) framework can only categorize the social-ecological effects of drivers, activities, and pressures; it cannot value them or prioritize them. Furthermore, depending on one's culture and belief system, state and impact categories cannot even be teased apart. For the Inuit, and perhaps other nonindigenous people who view themselves as a part of their environment, a combined "state changes-impacts on welfare" category (number 7 in the table) makes far more sense for developing management practices. Responses through governance, or management measures that are government-formed regulations for Arctic coastline are primarily developed to the South, away from the coastal margins. See Robards and Lovecraft (2010) for an extensive review of how measures can lack "fit" between policy goals and operational outcomes for the people affected

TABLE 1 The DAPSI(W)R(M) Framework Explained in Pan-Arctic Coastal Context

DAPSI(W)R(M) Framework	Pan-Arctic Coastal System
1. Drivers (features of human wants in the globalized Anthopocene). The systems themselves are unregulable and have no singular fixed institution or cause.	Identity formation Spirituality-ethics Caloric intake (living existence) Markets (material acquisition) Governance (social security) Geopolitical change Technology Planetary processes
2. Activities (what is regulable). Activities that can be causally, in part, located and identified by institutions.	Industrial development and the spread of infrastructure. Development is primarily extractive either through non-renewables (e.g., fossil fuels, minerals) or renewables (e.g., energy, fish, reindeer, tourist experience) Transport, shipping, and associated infrastructure Infrastructure such as cables, satellite technology, oilfield development, buildings, and ports. Development of human capital through educational processes Science production through research processes
3. Pressures (the results from Activities that will result in State changes)	Climate forcers Land fragmentation Local permafrost degradation Marine and terrestrial pollution (includes sound and light) Ocean acidification Ocean warming Slow population increase Expansion of capitalist markets Global commodity prices Feedbacks to global system
4. State Changes (Environmental)	Cryosphere changes in seasonality, thaw, and melt Ecosystem changes, biome shifts on longer time frame Phenology changes Changing climate patterns Regional/polar permafrost degradation Loss of sea ice platforms Atmospheric CO_2 increase
5. Impacts (Human Well-being)	Loss of fate control Demographic change Industrialization Loss of time on the land (tied with desk jobs) Pollution and health effects Privatization of the commons Increased shipping brings opportunities and risks More time spent in bureaucratic and geopolitical meetings steering development Reduced access to subsistence foods in/near industry Greater access to healthcare
6. Responses (M)	Global policies: U.N. Declaration on the Rights of Indigenous Peoples Pan-Arctic responses: The International Maritime Organization created the Polar Code (International Code for Ships Operating in Polar Waters) to regulate ship traffic, safety, and environmental pollution. It came into force January 1, 2017 (http://www.imo.org/) National responses (e.g. treaty obligations): Agreement on the Conservation of Polar Bears Treaty of 1973 Subnational responses: regulations from states, territories, provinces, and other locations in the Arctic
7. Combined 4+5	Stresses on the human-environmental coupled relationships: human-walrus; human-salmon, human-rangifer, human-whale complex

Sources: Arctic Council (2016). Arctic Resilience Assessment. M. Carson and G. Peterson (Eds.), Stockholm Environment Institute and Stockholm Resilience Centre, Stockholm. Available from http://www.Arctic-council.org/arr.; Arctic Council (2013). Arctic Resilience Interim Report 2013. Stockholm Environment Institute and Stockholm Resilience Centre, Stockholm. Available from http://www.Arctic-council.org/arr.; Larsen, J.N. and G. Fondahl (eds). (2015). Arctic human development report: regional processes and global linkages. Copenhagen. https://doi.org/10.6027/TN2014-567

by such measures. In recent decades, scientific literature has recognized how relationships between humans and animals affect the governance of industrial development. We cannot review how each nation-state balances indigenous vs industrial resource uses, but we want to call attention to several sets of relationships, or complexes, in the Arctic, including walrus-human and seal-human relationships. Subsistence harvesting and occupation along Arctic coastlines have been documented to stretch far back thousands of years. In particular, the large animals or herds that require multi-person or community efforts to catch, butcher, process, and share, have in turn shaped human relationships both to those animals as well as between members of those societies. As the Eskimo Walrus Commission in Alaska has noted, these relationships with animals, and their ecosystems, are about more than just collecting food (Eskimo Walrus Commission (EWC), 2003). Governments who recognize the UNDRIP and generally recognize human rights of indigenous peoples must make sure that industrial development does not interfere with these state variables made up of webs of relationships, resources, and communities. This is the vision of Arctic sustainable development adopted by the Arctic Council.

3.2 Measures and the Advent of the Arctic Council

We use the creation of the Arctic Council as the place in time to gauge the Arctic system with DAPSI(W)R(M). This framework analysis cannot exist outside of time, but for practical management purposes in the Arctic there are only a handful of significant pan-Arctic coastal system measures that precede the existence of the Arctic Council. This is due to the technological, climatic, and geographical inability for the Arctic to be its own entity (with any degree of accuracy or legal definitions), it isn't that the Arctic did not play a role in the human imagination as a place. Again, while not a governing body, the Arctic Council informs management of Arctic Coasts and serves as the locus for national and subnational management research and development. The Arctic Council itself is a measure—originating from pollution fears. But it is a measure that demonstrates the interactive and interdependent elements of DAPSI(W)R(M). It is an institution born of drivers related to activities that created pressures and led to state changes that eight nations felt were significant enough to warrant the formation, the measure, of a significant institution related to environmental management.

4 THE POSSIBLE FUTURES OF ARCTIC COASTAL ENVIRONMENTAL MANAGEMENT

The Arctic as a system of nation-states is in fact an ocean surrounded by these nations. This means different arctic measures will reflect governance measures of the nations and their populations. Nonetheless more general goals are shared, debated, researched, and put into place both at Arctic Council level, domestically, and in bilateral and multilateral agreements. While national-level priorities and actions set the stage for the behavior of governments in the North, they do not always determine government activities, largely due to pressure groups that can flourish in democracies. In other words, much of what is "Arctic Policy"—or actual governance in Arctic and sub-Arctic locations, is determined by national legislatures, state/territorial/provincial governments, comanagement regimes with indigenous peoples, local-scale legislatures, and even the activities of major city councils (e.g., zoning, education, and taxation). While this will be true in any Arctic nation, three very brief examples from the United States demonstrate how the Arctic as a location of measure-making may sink or rise on the national agenda depending on budgets, partisanship, and perceptions of risk. First, the US Senate's repeated, for decades, refusal to ratify UN Convention on the Law of the Sea (UNCLOS) in spite of requests from multiple presidential administrations and government officials from Alaska as well as international pressure. But second, on the other hand, President Bush signed a joint resolution with Congress in 2008 that resulted in the 2009 Arctic Fishery Management Plan, the closure of the US Arctic Management Area to commercial fishing pending more research on the marine environment. And third, across administrations, and despite military and scientific pressures, the United States possesses only a single icebreaker in commission, the *USCG Healy*. These policy paradoxes—of a G7 nation sometimes adopting and other times rejecting the governance conventions of arctic countries, seemingly to its detriment—are due to national and state politics.

4.1 Managing Change or Changing Management?

In this chapter, we addressed the DAPSI(W)R(M) from the perspective of its application to the pan-Arctic coasts with examples tied to management possibilities and realities. We feel strongly that as change is managed in the Arctic in new ways due to the rapidity and uncertainty surrounding climate, the cryosphere, and development, the opportunity to change management itself arises. This opportunity for change should reflect growing evidence and awareness of the importance of indigenous knowledge, the interrelationships between state variable change and impacts to human well-being, the diversity of measures at national and subnational levels that affect coastlines, and how political forces at these levels can dramatically shift the goals and practices of environmental management.

In summation, for Arctic coasts, exogenous pressures are outpacing endogenous pressures and governance capacity in the Arctic is scrambling to adapt. Concurrently, people and corporations within this coastal system want to capture the ability to be industrial actors or gain from the development of the region. In light of this, one should keep three facts in mind. First, there has not been a "gold rush" in the Artic as predicted only a decade ago in popular journalism. The unpredictability of climate change and enormous up-front investment from companies to operate effectively in the Arctic Ocean has kept massive fast development at bay. The resulting steady pressure may allow better adaptation and the review of many different forms of environmental management for goodness of policy fit. Second, building on the first fact, there is a need for nimble governance very different from what came before. This is where local scale practices and comanagement can offer real benefits because climate and other models are not yet able to fully downscale or predict a variety of events. The development of flexible adaptive management, we argue, should not split the state changes and impacts but examine them as integrated features of Arctic coastal life and manage them as such. Lastly, while there is much regulation of coastlines across the Arctic, they are a vast and complicated feature of ocean systems. Along with better "bottom-up" driven management, there needs to be a renewed focus on "top-down" research, debate, and discussion at the Arctic Council level in order to offer regulatory predictability and thus economic development that is guided by sound scientific principles informed by the Arctic nations and permanent participants.

5 CONCLUSION

Since the Illulisat Declaration in 2008, signed by the five Arctic Ocean littoral nations at the Arctic Ocean Conference, the United States has sent two policy signals about the future of its Arctic policy. First, the national government has committed itself to no new regime formations and to using the current international legal process (UNCLOS) to resolve disputes. It is worth quoting a portion of the declaration, although from a document only two pages in length.

> *In this regard, we recall that an extensive international legal framework applies to the Arctic Ocean ... Notably, the law of the sea provides for important rights and obligations concerning the delineation of the outer limits of the continental shelf, the protection of the marine environment, including ice-covered areas, freedom of navigation, marine scientific research, and other uses of the sea. We remain committed to this legal framework and to the orderly settlement of any possible overlapping claims. This framework provides a solid foundation for responsible management by the five coastal States and other users of this Ocean through national implementation and application of relevant provisions. We therefore see no need to develop a new comprehensive international legal regime to govern the Arctic Ocean. (italics added)*

To some extent, this has put to rest the discussion about the applicability of the Antarctica Treaty as a model for the Arctic. While depth of understanding of the former is insightful for shaping viewpoints and policies for the latter, for example related to issues of importance for both poles (e.g., scientific endeavors, tourism, and species preservation), it is unlikely to serve as a model of governance (Koivurova, 2010). The following comparison is not novel, but bears consideration. Antarctica is an island, a continental land mass surrounded by water with no indigenous human inhabitants, and a climate that dramatically restricts human endeavors. The Arctic is primarily an ocean surrounded by sovereign nations and inhabited by the Inuit and other indigenous groups for millennia. While debates may remain over transformative Arctic governance, it seems unlikely that any of the Arctic nations will take dramatic action altering current governance in the coming decades.

Second, Secretary Clinton's attendance at the 2011 biannual Arctic Council Ministerial meeting in Nuuk, Greenland signaled that the United States would play a greater role in Arctic institutions of governance at the global level and continue its internal elevation of Arctic issues in its bureaucracy (Myers, 2011). It did this through President Obama's term and his historic visit to Alaska in 2015. The current executive administration has yet to signal any specific new strategy for the Arctic as a whole, except for a pro-extraction development stance. Onshore this is seen through the US national government vote in 2017 to open the Arctic National Wildlife Refuge to oil exploration and the 2018 Bureau of Land Management's following notice of intent for environmental impact analysis. The latter is significant for offshore drilling in coastal regions (Fig. 6) and demonstrates how measures, including long-standing ones guiding entire socio-legal regimes, can be changed, even reversed, through politics depending on the members of any particular government.

On a 5-year cycle, the US Department of the Interior develops a plan determining where oil and gas companies can purchase leases for offshore drilling. The most recent plan was completed under the Obama administration in January 2017 and it excluded any new oil and gas leasing in Alaska's offshore Arctic waters through 2022. However, under the Trump administration, which came into office late in January 2017, there is a proposal announced and drafted that covers 2019–24 and promotes the removal of Arctic protections. While this plan will take possibly years to complete, in March 2018 the US Bureau of Ocean Energy Management (BOEM) formally asked where companies may want to drill in advance of leasing

Conceptual model of arctic oil spill exposure and injuries

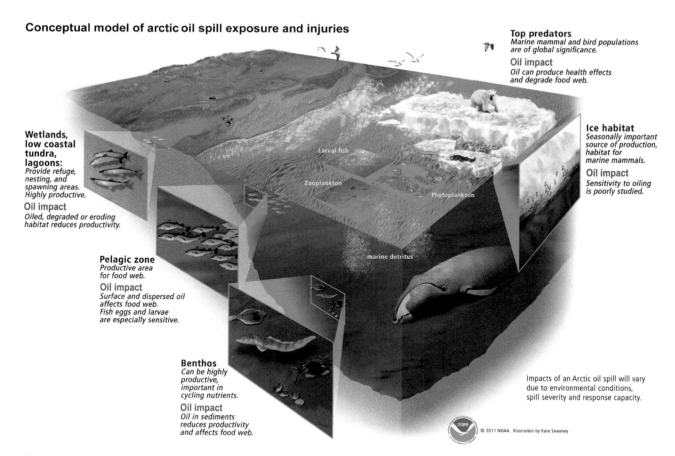

Top predators
Marine mammal and bird populations are of global significance.
Oil impact
Oil can produce health effects and degrade food web.

Ice habitat
Seasonally important source of production, habitat for marine mammals.
Oil impact
Sensitivity to oiling is poorly studied.

Wetlands, low coastal tundra, lagoons:
Provide refuge, nesting, and spawning areas. Highly productive.
Oil impact
Oiled, degraded or eroding habitat reduces productivity.

Larval fish
Zooplankton
Phytoplankton

marine detritus

Pelagic zone
Productive area for food web.
Oil impact
Surface and dispersed oil affects food web. Fish eggs and larvae are especially sensitive.

Benthos
Can be highly productive, important in cycling nutrients.
Oil impact
Oil in sediments reduces productivity and affects food web.

Impacts of an Arctic oil spill will vary due to environmental conditions, spill severity and response capacity.

© 2011 NOAA. Illustration by Kate Sweeney

FIG. 6 Potential of impacts of oil spill in Arctic waters, NOAA 2011. *Illustration by Kate Sweeney. Online: https://response.restoration.noaa.gov/multimedia/infographics.*

sales. This commenting period ended in April and the BOEM is now analyzing the feedback. There is currently a lawsuit filed in federal district court challenging the executive order of Trump administration to jump-start the drilling in Arctic and Atlantic Oceans. To this date, the US District court in Anchorage (where the lawsuit was filed) has denied the executive's effort to have the case dismissed.

If one thinks back to Fig. 2 of ecosystem services, DAPSI(W)R(M) provides a solid tool of analysis to consider the interrelated scales of social-ecological activity that produce human well-being. And yet, as humans choose measures to enact in the Arctic one must be sure that the measures, such as choices about who regulates what and why, themselves do not minimize or destroy the constituents of well-being through a narrow standpoint of what the material life, good social relations, and health mean for indigenous people and others for whom the coasts are a place of subsistence and livelihoods.

REFERENCES

ACIA, 2004. Impacts of a Warming Arctic: Arctic Climate Impact Assessment. Cambridge University Press, Cambridge.

Atkinson, D., 2011. The physical environment of Alaska's coasts. In: Lovecraft, A.L., Eicken, H. (Eds.), North by 2020: Perspectives on Alaska's Changing Social-Ecological Systems. University of Alaska Press, Fairbanks, AK, pp. 229–252.

Brattland, C., 2010. Mapping rights in coastal Sami seascapes. Arctic Rev. Law Pol. 1 (1/2010), 28–53.

Brondizio, E.S., Ostrom, E., Young, O.R., 2009. Connectivity and the governance of multilevel social-ecological systems: the role of social capital. Annu. Rev. Environ. Resour. 34, 253–278.

Carr, J.R., Stokes, C.R., Vieli, A., 2017. Threefold increase in marine-terminating outlet glacier retreat rates across the Atlantic Arctic: 1992–2010. Ann. Glaciol. 58 (74), 72–91. https://doi.org/10.1017/aog.2017.3.

Carson, M., Peterson, G. (Eds.), 2016. Arctic Resilience Assessment. Stockholm Environment Institute and Stockholm Resilience Centre. Stockholm. Available from: http://www.Arctic-council.org/arr.

Chapin, III, F.S., Kofinas, G.P., Folke, C. (Eds.), 2009. Principles of Ecosystem Stewardship: Resilience-Based Natural Resource Management in a Changing World. Springer Science & Business Media, Springer, New York.

Chapin, F.S., Lovecraft, A.L., Zavaleta, E.S., et al., 2006. Policy strategies to address sustainability of Alaskan boreal forests in response to a directionally changing climate. Proc. Natl. Acad. Sci. 103 (45), 16637–16643.

Cohen, J., Zhang, X., Francis, J., Jung, T., Kwok, R., Overland, J., Tayler, P.C., Lee, S., Laliberte, F., Feldstein, S., Maslowski, W., Henderson, G., Stroeve, J., Coumou, D., Handorf, D., Semmler, T., Ballinger, T., Hell, M., Kretschmer, M., Vavrus, S., Wang, M., Wang, S., Wu, Y., Vihma, T., Bhatt, U., Ionita, M., Linderholm, H., Rigor, I., Routson, C., Singh, D., Wendisch, M., Smith, D., Screen, J., Yoon, J., Peings, Y., Chen, H., Blackport, R., 2018. Arctic change and possible influence on mid-latitude climate and weather. US CLIVAR Report 41. https://doi.org/10.5065/D6TH8KGW. US CLIVAR Report 2018–1.

Crutzen, P.J. (2002). Geology of mankind. Nature. (415, 2 January), page 23.

Daniel, R. and C. Behe. (2017). Co-production of knowledge: an Inuit indigenous knowledge perspective. Abstract #C13H-04 given at the Fall Meeting of the American Geophysical Union.

Díaz, S., Demissew, S., Carabias, J., et al., 2015. The IPBES conceptual framework—connecting nature and people. Curr. Opin. Environ. Sustain. 14, 1–16.

Dietz, T., Ostrom, E., Stern, P.C., 2003. The struggle to govern the commons. Science 302. 12 December, 1907–1912.

Elliott, M., Burdon, D., Atkins, J.P., Borja, A., Comier, R., DeJonge, V.N., Turner, R.K., 2017. "And DPSIR begat DAPSI(W)R(M)!" – A unifying framework for marine environmental management. Marine Pollution Bulletin 18 (1–2), 27–40.

Eicken, H., Lovecraft, A.L., Druckenmiller, M., 2009. Sea-ice system services: a framework to help identify and meet information needs relevant for Arctic observing networks. Arctic 62, 119–136.

Eskimo Walrus Commission (EWC), 2003. Conserving Our Culture Through Traditional Management. Kawerak, Nome, Alaska, USA.

Fischer, J., Gardner, T.A., Bennett, E.M., Balvanera, P., Biggs, R., Carpenter, S., Daw, T., Folke, C., Hill, R., Hughes, T.P., Luthe, T., Maass, M., Meacham, M., Nordstrom, A.V., Peterson, G., Queiroz, C., Seppelt, R., Speirenburg, M., Tenhunen, J., 2015. Advancing sustainability through mainstreaming a social-ecological systems perspective. Environ. Sustain. 14, 144–149.

Francis, J., Vavrus, S.J., Cohen, J., 2017. WIREs. Clim. Chang. e474. https://doi.org/10.1002/wcc.474.

Harding, S., 2004. Rethinking standpoint epistemology: What is "strong objectivity"? In: Harding, S. (Ed.), The Feminist Standpoint Theory Reader: Intellectual and Political Controversies. Routledge, New York, pp. 127–140.

Inuit Circumpolar Council - Alaska, 2015. Alaskan Inuit Food Security Conceptual Framework: How to Assess the Arctic from an Inuit perspective. Technical Report. Anchorage, ICC-Alaska. Available from: www.iccalaska.org.

Koivurova, T. (2010). Environmental protection in the Arctic and AntArctica. In: Loukacheva (Ed.), Polar Law Textbook. Copenhagen: Nordic Council of Ministers. Ch 2.

Kumpula, T., Pajunen, A., Kaarlejärvi, E., Forbes, B.C., Stammler, F., 2011. Land use and land cover change in Arctic Russia: ecological and social implications of industrial development. Glob. Environ. Chang. 2 (2), 550–562.

Larour, E., Ivins, E.R., Adhikari, S., 2017. Should coastal planners have concern over where land ice is melting? Sci. Adv. 3 (11), e1700537. https://doi.org/10.1126/sciadv.1700537.

Larsen, J.N., Fondahl, G. (Eds.), 2015. Arctic human development report: regional processes and global linkages. Copenhagen. https://doi.org/10.6027/TN2014-567.

Latsch, A. (2012). Coastal Sami revitalization and rights claims in Finnmark (North Norway)—two aspects of one issue? Senter for samiske studier, skriftserie nr. 18. Available from: septentrio.uit.no/index.php/samskrift/article/download/2356/2177 (online text only Accessed 20 April 2018).

Lovecraft, A.L. and H. Eicken. (2009). The sea ice system services framework: development and application, Chapter 2 In H. Eicken, R. Gradinger, M. Salganek, K. Shrasawa, D. Perovich, M. Lepparanta (Eds.), Handbook on Field Techniques in Sea-Ice Research: A Sea-Ice System Services Approach. University of Alaska Press: Alaska, pp. 9-24.

Marino, E., 2012. The long history of environmental migration: assessing vulnerability construction and obstacles to successful relocation in Shishmaref, Alaska. Glob. Environ. Chang. 22, 374–381.

Maslow, A.H., 1943. A theory of human motivation. Psychol. Rev. 50 (4), 370–396.

Maslow, A.H., 1970. Motivation and Personality. Harper & Row, New York.

Meier, W.N., Hovelsrud, G.K., van Oort, B.E.H., Key, J.R., Kovacs, K.M., Michel, C., Haas, C., Granskog, M.A., Gerland, S., Perovich, D.K., Makshtas, A., Reist, J.D., 2014. Arctic sea ice in transformation: a review of recent observed changes and impacts on biology and human activity. Rev. Geophys. 52, 185–217. https://doi.org/10.1002/2013RG000431.

Metcalf, V., Robards, M., 2008. Sustaining a healthy human–walrus relationship in a dynamic environment: challenges for comanagement. Ecol. Appl. 18 (sp2).

Millennium Ecosystem Assessment, 2005. Ecosystems and Human Well-being: Synthesis. Island Press, Washington, DC.

Myers, S.L., 2011. Hillary Clinton Takes Seat at Arctic Council. New York Times. Available from: http://green.blogs.nytimes.com/2011/05/12/hillary-clinton-takes-seat-at-Arctic-council/.

Natcher, D.C., 2009. Subsistence and the Social Economy of Canada's Aboriginal North. Northern Review 30, 83–98. ISSN 1929–6657.

Post, E., Forchhammer, M.C., Syndonia Bret-Harte, M., Callaghan, T.V., Christensen, T.R., Elberling, B., Fox, A.D., Gilg, O., Hik, D.S., Høye, T.T., Ims, R.A., Jeppesen, E., Klein, D.R., Madsen, J., McGuire, A.D., Rysgaard, S., Schindler, D.E., Stirling, I., Tamstorf, M.P., Tyler, N.J.C., van der Wal, R., Welker, J., Wookey, P.A., Schmidt, N.M., Aastrup, P., 2009. Ecological dynamics across the Arctic associated with recent climate change. Science 325 (5946), 1355–1358.

Rasmussen, R. O. (Ed). 2011. Megatrends. Copenhagen: Nordic Council of Ministers as cited in Stockholm environment institute, Arctic Resilience Interim Report: pp. 71.

Robards, M.D., Lovecraft, A.L., 2010. Evaluating co-management for social-ecological fit: Indigenous priorities and agency mandates for Pacific Walrus. Policy Stud. J. 38 (2), 257–279.

Rochefort, D.A., Cobb, R.W., 1993. Problem definition, agenda access, and Policy Choice. Policy Stud. J. 21 (1), 56–71.

Romanovsky, V.E., Isaksen, K., Drozdov, D., et al. (2017). Changing permafrost and its impacts. pp 65-102 in AMAP, 2017. Snow, Water, Ice and Permafrost in the Arctic (SWIPA) 2017. Arctic Monitoring and Assessment Programme (AMAP), Oslo, Norway.

Scharin, H., Ericsdotter, S., Elliott, M., Turner, R.K., Niiranen, S., Blenckner, T., Hyytiainen, K., Ahlvik, L., Ahtiainen, H., Artell, J., Hasselström, L., Söderqvist, T., Rockström, J., 2016. Processes for the sustainable stewardship of marine environments. Ecological Economics 128, 55–67.

Sarker, A., Ross, H., Shrestha, K.K., 2008. Interdependence of common-pool resources: lessons from a set of nested catchments in Australia. Hum. Ecol. 36, 821–834.

Schweitzer, P., 2011. Humans in the coastal zone of the circumpolar north. In: Lovecraft, A.L., Eicken, H. (Eds.), North by 2020: Perspectives on Alaska's Changing Social-Ecological Systems. University of Alaska Press, Fairbanks, AK, pp. 253–260.

Smith, C.J., Papadopoulou, K.-N., Barnard, S., Mazik, K., Elliott, M., Patrício, J., Solaun, O., Little, S., Bhatia, N., Borja, A., 2016. Managing the marine environment, conceptual models and assessment: considerations for the European marine Strategy framework directive. Frontiers in Marine Science 3, 144. https://doi.org/10.3389/fmars.2016.00144.

Solomon, S., Plattner, G.-K., Knutti, R., Friedlingstein, P., 2009. Irreversible climate change due to carbon dioxide emissions. PNAS 106, 1704–1709. https://doi.org/10.1073/pnas.0812721106.

Stone, D., 2012. Policy Paradox: The Art of Political Decision Making. W.W. Norton and Company, New York.

Tadesco, M., Box, J.E., Cappelen, J., Fausto, R.S., Fettweis, X., Hansen, K., Mote, T., Smeets, C.J.P.P., vanAs, D., van de Wal, R.S.W., et al., 2015. Greenland ice sheet. In: Arctic Report Card 2015. Available from: http://www.Arctic.noaa.gov/reportcard. (Accessed December 14, 2015).

Tay, L., Diener, E., 2011. Needs and subjective well-being around the world. J. Pers. Soc. Psychol. 101 (2), 354–356. https://doi.org/10.1037/a00.

The World Factbook, 2018. Central Intelligence Agency. Washington, DC. https://www.cia.gov/library/publications/the-world-factbook/index.html.

Waters, C.N., Zalasiewicz, J., Summerhayes, C., Barnosky, A.D., Poirier, C., Gałuszka, A., Cearreta, A., Edgeworth, M., Ellis, E.C., Ellis, M., Jeandel, C., Leinfelder, R., McNeill, J.R., Richter, D.dB., Steffen, W., Syvitski, J., Vidas, D., Wagreich, M., Williams, M., Zhisheng, A., Grinevald, J., Odada, E., Oreskes, N., Wolfe, A.P., 2016. The Anthropocene is functionally and stratigraphically distinct from the Holocene. Science 351 (6269), 137.

Young, O.R., 2002. The Institutional Dimensions of Environmental Change: Fit Interplay and Scale. MIT Press, Cambridge, MA. 237 pp.

Index

Note: Page numbers followed by *f* indicate figures, *t* indicate tables, and *b* indicate boxes.